HANDBOOK OF NEURO-ONCOLOGY NEUROIMAGING

HANDBOOK OF NEURO-ONCOLOGY NEUROIMAGING

Chief Editors

Herbert B. Newton, M.D., FAAN

Professor of Neurology & Oncology, Director, Division of Neuro-Oncology
Esther Dardinger Chair in Neuro-Oncology
Co-Director, Dardinger Neuro-Oncology Center, Ohio State University Medical Center,
James Cancer Hospital & Solove Research Institute

Ferenc A. Jolesz, M.D., Ph.D.

B. Leonard Holman Professor of Radiology, Vice Chairman for Research Director,
Division of MRI and Image Guided Therapy Program
Department of Radiology Brigham and Women's Hospital Harvard Medical School, Boston, MA

Associate Editors

Mark G. Malkin, M.D., FRCPC, FAAN

Medical College of Wisconsin, Milwaukee, WI

Eric Bourekas, M.D., Ph.D.

Ohio State University Medical Center, Columbus, OH

Gregory Christoforidis, M.D., Ph.D.

Ohio State University Medical Center, Columbus, OH

ELSEVIER

AMSTERDAM • BOSTON • HEIDELBERG • LONDON • NEW YORK • OXFORD
PARIS • SAN DIEGO • SAN FRANCISCO • SINGAPORE • SYDNEY • TOKYO
Academic Press is an imprint of Elsevier

Academic Press is an imprint of Elsevier
360 Park Avenue South, New York, NY 10010-1710, USA
84 Theobald's Road, London WC1X 8RR, UK
30 Corporate Drive, Suite 400, Burlington, MA 01803, USA
525 B Street, Suite 1900, San Diego, CA 92101-4495, USA

First edition 2008

British Library Cataloguing in Publication Data
A catalogue record for this book is available from the British Library

Library of Congress Cataloging in Publication Data
A catalog record for this book is available from the Library of Congress

ISBN: 978-0-12-370863-2

For information on all Academic Press publications
visit our web site at http://books.elsevier.com

Typeset by Charon Tec Ltd (A Macmillan Company), Chennai, India
www.charontec.com

Printed and bound in Canada

08 09 10 11 10 9 8 7 6 5 4 3 2 1

I would like to thank my wife, Cheryl, and my two children, Alex and Ashley, for their love, patience, and support while I worked on this book project.

I would also like to thank my Neuro-Oncology patients and their families for their courage and strength in the face of adversity, which was a constant inspiration.

Herbert B. Newton

I would like to dedicate this book to my wife Anna, who had to sacrifice her professional life because of my dedication to my profession.

I would also like to thank the great many colleagues who have given unselfishly of their time and talents to make this book possible.

Ferenc A. Jolesz

Contents

Contributors

William Ankenbrandt
Assistant Professor of Clinical
Radiology, Department of Radiology,
Feinberg School of Medicine, North-
western University, Chicago IL

Harman P.S. Bajwa
Department of Neurology,
University of Virginia Health System,
Charlottesville, VA

Peter M. Black
Department of Neurosurgery,
Children's Hospital Boston/Harvard
Medical School, Boston, MA

Eric C. Bourekas
Department of Radiology, Ohio State
University Medical Center,
Columbus, OH

Nicole Petrovich Brennan
Department of Radiology, Memorial
Sloan-Kettering Cancer Center, Weill
Medical College of Cornell University,
New York, NY

Marc Bussiere
Department of Radiation Oncology,
Massachusetts General Hospital and
Harvard Medical School, Boston, MA

Marc C. Chamberlain
Neuro-Oncology Program/MCC
NeurProg, H. Lee Moffitt Cancer Center
and Research Institute, Tampa, FL

Paul H. Chapman
Department of Neurosurgery,
Massachusetts General Hospital
and Harvard Medical School,
Boston, MA

Clark C. Chen
Department of Neurosurgery,
Massachusetts General Hospital,
Boston, MA

D. Chourmouzi
Department of Radiology, Interbalkan
Medical Center, Thessaloniki, Greece

Gregory A. Christoforidis
Department of Radiology, Ohio State
University Medical Center,
Columbus, OH

L. Celso Hygino Cruz, Jr
Multi-Imagem Ressonância Mag-
nética, Centro Médico Barrashopping,
CDPI – Clínica de Diagnóstico por
Imagem, Rio de Janeiro, Brazil

Eric Davis
Neuroradiology Service, Memorial
Sloan-Kettering Cancer Center,
New York, NY

Romeu C. Domingues
Multi-Imagem Ressonância
Magnética, Centro Médico Barrashopping, CDPI – Clínica de Diagnóstico
por Imagem, Rio de Janeiro, Brazil

A. Drevelegas
Department of Radiology, Aristotele
University School of Medicine, AHEPA
University Hospital, Thessaloniki,
Greece

Shehanaz Ellika
Department of Neuroradiology, Henry
Ford Health System, Detroit, MI

Mark A. Ferrante
Department of Psychiatry and
Neurology, Tulane University School
of Medicine, New Orleans, LA

Jens H. Figiel
Department of Radiology, College of
Medicine, Philipps University
Marburg, Marburg, Germany

Alexandra Golby
Department of Neurosurgery, Brigham
and Women's Hospital, Harvard
Medical School, Boston, MA

R. Gilberto Gonzalez
Department of Radiology,
Massachusetts General Hospital,
Boston, MA

Jonathan P. Gordon
Division of Neuroradiology,
Department of Radiology, University
of Virginia Health Sciences Center,
Charlottesville, VA

Gordon J. Harris
Center for Neuro-Oncology,
Massachusetts General Hospital Can-
cer Center, Boston, MA

Nobuhiko Hata
Department of Radiology, Brigham
and Women's Hospital/Harvard
Medical School, Boston, MA

D. Hearshen
Diagnostic Radiology, Henry Ford
Hospital, Detroit, MI

John W. Henson
Pappas Brain Tumor Imaging Research
Program, Division of Neuroradiology,
Massachusetts General Hospital,
Boston, MA

Johannes T. Heverhagen
Department of Radiology, College of
Medicine, Philipps University
Marburg, Marburg, Germany

Andrei Holodny
Department of Radiology, Memorial
Sloan-Kettering Cancer Center, Weill
Medical College of Cornell University,
New York, NY

Liangge Hsu
Department of Neuroradiology,
Brigham and Women's Hospital,
Harvard Medical School, Boston, MA

Tudor H. Hughes
Department of Radiology, University
of California at San Diego Medical
Center, San Diego, CA

Masanori Ichise
Department of Radiology, College of
Physicians & Surgeons, Columbia
University, New York, NY

Rajan Jain
Department of Radiology, Henry Ford
Health System, Detroit, MI

Ferenc A. Jolesz
Division of MRI and Image Guided
Therapy Program, Department of
Radiology, Brigham and Women's
Hospital/Harvard Medical School,
Boston, MA

Daniel Kacher
Department of Radiology, Brigham
and Women's Hospital/Harvard
Medical School, Boston, MA

Alayar Kangarlu
Columbia University and New York
State Psychiatric Institute,
New York, NY

Boris R. Keil
Department of Radiology, College of
Medicine, Philipps University
Marburg, Marburg, Germany

Marie Foley Kijewski
Department of Radiology, Brigham
and Women's Hospital, Boston, MA

John T. Kissel
Division of Neuromuscular Disease,
Department of Neurology, Ohio State
University Hospital, Columbus, OH

Michael V. Knopp
Department of Radiology, Ohio State
University Medical Center,
Columbus, OH

George Krol
Chief, Neuroradiology Service,
Memorial Sloan-Kettering Cancer
Center, New York, NY

Michael H. Lev
Director, Emergency Neuroradiology,
Harvard Medical School Massachusetts
General Hospital, Boston, MA

E. Paul Lindell
Department of Diagnostic Radiology,
Mayo Clinic, Rochester, MN

Jay S. Loeffler
Chief, Radiation Oncology, Northeast
Proton Therapy Center, Massachusetts
General Hospital, Boston, MA

Stephan E. Maier
Department of Radiology, Brigham
and Women's Hospital, Boston, MA

Mark G. Malkin
Division of Neuro-Oncology,
Department of Neurology, Medical
College of Wisconsin, Milwaukee, WI

Hatsuho Mamata
Department of Radiology, Brigham
and Women's Hospital, Boston, MA

T. Mikkelsen
Co-Director, Department of
Neurology & Neurosurgery, Hermelin
Brain Tumor Center, Henry Ford
Hospital, Detroit, MI

Michelle Monje
Department of Neurology, Brigham
and Women's Hospital, Boston, MA

Robert V. Mulkern
Department of Radiology, Children's
Hospital Boston/Harvard Medical
School, Boston, MA

Michelle J. Naidich
Department of Radiology, Feinberg
School of Medicine, Northwestern
University, Chicago, IL

Herbert B. Newton
Division of Neuro-Oncology, Depart-
ment of Neurology, Ohio State
University Medical Center,
Columbus, OH

Erik B. Nine
Department of Radiology, The Ohio
State University Medical Center,
Columbus, OH

Alan B. Packard
Department of Radiology, Children's
Hospital Boston/Harvard Medical
School, Boston, MA

Nina A. Paleologos
Department of Neurology, Evanston
Memorial Hospital, Evanston, IL

N. Papanicolaou
Department of Radiology, University
Hospital of Heraklion, Medical School
of Crete, Crete, Greece

C. Douglas Phillips
Division of Neuroradiology,
Department of Radiology, University
of Virginia Health Sciences Center,
Charlottesville, VA

James M. Provenzale
Department of Radiology, Duke
University Medical College,
Durham, NC

Jeffrey J. Raizer
Director, Medical Neuro-Oncology,
Feinberg School of Medicine,
Northwestern University, Chicago, IL

Abhik Ray-Chaudhury
Division of Neuropathology,
Department of Pathology, Ohio
State University Medical Center,
Columbus, OH

Haricharan Reddy
Department of Radiology, Johns
Hopkins University Medical Center,
Baltimore, MD

Richard L. Robertson
Department of Radiology, Children's
Hospital Boston/Harvard Medical
School, Boston, MA

Lisa R. Rogers
Professor of Neurology, University of
Michigan Health Systems, Ann
Arbor, MI

Pamela W. Schaefer
Associate Director of Neuroradiology,
Massachusetts General Hospital,
Boston, MA

David Schiff
Division of Neurology, Department
of Neurology, University of
Virginia Health Sciences Center,
Charlottesville, VA

Kathleen Schmainda
Associate Professor, Radiology and
Biophysics, Department of Biophysics,
Medical College of Wisconsin,
Milwaukee, WI

Karl F. Schmidt
McLean Hospital, Harvard Medical
School, Belmont, MA

Lubdha M. Shah
Division of Neuroradiology,
Department of Radiology, University
of Virginia Health Sciences Center,
Charlottesville, VA

Xia Shuang
First Central Hospital, Nankai District,
Tianjin, China

V. Michelle Silvera
Children's Hospital Boston, Harvard
Medical School, Boston, MA

Thinesh Sivapatham
Department of Radiology, The Ohio
State University Medical Center,
Columbus OH

H. Wayne Slone
Department of Radiology,
Ohio State University Medical Center,
Columbus, OH

Aaron Sodickson
Department of Neurology, Brigham
and Women's Hospital, Boston, MA

Lilja Solnes
Department of Radiology,
Ohio State University Medical Center,
Columbus, OH

A. Gregory Sorensen
Department of Neuroradiology,
Massachusetts General Hospital,
Boston, MA

Yanping Sun
Department of Radiology,
Brigham and Women's Hospital,
Boston, MA

Ion-Florin Talos
Department of Radiology, Brigham
and Women's Hospital/Harvard
Medical School, Boston, MA

Suzanne Tharin
Department of Neurosurgery, Brigham
and Women's Hospital, Harvard
Medical School, Boston, MA

Stephan Ulmer
3D Imaging Service, Department of
Radiology, Massachusetts General
Hospital, Boston, MA

Ronald L. Van Heertum
Director, Columbia Kreitchman PET
Center, Department of Radiology,
College of Physicians and Surgeons,
Columbia University, New York, NY

Steven Vernino
Department of Neurology, UT
Southwestern Medical Center,
Dallas, TX

Arastoo Vossough
Department of Radiology, Harvard
Medical School, Massachusetts
General Hospital, Boston, MA

Simon K. Warfield
Department of Radiology, Brigham
and Women's Hospital/Harvard
Medical School, Boston, MA

Patrick Y. Wen
Center for Neuro-Oncology,
Dana-Farber/Brigham and Women's
Cancer Center, Boston, MA

E. Xinou
Department of Radiology, Interbalkan
Medical Center, Thessaloniki, Greece

Geoffrey S. Young
Department of Radiology, Brigham
and Women's Hospital, Boston, MA

Tina Young Poussaint
Division of Neuroradiology,
Department of Radiology, Children's
Hospital Boston, Boston, MA

Amir A. Zamani
Department of Radiology, Brigham
and Women's Hospital,
Boston, MA

Preface

The inspiration for this book evolved over the past decade, as we witnessed the dramatic growth in the field of Neuro-Oncology and realized the important contributions that Neuroimaging was making to this maturing discipline. Remarkable progress has taken place in Neuro-Oncology because of the increased utilization of advanced imaging technologies in clinical practice. With the introduction of computed tomography (CT), magnetic resonance imaging (MRI), and positron emission tomography (PET), anatomical, functional, and metabolic imaging have improved the visualization and localization of tumors, and provided target definition for various therapeutic modalities. The further refinements in MRI and CT technology, and the addition of newer imaging methods, such as MR spectroscopy, functional MRI, and diffusion MRI have allowed brain tumor patients to be diagnosed much earlier in the course of their illness and to be followed more carefully during the rigors of treatment. We felt there was a growing need for a single source, comprehensive, reference handbook that would encompass the most up to date clinical and technical information regarding the application of Neuroimaging techniques to brain tumor and Neuro-Oncology patients. With the timely publication of this book, an important void in the literature has now been filled.

This book should have broad appeal to anyone interested in the fields of Neuro-Oncology and Neuroimaging, and especially to those who are involved in the care of patients with brain tumors. It will satisfy clinicians that require in-depth overviews of various Neuro-Oncology topics, along with "cutting edge" information regarding how to apply Neuroimaging techniques to those areas of interest. In addition, the book can serve as a resource of background information to Neuroimaging researchers and basic scientists with an interest in brain tumors and Neuro-Oncology. We sincerely hope that the world-class group of authors assembled herein can assist you in providing the best oncologic care for your patients.

We are very thankful to the Publisher for providing the opportunity to introduce this extensive body of clinically relevant information to a larger professional audience. We would also like to thank all of our colleagues who contributed to this book and provided insight into the use of imaging methods in Neuro-Oncology.

Herbert B. Newton, M.D., FAAN
Columbus, OH 2007

Ferenc A. Jolesz, M.D., Ph.D.
Boston, MA 2007

SECTION I

Overview of Neuro-Oncological Disorders

1. Overview of Brain Tumor Epidemiology
2. Overview of Pathology and Treatment of Primary Brain Tumors
3. Overview of Pathology and Treatment of Metastatic Brain Tumors
4. Overview of Spinal Cord Tumor Epidemiology
5. Overview of Pathology and Treatment of Primary Spinal Cord Tumors
6. Overview of Pathology and Treatment of Intramedullary Spinal Cord Metastases
7. Epidural Spinal Cord Compression in Adult Neoplasms
8. Neoplastic Meningitis
9. Vascular Disorders: Epidemiology
10. Intracranial Hemorrhage in Cancer Patients
11. CNS Infarction
12. Intracranial Veno-Occlusive Disease
13. Paraneoplastic Syndromes
14. Neoplastic Plexopathies
15. Neurological Complications of Oncological Therapy

CHAPTER 1

Overview of Brain Tumor Epidemiology

Herbert B. Newton and Mark G. Malkin

INTRODUCTION

Brain tumors remain a significant health problem in the USA and worldwide. Overall, they comprise some of the most malignant tumors known to affect human beings and are generally refractory to all modalities of treatment. It is estimated that between 30 000 and 35 000 new cases of primary brain tumors (PBT) will be diagnosed each year in the USA (1–2% of newly diagnosed cancers overall) [1–6]. Metastatic brain tumors (MBT) are even more common and affect between 150 000 and 170 000 new patients each year in the USA [7–11]. Although significant advances have been made in our understanding of the molecular biology of brain tumors, further research is needed to improve our knowledge of the etiology and natural history of this disease.

EPIDEMIOLOGY OF PRIMARY BRAIN TUMORS

Most studies suggest that approximately 14 per 100 000 people in the USA will be diagnosed with a PBT each year [2–6]. Among this cohort with newly diagnosed tumors, 6–8 per 100 000 will have a high-grade neoplasm. Contemporary epidemiological studies suggest an increasing incidence rate for the development of PBT in children less than 14 years of age and in patients 70 years or older [12]. For people in the 15- to 44-year-old age group the overall incidence rates have remained fairly stable in recent years. The cause of the increased incidence of PBT in some age groups remains unclear, but may be due to improvements in diagnostic neuroimaging, such as magnetic resonance imaging (MRI), greater availability of specially qualified neurosurgeons and neuropathologists, improved access to medical care for children and elderly patients and more aggressive approaches to health care for elderly patients [5,12]. In other words, the increase in PBT incidence may be more apparent than real due to ascertainment bias.

The prognosis and survival of patients with PBT remains poor [1–6]. Although uncommon neoplasms, they rank among the top 10 causes of cancer-related deaths in the USA and account for a disproportionate 2.4% of all yearly cancer-related deaths [13]. The median survival for a patient with glioblastoma multiforme (GBM) is approximately 12–14 months, a figure which has not improved substantially over the past 30 years. For patients with a low-grade astrocytoma or oligodendroglioma, the median survival is still significantly curtailed and is about 6–10 years. For PBT patients in the USA as a whole, across all age groups and tumor types, the 5-year survival rate is 20% [3]. If a patient with a PBT survives for an initial 2 years, the probability of surviving another 3 years is 76.2%. In general, for any given tumor type, survival is better for younger patients than for older patients. The only exception to this generalization is for children with medulloblastoma and embryonal tumors, in which patients under 3 years of age have poorer survival rates than children between 3 and 14 years of age [14]. The 5-year survival rate for all children less than 14 years of age with a malignant PBT is 72%.

The median age at diagnosis for PBT is between 54 and 58 years [1–6]. Among different histological varieties of PBT, there is significant variability in the age of onset. A small secondary peak is also present in the pediatric age group, in children between the ages of 4 and 9. Overall, PBT are more common in males than females, with the exception of meningiomas, which are almost twice as common in females. Tumors of the sellar region, and of the cranial and spinal nerves, are almost equally represented among males and females. In the USA, gliomas are more commonly diagnosed in whites than blacks, while the incidence of meningiomas is relatively equal between the two groups.

Numerous epidemiological studies have been performed in an attempt to define risk factors involved in the development of brain tumors (Table 1.1) [2–6]. The vast majority of these potential risk factors have not been associated with any significant predisposition to brain tumors. One risk factor that has proven to be important is the presence of a hereditary syndrome with a genetic predisposition for developing tumors, some of which can affect the nervous system [4,5,15]. Several hereditary syndromes are associated with PBT, including tuberous sclerosis, neurofibromatosis types 1 and 2, nevoid basal cell carcinoma

TABLE 1-1 Risk factors that have been investigated in epidemiological studies of primary brain tumors

Hereditary syndromes (proven): tuberous sclerosis, neurofibromatosis types 1 and 2, nevoid basal cell carcinoma syndrome, Turcot's syndrome and Li-Fraumeni syndrome

Family history of brain tumors

Constitutive polymorphisms: glutathione transferases, cytochrome P-450 2D6 and 1A1, N-acetyltransferase and other carcinogen metabolizing, DNA repair and immune function genes

History of prior cancer

Exposure to infectious agents

Allergies (possible reduced risk)

Head trauma

Drugs and medications

Dietary history: N-nitroso compounds, oxidants, antioxidants

Tobacco usage

Alcohol consumption

Ionizing radiation exposure (proven)

Occupational and industrial chemical exposures: pesticides, vinyl chloride, synthetic rubber manufacturing, petroleum refining and production, agricultural workers, lubricating oils, organic solvents, formaldehyde, acrylonitrile, phenols, polycyclic aromatic hydrocarbons

Cellular telephones

Power frequency electromagnetic field exposure

Data adapted from references [2–6,14–26]

syndrome, Li-Fraumeni syndrome and Turcot's syndrome. However, it is estimated that hereditary genetic predisposition may be involved in only 2–8% of all cases of PBT. Familial aggregation of brain tumors has also been studied, with conflicting results [5,15]. The relative risk for developing a tumor among family members of a patient with a PBT is quite variable, ranging from 1 to 10. One study that performed a segregation analysis of families of more than 600 adult glioma patients showed that a polygenic model most accurately explained the inheritance pattern [16]. A similar analysis of 2141 first-degree relatives of 297 glioma families did not reject a multifactorial model, but concluded that an autosomal recessive model fitted the inheritance pattern more accurately [17]. Critics of these studies suggest that the common exposure of a family to a similar pattern of environmental agents could lead to a similar clustering of tumors. Other investigators have focused on genetic polymorphisms that might influence genetic and environmental factors to increase the risk for a brain tumor [4,5]. Alterations in genes involved in oxidative metabolism, detoxification of carcinogens, DNA stability and repair and immune responses might confer a genetic predisposition to tumors. For example, Elexpuru-Camiruaga and colleagues demonstrated that cytochrome P-4502D6 and glutathione transferase theta were associated with an increased risk for

brain tumors [18]. Other studies have not supported these results, but have found an increased risk for rapid N-acetyl-transferase acetylation and intermediate acetylation [19]. In general, further studies with larger cohorts of patients will be necessary to determine if genetic polymorphisms of key metabolic enzyme systems play a significant role in the risk for developing a brain tumor.

Cranial exposure to therapeutic ionizing radiation is a potent risk factor for subsequent development of a brain tumor and is known to occur after a wide range of exposures [1–6]. Application of low doses of irradiation (1000–2000 cGy), such as were prescribed in the past for children with tinea capitis or skin hemangiomas, have been associated with relative risks of 18 for nerve sheath tumors, 10 for meningiomas and 3 for gliomas [5,20]. Gliomas and other PBTs are also known to occur after radiotherapy for diseases such as leukemia, lymphoma and head and neck cancers [5,21,22]. In addition, alternative methods of radiation exposure, such as nuclear bomb blasts and employment at nuclear production facilities, have also been implicated as significant risk factors for the development of brain tumors [23,24].

Many other risk factors have been evaluated for their potential role in the genesis of brain tumors [1–6]. The majority of these factors have been proven to have little, if any, relationship to brain tumor development, or to have an indeterminate association due to a mixture of positive and negative studies. Factors in this category include the history of a prior primary systemic malignancy, head injury, prenatal or premorbid ingestion of various types of medications, exposure to viruses and other types of infection (except for the human immunodeficiency virus, which is known to be associated with brain lymphoma), dietary history (i.e. ingestion of N-nitroso compounds, oxidants and antioxidants), alcohol ingestion, smoking tobacco, residential chemical exposures and proximity to electromagnetic fields. The relationship between industrial and occupational chemical exposures and brain tumors is very complex and remains unclear [2,4,5]. Workers are exposed to chemicals that are potentially carcinogenic or neurotoxic, or both, including lubricating oils, organic solvents, formaldehyde, acrylonitrile, phenols and phenolic-based compounds, vinyl chloride and polycyclic aromatic hydrocarbons. Preclinical studies have proven the ability of vinyl chloride to induce brain tumors in rat models and some studies suggest an increased risk for chemical workers that handle this compound [25]. However, more recent and extensive analyses suggest that the relationship between vinyl chloride exposure and brain tumors remains inconclusive [26]. Similar inconclusive results for other chemicals are common in the epidemiology literature and demonstrate the difficulty of proving an association between workplace exposures and an uncommon form of cancer. At this time, no definitive associations have been proven between brain tumors and any specific chemicals found in the occupational or industrial setting, including those that are known to be definite or putative carcinogens.

Several large studies have evaluated the possibility of a link between the use of handheld cellular telephones and brain cancer, as well as other tumors of the head and neck region. Researchers from Denmark performed a nationwide review of 420 095 cell phone users and determined that the overall incidence of cancer was not elevated (OR = 0.89) in comparison to controls including brain tumors, salivary gland tumors and leukemias [27]. Other studies focusing on the incidence of high-grade gliomas in cellular telephone users have not been able to substantiate an increased incidence [28–30]. Several reports have focused on the use of cellular telephones and the incidence of acoustic schwannomas [31,32]. Neither study was able to discern a relationship between the duration of use, lifetime cumulative hours of use, or frequency of use of a cellular telephone and the risk of developing an acoustic schwannoma. The only positive report to date was a population-based, case-control study from Germany that evaluated 366 glioma, 381 meningioma and 1494 control patients [33]. In this study, the overall risk for a brain tumor was not associated with the use of a cellular telephone. However, there was a small increased risk of glioma (OR = 2.20), but not meningioma (OR = 1.09), in patients that had used a cellular telephone for 10 years or more.

More recent molecular epidemiological studies in adult patients with high-grade glioma are beginning to show promise for further research efforts [34]. In a study of the association between human leukocyte antigens (HLA) and related polymorphisms (HLA-A, -B, -C, -DRB1) and the onset and prognosis of GBM, 155 GBM patients and 157 controls were studied in the San Francisco area [35]. During multivariate logistical regression analysis, the HLA-B*13 and the HLA-B*07-Cw*07 haplotype were positively associated with the occurrence of GBM ($P = 0.01$, $P < 0.001$, respectively). The Cw*01 variant had a negative association with the occurrence of GBM ($P = 0.05$). In addition, progression to death among GBM patients was slower in patients with HLA-A*32 (HR = 0.45, $P < 0.01$) and faster in those with HLA-B*55 (HR = 2.27, $P < 0.01$). In a study of polymorphisms of ERCC1 and ERCC2, genes that are important for DNA nucleotide excision repair, 450 adult glioma patients and 500 controls were analyzed [36]. Overall, the presence of ERCC1 and ERCC2 were not associated with an increased risk for GBM. However, among whites, glioma patients were significantly more likely than controls to be homozygous for variants in ERCC1 C8092A and ERCC2 K751Q (OR = 3.2). In a similar study, 556 astrocytic tumors were analyzed for the expression of p53, epidermal growth factor receptor (EGFR), MDM2 and O^6-methylguanine-DNA-methyltransferase (MGMT) and then correlated with clinical parameters and risk factors [37]. The data confirmed the previously noted inverse relationship between p53 mutation and MDM2 ($P = 0.04$) or EGFR ($P = 0.004$) amplification.

In addition, the presence of p53 mutations was more likely to occur in younger patients ($P < 0.001$). EGFR gene amplification was more likely to occur in older patients (mean 63 years old amplified versus mean 48 years old non-amplified; $P = 0.005$). p53 mutations were more likely to occur in GBM among non-white patients than white patients ($P = 0.004$). Patient carriers of the MGMT variant 84Phe allele were significantly less likely to have tumors with p53 overexpression (OR = 0.30) and somewhat less likely to have tumors with p53 mutations (OR = 0.47). The authors concluded that these molecular data demonstrated ethnic variation in the pathogenesis of glioma.

Of all the potential risk factors studied, the only one that might be associated with a protective effect for developing a brain tumor is the presence of an allergy [38]. The presence of any form of allergy was inversely associated with the development of a glioma (OR = 0.7), but not with meningiomas or acoustic neuromas. Similar inverse associations were noted for the presence of autoimmune diseases and the presence of both gliomas and meningiomas. The authors suggested that allergy-related immunological factors might play a protective role in the genesis of certain brain tumors. As a follow-up to this initial study, Schwartzbaum and colleagues performed a population-based case-control evaluation of 111 GBM patients and 422 controls, using germ line polymorphisms associated with asthma and inflammation as biomarkers [39]. Self-reported asthma and eczema were inversely related to the incidence of GBM (OR=0.64). In addition, IL-4RA Ser478Pro TC, CC and IL-4RA Gln551ArgAG, AA were positively associated with GBM (OR=1.64), while IL-13-1,112CT, TT was negatively associated with GBM (OR = 0.56). The authors suggested that associations existed between IL-4RA, IL-13 and GBM that were independent of their role in allergic conditions.

EPIDEMIOLOGY OF METASTATIC BRAIN TUMORS

Brain metastases (MBT) are the most common complication of systemic cancer, with estimated incidence rates of 8.3 to 11 cases per 100 000 population [7–11,40]. Hospital and autopsy-based studies estimate that these tumors develop in 20–40% of all adult cancer patients, which corresponds to approximately 150 000 to 170 000 new cases per year in the USA. More recent data using population-based estimates would suggest a lower incidence of MBT, in the range of 10% [41]. The presence of an MBT does not always correlate with clinical sequelae; it is estimated that only 60–75% of patients with an MBT will become symptomatic. The frequency of MBT appears to be rising due to more successful systemic treatment and longer patient survival, earlier detection and implementation of therapy and improved imaging techniques. MBT most often arise from primary tumors of the lung (50–60%), breast (15–20%), melanoma

(5–10%) and gastrointestinal tract (4–6%) [7–11]. Empiric screening of patients with newly diagnosed non-small cell lung cancer identify MBT in 3–10% of cases [40]. However, MBT can develop from virtually any systemic malignancy, including primary tumors of the prostate, ovary and female reproductive system, kidney, esophagus, soft tissue sarcoma, bladder and thyroid [42–51]. In addition, between 10 and 15% of patients will develop MBT from an unknown primary [8,52]. Autopsy studies in adults would suggest that melanoma (20–45% of patients) has the most neurotropism of all primary tumors; however, small cell lung carcinoma, renal carcinoma, breast and testicular carcinoma also have a strong propensity for spread to the brain [8]. Tumors with a low degree of neurotropism include prostate, gastrointestinal tract, ovarian and thyroid malignancies. In children and young adults, MBT arise most often from sarcomas (e.g. osteogenic, Ewing's), germ cell tumors and neuroblastomas [7–11,53]. In 65–75% of patients, two or more metastatic tumors will develop simultaneously and be present at the time of cancer diagnosis. Single brain metastases are less common and are most often noted in patients with breast, colon and renal cell carcinoma. Patients with malignant melanoma and lung carcinoma are more likely to have multiple metastatic lesions.

The prognosis for patients with MBT is quite poor and is dependent on the histological tumor type, number and size of the metastatic lesions, neurological status and degree of systemic involvement. Overall, the presence of an MBT is associated with high morbidity and mortality, with approximately one third of all patients dying from the brain tumor [10]. The natural history is such that, left untreated, patients with MBT will usually die of neurological deterioration within 4 weeks. The addition of steroids will typically extend survival to 8 weeks. External beam radiotherapy, the most common modality of treatment, can further extend survival to 12–20 weeks in many patients [7–11]. However, survival is also dependent on the type of primary malignancy, as shown in a recent report by Hall and colleagues [54]. In their study, the overall 2-year survival rate for patients with MBT was 8.1%, with a range from 1.7% in patients with small cell lung carcinoma, up to 23.9% for those with ovarian cancer. Several studies have assessed how various prognostic factors relate to MBT patients at the time of diagnosis. A recent recursive partitioning analysis (RPA) of three RTOG trials evaluated a wide range of prognostic factors and their impact on patient survival [55]. The most important favorable factors were younger age (younger versus older than 65 years; $P < 0.0001$), higher Karnofsky Performance Status (KPS) score (greater or less than 70; $P < 0.0001$) and limited extent of systemic disease (controlled versus widespread disease; $P < 0.0001$). Using these criteria, patients could be grouped into three distinct classes. Class 1 included patients who were less than 65 years of age, had KPS scores greater than 70 and had well controlled systemic disease; class 3 consisted of all patients with KPS scores less than 70; while

class 2 included all other patients who did not fit into class 1 or class 3. The median overall survival varied significantly between groups: 28.4 weeks for patients in class 1; 16.8 weeks for those in class 2; and 9.2 weeks for class 3 patients. In addition, by univariate analysis, patients with multiple MBT had a significantly reduced survival compared to that of those with solitary lesions ($P = 0.021$).

In a similar study by Nussbaum and colleagues, the number of metastatic lesions present at diagnosis was found to correlate with overall survival [56]. They noted a significant difference ($P = 0.0001$) in median survival between patients with solitary brain metastases and those with multifocal disease: 5 months versus 3 months, respectively.

The molecular events that lead to the metastatic phenotype in a given primary tumor, with subsequent metastases to systemic organs and to the brain, remain unclear. Over the past few decades, the predominant theory postulated that somatic mutations in rare cells of the primary tumor (i.e. less than one in ten million) would lead to an acquired increase in metastatic capacity, with the ability to migrate through tissues, survive in blood and lymphatic fluid, invade distant organs and establish metastatic nodules [57]. Although this theory was supported somewhat by animal models, there were no data to verify this process in human tumors. More recent evidence, based on expression micro-array analyses of primary and metastatic tumors, supports the concept that metastatic potential is related to the intrinsic molecular biological state of the primary tumor as a whole, rather than to the emergence of a few rare cells [58,59]. The metastatic gene-expression signature consisted of a subset of eight genes that were upregulated (e.g. SNRPF, EIF4EL3, PTTG1) and a subset of nine genes that were downregulated (e.g. MHC class II DP-β1, RUNX1) in the primary cancer [58]. None of the genes were individual markers of the metastatic phenotype; they were only predictive when analyzed as a whole group. Patients with primary cancers that expressed the metastatic phenotypic signature had significantly shorter survival times in comparison to patients whose tumors did not express it ($P=0.009$).

In a related study, Milas and colleagues attempted to identify biological markers that could predict brain MBT and treatment outcome in patients with non-small cell lung cancer (NSCLC) [60]. Twenty-nine patients with MBT and matched controls without MBT were analyzed using immunohistochemical techniques. Primary cancer and brain tumor tissue samples were analyzed for the expression of EGFR, cyclooxygenase-2 (COX-2) and BAX. Expression of COX-2 in brain lesions correlated with expression in primary cancers ($P = 0.023$), while the expression of BAX was lower in the MBT in comparison to the primary cancer ($P = 0.045$). However, the overall expression of EGFR, COX-2 and BAX in primary NSCLC tumors did not differ between patients with MBT and those without MBT. Therefore, this set of molecular markers cannot be used to predict the likelihood of MBT in patients with NSCLC.

ACKNOWLEDGMENTS

The authors would like to thank Julia Shekunov for research assistance. Dr Newton was supported in part by National Cancer Institute grant, CA 16058 and the Dardinger Neuro-Oncology Center Endowment Fund.

REFERENCES

1. Newton HB (1994). Primary brain tumors: review of etiology, diagnosis, and treatment. Am Fam Phys 49:787–797.
2. Preston-Martin S (1996). Epidemiology of primary CNS neoplasms. Neurol Clin 14:273–290.
3. Davis FG, McCarthy BJ (2001). Current epidemiological trends and surveillance issues in brain tumors. Expert Rev Anticancer Ther 1:395–401.
4. Osborne RH, Houben MPWA, Tijssen CC, Coebergh JWW, van Duijn CM (2001). The genetic epidemiology of glioma. Neurology 57:1751–1755.
5. Wrensch M, Minn Y, Chew T, Bondy M, Berger MS (2002). Epidemiology of primary brain tumors: current concepts and review of the literature. Neuro-Oncology 4:278–299.
6. Hess KR, Broglio DR, Bondy ML (2004). Adult glioma incidence trends in the United States, 1977–2000. Cancer 101:2293–2299.
7. Newton HB (1999). Neurological complications of systemic cancer. Am Fam Phys 59:878–886.
8. Soffietti R, Ruda R, Mutani R (2002). Management of brain metastases. J Neurol 249:1357–1369.
9. Langer CJ, Mehta MP (2005). Current management of brain metastases, with a focus on systemic options. J Clin Oncol 23:6207–6219.
10. Bajaj GK, Kleinberg L, Terezakis S (2005). Current concepts and controversies in the treatment of parenchymal brain metastases: improved outcomes with aggressive management. Cancer Invest 23:363–376.
11. Lassman AB, DeAngelis LM (2003). Brain metastases. Neurol Clin N Am 21:1–23.
12. Legler JM, Ries LA, Smith MA et al (1999). Cancer surveillance series [corrected]: brain and other central nervous system cancers: recent trends in incidence and mortality. J Natl Cancer Inst 91:1382–1390.
13. ACS (American Cancer Society) (2002). Cancer Facts and Figures 2002. American Cancer Society, Atlanta.
14. Grovas A, Fremgen A, Rauck A et al (1997). The National Cancer Data Base report on patterns of childhood cancers in the United States. Cancer 80:2321–2332.
15. Bondy M, Wiencke J, Wrensch M, Kryitsis AP (1994). Genetics of primary brain tumors: a review. J Neuro-Oncol 18:69–81.
16. de Andrade M, Barnholtz JS, Amos CI, Adatto P, Spencer C, Bondy ML (2001). Segregation analysis of cancer in families of glioma patients. Genet Epidemiol 20:258–270.
17. Malmer B, Gronberg H, Bergenheim AT, Lenner P, Henriksson R (1999). Familial aggregation of astrocytoma in northern Sweden: an epidemiological cohort study. Int J Cancer 81:366–370.
18. Elexpuru-Camiruaga J, Buxton N, Kandula V et al (1995). Susceptibility to astrocytoma and meningioma: influence of allelism at glutathione S-transferase (GSTT1 and GSTM1) and cytochrome P-450 (CYP2D6) loci. Cancer Res 55:4237–4239.
19. Trizna Z, de Andrade M, Kyritsis AP et al (1998). Genetic polymorphisms in glutathione S-transferase mu and theta, N-acetyltransferase, and CYP1A1 and risk of gliomas. Cancer Epidemiol Biomarkers Prev 7:553–555.
20. Karlsson P, Holmberg E, Lundell M, Mattsson A, Holm LE, Wallgren A (1998). Intracranial tumors after exposure to ionizing radiation during infancy: a pooled analysis of two Swedish cohorts of 28 008 infants with skin hemangioma. Radiat Res 150:357–364.
21. Salvatai M, Aratico M, Caruso R, Rocchi G, Orlando EER, Nucci F (1991). A report on radiation-induced gliomas. Cancer 67:392–397.
22. Yeh H, Matanoski GM, Wang N, Sandler DP, Comstock GW (2001). Cancer incidence after childhood nasopharyngeal radium irradiation: a follow-up study in Washington County, Maryland. Am J Epidemiol 153:749–756.
23. Shintani T, Hayakawa N, Hoshi M et al (1999). High incidence of meningioma among Hiroshima atomic bomb survivors. J Radiat Res 40:49–57.
24. Loomis DP, Wolf SH (1996). Mortality of workers at a nuclear materials production plant at Oak Ridge, Tennessee, 1947–1990. Am J Ind Med 29:131–141.
25. Wong O, Whorton MD, Follart DE, Ragland D (1991). An industry-wide epidemiologic study of vinyl choride workers, 1942–1982. Am J Ind Med 20:317–334.
26. McLaughlin JK, Lipworth L (1999). A critical review of the epidemiologic literature on health effects of occupational exposure to vinyl chloride. J Epidemiol Biostat 4:253–275.
27. Johansen C, Boice JD, McLaughlin JK, Olsen JH (2001). Cellular telephones and cancer – a nationwide cohort study in Denmark. J Natl Cancer Inst 93:203–207.
28. Muscat JE, Malkin MG, Thompson S et al (2000). Handheld cellular telephone use and risk of brain cancer. J Am Med Assoc 284:3001–3007.
29. Inskip PD, Tarone RE, Hatch EE et al (2001). Cellular-telephone use and brain tumors. N Engl J Med 344:79–86.
30. Christensen HC, Schüz J, Kosteljanetz M (2005). Cellular telephones and risk for brain tumors. A population-based, incident case-control study. Neurology 64:1189–1195.
31. Muscat JE, Malkin MG, Shore RE et al (2002). Handheld cellular telephones and risk of acoustic neuroma. Neurology 58:1304–1306.
32. Schoemaker MJ, Swerdlow AJ, Ahlbom A et al (2005). Mobile phone use and risk of acoustic neuroma: results of the Interphone case-control study in five north European countries. Br J Cancer 93:842–848.

33. Schuz J, Bohler E, Berg G et al (2006). Cellular phones, cordless phones, and the risks of glioma and meningioma (Interphone Study Group, Germany). Am J Epidemiol 163:512–520.

34. Wrensch M, Fisher JL, Schwartzbaum JA, Bondy M, Berger M, Aldape KD (2005). The molecular epidemiology of gliomas in adults. Neurosurg Focus 19:E5.

35. Tang J, Shao W, Dorak MT et al (2005). Positive and negative associations of human leukocyte antigen variants with the onset and prognosis of adult glioblastoma multiforme. Cancer Epidemiol Biomarkers Prev 14:2040–2044.

36. Wrensch M, Kelsey KT, Liu M et al (2005). ERCC1 and ERCC2 polymorphisms and adult gliomas. Neuro-Oncology 7:495–507.

37. Wiencke JK, Aldape K, McMillan A et al (2005). Molecular features of adult glioma associated with patient race/ethnicity, age, and a polymorphism in O^6-methylguanine-DNA-methyltransferase. Cancer Epidemiol Biomarkers Prev 14:1774–1783.

38. Brenner AV, Linet MS, Fine HA et al (2002). History of allergies and autoimmune diseases and risk of brain tumors in adults. Int J Cancer 99:252–256.

39. Schwartzbaum J, Ahlbom A, Malmer B et al (2006). Polymorphisms associated with asthma are inversely related to glioblastoma multiforme. Cancer Res 65:6459–6465.

40. Gavrilovic IT, Posner JB (2005). Brain metastases: epidemiology and pathophysiology. J Neuro-Oncol 75:5–14.

41. Barnhholtz-Sloan JS, Sloan AE, Davis FG, Vigneau FD, Lai P, Sawaya RE (2004). Incidence proportions of brain metastases in patients diagnosed (1973 to 2001) in the metropolitan Detroit Cancer Surveillance System. J Clin Oncol 22:2865–2872.

42. Leroux PD, Berger MS, Elliott JP, Tamimi HK (1991). Cerebral metastases from ovarian carcinoma. Cancer 67:2194–2199.

43. Martinez-Manas RM, Brell M, Rumia J, Ferrer E (1998). Case report. Brain metastases in endometrial carcinoma. Gyn Oncol 70:282–284.

44. Mccutcheon IE, Eng DY, Logothetis CJ (1999). Brain metastasis from prostate carcinoma. Antemortem recognition and outcome after treatment. Cancer 86:2301–2311.

45. Lowis SP, Foot A, Gerrard MP et al (1998). Central nervous system metastasis in Wilms' tumor. A review of three consecutive United Kingdom trials. Cancer 83:2023–2029.

46. Culine S, Bekradda M, Kramar A et al (1998). Prognostic factors for survival in patients with brain metastases from renal cell carcinoma. Cancer 83:2548–2553.

47. Qasho R, Tommaso V, Rocchi G et al (1999). Choroid plexus metastasis from carcinoma of the bladder: case report and review of the literature. J Neuro-Oncol 45:237–240.

48. Salvati M, Frati A, Rocchi G et al (2001). Single brain metastasis from thyroid cancer: report of twelve cases and review of the literature. J Neuro-Oncol 51:33–40.

49. Ogawa K, Toita T, Sueyama H et al (2002). Brain metastases from esophageal carcinoma. Natural history, prognostic factors, and outcome. Cancer 94:759–764.

50. Espat NJ, Bilsky M, Lewis JJ et al (2002). Soft tissue sarcoma brain metastases. Prevalence in a cohort of 33829 patients. Cancer 94:2706–2711.

51. Schouten LJ, Rutten J, Huveneers HAM, Twijnstra A (2002). Incidence of brain metastases in a cohort of patients with carcinoma of the breast, colon, kidney, and lung and melanoma. Cancer 94:2698–2705.

52. Ruda R, Borgognone M, Benech F, Vasario E, Soffietti R (2001). Brain metastases from unknown primary tumour. A prospective study. J Neurol 248:394–398.

53. Kebudi R, Ayan I, Görgün O, Agaoglu FY, Vural S, Darendeliler E (2005). Brain metastasis in pediatric extracranial solid tumors: survey and literature review. J Neuro-Oncol 71:43–48.

54. Hall WA, Djalilian HR, Nussbaum ES et al (2000). Long-term survival with metastatic cancer to the brain. Med Oncol 17:279–286.

55. Gaspar L, Scott C, Rotman M et al (1997). Recursive partitioning analysis (RPA) of prognostic factors in three radiation therapy oncology group (RTOG) brain metastases trials. Int J Rad Oncol Biol Phys 37:745–751.

56. Nussbaum ES, Djalilian HR, Cho KH, Hall WA (1996). Brain metastases: histology, multiplicity, surgery, and survival. Cancer 78:1781–1788.

57. Poste G, Fidler IJ (1980). The pathogenesis of cancer metastasis. Nature 283:139–146.

58. Bernards R, Weinberg RA (2002). Metastasis genes: a progression puzzle. Nature 418:823.

59. Ramaswamy S, Ross KN, Lander ES, Golub TR (2003). A molecular signature of metastasis in primary solid tumors. Nat Genet 33:49–54.

60. Milas I, Komaki R, Hachiya T et al (2003). Epidermal growth factor receptor, cyclooxygenase-2, and BAX expression in the primary non-small cell lung cancer and brain metastases. Clin Cancer Res 9:1070–1076.

CHAPTER 2

Overview of Pathology and Treatment of Primary Brain Tumors

Herbert B. Newton, Abhik Ray-Chaudhury and Mark G. Malkin

INTRODUCTION

The epidemiology of primary brain tumors (PBT) was reviewed in detail in Chapter 1. In this chapter, we will provide an overview of the classification, pathology and treatment of the common PBT. Primary brain tumors will be diagnosed in approximately 30 000 to 35 000 patients in the USA this year and are associated with significant morbidity and mortality [1–6]. Of the estimated 14 patients per 100 000 population that will develop a PBT this year, 6–8 per 100 000 will have a high-grade neoplasm, usually some form of glioma such as glioblastoma multiforme (GBM) or anaplastic astrocytoma (AA).

PATHOLOGY OF SELECTED PRIMARY BRAIN TUMORS

The application of appropriate therapeutic strategies is dependent upon knowing the type of tumor affecting a given patient. In addition to assisting with treatment decisions, the tumor classification and grade provide important information regarding prognosis. This chapter will follow the World Health Organization (WHO) classification that separates nervous system tumors into different nosological entities and assigns a grade of I to IV to each lesion (Table 2.1), with grade I being biologically indolent and grade IV being biologically most malignant and having the worst

prognosis [7,8]. Within the WHO classification, tumors of neuroepithelial and meningeal origin contain the two largest and most clinically relevant groups of neoplasms.

Tumors of neuroepithelial origin comprise a large and diverse group of neoplasms, with a mixture of slowly growing and malignant tumor types (Table 2.2) [7–9]. Gliomas (e.g. GBM, AA, oligodendrogliomas, medulloblastoma) are the largest subgroup within the neuroepithelial class of neoplasms and are also the most common type of PBT. Tumors of neuroepithelial origin, and gliomas in particular, can grow diffusely within the brain or be more circumscribed. Diffusely growing tumors are most common and include the astrocytomas, oligodendrogliomas and mixed oligoastrocytomas. Any of these subtypes can undergo malignant transformation and degenerate into the most aggressive form of glioma, the GBM.

Diffuse astrocytoma

The current WHO classification divides astrocytomas into diffuse and localized varieties (Table 2.3) [7–9]. The diffuse astrocytomas are intrinsically invasive and often travel along white matter tracts deep into normal brain. There are three groups of diffuse astrocytic neoplasms: astrocytoma (WHO grade II; peak age of 30–39 years), AA (WHO grade III; peak age of 40–49 years) and GBM (WHO grade

TABLE 2-1 WHO Classification: tumors of the central nervous system
Tumors of neuroepithelial tissue
Tumors of cranial nerves and spinal nerves
Tumors of the meninges
Lymphomas and hemopoietic neoplasms
Germ cell tumors
Tumors of the sellar region
Cysts and tumor-like lesions
Metastatic tumors

TABLE 2-2 WHO Classification: tumors of neuroepithelial tissue
Astrocytic tumors
Oligodendroglial tumors
Ependymal tumors
Mixed gliomas
Choroid plexus tumors
Neuronal and mixed neuronal-glial tumors
Pineal parenchymal tumors
Neuroepithelial tumors of uncertain origin
Embryonal tumors

TABLE 2-3	WHO Classification: astrocytic tumors

Diffuse astrocytomas
    Astrocytoma (WHO grade II)
        Fibrillary
        Protoplasmic
        Gemistocytic
    Anaplastic astrocytoma (WHO grade III)
    Glioblastoma multiforme (WHO grade IV)
        Giant cell glioblastoma
        Gliosarcoma
Localized astrocytomas (WHO grade I)
    Pilocytic astrocytoma
    Pleomorphic xanthoastrocytoma
    Subependymal giant cell astrocytoma

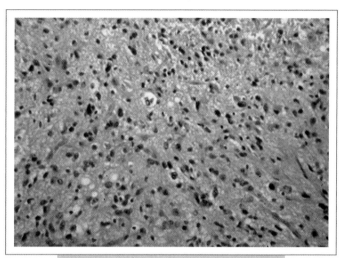

FIG. 2.2. WHO grade III fibrillary astrocytoma (AA). The tumor is more densely cellular than grade II, with significant cellular and nuclear pleomorphism and atypia. Mitotic figures are evident. H&E @ 200×.

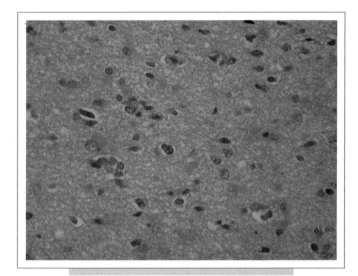

FIG. 2.1. WHO grade II fibrillary astrocytoma. Note the neoplastic astrocytes in a fibrillary matrix, with mildly increased cellularity and pleomorphism. No mitoses or hypervascularity are present. H&E @ 200×.

IV; peak age of 50–69 years). Diffuse astrocytic tumors can be divided into fibrillary, protoplasmic and gemistocytic forms, with the fibrillary form being most common. The presence of gemistocytic and protoplasmic cellular variations are most often seen in WHO grade II tumors. WHO grade II astrocytomas are considered low-grade tumors and usually occur in the cerebral white matter. These tumors are characterized by a relatively uniform population of proliferating neoplastic astrocytes in a fibrillary matrix, with minimal cellular and nuclear pleomorphism or atypia (Figure 2.1). Tumor margins are poorly delineated and suggest significant infiltration into surrounding brain. Mitotic figures are absent and there is no evidence for vascular hyperplasia. Microcystic change is commonly noted in all variants of grade II astrocytoma. The Ki-67 labeling index of WHO grade II astrocytomas is typically less than 4%, with a mean of approximately 2.0–2.5%.

Higher-grade diffuse astrocytomas include AA (WHO grade III) and GBM (WHO grade IV), as well as the GBM variants giant cell glioblastoma and gliosarcoma (WHO grade IV) (see Table 2.3) [7–10]. Anaplastic astrocytomas are similar to grade II tumors, except for the presence of more prominent cellular and nuclear pleomorphism and atypia and mitotic activity (Figure 2.2). In addition, grade III and IV tumors usually do not stain as intensely or as homogeneously with glial fibrillary acidic protein (GFAP). According to WHO criteria, the critical feature that upgrades a grade II tumor to an AA is the presence of mitotic activity, with anaplastic tumors having Ki-67 indices in the range of 5–10% in most cases. Other features of anaplasia can be present, such as multinucleated tumor cells, abnormal mitotic figures and regions of vascular proliferation. Necrosis is absent in grade III astrocytomas.

Glioblastoma multiforme is classified as a WHO grade IV tumor and has similar histological features to AA, but with more pronounced anaplasia (Figure 2.3A) [7–10]. The presence of microvascular proliferation and/or necrosis in an otherwise malignant astrocytoma upgrades the tumor to a GBM. Vascular proliferation is defined as blood vessels with 'piling up' of endothelial cells, including the formation of glomeruloid vessels (Figure 2.3B). The glomeruloid vessels can form undulating garlands that surround necrotic zones in some cases. Necrosis can be noted in large amorphous areas, which appear ischemic in nature, or can appear as more serpiginous regions with surrounding palisading tumor cells (i.e. perinecrotic pseudopalisading; Figure 2.3C). Necrosis with nuclear pseudopalisading is essentially pathognomonic for GBM. Other features of GBM that are typically prominent include marked cellular and nuclear pleomorphism and atypia, mitotic figures

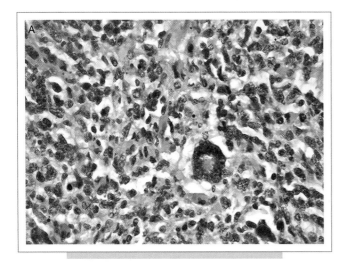

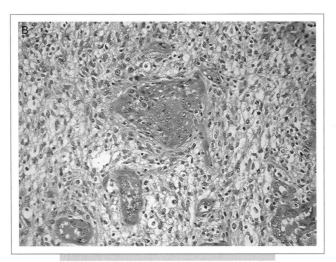

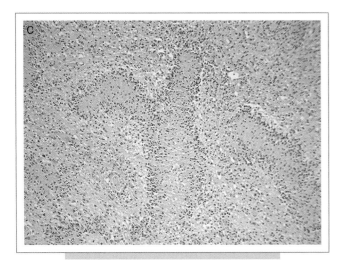

FIG. 2.3. WHO grade IV fibrillary astrocytoma (GBM). **(A)** A highly cellular tumor with marked cellular and nuclear pleomorphism, numerous mitoses, giant cells (H&E @ 400×); **(B)** dense vascular proliferation (H&E @ 200×) and **(C)** regions of necrosis with pseudopallisading tumor nuclei (H&E @ 100×).

and multinucleated giant cells, and pronounced infiltrative capacity into surrounding brain. Labeling indices with Ki-67 are usually in the range of 15–20%, but can be much higher in some tumors.

Localized astrocytomas

In the WHO classification, the localized astrocytomas include the pilocytic astrocytoma (WHO grade I), pleomorphic xanthoastrocytoma (PXA; WHO grade II) and the subependymal giant cell astrocytoma (WHO grade I) [7–10]. Pilocytic astrocytomas are slow growing, relatively circumscribed tumors that usually occur in children (peak age 10–12 years) and young adults. These tumors have a predilection for the cerebellum, optic nerves and optic pathways, and hypothalamus. The distinctive histological feature is the presence of cells with slender, elongated nuclei and thin, hair-like (i.e. piloid), GFAP-positive, bipolar processes. These cells are found in a biphasic background, which consists of dense fibrillary regions alternating with loose, microcystic areas. Labeling index studies with Ki-67 report values of 0.5–1.5% in most tumors. The PXA is a supratentorial tumor with a predilection for the superficial temporal lobes that usually occurs in younger patients (mean age 15–18 years) with a longstanding history of seizure activity [7–10]. On histological examination, PXA demonstrates significant pleomorphism, with numerous atypical giant cells and astrocytes with prominent nucleoli [11]. Also present are large foamy (xanthomatous) cells with lipidized cytoplasm that express GFAP. Subependymal giant cell astrocytoma is an indolent, slowly growing tumor that typically arises in the walls of the lateral ventricles and is almost invariably associated with tuberous sclerosis [7–10].

Oligodendrogliomas and oligoastrocytomas

Oligodendrogliomas are a form of diffuse glioma that can be of pure or mixed histology and are classified as WHO grade II or III [7–10]. They typically occur in young to middle-aged adults (peak age 35–45 years) with a history of seizures, within the white matter of the frontal and temporal lobes. Pure low-grade oligodendroglial tumors (WHO grade II) are characterized histologically by a moderately cellular, monotonous pattern of cells with round nuclei and perinuclear halos (the classic 'fried egg' appearance; Figure 2.4) [12]. The perinuclear halos are an artifact of the formalin fixation process of the tumor tissue. Foci of calcification are frequent and can be quite dense in some cases. Delicately branching blood vessels are prominent (i.e. 'chicken-wire' vasculature), but do not display endothelial proliferation. Oligodendrogliomas have a pronounced invasive capacity and are known to invade the gray and white matter diffusely, with a strong tendency to form secondary structures of Scherer, in particular perineuronal satellitosis. Mitoses are absent or rare and necrosis is not present. Labeling studies with Ki-67 usually demonstrate indices less than 5%, with a mean of approximately 2%. The

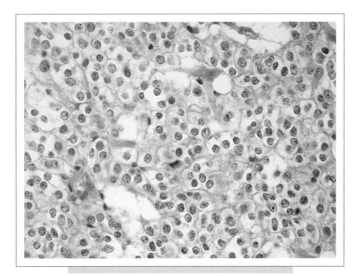

FIG. 2.4. WHO grade II oligodendroglioma. Demonstrates the classic features of typical oligodendroglioma, with moderate cellularity and numerous round cells with the 'fried egg' pattern of perinuclear halos and delicate 'chicken wire' vasculature. H&E @ 400×.

diagnosis of an anaplastic oligodendroglioma (WHO grade III) requires the presence of additional histologic features, including a higher degree of cellularity and mitotic activity, vascular endothelial hyperplasia, nuclear pleomorphism and regions of necrosis (Figures 2.5A and B). These tumors behave in a more aggressive fashion, with a higher proliferative rate (Ki-67 labeling index >5%) and capacity for invasion of surrounding brain. Mixed oligoastrocytomas can be classified as WHO grade II or III tumors [7–10]. Distinct populations of neoplastic oligodendroglial cells and astrocytes can be identified within the mass that have similar features to pure versions of the tumor. The percentage of each cell population can be quite variable, with an even mixture of cell types or with one cell type predominating.

Advances in molecular neuropathology have begun to clarify the biological underpinnings of variability in response to treatment of oligodendrogliomas [12–14]. The majority of tumors demonstrate genetic losses on chromosome 1p (40–92%) and/or 19q (50–80%). There is a strong predilection for deletions of 1p and 19q to occur together but, in some tumors, they can be singular events. Patients with oligodendrogliomas that contain deletions of 1p and 19q are consistently more responsive to irradiation and chemotherapy and have an overall median survival of 8–10 years. In contrast, patients with tumors that do not have deletion of 1p and 19q are more resistant to all forms of therapy and have an overall median survival of only 3–4 years.

Medulloblastoma and other embryonal tumors

Embryonal tumors are a group of aggressive, malignant neoplasms that usually affect children. They are classified by the WHO as grade IV in all cases (Table 2.4) [7–10]. All

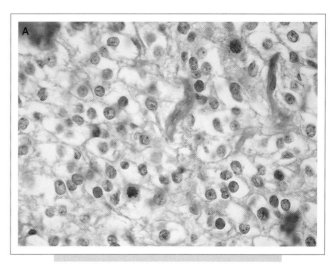

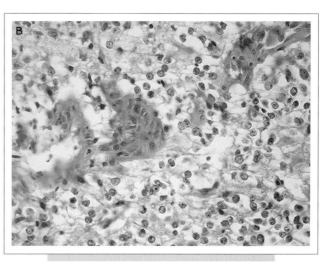

FIG. 2.5. WHO grade III oligodendroglioma. A more densely cellular tumor with **(A)** prominent cellular and nuclear pleomorphism, mitotic activity and **(B)** increased vascularity. H&E @ 400×.

TABLE 2-4 WHO Classification: embryonal tumors
Medulloepithelioma
Ependymoblastoma
Medulloblastoma
Desmoplastic medulloblastoma
Large cell medulloblastoma
Medullomyoblastoma
Melanotic medulloblastoma
Supratentorial PNET
Neuroblastoma
Ganglioneuroblastoma
Atypical teratoid/rhabdoid tumor

embryonal tumors share the common features of high cellularity, frequent mitoses, regions of necrosis and a propensity for metastases along CSF pathways. Medulloblastoma is the most common of the embryonal tumors and is

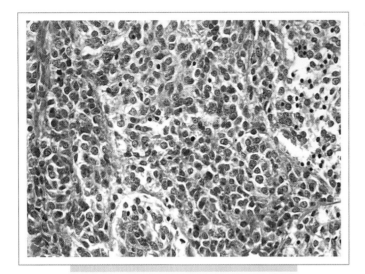

FIG. 2.6. WHO grade IV medulloblastoma. Note the dense cellularity and presence of undifferentiated cells with hyperchromatic, oval to carrot-shaped nuclei with scant cytoplasm. The nuclei have a tendency to mold against one another. H&E @ 200×.

TABLE 2-5 WHO Classification: tumors of the meninges
Tumors of meningothelial cells
Meningioma
Meningothelial
Fibrous (fibroblastic)
Transitional (mixed)
Psammomatous
Angiomatous
Microcystic
Secretory
Lymphoplasmacyte-rich
Metaplastic
Clear cell
Chordoid
Atypical
Papillary
Rhabdoid
Anaplastic meningioma
Mesenchymal, non-meningothelial tumors
Lipoma
Angiolipoma
Hibernoma
Liposarcoma (intracranial)
Solitary fibrous tumor
Fibrosarcoma
Malignant fibrous histiocytoma
Leiomyoma
Leiomyosarcoma
Rhabdomyoma
Rhabdomyosarcoma
Chondroma
Chondrosarcoma
Osteoma
Osteosarcoma
Osteochondroma
Hemangioma
Epithelioid hemangioendothelioma
Hemangiopericytoma
Angiosarcoma
Kaposi sarcoma
Primary melanocytic lesions
Diffuse melanocytosis
Melanocytoma
Malignant melanoma
Meningeal melanomatosis

considered a primitive neuroectodermal tumor (PNET) of the cerebellum. It usually arises in the midline in children, within the cerebellar vermis, while in adults it is more likely to have an off-center location within the cerebellar hemispheres. The typical medulloblastoma is densely cellular and composed of undifferentiated cells with hyperchromatic, oval to carrot-shaped nuclei with scant cytoplasm (Figure 2.6) [9,15]. The nuclei have a tendency to mold against one another. Mitoses and single cell necrosis are frequently present. Evidence of anaplasia is variable and may include increased nuclear size, abundant mitoses and the presence of large-cell or similar aggressive cellular morphology. Some tumors may display immunohistochemical and morphological evidence for differentiation along neuronal, glial or mesenchymal lines. Medulloblastomas are highly proliferative tumors, with Ki-67 labeling indices ranging from 15 to 50%.

Meningioma and other tumors of the meninges

Tumors of the meninges comprise a large and diverse group of neoplasms that mostly have meningothelial or mesenchymal, non-meningothelial origins (Table 2.5) [8,9]. The most common primary tumor of this group is the meningioma (18–20% of intracranial tumors), which has meningothelial cell origins and is composed of neoplastic arachnoidal cap cells of the arachnoidal villi and granulations. Meningiomas can occur anywhere within the intracranial cavity, but favor the sagittal area along the superior longitudinal sinus, over the lateral cerebral convexities, at the tuberculum sellae and parasellar region, the sphenoidal ridge and along the olfactory grooves. Numerous histologic variants of meningioma are described and recognized by the WHO (Table 2.5). However, the histopathological

description of most of these variants has no bearing upon the clinical behavior of the tumor. Meningioma subtypes that have a more indolent nature and low risk for aggressive growth or recurrence are classified as WHO grade I and include the meningothelial, fibrous/fibroblastic, transitional (mixed), secretory, psammomatous, angiomatous, microcystic, lymphoplasmocyte-rich and metaplastic variants [16,17]. Of this group, the meningothelial, fibrous and transitional variants are most frequently diagnosed. The

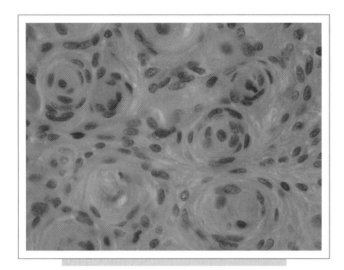

FIG. 2.7. WHO grade II meningioma. The tumor demonstrates a moderately dense, uniform pattern of cells with oval shaped nuclei and the presence of many cellular whorl patterns. H&E @ 200×.

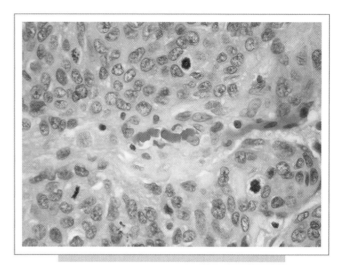

FIG. 2.8. WHO grade III anaplastic meningioma. This higher power view demonstrates increased nuclear pleomorphism and frequent mitoses. H&E @ 400×.

histological features common to most low-grade meningiomas are the presence of whorls (tightly wound, rounded collections of cells), psammoma bodies (concentrically laminated mineral deposits that often begin in the center of whorls), intranuclear pseudoinclusions (areas in which pink cytoplasm protrudes into a nucleus to produce a hollowed-out appearance) and occasional pleomorphic nuclei and mitoses (Figure 2.7) [16,17]. Meningothelial meningiomas are composed of lobules of typical meningioma cells, with minimal whorl formation. The tumor cells are uniform in shape, with oval nuclei that may show central clearing. Fibrous variants have spindle-shaped cells resembling fibroblasts that form parallel and interlacing bundles within a matrix of collagen and reticulin.

Meningioma subtypes that are more likely to display aggressive clinical behavior and to recur are classified by the WHO as grade II (atypical, clear cell, chordoid) and grade III (rhabdoid, papillary, anaplastic) [8,9,16,17]. On histological examination, all of the grade II tumors are likely to demonstrate increased cellularity, more frequent mitoses, diffuse or sheet-like growth, nuclear pleomorphism and atypia and evidence for micronecrosis. Grade III tumors, such as anaplastic meningioma, show features consistent with frank malignancy, including a high mitotic rate, advanced cytological atypia, nuclear pleomorphism and necrosis (Figure 2.8). Invasion of underlying brain is frequently noted in grade III meningiomas, but can also occur in lower grade variants. Proliferation studies using Ki-67 demonstrate labeling indices ranging from 8 to 15%.

Primary central nervous system lymphoma

Primary CNS lymphomas (PCNSL) are malignant tumors classified as WHO grade IV, that affect adults in the

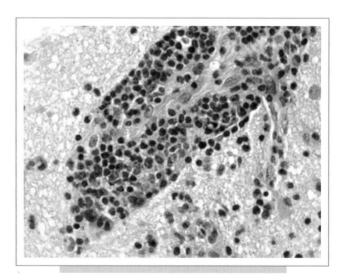

FIG. 2.9. WHO grade IV primary CNS lymphoma (PCNSL). Note the presence of neoplastic lymphocytes in an angiocentric growth pattern, with nuclear pleomorphism and mitoses. H&E @ 100×.

sixth and seventh decades of life [8,9]. They are often multifocal and usually arise in the deep supratentorial white matter, with a predilection for the periventricular region and basal ganglia. PCNSL are composed of a clonal expansion of neoplastic lymphocytes, typically of the diffuse, large cell or immunoblastic variety. In 95% of the tumors, the cells have a B-cell lineage, often with monoclonal IgM kappa production. On histological examination, PCNSL display a perivascular cellular orientation, with expansion of vessel walls and reticulin deposition (Figure 2.9) [18].

Regions of necrosis are common, especially if steroids have been administered prior to the biopsy. The lymphomatous cells are non-cohesive and usually have large, irregular nuclei, prominent nucleoli and scant cytoplasm. From the perivascular region, tumor cells are noted to invade the surrounding brain parenchyma, either in compact cellular aggregates or as singly infiltrating tumor cells. PCNSL are highly proliferative tumors, with Ki-67 labeling indices ranging from 20 to 50% in most studies. The diagnosis can be confirmed by immunohistochemical positivity for leukocyte common antigen (CD45) and specific B-cell markers (CD19, CD20, and CD79a).

SURGICAL THERAPY OF PRIMARY BRAIN TUMORS

Surgical intervention is the most common form of treatment for PBT and is an important aspect of initial therapy in most patients. Indications for surgery include reduction of tumor burden, alleviation of mass effect, control of seizures and reversal of neurological deficit, confirmation of the histological diagnosis, diversion of cerebrospinal fluid (CSF) by shunting procedures in selected cases, and the introduction of local antineoplastic agents [1,19,20]. Recent advances in neurosurgical technology offer new approaches to tumor removal, such as frame-based and frameless stereotactic biopsy, preoperative functional magnetic resonance imaging (MRI) and intraoperative cortical mapping, neuronavigation and tumor resection in the awake patient and the use of intraoperative MRI [21–23]. These techniques allow the surgeon to delineate more carefully tumor margins and to preserve surrounding regions of eloquent brain (e.g. Broca's area, primary motor cortex) and delicate vascular structures, while performing a more aggressive and thorough tumor resection. Complete removal of benign tumors such as meningioma, pilocytic astrocytoma and schwannomas can be curative. For malignant tumors (i.e. GBM, AA), although the lack of a prospective, controlled, randomized clinical trial still fosters debate in the literature, most neurosurgeons recommend a near total or gross total resection, whenever possible, of all enhancing tumor volume and regionally infiltrated brain as defined on T2-weighted or fluid attenuated inversion-recovery (FLAIR) MRI sequences. Gross total tumor resection is not curative for these tumor types, but has been associated with longer overall and progression-free survival in several studies, as well as improved neurological quality of life [24,25]. For tumors that are diffusely infiltrative or multifocal, a stereotactic biopsy is more likely to preserve neurological function than an attempt at resection and, in most cases, will be able to provide a histological diagnosis to guide further treatment. The accuracy of stereotactic biopsy is further improved when the region of interest is defined by contrast enhancement on MRI or abnormal signal on MRI spectroscopy or positron emission tomography (PET).

RADIATION THERAPY OF PRIMARY BRAIN TUMORS

External beam fractionated radiation therapy is an appropriate form of treatment for virtually all patients with high-grade gliomas (i.e. GBM, AA, AO, medulloblastoma), as well as for selected low-grade PBT that are surgically inaccessible or have progressed following initial resection [1,26–29]. Numerous randomized controlled trials have demonstrated a survival benefit for high-grade glioma patients receiving surgical resection and irradiation in comparison to resection alone (approximately 34–38 weeks versus 14–18 weeks, respectively). The mechanism of cell death appears to be the production of DNA strand damage by ionizing radiation and the generation of highly reactive oxygen radicals that induce further DNA damage and disrupt cellular processes. Sublethal or mortal damage to endothelial cells in tumor vessels may also be of importance. The standard approach is administered in the early postoperative phase and initially uses conformal radiation ports that encompass the T2-weighted target with a margin of 1–3 cm, using a dose of approximately 4500–4700 cGy in 180–200 cGy daily fractions. After this portion has been completed, a 'cone down' is performed, targeting the T1-weighted contrast-enhancing volume of the tumor with a 1–3 cm margin, bringing the total dose to approximately 6000 cGy. Irradiation is performed over the course of 6–7 weeks, with the patient receiving treatment 5 days per week. Radiation therapy schedules can sometimes be modified with hypofractionation and/or an abbreviated treatment course for elderly patients or for those with a low performance status, while maintaining a similar level of toxicity and overall survival [30,31]. More aggressive approaches to irradiation using hyperfractionation schemes have not been shown to improve tumor control and, in some reports, have been associated with worse outcomes [28]. Other techniques to increase localized radiation doses to the tumor resection cavity, such as brachytherapy with permanent or temporary radioactive seeds, have also had disappointing results in controlled trials [32]. In addition to the cranial dosage, spinal-axis RT is necessary for tumors that often seed the meninges, such as medulloblastoma, pineoblastoma and anaplastic ependymoma.

Stereotactic radiosurgery (SRS), using a linear accelerator-based system (e.g. Cyberknife®) or a Co^60-based system (e.g. Gamma Knife®) to deliver a single (or a few) high-dose radiation fraction(s) to a defined volume using stereotactic localization, is another method to boost radiation doses in the tumor bed of a newly diagnosed or recurrent glioma, while sparing normal surrounding tissues [33–35]. Because of the diffuse, infiltrative nature of the growth pattern of these tumors, the application of focal treatment modalities such as radiosurgery remains controversial. Retrospective and single-armed, uncontrolled prospective trials suggest an improvement in local tumor control rates and survival when using either radiosurgical system. However, these results were not confirmed in the randomized, controlled trial of radiosurgery for GBM recently reported by the Radiation Therapy Oncology Group (RTOG 93-05) [35,36]. In this study, 203 GBM patients were randomized to receive radiosurgery followed by conventional irradiation (60 Gy) and intravenous carmustine chemotherapy (80 mg/m² /day × 3 days every 8 weeks) or irradiation plus chemotherapy alone. The median survival for the radiosurgical and conventional treatment groups were 13.5 months and 13.6 months, respectively ($P = 0.5711$). In addition, the 2- and 3-year survival rates and patterns of failure were similar between groups. There was no difference in general quality of life or retention of cognitive function between groups.

CHEMOTHERAPY OF PRIMARY BRAIN TUMORS

Chemotherapy is used as an adjunctive treatment for malignant PBT (i.e. mostly high-grade gliomas – GBM, AA) and for selected low-grade gliomas that progress through initial surgical resection and irradiation [1,37–39]. The addition of chemotherapy has resulted in modest improvements in survival of patients with malignant glioma, as demonstrated by two detailed meta-analyses [40,41]. Over the past two decades and until recently, nitrosourea alkylating drugs such as carmustine and lomustine (BCNU and CCNU, respectively), were considered the most effective chemotherapeutic agents for these tumors [37]. Other agents with mild activity included procarbazine (administered alone or in combination with CCNU and vincristine; i.e. PCV), cisplatin, etoposide, carboplatin and cyclophosphamide. For the treatment of primary CNS lymphoma, methotrexate has been shown to be the most active agent, either alone or in combination with other drugs (e.g. cytarabine, rituximab) [42].

More recent studies have focused on the second generation alkylating agent, temozolomide (TZM), which has an activity profile superior to nitrosoureas and other agents. Temozolomide is an imidazotetrazine derivative of the alkylating agent dacarbazine, with activity against systemic and CNS malignancies [37,43–46]. The drug undergoes chemical conversion at physiological pH to the active species 5-(3-methyl-1-triazeno)imidazole-4-carboxamide (MTIC). Temozolomide exhibits schedule dependent antineoplastic activity by interfering with DNA replication through the process of methylation. The methylation of DNA is dependent upon formation of a reactive methyldiazonium cation, which interacts with DNA at the following sites: N^7-guanine (70%), N^3-adenine (9.2%) and O^6-guanine (5%). Because TZM is stable at acid pH, it can be taken orally in capsules. Oral bioavailability is approximately 100%, with rapid absorption of the drug. In addition, TZM has excellent penetration of the blood–brain barrier and brain tumor tissue.

Initial studies of TZM were in patients with recurrent AA and GBM, and suggested significant activity [37,43–46].

Subsequent larger studies demonstrated unequivocal efficacy in the recurrent setting. The first study evaluated the use of TZM (150–200 mg/m² /day × 5 days every 28 days) in a series of 162 patients with recurrent malignant gliomas, including 97 patients with AA [47]. In the AA cohort, there were six patients with complete response (CR), 27 with partial response (PR) and another 31 with stable disease (SD) (CR+PR+SD = 66%). The response rate was similar in patients that had failed prior chemotherapy or were chemotherapy-naïve. Median overall progression free survival (PFS) was 5.4 months, with 6- and 12-month PFS rates of 46% and 24%, respectively. The median overall survival was 13.6 months, with 6- and 12-months survival rates of 75% and 56%, respectively. A similar comparative phase II trial evaluated the activity of TZM versus procarbazine (125–150 mg/m² /day × 28 days every 8 weeks) in a cohort of 225 patients with GBM at first relapse [48]. Overall response rates (PR+SD) were significantly higher for patients in the TZM cohort (45.6% versus 32.7%; $P = 0.049$). Treatment with TZM resulted in a significant improvement in median PFS (12.4 weeks versus 8.32 weeks; $P = 0.0063$) and 6-month PFS (21% versus 8%; $P = 0.008$) in comparison to procarbazine. In addition, the 6-month overall survival rate was significantly higher for patients in the TZM arm of the study (60% versus 44%; $P = 0.019$).

Temozolomide has also been applied to GBM patients in the 'up-front' setting by Stupp and colleagues in a set of phase II and III studies, in combination with standard external beam irradiation and monthly adjuvant chemotherapy [37,49,50]. For the phase III study, a total of 573 patients were randomly assigned to receive radiation alone (6000 cGy; 200 cGy/day × 5 days/week for 6 weeks) or radiotherapy in combination with daily TZM (75 mg/m² /day × 7 days/week for 6 weeks) [50]. After the completion of irradiation, each patient in the chemotherapy arm went on to receive six cycles of adjuvant single-agent TZM (200 mg/m² /day × 5 days, every 28 days). The overall median survival was 14.6 months for the radiotherapy plus TZM cohort and 12.1 months for the cohort that received irradiation alone, for an overall median survival benefit of 2.5 months. The unadjusted hazard ratio for death due to the GBM for the radiotherapy plus TZM cohort was 0.63 ($P < 0.001$, log-rank test). The 2-year survival rate was 26.5% for the chemoradiation cohort and 10.4% for the cohort receiving radiotherapy alone. Chemoradiation using low-dose TZM, followed by adjuvant monthly TZM, has become the 'standard of care' for newly diagnosed GBM patients.

MOLECULAR OR 'TARGETED' TREATMENT

As noted above, conventional chemotherapeutic approaches to treatment for malignant glioma are not predicated on the biology of the malignant phenotype. It has become apparent that the transformed phenotype of brain tumor cells is highly complex and results from the dysfunction of a variety of inter-related regulatory pathways [51–53].

The transformation process involves amplification or overexpression of oncogenes in combination with loss or lack of expression of tumor suppressor genes. Oncogenes and signal transduction molecules that have been demonstrated to be important for gliomagenesis include platelet-derived growth factor and its receptor (PDGF, PDGFR), epidermal growth factor and its receptor (EGF, EGFR), CDK4, mdm-2, Ras, phosphoinositol-3 kinase (PI3K), Akt, and mTOR (mammalian target of rapamycin). Tumor suppressor genes of importance in glial transformation include p53, retinoblastoma (Rb), p16 and p15 (i.e. INK4a, INK4b), DMBT1 and PTEN. Most of these tumor suppressor genes function as negative regulators of the cell cycle, while others are inhibitors of important internal signal transduction pathways. The net effect of these acquired abnormalities is dysregulation of, and an imbalance between, the activity of the cell cycle and apoptotic pathways.

Because the survival of patients with high-grade gliomas has remained so poor using conventional chemotherapeutic approaches, new treatment modalities are being investigated that have a more molecular, 'targeted' mechanism of action, with the ability to overcome the transformed phenotype [54–58]. Recent advances in growth factor and signal transduction biology are now providing the background for the development of 'molecular therapeutics', a new class of drugs that manipulate and exploit these pathways [59]. Molecular drugs targeting critical signal transduction pathway effectors, such as PDGFR, EGFR, Ras, PI3K and mTOR, have entered early clinical trials in brain tumor patients [54–58]. Preliminary results suggest only modest activity against recurrent high-grade gliomas when used as single agents. Subsequent clinical trials will investigate using molecular drugs in combination with conventional chemotherapeutic agents (e.g. TZM, hydroxyurea), with other molecular drugs that target different signal transduction pathways, and with irradiation.

ACKNOWLEDGMENTS

The authors would like to thank Julia Shekunov for research assistance. Dr Newton was supported in part by National Cancer Institute grant, CA 16058 and the Dardinger Neuro-Oncology Center Endowment Fund.

REFERENCES

1. Newton HB (1994). Primary brain tumors: review of etiology, diagnosis, and treatment. Am Fam Phys 49:787–797.
2. Preston-Martin S (1996). Epidemiology of primary CNS neoplasms. Neurol Clin 14:273–290.
3. Davis FG, McCarthy BJ (2001). Current epidemiological trends and surveillance issues in brain tumors. Expert Rev Anticancer Ther 1:395–401.
4. Osborne RH, Houben MPWA, Tijssen CC, Coebergh JWW, van Duijn CM (2001). The genetic epidemiology of glioma. Neurology 57:1751–1755.
5. Wrensch M, Minn Y, Chew T, Bondy M, Berger MS (2002). Epidemiology of primary brain tumors: current concepts and review of the literature. Neuro-Oncology 4:278–299.
6. Hess KR, Broglio DR, Bondy ML (2004). Adult glioma incidence trends in the United States, 1977–2000. Cancer 101: 2293–2299.
7. Kleihues P, Cavanee WK (2000). *Pathology and Genetics of Tumours of the Nervous System*. International Agency for Research on Cancer, Lyon.
8. Kleihues P, Louis DN, Scheithauer BW et al (2002). The WHO classification of tumors of the nervous system. J Neuropathol Exp Neurol 61:215–225.
9. McLendon RE, Enterline DS, Tien RD, Thorstad WL, Bruner JM (1998). Tumors of central neuroepithelial origin. In *Russell & Rubinstein's Pathology of Tumors of the Nervous System*, 6th edn, Volume 1. Bigner DD, McLendon RE, Bruner JM (eds). Arnold, London. 9:307–571.
10. von Deimling A, Louis DN, Wiestler OD (1995). Molecular pathways in the formation of gliomas. Glia 15:128–138.
11. Kepes JJ (1993). Pleomorphic xanthoastrocytoma: the birth of a diagnosis and a concept. Brain Pathol 3:269–274.
12. Engelhard HH, Stelea A, Cochran EJ (2002). Oligodendrglioma: pathology and molecular biology. Surg Neurol 58:111–117.
13. Reifenberger G, Louis DN (2003). Oligodendroglioma: toward molecular definitions in diagnostic neuro-oncology. J Neuropathol Exp Neurol 62:111–126.
14. Jeuken JWM, von Deimling A, Wesseling P (2004). Molecular pathogenesis of oligodendroglial tumors. J Neuro-Oncol 70:161–181.
15. Eberhart CG, Burger PC (2003). Anaplasia and grading in medulloblastomas. Brain Pathol 13:376–385.
16. Lamszus K (2004). Meningioma pathology, genetics, and biology. J Neuropathol Exp Neurol 63:275–286.
17. Perry A, Gutmann DH, Reifenberger G (2004). Molecular pathogenesis of meningiomas. J Neuro-Oncol 70:183–202.
18. Jellinger KA, Paulus W (1992). Primary central nervous system lymphomas – an update. J Cancer Res Clin Oncol 119:7–27.
19. Salcman M (1990). Malignant glioma management. Neurosurg Clin N Am 1:49–64.
20. Dunn IF, Black PM (2003). The neurosurgeon as local oncologist: cellular and molecular neurosurgery in malignant glioma therapy. Neurosurg 52:1411–1424.

21. Matz PG, Cobbs C, Berger MS (1999). Intraoperative cortical mapping as a guide to the surgical resection of gliomas. J Neuro-Oncol 42:233–245.

22. Lemole GM, Henn JS, Riina HA, Spetzler RF (2001). Cranial application of frameless stereotaxy. BNI Quart 17:16–24.

23. Nimsky C, Ganslandt O, Kober H, Buchfelder M, Fahlbusch R (2001). Intraoperative magnetic resonance imaging combined with neuronavigation: a new concept. Neurosurgery 48:1082–1091.

24. DeVaux BC, O'Fallon JR, Kelly PJ (2001). Resection, biopsy, and survival in malignant glial neoplasms. A retrospective study of clinical parameters, therapy, and outcome. J Neurosurg 78:767–775.

25. Lacroix M, Abi-Said D, Fourney DR et al (2001). A multivariate analysis of 416 patients with glioblastoma multiforme: prognosis, extent of resection, and survival. J Neurosurg 95:190–198.

26. Shrieve DC, Loeffler JS (1995). Advances in radiation therapy for brain tumors. Neurol Clin 13:773–794.

27. Miyamoto C (2001). Radiation therapy principles for high-grade gliomas. In *Combined Modality Therapy of Central Nervous System Tumors*. Petrovich Z, Brady LW, Apuzzo ML, Bamberg M (eds). Springer, Berlin. 18:345–364.

28. Laperriere N, Zuraw L, Cairncross JG (2002). Radiotherapy for newly diagnosed malignant glioma in adults: a systematic review. Radiother Oncol 64:259–273.

29. Kortmann RD, Jeremic B, Bamberg M (2001). Radiotherapy in the management of low-grade gliomas. In *Combined Modality Therapy of Central Nervous System Tumors*. Petrovich Z, Brady LW, Apuzzo ML, Bamberg M (eds). Springer, Berlin. 16:317–326.

30. Chang EL, Yi W, Allen PK, Levin VA, Sawaya RE, Maor MH (2003). Hypofractionated radiotherapy for elderly or younger low-performance status glioblastoma patients: outcome and prognostic factors. Int J Rad Oncol Biol Phys 56:519–528.

31. Roa W, Brasher PMA, Bauman G et al (2004). Abbreviated course of radiation therapy in older patients with glioblastoma multiforme: a prospective randomized clinical trial. J Clin Oncol 22:1583–1588.

32. Selker RG, Shapiro WR, Burger P et al (2002). The Brain Tumor Cooperative Group NIH trial 87-01: a randomized comparison of surgery, external radiotherapy, and carmustine versus surgery, interstitial radiotherapy boost, external radiation therapy, and carmustine. Neurosurgery 51:343–357.

33. Nwokedi E, DiBiase SJ, Jabbour S, Herman J, Amin P, Chin LS (2002). Gamma knife stereotactic radiosurgery for patients with glioblastoma multiforme. Neurosurgery 50:41–47.

34. Prisco FE, Weltman E, Hanriot RM, Brandt RA (2002). Radiosurgical boost for primary high-grade gliomas. J Neuro-Oncol 57:151–160.

35. Tsao MN, Mehta MP, Whelan TJ et al (2005). The American Society for Therapeutic Radiology and Oncology (ASTRO) evidence-based review of the role of radiosurgery for malignant glioma. Int J Rad Oncol Biol Phys 63:47–55.

36. Souhami L, Seiferheld W, Brachman D et al (2004). Randomized comparison of stereotactic radiosurgery followed by conventional radiotherapy with carmustine to conventional radiotherapy with carmustine for patients with glioblastoma multiforme: report of Radiation Therapy Oncology Group 93-05 protocol. Int J Rad Oncol Biol Phys 60:853–860.

37. Newton HB (2006). Chemotherapy of high-grade astrocytomas. In *Handbook of Brain Tumor Chemotherapy*. Newton HB (ed.). Elsevier Medical Publishers-Academic Press, London. 24:347–363.

38. Lesser GJ (2001). Chemotherapy of low-grade gliomas. Sem Radiat Oncol 11:138–144.

39. Malkin MG (2006). Chemotherapy of low-grade astrocytomas. In *Handbook of Brain Tumor Chemotherapy*. Newton HB (ed.). Elsevier Medical Publishers-Academic Press, London. 25:364–370.

40. Fine HA, Dear KBG, Loeffler JS, Black PM, Canellos GP (1993). Meta-analysis of radiation therapy with and without adjuvant chemotherapy for malignant gliomas in adults. Cancer 71:2585–2597.

41. Steward LA, Burdett S, Souhami RL, Stenning S (2002). Chemotherapy in adult high-grade glioma: a systematic review and meta-analysis of individual patient data from 12 randomised trials. Lancet 359:1011–1018.

42. El Kamar FG, Abrey LE (2006). Chemotherapy for primary central nervous system lymphoma. In *Handbook of Brain Tumor Chemotherapy*. Newton HB (ed.). Elsevier Medical Publishers-Academic Press, London. 28:395–406.

43. Stevens MFG, Newlands ES (1993). From triazines and triazenes to temozolomide. Eur J Cancer 20A:1045–1047.

44. Newlands ES, O'Reilly SM, Glaser MG et al (1996). The Charing Cross Hospital experience with temozolomide in patients with gliomas. Eur J Cancer 32A:2236–2241.

45. Newlands ES, Stevens MFG, Wedge SR, Wheelhouse RT, Brock C (1997). Temozolomide: a review of its discovery, chemical properties, pre-clinical development and clinical trials. Cancer Treat Rev 23:35–61.

46. Hvizdos KM, Goa KL (1999). Temozolomide. CNS Drugs 12:237–243.

47. Yung WKA, Prados MD, Yaya-Tur R et al (1999). Multicenter phase II trial of temozolomide in patients with anaplastic astrocytoma or anaplastic oligoastrocytoma at first relapse. J Clin Oncol 17:2762–2771.

48. Yung WKA, Albright RE, Olson J et al (2000). A phase II study of temozolomide vs. procarbazine in patients with glioblastoma multiforme at first relapse. Br J Cancer 83:588–593.

49. Stupp R, Dietrich PY, Kraljevic SO et al (2002). Promising survival for patients with newly diagnosed glioblastoma treated with concomitant radiation plus temozolomide followed by adjuvant temozolomide. J Clin Oncol 20:1375–1382.

50. Stupp R, Mason WP, van den Bent MJ et al (2005). Radiotherapy plus concomitant and adjuvant temozolomide for glioblastoma. New Engl J Med 352:987–996.

51. von Deimling A, Louis DN, Wiestler OD (1995). Molecular pathways in the formation of gliomas. Glia 15:328–338.

52. Shapiro JR, Coons SW (1998). Genetics of adult malignant gliomas. BNI Quarterly 14:27–4

53. Maher EA, Furnari FB, Bachoo RM et al (2001). Malignant glioma: genetics and biology of a grave matter. Genes Dev 15:1311–1333.

54. Newton HB (2003). Molecular neuro-oncology and the development of 'targeted' therapeutic strategies for brain tumors. Part 1 – growth factor and ras signaling pathways. Expert Rev Anticancer Ther 3:595–614.

55. Newton HB (2004). Molecular neuro-oncology and the development of 'targeted' therapeutic strategies for brain tumors.

Part 2 – PI3K/Akt/PTEN, mTOR, SHH/PTCH, and angiogenesis. Expert Rev Anticancer Ther 4:105–128.

56. Newton HB (2005). Molecular neuro-oncology and the development of 'targeted' therapeutic strategies for brain tumors. Part 5 – apoptosis and cell cycle. Expert Rev Anticancer Ther 5:355–378.

57. Mischel PS, Cloughesy TF (2004). Targeted molecular therapy of glioblastoma. Brain Pathol 13:52–61.

58. Rich JN, Bigner DD (2004). Development of novel targeted therapies in the treatment of malignant glioma. Nature Rev Drug Discov 3:430–446.

59. Adjei AA, Hidalgo M (2005). Intracellular signal transduction pathway proteins as targets for cancer therapy. J Clin Oncol 23:5386–5403.

CHAPTER 3

Overview of Pathology and Treatment of Metastatic Brain Tumors

Herbert B. Newton, Abhik Ray-Chaudhury and Mark G. Malkin

INTRODUCTION

The epidemiology of metastatic brain tumors (MBT) was reviewed in detail in Chapter 1. In this chapter, we will provide an overview of the classification, pathology and treatment of the common MBT. Metastatic brain tumors arise in 20–40% of all adult cancer patients and are the most common complication of systemic neoplastic disease [1–5]. They will be diagnosed in approximately 150 000–170 000 patients this year in the USA and are associated with significant morbidity and mortality. Of the estimated 8.3–11 patients per 100 000 population that will develop an MBT this year, more than 75% will have underlying primary tumors of the lung, breast and skin (i.e. melanoma) (Table 3.1). However, virtually any primary tumor has the potential to metastasize to the brain.

PATHOLOGY OF METASTATIC BRAIN TUMORS

Systemic tumor cells usually travel to the brain by hematogenous spread through the arterial circulation, often after genetic alterations that produce a more motile and aggressive phenotype [6–11]. The metastasis most often originates from the lung, either from a primary lung tumor or from a pulmonary metastasis. Occasionally, cells reach the brain through Batson's paravertebral venous plexus or by direct extension from adjacent structures (e.g. sinuses, skull). The distribution of brain metastases follows the relative volume of blood flow to each area, so that 80% of tumors arise in the cerebral hemispheres, 15% in the cerebellum and 5% in the brainstem. Tumor cells typically lodge in small vessels at the gray–white junction and then spread into the brain parenchyma, where they proliferate and induce their own blood supply by neoplastic angiogenesis [9]. Expansion of the MBT disrupts the function of adjacent neural tissue through several mechanisms, including direct displacement of brain structures, perilesional edema, irritation of overlying gray matter and compression of arterial and venous vasculature.

The metastatic phenotype is the result of a complex alteration of gene expression that affects tumor cell adhesion, motility, protease activity and internal signaling pathways [8–10]. Initial changes involve downregulation of surface adhesion molecules, such as integrins and cadherins, which reduces cell-to-cell interactions and allows easier mobility through the surrounding extracellular matrix (ECM). Cell motility is also accelerated in response to specific ligands, such as scatter factor and autocrine motility factor [7–10]. Several oncogenes and signal transduction pathways are also commonly activated in these aggressive cells, including members of the *Ras* family, *Src*, *Met*, and downstream molecules such as Raf, MAPK 1/2, Rac/Rho, PI3-kinase and focal adhesion kinase. Cellular invasive capacity is augmented in the metastatic phenotype by increased tumor cell secretion of matrix metalloproteinases (e.g. collagenases, gelatinases) and other enzymes that degrade the ECM [9,10]. In addition, metastatic cells often have downregulated secretion of tissue inhibitors of metalloproteinases (i.e. TIMP-1, TIMP-2), which further enhances their invasive potential and access to the vasculature.

TABLE 3-1 Primary sites of metastatic brain tumors	
Primary tumor	*Percentage (%)*
Lung	50–60
squamous cell	25–30
adenocarcinoma	12–15
small cell	10–13
large cell	2
Breast	15–20
Melanoma	5–10
Gastrointestinal	4–6
Genitourinary	3–5
Unknown	4–8
Other	3–5

Data compiled from references [1–5]

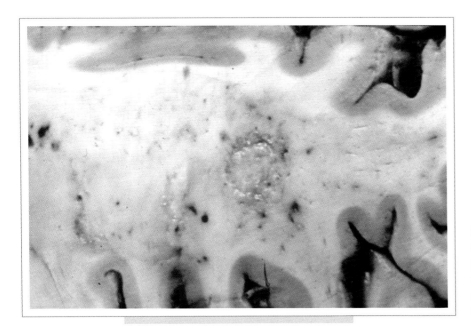

FIG. 3.1. Gross specimen of brain demonstrating a metastatic brain tumor from a primary lung carcinoma, located at the gray–white junction. Note the well circumscribed nature of the lesion, with little infiltration into surrounding brain.

Loss of certain metastasis-suppressor genes has also been implicated in the metastatic phenotype, including *nm23, KA11, KiSS1, PTEN, Maspin* and others [9–11]. Reduced expression of these genes removes inhibitory control over the formation of macroscopic metastases. A recent case control study of non-small cell lung cancer patients, with and without MBT, attempted to correlate the expression of epidermal growth factor receptor (EGFR), cyclooxygenase-2 and Bax with the risk of developing brain metastases [12]. It was found that expression of the biomarkers was similar for patients with and without MBT, and could not be used to predict the potential for developing an MBT. In addition, expression levels of EGFR, cyclooxygenase-2 and Bax did not correlate with patient survival in multivariate analysis.

Once the metastatic bolus of cells has traveled to the nervous system and has lodged within the brain, neoplastic angiogenesis is required for the tumor to grow to a clinically relevant size [9,10,13,14]. The angiogenic phenotype requires upregulation of angiogenic promoters such as vascular endothelial growth factor (VEGF), fibroblast growth factors (basic FGF, acidic FGF), angiopoietins (Ang-1, Ang-2), platelet-derived growth factor (PDGF), epidermal growth factor (EGF), transforming growth factors (TGFα and TGFβ), interleukins (IL-6, IL-8) and the various growth factor receptors (e.g. VEGFR, PDGFR, EGFR) [13,14]. During the 'angiogenic switch' to the metastatic phenotype, tumor cells also reduce secretion of angiogenesis inhibitors, such as thrombospondin-1, platelet factor-4 and interferons α and β [13]. This reduced concentration of inhibitory factors further 'tips the balance' in the local environment to permit angiogenic activity within and around the tumor mass.

On macroscopic evaluation, MBT usually form rounded, discrete deposits in the brain parenchyma that are well circumscribed and demarcated from surrounding neural

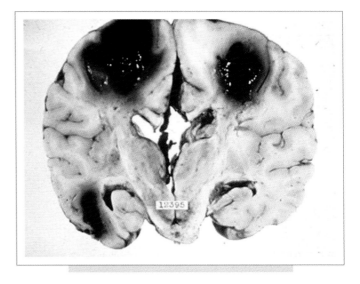

FIG. 3.2. Gross specimen of brain demonstrating several metastatic brain tumors from malignant melanoma. Note the hemorrhagic nature of the lesions, along with significant surrounding edema and mass effect.

tissues (Figures 3.1 and 3.2) [15,16]. The most common locations for metastases are the frontal and temporal lobes, other lobes of the cerebrum, the cerebellum and diencephalic region. The lesions can be single (25–35% of cases) or multiple (65–75% of cases) and may even present as a miliary pattern of numerous tiny masses. Primary tumors most likely to cause multifocal MBT include small cell and adenocarcinoma of the lung, melanoma and choriocarcinoma. Single metastatic deposits are more likely to arise from renal cell, gastrointestinal, breast, prostatic and

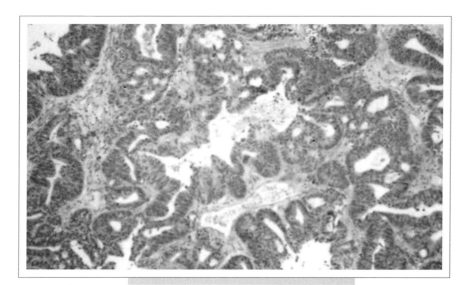

FIG. 3.3. Microscopic preparation of tissue from a metastatic adenocarcinoma of the lung. Note that the metastatic tissue in the brain maintains the ability to form complete glandular structures. H&E @ 200×.

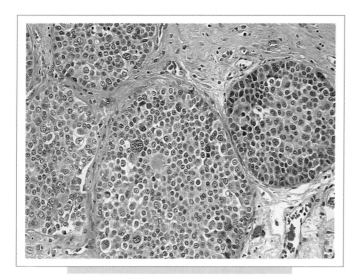

FIG. 3.4 Microscopic preparation of tissue from a metastatic ductal carcinoma of the breast. Note tumor nodules are sharply demarcated from surrounding brain parenchyma, with no infiltration. H&E @ 400×.

uterine carcinomas. The tumor deposits may have areas of hemorrhage or necrosis, particularly in the center of large lesions. Primary tumors most likely to cause hemorrhagic brain metastases include melanoma, choriocarcinoma, lung carcinoma and renal cell carcinoma. With or without hemorrhage, the tumor is usually surrounded by an extensive amount of vasogenic edema, which often seems out of proportion to the size of the mass and contributes to regional mass effect.

On microscopic examination, the histological features of the MBT are usually similar, if not identical, to those of the primary neoplasm (Figures 3.3 and 3.4) [15,16]. In some cases, there may be a vigorous angiogenic response, with more prominent vascular proliferation and the formation of glomeruloid structures. In other tumors, there may be extensive necrosis, with only small regions of recognizable neoplastic tissue at the periphery of the lesion or adjacent to blood vessels. However, unlike glioblastoma multiforme, pseudopalisading of tumor nuclei around necrotic foci is very uncommon. The tumor mass will usually have well defined borders, tending to displace adjacent brain parenchyma without significant infiltration. Areas of hemorrhage and gliosis are often noted. Initial review of the tissue morphology can often identify a major tumor category, such as metastatic carcinoma, melanoma or lymphoma. For a more detailed determination of cellular differentiation and assignment to a specific histological category, immunocytochemical analysis is required [15–17]. The tissue is usually screened with a detailed antibody panel, which includes numerous cell- and tumor-specific markers (Table 3.2). In some cases, further investigation with electron microscopy or molecular genetic techniques may be necessary to finalize the diagnosis.

SURGICAL THERAPY OF METASTATIC BRAIN TUMORS

In the modern era of neurosurgery, there is now an important role for surgical resection of MBT, in carefully selected patients [18–21]. Surgical removal should be considered in all patients with a magnetic resonance imaging (MRI) documented solitary metastasis. Unfortunately, this constitutes only 25–35% of all patients. Among those patients with solitary lesions, only half will be appropriate for surgery because of factors such as inaccessibility of the tumor (e.g. brainstem, eloquent cortex), extensive systemic tumor burden, or other medical problems (e.g. cardiac ischemia, pulmonary insufficiency). Using second generation image-guided, neuronavigation systems with frameless stereotaxy,

TABLE 3-2 Immunocytochemical staining techniques used in the diagnosis of metastatic brain tumors

Initial screening panel
  epithelial membrane antigen
  cytokeratins
  glial fibrillary acidic protein

Cell-specific markers
  lung cancer: cytokeratin 7, surfactant
  breast cancer: cytokeratin 7, estrogen and progesterone
    receptors
  gastrointestinal cancer: cytokeratin 20
  ovarian cancer: CA 125
  neuroendocrine: chromogranins, peptides
  thyroid cancer: thyroglobulin
  prostate cancer: prostate specific antigen, prostatic acid
    phosphatase

Germ cell tumors
  placental alkaline phosphatase

Sarcomas
  desmin
  smooth-muscle actin
  S-100

Malignant melanoma
  S-100
  HMB45
  MART-1

Lymphoma
  CD45
  CD3
  CD20

Data derived from references [14–16]

patients with MBT can undergo aggressive surgical resection with significantly less risk for neurological injury [22]. In a review of 49 patients by Tan and Black, the use of image-guided craniotomy allowed for a gross total resection of the tumor and complete resolution of symptoms in 96% and 70% of the cohort, respectively [22]. Neurological deterioration was noted in only two patients (3.6%), in whom significant deficits were present prior to surgery. The median survival for the entire group was 16.2 months, with a local recurrence rate of 16%. When neuronavigation and image-guidance is integrated with intraoperative magnetic resonance imaging (iMRI), the extent of surgical resection can be monitored and maximized in the operating room [23,24]. This often allows for a more complete resection of tumor and the potential for improved local control.

Class I evidence from two phase III trials is available to support the use of surgical resection in MBT patients [18–21,25,26]. In the seminal study by Patchell and colleagues, 48 patients with solitary MBT were randomly assigned to receive surgical resection plus irradiation versus irradiation alone [25]. Local recurrence at the site of the original metastasis was significantly less frequent in the surgical cohort in comparison to the irradiation alone

cohort (20% versus 52%; $P < 0.02$). Overall survival was significantly longer in the surgical group (median 40 weeks versus 15 weeks; $P < 0.01$). In addition, functional independence was maintained longer in the surgical cohort (median 38 weeks versus 8 weeks; $P < 0.005$). In a similar European phase III trial, 63 evaluable patients with solitary MBT were randomized to receive hyperfractionated irradiation (200 cGy × 2 per day; total of 4000 cGy) with or without surgical resection [26]. The overall survival was significantly longer in the surgical cohort (median 10 months versus 6 months; $P = 0.04$). A survival advantage was also noted for the surgical group in the 12-month (41% versus 23%) and 24-month (19% versus 10%) overall survival rates. The effect of the surgical procedure on survival was most pronounced in the patient cohort with stable systemic disease, with significant differences in overall survival (median 12 months versus 7 months; $P = 0.02$), 12-month survival rate (50% versus 24%) and 24-month survival rate (27% versus 10%). For patients with active systemic disease, the surgical resection and irradiation alone cohorts had the same median overall survival (5 months). One negative phase III trial has been reported by Mintz and co-workers, in their review of 84 patients randomized to receive irradiation with or without surgical resection [27]. The overall survival was similar between the surgical and irradiation alone groups (median 5.6 months versus 6.3 months; $P = 0.24$). There was also no difference between treatment cohorts in the ability of patients to maintain Karnofsky Performance Status equal to or above 70%. However, it should be mentioned that this study had several methodological shortcomings, including the fact that 73% of all patients had poorly controlled systemic disease, there was an unequal distribution of primary pathologies between treatment cohorts (i.e. more radio-resistant colorectal cancer in the surgical group and more radiosensitive breast cancer in the irradiation alone group) and non-uniform calculation of survival times [21].

There is also class II and III evidence to support the use of surgical resection for selected patients with a solitary MBT, mainly reflecting individual institutional experience [18–22,28–32]. This has been demonstrated in patients with solitary MBT from various types of primary tumors, including those from lung, breast, colon and rectum, melanoma, renal cell and others. In general, these studies also demonstrate improved local control rates and longer survival in patients with solitary, accessible MBT that receive surgical resection followed by external beam irradiation.

For patients with multiple MBT, the use of surgical treatment is more controversial and remains unclear [18–21]. Some authors advocate the removal of all metastatic tumors, if the lesions are accessible and not located in eloquent regions of brain [33]. Using this approach with carefully selected patients, the survival can be similar to that of patients undergoing surgery for solitary metastases. Other authors suggest limiting the use of surgical resection for the 'dominant or symptomatic' lesion, if it is accessible

[20,32]. The smaller and less symptomatic tumors can then be controlled by postoperative irradiation.

RADIATION THERAPY OF METASTATIC BRAIN TUMORS

Whole brain external beam irradiation (WBRT) remains the primary form of therapy for the majority of patients with brain metastases [1–5,34–36]. It is still the treatment of choice for tumors that are located in eloquent cortex or are too large or too numerous for surgical resection or radiosurgical approaches. Early randomized trials in the 1970s and 1980s by the Radiation Therapy Oncology Group (RTOG) and others evaluated variable dosing (10–54.4 Gy) and fractionation (1 to 34 fractions) schemes, in an attempt to determine the optimal therapeutic regimen [35,36]. The median survival across all studies ranged from 2.4 to 4.8 months, thereby proving that differences in dosing, timing and fractionation schedules did not significantly influence the results in MBT patients. Objective tumor responses (i.e. complete response (CR), partial response (PR), minor response (MR)) were noted in approximately 60% of patients in the randomized RTOG trials. The most widely used WBRT regimen delivers a total of 30 Gy in ten 3 Gy fractions over 2 weeks. Although this dose has limited potential for long-term tumor control, it is well tolerated and designed to minimize the neurotoxicity associated with WBRT. An analysis of RTOG clinical trial data suggests that this regimen can provide control of disease in roughly 50% of patients at 6 months. After receiving WBRT, most MBT patients note an improvement or stabilization of neurologic symptoms, including headache, seizures, impaired mentation, cerebellar dysfunction and motor deficits [35].

A randomized trial has also evaluated the utility of WBRT in the context of patients with a solitary MBT that have undergone surgical resection [37]. In this study, 95 patients with solitary MBT were treated with complete surgical resection and then randomized into a postoperative radiotherapy group or an observation group. The overall recurrence rate of MBT anywhere in the brain was significantly reduced in the radiotherapy group (18% versus 70%; $P < 0.001$). Postoperative WBRT was able to reduce the rate of MBT recurrence at the site of the original metastasis (10% versus 46%; $P < 0.001$) and at distant sites in the brain (14% versus 37%; $P < 0.01$). In addition, patients in the radiotherapy cohort were less likely to die of neurological causes than patients in the observation group (14% versus 44%; $P = 0.003$). However, there was no significant difference between groups in terms of the overall length of survival or the length of time that patients were able to maintain functional independence. This is not surprising since one would not expect WBRT to have any effect on the course of the systemic cancer.

Prophylactic cranial irradiation (PCI) is an 'up-front' application of WBRT that is only appropriate for consideration in selected patients with lung cancer. The efficacy of PCI was first demonstrated in patients with small cell lung cancer (SCLC), especially those with well-controlled systemic disease [38,39]. Initial reports demonstrated a survival benefit of 5.4% at 3 years, with a 25.3% reduction in the cumulative incidence of MBT in the cohort of patients achieving a complete systemic remission with chemotherapy [38]. A subsequent analysis of 505 patients that had participated in randomized trials has further characterized the benefit of PCI in SCLC patients [39]. The 5-year cumulative incidence of MBT as an isolated first site of relapse was 20% in the PCI cohort and 37% in control patients ($P < 0.001$). The overall 5-year incidence of MBT for the PCI and control groups was 43% and 59%, respectively (relative risk [RR] 0.50; $P < 0.001$). However, the effect on overall survival was modest, with 5-year rates for the PCI and control groups of 18% and 15%, respectively (RR 0.84; $P = 0.06$). Presumably, this is because the majority of SCLC patients ultimately die of systemic metastases, an issue not addressed by PCI. Prophylactic cranial irradiation has also been investigated in patients with non-small cell lung cancer (NSCLC), but with less compelling evidence of benefit [40,41]. Although there does appear to be a reduction in the incidence of MBT in the PCI cohorts, no survival benefit has been observed. This view is consistent with a recent Cochrane Review of the use of PCI in NSCLC patients [42]. The authors concluded that there was insufficient evidence at this time to recommend the use of PCI in clinical practice and that it should only be offered in the context of a clinical trial.

Stereotactic radiosurgery (SRS) is a method of delivering focused irradiation to the boundaries of a tumor (i.e. conformal dosing), in a single or few fractions, using great precision [34–36,43–47]. SRS has become an important therapeutic option for brain metastases for several reasons, including the fact that most MBT are spherical and small at the time of diagnosis, the degree of infiltration into surrounding brain is usually quite limited, the gray–white matter junction is considered a relatively 'non-eloquent' area of the brain and improved local control in the brain may extend patient survival. The treatment is most often administered using a Gamma Knife® (i.e. Co^{60} sources), however, linear accelerator (e.g. Cyberknife®) and proton beam units are also used and demonstrate comparable local control and complication rates. SRS is most effective for tumors less than or equal to 3 cm in diameter. However, some authors recommend treatment of tumors up to 4 cm in diameter. Typical doses are in the range of 15–20 Gy to the margins of the tumor, with higher doses administered at the center of the mass. Optimal dosing will depend on the size of the tumor, previous exposure to irradiation and proximity to delicate neural structures (e.g. optic chiasm).

There are two reports that provide class I evidence for the efficacy of SRS in the context of a boost to WBRT [48,49]. In the first study from the University of Pittsburgh,

27 patients with two to four MBT were randomized to receive WBRT (30 Gy over 12 fractions) plus SRS (tumor margin dose of 16 Gy) or WBRT alone [48]. Local control was improved by the use of the SRS boost, with local failure rates at 1 year of 8% for the combined treatment group and 100% for the WBRT alone group. The median time to local failure was 36 months for the WBRT plus SRS cohort and 6 months for the WBRT alone group ($P = 0.0005$). In addition, median time to overall brain failure (local or distant) was longer for the combined treatment cohort in comparison to the WBRT alone group (34 months versus 5 months; $P = 0.002$). However, the addition of the SRS boost did not significantly influence overall survival between the two groups (11 months versus 7.5 months, respectively; $P = 0.22$). Again, this lack of effect on overall survival could simply reflect the effect of systemic metastases in these patients. In a similar study by the RTOG (RTOG 9508), 333 patients with one to three MBT were randomized to receive either WBRT (37.5 Gy over 15 fractions) or WBRT plus an SRS boost of 15–24 Gy, depending on tumor size [49]. Local control at 1 year was significantly better for the SRS group in comparison to the WBRT alone group (82% versus 71%; $P = 0.01$). In addition, time to local progression was extended in the combined treatment cohort ($P = 0.0132$). Overall median survival was similar between groups, however, for patients with a single MBT, median survival was longer in the WBRT plus SRS cohort (6.5 months versus 4.9 months; $P = 0.0393$). The Karnofsky Performance Status (KPS) was more likely to be stable or improved at 6 months follow-up in the WBRT plus SRS group (43% versus 27%; $P = 0.03$). This is consistent with the multivariate analysis, which demonstrated improved survival in patients with RPA class 1 disease ($P < 0.0001$).

There are numerous reports in the literature describing class II and III evidence supporting the use of SRS for treatment of MBT [34–36,43–47]. A review of the larger trials (i.e. 100 or more patients) would suggest that SRS is as effective as, if not more effective than, WBRT [50–61]. In most of the studies, the median survival ranged between 5.5 and 13.5 months, with overall local control rates of 85–95%. The increase in local control rates did not translate into an improvement in survival, with most patients dying of systemic disease progression. Several factors have been found to influence the degree of local control, including primary tumor histology (e.g. melanoma versus lung carcinoma), tumor volume, tumor location, presentation (e.g. new versus recurrent) and pattern of MRI enhancement (e.g. homogeneous versus heterogeneous versus ring). Some authors are recommending the use of SRS as the primary, 'up-front' mode of irradiation in high performance patients with well-controlled systemic disease, instead of WBRT [50–61]. However, this view is not supported by the conclusions of a recent ASTRO meta-analysis of SRS treatment of MBT [62]. The ASTRO recommendations are to advise an SRS boost to WBRT in selected patients with

one to four newly diagnosed MBT. The omission of WBRT results in significantly lower rates of local and distant brain control.

CHEMOTHERAPY OF METASTATIC BRAIN TUMORS

Chemotherapy has become a more viable option for the treatment of MBT in recent years, especially for recurrent disease [63–69]. The prior reluctance to use chemotherapy stemmed from concerns about the ability of chemotherapy drugs to cross the blood–brain barrier (BBB) and penetrate tumor cells, intrinsic chemoresistance of metastatic disease and the high probability of early death from systemic progression. However, recent animal data suggest that metastatic tumors that strongly enhance on computed tomography (CT) or MRI have an impaired blood–brain barrier and will allow entry of chemotherapeutic drugs [63,65]. In addition, systemic resistance to a given drug does not always preclude sensitivity of the metastasis within the brain [63]. Several types of metastatic brain tumors are relatively chemosensitive and may respond, including breast cancer, small cell lung cancer, non-small cell lung cancer, germ cell tumors and ovarian carcinoma.

The most common approach to chemotherapy for brain metastases is to administer it 'up-front', before or during conventional WBRT or SRS [70–79]. Several authors have demonstrated that combination regimens given intravenously can be active in this context. The most frequently used agents included cisplatin (CDDP), etoposide (VP16) and cyclophosphamide (CTX). In a series of 19 patients with small cell lung cancer and brain metastases, Twelves and co-workers used intravenous (IV) CTX, vincristine and VP16 every three weeks before any form of irradiation [70]. Ten of the 19 patients (53%) had a radiological or clinical response. In nine patients, there was CT evidence of tumor shrinkage, while in one patient there was neurological improvement, without neuroimaging follow-up. The mean time to progression (TTP) was 22 weeks, with a median overall survival of 28 weeks. Cocconi and colleagues used up-front IV cisplatin and etoposide every 3 weeks for 22 evaluable patients with MBT from breast carcinoma [71]. There were 12 objective responses, for an overall objective response rate of 55%. The median TTP was 25 weeks overall and 40 weeks in the objective response cohort. Overall median survival was 58 weeks. The same authors have expanded their series to include 89 patients with MBT from breast, non-small cell lung carcinoma and malignant melanoma [72]. Objective responses were noted in the breast and lung cohorts. None of the patients with melanoma had objective responses. The overall objective response rate was 30% (34/89). Median TTP was 15 weeks, with a median survival for the cohort of 27 weeks. Similar responses have been noted in series of patients with MBT from lung and breast carcinoma [73–79]. However, although objective responses were noted in many of these studies, they did not translate into improvements in patient survival.

Topotecan is a semisynthetic camptothecan derivative that selectively inhibits topoisomerase I in the S phase of the cell cycle [80]. It demonstrates excellent penetration of the blood–brain barrier in primate animal models and humans. Summating the data of more than 60 patients in several European studies of single agent topotecan, the objective response rates have been encouraging, with 30–60% of patients demonstrating a complete response (CR) or partial response (PR) [81–84]. Topotecan is also being investigated in combination with radiotherapy and other cytotoxic chemotherapy agents, such as temozolomide. A recent phase I trial has evaluated the tolerability of temozolomide (50–$200\,mg/m^2$) and topotecan (1–$1.5\,mg/m^2$), given daily for 5 days every 28 days [85]. Twenty-five patients with systemic solid tumors were treated. Toxicity was mainly hematological, with frequent neutropenia and thrombocytopenia. Three patients were noted to have a PR.

Temozolomide is an imidazotetrazine derivative of the alkylating agent dacarbazine with activity against systemic and CNS malignancies [65,86–88]. The drug undergoes chemical conversion at physiological pH to the active species 5-(3-methyl-1-triazeno)imidazole-4-carboxamide (MTIC). Temozolomide exhibits schedule dependent antineoplastic activity by interfering with DNA replication through the methylation of DNA at the following sites: N^7-guanine (70%), N^3-adenine (9.2%) and O^6-guanine (5%). Several reports have suggested activity of single agent temozolomide against MBT, with occasional objective responses [89,90]. Temozolomide is also under investigation as a radiation sensitizer, including a randomized phase II trial by Antonadou and associates [91]. In this study, 52 newly diagnosed MBT patients (lung and breast) were treated with either WBRT alone (40 Gy) or WBRT plus conventional temozolomide. The addition of temozolomide improved the objective response rate when compared to WBRT alone (CR 38%, PR 58% versus CR 33%, PR 33%). In addition, neurologic improvement during treatment was more pronounced in the cohort of patients receiving chemotherapy. A similar randomized phase II trial by Verger and colleagues treated 82 patients with MBT (mostly lung and breast) using combined WBRT (30 Gy) and temozolomide ($75\,mg/m^2/day$ during irradiation, plus two cycles of conventional adjuvant dosing) versus WBRT alone [92]. The objective response rate and overall survival were similar between treatment groups. However, there was a significantly higher rate of progression-free survival at 90 days in the combined treatment cohort (72% versus 54%, $P = 0.03$). In addition, the percentage of patients dying from the MBT was lower in the chemotherapy arm (41% versus 69%; $P = 0.03$). Temozolomide has also been shown to have activity, as a single agent and in combination with other drugs (e.g. cisplatin, doxetaxel, thalidomide), against MBT from malignant melanoma [93–96].

In an effort to improve dose intensity to MBT, some authors have given some or all of the chemotherapy drugs by the intra-arterial (IA) route [65,97–101]. There are several advantages to administering chemotherapy IA instead of by the conventional IV route, including augmentation of the peak concentration of drug in the region of the tumor and an increase in the local area under the concentration-time curve [97]. Pathologically, metastatic brain tumors are excellent candidates for IA approaches, because they tend to be well circumscribed and non-infiltrative [1]. In addition, MBT almost always enhance on MRI imaging, indicating excellent arterial vascularization and impairment of the blood–tumor barrier. Pharmacologic studies using animal models of IA and IV drug infusion have shown that the IA route can increase the intratumoral concentration of a given agent by at least a factor of three- to five-fold [102,103]. For chemosensitive tumors, improving the intratumoral concentrations of drug should augment tumor cell kill and the ability to achieve objective responses [97]. Initial applications of IA chemotherapy to MBT involved the use of BCNU and cisplatin [98–101]. Although objective responses were noted in patients with lung and breast tumors, significant neurotoxicity occurred (e.g. seizures, confusion). More recent reports have used carboplatin as the primary IA agent and have resulted in similar objective response rates, with significantly less neurotoxicity [104–106].

The recent expansion of knowledge regarding the molecular biology of neoplasia and the metastatic phenotype has led to intense development of therapeutic strategies designed to exploit this new information [107]. Several targets of therapeutic intervention have been developed, including growth factor receptors and their tyrosine kinase activity, disruption of aberrant internal signal transduction pathways, inhibition of excessive matrix metalloproteinase activity, downregulation of cell cycle pathways and manipulation of the apoptosis pathways. The most promising approach thus far has been the development of small molecule drugs or monoclonal antibodies to the major growth factor receptors (e.g. PDGFR, EGFR, Her2, CD20) [108–112]. Monoclonal antibody agents such as rituximab (i.e. Rituxan®) and trastuzumab (i.e. Herceptin®) have proven to be clinically active against non-Hodgkin's lymphoma and breast cancer, respectively. Several small-molecule inhibitors of the tyrosine kinase activity of the EGFR (e.g. geftinib, erlotinib) are under clinical evaluation in phase I trials of patients with solid tumors [109–111]. Similar efforts are underway to develop agents that can target the tyrosine kinase activity of PDGFR and the ras signaling pathway [111,112]. Other agents under development will be designed to target downstream effectors, such as Raf, MAPK, Rac/Rho and angiogenesis. An initial report using imatinib, a tyrosine kinase inhibitor with activity against C-KIT and PDGFR, describes a 75-year-old male with a C-KIT positive GI stromal tumor who developed neurological deterioration and gait difficulty [113]. An MRI scan demonstrated leptomeningeal disease with brain infiltration and edema. After treatment with imatinib mesylate (400 mg bid) for 2 months, his neurological function and gait improved. A follow-up MRI scan revealed complete resolution of the meningeal and intra-parenchymal abnormalities. Several authors have recently described case reports of the use of gefitinib, an oral tyrosine kinase inhibitor of EGFR, in patients with MBT from NSCLC [114–118]. A few of these initial patients had objective responses, including CR, that were quite durable. These early reports led Ceresoli and colleagues to perform a prospective phase II trial of gefitinib in patients with MBT from NSCLC [119]. Forty-one consecutive patients were treated with gefitinib (250 mg/day); 37 had received prior chemotherapy and 18 had undergone WBRT. There were four patients with a PR and seven with SD. The overall progression-free survival was only 3 months. However, the median duration of responses in the patients with a PR was an encouraging 13.5 months.

ACKNOWLEDGMENTS

The authors would like to thank Julia Shekunov for research assistance. Dr Newton was supported in part by National Cancer Institute grant, CA 16058 and the Dardinger Neuro-Oncology Center Endowment Fund.

REFERENCES

1. Newton HB (1999). Neurological complications of systemic cancer. Am Fam Phys 59:878–886.
2. Soffietti R, Ruda R, Mutani R (2002). Management of brain metastases. J Neurol 249:1357–1369.
3. Langer CJ, Mehta MP (2005). Current management of brain metastases, with a focus on systemic options. J Clin Oncol 23:6207–6219.
4. Bajaj GK, Kleinberg L, Terezakis S (2005). Current concepts and controversies in the treatment of parenchymal brain metastases: improved outcomes with aggressive management. Cancer Investig 23:363–376.
5. Lassman AB, DeAngelis LM (2003). Brain metastases. Neurol Clin N Am 21:1–23.
6. Liotta LA, Stetler-Stevenson WG, Steeg PS (1991). Cancer invasion and metastasis: positive and negative regulatory elements. Cancer Investig 9:543–551.
7. Reilly JA, Fidler IJ (1993). The biology of metastasis. Contemp Oncol November:32–46.
8. Aznavoorian S, Murphy AN, Stetler-Stevenson WG, Liotta LA (1993). Molecular aspects of tumor cell invasion and metastasis. Cancer 71:1368–1383.
9. Webb CP, Vande Woude GF (2000). Genes that regulate metastasis and angiogenesis. J Neuro-Oncol 50:71–87.
10. Nathoo N, Chalhavi A, Barnett GH, Toms SA (2005). Pathobiology of brain metastases. J Clin Pathol 58:237–242.
11. Yoshida BA, Sokoloff MM, Welch DR, Rinker-Schaeffer CW (2000). Metastasis-suppressor genes: a review and perspective on an emerging field. J Natl Cancer Inst 92:1717–1730.
12. Milas I, Komaki R, Hachiya T et al (2003). Epidermal growth factor receptor, cyclooxygenase-2, and BAX expression in the primary non-small cell lung cancer and brain metastases. Clin Cancer Res 9:1070–1076.
13. Hanahan D, Folkman J (1996). Patterns of emerging mechanisms of the angiogenic switch during tumorigenesis. Cell 86:353–364.
14. Beckner ME (1999). Factors promoting tumor angiogenesis. Cancer Investig 17:594–623.
15. Bruner JM, Tien RD (1998). Secondary tumors. In *Russell & Rubinstein's Pathology of Tumors of the Nervous System*, 6th edn, Volume 2. Bigner DD, McLendon RE, Bruner JM (eds). Arnold, London. 17:419–450.
16. Ironside JW, Moss TH, Louis DN, Lowe JS, Weller RO (eds) (2002). Metastatic tumours. I *Diagnostic Pathology of Nervous System Tumours*. Churchill Livingstone, London. 12:319–341.
17. Perry A, Parisi JE, Kurtin PJ (1997). Metastatic adenocarcinoma to the brain: an immunohistochemical approach. Hum Pathol 28:938–943.
18. Sawaya R, Ligon BL, Bindal AK, Bindal RK, Hess KR (1996). Surgical treatment of metastatic brain tumors. J Neuro-Oncol 27:269–277.
19. Lang FF, Sawaya R (1996). Surgical management of cerebral metastases. Neurosurg Clin N Am 7:459–484.
20. Brem S, Panattil JG (2005). An era of rapid advancement: diagnosis and treatment of metastatic brain cancer. Neurosurgery 57:S4-5–S4-9.
21. Vogelbaum MA, Suh JH (2006). Resectable brain metastases. J Clin Oncol 24:1289–1294.
22. Tan TC, Black PM (2003). Image-guided craniotomy for cerebral metastases: techniques and outcomes. Neurosurgery 53:82–90.
23. Albayrak B, Samdani AF, Black PM (2004). Intra-operative magnetic resonance imaging in neurosurgery. Acta Neurochir (Wien) 146:543–556.
24. Nimsky C, Ganslandt O, von Keller B, Romstöck J, Fahlbusch R (2004). Intraoperative high-field-strength MR imaging: implementaton and experience in 200 patients. Radiology 233:67–78.
25. Patchell RA, Tibbs PA, Walsh JW et al (1990). A randomized trial of surgery in the treatment of single metastases to the brain. New Engl J Med 322:494–500.
26. Vecht CJ, Haaxma-Reiche H, Noordijk EM et al (1993). Treatment of single brain metastasis: radiotherapy alone or combined with neurosurgery? Ann Neurol 33:583–590.

27. Mintz AH, Kestle J, Rathbone MP et al (1996). A randomized trial to assess the efficacy of surgery in addition to radiotherapy in patients with a single cerebral metastasis. Cancer 78:1470–1476.

28. Arbit E, Wronski M, Burt M, Galicich JH (1995). The treatment of patients with recurrent brain metastases. A retrospective analysis of 109 patients with non-small cell lung cancer. Cancer 76:765–773.

29. Wronski M, Arbit E, McCormick B (1997). Surgical treatment of 70 patients with brain metastases from breast cancer. Cancer 80:1746–1754.

30. Wronski M, Arbit E (1999). Resection of brain metastases from colorectal carcinoma in 73 patients. Cancer 85:1677–1685.

31. O'Neill BP, Iturria NJ, Link MJ, Pollock BE, Ballman KV, O'Fallon JR (2003). A comparison of surgical resection and stereotactic radiosurgery in the treatment of solitary brain metastases. Int J Rad Oncol Biol Phys 55:1169–1176.

32. Paek SH, Audu PB, Sperling MR, Cho J, Andrews DW (2005). Reevaluation of surgery for the treatment of brain metastases: review of 208 patients with single or multiple brain metastases treated at one institution with modern neurosurgical techniques. Neurosurgery 56:1021–1034.

33. Bindal RK, Sawaya R, Leavens ME, Lee JJ (1993). Surgical treatment of multiple brain metastases. J Neurosurg 79:210–216.

34. Berk L (1995). An overview of radiotherapy trials for the treatment of brain metastases. Oncology 9:1205–1212.

35. Sneed PK, Larson DA, Wara WM (1996). Radiotherapy for cerebral metastases. Neurosurg Clin N Am 7:505–515.

36. Khuntia D, Brown P, Li J, Mehta MP (2006). Whole-brain radiotherapy in the management of brain metastasis. J Clin Oncol 24:1295–1304.

37. Patchell RA, Tibbs PA, Regine WF et al (1998). Postoperative radiotherapy in the treatment of single metastases to the brain: a randomized trial. J Am Med Assoc 280:1485–1489.

38. Aupérin A, Arriagada R, Pignon JP et al (1999). Prophylactic cranial irradiation for patients with small-cell lung cancer in complete remission. Prophylactic Cranial Irradiation Overview Collaborative Group. New Engl J Med 341:476–484.

39. Arriagada R, Le Chevalier T, Rivière A et al (2002). Patterns of failure after prophylactic cranial irradiation in small cell lung cancer: analysis of 505 randomized patients. Ann Oncol 13:748–754.

40. Laskin JJ, Sandler AB (2003). The role of prophylactic cranial radiation in the treatment of non-small cell lung cancer. Clin Adv Hem Oncol 1:731–740.

41. Gore EM (2003). Prophylactic cranial irradiation for patients with locally advanced non-small cell lung cancer. Oncology 17:775–779.

42. Lester JF, MacBeth FR, Coles B (2005). Prophylactic cranial irradiation for preventing brain metastases in patients undergoing radical treatment for non-small cell lung cancer: a Cochrane Review. Int J Rad Oncol Biol Phys 63:690–694.

43. Loeffler JS, Alexander E (1990). The role of stereotactic radiosurgery in the management of intracranial tumors. Oncology 4:21–31.

44. Boyd TS, Mehta MP (1999). Stereotactic radiosurgery for brain metastases. Oncology 13:1397–1409.

45. Sheehan J, Niranjan A, Flickinger JC, Kondziolka D, Lunsford LD (2004). The expanding role of neurosurgeons in the management of brain metastases. Surg Neurol 62:32–41.

46. McDermott MW, Sneed PK (2005). Radiosurgery in metastatic brain cancer. Neurosurgery 57:S4-45–S4-53.

47. Bhatnagar AK, Flickinger JC, Kondziolka D, Lunsford LD (2006). Stereotactic radiosurgery for four or more intracranial metastases. Int J Rad Oncol Biol Phys 64:898–903.

48. Kondziolka D, Patel A, Lunsford LD, Kassam A, Flickinger JC (1999). Stereotactic radiosurgery plus whole brain radiotherapy versus radiotherapy alone for patients with multiple brain metastases. Int J Rad Oncol Biol Phys 45:427–434.

49. Andrews DW, Scott CB, Sperduto PW et al (2004). Whole brain radiation therapy with or without stereotactic radiosurgery boost for patients with one to three brain metastases: phase III results of the RTOG 9508 randomised trial. Lancet 363:1665–1672.

50. Flickinger JC, Kondziolka D, Lunsford LD et al (1994). A multi-institutional experience with stereotactic radiosurgery for solitary brain metastasis. Int J Rad Oncol Biol Phys 28:797–802.

51. Alexander E, Moriarty TM, Davis RB et al (1995). Stereotactic radiosurgery for the definitive, noninvasive treatment of brain metastases. J Natl Cancer Inst 87:34–40.

52. Gerosa M, Nicolato A, Severi F et al (1996). Gamma knife radiosurgery for intracranial metastases: from local control to increased survival. Stereotact Funct Neurosurg 66:184–192.

53. Joseph J, Adler JR, Cox RS, Hancock SL (1996). Linear accelerator-based stereotactic radiosurgery for brain metastases: the influence of number of lesions on survival. J Clin Oncol 14:1085–1092.

54. Pirzkall A, Debus J, Lohr F et al (1998). Radiosurgery alone or in combination with whole-brain radiotherapy for brain metastases. J Clin Oncol 16:3563–3569.

55. Chen JC Petrovich Z, O'Day S et al (2000). Stereotactic radiosurgery in the treatment of metastatic disease to the brain. Neurosurgery 47:268–279.

56. Hoffman R, Sneed PK, McDermott MW et al (2001). Radiosurgery for brain metastases from primary lung carcinoma. Cancer J 7:121–131.

57. Gerosa M, Nicolato A, Foroni R et al (2002). Gamma knife radiosurgery for brain metastases: a primary therapeutic option. J Neurosurg 97:515–524.

58. Petrovich Z, Yu C, Giannotta SL, O'Day S, Apuzzo MLJ (2002). Survival and pattern of failure in brain metastases treated with stereotactic gamma knife radiosurgery. J Neurosurg 97:499–506.

59. Hasegawa T, Kondziolka D, Flickinger JC, Germanwala A, Lunsford LD (2003). Brain metastases treated with radiosurgery alone: an alternative to whole brain radiotherapy? Neurosurgery 52:1318–1326.

60. Lutterbach J, Cyron D, Henne K, Ostertag CB (2003). Radiosurgery followed by planned observation in patients with one to three brain metastases. Neurosurgery 52:1066–1073.

61. Muacevic A, Kreth FW, Tonn JC, Wowra B (2004). Stereotactic radiosurgery for multiple brain metastases from breast carcinoma. Cancer 100:1705–1711.

62. Mehta MP, Tsao MN, Whelan TJ et al (2005). The American Society for Therapeutic Radiology and Oncology (ASTRO) evidence-based review of the role of radiosurgery for brain metastases. Int J Rad Oncol Biol Phys 63:37–46.

63. Lesser GJ (1996). Chemotherapy of cerebral metastases from solid tumors. Neurosurg Clin N Am 7:527–536.

64. Newton HB (2000). Novel chemotherapeutic agents for the treatment of brain cancer. Expert Opin Investig Drugs 12:2815–2829.

65. Newton HB (2002). Chemotherapy for the treatment of metastatic brain tumors. Expert Rev Anticancer Ther 2:495–506.

66. Tosoni A, Lumachi F, Brandes AA (2004). Treatment of brain metastases in uncommon tumors. Expert Rev Anticancer Ther 4:783–793.

67. van den Bent MJ (2003). The role of chemotherapy in brain metastases. Eur J Cancer 39:2114–2120.

68. Schuette W (2004). Treatment of brain metastases from lung cancer: chemotherapy. Lung Cancer 45(suppl 2):S253–S257.

69. Bafaloukos D, Gogas H (2004). The treatment of brain metastases in melanoma patients. Cancer Treatment Rev 30:515–520.

70. Twelves CJ, Souhami RL, Harper PG et al (1990). The response of cerebral metastases in small cell lung cancer to systemic chemotherapy. Br J Cancer 61:147–150.

71. Cocconi G, Lottici R, Bisagni G et al (1990). Combination therapy with platinum and etoposide of brain metastases from breast carcinoma. Cancer Investig 8:327–334.

72. Franciosi V, Cocconi G, Michiarava M et al (1999). Front-line chemotherapy with cisplatin and etoposide for patients with brain metastases from breast carcinoma, non-small cell lung carcinoma, or malignant melanoma. A prospective study. Cancer 85:1599–1605.

73. Bernardo G, Cuzzoni Q, Strada MR et al (2002). First-line chemotherapy with vinorelbine, gemcitabine, and carboplatin in the treatment of brain metastases from non-small cell lung cancer: a phase II study. Cancer Investig 20:293–302.

74. Rosner D, Nemoto T, Lane WW (1986). Chemotherapy induces regression of brain metastases in breast carcinoma. Cancer 58:832–839.

75. Boogerd W, Dalesio O, Bais EM, Van Der Sande JJ (1992). Response of brain metastases from breast cancer to systemic chemotherapy. Cancer 69:972–980.

76. Robinet G, Thomas P, Breton JL et al (2001). Results of a phase III study of early versus delayed whole brain radiotherapy with concurrent cisplatin and vinorelbine combination in inoperable brain metastasis of non-small cell lung cancer: Groupe Français de Pneumo-Cancérologie (GFPC) protocol 95-1. Ann Oncol 12:59–67.

77. Postmus PE, Haaxma-Reiche H, Smit EF et al (2000). Treatment of brain metastases of small cell lung cancer: comparing teniposide and teniposide with whole-brain radiotherapy – a phase III study of the European Organization for the Research and Treatment of Cancer Lung Cancer Cooperative Group. J Clin Oncol 18:3400–3408.

78. Ushio Y, Arita N, Hayakawa T et al (1991). Chemotherapy of brain metastases from lung carcinoma: a controlled randomized study. Neurosurgery 28:201–205.

79. Guerrieri M, Wong K, Ryan G, Millward M, Quong G, Ball DL (2004). A randomized phase III study of palliative radiation with concomitant carboplatin for brain metastases from non-small cell carcinoma of the lung. Lung Cancer 46:107–111.

80. Slichenmyer WJ, Rowinsky EK, Donehower RC, Kaufmann SH (1993). The current status of camptothecin analogues as antitumor agents. J Natl Cancer Inst 85:271–291.

81. Ardizzoni A, Hansen H, Dombernowsy P et al (1997). Topotecan, a new active drug in the second-line treatment of small cell lung cancer: a phase II study in patients with refractory and sensitive disease. J Clin Oncol 15:2090–2096.

82. Korfel A, Oehm C, von Pawel J et al (2002). Response to topotecan of symptomatic brain metastases of small cell lung cancer also after whole-brain irradiation: a multicentre phase II study. Eur J Cancer 38:1724–1729.

83. Oberhoff C, Kieback DG, Würstlein R et al (2001). Topotecan chemotherapy in patients with breast and brain metastases: results of a pilot study. Onkologie 24:256–260.

84. Wong ET, Berkenblit A (2004). The role of topotecan in the treatment of brain metastases. Oncologist 9:68–79.

85. Eckardt JR, Martin KA, Schmidt AM, White LA, Greco AO, Needles BM (2002). A phase I trial of IV topotecan in combination with temozolomide daily times 5 every 28 days. Proc ASCO 21:83b.

86. Newlands ES, Stevens MFG, Wedge SR et al (1997). Temozolomide: a review of its discovery, chemical properties, pre-clinical development and clinical trials. Cancer Treat Rev 23:35–61.

87. Hvisdos KM, Goa KL (1999). Temozolomide. CNS Drugs 12:237–243.

88. Stupp RK, Gander M, Leyvraz S, Newlands E (2001). Current and future developments in the use of temozolomide for the treatment of brain tumors. Lancet Oncol 2:552–560.

89. Abrey LE, Olson JD, Raizer JJ et al (2001). A phase II trial of temozolomide for patients with recurrent or progressive brain metastases. J Neuro-Onc 53:259–265.

90. Christodoulou C, Bafaloukos D, Kosmidos P et al (2001). Phase II study of temozolomide in heavily pretreated cancer patients with brain metastases. Ann Oncol 12:249–254.

91. Antonadou D, Paraskaveidis M, Sarris N et al (2002). Phase II randomized trial of temozolomide and concurrent radiotherapy in patients with brain metastases. J Clin Oncol 20:3644–3650.

92. Verger E, Gil M, Yaya R et al (2005). Temozolomide and concomitant whole brain radiotherapy in patients with brain metastases: a phase II randomized trial. Int J Rad Oncol Biol Phys 61:185–191.

93. Biasco G, Pantaleo MA, Casadei S (2001). Treatment of brain metastases of malignant melanoma with temozolomide. New Engl J Med 345:621–622.

94. Agarwala SS, Kirdwood JM, Gore M et al (2004). Temozolomide for the treatment of brain metastases associated with metastatic melanoma: a phase II study. J Clin Oncol 22:2101–2107.

95. Bafaloukos D, Tsoutsos D, Fountzilas G et al (2004). The effect of temozolomide-based chemotherapy in patients with cerebral metastases from melanoma. Melanoma Res 14:289–294.

96. Hwu WJ, Raizer JJ, Panageas KS, Lis E (2001). Treatment of metastatic melanoma in the brain with temozolomide and thalidomide. Lancet Oncol 2:634–635.

97. Stewart DJ (1989). Pros and cons of intra-arterial chemotherapy. Oncology 3:20–26.

98. Yamada K, Bremer AM, West CR et al (1979). Intra-arterial BCNU therapy in the treatment of metastatic brain tumor from lung carcinoma. A preliminary report. Cancer 44:2000–2007.

99. Madajewicz S, West CR, Park HC et al (1981). Phase II study – Intra-arterial BCNU therapy for metastatic brain tumors. Cancer 47:653–657.

100. Cascino TL, Byrne TN, Deck MDF, Posner JB (1983). Intra-arterial BCNU in the treatment of metastatic tumors. J Neuro-Oncol 1:211–218.

101. Madajewicz S, Chowhan N, Iliya A et al (1991). Intracarotid chemotherapy with etoposide and cisplatin for malignant brain tumors. Cancer 67:2844–2849.

102. Barth RF, Yang W, Rotaru JH et al (1997). Boron neutron capture therapy of brain tumors: enhanced survival following intracarotid injection of either sodium borocaptate or boronophenylalanine with or without blood–brain barrier disruption. Cancer Res 57:1129–1136.

103. Kroll RA, Neuwelt EA (1998). Outwitting the blood–brain barrier for therapeutic purposes: osmotic opening and other means. Neurosurgery 42:1083–1100.

104. Gelman M, Chakares D, Newton HB (1999). Brain tumors: complications of cerebral angiography accompanied by intra-arterial chemotherapy. Radiology 213:135–140.

105. Newton HB, Stevens C, Santi M (2001). Brain metastases from fallopian tube carcinoma responsive to intra-arterial carboplatin and intravenous etoposide: a case report. J Neuro-Oncol 55:179–184.

106. Newton HB, Snyder MA, Stevens C et al (2003). Intra-arterial carboplatin and intravenous etoposide for the treatment of brain metastases. J Neuro-Oncol 61:35–44.

107. Garrett MD, Workman P (1999). Discovering novel chemotherapeutic drugs for the third millennium. Eur J Cancer 35:2010–2030.

108. Livitzki A, Gazit A (1995). Tyrosine kinase inhibition: an approach to drug development. Science 267:1782–1788.

109. Gibbs JB (2000). Anticancer drug targets: growth factors and growth factor signaling. J Clin Investig 105:9–13.

110. Dillman RO (2001). Monoclonal antibodies in the treatment of malignancy: basic concepts and recent developments. Cancer Investig 19:833–841.

111. Hao D, Rowinsky EK (2002). Inhibiting signal transduction: recent advances in the development of receptor tyrosine kinase and ras inhibitors. Cancer Investig 20:387–404.

112. Newton HB (2003). Molecular neuro-oncology and the development of 'targeted' therapeutic strategies for brain tumors. Part 1 – growth factor and ras signaling pathways. Expert Rev Anticancer Ther 3:595–614.

113. Brooks BJ, Bani JC, Fletcher CDM, Demeteri GD (2002). Response of metastatic gastrointestinal stromal tumor including CNS involvement to imatinib mesylate (STI-571). J Clin Oncol 20:870–872.

114. Cappuzzo F, Ardizzoni A, Soto-Parra H et al (2003). Epidermal growth factor receptor targeted therapy by ZD 1839 (Iressa) in patients with brain metastases from non-small cell lung cancer (NSCLC). Lung Cancer 41:227–231.

115. Cappuzzo F, Calandri C, Bartolini S, Crinò L (2003). ZD 1839 in patients with brain metastases from non-small cell lung cancer (NSCLC): report of four cases. Br J Cancer 89:246–247.

116. Poon ANY, Ho SSM, Yeo W, Mok TSK (2004). Brain metastases responding to gefitinib alone. Oncology 67:174–178.

117. Ishida A, Kanoh K, Nishisaka T et al (2004). Gefitinib as a first line of therapy in non-small cell lung cancer with brain metastases. Intern Med 43:718–720.

118. Katz A, Zalewski P (2003). Quality-of-life benefits and evidence of antitumor activity for patients with brain metastases treated with gefitinib. Br J Cancer 89:S15–S18.

119. Ceresoli GL, Cappuzzo F, Gregorc V, Bartolini S, Crinò L, Villa E (2004). Gefitinib in patients with brain metastases from non-small cell lung cancer: a prospective trial. Ann Oncol 15:1042–1047.

CHAPTER 4

Overview of Spinal Cord Tumor Epidemiology

Herbert B. Newton and Mark G. Malkin

INTRODUCTION

Primary spinal cord tumors (SCT) are a group of neoplasms that arise from the parenchyma of the spinal cord or from tissues that comprise or are contained within the surrounding spinal canal, including meninges, nerves, fat, bone and blood vessels [1–3]. Secondary SCT arise as metastatic deposits, within the spinal cord parenchyma or surrounding tissues, from a primary lesion outside of the central nervous system (CNS; see Chapters 6 and 7). Because of their location near delicate spinal cord structures that mediate motor control and sensation, even slowly growing benign SCT may cause severe neurologic sequelae. This chapter will provide an overview of the epidemiology of primary and metastatic tumors of the spinal cord, with a review of pathology and treatment to follow in the next two chapters.

EPIDEMIOLOGY OF PRIMARY SPINAL CORD TUMORS

Primary SCT are classified according to their cell of origin and the location of the mass in relationship to the spinal cord parenchyma: either intramedullary or extramedullary (i.e. developing from cells that are intrinsic or extrinsic, respectively, to the spinal cord). Extramedullary SCT are further subdivided into intradural and extradural (i.e. developing inside or outside, respectively, the fibrous dural covering of the spinal cord; while remaining extrinsic to the spinal cord itself). In adults, more than 85% of SCT are extramedullary (Table 4.1). Of this group, the intradural tumors are most frequent, with meningiomas and schwannomas comprising 55–60% of the total (Table 4.2) [1–3]. Less common extramedullary tumors are sarcomas, exophytic ependymomas and astrocytomas, epidermoids and dermoids. Intramedullary tumors are much less common, accounting for roughly 10–12% of all SCT in adults (see Table 4.1) [1–3]. Ependymomas and astrocytomas are responsible for more than 80% of these cases (see Table 4.2). In children, the distribution of SCT is different from

TABLE 4-1 Anatomic distribution of primary spinal cord tumors in adults	
Location	%
Intramedullary	10–12
Extramedullary	88–90
extradural	25
intradural	60
cauda equina	5

Adapted from References [1–3]

TABLE 4-2 Clinically important primary spinal cord tumors in adults			
Extramedullary	%	Intramedullary	%
Schwannoma	29	Ependymoma	55
Meningioma	25	Astrocytoma	30
Exophytic ependymoma	13	Hemangioblastoma	5
Sarcoma	12	Other	5–10
Exophytic astrocytoma	6	mixed glioma oligodendroglioma dermoid	
Other epidermoid dermoid teratoma lipoma vascular tumors chordoma	10–15	epidermoid teratoma Total	100
Total	100		

Adapted from References [1–3]

that in adults, with a greater percentage of intramedullary tumors, especially astrocytomas and ependymomas (Table 4.3) [4–6]. Of the extramedullary tumors that do develop in children, neuroblastomas are common; meningiomas and neurofibromas are much less frequently diagnosed.

TABLE 4-3 Clinically important primary spinal cord tumors in children

Extramedullary	%	Intramedullary	%
Extradural		Astrocytoma	15
Neuroblastoma	16	Ependymoma	13
Ganglioneuroma	3	Hemangioma	3
Osseous	2	Ganglioglioma	1
Other	5	Other	5
Intradural		lipoma	
Dermoid	9	oligodendroglioma	
Teratoma	8	epidermoid	
Neurofibroma	7	dermoid	
Lipoma	4		
Meningioma	3		
Schwannoma	2		

Adapted from References [4–6]

Spinal cord tumors are relatively rare, comprising 0.5% of all newly diagnosed tumors and 5–12% of all primary CNS neoplasms, with an annual incidence of approximately 1 case per 100 000 population per year [1–3,7–12]. The majority of SCT occur between the ages of 20 and 50, but can arise at any age. When all cases are taken into account (i.e. ante-mortem and post-mortem), the incidence of SCT appears to increase with age, such that the age groups of 0 to 24, 25 to 44, 45 to 64 and 65 years and older have rates of 0.7, 1.0, 2.8 and 3.6 per 100 000 person-years, respectively [7]. If only ante-mortem cases are included, the incidence drops off dramatically for patients over 60 years of age. The prevalence of SCT is much lower than that of primary brain tumors, with a ratio of 1:4 to 1:8. Men and women develop SCT with equal overall incidence, although specific tumor types are known to be more common in women (i.e. meningiomas) and men (i.e. ependymomas). The incidence of SCT does not appear to be related to social class, occupation or marital status [10]. However, among non-Hispanic white patients, an increased incidence has been observed in Jewish females and those of either gender born in Eastern Europe [10]. The anatomical distribution of primary SCT is uneven, with 50–55% of tumors arising in the thoracic spinal canal, 25–30% involving the lumbosacral region and the remaining 15–25% affecting the cervical spinal canal and foramen magnum.

The most common extramedullary tumors are peripheral nerve sheath tumors, which include schwannomas and neurofibromas, and meningiomas [1–3,7–10]. In patients with sporadic disease (i.e. do not have neurofibromatosis [NF1, NF2]), schwannomas are the most frequently diagnosed peripheral nerve sheath tumors of the spine [13–15]. They present as a solitary lesion attached to dorsal sensory branches. When schwannomas occur as a component of NF2, they are more likely to be multifocal and spread throughout the neuraxis. Schwannomas have an incidence rate similar to or slightly higher than meningiomas; however, some reports suggest they are significantly more common than meningiomas [14]. Schwannomas account for approximately one quarter of all SCT and 25–35% of all intradural, extramedullary tumors. Specific incidence estimates range between 0.3 and 0.5 cases per 100 000 population per year. In most studies, the incidence is similar for males and females, although a few scattered reports suggest these tumors are more common in males (61% versus 39%) [13].

Meningiomas of the spine account for 12–15% of all meningiomas and for 20–25% of all primary SCT, making them the second most common spinal neoplasm [1,2,7–10,16–18]. They are usually noted in the thoracic region (80%), but may occur in the cervical (17%) and lumbosacral (3%) spine as well. The tumor is intradural and often noted near the nerve root exit sites, but does not directly involve the nerve root. In up to 15% of cases, there can be an extradural component of the mass. The mass generally presents as a solitary lesion; multiple tumors are only noted in 1–2% of patients, sometimes in association with neurofibromatosis. Spinal meningiomas have a strong predilection for female patients, with a female to male ratio of up to 5:1 in some series. However, the age of onset and age distribution is equivalent between males and females. The median age at diagnosis is in the mid-50s, with the majority of cases occurring between 40 and 70 years of age. Although spinal meningiomas are rare in children, younger patients can develop these tumors and often have a more aggressive course [19]. In this younger cohort, the tumors have a stronger predilection for the cervical spine and are more often associated with predisposing factors (e.g. NF2, radiation exposure). Spinal trauma and exposure to ionizing radiation are both considered to be risk factors for developing a spinal meningioma [17]. However, the risk has been more firmly established in patients with intracranial meningiomas.

Spinal ependymomas most often arise as an intraparenchymal mass or as an intradural, extramedullary lesion in the lumbosacral spine [1,2,7–10]. They comprise approximately 35% of all CNS ependymomas and 13–15% of all spinal tumors. Ependymomas are the most common intramedullary SCT in adults and account for 55–60% of all cases [20–22]. The tumor can develop at any age, but has a typical onset in the middle adult years. Although some studies show an equal distribution between male and female patients, the majority of reports suggest a male predominance. Intramedullary ependymomas have a predilection for the cervical spinal cord, such that 60–70% of tumors arise from or extend into this region. Syringomyelia is associated with ependymomas of the spinal parenchyma in two-thirds of cases, especially those arising in the cervical cord. In 40–45% of patients, intradural ependymomas originate from the conus medullaris (30%) or filum terminale (65%) within the lumbosacral thecal sac, and are considered to be extramedullary tumors [2,20,23].

The vast majority of these lesions are of the myxopapillary histological subtype. Myxopapillary ependymomas account for approximately 15–18% of all spinal ependymomas and tend to affect young adults, with a mean age of onset of 36 years. Similar to their intramedullary counterparts, ependymomas of the cauda equina region tend to have a male predominance, with a male to female ratio of 1.7:1 to 2:1 [23].

Spinal astrocytomas are relatively uncommon and only account for 3–4% of all CNS astrocytomas, with an incidence of 0.8–2.5 cases per 100 000 population per year [1,2,11,12]. In adults, they comprise 7–10% of all primary SCT and 30–35% of all intramedullary SCT. In children, they tend to be more common and are the most frequently diagnosed intramedullary tumor, comprising 90% or more of primary SCT in patients less than 10 years of age, and up to 60% of primary SCT in adolescents [4,6]. Although spinal astrocytomas can occur at any age, they usually affect young adults, with an onset of symptoms between 30 and 35 years of age. Gender distribution between male and female patients is fairly even in some series, while others would suggest a slight male predominance [24]. The tumor arises in the cervical spinal cord or cervicothoracic region in approximately 60% of patients. Lesions of the lower thoracic and lumbosacral cord, conus medullaris and filum terminale are less common. Tumors spanning the entire spinal cord (i.e. holocord tumors) can occur, but are very rare. In up to 40% of cervical and upper thoracic cases, there may be an associated syringomyelia.

Hemangioblastomas are slowly growing, highly vascular tumors of the spinal cord that can occur sporadically (75–80%) or in association with von-Hippel-Lindau syndrome (VHL; 20–25%) [2,12,25,26]. The tumor is usually a solitary mass in sporadic cases, but can be multifocal in patients with VHL. Hemangioblastomas account for 2–8% of all intramedullary SCT and are the third most common intramedullary tumor. They can arise anywhere within the spinal cord, but have a predilection for the cervical and thoracic regions. This predilection for the cervical cord is more pronounced in patients with VHL. The mass is entirely intramedullary in 30% of cases, but more often will have intramedullary and extramedullary components (50–55%). In 15–20% of cases the lesion is located completely within the extramedullary compartment. The tumor usually affects young adults, with an age of onset between 35 and 40 years; it is relatively uncommon in pediatric patients. In patients with VHL, the age of onset tends to be 5–10 years earlier. Males are affected more often than females, with a male-to-female ratio of 1.7:1.

Spinal cord oligodendrogliomas are uncommon intramedullary tumors that represent 1–2% of all primary SCT and 1.5–2% of all CNS oligodendrogliomas [1–6,27–29]. They usually arise in the thoracic (30%) and cervical (25%) portions of the cord, but can involve multiple regions (e.g. thoraco-lumbar, 15%; holocord, 7.5%) and the lumbar spine (5%). The tumor is most often diagnosed in young

adults, however, children and adolescents can be affected. There appears to be an even distribution between male and female patients.

Other less common neoplasms that may involve the spine include lipomas, maldevelopmental tumors (e.g. teratoma, dermoids, epidermoids), paragangliomas, chordomas, chondrosarcomas, sarcomas and lymphomas [1–6,10,30,31]. Lipomas account for 1% of all intramedullary tumors and are the most common dysembryogenic lesion of the spine [32]. They are not true neoplasms and most likely arise from an inclusion of mesenchymal tissue within the spine. Dermoids, epidermoids and teratomas can be intramedullary or extramedullary, but tend to involve the thoracolumbar and lumbar spine. Epidermoid tumors are more common in male patients. Paragangliomas are tumors of neural crest origin that usually arise from the filum terminale or cauda equina. Chordomas of the spine account for 4% of all spinal tumors and 11–15% of all CNS chordomas [33]. They are most often diagnosed in the fourth decade of life and have a predilection for males, with a male-to-female ratio of 2:1. Chondrosarcomas of the spine account for 4–8% of all chondrosarcomas and have a male-to-female ratio of 1.5:1 [34]. They involve the true vertebrae in 30–35% of patients (usually thoracic) and affect the sacrum in another third. The age at presentation ranges between 45 and 50 years in most series. Primary sarcomas of the spine are very rare and account for 0.7% of all CNS tumors and 5.6% of all SCT [35]. Lymphomas of the spine are usually of B-cell origin and can arise within the spinal cord parenchyma or extradurally; 13% will have bony involvement [31]. The onset of symptoms is typically in the fifth to seventh decade.

EPIDEMIOLOGY OF INTRAMEDULLARY SPINAL CORD METASTASES

Intramedullary spinal cord metastases (ISCM) are an uncommon neurological complication of systemic cancer [36]. In comparison to neoplastic epidural spinal cord compression (see Chapter 7), ISCM are diagnosed much less frequently, with an incidence approximately one sixteenth as high [36,37]. ISCM are estimated to affect between 0.5 and 2.0% of all patients with cancer, representing only 8.5% of all CNS metastases, and comprise 1–3% of all intramedullary SCT [37–40]. However, the true incidence may be somewhat higher, since ISCM are noted in 2–4% of all cancer patients studied at autopsy, suggesting that many patients with this complication never become symptomatic.

ISCM can arise from any primary systemic neoplasm, but are due to lung cancer in 50–60% of cases (Table 4.4) [37–40]. Small cell lung carcinoma (SCLC) accounts for approximately half of these cases, but any of the other histological subtypes can be involved. Overall, ISCM are estimated to occur in up to 5% of patients with SCLC, but only 1% or so in patients with non-SCLC [39,41,42]. Other primary malignancies with a tendency to metastasize to

TABLE 4-4 Percentage of intramedullary spinal cord metastases by primary tumor site

Primary type	%
Lung	50–60
SCLC	50–55
Non-SCLC	45–50
Breast	10–13
Melanoma	9
Renal cell	5–9
Lymphoma	5
Colorectal & GI	3
Thyroid	2
Ovarian	1–2

Adapted from references [37–46]. SCLC: small cell lung carcinoma, GI: gastrointestinal

the spinal cord include breast cancer (10–13% of all ISCM), melanoma (9%), renal cell carcinoma (5–9%), lymphoma (5%), colorectal cancer and other gastrointestinal malignancies (3%), thyroid carcinoma (2%) and ovarian cancer (1–2%) [43–46]. The presence of ISCM suggests an advanced form of cancer and biologically aggressive disease. For example, at the time of diagnosis of ISCM, 55–60% of patients will already have been diagnosed with brain metastases and another 25% will have leptomeningeal metastases [37].

The mean age at presentation of ISCM in the literature is 58 years, with a range of 38 to 78 years [37–40]. Males are affected more often than females (64% versus 46%), but there does not appear to be any predilection for specific ethnic or racial groups. ISCM are distributed fairly evenly along the spinal cord parenchyma in most studies, affecting the cervical, thoracic and lumbar regions in 24%, 22% and 28% of cases, respectively. However, some reports would suggest a mild preponderance of involvement in the conus medullaris as, for example, the study by Schiff and O'Neill, in which the conus accounted for over half the cases [37]. The metastatic deposits are solitary in the majority of patients, with multifocal disease noted in approximately 15% [39]. ISCM usually occurs in the context of a well-established diagnosis of cancer. However, in some patients, ISCM can be the initial presentation of systemic cancer. This has been noted in up to 22.5% of ISCM patients in some reports [37].

ACKNOWLEDGMENTS

The authors would like to thank Julia Shekunov for research assistance. Dr Newton was supported in part by National Cancer Institute grant, CA 16058 and the Dardinger Neuro-Oncology Center Endowment Fund.

REFERENCES

1. Newton HB, Newton CL, Gatens C, Hebert R, Pack R (1995). Spinal cord tumors: Review of etiology, diagnosis, and multi-disciplinary approach to treatment. Cancer Pract 3:207–218.
2. Stieber VW, Tatter SB, Shaffrey ME, Shaw EG (2005). Primary spinal tumors. In *Principles of Neuro-Oncology*. Schiff D, O'Neill BP (eds). McGraw Hill Publishers, New York. 24:501–531.
3. Bhattacharyya AK, Guha A (2005). Spinal root and peripheral nerve tumors. In *Principles of Neuro-Oncology*. Schiff D, O'Neill BP (eds). McGraw Hill Publishers, New York. 25:533–549.
4. Townsend N, Handler M, Fleitz J, Foreman N (2004). Intramedullary spinal cord astrocytomas in children. Pediatr Blood Cancer 43:629–632.
5. Baysefer A, Akay KM, Izci Y, Kayali H, Timurkaynak E (2004). The clinical and surgical aspects of spinal tumors in children. Pediatr Neurol 31:261–266.
6. Auguste KI, Gupta N (2006). Pediatric intramedullary spinal cord tumors. Neurosurg Clin N Am 17:51–61.
7. Sasanelli F, Beghi E, Kurland KT (1983). Primary intraspinal neoplasms in Rochester, Minnesota 1935–1981. Neuroepidemiology 2:156–163.
8. Fogelholm R, Uutela T, Murros K (1984). Epidemiology of central nervous system neoplasms. A regional survey in central Finland. Acta Neurol Scand 69:129–136.
9. Helseth A, Mork SJ (1989). Primary intraspinal neoplasms in Norway, 1955 to 1986. A population-based survey of 467 patients. J Neurosurg 71:842–845.
10. Preston-Martin S (1990). Descriptive epidemiology of primary tumors of the spinal cord and spinal meninges in Los Angeles County 1972–1985. Neuroepidemiology 9:106–111.
11. Preston-Martin S (1996). Epidemiology of primary CNS neoplasms. Neurol Clin 14:273–290.
12. Tihan T, Chi JH, McCormick PC, Ames CP, Parsa AT (2006). Pathologic and epidemiologic findings of intramedullary spinal cord tumors. Neurosurg Clin N Am 17:7–11.
13. Conti P, Pansini G, Mouchaty H, Capuano C, Conti R (2004). Spinal neurinomas: retrospective analysis and long-term outcome of 179 consecutively operated cases and review of the literature. Surg Neurol 61:35–44.
14. Jinnai T, Hoshimaru M, Koyama T (2005). Clinical characteristics of spinal nerve sheath tumors: analysis of 149 cases. Neurosurgery 56:510–515.
15. Waldron J, Weinstein RP (2005). Nerve sheath tumors of the spine. In *Textbook of Neuro-Oncology*. Berger MS, Prados MD (eds). Elsevier Saunders, Philadelphia. 63:485–488.
16. Gezen F, Kahraman S, Canakci Z, Beduk A (2000). Review of 36 cases of spinal cord meningioma. Spine 15:727–731.

17. Cohen-Gadol AA, Krauss WE (2005). Spinal meningiomas. In *Textbook of Neuro-Oncology*. Berger MS, Prados MD (eds). Elsevier Saunders, Philadelphia. 64:489–492.

18. Peker S, Cerci A, Ozgen S, Isik N, Kalelioglu M, Pamir MN (2005). Spinal meningiomas: evaluation of 41 patients. J Neurosurg Sci 49:7–11.

19. Cohen-Gadol AA, Zikel OM, Koch CA, Scheithauer BW, Krauss WE (2003). Spinal meningiomas in patients younger than 50 years of age: a 21-year experience. J Neurosurg 98:258–263.

20. Schwartz TH, McCormick PC (2000). Intramedullary ependymomas: clinical presentation, surgical treatment strategies and prognosis. J Neuro-Oncol 47:211–218.

21. Chang UK, Choe WJ, Chung SK, Chung CK, Kim HJ (2002). Surgical outcome and prognostic factors of spinal intramedullary ependymomas in adults. J Neuro-Oncol 57:133–139.

22. Schwartz TH, Parsa AT, McCormick PC (2005). Intramedullary ependymomas. In *Textbook of Neuro-Oncology*. Berger MS, Prados MD (eds). Elsevier Saunders, Philadelphia. 66:497–500.

23. Parney IF, Parsa AT (2005). Myxopapillary ependymomas. In *Textbook of Neuro-Oncology*. Berger MS, Prados MD (eds). Elsevier Saunders, Philadelphia. 65:493–496.

24. Minehan KJ, Shaw EG, Scheithauer BW et al (1995). Spinal cord astrocytoma: pathological and treatment considerations. J Neurosurg 84:590–595.

25. Lonser RR, Oldfield EH (2006). Spinal cord hemangioblastomas. Neurosurg Clin N Am 17:37–44.

26. Wanebo JE, Lonser RR, Glenn GM, Oldfield EH (2003). The natural history of hemangioblastomas of the central nervous system in patients with von Hippel-Lindau disease. J Neurosurg 98:82–94.

27. Fortuna A, Celli P, Palma L (1980). Oligodendrogliomas of the spinal cord. Acta Neurochir 52:305–329.

28. Pagni CA, Canavero S, Gaidolfi E (1991). Intramedullary 'holocord' oligodendroglioma: case report. Acta Neurochir 113:96–99.

29. Fountas KN, Karampelas I, Nikolakakos LG, Troup EC, Robinson JS (2005). Primary spinal cord oligodendroglioma: case report and review of the literature. Childs Nerv Syst 21:171–175.

30. Waldron J, Ames C (2005). Benign tumors of the spine. In *Textbook of Neuro-Oncology*. Berger MS, Prados MD (eds). Elsevier Saunders, Philadelphia. 65:511–516.

31. Chou D, Gökaslan Z (2005). Malignant primary tumors of the vertebral column. In *Textbook of Neuro-Oncology*. Berger MS, Prados MD (eds). Elsevier Saunders, Philadelphia. 70:517–520.

32. Lee M, Rezai AR, Abbott R et al (1995). Intramedullary spinal cord lipomas. J Neurosurg 82:394–400.

33. Bjornsson J, Wold LE, Ebersold MJ, Laws ER (1993). Chordoma of the mobile spine. A clinicopathologic analysis of 40 patients. Cancer 71:735–740.

34. York JE, Berk RH, Fuller GN et al (1997). Chondrosarcoma of the spine: 1954 to 1997. J Neurosurg 90:73–78.

35. Merimsky O, Lepechoux C, Terrier P et al (2000). Primary sarcomas of the central nervous system. Oncology 58:210–214.

36. Newton HB (1999). Neurological complications of systemic cancer. Am Fam Phys 59:878–886.

37. Schiff D, O'Neill BP (1996). Intramedullary spinal cord metastases: clinical features and treatment outcome. Neurology 47:906–912.

38. Mut M, Schiff D, Shaffrey ME (2005). Metastases to nervous system: spinal epidural and intramedullary metastases. J Neuro-Oncol 75:43–56.

39. Chi JH, Parsa AT (2006). Intramedullary spinal cord metastasis: clinical management and surgical considerations. Neurosurg Clin N Am 17:45–50.

40. Kalayci M, Cagavi F, Gul S et al (2004). Intramedullary spinal cord metastases: diagnosis and treatment – an illustrated review. Acta Neurochir (Wien) 146:1347–1354.

41. Nikolaou M, Koumpou M, Mylonakis N, Karabelis A, Pectasides D, Kosmas C (2006). Intramedullary spinal cord metastases from atypical small cell lung cancer: a case report and literature review. Cancer Investig 24:46–49.

42. Potti A, Abdel-Raheem M, Levitt R et al (2001). Intramedullary spinal cord metastases (ISCM) and non-small cell lung carcinoma (NSCLC): clinical patterns, diagnosis and therapeutic considerations. Lung Cancer 31:319–323.

43. Villegas AE, Guthrie TH (2004). Intramedullary spinal cord metastasis in breast cancer: clinical features, diagnosis, and therapeutic consideration. Breast J 10:532–535.

44. Conill C, Sanchez M, Puig S et al (2004). Intramedullary spinal cord metastases of melanoma. Melanoma Res 14:431–433.

45. Taniura S, Tatebayashi K, Watanabe K, Watanabe T (2000). Intramedullary spinal cord metastasis from gastric cancer. Case report. J Neurosurg (Spine 1) 93:145–147.

46. Fakih M, Schiff D, Erlich R, Logan TF (2001). Intramedullary spinal cord metastasis (ISCM) in renal cell carcinoma: A series of six cases. Ann Oncol 12:1173–1177.

CHAPTER 5

Overview of Pathology and Treatment of Primary Spinal Cord Tumors

Herbert B. Newton, Abhik Ray-Chaudhury and Mark G. Malkin

INTRODUCTION

The epidemiology of primary spinal cord tumors (SCT) was reviewed in detail in Chapter 4. In this chapter, we will provide an overview of the classification, pathology and treatment of the common primary SCT. Primary SCT will be diagnosed in approximately 1500–2000 new patients this year and are associated with significant morbidity and mortality [1–6]. Of the estimated one patient per 100 000 population that will develop a primary SCT this year in the USA, the majority will have an intradural, extramedullary, low-grade neoplasm such as a schwannoma, meningioma or neurofibroma. Less often, the tumor will be intramedullary and arise from the spinal cord parenchyma, such as an ependymoma or astrocytoma. Although high-grade SCT derived from parenchymal cells or surrounding tissues are occasionally noted, they are infrequent in comparison to their counterparts in the brain.

PATHOLOGY OF SELECTED PRIMARY SPINAL CORD TUMORS

Similar to the situation discussed in Chapter 2 for primary brain tumors (PBT), specific knowledge of SCT pathology is necessary before therapeutic strategies can be finalized. In addition to assisting with treatment decisions, tumor classification and grade can provide important prognostic information for patients and families. This chapter will follow the World Health Organization (WHO) classification that separates nervous system tumors into different nosological entities and assigns a grade of I to IV to each lesion (Table 5.1), with grade I being biologically benign and grade IV being biologically most malignant and having the worst prognosis [7,8]. Within the WHO classification, tumors of neuroepithelial and meningeal origin, as well as tumors of the cranial and spinal nerves, are the most clinically relevant groups when considering primary SCT pathology.

In general, the histologic features of primary SCT are similar to the same type of tumor when it occurs in the

TABLE 5-1 WHO classification: tumors of the central nervous system
Tumors of neuroepithelial tissue
Tumors of cranial nerves and spinal nerves
Tumors of the meninges
Lymphomas and hemopoietic neoplasms
Germ cell tumors
Tumors of the sellar region
Cysts and tumor-like lesions
Metastatic tumors

brain or intracranial cavity [7,8]. However, the distribution of histological variants and grade may be different for a given tumor type when it develops in the spine as opposed to the brain. Tumors arising in the extramedullary compartment are most common and represent 88–90% of all primary SCT. Of this large and diverse group, schwannomas, meningiomas, neurofibromas, myxopapillary ependymomas, chordomas and hemangioblastomas are most frequently diagnosed.

Schwannomas

On gross pathological inspection, spinal schwannomas appear as discrete, rounded, firm, encapsulated masses arising from a nerve fascicle. The tumors may have variable amounts of cyst formation, yellowish areas of xanthomatous changes and hemorrhage. During the early 'intraneural' phase of growth, the tumor is fusiform in shape, similar to neurofibromas [3,7–10]. As the tumor enlarges, the adjacent nerve fascicles are compressed and displaced eccentrically. The nerve fascicles are usually not infiltrated by, or encased within, the mass, although they may be incorporated superficially into the tumor capsule. On microscopic examination, 'classic' schwannomas are composed of a heterogeneous, biphasic architecture that contains two distinct regions: Antoni A and Antoni B (Figure 5.1) [9].

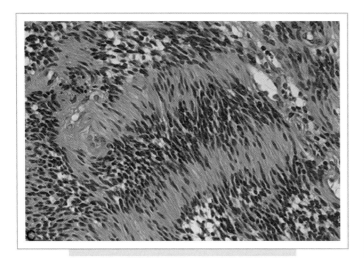

FIG. 5.1. Spinal schwannoma. Classic schwannoma demonstrating biphasic architecture composed of Antoni A and Antoni B regions. Antoni A regions are characterized by dense, compact rows of elongated, spindle-shaped cells with rod-shaped nuclei and eosinophilic cytoplasm. Antoni B regions are loosely organized and contain large, vacuolated, stellate cells with areas of microcystic change and lipid accumulation. H&E @ 400×.

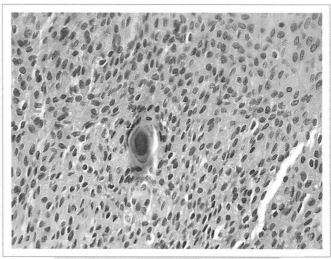

FIG. 5.2. WHO grade I spinal meningioma. High-power view of classic spinal meningioma, demonstrating numerous oval to elongated meningothelial cells in a syncytial pattern. A solitary psammoma body is present in the center of the field. H&E @ 400×.

In most tumors, the Antoni A regions predominate and are organized into dense, compact rows or arrays of elongated, spindle-shaped cells that have hyperchromatic, rod-shaped nuclei and eosinophilic cytoplasm. The nuclei are often aligned into palisades that alternate with dense, anuclear zones of fibrillar eosinophilic material; these structures are called Verocay bodies. In some tumors, the cell bundles form whorls of various sizes. The Antoni B regions are loosely organized and composed of large, vacuolated, pleomorphic stellate cells with pyknotic or irregular nuclei. Areas of microcystic change, hyalinization of blood vessels, hemorrhage with perivascular hemosiderin deposition, lipid accumulation and nuclear degenerative atypia are common. Mitoses and nuclear pleomorphism can be seen on occasion, but do not imply malignant potential. In addition to the 'classic' microscopic appearance of these tumors, several less common forms exist, including the cellular, ancient, plexiform, melanotic and malignant schwannoma variants [9].

Meningiomas

Spinal meningiomas have meningothelial cell origins and are derived from neoplastic arachnoidal cap cells embedded within the dura of the spinal canal [7,8,11]. The tumors usually present as rounded or globoid masses, with a variable consistency (i.e. soft to firm), that are attached to the dura and arise near the nerve root sleeve. Less often, the mass will have a carpet-like or 'en plaque' configuration, extending along the dura. Cervical meningiomas are most likely to have an anterior location, while thoracic cases tend

to occur more laterally. The microscopic features of spinal meningiomas are similar to meningiomas of the intracranial cavity (see Chapter 2), with the same spectrum of histological variants as recognized by the WHO. Meningioma subtypes that have a more indolent nature and low risk for aggressive growth or recurrence are classified as WHO grade I and include the meningothelial, fibrous/fibroblastic, transitional (mixed), secretory, psammomatous, angiomatous, microcystic, lymphoplasmocyte-rich and metaplastic variants [12,13]. Highly calcified, psammomatous lesions are the most frequent subtype noted in the spine. The histological features common to most low-grade meningiomas are the presence of whorls (tightly wound, rounded collections of cells), psammoma bodies (concentrically laminated mineral deposits that often begin in the center of whorls), intranuclear pseudoinclusions (areas in which pink cytoplasm protrudes into a nucleus to produce a hollowed-out appearance) and rare pleomorphic nuclei and mitoses (Figure 5.2). In general, high-grade meningiomas, especially atypical and anaplastic varieties (WHO grades II and III, respectively), are rare in the spinal meninges. However, the clear cell variant (WHO grade II) is often noted in the lumbar region, an unusual location for other subtypes.

Neurofibromas

On gross pathological examination, spinal neurofibromas are soft, well-circumscribed, pedunculated and unencapsulated gelatinous masses of a whitish or opalescent color [3,7,8,10]. Regions of cyst formation, xanthomatous changes and hemorrhage are not seen as commonly as they are with schwannomas. During initial phases of growth, the tumor infiltrates the parent nerve, causing a localized,

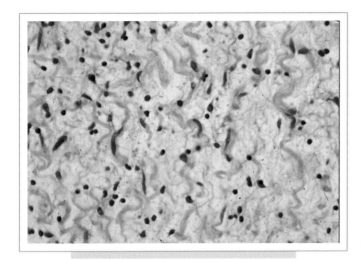

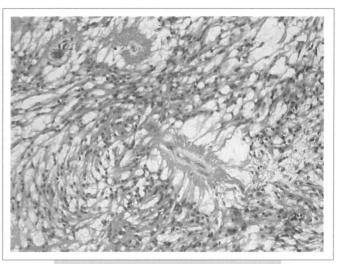

FIG. 5.3. Spinal neurofibroma. Typical neurofibroma demonstrating interlacing bundles of fusiform Schwann cells with wavy nuclei. H&E @ 200×.

FIG. 5.4. WHO grade I myxopapillary ependymoma. Medium-power view of classic myxopapillary ependymoma demonstrating the tumor cells forming perivascular radial arrangements (i.e. rosettes; at the center), within a background mucoid matrix. Note that some of the blood vessel walls (center and top left) display mucoid degeneration. H&E @ 200×.

fusiform swelling. As the tumor enlarges, the parent nerve and those nerves around it may develop gross alterations of shape (e.g. 'bag of worms') and can become encased within the mass. On microscopic histological examination, the typical spinal neurofibroma has interlacing bundles of fusiform Schwann cells with wavy nuclei within a matrix of collagen-rich and mucopolysaccharide-rich material (Figure 5.3). Because of the infiltrative growth pattern of neurofibromas, axons of nerve fibers are easily demonstrated after silver impregnation of tumor tissue. Features common to schwannomas such as palisading, Verocay bodies and whorling are noted infrequently. The architecture and cellular appearance can vary, producing several subtypes: storiform perineural fibroma, pacinian neurofibroma, epithelioid neurofibroma and pigmented neurofibroma [7,8].

Myxopapillary ependymomas

Myxopapillary ependymomas arise predominantly in the region of the filum terminale and cauda equina and present as elongated, sausage-shaped masses with a smooth, lobulated surface [7,8,14,15]. The tumor is usually well defined and almost appears encapsulated, displacing or compressing the nerve roots of the cauda equina, with only a mild tendency to infiltrate or envelope neural structures. Some tumors can be quite large at presentation (e.g. 10 cm or greater in length) and are known to cause erosion and scalloping of the surrounding vertebral bodies and sacrum. On microscopic histological examination, myxopapillary ependymomas are classified as WHO grade 1. Characteristic features include the presence of numerous small papillary structures, each surrounded by well-defined cuboidal or columnar cells, usually in a single layer (Figure 5.4) [15]. These cells have rounded, ependymal

type nuclei with distinct margins and a delicate chromatin meshwork, without obvious cytoplasmic processes. The cores of the papillae have a myxoid appearance and consist of a central blood vessel surrounded by a mucinous matrix, or are entirely filled by the mucinous material. In papillae with central blood vessels, extensive thickening and hyalinization of the vessel wall may be seen. Other regions of the tumor show a looser structure, with tumor cell nuclei embedded in a meshwork of fine cytoplasmic processes. The processes are often arranged in a radial fashion around the basement membranes of blood vessels and enclose numerous microcystic spaces. On occasion, more compact gliofibrillar areas resembling typical ependymomas can be noted, with the presence of perivascular pseudorosettes or even true ependymal rosettes. Degenerative features, such as microvascular hyalinization, thrombosis, hemorrhage and hemosiderin, are often present. At the edge of the tumor, there is usually a sharp margin between neoplastic cells and normal tissues, although in some cases nerve roots can be enclosed. Mitotic figures and other features of anaplasia are not usually seen.

Chordomas

Chordomas of the spine present as slow-growing, unencapsulated neoplasms that are locally invasive within bone and, in some cases, regional soft tissues [7,8,16–18]. A pseudocapsule may be noted around tumors that grow into soft tissues or the dura mater. As the tumors enlarge, they often stretch spinal nerves and displace blood vessels. Grossly, the tumors are usually reddish or purple in color,

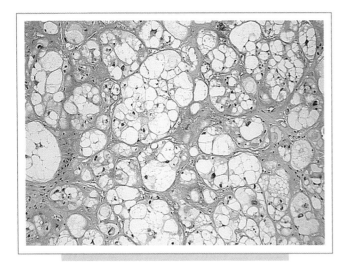

FIG. 5.5. Spinal chordoma. Medium-power view of classic or typical chordoma, demonstrating physaliphorous (bubble-bearing) cells. Note the large size, vacuolization and eccentric nuclei. Mitoses, spindle cells and regions of necrosis are absent. H&E @ 100×.

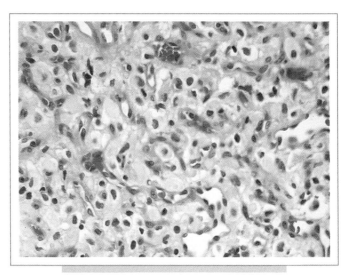

FIG. 5.6. Spinal hemangioblastoma. High-power view of typical spinal hemangioblastoma, demonstrating the rich capillary network and large, vacuolated (i.e. lipid laden) stromal cells. H&E @ 400×.

with a nodular appearance to the surface. Internally, the mass is frequently gelatinous and soft; regions that contain cartilage or calcium are more firm. Foci of hemorrhage may be present and can be small or extensive. The size of the lesion can be quite variable, with lumbosacral tumors often becoming very large. On microscopic examination, chordomas can be grouped into several different histological categories, including a typical pattern, a chondroid pattern and tumors with features of malignant degeneration [16–18]. The typical or classic pattern of chordoma (65–80% of all cases) is distinguished by a lobular arrangement, with the neoplastic cells disposed in solid sheets or irregular intersecting cords (Figure 5.5). The sheets and cords of cells are set in a stroma that contains an abundant mucinous matrix. The individual cells are large, often with vacuolated eosinophilic cytoplasm and contain variable amounts of mucin. The cell type considered diagnostic for chordomas is called physaliphorous (i.e. bubble-bearing). These cells are distinctively large and vacuolated, with eccentric nuclei (see Figure 5.5). Nuclei tend to be hyperchromatic, with prominent nucleoli, and rarely demonstrate atypia. Potentially aggressive features such as mitoses, necrosis, hypervascularity and spindle cells (i.e. sarcomatous degeneration) are typically absent or rare. Some authors contend that the chondroid pattern (15–30% of all cases) is a separate histological variant of chordoma, although this is controversial [16–19]. On histological examination, chondroid chordomas contain regions of typical chordoma with physaliphorous cells, against a background of areas characterized by cartilaginous matrix that have stellate tumor cells occupying lacunar spaces (resembling chondrocytes) [19]. As in typical chordoma, anaplastic or aggressive features

such as mitoses, necrosis, hypervascularity, and spindle cells are typically absent or rare.

Hemangioblastomas

Hemangioblastomas are red-brown tumors that are well circumscribed and can arise entirely within the spinal cord parenchyma, or be partly or entirely extramedullary [1–3,7,8,20,21]. Intra-parenchymal tumors often have a superficial margin at the pial surface, which is usually covered by ectatic and tortuous blood vessels. Extra-axial tumors are usually well-defined masses growing out over the posterior surface of the cord, sometimes with a projection of neoplastic tissue into the cord parenchyma. In some cases, the tumor may appear as discrete nodules attached to the posterior nerve roots. Between 60 and 70% of hemangioblastomas are associated with a large cystic cavity, with the solid portion of the tumor embedded in the cyst wall as a mural nodule. In the spinal cord, the cysts are elongated syrinx cavities, which can extend over several levels above and below the solid portion of the tumor. The cyst usually contains yellow or rusty brown proteinaceous fluid and has a lining that is smooth and opaque. Cystic or syrinx cavities are not formed from tumor tissue; the lining is composed of a dense layer of gliotic cells that often incorporates Rosenthal fibers and hemosiderin pigment. On microscopic examination, the predominant features of hemangioblastomas are an extensive meshwork of delicate blood vessels and larger, sinusoidal, vascular spaces (Figure 5.6) [7,8,20,21]. All of these vessels have a capillary structure, with thin walls formed by plump endothelial cells. In between the vascular channels are variable numbers of intervascular or stromal cells arranged in strands and sheets. The cells are large and rounded or polygonal in shape, with abundant cytoplasm and small, rounded, central or eccentrically placed nuclei. The nuclei may occasionally have a vacuolated appearance. In most cells, the cytoplasm is somewhat foamy in appearance and swollen by neutral lipid. Although pleomorphic, multinucleated or giant stromal cells may be noted in some cases, mitotic figures are extremely rare. Overall, these tumors are considered to be low-grade neoplasms (i.e. WHO grade I) and do not display intrinsic tumor necrosis or cytological evidence of malignancy.

Tumors arising in the intramedullary compartment are less common and represent 10–12% of all primary SCT in adults [1–3]. They are histologically similar to their counterparts in the brain, but are often of lower grade [7,8,22]. Of this group, ependymomas, astrocytomas, hemangioblastomas, maldevelopmental tumors (e.g. epidermoid) and other glial tumors (e.g. oligodendroglioma) are most often diagnosed.

Ependymomas

Intramedullary ependymomas are derived from the ependymal lining cells of the central canal and arise as well circumscribed masses within the spinal cord parenchyma (usually in the cervical cord or at the cervico-thoracic junction), with very little tendency to infiltrate adjacent neural tissue [7,8,22–24]. Spinal cord ependymomas are commonly divided into typical (WHO grade II) or anaplastic (WHO grade III) varieties. In addition, the WHO recognizes four histological variants of the typical ependymoma (all WHO grade II), including the cellular, papillary, clear cell and tanycytic subtypes, as well as two low-grade (grade I) forms, myxopapillary ependymoma and subependymoma. On histological examination, WHO grade II ependymomas are densely cellular and composed of oval to carrot-shaped cells, with a dense speckled nucleus and tapering eosinophilic cytoplasm (Figure 5.7) [22–25]. Some tumors have cells with a more glial appearance and more background fibrillarity or more epithelioid cells and architecture. Perivascular pseudorosettes, which are commonly observed, are circular arrangements of tumor cells that send processes towards vessel walls, creating a perivascular 'nuclear-free zone' that can be noted at low-power. Less commonly, true ependymal rosettes, surrounding a true lumen, can be observed. Ependymomas are usually glial fibrillary acidic protein (GFAP) positive. Anaplastic ependymomas (WHO grade III) are less common in the spinal cord and have additional features such as increased cellularity, mitotic activity, pleomorphic nuclei, vascular hyperplasia, nuclear atypia and necrosis.

Astrocytomas

Astrocytomas of the spinal cord tend to be of lower grade than their counterparts in the brain [1–6]. However, they remain infiltrative lesions that often span four to six spinal cord segments at the time of diagnosis. Approximately 50% of spinal astrocytomas are WHO grade I, 22% WHO

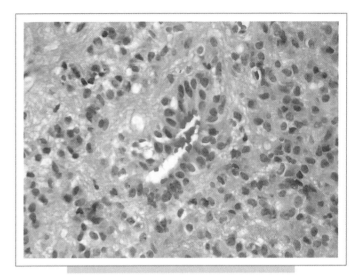

FIG. 5.7. WHO grade II intramedullary ependymoma. High-power view of typical intramedullary ependymoma, demonstrating a perivascular pseudorosette, surrounded by groups of cells with oval nuclei and scant cytoplasm. H&E @ 400×.

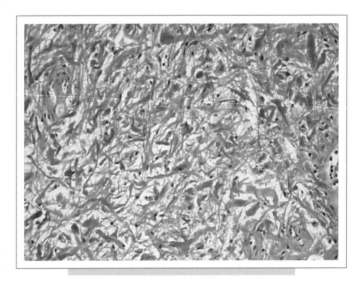

FIG. 5.8. WHO grade I spinal pilocytic astrocytoma. Medium-power view of a classic spinal pilocytic astrocytoma, demonstrating numerous eosinophilic Rosenthal fibers and astrocytic cells with thin bipolar processes which crisscross within the background. H&E @ 200×.

grade II, 20% WHO grade III and 8% WHO grade IV [1,4,26,27]. Grade I tumors (i.e. pilocytic astrocytomas) are more common in children, but can occur in young adults. They typically present as a homogeneous, well-demarcated mass, with a gliotic surrounding margin, or as an enhancing mural nodule associated with a large cyst [7,8,26,27]. The distinctive histological feature is the presence of cells with slender, elongated nuclei and thin, hair-like (i.e. piloid), GFAP-positive, bipolar processes (Figure 5.8). These cells are found

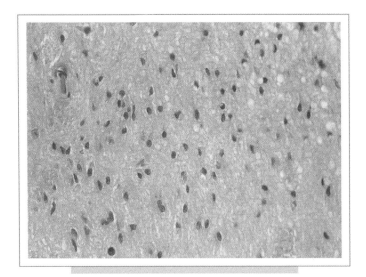

FIG. 5.9. WHO grade II spinal astrocytoma. High-power view of a typical grade II spinal astrocytoma, demonstrating oval to elongated, mildly pleomorphic and hyperchromatic astrocytic nuclei infiltrating into the neuropil. H&E @ 400×.

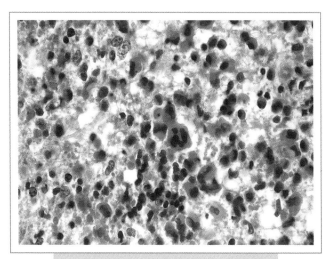

FIG. 5.10. WHO grade III spinal anaplastic astrocytoma. High-power view of a grade III spinal astrocytoma, demonstrating increased cellularity, nuclear pleomorphism, multinucleated cells and mitotic figures. No necrosis or vascular proliferation was present. H&E @ 600×.

in a biphasic background, which consists of dense fibrillary regions alternating with loose, microcystic areas. Also frequently noted are Rosenthal fibers, which are eosinophilic, refractile, corkscrew-shaped deposits. Spinal astrocytomas of higher grade (WHO grade II–IV) are considered diffuse tumors and are more infiltrative [7,8,26,27]. WHO grade II spinal astrocytomas are characterized by a relatively uniform population of proliferating neoplastic astrocytes in a fibrillary matrix, with minimal cellular and nuclear pleomorphism or atypia (Figure 5.9). Grade II tumors may display diffuse and intense staining with GFAP. Mitotic figures are absent and there is no evidence for vascular hyperplasia. Spinal anaplastic astrocytomas (AA; WHO grade III) are similar to grade II tumors, except for the presence of more prominent cellular and nuclear pleomorphism and atypia and mitotic activity (Figure 5.10). According to WHO criteria, the critical feature that upgrades a grade II tumor to an AA is the presence of mitotic activity, with anaplastic tumors having Ki-67 indices in the range of 5–10% in most cases. Other features of anaplasia can be present, such as multinucleated tumor cells and abnormal mitotic figures. Spinal glioblastoma multiforme (GBM) is classified as a WHO grade IV tumor and has similar histological features to AA, but with more pronounced anaplasia (Figure 5.11) [7,8]. The presence of microvascular proliferation and/or necrosis in an otherwise malignant astrocytoma upgrades the tumor to a GBM. Vascular proliferation is defined as blood vessels with 'piling up' of endothelial cells, including the formation of glomeruloid vessels. Necrosis can be noted in large amorphous areas, which appear ischemic in nature, or can appear as more serpiginous regions with surrounding palisading tumor

cells (i.e. perinecrotic pseudopalisading). Necrosis with nuclear pseudopalisading is essentially pathognomonic for GBM. Other features of GBM that are typically prominent include marked cellular and nuclear pleomorphism and atypia, mitotic figures and multinucleated giant cells, and pronounced infiltrative capacity into surrounding spinal cord.

SURGICAL THERAPY OF PRIMARY SPINAL CORD TUMORS

Once it is established that a patient has a lesion that is likely to be a primary SCT, surgical intervention is usually required to provide a definitive diagnosis and, if possible, to totally remove the mass [1–6]. Recent technical advances have improved the results of operative therapy for SCT. These advances include the use of the operating microscope and microneurosurgical instrumentation, bipolar coagulation, surgical lasers, ultrasonic aspirators, intraoperative ultrasound localization and the use of intraoperative spinal cord evoked potential monitoring. Intraoperative monitoring of sensory and motor evoked potentials should be considered for all patients undergoing surgery for an intramedullary spinal cord tumor [2,28]. A 50% decline in evoked potential amplitude during the procedure may be an indication of a new, possibly permanent, postoperative motor deficit.

The majority of intradural extramedullary tumors are well circumscribed, non-infiltrative, benign lesions that are amenable to complete surgical resection [1–3]. They typically have clear cleavage planes that allow separation from surrounding structures. The typical surgical approach

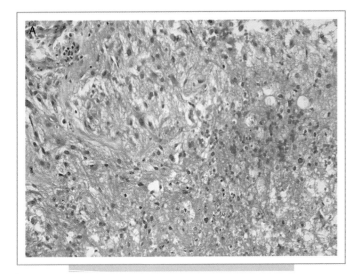

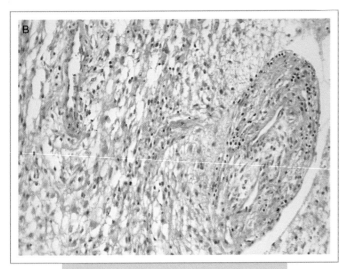

FIG. 5.11. WHO grade IV spinal GBM. Medium-power views of a grade IV spinal astrocytoma, demonstrating increased cellularity, cellular and nuclear pleomorphism, areas of necrosis **(A)** and microvascular proliferation **(B)**. H&E @ 200×.

for spinal schwannomas and neurofibromas is to attempt complete resection, which can be curative in most cases. The operative approach is usually a midline partial or total laminectomy [29–31]. The tumors are always attached to at least one nerve root. Schwannomas can be completely resected in 85–90% of cases without sacrifice of the parent nerve root [30,31]. Radical resection is also possible for large, invasive tumors that extend into surrounding bones and soft tissues [32]. Neurofibromas can be completely resected in 90% of cases [29]. However, in 80–90% of these patients, the parent nerve root must be sacrificed during tumor removal. Recurrences are rare for schwannomas and neurofibromas after complete resection. In patients with neurofibromatosis and multiple neurofibromas, surgical resection is restricted to the neurologically

symptomatic tumors [1,3,33]. For most meningiomas, a posterior approach with laminectomy is adequate to visualize and access the tumor for resection [1–3,11,34–36]. Tumors placed more ventrally may require a posterolateral approach or pediculectomy to allow safe tumor removal without cord retraction. Regions of the dura infiltrated with tumor should be resected when technically feasible. If complete removal of the dura cannot be achieved, extensive electrocautery of the remaining dura should be performed. Cauterization of the dura instead of resection, in selected cases, has not been associated with significantly higher recurrence rates [11]. Patients with schwannomas, meningiomas and neurofibromas that compress the spinal cord and cause myelopathy do well after surgical resection, often with various degrees of improvement in neurological status (50–80% in large series) [1–6,11]. It is theorized that because the tumors grow so slowly, the damage to the spinal cord is mainly compressive and somewhat reversible, which would not be the case were the damage due to severe demyelination, loss of axons and ischemia. Intradural, extramedullary epidermoids, dermoids, lipomas and teratomas can usually be removed completely.

The common intramedullary tumors, especially astrocytomas and ependymomas, are not as easily resected as the extramedullary tumors, although some authors advocate an aggressive approach [1–6,37–39]. Because these tumors usually lack a capsule and are more infiltrative, it is more difficult to discern a clear cleavage plane to separate tumor from surrounding normal tissues. Ependymomas of the cauda equina and conus medullaris are often adherent to surrounding neural structures and may erode into bone, making complete removal difficult without severe neurologic sequelae [1,14,15,40]. Gross total resection is recommended whenever feasible, but is only achieved in 22–42% of cases. In most patients, subtotal or incomplete piecemeal removal is performed and results in frequent recurrences and a compromise in survival. Spinal stabilization surgery (i.e. instrumentation and fusion) may also be of benefit for selected patients in whom tumor involvement with spinal elements has resulted in spinal instability. Intrinsic ependymomas can be completely resected in a greater percentage of patients because they tend to be more circumscribed than ependymomas of the filum terminale and cauda equina [1,2,23,24,40–43]. The majority of intramedullary ependymomas are low-grade, with very clear cleavage planes between tumor and surrounding normal cord tissue. For adequate visualization of the degree of infiltration of the mass and the clarity of the cleavage plane, an adequate myelotomy must be performed that extends over the entire rostral-caudal extent of the tumor. If a clear cleavage plane is identified, the goal should be gross total removal of all tumor tissue within the cord. Large tumors may require internal decompression with an ultrasonic aspirator or laser during the procedure. If the tumor is more infiltrative and a cleavage plane cannot be identified, then a biopsy should be obtained to establish a histologic diagnosis. Astrocytomas of the spinal cord vary in their degree of histologic malignancy and infiltrative capacity. In general, they tend to be of lower grade than similar tumors in the brain. Low-grade tumors that do not demonstrate severe infiltration into surrounding spinal segments can be aggressively resected, with good preservation of function and length of survival [1,2,4–6,26,27,37,38, 44–46]. The surgical approach would require a laminectomy in adult patients and an osteoplastic laminotomy in children, including verification of the location of the tumor and any associated cysts using ultrasound guidance, followed by a midline myelotomy. The tumor is resected using a combination of suction aspiration and the ultrasonic aspirator, working from the inside outward. Low-grade astrocytomas with significant infiltration and all high-grade tumors cannot be completely resected without risk of significant injury to the patient. In these tumors, a biopsy or partial resection that does not further compromise neurologic function is appropriate [47–49]. Microsurgical resection is the primary treatment for hemangioblastomas of the spinal cord, but the approach will vary depending on whether the tumor is sporadic or associated with von Hippel-Lindau syndrome (VHLS) [1,2,20,21,50–52]. For patients with sporadic tumors, resection of the mass is often required to confirm the diagnosis and is appropriate before the onset of significant symptoms. In patients with VHL, the indications for surgery are based on the presence of significant signs and symptoms attributable to the hemangioblastoma and its associated syrinx. Asymptomatic tumors should be followed clinically and considered for resection only when the patient develops significant signs and symptoms. For most patients, a direct posterior approach to the tumor will be adequate, including laminectomies that provide exposure 1–2 cm above and below the rostral and caudal margins of the tumor [20,52]. Ultrasound guidance is usually necessary to locate the mass and allow proper placement of the midline dural incision. The tumor is then resected in a circumferential manner, separating the tumor capsule from surrounding spinal cord and coagulating all of the individual small blood vessels associated with the capsule. The majority of dermoids, epidermoids and teratomas can be partially or completely resected using standard microneurosurgical techniques [1–3].

The surgical approach to extradural SCT, such as neuroblastoma, spinal chordoma and osteosarcoma, is to completely resect the tumor, if possible, or at least debulk the mass and relieve pressure on the thecal sac and spinal cord [1–3].

In general, the surgical approach to primary SCT in pediatric patients is similar to that of adults with the same type of tumor [4–6,45]. A complete surgical resection should be attempted in the majority of intradural, extramedullary SCT and virtually all extradural tumors. For intramedullary tumors, the preferred technique is osteoplastic laminotomy, with removal of the laminar roof in one piece and replacement of the entire block of tissue after resection of the mass. This technique allows for preservation of the posterior tension band, restores normal spinal anatomy and may result in bony fusion of the reapproximated lamina. In addition, there appears to be a reduced incidence of postoperative spinal deformity.

RADIATION THERAPY OF PRIMARY SPINAL CORD TUMORS

The majority of patients with primary SCT will not require external beam radiation therapy (RT) after surgical resection. Patients with completely resected extramedullary or extradural tumors (e.g. schwannoma, meningioma) should not receive RT, since it has not been shown to improve local tumor control or survival in this context [1–6,10,11,14,20,24,27,37,38,53,54]. These patients should be followed closely with regular neurologic examinations and MRI scans for evidence of SCT recurrence. At the time of recurrence, RT should be considered for those tumors not amenable to further surgical intervention [1–3,54]. Typical doses applied to the gross tumor volume are in the range of 45–55 Gy, with higher doses reserved for more histologically aggressive tumors. Other indications for RT are benign or low-grade SCT that cannot be completely resected, and all anaplastic or malignant tumors [1,2,53,54]. Radiation therapy is applied most often to incompletely resected or recurrent spinal ependymomas [1,2,14,24,55,56]. Although there are no controlled clinical trials, most authors agree that RT (50–55 Gy administered in 1.8 Gy daily fractions) improves local control and survival in properly selected spinal ependymoma patients. Patients with high-grade tumors and those who have undergone incomplete resection tend to have shorter disease-free intervals and survival [54–56]. An initial study of RT for spinal ependymoma reviewed 58 patients treated between 1950 and 1987, 43 of whom had received irradiation [55]. The 5- and 10-year cause-specific survival rates were 74% and 68%, respectively. Histological grade of the tumor was the only independent prognostic factor on multivariate analysis. However, RT was considered to be of benefit in the group of patients with incompletely resected tumors. In this cohort, the 5- and 10-year progression-free survival (PFS) rates were both 59%, and the cause-specific survival rates at 5 and 10 years were 69% and 62%, respectively. In a recent large multi-institutional review of spinal cord gliomas, 126 patients with ependymoma were treated, 59 of whom received postoperative RT [56]. Overall, the use of RT as compared to surgery alone did not improve PFS ($P = 0.11$). In addition, on univariate and multivariate analyses, the use of RT did not improve overall survival in comparison to surgery alone (adjusted hazard ratio 1.12, $P = 0.89$). These results are similar to those reported by other authors, such as Sgouros and colleagues, who also noted that RT did not confer a survival advantage

[57]. However, it must be pointed out that the apparent lack of efficacy of RT may be at least partially due to a selection bias in the data. The patients that received RT were less likely to undergo complete resections (20% versus 82%; $P < 0.01$) and had a significantly higher percentage of high-grade ependymomas (0% versus 21%; $P < 0.01$). Following the use of RT, spinal ependymomas generally tend to fail within the radiation port. Craniospinal RT (36–39.6 Gy) is only recommended for malignant ependymomas or tumors with documented leptomeningeal spread (i.e. cerebrospinal fluid evaluation with positive cytology or MRI with unequivocal non-bulky spinal meningeal enhancement). Higher doses of craniospinal RT are recommended (54 Gy) for patients that have MRI evidence for bulky leptomeningeal disease [2].

Spinal astrocytomas typically require postoperative RT because they are more infiltrative and cannot be completely resected as readily as ependymomas and other SCT [1,2,4,26,27,44–49]. For infiltrative low-grade tumors, the recommended RT dose is 50–54 Gy, administered in 1.8 Gy daily fractions. Anaplastic or malignant tumors require a higher dose, usually in the range of 55.8–59.4 Gy. The survival rates following RT are significantly affected by tumor grade. For example, in one older study, patients with low-grade astrocytomas had a 5-year actuarial survival rate of 89%, while patients with malignant tumors did not survive beyond 3 years [58]. A more recent study noted that median survival was significantly shorter for high-grade spinal astrocytomas in comparison to lower grade tumors, but did not differ between AA and GBM [49]. In this study, the median survival was 33 months for low-grade astrocytomas and 10 months for both AA and GBM. Similar results were noted in the large retrospective study cited above, in their cohort of 57 spinal astrocytoma patients [56]. The risk of disease progression in patients with high-grade tumors was more than twice that of patients with low- or moderate-grade tumors (hazard ratio 2.67; $P = 0.02$). In this same study, multivariate analysis demonstrated a significant risk reduction for patients with low- or moderate-grade tumors that received postoperative RT (adjusted hazard ratio 0.24; $P = 0.02$). However, postoperative RT did not appear to be beneficial in patients with high-grade astrocytomas ($P = 0.67$). Postoperative RT did not improve overall survival on univariate or multivariate analyses. Similar to the data for ependymomas cited above, a selection bias was noted in the RT group in terms of a lower percentage of patients receiving complete resections (13% versus 53%; $P < 0.01$). Relapse of spinal astrocytomas is almost always local, often with dissemination to the leptomeninges. Nonetheless, RT is recommended for the majority of spinal astrocytoma patients following resection, despite the poor results in high-grade tumors.

Radiation therapy is an important therapeutic consideration for many patients with locally invasive spinal chordomas that cannot be completely resected [59]. Unfortunately,

chordomas have proved to be relatively radioresistant tumors. The clinical results in most radiation therapy trials have demonstrated only modest improvements in local control, recurrence-free survival, and overall survival [59–62]. In general, the best results have been in patients that have undergone irradiation and an aggressive partial resection. For example, York and colleagues studied 27 patients with sacral chordomas and noted a significant difference in disease-free interval (24 months versus 8 months; $P < 0.02$) for patients receiving irradiation after subtotal resection [63]. In a study of 21 patients with chordomas of various sites, Keisch and colleagues concluded that irradiation prolonged the time to first relapse for tumors of the lower spine and sacrum, but not for tumors of the skull base [61]. The overall 5- and 10-year actuarial survival rates were 74% and 46%, respectively. Irradiation of chordomas with charged particles (i.e. protons, helium, neon, carbon) has shown promise as a more efficacious therapeutic option [59,64,65]. The high linear energy transfer of charged particles allows for a more defined and superior dose distribution (i.e. steeper fall-off in dose). Higher doses can be prescribed to the tumor volume with minimal risk of augmented toxicity to surrounding delicate structures (i.e. spinal cord). In a study using helium and neon particles, Berson and colleagues treated 25 patients with chordomas of the skull base and cervical spine and reported a 5-year local control rate of 55% [60]. For patients with unresectable sacral chordomas, carbon ion radiotherapy has also been effective [65]. The overall 5-year local control rate was 96%, with a 5-year overall survival rate of 52%.

A new technique being applied to selected primary SCT is the CyberKnife, an image-guided frameless stereotactic radiosurgery system [66]. In a series of 125 spinal tumors, including 17 extramedullary lesions (e.g. neurofibroma, meningioma), the mean dose was 14 Gy at the 80% isodose line (range 12–20 Gy). The treatment was well tolerated by patients. During follow-up evaluation that ranged from 6 to 27 months, all of the primary tumors remained stable.

The spinal cord is sensitive to the effects of radiation [1,2,54]. Overdosage must be avoided to reduce the risk of radiation-induced myelopathy, which is irreversible. The accepted spinal cord tolerance level is 5000–5500 cGy, administered over 5–6 weeks using 180–200 cGy daily fractions. This is the appropriate dose for most spinal ependymomas and astrocytomas. Above this dose, the therapeutic index falls sharply, so that the probability of myelopathy is 50% at 6500 cGy. The thoracic spinal cord segments are most radiation sensitive, while the cauda equina is least sensitive.

CHEMOTHERAPY OF PRIMARY SPINAL CORD TUMORS

There is a paucity of data to evaluate the efficacy of chemotherapy for the treatment of primary SCT. Because most of these tumors are histologically benign and respond well to surgical and/or radiation therapy, chemotherapy is rarely indicated. For gliomatous SCT that progress after surgery and RT, chemotherapy similar to that used for intracranial gliomas can be attempted [1,67–69]. However, due to the rarity of SCT that require chemotherapy, no phase III or large phase II trials have been reported. All of the available data have been from case reports or small institutional series that contain a handful of patients. Chemotherapy agents with potential activity include nitrosoureas (BCNU, CCNU), procarbazine, vincristine, etoposide, cyclophosphamide, temozolomide and carboplatin.

Patients with spinal cord astrocytomas have been treated with several forms of chemotherapy by a variety of authors [67–76]. The initial reports were in pediatric patients, including a 13-year-old girl with a thoracic AA that was treated with partial resection, multidrug chemotherapy and irradiation [70]. The chemotherapy regimen consisted of ifosfamide and etoposide alternating with cisplatin and cytarabine. Follow-up after RT demonstrated a complete response (CR) that was maintained for 16 months. Bouffet and co-workers, in their review of prognostic factors in spinal astrocytomas, reported a series of nine children that had received chemotherapy [72]. In three of these patients, chemotherapy was the sole mode of treatment, while in the other six patients it was used in combination with RT. Partial responses (PR) were noted in two of the patients that had only received chemotherapy. The same authors have also reported a 30-year-old patient with a grade II astrocytoma that was treated with neoadjuvant chemotherapy (carboplatin and vincristine) [71]. Treatment with the regimen over eleven cycles resulted in a CR, with improved neurological function, that was maintained during 14 months of follow-up. Allen and colleagues enrolled 13 pediatric patients with spinal astrocytomas (AA 8, GBM 4) into the Children's Cancer Group 945 protocol, which used two cycles of '8-drugs-in-1-day' chemotherapy before RT, and up to eight cycles afterwards [73]. The regimen included vincristine, carmustine, procarbazine, hydroxyurea, cisplatin, cytarabine, prednisone and cyclophosphamide. Pre-RT chemotherapy resulted in three objective responses (CR 1, PR 2) in evaluable patients. However, the 5-year PFS and overall survival were only 46% and 54%, respectively. The authors concluded that '8-drugs-in-1-day' chemotherapy did not offer a therapeutic advantage over the more traditional approach using CCNU and vincristine. Lewis and co-workers describe two pediatric patients with spinal cord astrocytomas that had excellent responses to multiagent chemotherapy [74]. The first was a 19-month-old child with a spinal AA, that progressed after partial resection. Chemotherapy was used instead of RT because of the young age, and consisted of carboplatin, vincristine, cyclophosphamide, methotrexate and cisplatin. The patient improved neurologically and developed a CR that remained durable at 29 months follow-up. An excellent

response was also noted in the other patient, a 4-year-old with a grade II astrocytoma, treated neoadjuvantly with carboplatin and vincristine. Similar responses have been noted in a report of three very young patients (26 to 41 months old) with low-grade gliomas of the spinal cord (pilocytic astrocytoma 2, ganglioglioma 1), treated with monthly single agent carboplatin ($560 \, mg/m^2$) [69]. Two of the patients developed PR, while the other had stable disease, all of which remained durable during 5 to 17 months follow-up. In another series of spinal cord gliomas in young children (age range 3 months to 9.5 years), Doireau and colleagues used an aggressive brain regimen consisting of carboplatin, procarbazine, vincristine, cisplatin, etoposide and cyclophosphamide [68]. After a 16-month course of chemotherapy, there were four patients with a CR, three with a PR, and one with stable disease. The overall survival rate was 87.5%, with an event-free survival rate at 2 and 4 years of 50%. Henson and co-workers described an adult patient with a low-grade astrocytoma of the conus medullaris that initially responded well to RT [75]. Two years later the patient had clinical and MRI progression and was placed on cisplatin and etoposide. However, after further progression he was placed on PCV (procarbazine, CCNU, vincristine), which induced tumor shrinkage on MRI and improvement in neurological function. The response was durable for 23 months (11 cycles) before disease progression. Temozolomide has been shown to be active in low-grade gliomas of the brain, and is now being applied to spinal gliomas. A recent report describes the use of temozolomide ($200 \, mg/m^2/day \times$ 5 days every 4 weeks) in two teenage patients with low-grade spinal astrocytomas [76]. Each patient was treated for a total of 10 cycles and had stabilization of disease clinically and by MRI scan.

There are very few published reports of patients with spinal ependymomas receiving chemotherapy [67]. The largest experience is from Chamberlain, who describes the use of oral etoposide ($50 \, mg/m^2/day \times 21$ days every 5 weeks) in a series of 10 adult patients with recurrent spinal cord ependymoma [77,78]. There were two patients with a PR and five patients with stable disease. The overall median response duration and median survival were 15 months and 17.5 months, respectively. A case report has also been described of a teenager with a progressive spinal myxopapillary ependymoma that was treated with the combination of tamoxifen ($60 \, mg/m^2/day$) and etoposide ($50 \, mg/m^2/day$), for three weeks each month [79]. The patient had a PR that was durable at 19 months follow-up.

MOLECULAR OR 'TARGETED' TREATMENT

There are very few published reports specifically examining the molecular biology of primary SCT [80–82]. However, it is likely that many, if not all, of the genetic alterations noted in intracranial gliomas are involved in the transformation and malignant phenotype of SCT (see Chapter 2). It has become apparent that the transformed phenotype of neoplastic glial tumor cells is highly complex and results from the dysfunction of a variety of interrelated regulatory pathways [83–85]. The transformation process involves amplification or overexpression of oncogenes in combination with loss or lack of expression of tumor suppressor genes. Oncogenes and signal transduction molecules that have been demonstrated to be important for gliomagenesis include platelet-derived growth factor and its receptor (PDGF, PDGFR), epidermal growth factor and its receptor (EGF, EGFR), CDK4, mdm-2, ras, phosphoinositol-3 kinase (PI3K), Akt, and mTOR (mammalian target of rapamycin). Tumor suppressor genes of importance in glial transformation include p53, retinoblastoma (Rb), p16 and p15 (i.e. INK4a, INK4b), DMBT1, NF2 and PTEN. Most of these tumor suppressor genes function as negative regulators of the cell cycle, while others are inhibitors of important internal signal transduction pathways. The net effect of these acquired abnormalities is dysregulation of, and an imbalance between, the activity of the cell cycle and apoptotic pathways.

Several reports have investigated the molecular genetics of spinal ependymomas [86–89]. In one study, five of seven patients were noted to have mutations of the NF2 gene, which resulted in the production of a truncated protein product [86]. Other investigators have also found mutations in NF2, as well as loss of DNA sequences from chromosomes 22q and 17p, in sporadic intramedullary spinal ependymomas [87]. Another study reviewed the molecular aspects of a large series of 62 ependymal tumors, including cases from the brain and spine [88]. Allelic loss was noted of chromosomes 10q (5 of 56 cases) and 22q (12 of 54 cases). In addition, somatic mutations of NF2 were present in six tumors, all of which were grade II intramedullary spinal ependymomas. A recent report on the methylation status of HIC-1, a putative tumor suppressor gene on chromosome 17p13.3, reviewed the findings in a series of 52 intracranial and spinal ependymomas [89]. There was a significant correlation between hypermethylation of HIC-1 and non-spinal localization of the tumor ($P = 0.019$).

Recent advances in growth factor and signal transduction biology are now providing the background for the development of 'molecular therapeutics', a new class of drugs that manipulate and exploit these pathways [90]. Molecular drugs targeting critical signal transduction pathway effectors, such as PDGFR, EGFR, ras, PI3K and mTOR, have entered early clinical trials in brain tumor patients [91–94]. In the near future, clinicians will have the opportunity to apply this new class of drugs to recurrent or progressive SCT.

ACKNOWLEDGMENTS

The authors would like to thank Julia Shekunov for research assistance. Dr Newton was supported in part by National Cancer Institute grant, CA 16058 and the Dardinger Neuro-Oncology Center Endowment Fund.

REFERENCES

1. Newton HB, Newton CL, Gatens C, Hebert R, Pack R (1995). Spinal cord tumors: review of etiology, diagnosis, and multidisciplinary approach to treatment. Cancer Pract 3:207–218.

2. Stieber VW, Tatter SB, Shaffrey ME, Shaw EG (2005). Primary spinal tumors. In *Principles of Neuro-Oncology*. Schiff D, O'Neill BP (eds). McGraw Hill Publishers, New York. 24:501–531.

3. Bhattacharyya AK, Guha A (2005). Spinal root and peripheral nerve tumors. In *Principles of Neuro-Oncology*. Schiff D, O'Neill BP (eds). McGraw Hill Publishers, New York. 25:533–549.

4. Townsend N, Handler M, Fleitz J, Foreman N (2004). Intramedullary spinal cord astrocytomas in children. Pediatr Blood Cancer 43:629–632.

5. Baysefer A, Akay KM, Izci Y, Kayali H, Timurkaynak E (2004). The clinical and surgical aspects of spinal tumors in children. Pediatr Neurol 31:261–266.

6. Auguste KI, Gupta N (2006). Pediatric intramedullary spinal cord tumors. Neurosurg Clin N Am 17:51–61.

7. Kleihues P, Cavanee WK (2000). *Pathology and Genetics of Tumours of the Nervous System*. International Agency for Research on Cancer, Lyon.

8. Kleihues P, Louis DN, Scheithauer BW et al (2002). The WHO classification of tumors of the nervous system. J Neuropathol Exp Neurol 61:215–225.

9. Kurtkaya-Yapicier O, Scheithauer B, Woodruff JM (2003). The pathobiologic spectrum of schwannomas. Histol Histopathol 18:925–934.

10. Waldron J, Weinstein RP (2005). Nerve sheath tumors of the spine. In *Textbook of Neuro-Oncology*. Berger MS, Prados MD (eds). Elsevier Saunders, Philadelphia. 63:485–488.

11. Cohen-Gadol AA, Krauss WE (2005). Spinal meningiomas. In *Textbook of Neuro-Oncology*. Berger MS, Prados MD (eds). Elsevier Saunders, Philadelphia. 64:489–492.

12. Lamszus K (2004). Meningioma pathology, genetics, and biology. J Neuropathol Exp Neurol 63:275–286.

13. Perry A, Gutmann DH, Reifenberger G (2004). Molecular pathogenesis of meningiomas. J Neuro-Oncol 70:183–202.

14. Parney IF, Parsa AT (2005). Myxopapillary ependymomas. In *Textbook of Neuro-Oncology*. Berger MS, Prados MD (eds). Elsevier Saunders, Philadelphia. 65:493–496.

15. Sonneland PR, Scheithauer BW, Onofrio BM (1985). Myxopapillary ependymoma. A clinicopathologic and immunocytochemical study of 77 cases. Cancer 56:883–893.

16. Rich TA, Schiller A, Suit HD, Mankin HJ (1985). Clinical and pathologic review of 48 cases of chordoma. Cancer 56:182–187.

17. Schoedel KE, Martinez AJ, Mahoney TM, Contis L, Becich MJ (1995). Chordomas: pathological features; ploidy and silver nucleolar organizing region analysis. A study of 36 cases. Acta Neuropathol 89:139–143.

18. Crapanzano JP, Ali SZ, Ginsberg MS, Zakowski MF (2001). Chordoma. A cytologic study with histologic and radiologic correlation. Cancer 93:40–51.

19. Ishida T, Dorfman HD (1994). Chondroid chordoma versus low-grade chondrosarcoma of the base of the skull: can immunohistochemistry resolve the controversy? J Neuro-Oncol 18:199–206.

20. Lonser RR, Oldfield EH (2006). Spinal cord hemangioblastomas. Neurosurg Clin N Am 17:37–44.

21. Lonser RR, Oldfield EH (2005). Spinal cord hemangioblastomas. In *Textbook of Neuro-Oncology*. Berger MS, Prados MD (eds). Elsevier Saunders, Philadelphia. 68:506–510.

22. McLendon RE, Enterline DS, Tien RD, Thorstad WL, Bruner JM (1998). Tumors of central neuroepithelial origin. In *Russell & Rubinstein's Pathology of Tumors of the Nervous System*, 6th edn. Volume 1. Bigner DD, McLendon RE, Bruner JM (eds). Arnold, London. 9:307–571.

23. Schwartz TH, McCormick PC (2000). Intramedullary ependymomas: clinical presentation, surgical treatment strategies and prognosis. J Neuro-Oncol 47:211–218.

24. Schwartz TH, Parsa AT, McCormick PC (2005). Intramedullary ependymomas. In *Textbook of Neuro-Oncology*. Berger MS, Prados MD (eds). Elsevier Saunders, Philadelphia. 66:497–500.

25. Rosenblum MK (1998). Ependymal tumors: a review of their diagnostic surgical pathology. Pediatr Neurosurg 28:160–165.

26. Houten JK, Cooper PR (2000). Spinal cord astrocytomas: presentation, management and outcome. J Neuro-Oncol 47:219–224.

27. Roonprapunt C, Houten JK (2006). Spinal cord astroctyomas: presentation, management, and outcome. Neurosurg Clin N Am 17:29–38.

28. Morota N, Deletis V, Constantini S, Kofler M, Cohen H, Epstein FJ (1997). The role of motor evoked potentials during surgery for intramedullary spinal cord tumors. Neurosurgery 41:1327–1336.

29. Seppala MT, Haltia MJ, Sankila RJ, Jaaskelainen JE, Heiskanen O (1995). Long-term outcome after removal of spinal neurofibroma. J Neurosurg 82:572–577.

30. Seppala MT, Haltia MJ, Sankila RJ, Jaaskelainen JE, Heiskanen O (1995). Long-term outcome after removal of spinal schwannoma: a clinicopathological study of 187 cases. J Neurosurg 83:621–626.

31. Conti P, Pansini G, Mouchaty H, Capuano C, Conti R (2004). Spinal neurinomas: retrospective analysis and long-term outcome of 179 consecutively operated cases and review of the literature. Surg Neurol 61:35–44.

32. Sridhar K, Ramamurthi R, Vasudevan MC, Ramamurthi B (2001). Giant invasive spinal schwannomas: definition and surgical management. J Neurosurg (Spine 2) 94:210–215.

33. Riccardi VM (1994). The neurofibromatoses. Hematol Oncol Ann 2:119–128.

34. Gezen F, Kahraman S, Çanakci Z, Bedük A (2000). Review of 36 cases of spinal cord meningioma. Spine 25:727–731.

35. Cohen-Gadol AA, Zikel OM, Koch CA, Scheithauer BW, Krauss WE (2003). Spinal meningiomas in patients younger than 50 years of age: a 21-year experience. J Neurosurg (Spine 3) 98:258–263.

36. Peker S, Cerci A, Ozgen S et al (2005). Spinal meningiomas: evaluation of 41 patients. J Neurosurg Sci 49:7–11.

37. Cooper PR, Epstein F (1985). Radical resection of intramedullary spinal cord tumors in adults. Recent experience in 29 patients. J Neurosurg 63:492–499.

38. Cooper PR (1989). Outcome after operative treatment of intramedullary spinal cord tumors in adults: intermediate and long-term results in 51 patients. Neurosurgery 25:855–859.

39. Constantini S, Miller DC, Allen JC, Rorke LB, Freed D, Epstein FJ (2000). Radical excision of intramedullary spinal cord tumors: surgical morbidity and long-term follow-up evaluation in 164 children and young adults. J Neurosurg 93:183–193.

40. Asazuma T, Toyama Y, Suzuki N, Fujimura Y, Hirabayshi K (1999). Ependymomas of the spinal cord and cauda equina: an analysis of 26 cases and a review of the literature. Spinal Cord 37:753–759.

41. McCormick PC, Torres R, Post KD, Stein BM (1990). Intramedullary ependymoma of the spinal cord. J Neurosurg 72:523–532.

42. Chang UK, Choe WJ, Chung SK, Chung CK, Kim HJ (2002). Surgical outcome and prognostic factors of spinal intramedullary ependymomas in adults. J Neuro-Oncol 57:133–139.

43. Hanbali F, Fourney DR, Marmor E et al (2002). Spinal cord ependymoma: radical surgical resection and outcome. Neurosurgery 51:1162–1174.

44. Epstein FJ, Farmer J, Freed D (1992). Adult intramedullary astrocytomas of the spinal cord. J Neurosurg 77:355–359.

45. Rossitch E, Zeidman SM, Burger PC, Curnes JT et al (1990). Clinical and pathological analysis of spinal cord astrocytomas in children. Neurosurgery 27:193–196.

46. Robinson CG, Prayson RA, Hahn JF et al (2005). Long-term survival and functional status of patients with low-grade astrocytoma of spinal cord. Int J Rad Oncol Biol Phys 63:91–100.

47. Cohen AR, Wisoff JH, Allen JC, Epstein F (1989). Malignant astrocytomas of the spinal cord. J Neurosurg 70:50–54.

48. Ciappetta P, Salvati M, Capoccia G, Artico M et al (1991). Spinal glioblastomas: report of seven cases and review of the literature. Neurosurgery 28:302–306.

49. Santi M, Mena H, Wong K, Koeller K, Olsen C, Rushing EJ (2003). Spinal cord malignant astrocytomas. Clinicopathologic features in 36 cases. Cancer 98:554–561.

50. Lonser RR, Weil RJ, Wanebo JE, DeVroom HL, Oldfield EH (2003). Surgical management of spinal cord hemangioblastomas in patients with von Hippel-Lindau disease. J Neurosurg 98:106–116.

51. Huang JS, Chang CJ, Jeng CM (2003). Surgical management of hemangioblastomas of the spinal cord. J Formos Med Assoc 102:868–875.

52. Lonser RR, Oldfield EH (2005). Microsurgical resection of spinal cord hemangioblastomas. Neurosurg (ONS Suppl 3) 57: ONS-372–ONS-376.

53. Linstadt DE, Wara WM, Leibel SA, Gutin PH et al (1989). Postoperative radiotherapy of primary spinal cord tumors. Int J Rad Oncol Biol Phys 16:1397–1403.

54. Petrovich Z, Liker M, Jozsef G (2001). Radiotherapy for tumors of the spine. In *Combined Modality Therapy of Central Nervous System Tumors*. Petrovich Z, Brady LW, Apuzzo ML, Bamberg M (eds). Springer, Berlin. 31:547–561.

55. Whitaker SJ, Bessell EM, Ashley SE, Bloom HJG et al (1991). Postoperative radiotherapy in the management of spinal cord ependymoma. J Neurosurg 74:720–728.

56. Abdel-Wahab M, Etuk B, Palermo J et al (2006). Spinal cord gliomas: a multi-institutional retrospective analysis. Int J Rad Oncol Biol Phys 64:1060–1071.

57. Sgouros S, Malluci CL, Jackowski A (1996). Spinal ependymomas – the value of postoperative radiotherapy for residual disease control. Br J Neursurg 10:559–566.

58. Kopelson G, Linggood RM (1982). Intramedullary spinal cord astrocytoma versus glioblastoma. The prognostic importance of histologic grade. Cancer 50:732–735.

59. Newton HB (2006). Chordoma. In *Textbook of Uncommon Cancers*, 3rd edn. Raghavan D, Brecher M, Johnson DH et al (eds). Wiley, Chichester. 56:614–625.

60. Berson AM, Castro JR, Petti P et al (1988). Charged particle irradiation of chordoma and chondrosarcoma of the base of skull and cervical spine: the Lawrence Berkeley laboratory experience. Int J Radiat Oncol Biol Phys 15:559–565.

61. Keisch ME, Garcia DM, Shibuya RB (1991). Retrospective long-term follow-up analysis in 21 patients with chordomas of various sites treated at a single institution. J Neurosurg 75:374–377.

62. Thieblemont C, Biron P, Rocher F et al (1996). Prognostic factors in chordoma: role of postoperative radiotherapy. Eur J Cancer 31:2255–2259.

63. York JE, Kaczaraj A, Abi-Said D et al (1999). Sacral chordoma: 40-year experience at a major cancer center. Neurosurgery 44:74–80.

64. Noel G, Feuvret L, Calugaru V et al (2005). Chordomas of the base of the skull and upper cervical spine. One hundred patients irradiated by a 3D conformal technique combining photon and proton beams. Acta Oncol 44:700–708.

65. Imai R, Kamada T, Tsuji H et al (2004). Carbon ion radiotherapy for unresectable sacral chordomas. Clin Cancer Res 10:5741–5746.

66. Gerszten PC, Ozhasoglu C, Burton SA et al (2004). CyberKnife frameless stereotactic radiosurgery for spinal lesions: clinical experience in 125 cases. Neurosurgery 55:89–99.

67. Balmaceda C (2000). Chemotherapy for intramedullary spinal cord tumors. J Neuro-Oncol 47:293–307.

68. Doireau V, Grill J, Zerah M et al (1999). Chemotherapy for unresectable and recurrent intramedullary glial tumours in children. Br J Cancer 81:835–840.

69. Hassall TEG, Mitchell AE, Ashley DM (2001). Carboplatin chemotherapy for progressive intramedullary spinal cord low-grade gliomas in children: three case studies and a review of the literature. Neuro-Oncology 3:251–257.

70. Weiss E, Klingbiel T, Kortmann RD, Hess CF, Bamberg M (1997). Intraspinal high-grade astrocytoma in a child – rationale for chemotherapy and more intensive radiotherapy? Child's Nerv Syst 13:108–112.

71. Bouffet E, Amat D, Devaux Y, Desuzinges C (1997). Chemotherapy for spinal cord astrocytoma. Med Pediatr Oncol 29:560–562.

72. Bouffet E, Pierre-Kahn A, Marchal JC et al (1998). Prognostic factors in pediatric spinal cord astrocytoma. Cancer 83:2391–2399.

73. Allen JC, Aviner S, Yates AJ et al (1998). Treatment of high-grade spinal cord astrocytoma of childhood with '8-in-1' chemotherapy and radiotherapy: a pilot study of CCG-945. J Neurosurg 88:215–220.

74. Lowis SP, Pizer BL, Coakham H, Nelson RJ, Bouffet E (1998). Chemotherapy for spinal cord astrocytoma: can natural history be modified? Child's Nerv Syst 14:317–321.

75. Henson JW, Thornton AF, Louis DN (2000). Spinal cord astrocytoma: response to PCV chemotherapy. Neurology 54:518–520.

76. Chamoun RB, Alaraj AM, Al Kutoubi AO, Abboud MR, Haddad GF (2006). Role of temozolomide in spinal cord low grade astrocytomas: results in two pediatric patients. Acta Neurochir (Wien) 148:175–180.

77. Chamberlain MC (2002). Etoposide for recurrent spinal cord ependymoma. Neurology 58:1310–1311.

78. Chamberlain MC (2002). Salvage chemotherapy for recurrent spinal cord ependymoma. Cancer 95:997–1002.

79. Madden JR, Fenton LZ, Weil M, Winston KR, Partington M, Foreman NK (2001). Experience with tamoxifen/etoposide in the treatment of a child with myxopapillary ependymoma. Med Pediatr Oncol 37:67–69.

80. Parsa AT, Fiore AJ, McCormick PC, Bruce JN (2000). Genetic basis of intramedullary spinal cord tumors and therapeutic implications. J Neuro-Oncol 47:239–251.

81. Parsa AT, Chi JH, Acosta FL, Ames CP, McCormick PC (2005). Intramedullary spinal cord tumors: molecular insights and surgical innovation. Clin Neurosurg 52:1–9.

82. Chi JH, Cachola K, Parsa AT (2006). Genetics and molecular biology of intramedullary spinal cord tumors. Neurosurg Clin N Am 17:1–5.

83. von Deimling A, Louis DN, Wiestler OD (1995). Molecular pathways in the formation of gliomas. Glia 15:328–338.

84. Shapiro JR, Coons SW (1998). Genetics of adult malignant gliomas. BNI Quarterly 14:27–34.

85. Maher EA, Furnari FB, Bachoo RM et al (2001). Malignant glioma: genetics and biology of a grave matter. Genes Dev 15:1311–1333.

86. Birch BD, Johnson JP, Parsa AT et al (1996). Frequent type 2, neurofibromatosis gene transcript mutations in sporadic intramedullary spinal cord ependymomas. Neurosurgery 39:135–140.

87. von Haken MS, White EC, Daneshvar-Shyesther L et al (1996). Molecular genetic analysis of chromosome arm 17p and chromosome arm 22q DNA sequences in sporadic pediatric ependymomas. Genes Chromosomes Cancer 17:37–44.

88. Ebert C, von Haken MS, Meyer-Puttlitz B et al (1999). Molecular genetic analysis of ependymal tumors: NF2 mutations and chromosome 22q loss occur preferentially in intramedullary spinal ependymomas. Am J Pathol 155:627–632.

89. Waha A, Koch A, Hartmann W et al (2004). Analysis of HIC-1 methylation and transcription in human ependymomas. Int J Cancer 110:542–549.

90. Adjei AA, Hidalgo M (2005). Intracellular signal transduction pathway proteins as targets for cancer therapy. J Clin Oncol 23:5386–5403.

91. Newton HB (2003). Molecular neuro-oncology and the development of 'targeted' therapeutic strategies for brain tumors. Part 1 – growth factor and ras signaling pathways. Expert Rev Anticancer Ther 3:595–614.

92. Newton HB (2004). Molecular neuro-oncology and the development of 'targeted' therapeutic strategies for brain tumors. Part 2 – PI3K/Akt/PTEN, mTOR, SHH/PTCH, and angiogenesis. Expert Rev Anticancer Ther 4:105–128.

93. Newton HB (2005). Molecular neuro-oncology and the development of 'targeted therapeutic strategies for brain tumors. Part 5 – apoptosis and cell cycle. Expert Rev Anticancer Ther 5:355–378.

94. Rich JN, Bigner DD (2004). Development of novel targeted therapies in the treatment of malignant glioma. Nature Rev Drug Discov 3:430–446.

CHAPTER 6

Overview of Pathology and Treatment of Intramedullary Spinal Cord Metastases

Herbert B. Newton, Abhik Ray-Chaudhury and Mark G. Malkin

INTRODUCTION

The epidemiology of intramedullary spinal cord metastases (ISCM) was reviewed in detail in Chapter 4. In this chapter, we will provide an overview of the classification, pathology and treatment of ISCM. Metastases to the spinal cord parenchyma are very uncommon in comparison to primary spinal cord tumors (SCT) and neoplastic epidural spinal cord compression (see Chapter 7), affecting only 0.5–2.0% of all cancer patients [1–5]. However, they are often associated with significant morbidity and mortality, and often herald the onset of a more biologically aggressive and advanced form of the primary malignancy [6,7]. ISCM represent 8.5% of all CNS metastases and only 1–3% of all intramedullary SCT. Although ISCM can arise from virtually any primary systemic neoplasm, the vast majority are due to tumors of the lung and breast, malignant melanoma and renal cell carcinoma (Table 6.1).

TABLE 6-1 Percentage of intramedullary spinal cord metastases by primary tumor site	
Primary type	%
Lung	50–60
SCLC	50–55
Non-SCLC	45–50
Breast	10–13
Melanoma	9
Renal cell	5–9
Lymphoma	5
Colorectal & GI	3
Thyroid	2
Ovarian	1–2

Adapted from references [2–5]. SCLC: small cell lung carcinoma, GI: gastrointestinal

PATHOLOGY OF INTRAMEDULLARY SPINAL CORD METASTASES

ISCM are distributed fairly evenly along the entire length of the spinal cord, with a possible subtle predilection for the conus medullaris [2–5,8,9]. The mechanism of metastatic spread of tumor cells from the primary neoplasm to the spinal cord parenchyma remains unclear. In the majority of cases, hematogenous dissemination is thought to be involved, usually via an arterial route [4,8]. This is corroborated by the common coexistence of pulmonary and brain metastasis in patients with ISCM. In addition, an arterial mechanism is consistent with spinal necropsy data demonstrating that ISCM typically involve the posterior horns, which contain spinal gray matter and have dense capillary beds, with five times the arterial perfusion in comparison to the surrounding spinal white matter [4,10]. Hematogenous spread through the vertebral venous plexus (i.e. Batson's venous plexus) has also been suggested as another mechanism in some patients with ISCM. In these cases, tumor cells gain access to the spinal cord parenchyma via retrograde movement through the plexus. Tumor cells may also gain access to the spinal cord directly from cerebrospinal fluid in patients with pre-existing leptomeningeal metastases [8,9]. In this situation, tumor cells infiltrate the Virchow-Robin spaces of blood vessels entering the cord, and then penetrate the pial layer to gain access to spinal cord parenchyma. The last possible mechanism involves direct extension of tumor cells into the spinal cord from an adjacent primary neoplasm. The tumor cells are able to enter the cord parenchyma by extending directly through the dura or by perineural spread along nerve roots.

Once the metastatic bolus of cells has gained access to the spinal cord and has lodged within the cord parenchyma,

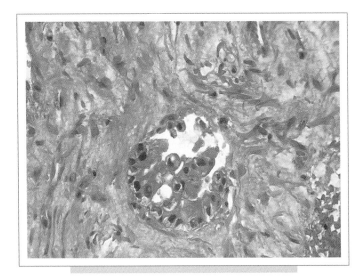

FIG. 6.1. Low power view of a bronchogenic carcinoma metastatic to the cervical spinal cord, demonstrating gliotic tissue surrounding the focus of tumor. H&E @ 200×.

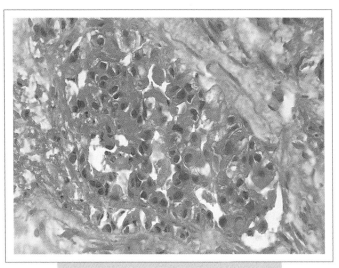

FIG. 6.2. High power view of the same intramedullary metastasis, revealing features of an adenocarcinoma, with the formation of complex glandular structures, as well as pleomorphic and hyperchromatic nuclei. H&E @ 400×.

neoplastic angiogenesis is required for the tumor to grow to a clinically relevant size [11–13]. The angiogenic phenotype requires upregulation of angiogenic promoters, such as vascular endothelial growth factor (VEGF), fibroblast growth factors (basic FGF, acidic FGF), angiopoietins (Ang-1, Ang-2), platelet-derived growth factor (PDGF), epidermal growth factor (EGF), transforming growth factors (TGFα and TGFβ), interleukins (IL-6, IL-8) and the various growth factor receptors (e.g. VEGFR, PDGFR, EGFR) [12,13]. Secretion of angiogenesis inhibitors, such as thrombospondin-1, platelet factor-4 and interferons α and β, is typically reduced in the ISCM, further 'tipping the balance' in the local environment to permit angiogenic activity within and around the tumor mass.

On macroscopic evaluation, ISCM usually form rounded, discrete deposits in the spinal cord parenchyma that are well circumscribed and demarcated from surrounding neural tissues (Figure 6.1) [4,8,9]. Foci of hemorrhage may be noted within the tumor and involvement of the posterior nerve roots is often present. On microscopic examination, the histological features of the ISCM are usually similar, if not identical, to those of the primary neoplasm (Figure 6.2) [14,15]. The tumor mass will usually have well defined borders, tending to displace adjacent spinal cord parenchyma without significant infiltration. Areas of hemorrhage and gliosis are frequently noted. Initial review of the tissue morphology can often identify a major tumor category, such as metastatic carcinoma or melanoma. However, in roughly 3% of ISCM cases, no known histopathological tissue pattern can be discerned [16]. For a more detailed determination of cellular differentiation and assignment to a specific histological category, immunocytochemical analysis may be required [14,15].

SURGICAL THERAPY OF INTRAMEDULLARY SPINAL CORD METASTASES

Until recently, surgical approaches to treatment of ISCM were not attempted, due to the frequent occurrence of neurological injury. Technical advances in neurosurgical technique have improved the results of operative therapy for ISCM in carefully selected patients [4,5,17,18]. These advances include the use of the operating microscope and microneurosurgical instrumentation, bipolar coagulation, surgical lasers, ultrasonic aspirators, intraoperative ultrasound localization and the use of intraoperative spinal cord evoked potential monitoring. Intraoperative monitoring of sensory and motor evoked potentials should be considered for all patients undergoing surgery for an ISCM [18,19]. A 50% decline in evoked potential amplitude during the procedure may be an indication of a new, possibly permanent, postoperative motor deficit.

Several recent reports suggest that carefully selected patients with ISCM can benefit from surgical resection [4,5,16–18]. Patients with a suspected ISCM that have an unknown primary tumor should be considered for a biopsy, to allow for a definitive pathological diagnosis. A more aggressive approach may not be possible for patients in whom the general medical condition is poor, with widespread systemic disease, or concomitant brain metastases and/or leptomeningeal disease. For the cohort of patients with higher functional status and well-controlled systemic and neurological disease, an open spinal procedure should be attempted, using a similar technique to what has been described for primary intramedullary spinal cord tumors (see Chapter 5) [4,18]. If there is a clear cleavage plane between the tumor margin and the surrounding normal spinal cord parenchyma, an aggressive gross total resection should be performed. If a clear cleavage plane cannot be discerned, a biopsy and decompressive laminectomy will be sufficient to allow for the development of an appropriate treatment plan. Using this treatment paradigm, Gasser and colleagues reviewed their experience with surgical therapy in a series of 13 patients with ISCM [18]. They were able to accomplish complete surgical removal of the tumor in six patients and an incomplete resection in the remaining seven patients. Eleven of the 13 patients were clinically stable in the postoperative period. The histology of the primary tumor appeared to affect the extent of resection and surgical outcome. A complete resection was possible for all ISCM from adenocarcinomas. However, tumors derived from sarcomas and poorly differentiated carcinomas could only be incompletely resected due to poor cleavage planes.

RADIATION THERAPY OF INTRAMEDULLARY SPINAL CORD METASTASES

The most common method of treatment for patients with ISCM has been a combination of dexamethasone and external beam irradiation of the spine [2–7,16,17]. Dexamethasone is able to acutely reduce peritumoral edema and can improve pain and neurological function in up to 85% of patients [17]. The use of dexamethasone provides neurological stabilization in the majority of patients while they are being prepared for more definitive therapy. Fractionated external beam radiotherapy (RT) is the treatment modality of choice for the vast majority of patients with ISCM, even if they have undergone an aggressive surgical resection [2–5,16,17,20]. Initial reports suggested that ISCM were often multiple and that RT of the entire spinal cord was necessary [7–9]. However, this approach often led to severe bone marrow toxicity and hematologic compromise. In the era of magnetic resonance imaging (MRI) of the spine, it is now known that ISCM are usually solitary lesions and that multifocal disease only occurs in approximately 12–15% of patients [2,4,21]. Therefore, focal RT to the site of the lesion, with a 1–2 cm margin above and below the mass, will be appropriate for most patients. Various RT treatment schedules have been used, but most authors suggest a total dose between 30 and 40 Gy, administered in daily fractions of 200–300 cGy, over 2–3 weeks [2–5,16,17].

The clinical response to RT is variable depending on several key factors, including the histology and radiosensitivity of the ISCM, the duration of the symptoms and the extent of the neurological deficit at the time of treatment initiation. Irradiation is very effective in radiosensitive tumor types, such as small cell lung carcinoma, lymphoma and germ cell tumors [2–5,16,17,22]. Clinical stabilization or improvement can also be noted in patients with less sensitive tumors, including those from breast, prostate, ovary and non-small cell lung carcinoma [2–5,16,17,23,24]. However, it must be noted that even some tumors that are traditionally considered to be insensitive to RT, such as renal cell carcinoma, can be stabilized if the treatment is initiated early in the course of ISCM [25]. Overall, it appears that early diagnosis and treatment of ISCM is even more important than sensitive histology [2,4,7,16,17]. The majority of patients, regardless of ISCM histology, will have the potential for clinical and neurological stabilization if RT (± surgical resection) and dexamethasone are utilized before the onset of significant spinal cord damage and weakness (i.e. myelopathy with paraparesis or tetraparesis). If RT and dexamethasone are initiated while the patient is still ambulatory, the prognosis for stabilization of function is very good. Regardless of the histologic subtype, if treatment of ISCM is started after the onset of severe myelopathy and limb weakness, the likelihood of any significant recovery is poor.

A new radiotherapy technique under evaluation for ISCM is image-guided, frameless stereotactic radiosurgery [26–28]. Several reports by Ryu and co-workers suggest that single-dose radiosurgery, given as a boost after RT (6–8 Gy) or as 'stand alone' treatment (10–60 Gy), can be safely delivered to the spine, is well tolerated, and often results in rapid pain control [26,27]. In their series, the incidence of tumor recurrence or progression into adjacent spinal regions was only 5%. Using the CyberKnife radiosurgery system, Gerszten and colleagues have described their experience in a series of 125 patients with spinal tumors, including 108 with metastatic lesions [28]. For the entire cohort, the mean dose was 14 Gy at the 80% isodose line (range 12–20 Gy). The treatment was well tolerated by patients. During follow-up evaluation that ranged from 6 to 27 months, most of the metastatic tumors remained stable.

CHEMOTHERAPY OF INTRAMEDULLARY SPINAL CORD METASTASES

Chemotherapy has occasionally been applied to ISCM as indicated by tumor staging and prior responses [2,4,5,9,16,25,29]. However, there are no chemotherapy protocols or drug strategies specific for ISCM. Several authors suggest adding chemotherapy to RT in cases with chemosensitive tumors (e.g. small cell lung carcinoma, lymphoma, germ cell tumors, breast carcinoma) [5,16,25]. In addition, intrathecal chemotherapy can be used for patients with concomitant ISCM and leptomeningeal disease [2,4]. In the review of survival and prognosis by Connolly and colleagues, the addition of chemotherapy to RT imparted a significant extension of survival in comparison to RT alone ($P < 0.05$) [16]. In their series, patients with melanoma had the best survival (43 weeks), while patients with lung and breast tumors only survived 19.7 and 13.0 weeks, respectively. Another report reviewed the results of chemotherapy as the sole treatment for two patients with ISCM from small cell lung carcinoma [30]. The first patient was treated with etoposide and nitrosoureas and had a partial response that lasted for 6 months. The second patient received cyclophosphamide, adriamycin, etoposide and cisplatin and had a complete response before the addition of focal RT. Several other patients with lung carcinoma were also reported to respond to similar chemotherapy regimens, usually in combination with RT [6,16,30].

ACKNOWLEDGMENTS

The authors would like to thank Julia Shekunov for research assistance. Dr Newton was supported in part by National Cancer Institute grant, CA 16058 and the Dardinger Neuro-Oncology Center Endowment Fund.

REFERENCES

1. Newton HB, Newton CL, Gatens C, Hebert R, Pack R (1995). Spinal cord tumors: review of etiology, diagnosis, and multidisciplinary approach to treatment. Cancer Pract 3:207–218.

2. Schiff D, O'Neill BP (1996). Intramedullary spinal cord metastases: clinical features and treatment outcome. Neurology 47:906–912.

3. Mut M, Schiff D, Shaffrey ME (2005). Metastases to nervous system: spinal epidural and intramedullary metastases. J Neuro-Oncol 75:43–56.

4. Chi JH, Parsa AT (2006). Intramedullary spinal cord metastasis: clinical management and surgical considerations. Neurosurg Clin N Am 17:45–50.

5. Kalayci M, Cagavi F, Gul S et al (2004). Intramedullary spinal cord metastases: diagnosis and treatment – an illustrated review. Acta Neurochir (Wien) 146:1347–1354.

6. Grem JL, Burgess J, Trump DL (1985). Clinical features and natural history of intramedullary spinal cord metastasis. Cancer 56:2305–2314.

7. Winkelman MD, Adelstein DJ, Karlins NL (1987). Intramedullary spinal cord metastasis. Diagnostic and therapeutic considerations. Arch Neurol 44:526–531.

8. Costigan DA, Winkelman MD (1985). Intramedullary spinal cord metastasis. A clinicopathological study of 13 cases. J Neurosurg 62:227–233.

9. Dunne JW, Harper CG, Pamphlett R (1986). Intramedullary spinal cord metastases: a clinical and pathological study of nine cases. Q J Med 61:1003–1020.

10. Qui MG, Zhu XH (2004). Aging changes of the angioarchitecture and arterial morphology of the spinal cord in rats. Gerontology 50:360–365.

11. Beckner ME (1999). Factors promoting tumor angiogenesis. Cancer Investig 17:594–623.

12. Webb CP, Vande Woude GF (2000). Genes that regulate metastasis and angiogenesis. J Neuro-Oncol 50:71–87.

13. Yoshida BA, Sokoloff MM, Welch DR, Rinker-Schaeffer CW (2000). Metastasis-suppressor genes: a review and perspective on an emerging field. J Natl Cancer Inst 92:1717–1730.

14. Bruner JM, Tien RD (1998). Secondary tumors. In *Russell & Rubinstein's Pathology of Tumors of the Nervous System*, 6th edn. Volume 2. Bigner DD, McLendon RE, Bruner JM (eds). Arnold, London. 17:419–450.

15. Ironside JW, Moss TH, Louis DN, Lowe JS, Weller RO (eds) (2002). Metastatic tumours. In *Diagnostic Pathology of Nervous System Tumours*. Churchill Livingstone, London. 12:319–341.

16. Connolly ES, Winfree CJ, McCormick PC, Cruz M, Stein BM (1996). Intramedullary spinal cord metastasis: report of three cases and review of the literature. Surg Neurol 46:329–337.

17. Gasser TG, Pospiech J, Stolke D, Schwechheimer K (2001). Spinal intramedullary metastases. Report of two cases and review of the literature. Neurosurg Rev 24:88–92.

18. Gasser T, Sandalcioglu IE, El Hamawi B, van de Nes JAP, Stolke D, Wiedemayer H (2005). Surgical treatment of intramedullary spinal cord metastases of systemic cancer: functional outcome and prognosis. J Neuro-Oncol 73:163–168.

19. Morota N, Deletis V, Constantini S, Kofler M, Cohen H, Epstein FJ (1997). The role of motor evoked potentials during surgery for intramedullary spinal cord tumors. Neurosurgery 41:1327–1336.

20. Petrovich Z, Liker M, Jozsef G (2001). Radiotherapy for tumors of the spine. In *Combined Modality Therapy of Central Nervous System Tumors*. Petrovich Z, Brady LW, Apuzzo ML, Bamberg M (eds). Springer, Berlin. 31:547–561.

21. Li MH, Holtas S (1991). MR imaging of spinal intramedullary tumors. Acta Radiol 32:505–513.

22. Nikolaou M, Kumpou M, Mylonakis N et al (2006). Intramedullary spinal cord metastases from atypical small cell lung cancer: a case report and literature review. Cancer Investig 24:46–49.

23. Potti A, Abdel-Raheem M, Levitt R, Schell DA, Mehdi SA (2001). Intramedullary spinal cord metastases (ISCM) and non-small cell lung carcinoma (NSCLC): clinical patterns, diagnosis, and therapeutic considerations. Lung Cancer 31:319–232.

24. Vellegas AE, Guthrie TH (2004). Intramedullary spinal cord metastasis in breast cancer: clinical features, diagnosis, and therapeutic consideration. Breast J 10:532–535.

25. Fakih M, Schiff D, Erlich R, Logan TF (2001). Intramedullary spinal cord metastasis (ISCM) in renal cell carcinoma: a series of six cases. Ann Oncol 12:1173–1177.

26. Ryu S, Fang YF, Rock J et al (2003). Image-guided and intensity-modulated radiosurgery for patients with spinal metastasis. Cancer 97:2013–2018.

27. Ryu S, Rock J, Rosenblum M, Kim JH (2004). Patterns of failure after single-dose radiosurgery for spinal metastasis. J Neurosurg 101:402–405.

28. Gerszten PC, Ozhasoglu C, Burton SA et al (2004). CyberKnife frameless stereotactic radiosurgery for spinal lesions: clinical experience in 125 cases. Neurosurgery 55:89–99.

29. Balmaceda C (2000). Chemotherapy for intramedullary spinal cord tumors. J Neuro-Oncol 47:293–307.

30. Ferrior JP, Cadranel J, Khalil A, Lebreton C, Contant S, Milleron B (1998). Intramedullary metastases of bronchogenic carcinoma. Two cases. Rev Neurol (Paris) 154:166–169.

CHAPTER 7

Epidural Spinal Cord Compression in Adult Neoplasms

Harman P.S. Bajwa and David Schiff

INTRODUCTION

Neoplasms can compromise spinal cord function resulting in devastating neurological deficits with significant limitations in daily functioning, including immobilizing pain and paresis. Neoplasm-related spinal cord dysfunction may arise from paraneoplastic involvement, complications of radiation and chemotherapy, leptomeningeal metastasis and intramedullary metastasis. Among all causes of myelopathies in cancer patients, epidural spinal cord compression (ESCC) is by far the most common, with significant morbidities. The best evidence from autopsy studies suggests that ESCC affects approximately 5% of all cancer deaths or about 25 000 Americans yearly in the USA [1]. Knowledge of pathophysiology, clinical features, diagnostic approaches and treatment options of ESCC is essential in decreasing the potential morbidities associated with this entity in cancer patients.

PATHOPHYSIOLOGY

ESCC usually arises as a complication of vertebral metastases. The majority of ESCC arise from lung, breast and prostate metastases to the vertebral column. Metastases generally show preference for the marrow-rich anterior vertebral body where their proliferation may eventually impact the thecal sac and epidural space that house the spinal cord blood supply. Metastasis to the marrow-destitute posterior vertebral arch, direct metastasis to the epidural space and intraspinal spread via the vertebral neural foramen are less common scenarios. About 60% of ESCC occurs in the thoracic spine, compared to about 30% in the lumbosacral spine [2]. This clear predilection for the thoracic region is directly related to reduced potential space available for tumor to expand in this region compared to other areas of the spine. The pathophysiology of ESCC is associated with the size and the location of the metastasis. Moreover, pathological fracture from metastasis-riddled vertebral bodies can cause destabilization of the spine with potential cord compression. In addition to direct compression of the cord and demyelination, animal and experimental models of ESCC suggest that the underlying pathophysiology of ESCC by vertebral metastases is also associated with venous congestion and vasogenic edema resulting from impingement of the epidural venous plexus [3]. This underlying phenomenon stresses the necessity of starting steroids early in the management of ESCC. If the impingement is not arrested, it ultimately leads to spinal cord infarction and irreversible loss of neurological function. The duration of symptoms of ESCC has been shown to be directly related to potential recovery after decompression. In clinical and animal studies, it has been shown that the rapid onset of ESCC is related to poor recovery of neurological dysfunction [4].

CLINICAL FEATURES

The cardinal clinical features of ESCC are pain, weakness, sensory deficits and autonomic dysfunction. Typically, the hallmark and sentinel symptom of ESCC is pain. As noted in Table 7.1, pain is present in approximately 83–96% of patients at the time of presentation [4]. Since early intervention can dramatically improve likelihood of preservation of neurological function, it is essential to consider ESCC in cancer patients presenting with back pain and radicular sensory complaints. Over time, the pain associated with ESCC usually intensifies, with change in distribution from localized to a radicular, sciatica-like quality, especially when involving the lumbosacral spine. Pain is usually reported to be worse with changes in intra-thoracic pressure, including coughing and straining or with position changes, likely an effect of further compression of the venous plexus, as seen especially in a state of recumbency.

Weakness is clinically the most apparent and objectifiable manifestation of ESCC. It is seen in 61–85% of patients at the time of presentation (see Table 7.1). Bilateral lower extremity weakness is frequently present and more severe with thoracic ESCC. As is seen with other myelopathies, ESCC

TABLE 7-1 Clinical presentation of malignant spinal cord compressions				
Patients	Pain (%)	Weakness (%)	Sensory deficits (%)	Autonomic dysfunction (%)
398	83	67	90	48
153	88	61	78	40
130	96	76	51	57
79	70	91	46	44
77	94	85	57	52

From Ref [4] Prasad D, Schiff D (2005). Malignant spinal-cord compression. Lancet Oncol 6:15–24.

preferentially affects the flexors of the lower extremities, including the iliopsoas muscles. A possible explanation of this could be that the naturally increased tonicity of the antigravity muscles is likely masking some of the underlying weakness in these muscle groups. Approximately two-thirds of all ESCC patients are non-ambulatory at the time of diagnosis [5].

Autonomic dysfunction, in the form of bowel and bladder involvement, is usually a late complication of ESCC. Urinary retention is typically present. With conservative estimates, about half of patients affected by ESCC are catheter dependent at the time of diagnosis [6]. This reflects the frustrating fact that most cases of ESCC are diagnosed fairly late in their course. Cauda equina involvement should be suspected if there is concomitant saddle anesthesia with bowel or bladder involvement.

DIAGNOSIS

With technological advances, specifically magnetic resonance imaging (MRI), radiographical evidence, rather than clinical localization, has become the cardinal diagnostic approach to ESCC [7]. With the ubiquitous presence of MRI scanners, the diagnosis of ESCC can now be made with confidence, with appropriate expeditious initiation of treatment. Other approaches, including myelography and computerized tomography (CT), are of secondary preference and are generally reserved for when MRI is contraindicated or as an adjunct to MRI for surgical planning. The standard MRI protocol for ESCC detection involves sagittal T1- and T2-weighted sequences with further axial images at the site of interest. The vertebral metastasis causing ESCC is usually hyperintense on T2 sequences with visible compression. Vertebral metastases generally show contrast enhancement, but this yields no further diagnostic utility and thus intravenous contrast administration may be unnecessary [8]. Approximately one-third of patients with metastatic epidural involvement have multiple spinal sites of disease [2]. This substantial number begets the need for imaging the entire spine in patients suspected of having ESCC. Surprisingly, even with the advent of such high yield diagnostic tools, there

continues to be a substantial lag between the presentation of symptoms and the eventual diagnosis of ESCC. One study reported that the median lag time between presentation of symptoms and actual diagnosis of ESCC was 14 days. The delay was shorter if presenting symptoms showed significant functional compromise, especially paralysis [5]. This underscores the importance of recognizing symptoms of ESCC quickly, with expedient utilization of intervening measures in a disease entity where early diagnosis greatly impacts treatment outcome.

PROGNOSIS

By definition, with ESCC of neoplastic etiology excluding primary spinal or paraspinal tumors, there is disseminated disease at the time of diagnosis, which itself is a poor prognosticator of overall survival. Likewise, the histological type and rate of dissemination of the primary cancer have a strong relationship with prognosis and development of ESCC, with primary lung tumors having the shortest time from diagnosis of the parent tumor to spinal cord compression [9]. As alluded to earlier, pretreatment functional status is the strongest prognostic indicator of post-treatment function, especially the maintenance of ambulation. One retrospective study found that, of the pretreatment ambulatory patients, 79% remained ambulatory post-treatment as opposed to only 21% of the non-ambulatory patients becoming ambulatory post-treatment [10]. The median survival time of patients with ESCC is estimated to be about 6 months, with survival being higher in the ambulatory cohort.

TREATMENT

The mainstays of treatment for ESCC are largely corticosteroids, radiotherapy and surgery. Chemotherapy is usually limited to lymphomas and germ cell tumors, given that most metastases that produce ESCC are not chemosensitive. Usually, the available treatment options are combined in the complete management of ESCC. In addition, ameliorating pain should always be a priority. Opiates, corticosteroids, localized radiation and surgical intervention are important tools in alleviating pain in ESCC. Moreover, nonsteroidal anti-inflammatories may be useful in decreasing bone pain from vertebral metastasis. As in many other neurological diseases, the pretreatment level of function is the best prognostic indicator of post-treatment status.

Corticosteroids

Consistent with the proposed vasogenic edema and inflammatory response associated with the pathophysiology of ESCC, systemic corticosteroids have been shown to be effective at reducing and delaying the onset of symptoms, including irreversible neurological deficits. Recommended doses in ESCC run the gamut from 16 g to 100 g of dexamethasone per day. Studies have suggested that there is likely equal efficacy at both ends and, as surmised, far less systemic steroidal side effects at lower doses [11]. With potential risk of gastric ulceration, gastrointestinal prophylaxis should be used with high dose steroids [12].

Radiotherapy

Radiation therapy has proven utility both in reducing pain and preserving neurological function in patients with ESCC. Intrinsic radiosensitivity of the tumor, the acuteness of onset of neurological deficits, the extent of pretreatment deficits and the extent of anatomic compression of the spinal cord are all factors in the efficacy of radiotherapy in ESCC [8]. As discussed earlier, multiple studies have shown that pretreatment function is directly correlated to post-treatment function after radiation therapy in ESCC. For example, radiotherapy has shown to preserve ambulation in 80–100% of those ambulating pretreatment [1]. In sharp contrast to this, only 2–6% of paraplegic patients from ESCC became ambulatory post radiation treatment.

Given that the general survival rates are higher for patients still ambulatory at the time of treatment, the importance of expeditious and accurate diagnosis cannot be overstated [13]. Traditionally, radiotherapy schedules for ESCC were fractionated to allow non-neoplastic cells, which have greater restorative capabilities than tumor cells, to recover between doses. However, fractionation must be weighed against delivering an adequate dose while achieving maximal treatment benefit with minimization of toxicity, mainly myelopathy, in patients with already low expected survival. Interestingly, recent studies suggest that short hypofractionation radiation schemes are similar in functional outcome, with decreased treatment time and acceptable toxicity to more protracted courses of radiotherapy [14]. Multiple radiation schedules are in use, from single fractions of 8 Gy, up to 30 Gy in 10 fractions [15]. Overall, radiotherapy in ESCC is usually well tolerated. However, cytopenias, gastrointestinal mucositis and diarrhea can all be complications, depending on the radiation dose, fractionalization schedule and size of the delivery field. Recent advances in technology, such as intensity-modulated radiation therapy, continue to improve on the delivery of precision-guided radiation to the metastases, limiting potential damage to healthy tissue.

Surgery

Typical indications for surgery for ESCC include pathological fracture resulting in bone fragments impinging on the spinal cord, radioresistant tumor and need for tissue for diagnosis. Surgical decompression is dependent in large part on the severity and location of the compression. Initial surgical decompressive interventions in ESCC are generally done with laminectomies and vertebral corpectomies, however, followed inevitably by radiotherapy for residual tumor. As discussed earlier, given that most bone metastases in ESCC usually affect vertebral bodies, surgical intervention can further destabilize the already weakened anterior column. Nowadays, specialized cements and instrument fixation techniques help to limit such complications. Nevertheless, a large retrospective study found complication rates of approximately 48% and a 30-day mortality of about 10% in its surgical patients [16]. On the other hand, a recent randomized, non-blinded, multi-institutional clinical trial established the potential role for aggressive decompressive resections combined with radiotherapy in ESCC for patients that could undergo a surgical procedure [17]. This study of 101 patients supported that patients with cord compression treated surgically with postoperative radiation were ambulatory longer than patients treated with radiotherapy alone. With eligibility criteria that excluded patients with multiple sites of cord compression, total paraplegia longer than 48 hours, or a life expectancy of less than 3 months, the generalizability of this trial to the entire ESCC population remains uncertain. Given the recurrence rate and median survival of patients with ESCC, surgery has to be considered on a case-by-case basis to provide the most complete care, yet with a minimization of risk in patients with short life expectancies.

RECURRENCE

Unfortunately, regardless of the potential treatment options available, local recurrence is estimated at about 10% in all patients with ESCC. In fact, one prospective study found that ESCC recurred in 20% of patients with a median interval of at least 7 months, with essentially all surviving patients having recurrence at 3 years. Additionally, ESCC recurred at the same spinal level about 55% of the time [18]. Treatment options as discussed above, including radiotherapy, chemotherapy and surgery, may be revisited as appropriate [19]. The concern of whether repeated cumulative irradiation may exceed the estimated spinal tolerance has to be carefully weighed against the potential benefits and usual short survival of these unfortunate patients.

REFERENCES

1. Bach F, Larsen BH, Rohde et al (1990). Metastatic spinal cord compression. Occurrence, symptoms, clinical presentations and prognosis in 398 patients with spinal cord compression. Acta Neuro Chir 107(1–2):37–43.

2. Schiff D, O'Neill BP, Wang CH, O'Fallon JR (1998). Neuroimaging and treatment implications of patients with multiple epidural spinal metastases. Cancer 83:1593–1601.

3. Ushio Y, Posner R, Posner JB et al (1977). Experimental spinal cord compression by epidural neoplasm. Neurology 27:422–429.

4. Prasad D, Schiff et al (2005). Malignant spinal-cord compression. Lancet Oncol 6:15–24.

5. Husband DJ (1998). Malignant spinal cord compression: prospective study of delays in referral and treatment. Br Med J 317:18–21.

6. Schiff, D (2003). Spinal cord compression. Neurol Clin N Am 21:67–86.

7. Husband DJ et al (2001). MRI in the diagnosis and treatment of suspected malignant spinal cord compression. Br J Radiol 74:15–23.

8. Schiff D (2005). Metastatic spinal cord disease. Continuum 11(5):30–46.

9. Helwig-Larsen S, Sorensen PS (1994). Symptoms and signs in metastatic spinal cord compression; a study of progression from first symptom until diagnosis in 153 pateints. Eur J Cancer 30A:396–398.

10. Sorensen S et al (1990). Metastatic epidural spinal cord compression. Results of treatment and survival. Cancer 65(7):1502–1508.

11. Heimdal K, Hirschberg H, Slettebo H et al (1992). High incidence of serious side effects of high-dose dexamethasone treatment in patients with epidural spinal cord compression. J Neuro-Oncol 12:141–144.

12. Hernández-Díaz S, Rodríguez L (2001). Steroids and risk of upper gastrointestinal complications. Am J Epidemiol 153:1089–1093.

13. Helwig-Larsen S, Sorensen PS, Kreiner S (2000). Prognostic factors in metastatic spinal cord compression: a prospective study using multivariate analysis of variables influencing survival and gait function in 153 pateints. Int J Radiat Oncol Biol Phys 46:1163–1169.

14. Maranzano EP et al (2005). Short course radiotherapy in metastatic spinal cord compression: results of a phase III, randomized, multicenter trial. J Clin Oncol 23:3358–3365.

15. Rades D et al (2004). A prospective evaluation of two radiotherapy schedules with 10 versus 20 fractions for the treatment of metastatic spinal cord compression: final results of a multicenter study. Cancer 101:2687–2692.

16. Sundaresan N, Sachdev VP, Holland JF et al (1995). Surgical treatment of spinal cord compression from epidural metastasis. J Clin Oncol 13:2330–2335.

17. Patchell R et al (2005). Direct decompressive surgical resection in the treatment of spinal cord compression caused by metastatic cancer: a randomized trial. Lancet 366:643–648.

18. Van der Sande JJ, Boogerd W, Kroger R, Kappelle AC (1999). Recurrent spinal epidural metastases: a prospective study with a complete follow up. J Neurol Neurosurg Psychiatr 66(5):623–627.

19. Schiff D, Shaw EG, Cascino TL (1995). Outcome after spinal reirradiation for malignant epidural spinal cord compression. Ann Neurol 37(5):583–589.

CHAPTER 8

Neoplastic Meningitis

Marc C. Chamberlain

EPIDEMIOLOGY

Neoplastic meningitis (NM) is diagnosed in 4–15% of patients with solid tumors (in which case it is termed carcinomatous meningitis), 5–15% of patients with leukemia and lymphoma (termed leukemic or lymphomatous meningitis, respectively) and 1–2% of patients with primary brain tumors [1–5]. Autopsy studies show that 19% of patients with cancer and neurologic signs and symptoms have evidence of meningeal involvement [6]. Adenocarcinoma is the most frequent histology and breast, lung and melanoma are the most common primary sites to metastasize to the leptomeninges (Table 8.1) [3,7,8]. Although small cell lung cancer and melanoma have the highest rates of spread to the leptomeninges (11% and 20% respectively [9,10]), because of the higher incidence of breast cancer (with a 5% rate of spread [11]), the latter accounts for most cases in large series of the disorder [1,7].

NM usually presents in patients with widely disseminated and progressive systemic cancer (>70%), but it can present after a disease-free interval (20%) and even be the first manifestation of cancer (5–10%), occasionally in the absence of other evidence of systemic disease [7,14–16].

PATHOGENESIS

Cancer cells reach the meninges by various routes:

1 hematogenous spread, either through the venous plexus of Batson or by arterial dissemination
2 direct extension from contiguous tumor deposits
3 through centripetal migration from systemic tumors along perineural or perivascular spaces [17–19].

Once cancer cells have entered the subarachnoid space, they are transported by CSF flow resulting in disseminated and multifocal neuraxis seeding of the leptomeninges. Tumor infiltration is most prominent in the base of brain (specifically the basilar cisterns) and dorsal surface of the spinal cord (in particular the cauda equina) [5,20]. Hydrocephalus or impairment of CSF flow may occur due to ependymal nodules or tumor deposits obstructing

TABLE 8-1 Most frequent primary tumors [3,7,8,12,13]	
Primary site of cancer	*%*
Breast cancer	27–50
Lung cancer	22–36
Adenocarcinoma	50–56
Squamous cell carcinoma	26–36
Small cell carcinoma	13–14
Malignant melanoma	12
Genitourinary	5
Head and neck	2
Adenocarcinoma of unknown primary	2

CSF outflow, particularly at the level of the fourth ventricle, basal cisterns, cerebral convexity or spinal subarachnoid space.

CLINICAL FEATURES

Leptomeningeal carcinomatosis classically presents with pleomorphic clinical manifestations encompassing symptoms and signs in three domains of neurological function:

1 the cerebral hemispheres
2 the cranial nerves
3 the spinal cord and associated roots.

Signs on examination generally exceed patient reported symptoms.

The most common manifestations of cerebral hemisphere dysfunction are headache and mental status changes. Other signs include confusion, dementia, seizures and hemiparesis. Diplopia is the most common symptom of cranial nerve dysfunction with the cranial nerve VI being the most frequently affected, followed by cranial nerves III and IV. Trigeminal sensory or motor loss, cochlear dysfunction and optic neuropathy are also common findings. Spinal signs and symptoms include weakness (lower extremities more often than upper), dermatomal or segmental sensory loss and pain in the neck, back, or following radicular patterns. Nuchal rigidity is only present in 15% of cases [3,7,8,15,21].

A high index of suspicion needs to be entertained in order to make the diagnosis of NM. The finding of multifocal neuraxis disease in a patient with known malignancy is strongly suggestive of NM, but it is also common for patients with NM to present with isolated syndromes such as symptoms of raised intracranial pressure, cauda equina syndrome or cranial neuropathy.

New neurological signs and symptoms may represent progression of NM, but must be distinguished from the manifestations of parenchymal disease (30–40% of patients with NM will have coexistent parenchymal brain metastases), from side effects of chemotherapy or radiation used for treatment and rarely from paraneoplastic syndromes. At presentation, NM must also be differentiated from chronic meningitis due to tuberculosis, fungal infection or sarcoidosis, as well as from metabolic and toxic encephalopathies in the appropriate clinical setting [7,22].

DIAGNOSIS

CSF examination

The most useful laboratory test in the diagnosis of NM is the CSF exam. Abnormalities include increased opening pressure (>200 mm of H_2O), increased leukocytes ($>4/mm^3$), elevated protein (>50 mg/dl) or decreased glucose (<60 mg/dl) which, though suggestive of NM, are not diagnostic. The presence of malignant cells in the CSF is diagnostic of NM but, in general, as is true for most cytological analysis, assignment to a particular tumor is not possible [23].

In patients with positive CSF cytology (see below), up to 45% will be cytologically negative on initial examination [6]. The yield is increased to 80% with a second CSF examination, but little benefit is obtained from repeat lumbar punctures after two lumbar punctures [7]. Of note, a series including lymphomatous and leukemic meningitis by Kaplan et al [3] observed the frequent dissociation between CSF cell count and malignant cytology (29% of cytologically positive CSF had concurrent CSF counts of less than $4/mm^3$). Murray et al [24] showed that CSF levels of protein, glucose and malignant cells [25] vary at different levels of the neuraxis even if there is no obstruction of the CSF flow. This finding reflects the multifocal nature of neoplastic meningitis and explains that CSF obtained from a site distant to that of the pathologically involved meninges may yield a negative cytology.

Of the 90 patients reported by Wasserstrom et al [7], 5% had positive CSF cytology only from either the ventricles or cisterna magna. In a series of 60 patients with NM, positive lumbar CSF cytology at diagnosis and no evidence of CSF flow obstruction, ventricular and lumbar cytologies obtained simultaneously were discordant in 30% of

cases [26]. The authors observed that in the presence of spinal signs or symptoms, the lumbar CSF was more likely to be positive and, conversely, in the presence of cranial signs or symptoms, the ventricular CSF was more likely to be positive. Not obtaining CSF from a site of symptomatic or radiographically demonstrated disease was found to correlate with false negative cytology results in a prospective evaluation of 39 patients, as did withdrawing small CSF volumes (<10.5 ml), delayed processing of specimens and obtaining less than two samples [27]. Even after correcting for these factors there remains a substantial group of patients with NM and persistently negative CSF cytology. Glass reported on a post-mortem evaluation evaluating the value of premortem CSF cytology and demonstrated that up to 40% of patients with clinically suspected NM proven at the time of autopsy are cytologically negative [6]. This figure increased to greater than 50% in patients with focal NM.

The low sensitivity of CSF cytology makes it difficult not only to diagnose NM, but also to assess the response to treatment. Biochemical markers, immunohistochemistry and molecular biological techniques applied to CSF have been explored in an attempt to find a reliable biological marker of disease.

Numerous biochemical markers have been evaluated but, in general, their use has been limited by poor sensitivity and specificity. Particular tumor markers such as CEA (carcinoembryogenic antigen) from adenocarcinomas, and AFP (α-fetoprotein) and β-HCG (β-human chorionic gonadotropin) from testicular cancers and primary extragonadal CNS tumors can be relatively specific for NM when elevated in CSF in the absence of markedly elevated serum levels [18,28]. Non-specific tumor markers such as CK-BB (creatine-kinase BB isoenzyme), TPA (tissue polypeptide antigen), $\beta_2$microglobulin, β-glucuronidase, LDH isoenzyme-5 and more recently VEGF (vascular endothelial growth factor) can be strong indirect indicators of NM, but none are sensitive enough to improve the cytological diagnosis [29–34]. The use of these biochemical markers can be helpful as adjunctive diagnostic tests and, when followed serially, to assess response to treatment. Occasionally, in patients with clinically suspected NM and negative CSF cytology, they may support the diagnosis of NM [35].

Use of monoclonal antibodies for immunohistochemical analysis in NM does not significantly increase the sensitivity of cytology alone [36–38]. However, in the case of leukemia and lymphoma, antibodies against surface markers can be used to distinguish between reactive and neoplastic lymphocytes in the CSF [39].

Cytogenetic studies have also been evaluated in an attempt to improve the diagnostic accuracy of NM. Flow cytometry and DNA single cell cytometry, techniques that measure the chromosomal content of cells, and fluorescent in situ hybridization (FISH), that detects numerical and structural genetic aberrations as a sign of malignancy, can give additional diagnostic information, but still have a low sensitivity [40–42]. Polymerase-chain reaction (PCR) can establish a correct diagnosis when cytology is inconclusive, but the genetic alteration of the neoplasia must be known for it to be amplified with this technique, and this is generally not the case, particularly in solid tumors [43].

In cases where there is no evidence of systemic cancer and CSF exams remain inconclusive, a meningeal biopsy may be diagnostic. The yield of this test increases if the biopsy is taken from an enhancing region on magnetic resonance imaging (MRI) (see below) and if posterior fossa or pterional approaches are used [44].

Neuroradiographic studies

Magnetic resonance imaging with gadolinium enhancement (MR-Gd) is the technique of choice to evaluate patients with suspected leptomeningeal metastasis [45]. Because NM involves the entire neuraxis, imaging of the entire CNS is required in patients considered for further treatment. T1-weighted sequences, with and without contrast, combined with fat suppression T2-weighted sequences constitute the standard examination [46]. MRI has been shown to have a higher sensitivity than cranial contrast enhanced computed tomography (CE-CT) in several series [45,47], and is similar to computerized tomographic myelography (CT-M) for the evaluation of the spine, but significantly better tolerated [48,49].

Any irritation of the leptomeninges (i.e. blood, infection, cancer) will result in their enhancement on MRI, which is seen as a fine signal-intense layer that follows the gyri and superficial sulci. Subependymal involvement of the ventricles often results in ventricular enhancement. Some changes such as cranial nerve enhancement on cranial imaging and intradural extramedullary enhancing nodules on spinal MR (most frequently seen in the cauda equina) can be considered diagnostic of NM in patients with cancer [50]. Lumbar puncture itself can rarely cause a meningeal reaction leading to dural-arachnoidal enhancement, so imaging should be obtained preferably prior to the procedure [51]. MR-Gd still has a 30% incidence of false negative results so that a normal study does not exclude the diagnosis of NM. On the other hand, in cases with a typical clinical presentation, abnormal MR-Gd alone is adequate to establish the diagnosis of NM [35,50].

Radionuclide studies, using either [111]indium-diethylenetriamine pentaacetic acid or ^{99}Tc macro-aggregated albumin, constitute the technique of choice to evaluate CSF flow dynamics [19,51]. Abnormal CSF circulation has been demonstrated in 30–70% of patients with NM, with blocks commonly occurring at the skull base, the spinal canal and over the cerebral convexities [49,52,53]. Patients with interruption of CSF flow demonstrated by radionuclide ventriculography have been shown in three clinical series to have decreased survival when compared to those with normal CSF flow [52,54,55]. Involved-field radiotherapy to the site of CSF flow obstruction restores flow in 30% of patients

with spinal disease and 50% of patients with intracranial disease [56]. Re-establishment of CSF flow with involved-field radiotherapy followed by intrathecal chemotherapy led to longer survival, lower rates of treatment-related morbidity and lower rate of death from progressive NM, compared to the group that had persistent CSF blocks [52,54]. These findings may reflect that CSF flow abnormalities prevent homogeneous distribution of intrathecal chemotherapy, resulting in:

1 protected sites where tumor can progress; and
2 accumulation of drug at other sites leading to neurotoxicity and systemic toxicity.

Based on this, many authors recommend that intrathecal chemotherapy be preceded by a radionuclide flow study and, if a block is found, that radiotherapy be administered in an attempt to re-establish normal flow [18,57].

STAGING

In summary, patients with suspected NM should undergo one or two lumbar punctures, cranial MR-Gd, spinal MR-Gd and a radioisotope CSF flow study to rule out sites of CSF block. If cytology remains negative and radiological studies are not definitive, consideration may be given to ventricular or lateral cervical CSF analysis based on suspected site of predominant disease. If the clinical scenario or radiological studies are highly suggestive of NM, treatment is warranted despite persistently negative CSF cytologies.

PROGNOSIS

The median survival of untreated patients with NM is 4–6 weeks and death generally occurs due to progressive neurological dysfunction [7]. Treatment is intended to improve or stabilize the neurological status and to prolong survival. Fixed neurological defects are rarely improved with treatment [18], but progression of neurological deterioration may be halted in some patients and median survival can be increased to 4–6 months. Of the solid tumors (Table 8.2), breast cancer responds best, with median survivals of 6 months and 11–25% 1-year survivals [22,28,58]. Numerous prognostic factors for survival and response have been evaluated (e.g. age, gender, duration of signs of

TABLE 8-2 Prognosis by tumor histology		
Tumor histology	*Median survival (months)*	*Range (months)*
Breast (n = 32) [22]	7.5	1.5–16
Non-small cell lung cancer (n = 32) [59]	5	1–12
Melanoma (n = 16) [54]	4	2–8
High-grade glioma (n = 20) [61]	3.5	1–6

NM, increased protein or low glucose in CSF, ratio of lumbar/ventricular CEA), but many remain controversial [58]. It is commonly accepted, however, that patients will do poorly with intensive treatment of NM if they have poor performance status, multiple fixed neurologic deficits, bulky CNS disease, coexistent carcinomatous encephalopathy and CSF flow abnormalities demonstrated by radionuclide ventriculography. In general, patients with widely metastatic aggressive cancers that do not respond well to systemic chemotherapies, are also less likely to benefit from intensive therapy [18,52,59,60]. What appears clear is that, optimally, NM should be diagnosed in the early stages of disease to prevent progression of disabling neurological deficits, analogous to the clinical situation of epidural spinal cord compression.

TREATMENT

The evaluation of treatment of NM is complicated by the lack of standard treatments, the difficulty of determining response to treatment given the suboptimal sensitivity of the diagnostic procedures and that most patients will die of systemic disease, and the fact that most studies are small, non-randomized and retrospective [58]. However, it is clear that treatment of NM can provide effective palliation and, in some cases, result in prolonged survival. Treatment requires the combination of surgery, irradiation and chemotherapy in most cases. Figure 8.1 outlines a treatment algorithm for NM.

Surgery

Surgery is used in the treatment of NM for the placement of:

1 intraventricular catheter and subgaleal reservoirs for administration of cytotoxic drugs
2 ventriculoperitoneal shunt in patients with symptomatic hydrocephalus.

Drugs can be instilled into the subarachnoid space by lumbar puncture or via an intraventricular reservoir system. The latter is the preferred approach because it is simpler, more comfortable for the patient and safer than repeated lumbar punctures. It also results in a more uniform distribution of the drug in the CSF space and produces the most consistent CSF levels. In up to 10% of lumbar punctures drug is delivered to the epidural space, even if there is CSF return after placement of the needle, and drug distribution has been shown to be better after drug delivery through a reservoir [62].

There are two basic types of reservoirs: the *Rickham style reservoir*, a flat rigid reservoir placed over a burr hole, and the *Ommaya reservoir*, a dome-shaped reservoir that can be palpated easily. They are generally placed over the right (non-dominant) frontal region using a small C-shaped incision. The catheter is placed in the frontal horn of the lateral ventricle or close to the foramen of Monroe through a standard ventricular puncture. In most cases, anatomical landmarks suffice, but ultrasonographic or CT guidance can be helpful in some situations [63]. It is very important to be sure that the tip and the side perforations of the catheter be inserted completely into the ventricle to avoid drug instillation into the brain parenchyma. Correct placement of the catheter should be checked by non-contrast CT prior to its use for drug administration and frequently it will show a small amount of air in both frontal horns [64].

NM often causes communicating hydrocephalus leading to symptoms of raised intracranial pressure. Relief of sites of CSF flow obstruction with involved-field radiation should be attempted to avoid the need for CSF shunting. If hydrocephalus persists, a ventriculoperitoneal shunt should be placed to relieve the pressure, because relief of pressure often results in clinical improvement. If possible an in-line on/off valve and reservoir should be used to permit the administration of intra-CSF chemotherapy, although some patients cannot tolerate having the shunt turned off to allow the circulation of the drug [19].

Furthermore, in patients with persistent blockage of ventricular CSF, a lumbar catheter and reservoir can be used in addition to a ventricular catheter, to allow treatment of the spine with intrathecal chemotherapy, although as discussed earlier, patients with persistent CSF flow blocks after radiation are probably best managed by supportive care alone.

Finally, occasional patients may undergo a meningeal biopsy, to confirm pathologically neoplastic meningitis. However, since most patients demonstrate MR leptomeningeal abnormalities, an abnormal CSF profile or a clinical examination consistent with NM, meningeal biopsies are rarely performed.

Radiotherapy

Radiotherapy is used in the treatment of NM for:

1 palliation of symptoms, such as a cauda equina syndrome
2 to decrease bulky disease such as coexistent parenchymal brain metastases; and
3 to correct CSF flow abnormalities demonstrated by radionuclide ventriculography.

Patients may have significant symptoms without radiographic evidence of bulky disease and still benefit from radiation. For example, patients with low back pain and leg weakness should be considered for radiation to the cauda equina, and those with cranial neuropathies should be offered whole-brain or base of skull radiotherapy [28].

Radiotherapy of bulky disease is indicated as intra-CSF chemotherapy is limited by diffusion to 2–3 mm penetration into tumor nodules. In addition, involved-field radiation can correct CSF flow abnormalities and this has been shown to improve patient outcome as discussed above.

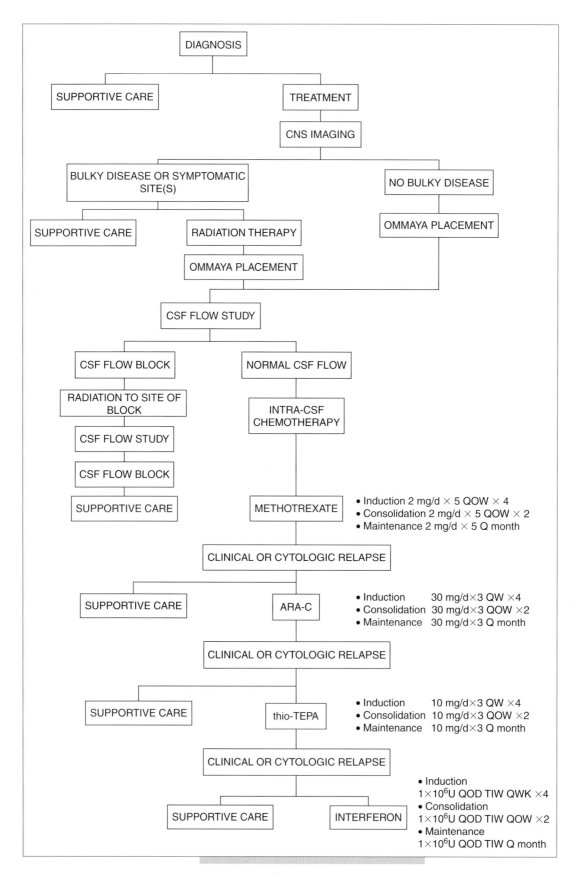

FIG. 8.1. Treatment algorithm of neoplastic meniningitis.

Whole neuraxis radiation is rarely indicated in the treatment of NM from solid tumors because it is associated with significant systemic toxicity (severe myelosuppression and mucositis among other complications) and is not curative.

Chemotherapy

Chemotherapy is the only treatment modality that can treat the entire neuraxis. Chemotherapy may be administered systemically or intrathecally.

Intrathecal chemotherapy is the mainstay of treatment for NM. Retrospective analysis or comparison to historical series suggest that the administration of chemotherapy to the CSF improves the outcome of patients with NM [1,22,54,65,66]. However, it is noted that most series will exclude patients that are too sick to receive any treatment, which may be up to one-third of patients with NM [67]. Three agents are routinely used: methotrexate, cytarabine (including liposomal cytarabine or DepoCyt®) and thio-TEPA. No difference in response has been seen when comparing single agent methotrexate with thio-TEPA [18] or when using multiple agent (methotrexate, thio-TEPA and cytarabine or methotrexate and cytarabine) versus single agent methotrexate in adult randomized studies of NM [68–70]. Table 8.3 outlines the common treatment regimens for these drugs. A sustained-release liposomal form of cytarabine (DepoCyt®) results in cytotoxic cytarabine levels in the CSF for $\geq$10 days, and when given bimonthly and compared to biweekly methotrexate resulted in longer time to neurological progression in patients with NM due to solid tumors [71]. Furthermore, quality of life and cause of death favored DepoCyt® over methotrexate. These findings were confirmed in a study of lymphomatous meningitis and in an open label study suggesting that DepoCyt® should be considered the drug of first choice in the treatment of NM when experimental therapies are unavailable [72,73].

Complications of intrathecal chemotherapy include those related to the ventricular reservoir and those related to the chemotherapy administered. The most frequent complications of ventricular reservoir placement are malposition (rates reported between 3 and 12%), obstruction and infection (usually skin flora). CSF infection occurs in 2–13% of patients receiving intrathecal chemotherapy. It commonly presents with headache, changes in neurologic status, fever and malfunction of the reservoir. CSF pleocytosis is commonly encountered. The most frequently isolated organism is *Staphylococcus epidermidis*. Treatment requires intravenous antibiotics, with or without oral and intraventricular antibiotics. Some authors advocate the routine removal of the ventricular reservoir, while others reserve device removal for those that do not clear with antibiotic therapy. Routine culture of CSF is not recommended because of the high rate of contamination with skin flora in the absence of infection [1,5,53,64,75]. Myelosuppression can occur

TABLE 8-3 Regional chemotherapy for neoplastic meningitis						
Drugs	*Induction regimens*		*Consolidation regimen*		*Maintenance regimen*	
	Bolus regimen	*CxT regimen*	*Bolus regimen*	*CxT regimen*	*Bolus regimen*	*CxT regimen*
Methotrexate	10–15 mg twice weekly (total 4 weeks)	2 mg/day for 5 days every other week (total 8 weeks)	10–15 g once weekly (total 4 weeks)	2 g/day for 5 days every other week (total 4 weeks)	10–15 mg once a month	2 mg/day for 5 days once a month
Cytarabine [71,74]	25–100 mg 2 or 3 times weekly (total 4 weeks)	25 mg/day for 3 days weekly (total 4 weeks)	25–100 mg once weekly (total 4 weeks)	25 mg/day for 3 days every other week (total 4 weeks)	25–100 mg once a month	25 mg/day for 3 days once a month
DepoCyt® [71,74]	50 mg every 2 weeks (total 8 weeks)		50 mg every 4 weeks (total 24 weeks)			
Thio-TEPA	10 mg 2 or 3 times weekly (total 4 weeks)	10 mg/day for 3 days weekly (total 4 weeks)	10 mg once weekly (total 4 weeks)	10 mg/day for 3 days every other week (total 4 weeks)	10 mg once a month	10 mg/day for 3 days once a month
α-Interferon	1 × 10^6 U 3 times weekly (total 4 weeks)		1 × 10^6 U 3 times weekly every other week (total 4 weeks)		1 × 10^6 U 3 times weekly one week per month)	

after administration of intrathecal chemotherapies and it is recommended that folinic acid rescue (10 mg every 6 hours for 24 hours) be given orally after the administration of methotrexate to mitigate this complication. Chemical aseptic meningitis occurs in nearly half of patients treated by intraventricular administration and is manifested by fever, headache, nausea, vomiting, meningismus and photophobia. In the majority of patients, this inflammatory reaction can be treated in the outpatient setting with oral antipyretics, antiemetics and corticosteroids. Rarely, treatment-related neurotoxicity occurs and may result in a symptomatic subacute leukoencephalopathy or myelopathy. However, in patients with NM and prolonged survival, the combination of radiotherapy and chemotherapy frequently results in a late leukoencephalopathy evident on neuroradiographic studies, and is occasionally symptomatic [1,5,76].

The rationale to give intrathecal chemotherapy is based on the presumption that most chemotherapeutic agents when given systemically have poor CSF penetration and do not reach therapeutic levels. Exceptions to this would be systemic high-dose methotrexate, cytarabine and thio-TEPA, all of which result in cytotoxic CSF levels. Their systemic administration, however, is limited by systemic toxicity and the difficulty to integrate these regimens into other chemotherapeutic programs being used to manage systemic disease. Some authors argue that intrathecal chemotherapy does not add to improved outcome in the treatment of NM, since systemic therapy can obtain access to the subarachnoid deposits through their own vascular supply [67]. In a retrospective comparison of patients treated with systemic chemotherapy and radiation to involved areas, plus or minus intrathecal chemotherapy, Bokstein et al [77] did not find significant differences in response rates, median survival or proportion of long-term survivors among the two groups but, of course, the group that did not receive the intrathecal treatment was spared the complications of this modality. Glantz et al [78] treated 16 patients with high dose intravenous methotrexate and compared their outcome with a reference group of 15 patients treated with intrathecal methotrexate. They found response rates and survival were significantly better in the group treated with intravenous therapy. Finally, a recent report describes two patients with breast cancer in whom NM was controlled with systemic hormonal treatment [79].

Nonetheless, intrathecal chemotherapy remains the preferred treatment route for NM at this time. New drugs are being explored to try to improve the efficacy, these include mafosphamide, diaziquone, topotecan [80], gemcitabine, interferon-α and temozolomide [81] which are being evaluated for intrathecal administration. Immunotherapy, using interleukin-2 and interferon-α [82], ^{131}I-radiolabelled monoclonal antibodies [83] and gene therapy [84] are other modalities that are being explored in clinical trials.

Supportive care

Not all patients with NM are candidates for the aggressive treatment outlined above. Most authors agree that combined-modality therapy should be offered to patients with life expectancy greater than 3 months and a Karnofsky performance status of greater than 60%.

Supportive care that should be offered to every patient, regardless of whether they receive NM directed therapy, include anticonvulsants for seizure control (seen in 10–15% of patients with NM), adequate analgesia with opioid drugs, as well as antidepressants and anxiolytics if necessary. Corticosteroids have a limited use in NM-related neurological symptoms, but can be useful to treat vasogenic edema associated with intraparenchymal or epidural metastases, or for the symptomatic treatment of nausea and vomiting together with routine antiemetics. Decreased attention and somnolence secondary to whole brain radiation can be treated with psychostimulants [5].

CONCLUSIONS

NM is a complicated disease for a variety of reasons. First, most reports concerning NM treat all subtypes as equivalent with respect to CNS staging, treatment and outcome. However, clinical trials in oncology are based on specific tumor histology. Comparing responses in patients with carcinomatous meningitis due to breast cancer to patients with non-small cell lung cancer outside of investigational new drug trials may be misleading. A general consensus is that breast cancer is inherently more chemosensitive than non-small cell lung cancer or melanoma, and therefore survival following chemotherapy is likely to be different. This observation has been substantiated in patients with systemic metastases, though comparable data regarding CNS metastases and, in particular, NM is meager.

A second feature of NM, which complicates therapy, is deciding whom to treat. Not all patients necessarily warrant aggressive CNS-directed therapy, however, few guidelines exist permitting appropriate choice of therapy. Based on the prognostic variables determined clinically and by evaluation of extent of disease, a sizable minority of patients will not be candidates for aggressive NM-directed therapy. Therefore supportive comfort care (radiotherapy to symptomatic disease, antiemetics and narcotics) is reasonably offered to patients with NM considered poor candidates for aggressive therapy, as seen in Figure 8.1.

Thirdly, optimal treatment of NM remains poorly defined. Given these constraints, the treatment of NM today is palliative and rarely curative with a median patient survival of 2–3 months based on data of the four prospective randomized trials in this disease. However, palliative therapy of NM often affords the patient protection from further neurological deterioration and consequently an improved neurologic quality of life. No studies to date have attempted an economic assessment of the treatment of NM and therefore no information is available regarding a cost-benefit analysis as has been performed for other cancer directed therapies.

Finally, in patients with NM, the response to treatment is primarily a function of CSF cytology and secondarily of clinical improvement of neurologic signs and symptoms. Aside from CSF cytology and perhaps biochemical markers, no other CSF parameters predict response. Furthermore, because CSF cytology may manifest a rostra-caudal disassociation, consecutive negative cytology samples (defined as a complete response to treatment) require confirmation by both ventricular and lumbar CSF cytologies. In general, only pain related neurologic symptoms improve with treatment. Neurologic signs such as confusion, cranial nerve deficit(s), ataxia and segmental weakness minimally improve or stabilize with successful treatment.

REFERENCES

1. Shapiro WR, Posner JB, Ushio Y et al (1977). Treatment of meningeal neoplasms. Cancer Treat Rep 61(4):733–743.
2. Nugent JL, Bunn PA, Jr, Matthews MJ et al (1979). CNS metastases in small cell bronchogenic carcinoma: increasing frequency and changing pattern with lengthening survival. Cancer 44(5):1885–1893.
3. Kaplan JG, DeSouza TG, Farkash A et al (1990). Leptomeningeal metastases: comparison of clinical features and laboratory data of solid tumors, lymphomas and leukemias. J Neuro-Oncol 9(3):225–229.
4. Siegal T, Lossos A, Pfeffer MR (1994). Leptomeningeal metastases: analysis of 31 patients with sustained off-therapy response following combined-modality therapy. Neurology 44(8):1463–1469.
5. Chamberlain MC (1997). Carcinomatous meningitis. Arch Neurol 54(1):16–17.
6. Glass JP, Melamed M, Chernik NL et al (1979). Malignant cells in cerebrospinal fluid (CSF): the meaning of a positive CSF cytology. Neurology 29(10):1369–1375.
7. Wasserstrom WR, Glass JP, Posner JB (1982). Diagnosis and treatment of leptomeningeal metastases from solid tumors: experience with 90 patients. Cancer 49(4):759–772.
8. Little JR, Dale AJ, Okazaki H (1974). Meningeal carcinomatosis. Clinical manifestations. Arch Neurol 30(2):138–143.
9. Rosen ST, Aisner J, Makuch RW et al (1982). Carcinomatous leptomeningitis in small cell lung cancer: a clinicopathologic review of the National Cancer Institute experience. Medicine (Baltimore) 61(1):45–53.
10. Amer MH, Al Sarraf M, Baker LH et al (1978). Malignant melanoma and central nervous system metastases: incidence, diagnosis, treatment and survival. Cancer 42(2):660–668.
11. Yap HY, Yap BS, Tashima CK et al (1978). Meningeal carcinomatosis in breast cancer. Cancer 42(1):283–286.
12. Theodore WH, Gendelman S (1981). Meningeal carcinomatosis. Arch Neurol 38(11):696–699.
13. Olson ME, Chernik NL, Posner JB (1974). Infiltration of the leptomeninges by systemic cancer. A clinical and pathologic study. Arch Neurol 30(2):122–137.
14. van Oostenbrugge RJ, Twijnstra A (1999). Presenting features and value of diagnostic procedures in leptomeningeal metastases. Neurology 53(2):382–385.
15. Balm M, Hammack J (1996). Leptomeningeal carcinomatosis. Presenting features and prognostic factors. Arch Neurol 53(7):626–632.
16. Sorensen SC, Eagan RT, Scott M (1984). Meningeal carcinomatosis in patients with primary breast or lung cancer. Mayo Clin Proc 59(2):91–94.
17. Gonzalez-Vitale JC, Garcia-Bunuel R (1976). Meningeal carcinomatosis. Cancer 37(6):2906–2911.
18. Grossman SA, Krabak MJ (1999). Leptomeningeal carcinomatosis. Cancer Treat Rev 25(2):103–119.
19. Chamberlain MC (1998). Radioisotope CSF flow studies in leptomeningeal metastases. J Neuro-Oncol 38(2–3):135–140.
20. Boyle R, Thomas M, Adams JH (1980). Diffuse involvement of the leptomeninges by tumour – a clinical and pathological study of 63 cases. Postgrad Med J 56(653):149–158.
21. Wolfgang G, Marcus D, Ulrike S (1998). Leptomeningeal carcinomatosis; clinical syndrome in different primaries. J Neuro-Oncol 38:103–110.
22. Chamberlain MC, Kormanik PR (1997). Carcinomatous meningitis secondary to breast cancer: predictors of response to combined modality therapy. J Neuro-Oncol 35(1):55–64.
23. Kolmel HW (1998). Cytology of neoplastic meningosis. J Neuro-Oncol 38(2–3):121–125.
24. Murray JJ, Greco FA, Wolff SN et al (1983). Neoplastic meningitis. Marked variations of cerebrospinal fluid composition in the absence of extradural block. Am J Med 75(2):289–294.
25. Rogers LR, Duchesneau PM, Nunez C et al (1992). Comparison of cisternal and lumbar CSF examination in leptomeningeal metastasis. Neurology 42(6):1239–1241.
26. Chamberlain MC, Kormanik PA, Glantz MJ. A comparison between ventricular and lumbar cerebrospinal fluid cytology in adult patients with leptomeningeal metastases. Neuro-Oncol 3(1):42–45.
27. Glantz MJ, Cole BF, Glantz LK et al (1998). Cerebrospinal fluid cytology in patients with cancer: minimizing false-negative results. Cancer 82(4):733–739.
28. DeAngelis LM (1998). Current diagnosis and treatment of leptomeningeal metastasis. J Neuro-Oncol 38(2–3):245–252.
29. Wasserstrom WR, Schwartz MK, Fleisher M et al (1981). Cerebrospinal fluid biochemical markers in central nervous system tumors: a review. Ann Clin Lab Sci 11(3):239–251.
30. van Zanten AP, Twijnstra A, Hart AA et al (1986). Cerebrospinal fluid lactate dehydrogenase activities in patients with central nervous system metastases. Clin Chim Acta 161(3): 259–268.
31. Klee GG, Tallman RD, Goellner JR et al (1986). Elevation of carcinoembryonic antigen in cerebrospinal fluid among patients with meningeal carcinomatosis. Mayo Clin Proc 61(1):9–13.
32. Twijnstra A, van Zanten AP, Hart AA et al (1987). Serial lumbar and ventricle cerebrospinal fluid lactate dehydrogenase activities in patients with leptomeningeal metastases from solid and haematological tumours. J Neurol Neurosurg Psychiatr 50(3):313–320.

33. Twijnstra A, Ongerboer d, V, van Zanten AP et al (1989). Serial lumbar and ventricular cerebrospinal fluid biochemical marker measurements in patients with leptomeningeal metastases from solid and hematological tumors. J Neuro-Oncol 7(1):57–63.

34. Stockhammer G, Poewe W, Burgstaller S et al (2000). Vascular endothelial growth factor in CSF: a biological marker for carcinomatous meningitis. Neurology 54(8):1670–1676.

35. Chamberlain MC (1998). Cytologically negative carcinomatous meningitis: usefulness of CSF biochemical markers. Neurology 50(4):1173–1175.

36. Garson JA, Coakham HB, Kemshead JT et al (1985). The role of monoclonal antibodies in brain tumour diagnosis and cerebrospinal fluid (CSF) cytology. J Neuro-Oncol 3(2):165–171.

37. Hovestadt A, Henzen-Logmans SC, Vecht CJ (1990). Immunohistochemical analysis of the cerebrospinal fluid for carcinomatous and lymphomatous leptomeningitis. Br J Cancer 62(4):653–654.

38. Boogerd W, Vroom TM, van Heerde P et al (1988). CSF cytology versus immunocytochemistry in meningeal carcinomatosis. J Neurol Neurosurg Psychiatr 51(1):142–145.

39. van Oostenbrugge RJ, Hopman AH, Ramaekers FC et al (1998). In situ hybridization: a possible diagnostic aid in leptomeningeal metastasis. J Neuro-Oncol 38(2–3):127–133.

40. Cibas ES, Malkin MG, Posner JB et al (1987). Detection of DNA abnormalities by flow cytometry in cells from cerebrospinal fluid. Am J Clin Pathol 88(5):570–577.

41. Biesterfeld S, Bernhard B, Bamborschke S et al (1993). DNA single cell cytometry in lymphocytic pleocytosis of the cerebrospinal fluid. Acta Neuropathol (Berl) 86(5):428–432.

42. van Oostenbrugge RJ, Hopman AH, Arends JW et al (1998). The value of interphase cytogenetics in cytology for the diagnosis of leptomeningeal metastases. Neurology 51(3):906–908.

43. Rhodes CH, Glantz MJ, Glantz L et al (1996). A comparison of polymerase chain reaction examination of cerebrospinal fluid and conventional cytology in the diagnosis of lymphomatous meningitis. Cancer 77(3):543–548.

44. Cheng TM, O'Neill BP, Scheithauer BW et al (1994). Chronic meningitis: the role of meningeal or cortical biopsy. Neurosurgery 34(4):590–595.

45. Chamberlain MC, Sandy AD, Press GA (1990). Leptomeningeal metastasis: a comparison of gadolinium-enhanced MR and contrast-enhanced CT of the brain. Neurology 40(3 Pt 1):435–438.

46. Schumacher M, Orszagh M (1998). Imaging techniques in neoplastic meningiosis. J Neuro-Oncol 38(2–3):111–120.

47. Sze G, Soletsky S, Bronen R et al (1989). MR imaging of the cranial meninges with emphasis on contrast enhancement and meningeal carcinomatosis. Am J Roentgenol 153(5):1039–1049.

48. Schuknecht B, Huber P, Buller B et al (1992). Spinal leptomeningeal neoplastic disease. Evaluation by MR, myelography and CT myelography. Eur Neurol 32(1):11–16.

49. Chamberlain MC (1995). Comparative spine imaging in leptomeningeal metastases. J Neuro-Oncol 23(3):233–238.

50. Freilich RJ, Krol G, DeAngelis LM (1995). Neuroimaging and cerebrospinal fluid cytology in the diagnosis of leptomeningeal metastasis. Ann Neurol 38(1):51–57.

51. Mittl RL, Jr., Yousem DM (1994). Frequency of unexplained meningeal enhancement in the brain after lumbar puncture. Am J Neuroradiol 15(4):633–638.

52. Glantz MJ, Hall WA, Cole BF et al (1995). Diagnosis, management, and survival of patients with leptomeningeal cancer based on cerebrospinal fluid-flow status. Cancer 75(12):2919–2931.

53. Trump DL, Grossman SA, Thompson G et al (1982). CSF infections complicating the management of neoplastic meningitis. Clinical features and results of therapy. Arch Intern Med 142(3):583–586.

54. Chamberlain MC, Kormanik PA (1996). Prognostic significance of [111]indium-DTPA CSF flow studies in leptomeningeal metastases. Neurology 46(6):1674–1677.

55. Mason WP, Yeh SD, DeAngelis LM (1998). [111]Indium-diethylenetriamine pentaacetic acid cerebrospinal fluid flow studies predict distribution of intrathecally administered chemotherapy and outcome in patients with leptomeningeal metastases. Neurology 50(2):438–444.

56. Chamberlain MC, Corey-Bloom J (1991). Leptomeningeal metastases: [111]indium-DTPA CSF flow studies. Neurology 41(11):1765–1769.

57. Chamberlain MC, Kormanik P, Jaeckle KA et al (1999). [111]Indium-diethylenetriamine pentaacetic acid CSF flow studies predict distribution of intrathecally administered chemotherapy and outcome in patients with leptomeningeal metastases. Neurology 52(1):216–217.

58. Hildebrand J (1998). Prophylaxis and treatment of leptomeningeal carcinomatosis in solid tumors of adulthood. J Neuro-Oncol 38(2–3):193–198.

59. Chamberlain MC, Kormanik PA (1997). Prognostic significance of coexistent bulky metastatic central nervous system disease in patients with leptomeningeal metastases. Arch Neurol 54(11):1364–1368.

60. Chamberlain MC (2003). Neoplastic meningitis-related encephalopathy: prognostic significance. Neurology 60(5): A17–18.

61. Chamberlain MC (2003). Combined modality treatment of leptomeningeal gliomatosis. Neurosurgery 52:324–330.

62. Shapiro WR, Young DF, Mehta BM (1975). Methotrexate: distribution in cerebrospinal fluid after intravenous, ventricular and lumbar injections. N Engl J Med 293(4):161–166.

63. Berweiler U, Krone A, Tonn JC (1998). Reservoir systems for intraventricular chemotherapy. J Neuro-Oncol 38(2–3): 141–143.

64. Sandberg DI, Bilsky MH, Souweidane MM et al (2000). Ommaya reservoirs for the treatment of leptomeningeal metastases. Neurosurgery 47(1):49–54.

65. Sause WT, Crowley J, Eyre HJ et al (1988). Whole brain irradiation and intrathecal methotrexate in the treatment of solid tumor leptomeningeal metastases – a Southwest Oncology Group study. J Neuro-Oncol 6(2):107–112.

66. Chamberlain MC, Kormanik P (1998). Carcinoma meningitis secondary to non-small cell lung cancer: combined modality therapy. Arch Neurol 55(4):506–512.

67. Siegal T (1998). Leptomeningeal metastases: rationale for systemic chemotherapy or what is the role of intra-CSF-chemotherapy? J Neuro-Oncol 38(2–3):151–157.

68. Giannone L, Greco FA, Hainsworth JD (1986). Combination intraventricular chemotherapy for meningeal neoplasia. J Clin Oncol 4(1):68–73.

69. Hitchins RN, Bell DR, Woods RL et al (1987). A prospective randomized trial of single-agent versus combination

chemotherapy in meningeal carcinomatosis. J Clin Oncol 5(10):1655–1662.

70. Bleyer WA, Drake JC, Chabner BA (1973). Neurotoxicity and elevated cerebrospinal-fluid methotrexate concentration in meningeal leukemia. N Engl J Med 289(15):770–773.

71. Glantz MJ, Jaeckle KA, Chamberlain MC et al (1999). A randomized controlled trial comparing intrathecal sustained-release cytarabine (DepoCyt) to intrathecal methotrexate in patients with neoplastic meningitis from solid tumors. Clin Cancer Res 5(11):3394–3402.

72. Glantz MJ, LaFollette S, Jaeckle KA et al (1999). Randomized trial of a slow release versus a standard formulation of cytarabine for the intrathecal treatment of lymphomatous meningitis. J Clin Oncol 17:3110–3116.

73. Chamberlain MC, Kormanik PA, Barba D (1997). Complications associated with intraventricular chemotherapy in patients with leptomeningeal metastases. J Neurosurg 87:694–699.

74. Jaeckle KA, Batchelor T, O'Day SJ et al (2002). An open label trial of sustained-release cytarabine (DepoCyt) for the intrathecal treatment of solid tumor neoplastic meningitis. J Neuro-Oncol 57(3):231–239.

75. Siegal T, Pfeffer MR, Steiner I (1988). Antibiotic therapy for infected Ommaya reservoir systems. Neurosurgery 22(1 Pt 1):97–100.

76. Kerr JZ, Berg S, Blaney SM (2001). Intrathecal chemotherapy. Crypt Rev Oncol Hematol 37(3):227–236.

77. Bokstein F, Lossos A, Siegal T (1998). Leptomeningeal metastases from solid tumors: a comparison of two prospective series treated with and without intra-cerebrospinal fluid chemotherapy. Cancer 82(9):1756–1763.

78. Glantz MJ, Cole BF, Recht L et al (1998). High-dose intravenous methotrexate for patients with nonelukemic leptomeningeal cancer: is intrathecal chemotherapy necessary? J Clin Oncol 16(4):1561–1567.

79. Boogerd W, Dorresteijn LDA, van der Sande JJ et al (2000). Response of leptomeningeal metastases from breast cancer to hormonal therapy. Neurology 55:117–119.

80. Blaney SM, Poplack DG (1998). New cytotoxic drugs for intrathecal administration. J Neuro-Oncol 38:219–223.

81. Sampson JH, Archer GE, Villavicencio AT et al (1999). Treatment of neoplastic meningitis with intrathecal temozolomide. Clin Cancer Res 5:1183–1188.

82. Herrlinger U, Weller M, Schabet M (1998). New aspects of immunotherapy of leptomeningeal metastasis. J Neuro-Oncol 38:233–239.

83. Coakham HB, Kemshead JT (1998). Treatment of neoplastic meningitis by targeted radiation using ^{131}I-radiolabelled monoclonal antibodies. J Neuro-Oncol 38:225–232.

84. Vrionis FD (1998). Gene therapy of neoplastic meningiosis. J Neuro-Oncol 38:241–244.

CHAPTER 9

Vascular Disorders: Epidemiology

Lisa R. Rogers

INTRODUCTION

Central nervous system vascular disease is a well-recognized complication of cancer. It may result from the direct or indirect effects of cancer and cancer treatment on blood vessels [1,2]. Until recent years, most clinical and pathology data relevant to these conditions were available only from retrospective reviews or case reports. Recent prospective studies have focused on the role of cancer treatment-related vascular injury and provide new information regarding the risk factors and incidence of these syndromes.

COAGULOPATHY

Cancer can be complicated by alterations of the coagulation system that result in intravascular thrombosis or hemorrhage. The coagulation disturbance is due to a complex interplay of tumor biologic factors and the host response. Malignant cells can express a variety of procoagulant molecules. The most widely studied of these is tissue factor, a cell surface glycoprotein that is capable of initiating the extrinsic pathway of blood coagulation, ultimately generating thrombin, which converts fibrinogen into fibrin. Tissue factor also activates platelets. It is speculated that an imbalance between tissue factor and its main inhibitor, tissue factor pathway inhibitor, contributes to clinically apparent coagulopathy in cancer patients. Another procoagulant that is expressed by tumor cells, termed cancer procoagulant, is a cysteine protease that is capable of initiating the coagulation process via a separate pathway from tissue factor. In addition, cancer cells can also alter coagulation through expression or suppression of fibrinolytic activity. Some cancer patients develop resistance to activated protein C which contributes to a coagulopathy.

Altered coagulation function in the cancer patient can result in vessel thrombosis, hemorrhage or a combination. Brain hemorrhage is most common in acute myelogenous leukemias, especially acute promyelocytic leukemia, early in its treatment. In this setting, disseminated intravascular coagulation is due to the release of procoagulant substances associated with tumor lysis. There is also a high incidence of hemorrhage associated with bone marrow transplant to treat leukemia; in one study by Blegi-Torres et al, intracranial hemorrhage was observed in 32% of 180 patients [3]. In addition to disseminated intravascular coagulation, liver dysfunction, thrombocytopenia or lumbar puncture can be precipitating factors in patients treated for leukemia or brain tumors [4–6].

Cerebral infarction in cancer patients is most commonly due to a thrombotic coagulopathy associated with the tumor or its treatment [1,2]. A recent retrospective review of cerebral ischemic events in 96 cancer patients identified coagulation disorders with cerebral embolism as the most common etiologic factor. Atherosclerosis accounted for only 22% of strokes [2]. Non-bacterial thrombotic endocarditis (NBTE) is one manifestation of the thrombotic coagulopathy that develops in cancer patients, especially those with adenocarcinomas. Autopsy studies indicate that multifocal cerebral infarctions are the result of cerebral embolization of cardiac vegetations or of intravascular thrombosis from the coagulation disorder [7]. Non-bacterial thrombotic endocarditis typically develops in patients with widely disseminated cancer, but stroke from NBTE can also be the presenting sign of cancer.

Cerebral venous thrombosis, typically involving the superior sagittal sinus, can also result from the thrombotic coagulopathy associated with cancer [8]. More often, it occurs secondary to cancer treatment, especially L-asparaginase administered for leukemia and lymphoma.

INTRACRANIAL TUMOR

Tumor in the skull or dura can result in epidural, subdural or subarachnoid hemorrhage [9]. The most common syndrome is subdural hemorrhage resulting from dural metastasis of carcinomas originating in the breast, lung or prostate or of lymphoma. The mechanism of hemorrhage is speculated to be congestion or stretching of vessels in the dura. Primary dural tumors, typically meningiomas, also bleed, but this is uncommon [10].

Hemorrhage into a tumor that is metastatic to the brain parenchyma occurs more commonly than in primary parenchymal tumors. The most common metastatic tumors are of melanoma, thyroid, renal cell, hepatocellular or lung origin. The mechanism of hemorrhage is presumed to be rapid tumor growth with rupture of neoplastic vessels or invasion of adjacent cerebral vessels. Jung et al recently demonstrated higher vascular endothelial growth factor expression and neovascularity in hemorrhagic, rather than non-hemorrhagic, brain metastases [11]. Overexpression of matrix metalloproteinases may also play a role. The most common primary brain tumors to be associated with hemorrhage are pituitary adenomas, meningiomas and gliomas. Rarely, arterial tumor emboli result in damage to cerebral vessels, resulting in neoplastic aneurysms with subsequent rupture and parenchymal or subarachnoid hemorrhages.

Tumors can also encase, compress or invade cerebral venous structures. Such tumors can be primary or metastatic and are typically located in the skull or dura. Arterial compression by tumors is rare. Komotar et al retrospectively reviewed the records of 1617 patients with meningiomas and identified only three patients with neurological symptoms attributed to internal carotid artery compression [12]. Rarely, primary or metastatic tumor in the leptomeninges surrounds and/or compresses vessels in the Virchow-Robin space, resulting in infarction [13,14]. Vascular occlusion in association with parenchymal brain tumors is rare, but is reported in glioblastoma [15].

Intravascular lymphomatosis is a rare variant of diffuse large cell lymphoma. Malignant cells proliferate within the vessels of systemic organs and the brain, causing diffuse thrombosis [16]. Thrombosis from tumor cell proliferation can also occur in leukemia when the leukemic cell count is very high (hyperleukocytosis). This type of vascular thrombosis can result in cerebral infarction or hemorrhage.

When systemic tumor metastasizes to the brain via the arterial circulation, the tumor embolus may be large enough to cause obstruction of a large- or medium-size vessel and result in transient or permanent cerebral ischemia [17]. The underlying tumor almost always originates in the heart or lung. Sarcoma is the most common cardiac tumor to cause symptomatic cerebral embolization. Lung tumors may be primary or metastatic. There are rare reports of brain embolism occurring from manipulation of the lung at the time of surgical removal of a lung tumor.

TREATMENT EFFECTS

Chemotherapy can increase the risk of systemic and cerebral thrombosis or hemorrhage. Some chemotherapies exaggerate the hypercoagulable state, whereas others may injure the endothelium or cause vasospasm. In other instances, thromboses are due to the release of tumor-derived coagulation factors into the bloodstream.

L-asparaginase is an enzymatic inhibitor of protein synthesis which is often used in the induction therapy of acute leukemia. It depletes plasma proteins involved in coagulation and fibrinolysis. It is most commonly associated with cerebral venous sinus thrombosis but, in a small percentage of patients, is associated with intraparenchymal thrombosis or hemorrhage.

In addition, inherited or acquired coagulation defects may render patients more susceptible to thrombotic complications of L-asparaginase. In a retrospective review of 19 pediatric patients with leukemia or lymphoma administered L-asparaginase-containing regimens, low levels of coagulation factors in association with increased plasma D-dimer levels during or after L-asparaginase administration, combined with fresh frozen plasma infusion, were associated with thrombosis [18].

The hemolytic uremic syndrome is a rare complication of chemotherapy, typically occurring in patients with carcinoma. It can result in brain hemorrhage [19]. Tamoxifen, used most commonly in the treatment of breast cancer, is associated with an increased risk of systemic venous thrombosis. Recent studies have sought to determine the risk of cerebral thrombosis. Although meta-analyses demonstrate that there is a slightly increased risk of stroke in breast cancer patients overall, as compared to control populations, the contribution of tamoxifen appears to be very small [20,21].

Radiotherapy can produce or accelerate atherosclerosis [22]. It is a causative factor in transient or permanent cerebral ischemia in some patients who are treated for head and neck cancer. Because of the radiation treatment portals, extensive areas of the common carotid artery and its branches in the neck are involved. In a report of stroke in patients younger than 60 years of age who received therapeutic neck radiation for head and neck tumors, Dorresteijn et al reported a 12% 15-year cumulative risk of stroke [23]. Other factors may also contribute to the development of carotid stenosis. In one study of duplex ultrasonography in patients treated for head and neck cancers, independent risk factors for severe post-radiation carotid stenosis included smoking history, interval from radiation of more than 5 years, no prior oncological surgery and cerebrovascular symptoms [24]. Survivors of childhood Hodgkin's disease are also at risk for stroke if they received mantle radiation exposure. Cerebral ischemia may be related to carotid artery disease or to cardiac valve disease [25]. A recent review of supraclavicular radiation for breast cancer did not identify an increase in the 10-year risk of stroke [26]. The lack of strokes in these patients might be explained by the lower radiation dose administered and the smaller volume of irradiated vessel, as compared to radiation for head and neck cancer.

Radiation-induced vasculopathy, resulting in multifocal vessel occlusions within the brain and sometimes resulting in hemorrhage (by a mechanism that is not yet identified) is a less common toxicity, and is reported in children who received therapeutic brain irradiation [6,27]. In addition, Bowers et al recently documented the incidence of and risk factors for strokes occurring in long-term survivors of childhood leukemia [28]. These patient groups were compared with a random sampling of siblings of cancer survivors. The relative risk of stroke for leukemia survivors as compared with the siblings was 64 (95% CI, 3.0–13.8; $P < 0.0001$) and for brain tumor survivors was 29 (95% CI, 13.8–60.6; $P < 0.0001$). An increased risk of stroke was found with a mean cranial radiation therapy dose of ≥ 30 Gy.

A variety of miscellaneous and rare vascular disorders attributed to radiation of brain tumors is reported, including pseudoaneurysms and cavernous malformations [29]. Reversible cerebral symptoms and signs associated with migraine headaches and with cortical enhancement on magnetic resonance imaging have been reported in a small number of patients years after brain radiation. The mechanism is not known, but speculated to be a delayed manifestation of vascular injury from radiation [30,31].

INFECTION

Immunosuppression in cancer patients renders them susceptible to opportunistic infections, especially fungal sepsis. Cerebral septic emboli are typically due to Aspergillus, Candida or Mucormycosis infections, and occur most often in patients with leukemia, lymphoma, or after bone marrow transplant [4,32].

OTHER

Rarely, CNS vasculitis is the presenting sign of non-Hodgkin's lymphoma, Hodgkin's disease or other hematologic malignancies [33]. The mechanism for this association is not known. Idiopathic thrombocytopenic purpura, predisposing to intracranial hemorrhage, can complicate chronic leukemia or lymphoma.

REFERENCES

1. Graus F, Rogers LR, Posner JB (1985). Cerebrovascular complications in patients with cancer. Medicine 64:16–35.
2. Cestari DM, Weine DM, Panageas KS, Segal AZ, DeAngelis LM (2004). Stroke in patients with cancer incidence and etiology. Neurology 62:2025–2030.
3. Blegi-Torres LF, Werner B, Gasparetto EL, de Medeiros BC, Pasquini R, de Medeiros CR (2002). Post-transplant complications. Intracranial hemorrhage following bone marrow transplantation: an autopsy study of 58 patients. Bone Marrow Transplant 29:29–32.
4. Coplin WM, Cochran MS, Levine SR, Crawford SW (2001). Stroke after bone marrow transplantation – frequency, aetiology and outcome. Brain 124:1043–1051.
5. Graus F, Saiz A, Sierra J et al (1996). Neurologic complications of autologous and allogeneic bone marrow transplantation in patients with leukemia: a comparative study. Neurology 46:1004–1009.
6. Kyrnetskiy EE, Kun LE, Boop FA, Sanford RA, Khan RB (2005). Types, causes, and outcome of intracranial hemorrhage in children with cancer. J Neurosurg 102(1 Suppl):31–35.
7. Rogers LR, Cho E-S, Kempin S, Posner JB (1987). Cerebral infarction from nonbacterial thrombotic endocarditis: clinical and pathological study including the effects of anticoagulation. Am J Med 83:746–756.
8. Raizer JJ, DeAngelis LM (2000). Cerebral sinus thrombosis diagnosed by MRI and MR venography in cancer patients. Neurology 54:1222–1226.
9. Minette SE, Kimmel DW (1989). Subdural hematoma in patients with systemic cancer. Mayo Clin Proc 64:637–642.
10. Bosnjak R, Derham C, Popovic M, Ravnik J (2005). Spontaneous intracranial meningioma bleeding: clinicopathological features and outcome. J Neurosurg 103:473–484.
11. Jung S, Moon K-S, Jung T-Y et al (2006). Possible pathophysiological role of vascular endothelial growth factor (VEGF) and matrix metalloproteinases (MMPs) in metastatic brain tumor-associated intracerebral hemorrhage. J Neuro-Oncol 76:257–263.
12. Komotar RJ, Keswani SC, Wityk RJ (2003). Meningioma presenting as stroke: report of two cases and estimation of incidence. J Neurol Neurosurg Psychiatr 74:136–137.
13. Herman C, Kupsky WJ, Rogers L, Duman R, Moore P (1995). Leptomeningeal dissemination of malignant glioma simulating cerebral vasculitis: case report with angiographic and pathological studies. Stroke 26:2366–2370.
14. Klein P, Haley EC, Wooten GF et al (1989). Focal cerebral infarctions associated with perivascular tumor infiltrates in carcinomatous leptomeningeal metastases. Arch Neurol 46:1149–1152.
15. Rojas-Marcos I, Martin-Duverneuil N, Laigle-Donadey, F, Taillibert S, Delattre JY (2005). Ischemic stroke in patients with glioblastoma multiforme. J Neurol 252:488–489.
16. Calamia KT, Miller A, Shuster EA, Perniciaro C, Menke DM (1999). Intravascular lymphomatosis: a report of ten patients with central nervous system involvement and a review of the disease process. Adv Exp Med Biol 455:249–265.
17. O'Neill BP, Dinapoli RP, Okazaki H (1987). Cerebral infarction as a result of tumor emboli. Cancer 60:90–95.
18. Kiyosawa N, Kano G, Yoshioka H, Sugimoto T, Imashuku S (2005). Cerebral thrombotic complications in adolescent leukemia/lymphoma patients treated with L-asparaginase-containing chemotherapy. Leukemia Lymphoma 46:729–735.
19. Antman KH, Skarin AT, Mayer RJ et al (1979). Microangiopathic hemolytic anemia and cancer: a review. Medicine 58:377–384.
20. Geiger AM, Fischberg GM, Chen W, Bernstein L (2004). Stroke risk and tamoxifen therapy for breast cancer. J Natl Cancer Inst 96:1528–1536.
21. Bushnell D, Goldstein LB (2004). Risk of ischemic stroke with tamoxifen treatment for breast cancer. A meta-analysis. Neurology 63:1230–1233.
22. Muzaffar K, Collins SL, Labropoulos N, Baker WH (2000). A prospective study of the effects of irradiation on the carotid artery. Laryngoscope 110:1811–1814.
23. Dorresteijn LD, Kappelle AC, Boogerd W et al (2002). Increased risk of ischemic stroke after radiotherapy on the neck in patients younger than 60 years. J Clin Oncol 20:282–288.

24. Cheng SW, Ting AC, Lam LK, Wei WI (2000) Carotid stenosis after radiotherapy for nasopharyngeal carcinoma. Arch Otolaryngol Head Neck Surg 126:517–521.

25. Bowers DC, McNeil DE, Liu Y et al (2005). Stroke as a late treatment effect of Hodgkin's disease: a report from the Childhood Cancer Survivor Study. J Clin Oncol 23:6508–6515.

26. Woodward WA, Giordano SH, Duan Z, Hortobagyi GN, Buchholz TA (2006). Supraclavicular radiation for breast cancer does not increase the 10-year risk of stroke. Cancer 106: 2556–2562.

27. Bowers DC, Mulne AF, Reisch JS et al (2002). Nonperioperative strokes in children with central nervous system tumors. Cancer 94:1094–1101.

28. Bowers DC, Liu Y, Leisenring W et al (2006). Late-occurring stroke among long-term survivors of childhood leukemia and brain tumors: a report from the Childhood Cancer Survivor Study. J Clin Oncol 24:5277–5282.

29. Lau WY, Chow CK (2005). Radiation-induced petrous internal carotid artery aneurysm. Ann Otol Rhinol Laryngol 114:939–940.

30. Bartleson JD, Krecke KN, O'Neill BP, Brown PD (2003). Reversible, strokelike migraine attacks in patients with previous radiation therapy. J Neuro-Oncol 5:121–127.

31. Pruitt A, Dalmau J, Detre J, Alavi A, Rosenfeld MR (2006). Episodic neurologic dysfunction with migraine and reversible imaging findings after radiation RRH. Neurology 67:676–678.

32. Kawanami T, Kurita K, Yamakawa M, Omoto E, Kato T (2002). Cerebrovascular disease in acute leukemia: a clinicopathological study of 14 patients. Intern Med 41:1130–1134.

33. Delobel P, Brassat D, Danjoux M et al (2004). Granulomatous angiitis of the central nervous system revealing Hodgkin's disease. J Neurol 251:611–612.

CHAPTER 10

Intracranial Hemorrhage in Cancer Patients

Lisa R. Rogers

COAGULOPATHY

Brain hemorrhage secondary to a systemic coagulation disorder is most common in acute myelogenous leukemias, especially early in the treatment of acute promyelocytic leukemia (APML). In APML undergoing chemotherapy, acute disseminated intravascular coagulation (DIC) is caused by procoagulant substances in the leukemic cells released during tumor lysis. Another common clinical setting is after bone marrow transplant to treat leukemia; in one study by Blegi-Torres and colleagues, intracranial hemorrhage was observed in 32% of 180 patients [1]. Graus et al identified that subdural hemorrhages, rather than parenchymal hemorrhages, are the most common site of bleeding in leukemic patients who have undergone bone marrow transplant [2]. In addition to DIC, liver dysfunction, thrombocytopenia, or lumbar puncture can be precipitating factors in patients treated for leukemia or for brain tumors [2–4]. Idiopathic thrombocytopenic purpura, complicating chronic leukemia or lymphoma, is a rare cause of intracranial hemorrhage.

Clinical

As in patients without cancer who develop parenchymal or subdural hemorrhage, signs may be acute or subacute and include headache, encephalopathy and vomiting. In acute DIC there will usually be signs of systemic thrombosis or hemorrhage.

Diagnosis

Acute DIC is diagnosed by severe abnormalities of coagulation function. Brain computed tomography (CT) or magnetic resonance imaging (MRI) scans often show a single parenchymal or subdural hemorrhage. There may be multiple hemorrhages when there is severe thrombocytopenia.

Treatment

Treatment of DIC remains controversial, as there is a need to control the underlying thrombotic disorder and also to replace clotting factors. Resection of a parenchymal or subdural hemorrhage due to DIC may be impossible, because of ongoing bleeding. Steroids can aid in temporarily reducing mass effect from the hemorrhage.

TUMOR-RELATED

In patients with solid tumors, hemorrhage into a metastatic central nervous system tumor is the most common cause of brain hemorrhage. The mechanism of parenchymal hemorrhage is speculated to be rapid tumor growth that results in rupture of neoplastic vessels or adjacent cerebral vessels. Vascular endothelial growth factor may play a role: Jung et al recently demonstrated higher vascular endothelial growth factor expression and neovascularity in hemorrhagic, rather than non-hemorrhagic, brain metastases [5]. Overexpression of matrix metalloproteinases may also play a role. When tumor metastasizes to the dura, congestion or stretching of vessels in the dura are speculated to be the reasons for subdural hemorrhage. The epidural or leptomeningeal spaces are unusual sites of hemorrhage associated with metastasis [6].

Hyperleukocytosis (peripheral blast count >100 000/ mm^3) in acute myelogenous leukemia can result in brain hemorrhage because of leukostasis (plugging of small cerebral vessels) or by the growth of leukemic nodules which invade blood vessels.

Clinical

The most common metastatic brain tumors associated with hemorrhage are of melanoma, thyroid, renal cell, hepatocellular or lung origin. The onset of neurological signs is usually acute, with headache, decline in consciousness, or a seizure. In some instances, however, the onset of signs is chronic and indistinguishable from those of a growing tumor. Depending on the location of the bleed, there may also be focal neurologic signs.

The most common skull or dural tumors associated with subdural hemorrhage are breast, lung, prostate or lymphoma metastasis. Primary dural tumors, typically meningiomas, also bleed, but this is uncommon [7]. Subdural hemorrhage usually causes subacute signs of headache, confusion and lethargy and there can be superimposed focal signs, depending on the location of the hemorrhage. The most common primary brain tumors to be associated with hemorrhage are pituitary adenomas, meningiomas and gliomas.

Rarely, arterial tumor emboli from systemic cancer result in damage to cerebral vessels, resulting in neoplastic aneurysms which rupture and cause bleeding into the brain or leptomeninges.

Method of diagnosis

Brain imaging with CT or MRI is the test of choice to diagnose intratumoral hemorrhage. The lesions are often multiple when the hemorrhage is into metastatic tumor. The early presence of edema and enhancement associated with the hemorrhage suggest intratumoral hemorrhage in patients who are not known to have cancer.

If there is no evidence of systemic cancer, biopsy or resection of the hematoma may be necessary to diagnose the histology of the lesion underlying the hemorrhage.

Subdural hemorrhages associated with primary or metastatic tumors to the skull or dura can be recognized by the skull expansion or destruction, if skull metastasis is present, and by dural enhancement. In some instances, however, the metastasis is microscopic and tumor can be identified only by cytologic examination of the subdural fluid or histologic examination of the dura.

Neoplastic aneurysms can be detected on cerebral angiography.

Treatment

Resection of a tumor-related hemorrhage may be indicated when the hemorrhage is life threatening but, in many instances, surgery is not required and therapy should be directed to the tumor, typically radiation, with or without chemotherapy. A neoplastic subdural hemorrhage may be large enough to require drainage. Definitive therapy for skull or dural metastasis is radiation.

Therapy for the rare occurrence of bleeding from a neoplastic aneurysm should be radiation, with or without chemotherapy. There is no published evidence that surgical repair of the aneurysm is indicated.

TREATMENT EFFECTS

Chemotherapy can increase the risk of systemic and cerebral hemorrhage. L-asparaginase, which is typically associated with cerebral venous thrombosis, can rarely result in brain hemorrhage. Hemorrhage from thrombocytopenia induced by chemotherapy is relatively rare. The hemolytic uremic syndrome is a rare complication of chemotherapy, typically occurring in patients with carcinoma, that can result in brain hemorrhage [8]. In addition, a variety of miscellaneous and rare vascular disorders attributed to radiation of brain tumors is reported, including pseudoaneurysms and cavernous malformations [9].

REFERENCES

1. Blegi-Torres LF, Werner B, Gasparetto EL, de Medeiros BC, Pasquini R, de Medeiros CR (2002). Post-transplant complications. Intracranial hemorrhage following bone marrow transplantation: an autopsy study of 58 patients. Bone Marrow Transplant 29:29–32.
2. Graus F, Saiz A, Sierra J et al (1996). Neurologic complications of autologous and allogeneic bone marrow transplantation in patients with leukemia: a comparative study. Neurology 46:1004–1009.
3. Coplin WM, Cochran MS, Levine SR, Crawford SW (2001). Stroke after bone marrow transplantation – frequency, aetiology and outcome. Brain 124:1043–1051.
4. Kyrnetskiy EE, Kun LE, Boop FA, Sanford RA, Khan RB (2005). Types, causes, and outcome of intracranial hemorrhage in children with cancer. J Neurosurg 102(1 Suppl):31–35.
5. Jung S, Moon K-S, Jung T-Y et al (2006). Possible pathophysiological role of vascular endothelial growth factor (VEGF) and matrix metalloproteinases (MMPs) in metastatic brain tumor-associated intracerebral hemorrhage. J Neuro-Oncol 76:257–263.
6. McIver JI, Scheithauer BW, Rydberg CH, Atkinson JLD (2001). Metastatic hepatocellular carcinoma presenting as epidural hematoma: case report. Neurosurgery 49:447–449.
7. Bosnjak R, Derham C, Popovic M, Ravnik J (2005). Spontaneous intracranial meningioma bleeding: clinicopathological features and outcome. J Neurosurg 103:473–484.
8. Antman KH, Skarin AT, Mayer RJ et al (1979). Microangiopathic hemolytic anemia and cancer: a review. Medicine 58:377–384.
9. Lau WY, Chow CK (2005). Radiation-induced petrous internal carotid artery aneurysm. Ann Otol Rhinol Laryngol 114:939–940.

CHAPTER 11

CNS Infarction

Arastoo Vossough and Michael H. Lev

INTRODUCTION

Symptomatic cerebral infarction or stroke affects more than 700 000 Americans each year [1]. Stroke is the third leading cause of mortality in the USA, with approximately 157 000 deaths per year [2]. Stroke is a leading cause of disability and the estimated direct and indirect cost of stroke for 2006 was $57.9 billion [1]. Approximately 85% of strokes are bland and 15% hemorrhagic. In western populations, the major causes of CNS infarcts include emboli from large vessels (e.g. carotid occlusive disease), emboli from the heart, small vessel (lacunar) disease and other causes, including hypercoagulable states, intracranial atherosclerosis, extracranial and intracranial arterial dissection, vasculitis, other vasculopathies and various causes of venous infarcts. In some cases, the etiology of stroke is undetermined.

IMAGING EVALUATION OF CNS INFARCTION

Advanced neuroimaging of acute stroke should answer four key issues [3]:

1. whether hemorrhage is present
2. whether there is intravascular clot that can be targeted for thrombolysis
3. whether a core of critically ischemic or irreversibly infarcted tissue is present
4. whether a penumbra of ischemic, but potentially salvageable, tissue is present.

To answer these and related questions, a variety of imaging techniques are used in the evaluation of CNS infarcts. Currently, the standard of care is thought to be non-contrast head computed tomography (CT) scan. Newer methods are increasingly being used as more physiologic and more powerful means of patient triage in the setting of acute stroke [4].

Computed Tomography

Non-Contrast Computed Tomography (NCCT)

CT is used to exclude hemorrhage in acute stroke, but can also show early ischemic changes, which include the following [5–7].

Vessel Hyperdensity

This finding is due to the presence of occlusive thrombus in a circle of Willis artery, most commonly in the M1 segment of the middle cerebral artery (MCA) known as the hyperdense MCA sign (Figure 11.1A) [8]. This finding is associated with a poorer prognosis [9]. Dense vessels may also be seen in cases of basilar artery or posterior cerebral artery occlusive thrombus. Arterial microcalcifications and high hematocrit levels can cause false positives.

Hypodensity

Early parenchymal hypoattenuation and loss of gray–white differentiation are the earliest parenchymal signs of CNS infarct. They may manifest as hypodensity in the basal ganglia, loss of the normal insular ribbon, or loss of cortex-white matter differentiation in the cerebral hemispheres (Figure 11.1B). The mechanism of hypodensity on CT is thought to represent cytotoxic edema or decreased blood volume [5]. The low density region seen on NCCT represents infarcted tissue. Reported sensitivity of CT in hyperacute stroke in the first 6 hours is in the range of 38–45% [10,11]. It has been shown that use of soft copy narrow window settings can improve the conspicuity of subtle ischemic hypodensity (Figure 11.1B,C) [12]. Hypodensity in greater than one-third of the MCA territory has been reported as a risk factor for hemorrhagic transformation [5].

Subtle effacement of cortical sulci and narrowing of the sylvian fissure and effacement of the ventricles and basal cisterns may also been seen.

CT Angiography (CTA)

CTA in the setting of ischemic stroke is used to evaluate the presence of arterial thrombosis, arterial stenosis and potentially evaluate the collateral pathways (see Figure 11.1). CTA is performed by dynamic intravenous injection of contrast with optimized timing to image the arterial vessels at their peak enhancement. For acute stroke, imaging of both the intracranial vasculature and the cervical vessels from the aortic arch should be performed. Various protocols are in use at different institutions but, in all cases, the CTA

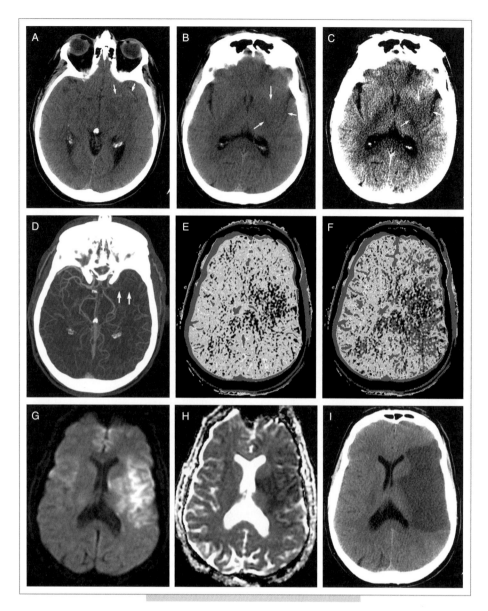

FIG. 11.1. Acute left MCA territory infarct. **(A)** NCCT shows dense MCA secondary to thrombus in the left MCA (arrows). **(B)** CT scan shown in standard brain windows shows subtle left basal ganglia hypodensity and loss of insular ribbon (arrows). **(C)** Same CT slice as in (B), but with narrow window settings, shows a much more conspicuous appearance of the left basal ganglia hypodensity and insular ribbon loss (arrows). **(D)** Maximum intensity projection of CTA shows no flow in the proximal left MCA (arrows). **(E)** CT perfusion color-coded CBV map demonstrating an area of decreased cerebral blood volume in the left MCA territory. **(F)** CT perfusion color-coded CBF map in the same slice as in (E), demonstrating a much larger area of decreased cerebral blood flow compared to the area of decreased blood volume (perfusion mismatch). **(G)** Diffusion weighted MRI image performed two hours later shows high DWI signal in the left MCA territory. **(H)** ADC map corresponding to DWI slice in (G), showing decreased signal and confirming restricted diffusion in the left MCA territory. **(I)** Follow-up CT scan 7 days later demonstrates the final area of infarct in the left MCA territory, which in this patient corresponds more closely with the area of decreased CBF. This patient was not a thrombolysis candidate at the time of presentation.

protocols should be optimized to find a balance between image quality and noise, slice thickness, pitch, table speed, gantry rotation rate, contrast dose, contrast flow rate and radiation dose. Use of newer multidetector scanners can help reduce contrast dose, reduce scanning time and potentially reduce radiation dose by automated tube modulation. Use of a saline chaser can help reduce contrast dose and decrease streak artifact from dense venous contrast in the thoracic inlet. Use of multiphasic injection rates can also help optimize contrast enhancement in the vessels and decrease contrast use. Use of bolus tracking or a test bolus prior to the actual scan can help optimize timing of the contrast bolus.

Once the CTA scan is performed, the source images can be viewed on the scanner to look for obvious areas of vessel occlusion or stenosis. Various 3D reconstruction methods can be utilized to visualize the vessels. These include maximum intensity projections (MIP), multiplanar reformats (MPR), curved reformats in which the vessels are straightened via a computer algorithm, shaded surface display (SSD) and, finally, volume rendering (VR). Some of these methods are time consuming and may require work on a dedicated workstation. In urgent situations, the CT technologists can produce triplanar thick slab MIP images (e.g. 30 mm thick slabs at 5 mm overlapping intervals) on the scanner console rapidly for quick interpretation before the detailed 3D images are later reconstructed (see Figure 11.1D).

The main advantages of CTA include widespread availability, speed of acquisition, higher accuracy compared to magnetic resonance angiography (MRA) and lower risk compared to catheter angiography. The relative disadvantages of CTA include iodinated contrast risk, radiation, limited field of view on older scanners, long final postprocessing times, beam hardening artifact from dental fillings and implants and lack of physiological data such as flow velocity and directionality that could be obtained by magnetic resonance imaging (MRI). Another limitation of CTA is that in areas of dense or circumferential calcification, there is substantial beam hardening artifact which limits accurate measurement of vessel lumen diameter.

CTA has been shown to be 98% sensitive and specific for proximal intracranial thrombus detection [13]. CTA has been shown to be quite accurate and superior to ultrasonography in assessing degree of stenosis, presence of hairline vessel lumen and atherosclerotic plaque ulceration [14]. Owing to its superior accuracy compared to MRA, CTA has become the first line neurovascular test in the acute setting for neurovascular disease. MRA remains an important screening test for non-acute or asymptomatic patients. CTA source images (CTA-SI) provide relevant data concerning tissue level perfusion and it has been shown that CTA-SI are blood volume weighted by theoretical modeling [15] and hence can be used to define infarct core.

CT Perfusion (CTP)

Perfusion CT is utilized to measure the ischemic penumbra in stroke, i.e. severely ischemic tissue which is not irreversibly damaged and potentially salvageable. Use of perfusion imaging to define physiologically the penumbra can be utilized potentially to extend the thrombolysis window beyond 3 hours for i.v. and 6 hours for anterior circulation i.a. thrombolysis [16]. Perfusion CT measures tissue level capillary blood flow in the brain [17]. Cerebral perfusion can be described by three parameters: cerebral blood volume (CBV) is the total volume of blood in a given unit volume of brain tissue; cerebral blood flow (CBF) is the volume of blood moving through a given unit volume of brain per unit time; mean transit time (MTT) is the average of the transit time of blood through a region of brain parenchyma. These parameters are related to each other by the 'central volume' formula MTT = CBV/CBF.

Quantitative CTP is often performed by continuous cine CT acquisition of a fixed brain region during dynamic intravenous contrast administration at a rate of 4–7 ml/s [18]. Currently, only a limited number of slices can be covered, but larger volumes of the brain may be examined with faster scanners. Once the data have been acquired, various post-processing methods are utilized to produce perfusion parameter (CBV, CBF, MTT) maps using a variety of deconvolution and non-deconvolution methods (see Figure 11.1).

The advantages of CTP include widespread availability of CT, lower costs, relative ease and higher speed of scanning, especially in very ill patients, and use in patients with contraindications to MRI. CTP also has a higher resolution, provides quantitative perfusion information compared to MR perfusion and is not limited by susceptibility effects from adjacent structures or large vessel contamination [18]. The disadvantages of CTP include use of ionizing radiation, the risks associated with iodinated contrast use, limited brain coverage with current scanner technology and more complex image post-processing requirements.

The operational ischemic penumbra in CTP is the area of CBV-CBF mismatch (see Figure 11.1E,F). The region of CBV abnormality is the core of the infarcted tissue and the area of CBV-CBF mismatch is the surrounding tissue that is hypoperfused, but potentially salvageable. Untreated or unsuccessfully treated patients with large CBV-CBF mismatch will show substantial growth of the infarcted tissue on follow-up (see Figure 11.1I). Patients with no mismatch or with early complete recanalization will show no or little lesion progression [18]. Therefore, the presence of a large mismatch can be used as a physiological triage tool for thrombolysis candidate selection. There is considerable ongoing research in better defining various threshold models to define and characterize more accurately the infarct core and ischemic penumbra and their relationship to clinical outcome and hemorrhagic risk.

Magnetic Resonance Imaging (MRI)
Conventional MRI

In the hyperacute stage of stroke (0–6 hours), there may be loss of the normal vessel flow voids on T2-weighted images. On FLAIR (fluid attenuation inversion recovery) sequences, there may be high intravascular signal within the vessels, a finding that can be seen in up to 65% of cases [19]. The sensitivity of parenchymal signal changes on T2-weighted or FLAIR images in the first 6 hours of stroke are only 18–29% [11,20,21]. Gradient echo T2*-weighted susceptibility images may also show the intravascular thrombus as a linear region of low signal susceptibility, due to the presence of deoxyhemoglobin within the clot. In one study, this finding had 83% sensitivity compared to 52% sensitivity of the dense vessel sign on non-contrast CT [22]. Contrast-enhanced T1-weighted images may show arterial enhancement without parenchymal enhancement in 50% of hyperacute strokes, believed to represent collateral flow, slow flow or hyperperfusion after early recanalization [23,24].

In the acute stage of stroke (6–24 hours), vasogenic edema causes increase in water content and the area of infarct is seen as hyperintensity on T2-weighted and FLAIR images with a sensitivity approaching 90% at 24 hours [24].

In the subacute stage of stroke (24 hours to 2 weeks), T2 and FLAIR hyperintensity persist and T1 hypointensity starts to develop. Brain swelling increases and peaks at day 3, manifesting as effacement of sulci, cisterns and ventricles, gyral thickening, midline shift and various forms of herniation. Arterial enhancement and meningeal enhancement are seen, which often resolve by 1 week. Gyriform parenchymal enhancement occurs at this stage and may persist up to 8 weeks [23].

In the chronic stage of stroke, the edema resolves and gliosis and tissue loss develops, characterized by T2 hyperintensity, T1 hypointensity, focal volume loss, and possibly cystic encephalomalacia. After 8 weeks, no parenchymal, vascular or meningeal enhancement is seen. With larger infarcts, Wallerian degeneration develops, manifested by volume loss and T2 hyperintensity in the corticospinal tract pathway.

Hemorrhagic transformation of brain infarction can range from small petechial bleeds to large parenchymal hematomas. It has a cumulative incidence of up to 43% in the first month and overall risk factors include embolic stroke etiology, hypertension, high glucose levels, higher NIH Stroke Scale score, reperfusion, good collateral circulation, longer time to recanalization after thrombolysis, anticoagulant therapy and thrombolytic therapy [25–28]. A common grading scheme uses the classification of hemorrhagic infarcts (HI1 and HI2) and parenchymal hematoma (PH1 and PH2) [29]. Gradient echoT2*-weighted sequences are best utilized in detecting these areas of intraparenchymal hemorrhage and it has been shown to have equivalent sensitivity compared to CT [30,31]. As time passes and blood products pass through the methemoglobin stage, they may manifest as areas of hyperintensity on T1-weighted images.

Conventional MRI is an adjunct to CTA or MRA in the diagnosis of cervicocranial vessel dissection. Fat-saturated T1-weighted images may show bright methemoglobin signal in the vessel wall hematoma caused by dissection. The caveat in the use of this technique is that the hematoma may not be seen in very early cases and also in chronic dissections.

MR Angiography (MRA)

MRA is used to evaluate the intracranial and cervical vascular supply to the brain. Both non-contrast (time-of-flight and phase-contrast) and contrast-enhanced (CE MRA) techniques are available. Images obtained from these various techniques are often post-processed via maximum intensity projections (MIP) to show the vascular tree anatomy as a rotating three-dimensional structure.

Time-of-flight (TOF) MRA

TOF MRA is a gradient echo sequence that is performed by repeatedly applying a radiofrequency (RF) pulse, followed by dephasing and rephasing gradients. Stationary tissues become saturated and lose signal by the repeated RF pulse, whereas flowing blood is bright as it brings unsaturated spins into the imaging volume [25]. This is known as 'flow-related enhancement'. Saturation bands are applied to suppress signal from incoming veins in the other direction. TOF MRA can be performed in a 2D or 3D acquisition. 2D TOF MRA is usually used in evaluation of neck vessels. It is generally reserved for instances where contrast enhanced MRA of the neck cannot be performed or has failed, or when flow directionality information is important. 3D TOF MRA is typically used to evaluate the intracranial circulation. 2D TOF is often the technique used in magnetic resonance venography (MRV) of the dural venous sinuses.

Phase-Contrast (PC) MRA

This is a gradient echo sequence which applies a pair of equal but opposing direction dephasing and rephasing RF pulses to the imaging volume. Stationary tissues have no net change in phase because they encounter equal but opposite direction dephasing and rephasing pulses, but moving blood acquires a phase shift. The amount of phase shift is proportional to the velocity of moving blood. PC MRA can also be performed in a 2D or 3D fashion. PC MRA is usually only reserved for instances where there is a concern that a venous or arterial subacute clot (intrinsically bright T1 signal) is mimicking flow-related enhancement.

Contrast-Enhanced (CE) MRA

CE MRA uses a 3D fast gradient-echo sequence in conjunction with a dynamic bolus injection of gadolinium. The acquisition is carefully timed during passage of the bolus of gadolinium through the arteries of interest. A preliminary test injection of gadolinium can be done to time the arrival of contrast into the arteries. With newer scanners, a fluoro technique can be used to monitor the region of interest in real time and start the acquisition when contrast is first seen or even use automated bolus detection. Due to a very low echo time (TE), there is good background fat suppression. Acquisition time can generally be less than 30 seconds. Similar techniques can be used for CE MRV of the venous sinuses [32].

Clinical Utility of MRA

CE MRA of the neck is reported to have a sensitivity of 94–97% and specificity of 81–95% compared to angiography (gold standard) for stratifying surgical from non-surgical cases of carotid stenosis and correlation of r = 0.94 to 0.96 [33–35]. 3D TOF MRA of the intracranial circulation is reported to have a sensitivity of 88–100% and specificity of 95–97% compared to digital subtraction angiography (DSA) in diagnosis of intracranial occlusion and MCA disease, respectively [36,37].

Diffusion Weighted Imaging (DWI)

Diffusion imaging techniques are discussed in more detail in Chapters 25 and 26. On diffusion imaging, acute infarcts present as bright areas on DWI images and dark on apparent diffusion contrast (ADC) maps (see Figure 11.1G,H). DWI is highly sensitive and specific in detection of hyperacute and acute stroke, ranging from 86 to 100% [38]. False-positive infarcts on DWI can be seen in patients with subacute or chronic infarcts with T2-shine through, cerebral abscesses, some tumors, venous infarcts, acute demyelinating lesions, hemorrhage, infections such as herpes encephalitis, Creutzfeldt-Jacob disease, status epilepticus, hemiplegic migraine, transient global amnesia, reversible posterior encephalopathy and diffuse axonal injury. False-negative infarcts on DWI often are seen in punctate brainstem, basal ganglia, or lacunar infarctions. In these cases, persistence of the specific neurological deficit on examination may prompt early follow-up imaging which often shows the abnormality not seen on the initial scan.

DWI is particularly helpful in detection of acute infarcts in the setting of subacute infarcts, where differentiation of new and older infarcts is very difficult. The area of diffusion abnormality represents the ischemic infarct core. In the absence of thrombolysis, reversibility of DWI lesions is quite rare and often seen in cases of transient ischemic attacks (TIA), transient global amnesia, status epilepticus, and venous sinus thrombosis [39]. With thrombolysis, a portion of the DWI abnormality may be reversible, especially in the white matter. Lower ADC values in areas of infarct are associated with higher risk of hemorrhagic transformation and larger areas of diffusion abnormality are associated with worse clinical outcome [38].

MR Perfusion Imaging

MR perfusion weighted imaging (PWI) techniques are discussed in more detail in Chapter 29. Like CT perfusion, MR perfusion provides CBV-CBF-MTT maps. The area of CBV abnormality on PWI is often matched with the DWI abnormality and represents the ischemic infarct core. The difference between CBV and CBF images constitute the operational ischemic penumbra. The advantages of MR perfusion imaging in stroke include whole brain coverage, performance of diffusion imaging at the same time and simpler post-processing with rapid perfusion map construction by the technologists at the scanner. The disadvantages of dynamic contrast MR PWI include susceptibility artifact due to metallic objects and near bone and air interfaces, need for high flow rate injections, sensitivity to patient motion, unreliability for calculating absolute values of perfusion parameters, unavailability of MR in all acute settings, and the possibility that the patient has a contraindication to MRI.

CBF is reportedly the best current estimate of whether the penumbra will infarct or survive [40]. Both MTT and CBF usually overestimate the final infarct volume. Very low CBF on MR perfusion has been associated with hemorrhagic transformation. The size of DWI, CBV, CBF, MTT abnormalities, as well the size of the diffusion–perfusion mismatch all correlate with clinical outcome [40]. The amount of decrease in size of MTT abnormality volume following intravenous thrombolysis correlates with clinical outcome.

Catheter Angiography

Catheter angiography no longer has a significant diagnostic role in evaluation of ischemic CNS infarction, given the increased diagnostic utility of less invasive imaging techniques. The exception is in the evaluation of vasculopathies, vasospasm, and rarely for confirmation of arterial dissection. Its diagnostic role in evaluation of aneurysms and arteriovenous malformations has also relatively decreased, but it remains an important modality in cases in which less invasive techniques do not provide adequate information. Nevertheless, endovascular techniques play a crucial role in the management of stroke patients for intra-arterial thrombolysis, mechanical thrombolysis and vasospasm treatment. Endovascular techniques are also employed in carotid, vertebral, and intracranial vessel angioplasty/stenting for the prevention and treatment of stroke.

CNS INFARCTION IN NEURO-ONCOLOGY

Neuro-oncologic patients are at increased risk for CNS infarcts, whether due to the direct effects of CNS tumors on the brain and spinal cord or as a result of treatment for these neoplasms. CNS infarcts also occur in higher frequency in patients with systemic neoplasia (see Chapter 55).

CNS Infarction Caused by Brain Tumors

Mass effect from brain tumors can cause displacement, kinking and resultant narrowing or occlusion of cerebral vessels, leading to infarction. Examples include cases of subfalcine herniation causing compression of the anterior cerebral arteries (ACAs) against the falx (Figure 11.2A–D). Also mass effect or acute hydrocephalus causing transtentorial herniation may lead to compression of the posterior cerebral arteries against the tentorium and resultant occipital lobe infarction (Figure 11.2E) or compression of the superior cerebellar arteries [41].

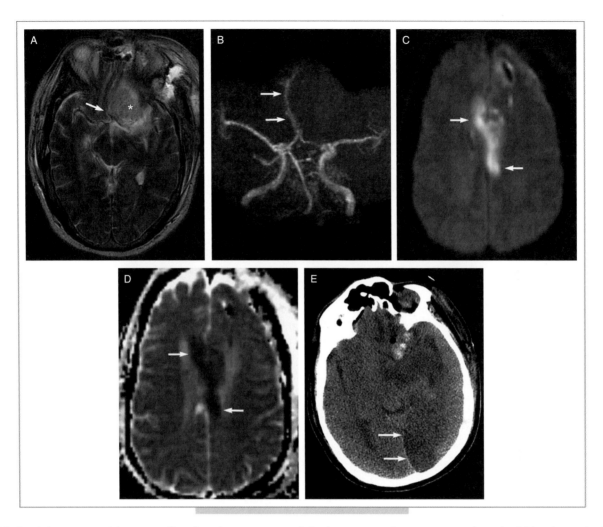

FIG. 11.2. Infarcts caused by mass effect from brain tumor and displacement and compression of cerebral blood vessels. **(A)** Left frontal glioblastoma multiforme (asterisk) causing rightward subfalcine herniation and displacement of the anterior cerebral arteries (arrows). **(B)** Rightward displacement of ACAs. **(C)** DWI image showing bilateral ACA territory infarcts (arrows). **(D)** ADC map confirming bilateral ACA territory infarcts (arrows). **(E)** CT scan from a few days later in same patient showing left occipital lobe infarct (arrows) caused by transtentorial herniation and compression of the posterior cerebral artery against the tentorium.

Primary or metastatic tumors of the brain, meninges or skull can infiltrate and compress the arterial supply or venous drainage systems and cause arterial or venous infarcts [42]. Examples include convexity meningiomas compressing or infiltrating the venous sinuses and also sellar/parasellar tumors infiltrating and occluding the carotid arteries or parts of the circle of Willis in the cavernous sinus, or suprasellar cisterns, respectively. However, these vascular occlusions may not necessarily lead to infarcts since the associated neoplasms frequently grow at a slow rate, often providing adequate time for the development of collateral pathways in the vascular supply or drainage systems.

Tumor infiltration in the Virchow-Robin perivascular spaces can cause vessel compression, spasm, or thrombosis leading to infarction, for example in diffuse leptomeningeal gliomatosis [43]. Intravascular lymphomatosis is a rare entity that often involves the CNS and can cause infarcts (Figure 11.3).

CNS Infarction as a Result of Treatment of Brain Tumors

After surgical resection of gliomas, focal areas of restricted diffusion and infarct adjacent to the resection cavity have been reported in 64–70% of cases [44,45]. These focal infarcts may or may not be associated with neurological deficits (Figure 11.4). Injury to the internal carotid arteries can occur during trans-sphenoidal surgery for pituitary/sellar tumors. Even in the absence of direct carotid injury, post-excision packing of the tumor cavity,

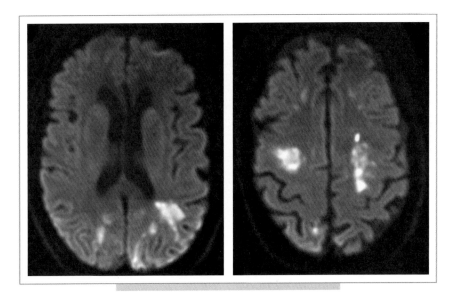

FIG. 11.3. Cerebral infarcts in a patient with proven intravascular lymphomatosis. Diffusion weighted images show multiple scattered bilateral areas of high DWI signal, in keeping with infarcts. None of these areas demonstrated contrast enhancement at this time (not shown).

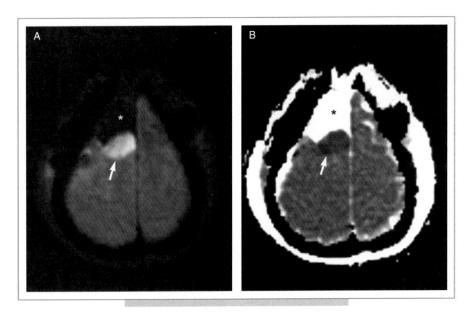

FIG. 11.4. Acute infarct developing in right frontal lobe adjacent to surgical resection cavity (asterisks), after resection of right frontal brain tumor. **(A)** DWI image showing high signal area (arrow) posterior to resection cavity. **(B)** ADC map showing low diffusion coefficient (arrow) in same area, confirming infarct.

intrasellar postoperative hemorrhage, and massive postoperative tumor swelling after partial resection have all been reported to cause infarcts, presumably due to compression of the internal carotid arteries [46]. Cortical venous infarcts can occur after surgery in cases of firmly adherent extra-axial tumors to the cortex [47].

Radiation therapy for treatment of head and neck cancer can lead to accelerated atherosclerosis of the cervical arterial supply to the brain and cause premature infarcts

[42]. Radiation therapy for brain tumors can also cause a similar phenomenon in intracranial vessels.

Spinal Cord Infarction in Neuro-oncology

Spinal cord infarcts are rare. They are often caused by interruption of the spinal cord blood supply secondary to thromboemboli or iatrogenic interruption of the supply after aortic or retroperitoneal surgery, or endovascular procedures. Neuro-oncologic patients may be at higher risk for spinal cord infarcts. Spinal cord infarction has been reported in acute hydrocephalus due to a colloid cyst of the third ventricle and downward brain herniation, presumably by pressure on the spinal artery [48]. Intravascular lymphomatosis can also cause spinal cord infarcts [49]. Spinal cord infarcts have also been reported as rare complications of posterior fossa and pineal surgery in the sitting position [50].

Application of regular diffusion MRI to imaging of the spinal cord is limited secondary to susceptibility artifact from surrounding structures, motion from CSF pulsation, respiratory motion, carotid-vertebral pulsations and the small size of the spinal cord. Many different methods for early detection of spinal cord infarcts have been proposed [51]. One particularly useful technique is line scan diffusion (LSD), with minimal susceptibility artifact, without the need for cardiac or respiratory gating, or flow compensation, all in a reasonable clinical timeframe [52].

CONCLUSION

Neuro-oncologic patients are at increased risk for CNS infarcts. A variety of imaging modalities are available for rapid evaluation and diagnosis of CNS infarcts in order to optimize the triage and management of these patients.

REFERENCES

1. Thom T (2006). *Heart Disease and Stroke Statistics – 2006 Update*. American Heart Association, Dallas.
2. Hoyert DL et al (2006). Deaths: final data for 2003. Natl Vital Stat Rep 54(13):1–120.
3. Shetty SK, Lev MH (2005). CT Perfusion in acute stroke. Semin Ultrasound CT MR 26(6):404–421.
4. Gonzalez RG (2006). Imaging-guided acute ischemic stroke therapy: From 'time is brain' to 'physiology is brain'. Am J Neuroradiol 27(4):728–735.
5. Vu D, Lev MH (2005). Noncontrast CT in acute stroke. Semin Ultrasound CT MR 26(6):380–386.
6. Kucinski T (2005). Unenhanced CT and acute stroke physiology. Neuroimaging Clin N Am 15(2):397–407, xi–xii.
7. Camargo EC et al (2006) Unenhanced computed tomography. In *Acute Ischemic Stroke: Imaging and Intervention*. Gonzalez RG et al (eds). Springer-Verlag: Berlin. 41–56.
8. Pressman BD, Tourje EJ, Thompson JR (1987) An early CT sign of ischemic infarction: increased density in a cerebral artery. Am J Roentgenol 149(3):583–586.
9. Tomsick T et al (1996). Prognostic value of the hyperdense middle cerebral artery sign and stroke scale score before ultraearly thrombolytic therapy. Am J Neuroradiol 17(1):79–85.
10. Mohr JP et al (1995). Magnetic resonance versus computed tomographic imaging in acute stroke. Stroke 26(5):807–812.
11. Gonzalez RG et al (1999). Diffusion-weighted MR imaging: diagnostic accuracy in patients imaged within 6 hours of stroke symptom onset. Radiology 210(1):155–162.
12. Lev MH et al (1999). Acute stroke: improved nonenhanced CT detection – benefits of soft-copy interpretation by using variable window width and center level settings. Radiology 213(1):150–155.
13. Lev MH et al (2001). CT angiography in the rapid triage of patients with hyperacute stroke to intraarterial thrombolysis: accuracy in the detection of large vessel thrombus. J Comput Assist Tomogr 25(4):520–528.
14. Sheikh SF, Gonzalez RG, Lev MH (2006). Stroke CT angiography, In *Acute Ischemic Stroke: Imaging and Intervention*. Gonzalez RG et al (eds). Springer-Verlag: Berlin. 57–86.
15. Hunter GJ et al (1998). Assessment of cerebral perfusion and arterial anatomy in hyperacute stroke with three-dimensional functional CT: early clinical results. Am J Neuroradiol 19(1):29–37.
16. Rother J (2003). Imaging-guided extension of the time window: ready for application in experienced stroke centers? Stroke 34(2):575–583.
17. Villringer A et al (1988). Dynamic imaging with lanthanide chelates in normal brain: contrast due to magnetic susceptibility effects. Magn Reson Med 6(2):164–174.
18. Shetty SK, Lev MH (2005). CT perfusion in acute stroke. Neuroimaging Clin N Am 15(3):481–501, ix.
19. Maeda M et al (2001). Time course of arterial hyperintensity with fast fluid-attenuated inversion-recovery imaging in acute and subacute middle cerebral arterial infarction. J Magn Reson Imaging 13(6):987–990.
20. Perkins CJ et al (2001). Fluid-attenuated inversion recovery and diffusion- and perfusion-weighted MRI abnormalities in 117 consecutive patients with stroke symptoms. Stroke 32(12):2774–2781.
21. Shimosegawa E et al (1993). Embolic cerebral infarction: MR findings in the first 3 hours after onset. Am J Roentgenol 160(5):1077–1082.
22. Flacke S et al (2000). Middle cerebral artery (MCA) susceptibility sign at susceptibility-based perfusion MR imaging: clinical importance and comparison with hyperdense MCA sign at CT. Radiology 215(2):476–482.
23. Crain MR et al (1991). Cerebral ischemia: evaluation with contrast-enhanced MR imaging. Am J Neuroradiol 12(4):631–639.
24. Yuh WT et al (1991). MR imaging of cerebral ischemia: findings in the first 24 hours. Am J Neuroradiol 12(4):621–629.
25. Vu D, Gonzalez RG, Schaefer PW (2006). Conventional MRI and MR angiography of stroke. In *Acute Ischemic Stroke: Imaging and Intervention*. Gonzalez RG et al (eds). Springer-Verlag, Berlin. 115–137.
26. Kidwell CS et al (2002). Predictors of hemorrhagic transformation in patients receiving intra-arterial thrombolysis. Stroke 33(3):717–724.
27. Hornig CR, Dorndorf W, Agnoli AL (1986). Hemorrhagic cerebral infarction – a prospective study. Stroke 17(2):179–185.

28. Hakim AM, Ryder-Cooke A, Melanson D (1983). Sequential computerized tomographic appearance of strokes. Stroke 14(6):893–897.

29. Fiorelli M et al (1999). Hemorrhagic transformation within 36 hours of a cerebral infarct: relationships with early clinical deterioration and 3-month outcome in the European Cooperative Acute Stroke Study I (ECASS I) cohort. Stroke 30(11):2280–2284.

30. Hermier M et al (2001). MRI of acute post-ischemic cerebral hemorrhage in stroke patients: diagnosis with T2*-weighted gradient-echo sequences. Neuroradiology 43(10):809–815.

31. Kidwell CS et al (2004). Comparison of MRI and CT for detection of acute intracerebral hemorrhage. J Am Med Assoc 292(15):1823–1830.

32. Farb RI et al (2003). Intracranial venous system: gadolinium-enhanced three-dimensional MR venography with auto-triggered elliptic centric-ordered sequence – initial experience. Radiology 226(1):203–209.

33. Nederkoorn PJ et al (2003). Carotid artery stenosis: accuracy of contrast-enhanced MR angiography for diagnosis. Radiology 228(3):677–682.

34. JM, UK-I et al (2004). Contrast-enhanced MR angiography vs intra-arterial digital subtraction angiography for carotid imaging: activity-based cost analysis. Eur Radiol 14(4):730–735.

35. Alvarez-Linera J et al (2003). Prospective evaluation of carotid artery stenosis: elliptic centric contrast-enhanced MR angiography and spiral CT angiography compared with digital subtraction angiography. Am J Neuroradiol 24(5):1012–1019.

36. Korogi Y et al (1994). Intracranial vascular stenosis and occlusion: diagnostic accuracy of three-dimensional, Fourier transform, time-of-flight MR angiography. Radiology 193(1):187–193.

37. Stock KW et al (1995). Intracranial arteries: prospective blinded comparative study of MR angiography and DSA in 50 patients. Radiology 195(2):451–456.

38. Schaefer PW et al (2006). Diffusion MRI of acute stroke. In *Acute Ischemic Stroke: Imaging and Intervention*. Gonzalez RG et al (eds). Springer-Verlag, Berlin. 139–171.

39. Grant PE et al (2001). Frequency and clinical context of decreased apparent diffusion coefficient reversal in the human brain. Radiology 221(1):43–50.

40. Schaefer PW, Copen WA, Gonzalez RG (2006). Perfusion MRI of acute stroke. In *Acute Ischemic Stroke: Imaging and Intervention*. Gonzalez RG et al (eds). Springer-Verlag, Berlin. 173–198.

41. Wang KC et al (1995). Infarction of the territory supplied by the contralateral superior cerebellar artery in a case of descending transtentorial herniation. Childs Nerv Syst 11(7):432–435.

42. Rogers LR (2004). Cerebrovascular complications in patients with cancer. Semin Neurol 24(4):453–460.

43. Singh M et al (1998). Diffuse leptomeningeal gliomatosis associated with multifocal CNS infarcts. Surg Neurol 50(4):356–362; discussion 362.

44. Smith JS et al (2005). Serial diffusion-weighted magnetic resonance imaging in cases of glioma: distinguishing tumor recurrence from postresection injury. J Neurosurg 103(3):428–438.

45. Ulmer S et al (2006). Clinical and radiographic features of peritumoral infarction following resection of gioblastoma. Neurology 67(9):1668–1670.

46. Kurschel S et al (2005). Rare fatal vascular complication of transsphenoidal surgery. Acta Neurochir (Wien)147(3):321–325; discussion 325.

47. Kiya K et al (2001). Postoperative cortical venous infarction in tumours firmly adherent to the cortex. J Clin Neurosci 8 Suppl 1:109–113.

48. Siu TL, Bannan P, Stokes BA (2005). Spinal cord infarction complicating acute hydrocephalus secondary to a colloid cyst of the third ventricle. Case report. J Neurosurg Spine 3(1):64–67.

49. Kinoshita T et al (2005). Intravascular malignant lymphomatosis: diffusion-weighted magnetic resonance imaging characteristics. Acta Radiol 46(3):246–249.

50. Morandi X et al (2004). Extensive spinal cord infarction after posterior fossa surgery in the sitting position: case report. Neurosurgery 54(6):1512–1515; discussion 1515–1516.

51. Schwartz ED et al (2002). Diffusion-weighted imaging of the spinal cord. Neuroimaging Clin N Am 12(1):125–146.

52. Maier SE et al (1998). Line scan diffusion imaging: characterization in healthy subjects and stroke patients. Am J Roentgenol 171(1):85–93.

CHAPTER 12

Intracranial Veno-Occlusive Disease

Lisa R. Rogers

NON-METASTATIC CEREBRAL VENOUS THROMBOSIS

Spontaneous thrombosis of cerebral venous sinuses, unrelated to tumor infiltration or compression, is a rare complication of systemic cancer. In a recent clinical series, cerebral venous thrombosis (CVT) accounted for only 0.3% of neurologic consultations at a large cancer center [1]. Santoro et al performed a multicenter retrospective analysis of ischemic stroke among 2318 children treated for acute lymphoblastic leukemia and identified a prevalence of 0.47%, all of which were due to CVT [2]. It presents important management considerations, particularly regarding the need for anticoagulation.

The most common cause of spontaneous CVT is a systemic coagulopathy that is associated with cancer or follows the administration of chemotherapy [1]. The most common chemotherapy to be associated with CVT is L-asparaginase that is administered for acute leukemia or lymphoma. L-asparaginase is an enzymatic inhibitor of protein synthesis and depletes plasma proteins involved in coagulation and fibrinolysis. Acquired or inherited coagulation defects may render patients more susceptible to the thrombotic complications of L-asparaginase. In a retrospective review of 19 pediatric patients with leukemia or lymphoma who were administered L-asparaginase-containing regimens, low levels of coagulation factors in association with increased plasma D-dimer levels during or after L-asparaginase administration, combined with fresh frozen plasma infusion, were associated with vascular thrombosis [3].

Clinical Features

Non-metastatic CVT, whether it occurs in adults or children, is predominantly associated with hematologic malignancies, especially acute leukemia [1,4]. The most common venous structure to be affected is the superior sagittal sinus. When CVT is associated with L-asparaginase therapy, it typically occurs during or shortly after the induction therapy, but there are instances of late thrombosis [5]. The clinical signs of spontaneous CVT, similar to patients without cancer who develop CVT, include headache, vomiting, papilledema and seizures. These may occur alone or in combination. Additionally, focal neurological signs or encephalopathy can occur if there is an associated cerebral infarction or hemorrhage. The symptoms usually develop acutely but, rarely, CVT can present with only chronic headache and papilledema identified on the physical examination.

Method of Diagnosis

Brain magnetic resonance imaging (MRI) is the procedure of choice to identify CVT. Unenhanced MRI allows for visualization of the thrombus and also parenchymal abnormalities that may also be present due to infarction or hemorrhage. Enlarged collateral veins may also be observed. The 'empty delta' sign may be observed following contrast injection, due to lack of filling in the area of thrombosis. The signal within the sinus may be difficult to interpret if the venous flow is slow, rather than obstructed, or if the occlusion is acute. In these instances, or if a CVT is suspected but is not visible on brain MRI or computed tomography (CT), magnetic resonance venography (MRV) or computed tomography venography is diagnostic of CVT.

Treatment and Prognosis

Spontaneous resolution or recanalization of CVT can occur when it is caused by a coagulopathy, especially when it occurs early in the course of cancer and when the tumor is responding to treatment. The prognosis is poor, however, when there is involvement of the deep venous sinuses [2]. The prognosis is also worse in patients with prothrombotic risk factors [6].

Treatment directed to the thrombosis should be considered for persistent and symptomatic cases. Anticoagulation, urokinase and endovascular thrombolysis are useful treatments when sinus occlusion occurs in patients without cancer [7,8]. Santoro and colleagues reported no complications from anticoagulation in children with acute leukemia and CVT [2].

METASTATIC CEREBRAL VENOUS THROMBOSIS

Cerebral venous sinuses or cortical veins may be compressed or infiltrated by tumors arising in or metastasizing to the skull or meninges. The altered venous flow may then result in stasis and thrombosis.

Clinical Features

Cancers, including solid tumors or lymphoma, that metastasize to the skull or dura can result in CVT due to adjacent venous compression or infiltration. In this setting, the cancer is usually widespread. Primary tumors of the skull or meninges, especially meningiomas, can also compress venous structures. Rarely, CVT is associated with metastasis to the leptomeninges [9]. The superior sagittal sinus is the most common venous structure affected by compression/infiltration, but other sites of compression can occur, depending upon the location of the tumor.

In contrast to spontaneous non-neoplastic CVT, thrombosis produced by skull or dural metastasis more often presents in a subacute fashion, with signs of increased intracranial pressure (e.g. headache, vomiting, papilledema). There may be focal neurologic signs or encephalopathy if the CVT results in venous infarction or hemorrhage [1].

Method of Diagnosis

Brain MRI or CT are useful to identify CVT as well as skull or dural tumor. As in non-metastatic CVT, venography by MRI or CT can be diagnostic of CVT if standard brain imaging is not.

Treatment and Prognosis

In contrast to non-metastatic CVT, the clinical course of metastatic CVT is generally progressive if left untreated and it can be fatal [10]. The thrombosis is less likely to spontaneously remit or recanalize. Depending on the type and location of tumor, radiation therapy or surgical intervention should be considered. There are no data to support the use of anticoagulation in this condition.

REFERENCES

1. Raizer JJ, DeAngelis LM (2000). Cerebral sinus thrombosis diagnosed by MRI and MR venography in cancer patients. Neurology 54:1222–1226.
2. Santoro N, Giordano P, Del Vecchio GC et al (2005). Ischemic stroke in children treated for acute lymphoblastic leukemia. A retrospective study. J Pediatr Hematol Oncol 27:153–157.
3. Kiyosawa N, Kano G, Yoshioka H, Sugimoto T, Imashuku S (2005). Cerebral thrombotic complications in adolescent leukemia/lymphoma patients treated with L-asparaginase-containing chemotherapy. Leukemia Lymphoma 46:729–735.
4. Reddingius RE, Patte C, Couanet D, Kalifa C, Lemerle J (1997). Dural sinus thrombosis in children with cancer. Med Pediatr Oncol 29:296–302.
5. Corso A, Castagnola C, Bernasconi C (1997). Thrombotic events are not exclusive to the remission induction period in patients with acute lymphoblastic leukemia: a report of two cases of cerebral sinus thrombosis. Ann Hematol 75:117–119.
6. Wermes C, Fleischhack G, Junker R et al (1999). Cerebral venous sinus thrombosis in children with acute lymphoblastic leukemia carrying the MTHFR TT677 genotype and further prothrombotic risk factors. Klin Padiatr 211:211–214.
7. de Bruijn SF, Stam J (1999). Randomized, placebo-controlled trial of anticoagulant treatment with low-molecular-weight heparin for cerebral sinus thrombosis. Stroke 30:484–488.
8. Soleau SW, Schmidt R, Stevens S, Osborn A, MacDonald JD (2003). Extensive experience with dural sinus thrombosis. Neurosurgery 52:534–544.
9. Akai T, Kuwayama N, Ogiichi T, Kurimoto M, Endo S, Takaku A (1997). Leptomeningeal melanoma associated with straight sinus thrombosis. Case report. Neurol Med Chir (Tokyo) 37:757–761.
10. López-Peláez MF, Millán JM, de Vergas J (2000). Fatal cerebral venous sinus thrombosis as major complication of metastatic cervical mass: computed tomography and magnetic resonance findings. J Laryngol Otol 114:798–801.

CHAPTER 13

Paraneoplastic Syndromes

Steven Vernino and E. Paul Lindell

INTRODUCTION

Paraneoplastic syndromes are complications of cancer that cannot be attributed to direct effects of the neoplasm or its metastases. Some of these disorders result from tissue-specific autoimmunity initiated by the immune response against cancer. The most dramatic examples of cancer-related autoimmunity are the paraneoplastic neurological disorders (PND). Certain cancers that express onconeural proteins (proteins that are usually restricted to the nervous system) are particularly prone to induce neurological autoimmunity. The pathogenesis of most PND is thought to involve cytotoxic T lymphocytes, which presumably recognize autoantigens in the context of MHC Class I on both tumor cells and neurons [1]. Antibodies against neuronal antigens may be produced as well. These paraneoplastic autoantibodies are useful diagnostic markers even if they are not directly pathogenic.

Paraneoplastic neurological disorders can involve any part of the nervous system and may affect multiple areas simultaneously (Table 13.1). Some PND have unique clinical characteristics and should be easily recognized (e.g. paraneoplastic cerebellar degeneration, sensory neuronopathy and limbic encephalitis). Others may be indistinguishable from more common neurological disorders (e.g. peripheral neuropathy, myasthenia gravis, motor neuron disease and myelitis). In these cases, the association with cancer may be under-recognized because of the absence of clinical suspicion.

Although the neurological symptoms vary between different disorders, most PND share some common clinical features. The neurological syndromes have a subacute onset and progressive course. Symptoms typically develop over weeks or months, but can progress more rapidly (over a few days) in some cases. The median age of onset is around 65 years, with a wide range. In American studies, female patients predominate (about 2:1) even when cases of gender specific tumors (breast, ovary and testes) are not considered [2,3]. The neurological illness precedes the diagnosis of cancer in the majority of cases. As a result, misdiagnosis or delay in diagnosis of PND is common. In patients with a previous history of cancer, the onset of neurological

TABLE 13-1 Paraneoplastic neurological syndromes
Brain and eye
Cerebellar degeneration
Limbic encephalitis [15]
Brainstem encephalitis [18]
Opsoclonus-myoclonus
Chorea [19]
Optic neuritis
Retinal degeneration
Spinal cord
Myelopathy
Myelitis with rigidity and spasms ('stiff-person' syndrome)
Motor neuronopathy
Nerve
Sensory neuronopathy (pure sensory neuropathy)
Sensorimotor peripheral neuropathy (subacute or chronic) [20]
Autonomic neuropathy, gastrointestinal dysmotility [2]
Neuromuscular junction/muscle
Lambert-Eaton myasthenic syndrome
Myasthenia gravis
Dermatomyositis
Neuromyotonia
Multifocal disorders (encephalomyeloneuropathies)
Combination of those above and others

symptoms may herald cancer recurrence. Even when the diagnosis of PND is suspected, the initial search for malignancy may be unrevealing. Tumors, when found, tend to be limited in stage [4]. Several reports suggest that patients with PND have a favorable cancer outcome (survival and treatment response) compared to those with identical tumors without PND [5–8]. The neurological disorder, on the other hand, may progress relentlessly despite treatment and patients are often left with significant neurological disability [5,9,10].

Overall, paraneoplastic neurological disorders are quite rare (estimated at 0.01% of cancer patients) [11]. Certain malignancies are more likely to be associated with PND. About 30% of patients with thymoma have some form of neurological autoimmunity, mostly myasthenia gravis [12]. Small cell carcinoma, most commonly arising in the lung, is associated with one or more PND in up to 3% of cases [13,14]. Other malignancies with definite PND associations include gynecological malignancies, arising from the breast, ovary, fallopian tube and peritoneum, Hodgkin's and non-Hodgkin's lymphoma, testicular cancer and neuroblastoma. PND occur at a much lower frequency in patients with non-small cell lung, renal, uterine and melanotic skin cancers. All PND are rare, but several classical syndromes are distinctive and common enough to warrant specific attention. Neuroimaging is important in the evaluation of PND of the central nervous system, but generally does not contribute to the diagnosis of PND restricted to the peripheral nervous system. Body imaging studies are very important in determining the presence and location of an underlying malignancy.

PARANEOPLASTIC CEREBELLAR DEGENERATION

Paraneoplastic cerebellar degeneration (PCD) presents as non-specific gait unsteadiness which progresses to a severe cerebellar ataxia over a few weeks or months. Occasionally, an acute onset may suggest brainstem or cerebellar stroke. More insidious onset may be confused with the inherited or degenerative ataxias. The typical clinical features are disabling incoordination of gait, trunk and limbs and an ataxic dysarthria. Often, severe vertigo with nausea, diplopia, nystagamus and oscillopsia are early complaints. Within a few months, most patients will lose the ability to walk or even sit independently, lose the ability to write or feed themselves and lose the ability to communicate effectively [8,9]. Dramatic tremors of the limbs and head (titubation) may occur. Symptoms often stabilize spontaneously, but usually leave the patient with severe ataxia and loss of independence. The classical association is with ovarian or breast carcinoma and the ataxia usually precedes the diagnosis of cancer. Autoantibodies reactive against the cytoplasm of cerebellar Purkinje cells (Purkinje cell antibody type I; PCA-1 or 'anti-Yo') may be found in the serum or cerebrospinal fluid (CSF) as markers of these tumors. PCD may also occur in the context of other malignancies (especially small cell lung carcinoma or Hodgkin's lymphoma) with different antibody markers (Table 13.2). Pathologically, PCD may be associated with lymphocytic infiltration in the cerebellum and progressive loss of cerebellar Purkinje cells.

LIMBIC ENCEPHALITIS

Paraneoplastic limbic encephalitis (PLE) is characterized by the triad of short-term memory impairment, temporal lobe seizures and psychiatric symptoms (commonly depression, psychosis or change in personality). Two-thirds of patients have overt seizures (usually complex partial temporal lobe seizures) which may be difficult to control [15]. PLE may stabilize or partially improve following

TABLE 13-2 Neuronal paraneoplastic autoantibodies

Antibody	Usual tumor	Commonly associated syndromes
Neuronal antibodies against nuclear or cytoplasmic antigens		
ANNA-1 (anti-Hu)*[2]	SCLC	Limbic encephalitis, ataxia, sensory neuronopathy, autonomic and sensorimotor neuropathies
CRMP-5 (anti-CV2)[22]	SCLC or thymoma	Encephalomyelitis, chorea, neuropathy, optic neuritis
PCA-1 (anti-Yo)	Ovarian or breast	Paraneoplastic cerebellar degeneration
Anti-Ma [18]	Lung, breast or testicular	Limbic and brainstem encephalitis
Amphiphysin	Lung or breast cancer	Encephalomyelitis, neuropathy, 'stiff-person syndrome'
PCA-2	SCLC	Encephalomyelitis
ANNA-2 (anti-Ri)	Lung or breast cancer	Ataxia, opsoclonus-myoclonus, neuropathy
PCA-Tr (anti-Tr)	Hodgkin's lymphoma	Paraneoplastic cerebellar degeneration
ANNA-3	SCLC	Encephalomyelitis
Anti-zic4	SCLC	Ataxia
Recoverin	SCLC	Cancer-associated retinopathy
Ion channel autoantibodies		
P/Q-type VGCC	SCLC (60%)	Lambert-Eaton syndrome, cerebellar degeneration
N-type VGCC	Lung or breast cancer	Encephalomyelitis, neuropathy
Muscle AChR	Thymoma (15%)	Myasthenia gravis
VGKC	Thymoma or SCLC	Neuromyotonia, limbic encephalitis
mGluR1	Hodgkin's lymphoma	Paraneoplastic cerebellar degeneration

* Alternate nomenclature is indicated in parentheses. SCLC: small cell lung carcinoma

treatment of the cancer or treatment with immunomodulatory therapies, but most patients are left with residual memory impairment and seizures.

At initial presentation, PLE must be distinguished from other forms of limbic encephalitis, including herpes simplex virus infection and various forms of non-paraneoplastic autoimmune encephalopathy [16]. The most common tumors associated with PLE are small cell lung carcinoma (SCLC), testicular cancer, breast cancer and thymoma. The neurological syndrome antedates the diagnosis of cancer in most cases. Several autoantibody markers have been associated with PLE including anti-Hu (ANNA-1) and anti-Ma (see Table 13.2). The autoantibody findings are very useful in directing the search for occult malignancy, but up to 30% of patients with PLE and cancer have negative antibody studies [15]. In those cases, a search for malignancy must be conducted according to the patient's individual cancer risk factors.

BRAINSTEM ENCEPHALITIS

This syndrome is characterized by prominent eye movement abnormalities (vertical gaze palsy, ophthalmoplegia, double vision, complex nystagamus or other involuntary eye movements) often associated with disorders of sleep and wakefulness (including excessive somnolence or central sleep apnea). Other cranial nerve findings include ptosis, facial weakness, flaccid dysarthria, dysphagia, subacute hearing loss and jaw or eyelid dystonia [17,18]. Symptoms of brainstem encephalitis may occur in combination with those of PLE, PCD, or opsoclonus/myoclonus. Several different cancers have been associated including SCLC, testicular and breast cancer.

OPSOCLONUS/MYOCLONUS

Opsoclonus refers to involuntary, chaotic, high-amplitude conjugate eye movements often associated with diffuse or focal myoclonus (involuntary brief muscle jerks of the trunk or limbs). This syndrome may occur in children with neuroblastoma. The neurological syndrome often improves with treatment of the neuroblastoma along with ACTH or prednisone, although many children are left with some degree of incoordination. In adults, less than half of patients with opsoclonus/myoclonus have cancer. Other etiologies include viral encephalitis, drug intoxication and idiopathic autoimmune causes. Patients without cancer may respond to immunomodulatory treatment and make a good recovery. Adults with paraneoplastic opsoclonus/myoclonus may also have features of PCD, PLE or brainstem encephalitis.

PARANEOPLASTIC CHOREA

Tremor and other movement disorders may occur in cancer patients, but are usually not manifestations of paraneoplastic autoimmunity [19]. However, the subacute onset of generalized or focal chorea in an adult may represent a paraneoplastic neurological disorder. Chorea refers to involuntary, random and coordinated but purposeless movements of one or more parts of the body. Chorea is a slow movement, often described as writhing or 'snake-like'. Chorea of the face (orofacial dyskinesia) consists of excessive pursing of the lips, grimacing or blinking and may be associated with a strained voice. Patients are often unconcerned by or even unaware of these movements. Observers may attribute the movements to restlessness or fidgeting. Chorea in adults can have diverse causes, but subacute onset in a patient over 60 years old and association with other neurological symptoms should strongly raise the possibility of a paraneoplastic disorder [19]. Paraneoplastic chorea may also present with significant side-to-side asymmetry. Based on pathological and radiological findings, paraneoplastic chorea appears to be the result of inflammation affecting extrapyramidal circuits including the caudate. Most patients have autoantibodies against the CRMP-5 protein (also known as 'anti-CV2' antibodies) in serum and CSF and the most commonly associated malignancy is SCLC.

PERIPHERAL NERVOUS SYSTEM DISORDERS

Signs and symptoms of a peripheral sensorimotor neuropathy (numbness in the feet and fingers or distal weakness) are very common in cancer patients. Neuropathy may present well in advance of the cancer diagnosis and is arguably the most common paraneoplastic neurological syndrome. One study estimated that 4.5% of patients with unexplained adult onset axonal sensorimotor neuropathy have a malignancy [20]. The exact incidence of neuropathy as a PND, however, remains uncertain. In many cases, the characteristics of a paraneoplastic peripheral neuropathy are those of a mixed sensory and motor length-dependent axonal neuropathy indistinguishable from the non-paraneoplastic neuropathies commonly encountered in the neurology clinic. A few clinical features should increase the suspicion of PND. The onset of paraneoplastic neuropathy tends to be more rapid with progression of symptoms, signs and electrophysiological changes over weeks or months. Pain is typical. Analysis of CSF may show mild abnormalities. Peripheral neuropathy has been associated with a number of cancers (small cell and non-small cell lung cancer, breast cancer and thymoma) and with several autoantibody markers (see Table 13.2). However, antibody studies are negative in many patients with paraneoplastic peripheral sensorimotor neuropathy [20].

Progressive neuropathy that exclusively affects the sensory nerves has been termed pure sensory neuropathy, sensory ganglionopathy or sensory neuronopathy. This disorder is more easily recognized as a PND. About 20% of cases of sensory neuronopathy are paraneoplastic; the remainder are either idiopathic or are associated with systemic autoimmune disease (notably Sjogren syndrome) or toxin exposure (including chemotherapy agents). Initial symptoms may commence in the upper or lower extremity and consist of distal pain, numbness and paresthesias which can be asymmetric. Because of marked loss of proprioception, clumsiness and gait unsteadiness develop (sensory ataxia). With eyes closed, the loss of balance and coordination becomes much worse (Romberg sign) and slow wandering movements of the digits or limbs (pseudoathetosis) may be seen. Muscle stretch reflexes are usually absent. Often, the disorder progresses relentlessly over weeks or months and leads to significant disability. Because of profound sensory loss, the patient may be unaware of serious injuries to the extremities. Any of several paraneoplastic antibodies may be found, but the typical correlation is with the anti-Hu (ANNA-1) antibody. SCLC is the most commonly associated tumor. Typically, the neurological syndrome precedes the diagnosis of cancer and the detection of cancer may be delayed by over a year despite intensive surveillance.

MYASTHENIA GRAVIS AND LAMBERT-EATON SYNDROME

These two disorders differ from the disorders described above in two important respects. First, antibodies against ion channels are not only a specific diagnostic marker, but also directly cause the disease. Secondly, patients often respond well to treatment. Myasthenia gravis (MG) is the prototypical autoimmune neurological disorder. Patients present with fatiguable weakness which usually affects the eyes (causing ptosis and diplopia). Antibodies against acetylcholine receptors at the neuromuscular junction cause the disease and can be detected in the serum in about 85% of MG patients. Up to 15% of patients with MG have thymoma. In Lambert-Eaton syndrome (LES), antibodies against P/Q-type voltage-gated calcium channels on the motor nerve terminal lead to inefficiency of neuromuscular transmission. LES is a paraneoplastic disorder associated with SCLC in about 60% of adult patients. Patients with paraneoplastic LES usually present after age 40 with complaints of generalized weakness and fatigue. When LES is associated with SCLC, features of other neurological syndromes may coexist and other paraneoplastic antibodies may be detected in addition to calcium channel antibodies.

LABORATORY FINDINGS

Since PND usually predates the diagnosis of cancer and routine laboratory tests in patients with these disorders are usually normal, diagnosis of PND depends on clinical suspicion. Analysis of CSF may be normal or show only mild lymphocytic pleocytosis and elevated protein. Oligoclonal bands and increased CSF IgG synthesis rate are seen in a minority of cases. The advent and expansion of testing for paraneoplastic neurological autoantibodies has been a great help in diagnosing PND.

Paraneoplastic antibodies are important as surrogate markers of a specific immune response to cancer. Neuronal nuclear and cytoplasmic antibodies are highly specific for the presence of cancer and also predictive of the cancer type. Nearly 90% of patients with ANNA-1 antibodies have SCLC and over 90% of patients with PCA-1 have cancer (76% ovarian or related peritoneal tumors and 13% breast cancer) [2,21]. Thus, finding a neuronal nuclear or cytoplasmic antibody in a patient with a neurological syndrome should mandate a thorough evaluation for occult malignancy and close oncological follow-up if cancer is not detected on the initial search. The antibody specificity helps direct the search for cancer by predicting the most likely cancer. Unfortunately, even comprehensive testing for all known paraneoplastic antibodies lacks the sensitivity to completely exclude a PND.

EVALUATION AND TREATMENT

The evaluation of a PND requires an initial clinical suspicion. These disorders usually precede a diagnosis of cancer and the tumors are typically limited in stage. Some clinical syndromes (including those described above) should immediately raise the possibility of PND and lead to a cancer evaluation regardless of serological results. Many other patients with PND have atypical, unusual or multifocal neurological complaints that are hard to characterize or localize neuroanatomically. Testing for paraneoplastic antibodies can help confirm the diagnosis of PND and direct the search for occult malignancy. Neurological imaging studies are an important part of the evaluation especially to help exclude alternative diagnoses, such as metastases, infection, demyelinating disease and cerebrovascular disorders.

Cancer evaluation starts with a complete medical history and examination. The search for malignancy then relies heavily on body imaging studies. Suspicious lesions on these imaging studies should be subjected to biopsy to confirm the cancer diagnosis and guide treatment.

Tumor therapy is the standard approach to treating PND. A complete oncological remission can be associated with stabilization or even improvement in the neurological syndrome [3]. Hence, early tumor diagnosis and prompt institution of therapy is critical. In addition to eliminating the tumor as the stimulus for the autoimmune syndrome, many chemotherapy regimens also provide direct immunosuppressive effects. Once a tumor remission is achieved, the serum titers of paraneoplastic antibodies tend to decline slowly over time, but may never normalize. A relapse of neurological symptoms or onset of a new unexplained neurological syndrome may herald tumor recurrence. A rise in the paraneoplastic antibody titer can also signal the return of cancer, but monitoring of paraneoplastic serology over time is not efficient or reliable for this purpose. In cases of PND where cancer cannot be identified despite an exhaustive search, or where the neurological symptoms progress despite cancer remission, immunomodulatory therapies can be considered [5].

SUMMARY

Paraneoplastic neurological syndromes represent uncommon immunological complications of certain malignancies. Clinical manifestations can be quite varied and multifocal in the nervous system. Several distinct clinical syndromes are recognized including sensory neuronopathy, cerebellar degeneration, limbic encephalitis and Lambert-Eaton myasthenic syndrome. These disorders are usually associated with a subacute onset and significant disability. Typically, the neurological presentation antedates the diagnosis of malignancy and the cancer, when found, tends to be localized and responsive to treatment. Diagnosis depends on clinical suspicion, serology for paraneoplastic antibodies and a focused search for cancer. Imaging studies are important in the diagnosis of central nervous system disorders and also for the detection of occult malignancy.

REFERENCES

1. Albert ML, Darnell JC, Bender A et al (1998). Tumor-specific killer cells in paraneoplastic cerebellar degeneration. Nature Med 4(11):1321–1324.
2. Lucchinetti CF, Kimmel DW, Lennon VA (1998). Paraneoplastic and oncologic profiles of patients seropositive for type 1 antineuronal nuclear autoantibodies. Neurology 50(3):652–657.
3. Candler PM, Hart PE, Barnett M et al (2004). A follow up study of patients with paraneoplastic neurological disease in the United Kingdom. J Neurol Neurosurg Psychiatr 75(10):1411–1415.
4. Graus F, Dalmou J, Rene R et al (1997). Anti-Hu antibodies in patients with small cell lung cancer: association with complete response to therapy and improved survival. J Clin Oncol 15(8):2866–2872.
5. Vernino S, O'Neill BP, Marks RS et al (2004). Immunomodulatory treatment trial for paraneoplastic neurological disorders. Neuro-Oncol 6:55–62.
6. Maddison P, Newsom-Davis J, Mills KR, Souhami RL (1999). Favourable prognosis in Lambert-Eaton myasthenic syndrome and small cell lung carcinoma. Lancet 353(9147):117–118.
7. Altman AJ, Baehner RL (1976). Favorable prognosis for survival in children with coincident opso-myoclonus and neuroblastoma. Cancer 37(2):846–852.
8. Hammack JE, Kimmel DW, O'Neill BP, Lennon VA (1990). Paraneoplastic cerebellar degeneration: a clinical comparison of patients with and without Purkinje cell cytoplasmic antibodies. Mayo Clin Proc 65(11):1423–1431.
9. Rojas I, Graus F, Keime-Guibert F et al (2000). Long-term clinical outcome of paraneoplastic cerebellar degeneration and anti-Yo antibodies. Neurology 55(5):713–715.
10. Dalmau J, Graus F, Rosenblum M, Posner J (1992). Anti-Hu associated paraneoplastic encephalomyelitis/sensory neuronopathy. A clinical study of 71 patients. Medicine (Baltimore) 71:59–72.
11. Darnell RB, Posner JB (2003). Paraneoplastic syndromes involving the nervous system. N Engl J Med 349(16):1543–1554.
12. Vernino S, Lennon VA (2004). Autoantibody profiles and neurological correlations of thymoma. Clin Cancer Res 10:7270–7275.
13. Elrington GM, Murray NM, Spiro SG, Newsom-Davis J (1991). Neurological paraneoplastic syndromes in patients with small cell lung cancer. A prospective survey of 150 patients. J Neurol Neurosurg Psychiatr 54(9):764–767.
14. Sculier JP, Feld R, Evans WK et al (1987). Neurologic disorders in patients with small cell lung cancer. Cancer 60(9): 2275–2283.
15. Lawn ND, Westmoreland BF, Kiely MJ et al (2003). Clinical, magnetic resonance imaging, and electroencephalographic findings in paraneoplastic limbic encephalitis. Mayo Clin Proc 78(11):1363–1368.
16. Thieben MJ, Lennon VA, Boeve BF et al (2004). Potentially reversible autoimmune limbic encephalitis with neuronal potassium channel antibody. Neurology 62(7):1177–1182.
17. Pittock SJ, Lucchinetti CF, Lennon VA (2003). Anti-neuronal nuclear autoantibody type 2: paraneoplastic accompaniments. Ann Neurol 53(5):580–587.
18. Voltz R, Gultekin SH, Rosenfeld MR et al (1999). A serologic marker of paraneoplastic limbic and brain-stem encephalitis in patients with testicular cancer. N Engl J Med 340(23):1788–1795.
19. Vernino S, Tuite P, Adler CH et al (2002). Paraneoplastic chorea associated with CRMP-5 neuronal antibody and lung carcinoma. Ann Neurol 51(5):625–630.
20. Antoine JC, Mosnier JF, Absi L et al (1999). Carcinoma associated paraneoplastic peripheral neuropathies in patients with and without anti-onconeural antibodies. J Neurol Neurosurg Psychiatr 67(1):7–14.
21. Pittock SJ, Kryzer TJ, Lennon VA (2004). Paraneoplastic antibodies coexist and predict cancer, not neurological syndrome. Ann Neurol 56(5):715–719.
22. Yu Z, Kryzer TJ, Griesmann GE et al (2001). CRMP-5 neuronal autoantibody: marker of lung cancer and thymoma-related autoimmunity. Ann Neurol 49(2):146–154.

CHAPTER 14

Neoplastic Plexopathies

Mark A. Ferrante and John T. Kissel

INTRODUCTION

The cervical, brachial, lumbar and sacral plexuses constitute the primary plexuses of the peripheral nervous system (PNS), with the lumbar and sacral plexuses typically discussed together as the lumbosacral (LS) plexus. Neoplastic processes may involve any of these plexuses with significant adverse effects on patients' quality of life. When misattributed to other orthopedic or neuromuscular conditions, therapeutic delay results which further adversely affects outcome. Most patients with metastatic involvement have advanced neoplastic spread and die from vital organ complications. To improve survival and quality of life, clinicians must possess an understanding of plexus anatomy, the clinical features associated with plexopathies and the diagnostic and therapeutic options available.

CLASSIFICATION

In 2000, the World Health Organization classified peripheral nerve tumors into four categories:

1 neurofibromas
2 schwannomas
3 perineuriomas (intraneural and extraneural)
4 malignant neural sheath tumors (NSTs).

The latter, which are also called neurogenic sarcomas, include malignancies of peripheral nerve origin or with nerve sheath differentiation, excluding tumors of epineurial or vascular origin [1]. A more practical approach uses two features, malignant status and cell of origin, to classify these tumors into four categories: benign NSTs; benign non-NSTs; malignant NSTs; and malignant non-NSTs. The latter category includes all malignant tumors of extraplexal origin that invade the plexus by direct extension or metastasis.

PLEXUS ANATOMY

Cervical Plexus

Upon exiting the intervertebral foramen, each spinal nerve gives off a posteriorly directed *posterior primary ramus* and then continues as the *anterior primary ramus*

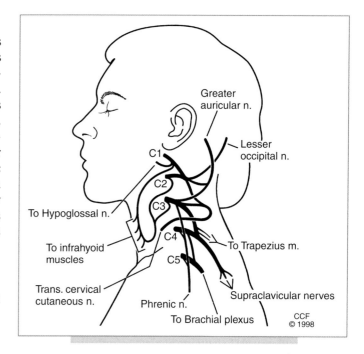

FIG. 14.1. The cervical plexus.

(APR). The cervical plexus is derived from the C1–C4 APR and is located superficial to the scalenus medius and levator scapulae and deep to the sternocleidomastoid (Figure 14.1). The sensory nerves from this plexus (lesser occipital, greater auricular, transverse cutaneous and supraclavicular nerves) supply sensation to the back of the head, neck and cape area. Its motor branches supply the diaphragm, infrahyoid, scalenus medius, trapezius, levator scapulae and C1–C4 paravertebral muscles. Postganglionic sympathetic fibers, derived from the superior cervical ganglion, also traverse this plexus [2,3].

Brachial Plexus

The brachial plexus (BP) is composed of five *roots* (C5–T1), three *trunks* (upper, middle, lower), six *divisions* (three

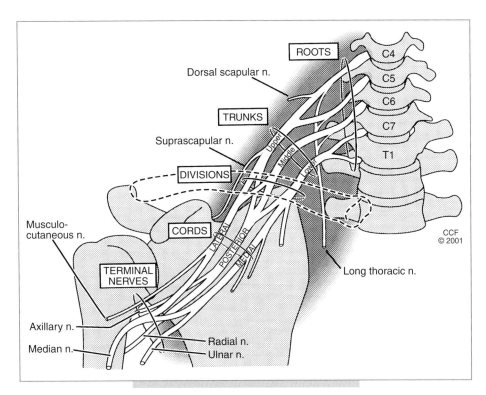

FIG. 14.2. The brachaial plexus.

anterior; three posterior), three *cords* (lateral, posterior, middle) and several *terminal nerves* (Figure 14.2). Nerve elements derived from the BP include motor branches to the phrenic, dorsal scapular and long thoracic nerves (via the C5 APR); branches to the long thoracic nerve (via C6–7 APR); the suprascapular nerve (via the upper trunk); the lateral pectoral and musculocutaneous nerves and the lateral head of the median nerve (via the lateral cord); the subscapular, thoracodorsal, axillary, and radial nerves (via the posterior cord); and the medial pectoral, medial brachial cutaneous, medial antebrachial cutaneous, and ulnar nerves and the medial head of the median nerve (via the medial cord) [2]. Conceptually, the supraclavicular plexus is divided into three parts: upper plexus (upper trunk; C5 and C6 roots), middle plexus (middle trunk; C7 root) and lower plexus (lower trunk; C8 and T1 roots). Many tumors associated with metastatic BP lesions drain to the lateral axillary lymph nodes near the lower trunk and its divisions, producing lower plexopathies. Since the upper trunk and its divisions are relatively free of lymph nodes, upper plexopathies are less commonly the onset site for metastatic disease arising from the lymphatic system [4].

Lumbosacral Plexus

The LS plexus lies in the retroperitoneal space (Figure 14.3). The lumbar plexus receives nerve fibers from the L1–L4 APR, and often T12 APR. Fibers from the L4 and L5 APR join to form the LS trunk which, in turn, joins the S1–S4 APR to form the sacral plexus. This plexus gives off the

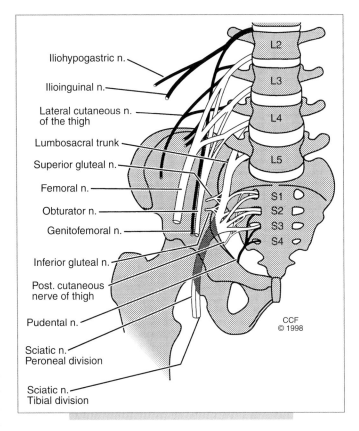

FIG. 14.3. The lumbosacral plexus.

iliohypogastric, ilioinguinal, genitofemoral, obturator, femoral, lateral femoral cutaneous, common peroneal, tibial, superior and inferior gluteal, pudendal and posterior femoral cutaneous nerves, as well as motor branches to the psoas and iliacus muscles. A portion of S4 APR contributes to the coccygeal plexus. The lumbar plexus is situated within the upper two-thirds of the psoas major. The iliohypogastric, ilioinguinal and genitofemoral nerves descend posterior to the iliac fascia and paraaortic and iliac lymph nodes. The sacral plexus lies within the pelvis adjacent to the rectum, colon and ureter. The lumbar and sacral plexuses receive their blood supply from lumbar arteries derived from the abdominal aorta and internal iliac artery, respectively [2].

EPIDEMIOLOGY AND RISK FACTORS

Although only 1% of cancer patients develop neoplastic plexopathies, approximately 15% of BP lesions are due to plexus tumors, most commonly neurofibromas and schwannomas [5–7]. Neurofibromas are more common in patients with neurofibromatosis 1 (NF1) [8,9], but only 10% of patients with solitary neurofibromas have NF1 [10]. Those tumors associated with NF1 tend to be larger, are more frequently multiple or plexiform, and manifest at an earlier age [7,11]. Sporadic neurofibromas are more frequent on the right and are more common in women [11]. The risk of malignant degeneration in solitary neurofibromas is low, but occurs in 5–10% of NF1 patients [12,13]. Schwannomas are more common in the head and neck and along major nerve trunks [14]. Their incidence peaks in the fourth and fifth decades and is higher in women [7,11,15,16].

The incidence of non-NSTs is lower than that of NSTs. Malignant NSTs, which account for 10% of sarcomas, are rare, with an incidence of 0.001%; this increases to 4.6% in NF1 patients [17,18]. Fifty percent of these tumors occur in NF1 patients, with a lifetime risk of 2–10% and an expected life reduction of 10–15 years [19,20,21]. NF1 patients with malignant NSTs present one to two decades earlier [22]. Reported figures on the incidence of metastatic plexopathies are suspect, since plexus evaluations may not be undertaken when study results would not affect management or when patients are too ill or refuse such evaluations. Overall, the most common metastatic lesion to involve peripheral nerve, and usually the BP, is breast carcinoma [8,15], which more frequently involves middle-aged to elderly females.

The incidence of primary neoplastic plexopathies is not uniform among the plexuses. Most involve the BP [23]. In adults, solitary neurofibromas commonly involve the supraclavicular plexus, especially the upper and middle plexuses [6,7,11,15,24,25]. This distribution is less pronounced among NF1 patients [6,7]. Schwannomas more commonly involve the upper plexus [6,7,15]. Less than 1% of patients with neoplasms have LS plexus involvement

[26,27]. Nonetheless, malignancy constitutes the most common diagnosis in patients presenting with LS plexopathies [28,29]. Risk factors for the primary neoplastic plexopathies include NF1, NF2 and prior radiation to the area. About 11% of malignant NSTs follow radiation, with a latency period ranging from 4 to 41 years (average, 15 years) [30,31].

PATHOGENESIS AND PATHOPHYSIOLOGY

Schwannomas and neurofibromas grow along a spectrum ranging from discrete, paraneural schwannomas to poorly delineated, plexiform neurofibromas. Schwannomas typically are solitary, benign, slow growing tumors arising anywhere along the neural sheath, especially on spinal roots at the myelination transition zone (where Schwann cells and oligodendrocytes abut). Larger lesions may compress adjacent nerves. Rarely, schwannomas are multiple, undergo malignant transformation, or occur in NF1 patients [32]. Solitary neurofibromas, which probably arise from perineural fibroblasts, involve the entire cross-section of the nerve [7,8,15,33]. Since plexiform neurofibromas do not have a capsule, they are infiltrative and typically many inches long at presentation [23]. Malignant NSTs arise commonly de novo or via the malignant transformation of a plexiform neurofibroma, less commonly from a solitary intraneural neurofibroma or ganglioneuroma, and rarely from a schwannoma [21,34]. Malignant NSTs spread via direct extension or blood-borne metastasis.

Secondary neoplastic plexopathies reflect invasion from extraplexal sources, such as adjacent primary or metastatic malignancies or lymph node metastases. Infiltration may also occur through the blood in hematological malignancies and via spinal fluid in both hematologic and certain solid malignancies (especially breast and lung cancer and melanoma) [35]. Intraneural metastasis is rare. Compression may disrupt plexus fiber integrity (e.g. metastatic axillary lymphadenopathy) [36–38]. When chemotherapy-induced immunosuppression results in thoracic outlet infection, indirect BP involvement may occur [39–41]. Leukemic infiltrates may produce vascular occlusion and plexus infarction.

Although nerve fibers can be damaged by neoplastic processes in a variety of ways, their pathologic responses are limited to axon loss (most common) and demyelination. Depending on the fiber type affected, axon loss results in conduction failure and weakness, sensory loss and dysautonomia. When the insult to myelinated fibers is insufficient to cause axon disruption, isolated demyelination may occur. Unlike axon loss, demyelination does not induce distal changes, producing either conduction block, when nerve impulses cannot traverse the lesion site, or conduction slowing. Demyelination can be observed with abrupt onset lesions, but even then, axon loss usually occurs.

CLINICAL FEATURES

The main clinical features associated with neoplasia include pain and sensorimotor dysfunction, depending on the plexus region and fiber types involved. Benign NSTs usually present as painless, palpable masses over the course of a nerve, often with Tinel's sign and lateral, but not longitudinal, mobility [23,42]. Due to the slow growth rate of these tumors, pain and neurologic deficit are infrequent, but more common with neurofibromas [8,15,34,42–44]. Although malignant NSTs can occur sporadically, they more commonly develop from pre-existing plexiform neurofibromas. They present as painful, rapidly expanding masses with progressive neurologic dysfunction [11,13]. Patients with metastatic plexopathies typically present with severe pain. Most of these patients are already known to harbor a malignancy, have already been treated, and have widely metastatic disease.

With metastatic cervical plexopathies, pain radiating to the neck, shoulder, or throat that worsens with swallowing or neck movement is the primary symptom [27]. Sensory disturbances are frequent. Motor involvement may go unrecognized when the affected muscle is multiply innervated or not easily assessed [3,27,45,46]. With intraspinal canal extension, respiratory paralysis (epidural spinal cord compression) and Horner's syndrome (sympathetic trunk involvement) may occur. Mass lesions or lymphadenopathy may be noted on general examination.

Most metastatic brachial plexopathies present with pain involving the shoulder or axilla with radiation along the medial aspects of the arm, forearm and hand [4]. As the malignancy spreads, progressive neurologic dysfunction occurs. With upper plexopathies (C5, C6), weakness may cause the arm to assume the 'waiter's tip' position (i.e. adduction and internal rotation of the shoulder, extension and pronation of the arm and forearm and posterolateral rotation of the palm). Sensory disturbances occur along the lateral aspects of the arm, forearm and hand and the biceps and brachioradialis reflexes may be lost. With middle plexopathies, weakness is confined to C7 muscles. Sensory abnormalities occur along the posterolateral arm, dorsal forearm and lateral hand. The triceps reflex may be affected. With lower plexus involvement, weakness involves C8/T1 muscles and sensory disturbances occur along the medial arm, forearm and hand. An ipsilateral Horner's sign occurs in 25% of cases [5]. The finger flexor reflex may be affected. The medial extent of the hand symptoms reflects the dorsal root ganglion (DRG) from which the affected sensory nerve fibers originate [47]. Lateral cord lesions may produce weakness of the biceps, brachialis, pronator teres and flexor carpi radialis and sensory disturbances in the lateral forearm or hand. Posterior cord lesions may cause weakness in muscles innervated by the subscapular, thoracodorsal, axillary and radial nerves. Sensory disturbances in the upper and lower lateral brachial cutaneous, posterior brachial and antebrachial cutaneous and superficial radial nerves

may occur. Medial cord lesions may generate weakness in C8/T1-ulnar, C8/T1-median or C8-radial nerve innervated muscles, as well as sensory disturbances in the medial brachial and antebrachial and ulnar nerve distributions. Terminal nerve involvement produces sensorimotor abnormalities restricted to the domains of the affected nerve.

Approximately 15% of patients with neoplastic LS plexopathies present without a known malignancy [26]. Pain is the most common and, often, the most disabling feature, typically preceding any other symptom by weeks to months [26,27,48,49]. With lumbar plexopathies, pain occurs in the lower torso, buttock, hip, anterolateral thigh and medial leg regions; it more commonly occurs in the posterolateral thigh, remainder of the leg and foot with sacral plexus involvement [26,50]. The pain may be elicited or exacerbated by reverse straight leg testing (lumbar plexus) or by straight leg testing (sacral plexus). Upper lumbar plexopathies are associated with sensory disturbances in the cutaneous distributions of the iliohypogastric, ilioinguinal and genitofemoral nerves. More inferior lesions affect fibers destined for the femoral (anteromedial thigh and medial leg sensation; hip flexor and knee extensor strength; quadriceps reflex), obturator (superomedial thigh sensation, thigh adductor strength, adductor reflex), or lateral femoral cutaneous nerve. When LS trunk or upper sacral plexus involvement affects fibers destined for the deep peroneal nerve, foot drop may result. Lower sacral plexopathies produce weakness in the thigh extensors and abductors, leg flexors and foot, sensory disturbances in the posterior thigh, leg (sparing the medial aspect) and foot and, when S1-derived nerve fibers are involved, ankle reflex changes. Patients with deep pelvic tumors often present with sensory disturbances in the perineum and perianal area prior to leg pain or weakness. Other features associated with pelvic malignancy include sciatic notch tenderness, anal sphincter tone abnormalities, urinary incontinence, leg or penile edema, thrombophlebitis, ascites, hepatosplenomegaly and generalized lymphadenopathy [27,48–51]. When retroperitoneal lymph node metastases involve the LS plexus, findings include vertebral body destruction, leg edema and features of ureteral obstruction [52,53].

ASSOCIATED NEOPLASTIC DISORDERS

Most neoplastic cervical plexopathies are due to the direct invasion of tumor from adjacent tissues; invasion via adjacent metastases is less frequent [27]. Breast and lung cancers account for 70% of the primary tumors involving the BP [4]. Apical lung cancers are the most common non-metastatic malignancy to involve the BP [54]. Other sources include lymphoma, sarcoma, melanoma, thyroid, testicular, bladder, gastrointestinal and head and neck cancer [5,15,43]. Intraneural metastasis is rare, but has been reported with carcinoid and hematologic malignancies [55–61]. Colorectal carcinoma, sarcoma, breast cancer, lymphoma and carcinoma of the cervix most usually affect the LS plexus [5,15].

Primary neoplastic plexopathies are infrequent and usually occur in the BP [8]. Of these, schwannomas and neurofibromas predominate, most of which are benign [6,7]. Benign non-NSTs include desmoid, lipoma, hamartoma, ganglioneuroma, myoblastoma, lymphangioma, myositis ossificans, osteochondroma and vascular tumors (e.g. venous angioma, hemangiopericytoma, glomus tumor hemangioblastoma) [6,7,62,63]. Malignant non-NSTs can directly extend or metastasize to nerve. Most malignant non-NSTs of the BP are accounted for by breast and lung cancer and melanoma.

While most secondary malignancies of the BP do not involve particular plexus regions [4], breast, head and neck and apical lung cancers are exceptions. Breast cancer favors the infraclavicular plexus when it metastasizes to the lateral axillary lymph nodes, head and neck cancer favors the upper plexus when it infiltrates from above and apical lung cancers favor the lower plexus with infiltration below. Other common BP malignancies include lymphoma, melanoma and esophageal and bladder cancer [64].

A variety of malignant carcinomas and lymphoreticular tumors involve the LS plexus. These may be restricted to the lumbar (31%), sacral (51%), or LS (18%) plexus [26]. Most secondary lesions reflect direct invasion from primary pelvic malignancies; less commonly, metastatic growth from regional lymph nodes or bone occurs [5]. Direct extension from an intra-abdominal source more commonly affects the lumbar plexus, whereas metastatic involvement more commonly involves the sacral plexus [26]. The most common neoplastic LS plexopathy is invasive colorectal carcinoma [65], while breast cancer is the most common cause of metastatic plexopathy [26]. Other common sources include uterine and cervical malignancies, retroperitoneal sarcoma, lymphoma, melanoma and myeloma, as well as lung, testicular, thyroid, renal and bladder carcinoma [5,26,27,29,65]. Malignant psoas syndrome is observed in patients with advanced solid tumors of the genitourinary tract, prostate and colorectum, a reflection of their patterns of spread.

Benign neoplasms, most commonly neurofibromas and schwannomas, infrequently affect the LS plexus [11,65–67] and are often incidentally discovered when individuals with sensorimotor disturbances are referred for electrodiagnostic (EDX) testing, which localizes the lesion and directs imaging. Compression from benign tumors (dermoid cyst, uterine leiomyoma) may affect the LS plexus [65].

Pancoast Syndrome

Apical lung cancers comprise 3% of all lung cancers. Unlike most lung cancers, they are predominantly extrapulmonary, slow growing, locally aggressive, infrequently metastatic and radiosensitive [68–70]. Most patients are elderly males with smoking histories that present with severe, unrelenting shoulder pain from pleural involvement. The pain usually extends into the axilla and medial arm (C8–T2), is worse at night and interferes with sleep. Medial scapular pain follows tumor extension into the superior mediastinum or posterior primary rami [71]. Lower plexus and stellate ganglion involvement produce C8–T1 sensory and motor loss and Horner's syndrome. Other features include supraclavicular fullness, venous distension and arm edema. Bone destruction of the upper thoracic ribs and, less often, the adjacent vertebrae occurs [72,73]. Although bronchogenic carcinoma is the most common cause of this disorder, it may follow any thoracic inlet disorder, including other tumors, inflammatory processes and infections [39–41,73–77].

DIFFERENTIAL DIAGNOSIS

A palpable mass with pain or paresthesias in a plexus distribution is suggestive of a primary neoplastic plexopathy. Other considerations include benign cysts, nerve hamartomas and traumatic neuromas. Neoplastic cervical plexopathies may be mimicked by neoplastic meningitis or epidural tumor spread involving cervical roots. When patients develop cervical plexus dysfunction following radical neck dissections, the relationship between symptoms and surgery helps differentiate an iatrogenic process from tumor recurrence. Cervical plexopathies may also follow radiation therapy [78].

The differential diagnosis of BP lesions among cancer patients is much more extensive. The two most common causes are metastatic plexopathy and post-radiation plexopathy (in those who received radiation). Other considerations include neoplastic advancement (e.g. epidural extension, neoplastic meningitis), treatment-related (e.g. subclavian artery occlusion, effects of chemotherapy), iatrogenic trauma (e.g. secondary to surgical intervention or anesthetic administration), orthopedic conditions (e.g. rotator cuff tear) and non-traumatic entities [4,24,64,79–81].

LS plexopathies must be differentiated from more proximal and more distal lesions. When patients with known malignancies develop LS plexopathies, metastatic spread to the plexus or cauda equine and neoplastic meningitis are considerations. When severe unilateral, local or radicular pain is present, disc herniation is a possibility. If the patient has received pelvic irradiation, post-radiation plexopathy is a concern. Treatment with chemotherapy via internal iliac artery injection may affect the LS plexus through ischemic or neurotoxic mechanisms. Intragluteal injections of analgesics can also damage the LS plexus, presumably due to vasospasm and thrombosis of the sacral plexus blood supply [29,82]. Patients with retroperitoneal hemorrhage (e.g. leukemia) may develop abnormalities in the distribution of the particular plexus fibers affected. When the bleeding is limited to the iliacus muscle, an isolated femoral neuropathy may develop. Extensive bleeding into the psoas muscle may involve lumbar plexus fibers or, less commonly, both lumbar and sacral plexus fibers [65,83,84]. Several features of diabetic amyotrophy and peripheral nerve vasculitis suggest malignancy (e.g. older age, progressive weight loss, persistent lower extremity pain). Normal imaging studies help exclude malignancy.

Post-radiation Plexopathy

Although peripheral nerve fibers are relatively radio-resistant [85], radiation therapy (XRT) may have toxic effects on nerve fibers and vasa nervorum, with resultant demyelination, impaired circulation and fibrosis (post-radiation plexopathy) [50,86–90]. The interval between the last dose of XRT and the onset of symptoms may vary from a few weeks to more than 30 years, although symptoms usually occur within 12–20 months [38,64,91]. The initial clinical features reflect demyelinating conduction block, with paresthesias, often in a lateral cord distribution. A presumed diagnosis of radiculopathy or carpal tunnel syndrome may result in unnecessary surgical procedures. Pain at onset is a major symptom in only a third of cases. Weakness and sensory loss follow, sometimes with limb edema. As the disorder progresses, the demyelinating conduction block converts to axon loss and the arm becomes atrophic, flail, insensate, edematous and painful [29,50]. Risk factors for this disorder include higher XRT doses, higher fraction sizes, the three-field technique and concomitant chemotherapy. Over the past few decades, risk factor modification has significantly reduced the incidence of this disorder [92–94]. Subclavian artery thrombosis with resultant plexus fiber ischemia and permanent, painless weakness and sensory loss, infrequently follows XRT. A lower motor neuron syndrome can occur 3–5 years after XRT to the cauda equina [64,65].

Differentiating post-radiation plexopathy from neoplastic plexopathy can be challenging. Features predicting tumor recurrence include shoulder pain, rapid progression, lower plexus involvement and metastatic disease in other body regions. Findings suggesting post-radiation plexopathy include painless paresthesias at onset, lateral cord distribution, slow progression and certain EDX features (i.e. demyelinating conduction block, fasciculations, grouped repetitive discharges and myokymic discharges) [4,38,91,95–97]. Bilateral, asymmetric, distal lower extremity weakness, often preceded by sensory disturbances, may follow radiation damage to the LS plexus, with EDX abnormalities similar to those described for the BP. Features of post-radiation plexopathy never exclude tumor recurrence, since the two may occur concomitantly. Surgical confirmation may be necessary, although a negative biopsy also does not exclude tumor recurrence [4].

EVALUATION

Accurate diagnosis of neoplastic plexopathies requires the coupling of historical and examination data with laboratory findings, imaging and EDX studies. Pertinent historical data include the growth rate of the lesion, the presence and progression of pain or neurologic deficits, the presence of other masses, a personal or family history of NF or cancer, previous cancer treatments and the presence of constitutional symptoms (e.g. fever; weight loss). On examination, the size and location of the lesion and the presence of tenderness, Tinel's sign, neurologic deficit, other tumors and NF1 stigmata are sought. Plain x-rays and computed tomography (CT) scans may identify bony neoplastic destruction or radiation exposure and spinal column instability, while radionuclide bone scans identify extraplexal metastases. Magnetic resonance imaging (MRI) is the procedure of choice for defining the location and margins of the tumor, its relationship to adjacent structures and lymphadenopathy [21,98,99]. On MRI, NSTs appear as fusiform or spherical enlargements; often, the capsule and the nerve from which it arises are visualized [98,100]. Multiple or plexiform lesions suggest NF1 [23]. When malignant cells track along the connective tissue of a plexus, visualization by MRI may be impossible. Radiographic studies do not distinguish neurofibromas from schwannomas or benign from malignant ones [101]. Heterogeneous signal and bone erosion are seen with benign NSTs and benign plexiform neurofibromas may have irregular infiltrative borders [102–104]. Positron emission tomography (PET) scanning may be useful for identifying neoplastic changes within a plexus and differentiating benign and malignant plexiform neurofibromas and magnetic resonance neurography to differentiate intraneural and perineural masses [5,105,106]. Angiography defines tumor vascularity and the relationship between the tumor and any neighboring vessels [21]. CSF examination is performed when neoplastic meningitis is a consideration. EDX studies localize and quantify the plexopathy, identify subclinical lesions, identify features of post-radiation plexopathy, direct imaging studies and assess progression. EDX studies are often normal until persistent sensorimotor deficits are present [23].

Accurate diagnosis requires histologic studies of tumor tissue or enlarged lymph nodes. Unfortunately, when benign plexus tumors are mistaken for non-neural lesions and improperly biopsied or excised, severe, permanent and unnecessary neurologic deficits and pain may result [7,15,43]. Consequently, whenever a mass lesion is associated with neurologic dysfunction or Tinel's sign, referral to an experienced neurosurgeon is indicated. When a malignant NST is suspected, a metastatic evaluation, including chest films, bone scan, chest and abdominal CT and MRI of the surrounding structures, is performed. Definitive diagnosis is made by biopsy [8]. When a malignant plexopathy is suspected and the MRI is normal, surgical exploration and studies to search for extraplexal lesions may be required. When the results are negative, re-exploration should be carried out 3–6 months later [4].

Other aspects of the evaluation reflect the plexus under consideration. When malignancy involves the cervical plexus, a thorough head and neck examination is mandatory, preferably by an otolaryngologist. Plain films and

CT scans of the neck identify bone erosion and neck instability and chest films may identify benign NSTs of the lower trunk or hemi-diaphragmatic paralysis. With the latter, determination of whether the lesion reflects phrenic nerve (more common) or cervical plexus involvement is needed [23,29,107]. When tumor infiltration of the cervical plexus does not produce a recognizable mass, a repeat scan or surgical exploration may be warranted [27]. With suspected neoplastic brachial plexopathies, the neck, supraclavicular fossa, shoulder region and axilla are palpated for masses. Horner's syndrome is sought because it is associated with epidural spinal cord compression among patients with metastatic lesions [38]. Plain films of the chest, cervical spine, clavicle, scapula, shoulder and humerus help identify lung lesions (e.g. superior sulcus masses) and bone abnormalities related to malignant erosion or radiation necrosis. Axial MRI images best assess the APR and trunk elements, while coronal images better assess the divisions and cords. A new MRI technique, *magnetic resonance myelography*, images proximal BP elements traversing the cerebrospinal fluid [108,109]. Following 3D reconstruction, it produces T2-weighted myelogram-like images of the CSF. With suspected neoplastic LS plexopathies, pelvic CT scanning can visualize nerve sheath tumors and is abnormal at presentation in up to 96% of patients with metastatic disease [26]. Apical lordotic chest films identify apical lung cancers, typically with extension through the visceral pleura into the parietal pleura and chest wall. CT scanning and MRI of the neck, chest and upper abdomen help stage the tumor (extent, epidural spread) and dictate appropriate surgical intervention [76,101]. Percutaneous transthoracic needle biopsy has a diagnostic yield of 95% [101]. Bone, liver and brain imaging assess for metastatic lesions [76]. Open biopsy to delineate the tumor cell type usually is unnecessary prior to treatment [76]. With neoplastic LS plexopathies, plain films of the spine and pelvis may reveal bony erosion or loss of the psoas shadow suggesting a retroperitoneal mass. Lower abdominal and pelvic CT scanning can identify solid tumors, calcifications, bony erosion, spinal column instability, bleeding and radiation osteitis [29,50]. Modern scanners permit visualization of plexus elements when oral, intravenous and urographic contrast agents are administered. Although 96% of patients presenting with neoplastic LS plexopathies have abnormal CT scans, CT cannot categorize lesions and does not recognize non-structural plexus lesions, such as vasculitis. CT-myelography may identify intraspinal neoplastic involvement of the cauda equina or epidural space [26].

MRI remains more sensitive for identifying plexus lesions [110]. Neoplastic invasion, NSTs and some metastatic lesions require gadolinium for visualization. Magnetic resonance myelography can delineate structures proximal to the termination site of the nerve root sleeve, such as thecal margins, nerve roots and their sheaths, and the cauda equina and conus medullaris. Other useful radiographic modalities include isotope bone scans (bony metastases), ultrasonography (lower abdominal and pelvic masses) and intravenous pyelography and barium enemas (distortion of bladder, ureter, or bowel).

EDX testing is useful for LS plexus assessment. Because of the high innervation ratio of lower extremity muscles, EDX testing can identify even minimal motor axon loss. Also, it frequently identifies abnormalities outside the region of the imaged mass, including contralaterally. Consequently, a normal EDX examination argues against a neoplastic plexopathy. Because of the limited number of reliable sensory nerve conduct studies (NCS) available to assess the LS plexus, normal sensory responses do not exclude a plexopathy and abnormal ones may reflect multiple nerve involvement. EDX features of post-radiation plexopathy may also be noted [111,112].

Surgical biopsy typically is required when a mass lesion of unknown primary is identified. Serological testing may be helpful, including serum sedimentation rate, testing for diabetes and specific antigen testing (e.g. PSA, CEA, CA-125). When patients with known primaries develop widely metastatic disease and present with LS plexus involvement, diagnostic testing may not be pursued, especially when the primary was known to neighbor the LS plexus.

MANAGEMENT

The diversity of neoplastic plexopathies requires individualized treatment plans that reflect tumor characteristics, sensitivity to radiation and chemotherapy, the degree of spread and the general health and wishes of the patient. Surgical resection is the treatment of choice for most non-metastatic plexus tumors, the goals of which include cure, restoration of function, symptom relief, palliation and cosmetic improvement. Excision types include intralesional (excision leaves visible tumor), marginal (excision includes surrounding capsule or reactive zone), wide (excision includes a cuff of normal surrounding tissue) or radical (includes the entire anatomical compartment) [21]. Intralesional and marginal resections usually are indicated for benign tumors, whereas wide and radical resections are reserved for malignant lesions. Chemotherapy and XRT typically are adjunctive treatments.

With benign tumors, surgical indications include pain, neurologic dysfunction, suspicion of malignant transformation (e.g. rapid growth) and disfigurement [15,23]. Schwannomas are encapsulated and, when symptomatic, often can be excised without deficit or recurrence. Since neurofibromas arise within the fascicle, their surgical excision requires sacrifice of non-functioning parent and adjacent fascicles. Intraoperative EDX studies permit total tumor resection without added deficit from sacrifice of functioning fascicles [15,23]. Postoperative neurologic deficits are more common among NF1 patients, occurring in one-third with solitary tumors [115]. The poor delineation, increased vascularity and infiltrative nature of plexiform neurofibromas increase the chance of postoperative neurologic dysfunction to one-half and reduce the likelihood of complete resection [8,23,113,114]. When nerve function cannot be preserved, nerve grafting may be performed. Benign non-NSTs that grow extrinsic to their parent fascicles and intraneural lipomas often can be excised without neurologic deficit [115].

Malignant NSTs have a high risk of local and distant spread. Survivability requires surgical excision coupled with adjuvant high-dose XRT [116,117]. Although postoperative XRT provides local control and delays recurrence, especially when resection is incomplete, better long-term survival has not been proven with this regimen and there is a risk of generating malignant NSTs among unaffected fibers [113,118]. The roles of brachytherapy and intraoperative electron beam irradiation are unresolved [113]. The role of chemotherapy is unclear. A meta-analysis showed no difference in survival despite improved progression-free survival and reduced 10-year relapse rates with chemotherapy [119]. Although chemotherapy has no role in the initial treatment of patients without metastatic disease when gross total tumor resection is accomplished, it may be considered when complete resection is impossible, when extremity amputation is to be avoided, with widely metastatic disease, as an alternative to radiotherapy or with treatment failures [22,113,117]. The surgical goal is complete resection with wide tumor margins [116]. Extremity amputation well above the lesion may be required [8,15], although even this may not prevent metastases or improve survival [9]. In selected patients, limb-sparing procedures, which include wide local resection, radiation and chemotherapy, have been performed with good to excellent survival rates [8]. Neural sacrifice usually is unavoidable. Unfortunately, the large area resected and the adverse effects of postoperative XRT on graft tissue preclude nerve grafting [22].

Metastatic plexopathies occur late in the course of the malignancy, are relatively resistant to chemotherapy and radiotherapy and typically are incurable. Treatment is palliative and responses modest and short-lived [5,8,27]. With metastatic breast carcinoma, however, the malignancy often does not invade beyond the epineurial level; thus, external neurolysis and tumor removal is an option. Likewise, with melanoma and lymphoma, tumor removal from epineurial attachments, followed by local radiation, may be beneficial [8].

The pain associated with neoplastic plexopathies usually is severe in degree, unrelenting in nature and difficult to control. Opioid therapy, the mainstay, is associated with multiple side effects and may be inefficacious. Dose-limiting side effects may respond to continuous delivery (transdermal, subcutaneous). Non-opioid analgesic (e.g. antiepileptic drugs [AEDs], tricyclic antidepressants [TCAs], dexamethasone) are also helpful. When these agents are started early at dosages adequate to achieve pain control, pain severity and adverse effects may be significantly diminished. Other approaches include transcutaneous electric nerve stimulation (TENS), local anesthesia (e.g. paravertebral nerve blocks) and implantable intrathecal pumps. Surgical procedures (e.g. DREZ procedure; contralateral percutaneous cordotomies at C1–C2; selective rhizotomies) may be required for severe intractable pain. Destructive procedures (e.g. limb amputations) are ineffective [4,27,29,38].

Occupational therapy consultation for assistive devices to maximize remaining neurologic function and measures to prevent neuromuscular complications (e.g. painful contractures, deep venous thromboses, ulcerations, compression neuropathies) should also be employed to optimize quality of life. Compressive devices and extremity elevation may benefit lymphedema.

The current treatment of Pancoast syndrome is surgical resection, with many patients, even some with extensive BP or spinal column involvement, achieving curative resections [110,120]. Surgical contraindications include extensive invasion, mediastinal lymphadenopathy and peripheral metastases, although palliative surgery may be employed for epidural compression, mechanical instability, or pain unresponsive to radiation [76,120]. Induction chemo radiotherapy increases the chance of performing a complete resection [121]. Most studies have failed to demonstrate a survival advantage using either postoperative XRT or intraoperative brachytherapy [121]. The pain associated with Pancoast syndrome can be unbearable and the relief following induction chemoradiotherapy typically is short-lived. Combinations of opioid and non-opioid analgesics, steroids, tricyclics, antiepileptics, ketamine and TENS may be ineffective. Continuous administration of local anesthetics into the BP region was successful in six patients [122]. Adequate pain relief may follow percutaneous cervical cordotomy, but the success rate of other invasive techniques (e.g. scalenotomy, spinothalamic tractotomy, myelotomy, cingulotomy, dorsal column stimulators and subarachnoid phenol or alcohol) is unclear [122].

There is no effective treatment for post-radiation plexopathy and care is symptomatic [5,11,29,64]. Physical therapy can prevent atrophy, capsulitis, contractures and lymphedema, and orthotics help maintain function. Medications, TENS and dorsal column stimulators may reduce pain [38]. External neurolysis can break up perineural adhesions and scar tissue. Although this may lessen pain, progression usually continues [5,15]. Because of its ability to maintain blood flow in vessels with radiation-induced endothelial damage, anticoagulation has been suggested as a potential treatment for post-radiation plexopathy, but currently is of unproven value [123,124].

PROGNOSIS AND FUTURE PERSPECTIVES

The prognosis for neurologic recovery primarily reflects the underlying malignancy. The prognosis is excellent for most benign NSTs and non-NSTs. Following tumor resection, worsening of pain or motor dysfunction is infrequent and the recurrence rate is under 5% [23,62,113,125], although desmoid tumors tend to recur despite thorough excision [15]. The total removal of plexiform tumors causes loss of function and subtotal removal permits progression of tumor growth. Thus, these tumors are associated with functional loss of the involved nerve [23]. Malignant NSTs have a high recurrence rate; about one-fourth of non-NF1 patients experience local recurrence within 2 years with an even higher rate among NF1 patients [13,124]. The 5-year survival rate for non-NF-1 patients is about two-thirds and for NF1 patients about 25% [9,21,32,113]. Poor prognostic factors include tumor size >5 cm, high histological grade, positive surgical margins and the presence of NF1. Patients with involvement of proximal PNS structures do worse because of incomplete resection and spread into the CNS [13].

The prognosis for patients with metastatic plexopathies is poor. Typically, the lesion expands, pain is difficult to control and widespread metastatic disease ensues [26,27,29]. Most patients die from neoplastic involvement of vital organs within two years with pain and motor dysfunction adversely affecting quality of life [26,27]. Patients with lymphoma, breast cancer and certain head and neck malignancies may achieve better pain control and functional improvement.

With Pancoast syndrome, the degree of local invasion at presentation determines the resectability of the tumor and survival [120]. Favorable features include absence of metastatic disease, lack of lymph node involvement and resolution of pain following preoperative XRT [69,70,126]. Conversely, extensive invasion of the BP, spine invasion and cord compression portend a poor prognosis [126]. Tumor cell type does not seem to influence survival [70]. Future studies, to determine the best combination of chemotherapy and biological agents and whether prophylactic cranial irradiation would lessen the occurrence of brain metastases following treatment, are needed [121].

The prognosis for patients with post-radiation plexopathy is poor because the disorder is inexorably progressive, spreading through plexus and converting from chronic demyelination to severe axon loss [27,47,50,64,84]. Many patients with arm involvement develop a useless, swollen, painful limb [64,91]. Even when the lesion is incomplete, sensory ataxia often severely limits useful function [29]. Radiologic innovations continue to improve the ability to diagnose plexus tumors and surgical advancements continuously improve functional outcome and reduce surgical morbidity. Advances in genetics and oncology should provide new therapeutic options that will extend and improve the quality of life of those affected. An excellent example of these advances relates to NF1, where tumorigenesis may reflect loss of neurofibromin function and resultant overactivity of the p21 ras signaling pathway [113]. Ongoing clinical trials include anti-ras drugs and, for plexiform neurofibromas, interferon-alpha2b (an anti-angiogenic agent thought to impede or prevent tumor growth) [116,127].

REFERENCES

1. Kleihues P, Cavenee WK (2000). *World Health Organization Classification of Tumours: Pathology and Gentics of Tumours of the Nervous System*. AZRC Press, Lyon.
2. Clemente CD (1985). *Gray's Anatomy*, 30th edn (American). Williams & Wilkins, Baltimore.
3. Brazis PW, Masdeu JC, Biller J (1990). *Localization in Clinical Neurology*, 2nd edn, Little Brown, Boston.
4. Kori SH, Foley KM, Posner JB (1981). Brachial plexus lesions in patients with cancer: 100 cases. Neurology 31:45–50.
5. Jaeckle KA (2004). Neurological manifestations of neoplastic and radiation-induced plexopathies. Semin Neurol 24:385–393.
6. Kim DH, Cho YJ, Tiel RL, Kline DG (2003). Outcomes of surgery in 1019 brachial plexus lesions treated at Louisiana State University Health Sciences Center. J Neurosurg 98:1005–1016.
7. Ganju A, Roosen N, Kline DG, Tiel RL (2001). Outcomes in a consecutive series of 111 surgically treated plexal tumors: a review of the experience at the Louisiana State University Health Sciences Center. J Neurosurg 95:51–60.
8. Chang SD, Kim DH, Hudson AR, Kline DG (2002). Peripheral nerve tumors. In *Neuromuscular Disorders in Clinical Practice*. Katirji B, Kaminski HJ, Preston DC, Ruff RL, Shapiro BE (eds). Butterworth Heinemann, Boston. 828–837.
9. Park JK (2003). Peripheral nerve tumors. In *Office Practice of Neurology*, 2nd edn. Samuels MA, Feske SK (eds). Churchill Livingstone, Philadelphia. 1118–1121.
10. Geschickter CF (1935). Tumors of peripheral nerve. Am J Cancer 25:377–410.
11. Kline DG, Hudson AR (1995). *Nerve Injuries*. Saunders, Philadelphia.
12. Ferner RE, O'Doherty MJ (2002). Neurofibroma and schwannoma. Curr Opin Neurol 15:679–684.
13. Baehring JM, Betensky RA, Batchelor TT (2003). Malignant peripheral nerve sheath tumor: the clinical spectrum and outcome of treatment. Neurology 61:696–698.
14. Moukarbel RV, Sabri AN (2005). Current management of head and neck schwannomas. Curr Opin Otolaryngol Head Neck Surg 13:117–122.
15. Lusk MD, Kline DG, Garcia CA (1987). Tumors of the brachial plexus. Neurosurgery 21:439–453.
16. Pilavaki M, Chourmouzi D, Kiziridou A, Skordalaki A, Zarampoukas T, Drevelengas A (2004). Imaging of peripheral nerve sheath tumors with pathologic correlation: pictorial review. Eur J Radiol 52:229–239.
17. Ducatman BS, Scheithauer BW, Piepgras DG, Reiman HM, Ilstrup DM (1986). Malignant peripheral nerve sheath tumors. A clinicopathologic study of 120 cases. Cancer 57:2006–2021.
18. Collin C, Godbold J, Hajdu S, Brennan M (1987). Localized extremity soft tissue sarcoma: an analysis of factors affecting survival. J Clin Oncol 5:601–612.

19. Zoller M, Rembeck B, Akesson H, Angervall L (1995). Life expectancy, mortality and prognostic factors in neurofibromatosis type 1. A twelve-year follow-up of an epidemiological study in Goteborg, Sweden. Acta Derm Venereol 75:136–140.

20. Korf BR (2003). The neurofibromatoses. In *Office Practice of Neurology*, 2nd edn. Samuels MA, Feske SK (eds). Churchill Livingstone, Philadelphia. 1076–1082.

21. Lee DH, Dick HM (1998). Management of peripheral nerve tumors. In *Management of Peripheral Nerve Problems*, 2nd edn. Omer GE Jr, Spinner M, Van Beek AL (eds). Saunders, Philadelphia.

22. Perrin RG, Guha A (2004). Malignant peripheral nerve sheath tumors. Neurosurg Clin N Am 15:203–216.

23. Donner TR, Voorhies RM, Kline DG (1994). Neural sheath tumors of major nerves. J Neurosurg 81:362–373.

24. Wilbourn AJ (2002). Brachial plexopathies. In *Neuromuscular Disorders in Clinical Practice*. Katirji B, Kaminski HJ, Preston DC, Ruff RL, Shapiro BE (eds). Butterworth Heinemann, Boston. 884–906.

25. Wilbourn AJ, Ferrante MA (2003). Clinical electromyography. In *Baker's Clinical Neurology on CD-ROM*. Joynt RJ, Griggs RC (eds). Lippincott, Philadelphia.

26. Jaeckle KA, Young DF, Foley KM (1985). The natural history of lumbosacral plexopathy in cancer. Neurology 35:8–15.

27. Jaeckle KA (1991). Nerve plexus metastases. Neurol Clin 9:857–866.

28. Mumenthaler M, Schliack H (1991). *Peripheral Nerve Lesions: Diagnosis and Therapy*. Thieme, New York.

29. Wilbourn AJ, Ferrante MA (2001). Plexopathies. In *Neuromuscular Diseases: Expert Clinicians' Views*. Pourmand R (ed.). Butterworth Heinemann, Boston. 493–527.

30. Foley KM, Woodruff JM, Ellis FT, Posner JB (1980). Radiation-induced malignant and atypical peripheral nerve sheath tumors. Ann Neurol 7:288–296.

31. Ducatman BS, Scheithauer BW (1983). Postirradiation neurofibrosarcoma. Cancer 51:1028–1033.

32. Ghosh BC, Ghosh L, Huvos AG, Fortner JG (1973). Malignant schwannoma: a clinicopathologic study. Cancer 31:184–190.

33. Enzinger FM, Weiss SW (1988). *Soft Tissue Tumors*, 2nd edn. CV Mosby, St Louis.

34. Harkin JC, Reed RJ (1969). Tumors of the peripheral nervous system. In *Atlas of Tumor Pathology*, 2nd series, fascicle 3. Armed Forces Institute of Pathology, Washington DC.

35. Mackinnon SE, Dellon AL (1988). Tumors of the peripheral nerve. In *Surgery of the Peripheral Nerve*. Thieme, Stuttgart. 535–548.

36. Van Echo DA, Sickles EA, Wiernik PH (1973). Thoracic outlet syndrome, supraclavicular adenopathy, Hodgkin's disease. Ann Int Med 78:608–609.

37. Felice KJ, Donaldson JO (1995). Lumbosacral plexopathy due to benign uterine leiomyoma. Neurology 45:1943–1944.

38. Kori SH (1995). Diagnosis and management of brachial plexus lesions in cancer patients. Oncology 9:756–765.

39. Winston DJ, Jordan MC, Rhodes J (1977). *Allescheria boydii* infections in the immunosuppressed host. Am J Med 63:830–835.

40. Simpson FG, Morgan M, Cooke NJ (1986). Pancoast's syndrome associated with invasive aspergillosis. Thorax 41:156–157.

41. Shamji FM, Leduc JR, Bormanis J, Sachs HJ (1988). Acute Pancoast's syndrome caused by fungal infection. Can J Surg 31:441–443.

42. Dodge HW Jr, Craig WM (1957). Benign tumors of peripheral nerves and their masquerade. Minn Med 40:294–301.

43. Dart LH Jr, MacCarty CS, Love JG, Dockerty MB (1970). Neoplasms of the brachial plexus. Minn Med 53:959–964.

44. Fisher RG, Tate HB (1970). Isolated neurilemmomas of the brachial plexus. J Neurosurg 32:463–467.

45. Haymaker W, Woodhall B (1953). *Peripheral Nerve Injuries: Principles of Diagnosis*. Saunders, Philadelphia.

46. Schaafsma SJ (1987). Plexus injuries. In *Handbook of Clinical Neurology, vol 7, Diseases of Nerves, Part 1*. Vinken PJ, Bruyn GW (eds). Elsevier, Amsterdam. 402–429.

47. Ferrante MA, Wilbourn AJ (1995). The utility of various sensory nerve conduction responses in assessing brachial plexopathies. Muscle Nerve 18:879–889.

48. Pettigrew LC, Glass JP, Moar M, Zornoza J (1984). Diagnosis and treatment of lumbosacral plexopathies in patients with cancer. Arch Neurol 41:1282–1285.

49. Chad DA, Bradley WG (1987). Lumbosacral plexopathy. Semin Neurol 7:97–107.

50. Thomas JE, Cascino TL, Earle JD (1985). Differential diagnosis between radiation and tumor plexopathy of the pelvis. Neurology 35:1–7.

51. McKinney AS (1973). Neurologic findings in retroperitoneal mass lesions. South Med J 66:862–864.

52. Van Nagell JR, Sprague AD, Roddick JW (1975). The effect of intravenous pyelography and cystoscopy on the staging of cervical cancer. Gynecol Oncol 3:87–91.

53. Saphner T, Gallion HH, Van Nagell JR, Kryscio R, Patchell RA (1989). Neurologic complications of cervical cancer. Cancer 64:1147–1151.

54. Vargo MM, Flood KM (1990). Pancoast tumor presenting as cervical radiculopathy. Arch Phys Med Rehabil 71:606–609.

55. Liang R, Kay R, Maisey MN (1985). Brachial plexus infiltration by non-Hodgkin's lymphoma. Brit J Radiol 58:1125–1127.

56. Grisold W, Jellinger K, Lutz D (1990). Human neurolymphomatosis in a patient with chronic lymphatic leukemia. Clin Neuropathol 9:224–230.

57. Diaz-Arrastia R, Younger DS, Hair L et al (1992). Neurolymphomatosis: a clinicopathological syndrome reemerges. Neurology 42:1136–1141.

58. Glass J, Hochberg FH, Miller DC (1993). Intravascular lymphomatosis: a systemic disease with neurologic manifestations. Cancer 71:3156–3164.

59. Levin KH, Lutz G (1996). Angiotrophic large cell lymphoma with peripheral nerve and skeletal muscle involvement: early diagnosis and treatment. Neurology 47:1009–1011.

60. van den Bent MJ, de Bruin HG, Bos GM, Brutel de la Rivere G, Sillevis Smitt PA (1999). Negative sural nerve biopsy in neurolymphomatosis. J Neurol 246:1159–1163.

61. Grisold W, Piza-Katzer H, Jahn R, Herczeg E (2000). Intraneural nerve metastasis with multiple mononeuropathies. JPNS 5:163–167.

62. Huang JH, Zaghloul K, Zager EL (2004). Surgical management of brachial plexus region tumors. Surg Neurol 61:372–378.

63. Todd M, Shah GV, Mukherji SK (2004). MR imaging of brachial plexus. Top Magn Reson Imaging 15:113–125.

64. Wilbourn AJ (2005). Brachial plexus lesions. In *Peripheral Neuropathy*, 4th edn. Dyck PJ, Thomas PK (eds). Elsevier Saunders, Philadelphia. 1339–1373.

65. Donaghy, M (2005). Lumbosacral plexus lesions. In *Peripheral Neuropathy*, 4th edn. Dyck PJ, Thomas PK (eds). Elsevier Saunders, Philadelphia. 1375–1390.

66. Benzel EC, Morris DM, Fowler MR (1988). Nerve sheath tumors of the sciatic nerve and sacral plexus. J Surg Oncol 39:8–16.

67. Hunter VP, Burke TW, Crooks LA (1988). Retroperitoneal nerve sheath tumors: an unusual cause of pelvic mass. Obstet Gynecol 71:1050–1052.

68. Komaki R, Roh J, Cox JD et al (1981). Superior sulcus tumors: results of irradiation in 36 patients. Cancer 48:1563–1568.

69. Kanner RM, Martini N, Foley KM (1982). Incidence of pain and other clinical manifestations of superior pulmonary sulcus (Pancoast) tumors. In *Advances in Pain Research and Therapy*, Vol 4. Bonica JJ, Ventafridda V, Pagni CA (eds). Raven Press, New York. 27–39.

70. Ricci C, Rendina EA, Venuta F et al (1989). Superior pulmonary sulcus tumors: radical resection and palliative treatment. Int Surg 74:175–179.

71. Hepper NGG, Herskovic T, Witten DM, Mulder DW, Woolner LB (1966). Thoracic inlet tumors. Ann Int Med 64:979–989.

72. Pancoast HK (1932). Superior pulmonary sulcus tumor. J Am Med Assoc 99:1391–1396.

73. Attar S, Miller JE, Satterfield J et al (1979). Pancoast's tumor: irradiation or surgery? Ann Thor Surg 28:578–586.

74. Stathatos C, Kontaxis AN, Zafiracopoulos P (1969). Pancoast's syndrome due to hydatid cysts of the thoracic outlet. J Thorac Cardiovasc Surg 58:764–768.

75. Omenn GS (1971). Pancoast syndrome due to metastatic carcinoma from the uterine cervix. Chest 60:268–270.

76. Urschel HC Jr (1988). Superior pulmonary sulcus carcinoma. Surg Clin 68:497–509.

77. Silverman MS, MacLeod JP (1990). Pancoast's syndrome due to staphylococcal pneumonia. Can Med Assoc J 142:343–345.

78. Westling P, Svensson H, Hele P (1972). Cervical plexus lesions following post-operative radiation therapy of mammary carcinoma. Acta Radiol Ther Phys Biol 11:209–216.

79. Jackson L, Keats AS (1965). Mechanisms of brachial plexus palsy following anesthesia. Anesthesiology 26:190–194.

80. Pezzimenti JF, Bruckner HW, DeConti RC (1973). Paralytic brachial neuritis in Hodgkin's disease. Cancer 31:626–629.

81. Thyagarajan D, Cascino T, Harms G (1995). Magnetic resonance imaging in brachial plexopathy of cancer. Neurology 45:421–427.

82. Stoehr M, Dichgans J, Dorstelmann D (1980). Ischaemic neuropathy of the lumbosacral plexus following intragluteal injection. J Neurol Neurosurg Psychiatr 43:489–494.

83. Emery S, Ochoa J (1978). Lumbar plexus neuropathy resulting from retroperitoneal hemorrhage. Muscle Nerve 1:330–334.

84. Stewart JD (2000). *Focal Peripheral Neuropathies*, 3rd edn. Lippincott Williams & Wilkins, Philadelphia.

85. Fajardo LF (1988). Vascular lesions following radiation. Pathol Annu 23:297–330.

86. Greenfield MM, Stark FM (1948). Post-irradiation neuropathy. Am J Radiol 60:617–622.

87. Stoll BA, Andrews JT (1966). Radiation-induced peripheral neuropathy. Br Med J 1:834–837.

88. Haymaker N, Lindgren M (1970). Nerve disturbances following exposure to ionizing radiation. In *Handbook of Clinical Neurology*. Vinken PJ, Bruyn GW (eds). North Holland, Amsterdam.

89. Burns RJ (1978). Delayed radiation-induced damage to the brachial plexus. Clin Exp Neurol 15:221–227.

90. Johannson S, Svensson H, Larson LG, Denekamp J (2000). Brachial plexopathy after postoperative radiotherapy of breast cancer patients. Acta Oncol 39:373–382.

91. Wilbourn AJ, Levin KH, Lederman RJ (1994). Radiation-induced brachial plexopathy: electrodiagnostic changes over 13 years (Abstract). Muscle Nerve 17:1108.

92. Pierce SM, Recht A, Lingos TI et al (1992). Long-term radiation complications following conservative surgery (CS) and radiation therapy (RT) in patients with early stage breast cancer. Int J Radiat Oncol Biol Phys 23:915–923.

93. Olsen NK, Pfeiffer P, Johanssen L, Schroder H, Rose C (1993). Radiation-induced brachial plexopathy: neurological follow-up in 161 recurrence-free breast cancer patients. Int J Radiat Oncol Biol Phys 26:43–49.

94. Schierle C, Winograd JM (2004). Radiation-induced brachial plexopathy: review. Complication without a cure. J Reconstr Microsurg 20:149–152.

95. Allen AA, Albers JW, Bastron JA, Daube JR (1977). Myokymic discharges following radiotherapy for malignancy (abstract). Electroenceph Clin Neurophysiol 43:148.

96. Lederman RJ, Wilbourn AJ (1984). Brachial plexopathy: recurrent cancer or radiation? Neurology 34:1331–1335.

97. Harper CM, Thomas JE, Cascino TL, Litchy WJ (1989). Distinction between neoplastic and radiation-induced brachial plexopathy, with emphasis on the role of EMG. Neurology 39:502–506.

98. Cerofolini E, Landi A, DeSantis G, Maiorana A, Canossi G, Romagnoli R (1991). MR of benign peripheral nerve sheath tumors. J Comput Assist Tomogr 15:593–597.

99. Rapoport S, Blair DN, McCarty SM, Desser TS, Hammers LW, Sostman HD (1988). Brachial plexus: correlation of MR imaging with CT and pathologic findings. Radiology 167:161–165.

100. Smith W, Amis JA (1992). Neurilemmoma of the tibial nerve: a case report. J Bone Joint Surg (Am) 74:443–444.

101. Mukherji SK, Castillo M, Wagle AG (1996). The BP. Sem US CT MRI 17:519–538.

102. Petasnick JP, Turner DA, Charters JR, Gritelis S, Zacharias CE (1986). Soft tissue masses of the locomotor system: comparison of MRI with CT. Radiology 160:125–133.

103. Levine E, Huntrakoon M, Wetzel LH (1987). Malignant nerve sheath neoplasms in neurofibromatosis: distinction from benign tumors by using imaging techniques. Am J Radiol 149:1059–1064.

104. Kransdorf MJ, Jelinek JS, Moser RP et al (1989). Soft tissue masses: diagnosis using MR imaging. Am J Radiol 153:541–547.

105. Howe FA, Filler AG, Bell BA, Griffiths JR (1992). Magnetic resonance neurography. Magn Reson Med 28:328–338.

106. Ferner RE, Lucas JD, O'Doherty MJ et al (2000). Evaluation of (18)Fluorodeoxyglucose positron emission tomography ((18)FDG PET) in the detection of malignant peripheral nerve sheath tumours arising from within plexiform neurofibromas in neurofibromatosis 1. J Neurol Neurosurg Psychiatr 68:353–357.

107. Gyhra A, Israel J, Santander C, Acuna D (1980). Schwannoma of the brachial plexus with intrathoracic extension. Thorax 35:703–704.

108. Krudy AG (1992). MR myelography using heavily T2-weighted fast spin-echo pulse sequences with fat presaturation. Am J Radiol 159:1315–1320.

109. Toshiyasu N, Yabe Y, Horiuchi Y, Takayama S (1997). Magnetic resonance myelography in brachial plexus injury. J Bone Joint Surg (Br) 79B:764–769.

110. Taylor BV, Kimmel DW, Krecke KN, Cascino TL (1997). Magnetic resonance imaging in cancer-related lumbosacral plexus. Mayo Clin Proc 72:823–829.

111. Aho K, Sainio K (1983). Late irradiation-induced lesions of the lumbosacral plexus. Neurology 33:953–955.

112. Albers JW, Allen AA Jr, Bastron JA, Daube JR (1981). Limb myokymia. Muscle Nerve 4:494–504.

113. Huang JH, Johnson VE, Zager EL (2006). Tumors of the peripheral nerves and plexuses. Curr Treat Options Neurol 8:299–308.

114. Tiel R, Kline D (2004). Peripheral nerve tumors: surgical principles, approaches, and techniques. Neurosurg Clin N Am 15:167–175.

115. Louis DS (1987). Peripheral nerve tumors in the upper extremity. Hand Clin 3:311–318.

116. Ferner RE, Gutmann DH (2002). International consensus statement on malignant peripheral nerve sheath tumors in neurofibromatosis. Cancer Res 62:1573–1577.

117. Rawal A, Yin Q, Roebuck M, Sinopidis C, Kalogrianitis S, Helliwell TR, Frostick S (2006). Atypical and malignant peripheral nerve-sheath tumors of the brachial plexus: report of three cases and review of the literature. Microsurgery 26:80–86

118. Khanfir K, Alzieu L, Terrier P et al (2003). Does adjuvant radiation therapy increase loco-regional control after optimal resection of soft-tissue sarcoma of the extremities? Eur J Cancer 39:1872–1880.

119. Tierney JF, Stewart LA, Parmar MKB et al (1997). Adjuvant chemotherapy for localised resectable soft tissue sarcoma of adults: meta-analysis of individual data. Sarcoma Meta-Analysis Collaboration. Lancet 350:1647–1654.

120. Bilsky MH, Vitaz TW, Boland PJ, Bains MS, Rajaraman V, Rusch VW (2002). Surgical treatment of superior sulcus tumors with spinal and brachial plexus involvement. J Neurosurg (Spine 3) 97:301–309.

121. Pitz, CC de la Riviere AB, van Swieten HA, Duurkens VA, Lammers JW, van den Bosch JM (2004). Surgical treatment of Pancoast Tumours. Eur J Cardiothorac Surg 26:202–208.

122. Vranken JH, Zuurmond WW, de Lange JJ (2000). Continuous brachial plexus block as treatment for the pancoast syndrome. Clin J Pain 16:327–333.

123. Glantz MJ, Burger PC, Friedman AH, Radtke RA, Massey EW, Schold SC Jr (1994). Treatment of radiation-induced nervous tissue injury with heparin and warfarin. Neurology 44:2020–2027.

124. Soto O (2005). Radiation-induced conduction block: resolution following anticoagulant therapy. Muscle Nerve 31:642–645.

125. Artico M, Cervoni L, Wierzbicki V, D'Andrea V, Nucci F (1997). Benign neural sheath tumours of major nerves: characteristics in 119 surgical cases. Acta Neurochir 139: 1108–1116.

126. Sundaresan N, Hilaris BS, Martini N (1987). The combined neurosurgical-thoracic management of superior sulcus tumors. J Clin Oncol 5:1739–1745.

127. NIH Clinical Research Studies (protocol number: 05-C-0232). A phase 1 trial of peginterferon Alfa-2b (Peg-Intron) for plexiform neurofibromas. http://clinicalstudies.info.nih.gov/detail/a_2005-c-0232.html.

CHAPTER 15

Neurological Complications of Oncological Therapy

Michelle Monje and Patrick Y. Wen

COMPLICATIONS OF RADIOTHERAPY

Radiation therapy is one of the most important causes of neurotoxicity in patients with cancer. It may affect the nervous system by:

1   direct injury to neural structures included in the radiation portal or
2   indirectly by damaging blood vessels or endocrine organs necessary for functioning of the nervous system or by producing tumors.

Radiation injury may occur acutely, but more commonly occurs after a delay of months or years. Many factors determine whether radiation injury will occur. These include the radiation dose, fraction size, duration of treatment, the volume treated, the length of survival following radiation therapy and the presence of other therapies and systemic diseases.

The complications of radiotherapy are classically categorized in terms of acute, early-delayed and late-delayed forms (Table 15.1) [1]. Radiation toxicity occurs in all parts of the nervous system: brain, spinal cord and peripheral nerve.

Brain

Acute (0–14 days)

Acute Encephalopathy

Following cranial radiotherapy, acute encephalopathy may occur. Symptoms include nausea/vomiting, drowsiness, headache, and/or possibly worsening of pre-existing neurological deficits. This syndrome may be associated with fever. The risk of acute encephalopathy increases with radiation fraction size exceeding 2 Gy. The etiology appears to be disruption of the blood–brain barrier, with resultant vasogenic edema and subsequently increased intracranial pressure (ICP) [2]. A transient increase in white matter edema may be seen on neuroimaging. The prognosis is generally excellent, however, herniation and death are possible if the ICP is already increased, especially in the

TABLE 15-1 SUMMARY OF NEUROLOGIC COMPLICATIONS IN NERVOUS SYSTEM
Neurologic complications of radiation therapy to the brain
Direct effects of radiation therapy
Acute reactions (hours or days)
Cerebral edema: headaches, nausea, vomiting, lethargy
Early delayed reactions (2 weeks to 4 months)
Drowsiness, increased cognitive dysfunction, exacerbation of neurologic deficits
Late delayed reactions (4 months to several years)
Mild to moderate cognitive decline
Dementia
Cerebral necrosis
Leukoencephalopathy
Normal pressure hydrocephalus
Indirect effects of radiation therapy
Cerebrovascular disorders
Large and small vessel disease, moyamoya, telangiectasias, cavernomas, angiomatous malformations, aneurysms
Radiation-induced neoplasms
Meningiomas, sarcomas, gliomas, schwannomas
Endocrine dysfunction
Hypothyroidism, hypogonadism, growth hormone deficiency, hypoadrenalism
Neurologic complications of radiation therapy to the spinal cord
Early delayed radiation myelopathy
Late delayed radiation myelopathy
Delayed radiation myelopathy (DRM)
Radiogenic lower motor neuron disease
Spinal vasculopathy and hemorrhage
Neurologic complications of radiation therapy to the brachial plexus
Early delayed brachial plexopathy
Late delayed brachial plexopathy
Neurologic complications of radiation therapy to peripheral nerves
Cranial neuropathies
Peripheral neuropathy
Nerve sheath tumor

case with posterior fossa or intraventicular tumors. Acute radiation-induced encephalopathy usually responds to corticosteroid therapy. Patients with large tumors and mass effect should receive corticosteroids prior to beginning radiotherapy to prevent acute encephalopathy.

Fatigue

A high percentage of patients undergoing cranial irradiation experience severe and progressive fatigue. This occurs partway through the course of radiotherapy, typically between weeks 3 and 6 of radiotherapy for primary brain tumors and near the end of radiotherapy, to 1–2 weeks after the completion of radiotherapy for metastatic disease [3]. In one study of 104 patients with metastatic disease to brain, the incidence of excessive fatigue was reported to be 85% in patients who underwent whole brain radiotherapy plus radiosurgery, as compared to 28% in patients undergoing radiosurgery alone [4]. Methylphenidate and modafinil may reduce the severity of fatigue [5].

Early Delayed (4 weeks–months)

Several neurological syndromes occur between 4 weeks to several months after radiation therapy. The precise etiology is unclear but may be related to transient demyelination and remyelination. These syndromes usually resolve spontaneously [2].

Somnolence Syndrome

The somnolence syndrome consists of drowsiness, lethargy, excessive sleep, headache, nausea and anorexia. It occurs several weeks after cranial irradiation and the course is usually monophasic, resolving within a few weeks [6,7]. The reported incidence varies greatly, depending on factors such as radiation dose, fractionation, tumor type and diagnostic criteria. EEG findings reveal diffuse slowing during the symptomatic period, with normalization of the EEG after symptoms resolve [6]. The somnolence syndrome was suggested to be a predictor of long-term cognitive outcome in one series, but this finding was not reproduced in subsequent studies [8–10]. Informing patients of the probability of somnolence for a period after radiotherapy greatly reduces anxiety.

Worsening of Pre-existing Deficits

A worsening of pre-existing focal deficits can occur after radiotherapy. This usually occurs about 2 months following treatment. Peritumoral edema and contrast enhancement is sometimes, but not always, seen on neuroimaging. Clinical and radiological improvement is usually seen within 4–8 weeks [11].

Transient Cognitive Impairment

A transient decline in cognition, distinct from the late onset progressive cognitive decline described below, may occur. This type of cognitive dysfunction, predominantly verbal memory dysfunction, carries a good prognosis and does not appear to predict development of the late-delayed cognitive syndrome [12,13].

Subacute Rhombencephalitis

A subacute rhombencephalitis may occur 1–3 months after treatment when radiation portals involve the brainstem, such as with ocular, pituitary or head and neck tumors. This syndrome is distinct from the brainstem radionecrosis that can occur later. Clinically, patients present with diplopia, nystagmus, ataxia, dysarthria and other brainstem symptoms and signs. CSF evaluation may show signs of inflammation. Neuroimaging reveals brainstem white matter abnormalities, with lesions that are T1 hypointense and T2 hyperintense; the lesions may enhance with gadolinium [14]. Prognosis is generally good, with progressive improvement over weeks to months, but death has been reported [15].

Late Delayed (months to years)

Radionecrosis

Radiation necrosis has been reported in the treatment of both intracranial and extracranial tumors, such as nasopharyngeal carcinoma (Figure 15.1). Radiation necrosis typically occurs 1–2 years after radiation, but latency as short as 3 months and as long as 30 years have been reported [16,17]. Radiographically, it can be difficult to distinguish from recurrent tumor.

Recognition of the risk factors for radiation necrosis has resulted in a decrease in incidence. Radiation total dose and fraction size are important risk factors and a total external beam dose of 55–60 Gy delivered in 1.8–2.0 Gy fractions constitutes the upper limits of 'safe' dose. Other risk factors include lesion volume and location, old age, associated chemotherapy and vascular risk factors such as diabetes. Therapies that attempt to increase the dose of radiation delivered to the tumor volume, such as interstitial brachytherapy and stereotactic radiosurgery, also result in an increased incidence of radiation necrosis [18].

Radiation necrosis can result in devastating clinical consequences. Patients present with focal neurological symptoms, often recapitulating the presenting symptoms of the patient's initial disease in the case of primary brain tumors, severe cognitive deficits, signs and symptoms of increased intracranial pressure and/or seizures. About one half of patients present with seizure as the first sign. Imaging characteristics of radiation necrosis are very similar to recurrent tumor. CT reveals hypodensity and variable contrast enhancement. MRI shows T1 hypointensity and T2 hyperintensity, predominantly involving white matter. Lesions frequently enhance with gadolinium [19]. Mass effect may be present. Positron emission tomography (PET) or single photon emission computerized tomography (SPECT) imaging and MR spectroscopy may be employed

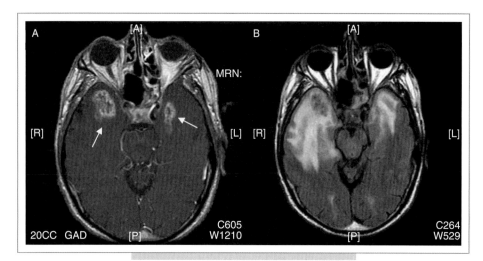

FIG. 15.1. A 49-year-old man with T4 squamous cell carcinoma of the paranasal sinuses treated with radiotherapy 6 years previously presenting with confusion and memory loss. **(A)** T1-weighted axial MRI with gadolinium showing enhancing necrotic areas in both temporal lobes (white arrows); **(B)** axial FLAIR MRI showing edema around necrotic areas. FDGPET scan showed no uptake, suggesting that the areas of enhancement were necrosis.

to differentiate radiation necrosis from malignancy [20]. In one series, a (201)T1-SPECT technique yielded diagnostic sensitivity of 0.88 and a specificity of 0.83, compared to routine neuroanatomincal imaging (CT and MRI), which was found to have a sensitivity of 0.63 and a specificity of 0.59 [20]. However, biopsy remains the only definitive means of diagnosis.

Histopathologically, the lesions of radiation necrosis predominantly affect white matter. Vascular changes include vessel wall hyalinization, thickening and fibrinoid necrosis, fibrinous exudates, vascular hemorrhage and thrombosis. These changes affect the small arteries and arterioles. Demyelination is also evident.

The etiology of radiation necrosis is not yet clear. Histopathology reveals vascular damage and demyelination, implicating the vascular cells and/or oligodendrocytes as the targets of radiation injury. Classically, radiation injury has been attributed to either the vascular hypothesis or the glial hypothesis. The *vascular hypothesis* states that radiation-induced vasculopathy and the resultant ischemic necrosis account for radiation injury. The *glial hypothesis* proposes that radiation-induced damage to oligodendrocytes and their precursors is the underlying cause of radiation injury. However, the lesions of radiation necrosis are not typical of either pure vascular or demyelinating disease, and neither hypothesis seems sufficient alone to account for radiation necrosis [21].

Treatment of radiation necrosis involves surgical excision and steroid therapy. Additional therapies have been proposed, including anticoagulants, hyperbaric oxygen and alpha-tocopherol, but their clinical utility has yet to be proven [22–24].

Leukoencephalopathy

Diffuse white matter changes may occur as a late-delayed effect of radiation alone, a combination of radiotherapy and chemotherapy or, more rarely, after chemotherapy alone. Clearly, radiation and chemotherapy have a synergistic effect. In addition to concomitant chemotherapy such as methotrexate, risk factors for radiation-induced leukoencephalopathy include higher radiation dose and age greater than 60 years [2,25]. This diffuse white matter injury is an entity distinct from radiation necrosis. As patients are surviving longer after cancer therapies, radiation-induced leukoencephalopathy is an increasingly important complication of treatment. Histopathology reveals rarefication of white matter, reactive astrogliosis and foci of necrosis [26,27]. In the most severe form, disseminated necrotizing leukoencephalopathy, necrotic foci become confluent and a prominent axonopathy is noted ultrastructurally [28]. Neuroimaging reveals hypodensity on CT scans and increased T2/FLAIR signal in the white matter. Diffuse atrophy is often seen. MR spectroscopy reveals loss of N-acetyl aspartate (NAA), choline and creatine, implying axonal and membrane damage in the abnormal-appearing white matter [29].

Dementia

Associated with diffuse leukencephalopathy, with or without radionecrosis, is a devastating dementia syndrome (Figure 15.2) [30]. The dementia is typically a subcortical dementia characterized by deficits in memory, attention and intellectual function. Gait disturbance, urinary incontinence and personality changes may occur. Cortical functions such apraxis and language are relatively spared. Typical onset is within two years of radiation exposure and the course is usually progressive. A large meta-analysis of the literature found an incidence of post-radiation dementia to be 12% [31]. Risk factors are the same as for leukencephalopathy, including radiation dose, fractions size, volume of brain irradiated, older age and concomitant chemotherapy. Methylphenidate and anticholinesterases such as donepezil are sometimes used for symptomatic relief [32,33]. As the survival of brain tumor patients improves, an increasing number develop impairment of normal cerebrospinal fluid reabsorption through the arachnoid granulations.

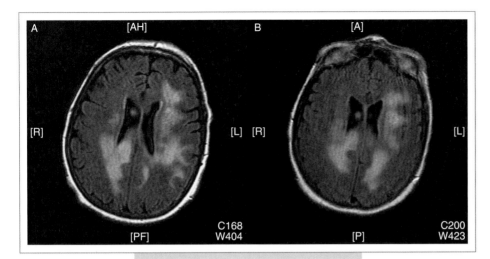

FIG. 15.2. A 70-year-old woman with CNS lymphoma treated with methotrexate and radiation therapy. Axial FLAIR MRI of the brain showing increased periventricular leucoencephalopathy.

This leads to a communicating hydrocephalus with cognitive impairment, gait unsteadiness and urinary symptoms. Some of these patients may benefit from placement of a ventriculo-peritoneal shunt [34].

Mild to Moderate Cognitive Impairment

Cognitive dysfunction, characterized by prominent dysfunction of short-term memory, is perhaps the most common sequela of radiotherapy. Cranial radiotherapy causes a debilitating cognitive decline in both children and adults [31,35–41]. Months to years after cranial radiation exposure, patients exhibit progressive deficits in short-term memory, spatial relations, visual motor processing, quantitative skills and attention [42]. Hippocampal dysfunction is a prominent feature of these neuropsychological sequelae. In fact, the severity of the cognitive deterioration appears to depend upon the radiation dosage delivered to the medial temporal lobes [43].

The incidence of treatment–induced impairment in cognition has been very well described in children. It is estimated that, when irradiated at age less than 7 years, nearly 100% of children require special education; after 7 years of age approximately 50% of children require special education. Some degree of memory dysfunction is thought to occur in the majority of children. The incidence of memory dysfunction in adult patients has been difficult to quantify, largely due to a lack of uniformity in neuropsychometric testing in the literature. However, as adults are surviving longer after treatment and the long-term consequences of radiation are becoming more important for this population, an extremely high rate of cognitive dysfunction of varying degrees has been recognized.

Mild to moderate cognitive dysfunction is inconsistently associated with radiological findings and frequently occurs in patients with normal-appearing neuroimaging [44]. Clinically significant memory deficit in the absence of radiological findings implicates damage to a subtle process with robust physiological consequences.

One such process is hippocampal neurogenesis. Studies in animal models have demonstrated that therapeutic doses of cranial irradiation virtually ablate neurogenesis and that this inhibition of neurogenesis correlates with impaired performance on hippocampal-dependent memory tests [21,45–48]. Surprisingly, irradiation does not simply deplete the stem cell population, but rather disrupts the microenvironment that normally supports hippocampal neurogenesis [46]. This microenvironmental perturbation is due largely to irradiation-induced microglial inflammation, and anti-inflammatory therapy with the non-steroidal anti-inflammatory agent indomethacin partially restores hippocampal neurogenesis and function [47]. Human trials are currently underway to evaluate the clinical utility of anti-inflammatory therapy during cranial radiotherapy.

Additional possible mechanisms underlying mild to moderate cognitive dysfunction include subtle white matter dysfunction and altered regional blood flow due to microvascular disease.

Radiotherapy-induced Tumors

Criteria for definition as a radiation-induced tumor include a long latency (years to decades) to occurrence of the second tumor and location of the tumor within the radiation portal. Although uncommon, secondary tumors do occur following cranial irradiation, particularly meningiomas, gliomas (Figure 15.3) and sarcomas. The relative proportion of secondary cranial tumors is roughly 70%, 20% and 10%, respectively. In an Israeli study of 10 834 patients exposed in childhood to an average dose of only 1.5 Gy for *Tinea capitis*, the relative risk of developing a neural tumor was found to be 6.9. In patients receiving 2.5 Gy, the relative risk was 20 [49]. After cranial irradiation for childhood leukemia, one series found a relative risk for secondary tumor of 22 [50].

More than 300 cases of radiation-induced meningiomas have been reported [51]. Authors frequently differentiate between low-dose (<10 Gy) irradiation and high-dose

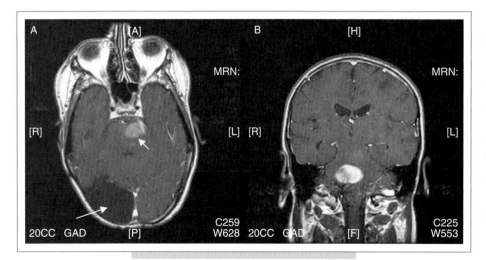

FIG. 15.3. A 25-year-old with a history of right parietal cerebral neuroblastoma at age 3 treated with surgical resection and radiotherapy who presented with a 2-month history of ataxia, headaches and drowsiness. **(A)** Axial and **(B)** coronal T1-weighted MRI with gadolinium shows enhancing brainstem lesion. The right parietal surgical cavity from his prior surgery for neuroblastoma is visible (longer arrow). Biopsy showed anaplastic astrocytoma, presumably induced by prior radiotherapy.

(>20 Gy) irradiation-induced meningiomas. The latency from time of treatment to meningioma development ranges from one to three decades [52,53]. There is an increased incidence of cellular atypia and aggressive subtypes following irradiation [53,54]. A cytogenetic study of radiation-induced meningiomas revealed consistent abnormalities involving chromosome 1p [55].

More than 100 case of secondary gliomas have been reported (see Figure 15.3); of these, approximately 40% are glioblastoma [2]. Gliomas arise after a mean latency of 9.6 years [56]. To date, four cases of gliomas at sites of previous radiosurgery have been also been reported [56,57]. Prognosis is often poor in secondary gliomas relative to spontaneous forms, due either to a more aggressive behavior or because treatment options are limited by previous exposures.

Rarely, sarcomas involving the skull base, calvaria or dura may occur. Osteosarcoma, fibrosarcoma and chondrosarcoma have been reported.

Vasculopathy

Large Vessel Atherosclerosis Radiation of head and neck cancers or lymphoma can cause accelerated atherosclerosis of the carotid arteries [58]. Intracranial accelerated atherosclerosis also occurs [59]. Treatment involves aggressive control of vascular risk factors such as hyperlipidemia and hypertension. Carotid endarterectomy may be pursued when indicated, but may be more complicated due to vascular fibrosis and post-radiation skin changes that may impair healing and increase risk of infection.

Moyamoya Pattern Progressive cerebral arterial occlusive disease, also known as moyamoya disease, is a stenosis or occlusion of large and intermediate cerebral arteries, abnormal netlike vessels and transdural anastomoses. Consequences include transient ischemic attacks (TIA), stroke and seizure. Cranial irradiation is a rare cause of

secondary moyamoya disease, particularly radiation for tumors of the optic chiasm, suprasellar region and brainstem during childhood [60,61].

Vascular Malformations Radiation-induced vascular malformations occur more frequently in children than in adults and carry a significant risk of hemorrhage. In one series, 20% of children who had undergone cranial irradiation exhibited radiological evidence of at least one new telangiectasia, defined as small low-signal intensity foci on T2 MR images [62]. In another series, 5 out of 20 patients with post-irradiation telangiectasias developed hematoma formation at the site of previously identified T2 shortening [63]. Post-mortem histopathology reveals thin-walled vessels surrounded by hemosiderin and gliosis [63]. Radiation-induced retinal telangeiectasias and microaneurysms have also been reported [64]. Cavernous angiomas also develop after cranial irradiation, particularly when radiation exposure occurs during childhood. The reported latency between irradiation and diagnosis ranges from 3 to 41 years [65,66]. Radiation-induced cavernomas may appear as contrast-enhancing masses and may mimic tumor radiographically [67]. These lesions are of great concern given their hemorrhagic potential.

Small Vessel Disease Radiation injury to the microvasculature has been extensively described. A mineralizing microangiopathy occurs more frequently in children than in adults, resulting in calcifications in the basal ganglia, subcortical white matter or dentate nuclei [68]. The etiology is radiation-induced intimal injury to small vessels, subsequent tissue ischemia and dystrophic calcification. Lacunar infarctions have also been reported to be a consequence of cranial irradiation. In one series of 421 children treated with cranial radiotherapy, 25 (6%) developed asymptomatic lacunes at a median latency of 2 years. The burden of lacunar disease increased with time from irradiation. The most significant risk factor for development of lacunar disease in this series was age less than 5 at the time of irradiation [69].

Neuroendocrine Dysfunction The hypothalamic–pituitary axis is exquisitely sensitive to irradiation [70]. Endocrine disorders can be the consequence of direct irradiation of an endocrine gland (e.g. 50% of patients developing hypothyroidism within 20 years following radiotherapy for Hodgkin's disease as a result of irradiation of the thyroid gland) or as a result of hypothalamic–pituitary dysfunction secondary to cranial irradiation [70–72]. In children, the most common endocrinopathy is growth hormone deficiency. Gonadotrophin deficiency and secondary and tertiary hypothyroidism occur less frequently. In adults, although growth hormone deficiency is common, it is rarely symptomatic. Approximately 67% of adult males experience sexual difficulties, usually decreased libido and impotence, within two years of radiotherapy. These problems are thought to result from gonadotrophin deficiency from hypothalamic damage. Hypothyroidism and hypoadrenalism occur less commonly and may require hormonal replacement. Hyperprolactinemia may also occur [73,74].

Spinal Cord

The spinal cord may be affected by radiotherapy for primary spinal tumors, epidural metastases, tumors of the head and neck or tumors near the spinal cord, such as Hodgkin's lymphoma.

Acute

Acute complications of the spinal cord are exceedingly rare and acute neurological worsening should lead to an evaluation for intratumoral hemorrhage or tumor progression [2].

Early-delayed Radiation Myelopathy

This complication occurs 6 weeks to 6 months after radiotherapy and generally presents as Lhermitte's sign, an unpleasant sensation of tingling, numbness and/or electrical-like feeling that travels down the spine and to the extremities following neck flexion. Symptoms typically resolve completely within 2–9 months after symptom onset [75,76]. There are no radiological findings. The etiology is thought to be a transient demyelination due to radiation-induced oligodendrocyte apoptosis; rebound of the oligodendrocyte precursor population several weeks after radiation exposure likely accounts for recovery [77,78].

Late-delayed Radiation Myelopathy

Late-delayed myelopathy takes two main forms – a progressive radiation myelopathy and a rare radiogenic lower motor neuron disease. In comparison to the early-delayed myelopathy, the late forms of myelopathy carry a poor prognosis.

Delayed Radiation Myelopathy (DRM)

This complication occurs 6 months to 10 years after radiation exposure. Risk factors include higher doses, larger fraction sizes, previous radiation exposure (therapeutic or incidental), old age, concurrent radiosensitizing chemotherapies and radiation sites in the lower thoracic and lumbar spine [75]. Clinical presentation may be acute or insidious in onset, progressing to paraparesis or quadriparesis with variable sensory disturbances, bowel and bladder dysfunction or diaphragmatic weakness in the case of high cervical lesions. Patients may present with a Brown-Sequard hemicord syndrome or a transverse myelitis. Pathology reveals findings similar to cerebral radionecrosis, with areas of focal necrosis, hyalinization and fibrinoid necrosis of the vasculature, demyelination and telangeictasias with focal hemorrhage. As in cerebral radiation injury, the etiology is thought to be either radiation-induced vascular and/or glial injury, but the pathogenesis is incompletely understood. MRI often shows cord edema and enhancement early in the course and atrophy late in the course, but imaging may also be normal at very early or intermediate stages [79]. Cyst formation has been reported [80]. Prognosis is poor. There is no effective treatment, although hyperbaric oxygen, anticoagulation and steroid therapies have been suggested [22,81,82]. To minimize risk of this terrible complication, radiation to the spinal cord is generally limited to 4500 cGy in 22–25 fractions.

Radiogenic Lower Motor Neuron Disease

A lower motor neuron syndrome is a rare complication of radiotherapy to the spinal cord or paraspinal tumors [83,84]. This complication predominately affects the lower extremities; in one series of 47 patients, only one had upper extremity weakness [85]. Patients tend to be adolescent to young adults and there is a male predominance [85]. Reported latency from irradiation to presentation ranges from 4 to 312 months [85]. There is no apparent dose dependence and authors have suggested that the etiology of radiogenic lower motor neuron disease may be multifactorial [85]. Based on electrophysiologic data and the absence of sensory findings, the anterior horn cell has been implicated as the target cell, however, damage to motor nerve roots is also a possibility [86].

Vascular Changes

Rarely, spinal cord hemorrhage may occur. This is likely to be related to formation of spinal telangiectesias or cavernous malformations [87].

Plexuses and Peripheral Nerve
Plexopathies

Irradiation of supraclavicular, infraclavicular or axillary lymph nodes can cause a brachial plexopathy, while irradiation of the pelvic and retroperitoneal lymph nodes can rarely result in a lumbosacral plexopathy. An early-delayed variant of brachial plexopathy presents several months after irradiation, occurring in 1–2% of patients irradiated for breast cancer [88]. Early-delayed brachial plexopathy is generally reversible. Late-delayed brachial plexopathy occurs years after irradiation, with a progressive course. Pathology of late-delayed plexopathy reveals perineural fibrosis and endoneural thickening, demyelination, axon loss and small vessel changes. It is sometimes difficult to distinguish a radiation-induced plexopathy from recurrent tumor. The presence of myokymia on electromyography is suggestive of radiation plexopathy. Neuroimaging studies such as MRI or PET may be helpful in differentiating tumor infiltration from radiation plexopathy.

Peripheral Nerve
Peripheral Neuropathy

The peripheral nerves are relatively radioresistant and radiation-induced injury is rare.

Nerve Sheath Tumors

Rarely, nerve sheath tumors can develop [89].

Cranial Neuropathies

The optic nerve (strictly speaking, part of the central nervous system) is the most frequently affected cranial nerve. Other cranial nerves affected by radiotherapy include the olfactory nerve (also part of the central nervous system), hypoglossal nerve and the spinal accessory nerve. Multiple cranial neuropathies are also possible. Involvement of the occulomotor, trochlear and abducens nerves are rare.

COMPLICATIONS OF CHEMOTHERAPY

Chemotherapy-induced Cognitive Impairment

Chemotherapy-induced cognitive impairment, known among cancer patients and oncologists as 'chemofog' or 'chemobrain', and characterized by deficits in memory function and concentration is an increasingly recognized complication [90–94]. A recent meta-analysis estimates that mild cognitive impairment occurs in 10–40% of breast cancer survivors [94]. In contrast to the cognitive dysfunction that follows cranial irradiation, chemotherapy-induced deficits in memory and concentration appear to be transient, but may resolve slowly over a number of years [94].

Additional complications of chemotherapy include acute encephalopathy, seizure, headaches, aseptic meningitis, acute cerebellar syndrome, vasculopathy and stroke, neuropathy, visual loss, myelopathy, reversible posterior leukoencephalopathy syndrome (RPLES) and dementia. The following section will focus on the drugs that frequently cause neurotoxicity. Neurologic complications of chemotherapy are reviewed in greater detail in several recent reviews (Table 15.2) [95].

Antimetabolites
Methotrexate

Methotrexate is a dihydrofolate reductase inhibitor used in the treatment of leukemia, lymphomas (including central nervous system lymphoma), choriocarcinoma, breast cancer and leptomeningeal metastases. The manifestations of methotrexate toxicity depend of the route of administration, dose and the use of other treatment modalities such as irradiation.

Intrathecal methotrexate produces aseptic meningitis in about 10% of patients. Symptoms of headache, nuchal rigidity, back pain, nausea, vomiting, fever and lethargy begin 2–4 hours after the drug is administered into the intrathecal space and persist for 12–72 hours. Transverse myelopathy is also possible, presenting as back or leg pain, followed by paraplegia, sensory loss and sphincter dysfunction. This can occur 30 minutes to up to 2 weeks later, but typically happens within 2 days. Variable recovery ensues after methotrexate-induced transverse myelopathy. Rarely, intrathecal methotrexate results in acute encephalopathy, subacute focal neurological deficits, neurogenic pulmonary edema and death. Accidental overdose (more than 500 mg) of intrathecal methotrexate usually results in death.

Low-dose methotrexate toxicity is associated with dizziness, headache and mild cognitive impairment. Symptoms resolve after discontinuation of methotrexate [96].

High-dose methotrexate toxicity can have acute, subacute and chronic manifestations. Acute toxicity produces somnolence, confusion and seizures within 24 hours of treatment. Subacutely, high-dose methotrexate can cause a subacute stroke-like syndrome, characterized by transient focal neurological deficits, confusion and sometimes seizures, that appears within 6 days of exposure and resolves completely within 72 hours. Chronically, methotrexate can cause leukoencephalopathy, either alone or in combination with radiotherapy (see Figure 15.2). The manifestations and consequences of leukoencephalopathy were discussed above.

5-Fluorouracil

5-Fluorouracil (5-FU) is a fluorinated pyrimidine that disrupts DNA synthesis and is used in treatment of many cancers, including colon and breast cancers. 5-FU crosses the blood–brain barrier and the highest concentration is found in the cerebellum, where it is toxic to Purkinje and granule cells [97]. An acute cerebellar syndrome occurs in approximately 5% of patients, characterized by acute onset of ataxia, dysmetria, dysarthria and nystagmus. Onset begins weeks to months after initiation of treatment and the drug should be discontinued when symptoms emerge. Complete recovery is the rule. Patients with decreased dihydropyrimidine dehydrogenase activity are at an increased risk for developing severe neurologic toxicity following 5-FU chemotherapy [98,99]. Other toxicities of 5-FU include encephalopathy, optic neuropathy, focal dystonia,

TABLE 15-2 NEUROLOGICAL COMPLICATIONS OF CHEMOTHERAPY, BIOLOGICAL RESPONSE MODIFIERS AND TARGETED MOLECULAR AGENTS

Acute encephalopathy
- Asparaginase
- 5-Azacytidine
- BCNU (IA or HD)
- Chlorambucil
- Cisplatin
- Corticosteroids
  - Cyclophosphamide
- Cytosine arabinoside (HD)
- Dacarbazine
- Doxorubicin
  - Etoposide (HD)
- Fludarabine
- 5-Fluorouracil (5-FU)
- Hexamethylmelamine
- Hydroxyurea
- Ibritumomab
  - Ifosfamide
  - Imatinib
  - Interferons
- Interleukins 1 and 2
- Mechloramine
- Methotrexate
- Misonidazole
- Mitomycin C
- Nelarabine
- Paclitaxel
- Pentostatin
- Procarbazine
- Tamoxifen
- Thalidomide
- Thiotepa (HD)
- Tumor necrosis factor
- Vinca alkaloids
- Zarnestra

Dementia
- BCNU (IA and HD)
- Corticosteroids
- Cytosine arabinoside
- Dacarbazine
- 5-FU + Levamisole
- Fludarabine
- Interferon-alpha
- Methotrexate

Cranial neuropathy
- BCNU (IA) (ototoxicity)
- Cisplatin (ototoxicity)
- Cytosine arabinoside
- Ifosfamide
- Methotrexate
- Nelarabine
- Vincristine (extraocular palsies)

Acute cerebellar syndrome
- Cytosine arabinoside
- 5-FU

- Hexamethylmelamine
- Ifosfamide
  - Interleukin-2
- Procarbazine
- Tamoxifen
- Thalidomide
- Vinca alkaloids
- Zarnestra

Leukoencephalopathy
- Capecitabine
- Cisplatin
- Cytarabine (IT)
- 5-FU with levamisole
- Cyclosporin-A
- Methotrexate (IT)
- Nelarabine
- Sunitinib malate

Headache
- Asparaginase
- Capecitabine
- Cetuximab
- Cisplatin
- Corticosteroids
- Cytosine arabinoside
- Danazol
- Estramustine
- Etoposide
- Fludarabine
- Gefitinib
- Hexamethylmelamine
- Interferons
- Interleukins 1, 2, 4
- Ibritumomab
- Levamisole
- Mechlorethamine
- Methotrexate (IT)
- Nelarabine
- Octreotide
- Oprelvekin
- Plicamycin
- Rituximab
- Retinoic acid
- SU5416
- Tamoxifen
- Temozolomide
- Thiotepa (IT)
- Topotecan
- Tositumomab
- Trastuzumab
- ZD1839

Aseptic meningitis
- Cytosine arabinoside (IT)
- Levamisole
- Methotrexate (IT)
- Thiotepa (IT)

Myelopathy
- Cisplatin
- Cladribine
- Corticosteroids
- Cytosine arabinoside
- Doxorubicin
  - DFMO
- Fludarabine
- Interferon alpha
- Methotrexate (IT)
- Mitoxantrone (IT)
- Docetaxel
- Thiotepa (IT)
- Vincristine (IT)

Vasculopathy and stroke
- Asparaginase
- Bevacizumab
- BCNU (IA)
- Bleomycin
- Carboplatin (IA)
- Cisplatin (IA)
- Doxorubicin
- Erlotinib
- Estramustine
- 5-FU
- Imatinib mesylate
- Methotrexate
- Nelarabine
- Tamoxifen

Seizures
- Amifostine
  - Asparaginase
- BCNU
- Busulphan (HD)
  - Chorambucil
- Cisplatin
- Corticosteroids
- Cytosine arabinoside
- Dacarbazine
- Etanercept
- Etoposide
  - 5-FU
- Ifosfamide
  - Interferon
  - Interleukin-2
- Letrozole
  - Levamisole
- Mechloramine
- Methotrexate
- Octreotide
- Paclitaxel
- Pentostatin
- Suramin
- Teniposide

- Thalidomide
  - Vinca alkaloids

Visual loss
- BCNU (IA)
- Cisplatin
- Etanercept
- Fludarabine
- Methotrexate (HD)
- Tamoxifen

Neuropathy
- 5-Azacytidine
- Bortezomib
- Capecitabine
- Carboplatin
- Cisplatin
- Cytosine arabinoside
- Docetaxel
- Etoposide
- 5-FU
- Gemcitabine
- Hexamethylmelamine
- Ifosphamide
- Interferon-alpha
- Misonidazole
- Nelarabine
- Oprelvekin
- Oxaliplatin
- Paclitaxel
- Pemetrexed
- Procarbazine
- Purine analogs (fludarabine, cladribine, pentostatin)
- Sorafenib
- Sunitinib malate
- Suramin
- Teniposide (VM-26)
- Thalidomide
- Tumor necrosis factor
- Vinca alkaloids
- Zarnestra

Syncope
- Bevacizumab
- Erlotinib
- Nelarabine

Adapted with permission from Medlink Neurology (Wen PY, Kesari S, Grier J. Neurologic complications of chemotherapy 2006)

Parkinsonian syndromes, eye movement abnormalities and cerebrovascular disorders [100–108]. The combination of 5-fluorouracil and levamisole used to treat colon cancer has been rarely associated with the development of a multifocal leucoencephalopathy [109].

Capecitabine, a pro-drug that is metabolized to 5-FU by the enzyme thymidine phosphorylase, is used to treat breast and gastrointestinal malignancies. Neurologic complications are uncommon, but some patients experience paresthesias, headaches and cerebellar symptoms. Several cases of neuropathies and multifocal leukoencephalopathy have been described [110,111].

Cytosine Arabinoside (cytarabine, ara-C)

Ara-C is a pyrimidine analogue that disrupts DNA synthesis and is used in the treatment of leukemias, lymphomas and leptomeningeal metastases. Conventional doses do not produce much neurotoxicity. However, at higher doses, an acute cerebellar syndrome may develop in 10–25% of patients, particularly those with renal impairment [112–114]. Widespread loss of Purkinje cells and cerebellar atrophy may ensue. Somnolence is another frequent symptom. Occasionally, high-dose Ara-C produces encephalopathy, seizures, reversible ocular toxicity, lateral rectus syndrome, bulbar and psuedobulbar palsy, Horner's syndrome, aseptic meningitis, anosmia and/or an extrapyramidal syndrome.

Alkylating Agents

Ifosfamide

Ifosfamide is an alkylating agent that causes cross-linking of DNA and is used in the treatment of many cancers, including a number of sarcomas and gynecological cancers. It produces encephalopathy in 20–30% of patients that presents hours to days after exposure. Clinical manifestations include confusion, cerebellar dysfunction, hallucinations, cerebellar dysfunction, seizures, cranial neuropathies, extrapyramidal signs and occasionally coma [115–118]. The etiology appears to be accumulation of chloracetaldehyde, a breakdown product of ifosfamide. The development of encephalopathy is a relative contraindication to future use of the drug. Methylene blue and benzodiazepines have been used to attenuate the symptoms of ifosfamide-induced encephalopathy [118–121]. The encephalopathy typically resolves without intervention, but rarely the deficits are irreversible.

Cisplatin

Cisplatin produces DNA cross-linking and is used to treat medulloblastoma, head and neck, ovarian, germ cell, cervical, lung and bladder cancers. Cisplatin produces a neuropathy that affects predominantly large myelinated sensory fibers at the level of the dorsal-root ganglion [122,123], causing proprioceptive loss, numbness and parasthesias [122,123]. Sural nerve biopsy shows demyelination and axonal loss. The neuropathy typically resolves with time after cessation of treatment. Cisplatin also commonly causes ototoxicity with high-frequency sensorineural hearing loss and tinnitus due to dose-related damage to the hair cells in the organ of Corti. Audiometric hearing loss is present in 74–88% of patients, with symptomatic hearing loss in 16–20% of patients. Irradiation increases the risk of hearing loss [124,125]. Lhermitte's sign occurs in 20–40% of patients receiving cisplatin. Rarely, cisplatin may cause encephalopathy, reversible posterior leukoencephalopathy, late vascular toxicity, taste disturbance and a myasthenic syndrome. Oxaliplatin is also associated with a high incidence of neuropathy, while neuropathies occur infrequently with carboplatin.

Taxanes (paclitaxel and docetaxel)

Taxanes inhibit microtubule function and are used in the treatment of many cancers including non-small cell lung, ovary and breast cancers. Paclitaxel (Taxol) produces a dose-limiting peripheral neuropathy in 60% of patients receiving $250 \, mg/m^2$. The neuropathy is predominantly sensory, affecting both small and large fibers and manifesting clinically as burning parasthesias of the hands and feet and loss of reflexes. It does not commonly progress. Motor and autonomic neuropathies have also been reported. Rarely, paclitaxel causes seizures and encephalopathy. Docetaxel causes neuropathy less frequently, but does occasionally cause Lhermittes's sign [126].

Vincristine

Vincristine is a vinca alkaloid that disrupts microtubules and is used to treat many cancers including leukemia, lymphoma, sarcomas and brain tumors. It causes an axonal neuropathy affecting both sensory and motor fibers in almost all patients. Small sensory fibers are particularly affected. Clinical manifestations include fingertip and foot parasthesias, muscle cramps, foot and wrist drop and sensory loss of varying degrees. Focal neuropathies and cranial neuropathies are also possible. In addition to the sensory and motor neuropathy, vincristine commonly causes an autonomic neuropathy, characterized by gastrointestinal, urinary and/or sexual dysfunction. Rarely, vincristine causes the syndrome of inappropriate anti-diuretic hormone (SIADH), resulting in hyponatremia leading to metabolic encephalopathy and seizures.

Vincristine should never be administered intrathecally and accidental administration of vincristine into the CSF produces a rapidly ascending myelopathy, coma and death [9,127].

Other vinca alkloids such as vinblastine and vinorlbine are associated with a lower incidence of neuropathies.

L-asparaginase

L-asparaginase is used in the treatement of acute lymphocytic leukemia and functions by catalyzing asparagine into aspartate and ammonia, thereby inhibiting protein synthesis. A hyperammoniemic encephalopathy can occur. The severity of the encephalopathy correlates with the amount of slow wave activity on the electroencephalogram, but not with the degree of hyperammoniemia [128,129].

L-asparaginase can cause a deficiency in anti-thrombin III, plasminogen and fibrinogen. This predisposes patients to thrombotic cerebrovascular disease, especially venous sinus thrombosis. Patients receiving L-asparaginase who present with confusion, obtundation, seizure or new focal neurological deficit should be evaluated for venous sinus thrombosis [95].

Anthracycline Antibiotics (daunorubicin, doxorubicin, mitoxantrone)

The anthracycline antibiotics exert their anti-neoplastic action by intercalating between DNA bases, causing uncoiling of the double helix structure of DNA. Side effects of the anthracylines on the nervous system are indirect. The anthracyclines can cause a cardiomyopathy and, subsequently, reduced ventricular contractility can promote intraventricular thrombus formation. These thrombi can cause cardioembolic strokes or transient ischemic attacks [130].

Bevacizumab (Avastin)

Avastin and other inhibitors of vascular endothelial growth factor (VEGF) and the VEGF receptor (VEGFR) are associated with reversible posterior leucoencephalopathy syndrome (RPLS). RPLS has been reported with various chemotherapeutic and immunosuppressive agents and typically presents as cortical blindness, headache and confusion. Hypertension may be present and seizures may occur. Imaging reveals non-enhancing subcortical leukoencephalopathy in a distal vascular distribution. Discontinuing the causative agent usually results in complete resolution of symptoms and signs. Recently, two cases of RPLS were reported with Bevacizumab use [131,132].

Glucocorticoids

Corticosteroids such as prednisone and dexamethasone are used for a number of reasons in oncological therapy, including cytolytic effect on neoplastic lymphocytes and reduction of peritumoral edema in patients with brain

TABLE 15-3 NEUROLOGIC COMPLICATIONS OF CORTICOSTEROIDS
Common
• Myopathy
• Behavioral changes
• Visual blurring
• Tremor
• Insomnia
• Reduced taste and olfaction
• Cerebral atrophy
Uncommon
• Psychosis
• Hallucinations
• Hiccups
• Dementia
• Seizures
• Dependence
• Epidural lipomatosis

tumors. Corticosteroids have a number of systemic and neurological side effects (Table 15.3). Common neurological side effects include steroid myopathy and alterations in mood. Steroid psychosis, steroid-induced dementia and cortical atrophy also occur. Corticosteroids may play an important role in cognitive dysfunction during and after cancer therapy. In animal models, corticosteroids are known to impair the physiology of the developing brain, including hippocampal neurogenesis. In humans, prednisone therapy has been demonstrated to impair verbal memory function and children treated for acute lymphoblastic leukemia exhibited more severe long-term cognitive dysfunction if their regimen included dexamethasone [133,134].

SUMMARY

Neurologic complications of oncologic therapies are occurring with increasing frequency in cancer patients. This is a result of the availability of a growing number of treatments associated with neurotoxicities and prolonged patient survival. Increased understanding of the underlying mechanisms for these neurologic complications and methods to prevent them will be an important challenge for the future and will hopefully lead to reduced neurologic morbidity in cancer patients.

ACKNOWLEDGEMENTS

We gratefully acknowledge the support of the James Haggerty Fund.

REFERENCES

1. Sheline GE, Wara WM, Smith V (1980). Therapeutic irradiation and brain injury. Int J Radiat Oncol Biol Phys 6:1215–1228.

2. Behin A, Delattre JY (2004). Complications of radiation therapy on the brain and spinal cord. Semin Neurol 24:405–417.

3. Cross NE, Glantz MJ (2003). Neurologic complications of radiation therapy. Neurol Clin 21:249–277.

4. Kondziolka D, Niranjan A, Flickinger JC, Lunsford LD (2005). Radiosurgery with or without whole-brain radiotherapy for brain metastases: the patients' perspective regarding complications. Am J Clin Oncol 28:173–179.

5. Wen PY, Schiff D, Kesari S, Drappatz J, Gigas DC, Doherty L (2006). Medical management of patients with brain tumors. J Neuro-Oncol 80(3):313–326.

6. Garwicz S, Aronson S, Elmqvist D, Landberg T (1975). Postirradiation syndrome and eeg findings in children with acute lymphoblastic leukaemia. Acta Paediatr Scand 64:399–403.

7. Littman P et al (1984). The somnolence syndrome in leukemic children following reduced daily dose fractions of cranial radiation. Int J Radiat Oncol Biol Phys 10:1851–1853.

8. Ch'ien LT et al (1980). Long-term neurological implications of somnolence syndrome in children with acute lymphocytic leukemia. Ann Neurol 8: 273–277.

9. Berg RA, Ch'ien LT, Lancaster W, Williams S, Cummins J (1983). Neuropsychological sequelae of postradiation somnolence syndrome. J Dev Behav Pediatr 4:103–107.

10. Trautman PD et al (1988). Prediction of intellectual deficits in children with acute lymphoblastic leukemia. J Dev Behav Pediatr 9:122–128.

11. Hoffman WF, Levin VA, Wilson CB (1979). Evaluation of malignant glioma patients during the postirradiation period. J Neurosurg 50:624–628.

12. Armstrong CL et al (2000). Radiotherapeutic effects on brain function: double dissociation of memory systems. Neuropsychiatr Neuropsychol Behav Neurol 13:101–111.

13. Vigliani MC, Sichez N, Poisson M, Delattre JY (1996). A prospective study of cognitive functions following covential radiotherapy for supratentorial gliomas in young adults: 4-year results. Int J Radiat Oncol Biol Phys 35: 527–533.

14. Creange A et al (1994). Subacute leukoencephalopathy of the rhombencephalon after pituitary radiotherapy. Rev Neurol (Paris) 150:704–708.

15. Lampert P, Tom MI, Rider WD (1959). Disseminated demyelination of the brain following Co60 (gamma) radiation. Arch Pathol 68:322–330.

16. Oppenheimer JH et al (1992). Radionecrosis secondary to interstitial brachytherapy: correlation of magnetic resonance imaging and histopathology. Neurosurgery 31:336–343.

17. Hoshi M, Hayashi T, Kagami H, Murase I, Nakatsukasa M (2003). Late bilateral temporal lobe necrosis after conventional radiotherapy. Neurol Med Chir (Tokyo) 43:213–216.

18. Gabayan AJ et al (2006). Glia site brachytherapy for treatment of recurrent malignant gliomas: a retrospective multi-institutional analysis. Neurosurgery 58:701–709.

19. Kumar AJ et al (2000). Malignant gliomas: MR imaging spectrum of radiation therapy- and chemotherapy-induced necrosis of the brain after treatment. Radiology 217:377–384.

20. Gomez-Rio M. et al (2004). (201)Tl-SPECT in low-grade gliomas: diagnostic accuracy in differential diagnosis between tumour recurrence and radionecrosis. Eur J Nucl Med Mol Imaging 31:1237–1243.

21. Monje ML, Palmer T (2003). Radiation injury and neurogenesis. Curr Opin Neurol 16:129–134.

22. Glantz MJ et al (1994). Treatment of radiation-induced nervous system injury with heparin and warfarin. Neurology 44:2020–2027.

23. Leber KA, Eder HG, Kovac H, Anegg U, Pendl G (1998). Treatment of cerebral radionecrosis by hyperbaric oxygen therapy. Stereotact Funct Neurosurg 70 Suppl 1:229–236.

24. Chan AS, Cheung MC, Law SC, Chan JH (2004). Phase II study of alpha-tocopherol in improving the cognitive function of patients with temporal lobe radionecrosis. Cancer 100:398–404.

25. Wassenberg MW, Bromberg JE, Witkamp TD, Terhaard CH, Taphoorn MJ (2001). White matter lesions and encephalopathy in patients treated for primary central nervous system lymphoma. J Neuro-Oncol 52:73–80.

26. Price RA, Jamieson PA (1975). The central nervous system in childhood leukemia. II. Subacute leukoencephalopathy. Cancer 35:306–318.

27. Wang AM, Skias DD, Rumbaugh CL, Schoene WC, Zamani A (1983). Central nervous system changes after radiation therapy and/or chemotherapy: correlation of CT and autopsy findings. Am J Neuroradiol 4:466–471.

28. Perry A, Schmidt RE (2006). Cancer therapy-associated CNS neuropathology: an update and review of the literature. Acta Neuropathol (Berl) 111:197–212.

29. Virta A et al (2000). Spectroscopic imaging of radiation-induced effects in the white matter of glioma patients. Magn Reson Imaging 18:851–857.

30. DeAngelis LM, Delattre JY, Posner JB (1989). Radiation-induced dementia in patients cured of brain metastases. Neurology 39:789–796.

31. Crossen JR, Garwood D, Glatstein E, Neuwelt EA (1994). Neurobehavioral sequelae of cranial irradiation in adults: a review of radiation-induced encephalopathy. J Clin Oncol 12:627–642.

32. Meyers CA, Weitzner MA, Valentine AD, Levin VA (1998). Methylphenidate therapy improves cognition, mood, and function of brain tumor patients. J Clin Oncol 16: 2522–2527.

33. Shaw EG et al (2006). Phase II study of donepezil in irradiated brain tumor patients: effect on cognitive function, mood, and quality of life. J Clin Oncol 24:1415–1420.

34. Thiessen B, DeAngelis LM (1998). Hydrocephalus in radiation leukoencephalopathy: results of ventriculoperitoneal shunting. Arch Neurol 55:705–710.

35. Roman DD, Sperduto PW (1995). Neuropsychological effects of cranial radiation: current knowledge and future directions. Int J Radiat Onco Biol Phys 31:983–998.

36. Anderson VA, Godber T, Smibert E, Weiskop S, Ekert H (2000). Cognitive and academic outcome following cranial irradiation and chemotherapy in children: a longitudinal study. Br J Cancer 82:255–262.

37. Moore BD III, Copeland DR, Ried H, Levy B (1992). Neurophysiological basis of cognitive deficits in long-term survivors of childhood cancer. Arch Neurol 49:809–817.

38. Abayomi OK (1996). Pathogenesis of irradiation-induced cognitive dysfunction. Acta Oncol 35:659–663.

39. Lee PW, Hung BK, Woo EK, Tai PT, Choi DT (1989). Effects of radiation therapy on neuropsychological functioning in patients with nasopharyngeal carcinoma. J Neurol Neurosurg Psychiatr 52:488–492.

40. Surma-aho O et al (2001). Adverse long-term effects of brain radiotherapy in adult low-grade glioma patients. Neurology 56:1285–1290.

41. Kramer JH et al (1997). Neuropsychological sequelae of medulloblastoma in adults. Int J Radiat Oncol Biol Phys 38:21–26.

42. Strother DR (2002). Tumors of the central nervous system. In *Principles and Practice of Pediatric Oncology*, 4th edn. Pizzo PA, Poplack DG (eds). Lippincott, Williams and Wilkins, Philadelphia. 751–824.

43. Abayomi OK (2002). Pathogenesis of cognitive decline following therapeutic irradiation for head and neck tumors. Acta Oncol 41:346–351.

44. Dropcho EJ (1991). Central nervous system injury by therapeutic irradiation. Neurol Clin 9:969–988.

45. Parent JM, Tada E, Fike JR, Lowenstein DH (1999). Inhibition of dentate granule cell neurogenesis with brain irradiation does not prevent seizure-induced mossy fiber synaptic reorganization in the rat. J Neurosci 19:4508–4519.

46. Monje ML, Mizumatsu S, Fike JR, Palmer TD (2002). Irradiation induces neural precursor-cell dysfunction. Nat Med 8:955–962.

47. Monje ML, Toda H, Palmer TD (2003). Inflammatory blockade restores adult hippocampal neurogenesis. Science 302:1760–1765.

48. Raber J et al (2004). Radiation-induced cognitive impairments are associated with changes in indicators of hippocampal neurogenesis. Radiat Res 162:39–47.

49. Ron E et al (1988). Tumors of the brain and nervous system after radiotherapy in childhood. N Engl J Med 319:1033–1039.

50. Neglia JP et al (1991). Second neoplasms after acute lymphoblastic leukemia in childhood. N Engl J Med 325:1330–1336.

51. Amirjamshidi A, Abbassioun K (2000). Radiation-induced tumors of the central nervous system occurring in childhood and adolescence. Four unusual lesions in three patients and a review of the literature. Childs Nerv Syst 16:390–397.

52. Nishio S et al (1998). Radiation-induced brain tumours: potential late complications of radiation therapy for brain tumours. Acta Neurochir (Wien) 140:763–770.

53. Strojan P, Popovic M, Jereb B (2000). Secondary intracranial meningiomas after high-dose cranial irradiation: report of five cases and review of the literature. Int J Radiat Oncol Biol Phys 48:65–73.

54. Regel JP et al (2006). Malignant meningioma as a second malignancy after therapy for acute lymphatic leukemia without cranial radiation. Childs Nerv Syst 22:172–175.

55. Zattara-Cannoni H et al (2001). Cytogenetic study of six cases of radiation-induced meningiomas. Cancer Genet Cytogenet 126:81–84.

56. Salvati M et al (2003). Radiation-induced gliomas: report of 10 cases and review of the literature. Surg Neurol 60:60–67.

57. McIver JI, Pollock BE (2004). Radiation-induced tumor after stereotactic radiosurgery and whole brain radiotherapy: case report and literature review. J Neuro-Oncol 66:301–305.

58. Murros KE, Toole JF (1989). The effect of radiation on carotid arteries. A review article. Arch Neurol 46:449–455.

59. Laplane D, Carydakis C, Baulac M, Elhadi D, Chiras J (1986). Intracranial artery stenoses 44 years after craniofacial radiotherapy. Rev Neurol (Paris) 142:65–67.

60. Kondoh T et al (2003). Moyamoya syndrome after prophylactic cranial irradiation for acute lymphocytic leukemia. Pediatr Neurosurg 39:264–269.

61. Bitzer M, Topka H (1995). Progressive cerebral occlusive disease after radiation therapy. Stroke 26:131–136.

62. Koike S et al (2004). Asymptomatic radiation-induced telangiectasia in children after cranial irradiation: frequency, latency, and dose relation. Radiology 230:93–99.

63. Gaensler EH et al (1994). Radiation-induced telangiectasia in the brain simulates cryptic vascular malformations at MR imaging. Radiology 193:629–636.

64. Bagan SM, Hollenhorst RW (1979). Radiation retinopathy after irradiation of intracranial lesions. Am J Ophthalmol 88:694–697.

65. Jain R et al (2005). Radiation-induced cavernomas of the brain. Am J Neuroradiol 26:1158–1162.

66. Duhem R, Vinchon M, Leblond P, Soto-Ares G, Dhellemmes P (2005). Cavernous malformations after cerebral irradiation during childhood: report of nine cases. Childs Nerv Syst 21:922–925.

67. Olivero WC, Deshmukh P, Gujrati M (2000). Radiation-induced cavernous angioma mimicking metastatic disease. Br J Neurosurg 14:575–578.

68. Shanley DJ (1995). Mineralizing microangiopathy: CT and MRI. Neuroradiology 37:331–333.

69. Fouladi M et al (2000). Silent lacunar lesions detected by magnetic resonance imaging of children with brain tumors: a late sequela of therapy. J Clin Oncol 18:824–831.

70. Constine LS et al (1993). Hypothalamic-pituitary dysfunction after radiation for brain tumors. N Engl J Med 328:87–94.

71. Illes A et al (2003). Hypothyroidism and thyroiditis after therapy for Hodgkin's disease. Acta Haematol 109:11–17.

72. Samaan NA et al (1987). Endocrine complications after radiotherapy for tumors of the head and neck. J Lab Clin Med 109:364–372.

73. Washburn LC, Carlton JE, Hayes RL (1974). Distribution of WR-2721 in normal and malignant tissues of mice and rats bearing solid tumors: dependence on tumor type, drug dose and species. Radiat Res 59:475–483.

74. Toogood AA (2004). Endocrine consequences of brain irradiation. Growth Horm IGF Res 14 Suppl A:S118-S124.

75. Rampling R, Symonds P (1998). Radiation myelopathy. Curr Opin Neurol 11:627–632.

76. Lewanski CR, Sinclair JA, Stewart JS (2000). Lhermitte's sign following head and neck radiotherapy. Clin Oncol (R Coll Radiol) 12:98–103.

77. Li YQ, Guo YP, Jay V, Stewart PA, Wong CS (1996). Time course of radiation-induced apoptosis in the adult rat spinal cord. Radiother Oncol 39:35–42.

78. Atkinson SL, Li YQ, Wong CS (2005). Apoptosis and proliferation of oligodendrocyte progenitor cells in the irradiated rodent spinal cord. Int J Radiat Oncol Biol Phys 62:535–544.

79. Wang PY, Shen WC, Jan JS (1992). MR imaging in radiation myelopathy. Am J Neuroradiol 13:1049–1055.

80. Shindo K, Nitta K, Amino A, Nagasaka T, Shiozawa Z (1995). A case of chronic progressive radiation myelopathy with cavity formation in the thoracic spinal cord. Rinsho Shinkeigaku 35:1012–1015.

81. Udaka F, Tsuji T, Shigematsu K, Kawanishi T, Kameyama M (1990). A case of chronic progressive radiation myelopathy

successfully treated with corticosteroid. Rinsho Shinkeigaku 30:439–443.

82. Feldmeier JJ, Lange JD, Cox SD, Chou LJ, Ciaravino V (1993). Hyperbaric oxygen as prophylaxis or treatment for radiation myelitis. Undersea Hyperb Med 20:249–255.

83. Lagueny A et al (1985). Post-radiotherapy anterior horn cell syndrome. Rev Neurol (Paris) 141:222–227.

84. De CP et al (1986). Isolated lower motoneuron involvement following radiotherapy. J Neurol Neurosurg Psychiatr 49:718–719.

85. Esik O, Vonoczky K, Lengyel Z, Safrany G, Tron L (2004). Characteristics of radiogenic lower motor neurone disease, a possible link with a preceding viral infection. Spinal Cord 42:99–105.

86. Bowen J, Gregory R, Squier M, Donaghy M (1996). The post-irradiation lower motor neuron syndrome neuronopathy or radiculopathy? Brain 119:1429–1439.

87. Jabbour P, Gault J, Murk SE, Awad IA. (2004). Multiple spinal cavernous malformations with atypical phenotype after prior irradiation: case report. Neurosurgery 55:1431.

88. Salner AL et al (1981). Reversible brachial plexopathy following primary radiation therapy for breast cancer. Cancer Treat Rep 65:797–802.

89. Foley KM, Woodruff JM, Ellis FT, Posner JB (1980). Radiation-induced malignant and atypical peripheral nerve sheath tumors. Ann Neurol 7:311–318.

90. Cull A et al (1996). What do cancer patients mean when they complain of concentration and memory problems? Br J Cancer 74:1674–1679.

91. Ganz PA (1998). Cognitive dysfunction following adjuvant treatment of breast cancer: a new dose-limiting toxic effect? J Natl Cancer Inst 90:182–183.

92. Brezden CB, Phillips KA, Abdolell M, Bunston T, Tannock IF (2000). Cognitive function in breast cancer patients receiving adjuvant chemotherapy. J Clin Oncol 18:2695–2701.

93. Jansen CE, Miaskowski C, Dodd M, Dowling G, Kramer J (2005). A metaanalysis of studies of the effects of cancer chemotherapy on various domains of cognitive function. Cancer 104:2222–2233.

94. Matsuda T et al (2005). Mild cognitive impairment after adjuvant chemotherapy in breast cancer patients – evaluation of appropriate research design and methodology to measure symptoms. Breast Cancer 12:279–287.

95. Plotkin SR, Wen PY (2003). Neurologic complications of cancer therapy. Neurol Clin 21:279–318.

96. Wernick R, Smith DL (1989). Central nervous system toxicity associated with weekly low-dose methotrexate treatment. Arthritis Rheum 32:770–775.

97. Riehl JL, Brown WJ (1964). Acute cerebellar syndrome secondary to 5-fluorouracil therapy. Neurology 14:961–967.

98. Takimoto CH et al (1996). Severe neurotoxicity following 5-fluorouracil-based chemotherapy in a patient with dihydropyrimidine dehydrogenase deficiency. Clin Cancer Res 2:477–481.

99. Johnson MR et al (1999). Life-threatening toxicity in a dihydropyrimidine dehydrogenase-deficient patient after treatment with topical 5-fluorouracil. Clin Cancer Res 5:2006–2011.

100. Greenwald ES (1976). Letter: organic mental changes with fluorouracil therapy. J Am Med Assoc 235:248–249.

101. Lynch HT, Droszcz CP, Albano WA, Lynch JF (1981). 'Organic brain syndrome' secondary to 5-fluorouracil toxicity. Dis Colon Rectum 24:130–131.

102. Liaw CC et al (1999). Risk of transient hyperammonemic encephalopathy in cancer patients who received continuous infusion of 5-fluorouracil with the complication of dehydration and infection. Anticancer Drugs 10:275–281.

103. Adams JW, Bofenkamp TM, Kobrin J, Wirtschafter JD, Zeese JA (1984). Recurrent acute toxic optic neuropathy secondary to 5-FU. Cancer Treat Rep 68:565–566.

104. Bixenman WW, Nicholls JV, Warwick OH (1977). Oculomotor disturbances associated with 5-fluorouracil chemotherapy. Am J Ophthalmol 83:789–793.

105. Gradishar W, Vokes E, Schilsky R, Weichselbaum R, Panje W (1991). Vascular events in patients receiving high-dose infusional 5-fluorouracil-based chemotherapy: the University of Chicago experience. Med Pediatr Oncol 19:8–15.

106. Brashear A, Siemers E (1997). Focal dystonia after chemotherapy: a case series. J Neuro-Oncol 34:163–167.

107. Pirzada NA, Ali II, Dafer RM (2000). Fluorouracil-induced neurotoxicity. Ann Pharmacother 34:35–38.

108. Cho HJ et al (2003). Conjugated linoleic acid inhibits cell proliferation and ErbB3 signaling in HT-29 human colon cell line. Am J Physiol Gastrointest Liver Physiol 284:G996–1005.

109. Luppi G et al (1996). Multifocal leukoencephalopathy associated with 5-fluorouracil and levamisole adjuvant therapy for colon cancer. A report of two cases and review of the literature. The INTACC. Intergruppo Nazionale Terpia Adiuvante Colon Carcinoma. Ann Oncol 7:412–415.

110. Niemann B, Rochlitz C, Herrmann R, Pless M (2004). Toxic encephalopathy induced by capecitabine. Oncology 66:331–335.

111. Saif MW, Wood TE, McGee PJ, Diasio RB (2004). Peripheral neuropathy associated with capecitabine. Anticancer Drugs 15:767–771.

112. Winkelman MD, Hines JD (1983). Cerebellar degeneration caused by high-dose cytosine arabinoside: a clinicopathological study. Ann Neurol 14:520–527.

113. Hwang TL, Yung WK, Estey EH, Fields WS (1985). Central nervous system toxicity with high-dose Ara-C. Neurology 35:1475–1479.

114. Herzig RH et al (1983). High-dose cytosine arabinoside therapy for refractory leukemia. Blood 62:361–369.

115. Meanwell CA, Kelly KA, Blackledge G (1986). Avoiding ifosfamide/mesna encephalopathy. Lancet 2:406.

116. Zalupski M, Baker LH (1988). Ifosfamide. J Natl Cancer Inst 80:556–566.

117. Pratt CB, Goren MP, Meyer WH, Singh B, Dodge RK (1990). Ifosfamide neurotoxicity is related to previous cisplatin treatment for pediatric solid tumors. J Clin Oncol 8:1399–1401.

118. Watkin SW, Husband DJ, Green JA, Warenius HM (1989). Ifosfamide encephalopathy: a reappraisal. Eur J Cancer Clin Oncol 25:1303–1310.

119. Aeschlimann C, Cerny T, Kupfer A (1996). Inhibition of (mono)amine oxidase activity and prevention of ifosfamide encephalopathy by methylene blue. Drug Metab Dispos 24:1336–1339.

120. Pelgrims J et al (2000). Methylene blue in the treatment and prevention of ifosfamide-induced encephalopathy: report of 12 cases and a review of the literature. Br J Cancer 82:291–294.

121. Simonian NA, Gilliam FG, Chiappa KH (1993). Ifosfamide causes a diazepam-sensitive encephalopathy. Neurology 43:2700–2702.

122. Roelofs RI, Hrushesky W, Rogin J, Rosenberg L (1984). Peripheral sensory neuropathy and cisplatin chemotherapy. Neurology 34:934–938.

123. Thompson SW, Davis LE, Kornfeld M, Hilgers RD, Standefer JC (1984). Cisplatin neuropathy. Clinical, electrophysiologic, morphologic, and toxicologic studies. Cancer 54:1269–1275.

124. Moroso MJ, Blair RL (1983). A review of cis-platinum ototoxicity. J Otolaryngol 12:365–369.

125. Schell MJ et al (1989). Hearing loss in children and young adults receiving cisplatin with or without prior cranial irradiation. J Clin Oncol 7:754–760.

126. Lee JJ, Swain SM (2006). Peripheral neuropathy induced by microtubule-stabilizing agents. J Clin Oncol 24:1633–1642.

127. Gaidys WG, Dickerman JD, Walters CL, Young PC (1983). Intrathecal vincristine. Report of a fatal case despite CNS washout. Cancer 52:799–801.

128. Leonard JV, Kay JD (1986). Acute encephalopathy and hyperammonaemia complicating treatment of acute lymphoblastic leukaemia with asparaginase. Lancet 1:162–163.

129. Moure JM, Whitecar JP Jr, Bodey GP (1970). Electroencephalogram changes secondary to asparaginase. Arch Neurol 23:365–368.

130. Schachter S, Freeman R (1982). Transient ischemic attack and adriamycin cardiomyopathy. Neurology 32:1380–1381.

131. Glusker P, Recht L, Lane B (2006). Reversible posterior leukoencephalopathy syndrome and bevacizumab. N Engl J Med 354:980–982.

132. Ozcan C, Wong SJ, Hari P (2006). Reversible posterior leukoencephalopathy syndrome and bevacizumab. N Engl J Med 354:980–982.

133. Hajek T, Kopecek M, Preiss M, Alda M, Hoschl C (2006). Prospective study of hippocampal volume and function in human subjects treated with corticosteroids. Eur Psychiatr 21:123–128.

134. Waber DP et al (2000). Cognitive sequelae in children treated for acute lymphoblastic leukemia with dexamethasone or prednisone. J Pediatr Hematol Oncol 22:206–213.

Section II

Physics and Basic Science of Neuroimaging

CHAPTER 16

Physics of CT: Scanning

Boris R. Keil and Johannes T. Heverhagen

HISTORICAL DEVELOPMENT

Translate/rotate geometry

The first head CT scanner was introduced in 1972 by Sir Godfrey Hounsfield and called EMI Mark 1. The principal mathematical basis behind this technology was already published in 1917 by J. Radon. The first CT scanner produced images with an 80-by-80 matrix of 3 mm pixel resolution and required approximately five minutes of scan time per pair of slices [1]. In this type of system, a pencil beam with only two x-ray detectors was used. The x-ray tube and the detector translate linearly across the patient at a particular angle. The system then rotates at an angle of one degree and translates back across the patient. During a total examination time of five minutes the raw data of one image pair were acquired with a total gantry rotation of 180 degrees. This system was the sole CT scanner that used parallel beam geometry [2]. The pencil beam geometry provides very efficient scatter reduction, but it was an inefficient utilization of x-ray source. In the second generation of CT scanners, the pencil beam expanded to a small, fan-shaped beam and the detector increased to a linear array of 30 detectors. The translate/rotate configuration was still used, but this new development reduced the acquisition time of 18 seconds per slice and made more efficient use of the x-ray beam.

Rotate/rotate geometry

The third generation scanners opened up the x-ray fan beam to include the entire patient in the fan. This geometry involves an expanded detector array with more than 800 detector elements. This type of scanner eliminated the time-consuming translation movement and allowed a fast image acquisition, approximately 2 seconds per slice. The geometry of third generation scanners is especially sensitive to detector mis-calibration, because the detectors are permanently irradiated. Because they are constantly moving and vibrating, their calibration could drift significantly.

Rotate/stationary geometry

In the next type of CT generation, only the x-ray source was rotating around the patient and the x-rays were detected using a stationary 360 degree ring of detectors. This allows very fast image acquisition, as short as half a second. These fourth-generation CT scanners require many detectors. Modern scanners use about 4800 detector elements for the whole ring. The fixed detector avoid calibration problems found in the third generation, because the beam always irradiates a newly calibrated detector.

Stationary/stationary geometry

In the next step, a so-called cine-CT scanner or electron beam scanner dedicated for cardiac tomography has been developed. It seems like a huge x-ray tube, which the patient enters. A large anode arc of tungsten encircles the patient semicircularly. A moving electron beam sweeps across the anode and generates the x-rays. By using multiple target layers in anode material, it is possible to produce up to eight fan beams simultaneously. Consequently, electron beam CT scanners are able to acquire eight slices within a very short time of 50 milliseconds. The production of such a CT scanner is double the cost of a conventional CT scanner. Today, only 120 of these scanners exist worldwide.

Helical CT

In the early 1990s, the newly developed slip-ring technology for x-ray tube and detector connections replaced the old CT scanner equipment, which worked with spooled cable technology. Now the gantry was able to rotate continuously around the patient during the entire examination. Implementation of the helical CT, or commonly called spiral CT, revolutionized CT development, as well as clinical practice [3]. During scanning, data acquisition is combined with continuous movement of the patient through the gantry. The path of the x-ray beam around the patient can be described as a spiral. Whereas conventional CT required a step-and-shoot acquisition procedure, helical

CT uses much faster and non-stop technology to acquire multiple transverse slices [4]. These advances have revolutionized CT scanning in clinical routine. Entire organs such as liver or lung can be scanned in a single breath-hold [5]. This approach enables imaging of continuous sections without gaps as a result of changes in respiration levels between acquisitions of individual slices. Such gaps could lead to masking of lesions hidden in those gaps. Due to the better resolution, it is also possible to detect smaller lesions. Furthermore, application of intravenous contrast agents can be better tailored to the acquisition protocol and thinner slices can be obtained. New applications and expansion of existing techniques have improved clinical diagnosis. The data obtained from spiral CT are truly three-dimensional and allow multiplanar image reconstruction as a consequence of the lack of motion mis-registration and the increased through plane resolution [6]. The ability to acquire volume data also paved the way for the development of three-dimensional image processing techniques.

The helical trajectory of the spiral data acquisition does not allow the direct reconstruction of a cross-section of the patient. Before reconstruction, the raw data from helical data set must be interpolated to the reconstruction plane of interest [7,8]. Helical scan protocols have introduced a new parameter, called pitch. The pitch is determined by table speed (in mm/s) divided by beam collimation (in mm) multiplied by the gantry rotation period:

$$Pitch = \frac{table\ movement\ per\ 360\text{-}degree\ gantry\ rotation}{collimator\ width}$$

In other words, the pitch factor is a ratio without units which provides information about table movement relative to beam collimation. A pitch of less than 1 means that the data are over-sampled, which may result in a better image quality but higher radiation dose for the patient. Therefore, pitch factors greater than 1 and up to 1.5 are commonly used in clinical routine. However, the key limitation is that slice thickness is still determined by the collimation of the x-ray beam. Opening up the collimator increases the slice thickness which improves x-ray beam utilization, but reduces the through plane resolution.

Multiple detector array CT

Multislice CT represents another major shift in CT technology. By using multiple detector arrays, the collimator spacing of the x-ray beam is wider and therefore more x-rays are used in producing image data. With the introduction of this approach, the slice thickness is determined by the detector size and not by the collimator [9]. This allows the acquisition of several images simultaneously [10]. Modern scanners are available with up to 128 detector rows (current prototypes use 256 detector rows). The main benefit is an improved scan speed and the availability of isotropic

images. In multislice CT, the beam is split into several slices and therefore the definition of the pitch factor is different than in single detector spiral CT scanners [11]. For example, a four detector array multislice scanner is acquiring 2.5 mm-thick slices at a tabletop speed of 15 mm per one-second gantry rotation. This results in a pitch of 6 (15 m/2.5 m) which corresponds to a single collimator pitch of 1.5.

$$Pitch = \frac{table\ movement\ per\ 360\text{-}degree\ gantry\ rotation}{collimator\ width \times number\ of\ detectors}$$

Dual source CT

Dual source CT scanner is the newest step of development in CT technology. This type of system uses two x-ray sources with two corresponding detector arrays which are mounted onto a rotating gantry with an offset of 90 degrees. The x-ray tubes can operate with two different energies simultaneously [12]. Alternatively, it is possible to acquire image data in half the time, unlike conventional technology. For example, a dual-source multidetector row scanner can acquire an entire cardiac study within a single 10-second breath hold.

IMAGE RECONSTRUCTION

Each registered beam is just a projection of attenuation characteristics of the irradiated tissue. The aim of the reconstruction algorithm is to estimate how the tissue absorptions are distributed along the x-ray path. This goal is not achievable using a single projection profile. Instead, it needs a large number of projections for many different angles. The acquired projection profile can be displayed as a sinogram (Figure 16.1). Sinograms are not used for clinical routine, but they are relevant for understanding tomographic principles. The horizontal axis of the sinogram represents the different projection profiles. The vertical axis in the sinograms corresponds to each angle of projections. Objects closer to the center of the field of view produce small sinusoid amplitudes in the sinogram, objects closer to the edge heighten sinusoidal amplitudes.

Reconstructing an image from the projection profiles is a classical inverse problem. The early attempts at CT reconstruction used an iterative approach called algebraic reconstruction algorithm (ART) [13]. This algorithm starts with an assumed image, computes projections from the image, compares the original projection data and updates the image based on errors between projections that would be obtained from the current pixels, values and actual projection. This method was very time consuming and computationally intensive. A faster CT reconstruction approach has been developed called filtered back-projection [14]. It became widely accepted because reconstruction of the images is completed as soon as the CT examination is finished.

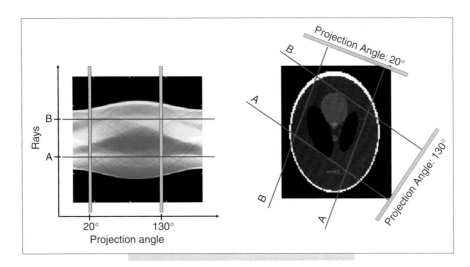

FIG. 16.1. The sinogram contains the raw data of a CT acquisition. The horizontal axis corresponds to each angle of acquired projection profiles. The vertical axis corresponds to the rays in each projection. Here, there are two rays (A and B) and two different projection profiles (20° and 130°) illustrated.

Figure 16.2 shows a simplified 3-by-3 matrix array containing nine pixels, each pixel having specific values of absorption coefficients. In such a way, a high attenuation pixel is in the center and lighter attenuation pixels are in the periphery. Three projection profiles pass through the array vertically, horizontally and diagonally. By using the direct back-projection procedure alone, the record measurement reading for each detector element is distributed evenly along the path of the x-ray. Subsequently, each pixel is set to the average of all angularity of beams that traverse that pixel. Now the measurement reading is back-projected and averaged into each pixel. The reconstructed image resembles the original, but is modified by a blurring map.

By virtue of the extensive blurring, the direct back-projection yielded no suitable outcomes. In practice, before back-transformation is carried out, each projection profile is filtered. The mathematical filtering step involves convolving the projection data with a convolution kernel. The filtered back-projection is easiest to implement in the frequency domain using a special Fourier transformation called the Fourier slice theorem. The Fourier slice theorem states that the one-dimensional Fourier transform of the projection profile is equal to the two-dimensional Fourier transform of the image evaluated on the radial line that the projection was taken on [15]. Once the two-dimensional Fourier transform space is filled from the individual one-dimensional Fourier transforms of the projection profiles, the image can be generated by applying an inverse two-dimensional Fourier transform to this space. Before the inverse transform is done, the user can apply different convolution kernels as filter procedure to create an appropriate character of the diagnostic image. The most common kernels are derived by a so-called Ram-Lak filter multiplied with a low pass filter [16]. In filter procedures, the shape of the kernel

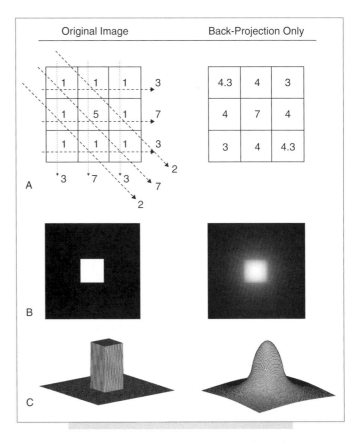

FIG. 16.2. Image reconstruction using a simple back-projection. Back-projection alone produces a blurred image. **(A)** The left 3-by-3 pixel array represents the original grid with a pixel of high attenuation in the center and pixels of weaker attenuation in the periphery. The right array shows the reconstructed grid by a simple back-projection. **(B)** Represents the corresponding images and **(C)** illustrates the same smearing effect by a relief figure.

would be impressed to each projection profile, so avoiding the blurring effects in reconstructed images (Figure 16.3).

After CT reconstruction but before storing and displaying, CT images are normalized and truncated to integer values. These values are called Hounsfield units or CT numbers and represent quantitative data with specific values for each tissue quality [17]. In general, they are related to the attenuation coefficient of water, because most tissues have linear attenuation coefficients very similar to that of water over a large photon-energy interval. This is the reason for introducing the dimensionless Hounsfield unit (HU):

$$HU = \frac{\mu_{tissue} - \mu_{water}}{\mu_{water}} \cdot 1000$$

Where μ is the linear attenuation coefficient of the material in the irradiated pixel and μ_w represents the attenuation coefficient of water. Each Hounsfield unit is equivalent to 0.1% of the attenuation of water [18]. These normalized results in Hounsfield units range typically from -1000 to $+3000$, where -1000 corresponds to air, soft tissues range from -100 to $+100$, water is 0, and dense bone and areas which are filled with contrast agent range up to $+3000$. Hounsfield units are quantitative and provide important information leading to more accurate diagnosis in many clinical settings. For example, the identification of calcified or non-calcified pulmonary nodules can be determined from CT images based on mean Hounsfield units of the nodule. CT scanners can measure bone density with good accuracy. Furthermore, the quantitative image pixels are used for radiotherapy dose planning. The accuracy of dose calibrations based on such CT data is partly determined by the precision of the calibration of CT Hounsfield units to relative electron density.

CT images typically possess 12 bits of gray scale for a total of 4096 shades of gray. This corresponds to the range of Hounsfield units -1000 to 3095. However, human eyes cannot resolve that many shades of gray in medical images. If the full value range (-1000 to $+3095$) were displayed, the particular tissues of interest would be limited to a relatively small range of Hounsfield units. Consequently, the visualization and medical diagnoses would be limited. The most common way to adjust the intensity values of the anatomic tissues of interest to the dynamic range of human eyes is applying a look-up table. A look-up table lists the relationship between stored Hounsfield units and their corresponding gray scale values of the output image. Contrast can be enhanced by assigning a narrow interval of Hounsfield units to the entire gray scale on the display monitor. This postprocessing procedure is called windowing and leveling. The window width determines the contrast of the image. A narrower window results in greater contrasts (Figure 16.4). The level is the Hounsfield unit at the center of the window and selects the structures in the image displayed on the gray scale, i.e. from black to white. For example, a head CT image

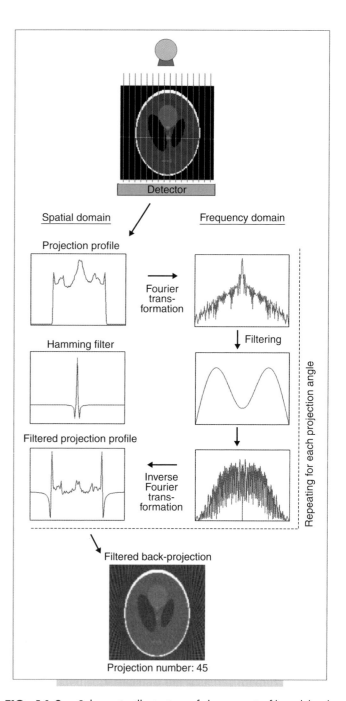

FIG. 16.3. Schematic illustration of the steps in filtered backprojection. Each single acquired projection profile will be converted to the frequency domain using Fourier-transformation. In the frequency domain, the mathematical filter operation (convolution) is easy to apply using multiplication with a frequency domain filter kernel. The resulting product will be reconverted to the spatial domain by applying the inverse Fourier transform and is ready to be back projected. These steps would be repeated for each angle. The illustration shows a reconstruction result of an original object using 45 projections.

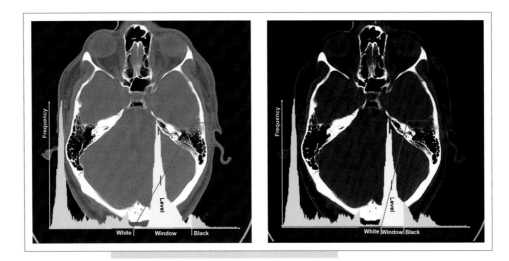

FIG. 16.4. The images show the effect of changes in window width. The *level* is the Hounsfield unit which corresponds to the center of the *window*. Only the Hounsfield units inside the window are displayed in the image. Structures outside the window width are displayed either completely black or white. The left image shows a window of 1400 HU by a level of 250 HU. The right image shows the same level but with a narrower window of 700 HU. This narrow window generates a very high contrast image.

viewed at a center of +50 and a window of 200 Hounsfield units is optimized for gray and white matter contrast. The center of +50 corresponds to the approximate Hounsfield unit of gray and white matter and it is these tissues that will be displayed with the mid-gray illumination. A window of 200 HU implies a range from −50 to +150. Consequently, all tissues having Hounsfield units of 150 or above will be displayed as white, while all tissues of −50 and below will be displayed as black. It is routine to display CT images using several different window and level settings for each image in order to enhance the depiction of various tissues, such as lung, soft tissues and bones.

REFERENCES

1. Beckmann EC (2006). CT scanning the early days. Br J Radiol 79:5–8.
2. Robb RA (1982). X-ray computed tomography: from basic principles to applications. Annu Rev Biophys Bioeng 11:177–201.
3. Kalender WA, Polacin A (1991). Physical performance characteristics of spiral CT scanning. Med Phys 18:910–915.
4. Kalender WA (1995). Thin-section three-dimensional spiral CT: is isotropic imaging possible? Radiology 197:578–580.
5. Kalender WA, Seissler W, Klotz E, Vock P (1990). Spiral volumetric CT with single-breath-hold technique, continuous transport, and continuous scanner rotation. Radiology 176:181–183.
6. Hu H (1999). Multi-slice helical CT: scan and reconstruction. Med Phys 26:5–18.
7. Flohr TG, Schaller S, Stierstorfer K, Bruder H, Ohnesorge BM, Schoepf UJ (2005). Multi-detector row CT systems and image-reconstruction techniques. Radiology 235:756–773.
8. Taguchi K, Aradate H (1998). Algorithm for image reconstruction in multi-slice helical CT. Med Phys 25:550–561.
9. Kohl G (2005). The evolution and state-of-the-art principles of multislice computed tomography. Proc Am Thorac Soc 2:470–500.
10. Flohr T, Stierstorfer K, Bruder H, Simon J, Schaller S (2002). New technical developments in multislice CT – Part 1: approaching isotropic resolution with sub-millimeter 16-slice scanning. Rofo 174:839–845.
11. Lewis MA (2001). Multislice CT: opportunities and challenges. Br J Radiol 74:779–781.
12. Flohr TG, McCollough CH, Bruder H et al (2006). First performance evaluation of a dual-source CT (DSCT) system. Eur Radiol 16:256–268.
13. Brooks RA, Di CG (1975). Theory of image reconstruction in computed tomography. Radiology 117:561–572.
14. Stark H, Woods J, Paul I, Hingorani R (1981). Direct Fourier reconstruction in computer tomography. Acoustics, speech, and signal processing, IEEE 29:237–245.
15. Edelheit LS, Herman GT, Lakshminarayanan AV (1977). Reconstruction of objects from diverging x-rays. Med Phys 4:226–231.
16. Ramachandran GN, Lakshminarayanan AV (1971). Three-dimensional reconstruction from radiographs and electron micrographs: application of convolutions instead of Fourier transforms. Proc Natl Acad Sci USA 68:2236–2240.
17. Kalender WA (2005). *Computer Tomography – Fundamentals, System Technology, Image Quality, Applications*, 2nd edn. Publicis Corporate Publishing.
18. Brooks RA, Di Chiro G (1976). Principles of computer assisted tomography (CAT) in radiographic and radioisotropic imaging. Phys Med Biol 21:689–732.

CHAPTER 17

Physics of CT: Contrast Agents

Jens H. Figiel and Johannes T. Heverhagen

HISTORY

While x-rays revolutionized modern medicine, many soft tissue structures in the human body remain invisible on x-ray images, including computed tomography (CT). Therefore, the value of contrast agents was recognized early on in the development of diagnostic x-ray applications. Consequently, development of x-ray contrast agents has made significant progress over the past century, in order to accommodate the diagnostic requirements and to design safely applicable products. Even though new generations of CT scanners as well as modalities with better soft tissue contrast, e.g. magnetic resonance imaging and ultrasound, improved the capabilities of native scans, contrast agents are employed routinely in CT. Their functions are to depict morphology by creating or increasing the contrast between different anatomic structures and to visualize function, e.g. perfusion, integrity of the blood–brain barrier or capacity of elimination processes (kidneys or liver).

GENERAL PRINCIPLE OF X-RAY CONTRAST AGENTS

Contrast in x-ray images is determined by the absorption of x-rays by the irradiated tissue depending on atomic number, concentration and volume of the absorbing material. In some regions of the body, different tissues provide enough inherent contrast, e.g. chest, but in other regions where the properties of the present organs are similar, e.g. abdomen, virtually no intrinsic contrast is present. Therefore, the introduction of materials that reduce (gases, negative contrast agents) or increase (iodine, barium, positive contrast agents) absorption and as a result enhance contrast is necessary.

In the 1950s, iodinated contrast agents based on triiodobenzene were established and continued to dominate the field until today. This is based on the physico-chemical properties of iodine (high density, firm binding to benzene and low toxicity) and the availability of positions 1, 3 and 5 for the introduction of side chains in order to modify the biological and chemical properties of the complex (Figure 17.1) [2–5].

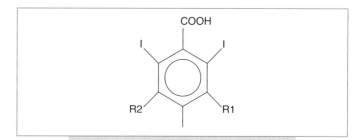

FIG. 17.1. Triiodianted contrast agent. Aromate = Parent substance; I = iodine, contrast enhancement; COOH = water solubility; R1, R2 = reduction of toxicity.

Soon, it became obvious that the side effects of these contrast agents were attributable to the high osmolality of the agents, and substances with lower osmolality (non-ionic contrast agents) were formulated [6]. Besides the lower osmolality, the non-ionic contrast agents have two distinct advantages over ionic ones:

1 The incidence of general reactions and of allergic reactions that can be life-threatening is markedly reduced [7–10]. The improved tolerance of non-ionic versus ionic substances is rooted in several physico-chemical properties such as the absence of any electrical charges or cations and significantly better shielding by hydrophilic side chains.
2 Neural tolerance improved due to the blood-isotonic character of the substances. Therefore, non-ionic contrast agents have replaced ionic ones virtually completely.

CURRENT APPLICATIONS OF CONTRAST AGENTS IN CT

Initially, it was thought that the high soft tissue contrast of CT would make the administration of contrast agents unnecessary. However, nowadays the use of contrast agents in CT is routine and utilized in a wide variety of applications. These applications range from mechanical filling of cavity structures in CT myelography (intrathecal administration) over purely anatomical depiction of perfusion defects in stroke (intravenous application) to functional assessment of the blood–brain barrier in the diagnosis of brain tumors (intravenous application).

If injected intrathecally the contrast agent mixes with the CSF, fills and opacifies the luminal area of spinal canal. Thereby, it unmasks luminal changes, such as stenoses or cavities [11]. Intravenous application can reveal perfusion defects caused by occlusion of a vessel and can depict vascular anatomy. The CT scan then not only shows the occluded vessel, but also the affected area of the brain by lack of contrast enhancement in that area. Intravenous application also demonstrates bleeding due to rupture of a vessel or an aneurysm. Here, contrast agent leaks out of the ruptured vessel, into the surrounding tissue and reveals the extent of hemorrhage in the brain. It also indicates intraluminal filling defects in sinus thrombosis and can enhance the entire vasculature in CT angiography. Leaking of the contrast through the blood–brain barrier into the parenchyma of the brain represents functional disruption of the blood–brain barrier by a tumor with immature, leaky vasculature [12].

Modern multislice spiral CT scanners allow the acquisition of serial, time-resolved images of the same slice, even of the same volume, after injection of the contrast agent (dynamic CT). These time series represent the first pass of contrast agent through the tissue allowing the density patterns in the observed volume to be followed over time. The individual time-density curve (arrival and wash out of the contrast agent) permits conclusions about the functional status and distribution of the contrast agent. In addition, pharmacologically induced functional changes can be observed [13].

SIDE EFFECTS AND TOXICITY

X-ray contrast agents are usually injected in high volumes and at high rates. Therefore, important prerequisites of contrast agents are low toxicity and safe application. A major improvement in patient safety has been the step from ionic to non-ionic contrast agents. For ionic agents, the incidence of side effects is high, depending on the patient's condition, the type of examination, the contrast agent, its dose and the circumstances the examination has been performed under, elective versus emergency. Of special interest are severe or fatal incidents. The numbers for these incidents vary greatly between 1 out of every 116 000 patients to 1 out of every 10 000 patients. Non-ionic contrast agents are being better tolerated in various ways (Table 17.1). It has been shown that the incidences of general reactions has been reduced using non-ionic contrast agents [7,14]. The frequency of severe reactions has also been reduced. However, up to now no conclusion can be drawn about the frequency of fatal incidents but, since life-threatening events are reduced by using non-ionic contrast agents, it is very likely that they also reduce the number of fatalities.

TABLE 17-1 Tolerance of ionic and non-ionic contrast agents in comparison

Reactions, side effects	Ionic contrast agents	Non-ionic contrast agents
General reactions	–	+
Osmolality-dependent effects	–	+
General renal tolerance (IV application)	0	0
Renal angiography (IA application)	–	+
Cardiodepression (due to calcium binding)	–	+
Neural tolerance	–	+

–: worse than; +: better than; 0: no difference

Nevertheless, it has to be stressed that only the frequency of side effects is reduced by non-ionic contrast agents and that the same kind of side effects do occur. Therefore, if administering contrast agents, even non-ionic ones, one has to be prepared to treat reactions. With the introduction of non-ionic contrast agents, delayed reactions were described, noticed hours to days following the administration of the agent. These reactions include rash, parotitis, headache and nausea. No differences between ionic and non-ionic contrast agents could be demonstrated for delayed reactions [15]. However, for non-ionic contrast agents, it has been demonstrated that delayed reactions were twice as common as early reactions occurring within the first 30 minutes after administration [14]. This should lead the attending physician to draw the patient's attention to possible delayed reactions, even though these reactions were usually mild in intensity.

While side effects cannot be generally attributed to one singular mechanism, they can be classified into two main reactions:

1 General and dose-independent anaphylactic reactions: these effects do not correlate with the osmolality or the amount of the injected contrast agent. Even small diluted or isotonic amounts of contrast agent can lead to general reactions. Several mechanisms have been discussed as a trigger for these incidents, including effects on the blood coagulation or the vascular endothelia, an effect on the central nervous system or a cross-reaction with antibodies against immunogenic substances. The reactions range from mild (urticaria, dizziness) to severe (cardiac arrest) reactions. The mortality with ionic contrast agents is reported to range from one in every 10 000 to 100 000 patients.

While these reactions can occur with ionic and non-ionic agents, they occur less frequently with non-ionic agents. Often, in patients who had repeated reactions to ionic contrast agents, non-ionic ones were tolerated without any symptoms [7,8,10,16]. In patients at risk, with a prior history of anaphylactic or allergic reactions or cardiopulmonary patients, the use of prophylactic medications reduces the frequency and severity of anaphylaxis. However, it does not rule out anaphylactic reactions and, moreover, it does not alleviate other reactions or side effects. Possible medications are the combination of H_1 and H_2 blockers or oral administration of 32 mg methylprednisolone twice (6 hours and 2 hours) before the injection. Patients without any risk factors do not benefit from prophylaxis [17,18].

2 Dose-dependent side effects: these effects can be related to the osmolality and pharmacological effects of the contrast agent. Examples for these effects are pain, cardiovascular effects, renal damage or a sensation of heat. Non-ionic contrast agents have replaced ionic ones to a considerable degree due to their lower osmolality. They have distinct advantages in pain intensive applications and angiography.

Certain conditions demand special attention and special measures before an iodinated contrast agent is applied. In addition, certain precautions have to be taken and arrangements have to be made in case a reaction occurs. In the following, some scenarios are described.

Pregnant and breast-feeding patients: x-ray contrast agents have not proven to be safe in pregnant patients. However, since exposure to x-rays should be avoided during pregnancy (and alternative investigations, such as magnetic resonance imaging (MRI) or ultrasound (US), are preferred), the application of x-ray contrast agents during pregnancy does not pose a real imminent problem. Contrast agents do not (or only minimally) enter the breast milk and therefore do not pose a threat to the infant.

Iodine-induced hyperthyroidism: in pathologically altered thyroid glands, the injection of iodine (diagnostic or therapeutic) can have serious metabolic side effects as severe as thyrotoxic crisis. This applies especially for patients with struma or hyperthyroidism and, therefore, hyperthyroidism has to be excluded prior to contrast agent injection. The risk of hyperthyroidism is exclusively determined by the injection of iodine. As a result, the risk of hyperthyroidism is not alleviated by the use of non-ionic contrast agents. It is the same for ionic and non-ionic ones. Iodine-induced hyperthyroidism occurs weeks or months after the iodine administration and patients have to be alerted to watch for symptoms.

In case of a mandatory application of iodinated contrast agent in a patient with hyperthyroidism, a double prophylactic medication should be administered. It consists of perchlorate (3×300 mg daily) for 2 days prior and 1 week after and of thiamizole (2×20 mg daily) for 2 days prior and 3 weeks after contrast agent application.

Renal damage: intravenously administered contrast agents in CT are eliminated by the kidneys. In patients without further risk factors, the possibility of impairment of renal function is not imminent [19,20]. Deterioration is defined as an increase in serum creatinine of at least 1 mg/dl. In patients with risk factors in addition to the application of contrast agents, such as renal insufficiency, insulin-dependent diabetes mellitus, dehydration, cardiac insufficiency or age > 70 years, impairment of the kidneys cannot be excluded. Therefore, alternative procedures such as MRI and US should be chosen. If the application of iodinated contrast agents cannot be avoided, various prophylactic measures are recommended. They include appropriate hydration, discontinuation of drugs which can compromise renal function, avoiding multiple examinations with contrast agent application and administration of acetylcysteine.

Metformin-induced lactic acidosis: care should be taken with the medication given to the patient. The administration of iodinated contrast medium in addition to oral antihyperglycemic agents containing metformin may put the patient at an additional risk of lactic acidosis. Even though a rare condition, it is a very severe one with a mortality of up to 50% [21]. The incidence is reported to be 9 per 100 000 patients per year of metformin intake, many of whom can be attributed to contraindications like congestive heart failure, renal failure, advanced age and states with tissue hypoxemia.

Impaired renal function is thought to lead to an accumulation of metformin, predisposing to a lactic acidosis. Only particular cases are reported which show an association of metformin with the administration of iodinated contrast-medium [22]. Most of these patients had renal insufficiency. These results have led to guidelines, which propose a regimen solely dependent on renal function. When a normal renal function is found, metformin should be stopped at the time of contrast medium administration and should be resumed at the earliest 48 h later given a normal renal function by monitoring serum creatinine. In renal dysfunction, metformin therapy should be withdrawn 48 h before and after the administration of contrast medium [23,24].

For guidelines for therapy and prophylaxis of adverse reactions to contrast agents, please also refer to Bush and Swanson [25]. Even though such reactions are rare, the administering physician always has to be prepared to react to an emergency and treat possible side effects.

REFERENCES

1. Speck U (1999). *Contrast Media – Overview, Use and Pharmaceutical Aspects*, 4th edn. Springer, Berlin.
2. Yamamoto Y, Satoh T, Sakurai M, Asari S, Sadamoto K (1982). Minimum dose contrast bolus in computed angio-tomography of the brain. J Comput Assist Tomogr 6: 575–585.
3. Alfidi RJ, Laval-Jeantet M (1976). AG 60.99: a promising contrast agent for computed tomography of the liver and spleen. Radiology 121:491.
4. Gerzof SG, Robbins AH, Pugatch RD, Gerson ES (1977). New applications of old radiographic techniques applied to computed tomography. Comput Tomogr 1:331–338.

5. Huang HK, Chamberlin K, Schellinger D, Raptopoulos V, Garnic JD (1977). A subtraction technique comparing pre- and post-contrast medium enhancement CT scans. Comput Tomogr 1:267–271.

6. Almen T (1969). Contrast agent design. Some aspects on the synthesis of water soluble contrast agents of low osmolality. J Theor Biol 24:216–226.

7. Katayama H, Yamaguchi K, Kozuka T, Takashima T, Seez P, Matsuura K (1990). Adverse reactions to ionic and nonionic contrast media. A report from the Japanese Committee on the Safety of Contrast Media. Radiology 175:621–628.

8. Rapoport S, Bookstein JJ, Higgins CB, Carey PH, Sovak M, Lasser EC (1982). Experience with metrizamide in patients with previous severe anaphylactoid reactions to ionic contrast agents. Radiology 143:321–325.

9. Zukiwski AA, David CL, Coan J, Wallace S, Gutterman JU, Mavligit GM (1990). Increased incidence of hypersensitivity to iodine-containing radiographic contrast media after interleukin-2 administration. Cancer 65:1521–1524.

10. Wolf GL, Arenson RL, Cross AP (1989). A prospective trial of ionic vs nonionic contrast agents in routine clinical practice: comparison of adverse effects. Am J Roentgenol 152:939–944.

11. Saifuddin A (2000). The imaging of lumbar spinal stenosis. Clin Radiol 55:581–594.

12. Haider MA, Milosevic M, Fyles A et al (2005). Assessment of the tumor microenvironment in cervix cancer using dynamic contrast enhanced CT, interstitial fluid pressure and oxygen measurements. Int J Radiat Oncol Biol Phys 62:1100–1107.

13. Rydberg J, Buckwalter KA, Caldemeyer KS et al (2000). Multisection CT: scanning techniques and clinical applications. Radiographics 20:1787–1806.

14. Christiansen C (2005). X-ray contrast media – an overview. Toxicology 209:185–187.

15. McCullough M, Davies P, Richardson R (1989). A large trial of intravenous Conray 325 and Niopam 300 to assess immediate and delayed reactions. Br J Radiol 62:260–265.

16. Holtas S (1984). Iohexol in patients with previous adverse reactions to contrast media. Invest Radiol 19:563–565.

17. Lasser EC, Berry CC (1991). Adverse reactions to contrast media. Ionic and nonionic media and steroids. Invest Radiol 26:402–403.

18. Reimann HJ, Tauber R, Kramann B, Gmeinwieser J, Schmidt U, Reiser M (1986). Premedication with H1- and H2-receptor antagonists for intravenous urography using contrast media. Rofo 144:169–173.

19. Cramer BC, Parfrey PS, Hutchinson TA et al (1985). Renal function following infusion of radiologic contrast material. A prospective controlled study. Arch Intern Med 145:87–89.

20. Heller CA, Knapp J, Halliday J, O'Connell D, Heller RF (1991). Failure to demonstrate contrast nephrotoxicity. Med J Aust 155:329–332.

21. Stang M, Wysowski DK, Butler-Jones D (1999). Incidence of lactic acidosis in metformin users. Diabetes Care 22:925–927.

22. Sirtori CR, Pasik C (1994). Re-evaluation of a biguanide, metformin: mechanism of action and tolerability. Pharmacol Res 30:187–228.

23. Jones GC, Macklin JP, Alexander WD (2003). Contraindications to the use of metformin. Br Med J 326:4–5.

24. Thomsen HS, Morcos SK (2006). ESUR guidelines on contrast media. Abdom Imaging 31:131–140.

25. Bush WH, Swanson DP (1991). Acute reactions to intravascular contrast media: types, risk factors, recognition, and specific treatment. Am J Roentgenol 157:1153–1161.

CHAPTER 18

Introductory MRI Physics

Aaron Sodickson

EXTERNAL MAGNETIC FIELD, PROTONS AND EQUILIBRIUM MAGNETIZATION

Much of the bulk of the magnetic resonance imaging (MRI) scanner apparatus is dedicated to producing an extremely strong magnetic field, denoted as B_0. Common present day whole-body clinical systems operate at field strengths of 1.5 Tesla (30000 times the strength of the earth's magnetic field), although both lower and higher field strength systems are in use.

The great majority of clinical MRI applications image protons, present in vast numbers throughout the body in all molecules containing hydrogen nuclei. These protons have a physical characteristic, the nuclear spin, which imparts magnetic properties to the nucleus and which is vital to MR image formation.

When placed into the strong B_0 magnetic field, a small excess of the nuclear spins align along rather than opposed to the field direction and their microscopic magnetizations sum into a net equilibrium magnetization vector denoted M_0 (Figure 18.1). The strength of M_0 (and subsequently the maximum available signal strength) scales with the size of B_0, providing the primary motivation for the development and use of higher B_0 field strength systems.

LARMOR PRECESSION AND MRI SIGNAL

Once M_0 is perturbed away from its equilibrium z-axis, it behaves as a classical bar magnet and precesses about B_0 at a characteristic frequency, the Larmor frequency $\omega_0 = \gamma B_0$. This precession frequency scales with the strength of the external magnetic field B_0, with a proportionality constant, the gyromagnetic ratio γ, that is an intrinsic property of the nucleus being interrogated and is the same for all protons throughout the body. At typical clinical MRI field strengths, the proton Larmor frequency is in the megaHertz (MHz) or radiofrequency range.

When surrounded by a loop of wire, 'the coil', the time-varying magnetic field created by the precessing magnetization vector induces a current, 'the signal', which forms the basis for image formation in MRI (Figure 18.2).

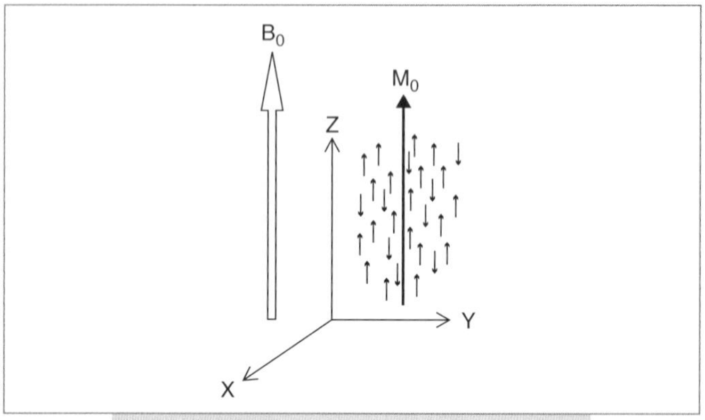

FIG. 18.1. The direction of the external static magnetic field B_0 defines the z-axis, or 'longitudinal' axis of the MRI apparatus. The perpendicular x-y plane is often denoted the 'transverse' plane. A vector sum of the microscopic nuclear spin magnetizations yields the net equilibrium magnetization vector M_0 oriented along the z-axis.

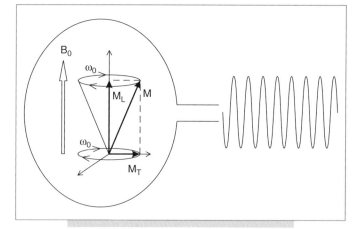

FIG. 18.2. Precessing magnetization vector M, separated into its longitudinal component M_L which remains fixed along the z-axis and its transverse component M_T which precesses about the z-axis in the x-y plane. The time varying magnetic field produced by the precessing magnetization induces an oscillating current in the coil surrounding the magnetization vector.

RADIOFREQUENCY ENERGY AND RESONANCE

In order to precess and induce a signal, M_0 must first be tilted away from its equilibrium 'longitudinal' z-axis orientation and into the 'transverse' x-y plane. In fact, the magnetization vector may best be thought of in terms of its longitudinal component M_L which remains fixed along the z-axis and its transverse component M_T which precesses about B_0 in the x-y plane (see Figure 18.2). MRI relies on the use of radiofrequency or 'RF' magnetic fields, denoted B_1, to convert longitudinal magnetization into transverse magnetization, or equivalently, to tilt the magnetization vector into the transverse plane.

The weak B_1 magnetic fields are aligned in the transverse plane, perpendicular to the z-axis. By its nature, M_0 will precess about any externally applied magnetic field, so this B_1 field will tilt the magnetization away from its equilibrium z-axis. However, the strength of the applied B_1 field is so tiny in comparison to B_0 that if its direction remained fixed, it would have a negligible net effect due to the rapid precession of the magnetization vector about the z-axis. MRI works in this setting by rotating the axis of the B_1 field about the z-axis exactly in concert with the precessing magnetization vector. When the direction of the applied B_1 field is altered in step with the Larmor precession of M_0, the effect of the weak B_1 field can accumulate over time and cause a net tilting of M_0 towards the transverse plane (Figure 18.3). This precise matching of the two frequencies is the 'resonance' criterion from which MRI borrows its name and it sets the B_1 frequency in the megaHertz radiofrequency range.

Resonant radiofrequency energy is typically applied as 'RF-pulses', whose combination of amplitude and duration produces a predictable net tilting of the magnetization vector away from the z-axis by a prescribed angle. A 90º pulse takes M_0 from the z-axis fully into the transverse plane, while a 180º pulse produces an excursion from +z to the −z axis (Figure 18.3).

RELAXATION

T1 and T2 Relaxation

After the net magnetization vector has been perturbed away from equilibrium, it will eventually return to its equilibrium state, with restoration of M_0 along the +z-axis. This occurs through simultaneous recovery of the longitudinal magnetization component M_L along the z-axis via T1 relaxation, and disappearance of the transverse magnetization component M_T via T2 relaxation. The two relaxation rates, denoted T1 and T2, are intrinsic features of the underlying tissue (related to tissue structure and microscopic proton motion) and vary with tissue type.

Figure 18.4 demonstrates the relaxation curves of the longitudinal and transverse magnetization components following a 90º RF pulse. The 90º pulse converts all of the longitudinal magnetization into transverse magnetization, leaving no residual M_L, and a maximal value of M_T. M_L then re-grows from zero to its equilibrium value M_0 along +z, recovering most rapidly for fat and most slowly for CSF. M_T decays away to zero at the T2 relaxation rate, with loss of M_T (and thus signal strength) occurring most rapidly for fat and most slowly for CSF.

T2* Relaxation

Signal strength actually decays at the T2 relaxation rate only in an idealized situation in which all spins experience exactly the same external magnetic field. In reality, any source of inhomogeneity in this field causes nearby spins to precess at different frequencies and to dephase more rapidly, resulting in more rapid signal loss with a shortened relaxation time, T2*. Common sources of these inhomogeneities

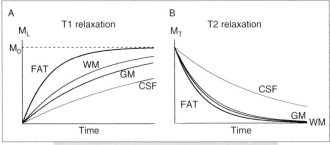

FIG. 18.4. **(A)** T1-relaxation curves describing recovery of longitudinal magnetization towards equilibrium following a 90° RF pulse. **(B)** T2-relaxation curves describing the decay of transverse magnetization towards zero following a 90° RF pulse. In sequence from shortest to longest relaxation times (fastest to slowest relaxation rates): fat, white matter (WM), gray matter (GM) and cerebrospinal fluid (CSF). As T1 relaxation represents recovery of longitudinal magnetization towards M_0, while T2 relaxation represents decay of transverse magnetization towards zero, the directions of the curves are reversed from (A) to (B).

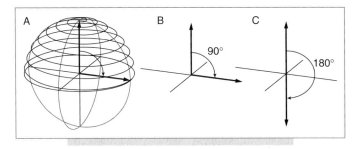

FIG. 18.3. **(A)** In the presence of the RF field, the magnetization vector spirals down from the z-axis into the transverse plane. **(B)** Net effect of a 90° RF-pulse. **(C)** Net effect of a 180° RF-pulse.

include the inability to engineer an absolutely uniform magnetic field and 'susceptibility effects' in which differences in the magnetic properties of nearby materials (air, metal, blood products) distort the magnetic field.

IMAGE CONTRAST MECHANISMS

Basic Pulse Sequence and Imaging Parameters TR and TE

Figure 18.5 outlines a simplified 'pulse sequence' diagram and defines two key variables, TE and TR, that may be adjusted to alter image contrast. The 'echo time', TE, may be thought of as the delay after the RF pulse during which the signal is gathered. The 'repetition time', TR, denotes the time interval between subsequent excitations by RF pulses. The process of creating an image requires numerous repetitions of this basic building block (see below).

During each TR, longitudinal magnetization is allowed to grow back along +z, to a degree depending on the tissue's T1 relaxation rate. Adjusting TR in this pulse sequence thus controls the extent of T1-weighted contrast introduced into the image (see below).

Immediately following the subsequent RF pulse, each tissue has its maximal amount of transverse magnetization. This transverse magnetization then decays towards zero during TE, to a degree depending on the tissue's T2 relaxation rate. Adjusting the length of the TE delay before collecting the induced signal thus controls the extent of T2-weighted contrast introduced into the image (see below).

Proton-Density, T1-Weighted and T2-Weighted Images

The 'proton density' refers to the number of hydrogen nuclei per unit volume. A region with a greater number of spins will contribute more magnetization and thus more signal than an equivalent region containing fewer spins. Proton density images are designed to show purely this underlying density variation. T1- and T2-weighted images then superimpose additional image contrast based on the differing relaxation times of the tissues, to degrees depending on the TR and TE values chosen for the scan. It must be emphasized that T1 and T2 are fixed parameters reflecting physical characteristics of the tissue, while TR and TE are imaging variables that are adjusted at the MR scanner to accentuate image contrast between tissues.

The magnitude of the signal collected for a given tissue type is determined by multiplying the underlying proton density distribution by the values of the T1 and T2 relaxation curves of Figure 18.5 for each particular tissue at the chosen TR and TE times. The appropriate choices of TR and TE values are contained in Table 18.1 and discussed below.

Proton-Density Images

Proton density images rely on a very long TR value to eliminate T1-weighted contrast and a very short TE value to eliminate T2-weighted contrast. The long TR allows recovery of full equilibrium magnetization for all tissues. The next RF pulse transfers these magnetizations to the transverse plane, where they precess and induce the signal that is recorded immediately (at a very short TE) before significant T2 relaxation can take place. The resulting image then reflects the spatial distribution of proton density across the patient, without additional T1- or T2-weighted contrast.

T1-Weighted Images

T1-weighted images rely on relatively short values of TR to introduce T1-weighted contrast and very short TE values to eliminate T2-weighting. The short value of TR establishes differences in longitudinal magnetization values for different tissue types (see Figure 18.5B). Following the subsequent RF pulse to transfer this magnetization to the transverse plane, the induced signal is collected rapidly

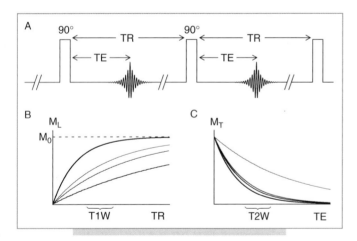

FIG. 18.5. **(A)** Schematic pulse sequence diagram. **(B)** T1 relaxation occurs during each repetition time (TR) between subsequent 90° RF pulses. Choice of TR thus determines the extent of T1 weighting in the image. **(C)** T2 relaxation occurs during each echo time (TE), which defines the timing of signal collection after each RF pulse. Choice of TE thus determines the amount of T2 weighting in the image. TR and TE ranges to produce T1-weighted and T2-weighted images are bracketed in (B) and (C).

TABLE 18-1	Appropriate choices of TR and TE to create T1-weighted, T2-weighted and proton density images		
	Proton density	*T1-weighted*	*T2-weighted*
TR	As long as possible	**Relatively short**	As long as possible
TE	As short as possible	As short as possible	**Relatively long**

(at short TE) to minimize T2-weighting. This combination produces a T1-weighted image while eliminating the confounding effects of T2-weighting.

T2-Weighted Images

T2-weighted images rely on a very long TR value to eliminate T1-weighting and a relatively long TE value to introduce T2-weighted contrast by creating differences in transverse magnetization between different tissue types. The long TR allows recovery of full equilibrium magnetization for all tissues. Following the subsequent RF pulse to transfer this magnetization to the transverse plane, signal is not collected until enough TE time has passed to highlight differences in transverse magnetization based on T2-relaxation differences (see Figure 18.5C). This combination produces a T2-weighted image while eliminating the confounding effects of T1-weighting.

Figure 18.6 shows examples of the different image types, achieved by appropriate adjustment of TR and TE.

Inversion Recovery

Inversion recovery sequences add an additional 180° RF pulse to the basic pulse sequence building block of Figure 18.5. Following this pulse, there is a delay time τ before the subsequent 90° pulse and signal collection at TE (Figure 18.7). TE and TR are chosen to produce T2-weighted images, but the additional τ imaging variable provides a mechanism to use T1 relaxation differences to uniformly suppress signal from a tissue of interest (water in FLAIR images and fat in STIR images).

FLAIR

In FLAIR (*fl*uid *a*ttenuated *i*nversion *r*ecovery) images, τ is chosen to correspond to the zero-crossing point in the T1 relaxation curve of water. The 90° RF pulse is applied at just the moment that the longitudinal magnetization of cerebrospinal fluid crosses through zero, so that there is no formation of transverse magnetization to produce signal from the CSF. Other tissue types with faster T1 relaxation

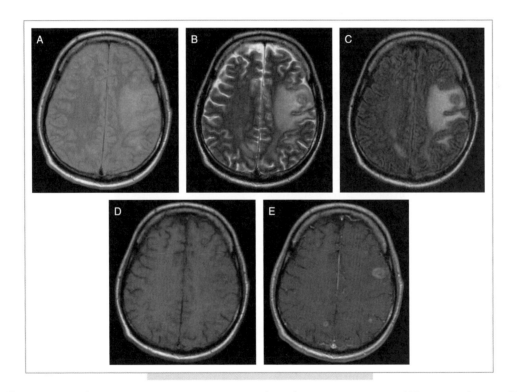

FIG. 18.6. Different types of image contrast in a patient with melanoma metastases. **(A)** Proton density, **(B)** T2-weighted, **(C)** FLAIR, **(D)** T1-weighted and **(E)** T1-weighted following intravenous gadolinium administration. Note the slightly lower signal intensity of white matter relative to gray matter on the T2-weighted and FLAIR images, as opposed to its slightly higher relative intensity on the T1-weighted image. In (E), the metastases enhance due to the T1-shortening effect of gadolinium in regions where it enters tissues across a disrupted blood–brain barrier. The shorter T1 time leads to greater recovery of longitudinal magnetization during each TR and produces greater signal after the subsequent RF pulse. In (C) the FLAIR image eliminates the signal from CSF, while continuing to demonstrate the vasogenic edema surrounding the metastases.

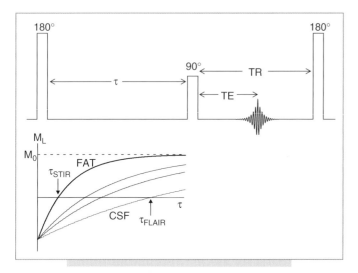

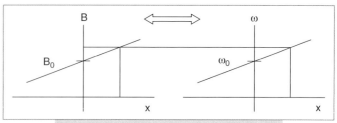

FIG. 18.8. In the presence of the magnetic field gradient, magnetic field strength B varies linearly with position. As a result, corresponding precession frequencies vary linearly with position.

FIG. 18.7. Immediately following the 180° inversion pulse, all of the equilibrium magnetization has been flipped to the −z−axis. The longitudinal magnetization recovers from −z to +z at the characteristic T1 relaxation rates of the different tissues. Appropriate selection of τ eliminates subsequent signal production from water in FLAIR images and from fat in STIR images.

have recovered significant portions of their equilibrium magnetization, so the 90º pulse does produce transverse magnetization and MR signal for these tissues. Figure 18.6C provides a sample image.

STIR

In STIR (*short tau inversion recovery*) images, τ is chosen to match the zero-crossing point in the T1 relaxation curve of fat. As a result, STIR provides a robust method of uniform fat suppression and is particularly helpful in accentuating soft tissue edema within fat-containing tissues such as bone marrow.

MAKING AN IMAGE

Spatial Localization by Magnetic Field Gradients

Spatial localization in MR images is primarily achieved through magnetic field gradients. These gradients, G, are spatially varying magnetic fields whose strength increases linearly with position along a given direction and are applied in addition to the uniform B_0 field that is present at baseline. For example, when an x-axis gradient G_x is applied, the total magnetic field strength is slightly weaker on one side of the imaging volume (−x locations) and stronger on the other (+x locations) (Figure 18.8). Because the Larmor precession frequency is directly proportional to magnetic field strength, the result is that signal from one

side of the patient oscillates at slightly slower frequencies than signal from the other side of the patient.

All of these signal components are simultaneously detected in the surrounding coil, but may be separated into the individual frequency components through use of the Fourier transform. The amount of signal oscillating at a given frequency corresponds to the amount of magnetization precessing at the corresponding magnetic field strength, which determines the spatial location of that magnetization along the gradient direction. In this way, magnetic field gradients create a mapping of frequency to position, a property which is crucial to MR image formation.

k-space

The concept of frequency-to-position mapping produced by magnetic field gradients is most intuitive in a single dimension, but is more challenging to conceptualize in multidimensional images. The 'k-space' formalism is a convenient way to describe gradient encoding along one or more dimensions.

Magnetic Field Gradients, Spatial Frequencies and k-space

As magnetic field gradients establish different precession frequencies across the imaging volume, evolution in a gradient over a specified time creates a spatial distribution of magnetization that varies along the gradient direction. Greater amounts of evolution in the gradient produce progressively tighter spatial oscillations across the patient, or oscillations of progressively higher 'spatial frequency' (Figure 18.9A).

The basic tenet of Fourier transform mathematics is that any given shape may be defined as an appropriate combination of spatially oscillating distributions. k-space tabulates the contributions of these spatial frequency components. It is the mathematical domain in which MRI signal is collected and is in fact related to the actual image by means of Fourier transformation.

Each data point of collected signal occupies a particular location in k-space and contains information about

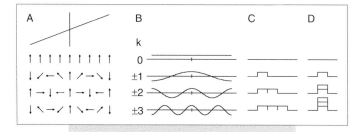

FIG. 18.9. **(A)** Creation of spatial distributions via gradient evolution. Relative to the center position, rightward magnetizations precess at increasing clockwise rates, leftward magnetization at increasing counterclockwise rates. The top row is a uniform distribution of magnetization prior to gradient evolution, lower rows show the spatial distribution after progressive steps of gradient evolution. **(B)** Corresponding spatial distributions and k-space values. (C) and (D) Equivalent gradient evolution steps: **(C)** frequency encoding approach, with fixed gradient strength and increasing evolution time intervals; **(D)** phase encoding approach, with fixed evolution times, but increasing gradient strengths.

the corresponding spatial frequency of magnetization distributed across the imaging volume. The k = 0 data point represents magnetization uniformly distributed across the imaging volume. The k = ±1 data points describe distributions of magnetization with a single cycle of oscillation from one end of the imaging volume to the other, k = ±2 with two cycles of oscillation, and so on (Figure 18.9B). A complete set of these data points specifies the amount of each spatial frequency component and uniquely determines the true spatial map of magnetization. This k-space 'map' of spatial-frequencies is converted via the Fourier transform to the 'spatial domain' to produce the image.

Each spatial frequency component is interrogated by creating the appropriate spatial oscillation via increasing intervals or 'steps' of gradient evolution. In practice, this may be done by allowing evolution for longer time intervals under a gradient of fixed strength – the 'frequency encoding' approach (Figure 18.9C), or by producing evolution for a fixed time interval under gradients of increasing strength – the 'phase encoding' approach (Figure 18.9D). The signal produced by each gradient evolution step reflects the size of the corresponding spatial frequency component and 'fills' the corresponding k-space position.

Frequency Encoding

During the signal collection phase of the pulse sequence, a frequency-encoding gradient or 'readout' gradient is applied. This gradient maps precession frequency to spatial position. Fourier transformation decomposes the collected signal into its component frequencies, whose magnitudes correspond to the amount of magnetization at each position, and yield a one-dimensional image.

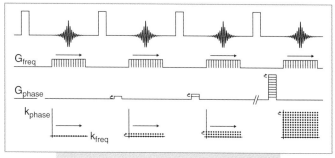

FIG. 18.10. Spatial encoding technique combining frequency and phase encoding. Signal collected during each frequency encoding gradient fills a horizontal line in k-space (straight arrows). Each phase encoding gradient step toggles to the next vertical k-space position (curved arrows). After a complete set of phase encode steps, a full plane of k-space data is filled, allowing 2D image reconstruction.

In the k-space description, gradient evolution 'steps' are defined by successive time intervals under a readout gradient of fixed strength. Each associated signal data point fills the corresponding k-space position (Figure 18.10). All of the data points needed to fill the complete 'line' of k-space may be acquired after a single RF pulse and Fourier transformation yields a one-dimensional image.

Phase Encoding

Phase encoding is used to encode spatial information in the second dimension and uses a gradient along an axis perpendicular to the frequency encoding direction. Unlike frequency encoding, 'phase-encoding steps' are achieved via stepwise changes in the strength of the phase-encoding gradient, with the evolution interval remaining fixed (see Figure 18.10).

Phase encoding is otherwise entirely analogous to frequency encoding, as each encoding step provides information about a different spatial frequency component and fills the corresponding position in k-space.

Each phase encoding evolution step must be performed after a new RF pulse, which requires numerous repetitions of the pulse sequence building block in order to collect all of the required signal data. Data collected after a full set of frequency and phase encoding steps then fills a full plane of k-space, which may be Fourier transformed into a two-dimensional image.

Slice Selection

In order to spatially encode the third dimension, a 'slice selection' step is often employed, in which the RF pulses are applied in the presence of a slice-selection magnetic field gradient. The RF pulses contain an adjustable range of frequencies. As discussed in the section on radiofrequency energy and resonance, the RF excitation frequency must match the spin precession frequency in order to tilt magnetization into the transverse plane. As a result, each RF pulse only excites spins with a matching range of precession frequencies, which corresponds to a predictable range of locations along the gradient direction (Figure 18.11).

Expanded Pulse Sequence Diagram and 2D Versus 3D Imaging Techniques

Figure 18.12 contains a simplified schematic pulse sequence diagram for two-dimensional imaging, in which slice selection, phase encoding and frequency encoding extract spatial information along each of the three perpendicular axes. For each slice selection step, in-plane spatial encoding is performed with a combination of phase and frequency encoding. Each slice requires a full set of phase encoding steps, each of which in turn requires its own TR.

It is primarily this need to repeat numerous excitations with different slice-selection and phase-encoding steps that accounts for the lengthy scan times of MRI (although in reality data acquisition from different slices may be interleaved to save some time).

Rather than performing slice selection as above, three-dimensional imaging techniques instead add an additional nested phase-encoding cycle to spatially encode the third dimension, again requiring repetition through multiple excitations.

Echoes

In actual practice, it is generally necessary to acquire signal data from both positive and negative sides of k-space. In order to do this, data are collected during symmetric 'echoes', with the center point of the echo corresponding to k = 0. There are two general techniques to do this, called, respectively, spin echo and gradient echo.

Spin Echoes

Spin echoes are formed by placing a 180° pulse in the center of the TE time (left out of the pulse sequence diagrams for simplicity). This additional 180° pulse has the effect of refocusing any dephasing of transverse magnetization caused by persistent magnetic field inhomogeneities as discussed above. The result is an echo height determined by T2 rather than T2*, which not only produces greater signal strength, but also better reflects intrinsic tissue properties. A commonly used analogy is to visualize runners on a track, who run in one direction for TE/2 before the sound of a whistle (the 180° pulse) causes them all to reverse directions but to maintain their individual speeds. After the second TE/2, they all rejoin one another in an 'echo' at the starting line before again beginning to 'de-phase' around the track in the other direction.

Gradient Echoes

Gradient echoes do not use any additional RF pulses during TE. Instead, the strength of the frequency encoding readout gradient is reversed to form an echo. The track runner analogy to a gradient echo is that after running for a length of time, the whistle (the gradient reversal) causes the fastest and slowest runners each to adopt the other's previous running speed. After an equivalent time interval, the two again meet, coming momentarily back 'in phase' with one another.

Because fixed magnetic field inhomogeneities are not corrected by gradient echo sequences, such sequences are more prone to susceptibility effects. These may be intended, as with use of gradient echo sequences to improve

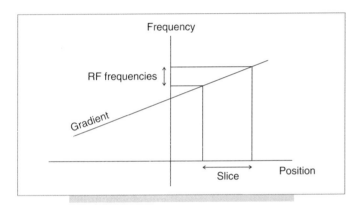

FIG. 18.11. Slice selection. The position of the slice is determined by the center frequency of the RF pulse and the slope of the magnetic field gradient, while the slice thickness is determined by the range of frequencies in the RF pulse and by the slope of the gradient.

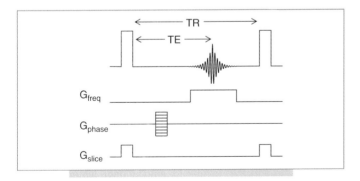

FIG. 18.12. Schematic two-dimensional imaging pulse sequence diagram (several practical details omitted). The slice-selection gradient is applied during the RF pulses. The phase encoding gradients step in amplitude during successive TR repetitions. The frequency encoding gradients are applied during signal readout.

detectability of subtle foci of hemorrhage, or undesired, in the form of susceptibility artifacts arising from dental or surgical metal, or from air/tissue interfaces in diffusion weighted images.

CONCLUSION

This chapter has introduced many of the key conceptual underpinnings of MRI physics. Several good references contain extensive sections that expand further upon these crucial concepts and typical imaging methods [1–4]. A basic understanding of these principles is vital to a firm grasp of clinical MR image content and forms the foundation for understanding more advanced MR imaging techniques.

REFERENCES

1. Stark DD, Bradley WG (1999). *Magnetic Resonance Imaging*, 3rd edn. Mosby, St Louis.
2. Edelman RR, Hesselink JR, Zlatkin M (2005). *Clinical Magnetic Resonance Imaging*, 3rd edn. Saunders, Philadelphia.
3. Hornak JP *The Basics of MRI* http://www.cis.rit.edu/htbooks/mri/
4. Higgins CB, Hricak H, Helms CA (1996). *Magnetic Resonance Imaging of the Body*, 3rd edn. Lippincott-Raven, Philadelphia.

CHAPTER 19

Advanced MR Techniques in Clinical Brain Tumor Imaging

Geoffrey S. Young and Shuang Xia

DIFFUSION-WEIGHTED IMAGING (DWI)

Introduction

DWI contrast reflects the mean distances traveled during the pulse sequence by free water protons in tissue due to Brownian motion – distances of the order of 10 microns. Diffusivity is characterized by the 'apparent diffusion coefficient' (ADC), a scalar parameter with units of mm^2/s [1]. Intracellular organelles and membranes result in an intracellular water diffusion path length that is an order of magnitude shorter than the diffusion path length of extracellular water. For this reason, the fraction of each voxel comprised of extracellular space, known as the extracellular volume fraction (ECF), is the primary determinant of ADC in that voxel.

Because routine DWI is produced from a T2*-weighted spin-echo echoplanar (SE-EPI) sequence with diffusion sensitization gradients added before and after the $180°$ pulse, the contrast reflects susceptibility weighting and T2-weighting as well as diffusion weighting. The magnitude and timing of each of the three roughly orthogonal diffusion sensitization gradients determines the degree of diffusion weighting, expressed in terms of a 'b-value' with units of s^2/mm. In current practice, b-values of $1000 s^2/mm$ are most common. In order to separate the contributions of diffusion weighting and of T2 weighting to the observed signal intensity (SI) on DWI images, ADC maps are calculated from the DWI and from an image set acquired with the diffusion gradient magnitude set to zero (B_0). Although b-values of 1500–2500 may allow more accurate estimates of cellularity, the lower signal-to-noise ratio (SNR), increased sensitivity to artifact and longer scan time associated with these stronger gradient settings have kept them from widespread clinical use [2].

Differential Diagnosis, Grading, Monitoring of Progression and Response

Very low ADC in the viscous contents of epidermoid cysts and abscesses can be used to distinguish these lesions from arachnoid cyst, tumefactive demyelination and necrotic tumor with accuracy greater than 90% [3–6]. Vasogenic edema and necrosis decrease net cellularity and thus increase ADC, whereas highly cellular tumors (primitive neuroectodermal tumor (PNET), glioblastoma multiforme (GBM), meningioma, lymphoma, metastases) increase cellularity, thus lowering ADC. Low ADC in an intra-axial lesion should suggest lymphoma or metastasis and, in a peripheral lesion, meningioma or dural metastasis more likely than glioma but, because low ADC is seen in a small number of glioblastoma, correlation with other advanced and conventional neuroimaging data is required for reliable clinical interpretation [7–11].

Minimum ADC (ADC_{min}) correlates well with histologic measures of tumor cellularity in high- and low-grade glioma, lymphoma, medulloblastoma, meningioma and metastases, among other tumors [7,12–16]. ADC_{min} below a threshold of 1.7–2.5 can contribute to the distinction of high-grade from low-grade glioma [17,18], but overlap between tumor grades makes correlation with other MRI data mandatory [10,17,19,20]. Wide variations in ADC_{min} and cellularity among GBM is likely in part due to variable necrosis, hemorrhage and calcification, but also may also reflect genotypic heterogeneity since these measures correlate well with radiation responsiveness [21,22]. Conversely, ADC cannot be used to distinguish accurately typical from atypical meningioma, since variation in cellularity between individual tumors is greater than the difference between the groups [15]. In the postoperative patient, low ADC at the resection margin or in the cavity can alter management by suggesting ischemia or abscess, respectively, when conventional MR sequences exclude artifact from hematoma and are consistent with these diagnoses [23,24]. Because increases or decreases in ADC from baseline are sensitive markers of cytoreductive response to radiation therapy (XRT)/chemotherapy or recurrence respectively, functional diffusion mapping (fDM) and other means of presenting longitudinal DWI data are under development (Figures 19.1 and 19.2) [13,14,25–29].

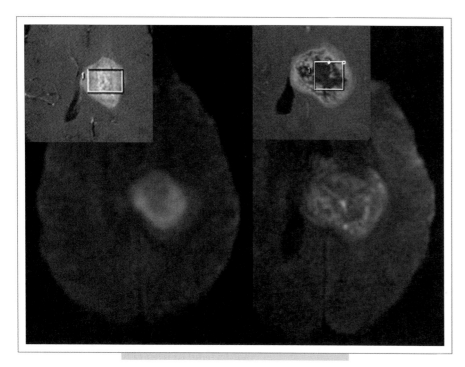

FIG. 19.1. Increasing ADC suggestive of decreasing cellularity in patient whose spectroscopy is presented in Figure 19.5 supports the impression that the enlargement of the Gd enhancing lesion is due to radiation necrosis rather than tumor recurrence.

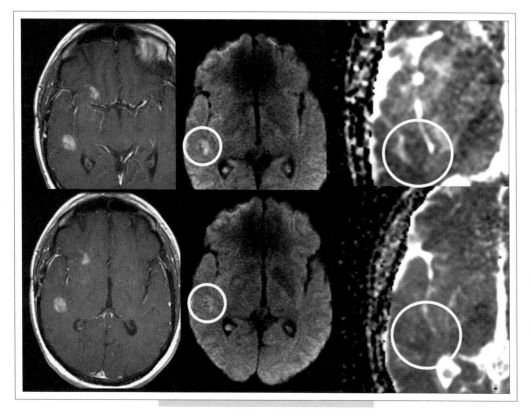

FIG. 19.2. Post-Gd T1WI (left), DWI (center) and ADC (right) at baseline (top row) and after one month of chemoradiation (bottom row) demonstrate marked decrease in cellularity which, despite unchanged enhancement, suggests good therapeutic response.

DWI sequences have been adapted to perform diffusion tensor imaging (DTI) by acquiring data in six or more directions. This yields sufficient data to define fully a 3D tensor (vector) matrix describing the direction and magnitude of water diffusion in each voxel. Increasing the number of diffusion gradient directions allows more precise estimation of the preferred direction of water diffusion. The degree to which water diffuses faster in one direction than another within a given voxel is referred to as the degree of anisotropy and can be characterized by a large number of derived scalar measures, the most commonly used of which is called 'fractional anisotropy' (FA) [30,31]. FA and other anisotropy measures reflect the existence of tissue microstructures – such as myelinated axonal bundles – and physiologic processes – such as axonal transport – that facilitate the diffusion of water in one or more directions and hinder it in other directions. DTI data can be used to generate a voxel-by-voxel map of anisotropy and of other scalar measures and to produce a map depicting the principal diffusion direction in each voxel as a vector superimposed on the anatomic image.

DTI for Preoperative Assessment, Surgical Guidance and Follow-up

Numerous mathematical 'tractography' algorithms have been published that connect the principal direction or directions of preferred water diffusion in each voxel to those of neighboring voxels to trace out 3D curves depicting the major white matter (WM) tracts in the brain. In normal white matter the high anisotropy produced by the tightly bundled myelin sheaths of the major tracts allows reliable tractography. Although more technically challenging, the depiction of eloquent WM tracts infiltrated and displaced by tumor has been shown to be useful in surgical planning and in prediction of postoperative dysfunction related to WM tract injury [32–36].

Initial experience suggests that DTI can contribute to the assessment of WM microinvasion by infiltrating glioma and thus assist in differentiating tumor recurrence from radiation necrosis, edema and gliosis [32,37,38,39,40]. Although published attempts to use FA and related anisotropy measures clinically to define the margin of glioma WM infiltration and to distinguish vasogenic edema around metastases from tumor infiltration around glioma have not been uniformly successful [22,38,41–46], work continues on a number of more promising and sophisticated analyses of white matter microstructure such as the 'fiber coherence index' [47].

MRS

$^1H_0$ MR spectroscopy (MRS) is essentially proton NMR spectroscopy modified by use of point resolved spin echo (PRESS) or stimulated echo (STEAM) based anatomic localization schemes performed on standard MRI systems. PRESS produces higher SNR and fewer artifacts, but cannot be performed at very short echo time (TE), so both sequences are in widespread use. Simultaneous localization of MRS from multiple voxels to increase spatial resolution is called magnetic resonance spectroscopic imaging (MRSI) or chemical shift imaging (CSI). At clinical scanner field strengths (1–2 orders of magnitude lower than analytic NMR spectrometers), employing standard chemical shift selective (CHESS) suppression of more than 99.99% of the free water proton signal (needed because the metabolites of interest are present in concentration four to five orders of magnitude

lower than water), with localization of MRSI data to $1 cm^3$ voxels, the available SNR is so low that current MRSI techniques require long acquisition times (more than 10 minutes in some cases) and are thus very sensitive to patient motion and other artifacts. Single voxel spectroscopy (SVS) of large regions of interest (ROI) is less time-consuming but yields far fewer useful data, since the large voxels may contain a mixture of necrotic tumor, high-grade tumor, low-grade tumor and surrounding brain tissue, so technical development of faster 3D-MRSI techniques with greater anatomic coverage and smaller voxel sizes continues [48–53].

$^1H_0$ MRS and NMR separate signals from chemically distinct hydrogen protons based on the slightly different 'chemical shift' of each $^1H_0$ resonant frequency due to differences in the charge density of the surrounding covalent electron-bond cloud. The resulting $^1H_0$ MR spectrum is graphed with the SI of each distinct species on the y-axis expressed in arbitrary units (AU) and the chemical shift of each species on the x-axis expressed in parts per million (ppm) of the resonant-frequency tetramethylsilane (TMS) protons. The use of ppm rather than Hz produces spectra that are similar regardless of scanner field strength. Because the SI detected from each $^1H_0$ species depends on a host of technical factors that vary from scan to scan, peak areas vary by 20% between serial scans even under ideal conditions [54]. Absolute quantitation of metabolite concentrations is generally reserved for the laboratory [55] and, instead, clinical spectra are presented scaled to the highest peak in the spectrum.

Because peak height is subject to less measurement error than peak area, clinical interpretation focuses on ratios of peak heights in the area of interest relative to each other and to normal-appearing voxels in the contralateral brain. The main peaks of interest in brain tumor MRS are: branched-chain amino acids (AA: 0.9–1.0 ppm); lipid (Lip: 0.9–1.5 ppm); lactate (Lac: 1.3 ppm); alanine (Ala: 1.5 ppm); n-acetyl aspartate (NAA: 2.0 ppm); choline (Cho: 3.2 ppm); creatine (Cr: 3.0 ppm and 3.9 ppm); and myo-inositol (mI: 3.6 ppm). The two creatine peaks and the broad Lip and AA peaks reflect the presence of multiple non-equivalent proton species in these molecules. Occasionally, when definite resolution of the overlapping AA, Lac and Lip peaks is clinically important, combinations of short (<45 ms), intermediate (130–140 ms) and/or long (270–280 ms) echo times are used to differentiate them based on the phase difference of AA, Lac and Ala with respect to Lip, NAA, Cr and Cho.

NAA, found only in neurons, is involved in lipid metabolism, nitrogen balance, neuronal-glial signaling and energy metabolism, among other processes. NAA varies in concentration with anatomic region and patient age and is decreased by all processes that impair neuronal function or kill neurons, including tumor, necrosis, demyelination, ischemia, neurodegeneration, etc. [56–58]. Creatine, an important high-energy phosphate buffer substrate for creatine kinase (CK), has an important role in maintaining vital cellular levels of ATP [59]. Because Cr varies in concentration with anatomic region, diet and pathophysiology, great care is needed when using it as an internal reference for evaluating other peak heights [60]. Cho-containing compounds play a key role in biosynthesis of phosphatidylcholine, other membrane lipids and acetyl choline. Brain Cho concentration varies with anatomic region, and dietary choline intake is very useful as a marker of cell-membrane synthesis and degradation in glial growth, injury and repair [61,62]. In contrast to Cr, NAA and Cho peaks present in normal brain spectra, Lip is only observed in disease states producing frank necrosis or partial volume averaging of fat in scalp, marrow, lipoma or teratoma. Lac, observed as a doublet in MRSI because of J-coupling between its protons, is present only in disease states producing anaerobic glycolysis. Because the lipid and lactate peaks are often observed together in pathologic processes producing anaerobic metabolism together with frank necrosis, distinction of the two is often diagnostically unimportant, in which case the term 'Lip-Lac' is used to refer to both overlapping peaks.

Because of the distinct T2s of Cr (150–160 ms), Cho (240–250 ms) and NAA (250–300 ms), different fractions of the total signal from each metabolite will have decayed at a given echo time and so both relative and absolute peak heights will vary considerably with TE. In addition, the T2 of these metabolites varies with their anatomic location [58,63], so careful reference to normal spectra acquired in the same anatomic location at the same TE are required to interpret spectroscopy correctly. The T2 of mI, among other metabolites including glutamate and glucose, is so short that short TE spectroscopy (by convention, less than 35 ms) is required to detect it.

MRS in Differential Diagnosis, Grading, Surgical Planning and Follow-up

Because Cho concentrations are increased by glial membrane injury in ischemia and demyelination and by glial division in gliosis and glial neoplasia, high Cho may be seen in the spectra of a wide variety of pathologic lesions. Similarly, demyelination, ischemia and glioma all injure neurons and decrease NAA. Finally, ischemia and neoplasia both may produce Lac and Lip since anaerobic glycolysis and cellular necrosis occur in both pathophysiologies. Predictably, numerous studies have demonstrated that MRS is of no significant benefit in differential diagnosis of brain tumor from non-neoplastic brain lesions [64–67]. The only major exception to this rule is the use of MRS in rim-enhancing lesions to distinguish bacterial abscess and parasitic cyst from necrotic tumor by demonstrating the presence of AA in the cystic contents. Because AA represent lysosomal breakdown products their detection implies the presence of activated polymorphonuclear leukocytes (PMN), a finding nearly pathognomonic of infection [4,68].

MRS has a similarly limited value for differentiating glial from non-glial neoplasms since both meningioma, which contain no neurons, and necrotic GBM, in which the neurons have died, produce very high Cho and little or no NAA. Although some meningioma have very high concentrations

of Ala, up to 80% of them contain only low levels similar to the concentrations detected in metastases and schwannoma. Finally, MRS has no role in the differentiation of metastasis from GBM, since both may have high Cho, low or no NAA and may have prominent Lac and Lip [69].

Although not useful in differential diagnosis, MRS estimation of the Cho/NAA ratio (CNR) is valuable for preoperative estimation of the grade of known or suspected glioma, since Cho increases and NAA decreases in proportion to increasing tumor grade from Gr I to Gr III. Although frequent necrosis in Gr IV glioma makes GBM CNR somewhat variable, either a CNR greater than 1.5 or the presence of Lac/Lip should suggest a Gr IV lesion [18,70–74]. Because the high blood volume observed in all grades of oligodendroglioma makes prediction of oligodendroglioma grade by perfusion unreliable, MRS remains especially useful in grading this glioma subtype [75–78]. MRS may also be more useful than perfusion in areas where radiation necrosis mixed with recurrent tumor lowers the apparent blood volume, but this remains to be proven [79]. Similarly, whole brain NAA (WBNAA) and CH_2/CH_3 ratios in normal-appearing brain have been proposed as alternative ways to measure glioma WM infiltration, but these measure have not yet been clinically validated (Figures 19.3 and 19.4) [80,81].

MRS targeting of foci with high CNR improves the accuracy of biopsy by reducing false negatives and undergrading and it has been used to guide radiosurgery and resection [82–86].

MRS in Assessment of Treatment Response

Because both recurrent GBM and radiation necrosis may produce Lac/Lip, high Cho and low NAA, MRS performed at a single time point is not useful in glioma follow-up, except in foci where florid necrosis produces a complete absence of all metabolites [87]. Longitudinal CNR decrease (suggesting necrosis) or increase (suggesting recurrence) over time has been demonstrated to be much more useful when performed by groups with the resources to apply consistently meticulous technical acquisition and post-processing standards to produce reliable data integrated with diffusion, perfusion and anatomic imaging [83,85,88–90]. However, outside a handful of major laboratories, the experience in routine clinical practice using currently-available hardware and software is quite disappointing, because the temporal variation in metabolite concentrations on sequential scans is often far smaller than the variations in spectra due to changes in voxel location, differences in acquisition parameters, scalp lipid averaging, susceptibility artifacts and other technical factors. In essence, reliable MRSI glioma follow-up can only be achieved when staffing and scan time are sufficient to allow a trained technical spectroscopy specialist to monitor each individual MRS scan under the direct supervision of the interpreting radiologist. This is impossible to achieve at most centers because MRS is currently not reimbursed and so the associated personnel and scanner-time costs of performing it are prohibitive (Figure 19.5).

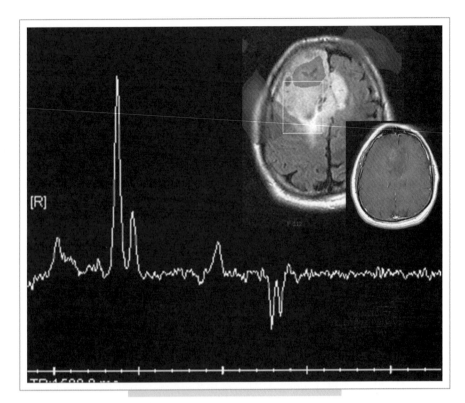

FIG. 19.3. Intermediate TE (144 ms) SV-MRS of anaplastic oligoastrocytoma reveals CNR >4 and prominent lactate peak consistent with high grade glioma.

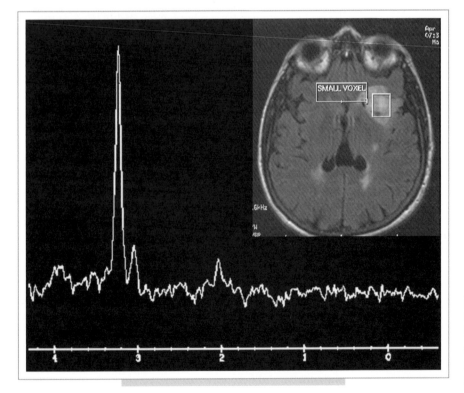

FIG. 19.4. Extremely high Cho peak and CNR in oligodendroglioma suggest high WHO grade. MRS is preferred for grading of oligodendroglioma because the elevated rCBV seen in all grades makes perfusion MR grading less reliable.

MICROVASCULAR IMAGING IN BRAIN TUMORS: PMR AND T1P

Imaging of tumor neovasculature directly assesses several of the cardinal features of high-grade glioma of great importance in assessment and monitoring of patients taking angiogenesis inhibitors. While much active investigation focuses on more advanced techniques, such as first pass permeability estimation, arterial spin labeling (ASL)

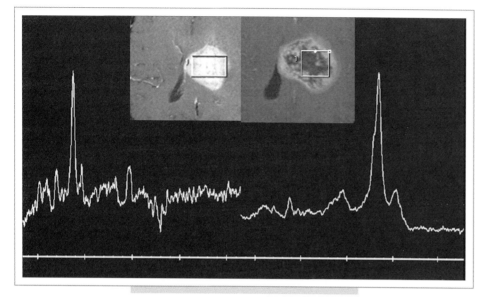

FIG 19.5 Despite the enlargement of the area of abnormal enhancement following chemoradiation, serial TE = 144 ms SV-MRS showing disappearance of choline and dramatic increase in lipid suggests radiation necrosis. Note that the presence of lactate, seen as a small inverted doublet on the earlier (left) spectrum, can no longer be assessed on the later spectrum (right) because of the dominant overlapping lipid peak.

and vascular morphologic imaging, these techniques have not been validated sufficiently for routine clinical use. Thus this discussion will focus on the widely accepted use of susceptibility-weighted first-pass techniques (PMR) to estimate blood volume and of T1-weighted techniques (T1P) to estimate capillary permeability.

PMR

Because the signal change exploited by PMR results from the susceptibility effect of a concentrated intravascular bolus of paramagnetic Gd contrast agent, production of reliable PMR maps requires high-pressure power injection through an 18–20 gauge antecubital IV catheter of a high contrast dose (typically 0.2 mmol/kg or twice the standard dose) at a high rate (typically 4 ml/s or higher). The data acquisition typically consists of a single-shot echoplanar (ssEPI or EPI) sequence covering the whole brain every 1–2 seconds which is repeated continuously for 40–60 seconds, starting about 10 seconds before the arrival of the bolus in the brain and ending 10–20 seconds after recirculation of the first pass of contrast agent. Among EPI sequences, SE-EPI remains the gold standard, despite the slightly higher contrast requirements of this sequence, because of its contrast sensitivity to 1–2 micron neovessels and its lower sensitivity to the susceptibility artifacts that remain a significant limitation of PMR [91,92]. Post-processing of the time-series data yields a time-signal intensity curve (TIC) for each voxel from which the relative capillary blood volume (rCBV) in the voxel can be derived by integrating the area under the signal intensity curve during the passage of the contrast bolus. rCBV is displayed as a rainbow color map, windowed by convention to display high gray matter (GM) rCBV in red and lower white matter (WM) rCBV in blue. These data are supplemented by inspection of TIC in the ROI within the lesion and a corresponding equally-sized ROI within normal-appearing (ideally, contralateral) brain and, when needed, calculation of relative rCBV (rrCBV) ratios between the two ROI [93]. Overlaying these maps on higher resolution anatomic images is critical for interpretation because, despite ongoing work to develop faster and higher resolution methods [92,94], the limit of resolution of clinical PMR using commercial hardware and software remains 2–4 mm in-plane voxel dimensions, 5–10 mm sections and 1–2 s temporal resolution.

PMR in Diagnosis, Grading, Operative Planning and Follow-up

Visual inspection of TIC and source images are essential for interpretation because the effect of patient motion, susceptibility, SNR, bolus rate and timing and partial-volume averaging artifacts cannot be assessed from color maps alone. Because the arterial input function is difficult to estimate accurately from signal loss techniques such as Gd bolus imaging, clinically available rCBV maps remain qualitative, rather than quantitative. Nevertheless, knowledge that the normal CBV of GM is approximately two to three times that of WM [95,96] allows effective clinical interpretation of tumor rCBV maps, since a maximum tumor rCBV two to three times that of normal-appearing WM (NAWM) has been demonstrated to be a strong predictor of the presence of high grade glioma [97,98]. Thus, correctly thresholded and windowed rCBV maps, corroborated by inspection of ROI, TIC and rrCBV ratios, that show tumor blood volume similar to or higher than normal-appearing GM are strongly predictive of hypervascular and thus high-grade glioma. Low-grade tumor rCBV, on the other hand, tends to be less than or equal to that of NAWM. Focal areas of rCBV intermediate between NAWM and GM deserve careful attention to TIC, rrCBV ratios, prior comparison PMR and close interval follow-up. In all cases, comparison of rCBV maps to prior maps in order to detect longitudinal change and correlation of rCBV with independent advanced data and anatomic MR images is critical (Figure 19.6).

Inspection of TIC, as required for reliable interpretation, also allows detection of significant first-pass contrast leakage that violates the central volume theorem assumption of a non-diffusible tracer [99] and complicates calculation of CBV [100]. Although multiple curve fitting algorithms [101–106] have been reported to address this, standard commercial curve integration software still relies on simple interpolated and constant baseline methods. Since the former will systematically underestimate, and the latter systematically overestimate, CBV in high permeability lesions, inspection of both may be necessary in highly permeable lesions. While the GBM neovessel blood–brain barrier (BBB) is often impaired and may produce a TIC which returns as little as half way to baseline, detection of very high first pass leak with minimal or no return to baseline on lesion TIC should suggest a non-glial tumor such as meningioma, choroid plexus papilloma, metastases, or lymphoma, since capillaries within these lesions have essentially no BBB [105,107,108]. When interpreted in the context of DWI and conventional imaging, this visual TIC analysis can be quite useful in helping to distinguish such lesions from GBM and oligodendroglioma. In a rim-enhancing lesion, peripheral rCBV less than or equal to that of the surrounding white matter should suggest the possibility of abscess and indicates a need for careful correlation with DWI, MRS and anatomic images [5]. The usefulness of rCBV to distinguish glioma from metastases remains undefined. Published data shows that the rCBV of metastases varies with the vascularity of the primary tumor in a range overlapping that of GBM, but it is unclear how this applies to standard SE-EPI techniques [109–112]. Thus, although DWI and MRS remain the primary advanced techniques for differentiating abscess from cystic metastasis, SE-EPI PMR demonstration of peripheral high rCBV argues against abscess. In a solid enhancing intra-axial lesion, SE-EPI demonstration of low rCBV should suggest consideration of lymphoma or metastasis [10,107]. In a cortical or subcortical lesion with conventional imaging features suggestive of focal circumscribed low-grade glioma, demonstration of rCBV greater than or equal to GM strongly suggests oligodendroglioma (Figures 19.7 and 19.8).

Because 'chicken-wire' capillaries produce high rCBV in all grades of oligodendroglioma, MRS should be performed in grading of such lesions [75–78]. In astrocytoma, on the other hand, $rCBV_{max}$ is a strong predictor of grade, and rCBV greater than that of normal GM should raise concern for GBM, in which mean $rCBV_{max}$ values are often four to six times those of WM [98,113–118]. This strong correlation also forms the basis for the demonstrated efficacy of rCBV maps in targeting the high-grade areas of heterogeneous glioma for biopsy, ablation, or resection [118–120].

Pronounced variation of rCBV within and between patients makes it essential to obtain baseline PMR data using equivalent techniques if PMR is to be used in assessing treatment response [119]. After chemoradiation of glioma, increasing rCBV within an area of abnormal enhancement or T2 prolongation suggests recurrence, whereas low rCBV, especially when correlated with increasing ADC, suggests

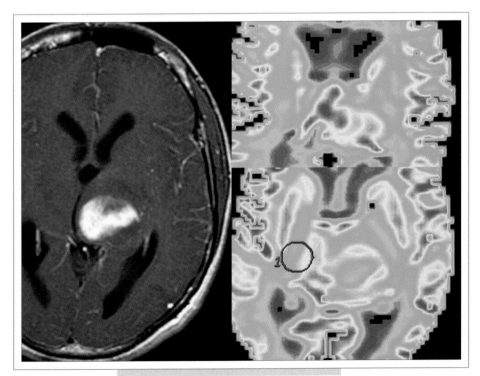

FIG. 19.6. Perfusion MR of this astrocytoma reveals rCBV higher than cortical GM at the periphery and just outside of the area of abnormal enhancement (upper right), but lower blood volume in the center and inferior portion of the lesion (lower right). This information about the heterogeneous spatial distribution of high-grade tumor can be useful in preventing false negative or undergraded biopsy results and in directing focal resection or ablation to areas of unsuspected hypervascular glioma.

radiation necrosis [27,121–123]. Good correlation between rCBV and expression of pro-angiogenic cytokines suggests that this will remain an important method of follow-up in the era of angiogenesis inhibition [115]. The pronounced effect of steroids on vascular permeability in glioma and the difficulty of completely deconvolving permeability from blood volume with current techniques may explain the mixed conclusions reached by studies examining the effects of steroid dose on CBV, and certainly suggests attention to steroid dosing during follow-up at present [124–126].

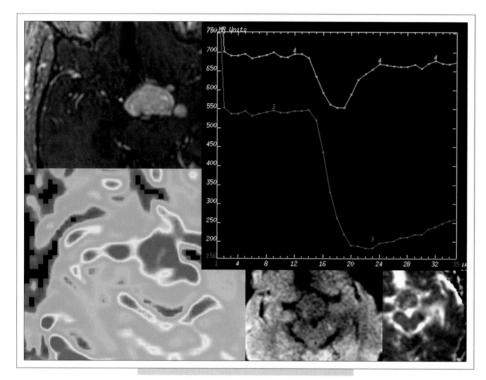

FIG. 19.7. Although the intermediate ADC of this enhancing lesion does not indicate sufficiently high cellularity to be definitive, the spin echo echoplanar perfusion imaging findings are characteristic of meningioma. The time-intensity curve (TIC) within the lesion (purple curve) demonstrates near complete leak of contrast during first pass consistent with a lesion of non-glial origin and both the TIC and rCBV color map (left) demonstrate very high blood volume compared to a reference ROI (green curve).

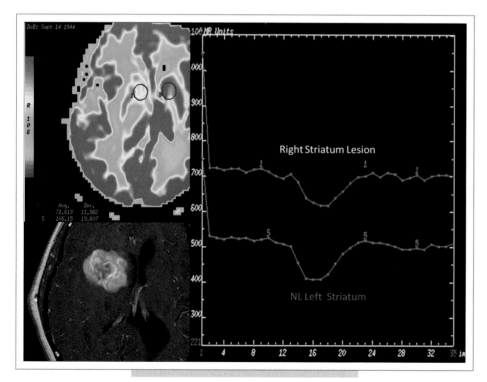

FIG. 19.8. rCBV color map (upper left) demonstrates lower blood volume in the right basal ganglia lesion region of interest than in the contralateral anatomically appropriate left basal ganglia region of interest. In this prominently enhancing solitary right basal ganglia periventricular lesion (lower left) this PMR is suggestive of CNS lymphoma. The relatively little first pass leak seen in the TIC (right) in this lesion may be due to prior high dose steroid therapy.

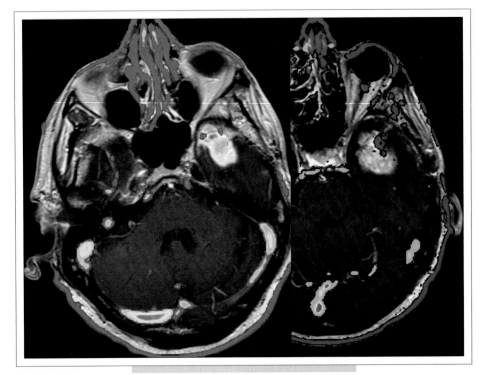

FIG. 19.9. Despite differences in windowing and slice registration, a 4 week follow-up permeability Kps map (right) superimposed on post-Gd T1WI demonstrates increasing permeability (similar to systemic scalp capillaries) in the rapidly expanding lateral aspect of the left temporal recurrent high-grade glioma. This type of data holds promise for distinguishing recurrent tumor from necrosis, but clinical validation and routine application await improvements in post-processing and display technique to achieve practical longitudinally comparable maps. (Post-processing software courtesy Timothy Roberts PhD, Children's Hospital of Philadelphia.)

T1P

Because increasing secretion of pro-angiogenic factors correlates with tumor grade, both permeability of co-opted native capillaries and formation of fenestrated neocapillaries increases with tumor grade [127]. These different degrees of abnormal permeability are difficult to differentiate on delayed post-Gd T1-weighted imaging (Gd-T1WI), likely accounting for its imperfect performance in distinguishing low- and high-grade glioma [128].

Dynamic Gd-enhanced T1-weighted permeability imaging (T1P), a semiquantitative dynamic adaptation of Gd-T1WI, allows a much more sensitive assessment of microvessel permeability by acquiring data continuously throughout the first few passes of Gd recirculation. Serial fast Gd-T1WI, usually acquired with radiofrequency spoiled gradient echo (SPGR) technique, is performed continuously throughout the first 2–5 minutes after bolus Gd injection. TIC calculated in T1P imaging demonstrate an increase in available signal – due to Gd augmentation of the T1 relaxation rate of adjacent water protons – that is far more easily quantifiable than the signal loss seen in DSC PMR. Because the TIC reflects several passes through the brain at a somewhat lower temporal resolution than PMR, curve fitting, deconvolution of the initial first-pass effect and one- or two-compartment pharmacokinetic modeling must be applied to calculate parameters that reliably reflect the degree of BBB impairment. The most widely used parameters are derived from the slope of the delayed plateau phase of the TIC. When appropriately corrected for first-pass effects, T2* effects and

large vessel concentrations, this slope should provide a good estimate of the net forward volume transfer constant (K^{trans}), reflecting the net rate of passage of Gd from the intravascular space to the extravascular extracellular space. The validity of the assumptions underlying the mathematics of K^{trans} calculation depend heavily on the repetition time (TR) and TE of the sequence, since Gd has T2 and T2* shortening as well as T1 shortening effects. T1 mapping can be performed prior to T1P in order to allow better quantitation, if required by the pulse sequence parameters and analysis employed. Even so, the rate of Gd diffusion across the BBB depends on variable CBF, variable intravascular Gd concentration, capillary surface area (which is abnormal in neocapillaries) and pressure gradients across the BBB that result from a combination of tumor interstitial oncotic pressure and intravascular and tissue hydrostatic pressure, as well as BBB permeability. Thus, although in general the T1 method described here is far less controversial than more-recently reported T2*-based first-pass methods, the calculation of a true K^{trans} is not trivial. Because controversy persists over the degree to which parameters calculated from T1P reflect pure K^{trans} or an admixture of K^{trans} with CBF, surface area and other factors, and over the relative advantages of K^{trans} versus other compound metrics such as permeability surface area (K_{ps}), the term 'measures of permeability' is used below in order to recognize that most of the large variety of published delayed phase measures depend heavily, but not exclusively, on K^{trans}. In addition, a number of measures reflecting aspects of permeability dynamics other than the slope of the delayed

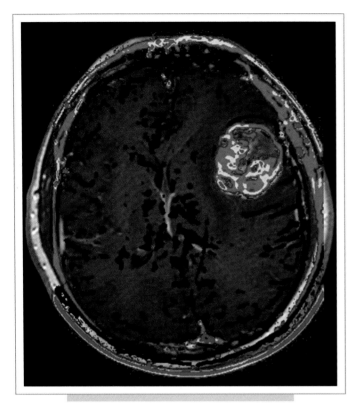

FIG. 19.10. Permeability Kps color map superimposed on post-Gd T1WI depicts heterogeneous permeability within a lung cancer metastasis. Permeability similar to systemic capillaries in scalp is seen in a portion of the lesion. (Post-processing software courtesy Timothy Roberts PhD, Children's Hospital of Philadelphia.)

phase have been devised in the quest for a practical, reproducible and informative metric, including the maximum rate of enhancement during the first pass (max dI/dt, rR, etc.) [129–132]. It seems likely that these two types of metrics each reflect different aspects of the underlying endothelial abnormality.

T1P in Diagnosis, Grading and Follow-up

Assuming reliable T1P technique, increasing permeability correlates well with increasing tumor grade [129,131–134]. Although some reports suggest that rCBV may be slightly more predictive, the two measures are independent but strongly correlated in high-grade glioma, suggesting that some combination of both may prove ideal [97,135,136]. Like rCBV, T1P can be used to aid the differentiation of recurrent glioma from radiation necrosis [130,133], but it is much less widely used due to a combination of imaging time, profusion of reported permeability measures and lack of robust and user-friendly post-processing software. As these technical needs are answered over the next few years, T1P will likely become a significant clinical tool (Figures 19.9 and 19.10).

CONCLUSION

The techniques discussed above can produce data on brain tumor cellularity, tissue invasion, metabolism and microvasculature that, when appropriately integrated with anatomic MRI in a multidimensional brain tumor imaging protocol, can contribute significantly to differential diagnosis, estimation of tumor grade, planning of biopsy and focal resection/ablation and therapeutic monitoring. Meticulous attention to acquisition and post-processing technique, reference to source images and graphical data and comparison with identically-acquired preoperative baseline and serial follow-up data are essential for the safe and effective clinical use of all of these techniques. Although space limitations have prevented more than a cursory allusion to the potential pitfalls of advanced imaging, the importance of integrating all available advanced MRI data and interpreting them in the context of the anatomic imaging should be clear. Similarly, now that such a wealth of tumor-specific information is available, the importance of correlation with accurate history regarding tumor histology, radiation therapy, steroid administration and cytotoxic and anti-angiogenic chemotherapy cannot be overstated.

REFERENCES

1. Provenzale J, Mukundan S, Barboriak DP (2006). Diffusion-weighted and perfusion MR imaging for brain tumor characterization and assessment of treatment response. Radiology 239:632–649.
2. Mardor Y, Pfeffer R, Spiegelmann R et al (2003). Early detection of response to radiation therapy in patients with brain malignancies using conventional and high b-value diffusion-weighted magnetic resonance imaging. J Clin Oncol 21:1094–1100.
3. Reddy JS, Mishra AM, Behari S et al (2009). The role of diffusion-weighted imaging in the differential diagnosis of intracranial cystic mass lesions: a report of 147 lesions. Surg Neurol 66:246–250.
4. Mishra AM, Gupta RK, Jaggi RS et al (2004). Role of diffusion-weighted imaging and in vivo proton magnetic resonance spectroscopy in the differential diagnosis of ring-enhancing intracranial cystic mass lesions. J Comput Assist Tomogr 28:540–547.
5. Erdogan C, Hakyemez B, Yildirim N, Parlak M (2005). Brain abscess and cystic brain tumor: discrimination with dynamic susceptibility contrast perfusion-weighted MRI. J Comput Assist Tomogr 29:663–667.
6. Tsui EY, Leung WH, Chan JH, Cheung YK, Ng SH (2002). Tumefactive demyelinating lesions by combined perfusion-weighted and diffusion-weighted imaging. Comput Med Imaging Graph 26:343–346.

7. Guo AC, Cummings TJ, Dash RC, Provenzale JM (2002). Lymphomas and high-grade astrocytomas: comparison of water diffusibility and histologic characteristics. Radiology 224:177–183.

8. Okamoto K, Ito J, Ishikawa K, Sakai K, Tokiguchi S (2000). Diffusion-weighted echo-planar MR imaging in differential diagnosis of brain tumors and tumor-like conditions. Eur Radiol 10:1342–1350.

9. Toh CH, Chen YL, Hsieh TC, Jung SM, Wong HF, Ng SH (2006). Glioblastoma multiforme with diffusion-weighted magnetic resonance imaging characteristics mimicking primary lymphoma. Case report. J Neurosurg 105:132–135.

10. Calli C, Kitis O, Yunten N, Yurtseven T, Islekel S, Akalin T (2006). Perfusion and diffusion MR imaging in enhancing malignant cerebral tumors. Eur J Radiol 58:394–403.

11. Krabbe K, Gideon P, Wagn P, Hansen U, Thomsen C, Madsen F (1997). MR diffusion imaging of human intracranial tumours. Neuroradiology 39:483–489.

12. Kotsenas AL, Roth TC, Manness WK, Faerber EN (1999). Abnormal diffusion-weighted MRI in medulloblastoma: does it reflect small cell histology? Pediatr Radiol 29:524–526.

13. Chenevert TL, Stegman LD, Taylor JM (2000). Diffusion magnetic resonance imaging: an early surrogate marker of therapeutic efficacy in brain tumors. J Natl Cancer Inst 92:2029–2036.

14. Chenevert TL, McKeever PE, Ross BD (1997). Monitoring early response of experimental brain tumors to therapy using diffusion magnetic resonance imaging. Clin Cancer Res 3:1457–1466.

15. Filippi CG, Edgar MA, Ulug AM, Prowda JC, Heier LA, Zimmerman RD (2001). Appearance of meningiomas on diffusion-weighted images: correlating diffusion constants with histopathologic findings. Am J Neuroradiol 22:65–72.

16. Hayashida Y, Hirai T, Morishita S et al (2006). Diffusion-weighted imaging of metastatic brain tumors: comparison with histologic type and tumor cellularity. Am J Neuroradiol 27:1419–1425.

17. Sugahara T, Korogi Y, Kochi M et al (1999). Usefulness of diffusion-weighted MRI with echo-planar technique in the evaluation of cellularity in gliomas. J Magn Reson Imaging 9:53–60.

18. Catalaa I, Henry R, Dillon WP et al (2006). Perfusion, diffusion and spectroscopy values in newly diagnosed cerebral gliomas. NMR Biomed 19:463–475.

19. Yang D, Korogi Y, Sugahara T et al (2002). Cerebral gliomas: prospective comparison of multivoxel 2D chemical-shift imaging proton MR spectroscopy, echoplanar perfusion and diffusion-weighted MRI. Neuroradiology 44:656–666.

20. Bulakbasi N, Kocaoglu M, Ors F, Tayfun C, Ucoz T (2003). Combination of single-voxel proton MR spectroscopy and apparent diffusion coefficient calculation in the evaluation of common brain tumors. Am J Neuroradiol 24:225–233.

21. Mardor Y, Roth Y, Ochershvilli A et al (2004). Pretreatment prediction of brain tumors' response to radiation therapy using high b-value diffusion-weighted MRI. Neoplasia 6:136–142.

22. Castillo M, Smith JK, Kwock L, Wilber K (2001). Apparent diffusion coefficients in the evaluation of high-grade cerebral gliomas. Am J Neuroradiol 22:60–64.

23. Khan RB, Gutin PH, Rai SN, Zhang L, Krol G, DeAngelis LM (2006). Use of diffusion-weighted magnetic resonance imaging in predicting early postoperative outcome of new neurological deficits after brain tumor resection. Neurosurgery 59:60–66.

24. Schaefer PW, Grant PE, Gonzalez RG (2000). Diffusion-weighted MR imaging of the brain. Radiology 217:331–345.

25. Hein PA, Eskey CJ, Dunn JF, Hug EB (2004). Diffusion-weighted imaging in the follow-up of treated high-grade gliomas: tumor recurrence versus radiation injury. Am J Neuroradiol 25:201–209.

26. Chan YL, Yeung DK, Leung SF, Chan PN (2003). Diffusion-weighted magnetic resonance imaging in radiation-induced cerebral necrosis. Apparent diffusion coefficient in lesion components. J Comput Assist Tomogr 27:674–680.

27. Tsui EY, Chan JH, Ramsey RG et al (2001). Late temporal lobe necrosis in patients with nasopharyngeal carcinoma: evaluation with combined multi-section diffusion weighted and perfusion weighted MR imaging. Eur J Radiol 39:133–138.

28. Moffat BA, Chenevert TL, Lawrence TS et al (2005). Functional diffusion map: a noninvasive MRI biomarker for early stratification of clinical brain tumor response. Proc Natl Acad Sci USA 102:5524–5529.

29. Hamstra DA, Chenevert TL, Moffat BA et al (2005). Evaluation of the functional diffusion map as an early biomarker of time-to-progression and overall survival in high-grade glioma. Proc Natl Acad Sci USA 102:16759–16764.

30. Inoue T, Ogasawara K, Beppu T, Ogawa A, Kabasawa H (2005). Diffusion tensor imaging for preoperative evaluation of tumor grade in gliomas. Clin Neurol Neurosurg 107:174–180.

31. Field AS, Wu YC, Alexander AL (2005). Principal diffusion direction in peritumoral fiber tracts: color map patterns and directional statistics. Ann NY Acad Sci 1064:193–201.

32. Goebell E, Fiehler J, Ding XQ et al (2006). Disarrangement of fiber tracts and decline of neuronal density correlate in glioma patients – a combined diffusion tensor imaging and 1H-MR spectroscopy study. Am J Neuroradiol 27:1426–1431.

33. Yu CS, Li KC, Xuan Y, Ji XM, Qin W (2005). Diffusion tensor tractography in patients with cerebral tumors: a helpful technique for neurosurgical planning and postoperative assessment. Eur J Radiol 56:197–204.

34. Nimsky C, Grummich P, Sorensen AG, Fahlbusch R, Ganslandt O (2005). Visualization of the pyramidal tract in glioma surgery by integrating diffusion tensor imaging in functional neuronavigation. Zentralbl Neurochir 66:133–141.

35. Lazar M, Alexander AL, Thottakara PJ, Badie B, Field AS (2006). White matter reorganization after surgical resection of brain tumors and vascular malformations. Am J Neuroradiol 27:1258–1271.

36. Schonberg T, Pianka P, Hendler T, Pasternak O, Assaf Y (2006). Characterization of displaced white matter by brain tumors using combined DTI and fMRI. Neuroimage 30:1100–1111.

37. Stieltjes B, Schlüter M, Didinger B et al (2006). Diffusion tensor imaging in primary brain tumors: reproducible quantitative analysis of corpus callosum infiltration and contralateral involvement using a probabilistic mixture model. Neuroimage 31:531–542.

38. Lu S, Ahn D, Johnson G, Law M, Zagzag D, Grossman RI (2004). Diffusion-tensor MR imaging of intracranial neoplasia and associated peritumoral edema: introduction of the tumor infiltration index. Radiology 232:221–228.

39. van Westen D, Latt J, Englund E, Brockstedt S, Larsson EM (2006). Tumor extension in high-grade gliomas assessed with diffusion magnetic resonance imaging: values and lesion-to-brain ratios of apparent diffusion coefficient and fractional anisotropy. Acta Radiol 47:311–319.

40. Sundgren PC, Fan X, Weybright P et al (2006). Differentiation of recurrent brain tumor versus radiation injury using diffusion tensor imaging in patients with new contrast-enhancing lesions. Magn Reson Imaging 24:1131–1142.

41. Provenzale JM, McGraw P, Mhatre P, Guo AC, Delong D (2004). Peritumoral brain regions in gliomas and meningiomas: investigation with isotropic diffusion-weighted MR imaging and diffusion-tensor MR imaging. Radiology 232:451–460.

42. Price SJ, Burnet NG, Donovan T et al (2003). Diffusion tensor imaging of brain tumours at 3T: a potential tool for assessing white matter tract invasion? Clin Radiol 58:455–462.

43. Lu S, Ahn D, Johnson G, Cha S (2003). Peritumoral diffusion tensor imaging of high-grade gliomas and metastatic brain tumors. Am J Neuroradiol 24:937–941.

44. Chiang IC, Kuo YT, Lu CY et al (2004). Distinction between high-grade gliomas and solitary metastases using peritumoral 3-T magnetic resonance spectroscopy, diffusion, and perfusion imagings. Neuroradiology 46:619–627.

45. Kono K, Inoue Y, Nakayama K et al (2001). The role of diffusion-weighted imaging in patients with brain tumors. Am J Neuroradiol 22:1081–1088.

46. Stadnik TW, Chaskis C, Michotte A et al (2001). Diffusion-weighted MR imaging of intracerebral masses: comparison with conventional MR imaging and histologic findings. Am J Neuroradiol 22:969–976.

47. Zhou XJ, Engelhard HH, Leeds NE, Villano JL (2007). Studies of glioma infiltration using a fiber coherence index. Magn Reson Med (in press).

48. Star-Lack J, Vigneron DB, Pauly J, Kurhanewicz J, Nelson SJ (1997). Improved solvent suppression and increased spatial excitation bandwidths for 3-D PRESS CSI using phase-compensating spectral/spatial spin-echo pulses. J Magn Reson Imaging 7:745–757.

49. Tran T-KC, Vigneron DB, Sailasuta N et al (2000). Very selective suppression pulses for clinical MRSI studies of brain and prostate cancer. Magn Reson Med 43:23–33.

50. Duyn JH, Gillen J, Sobering G, van Zijl PC, Moonen CT (1993). Multisection proton MR spectroscopic imaging of the brain. Radiology 188:277–282.

51. Duyn J, Moonen CT (1993). Fast proton spectroscopic imaging of human brain using multiple spin-echoes. Magn Res Med 30:409–414.

52. Adelsteinsson E, Irarrazabal P, Spielman DM, Macovski A (1995). Three-dimensional spectroscopic imaging with time-varying gradients. Magn Res Med 33:461–466.

53. Posse S, Tedeschi G, Risinger R, Ogg R, Le Bihan D (1995). High speed 1H spectroscopic imaging in human brain by echo planar spatialspectral encoding. Magn Res Med 33:34–40.

54. Marshall I, Wardlaw J, Cannon J, Slattery J, Sellar RJ (1996). Reproducibility of metabolite peak areas in 1H MRS of brain. Magn Reson Imaging 14:281–292.

55. Calvar JA (2006). Accurate (1)H tumor spectra quantification from acquisitions without water suppression. Magn Reson Imaging 24:1271–1279.

56. Birken DL, Oldendorf WH (1989). N-acetyl-L-aspartic acid: a literature review of a compound prominent in 1H-NMR spectroscopic studies of brain. Neurosci Biobehav Rev 13:23–31.

57. Moffett JR, Ross B, Arun P, Madhavarao CN, Namboodiri AM (2007). N-Acetylaspartate in the CNS: from neurodiagnostics to neurobiology. Prog Neurobiol 81:89–131.

58. Christiansen P, Toft P, Larsson HB, Stubgaard M, Henriksen O (1993). The concentration of N-acetyl aspartate, creatine + phosphocreatine, and choline in different parts of the brain in adulthood and senium. Magn Reson Imaging 11:799–806.

59. Wyss M, Kaddurah-Daouk R (2000). Creatine and creatinine metabolism. Physiol Rev 80:1107–1213.

60. Degaonkar MN, Pomper MG, Barker PB (2005). Quantitative proton magnetic resonance spectroscopic imaging: regional variations in the corpus callosum and cortical gray matter. J Magn Reson Imaging 22:175–179.

61. Kent C (205). Regulatory enzymes of phosphatidylcholine biosynthesis: a personal perspective. Biochim Biophys Acta 1733:53–66.

62. Babb SM, Ke Y, Lange N, Kaufman MJ, Renshaw PF, Cohen B (2004). Oral choline increases choline metabolites in human brain. Psychiatr Res 130:1–9.

63. Brief EE, Whittall KP, Li DK, MacKay A (2003). Proton T1 relaxation times of cerebral metabolites differ within and between regions of normal human brain. NMR Biomed 16:503–509.

64. Gajewicz W, Papierz W, Szymczak W, Goraj B (2003). The use of proton MRS in the differential diagnosis of brain tumors and tumor-like processes. Med Sci Monit 9:MT97–105.

65. Preul MC, Caramanos Z, Collins DL et al (1996). Accurate, noninvasive diagnosis of human brain tumors by using proton magnetic resonance spectroscopy. Nat Med 2:323–325.

66. Del Sole A, Falini A, Ravasi L et al (2001). Anatomical and biochemical investigation of primary brain tumours. Review. Eur J Nucl Med 28:1851–1872.

67. Delorme S, Weber MA (2006). Applications of MRS in the evaluation of focal malignant brain lesions. Cancer Imaging 22:95–99.

68. Lai PH, Ho JT, Chen WL et al (2002). Brain abscess and necrotic brain tumor: discrimination with proton MR spectroscopy and diffusion-weighted imaging. Am J Neuroradiol 23:1369–1377.

69. Cho YD, Choi GH, Lee SP, Kim JK (2003). (1)H-MRS metabolic patterns for distinguishing between meningiomas and other brain tumors. Magn Reson Imaging 21:663–672.

70. Devos A, Lukas L, Suykens JA et al (2004). Classification of brain tumours using short echo time 1H MR spectra. J Magn Reson 170:164–175.

71. Li X, Vigneron DB, Cha S et al (2005). Relationship of MR-derived lactate, mobile lipids, and relative blood volume for gliomas in vivo. Am J Neuroradiol 6:760–769.

72. Law M, Yang S, Wang H et al (2003). Glioma grading: sensitivity, specificity, and predictive values of perfusion MR imaging and proton MR spectroscopic imaging compared with conventional MR imaging. Am J Neuroradiol 24:1989–1998.

73. Fayed N, Morales H, Modrego PJ, Pina MA (2006). Contrast/Noise ratio on conventional MRI and choline/creatine ratio on proton MRI spectroscopy accurately discriminate low-grade from high-grade cerebral gliomas. Acad Radiol 13:728–737.

74. Chen J, Huang SL, Li T, Chen XL (2006). In vivo research in astrocytoma cell proliferation with 1H-magnetic resonance spectroscopy: correlation with histopathology and immunohistochemistry. Neuroradiology 48:312–318.

75. Jenkinson MD, Smith TS, Joyce K et al (2005). MRS of oligodendroglial tumors: correlation with histopathology and genetic subtypes. Neurology 64:2085–2089.

76. White ML, Zhang Y, Kirby P, Ryken TC (2005). Can tumor contrast enhancement be used as a criterion for differentiating

tumor grades of oligodendrogliomas? Am J Neuroradiol 26:784–790.

77. Lev MH, Ozsunar Y, Henson JW et al (2004). Glial tumor grading and outcome prediction using dynamic spin-echo MR susceptibility mapping compared with conventional contrast-enhanced MR: confounding effect of elevated rCBV of oligodendrogliomas. Am J Neuroradiol 25:214–221.

78. Xu M, See SJ, Ng WH et al (2005). Comparison of magnetic resonance spectroscopy and perfusion-weighted imaging in presurgical grading of oligodendroglial tumors. Neurosurgery 56:919–924.

79. Chernov M, Hayashi M, Izawa M et al (2005). Differentiation of the radiation-induced necrosis and tumor recurrence after gamma knife radiosurgery for brain metastases: importance of multi-voxel proton MRS. Minim Invasive Neurosurg 48:228–234.

80. Matulewicz L, Sokol M, Wydmanski J, Hawrylewicz L (2006). Could lipid CH_2/CH_3 analysis by in vivo 1H MRS help in differentiation of tumor recurrence and post-radiation effects? Folia Neuropathol 44:116–124.

81. Cohen BA, Knopp EA, Rusinek H, Babb JS, Zagzag D, Gonen O (2005). Assessing global invasion of newly diagnosed glial tumors with whole-brain proton MR spectroscopy. Am J Neuroradiol 26:2170–2177.

82. Gajewicz W, Grzelak P, Gorska-Chrzastek M, Zawirski M, Kusmierek J, Stefanczyk L (2006). The usefulness of fused MRI and SPECT images for the voxel positioning in proton magnetic resonance spectroscopy and planning the biopsy of brain tumors: presentation of the method. Neurol Neurochir Pol 40:284–290.

83. Hall WA, Martin A, Liu H, Truwit CL (2001). Improving diagnostic yield in brain biopsy: coupling spectroscopic targeting with real-time needle placement. J Magn Reson Imaging 13:12–15.

84. Payne GS, Leach MO (2006). Applications of magnetic resonance spectroscopy in radiotherapy treatment planning. Br J Radiol 79 (Special Issue 1):S16–S26.

85. Graves EE, Nelson SJ, Vigneron DB et al (2000). A preliminary study of the prognostic value of 1H-spectroscopy in gamma knife radiosurgery of recurrent malignant gliomas. Neurosurgery (Baltimore) 46:319–328.

86. Graves EE, Pirzkall A, Nelson SJ, Verhey L, Larson D (2001). Registration of magnetic resonance spectroscopic imaging to computed tomography for radiotherapy treatment planning. Med Phys 28:2489–2496.

87. Chan YL, Yeung DK, Leung SF, Cao G (1999). Proton magnetic resonance spectroscopy of late delayed radiation-induced injury of the brain. J Magn Reson Imaging 10:130–137.

88. Graves EE, Nelson SJ, Vigneron DB et al (2001). Serial proton MR spectroscopic imaging of recurrent malignant gliomas after gamma knife radiosurgery. Am J Neuroradiol 2:613–624.

89. Plotkin M, Eisenacher J, Bruhn H et al (2004). 123I-IMT SPECT and 1H MR-spectroscopy at 3.0 T in the differential diagnosis of recurrent or residual gliomas: a comparative study. J Neuro-Oncol 70:9–58.

90. Hollingworth W, Medina LS, Lenkinski RE et al (2006). A systematic literature review of magnetic resonance spectroscopy for the characterization of brain tumors. J Neuroradiol 27:1404–1411.

91. Sugahara T, Korogi Y, Kochi M, Ushio Y, Takahashi M (2001). Perfusion-sensitive MR imaging of gliomas: comparison between gradient-echo and spin-echo echo-planar imaging techniques. Am J Neuroradiol 22:1306–1315.

92. Cha S (2006). Dynamic susceptibility-weighted contrast-enhanced perfusion MR imaging in pediatric patients. Neuroimaging Clin N Am 6:137–147.

93. Cha S, Knopp EA, Johnson G, Wetzel SG, Litt AW, Zagzag D (2002). Intracranial mass lesions: dynamic contrast-enhanced susceptibility-weighted echo-planar perfusion MR imaging. Radiology 223:11–29.

94. Lupo JM, Lee MC, Han ET et al (2006). Feasibility of dynamic susceptibility contrast perfusion MR imaging at 3T using a standard quadrature head coil and eight-channel phased-array coil with and without SENSE reconstruction. J Magn Reson Imaging 24:520–529.

95. Nakagawa T, Tanaka R, Takeuchi S, Takeda N (1998). Haemodynamic evaluation of cerebral gliomas using XeCT. Acta Neurochir (Wien) 140:223–233.

96. Muizelaar JP, Fatouros PP, Schroder ML (1997). A new method for quantitative regional cerebral blood volume measurements using computed tomography. Stroke 28:1998–2005.

97. Law M, Young R, Babb J et al (2006). Comparing perfusion metrics obtained from a single compartment versus pharmacokinetic modeling methods using dynamic susceptibility contrast-enhanced perfusion MR imaging with glioma grade. Am J Neuroradiol 27:1975–1982.

98. Hakyemez B, Erdogan C, Ercan I, Ergin N, Uysal S, Atahan S (2005). High-grade and low-grade gliomas: differentiation by using perfusion MR imaging. Clin Radiol 60:493–502.

99. Stewart GN (1893). Researches on the circulation time in organs and on the influences which affect it: Parts I–III. J Physiol (London) 15:1–89.

100. Lupo JM, Cha S, Chang SM, Nelson SJ (2005). Dynamic susceptibility-weighted perfusion imaging of high-grade gliomas: characterization of spatial heterogeneity. Am J Neuroradiol 26:1446–1454.

101. Uematsu H, Maeda M (2006). Double-echo perfusion-weighted MR imaging: basic concepts and application in brain tumors for the assessment of tumor blood volume and vascular permeability. Eur Radiol 16:180–186.

102. Vonken EJ, van Osch MJ, Bakker CJ, Viergever MA (1999). Measurement of cerebral perfusion with dual-echo multi-slice quantitative dynamic susceptibility contrast MRI. J Magn Reson Imaging 10:109–117.

103. Vonken EP, van Osch MJ, Bakker CJ, Viergever MA (2000). Simultaneous quantitative cerebral perfusion and Gd-DTPA extravasation measurement with dual-echo dynamic susceptibility contrast MRI. Magn Reson Med 43:820–827.

104. Daldrup-Link HE, Brasch RC (2003). Macromolecular contrast agents for MR mammography: current status. Eur Radiol 13:354–365.

105. Yang S, Law M, Zagzag D (2003). Dynamic contrast-enhanced perfusion MR imaging measurements of endothelial permeability: differentiation between atypical and typical meningiomas. Am J Neuroradiol 24:1554–1559.

106. Boxerman JL, Schmainda KM, Weisskoff RM (2006). Relative cerebral blood volume maps corrected for contrast agent extravasation significantly correlate with glioma tumor grade, whereas uncorrected maps do not. Am J Neuroradiol 27:859–867.

107. Hartmann M, Heiland S, Harting I et al (2003). Distinguishing of primary cerebral lymphoma from high-grade gliomas with perfusion-weighted magnetic resonance imaging. Neurosci Lett 338:119–122.

108. Rollin N, Guyotat J, Streichenberger N, Honnorat J, Tran Minh VA, Cotton F (2006). Clinical relevance of diffusion and perfusion magnetic resonance imaging in assessing intra-axial brain tumors. Neuroradiology 48:150–159.

109. Essig M, Waschkies M, Wenz F, Debus J, Hentrich HR, Knopp MV (2003). Assessment of brain metastases with dynamic susceptibility-weighted contrast-enhanced MR imaging: initial results. Radiology 228:193–199.

110. Kremer S, Grand S, Berger F et al (2003). Dynamic contrast-enhanced MRI: differentiating melanoma and renal carcinoma metastases from high-grade astrocytomas and other metastases. Neuroradiology 45:44–49.

111. Kremer S, Grand S, Remy C et al (2004). Contribution of dynamic contrast MR imaging to the differentiation between dural metastasis and meningioma. Neuroradiology 46:642–648.

112. Calli C, Kitis O, Yunten N, Yurtseven T, Islekel S, Akalin T (2006). Perfusion and diffusion MR imaging in enhancing malignant cerebral tumors. Eur J Radiol 58:394–403.

113. Knopp EA, Cha S, Johnson G et al (1999). Glial neoplasms: dynamic contrast-enhanced T2*-weighted MR imaging. Radiology 211:791–798.

114. Sugahara T, Korogi Y, Kochi M et al (1998). Correlation of MR imaging-determined cerebral blood volume maps with histologic and angiographic determination of vascularity of gliomas. Am J Roentgenol 171:1479–1486.

115. Maia AC, Malheiros SM, da Rocha AJ et al (2005). MR cerebral blood volume maps correlated with vascular endothelial growth factor expression and tumor grade in nonenhancing gliomas. Am J Neuroradiol 26:777–783.

116. Shin JH, Lee HK, Kwun BD et al (2002). Using relative cerebral blood flow and volume to evaluate the histopathologic grade of cerebral gliomas: preliminary results. Am J Roentgenol 179:783–789.

117. Aronen HJ, Gazit IE, Louis DN et al (1994). Cerebral blood volume maps of gliomas: comparison with tumor grade and histologic findings. Radiology 191:41–51.

118. Chaskis C, Stadnik T, Michotte A, Van Rompaey K, D'Haens J (2006). Prognostic value of perfusion-weighted imaging in brain glioma: a prospective study. Acta Neurochir (Wien) 148:277–285.

119. Lupo JM, Cha S, Chang SM, Nelson SJ (2005). Dynamic susceptibility-weighted perfusion imaging of high-grade gliomas: characterization of spatial heterogeneity. Am J Neuroradiol 26:1446–1454.

120. Maia AC, Malheiros SM, da Rocha AJ et al (2004). Stereotactic biopsy guidance in adults with supratentorial nonenhancing gliomas: role of perfusion-weighted magnetic resonance imaging. J Neurosurg 101:970–976.

121. Tsui EY, Chan JH, Leung TW et al (2000). Radionecrosis of the temporal lobe: dynamic susceptibility contrast MRI. Neuroradiology 42:149–152.

122. Siegal T, Rubinstein R, Tzuk-Shina T, Gomori JM (1997). Utility of relative cerebral blood volume mapping derived from perfusion magnetic resonance imaging in the routine follow up of brain tumors. J Neurosurg 86:22–27.

123. Sugahara T, Korogi Y, Tomiguchi S et al (2000). Posttherapeutic intraaxial brain tumor: the value of perfusion-sensitive contrast-enhanced MR imaging for differentiating tumor recurrence from nonneoplastic contrast-enhancing tissue. Am J Neuroradiol 21:901–909.

124. Ostergaard L, Hochberg FH, Rabinov JD et al (1999). Early changes measured by magnetic resonance imaging in cerebral blood flow, blood volume, and blood-brain barrier permeability following dexamethasone treatment in patients with brain tumors. J Neurosurg 90:300–305.

125. Bastin ME, Carpenter TK, Armitage PA, Sinha S, Wardlaw JM, Whittle IR (2006). Effects of dexamethasone on cerebral perfusion and water diffusion in patients with high-grade glioma. Am J Neuroradiol 27:402–408.

126. Wilkinson ID, Jellineck DA, Levy D et al (2006). Dexamethasone and enhancing solitary cerebral mass lesions: alterations in perfusion and blood-tumor barrier kinetics shown by magnetic resonance imaging. Neurosurgery 58:640–646.

127. Liebner S, Fischmann A, Rascher G et al (2000). Claudin-1 and claudin-5 expression and tight junction morphology are altered in blood vessels of human glioblastoma multiforme. Acta Neuropathol (Berl) 100:323–331.

128. Ginsberg LE, Fuller GN, Hashmi M, Leeds NE, Schomer DF (1998). The significance of lack of MR contrast enhancement of supratentorial brain tumors in adults: histopathological evaluation of a series. Surg Neurol 49:436–440.

129. Roberts HC, Roberts TP, Ley S, Dillon WP, Brasch RC (2002). Quantitative estimation of microvascular permeability in human brain tumors: correlation of dynamic Gd-DTPA-enhanced MR imaging with histopathologic grading. Acad Radiol 9(Suppl 1):S151–S155.

130. Hazle JD, Jackson EF, Schomer DF, Leeds NE (1997). Dynamic imaging of intracranial lesions using fast spin-echo imaging: differentiation of brain tumors and treatment effects. J Magn Reson Imaging 7:1084–1093.

131. Uematsu H, Maeda M, Sadato N et al (2000). Vascular permeability: quantitative measurement with double-echo dynamic MR imaging-theory and clinical application. Radiology 214:912–917.

132. Roberts HC, Roberts TPL, Brasch RC, Dillon WP (2000). Quantitative measurement of microvascular permeability in human brain tumors achieved using dynamic contrast-enhanced MR imaging: correlation with histologic grade. Am J Neuroradiol 21:891–899.

133. Provenzale JM, Wang GR, Brenner T, Petrella JR, Sorensen AG (2002). Comparison of permeability in high-grade and low-grade brain tumors using dynamic susceptibility contrast MR imaging. Am J Roentgenol 178:711–716.

134. Law M, Yang S, Babb JS et al (2003). Comparison of cerebral blood volume and vascular permeability from dynamic susceptibility contrast enhanced perfusion MR imaging with glioma grade. Am J Neuroradiol 25:746–755.

135. Jackson A, Kassner A, Annesley-Williams D, Reid H, Zhu XP, Li KL (2002). Abnormalities in the recirculation phase of contrast agent bolus passage in cerebral gliomas: comparison with relative blood volume and tumor grade. Am J Neuroradiol 23:7–14.

136. Provenzale JM, York G, Moya MG et al (2006). Correlation of relative permeability and relative cerebral blood volume in high-grade cerebral neoplasms. Am J Roentgenol 187:1036–1042.

CHAPTER **20**

Contrast Agents in Neuroradiological MRI: Current Status

Robert V. Mulkern, Richard L. Robertson and Alan B. Packard

INTRODUCTION

New contrast agents and new ideas for their use in magnetic resonance imaging (MRI) continue to be explored and developed and there are several reviews available [1–4]. Tagging plasma proteins with paramagnetic moieties [5–7] as 'blood pool' agents, labeling stem cells with superparamagnetic nanoparticles for tracking their migration in vivo [8–10], targeting cell surface receptors with specially designed paramagnetic probes [11], achieving paramagnetic probes which can be 'activated' in response to specific in vivo conditions such as pH or temperature [12], and even loading tumors with gadolinium-based agents for enhancing the effectiveness of neutron capture therapy [13] are all fascinating works in progress at this time. Some of these methods may achieve useful clinical status while some will remain exclusively research tools. Our goal here is not to describe these developments or attempt to predict the future utility of any approach, but rather to focus on the current status of the MRI contrast agents in clinical use. These are the low molecular weight (600–800 dalton) gadolinium (Gd) complexes that have been used on a daily basis in neuroradiological exams for nearly two decades. We briefly review the safety aspects, relaxation properties, biodistributions and the established clinical and research applications associated with the use of these 'garden variety' MRI contrast agents in neuroradiology.

The chemical and physical aspects of the common gadolinium-based MRI contrast agents, safety, biodistributions and relaxation effects have been well-documented [14–19] and there is an enormous body of literature regarding their application in neuroradiological MR examinations. Essentially there are three approved Gd contrast agents, each prepared at 500 mM Gd/kg concentration, currently in use in the USA [16]. The practical differences between these three agents are marginal. Their on-label uses, found in the package inserts, are similar. As pointed out by Runge and Knopp [20], however, package insert information for the same product can differ widely depending on the country in which the product is sold. In neuroradiological MRI examinations, venous injection of some 20 ml of the agent, typically resulting in in vivo concentrations of approximately 0.1 mM/kg, is routine. Power injectors for controlled bolus injections at rates from 1 to 10 ml/s are in common use and are recommended when using the agent for 'first pass perfusion' studies in the brain (see below).

SAFETY

The clinically established MRI contrast agents are all based on the magnetic properties of the gadolinium (III) ion (Gd^{+++}) which shortens the T1 and T2 relaxation times of water molecules with which the ion comes in contact or in close proximity. Gadolinium (III) is toxic and so must be trapped within a molecular cage, or 'chelated' so that it can be safely excreted from the body [16]. Though many Gd complexes have been created and tested [14], the three agents approved by the US Food and Drug Administration (FDA) and most commonly used in the USA are ProHance (gadoteridol, Bracco Diagnostics, Princeton, NJ), Omniscan (gadodiamide, Amersham Health, Princeton, NJ) and Magnevist (gadopentatate dimeglumine, Berlex Laboratories, Wayne, NJ). Omniscan and ProHance are both non-ionic while Magnevist is an ionic compound. In the healthy human body, excretion through the kidneys occurs with half-lives on the order of 90 minutes – nearly 500 times shorter than free Gd [16]. The minor concern associated with Gd complexes administered intravenously is that other ions, such as zinc or copper, can displace the Gd ion from its 'cage'. Thus, the longer half-lives of the complexes in patients with renal insufficiencies or increased copper levels, as in patients with Wilson's disease, have raised concerns, though clinical manifestations of Gd toxicity in such patient populations have not been established [16]. Common dose levels of 0.1 mM Gd/kg, administered intravenously, appear to be well tolerated with triple dosing and repeat administration on a monthly basis, yielding no apparent toxicity problems [18]. Lethal dose levels

(LD_{50}) for rodents are greater than 6 mM/kg, some 20 to 60 times higher than the doses administered clinically [19]. Intrathecal administration is considered dangerous [20–22]. Rare adverse reactions to clinical doses in humans, less than 5% [16], include hives, emesis, nausea, headache, irritation or hot/cool sensations at the injection site and, very rarely, hypocalcemia reactions [23,24]. It is generally understood that the standard gadolinium-based contrast agents are safer than iodinated contrast agents used for CT. Of course, from an environmental perspective, the administration of millions of doses of labile Gd contrast agents to patients is ultimately a significant source of heavy metal pollution, affecting first sewerage and then groundwater [25]. As such, this aspect of Gd contrast agent use, which tends to be very little discussed, should probably be considered when truly pondering the overall effects of these agents on the many varied species inhabiting our planet.

RELAXATION EFFECTS: WHY THE FOCUS ON T1 AND NOT T2?

The classic radiological sign associated with venous administration of a Gd agent is signal enhancement on T1-weighted images so that T1-weighted sequences, both gradient and spin-echo, are traditionally employed to query the effects of the agent. The Gd complex causes a reduction in T1 that is related to the concentration of the agent in any given voxel. The size of the T1 reduction is related to the T1-relaxivity of the agent. The relaxivity refers to the slope of the relaxation rate, $R1 = 1/T1$, versus mM Gd concentration, and is on the order of $5 s^{-1}/mM$ Gd, as measured in water or plasma, for the common Gd agents [14,15]. Of course, there are also T2-reductions associated with the T2-relaxivities. The T2-relaxivities are slightly higher than the T1-relaxivities, on the order of $6–7 s^{-1}/mM$ Gd and, in fact, Gd-related T2-shortening generally works against the T1-related signal enhancement. In typical T1-weighted imaging sequences with short echo times (TEs) and short repetition times (TRs), however, the dominant effect is from the T1 change and it is fairly straightforward to see why. Consider a typical tissue T1 value which, in the absence of Gd, is on the order of 1 s, or equivalently a non-enhanced relaxation rate, $R1 = 1/T1$, of $1 s^{-1}$. At a tissue concentration of 0.1 mM Gd/kg, some $0.5 s^{-1}$ (relaxivity × mM Gd) is added to the native relaxation rate so that the Gd-associated R1 becomes $1.5 s^{-1}$, yielding a T1 of 667 ms. This translates to a substantial 33% decrease in T1. Typical T2 and $R2 = 1/T2$ values in brain parenchyma are on the order of 80 ms and $12.5 s^{-1}$, respectively. Adding the $0.6 s^{-1}$ associated with a 0.1 mM Gd concentration to the native T2 relaxation rate yields only a 4% decrease in T2, from 80 ms to 77 ms. Thus, the T2 effect from Gd is generally negligible compared to T1 changes, though T2-darkening of signal can be found in regions of high Gd concentration, such as the renal medulla, as the Gd complexes are extracted from the blood via the kidneys. Finally, when reporting results from 'routine' clinical studies of, for example, magnetic resonance spectroscopy or diffusion imaging, it is prudent to include information regarding whether the exams were performed before or after the administration of Gd contrast, as these can confound interpretations (26).

BIODISTRIBUTION AND APPLICATIONS: CLINICAL AND RESEARCH

Following intravenous injection, the Gd contrast agents distribute themselves first through the vasculature and then, with the exception of neural tissue with an intact blood–brain barrier (BBB), throughout the extracellular space. Neither the ionic nor non-ionic Gd agents currently in use cross cell membranes. The wash-in phase to the extracellular (extravascular but not intracellular, sometimes referred to as 'interstital') space is followed by a wash-out phase as the Gd agents are cleared from the blood by the kidneys and, by extension, from the extracellular spaces back to the blood for renal excretion. Neural tissue is special in regard to these Gd agents in that an intact BBB truly minimizes the effect of the agent in typical T1-weighted scans. For example, Steen et al measured T1 in normal-appearing gray and white matter of brain tumor patients before and after injection of gadodiamide [27]. They found T1 reductions of less than 2% in the normal brain tissues despite obvious T1 reductions (signal enhancement) in the tumors. Aronen et al measured T1 in brain tumors pre- and post- 0.1 mM/kg and 0.2 mM/kg injections of Gd contrast and reported T1 reductions on the order of 50% in the tumors with only marginal effects on gray and white matter T1 values [28]. Thus, the intact BBB is a key factor in how Gd agents affect brain signal intensity on T1-weighted images. Even with an intact BBB, however, so-called T2* effects are observable when the agent first appears at full strength in the brain vasculature. This effect can be gainfully employed for 'first pass' perfusion studies as discussed below, following a brief review of the standard post-Gd T1-weighted imaging findings.

TYPICAL GD CONTRAST AGENT ENHANCEMENT FEATURES IN CLINICAL PRACTICE

Fairly non-specific signal enhancement is observed in many tumors [29–31], 'active' multiple sclerosis lesions [18,32–34], inflammation [35–38] and radiation myelopathy [39–41]. To varying degrees, the use of Gd agents helps demarcate the extent and pattern of such pathologies and can make small lesions, like acoustic neuromas, more conspicuous. In general, however, the enhancement must be considered more sensitive than specific. Some normal tissues encountered in brain imaging, such as the nasal/sinus mucosa, choroid plexus, cranial nerves, cavernous sinus, infundibulum and pituitary gland will also show varying degrees of enhancement [29,30,42–45] which will depend on the subject's age. Figure 20.1 shows normal Gd

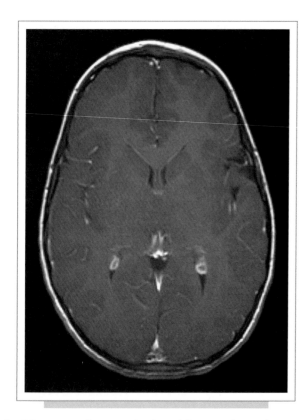

FIG. 20.1. Axial T1-weighted gadolinium-enhanced image in a 14-year-old boy shows prominent high signal in the glomus of the choroid plexus bilaterally. Enhancement of intracranial vessels is also apparent.

enhancement of the glomus of the choroid plexus, as well as several intracranial vessels, in a 14-year-old patient in whom no abnormal enhancements were otherwise observed. Considerable meningeal enhancement is often observed in cases of low cerebrospinal fluid (CSF) pressure and/or leaks, as in the case of intracranial hypotension syndrome and orthostatic headaches [46,47]. Subjects with diabetes [48] and renal failure [49,50] have shown increased Gd enhancement in brain structures and other tissues naturally imaged when performing brain imaging, like orbital vitreous, presumably due to a compromised BBB in these areas. A full accounting of the numerous findings associated with signal enhancement from the common Gd agents is impossible within the scope of a brief review and the references cited in this work are to be understood as a representative, not comprehensive, sample. Since the use of these agents is pervasive and presumably useful, the practicing MRI radiologist should become familiar with the various benefits and limitations of the information gleaned from Gd-associated signal enhancement, or non-enhancement, as for example often encountered in necrotic regions of tumor. It should also be taken into account that studies designed to prove the efficacy of Gd contrast agent media have not been as rigorous as one would desire. In a critical examination of the

literature on Gd contrast agents in neuroimaging published several years ago, Breslau et al concluded that although Gd contrast agents '…play an essential role in lesion detection and confidence of interpretation, no rigorous studies exist to establish valid sensitivity and specificity estimates for their application' [51], an observation suggesting caution when interpreting Gd enhancement features in general.

FIRST PASS PERFUSION STUDIES AND GD CONTRAST AGENTS

Aside from direct signal enhancement in T1-weighted scans, the second most common use of Gd agents in neuroimaging is probably the so-called 'first pass' perfusion studies in which estimates, relative and/or absolute, of cerebral blood volume (CBV) and cerebral blood flow (CBF) may be determined. As stated above, an intact BBB keeps Gd agents within the vasculature and mitigates T1 effects. The presence of Gd in the blood, however, causes a change in the magnetic susceptibility of the vessels. This in turn increases the susceptibility difference between brain parenchyma and vessels, generating microscopic magnetic field gradients which reach beyond the vessel walls and cause dephasing of spins within brain parenchyma [52]. This dephasing is formally described as a reduction in the transverse relaxation times T2 and T2* of the extravascular protons. Since the Gd within the blood rapidly escapes into the extracellular spaces throughout the rest of the body, this dephasing effect is only transient and primarily observed during the 'first pass' of the agent through the brain. Thus, the phenomenon must be accessed using the high temporal resolution imaging afforded by echo planar imaging (EPI) sequences. Typically, either gradient echo EPI sequences sensitive to T2* changes, or spin echo EPI sequences sensitive to T2 changes, are used to monitor the transient dephasing during the first pass of the agent through the brain vasculature. These sequences allow for the acquisition of some 20 brain slices every 2 or 3 seconds, with data collection commencing some 10–15 seconds prior to a controlled bolus intravenous injection. Data collection generally continues for up to 2 minutes or more after injection so that the time course of the brain signal can be measured within each voxel at 2–3 second intervals. What one looks for in the resulting signal versus time curves are 'dips' associated with the first pass of the agent through the brain. The overall shape of the dip depends on a number of physiological parameters including regional CBV and CBF. A thorough mathematical modeling based on standard physiological models can be found in the literature [53,54]. In practical terms, most manufacturers provide post-processing image analysis tools in which images of the area of the dip, numerically integrated over operator-defined time periods, reflect local CBV values (big dip, big CBV). They also generally provide time-to-peak (TTP) images and images of CBF, though the latter generally requires estimating the arterial input function from a conspicuous blood vessel like the middle cerebral artery (MCA). Though the standard Gd agents are used routinely for this application, a newer agent, gadobutrol, similar to the others but prepared at 1 M rather than the 500 mM Gd concentration and approved for use in Europe, may offer some advantages in terms of tighter boluses [55]. There is now a fairly large body of literature regarding the use of first pass 'perfusion' methods performed with Gd agents for clinical applications in stroke [56–60] and tumors [61–65]. In stroke, it is becoming clear that first pass perfusion methods can help identify the extent of 'ischemic penumbra' in relation to the infarct 'core' as defined by diffusion imaging methods. This so-called 'perfusion/diffusion mismatch' may ultimately help with patient management decisions as recently reviewed by Gonzales [60]. In tumors, the disruption of the BBB results in leakage of the Gd into the tumor, a phenomenon that complicates justification of some of the standard 'first pass' perfusion assumptions while offering additional information regarding tumor vascularity and function which may prove useful in characterizing tumors [65]. Though the first pass Gd perfusion studies based on T2* sequences can be used for assessing blood volume and flow dynamics; slower imaging methods, which monitor the T1 changes with time as Gd washes in and out of tissues, are also used for measuring aspects of vascular permeability and compartmental volumes. This topic, as yet more in the research as opposed to the clinical realm, is discussed further below following a discussion of Gd contrast agents for magnetic resonance angiography examinations.

MAGNETIC RESONANCE ANGIOGRAPHY AND GD CONTRAST AGENTS

The two primary imaging methods used in MRI to highlight blood vessels, magnetic resonance angiography (MRA), are the time-of-flight (TOF) technique and the phase contrast (PC) technique. To some degree, judicious use of Gd contrast can enhance the value of either MRA method by essentially allowing better visualization of the smaller distal branches due to the (transient) shortened T1 of blood [66–71]. With TOF for example, rapid radiofrequency (RF) pulsing of selected tissue sections serves to T1 saturate static spins within the section while spins flowing into the section have seen none, or only a few, of the RF pulses. These spins are thus largely T1 unsaturated and so appear bright, allowing the separation of flowing blood within vessels from the surrounding tissue. The bright vessels are most often displayed using maximum intensity projection (MIP) algorithms to generate MR angiograms. TOF methods inherently utilize T1 saturation effects so that lowering blood T1 values with Gd contrast helps the blood remain unsaturated, even for slow flowing blood in small vessels which may experience many RF pulses. Similarly, phase contrast techniques, where the term 'contrast' here refers to the contrast generated by the

phase as opposed to magnitude of the MR signal, and not to the 'contrast' associated with the Gd agent contrast material, primarily utilize rapid gradient echo sequences whose signals will generally benefit from shortened blood T1 values. Thus, at appropriate doses, Gd contrast can enhance the value of TOF and PC MRA, though at costs including enhancement of some surrounding tissues and additional enhancement of the venous system that can confound the interpretation of arterial signals which may be of greater interest. As both 2D and 3D TOF acquisitions have become faster, timing of bolus injections of Gd contrast is more and more often coordinated with the actual acquisitions so that the various phases of arterial and venous enhancement can be separately visualized. Figures 20.2 A–C are inverted MIP images from a rapid 3D TOF acquisition from a 6-year-old

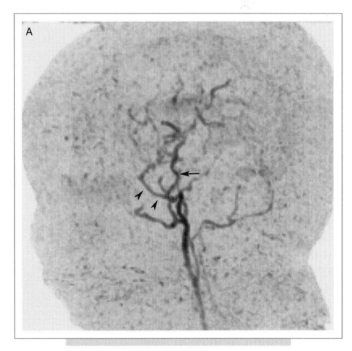

FIG. 20.2. Time resolved MR angiography in a 6-year-old female with moyamoya disease. The exam was performed one year after surgical revascularization with pial synangiosis. The pulse sequence was a 3D rapid gradient echo sequence with TR/TE values of 4.2/1.3 ms/ms, and basically acquired 16 slice 3D data sets at approximately 5 s intervals for 75 s after administration of 0.1 mM Gd contrast. The maximum intensity projection (MIP) images are displayed with inverted gray scale. **(A)** Lateral projection of early arterial phase showing a prominent superficial temporal artery (arrow) and middle meningeal artery (arrowheads). **(B)** Lateral projection of mid-arterial/parenchymal phase demonstrating numerous parenchymal branches and early filling of the sagittal sinus (arrows). **(C)** Lateral projection venous phase shows washout of the arterial structures and opacification of the dural venous sinuses.

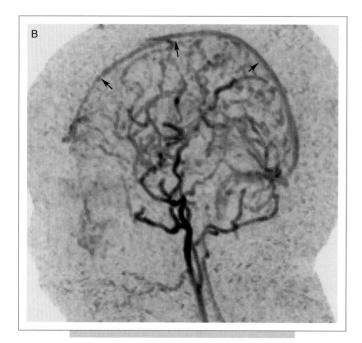

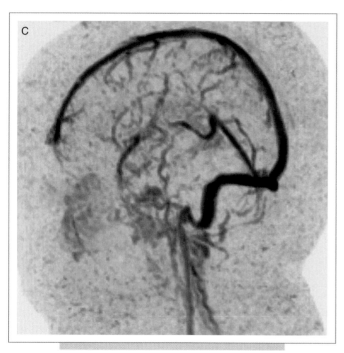

FIG. 20.2. (*Continued*)

female with moyamoya disease demonstrating how the angiograms change with time following Gd administration, first showing arterial enhancement followed by venous enhancement. The use of Gd contrast agents in MRA is not necessary and can add to the cost and complexity of the exam but, when used properly, can also enhance the utility of a given MRA study.

DYNAMIC CONTRAST AGENT STUDIES FOR VASCULAR PERMEABILITY AND EXTRACELLULAR VOLUME CHARACTERIZATION

In brain tumors, active MS lesions and other pathologies where a compromised BBB occurs, varying levels of T1-related signal enhancement will occur over time as the Gd washes in and then out of the tissue. Current standard of care generally involves documenting regions and patterns of enhancement on post-Gd images acquired several minutes (usually an unspecified number of minutes) after contrast agent administration. This approach largely ignores the precise temporal course of the enhancement that, in fact, contains a great deal of information. In so-called dynamic contrast enhanced (DCE) studies, a series of T1-weighted images are collected over time after Gd administration in order to extract this information. Typically, time resolutions of 5 to10 seconds for several slices are achieved with rapid gradient echo sequences applied for several minutes. From such data sets, the time course of the signal changes, which are directly related to Gd concentrations, are recorded from voxels of interest. Three general features help characterize these signal versus time curves. These are:

1 how quickly enhancement takes place during the wash-in phase
2 the peak enhancement level achieved; and
3 how quickly the enhancement recedes during the wash-out phase.

Pharmacokinetic models, similar to the compartmental models used for nuclear medicine tracer studies [72,73], are used to extract information regarding aspects of abnormal capillary leakage and the extravascular extracellular space (EES) which may be used to characterize lesions. In the most quantitative analyses of such studies, signal from a large blood vessel is used to estimate plasma Gd concentration which spikes rapidly to a maximum and generally follows a two phase wash-out, or bi-exponential signal decay, from the vasculature [74–76]. Complete modeling also requires a baseline measurement of the T1 value prior to Gd administration, as well as knowledge of the T1 relaxivity of the agent. Coupling these data with the basic signal versus time curve from the lesion allows one to extract three primary physiological parameters. These are: K^{trans}, the volume transfer constant between plasma and the EES in min^{-1}; v_e, the volume fraction of the extravascular extracellular space (EES) which is dimensionless and is also referred to as the interstitial space; and the so-called rate constant k_{ep}, which is the ratio K_{trans}/v_e. The latter rate constant is the simplest to extract as it can be estimated from the shape of the signal versus time curve alone, while the other two parameters require complete measurements including baseline T1 values, in vivo relaxivities and plasma concentrations of Gd. In the limit of low permeability, K^{trans}

may be interpreted as the product of the permeability P times the surface area S per unit mass of tissue and has units of ml g^{-1} min^{-1}. Tofts et al have provided an extensive discussion of the numerous symbols and units associated with the MRI based Gd dynamic contrast agent studies [75] which, though generally associated with T1 based signal changes, can also be adapted to T2 or T2* based studies. Despite the complexity of the modeling and the more demanding experimental protocols, fairly interesting and promising assessments of the potential for improved tissue characterization have been published, particularly in the grading and differentiation of brain tumors [61–65,77–80]. Such studies will no doubt continue to be assessed and hold significant potential for characterizing elements of the complex vasculature and abnormal capillary permeabilities associated with various disease processes. In this regard,

advanced modeling of the signal intensities from both spin echo and gradient echo sequences following Gd administration has been forwarded for assessing aspects of angiogenesis, including estimates of mean vessel size [81,82], a subject less well-developed than the tracer dynamic field but of considerable interest and potential for expanded use of the common Gd contrast agents.

Post-script: The discussion of safety aspects for the common gadolinium chelates in this chapter is based on literature reviews performed prior to recently reported observations linking nephrogenic systemic fibrosis (NSF) to gadolinium contrast use in patients with renal insufficiency. Readers are alerted to ongoing investigations [83,84] and concerns expressed by the FDA [85] regarding this potential aspect of the toxicity of the gadolinium contrast agents currently in use.

REFERENCES

1. Jasanoff A (2005). Functional MRI using molecular imaging agents. Trend Neurosci 28:120–126.
2. Benaron DA (2002). The future of cancer imaging. Cancer Metastasis Rev 21:45–78.
3. Roberts TPL, Chuang N, Roberts HC (2000). Neuroimaging: do we really need new contrast agents for MRI? Eur J Radiol 34:166–178.
4. Knopp MV, Tengg-Kobligk H von, Floemer F, Schoenberg SO (1999). Contrast agents for MRA: future directions. J Magn Reson Imaging 10:314–316.
5. Adzamli K, Yablonskiy DA, Chicoine MR et al (2003). Albumin-binding MR blood pool agents as MRI contrast agents in an intracranial mouse glioma model. Magn Reson Med 49:586–590.
6. Schwarzbauer C, Morrissey SP, Deichmann R et al (1997). Quantitative magnetic resonance imaging of capillary water permeability and regional blood volume with an intravascular MR contrast agent. Magn Reson Med 37:769–777.
7. Eldredge HB, Spiller M, Chasse JM, Greenwood MT, Caravan P (2006). Species dependence on plasma protein binding and relaxivity of the gadolinium-based MRI contrast agent MS-325. Invest Radiol 41:229–243.
8. Zhang ZG, Jiang Q, Jiang F et al (2004). In vivo magnetic resonance imaging tracks adult neural progenitor cell targeting of brain tumor. Neuroimage 23:281–287.
9. Anderson SA, Glod J, Arbab AS (2005). Noninvasive MR imaging of magnetically labeled stem cells to directly identify neovasculature in a glioma model. Blood 105:420–425.
10. Modo M, Roberts TJ, Sandhu JK, Williams SCR (2004). In vivo monitoring of cellular transplants by magnetic resonance imaging and positron emission tomography. Expert Opin Biol Ther 4:145–155.
11. Artemov D, Bhujwalla ZM, Bulte JWM (2004). Magnetic resonance imaging of cell surface receptors using targeted contrast agents. Curr Pharm Biotechnol 5:485–494.
12. Lowe MP (2004). Activated MR contrast agents. Curr Pharm Biotechnol 5:519–528.
13. Matsumura A, Zhang T, Yamamoto T et al (2003). In vivo gadolinium neutron capture therapy using a potentially effective compound (Gd-BOPTA). Anticancer Res 23:2451–2456.
14. Caravan P, Ellison JJ, McMurry TJ, Lauffer RB (1999). Gadolinium (III) chelates as MRI contrast agents: structure, dynamics, and applications. Chem Rev 99:2293–2352.
15. Gore JC, Joers JM, Kennan RP (2002). Contrast agents and relaxation effects. In *Magnetic Resonance Imaging in the Brain and Spine*, 3rd edn. Atlas SW (ed.). Lippincott Williams & Wilkins, Philadelphia. 79–99.
16. Shellock FG, Kanal E (1999). Safety of magnetic resonance imaging contrast agents. J Magn Reson Imaging 10:477–484.
17. Yoshikawa K, Davies A (1997). Safety of ProHance in special populations. Eur Radiol 7(suppl 5):246–250.
18. Filippi M, Rovaris M, Capra R et al (1998). A multi-centre longitudinal study comparing the sensitivity of monthly MRI after standard and triple dose gadolinium-DTPA for monitoring disease activity in multiple sclerosis. Implications for phase II clinical trials. Brain 121:2011–2020.
19. Oksendal A, Hals P (1993). Biodistribution and toxicity of MR imaging contrast media. J Magn Reson Imaging 3:157–165.
20. Runge VM, Knopp MV (1999). Off-label use and reimbursement of contrast media in MR. J Magn Reson Imaging 10:489–495.
21. Siebner HR, Grafin von Einsiedel H, Conrad B (1997). Magnetic resonance ventriculography with gadolinium DTPA: report of two cases. Neuroradiology 39:418–422.
22. Ray DE, Cavanagh JB, Nolan CC, Williams SC (1996). Neurotoxic effects of gadopentetate dimeglumine: behavioral disturbance and morphology after intracerebroventricular injection in rats. Am J Neuroradiol 17:365–373.
23. Davenport A, Whiting S (2006). Profound pseudohypocalcemia due to gadolinium (Magnevist) contrast in a hemodialysis patient. Am J Kidney Dis 47:350–352.
24. Williams SF, Meek SE, Moraghan TJ (2005). Spurious hypocalcemia after gadodiamide administration. Mayo Clinic Proc 80:1655–1657.

25. Kummerer K, Helmers E (2000). Hospital effluents as a source of gadolinium in the aquatic environment. Environ Sci Technol 34:573–577.

26. Sijens PE, vandenBent MJ, Nowak PCJM, vanDijk P, Oudkerk M (1997). H-1 chemical shift imaging reveals loss of brain tumor choline signal after administration of Gd-contrast. Magn Reson Med 37:222–225.

27. Steen RG, Reddick WE, Ogg RJ, Langston JW (1999). Effect of a gadodiamide contrast agent on the reliability of brain tissue T1 measurements. Magn Reson Imaging 17:229–235.

28. Aronen HJ, Niemi P, Kwong KK, Pardo FS, Davis TL (1998). The effect of paramagnetic contrast media on T1 relaxation times in brain tumors. Acta Radiol 39:474–481.

29. Stack JP, Antoun NM, Jenkins JPR, Metcalfe R, Isherwood I (1988). Gadolinium-DTPA as a contrast agent in magnetic resonance imaging of the brain. Neuroradiology 30:145–154.

30. Ge HL, Hirsch WL, Wolf GL, Rubin RA, Hackett RK (1992). Diagnostic role of gadolinium-DTPA in pediatric neuroradiology: a retrospective review of 655 cases. Neuroradiology 34:122–125.

31. Forsyth PAJ, Petrov E, Mahallati H et al (1997). Prospective study of postoperative magnetic resonance imaging in patients with malignant gliomas. J Clin Oncol 15:2076–2081.

32. McFarland HF, Frank JA, Albert PS et al (1992). Using gadolinium-enhanced magnetic resonance imaging lesions to monitor disease activity in multiple sclerosis. Ann Neurol 32:758–766.

33. Miller DH, Barkhof F, Nauta JJP (1993). Gadolinium enhancement increases the sensitivity of MRI in detecting disease activity in multiple sclerosis. Brain 116:1077–1094.

34. Inglese M, Grossman RI, Filippi M (2005). Magnetic resonance imaging of multiple sclerosis lesion evolution. J Neuroimaging 15:22S–29S.

35. Perry JR, Fung A, Poon P, Bayer N (1994). Magnetic resonance imaging of nerve root inflammation in the Guillain-Barre syndrome. Neuroradiology 36:139–140.

36. Fitzgerald DC, Mark AS (1996). Endolymphatic duct/sac enhancement on gadolinium magnetic resonance imaging of the inner ear: preliminary observations and case reports. Am J Otol 17:603–606.

37. Donovan MJD, Sze G, Quencer RM, Eismont FJ, Green BA, Gahbauer H (1990). Gadolinium-enhanced MR in spinal infection. J Comput Assist Tomogr 14:721–729.

38. Casselman JW, Kuhweide R, Ampe W, Meeus L, Steyaert L (1993). Pathology of the membranous labyrinth: comparison of T1- and T2-weighted and gadolinium-enhanced spin-echo and 3DFT-CISS imaging. Am J Neuroradiol 14:59–69.

39. Michikawa M, Wada Y, Sano M et al (1991). Radiation myelopathy: significance of gadolinium-DTPA enhancement in the diagnosis. Neuroradiology 33:286–289.

40. Rampling R, Symonds P (1998). Radiation myelopathy. Curr Opin Neurol 11:627–632.

41. Martin D, Delacollette D, Collignon J et al (1994). Radiation-induced myelopathy and vertebral necrosis. Neuroradiology 36:405–407.

42. Kilgore DP, Breger RK, Daniels DL, Pojunas KW, Williams AL, Haughton VM (1986). Cranial tissues: normal MR appearance after intravenous injection of Gd-DTPA. Radiology 160:757–761.

43. Sakamoto Y, Takahashi M, Korogi Y, Bussaka H, Ushio Y (1991). Normal and abnormal pituitary glands: gadopentetate dimeglumine-enhanced MR imaging. Radiology 178:441–445.

44. Holodny AI, George AE, Golomb J, de Leon MJ, Kalnin AJ (1998). The perihippocampal fissures: normal anatomy and disease states. Radiographics 18:653–665.

45. Elster AD (1990). Cranial MR imaging with Gd-DTPA in neonates and young infants: preliminary experience. Radiology 176:225–230.

46. Mokri B, Piepgras DG, Miller GM (1997). Syndrome of orthostatic headaches and diffuse pachymeningeal gadolinium enhancement. Mayo Clin Proc 72:400–413.

47. Brightbill TC, Goodwin RS, Ford RG (2000). Magnetic resonance imaging of intracranial hypotension syndrome with pathophysiological correlation. Headache 40:292–299.

48. Starr JM, Wardlaw J, Ferguson K, MacLullich A, Deary IJ, Marshall I (2003). Increased blood–brain barrier permeability in type II diabetes demonstrated by gadolinium magnetic resonance imaging. J Neurol Neurosurg Psychiatr 74:70–76.

49. Lev MH, Schafer PW (1999). Subarachnoid gadolinium enhancement mimicking subarachnoid hemorrhage on FLAIR MR images. Am J Roentgenol 173:1414–1415.

50. Kanamalla US, Boyko OB (2002). Gadolinium diffusion into orbital vitreous and aqueous humor, perivascular space, and ventricles in patients with chronic renal disease. Am J Roentgenol 179:1350–1352.

51. Breslau J, Jarvik JG, Haynor DR, Longstreth WT, Kent DL, Maravilla KR (1999). MR contrast media in neuroimaging: a critical review of the literature. Am J Neuroradiol 20:670–675.

52. Villringer A, Rosen BR, Belliveau JW et al (1988). Dynamic imaging with lanthanide chelates in normal brain: contrast due to magnetic susceptibility effects. Magn Reson Med 6:164–174.

53. Ostergaard L, Weisskoff RM, Chesler DA, Gyldensted C, Rosen BR (1996). High resolution measurement of cerebral blood flow using intravascular tracer bolus passages. Part I: Mathematical approach and statistical analysis. Magn Reson Med 36:715–725.

54. Ostergaard L, Sorenson AG, Kwong KK, Weisskoff RM, Gyldensted C, Rosen BR (1996). High resolution measurement of cerebral blood flow using intravascular tracer bolus passages. Part II: Experimental comparison and preliminary results. Magn Reson Med 36:726–736.

55. Tombach B, Heindel W (2002). Value of 1.0-M gadolinium chelates: review of preclinical and clinical data on gadobutrol. Eur Radio 12:1550–1556.

56. Ueda T, Yuh WTC, Taoka T (1999). Clinical application of perfusion and diffusion MR imaging in acute ischemic stroke. J Magn Reson Imaging 10:305–309.

57. Parsons MW, Barber A, Chalk J et al (2001). Diffusion- and perfusion-weighted MRI response to thrombolysis in stroke. Ann Neurol 51:28–37.

58. Karonen JO, Liu Y, Vanninen RL et al (2000). Combined perfusion- and diffusion-weighted MR imaging in acute stroke during the 1st week: A longitudinal study. Radiology 217:886–894.

59. Grandin CB, Duprez TP, Smith AN et al (2002). Which MR-derived perfusion parameters are the best predictors of infarct growth in hyperacute stroke? Comparative study between relative and quantitative measurements. Radiology 223:361–370.

60. Gonzales RG (2006). Imaging-guided acute ischemic stroke therapy: from 'Time is brain' to 'Physiology is brain'. Am J Neuroradiol 27:728–735.

61. Law M, Yang S, Babb JS et al (2004). Comparison of cerebral blood volume and vascular permeability from dynamic susceptibility contrast-enhanced perfusion MR imaging with glioma grade. Am J Neuroradiol 25:746–755.

62. Li KL, Zhu XP, Waterton J, Jackson A (2000). Improved 3D quantitative mapping of blood volume and endothelial permeability in brain tumors. J Magn Reson Imaging 12:347–357.

63. Lupo JM, Cha SM, Chang SM, Nelson SJ (2005). Dynamic susceptibility-weighted perfusion imaging of high-grade gliomas: characterization of spatial heterogeneity. Am J Neuroradiol 26:1446–1454.

64. Cha S, Tihan T, Crawford F et al (2005). Differentiation of low-grade oligodendrogliomas from low-grade astrocytomas by using quantitative blood-volume measurements derived from dynamic susceptibility contrast-enhanced MR imaging. Am J Neuroradiol 26:266–273.

65. Patankar TF, Haroon HA, Mills SJ et al (2005). Is volume transfer coefficient (K^{trans}) related to histologic grade in human gliomas? Am J Neuroradiol 26:2455–2465.

66. Ozsarlak O, Van Goethem JW, Maes M, Parizel PM (2004). MR angiography of the intracranial vessels: technical aspects and clinical applications. Neuroradiology 46:955–972.

67. Mathews VP, Elster AD, King JC, Ulmer JL, Hamilton CA, Strottmann JM (1995). Combined effects of magnetization transfer and gadolinium in cranial MR imaging and angiography. Am J Roentgenol 164:169–172.

68. Marchal G, Michiels J, Bosmans H, Van Hecke P (1992). Contrast enhanced MRA of the brain. J Comput Assist Tomogr 16:25–29.

69. Jager HR, Ellamushi H, Moore EA, Grieve JP, Kitchen ND, Taylor WJ (2000). Contrast-enhanced MR angiography of intracranial giant aneurysms. Am J Neuroradiol 21:1900–1907.

70. Ozsarlak O, Parizel PM, Van Goethem JW (2004). Low-dose gadolinium enhanced 3D time-of-flight MR angiography of the intracranial vessels using PAT optimized phased array 8-channel head coil. Eur Radiol 14:2067–2071.

71. Jung HW, Chang KH, Choi DS, Han MH, Han MC (1995). Contrast-enhanced MR angiography for the diagnosis of intracranial vascular disease: optimal dose of gadopentate dimeglumine. Am J Roentgenol 165:1251–1255.

72. Ohno K, Pettigrew KD, Rapoport SI (1978). Lower limits of cerebro-vascular permeability to nonelectrolytes in the conscious rat. Am J Physiol 235:H299–307.

73. Iannotti F, Fieschi C, Alfano B et al (1987). Simplified noninvasive PET measurement of blood-brain barrier permeability. J Comput Assist Tomogr 11:390–397.

74. Tofts PS (1997). Modeling tracer kinetics in dynamic Gd-DTPA MR imaging. J Magn Reson Imaging 7:91–101.

75. Tofts PS, Brix G, Buckley DL et al (1999). Estimating kinetic parameters from dynamic contrast-enhanced T1-weighted MRI of a diffusible tracer: Standardized quantities and symbols. J Magn Reson Imaging 10:223–232.

76. Tofts PS, Kermode AG (1991). Measurement of the blood–brain barrier permeability and leakage space using dynamic MR imaging. 1. Fundamental concepts. Magn Reson Med 17:357–367.

77. Wong ET, Jackson EF, Hess KR et al (1998). Correlation between dynamic MRI and outcome in patients with malignant gliomas. Neurology 50:777–781.

78. Hawighorst H, Knopp MV, Debus J et al (1998). Pharmacokinetic MRI for assessment of malignant glioma response to stereotactic radiotherapy: Initial results. J Magn Reson Imaging 8:783–788.

79. Roberts HC, Roberts TPL, Brasch RC, Dillon WP (2000). Quantitative measurement of microvascular permeability in human brain tumors achieved using dynamic contrast-enhanced MR imaging: correlation with histologic grade. Am J Neuroradiol 21:891–899.

80. Knopp MV, von Tengg-Kobligk H, Choyke PL (2003). Functional magnetic resonance imaging in oncology for diagnosis and therapy monitoring. Molec Cancer Ther 2:419–426.

81. Schmainda KM, Rand SD, Joseph AM et al (2004). Characterization of a first-pass gradient-echo spin-echo method to predict brain tumor grade and angiogenesis. Am J Neuroradiol 25:1524–1532.

82. Kiselev VG, Strecker R, Ziyeh S, Speck O, Hennig J (2005). Vessel size imaging in humans. Magn Reson Med 53:553–563.

83. Akgun H, Gonlusen G, Cartwright J Jr, Suki WN, Truong LD (2006). Are gadolinium-based contrast media nephrotoxic? A renal biopsy study. Arch Pathol Lab Med 130:1354–1357.

84. Grobner T (2006). Gadolinium: a specific trigger for development of nephrogenic fibrosing dermopathy and nephrogenic systemic fibrosis? Nephrol Dial Transplant 21:1104–1108.

85. US Federal Drug Administration (FDA). Public health advisory: update on MRI contrast agents containing gadolinium and nephrogenic fibrosing dermopathy. Created June 8, 2006. Updated December 22, 2006 and May 23, 2007. Available at: http://www.fda.gov/cder/drug/advisory/gadolinium_agents.htm.

CHAPTER 21

Physics of High Field MRI and Applications to Brain Tumor Imaging

Alayar Kangarlu

INTRODUCTION

Magnetic resonance imaging (MRI) has launched a revolution in diagnostic medicine since its introduction in the early 1980s. Since the beginning, it has witnessed an unremitting drive toward higher magnetic fields [1–5]. The primary obstacles against this drive have been magnet technology and patient safety [6,7]. The whole body scanners use superconducting wires wound around cylinders with diameters about 1 meter (m) and a length of about 2 m. The magnetic field strength of the majority of clinical scanners presently is 1.5 Tesla. In the early 1990s, a few 4T research whole body scanners were introduced by the major manufacturers and these units established the bases for a gradual introduction of higher fields into clinical settings [1–4]. Today, 3T whole body scanners are appearing in clinics and hospitals at a rapid rate. In 1997, an 8T whole body scanner was debuted by Ohio State University [5], whose proof of safe exposure of human subjects [6,7] prepared the stage for manufacturers to follow suit and consider the feasibility of high field (HF) MRI for medical imaging. Today, more than twenty academic institutions and research laboratories around the world operate 7T whole body scanners. This has changed the classification of MRI scanners as low field (LF), high field (HF), or ultra high field (UHF) as they have been transformed every decade. While as early as last decade a 3T scanner was considered to be a UHF magnet for research, today this field strength is viewed simply as high field for clinical applications and low field or medium field strength for research applications. And 7T and higher field scanners, which were considered ultra high field at their debut, are now simply considered to be HF scanners. Such rapid change in taxonomy and proliferation of HF technology was achieved due to the feasibility and stringent safety work that was demonstrated by the 8T scanner. Another byproduct of this rapid increase in the number of HF scanners is the interest of many scientists and engineers in its potential leading to developments such as parallel imaging in hardware [8,9],

diffusion tensor imaging (DTI) in pulses sequences and molecular imaging in the area of contrast agents.

In this review, we present a summary of the physics of MRI, in which the advantages of HF MRI are identified. In addition, the inevitable disadvantages are described, and opportunities for engineering innovations for their minimization or elimination are surveyed. Considering that HF scanners are approaching the limits of existing superconducting materials' tolerance for magnetic flux, ways of understanding these limits and working with or around them are also illustrated.

MAGNETIC RESONANCE IMAGING PHYSICS

In order to better understand the role of the latest developments in enhancing sensitivity and specificity of MRI in augmenting its diagnostic capabilities, a brief review of the process of signal generation by magnetic resonance is necessary. The underlying process in this technique is nuclear magnetization, which all elements in nature exhibit. Hydrogen (^1H) is the most abundant element in biological tissues that has a large gyromagnetic ratio (γ). γ represents the nuclear magnetic dipole (μ) and its rotational dynamic that governs the ways which individual μ's from spin and orbital motion of different protons and neutrons add to constitute nuclear magnetic moment, ultimately causing a detectable signal from outside the body [10]. All elements with odd numbers of protons or neutrons in their nucleus have a non-zero angular momentum and magnetic moment. In this regard, ^1H is a unique element with the smallest odd number of protons. By the same token, other nuclei with odd number nucleons (protons and neutrons) exhibit intrinsic magnetic moment which is the resultant of individual magnetic moments. In the same manner that tiny magnetic moments (fields) of the nucleons of an individual nucleus add to make up the elemental μ such as μ (^1H), the magnetic fields generated by each atom of a nucleus within an imaging voxel add to make up the magnetization vector associated with that voxel.

Upon exposure of an object to a strong static magnetic field (B_0), it interacts with the spin motion splitting their randomly distributed population into two groups of parallel (N^-) and antiparallel (N^+) protons with respect to the direction of B_0. The population difference between the two groups is governed by the Boltzman distribution function:

$$\frac{N^-}{N^+} = e^{\frac{\Delta E}{kT}}$$

where $\Delta E = /2\pi h\gamma B_0/2\pi$, T is the room temperature and k is the Boltzman constant ($1.38 * 10^{-23} J/K$).

For ^1H with a $\gamma = 42.575 MHz/T$ a field strength of $B_0 = 1T$ will split the ^1H population with two groups whose energies are $\Delta E = 2.82 \times 10^{-26}$ Joules (J) apart. Considering that human MRI is performed at room temperature, $kT = 4.01 \times 10^{-21}$ J or about 25 meV, the ratio of $\Delta E/kT$ becomes an infinitesimally small number, of the order of $\approx 10^{-6}$. For such a small number, exponential terms on the right side of the equation could be approximated by $\approx 1 + \Delta E/kT$ or expressed in terms of the excess proton or population difference of $\Delta N = N^- - N^+$ which, for such fields and temperatures, amounts to a $\Delta N/N$ of 3 parts per million (ppm). This simple analysis implies that a magnetic field of 1T will divide every group of 1 million water molecules, with 2 million hydrogen atoms, into two almost equal groups of about $1\,000\,000 \pm 2$ hydrogen atoms. In other words, population difference caused by a 1T magnetic field amounts to only 3 more parallel protons (N^-) than antiparallel ones (N^+). This tiny population difference for such a relatively large external field is the reason for MRI's insensitivity. Were it possible to probe every hydrogen atom of the water molecules within the body, signal-to-noise ratio (SNR) of MRI would have been more than $100\,000$ times stronger. Such increase in SNR would have easily placed cellular and even molecular imaging within the reach of clinical magnets operating at 1.5T. But this is not the case, and we had to work around this inherent insensitivity of MRI. So, once again, resorting to HF remains the only way to increase SNR.

On the other hand, given the high concentration of water in the body, each $1\,\mu l$ or $1\,mm^3$ of biological tissues contains about 10^{15} excess protons, ΔN. Such a population of uncompensated protons will generate an SNR ≈ 10 compared to the background noise. This is the reason why, in spite of the NMR insensitivity, we can still generate images representing voxels of $\approx 1\,\mu l$. But, not much smaller! This warrants a discussion on how HF MRI can be further sensitized to become a cancer-specific diagnostic imaging tool. Following, we will discuss the principles of operation of critical technologies in HF MRI and those areas of highest potential for increasing MRI

sensitivity towards the limits needed to visualize pre-angiogenic tumor pathology [11].

MRI OF BRAIN TUMORS – STATE OF THE ART

In spite of the availability of whole body HF scanners for a decade, few applications in cancer have surfaced to date [12,13]. Aside from the HF, the state of the art in brain tumor imaging is incorporation of fast techniques and new contrast agents in visualizing the tumor. Yet, for most tumors and their metastases, the current state of the art in clinical MRI is unable to detect the tumor until after it has developed neovasculature [14]. While the ability for visualization of microstructures by clinical scanners based on the indigenous contrast of deoxyhemoglobin is low, administration of contrast agents in animals have shown some promising results. Reliable vascular density quantification by contrast agents will help make routine MRI a better indicator of tumor activity. Correlations between microvascular density and cell proliferation in human glioma found by imunohistochemical methods underlines the significance of such imaging capability for human cerebral gliomas [14]. This also emphasizes the significance of HF in vivo detection of microvessels of about $100\,\mu m$ diameter which can not be easily detected by 1.5T scanners.

INDIRECT MEASURE OF PATHOLOGY

The primary collateral damage of cancer to brain tissues detectable by routine MRI are edema and hemorrhage. Also, due to the ability of diffusion imaging at 1.5T to distinguish fiber structures, such techniques can be used to detect any damage to myelinated axons. But, besides these indirect measures of tumor etiology, clinical MRI lacks specificity for prediction of underlying pathology or histologic grade of tumors. So, one mission for MRI could be enhancement of such specificity. In therapeutics, MRI could develop into a tool to make a distinction between tumor recurrence and consequences of radiation therapy, such as tumor necrosis. Presently, all means for assessment of tumor activity offered by MRI, such as vascular, metabolic and physiological, are indirect. Metabolic quantification offered by magnetic resonance spectroscopy (MRS) enables measurement of choline which is a good marker of membrane turnover that could precede the formation of necrotic lesions. High resolution ^1H MRS will provide a measure of this activity. MRI studies to date have visualized brain tumors in contrast with normal tissues. The mechanism of signal generation in MRI has not been proven to correlate with the underlying pathology of cancer, although consequences of cancer, such as breakdown of the blood–brain barrier (BBB), necrosis, hemorrhage, edema and inflammation, have MRI implications that have been used in the past for visualization of tumors. However, correlation between these measures and clinical manifestations of the disease has not yet been established.

A DIRECT MEASURE OF PATHOLOGY

The ideal feature in MRI for cancer studies would be an ability to monitor directly the pathologic processes associated with cancer. Imaging techniques could be sensitized for particular structural or physiological characteristics which best represent a particular pathology. In cancer, angiogenesis lends itself well to MRI optimizations for cancer imaging. For example, routine MRI operating at 1.5T has come a long way since its early days in imaging brain tumors. Today's 1.5T images have higher SNR, contrast to noise ratio (CNR) and fewer artifacts than their earlier counterparts. This level of progress has enabled clinicians to use spectroscopy, perfusion, DTI, and soon functional magnetic resonance imaging (fMRI), at 1.5T as approved tools of medical diagnosis. But, in spite of these many innovations to improve the quality of images and speed up acquisition, the underlying insensitivity of MRI has prevented this modality from being able directly to detect tumors at their inception, based on its etiology and not collateral consequences. Beyond the HF increase of SNR, affiliated technologies such as radiofrequency (RF) coils, pulse sequences and gradient technology have reached the level of maturity that engineering and safety limits allow. Further enhancement in MRI ability of tumor detection requires extension of its reach into microscopic aspects of tissue architecture. One active area for sensitization of MRI to cancer pathology is with application of contrast agents. A new class of contrast enhancing agents with the ability to target directly pathological sites has raised hopes for achieving molecular level detection of cancer with whole body MRI scanners. When this happens, it will amount to a second revolution in diagnostic imaging, the first being the appearance of MRI as a routine radiological device.

HOW MRI COMPARES WITH OTHER IMAGING TECHNIQUES

One of the major disadvantages of MRI is its long acquisition time compared to optical imaging, computer assisted tomography (CT) and positron emission tomography (PET). But all of these other techniques are beset by difficulties of their own. CT scanning involves the use of ionizing radiation. Optical imaging suffers from very small penetration depth within the human body. PET images can only be produced at low resolution (5 mm) and are invasive. MR images produce a high soft tissue contrast at a high resolution. But the primary contrast mechanisms are magnetization relaxation times, T1 and T2. Considering that T1s are of the order of 1 second and T2s of 0.1 second, and spatial encoding of the resonant spins requires phase encoding in at least one dimension of a 2D image, the acquisition times become long, on the order of 5 minutes per pulse sequence. Since its inception, MRI has constantly witnessed faster imaging sequences to deal with this drawback. Imaging in steady-state, single shot excitation techniques such as echoplanar imaging (EPI), multiple refocusing techniques such as fast spin echo (FSE), etc. have considerably reduced imaging time, but not down to real time yet.

HIGH FIELD IMAGING: PROS AND CONS

In assessing the role of HF MRI for cancer, the underlying pathology should be kept in perspective. Existing imaging techniques have been proven to be insensitive for the detection of most tumors or their metastases at their early stages. Subsequent to tumor neovascularization, MRI can use the contrast generated by the presence of extra paramagnetic enhancement to visualize neovasculature or other concurring or ensuing events such as edema. Dependence of 1.5T MRI on collateral events such as edema for visualization of tumors has set a ceiling for the utility of this tool for non-invasive monitoring of the progression of the disease and therapeutic efficacy. This is why the search for techniques capable of accessing the microstructure of the brain tissues is at full throttle. Few alternatives are available to sensitize MR to cancer pathology at a microscopic level. High magnetic field (B_0) is one such option, representing the ultimate limit on the SNR of images. SNR is directly proportional to B_0 and is the primary determinant of the image resolution [15,16]. HF MRI is also critically dependent on RF coil technology and gradient design. Particularly, RF plays an important role on the quality of HF images. This issue and ways of improving it are analyzed in this chapter. Improvement of the image quality to the extent of microscopic resolution in MRI is a significant goal for cancer and must be rigorously pursued. In this regard, while advantages of HF MRI in its application to cancer are well known [17], the disadvantages and challenges in materializing its goals must also be fully scrutinized. Understanding of these challenges will translate them into engineering topics for research that will ultimately help improve the technology and sharpen its usefulness [18].

HIGH RESOLUTION IMAGING

In further enhancing the visibility of cancer pathology to HF MRI, the visualization of the normal tissues and the etiology of cancer must be considered. The most striking aspect of images published by HF whole body MRI scanners provide evidence that acquisition of microscopic resolution from the human body within clinically acceptable time is possible, as shown in Figure 21.1. Such resolution is achieved [17] while staying within regulatory guidelines concerning RF power deposition and gradient switching. In addition, each pulse sequence has an RF energy associated with it which is a function of sequence parameters. These sequence parameters could be adjusted to prevent most sequences from posing a safety hazard at high field strengths [7]. For example, gradient echo (GRE) contains less RF energy than does spin echo. This makes it ideal for high field applications. This is particularly true, since GRE

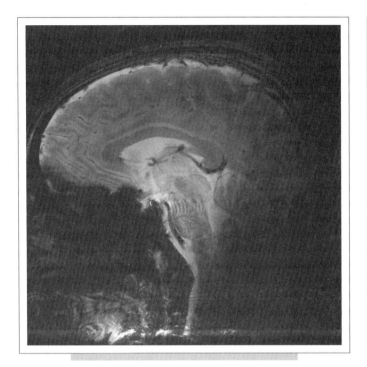

FIG. 21.1. A sagittal GRE image of a human brain at 8 T. A number of artifacts appear on this image including (A) strong signal void due to the tissue–air interface in the prefrontal cortex around the sinuses and orbitals; (B) loss of all tissues in the oral cavity; (C) RF ripple artifact originating from multiple flip angle, the off-center placement of the head in the coil causing signal loss in posterior region and inadequate insertion of the head in the coil subjecting the lower portion of the image to inhomogeneous B1 causing strong distortion in tissue representation.

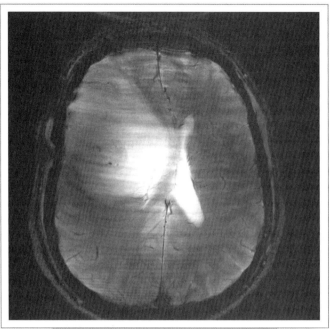

FIG. 21.2. An axial GRE image of a brain tumor at 8 T. Distinct features are (A) high SNR as translated into 200 μm in-plane resolution; (B) bright center or dielectric resonances; (C) no signal from the skull due to its short T2* and high susceptibility; (D) horizontal striped artifacts due to a repetitive motion caused by pulsatile motions in the disproportionate amount of CSF within the slice plane.

has high sensitivity for the detection of microvasculature, such as capillaries and venules [17]. The high paramagnetic effect produced by deoxyhemoglobin will generate an inhomogeneous magnetic field in the near vicinity of these vessels, causing rapid dephasing in pressesional coherence, i.e. drastically reducing T2*. Such a dephasing effect is a function of the magnetic field, giving rise to a more striking signature for these vessels at higher fields as is shown in Figures 21.1 and 21.2. At HF, GRE gains additional sensitivity that is particularly useful for visualization of neovasculature around and within the tumor. While 100 μm resolution images of the whole head from an awake subject is possible, in the absence of ultrafast techniques, acquisition of these images is long. However, given the prevalence of 200–300 μm vessels in most brain tumors, GRE could be further sensitized routinely to detect neovasculature in vivo. This aspect of HF MRI lends itself to the development of new non-invasive tools for monitoring tumor progression and identifying and validating biological endpoints whereby novel anti-angiogenesis agents can be more rapidly evaluated regarding their angiostatic efficacy and clinical relevance. Other sequences have the potential

for cancer specific enhancement. These include diffusion-based images and spectroscopic techniques. While diffusion as a thermally induced process is not field dependent, the high SNR of high field will generate a high contrast to noise ratio (CNR) that enhances conspicuity of the diffusion-modifying pathology. Edema, in particular, has a strong diffusion signature. Even with diffusion-weighted imaging (DWI) and a spatial resolution of $250 \times 250 \times 1000\,\mu m$, images of regional morphological and functional heterogeneity in the tumor could be obtained at HF. This resolution, however, is not sufficient to resolve the microvasculature. Reports of sequences visualizing cortical microanatomy in brain samples have shown a contrast superior to any clinical imaging techniques that rival histology [18]. Others have been able to optimize resolution of 3 T scanners to observe cortical layers in human brain [19]. Observation of multilaminar structures in the cortical gray matter is a milestone in in vivo detection of brain microstructures and they can be clearly seen in images from today's HF scanners. While high SNR in HF MRI has already found many applications, its help in cancer diagnosis and therapy has not been immediate. But it holds promises that should be further explored, while considering other means of image enhancement, such as data acquisition, RF coils, relaxation effects, susceptibility contrast and contrast agents. Innovations in these areas, just within the short time since the 8T debut, have made great strides in making HF scanners a viable investment in radiology. Scientists are actually designing experiments with their 7T whole body scanners that were inconceivable with clinical imaging at 1.5T. Development of MR microscopy (MRM) for in vivo application neuroimaging has the potential to produce results that would correlate the clinical deficit with presence or absence of different types of tumors. Development of in vivo human MRM with a 50–100 μm resolution is not far from reach and will further inform tumor studies. Our ability to routinely acquire images with a 200 μm resolution have set the stage for further innovations in lesion microscopy [17]. In vivo human MRM will be the precursor to in vivo human cellular imaging. The technical issues on the path to full realization of in vivo MRM are presently the subject of many engineering studies in MRI.

MAGNETIC SUSCEPTIBILITY

Inhomogeneity of living tissues combined with compartmentalized organs and cavities introduce a challenge in MRI. This is due to the necessity in MRI for a highly homogeneous magnetic field at the imaged site. This stringent homogeneity requirement, i.e. about 0.1 ppm/cm, is the primary reason for the size and price of the MRI magnets. Tissue heterogeneity modifies B_0 homogeneity and must be compensated for by shimming schemes, pulse sequence design, or post-processing. In humans, tissue contents vary from solid to gaseous and various consistencies of liquid in between. This drastic variation in consistencies has profound contrast implications, particularly on HF MRI images due to magnetic susceptibility (χ) effects on magnetization (M) of tissue within each voxel. As $\mathbf{B_0} = \mu_0(1 + \chi)\mathbf{H}$, where $\mathbf{M} = \chi\mathbf{H}$, μ_0 is the permeability of air, and $\mathbf{H}$ is the magnetic flux, variation of χ causes change in B_0 which deteriorates B_0 homogeneity over the sample. In spite of the fact that susceptibility causes loss of signal, an accurate account of its magnitude could enable counter measures to minimize its adverse effects and enable one to use its T2* contrast potentials to their advantage [20]. At high field, susceptibility induced T2* effects could be used as a source of contrast to achieve very high resolution to enhance MRI images to the limit of microscopy. While magnetic susceptibility causes artifacts causing signal loss in brain near air-filled sinuses, the T2*-based gradient recalled echo (GRE) sequence has shown much potential on high resolution images at 7T and higher. GRE uses a gradient reversal to rephase the transverse magnetization (**M**) in contrast to spin echo (SE) sequence that uses a 180° RF pulse [21,22]. As gradients are unable to rephase the dephasing caused by local susceptibility-induced inhomogeneities, the heterogeneous nature of the tissues drastically reduces the T2* at high field making it suitable for microstructure depiction. The 8T GRE images [14,17] made the case for

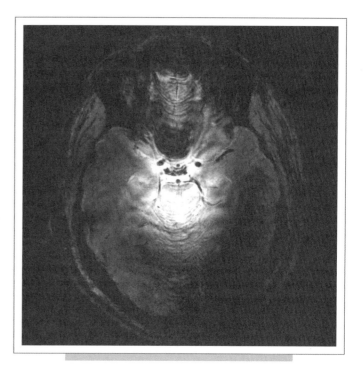

FIG. 21.3. An axial GRE view of a human head image acquired at 8 T. This image has a strong dielectric resonance at the central regions and a strong susceptibility source inferior to the slice location. The susceptibility inhomogeneity caused in the inferior regions of this slice extends over a large region including superior occipital gyrus, calcarine sulcus and fusiform gyrus.

the utility of GRE in neurology applications, such as brain tumor studies, by highlighting its unique capability in showing microvasculature. However, GRE images bear testimony to artifacts that go beyond voxel point spread function as seen in Figures 21.1 and 21.3. Any large source of magnetic susceptibility will cast a large dark shadow on the images. Susceptibility-induced artifacts due to sinuses, vestibular tube or dental fillings are particularly pronounced at HF and could wipe out signal in a large portion of the image. GRE artifacts can also lower the efficiency of fat suppression as fat and water resonances are only 3.5 ppm apart. Susceptibility could also have a detrimental effect on diffusion-based contrasts such as diffusion weighted imaging (DWI), apparent diffusion constant (ADC) and diffusion tensor imaging (DTI). The effect of magnetic susceptibility on diffusion is due to diffusion theory assumption that signal loss from various molecules is a function of their state of motion alone. Magnetic susceptibility and its variation across the sample add complications to this picture that should be taken into account for diffusion image interpretation. The initial studies at 8T highlighted the contrast between strong T2* of GRE and T2 of SE by investigating the feasibility of fast spin echo (FSE) sequences, such as rapid acquisition by relaxation enhancement (RARE) [23]. The use of FSE sequences was viewed with caution for HF applications due to their high RF power content caused by RF refocusing. But the ability of FSE for suppression of T2* mechanism, thereby increasing SNR and enabling the use of T2 mechanism for brain studies, provided motivation for investigation into the health and feasibility consequences of their use at HF [6,7,23,24].

As was mentioned above, the T2* could also be used as a source of contrast. Such is the case in blood oxygen level dependent (BOLD) mechanisms that generate a T2*-based change in SNR, also called activation SNR, in functional MRI studies. BOLD SNR increases proportional to T2* generating functional contrast in brain tissues based on the difference in susceptibility $\Delta\chi$ of oxyhemoglobin and deoxyhemoglobin. Such $\Delta\chi$ in tissue translates into ΔR^* or difference in relaxativity which is a measure of hemodynamic response function (HRF) to an external neurological stimulation (NS). Establishment of a mathematical relationship between HRF and NS will advance our knowledge of the inner workings of the brain. Such relationships could in turn be used as a measure of functional performance of the brain which could be useful in evaluation of the damage to brain tissue outside the visible tumor boundaries. fMRI maps have shown great potential for pre-surgery planning and refinement of their results will strengthen the neurosurgeon's hand in the process of tumor resection.

In other applications, susceptibility-induced signal dephasing is used to advantage. The so-called iron oxide based contrast agents, such as ultra small superparamagnetic iron oxide particles (USPIO), with their ability to kill MRI signal in their near vicinity, are a promising candidate for negative contrast. In addition, gadolinium-based molecules, such as Gd-DTPA, through their ability for T1 relaxativity enhancement, are used as positive contrast agents. This means that negative agents or T2 agents cause darkening of images on their sites and positive agents or T1 agents brighten the images in the regions where they reside. The role of molecular dynamics in determination of Lanthanide-based contrast agents has given rise to simulation research to improve our understanding of the bonding dynamics and dipole–dipole interaction that will enable designing of cancer-specific contrast agents with higher detectabilities. While both of these classes of contrast agents, i.e. T2-reducing and T1-reducing contrast agents, are used as passive agents for detection of abnormalities of the BBB, their potential for active uses have found considerable interest recently [25]. Both USPIOs and Gd-contrast agents could also be used as targeted agents through conjugation with ligand mediated pathology-specific molecules such as vascular endothelial growth factor (VEGF). Application of targeted molecular imaging at high fields has great potential that could be explored with HF and UHF animal magnets on animal models of cancer. Strong susceptibility artifacts and dipole–dipole interactions, freedom with to use higher doses than known to be safe for humans have made UHF animal magnets a rapidly proliferating research tool in cancer research.

STATIC MAGNETIC FIELD: SNR

The most fundamental parameter in enhancing sensitivity of MRI to cancer pathology is SNR. As 8T MRI has demonstrated through producing high SNR that depiction of microvasculature in in vivo human subjects is feasible (see Figures 21.1 and 21.2) Regarding its benefits to cancer research, high SNR could be used in three major areas:

1 visualization of neovasculature
2 tumor properties based contrast
3 localization and demarcation of tumor.

The first application is one of the most important potentials of high SNR that has not yet been fully realized [13,14]. Enhancing the visibility of blood vessels enables a clearer depiction of high tortuosity, irregular vascular patterns and longer vessels within and around the tumor. The ability to visualize and map out these features of the mass, in comparison with normal blood vessels and other landmarks, could provide an independent measure of in vivo staging of brain tumors. High SNR's capability for microstructure depiction can be used for imaging-based distinction of neovasculature from mature blood vessels. This is an important diagnostic capability that radiology presently lacks. Its availability will offer a significant capability for anti-angiogenic therapy. Results of microvascular density could also be used, along with dynamic contrast enhanced (DCE) MRI to get an account of vascular leakiness. High SNR and high susceptibility have the potential in HF MRI to enable detection of neovasculature in a longitudinal study that keeps track of vascular density making them distinct from mature blood vessels. 8T images have shown that susceptibility enhances the contrast between microvasculature and brain parenchyma to the point of visualization of subvoxel structures. Histopathalogic findings have highlighted the importance of in vivo distinction between neovasculature and mature blood vessels in the early stages of the disease [11]. More work is needed to show how high SNR can help to achieve such a distinction.

To enhance the role of SNR in brain tumor imaging beyond vasculature depiction, other tumor properties such as microstructure and physiology could be used. Ability to stage tumors, as is done by histologic methods, could be a useful role for HF MRI. High SNR will help the use of other tumor properties such as hypoxia and acidity to enable MRI to determine histologic type and grade in brain tumors non-invasively. Hypoxic mapping by BOLD contrast in HF MRI using echoplanar imaging (EPI) could provide an estimation of the pathophysiology of a tumor and a measure of therapeutic efficacy. Acidity could be obtained from MRS detection of lactic peaks. At HF, higher SNR is combined with larger chemical shift dispersion, enabling spectroscopy (MRS) peaks to become both stronger and further from each other. Both of these HF effects contribute to a better resolution of the metabolite peaks. Chemical shift imaging (CSI) maps offer measures of n-acetyl aspartate (NAA), choline, creatine and lactate in the brain. High SNR will make more peaks accessible and measures such as acidity and choline content could help with the determination of malignancy. The SNR at 1.5 T is sufficient for relative quantification of metabolites like creatine/NAA, while high SNR could allow absolute quantification of maps of all these metabolites at HF. Such capabilities will elevate MRI to a true non-invasive neuropathologic method.

The relationship between high SNR, and the ability to detect tumor boundaries more reliably is not clear. In general, enhanced sensitivity will enhance the capability for signal modulation as a function of tissue microstructure and physiology. Independence of relaxation mechanisms in creating contrast for tumor detection from SNR, and its ability to detect signals from smaller voxels, increases the possibility of a better demarcation of the tumor boundaries due to a less severe partial volume effect at the tumor boundaries. Furthermore, events such as increased cellularity and diffuse calcification have the potential for additional contrast generation. Both processes will become visible earlier as their contribution to the determination of voxel SNR becomes more dependent on the fraction of voxel occupied by calcificied particles and higher cellular density.

As mentioned in the susceptibility section, small animal models could be used in UHF MRI scanners to allow the testing of hypothesis-driven experiments to determine the viability of the model to enhance tumor visibility and to evaluate the therapeutic efficacy of new drugs. The high SNR of UHF animal scanners provides a powerful tool to push the boundaries of resolution towards microscopy and contrast-free molecular imaging. Since the results of animals studies, as well as new MRI techniques, are directly applicable to human scanners, they reduce the bench-to-bed development time and offer a realistic alternative to research involving human subjects.

RF PENETRATION

The Larmor equation is given by $\omega_0 = \gamma B_0$, and determines the presessional frequency of a proton within a magnetic field B_0. As B_0 increases, so does the rotational frequency of protons, ω_0. Resonance effect in MRI requires irradiation of the subject with an electromagnetic wave of the same frequency as ω_0 in order to deflect proton magnetic moments away from B_0, i.e. excitation. Considering the $^1H_\gamma$ and the range of B_0 used in clinical MRI, the frequency of electromagnetic fields needed in MRI falls in the 63.5 MHz (at 1.5 T) to 128 MHz (at 3 T) range, which falls in the RF category. For whole body research magnets, RF goes up to 400 MHz at 9.4 T. This range (>300 MHz) corresponds to the UHF band used for TV broadcasting. With the increase in RF frequency, the electrodynamic properties of tissues will vary proportional to ω_0. Tissue conductivity is a function of frequency which increases at high fields

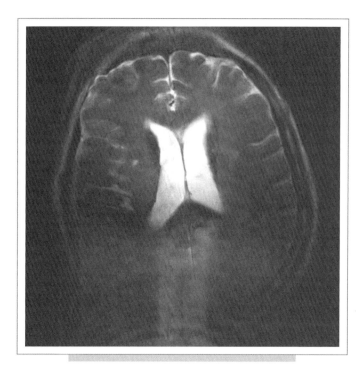

FIG. 21.4. A RARE image obtained with a 512 × 512 matrix size. This image is heavily T2 weighted with a CSF flow artifact extended along the phase encoding direction. Also seen here are motion artifacts and strong loss of signal in the posterior aspect of the slice.

and, as a result, increases the RF absorption by the tissue. This extra RF absorption requires higher power to achieve the same deflection angle at HF compared to lower fields. So, RF power content of pulse sequences increases at HF. The electric field (E or E_1) associated with the RF magnetic field (B1) that is used to excite the spins causes tissue heating [27]. Delivery of RF to a voxel occupied by a tissue of conductivity σ (ohm^{-1}) and density ρ (kg/m^3) requires an electric field E_1 (V/m) which will cause RF absorption or specific absorption rate (SAR(Watt/kg)) given by the equation $SAR = \sigma E^2/2\rho$. SAR is a standard measure of tissue heating and regulatory bodies such as the Food and Drug Administration (FDA) provide guidelines for RF limits used in human applications [7]. Present values for SAR suggested by the FDA since July 14, 2003 stand at SAR = 3W/kg averaged overhead for 10 minutes. While SAR implies absorption of RF by the tissue in a uniform manner, RF distribution across the body parts, e.g. the human head, is by no means uniform. The heterogeneous nature of the biological tissues causes this non-uniformity that leads to a non-uniform local SAR across the head [28]. Other factors, such as head size and coil size, also affect the extent of non-uniform distribution of RF within the brain [27,28]. While higher absorption at HF attenuates the RF penetration into the tissue, other properties of the human body, i.e. high dielectric constant, cause opposite effects as is discussed in the following section [26,29]. In fact, the dielectric effects are so strong that they dominate the B1 distribution generating the bright center, a trade mark of high field images as seen on 8T and 7T images of the human head shown in Figures 21.2–21.4. Aside from the safety considerations of inhomogeneous RF distribution, penetration effects must be addressed as an image quality issue as well. Such RF distribution has been addressed by offering solutions that use innovative coil design and excitation schemes [27].

OTHER RF ISSUES

Biological tissues have a high dielectric constant, ε, of about 50. This is due to the high water content and ε (water) is around 80. The unique molecular structure of water is the primary determinant of its high dielectric properties. Within the human body, as in water, electromagnetic waves experience pronounced changes in their wave properties, such as wavelength, that impacts their propagation and distribution. The dimensions of imaged objects also play in as a factor in the formation of dielectric effects. The typical dimension, d, of a human head is about 10 cm while the same d for the human torso is about 25 cm. At the same time, after RF enters the body its wavelength within the tissue changes and is given by $\lambda_t = \lambda_0 / \sqrt{\varepsilon_r}$ where λ_t is the RF wavelength inside the tissue, λ_0 is the RF wavelength in air, and ε_r is the relative permittivity of the tissue. Evidence for dielectric effects begin to appear as λ_t approaches d from above which is caused by B_0 increase, and becomes more severe as λ_t becomes smaller than d.

Dielectric resonance, or center brightness effects, begin to appear on MRI images at around 2T when the patterns of inhomogeneity across MRI images grow in severity and complexity [29]. At 8T, as seen in Figures 21.2 and 21.3, the dielectric patterns clearly cause a significantly stronger B1 at the central regions making these regions of the image considerably brighter than peripheral regions. While undesirable in general, this feature of HF MRI lends itself to the studies involving the center of the brain as it enables signal to be focused selectively in those regions. For all other applications requiring a homogeneous distribution of image intensity, new techniques in coil design and pulse sequences have shown to be able largely to suppress this effect [8,9].

For magnetic fields above 7T, a unique coil design called a transverse electromagnetic or TEM resonator is used [30,31]. HF MRI at 4T and above has benefited from the electromagnetic characteristics of TEM coil design [32]. TEM coils incorporate a distributed capacitor concept in their construction, unlike the low field RF coil's birdcage-like designs, which use a lumped element concept for capacitor/inductor use. While unable to suppress the artifacts such as center brightness at high field, as seen in Figures 21.2 and 21.3, the TEM resonator has been able to address some fundamental issues, such as high electric field concentration around capacitors of lumped element coils and radiation damping that becomes severe in birdcage design at HF. The issue of radiation damping in birdcage RF coils [28,32] refers to the leaking of RF power as radiation in an outwardly direction that takes energy away from the internal regions of the coil where that energy is needed for spin excitation. This effect in birdcage coils degrades image homogeneity at proton frequencies due to a need for more RF power that introduces more electric noise into the body, reducing SNR and flip angle uniformity across the sample.

One remedy for dielectric effects is the choice of phase array volume coils and multichannel coils in a scheme called parallel imaging [8,9,28]. Phase array coils and multichannel coils are made up of a set of surface coils in tandem wrapped around the imaged body. Multichannel coils and phase array coils take advantage of their surface elements that generate stronger signal in the near proximity of the coil. This characteristic of multichannel coils offers higher SNR in peripheral regions compared to deep tissues. This effect acts opposite to dielectric effects and compensates the center brightness, thereby restoring the intensity homogeneity to HF MRI images. Multichannel coils could be comprised of traditional surface coil elements or structures such as TEM coils could be used in a way that their individual elements receive signals independently within a parallel receive mode. Multichannel coils have been used in parallel receive mode (pRX) in different partial parallel imaging schemes such as *sen*sitivity *en*coding (SENSE) [8], and *sim*ultaneous *a*cquisition of *s*patial *h*armonics (SMASH) [9], which are two variants of radiofrequency coil array reception techniques. Both SENSE and SMASH have the dual advantage of homogeneity and fast acquisition. The fast imaging is achieved by shortening the readout duration as it is for individual surface coils ability to skip k-space lines. At lower fields, parallel imaging is used to reduce scan time for dynamic imaging studies and others in need of higher temporal resolution. As far as the use of parallel imaging for restoration of homogeneity of RF (B1) distribution is concerned, there are two opportunities for reinstating image homogeneity, one during transmit and the other during receive, which leads into a multitude of configurations for parallel imaging. These configurations include the use of (a) single channel transceiver (Tx/Rx) volume coil; (b) single channel transmit/parallel receive (Tx/pRx); (c) parallel transmit/parallel receive (pTx/pRx) also called transmit SENSE/receive SENSE. These configurations have different advantages and disadvantages in terms of RF power, SAR, image uniformity and SNR.

Parallel imaging has another benefit related to the economics of RF power generation and delivery technology. As RF power per unit flip angle excitation increases at HF, it becomes increasingly more expensive and difficult to incorporate RF amplifiers with the capability for high fidelity pulse generation needed for spin excitation in HF scanners. A typical RF power amplifier for a 7T scanner should have

10–30 kW of power for delivery for various applications. This makes pTx/pRx even more attractive. It is more economical to use multiple smaller transmitters feeding individual coil elements than delivering massive amounts of RF power to an individual volume coil. This means that at HF one could use the smallest possible Tx coils to their advantage, since they can produce larger B1 with the same amount of RF power compared to volume coils. On the other hand, for each unit of RF power, smaller elements of pRx coils can produce a more intense and shorter pulse which has a high bandwidth (BW). Such high BW pulses allow more versatile sequence designs with better image contrast. Shorter echo times and higher sensitivity are achieved by pRx, in addition to high SNR advantages of pRx mode, making this technology more favored for HF MRI studies.

Small HF and UHF animal scanners have already made a foothold for imaging of CNS tumor models. These scanners operating at fields of 7–14T have demonstrated the gain in most of the fundamental MRI parameters like SNR, CNR, resolution, etc. The ability to generate a 3D image of a tumor at $(0.1\,\text{m})^3$ will open the path for conducting the most basic evaluation of therapeutic efficacy in humans, as is now done in animal models of brain tumors. Examples of in vivo anatomical imaging obtained at 8T (see Figures 21.1–21.3) have voxels about ten times bigger than $(0.1\,\text{mm})^3$. Other chapters in this book also illustrate 8T images with excellent SNR characteristics, raising hope that HF is truly advancing and will eventually bring anatomical, functional and molecular imaging with MRI possible at submillimeter resolution. Another quantum leap in this technology will be made upon optimization of parallel imaging in the next generation of MRI scanners. In addition, targeted contrast agents should make it possible to achieve resolutions of about $(0.1\,\text{mm})^3$

throughout the human brain in reasonable scan times. This will lead to a reassertion of MRI as the most effective tool in the diagnostic and therapeutic aspects of cancer care.

CONCLUSION

MRI has achieved an order of magnitude improvement in image quality in the last two decades since its appearance as a medical imaging modality. During this time, HF has remained its trusted path to development. HF has shown that it has much to offer to cancer diagnosis and monitoring of therapy. As field strength increases, so does the sensitivity, magnetic susceptibility, T1 relaxation and chemical shift. Gain in SNR has the most profound effect on new capabilities which open new applications to MRI. HF MRI has already shown that it is capable of acquisition of images with submillimeter resolution, microvascular visualization, reliable brain metabolite mapping and high temporal resolution dynamic imaging. fMRI is another feature that MRI has offered to cancer research with a great deal to offer at high field in the arena of neurosurgery planning. Other developments such as tumor-specific contrast agents, parallel imaging, diffusion tensor imaging and complex new pulse sequences are further empowering HF MRI, helping its march toward a true quantitative imaging technique with potential to develop specificity for imaging-based tumor staging. Within less than a decade since its appearance, parallel imaging has increased the number of channels for simultaneous acquisition to 128 channels, which makes single shot k-space read-out acquisition possible. Such rapid developments will substantially facilitate HF MRI's ascension to a non-invasive neuropathological tool, with the potential for real-time therapeutic monitoring as well.

ACKNOWLEDGMENT

I wish to thank Gregory Christoforidis of Ohio State University, Department of Radiology, for providing images for this chapter.

REFERENCES

1. Vetter J, Ries G, Reichert (1988). A 4-Tesla superconducting whole-body magnet for MR imaging and spectroscopy. IEEE Trans Magn 24:1285–1287.
2. Barfuss H, Fischer H, Hentschel D, Ladebeck R, Vetter J (1988). Whole-body MR imaging and spectroscopy with a 4-T System. Radiology 169:811–816.
3. Barfuss H, Fischer H, Hentschel D et al (1990). In vivo magnetic resonance of humans with a 4T whole-body magnet. NMR Biomed 1:31–45.
4. Bomsdorf H, Helzel T, Kunz D, Roschmann P, Tschendel O (1988). Spectroscopy and imaging with a 4 Tesla whole-body MR system. NMR Biomed 1:151–156.
5. Robitaille PML, Abduljalil AM, Kangarlu A et al (1998). Human magnetic resonance imaging at eight Tesla. NMR Biomed 11:263–265.
6. Kangarlu A, Zhu H, Burgess, RE, Hamlin RL, Robitaille, PML (1999). Cognitive, cardiac, and physiological safety studies in

ultra high field magnetic resonance imaging. J Magn Reson Imaging 17:1407–1416.

7. Kangarlu A, Robitaille P (2000). Biological effects and health implications in magnetic resonance imaging. Concep Magn Reson 12:321–359.

8. Prussemann KP, Weiger M, Scheidegger MB, Boesiger P (1999). SENSE: sensitivity encoding for fast MRI. Magn Reson Med 42:952–962.

9. Sodickson DK, Manning WJ (1997). Simultaneous acquisition of spatial harmonics (SMASH): fast imaging with radiofrequency coil arrays. Magn Reson Med 38:591–603.

10. Abragam A (1994). *Principles of Nuclear Magnetism*. Oxford Science Publication, Oxford.

11. McDonald DM, Choyke PL (2003). Imaging of angiogenesis: from microscope to clinic. Nat Med 9:713–725.

12. Christoforidis GA, Kangarlu A, Abduljalil AM et al (2004). Susceptibility-based imaging of glioblastoma microvascularity at 8T: correlation of MR imaging and postmortem pathology. Am J Neuroradiol 25:756–760.

13. Christoforidis GA, Kangarlu A, Abduljalil AM et al (2002). Visualization of microvascularity in glioblastoma multiforme with 8-T high-spatial-resolution MR imaging. Am J Neuroradiol 23:1553–1556.

14. Tynninen O, Aronen HJ, Ruhala M et al (1999). MRI enhancement and microvascular density in gliomas: correlation with tumor cell proliferation. Invest Radiol 34:427–34.

15. Chen CN, Sank VJ, Cohen SM, Hoult DI (1986). The field dependence of NMR imaging: I. Laboratory assessment of signal-to-noise ratio and power deposition. Magn Reson Med 3:722–729.

16. Hoult DI, Chen CN, Sank VJ (1986). The field dependence of NMR imaging: II. Arguments concerning an optimal field strength. Magn Reson Med 3:730–746.

17. Robitaille PML, Abduljalil AM, Kangarlu A (2000). Ultra high resolution imaging of the human head at 8 Tesla:2K×2K for Y2K. J Com Assist Tomog 24:2–8.

18. Kangarlu A, Bourekas BC, Ray-Chaudhry A, Rammohan KW (2007). Cerebral cortical lesions in multiple sclerosis detected by magnetic resonance imaging at 8 Tesla. Am J Neuroradiol 28(2):262–266.

19. Koretsky AP (2004). New developments in magnetic resonance imaging of the brain. NeuroRx 1:155–164.

20. Schenck JF (1996). The role of magnetic susceptibility in magnetic resonance imaging: MRI, magnetic compatibility of the first and second kinds. Med Phys 23:815–850.

21. Abduljalil AM, Robitaille PML (1999). Macroscopic susceptibility in ultra high field MRI. J Com Assist Tomog 23:823–841.

22. Abduljalil AM, Kangarlu A, Robitaille PML (1999). Macroscopic susceptibility in ultra high field MRI-II: Acquisition of spin echo images from the human head. J Com Assist Tomogr 23:842–844.

23. Kangarlu A, Abduljalil A, Norris DG, Schwartzbauer C, Robitaille P (1999). Human RARE imaging at 8 tesla: RF intense imaging without SAR violation. Magn Reson Mater Phys Biol Med 9:81–84.

24. Norris DG, Kangarlu A, Abduljalil A, Schwartzbauer C, Robitaille P (1999). Human MDEFT Imaging at 8 tesla. Magn Reson Mater Phys Biol Med 9:92–96.

25. Lubbe AS, Bergemann C, Brook J, McClure DG (1999). Physiological aspects in magnetic drug targeting. J Magn Magn Mater 194:149–156.

26. Robitaille P, Kangarlu A, Abduljalil A (1999). RF penetration in ultra high field magnetic resoance imaging (UHF MRI): challenges in visualizing details within the center of the human brain. J Com Assist Tomogr 23:845–849.

27. Kangarlu A, Ibrahim TS, Shellock F (2005). Effects of RF coil design and field polarization on RF heating inside of a head phantom. Magn Reson Imaging 23:53–60.

28. Ibrahim TS, Lee R, Baertlein BA, Kangarlu A, Robitaille P (2000). Application of finite difference time domain method in RF coil design using multi-port excitation. Magn Reson Imaging 18:733–742.

29. Kangarlu A, Baertlein BA, Lee R, Ibrahim T, Abduljalil AM, Robitaille PML (1999). Dielectric resonance phenomena in ultra high field magnetic resonance imaging. J Comp Assist Tomogr 23:820–831.

30. Roschmann P (1987). Radiofrequency penetration and absorption in the human body: limitations to high-field whole-body nuclear magnetic resonance imaging. Med Phys 14:922–931.

31. Vaughn JT, Hetherington HP, Otu JO, Pan JW, Pohost GM (1994). High frequency volume coils for clinical NMR imaging and spectroscopy. Magn Reson Med 32:206–218.

32. Baertlein BA, Ozbay O, Ibrahim T et al (2000). Theoretical model for a MRI radio frequency resonator. IEEE Trans Biomed Eng 47:535–546.References

SECTION III

Advances in Neuroimaging of Brain Tumors

CHAPTER 22

Magnetic Resonance Image Guided Neurosurgery

Ferenc A. Jolesz, Ion-Florin Talos, Simon K. Warfield, Daniel Kacher,
Nobuhiko Hata, Nicholas Foroglou and Peter M. Black

INTRODUCTION

Since its introduction as a diagnostic tool in the mid-1980s, magnetic resonance imaging (MRI) has evolved into the premier neuroimaging modality. With the introduction of higher field magnets, we are now able to achieve spatial resolutions of such superb quality that even the most exquisite details of the brain anatomy can be visualized. Due to these advantages, MRI has had a great impact on the diagnosis of neurosurgical conditions and, with the implementation of intraoperative MRI, on surgical therapy as well. With intraoperative image updates, brain deformations throughout the course of surgery are detected in near real-time and the resection progress is readily monitored. This enables accurate neuronavigation. Intraoperative MRI (iMRI) is approaching the point when it will be used not only to localize, target and guide resection of brain tumors and other intracranial lesions, but also to provide a comprehensive picture of the surrounding functional anatomy. In addition to new imaging methods such as diffusion tensor imaging, multivoxel spectroscopy (MRS) and functional MRI (fMRI), novel therapeutic approaches have been added to the arsenal of MR-guided therapy. Among these, the promising results in MRI-guided focused ultrasound surgery, in which the non-invasive thermal ablation of tumors is monitored and controlled by MRI, are especially encouraging. With the clinical introduction of these advances, intraoperative MRI is changing the face of neurosurgery today.

Modern neurosurgery requires ever more comprehensive integration of sophisticated morphologic and functional imaging techniques into the surgical flow. This integration process includes the incorporation of interactive dynamic imaging, high performance computing and real-time image processing in the operating room.

The evolution of novel intraoperative imaging techniques has driven the development of several revolutionary image-guided therapy methods currently in the process of implementation into the neurosurgical practice.

In the following sections, we will present the various intraoperative systems currently in use. We will then review the current state of image-guided brain tumor surgery, including the role of multimodal imaging for guiding tumor resection and methods employed for image fusion and visualization. We will conclude with a brief presentation of novel, minimally-invasive therapeutic approaches, including high-intensity focused ultrasound (HIFUS).

IMAGE-GUIDED THERAPY SYSTEMS

Due to the ever growing need for fast, high-resolution morphologic and functional imaging, competitors in the MRI market have begun to pursue higher field MR scanner solutions. Closed (cylindrical) bore MR systems are marketed at a field strength of 3T, while open systems are now being introduced at 1.0 Tesla. The migration to higher field solutions is expected to continue, since the dominant competitors are currently pursuing even higher field solutions for the research community that range from 4T to 9.4T. Recently, the US Food and Drug Administration (FDA) designated 7T as non-significant risk.

One of the implications of this migration is the recalibration of field strength definitions [1]. Demand for the 3T systems has been so great that a sub-segment has developed to describe more accurately the change that is taking place. For the purposes of this review, the 3T market will be designated as the very high field. The high field segment will include only those scanners with a field strength of 1.5T. Mid-field scanners will now include field strengths from 0.5T to 1T, while low-field scanners are less than 0.5T. A second trend in MRI scanner design includes decreasing bore length and increasing bore diameter, which greatly benefits access during interventional procedures and provides greater flexibility in patient positioning.

Choice of MRI Scanner and Procedural flow

At the time of this publication, scanners ranging from 0.12T to 3T are in use for MRI-guided neurosurgery with a range of configurations (Figure 22.1). The choice of scanner may strongly influence operating room architecture and layout, as well as the manner in which imaging and intervention are performed. Solutions for iMRI include: operating within a vertically open scanner, which eliminates the need to move the patient in between imaging cycles (Figure 22.1A,B); use of a dedicated portable scanner (Figure 22.1D); horizontally open scanners (Figure 22.1C); or conventional closed bore scanners (Figure 22.1E,F), which require that the patient be moved for intraoperative image updates, by translating or pivoting a table up to 180 degrees (with this configuration, the surgery takes place outside the 5 gauss line); mobile closed bore scanners, which move to the patient (Figure 22.1G,H).

Different procedures are enabled by certain scanner capabilities. Greater signal-to-noise ratio and faster scan times, characteristic of the higher field, closed bore scanners, enable intraoperative functional MRI (fMRI), diffusion tensor MRI (DT-MRI) and MR spectroscopy (MRS) acquisitions, which may facilitate tumor resections. The high field scanners, moreover, have greater sensitivity to temperature changes during thermal ablations, which is ideal for minimally-invasive procedures, such as laser ablation and focused ultrasound ablation [2,3]. Lower field open scanners offer greater patient access for biopsy and therapeutic probe placement. The tradeoffs of these scanners are listed in Table 22.1.

Safety

Complexity of controlling access increases in situations where a single iMRI room services multiple operating rooms. Depending on layout, more entrances into the iMRI room may be necessary. Moreover, moving an intubated patient between the surgical station and the scanner poses a clinical safety concern. When the surgery occurs within the 5 gauss line, the use of custom-made, non-ferrous instruments is mandatory (see Figure 22.1A,F). Solutions where surgery occurs outside the 5 gauss line enable the use of conventional, more readily available instruments. However, great pains must be taken to prevent instruments from coming too close to the scanner where they may become projectiles. This point is critical to the safety of the patient and staff.

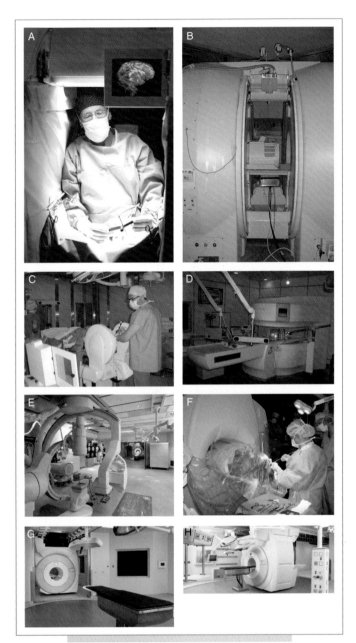

FIG. 22.1. Solutions for MRI-guided neurosurgery. **(A)** Vertical open superconducting 0.5 T interventional scanner with integrated optical tracking system for neuronavigation; **(B)** same system shown with integrated fluoro x-ray system (courtesy Rebecca Farhig); **(C)** portable dedicated low field system (courtesy Medtronics); **(D)** horizontal open permanent magnet low field system (courtesy Roberto Blanco); **(E)** two room suite with closed bore 3 T scanner and bi-plane fluoro x-ray system (courtesy Walter Kucharcyzk); **(F)** one room solution with 3 T scanner; surgery occurs at the back of the scanner to minimize excursion between surgical location and imaging location; **(G,H)** mobile 1.5 T scanner; subsequent to image acquisition, scanner is moved to garage.

TABLE 22-1 Possible MRI configurations for intraoperative and interventional imaging			
Configuration	Field	Direct access	Advantages
Long bore (closed)	1–3 T	No	Highest image quality
Short/wide bore (closed)	1.5 T	Limited	Moderate homogeneity and gradient speed
Horizontal open gap	0.02–1.0 T	Limited	Safety and instrument compatibility
Vertical open gap	0.5 T	Full	Intraoperative imaging; flexible positioning

Patient Positioning

Many iMRI procedures require general anesthesia. It is essential the patient is positioned properly and all pressure points must be appropriately padded to maintain good circulation. When the head has to be pinned in a head holder, as is the case with most neurosurgical procedures, centering the head in the sweet spot of the imaging volume may interfere with achieving the ideal surgical position. These requirements may sometimes preclude fitting of a large patient into the bore of the scanner.

BRAIN TUMOR RESECTION UNDER INTRAOPERATIVE MRI GUIDANCE

Rationale

Currently, there is no generally accepted management standard for primary brain tumors. With the exception of patients with intractable epilepsy and symptomatic mass effect, the role of surgery remains controversial. However, an increasing number of studies suggest a beneficial effect of gross total tumor resection on overall- and recurrence-free survival in patients with low- and high-grade gliomas [4–7].

The fundamental goal of neurosurgery is to target, access and remove intracranial lesions without causing damage to normal brain tissue or vascular structures. The overall concern is the preservation of neurological function, which requires precise delineation of functional anatomy and correct definition of the resection target volume.

Intraoperative MRI has become the method of choice for achieving the safest and most accurate resection of glial neoplasms. Visual inspection is often unreliable in distinguishing tumor from normal brain tissue. Compared with visual inspection and other imaging modalities (CT, ultrasound), MRI has a much higher sensitivity in detecting these lesions. However, the usefulness of preoperatively acquired MR images for surgical guidance is limited by changes in brain configuration, which inevitably occur during surgery. At the outset of any given procedure, surgical manipulations and anesthesia result in a change in the anatomic position of brain structures and intracerebral tumors. For example, the initial loss of cerebrospinal fluid (CSF) following a craniotomy, as well as retraction and resection of tissue, tissue hydration state, partial arterial CO_2 pressure, all play a role in the intraoperative brain shift [8,9]. Hence, localization and targeting of brain tumors based on preoperatively acquired images is unreliable. In fact, literature data show that, in the absence of intraoperative image updates, visual inspection of the surgical field alone fails to detect residual tumor otherwise accessible for resection [7,10–12].

Another major challenge when attempting gross total tumor resection is determining the relationship between lesion and functionally critical cortical areas and white matter tracts. A recent analysis of a large patient series harboring supratentorial gliomas, operated on with intraoperative anatomic MRI guidance, has shown that the main predictors for incomplete tumor resection are, besides large tumor volume and oligodendroglioma histopathology, tumor involvement of eloquent cortex and large fiber tracts, such as the corticospinal tract [13]. These findings underscore the importance of multimodal imaging, such as fMR and DT-MRI for surgical guidance.

Multimodality Imaging for Guiding Brain Tumor Resection

Functional magnetic resonance imaging provides information about cortical function by non-invasive means. Such functional information is of paramount value for surgical planning in lesions located in or around eloquent cortex.

Basically, neural activation is accompanied by a hemodynamic response, i.e. increased supply of arterial (oxygenated) blood to the activated cortical regions. This results in a local increase of the oxyhemoglobin-to-deoxyhemoglobin ratio. Given the fact that oxyhemoglobin is diamagnetic and deoxyhemoglobin is paramagnetic, the hemodynamic response during activation leads to an alteration of local magnetic susceptibility, which can be readily detected with appropriate MRI sequences, such as echoplanar gradient echo imaging [14].

The interpretation of fMRI data is based on statistical analysis of the temporal variations in voxel intensity distribution during rest and activation. The functional maps are obtained by displaying the distribution of the voxels activated by various physiological tasks (i.e. motor, sensory, visual). Although fMRI data do not alter the diagnosis, they can contribute substantially to surgical planning by providing critical information about optimal positioning of craniotomy and corticotomy sites, surgical excision margins and access routes to the lesions. fMRI based surgical planning also yields various potential strategies that enable the surgeon to avoid targeting the lesion too close to eloquent cortex or to other essential structures.

The relationship between intra-axial infiltrative brain tumors, such as low- and high-grade glial neoplasm and the functionally active neural tissue has not been entirely elucidated. Large-population MEG (magneto-encephalography) studies [15,16] demonstrated the presence of functionally active tissue within the tumor as defined by MRI in 18% of grade II, 17% of grade III and 8% of grade IV tumors. Thus, obtaining functional information well in advance of the surgical procedure is critical in defining the safest surgical strategy. Moreover, the incorporation of preoperative fMRI information into frameless, stereotactic systems is becoming routine at some institutions [17].

In our experience, preoperative fMRI data have proven very useful in guiding electrophysiological, cortical mapping. Furthermore, such information may be crucial to stereotactic biopsies, where, due to the small skull opening, cortical stimulation is not practicable. However, the intraoperative use of fMRI data is limited by the changes in brain's shape which inevitably occur during any neurosurgical procedure. Intraoperative fMRI acquisitions are not practical in low- and mid-field iMRI systems. The high-field systems, however, do not suffer from this limitation [18].

Diffusion tensor MRI can add important information about the proximity of tumors to critical white matter fiber tracts [19,20]. Precise knowledge of white matter fiber tract locations, as well as their topographic relationship with the lesion, is not only essential to accurate surgical planning, but also to understanding and predicting neurological function both before and after surgery. White matter fiber tracts can be depicted by displaying the eigenvectors of the diffusion tensor or by color-coding the directional information. With such information overlaid on anatomical images, fiber tract alterations such as displacement, infiltration or disruption can be readily observed. However, it is important to note the difficulty of identifying a tract in a two-dimensional plane.

During the past few years, several algorithms for three-dimensional tracking of water diffusion have been developed (3D-tractography) [21–23]. Unlike the task of detecting white matter fiber tracts within a two-dimensional plane, the 3D approach enables more precise tract identification and produces more reliable anatomic information. By registering multiple image data, such as functional MRI, diffusion-weighted tensor images and structural images including gadolinium contrast-enhanced images in three dimensions, and by rendering all information on to a 3D brain model, the surgeon can interpret the overall view of tumor and white matter fiber relationship with relative ease.

Rigid and Non-rigid Image Registration for Surgical Guidance

Recently developed, sophisticated image registration techniques have enabled the use of preoperatively acquired multimodal images for surgical navigation.

Registration has been extensively studied and excellent techniques are now available to achieve alignment of several different types of images. Different registration algorithms are distinguished by the alignment metric used to assess alignment quality and the nature of the transform that may be estimated. Different choices for these parameters lead to different performance characteristics, different level of robustness to signal variation, to the type of geometrical change that may be identified automatically and to changes in the capture range and speed of execution. Excellent reviews of recent work in image registration algorithms and strategies for registration validation are available [24–26].

In our work we have focused on robust methods, able to be executed within the time constraints of surgery [27] and fast implementations of methods suitable for multi-modality data [28]. Alignment of preoperative data can utilize transformation models suitable for the capture of rigid body motion alone.

Figure 22.2 illustrates the fusion of functional MRI, diffusion tensor MRI and structural MRI for the purpose of enhancing preoperative surgical planning. We applied

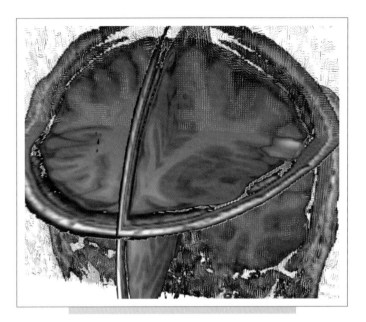

FIG. 22.2. Intrasubject alignment of preoperative structural MRI, functional MRI and diffusion tensor MRI. This type of visualization enables ready appreciation of the 3D relationship between anatomical structures, the tumor and critical fiber pathways.

a previously described segmentation strategy [29,30] to high resolution MRI (spoiled gradient echo (SPGR) at $0.9375 \times 0.9375 \times 1.3 \, \text{mm}^3$ acquired on a GE 3T scanner), and a T2W fast spin-echo ($0.46875 \times 0.46875 \times 5.0 \, \text{mm}^3$). We aligned the T2W scan, and also preoperative DT-MRI and fMRI to the SPGR scan, as shown in Figure 22.2. The distance between regions of activation in a language fMRI paradigm, critical white matter fiber tracts and the projected area of resection was carefully considered.

Rigid registration is able to compensate for translational and rotational differences, but non-rigid registration transformations are necessary to compensate for brain shift. A recent review [31] describes the motivation and methods for successfully capturing such intraoperative deformations and a robust, rapid and effective method for registering whole brain preoperative MRI to intraoperative MRI has recently been described [32].

One predictor of neurological deficit following neurosurgery is the distance between the resected tissue and activation observed with fMRI [33]. We have recently experimented with the alignment of preoperative fMRI to iMRI during neurosurgical cases. We repeatedly updated the alignment as new iMRI was acquired over the course of the surgery and utilized rigid followed by non-rigid registration [32] to project the fMRI and DT-MRI into alignment with the patient's brain as the surgery proceeded. Reorientation of the DT-MRI was carried out using the method of Ruiz-Alzola [34]. The surgeon was able to visualize the spatial relationship between preoperative fMRI, DT-MRI, the location of the tumor and the region of the tumor resection. This is illustrated in Figure 22.3.

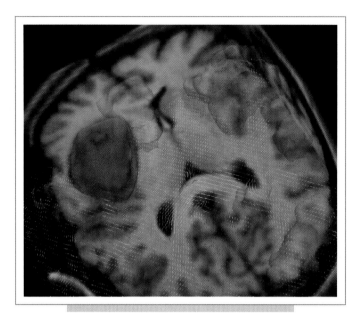

FIG. 22.3. Visualization of fusion of preoperative and intraoperative MRI. Functional MRI activation is represented in blue and DT-MRI is represented by line glyphs indicating the orientation of local white matter fibers. The preoperative appearance of the tumor is indicated by surface rendering in red and cross-sectional slices through intraoperative volumetric MRI are shown.

Intraoperative Dynamic Contrast Imaging

One charateristic of high-grade glial neoplasms is their capacity to induce formation of new, abnormal vessels (tumor angiogenesis). These abnormal vessels lack a blood–brain barrier and are inherently highly permeable. Dynamic contrast enhancement studies performed during image-guided resection procedures have demonstrated a higher rate and degree of enhancement at sites of tumor recurrence, compared with areas associated with radiation necrosis, which results from damage to existing vessels [35].

In this technique, an SPGR scan sequence is performed sequentially while gadolinium is injected intravenously (2D FSPGR, with echo time (TE) minimum, 45 degree flip angle, 1 NEX, sequential, multiphase, interleaved, variable bandwidth, and extended dynamic range, 10 phases/location; 22 cm FOV; 256 × 128; 1.87 s/image). This technique produces a series of images depicting regional rates of enhancement through the tumor volume.

A series of 24 patients who underwent dynamic MR imaging while in the intraoperative magnet was reported in Schwartz et al [35]. In patients *with* tumor recurrence, the signal intensity at first pass through active tumor had increased approximately 50% over baseline, while the maximal signal intensity change at first pass in patients without evidence of recurrent tumor was only approximately 15% over baseline. These data suggest that dynamic MR imaging can at once establish whether active tumor is likely present within a treated tumor bed and direct the surgeon to these sites; we now routinely use it for these purposes.

Schwartz et al demonstrated an accuracy of 90% for this method in distinguishing patients with recurrent tumor from those with radiation necrosis.

Intraoperative Detection of Complications

One of the great advantages of performing surgery within the intraoperative MR system is the ability of the operator to evaluate for intraoperative or perioperative hemorrhage. To this end, we routinely use a 'heme-susceptibility sequence' that consists of a series of gradient echo images with a time to recovery (TR) of 600 ms, a 30 degree flip angle, 1 NEX and TEs of 9, 20, 40 and 60 ms. These images are acquired as two sets of double echoes: one with TEs of 9 and 40 ms, and the other with TEs of 20 and 60 ms.

We have found that gradient echo sequences with long TEs (i.e. 40–60 ms) show dark signal ('blooming') due to hyper-acute blood collections, which is not evident on conventional imaging sequences or on shorter gradient-echo (GRE) sequences. This most likely reflects the greater sensitivity of the long TE GRE sequence to the presence of small amounts of deoxyhemoglobin that develop in the periphery of a hyper-acute blood collection. Atlas and Thulborn [36] have suggested that the low partial pressure of dissolved oxygen in the brain when compressed by the hematoma leads to deoxygenation of the blood at the hematoma–brain interface. Hence, deoxyhemoglobin can be visualized only on long TE GRE images and the attendant conspicuity of blood is directly correlated with increasing TE. It is important, however, to compare these images with conventional sequences to differentiate acute blood from postoperative air collections, which are visible as signal voids on all imaging sequences.

By understanding the changing appearance of the operative site on GRE images as a function of increasing TE, we can measure intraoperatively the amount of acute blood in a way that positively impacts the outcome of intracranial surgery performed in the intraoperative MR system. In 15 of the 18 cases studied, a small amount of hemorrhage was noted on the periphery of the operative cavity, but was considered too small to warrant any further action by either the radiologist or surgeon. In three cases, however, the collections produced significant mass effect on surrounding brain structures or were enlarging and were subsequently drained using MR guidance. In the most extreme case, where acute blood was exerting mass effect on surrounding brain and rapidly increasing in size, the collection was removed under MR guidance. Since diffusion-weighted MRI can detect ischemic injury, this method shows great potential for detecting white matter tracts, as well as avoiding surgical complications related to vascular spasm and occlusion. As a proof of concept, intraoperative line scan diffusion imaging (LSDI) on a 0.5 Tesla interventional MRI was performed during neurosurgery in three patients. Diffusion trace images were obtained in acute ischemic cases. Diagnosis of acutely developed vascular occlusion was confirmed with follow-up scans [37].

Instrument Tracking and Neuronavigation

Early work with brain stereotaxis has established the importance and value of image guidance through better determination of tumor margins, better localization of lesions and optimization of the targeting. Frame based and frameless stereotaxis using 2D and 3D images have shown that neurosurgery can be performed through less invasive approaches, with greater accuracy. Today, the field of computer-assisted, image-guided surgery has replaced stereotactic neurosurgery.

With the introduction of intraoperative MRI, the need for frame of reference transformation and registration has been eliminated. The lesion localization in the three-dimensional space is accomplished using image coordinates. Furthermore, with intraoperative image updates, the localization can be updated in a dynamic fashion. These features have dramatically enhanced the ability to obtain accurate biopsies and correct resection margins. Interactive use of MRI allows the selection of the optimal trajectories for various neurosurgical approaches. Brain surgeries can be carefully planned and then executed under MRI guidance, which allows the surgical exposure and the related damage to the normal brain to be minimized.

Interactive instrument tracking is the process by which interactive localization is possible in the patient's coordinate system. Tracking methods include articulated arms, optical tracking, passive systems, sonic digitizers and electromagnetic sensors. Active optical trackers use multiple video cameras to triangulate the 3D location of flashing light emitting diodes (LEDs) that may be mounted on any surgical instrument. Passive tracking systems use single or multiple video cameras to localize markers that have been placed on surgical instruments. Such passive systems do not require a power cable to be attached to the hand-held localizer. Unfortunately, both LED and passive vision localization systems require line of sight between (at least some of) the landmarks or emitters and the imaging sensor at all times when an object is to be tracked. Electromagnetic digitizers operate without such a restriction and offer the ability to track objects out of view or instruments placed inside the body.

Interventional imaging and display provide the neurosurgeon with an enhanced and dynamic view of the anatomy. Coupled with a stereotactic system, such as the optical tracking system described earlier, this allows the surgeon to navigate the surgical instruments within the surgical field, based on preoperative and/or intraoperative imaging information.

Efforts to refine the integration of neuronavigation tools into iMRI systems further are rather critical for close-bore and horizontal-gap MRI scanner configurations, where images and anatomy have to be correlated outside the scanner after the imaging is completed and the patient is taken out of the scanner. The neuronavigation system is then used to register the image updates to the patient's

frame of reference. After accomplishing this critical step, the neuronavigation system is used for guiding tumor resection.

Modern neuronavigation systems are not limited to tracking hand-held surgical instruments. Nimsky et al have reported 22 cases in which a surgical microscope was integrated with the tracking system and used for guiding the tumor resection. The authors found that the combination of intraoperative imaging with functional neuronavigation offers the opportunity for more radical resections and may help decrease surgical morbidity [38]. Iseki et al reported their setup of MRI-guided surgery suite with a horizontal gap scanner and in-house neuronavigation software [39]. Similar efforts to combine neuronavigation and open-MRI scanners can be also found in Mursch et al and Samset et al [40,41]. A unique approach of combining a neuronavigation system with a compact mobile MRI scanner designed for use in a conventional operating room has been described in Schulder et al [42].

The development of neuronavigation systems for use within the iMRI environment has been a constant research focus at our institution. As a result of these efforts, we have developed the 3D-Slicer software package. This is a modular image processing software package which features capabilities such as segmentation, registration, 3D-model generation, DT-tractography, mutlilayered display and multiplanar image reformatting, along with instrument tracking capabilities (Figure 22.4). The 3D-Slicer has become an integral part

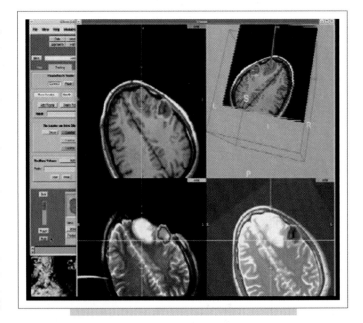

FIG. 22.4. Neuronavigation using the 3D Slicer. Preoperative fMRI registered and displayed on top of a preoperative T1 (top left), intraoperative T2 (bottom left) and preoperative T2 (bottom left) image. Note the virtual surgical probe (green bar) in the 3D view window (top right), which indicates the position of the surgical instrument used by the physician to navigate the multimodal image volume.

of our iMRI system and has been successfully used for guiding brain biopsies and brain tumor resections for the past seven years on a routine basis. Rigid and non-rigid fusion of preoperative, multimodal images with intraoperative image updates can be accomplished within the time constraints imposed by the ongoing surgery in this environment. The resulting updated multimodal data sets are employed for surgical guidance [43].

After registering the multimodal preoperative images with the intraoperative scans, the composite volume can be immediately used for surgical navigation ('virtual intraoperative scan').

MRI-GUIDED THERMAL ABLATIONS AND FOCUSED ULTRASOUND SURGERY

With the introduction of MRI as a monitoring method for thermal therapies, a novel mechanism for controlling energy deposition became available. It was recognized that many MRI parameters are sensitive to temperature changes, which makes MRI suitable for monitoring thermal ablations by non-invasive means. Furthermore, one can take advantage of diffusion MRI, which detects changes in water mobility and compartmentalization, to identify reversible as well as irreversible thermally induced tissue changes [44,45]. It became obvious, however, that MRI monitoring of thermal ablation is only feasible if the imaging and therapy delivering systems are integrated.

The role of MRI during thermal ablations is to monitor temperature levels and to ensure that the thermal coagulation is restricted to the targeted tissue volume. MRI can also detect irreversible tissue necrosis and demonstrate permanent changes within the treated tissue volume. Physiologic effects such as perfusion or metabolic response to elevated temperature can also be used for monitoring the ablation.

Both flow and tissue perfusion can affect the rate and extent of energy delivery and the size of the treated tissue volumes [46].

Since the original description of MRI monitoring and control of laser–tissue interactions, MRI-guided interstitial laser (ILT) has become a clinically tested and accepted minimally invasive treatment option. It is a relatively simple straightforward method, which can be well adapted to the interventional MRI environment.

Overall, early results suggest that ILT is a safe therapy method. Although no definitive conclusion can be drawn based on the currently available data, it appears that ILT can be of benefit in patients with low-grade gliomas (Figure 22.5) [47, 48]. In malignant gliomas, thermal therapy has been essentially unsuccessful, a predictable outcome since such tumors extend far beyond the area of MRI contrast enhancement.

Among currently developed thermal therapy methods, focused ultrasound (FUS) appears to be the most promising, since its use does not require any invasive intervention. The potential therapeutic use of ultrasound energy for intracranial pathology has long been acknowledged [49]. There is no more convincing example for the FUS benefits than in the brain, where deep lesions can be induced without any associated damage along the path of the acoustic beam [49,50]. In the brain, where most injuries have detectable functional consequences, it is extremely important to limit tissue damage to the targeted area. This necessitates the use of an imaging technique for localization, targeting and real-time intraoperative monitoring, and to control the spatial extent of heat deposition [51]. By combining FUS with MRI-based guidance and control, it may be possible to achieve complete tumor ablation without any associated structural injury or functional deficit.

Since the skull bone scatters and attenuates the propagation of the ultrasound beam, most clinical trials have been performed following craniotomy in order to provide an ultrasound window. However, the transcranial application of FUS, although challenging, is not impossible [49]. Although bone scatters and absorbs most of the acoustic energy, a small fraction can penetrate through the skull. Recent simulation and experimental studies have demonstrated the feasibility of accurately focusing ultrasound through the intact skull by using an array of multiple ultrasound transducers arranged over a large surface area [50]. To correct for beam distortion, the driving signal for the transducer elements of the array is individually adjustable, either based on measurements obtained with an invasive hydrophone probe, or better, based on detailed MRI.

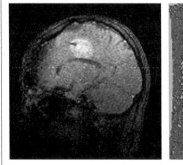  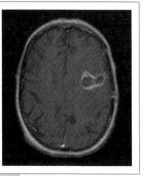

FIG. 22.5. **(A)** Intraoperative image guiding the insertion of the laser fiber into the tumor; **(B)** MR thermal mapping during laser deposition; **(C)** postoperative Gd-enhanced image.

Because of the large surface area, the ultrasound energy is distributed in such a manner as to avoid heating and consequent damage of skin, bone, meninges or surrounding normal brain parenchyma, while at the same time being able to coagulate the tissue at the focus. The experimental data are extremely promising and a clinical trial is in progress at Harvard Medical School. From the preliminary results we conclude that it appears possible to coagulate brain tumors thermally through the intact skull under MRI thermometry control using MR-compatible arrays.

By applying multiple, smaller transducers around the skull in a helmet-like phased-array system, sufficient amounts of energy can be deposited in the target tissue. Unfortunately, the skull thickness is uneven, causing variable delays of the acoustic waves originating from individual phased array elements. Phase incoherence can be corrected, however, if the skull thickness is known from preoperative x-ray computed tomography scans. In an experimental setup, successful focusing through the skull was achieved and verified by MRI, thus providing the foundation for developing the first human MRI-guided FUS system for brain tumor treatment.

Beyond thermal coagulation of tissue, FUS has various other effects that can be therapeutically exploited and thus open the way for potentially innovative vascular and functional neurosurgery applications and for targeted drug delivery to the central nervous system. Among the most important, the capability of occluding vessels could make FUS a therapeutic tool for the treatment of vascular malformations [52].

FUS can be used selectively to open the blood–brain barrier, without damaging the surrounding brain parenchyma [53,54]. For this effect to be achieved, preformed gas bubbles must be introduced into the vasculature, as is routinely done with ultrasound contrast agents. The gas bubbles implode and release cavitation-related energy, which transiently inactivates the tight junctions. As a consequence, large molecules can pass through the artificially created 'window' in the blood–brain barrier [55]. These large molecules can be chemotherapeutic or neuropharmacological agents. FUS-based targeted selective drug delivery to the brain could result in novel therapeutic interventions for movement and psychiatric disorders.

Such MRI-guided focal opening of the blood–brain barrier, combined with ultrasound technology that permits sonications through the intact skull, will open the way for new, non-invasive, targeted therapies. Specifically, it would provide targeted access for chemotherapeutic and gene therapy agents [56], as well as monoclonal antibodies, and could even provide a vascular route for performing neurotransplantations.

CONCLUSION

The role of imaging for both diagnosis and therapy has become accepted. The increasingly by favorable reception and dissemination of minimally invasive procedures resulted in the recognition of the feasibility of image-guided approaches. Although radiology has combined imaging with various novel therapeutic methods, the full utilization of advanced imaging technology has not yet been accomplished. Current research focuses on the development of therapy delivery systems in which advanced imaging modalities are closely linked with high performance computing. Obviously, the operating room of the future will accommodate various instruments, tools and devices, which are attached to the imaging systems and controlled by image-based feedback.

REFERENCES

1. *US MRI Scanners and Coils Markets A435–50.* (2003). Frost and Sullivan Marketing Report pp. 2–7.
2. Jolesz FA (2005). Future perspectives for intraoperative MRI. Neurosurg Clin N Am 16(1):201–213.
3. Jolesz FA, Hynynen K, McDannold N, Tempany C (2005). MR imaging-controlled focused ultrasound ablation: a noninvasive image-guided surgery. Magn Reson Imag Clin N Am 13(3):545–560.
4. Keles GE, Lamborn KR, Berger MS (2001). Low-grade hemispheric gliomas in adults: a critical review of extent of resection as a factor influencing outcome. J Neurosurg 95(5):735–745.
5. Lacroix M, Abi-Said D, Fourney DR et al (2001). A multivariate analysis of 416 patients with glioblastoma multiforme: prognosis, extent of resection, and survival. J Neurosurg 95(2):190–198.
6. Claus EB, Horlacher A, Hsu L et al (2005). Survival rates in patients with low-grade glioma after intraoperative magnetic resonance image guidance. Cancer 103(6):1227–1233.
7. Schneider JP, Trantakis C, Rubach M et al (2005). Intraoperative MRI to guide the resection of primary supratentorial glioblastoma multiforme – a quantitative radiological analysis. Neuroradiology 47(7):489–500.
8. Nabavi A, Black PM, Gering DT et al (2001). Serial intraoperative magnetic resonance imaging of brain shift. Neurosurgery 48(4):787–797; discussion 797–798.
9. Nimsky C, Ganslandt O, Cerny S et al (2000). Quantification of, visualization of, and compensation for brain shift using intraoperative magnetic resonance imaging. Neurosurgery 47(5):1070–1079; discussion 1079–1080.
10. Schneider JP, Schulz T, Schmidt F et al (2001). Gross-total surgery of supratentorial low-grade gliomas under intraoperative MR guidance. Am J Neuroradiol 22(1):89–98.
11. Black PM, Alexander E 3rd, Martin C et al (1999). Craniotomy for tumor treatment in an intraoperative magnetic resonance imaging unit. Neurosurgery 45(3):423–431; discussion 431–433.

12. Hadani M, Spiegelman R, Feldman Z et al (2001). Novel, compact, intraoperative magnetic resonance imaging-guided system for conventional neurosurgical operating rooms. Neurosurgery 48(4):799–807; discussion 807–809.

13. Talos IF, Zou KH, Ohno-Machado L et al (2006). Supratentorial low-grade glioma resectability: statistical predictive analysis based on anatomic MR features and tumor characteristics. Radiology 239(2):506–513.

14. Logothetis NK, Pfeuffer J (2004). On the nature of the BOLD fMRI contrast mechanism. Magn Reson Imaging 22(10):1517–1531.

15. Schiffbauer H, Berger MS, Ferrari P et al (2002). Preoperative magnetic source imaging for brain tumor surgery: a quantitative comparison with intraoperative sensory and motor mapping. J Neurosurg 97(6):1333–1342.

16. Schiffbauer H, Berger MS, Ferrari P et al (2003). Preoperative magnetic source imaging for brain tumor surgery: a quantitative comparison with intraoperative sensory and motor mapping. Neurosurg Focus 15(1):E7.

17. Rohlfing T, West JB, Beier J et al (2000). Registration of functional and anatomical MRI: accuracy assessment and application in navigated neurosurgery. Comput Aided Surg 5(6): 414–425.

18. Nimsky C, Ganslandt O, Kober H et al (1999). Integration of functional magnetic resonance imaging supported by magnetoencephalography in functional neuronavigation. Neurosurgery 44(6):1249–1255; discussion 1255–1256.

19. Witwer BP, Moftakhar R, Hasan KM et al (2002). Diffusion-tensor imaging of white matter tracts in patients with cerebral neoplasm. J Neurosurg 97(3):568–575.

20. Talos IF, O'Donnell L, Westin CF et al (2003). Diffusion tensor and functional MRI fusion with anatomical MRI for image-guided neurosurgery. Lect Notes Comp Sci 2878:407–415.

21. Basser PJ, Pajevic S, Pierpaoli C et al (2000). In vivo fiber tractography using DT-MRI data. Magn Reson Med 44(4):625–632.

22. Mori S, van Zijl PC (2002). Fiber tracking: principles and strategies – a technical review. NMR Biomed 15(7–8):468–480.

23. Westin CF, Maier SE, Mamata H et al (2002). Processing and visualization for diffusion tensor MRI. Med Image Anal 6(2):93–108.

24. Pluim JP, Maintz JB, Viergever MA (2003). Mutual-information-based registration of medical images: a survey. IEEE Trans Med Imaging 22(8):986–1004.

25. West J, Fitzpatrick JM, Wang MY et al (1997). Comparison and evaluation of retrospective intermodality brain image registration techniques. J Comput Assist Tomogr 21(4):554–566.

26. Maintz JB, Viergever MA (1998). A survey of medical image registration. Med Image Anal 2(1):1–36.

27. Warfield SK, Jolesz FA, Kikinis R (1998). A high performance computing approach to the registration of medical imaging data. Parallel Comput 24(9–10): 1345–1368.

28. Wells WM 3rd, Viola P, Atsumi H et al (1996). Multi-modal volume registration by maximization of mutual information. Med Image Anal 1(1):35–51.

29. Warfield SK (2005). Medical Image Analysis for Image Guided Therapy. DICTA.

30. Grau V, Mewes AU, Alcaniz M et al (2004). Improved watershed transform for medical image segmentation using prior information. IEEE Trans Med Imaging 23(4):447–458.

31. Warfield SK, Haker SJ, Talos IF et al (2005). Capturing intraoperative deformations: research experience at Brigham and Women's Hospital. Med Image Anal 9(2):145–162.

32. Clatz O, Delingette H, Talos IF et al (2005). Robust nonrigid registration to capture brain shift from intraoperative MRI. IEEE Trans Med Imaging 24(11):1417–1427.

33. Krishnan R, Raabe A, Hattingen E et al (2004). Functional magnetic resonance imaging-integrated neuronavigation: correlation between lesion-to-motor cortex distance and outcome. Neurosurgery 55(4):904–914; discusssion 914–915.

34. Ruiz-Alzola J, Westin CF, Warfield SK et al (2002). Nonrigid registration of 3D tensor medical data. Med Image Anal 6(2):143–161.

35. Schwartz RB, Hsu L, Kacher DF et al (1998). Intraoperative dynamic MRI: localization of sites of brain tumor recurrence after high-dose radiotherapy. J Magn Reson Imaging 8(5):1085–1089.

36. Atlas SW, Thulborn KR (1998). MR detection of hyperacute parenchymal hemorrhage of the brain. Am J Neuroradiol 19(8):1471–1477.

37. Mamata Y, Mamata H, Nabavi A et al (2001). Intraoperative diffusion imaging on a 0.5 Tesla interventional scanner. J Magn Reson Imaging 13(1):115–119.

38. Nimsky C, Ganslandt O, Kober H et al (2001). Intraoperative magnetic resonance imaging combined with neuronavigation: a new concept. Neurosurgery 48(5):1082–1089; discussion 1089–1091.

39. Iseki H, Muragaki Y, Nakamura R et al (2005). Intelligent operating theater using intraoperative open-MRI. Magn Reson Med Sci 4(3):129–136.

40. Mursch K, Gotthardt T, Kroger R et al (2005). Minimally invasive neurosurgery within a 0.5 tesla intraoperative magnetic resonance scanner using an off-line neuro-navigation system. Minim Invasive Neurosurg 48(4):213–217.

41. Samset E, Hogetveit JO, Cate GT, Hirschberg H (2005). Integrated neuronavigation system with intraoperative image updating. Minim Invasive Neurosurg 48(2):73–76.

42. Schulder M, Sernas TJ, Carmel PW (2003). Cranial surgery and navigation with a compact intraoperative MRI system. Acta Neurochir Suppl 85:79–86.

43. Nabavi A, Gering DT, Kacher DF et al (2003). Surgical navigation in the open MRI. Acta Neurochir Suppl 85:121–125.

44. Bleier AR, Jolesz FA, Cohen MS et al (1991). Real-time magnetic resonance imaging of laser heat deposition in tissue. Magn Reson Med 21(1):132–137.

45. MacFall J, Prescott DM, Fullar E, Samulski TV (1995). Temperature dependence of canine brain tissue diffusion coefficient measured in vivo with magnetic resonance echo-planar imaging. Int J Hyperthermia 11(1):73–86.

46. McDannold NJ, Jolesz FA (2000). Magnetic resonance image-guided thermal ablations. Top Magn Reson Imaging 11(3): 191–202.

47. Kahn T, Bettag M, Ulrich F et al (1994). MRI-guided laser-induced interstitial thermotherapy of cerebral neoplasms. J Comput Assist Tomogr 18(4):519–532.

48. Bettag M, Ulrich F, Schober R et al (1991). Stereotactic laser therapy in cerebral gliomas. Acta Neurochir Suppl (Wien) 52:81–83.

49. Hynynen K, Jolesz FA (1998). Demonstration of potential non-invasive ultrasound brain therapy through an intact skull. Ultrasound Med Biol 24(2):275–283.

50. Clement GT, White J, Hynynen K (2000). Investigation of a large-area phased array for focused ultrasound surgery through the skull. Phys Med Biol 45(4):1071–1083.

51. McDannold K, Hynynen D, Wolf G et al (1998). MRI evaluation of thermal ablation of tumors with focused ultrasound. J Magn Reson Imaging 8(1):91–100.

52. Hynynen K, Colucci V, Chung A, Jolesz F (1996). Noninvasive arterial occlusion using MRI-guided focused ultrasound. Ultrasound Med Biol 22(8):1071–1077.

53. Hynynen K, McDannold N, Sheikov NA et al (2005). Local and reversible blood-brain barrier disruption by noninvasive focused ultrasound at frequencies suitable for trans-skull sonications. Neuroimage 24(1):12–20.

54. McDannold N, Vykhodtseva N, Hynynen K (2006). Targeted disruption of the blood-brain barrier with focused ultrasound: association with cavitation activity. Phys Med Biol 51(4):793–807.

55. Kinoshita M, McDannold N, Jolesz FA, Hynynen K (2006). Targeted delivery of antibodies through the blood–brain barrier by MRI-guided focused ultrasound. Biochem Biophys Res Commun 340(4):1085–1090.

56. Kinoshita M, Hynynen K (2005). A novel method for the intracellular delivery of siRNA using microbubble-enhanced focused ultrasound. Biochem Biophys Res Commun 335(2):393–399.

CHAPTER 23

Neurosurgical Treatment Planning

Suzanne Tharin and Alexandra Golby

INTRODUCTION

The goal of neurosurgical resection of brain tumors is maximal excision with minimal permanent injury to surrounding normal brain tissue and, more importantly, no resultant neurologic deficit. A new deficit may be caused by damage to cortical areas immediately surrounding a lesion, as well as to the white matter at the depths of the lesion and to brain tissue involved in the surgical approach. Functional brain mapping using a variety of techniques may aid in neurosurgical planning by providing information on the localization of eloquent cortex supporting behavior such as movement and speech, as well as of white matter tracts connecting critical areas. Together with conventional imaging methods used to localize the lesion, functional mapping can be helpful in defining the relationship of the lesion to critical brain structures for operative planning. While neurosurgeons have long used invasive brain mapping techniques, recent technologic advances have led to several additional less invasive methods for mapping brain function. These methods differ not only in their methodology and physiologic basis but also in their level of spatial and temporal resolution, their invasiveness and their cost.

Functional brain mapping techniques may be used for preoperative planning, as well as for intraoperative assistance in navigation and decision-making. Given incomplete information inherent in any single technique, combining information from multiple methods may be the most useful in operative planning. The nature of the pre- and intraoperative functional mapping will be dependent on the eloquent cortex in the vicinity of the lesion. This chapter will briefly present the most common brain mapping techniques and will then discuss the role of functional imaging studies in operative planning according to the location of the brain tumor and critical functions of adjacent brain.

BRAIN MAPPING TECHNIQUES

Functional Magnetic Resonance Imaging (fMRI)

Functional MRI measures changes in cerebral blood flow as a surrogate for neuronal activity. The most commonly used fMRI method measures blood oxygen level-dependent (BOLD) changes in the MR signal. Neuronal activity results in increased blood flow through local capillaries. The increased perfusion outstrips the increased demand, resulting in an increase in the ratio of oxyhemoglobin (oxy-Hb) to deoxy-Hb [1]. The iron in deoxy-Hb is paramagnetic and reduces $T2^*$ signal. The relative increase in oxy-Hb concentration with neuronal activity results in an increase in $T2^*$ signal with neuronal activity, forming the basis of the BOLD signal [2]. For each voxel, the BOLD signal during performance of a task is compared to that during the resting or control state. BOLD signal change represents a ratio rather than an absolute physiologic measure. The change in signal is small, on the order of 0.5–5% [3]. To obtain an acceptable signal-to-noise ratio, the BOLD signal is averaged over multiple repeated trials and then subjected to statistical analysis. Finally, the BOLD map is superimposed on structural MRI.

Functional MRI is still being developed as a clinical tool and validated against methods such as Wada testing or intraoperative mapping that have the proven clinical track records (see below). fMRI has been validated against positron emission tomography (PET), transcranial magnetic stimulation (TMS) and electrocortical stimulation testing (ECS) for localization of the motor cortex in patients with lesions displacing the central sulcus, demonstrating overlapping (<1 cm distance) maps in the majority of cases [4]. The typical spatial resolution of fMRI is 2–5 mm. Advanced MRI techniques, including higher field strengths [5–7] and parallel imaging [8,9], can provide increased signal-to-noise, increased spatial resolution or shortened acquisition times. The latency of the observed signal change in BOLD imaging is several seconds, making the temporal resolution of fMRI poor when compared with techniques such as ECS or electroencephalography (EEG).

Functional MRI is non-invasive and therefore repeatable, both for many runs in a single session (unlike Wada testing, see below) and on multiple occasions to follow patients over time. Due to its non-invasiveness and safety, fMRI is also suitable for use in children. Unlike Wada testing, fMRI can provide localization and not merely lateralization of

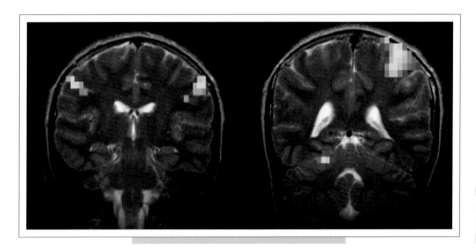

FIG. 23.1. Coronal functional MRI images show motor activation for lip pursing (left) demonstrating bilateral activation and hand clenching (right).

critical functions such as motor, sensory, language and memory. Finally, fMRI is able to demonstrate functional activations in the depths of cortical sulci not just at the cortical surface, an advantage over even the gold-standard ECS (Figure 23.1). Disadvantages of fMRI include sensitivity to motion-related artifacts, including those arising from heartbeat, breathing and head motion. Disruption of the signal due to air and face movement associated with vocalization have generally precluded the use of tasks involving overt spoken speech for language mapping. The technique is furthermore sensitive to signal from large draining veins, although this is less prevalent at higher field strengths [10].

Electrocorticography and Electrical Cortical Stimulation (ECS)

Electrocorticography is the direct recording of electrical potentials associated with brain activity from the cerebral cortex. A widely used application of electrocorticography is the localization of the central sulcus by means of phase reversal of somatosensory evoked potentials (SSEP-PR) [11]. Since the 1930s, direct electrocortical stimulation (ECS) testing has been the gold standard method for mapping brain function in preparation for surgical resection [12,13]. Invasive recordings from the human brain can provide unique opportunities to study fundamental processes at fine temporal and spatial resolution [14].

Cortical stimulation testing is limited due to its invasive nature requiring craniotomy and generally specialized surgical and anesthesia teams. ECS may also not be able to examine functional cortex within the sulcal depths that comprise as much as two-thirds of the cortical surface [15], the deep structures of the mesial temporal lobe or the underlying white matter. For example, it is not uncommon for patients who have undergone cortical mapping and resection respecting the boundaries of the eloquent cortex nevertheless to be left with neurological deficits secondary to damage to associated white matter tracts. To minimize such occurrences, a few groups have reported success

using white matter stimulation intraoperatively to define critical tracts [16].

Diffusion Tensor Imaging (DTI)

A second MRI-based technique useful in the planning of brain tumor resections, DTI is able to demonstrate white matter tracts by using MRI to measure the direction of diffusion of water molecules as a marker for the axis of these tracts. The technique is based upon the restriction of diffusion of water by axonal membranes and myelin. Water diffusion in axon tracts is direction dependent, or anisotropic. For physiologic reasons that are incompletely understood, water diffuses least in the direction perpendicular to fiber tracts and most in parallel to them [17]. In DTI, magnetic field gradients are applied in multiple orientations. The diffusion tensor is a mathematical model that is used to estimate the direction of maximum diffusivity of water molecules for every voxel. This direction of maximum diffusivity corresponds to the dominant axis of white mater tracts in that voxel [18]. Typical spatial resolution in DTI is a voxel size of $2 \times 2 \times 5\,mm^3$, although this is improving quickly [19]. An advantage of DTI is that it specifically examines white matter, which is not amenable to characterization by other functional techniques. One limitation is that it offers little specific information on the functional status or substrates of these tracts. The technique also suffers from poor signal-to-noise ratio [19] and is vulnerable to artifact from air spaces. Other difficulties include imaging crossing tracts, although acquisition and computational advances are continuously being made, providing solutions to these and other issues.

DTI may be used to visualize major white matter tracts including the cingulum, superior and inferior occipitofrontal fasciculi, uncinate fasciculus, arcuate fasciculus, occipitotemporal fasciculus, the corticospinal, corticobulbar and corticopontine tracts, the optic radiations, corpus callosum and anterior commissure [18]. DTI has been used to image the effect of neoplasms on the integrity and trajectory of

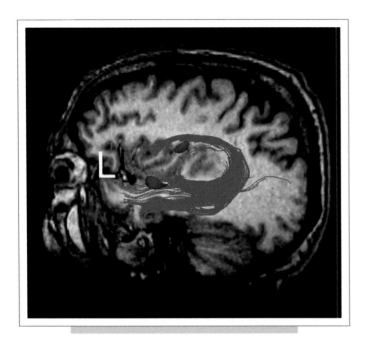

FIG. 23.2. Sagittal image shows coregistered fMRI for language activation and DTI data representing the arcuate fasciculus.

white matter tracts [20]. Four patterns of anisotropy have been observed: normal signal with altered position or direction, corresponding to tract displacement; decreased but present signal with normal direction and location, thought to correspond to vasogenic edema; decreased signal with disrupted direction maps, thought to correspond to infiltration; and loss of anisotropic signal corresponding to fiber tract obliteration or destruction [18]. Histopathologic studies for validation are still lacking, however, due to the difficulty in performing such studies. Information on white matter tracts may be useful in operative planning, to assist in the avoidance of both displaced tracts as well as those affected by edema or tumor, and to allow the surgeon greater confidence when working in areas of brain where critical white mater tracts might be expected, but where DTI reveals pre-existing tract destruction (Figure 23.2).

Magnetoencephalography (MEG)

Magnetoencephalography (MEG) is a non-invasive method of measuring brain activity by measuring the magnetic fields that accompany neuronal activity. MEG is similar to EEG but is based on magnetic field changes rather than voltage changes. Neural activity can be described as the generation and propagation of ion currents. The longitudinal current flow generated by several thousands of neurons firing synchronously can be detected at the scalp surface using a biomagnetometer. MEG data are gathered using a biomagnetometer made up of wire induction coils arranged in an array covering the entire head. The magnetic fields produced by neural activity induce electric

currents in these coils and can be used to reconstruct an image of the distribution of evoked neural electrical activity of brain function in real time. Superconducting quantum interference devices (SQUIDs) are used to record the conductance change caused by the small magnetic fields associated with neuronal activity. MEG source localization is generally modeled as the equivalent current dipole. During MEG functional mapping a time-locked stimulus task is presented multiple times. The resulting evoked neuromagnetic signals are then averaged over many trials to separate the signal produced by a focal population of active neurons from background activity. The use of MEG at this time is mainly restricted to centers pursuing research programs, though MEG for mapping both seizure foci and functional cortex is a clinically reimbursable procedure.

Although MEG is currently used primarily to localize epileptogenic foci for resection [21], its use has more recently expanded to stereotactic and image-guided surgery to aid in the safe resection of lesions adjacent to eloquent cortex [22,23]. Magnetic source imaging is the coregistration of MEG data to a structural image to facilitate the anatomo-functional correlations and to incorporate this information into stereotactic neuronavigation systems [24]. Several studies report good correlation between preoperative MEG functional data and intraoperative maps of sensory and motor evoked potentials and electrocortical mapping [25–29]. Several groups have merged functional MEG data with anatomic data in order to locate key functional cortex near and within cortical lesions including gliomas and brain metastases. Rezai et al have reported a technique of integrating MEG functional mapping data for both motor and sensory tasks into a stereotactic database for use intraoperatively, as well as for preoperative planning [22]. Similarly McDonald et al report the successful combination of both fMRI and MEG data into a frameless stereotactic system that also incorporates digital registration of cortical stimulation sites [30].

Wada Testing

The intra-carotid amytal, or Wada, test is a method in which sodium amytal is injected into the internal carotid artery (ICA) temporarily anesthetizing the brain in the territory of that ICA. Following confirmation of clinical effect by the onset of contralateral hemiparesis and EEG changes, a battery of behavioral tests is administered. Wada testing was initially developed to lateralize language dominance in patients undergoing electroconvulsive therapy, but has long been used for language lateralization in preoperative patients [31,32]. The Wada test was subsequently modified to test lateralization of memory and to assess the risk of postoperative amnesia in patients undergoing temporal lobectomy for medial temporal lobe epilepsy [33]. The spatial resolution of the Wada test is usually hemispheric. However, Wada testing has been employed in a more highly localized manner with selective catheterization, for instance to investigate function of the mesial temporal structures in epilepsy patients [34].

Wada testing is limited in that the examiner has only a few minutes to test each hemisphere. The test is invasive, carrying a 0.6–1% risk of stroke [35]. Agitation or obtundation can preclude language testing in some individuals. Because of its invasive nature, Wada testing is not readily repeatable. Technically, the test may potentially be confounded by cross flow between hemispheres, resulting in anesthesia of both hemispheres from a unilateral injection. fMRI correlates well with and may soon substitute for Wada testing for the preoperative determination of language lateralization, and may indeed offer many advantages (discussed below).

Transcranial Magnetic Stimulation (TMS)

Transcranial magnetic stimulation involves the stimulation or inhibition of neuronal electrical activity via a magnetic field delivered at the scalp. Barker et al discovered that a magnetic field could be used to set up an electrical stimulus across the scalp and skull [36]. In TMS, a current is discharged into an electromagnetic coil held over the skull; this discharge creates a magnetic field that induces a perpendicular electric field. The field is conducted through the skin and skull and produces an electric current in the cortex, without causing pain in the patient [37]. TMS is relatively non-invasive though there is a risk of repetitive TMS causing seizures. Disadvantages include poor spatial resolution and very limited availability [37]. While there has been relatively little reported use of TMS in surgical planning, Krings et al have described the correlation of TMS with ECS in motor mapping in two patients with tumors near the central sulcus [38]. Neggers et al have recently developed a frameless stereotactic navigation system for delivering TMS and validated their results showing correlation with fMRI and ECS to within 5 mm in motor cortex [39]. Despite one study reporting that TMS does not replicate Wada testing [40], the above-mentioned reports suggest tremendous promise for the technique in operative planning in the future.

IMAGING SPECIFIC FUNCTIONS

Motor Mapping

Voluntary movement is associated with activation of primary motor cortex (M1), premotor area (PMA), supplementary motor area (SMA) and superior parietal lobule (SPL). Motor cortex may be localized intraoperatively by stimulating the precentral and post-central gyri, eliciting motor or somatosensory responses respectively. ECS for motor mapping [13,41] may be performed under general anesthesia as long as muscle relaxants are not used. Low or high frequency stimulation delivered to the motor cortex causes contralateral muscular contractions. Preoperatively,

fMRI is used in many centers to delineate brain regions activated by motor tasks.

Several groups have validated fMRI against ECS in localizing motor areas [42–45]. Majos et al studied 33 patients with tumors near the central sulcus and found 84% agreement between preoperative fMRI and intraoperative ECS in identification of motor areas and 83% agreement in the identification of sensory centers between the two techniques. When sensory and motor data are combined, the agreement between fMRI and ECS increased to 98% [43]. Roessler et al recently used 3-Tesla fMRI and ECS to guide resection of gliomas in motor cortex [46]. In those patients for whom the two modalities could be correlated (17 of 22), there was 100% agreement between fMRI and ECS data within 1 cm.

Berman et al have combined motor ECS with DTI of motor fiber tracts to visualize motor areas as well as their descending axons in patients with gliomas [47]. A few groups have described the use of DTI with fMRI to evaluate the motor cortex and its descending (pyramidal) tracts in patients with tumors near motor cortex [20,48–51]. Reliable integration of DTI datasets into standard neuronavigation systems, by co-registration of fiber tract data with standard anatomic data, will greatly facilitate the routine use of this technology. Nimsky et al have described such integration of DTI data into a standard intraoperative neuronavigation system [52,53] and have recently used this technique to visualize the pyramidal tracts and optic radiations during surgery in 16 patients [53]. Combining DTI with fMRI allows imaging of a particular functional area as well as its connections to other areas, non-invasively, prior to surgery.

Single pulse TMS (spTMS) has also been used to map motor cortex [38,54]. This technique, however, has not been widely used to date.

Several studies have shown altered motor maps in patients with lesions near motor areas. Yousry et al have found alterations in the cortical representation of the motor hand area in neurologically intact patients with space-occupying lesions near the central sulcus [45]. Using fMRI to study patients with tumors located in the vicinity of the motor strip, Krings et al found decreasing activation in primary motor cortex in proportion to the degree of preoperative hemiparesis [48]. They also noted increasing activation in secondary motor areas with increasing preoperative weakness. Motor cortex may also be frequently (and unpredictably) shifted by mass lesions. These accounts of the alterations of motor maps illustrate the potential but limited plasticity of the adult brain.

Language Lateralization

fMRI and Wada are used for language lateralization in patients with tumors of the frontal and temporal lobes. Wada testing with comprehensive language assessment [55] reliably lateralizes language function. Benbadis et al have demonstrated poor correlation between language lateralization based solely on speech arrest, when compared to lateralization by Wada testing based on comprehensive language assessment and lateralization by fMRI in 12 patients with intractable epilepsy, suggesting that speech arrest alone is an unreliable method of preoperative language lateralization [56]. Woerman et al compared language lateralization by fMRI and Wada in 100 patients with temporal lobe epilepsy (TLE) or extratemporal epilepsy and found 91% concordance in the results of the two tests [57]. In their study, fMRI falsely categorized language lateralization in only 3% of left-sided TLE patients, but in 25% of left-sided extratemporal epilepsy patients. This large study using a simple word-generation task and a rapid (15 minute) acquisition time suggests that fMRI may reduce the need for Wada testing in TLE, but is less useful for language lateralization in extratemporal epilepsy and hence may be less useful for language lateralization in patients with brain tumors.

Lehericy et al have demonstrated that frontal, not temporal, asymmetry of language dominance correlated with Wada testing [58]. Further validation of fMRI for preoperative language lateralization comes from studies demonstrating atypical (bilateral or right-dominant) language dominance in 22–24% of non-right-handed subjects [59,60], in agreement with earlier findings using Wada testing of left-handed epileptic patients [61]. Fernandez et al have recently demonstrated high within-test and test-retest intrasubject reproducibility for fMRI language lateralization in patients with epilepsy [62]. Binder et al noted a linear relationship between the intensity of right hemispheric activation on fMRI and the severity of language deficits

observed on Wada testing of right-dominant individuals [63], suggesting a role for fMRI in predicting the severity of language deficits in right hemisphere surgery in this population and suggesting a possible advantage of fMRI with its graded output over the Wada test with binary output [64]. In addition to carrying fewer risks, fMRI has the advantage of taking less time and costing less than Wada testing [65].

Repetitive TMS (rTMS) is used to disrupt language processing [66,67] and may be used for the determination of language lateralization [68]. However, language lateralization by rTMS does not correlate completely with Wada results [69].

Language Localization

Specific frontal and temporal language areas are essential for language function. While ECS can cause various speech disruptions with stimulation of both frontal and temporal cortex, different non-invasive functional mapping paradigms will preferentially demonstrate some language areas. Some of these areas can be preferentially activated with different protocols, such as frontal regions with word generation paradigms [58,60]. An observational technique like fMRI cannot distinguish essential from participating, but non-essential, areas. fMRI has not yet been well correlated with ECS in language localization studies [70] and so its use remains an adjunct as research efforts to improve its agreement with standard techniques continue. MEG has also been used for language localization; interestingly, whereas fMRI most commonly highlights frontal language areas, MEG tends to demonstrate most robustly temporal language areas (Figure 23.3). This is an example

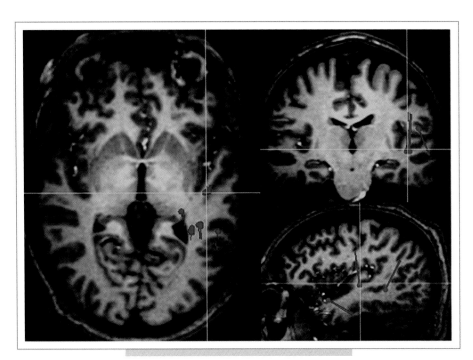

FIG. 23.3. MEG maps obtained during a language task. Note location of dipoles primarily in the left temporal region.

of why combining complementary techniques may be particularly useful.

ECS language mapping requires that the patient be able to cooperate in performing language tasks during an awake craniotomy [71]. Awake craniotomy for language mapping is typically performed using a combination of local anesthetic field block and short acting general agents. Once the scalp, skull and dura are opened the sedation is allowed to wear off so that the patient may cooperate with behavioral testing. During testing, the patient is awake and asked to perform language tests such as counting or naming while the surgeon stimulates the cortical surface. Areas where cortical stimulation induces speech arrest or paraphasic errors are considered essential for language function. In this method, ECS is employed as an inhibition technique causing disruption in normal neuronal firing. Haglund et al reported that, for 40 patients with dominant temporal gliomas without language deficits preoperatively, 87% had no deficits postoperatively using the above methods [72]. Even cooperative patients may have trouble maintaining task performance over the course of the investigation. Awake craniotomy generally requires dedicated neuroanesthesia support and a sufficient caseload to provide training and expertise and hence may not be available in many centers.

While fMRI protocols in current use have demonstrated excellent concordance with Wada testing for language lateralization, correlation between pre- and postoperative fMRI and intraoperative cortical mapping remains inconsistent in studies. In a study of 14 right-handed patients with left hemisphere tumors, Roux et al [70] compared language areas activated by naming and verb generation tasks with intraoperative ECS speech mapping. Twenty-two language sites were identified with cortical stimulation, five of which were concordant with sites identified by fMRI, but 17 of which were not associated with fMRI signal. Based on these findings they conclude that while fMRI is a helpful adjunct, its failure to identify speech areas identified by ECS in the operating room makes it insufficient to form the basis of critical preoperative decision-making prior to resection. Postoperative fMRI of a subset of their patients identified language foci that correlated with those identified intraoperatively with ECS in only six of eight patients examined, with complete agreement in only three of eight patients [70]. The authors note that sensitivity and specificity of fMRI can be improved by combining data from both tasks, suggesting that modification of the behavioral paradigms used could improve correlation with ECS data. fMRI does generally identify a greater number of language-associated areas, partly due to limited coverage of cortex by electrodes, and partly due to the fundamental differences between fMRI, an observational technique, and ECS, an inhibition technique [73]. This finding suggests that, in addition to being less sensitive, fMRI as presently performed may also be less specific for language mapping.

Memory Lateralization

Several groups have used fMRI to determine memory lateralization in patients with medically refractory medial temporal lobe (MTL) epilepsy and have also demonstrated good agreement with Wada testing [74–76]. This approach may also be helpful when planning resections of tumors in the medial temporal lobe, although this has not been studied.

Other Functional Localization

Both primary visual cortex and the visual pathways may be interrupted by surgery causing permanent visual field loss. Tummala et al have used DTI of the optic radiations to facilitate complete resection of tumors with no postoperative visual field deficit in two pediatric patients [77].

CONSIDERATIONS FOR SPECIFIC TUMOR TYPES

Many patients with primary tumors of the CNS have infiltrative lesions, so resections are performed to obtain tissue diagnosis and for reduction of mass effect and disease burden, but generally not for cure. Thus, minimizing resultant neurologic deficits and associated reduction in quality of life is paramount. Similarly, patients with brain metastases usually have limited life expectancy, so palliative surgery must not worsen existing symptoms or introduce new deficits.

Low-grade Gliomas

There are data to suggest that the degree of resection of low-grade gliomas correlates with long-term survival [78–82], although the evidence remains controversial [83,84]. Because low-grade gliomas can have a slow clinical course, when they are located within or near eloquent brain, the need to balance aggressive resection with postoperative morbidity, particularly in the neurologically intact patient, is heightened.

Preoperative fMRI has been widely used either alone [48,85–88] or in combination with ECS [4,42,89] or intraoperative MRI [86] to map eloquent cortex in the vicinity of low-grade gliomas prior to resection. Due to their infiltrative nature, low-grade gliomas may have functional tissue within the tumor [90,91]. There is also some evidence that low-grade tumors may contain white mater tracts within the boundaries of the tumor [90]. Russel et al have found that resection of low-grade gliomas involving the supplementary motor area (SMA) results in a higher incidence of transient weakness (SMA syndrome) than resection of high-grade gliomas in the same region, presumably owing to the presence of more functional SMA cortex within the lower grade lesions [92]. Detailed and accurate functional imaging is therefore particularly important for planning their resection.

High-grade Gliomas

Tumors of higher grade present a somewhat different set of challenges. There is inconsistent evidence that degree of resection of glioblastoma multiforme (GBM) correlates with time to progression and median survival [93]. However, with the best median survival times on the order of 1–2 years [93,94], time for postoperative recovery and duration of even transient deficits are particularly relevant for GBM patients. The transient hemiparesis of the SMA syndrome, for instance [92], which usually resolves within weeks to months, could represent an unacceptable morbidity in a patient with such a limited life expectancy.

The BOLD effect of fMRI may be affected by altered regional blood flow as is found in high-grade tumors of the CNS [95]. Higher-grade tumors are associated with hemodynamic-neural uncoupling; fMRI data must accordingly be interpreted with this in mind. Perhaps owing in part to this effect, Liu et al have reported that BOLD fMRI activation volume in the SMA is affected by both tumor type (intra- versus extra-axial) and distance from motor cortex [96]. Edema surrounding the tumor can also affect the MRI signal. While a potential source of error in the interpretation of fMRI, this effect has been capitalized upon in DTI to distinguish between primary and metastatic brain tumors: in the area of edema surrounding a metastasis, there in an increase in the diffusion of water along white matter (WM) tracts, whereas a decrease in diffusion is generally noted in primary brain tumors (see above), presumably owing to their infiltrative nature [97–99]. DTI may also be used to image the effect of tumor on fiber tracts (see above), as well as to measure early response to therapy in densely cellular tumors [100]. Metabolic imaging using FDG-PET or single photon emission computerized tomography (SPECT) are commonly used to distinguish radiation necrosis from tumor recurrence where MRI data are ambiguous [101].

Functional imaging may also be useful in monitoring the response of tumors to treatment. Henson et al [100] have suggested that the use of BOLD fMRI to measure oxygen levels within tumors [102] could be used in monitoring the efficacy of agents that increase DNA damage from chemotherapy or radiation, currently under investigation in animal models [103].

INTRAOPERATIVE IMAGE-GUIDED SURGERY

The recent and rapid adoption of intraoperative neuronavigation systems into neurosurgical practice has made image-guided intracranial surgery widely accessible. Extending the capabilities of these systems so that various functional mapping data can be integrated into the operating room environment will significantly increase the clinical utility of these types of studies. In addition, the intraoperative 3-dimensional display of multimodal preoperatively obtained

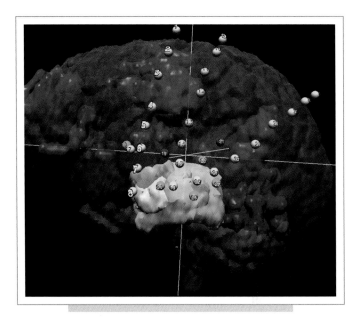

FIG. 23.4. Intraoperative view of integrated images including segmented tumor in the left temporal lobe, fMRI activations (blue) and intraoperative recording sites (yellow spheres). Negative intraoperative stimulation sites are represented by green spheres whereas positive sites are shown in red.

functional information may be integrated with intraoperative findings from cortical mapping (Figure 23.4), allowing for further validation of the non-invasive techniques.

A problem common to the intraoperative use of preoperative mapping techniques is the failure to take into account brain shift following opening of the craniotomy flap and dura. Shifts of up to 24 mm at the cortical surface have been described [104]. Causes of brain shift include patient positioning and gravity, edema, administration of osmotic diuretics, drainage of CSF, retraction and resection [104–108]. Investigations of brain shift have furthermore revealed that displacement of the cortical surface and of deeper structures are uncorrelated [104,109]. Because of this, and because of the heterogeneity among lesions encountered in the operating room, intraoperative MRI (ioMRI) is necessary to address this problem by providing images of the brain during surgery [110]. Recently, multiple groups have put forward algorithms to update neuronavigation systems with intraoperative MRI images following brain shift [104,105]. Continued advances in imaging and/or modeling intraoperative brain shift and the application of these transformations to preoperative functional datasets will allow more accurate intraoperative information to be available to the surgeon.

In addition to adjusting preoperatively acquired images to compensate for brain shift, future prospects will include the intraoperative acquisition of functional data, such as intraoperative DTI and fMRI. Nimsky et al have demonstrated the feasibility of intraoperative DTI to assess shift of white matter tracts [111]. By comparing preoperatively and intraoperatively acquired DTI data, the authors describe variable and sometimes marked shifting (from 8 mm inward to 15 mm outward) of white matter tracts during resection of adjacent mass lesions [111,112]. Schulder et al have described the use of low-field intraoperative fMRI of motor cortex [113]. Intraoperative fMRI would be facilitated by the use of high field strength systems (≥1.5 Tesla) [114]. Gasser et al have demonstrated the feasibility of high field intraoperative fMRI. The authors used peripheral sensory stimulation and measured fMRI activation of sensory cortex, which was verified with phase reversal of SSEPs [115]. The use of high field strength ioMRI may allow the intraoperative acquisition of functional data, though this approach suffers from some of the limitations of intraoperative ECS such as invasiveness, lack of preoperative access to information and limited testing time. On the other hand, such a set-up will provide an unprecedented opportunity for brain functional mapping modalities.

Current functional brain mapping techniques suffer from limitations in either temporal or spatial resolution, or are highly invasive. It follows that resolution and quality of data can be improved by integrating, or coregistering, data from multiple complementary sources [116], including conventional imaging data and neuronavigation systems [117,118]. While pre- and intraoperative coregistration of data from multiple functional mapping techniques has yet to find its way into mainstream use, an increasing number of studies suggesting its utility may be found in the literature. For instance, Roessler et al recently used 3-Tesla fMRI and ECS to maximize resection of gliomas in motor cortex with no permanent morbidity [46]. Neuronavigation-integrated fMRI has been combined with intraoperative ECS to predict and minimize risk of new postoperative neurologic deficit in resection of tumors near motor cortex [119].

There will likely be an increased role for functional imaging in guidance of minimally invasive techniques such as focused ultrasound (FUS), laser thermal ablation, stereotactic radiosurgery (SRS), and other emerging techniques in patients with brain tumors.

New minimally invasive functional imaging techniques, such as functional transcranial Doppler ultrasound imaging (fTCDs), are finding increasing use in preoperative planning. Rihs et al have recently validated fTCDs for determination of language lateralization against Wada testing and demonstrated the utility of fTCDs when patients may be unable to undergo Wada testing [120]. fTCDs have also been used to map areas involved in visual [121] and auditory [122] perception. Studies comparing the reliability of less invasive methods to that of the more invasive, gold standard, techniques in current use will pave the way for increasingly non-invasive testing. As fMRI has revealed, non-invasive studies carry the advantage of applicability to more patient populations including children, repeatability during a session, repeatability over the patient's clinical course, and whole-brain coverage. In addition, the ability to prepare pre-procedure non-invasive functional brain maps will be particularly important as more minimally invasive treatments are developed.

The combination of functional imaging with conventional imaging, image-guidance and intraoperative imaging systems will lead to our ability to perform more complete and precise resections of lesions while preserving neurological functions. Functional brain mapping stands eventually to change the way intracranial processes are treated by creating a road map of brain function that not only defines eloquent 'no-go' areas, but also illuminates potential functional targets. Such a non-invasively obtained road map will be critical to the development of minimally invasive therapies.

REFERENCES

1. Fox PT. Raichle ME (1986). Focal physiological uncoupling of cerebral blood flow and oxidative metabolism during somatosensory stimulation in human subjects. Proc Natl Acad Sci USA 83:1140–1144.

2. Ogawa S et al (1990). Brain magnetic resonance imaging with contrast dependent on blood oxygenation. Proc Natl Acad Sci 87:9868–9872.

3. Bandettini PA et al (1992). Time course EPI of human brain function during task activation. Magn Reson Med 25(2): 390–397.

4. Krings T et al (2001). Metabolic and electrophysiological validation of functional MRI. J Neurol Neurosurg Psychiatr 71(6):762–771.

5. Kim DS, Garwood M (2003). High-field magnetic resonance techniques for brain research. Curr Opin Neurobiol 13(5): 612–619.

6. Pfeuffer J et al (2002). Perfusion-based high-resolution functional imaging in the human brain at 7 Tesla. Magn Reson Med 47(5):903–911.

7. Yacoub E et al (2001). Imaging brain function in humans at 7 Tesla. Magn Reson Med 45(4):588–594.

8. Sodickson DK et al (2005). Rapid volumetric MRI using parallel imaging with order-of-magnitude accelerations and a 32-element RF coil array: feasibility and implications. Acad Radiol 12(5):626–635.

9. Lin FH et al (2005). Functional MRI using regularized parallel imaging acquisition. Magn Reson Med 54(2):343–353.

10. Gati J et al (1997). Experimental determination of the BOLD field strength dependence in vessels and tissue. Magn Reson Med 38:296–302.

11. Wood CC et al (1988). Localization of human sensorimotor cortex during surgery by cortical surface recording of somatosensory evoked potentials. J Neurosurg 68(1):99–111.

12. Penfield W (1937). The cerebral cortex and consciousness. In Harvey Lecture. 1937.

13. Ojemann G et al (1989). Cortical language localization in left, dominant hemisphere. An electrical stimulation mapping investigation in 117 patients. J Neurosurg 71(3):316–326.

14. Engel AK et al (2005). Invasive recordings from the human brain: clinical insights and beyond. Nat Rev Neurosci 6(1):35–47.

15. Cosgrove GR, Buchbinder BR, Jiang H (1996). Functional magnetic resonance imaging for intracranial navigation. Neurosurg Clin N Am 7(2):313–322.

16. Keles GE et al (2004). Intraoperative subcortical stimulation mapping for hemispherical perirolandic gliomas located within or adjacent to the descending motor pathways: evaluation of morbidity and assessment of functional outcome in 294 patients. J Neurosurg 100(3):369–375.

17. Moseley ME et al (1990). Diffusion-weighted MR imaging of anisotropic water diffusion in cat central nervous system. Radiology 176(2):439–445.

18. Jellison BJ et al (2004). Diffusion tensor imaging of cerebral white matter: a pictorial review of physics, fiber tract anatomy, and tumor imaging patterns. Am J Neuroradiol 25(3):356–369.

19. Hunsche S et al (2001). Diffusion-tensor MR imaging at 1.5 and 3.0 T: initial observations. Radiology 221(2):550–556.

20. Witwer BP et al (2002). Diffusion-tensor imaging of white matter tracts in patients with cerebral neoplasm. J Neurosurg 97(3):568–575.

21. Knowlton RC, Shih J (2004). Magnetoencephalography in epilepsy. Epilepsia 45 Suppl 4:61–71.

22. Rezai AR et al (1996). The interactive use of magnetoencephalography in stereotactic image-guided neurosurgery. Neurosurgery 39(1):92–102.

23. Orrison WW Jr (1999). Magnetic source imaging in stereotactic and functional neurosurgery. Stereotact Funct Neurosurg 72(2–4):89–94.

24. Fagaly R (1990). Neuromagnetic instrumentation. In *Advances in Neurology: Magnetoencephalography*. Sato S (ed.). Raven Press: New York. 11–32.

25. Gallen CC, Bucholz R, Sobel DF (1994). Intracranial neurosurgery guided by functional imaging. Surg Neurol 42(6):523–530.

26. Gallen CC et al (1995). Presurgical localization of functional cortex using magnetic source imaging. J Neurosurg 82(6):988–994.

27. Kamada K et al (1993). Functional neurosurgical simulation with brain surface magnetic resonance images and magnetoencephalography. Neurosurgery 33(2):269–272; discussion 272–273.

28. Rezai AR et al (1995). Introduction of magnetoencephalography to stereotactic techniques. Stereotact Funct Neurosurg 65(1–4):37–41.

29. Sutherling WW et al (1988). The magnetic and electric fields agree with intracranial localizations of somatosensory cortex. Neurology 38(11):1705–1714.

30. McDonald J et al (1999). Integration of preoperative and intraoperative functional brain mapping in a frameless stereotactic environment for lesions near eloquent cortex. J Neurosurg 90(3):591–598.

31. Wada J, Rasmussen T (1960). Intracarotid injection of sodium amytal for the lateralization of cerebral speech dominance: experimental and clinical observations. J Neurosurg 17:226–282.

32. Wada J (1949). A new method for determination of the side of cerebral speech dominance: a preliminary report on the intracarotid injection of sodium amytal in man. Igaku to Seibutsugaki 14:221–222.

33. Milner B, Branch C, Rasmussen T (1962). Study of short-term memory after intracarotid injection of sodium amytal. Transact Am Neurol Assoc 87:224–226.

34. Brassel F et al (1996). Superselective intra-arterial amytal (Wada test) in temporal lobe epilepsy: basics for neuroradiological investigations. Neuroradiology 38(5):417–421.

35. Hankey GJ, Warlow CP, Sellar RJ (1990). Cerebral angiographic risk in mild cerebrovascular disease. Stroke 21(2):209–222.

36. Barker AT, Jalinous R, Freeston IL (1985). Non-invasive magnetic stimulation of human motor cortex. Lancet 1(8437):1106–1107.

37. Sack AT, Linden DE (2003). Combining transcranial magnetic stimulation and functional imaging in cognitive brain research: possibilities and limitations. Brain Res Rev 43(1):41–56.

38. Krings T et al (1997). Stereotactic transcranial magnetic stimulation: correlation with direct electrical cortical stimulation. Neurosurgery 41(6):1319–1325; discussion 1325–1326.

39. Neggers SF et al (2004). A stereotactic method for image-guided transcranial magnetic stimulation validated with fMRI and motor-evoked potentials. Neuroimage 21(4):1805–1817.

40. Epstein CM et al (2000). Repetitive transcranial magnetic stimulation does not replicate the Wada test. Neurology 55(7):1025–1027.

41. Berger MS et al (1989). Brain mapping techniques to maximize resection, safety, and seizure control in children with brain tumors. Neurosurgery 25(5):786–792.

42. Roux FE et al (1999). Usefulness of motor functional MRI correlated to cortical mapping in Rolandic low-grade astrocytomas. Acta Neurochir 141(1):71–79.

43. Majos A et al (2005). Cortical mapping by functional magnetic resonance imaging in patients with brain tumors. Eur Radiol 15(6):1148–1158.

44. Roux FE et al (2000). Functional MRI and intraoperative brain mapping to evaluate brain plasticity in patients with brain tumours and hemiparesis. J Neurol Neurosurg Psychiatr 69(4):453–463.

45. Yousry TA et al (1995). Topography of the cortical motor hand area: prospective study with functional MR imaging and direct motor mapping at surgery. Radiology 195(1):23–29.

46. Roessler K et al (2005). Evaluation of preoperative high magnetic field motor functional MRI (3 Tesla) in glioma patients by navigated electrocortical stimulation and postoperative outcome. J Neurol Neurosurg Psychiatr 76(8):1152–1157.

47. Berman JI et al (2004). Diffusion-tensor imaging-guided tracking of fibers of the pyramidal tract combined with intraoperative cortical stimulation mapping in patients with gliomas. J Neurosurg 101(1):66–72.

48. Krings T et al (2002). Activation in primary and secondary motor areas in patients with CNS neoplasms and weakness. Neurology 58(3):381–390.

49. Kamada K et al (2003). Visualization of the eloquent motor system by integration of MEG, functional, and anisotropic diffusion-weighted MRI in functional neuronavigation. Surg Neurol 59(5):352–361; discussion 361–362.

50. Parmar H, Sitoh YY, Yeo TT (2004). Combined magnetic resonance tractography and functional magnetic resonance imaging in evaluation of brain tumors involving the motor system. J Comp Assist Tomogr 28(4):551–556.

51. Moller-Hartmann W et al (2002). Preoperative assessment of motor cortex and pyramidal tracts in central cavernoma

employing functional and diffusion-weighted magnetic resonance imaging. Surg Neurol 58(5):302–307; discussion 308.

52. Nimsky C et al (2005). Visualization of the pyramidal tract in
glioma surgery by integrating diffusion tensor imaging in functional neuronavigation. Zentralbl Neurochir 66(3):133–141.

53. Nimsky C, Ganslandt O, Fahlbusch R (2006). Implementation
of fiber tract navigation. Neurosurgery 58(4 Suppl 2):ONS-
292-303; discussion ONS-303-304.

54. Maldjian J et al (1996). Functional magnetic resonance imaging of regional brain activity in patients with intracerebral
arteriovenous malformations before surgical or endovascular
therapy. J Neurosurg 84(3):477–483.

55. Loring DW et al (1993). Wada memory testing and hippocampal volume measurements in the evaluation for temporal
lobectomy. Neurology 43(9):1789–1793.

56. Benbadis SR et al (1998). Is speech arrest during wada testing
a valid method for determining hemispheric representation
of language? Brain Lang 65(3):441–446.

57. Woermann FG et al (2003). Language lateralization by Wada
test and fMRI in 100 patients with epilepsy. Neurology
61(5):699–701.

58. Lehericy S et al (2000). Functional MR evaluation of temporal and frontal language dominance compared with the Wada
test. Neurology 54(8):1625–1633.

59. Szaflarski J et al (2002). Language lateralization in left-handed
and ambidextrous people. Neurology 59(2).

60. Pujol J et al (1999). Cerebral lateralization of language in normal left-handed people studied by functional MRI. Neurology
52(5):1038–1043.

61. Rasmussen T, Milner B (1977). The role of early left-brain
injury in determining lateralization of cerebral speech functions. Ann NY Acad Sci. 299:355–369.

62. Fernandez G et al (2003). Intrasubject reproducibility of presurgical language lateralization and mapping using fMRI.
Neurology 60(6):969–975.

63. Binder JR et al (1996). Determination of language dominance
using functional MRI: a comparison with the Wada test.
Neurology 46(4):978–984.

64. Springer JA et al (1999). Language dominance in neurologically normal and epilepsy subjects: a functional MRI study.
Brain 122(Pt 11):2033–2046.

65. Medina LS et al (2004). Functional MR imaging versus Wada
test for evaluation of language lateralization: cost analysis.
Radiology 230(1):49–54.

66. Knecht S et al (2002). Degree of language lateralization determines susceptibility to unilateral brain lesions. Nat Neurosci
5(7):695–699.

67. Pascual-Leone A, Gates JR, Dhuna A (1991). Induction of
speech arrest and counting errors with rapid-rate transcranial
magnetic stimulation. Neurology 41(5):697–702.

68. Jennum P et al (1994). Speech localization using repetitive
transcranial magnetic stimulation. Neurology 44(2):269–273.

69. Epstein CM et al (2000). Repetitive transcranial magnetic
stimulation does not replicate the Wada test. Neurology 55(7):
1025–1027.

70. Roux FE et al (2003). Language functional magnetic resonance imaging in preoperative assessment of language areas:
correlation with direct cortical stimulation. Neurosurgery
52(6):1335–1345; discussion 1345–1347.

71. Meyer FB et al (2001). Awake craniotomy for aggressive resection of primary gliomas located in eloquent brain. Mayo Clin
Proc 76(7):677–687.

72. Haglund MM et al (1994). Cortical localization of temporal
lobe language sites in patients with gliomas. Neurosurgery
34(4):567–576; discussion 576.

73. Carpentier A et al (2001). Functional MRI of language processing: dependence on input modality and temporal lobe epilepsy. Epilepsia 42(10):1241–1254.

74. Golby A et al (2002). Memory lateralization in medial temporal lobe epilepsy assessed by functional MRI. Epilepsia
43(8):855–863.

75. Binder J et al (1996). Functional MRI demonstrates left medial
temporal lobe activation during verbal episodic memory
encoding. Neuroimage 3:S530.

76. Detre JA et al (1998). Functional MRI lateralization of memory
in temporal lobe epilepsy. Neurology 50(4):926–932.

77. Tummala RP et al (2003). Application of diffusion tensor
imaging to magnetic-resonance-guided brain tumor resection. Pediatr Neurosurg 39(1):39–43.

78. Berger MS et al (1994). The effect of extent of resection on
recurrence in patients with low grade cerebral hemisphere
gliomas. Cancer 74(6):1784–1791.

79. Lo SS et al (2001). Does the extent of surgery have an impact
on the survival of patients who receive postoperative radiation therapy for supratentorial low-grade gliomas? Int J
Cancer 96 Suppl:71–78.

80. Philippon JH et al (1993). Supratentorial low-grade astrocytomas in adults. Neurosurgery 32(4):554–559.

81. Wisoff JH et al (1998). Current neurosurgical management
and the impact of the extent of resection in the treatment of
malignant gliomas of childhood: a report of the Children's
Cancer Group trial no. CCG-945. J Neurosurg 89(1):52–59.

82. Claus EB et al (2005). Survival rates in patients with low-grade
glioma after intraoperative magnetic resonance image guidance. Cancer 103(6):1227–1233.

83. Keles GE, Lamborn KR, Berger MS (2001). Low-grade
hemispheric gliomas in adults: a critical review of extent
of resection as a factor influencing outcome. J Neurosurg
95(5):735–745.

84. McCormack BM et al (1992). Treatment and survival of low-
grade astrocytoma in adults – 1977–1988. Neurosurgery
31(4):636–642; discussion 642.

85. Achten E et al (1999). Presurgical evaluation of the motor hand
area with functional MR imaging in patients with tumors and
dysplastic lesions. Radiology 210(2):529–538.

86. Hall WA, Liu H, Truwit CL (2005). Functional magnetic resonance imaging-guided resection of low-grade gliomas. Surg
Neurol 64(1):20–27; discussion 27.

87. Krings T et al (2001). Functional MRI for presurgical planning: problems, artefacts, and solution strategies. J Neurol
Neurosurg Psychiatr 70(6):749–760.

88. Wilkinson ID et al (2003). Motor functional MRI for preoperative and intraoperative neurosurgical guidance. Br J
Radiol 76(902):98–103.

89. Hirsch J et al (2000). An integrated functional magnetic resonance imaging procedure for preoperative mapping of cortical areas associated with tactile, motor, language, and visual
functions. Neurosurgery 47(3):711–721; discussion 721–722.

90. Skirboll SS et al (1996). Functional cortex and subcortical white matter located within gliomas. Neurosurgery 38(4):678–684; discussion 684–685.

91. Ojemann JG, Miller JW, Silbergeld DL (1996). Preserved function in brain invaded by tumor. Neurosurgery 39(2):253–258; discussion 258–259.

92. Russell SM, Kelly PJ (2003). Incidence and clinical evolution of postoperative deficits after volumetric stereotactic resection of glial neoplasms involving the supplementary motor area. Neurosurgery 52(3):506–515; discussion 515–516.

93. Keles GE, Anderson B, Berger MS (1999). The effect of extent of resection on time to tumor progression and survival in patients with glioblastoma multiforme of the cerebral hemisphere. Surg Neurol 52(4):371–379.

94. Lacroix M et al (2001). A multivariate analysis of 416 patients with glioblastoma multiforme: prognosis, extent of resection, and survival. J Neurosurg 95(2):190–198.

95. Bogomolny DL et al (2004). Functional MRI in the brain tumor patient. Top Magn Reson Imaging 15(5):325–335.

96. Liu WC et al (2005). The effect of tumor type and distance on activation in the motor cortex. Neuroradiology 47:813–819.

97. Lu S et al (2003). Peritumoral diffusion tensor imaging of high-grade gliomas and metastatic brain tumors. Am J Neuroradiol 24(5):937–941.

98. Lu S et al (2004). Diffusion-tensor MR imaging of intracranial neoplasia and associated peritumoral edema: introduction of the tumor infiltration index. Radiology 232(1):221–228.

99. Provenzale JM et al (2004). Peritumoral brain regions in gliomas and meningiomas: investigation with isotropic diffusion-weighted MR imaging and diffusion-tensor MR imaging. Radiology 232(2):451–460.

100. Henson JW, Gaviani P, Gonzalez RG (2005). MRI in treatment of adult gliomas. Lancet Oncol 6(3):167–175.

101. Buchpiguel CA et al (1995). PET versus SPECT in distinguishing radiation necrosis from tumor recurrence in the brain. J Nuc Med 36(1):159–164.

102. Hsu YY et al (2004). Blood oxygenation level-dependent MRI of cerebral gliomas during breath holding. J Magn Reson Imaging 19(2):160–167.

103. Hou H et al (2004). Effect of RSR13, an allosteric hemoglobin modifier, on oxygenation in murine tumors: an in vivo electron paramagnetic resonance oximetry and bold MRI study. Int J Radiat Oncol Biol Phys 59(3):834–843.

104. Hastreiter P et al (2004).Strategies for brain shift evaluation. Med Image Anal 8(4):447–464.

105. Clatz O et al (2005). Robust nonrigid registration to capture brain shift from intraoperative MRI. IEEE Transact Med Imaging 24(11):1417–1427.

106. Platenik LA et al (2002). In vivo quantification of retraction deformation modeling for updated image-guidance during neurosurgery. IEEE Transact Biomed Eng 49(8):823–835.

107. Nimsky C et al (2000). Quantification of, visualization of, and compensation for brain shift using intraoperative magnetic resonance imaging. Neurosurgery 47(5):1070–1079; discussion 1079–1080.

108. Hartkens T et al (2003). Measurement and analysis of brain deformation during neurosurgery. IEEE Transact Med Imaging 22(1):82–92.

109. Reinges MH et al (2004). Course of brain shift during microsurgical resection of supratentorial cerebral lesions: limits of conventional neuronavigation. Acta Neurochir 146(4):369–377; discussion 377.

110. Hall WA et al (2000). Safety, efficacy, and functionality of high-field strength interventional magnetic resonance imaging for neurosurgery. Neurosurgery 46(3):632–641; discussion 641–642.

111. Nimsky C et al (2005). Intraoperative diffusion-tensor MR imaging: shifting of white matter tracts during neurosurgical procedures – initial experience. Radiology 234(1):218–225.

112. Nimsky C et al (2005). Preoperative and intraoperative diffusion tensor imaging-based fiber tracking in glioma surgery. Neurosurgery 56(1):130–137; discussion 138.

113. Schulder M, Azmi H, Biswal B (2003). Functional magnetic resonance imaging in a low-field intraoperative scanner. Stereotact Funct Neurosurg 80(1–4):125–131.

114. Nimsky C, Ganslandt O, Fahlbusch R (2005). Comparing 0.2 tesla with 1.5 tesla intraoperative magnetic resonance imaging analysis of setup, workflow, and efficiency. Acad Radiol 12(9):1065–1079.

115. Gasser T et al (2005). Intraoperative functional MRI: implementation and preliminary experience. Neuroimage 26(3):685–693.

116. Dale AM, Halgren E (2001). Spatiotemporal mapping of brain activity by integration of multiple imaging modalities. Curr Opin Neurobiol 11(2):202–208.

117. Roux FE et al (2001). Methodological and technical issues for integrating functional magnetic resonance imaging data in a neuronavigational system. Neurosurgery 49(5):1145–1156; discussion 1156–1157.

118. O'Shea JP, Branco WS, Petrovich DM, Knierim N, Golby K (2006). Integrated image and function-guided surgery in eloquent cortex: a technique report. Int J Med Robot Comput Assist Surg 2:75–83.

119. Krishnan R et al (2004). Functional magnetic resonance imaging-integrated neuronavigation: correlation between lesion-to-motor cortex distance and outcome. Neurosurgery 55(4):904–914; discusssion 914–915.

120. Rihs F et al (1999). Determination of language dominance: Wada test confirms functional transcranial Doppler sonography. Neurology 52(8):1591–1596.

121. Aaslid R (1987). Visually evoked dynamic blood flow response of the human cerebral circulation. Stroke 18(4):771–775.

122. Matteis M et al (1997). Transcranial doppler assessment of cerebral flow velocity during perception and recognition of melodies. J Neurol Sci 149(1):57–61.

CHAPTER **24**

Stereotactic Radiosurgery: Basic Principles, Delivery Platforms and Clinical Applications

Clark C. Chen, Paul H. Chapman, Marc Bussiere and Jay S. Loeffler

INTRODUCTION

The clinical application of ionizing radiation (IR) was recognized soon after the discovery of x-rays. Since then, the discipline of radiation oncology has become an indispensable part of modern oncology. The tumoricidal effect of ionizing radiation is derived primarily from the induction of DNA damage beyond the cellular capacity for repair. In its initial clinical applications, radiation was delivered to both the normal and cancer tissues. The therapeutic efficacy was based on the increased DNA repair capacity of normal cells relative to tumor cells when repeatedly exposed to small doses of radiation. This form of radiation treatment became known as radiotherapy (Figure 24.1A).

With advances in stereotactic neurosurgery, non-invasive neuroimaging and radiation physics, it has become possible selectively to irradiate a defined tumor volume with relative sparing of surrounding normal tissues. This is achieved by converging multiple, non-parallel radiation beams (Figure 24.1B). Because the method was largely developed

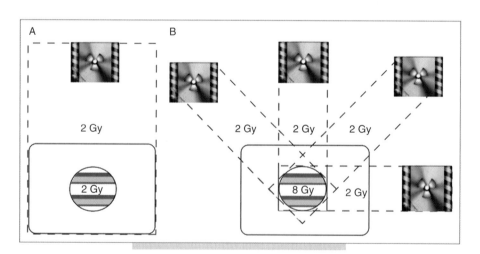

FIG. 24.1. Radiotherapy and radiosurgery. **(A)** In radiotherapy, radiation is delivered to both normal and cancer tissues. The therapeutic efficacy is derived from the increased DNA repair capacity of normal cells relative to tumor cells when repeatedly exposed to small doses of radiation. In the illustration, the dashed line represents the radiation field. The solid square represents the intracranial space. The striped circle represents the tumor volume. In this theoretic illustration, both the tumor and the normal tissue receive 2 Gy of ionization radiation. **(B)** In radiosurgery, multiple, non-parallel radiation beams are converged on the tumor volume. The radiation received by the normal tissue in each beam path is minimal relative to the point of beam convergence. In this illustration, the dashed line represents the radiation field. The solid square represents the intracranial space. The striped circle represents the tumor volume. The tumor volume receives 8 Gy of radiation from four radiation sources while the surrounding tissues receive significantly lower doses of radiation.

by neurosurgeons and because the expertise required was analogous to the skills required for microsurgery, this practice became known as radiosurgery.

The following chapter will review the principles of radiosurgery. Different methods of administering radiosurgery (gamma knife, linear accelerator (LINAC) based radiosurgery and proton beam radiosurgery) are also reviewed. Clinical applications of radiosurgery (cerebral metastasis, meningiomas, acoustic neuromas, pituitary adenomas and gliomas) will be discussed. Finally, principles of neuroimaging as they relate to radiosurgery will be reviewed.

DEFINITION OF RADIOSURGERY

Conventional Radiotherapy

In conventional radiotherapy, small doses of radiation are administered in daily sessions to both the normal and cancerous cells. The therapeutic efficacy of radiotherapy is derived from the increased radiation resistance of the normal tissues relative to the cancer cells. Typically, 1.5–2.0 Gy in daily sessions with total cumulative doses of 30–60 Gy are administered. Daily doses exceeding 2.0 Gy are associated with increased radiation toxicity [1].

Radiosurgery

Radiosurgery, on the other hand, refers to the delivery of a single, high dose radiation dose to a defined lesion. As originally proposed by Dr Lars Leksell [2], this is achieved by converging multiple, non-parallel radiation beams on the target and minimizing radiation received in each individual beam path. The therapeutic efficacy is derived from the sharp dose fall off immediately outside of the treatment volume. Typically, the marginal dose delivered in a single session ranges from 11 Gy (for the treatment of benign lesions) to 70 Gy (for thalamotomy in the treatment of movement disorders) [3,4].

Hypofractionated Radiotherapy

Radiosurgery and conventional radiotherapy define two extremes in a continuum of potential fractionation schemes. This continuum, where doses >2.0 Gy are administered over <25 sessions, are termed hypofractionated radiotherapy. The rationale for the various hypofractionated regimens is based on mathematical extrapolations and the efficacy of these regimens awaits clinical confirmation. These regimens range from weekly doses of 4–5 Gy with a cumulative dose of 20 Gy [5] to doses of 8 Gy separated by 8 hours summing to 24 Gy in one day [6].

Hypofractionated radiotherapy needs to be distinguished from multistage radiosurgery. In the latter, a different region of the target is irradiated in each session and the sessions are separated by time intervals of 3 to 12 months [7]. In the former, the entire target volume is repeatedly radiated during treatment sessions separated by hours to weeks.

PRINCIPLES OF RADIOSURGERY

General Introduction

Radiosurgery is administered by converging multiple, non-parallel beams of radiation on the target [2]. The radiation received by the normal tissue in each beam path is minimal relative to the point of beam convergence. As an analogy, if each person in an unlit, crowded stadium holds a flashlight and points the light such that they converge at a single point, the point of convergence will be bright while the entire stadium remains dark.

Prerequisites for radiosurgery include the ability to delineate precisely the diseased tissue from the surrounding parenchyma, a detailed understanding of the pertinent neuroanatomy and the technology to deliver reliably highly conformation radiation. These prerequisites mandate a close collaboration between radiation oncologists, neurosurgeons and medical physicists in the planning and administration of radiosurgery.

Principles Guiding Clinical Practice of Radiosurgery

The practice of radiosurgery is guided by three general principles. These principles involve evaluating the target lesion with regard to its size, relationship to radiosensitive cranial nerves and its location in terms of regional anatomy. These principles are discussed below.

Size of the Lesion

One of the key principles of radiosurgery is that as the size of the irradiated target increases, the undesired radiation of the surrounding non-target tissue also increases. This phenomenon can be illustrated using the analogy of flashlights in a dark stadium. Recall, illumination of a small area in an unlit stadium can be achieved by converging multiple sources of flashlights. To illuminate a larger area, either more intense or larger flashlights will be needed. Consequently, areas outside of the target will receive more illumination. Similarly, undesired radiation of non-target tissues increases with radiosurgery of larger lesions.

The clinical impact of this geometric inevitability is magnified by the higher doses of radiation delivered in radiosurgery. A dose escalation study by the Radiation Therapy Oncology Group (RTOG 90-05) defined the maximally tolerated radiosurgical dose in the treatment of cerebral metastasis as a function of the tumor size [8]. This study recommended doses of 24, 18 and 15 Gy for lesions less than or equal to 2 cm, 3 cm and 4 cm in the largest diameter, respectively. These recommendations limited treatment-related morbidity to <20%. Radiosurgery of lesions >4 cm is associated either with an unacceptable level of radiation toxicity or an ineffectual radiation dose.

Proximity to Cranial Nerves

Another principle of radiosurgery is that close proximity of the target to radiosensitive cranial nerves may expose these structures to sufficient damage and cause cranial neuropathy. In general, cranial nerves II and VIII appear highly sensitive to radiation relative to the other cranial nerves. Radiotherapy should be given consideration in cases where radiosurgery may jeopardize the functional status of the radiosensitive nerves [3,9–11].

Radiation sensitivity of the optic apparatus is likely a complex interaction between the intrinsic condition of the nerve (i.e. whether it had incurred damage from previous surgery or radiation), the total volume irradiated and the dose received [12,13]. As a first step to understand this complex interaction, several studies have focused on identifying the maximal radiosurgery dose tolerated by the optic apparatus. Tishler et al reported that in a series of 64 patients treated with radiosurgery for cavernous sinus lesions, four out of 17 patients (24%) receiving >8 Gy to any part of the optic apparatus developed visual complications, as compared to none out of 35 receiving <8 Gy [14]. In another series of 215 patients, Stafford et al reported four patients who developed optic neuropathy after radiosurgery of cavernous sinus lesions [11]. Of these patients, only one had not received prior radiotherapy or radiosurgery. This patient received 12.8 Gy to the optic apparatus. The remaining three received prior radiotherapy in addition to radiosurgery of 7–10 Gy (59 Gy + 7 Gy; 45 Gy + 9 Gy; and 50.4 Gy + 9 Gy + 12 Gy). Finally, Pollock et al [15] reported the incidence of optic neuropathy after 8 Gy to the optic chiasm to be 3% (1/38). Extrapolating from these data, radiosurgery is generally avoided when radiation to any part of the optic apparatus exceeds 10 Gy.

Assessment of the radiosensitivity of cranial nerve VIII is highly problematic since it is almost always radiated in the context of acoustic neuroma treatment. As such, it is impossible to distinguish between damage intrinsic to the disease process, induced by the radiation treatment, or some combination thereof. Further, most radiotherapy and radiosurgery series fail to document detailed audiologic assessments prior to and after treatment. To our knowledge, only three studies carried out such careful audiologic assessment [9,10,16]. Ito et al reported that 69% of patients who underwent radiosurgery (16.8 Gy median dose) for acoustic neuroma exhibited pure tone average (PTA) elevation of >20 decibels (dB), suggesting hearing loss after treatment [9]. Similarly, Paek et al [16], reported that 64% of patients (16/25) with serviceable hearing pretreatment suffered hearing loss of >20 dB after radiosurgery (12 Gy at 50% isodose). These results are generally consistent with other studies where qualitative assessments of hearing preservation were performed [17–19]. In sum, the probability of cranial VIII dysfunction (as measured by loss of hearing) after radiosurgery (12 Gy marginal dose) for acoustic neuroma is approximately 60%.

Other cranial nerves appear more resilient to radiosurgery. Neuropathy of cranial nerves III–VI has not been reported for doses <15 Gy delivered to the cavernous sinus [14]. Doses of 15–40 Gy to the cavernous sinus are associated with a 10–15% incidence of cranial nerve III–VI neuropathy. Cranial nerve VII routinely receives doses of 11–15 Gy in acoustic neuroma radiosurgery. In a series of 829 patients followed for 10 years after acoustic neuroma radiosurgery, the incidence of facial neuropathy is less than 1% [20]. Zero to 2% cranial neuropathy of cranial nerves IX–XI has been reported after radiosurgery (8–12 Gy) of jugular foramen schwannomas or skull base meningiomas [21–24]. In sum, lesions in proximity to cranial nerves III, IV, V, VI, VII, IX, X and XI, bear less consideration in radiosurgery planning.

Regional Cerebral Anatomy

The final principle of radiosurgery is that post-treatment morbidity is a function of the regional anatomy. Flickinger et al reviewed 332 patients with arteriovenous malformation (AVM) treated with radiosurgery and correlated the risk of post-treatment neurologic injury to the volume of irradiation, as well as the location of the lesion [25]. It was not surprising that the volume of parenchyma receiving 12 Gy radiation directly correlated with the risk of post-treatment neurologic deficit. More importantly, the study demonstrated that the risk of radiosurgical morbidity depended on the region irradiated. For instance, 12 Gy irradiation of the basal ganglia/thalamus region was associated with approximately 20% morbidity. The same irradiation of the frontal, temporal or parietal lobe was associated with a <1% morbidity. Of note, the various regions at high risk to radiosurgery can be safely irradiated by stereotactic radiotherapy.

As another illustration of the importance of regional anatomy, radiosurgery of lesions in proximity of the cerebrospinal fluid (CSF) space is associated with an increased risk of post-treatment hydrocephalus. For instance, roughly 10% of patients with vestibular schwannoma develop communicating hydrocephalus after radiosurgery [26]. The etiology of the hydrocephalus is unclear, though many investigators attribute the phenomenon to CSF malabsorption secondary to tumor necrosis. Understanding the regional anatomy as it pertains to the risk of hydrocephalus is a prerequisite for timely neurosurgical intervention.

As a final example, irradiation of the pituitary/cavernous sinus region often involves the intracavernous segment of the internal carotid artery (ICA). The risk of ICA stenosis after radiosurgery is small but finite. In the over 1400 reported cases of pituitary/cavernous sinus lesions treated with radiosurgery, four cases of ICA stenosis have been noted [15,27,28]. To avoid such complications, Pollock et al recommended that the prescription dose be limited to <50% of the ICA vessel diameter [15]. Shin et al recommended restricting the dose received by the ICA to <30 Gy [29]. It should be noted that these recommendations are somewhat arbitrary. Nevertheless, caution with regard to the exposure of the ICA is warranted, especially in cases where the ICA is compromised by mass compression. In contrast, the cerebral venous structures appear highly resistant to the effects of radiation. No cases of post-radiosurgery sinus scarring/thrombosis have been reported to date.

PLATFORMS FOR RADIOSURGERY

Photon and Proton Radiosurgery

Radiosurgery can be administrated using photon or proton as the element of energy transfer. The following sections will review the basic physical properties of these elementary particles, as well as the various platforms designed to deliver them.

Modern physics revealed the energy in gamma or x-irradiation could be conceptualized as discrete packets (the complexity of the wave particle duality will not be discussed in this chapter). These energy packets are termed photons. Since both gamma- and x-rays are made up of photons, they share identical physical properties. The

difference between these two forms of radiation lies in the method by which the photons are produced. The photons in gamma rays are derived from the natural decay of unstable nuclei, whereas the photons in x-rays are produced by linear accelerators. Gamma rays are used in Gamma Knife radiosurgery, whereas x-rays are used in LINAC radiosurgery. Since photons are used as the source of radiation in both Gamma Knife and LINAC radiosurgery, it is not surprising that the clinical efficacies of these modalities are comparable [30].

The physical properties of protons, on the other hand, contrast those of photons. A proton beam is generated by stripping an atom of its electron and accelerating the residual proton in the magnetic field of a cyclotron or a synchrocyclotron [31]. The energy distribution of a proton beam consists of a slowly rising dose, a rapid rise to a maximum (Bragg peak) and a fall to near zero (Figure 24.2). This contrasts the exponential decay of photons.

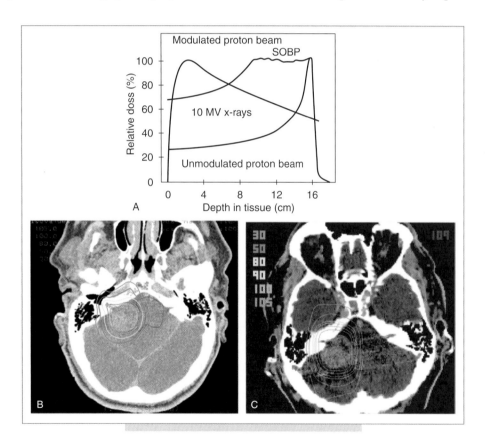

FIG. 24.2. Physical characteristics of proton and photons. **(A)** Depth dose curve for an unmodulated proton beam, 10 MeV photon radiation and *spread-out Bragg peak* (SOBP) proton radiation. The dose distribution of a proton beam consists of a slowly rising dose, a rapid rise to a maximum (also known as the Bragg peak), followed by a fall to near zero. As a result, a significant portion of the beam energy is deposited in a small volume. In contrast, the energy deposition of photon irradiation is dictated by the laws of exponential decay and is distributed over a larger volume. Because of the sharpness of the Bragg peak, a single monoenergitic proton beam will only irradiate an area approximately the size of a pituitary gland. Irradiation of larger targets is achieved by superimposing proton beams of different energies. The end result of this superposition is known as the *spread-out Bragg peak* (SOBP). **(B, C)** Comparative treatment plans for a large acoustic neuroma using proton radiosurgery **(B)** and conformal photon radiosurgery **(C)**. The various isodose lines are as indicated in the right upper corner of each panel. The comparison reveals that proton radiosugery minimizes irradiation of the non-target tissues relative to photon radiosurgery.

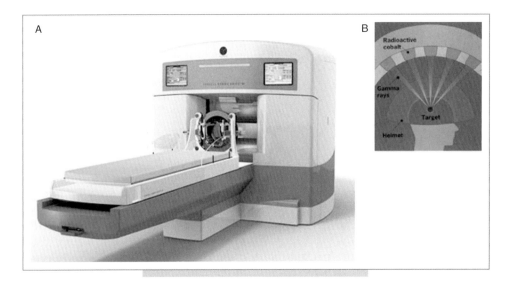

FIG. 24.3. Gamma Knife radiosurgery. **(A)** The Gamma Knife device (Electa™) consists of an 18 000 kg shield surrounding a hemispheric array of 201 cobalt 60 sources. **(B)** The sources are oriented such that all 201 beams converge at a single point (also known as the isocenter of the Gamma Knife).

Because of the sharpness of the Bragg peak, a single monoenergitic proton beam will only irradiate an area approximately the size of a pituitary gland. Irradiation of larger targets is achieved by superimposing proton beams of different energies. By doing so, the Bragg peaks can be manipulated to generate a moderate entrance dose, a uniformly high dose within the target tissue and a zero dose beyond the target (see Figure 24.2). Clinically, the unique property of protons translates into less integral doses to the patient when compared to photon-based treatment, particularly in the treatment of larger sized lesions [32].

Relative to photon radiosurgery, proton radiosurgery represents a recent technological innovation. As such, access to proton radiosurgery remains limited. For instance, there are currently 23 clinical proton radiotherapy facilities worldwide, with three centers available in the USA. Of these centers, less than half offer radiosurgery as a therapeutic option. In contrast, there are approximately 200 gamma knife centers worldwide and many more LINAC based radiosurgery programs [33].

Gamma Knife

The Gamma Knife system is a platform for the delivery of photon-based radiosurgery. The device consists of an 18 000 kg shield surrounding a hemispheric array of 201 cobalt 60 sources (Figure 24.3). The sources are oriented such that all 201 beams converge at a single point (also known as the isocenter of the Gamma Knife). This array produces a target accuracy of 0.1–1 mm [34], which is at least as good as the best possible lesion delineation with the current imaging technology. Prior to treatment, a stereotactic head frame is placed on the patient to guide the radiosurgery planning, as well as to achieve patient immobilization during the treatment session.

Linear Accelerator (LINAC) Radiosurgery

LINAC is another platform that delivers photon-based radiosurgery. Instead of using an array of fixed cobalt sources, LINAC radiosurgery utilizes multiple non-coplanar arcs of radiation that intersect at the target volume (Figure 24.4). The general principle of LINAC is otherwise identical to that of Gamma Knife radiosurgery. LINAC based devices achieve target accuracy of 0.1–1 mm [34]. Similar to Gamma Knife radiosurgery, a stereotactic head frame is placed on the patient prior to LINAC radiosurgery.

CyberKnife

The CyberKnife device combines a mobile linear accelerator with an image-guided robotic system (Figure 24.5). Thus, it is yet another platform for the delivery of photon-based radiosurgery. In this system, the placement of a stereotactic head frame is not required. Prior to the CyberKnife treatment, CT images are taken to define the spatial relationship between the patient's bony anatomy and the target volume. During the actual treatment, patient movements are monitored by the system's x-ray cameras. The images from these cameras are compared to the pre-treatment CT scan. Based on calculations derived from these comparisons, the mobile linear accelerator is maneuvered in response to changes in patient position. Less than 1 mm accuracy is achieved using this system [35].

Proton Beam Radiosurgery

Proton delivery systems are not compact like conventional LINAC or Gamma-knife units. Until recently, medicinal applications of protons were done using retrofitted equipments and fixed beam portals [36]. At the Massachusetts General Hospital (MGH), we have developed two devices designed specifically for proton radiosurgery. The first is

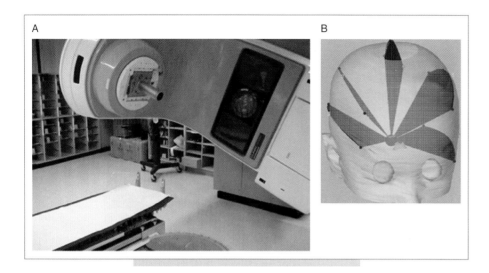

FIG. 24.4. LINAC radiosurgery. **(A)** A LINAC apparatus. The source of radiation can be rotated relative to the treatment couch. **(B)** Schematic depiction of multiple non-coplanar arcs of radiation that intersect at the target volume.

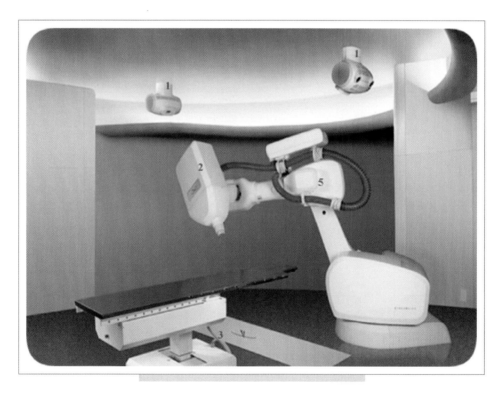

FIG. 24.5. Cyberknife radiosurgery. The CyberKnife device (Accuray™) combines a mobile linear accelerator with image-guided robotic system. During the treatment, patient movement is monitored with minimal time lag by the system's low-dose x-ray cameras. These planar images are compared to digitally reconstructed radiographs derived from the pretreatment CT scan. Based on calculations derived from these comparisons, the computer-controlled robotic arm maneuvers the mobile linear accelerator in response to any changes in patient position during treatment. 1: Diagnostic x-ray tubes; 2: mobile LINAC; 3: camera; 4: couch; 5: robotic arm.

the STAR device (*s*tereotactic *a*lignment for *r*adiosurgery apparatus; Figure 24.6). With the STAR device, the patient is placed in an immobilizing head frame attached to a couch apparatus that can be rotated relative to a fixed beam portal. To mobilize the beam source, we have also developed a device where a beam source is mounted on a rotating gantry (Figure 24.7). In both the STAR and gantry device, the target accuracy is on the order of 0.1 mm.

Because of the small number of facilities available, clinical judgment must be exercised in patient selection for proton radiosurgery. In general, treatments of small (<10 ml in volume), spherically shaped lesions that are not in proximity to critical anatomic structures, in eloquent regions, in deep subcortical areas, or in previously radiated volumes do not require proton radiosurgery. In most of these cases, equally efficacious results can be attained with photon

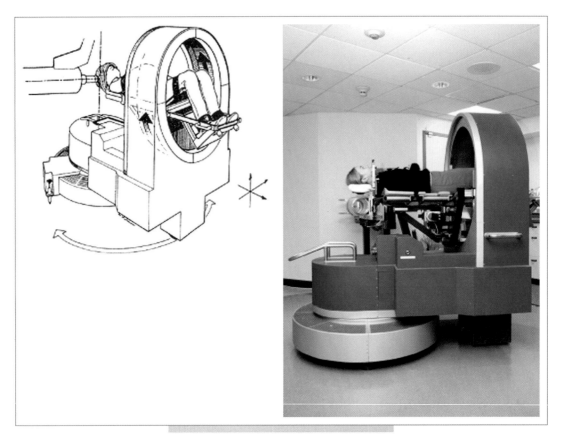

FIG. 24.6. The STAR (*stereotactic alignment radiosurgery*) system at the MGH Northeast Proton Therapy Center (MGH-NPTC). An isocentric patient positioning device designed to rotate the patient around a fixed proton beam line. The unit has five degrees of freedom, three linear and two rotations. Using these axes of movement, the patient is positioned such that the target lesion is at the center of the beam isocenter.

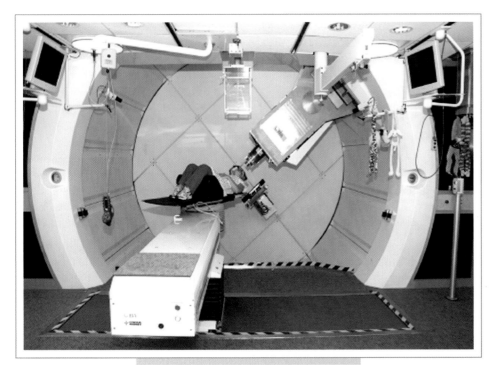

FIG. 24.7. The proton gantry system at the MGH Northeast Proton Therapy Center (MGH-NPTC). An isocentric gantry with a mobile proton beam source is used in conjunction with a robotic patient positioning treatment couch. The robotic unit has six degrees of freedom, three linear and three rotational. The patient is positioned to place the target lesion at the center of beam isocenter.

radiosurgery. Patients with poor expected survival are also unlikely to derive additional benefit from proton radiosurgery, since delayed radiation toxicity develops years after treatment.

CLINICAL EFFICACY OF RADIOSURGERY

In general, surgically accessible brain tumors with significant mass effect or peri-tumoral edema are best treated by resection. Radiotherapy and radiosurgery represent therapeutic options for poor surgical candidates or surgically inaccessible lesions. In the ensuing section, the common applications of radiosurgery in neuro-oncology, including the treatment of cerebral metastases, meningiomas, acoustic neuromas, pituitary adenomas and gliomas, will be reviewed. In selected cases, the efficacy of radiosurgery will be compared to that of radiotherapy.

Cerebral Metastases

Cerebral metastases constitute 30–40% of all intracranial tumors [37,38]. The natural history of cerebral metastases is one of aggressive and rapid growth. Despite the tremendous histologic heterogeneity, cerebral metastatic disease, as a whole, carries a very poor prognosis. Without intervention, mortality is expected 1–2 months after the diagnosis [39]. The addition of surgery, chemotherapy, and *w*hole *b*rain *r*adiation *t*herapy (WBRT) extends median survival by only 2–4 months [40].

The observation that metastases resistant to WBRT (e.g. sarcoma, melanoma and renal cell carcinoma) respond to radiosurgery [41,42] led to several retrospective series and three randomized trials investigating the use of radiosurgery as an adjuvant therapy to WBRT [30,43–45]. The sum of this body of work suggests that radiosurgery is an effective means of augmenting WBRT in preventing intracranial tumor growth. However, the patients treated eventually die of systemic disease progression. Consequently, while local control is improved by the combination therapy, patient survival remains unaffected (Table 24.1). The following represents a brief summary of the three randomized trials.

RTOG 95-08 [30] randomized 331 patients with one to three metastases that were <4 cm to WBRT (164 patients)

and combined WBRT/radiosurgery (167 patients). Chougule et al [43], randomized 73 patients with one to three brain metastases with tumor volume <30 ml and a minimum life expectancy of 3 months to WBRT (36 patients) and combined WBRT/radiosurgery (37 patients). Kondziolka et al [44] randomized 27 patients with two to four brain metastases <2.5 cm to WBRT (14 patients) and combined WBRT/radiosurgery (13 patients). In all three studies, no statistically significant difference in median survival was observed between the treatment arms. However, the local control rate at the 1-year follow-up was consistently superior in the WBRT/radiosurgery arm (82–92%) relative to the WBRT arm (0–71%, $P<0.05$). The discrepancy between patient survival and local control was due to death from systemic disease progression.

Additionally, the RTOG 95-08 study [30] revealed that patients undergoing WBRT/radiosurgery treatment were more likely to have stable/improved neurologic function, as well as decreased steroid use at the 6-month follow-up. In all three studies, no statistically significant difference in short-term toxicity was observed between the treatment arms. While WBRT is associated with cognitive impairment 7–10 years after treatment [46,47], very few patients with brain metastases achieve such survival. Only in this subset of patients (e.g. patients receiving tamoxifen for estrogen receptor (ER)+ breast cancer) will the ramifications of WBRT be pertinent.

In addition to its role as an adjuvant therapy to WBRT, radiosurgery is also proposed as a standalone therapy for cerebral metastases. In this regard, two randomized trials, a prospective study, as well as several retrospective series have been conducted [43,45,48–51]. These studies, in aggregate, suggest that radiosurgery is an effective means of controlling tumor growth in the volume treated. However, unlike WBRT, radiosurgery fails to address the micrometastasis outside of the treatment volume. As such, metachronous tumor growth is often observed in the absence of WBRT. Nevertheless, with diligent monitoring, timely salvage therapy can be administered (either in the form of WBRT or radiosurgery) to attain patient survival comparable to that achieved by up-front WBRT/radiosurgery.

TABLE 24-1 Randomized clinical trials of whole brain radiotherapy versus combined whole brain radiotherapy/radiosurgery treatment					
Studies	Number of patients	Number of metastases	Size of metastases (cm)	Median survival (months)	Local control (%)
RTOG 95-08, 2004 [30]	167/164	1–3	<4	6.5 versus 5.7 ($P=0.1$)	82 versus 71 ($P<0.05$)
Kondziolka et al, 1999 [44]	14/13	2–4	<2.5	11 versus 7 ($P=0.22$)	92 versus 0 ($P<0.005$)
Chougule et al, 2000 [43]	37/36	1–3	<3	7 versus 9 ($P>0.05$)	91 versus 62 ($P<0.05$)

TABLE 24-2 Studies comparing radiosurgery and combined whole brain radiotherapy/radiosurgery treatment					
Studies	Number of patients	Number of metastases	Size of metastases (cm)	Local control (%)	Freedom from new metastasis (%)
JROSG99-1, 2004 [48]	61/59	1–4	<3	70 versus 86 (P=0.01)	48 versus 82 (P=0.003)
Chougule et al, 2000 [43]	37/31	1–3	<3	87 versus 91 (P>0.05)	Not reported
Lutterbach et al, 2003* [49]	101	1–3	<3	91	51

*Single arm prospective study.

The two randomized trials that examined the efficacy of radiosurgery as a standalone treatment (Table 24.2) included the JROSG99-1 trial [48] (patients with KPS>70, <4 metastases <3 cm in diameter) and the study by Chougule et al [43], (patients with one to three brain metastases with tumor volume <30 ml and a minimum life expectancy of 3 months). In JROSG99-1, a slight superiority in the 1-year local control rate was observed in the WBRT/radiosurgery arm (86 versus 70%, $P = 0.01$). In the Chougule study, however, the radiosurgery and WBRT/radiosurgery arms yielded comparable 1-year local control rate (87 versus 91%). Overall, these results suggest that radiosurgery is effective in controlling the growth of cerebral metastases. This conclusion was further substantiated by a prospective single-arm study, where a 91% local control rate was achieved one year after radiosurgery, as well as by several other retrospective series [45,49].

Because micrometastasis, invisible on imaging modalities, cannot be treated without WBRT, the risk of metachronous cancer growth is significantly higher in patients receiving radiosurgery alone. In JROSG99-1, freedom from new intracranial metastases was 82% in the combined treatment arm versus 48% in the radiosurgery arm [48]. However, with diligent monitoring, timely salvage therapy can be administered (either in the form of WBRT or radiosurgery) to attain patient survival comparable to that achieved by upfront WBRT/radiosurgery [51]. Given the long-term cognitive impairment associated with WBRT [46], sequential radiosurgeries without WBRT may be appropriate for a subset of cancer patients.

Meningiomas

Meningiomas comprise 15–20% of all adult intracranial tumors. Excluding the rare atypical (WHO grade II) and malignant meningiomas (WHO grade III), the natural history of meningioma (WHO grade I) is one of slow growth over an extended period of time. While the kinetics of meningioma growth is likely complex and non-linear, extrapolation from retrospective series suggest an average growth rate of 0.8–5 ml per year [52]. On average, the time required for tumor volume doubling ranges from 1 to 140 years. Despite this broad range, most meningiomas grow at a very slow pace. Only 20–30% of all meningiomas exhibit doubling time of <5 years. These faster growing meningiomas tend to occur in younger patients [52,53]. Given the slow growth rate, observation is the treatment of choice for asymptomatic meningiomas without evidence of growth.

When surgically accessible, the treatment of choice for symptomatic or progressively enlarging meningiomas is resection. The efficacy of surgery is proportional to the extent of resection. This relationship is formalized by the Simpson grade [54]. Simpson grade 1 refers to complete tumor excision, including the affected dura and bone. In Simpson grade 2, complete excision of the tumor mass is achieved without removal of the affected bone. The affected dura is cauterized but not removed. In grade 3, no visible tumor remains after the resection, but the affected bone/dura is not removed or cauterized. Simpson grade 4 refers to cases where visible residual tumor remains after resection. Simpson grade 5 refers to surgical biopsies without intent of tumor resection. At 8–10-year follow-up, the proportion of patients suffering tumor recurrence after Simpson grade 1, 2, 3, 4 and 5 resections of WHO grade I meningiomas are approximately 5–10%, 20%, 30%, 40% and 90%, respectively [54–56]

Radiotherapy and radiosurgery are used for treatment of partially resected or unresectable meningiomas. Unlike surgery, where the mass effect can be reversed by debulking, the goal of radiation treatment is to arrest or delay tumor growth, though tumor shrinkage is occasionally observed [21,22,57–61]. The currently available data (Table 24.3) suggests that radiotherapy and radiosurgery are equally efficacious in controlling the growth of meningiomas. Moreover, the risk of post-treatment morbidity is comparable between these modalities. For surgically inaccessible or residual meningiomas (WHO grade I), conventional radiotherapy achieves local control rates of 76–98% at 5 years and 77–98% at 10 years. Complications following radiotherapy ranges are 3.3–10% [57–60]. With radiosurgery, local control rates are 93–98% at 5 years and 89–98% at 10 years [21,22,61]. Complication following radiosurgery ranges are 2.5–5.5%.

These studies suggest that the efficacy of radiotherapy and radiosurgery in the treatment of WHO grade I

TABLE 24-3 Studies of radiotherapy and radiosurgery as treatment for meningiomas					
Studies	Number of patients	Follow-up (years)	Actuarial control (%)	Dose (median) (Gy)	Complications (%)
Radiotherapy					
Goldsmith et al, 1994 [59]	140	5[a]	98	54	3.6
Glaholm et al, 1990 [58]	186	10[a]	77	53	2.2
Nutting et al, 1999 [60]	82	10[a]	83	57	5
Condra et al, 1999 [57]	21	15[a]	87	53	10
Radiosurgery[c]					
Kobayashi et al, 2001 [61]	54	5[b]	89	14.5	5.5
Eustacchio et al, 2002 [21]	121	6[b]	98	13	3.3
Kreil et al, 2004 [22]	200	8[b]	98	12	2.5

[a]Actuarial follow-up; [b]median follow-up; [c]only studies with >5 year median follow-up are included in this table.

meningiomas is comparable to that achieved by a Simpson grade 1 resection. As such, meningiomas located in regions where Simpson grade 1 resection cannot be safely achieved (e.g. meningiomas of the cavernous sinus or other skull base locations), radiosurgery/radiotherapy should be considered [56]. However, for surgically accessible meningiomas in the areas of the convexity, falx and parasagittal region, complete resection should be the treatment of choice since radiotherapy and radiosurgery of these regions are often associated with significant peritumoral edema and neurologic morbidity [3].

Acoustic Neuromas

Acoustic neuromas, also known as vestibular schwannomas, are benign tumors that arise from the Schwann cells associated with the eighth cranial nerve. They account for 6% of all intracranial tumors [37]. The natural history of acoustic neuromas, like that of meningiomas, is one of slow growth. The average growth rate of an acoustic neuroma is estimated to range from 0.6 to 7 mm per year [62–64]. Of note, cyclic fluctuations in size (e.g. growth followed by regression or vice versa) have been noted in up to 60% of the tumors [64]. However, tumor growth demonstrated by more than two sequential scans is highly predictive of future growth [63].

Historically, surgical resection has been the treatment of choice for acoustic neuromas. Two important goals of acoustic neuroma surgery are facial nerve and hearing preservation. The likelihood of achieving these goals is largely a function of the tumor size [65,66]. Ninety-three to ninety-seven percent of patients with small tumors (<2 cm) have satisfactory (normal or slight weakness on close inspection) facial nerve function postoperatively. Thirty-five to thirty-nine percent of patients with large tumor (>2 cm) have serviceable hearing preservation postoperatively [65,66]. Other surgical complications associated with acoustic neuroma resection include cerebrospinal fluid (CSF) leak (9–15%), trigeminal neuropathy (0–5%), lower cranial nerve

palsy (0–3%), wound or CSF infections (0–3%) and brainstem injury (0–1%). The incidence of tumor growth after resection ranges from 0–9% [65,66].

In recent years, radiosurgery has emerged as an alternative to microsurgical resection for small acoustic neuromas (<3 cm). In the largest series reported to date (827 patients), radiosurgery of acoustic schwannoma achieved a local control rate of 97% at the 10-year follow-up [20]. Other series have reported local control rates of 87–95% at the 10-year follow-up [17–19,67–71]. The incidence of facial neuropathy following radiosurgery ranges from 1–5%. Serviceable hearing preservation ranges from 13–40%. Trigeminal neuropathy ranges from 1–3%. The efficacy of surgical resection and radiosurgery, therefore, appears comparable [17,18], with one notable exception. Posttreatment hydrocephalus is rarely observed after microsurgical resection but is observed in 10–15% of the patients after radiosurgery [26].

Stereotactic radiotherapy is another option for the treatment of acoustic neuromas. The efficacy of radiotherapy in terms of local control is comparable to that of radiosurgery. In multiple series, the actuarial control of acoustic neuromas ranges from 87–98% at 5 years (Table 24.4). However, hearing appears better preserved by stereotactic radiotherapy relative to radiosurgery [19,26,40,72,73]. For instance, Andrews et al performed audiometric assessment of 17 patients undergoing radiosurgery and 34 patients undergoing stereotactic radiotherapy at the Thomas Jefferson University Hospital [19]. In the radiosurgery group, 67% of the patients lost serviceable hearing after treatment. In the radiotherapy group, only 19% of the patients suffered loss of serviceable hearing after treatment. Similarly, Sakamoto et al reported that 21% of patients who underwent stereotactic radiotherapy for acoustic neuroma treatment (36–44 Gy cumulative dose in 1.8–2.0 Gy fractions followed by a 4-Gy boost) suffered hearing loss as defined by pure tone average (PTA) elevation of >20 decibels (dB) [10]. This contrasts the report of Ito et al [9] where 69% of patients who underwent radiosurgery

TABLE 24-4 Studies of radiotherapy and radiosurgery as treatment for acoustic neuroma				
Studies	Actuarial tumor control (%)	Loss of serviceable hearing[a] (%)	Cranial nerve V neuropathy (%)	Cranial nerve VII neuropathy (%)
Radiotherapy				
Chan et al, 2005 [72]	98, 5 years	26	4	1
Sawamuara et al, 2003 [76]	91, 5 years	29	4	0
Combs et al, 2005 [73]	97, 5 years	4	3	2
Andrews et al, 2004 [19]	94, 5 years	19	7	2
Radiosurgery				
Lunsford et al, 2005 [20]	97, 12 years	23–50	3.1	1
Hasegawa et al, 2005 [69]	87, 10 years	63	5	5
Chung et al, 2005 [67]	95, 7 years	40	1	1
Andrews et al, 2004 [19]	98, 5 years	67	5	2

[a]Percentage of patients who retained serviceable hearing after treatment.

suffered hearing loss (PTA elevation of >20 dB). Using a questionnaire based assessment, Chan et al [72] reported that 19% of patients lost speech discrimination after radiotherapy (54 Gy in 1.8 Gy fractions). Using a similarly qualitative approach, Sawamura et al reported a 29% serviceable hearing loss following radiation therapy [26]. Granted, the studies discussed here consist of small retrospective series and are, therefore, subject to multiple study biases. However, the consistency among these reports suggests that hearing preservation may be better achieved by radiotherapy in comparison to radiosurgery.

Given the efficacy of radiosurgery and radiotherapy as treatment options for acoustic neuromas, the paradigm of surgery as the preferred intervention has been challenged. While surgery, no doubt, is the treatment of choice for symptomatic, large acoustic neuromas (>3 cm), for smaller neuromas, the decision of surgery versus radiation treatment is largely a function of the patient's age, preference and the available surgical expertise. In deciding which form of radiation treatment to pursue, consideration should be given to stereotactic radiotherapy as a means of hearing preservation.

Pituitary Adenomas

Pituitary adenomas constitute between 10 and 20% of all primary brain tumors [37,38]. Histologically, pituitary adenomas are divided into those that secrete an excessive amount of pituitary hormones, including prolactin, growth hormone (GH) and adrenal corticotrophic hormone (ACTH) and those that do not secrete biologically active hormones (non-functioning adenomas). Irrespective of histology, the natural history of pituitary adenomas is one of slow growth [74].

Based on autopsy series and MRI surveys, it is estimated that 10–20% of the normal adult population harbors pituitary abnormalities [74]. Given the prevalence and the benign nature of the pituitary adenomas, treatment should be restricted to lesions that are hormonally active, causing

mass effects or increasing in size. The role and efficacy of radiosurgery in treating the various subtypes of pituitary adenomas will be reviewed.

Prolactinomas

Prolactin-secreting adenomas typically present with the classic Forbes-Albright syndrome, consisting of amenorrhea/galactorrhea, reduced fertility, loss of libido, or erectile dysfunction. The diagnosis is made by a brain MRI and an elevated serum prolactin level. Once the diagnosis is made, medical therapy with dopamine agonists is the primary therapy. Surgery, radiotherapy and radiosurgery are restricted to patients who fail to tolerate or respond to medical therapy.

Because hormonal hypersecretion can be more quickly normalized by surgery than radiation treatment [75–77], it is considered the treatment of choice after medical therapy. In a landmark series of 889 surgically treated prolactinoma patients, 65% of patients remained in disease remission (without medical therapy) after 10 years. The likelihood of disease remission after surgery was inversely proportional to the size of the lesion (<1 cm: 87% remission; >1 cm: 56% remission at 10 years) [76]. Approximately 20–30% of patients developed hypopituitarism or diabetes insipidus after surgery [77,78].

Patients unlikely to tolerate surgery or who opt not to undergo surgery are treated with radiation therapy or radiosurgery. Without surgery or medical therapy, serum prolactin level normalization was achieved in 30–50% of the patients 7–11 years after radiotherapy (45–50 Gy) [79–81]. The mean time required for serum normalization was approximately 7.3 years. Radiation induced hypopituitarism range was 20–30% [81].

Comparable efficacy was reported for radiosurgery, with the caveat that long-term (>5 years) follow-up is unavailable at the present time. After radiosurgery, serum prolactin normalization was achieved in 20–60% of the patients [82]. This normalization was generally achieved within

2 years of treatment [82]. The risk of post-radiosurgery hypopituitarism was comparable to surgery or radiotherapy. The therapeutic effects of surgery and radiation appeared additive since tumor mass was one of the predictors of radiation response [83].

GH-secreting Adenomas

Clinically, GH-secreting adenomas present with acromegaly in adults and gigantism in children (before closure of the epiphyseal plate). The diagnosis is made by brain imaging, clinical examination and an elevated serum level of GH and insulin-like growth factor 1 (IGF-1). The therapeutic goal of treatment is to reduce serum IGF-1 levels to age- and sex-adjusted levels and nadir GH levels of <1 ng/ml during an oral glucose tolerance test (OGTT). The use of these criteria for evaluating disease remission after surgery or radiation treatment is problematic on several fronts. First, elevated, but medically non-suppressible levels of GH are frequently encountered in acromegalic patients after surgical or radiation treatment. If the patient is otherwise asymptomatic, close follow-up without further therapy is advised. Secondly, up to a third of the patients with acromegaly who undergo surgical or radiation treatment exhibit elevated IGF-1 level despite a nadir GH level of <1 ng/ml during OGTT. Given these caveats, most studies have arbitrarily adopted a basal GH level of <10 ng/ml without medical therapy as a proxy for endocrine normalization [82].

Medical therapy with long-acting somatostatin analogues normalizes serum IGF-1 level in up to two-thirds of acromegalics. Symptoms of acromegaly, including headaches, sweating and arthralgia, decrease with treatment. Gastrointestinal side effects such as colic, diarrhea or constipation combined with the financial burden of chronic therapy make long-term medical suppression prohibitive for many patients [84].

Because of the rapidity of endocrine correction after surgery, it remains the mainstay treatment for GH-secreting adenomas. Trans-sphenoidal resection for GH-secreting adenomas consistently achieves a remission rate of 70–74% using post-surgical GH level <10 ng/ml as the criterion [85–87]. The success rate drops to approximately 60% when the strict 'cure' criteria (OGTT GH<1 ng/ml and IGF-1 normalization) are applied. Biochemical normalization occurs immediately after surgery, but disease recurrence is observed in <10% of the surgical patients [86]. Surgical morbidities include: CSF leak (1–5%), septal perforation (0–5%) and pituitary insufficiency (20–30%).

Radiation treatment is a therapeutic option for patients who failed surgery or opted not to undergo surgery. Following radiotherapy, 60–80% of treated patients achieved GH <10 ng/ml after 10 years [88–90]. As was the case with prolactinomas, the therapeutic effects of surgery and radiation appeared additive [83].

The reported 'cure' for GH-secreting adenomas after radiosurgery ranged from 0–100% [82]. This wide range of values reflected the variability in the criteria used to define disease remission and differences in radiosurgery planning. As such, meaningful conclusions could not be drawn based on a broad review of the literature. As such, the discussion here will focus on a few well-carried-out studies. Landolt et al [91] compared the efficacy of radiosurgery (16 patients, 25 Gy marginal dose) and radiotherapy (50 patients, 40 Gy cumulative dose) in patients who suffered recurrence after surgical resection of GH-secreting adenoma. The same criteria of disease remission (IGF-1<50 mIU/l and GH<10 ng/ml) were applied to both groups of patients. In this study, the proportion of patients achieving remission was comparable between the two groups (70–80%). However, the mean time to cure was 1.4 years with radiosurgery and 7.1 years with radiotherapy. Similar results have been reported by others [92]. These results suggest comparable disease control by radiosurgery and radiotherapy. However, the latency of treatment effect is significantly reduced by the use of radiosurgery.

ACTH-secreting Adenomas

Adenomas that secrete ACTH cause Cushing's disease or Nelson's disease in patients with bilateral adrenalectomies. The best screening test for hypercortisolemia is the 24-hour urine-free cortisol (UFC) and the definitive test is the dexamethasone suppression of corticotrophin releasing hormone (CRH) test [93]. Long-term use of medical therapy such as ketoconazole is associated with serious side effects. Thus, surgery remains the mainstay of treatment. Patients who failed or opt not to undergo surgery are treated with radiotherapy or radiosurgery.

The biochemical definition of a 'cure' for Cushing's disease remains a topic of debate. Many favor the use of a 24-hour urine free cortisol level as the gold standard while others argue the importance of morning serum cortisol, basal serum cortisol, or ACTH level [82]. Most studies (including the ones discussed below) use a combination of normal urine free cortisol and resolution of clinical stigmata as definition of disease cure.

Microsurgery of ACTH-secreting adenomas is often curative. Thapar and Laws [94] reported curative excisions in 90% of microadenomas and 60% of macroadenomas – results consistent with other surgical series [95,96]. The overall surgical mortality ranged 0.5–2% and morbidity ranged 3.3–9.3% [94–97]. Twenty to thirty percent of all patients undergoing pituitary surgery developed treatment-related hypopituitarism and diabetes insipidus [77,78]. Diligent monitoring of endocrine status is warranted since up to 30% of the 'cured' patients suffered relapse [98].

Approximately 80% of the patients who failed surgical resection achieved endocrine normalization 2 years after radiotherapy [99,100]. In comparable radiosurgery series, 50–80% 'cure' rates have been reported [82]. In experienced centers, morbidity related to radiotherapy or radiosurgery appeared comparable. The median time to endocrine normalization, however, was one to three months shorter in the radiosurgery series [3,82,101] when compared to the radiotherapy series.

Non-secreting Adenomas

Non-secreting adenomas cause clinical manifestations as they expand in size, compressing critical anatomic structures in proximity of the sella turcica – particularly the optic chiasm. Surgery is the primary treatment strategy for these adenomas. For patients presenting with visual deterioration, 84–87% regain normal visual function after surgery [76,102]. Routine postoperative radiation treatment of the residual tumor is not necessary [103].

Radiation therapy and radiosurgery are indicated for recurring adenomas. Because of the slow growing nature of pituitary adenomas, long-term follow-up is required to establish efficacy. For instance, Flickinger et al reported on a series of 87 patients who received postoperative radiation for non-functioning adenomas. The progression-free survival was 97% at 5 years and 76% at 20 years [104]. Unfortunately, the current radiosurgery literature consists entirely of short-term follow-ups (92–100% control rate, median follow up of 2–5 years). While the long-term results are likely comparable to those of radiotherapy, this hypothesis awaits validation.

Asymptomatic but progressively enlarging non-functional adenomas in poor surgical candidates are often treated with radiation therapy. In such cases, the 10-year progression-free survival was approximately 55% [105]. In comparison, 75% of the patients who underwent both surgery and radiation therapy remain disease free at 10 years [105].

Summary

With the exception of prolactinomas, where dopamine agonists remain the mainstay therapy, microsurgery is the treatment of choice for pituitary adenomas. Radiation therapy and radiosurgery represent treatment options for residual/recurrent tumors or for poor surgical candidates. Radiotherapy and radiosurgery appear equally efficacious in inducing endocrine normalization of secretory adenomas (Table 24.5). However, the time to GH and prolactin normalization after radiosurgery is significantly shorter in comparison to radiotherapy. Though a similar trend is observed for the ACTH-secreting adenomas, the difference is less dramatic. Long-term follow-up data are needed to establish the efficacy of radiosurgery as a treatment for non-functioning pituitary adenomas.

Gliomas

Gliomas constitute 40–50% of all primary brain tumors [37,38]. The term glioma denotes any tumors derived from the neuroglia. As such, it refers to several tumor subtypes that exhibit distinct natural histories. These subtypes are commonly categorized using the World Health Organization (WHO) classification scheme.

TABLE 24-5 Summary of the efficacy of surgery, radiation therapy and radiosurgery in the treatment of pituitary adenomas

	Surgery (%)	Radiotherapy (%)	Radiosurgery (%)
Prolactinoma[a]	65	30–50	30–50
GH secreting[b]	60	60–80	60–80
ACTH secreting	80	80	80
Non-functional	90	55	92–100, 2–4 years[c]

[a]Time to endocrine normalization considerably shorter after radiosurgery; [b]no long term data available; [c]the numbers shown indicate the 10-year progression-free survival after treatment unless otherwise indicated.

Grade I Gliomas

WHO grade I gliomas, including pilocytic astrocytomas and subependymal giant cell astrocytomas, are focal lesions with very low malignant potential. These tumors are amenable to cure with complete excision. As such, surgery remains the primary therapy. The complete resection of a pilocytic astrocytoma is associated with a 100% disease-free survival at 10 years. The 10-year survival drops to 84% for subtotal resections and 44% for biopsies [106]. Radiotherapy and radiosurgery are highly effective in cases of unresectable or residual tumor, with 85–95% disease-free progression at 5 years [107–109]. Disease progression after the first radiation treatment can be successfully controlled with a second treatment [107]. While radiosurgery and radiotherapy exhibit comparable disease control, many pilocytic astrocytomas are located in regions (e.g. in the hypothalamic region) where a single high-dose radiation may be associated with significant toxicity [25]. The choice of therapy, therefore, is dictated by the regional anatomy.

Grade II Gliomas

Grade II gliomas, including protoplastic, fibrillary, gemistocytic and mixed astrocytomas, tend to be diffuse and infiltrative, with a propensity for progression to higher grades. The risk of disease recurrence, and thus the probability of patient survival, correlates with the extent of resection [110–112]. Thus, surgical resection remains the primary therapy. When complete resection is achieved, the 10-year survival rate is roughly 68% without additional therapy [113]. The 10-year survival drops to the range of 30–50% with subtotal resections (<50% tumor mass resected) [114–117]. When incomplete resection is combined with radiotherapy, the 5-year survival rate increases to the range of 43–75% [114–117].

Two randomized trials have been conducted to determine the optimal radiation dose for the treatment of low-grade gliomas. The European Organization for Research and Treatment of Cancer (EORTC) study 22844 randomized 379 low-grade glioma patients to 45 versus 59.4 Gy radiotherapy [118]. Similarly, the Eastern Cooperative Group

(ECOG) randomized 211 low-grade glioma patients to 50.4 versus 64.8 Gy radiotherapy [119]. There was no difference in the 5-year survival or progression-free survival between the dose groups in either study. With regard to the timing of radiation therapy, the EORTC study 22845 revealed no survival benefit in patients receiving immediate postoperative radiotherapy [120]. However, the patients irradiated immediately after surgery had an improved 5-year progression free survival (44% versus 37%; $P < 0.05$). Since the patients were treated with radiation therapy at the time of relapse, these results suggest that radiotherapy is effective both as an up-front and a salvage modality.

In contrast to the large body of literature supporting radiotherapy as a treatment for WHO grade II gliomas, the literature in support of radiosurgery suffers from short-term follow-up and small sample size [108,121–125]. To date, the only study with a sufficient sample size (n = 49) and adequate length of follow-up (5 years) reported progression-free survival of 37% 5 years after radiosurgery [108]. If confirmed by other studies, this result suggests that the efficacy of radiosurgery compares favorably to that of radiotherapy in the treatment of low-grade gliomas.

Grade III and IV Gliomas

Grade III (anaplastic astrocytoma) and IV (glioblastoma multiforme) gliomas are highly malignant neoplasms that are associated with very poor clinical prognoses. With maximal chemo-, radiation and surgical therapy, the median survival for patients affected with grade III glioma is approximately 50 months. For grade IV gliomas, the median survival is approximately 12 months [126].

Surgery remains the mainstay therapy for the management of malignant gliomas, serving to achieve definitive diagnosis as well as cytoreduction. As demonstrated by two randomized trials, the addition of radiotherapy prolongs survival in patients with malignant glioma [127,128]. The survival benefit requires a cumulative dose of 60 Gy (median survival of 12 months). While survival is further improved by dose escalation to 90 Gy using proton irradiation (median survival 18.6 months), almost all patients developed radiation necrosis within the 90 Gy volume and suffered neurologic deterioration [129]. Thus, most current regimens of radiotherapy restrict the cumulative dose to 60 Gy.

The majority of malignant gliomas recur within 2 cm of the enhancing edge of the original tumor [130]. This observation has led investigators to explore the use of radiosurgery as boost therapy to the tumor margins. Though the initial reports suggest that this treatment strategy improved patient survival [131–136], such benefit was not confirmed by the randomized trial RTOG 93-05 [137]. In this trial, patients were randomized to arms of radiosurgery boost + radiation therapy + BCNU and radiation therapy + BCNU. The median survival was not significantly different between the treatment arms (13.6 versus 13.5 months; $P = 0.5711$). This result prompted a re-examination of the earlier

retrospective series. The resulting analysis revealed that the previously reported benefits were largely due to patient selection bias [132,138].

Similarly, retrospective series describing radiosurgery as a salvage therapy of recurrent tumor have suggested potential benefits [139,140]. These results await interrogation by randomized trial methodologies.

Summary

The clinical efficacies of radiation therapy in the treatment of grade I, II, III and IV gliomas are well established. Though several case series suggest that radiosurgery may be beneficial for a subset of glioma patients, the use of radiosurgery for the treatment of glioma remains investigational at the present time.

Radiosurgery Associated Tumor Formation

Radiation associated carcinogenesis after radiotherapy is a well-described phenomenon. It has been observed after radiation treatment for tinea capitis, pituitary adenomas and lymphoblastic leukemias [141–143]. The latency period for radiation associated tumor formation ranges from 6 to 32 years. The cumulative risk of tumor formation 30 years after radiotherapy ranges from 0.8–1.9% [141–143]. Because of the smaller irradiated field and the steep dose fall-off in radiosurgery, the probability of radiation induced secondary cancer should be less than that of radiotherapy. To date, only six cases of radiosurgery associated tumors have been reported [144]. While the true incidence of radiosurgery associated carcinogenesis is likely extremely low, long-term surveillance is warranted. Further, patients should be warned of the potential risk during the informed consent process.

NEUROIMAGING IN RADIOSURGERY

In radiosurgery, the focal dose distribution allows for a positive therapeutic gain. The need for accurate anatomic visualization is magnified by the use of higher radiation dose delivered in radiosurgery as compared to radiotherapy. To achieve maximal spatial accuracy in radiosurgical planning, an understanding of the basic principles underlying neuroimaging as well as the limitations associated with each imaging modality is mandatory.

Computerized Tomography (CT) Imaging

Computerized tomography (CT) provides cross-sectional images of the body using mathematical reconstructions based on x-ray images taken circumferentially around the subject. In practice, x-ray transmissions through the subject from a rotating emitter are detected and digitally converted into a grayscale image. Because CT images are ultimately a compilation of x-ray transmissions, the physical principles underlying the two modalities are identical. That is, structural discrimination is made based on the relative atomic composition, and therefore the electron density, of the tissue imaged. However, CT images offer improved anatomic resolution because each image represents the synthesis of information from multiple x-ray images.

Besides improved anatomic delineation, CT imaging aids radiosurgical planning in another way. Because the pixel intensity on a CT image reflects the electron density of the tissues imaged, the pixel intensity can be mathematically converted into electron density maps (electrons per cm^3). This information can be used to define isodose lines in radiosurgical planning. Without this information, actual radiation dose delivered can deviate from the desired dose by as much as 20% as a result of tissue inhomogeneity [145].

Despite yielding improved anatomic resolution as well as electron density information, delineation of soft tissue structures by CT imaging is suboptimal, even with the aid of intravenous contrast agents. For the most part, delineation of soft tissue structures is achieved by the use of magnetic resonance imaging, especially for targets in the cranial base.

Magnetic Resonance (MR) Imaging

The human body consists primarily of fat and water, both having a high content of hydrogen atoms. Magnetic resonance (MR) imaging exploits the nuclear spin property of these hydrogen atoms as a means to attain soft tissue resolution. In MR imaging, a radiofrequency pulse is applied to the imaged subject. As a result, the nuclear spin states of these atoms shift from that of equilibrium to that of excitation. To return to their equilibrium state, the law of energy conservation dictates that an energy equal to that absorbed must be emitted. The energy release between nuclear spin state transitions can be measured and analyzed. Since the process of energy absorption and emission is affected by the local chemical environment, hydrogen atoms in soft tissues of varying chemical composition will absorb and emit differential energy. Mathematical transformation of this information yields fine resolution maps of soft tissue structures. Since tumor and normal tissues often differ in chemical composition [146], the same principle allows delineation of these tissue types.

Because of the complexity of the nuclear interactions involved in MR imaging, the modality is subject to many sources of error, resulting in distortion of the image obtained. One such source of error involves the imperfection of the input magnetic field. The input magnetic field in MR imaging is produced by electric currents passing through sets of mutually orthogonal coils. Ideally, the magnetic field generated should be uniform such that a linear relationship between space and resonance frequency can be established [147]. However, such uniform fields cannot be easily achieved in practice. This phenomenon is referred to as gradient field non-linearity and tends to escalate with increasing distance from the central axis of the main magnet. For the most part, gradient field non-linearity can be corrected computationally. Prior to correction, gradient field non-linearity can induce spatial distortions as large

as 4 mm. After computational correction, the distortion is minimized to <1 mm [148].

A more complex MR distortion that is more difficult to correct computationally involves electromagnetic interactions between the imaged tissue and the input magnetic field. This distortion is often referred to as resonance offset. Resonance offset occurs because hydrogen atoms carry with them an inherent magnetic field. Thus, placement of hydrogen-bearing tissues in a magnetic field necessarily induces a perturbation in the input magnetic field. This perturbation disrupts the linear relationship between space and resonance frequency to produce geometric distortions. The physics of this perturbation is complex since it depends on the inherent magnetic properties as well as the volume and shape of the imaged object. Resonance offset distortions tend to be largest at the interface of materials that differ in magnetic properties, such as at the air/water interface. In anatomic imaging, this translates into large distortions at the air/bone or air/tissue interfaces. Studies reveal that distortions at these interfaces can be as large as 2 mm [147,149,150].

While the development of higher strength magnets has allowed for improved resolution of soft tissue structures, as well as minimization of geometric distortions related to gradient field non-linearity, higher strength magnets do not address the issue of resonance offset. Since resonance offset is a product of the input magnetic field and the local field imposed by the imaged tissue, increasing the strength of the input magnetic field will magnify the effects of resonance offset [151].

The accuracy of MR as a standalone imaging modality has been determined by a number of investigators [152–156]. While most investigators report a localization uncertainty of 2–3 mm [153–156], maximal absolute errors of 7–8 mm have also been reported [152]. These studies reveal that error in fiducial localization is amplified by subsequent mathematical transformation. Though the degree of localization uncertainty varies between studies, the reported uncertainty consistently remains greater than one millimeter, failing to achieve the current radiosurgical standard set forth by the American Society of Therapeutic Radiology and Oncology (ASTRO) [157–159].

Another downside of MR imaging as it pertains to radiosurgical planning is the absence of electron density information (see section on CT). Contrary to CT imaging, where the image is derived based on differential electron density, pixel intensities in MR images bear no correlation to electron density. For radiosurgical planning using MR as the only imaging modality, image processing and assignment of hypothetical electron density values are required. Such strategies have led to suboptimal radiosurgical plans, especially when protons are used [160].

Motion artifact is another consideration affecting spatial accuracy in MR imaging. The prolonged duration required for image acquisition increases the potential for patient movement. Even with a cooperative patient, motion artifact occurs with breathing and internal physiologic motions. The resultant motion compromises the accuracy of spatial resolution.

Though MR imaging may be inadequate as a stand-alone modality in radiosurgical planning, combining MR and CT images has led to radiosurgical plans that are superior to plans derived from each modality alone [161–164]. For example, Shuman et al reported that the incorporation of MR information into CT-based radiotherapy plans resulted in better definition of tumor volume in 53% of the cases [163]. These observations have led to the development of algorithms for superimposing MR and CT images.

CT-MR Image Integration

The differences between CT and MR imaging illustrate the conceptual distinction between geometric and diagnostic accuracy. While CT imaging is geometrically accurate due to absence of spatial distortion effects, diseased tissues are often missed by this modality. As such, CT imaging is diagnostically inaccurate. On the other hand, due to enhanced soft tissue resolution, MR imaging affords enhanced diagnostic accuracy. However, the spatial accuracy is limited due to MR distortion effects.

Algorithms have been developed to utilize maximally the different types of information afforded by CT and MR imaging. Simple approaches to image integration involve manual superposition of equivalent views of MR and CT images, using bony landmarks as correlation points. Such approaches, however, are labor intensive and error-prone with uncertainties of up to 8 mm [165].

Advances in computational technology have allowed for the development of automated algorithms for superposition of CT and MR images in three-dimensional space. One way of integrating CT and MR images requires that the patient be placed in an immobilization device, such as the stereotactic frame. The immobilization device minimizes motion artifacts and ensures that the images are acquired in a predetermined manner. Fiduciary markers are used to establish the spatial relationship between the target and the head-frame. Additionally, they serve as coregistration points between the MR and CT images. Because image acquisition and correlative points are fixed in space in a predetermined way, this mode of image fusion is sometimes referred to as prospective image coregistration [158].

Alternatively, image coregistration can be done with images that are not acquired in a predetermined manner. This mode of image fusion is also known as retrospective coregistration. Retrospective image coregistration relies on matching corresponding anatomic landmarks instead of fiduciary markers. The CT and MR images are integrated on the basis of aligning these anatomic landmarks [166]. Various computational techniques, including point matching [167], line matching, iterative matching [168], have been developed for retrospective image superposition. Whether one method

is superior to another remains an area of research. In general, with proper training and quality control, most current algorithms will coregister MRI and CT images to an uncertainty of 1–2 mm using prospective registration and 2–3 mm using retrospective registration [158].

Contrast Administration

Contrast administration takes advantage of the observation that disease processes, such as tumor growth, often result in vascular encroachment or faulty angiogenesis [146]. These processes allow contrast material to escape the vasculature and preferentially accumulate in the diseased tissue. The accumulation of contrast material can be easily visualized on CT or MR imaging. In malignant gliomas for instance, contrast enhancement correlates with diseased tissue. Kelly et al evaluated 195 brain tumor biopsies acquired from various locations relative to the contrast-enhancing regions of CT or MRI scans and showed that the regions of contrast enhancement best correlated with regions of tumor burden [169].

Since contrast-enhancing volumes are used for radiosurgery target definition, diseased tissues without contrast enhancement often escape therapy. Investigators have used various functional imaging modalities to address this issue. While these modalities hold tremendous promise, they are limited by poor anatomic resolution. As such, functional imaging is most useful in conjunction with traditional anatomic imaging modalities. In many instances, the clinical applications of functional imaging remain investigational.

MR Spectroscopy (MRS)

Another type of functional imaging capitalizes on the ability of MRI to measure the levels of biochemical metabolites. Three metabolites commonly used to distinguish tumor and healthy tissue include choline, creatine and N-acetylaspartate (NAA). Choline is an essential component of the cell membrane. The level of choline reflects the rapidity of membrane turnover and is increased in rapidly proliferating tumors. Creatine is a metabolic intermediate for the synthesis of phosphocreatine, an energy source for cellular metabolism. The level of creatine corresponds to the level of cellular energy reserve, which is decreased in tumor tissues. NAA is a marker for neuronal differentiation and is decreased in tumors [146].

Using elevated choline and decreased NAA as criteria, Pirzkall et al compared the MRS defined tumor volume to that defined by contrast-enhanced MRI for malignant gliomas. The authors report that the MRS defined volume extended outside of the MRI defined volume by >2 cm in 88% of the patients [170]. In another study of 46 patients with malignant gliomas, patients with MRS abnormality outside of the MR defined target volume showed decreased median survival relative to those with MRS abnormality inside the MR defined tumor volume (10.7 months versus 17.4 months, $P=0.002$) [171]. Other studies have confirmed the correlation between untreated MRS abnormality and worse prognosis [172–176]. These studies suggest that MRS data should be taken into consideration in target volume determination for the treatment of gliomas.

MR Perfusion Imaging

Like levels of biochemical metabolites, perfusion parameters such as cerebral blood volume (CBV) can be measured using MR imaging techniques. CBV is measured by monitoring the transit of a rapid bolus of contrast with respect to time. This parameter is an indirect measure of tissue vascularity, a property often associated with tumor burden. It is, therefore, not surprising that MR-derived measurements of cerebral blood volume correlate with tumor grading and clinical outcome. In a series of 28 patients with gliomas, pretreatment high CBV intensity was associated with shorter median survival [146,177,178].

The use of CBV in radiosurgical planning is limited by several factors. CBV values in tumor volumes are often greater than CBVs of normal white matter, but comparable to CBVs of normal gray matter. So, distinguishing tumor and cortex is problematic, especially in the context of anatomic distortion caused by large tumors. Additionally, regions of increased CBV correlate well with regions of contrast enhancement. As such, incorporation of CBV information will only alter radiosurgical plans in a subpopulation of patients.

MR Diffusion Weighted Imaging (DWI)

The white matter in a normal cerebrum is organized into tracts that allow communication between cortical neurons. As a result of this high degree of organization, water molecules in the cerebral cortex diffuse in a highly directional manner. The extent of this directional diffusion can be estimated using specialized MR techniques and is referred to as apparent diffusion coefficient (ADC). Because gliomas often distort cerebral architecture, regions with altered ADC are expected to correlate to tumor burden. This expectation was demonstrated in several studies [179–181]. These studies revealed that patients with lower ADC values in the tumor volume showed shorter median survival than patients with normal or near normal ADC values (12 months versus 21.7 months). These studies suggest that ADC maps may be helpful in guiding radiosurgical planning, especially in cases where conventional MRI and ADC maps yield discordant information with regard to tumor volume.

Summary

Given the complexity of physiology and pathology underlying tumor biology, it is unlikely that any single imaging modality will allow perfect definition of the diseased volume [182]. The failure precisely to define tumor volume will result in inadequate or excessive radiation treatment and suboptimal clinical outcomes. Precise tumor volume definition likely requires a synthesis of information obtained from contrast enhanced CT, MRI, MRS, CBV and ADC in a meaningful way. The optimal algorithm for the synthesis of this information remains an area of active research.

CONCLUSION

Radiosurgery has emerged to become an important therapeutic option in the field of neuro-oncology. While radiosurgery has been applied to a wide spectrum of neuro-oncologic diseases, it is important to note that the long-term follow-up for some of these applications, including the treatment of non-functioning pituitary adenomas, remains sparse. In the treatment of gliomas, reports of radiosurgery efficacy are limited to small case series that are tainted with patient selection bias. As such, the use of radiosurgery in these clinical contexts remains investigational.

For the treatment of cerebral metastases, meningiomas, functioning pituitary adenomas and acoustic neuromas, the radiosurgery literature is sufficiently mature in demonstrating its clinical efficacy. Comparative analysis of this literature with the radiotherapy literature suggests specific clinical contexts in which one modality may be more suitable than the other. For instance, certain metastatic diseases, including sarcomas, renal cell carcinomas and melanomas, are more likely to respond to radiosurgery than radiotherapy. The latency of endocrine normalization following treatment of functional pituitary adenomas is also shorter when radiosurgery is employed. On the other hand, the likelihood of hearing preservation following acoustic neuroma treatment is higher with stereotactic radiotherapy. An understanding of the clinical contexts as they relate to the efficacy of radiosurgery and radiotherapy is a prerequisite for the optimization of patient treatment strategies.

REFERENCES

1. Halperin E, Schidt-Ulrich RP, CA, Luther W (2004). The discipline of radiation oncology. In *Principles and Practice of Radiation Oncology*. Perez C, Brady L, Halperin E, and Schmidt-Ullrich R (eds). Lippincott Williams & Wilkins, Philadelphia. 1–96.
2. Leksell L (1951). The stereotaxic method and radiosurgery of the brain. Acta Chir Scand 102:316–319.
3. Chan A, Cardinale R, Loeffler J (2004). Stereotactic irradiation In *Principles and Practice of Radiation Oncology*. Perez C, Brady L, Halperin E, and Schmidt-Ullrich R (eds). Lippincott Williams & Wilkins, Philadelphia. 410–428.
4. Duma C, Jacques D, Kopyov O (1999). The treatment of movement disorders using Gamma Knife stereotactic radiosurgery. Neurosurg Clin N Am 10:379–389.

5. Lederman G, Lowry J, Wertheim S et al (1997). Acoustic neuroma: potential benefits of fractionated stereotactic radiosurgery. Stereotact Funct Neurosurg 69(1–4): 75–182.

6. Poen JC, Golby AJ, Forster KM et al (1999). Fractionated stereotactic radiosurgery and preservation of hearing in patients with vestibular schwannoma: a preliminary report. Neurosurgery 45:1299–1305; discussion 1305–1307.

7. Pollock BE, Lunsford LD (2004). A call to define stereotactic radiosurgery. Neurosurgery 55:1371–1373.

8. Shiau CY, Sneed PK, Shu HK et al (1997). Radiosurgery for brain metastases: relationship of dose and pattern of enhancement to local control. Int J Radiat Oncol Biol Phys 37:375–383.

9. Ito K, Kurita H, Sugasawa K, Mizuno M, Sasaki T (1997). Analyses of neuro-otological complications after radiosurgery for acoustic neurinomas. Int J Radiat Oncol Biol Phys 39:983–988.

10. Sakamoto T, Shirato H, Sato N et al (1998). Audiological assessment before and after fractionated stereotactic irradiation for vestibular schwannoma. Radiother Oncol 49:185–190.

11. Stafford SL, Pollock BE, Leavitt JA et al (2003). A study on the radiation tolerance of the optic nerves and chiasm after stereotactic radiosurgery. Int J Radiat Oncol Biol Phys 55:1177–1181.

12. Leber KA, Bergloff J, Langmann G, Mokry M, Schrottner O, Pendl G (1995). Radiation sensitivity of visual and oculomotor pathways. Stereotact Funct Neurosurg 64:233–238.

13. Leber KA, Bergloff J, Pendl G (1998). Dose-response tolerance of the visual pathways and cranial nerves of the cavernous sinus to stereotactic radiosurgery. J Neurosurg 88:43–50.

14. Tishler RB, Loeffler JS, Lunsford LD et al (1993). Tolerance of cranial nerves of the cavernous sinus to radiosurgery. Int J Radiat Oncol Biol Phys 27: 215–221.

15. Pollock BE, Nippoldt TB, Stafford SL, Foote RL, Abboud CF (2002). Results of stereotactic radiosurgery in patients with hormone-producing pituitary adenomas: factors associated with endocrine normalization. J Neurosurg 97:525–530.

16. Paek SH, Chung HT, Jeong SS et al (2005). Hearing preservation after gamma knife stereotactic radiosurgery of vestibular schwannoma. Cancer 104:580–590.

17. Karpinos M, Teh B, Zeck O et al (2002). Treatment of acoustic neuroma: stereotactic radiosurgery vs. microsurgery. Int J Radiat Oncol Biol Phys 54:1410–1421.

18. Pollock B, Lunsford L, Kondziolka D et al (1995). Outcome analysis of acoustic neuroma management: a comparison of microsurgery and stereotactic radiosurgery. Neurosurgery 36:215–224.

19. Andrews DW, Suarez O, Goldman HW et al (2001). Stereotactic radiosurgery and fractionated stereotactic radiotherapy for the treatment of acoustic schwannomas: comparative observations of 125 patients treated at one institution. Int J Radiat Oncol Biol Phys 50:1265–1278.

20. Lunsford LD, Niranjan A, Flickinger JC, Maitz A, Kondziolka D (2005). Radiosurgery of vestibular schwannomas: summary of experience in 829 cases. J Neurosurg 102 Suppl:195–199.

21. Eustacchio S, Trummer M, Fuchs I, Schrottner O, Sutter B, Pendl G (2002). Preservation of cranial nerve function following Gamma Knife radiosurgery for benign skull base meningiomas: experience in 121 patients with follow-up of 5 to 9.8 years. Acta Neurochir Suppl 84:71–76.

22. Kreil W, Luggin J, Fuchs I, Weigl V, Eustacchio S, Papaefthymiou G (2005). Long term experience of gamma knife radiosurgery for benign skull base meningiomas. J Neurol Neurosurg Psychiatr 76:1425–1430.

23. Muthukumar N, Kondziolka D, Lunsford LD, Flickinger JC (1999). Stereotactic radiosurgery for jugular foramen schwannomas. Surg Neurol 52:172–179.

24. Zhang N, Pan L, Dai JZ, Wang BJ, Wang EM, Cai PW (2002). Gamma knife radiosurgery for jugular foramen schwannomas. J Neurosurg 97:456–458.

25. Flickinger JC, Kondziolka D, Lunsford LD et al (2000). Development of a model to predict permanent symptomatic postradiosurgery injury for arteriovenous malformation patients. Arteriovenous Malformation Radiosurgery Study Group. Int J Radiat Oncol Biol Phys 46:1143–1148.

26. Sawamura Y, Shirato H, Sakamoto T et al (2003). Management of vestibular schwannoma by fractionated stereotactic radiotherapy and associated cerebrospinal fluid malabsorption. J Neurosurg 99:685–692.

27. Lim YL, Leem W, Kim TS, Rhee BA, Kim GK (1998). Four years' experiences in the treatment of pituitary adenomas with gamma knife radiosurgery. Stereotact Funct Neurosurg 70 Suppl 1:95–109.

28. Muramatsu J, Yoshida M, Shioura H et al (2003). Clinical results of LINAC-based stereotactic radiosurgery for pituitary adenoma. Nippon Igaku Hoshasen Gakkai Zasshi 63:225–230.

29. Shin M, Kurita H, Sasaki T et al (2000). Stereotactic radiosurgery for pituitary adenoma invading the cavernous sinus. J Neurosurg 93 Suppl 3:2–5.

30. Andrews DW, Scott CB, Sperduto PW et al (2004). Whole brain radiation therapy with or without stereotactic radiosurgery boost for patients with one to three brain metastases: phase III results of the RTOG 9508 randomised trial. Lancet 363:1665–1672.

31. Lawrence E, Edlefsen N (1930). On the production of high speed protons. Science 72:376.

32. Miller D (1995). A review of proton beam radiation therapy. Med Phys 22:1943–1954.

33. Sisterson J (2005). Particle Therapy Co-Operative Group (PTCOG). Particles Newsletter 35.

34. Schwartz M (1998). Stereotactic radiosurgery: comparing different technologies. Cmaj 158:625–628.

35. Kuo JS, Yu C, Petrovich Z, Apuzzo ML (2003). The CyberKnife stereotactic radiosurgery system: description, installation, and an initial evaluation of use and functionality. Neurosurgery 53:1235–1239; discussion 12:39.

36. Kjellberg R, Koehler A, Preston W (1964). Intracranial lesions made by a proton beam. In *Response of the Nervous System to Ionizing Radiation*. Haley T and Haley RS (eds). Little, Brown and Co., Boston.

37. Mahaley MS Jr, Mettlin C, Natarajan N, Laws ER Jr, Peace BB (1990). Analysis of patterns of care of brain tumor patients in the United States: a study of the Brain Tumor Section of the AANS and the CNS and the Commission on Cancer of the ACS. Clin Neurosurg 36:347–352.

38. Osborn A (1994). *Diagnostic Neuroradiology*, Vol. MD, FACR. Mosby, St. Louis. 936.

39. Lang EF, Slater J (1964). Metastatic brain tumors. results of surgical and nonsurgical treatment. Surg Clin North Am 44:865–872.

40. Agboola O, Benoit B, Cross P et al (1998). Prognostic factors derived from recursive partition analysis (RPA) of Radiation Therapy Oncology Group (RTOG) brain metastases trials applied to surgically resected and irradiated brain metastatic cases. Int J Radiat Oncol Biol Phys 42:155–159.

41. Alexander E 3rd, Moriarty TM, Davis RB et al (1995). Stereotactic radiosurgery for the definitive, noninvasive treatment of brain metastases. J Natl Cancer Inst 87:34–40.

42. Brown PD, Brown CA, Pollock BE, Gorman DA, Foote RL (2002). Stereotactic radiosurgery for patients with 'radioresistant' brain metastases. Neurosurgery 51:656–665; discussion 665–667.

43. Chougule P, Burton-Williams M, Saris S (2000). Randomized treatment of brain metastasis with gamma knife radiosurgery. Int J Radit Oncol Biol Phys 48:114.

44. Kondziolka D, Patel A, Lunsford LD, Kassam A, Flickinger JC (1999). Stereotactic radiosurgery plus whole brain radiotherapy versus radiotherapy alone for patients with multiple brain metastases. Int J Radiat Oncol Biol Phys 45:427–434.

45. Mehta MP, Tsao MN, Whelan TJ et al (2005). The American Society for Therapeutic Radiology and Oncology (ASTRO) evidence-based review of the role of radiosurgery for brain metastases. Int J Radiat Oncol Biol Phys 63:37–46.

46. Surma-aho O, Niemela M, Vilkki J et al (2001). Adverse long-term effects of brain radiotherapy in adult low-grade glioma patients. Neurology 56:1285–1290.

47. Taphoorn MJ, Klein M (2004). Cognitive deficits in adult patients with brain tumours. Lancet Neurol 3:159–168.

48. Aoyama H, Shirato H, Nakagawa K (2004). Interim report of the JROSG99-1 multi-institutional randomized trial, comparing radiosurgery alone vs. radiosurgery plus whole brain irradiation for 1–4 brain metastases [abstract]. Proceedings from the 40th annual meeting of the American Society of Clinical Oncology 108S.

49. Lutterbach J, Cyron D, Henne K, Ostertag CB (2003). Radiosurgery followed by planned observation in patients with one to three brain metastases. Neurosurgery 52: 1066–1073; discussion 1073–1074.

50. McDermott M, Sneed P (2005). Radiosurgery in metastaic brain cancer. Neurosurgery 57:45–53.

51. Sneed PK, Lamborn KR, Forstner JM et al (1999). Radiosurgery for brain metastases: is whole brain radiotherapy necessary? Int J Radiat Oncol Biol Phys 43:549–558.

52. Nakamura M, Roser F, Michel J, Jacobs C, Samii M (2003). The natural history of incidental meningiomas. Neurosurgery 53:62–70; discussion 70–71.

53. Firsching RP, Fischer A, Peters R, Thun F, Klug N (1990). Growth rate of incidental meningiomas. J Neurosurg 73:545–547.

54. Simpson D (1957). The recurrence of intracranial meningiomas after surgical treatment. J Neurol Neurosurg Psychiatr 20:22–39.

55. Mathiesen T, Lindquist C, Kihlstrom L, Karlsson B (1996). Recurrence of cranial base meningiomas. Neurosurgery 39: 2–7; discussion 8–9.

56. Pollock BE, Stafford SL, Utter A, Giannini C, Schreiner SA (2003). Stereotactic radiosurgery provides equivalent tumor control to Simpson Grade 1 resection for patients with small- to medium-size meningiomas. Int J Radiat Oncol Biol Phys 55:1000–1005.

57. Condra KS, Buatti JM, Mendenhall WM, Friedman WA, Marcus RB Jr, Rhoton AL (1997). Benign meningiomas: primary treatment selection affects survival. Int J Radiat Oncol Biol Phys 39:427–436.

58. Glaholm J, Bloom H, Crow J (1990). The role of radiotherapy in the management of intracranial meningiomas: The Royal Marsden Hospital experience with 186 patients. Int J Rad Biol Phys 18:755–761.

59. Goldsmith B, Wara W, Wilson C, Larson D (1994). Postoperative irradiation for subtotally resected meningiomas. J Neurosurg 1994:195–201.

60. Nutting C, Brada M, Brazil L et al (1999). Radiotherapy in the treatment of benign meningioma of the skull base. J Neurosurg 90:823–827.

61. Kobayashi T, Kida Y, Mori Y (2001). Long-term results of stereotactic gamma radiosurgery of meningiomas. Surg Neurol 55:325–331.

62. Rosenberg SI, Silverstein H, Gordon MA, Flanzer JM, Willcox TO, Silverstein JA (1993). Comparison of growth rates of acoustic neuromas: nonsurgical patients vs. subtotal resection. Otolaryngol Head Neck Surg 109:482–487.

63. Shin YJ, Fraysse B, Cognard C et al (2000). Effectiveness of conservative management of acoustic neuromas. Am J Otol 21:857–862.

64. Tschudi DC, Linder TE, Fisch U (2000). Conservative management of unilateral acoustic neuromas. Am J Otol 21:722–728.

65. Gormley W, Sekhar L, Wright D, Kamerer D, Schessel D (1997). Acoustic neuromas: results of current surgical management. Neurosurgery 41:50–58.

66. Samii M, Matthies C (1997). Management of 1000 vestibular schwannomas (acoustic neuromas): the facial nerve – preservation and restitution of function. Neurosurgery 40:684–694.

67. Chung WY, Liu KD, Shiau CY et al (2005). Gamma knife surgery for vestibular schwannoma: 10-year experience of 195 cases. J Neurosurg 102 Suppl:87–96.

68. Harsh G, Thornton A, Chapman P, Bussiere M, Rabinov J, Loeffler J (2002). Proton beam stereotactic radiosurgery of vestibular schwannomas. Int J Radiat Oncol Biol Phys 54:35–44.

69. Hasegawa T, Kida Y, Kobayashi T, Yoshimoto M, Mori Y, Yoshida J (2005). Long-term outcomes in patients with vestibular schwannomas treated using gamma knife surgery: 10-year follow up. J Neurosurg 102:10–16.

70. Petit J, Hudes R, Chen T, Eisenberg H, Simard J, Chin L (2001). Reduced-dose radiosurgery for vestibular schwannomas. Neurosurgery 49:1299–1306.

71. Prasad D, Steiner M, Steiner L (2000). Gamma surgery for vestibular schwannoma. Neurosugery 92:745–759.

72. Chan AW, Black P, Ojemann RG et al (2005). Stereotactic radiotherapy for vestibular schwannomas: favorable outcome with minimal toxicity. Neurosurgery 57:60–70; discussion 60–70.

73. Combs SE, Volk S, Schulz-Ertner D, Huber PE, Thilmann C, Debus J (2005). Management of acoustic neuromas with fractionated stereotactic radiotherapy (FSRT): long-term results in 106 patients treated in a single institution. Int J Radiat Oncol Biol Phys 63:75–81.

74. Stieber V, Deguzman A, Shaw E (2004). Pituitary. In *Principles and Practice of Radiation Oncology*. Perez C, Brady L, Halperin E, and Schmidt-Ullrich R (eds). Lippincott Williams & Wilkins, Philadelphia. 839–859.

75. Hamilton DK, Vance ML, Boulos PT, Laws ER (2005). Surgical outcomes in hyporesponsive prolactinomas: analysis of patients with resistance or intolerance to dopamine agonists. Pituitary 8:53–60.

76. Jane JA Jr, Laws ER Jr (2001). The surgical management of pituitary adenomas in a series of 3,093 patients. J Am Coll Surg 193:651–659.

77. Webb SM, Rigla M, Wagner A, Oliver B, Bartumeus F (1999). Recovery of hypopituitarism after neurosurgical treatment of pituitary adenomas. J Clin Endocrinol Metab 84:3696–3700.

78. Sudhakar N, Ray A, Vafidis JA (2004). Complications after trans-sphenoidal surgery: our experience and a review of the literature. Br J Neurosurg 18:507–512.

79. Grossman A, Cohen BL, Charlesworth M et al (1984). Treatment of prolactinomas with megavoltage radiotherapy. Br Med J (Clin Res Ed) 288:1105–1109.

80. Tsagarakis S, Grossman A, Plowman PN et al (1991). Megavoltage pituitary irradiation in the management of prolactinomas: long-term follow-up. Clin Endocrinol (Oxf) 34:399–406.

81. Tsang RW, Brierley JD, Panzarella T, Gospodarowicz MK, Sutcliffe SB, Simpson WJ (1996). Role of radiation therapy in clinical hormonally-active pituitary adenomas. Radiother Oncol 41:45–53.

82. Sheehan JP, Niranjan A, Sheehan JM et al (2005). Stereotactic radiosurgery for pituitary adenomas: an intermediate review of its safety, efficacy, and role in the neurosurgical treatment armamentarium. J Neurosurg 102:678–691.

83. Volker S, Stolke D, Hubertus M, Hoffmann B (1994). Clinical and radiological evaluation of long-term results of stereotactic proton beam radiosurgery in patients with cerebral arteriovenous malformation. J Neurosurg 81:683–689.

84. McKeage K, Cheer S, Wagstaff AJ (2003). Octreotide long-acting release (LAR): a review of its use in the management of acromegaly. Drugs 63:2473–2499.

85. Hardy J, Somma M, Vezina J (1976). Treatment of acromegaly: radiation or surgery? In *Current Controversies in Neurosurgery*. Morley G (ed.). WB Saunders, Philadelphia. 261.

86. Kreutzer J, Vance ML, Lopes MB, Laws ER Jr (2001). Surgical management of GH-secreting pituitary adenomas: an outcome study using modern remission criteria. J Clin Endocrinol Metab 86:4072–4077.

87. Ogilvy KM, Jakubowski J, Shortland JR (1974). Letter: spinal subarachnoid spread of pituitary adenoma. J Neurol Neurosurg Psychiatr 37:1186.

88. Dowsett RJ, Fowble B, Sergott RC et al (1990). Results of radiotherapy in the treatment of acromegaly: lack of ophthalmologic complications. Int J Radiat Oncol Biol Phys 19:453–459.

89. Lamberg BA, Kivikangas V, Vartianen J, Raitta C, Pelkonen R (1976). Conventional pituitary irradiation in acromegaly. Effect on growth hormone and TSH secretion. Acta Endocrinol (Copenh) 82:267–281.

90. Sheline G, Wara W (1975). Radiation therapy of acromegaly and non-secretory chromophobe adenomas of the pituitary. In *Tumors of the Central Nervous System*. Seydel H (ed.). Wiley, New York. 119–131.

91. Landolt AM, Haller D, Lomax N et al (1998). Stereotactic radiosurgery for recurrent surgically treated acromegaly: comparison with fractionated radiotherapy. J Neurosurg 88:1002–1008.

92. Mitsumori M, Shrieve DC, Alexander E 3rd et al (1998). Initial clinical results of LINAC-based stereotactic radiosurgery and stereotactic radiotherapy for pituitary adenomas. Int J Radiat Oncol Biol Phys 42:573–580.

93. Yanovski JA, Cutler GB Jr, Chrousos GP, Nieman LK (1993). Corticotropin-releasing hormone stimulation following low-dose dexamethasone administration. A new test to distinguish Cushing's syndrome from pseudo-Cushing's states. J Am Med Assoc 269:2232–2238.

94. Thapar K, Laws E (2001). *Brain tumors: an encyclopedia approach*. Kaye A, Laws E (eds). Churchill Livingstone, London. 803–853.

95. De Tommasi C, Vance ML, Okonkwo DO, Diallo A, Laws ER Jr (2005). Surgical management of adrenocorticotropic hormone-secreting macroadenomas: outcome and challenges in patients with Cushing's disease or Nelson's syndrome. J Neurosurg 103:825–830.

96. Robert F, Hardy J (1991). Cushing's disease: a correlation of radiological, surgical and pathological findings with therapeutic results. Pathol Res Pract 187:617–621.

97. Semple PL, Laws ER Jr (1999). Complications in a contemporary series of patients who underwent transsphenoidal surgery for Cushing's disease. J Neurosurg 91:175–179.

98. Benveniste RJ, King WA, Walsh J, Lee JS, Delman BN, Post KD (2005). Repeated transsphenoidal surgery to treat recurrent or residual pituitary adenoma. J Neurosurg 102:1004–1012.

99. Estrada J, Boronat M, Mielgo M et al (1997). The long-term outcome of pituitary irradiation after unsuccessful transsphenoidal surgery in Cushing's disease. N Engl J Med 336:172–177.

100. Jennings AS, Liddle GW, Orth DN (1977). Results of treating childhood Cushing's disease with pituitary irradiation. N Engl J Med 297:957–962.

101. Sheehan JM, Vance ML, Sheehan JP, Ellegala DB, Laws ER Jr (2000). Radiosurgery for Cushing's disease after failed transsphenoidal surgery. J Neurosurg 93:738–742.

102. Jane JA Jr, Laws ER Jr (2002). Surgical management of pituitary adenomas. Singapore Med J 43:318–323.

103. Dekkers OM, Pereira AM, Roelfsema F et al (2006). Observation alone after transsphenoidal surgery for non-functioning pituitary macroadenoma. J Clin Endocrinol Metab 28:28.

104. Flickinger JC, Nelson PB, Martinez AJ, Deutsch M, Taylor F (1989). Radiotherapy of nonfunctional adenomas of the pituitary gland. Results with long-term follow-up. Cancer 63:2409–2414.

105. Urdaneta N, Chessin H, Fischer JJ (1976). Pituitary adenomas and craniopharyngiomas: analysis of 99 cases treated with radiation therapy. Int J Radiat Oncol Biol Phys 1:895–902.

106. Forsyth PA, Shaw EG, Scheithauer BW, O'Fallon JR, Layton DD Jr, Katzmann JA (1993). Supratentorial pilocytic astrocytomas. A clinicopathologic, prognostic, and flow cytometric study of 51 patients. Cancer 72:1335–1342.

107. Boethius J, Ulfarsson E, Rahn T, Lippitz B (2002). Gamma knife radiosurgery for pilocytic astrocytomas. J Neurosurg 97: 677–680.

108. Heppner PA, Sheehan JP, Steiner LE (2005). Gamma knife surgery for low-grade gliomas. Neurosurgery 57:132–139.

109. Kidd EA, Mansur DB, Leonard JR, Michalski JM, Simpson JR, Perry A (2006). The efficacy of radiation therapy in the management of grade I astrocytomas. J Neuro-Oncol 76:55–58. Epub 2005 Aug 11.

110. Berger MS, Deliganis AV, Dobbins J, Keles GE (1994). The effect of extent of resection on recurrence in patients with low grade cerebral hemisphere gliomas. Cancer. 74:1784–1791.

111. Claus EB, Horlacher A, Hsu L et al (2005). Survival rates in patients with low-grade glioma after intraoperative magnetic resonance image guidance. Cancer 103:1227–1233.

112. Piepmeier J, Christopher S, Spencer D et al (1996). Variations in the natural history and survival of patients with supratentorial low-grade astrocytomas. Neurosurgery 38:872–878; discussion 878–879.

113. Janny P, Cure H, Mohr M et al (1994). Low grade supratentorial astrocytomas. Management and prognostic factors. Cancer 73:1937–1945.

114. Fazekas J (1977). Treatment of grades I and II brain astrocytomas: the role of radiotherapy. Int J Rad Biol Phys 2:661–666.

115. Garcia D, Fulling K, Marks J (1985). The value of radiation therapy in addition to surgery for astrocytomas of the adult cranium. Cancer 55:919–927.

116. Medbery C, Straus K, Steinberg S, Cotelingam J, Fisher W (1988). Low-grade astrocytomas: treatment results and prognostic variables. Int J Rad Biol Phys 15:837–841.

117. Shaw E, Dumas-Duport C, Scheithauer B et al (1989). Radiation therapy in the management of low-grade supratentorial astrocytomas. J Neurosurg 70:853–861.

118. Karim AB, Maat B, Hatlevoll R et al (1996). A randomized trial on dose-response in radiation therapy of low-grade cerebral glioma: European Organization for Research and Treatment of Cancer (EORTC) Study 22844. Int J Radiat Oncol Biol Phys 36:549–556.

119. Shaw E, Arusell R, Scheithauer B et al (2002). Prospective randomized trial of low- versus high-dose radiation therapy in adults with supratentorial low-grade glioma: initial report of a North Central Cancer Treatment Group/Radiation Therapy Oncology Group/Eastern Cooperative Oncology Group study. J Clin Oncol 20:2267–2276.

120. van den Bent MJ, Afra D, de Witte O et al (2005). Long-term efficacy of early versus delayed radiotherapy for low-grade astrocytoma and oligodendroglioma in adults: the EORTC 22845 randomised trial. Lancet 366:985–990.

121. Barcia JA, Barcia-Salorio JL, Ferrer C, Ferrer E, Algas R, Hernandez G (1994). Stereotactic radiosurgery of deeply seated low grade gliomas. Acta Neurochir Suppl 62:58–61.

122. Fuchs I, Kreil W, Sutter B, Papaethymiou G, Pendl G (2002). Gamma Knife radiosurgery of brainstem gliomas. Acta Neurochir Suppl 84:85–90.

123. Grabb PA, Lunsford LD, Albright AL, Kondziolka D, Flickinger JC (1996). Stereotactic radiosurgery for glial neoplasms of childhood. Neurosurgery 38:696–701; discussion 701–702.

124. Hadjipanayis CG, Kondziolka D, Gardner P et al (2002). Stereotactic radiosurgery for pilocytic astrocytomas when multimodal therapy is necessary. J Neurosurg 97:56–64.

125. Hadjipanayis CG, Niranjan A, Tyler-Kabara E, Kondziolka D, Flickinger JC, Lunsford LD (2002). Stereotactic radiosurgery for well-circumscribed fibrillary grade II astrocytomas: an initial experience. Stereotact Funct Neurosurg 79:13–24.

126. Curran WJ Jr, Scott CB, Horton J et al (1993). Recursive partitioning analysis of prognostic factors in three Radiation Therapy Oncology Group malignant glioma trials. J Natl Cancer Inst 85:704–710.

127. Bleehen NM, Wiltshire CR, Plowman PN et al (1981). A randomized study of misonidazole and radiotherapy for grade 3 and 4 cerebral astrocytoma. Br J Cancer 43:436–442.

128. Sheline GE (1990). Radiotherapy for high grade gliomas. Int J Radiat Oncol Biol Phys 18:793–803.

129. Fitzek M, Thornton A, Rabinov J et al (1999). Accelerated fractionated proton/photon irradiation to 90 cobalt gray equivalent for glioblastoma multiforme: results of a phase II prospective trial. J Neurosurg 91:251–260.

130. Hochberg FH, Pruitt A (1980). Assumptions in the radiotherapy of glioblastoma. Neurology 30:907–911.

131. Buatti JM, Friedman WA, Bova FJ, Mendenhall WM (1995). Linac radiosurgery for high-grade gliomas: the University of Florida experience. Int J Radiat Oncol Biol Phys 32:205–210.

132. Larson DA, Gutin PH, McDermott M et al (1996). Gamma knife for glioma: selection factors and survival. Int J Radiat Oncol Biol Phys 36:1045–1053.

133. Nwokedi EC, DiBiase SJ, Jabbour S, Herman J, Amin P, Chin LS (2002). Gamma knife stereotactic radiosurgery for patients with glioblastoma multiforme. Neurosurgery 50:41–46; discussion 46–47.

134. Prisco FE, Weltman E, de Hanriot RM, Brandt RA (2002). Radiosurgical boost for primary high-grade gliomas. J Neurooncol 57:151–160.

135. Sarkaria JN, Mehta MP, Loeffler JS et al (1995). Radiosurgery in the initial management of malignant gliomas: survival comparison with the RTOG recursive partitioning analysis. Radiation Therapy Oncology Group. Int J Radiat Oncol Biol Phys 32:931–941.

136. Shrieve DC, Alexander E 3rd, Black PM et al (1999). Treatment of patients with primary glioblastoma multiforme with standard postoperative radiotherapy and radiosurgical boost: prognostic factors and long-term outcome. J Neurosurg 90:72–77.

137. Tsao MN, Mehta MP, Whelan TJ et al (2005). The American Society for Therapeutic Radiology and Oncology (ASTRO) evidence-based review of the role of radiosurgery for malignant glioma. Int J Radiat Oncol Biol Phys 63:47–55.

138. Ulm AJ 3rd, Friedman WA, Bradshaw P, Foote KD, Bova FJ (2005). Radiosurgery in the treatment of malignant gliomas: the University of Florida experience. Neurosurgery 57: 512–517; discussion 512–517.

139. Laing RW, Warrington AP, Graham J, Britton J, Hines F, Brada M (1993). Efficacy and toxicity of fractionated stereotactic radiotherapy in the treatment of recurrent gliomas (phase I/II study). Radiother Oncol 27:22–29.

140. Lederman G, Wronski M, Arbit E et al (2000). Treatment of recurrent glioblastoma multiforme using fractionated stereotactic radiosurgery and concurrent paclitaxel. Am J Clin Oncol 23:155–159.

141. Brada M, Ford D, Ashley S et al (1992). Risk of second brain tumour after conservative surgery and radiotherapy for pituitary adenoma. Br Med J 304:1343–1346.

142. Kimball Dalton VM, Gelber RD, Li F, Donnelly MJ, Tarbell NJ, Sallan SE (1998). Second malignancies in patients treated for childhood acute lymphoblastic leukemia. J Clin Oncol 16:2848–2853.

143. Sadetzki S, Flint-Richter P, Ben-Tal T, Nass D (2002). Radiation-induced meningioma: a descriptive study of 253 cases. J Neurosurg 97:1078–1082.

144. Loeffler JS, Niemierko A, Chapman PH (2003). Second tumors after radiosurgery: tip of the iceberg or a bump in the road? Neurosurgery 52:1436–1440; discussion 1440–1442.

145. Battista J, Rider W, Van DJ (1980). Computed tomography for radiotherapy planning. Int J Radiat Oncol Biol Phys 6:99–107.

146. Nelson SJ, Cha S (2003). Imaging glioblastoma multiforme. Cancer J 9:134–145.

147. Sumanaweera TS, Adler JR Jr, Napel S, Glover GH (1994). Characterization of spatial distortion in magnetic resonance imaging and its implications for stereotactic surgery. Neurosurgery 35:696–703; discussion 703–704.

148. Chang H, Fitzpatrick J (1992). A technique for accurate magnetic resonance imaging in the presence of field inhomogeneities. IEEE Trans Med Imaging 11:319–329.

149. Sumanaweera T, Glover G, Song S, Adler J, Napel S (1994). Quantifying MRI geometric distortion in tissue. Magn Reson Med 31:40–47.

150. Sumanaweera T, Glover G, Binford T, Adler J (1993). MR susceptibility misregistration correction. IEEE Trans Med Imaging 12:251–259.

151. Mack A, Wolff R, Scheib S et al (2005). Analyzing 3-tesla magnetic resonance imaging units for implementation in radiosurgery. J Neurosurg 102 Suppl:158–164.

152. Walton L, Hampshire A, Forster DM, Kemeny AA (1997). Stereotactic localization with magnetic resonance imaging: a phantom study to compare the accuracy obtained using two-dimensional and three-dimensional data acquisitions. Neurosurgery 41:131–137; discussion 137–139.

153. Walton L, Hampshire A, Forster DM, Kemeny AA (1996). A phantom study to assess the accuracy of stereotactic localization, using T1-weighted magnetic resonance imaging with the Leksell stereotactic system. Neurosurgery 38:170–176; discussion 176–178.

154. Orth RC, Sinha P, Madsen EL et al (1999). Development of a unique phantom to assess the geometric accuracy of magnetic resonance imaging for stereotactic localization. Neurosurgery 45:1423–1429; discussion 1429–1431.

155. Kondziolka D, Dempsey PK, Lunsford LD et al (1992). A comparison between magnetic resonance imaging and computed tomography for stereotactic coordinate determination. Neurosurgery 30:402–406; discussion 406–407.

156. Bednarz G, Downes MB, Corn BW, Curran WJ, Goldman HW (1999). Evaluation of the spatial accuracy of magnetic resonance imaging-based stereotactic target localization for gamma knife radiosurgery of functional disorders. Neurosurgery 45:1156–1161; discussion 1161–1163.

157. Shaw E, Kline R, Gillin M et al (1993). Radiation Therapy Oncology Group: radiosurgery quality assurance guidelines. Int J Radiat Oncol Biol Phys 27:1231–1239.

158. Khoo VS, Dearnaley DP, Finnigan DJ, Padhani A, Tanner SF, Leach MO (1997). Magnetic resonance imaging (MRI): considerations and applications in radiotherapy treatment planning. Radiother Oncol 42:1–15.

159. The American Society for Therapeutic Radiology and Oncology, Task Force on Stereotactic Radiosurgery and the American Association of Neurological Surgeons, Task Force on Stereotactic Radiosurgery (1994). Consensus statement on stereotactic radiosurgery quality improvement. Int J Radiat Oncol Biol Phys 28:527–530.

160. Schad LR, Gademann G, Knopp M, Zabel HJ, Schlegel W, Lorenz WJ (1992). Radiotherapy treatment planning of basal meningiomas: improved tumor localization by correlation of CT and MR imaging data. Radiother Oncol 25:56–62.

161. Ten Haken RK, Thornton AF Jr, Sandler HM et al (1992). A quantitative assessment of the addition of MRI to CT-based, 3-D treatment planning of brain tumors. Radiother Oncol 25:121–133.

162. Shuman WP, Griffin BR, Haynor DR et al (1987). The utility of MR in planning the radiation therapy of oligodendroglioma. Am J Roentgenol 148:595–600.

163. Shuman WP, Griffin BR, Haynor DR et al (1985). MR imaging in radiation therapy planning. Work in progress. Radiology 156:143–147.

164. Just M, Rosler HP, Higer HP, Kutzner J, Thelen M (1991). MRI-assisted radiation therapy planning of brain tumors – clinical experiences in 17 patients. Magn Reson Imaging 9:173–177.

165. De Salles AA, Asfora WT, Abe M, Kjellberg RN (1987). Transposition of target information from the magnetic resonance and computed tomography scan images to conventional X-ray stereotactic space. Appl Neurophysiol 50:23–32.

166. Pelizzari CA, Chen GT, Spelbring DR, Weichselbaum RR, Chen CT (1989). Accurate three-dimensional registration of CT, PET, and/or MR images of the brain. J Comput Assist Tomogr 13:20–26.

167. Judnick JW, Kessler ML, Fleming T, Petti P, Castro JR (1992). Radiotherapy technique integrates MRI into CT. Radiol Technol 64:82–89.

168. Evans PM, Gildersleve JQ, Morton EJ et al (1992). Image comparison techniques for use with megavoltage imaging systems. Br J Radiol 65:701–709.

169. Kelly PJ, Daumas-Duport C, Scheithauer BW, Kall BA, Kispert DB (1987). Stereotactic histologic correlations of computed tomography- and magnetic resonance imaging-defined abnormalities in patients with glial neoplasms. Mayo Clin Proc 62:450–459.

170. Pirzkall A, McKnight TR, Graves EE et al (2001). MR-spectroscopy guided target delineation for high-grade gliomas. Int J Radiat Oncol Biol Phys 50:915–928.

171. Chan A, Nelson S (2002). Use of 1H magnetic resonance spectroscopic imaging in managing recurrent glioma patients undergoing radiosurgery. ISMRM Workshop Proceedings on MR of Cancer 27.

172. Nelson SJ, Graves E, Pirzkall A, et al (2002). TR In vivo molecular imaging for planning radiation therapy of gliomas: an application of 1H MRSI. J Magn Reson Imaging 16:464–476.

173. McKnight TR, von dem Bussche MH, Vigneron DB et al (2002). Histopathological validation of a three-dimensional magnetic resonance spectroscopy index as a predictor of tumor presence. J Neurosurg 97:794–802.

174. Graves EE, Nelson SJ, Vigneron DB et al (2001). Serial proton MR spectroscopic imaging of recurrent malignant gliomas after gamma knife radiosurgery. Am J Neuroradiol 22:613–624.

175. Graves EE, Nelson SJ, Vigneron DB et al (2000). A preliminary study of the prognostic value of proton magnetic resonance spectroscopic imaging in gamma knife radiosurgery of recurrent malignant gliomas. Neurosurgery 46:319–326; discussion 326–328.

176. Chan AA, Lau A, Pirzkall A et al (2004). Proton magnetic resonance spectroscopy imaging in the evaluation of patients undergoing gamma knife surgery for Grade IV glioma. J Neurosurg 101:467–475.

177. Cha S (2004). Perfusion MR imaging of brain tumors. Top Magn Reson Imaging 15:279–289.

178. Cha S (2003). Perfusion MR imaging: basic principles and clinical applications. Magn Reson Imaging Clin N Am 11:403–413.

179. Nelson S (1999). Imaging of brain tumors after therapy. Neuro-imaging Clin N Am 9:801–819.

180. Moffat BA, Chenevert TL, Lawrence TS et al (2005). Functional diffusion map: a noninvasive MRI biomarker for early stratification of clinical brain tumor response. Proc Natl Acad Sci USA 102:5524–5529. Epub 2005 Apr 1.

181. Castillo M, Smith JK, Kwock L, Wilber K (2001). Apparent diffusion coefficients in the evaluation of high-grade cerebral gliomas. Am J Neuroradiol 22:60–64.

182. Benjamin R, Capparella J, Brown A (2003). Classification of glioblastoma multiforme in adults by molecular genetics. Cancer J 9:82–90.

CHAPTER 25

Diffusion Magnetic Resonance Imaging in Brain Tumors

L. Celso Hygino Cruz Jr, Romeu C. Domingues and A. Gregory Sorensen

INTRODUCTION

The prevalence of primary neoplasms of the central nervous system (CNS) is around 15 000 to 17 000 new cases annually in the USA and it is estimated that brain tumors cause the deaths of 13 000 patients every year. Although this incidence represents annually only 1.3% of all new cases of cancer, brain tumors correspond to 2.3% of all deaths caused by cancer. Brain metastases occur in approximately 30% of all patients with disseminated cancer [1]. When metastatic lesions are included, brain tumors are estimated to cause the death of 90 000 patients every year [2–4]. Gliomas are the leading cause of primary CNS tumor, accounting for 40–50% of cases [5] and 2–3% of all cancers [6]. Despite new techniques of treatment, patient survival still remains very low, varying between 16 and 53 weeks [7].

A variety of imaging techniques have been developed for assessing brain tumors. This task of assessment includes initial detection, diagnosis, guiding intervention, monitoring after therapy and overall prognosis. For more than 20 years, conventional magnetic resonance imaging (MRI), typically T1- and T2-weighted imaging, has been used for this purpose [8]. While in clinical practice nearly all detected lesions undergo some type of surgical intervention, such as biopsy or excision, and therefore histologic diagnosis is available, some cases will still require a determination of tumor grade. For example, surgical excision of the visible abnormality is often attempted, despite a consensus that conventional MRI tends to underestimate the extent of the tumor, which in turn many suspect leads to suboptimal treatment [9]. In part this is because most primary brain neoplasms infiltrate the brain parenchyma out beyond the enhancing portion of the lesion. High-grade gliomas have the ability to infiltrate the brain parenchyma. This infiltration appears to follow the main fiber bundles, possibly along the vessel channels, without disrupting the blood–brain barrier [10]. An ability to detect these

bundles, therefore, might allow better determination of tumor growth directions and extent. Many other imaging techniques are under development to attempt to answer these and other questions [11].

New functional MRI sequences, such as diffusion-weighted imaging (DWI), perfusion imaging and MR spectroscopy (MRS) have been widely used to determine the grade, heterogeneity and extent of these primary and secondary brain tumors. Diffusion weighted imaging first came into wide use because of its ability to detect hyperacute infarcts (Figure 25.1). In recent years, diffusion has been studied extensively in the evaluation of other pathologies, including brain tumors. Diffusion tensor MR imaging (DT-MRI or DTI) is a relatively new MR method that may be able to delineate more accurately the tumor versus the infiltrating tumor between the peritumoral edema and the normal brain parenchyma [12]. DTI has been widely used because of its apparent ability to illustrate the relationship of a tumor with the nearby main fiber tracts, though this is difficult to validate. Because of these early findings, some have suggested that DTI might also be used to aid in surgical planning [13], possibly in radiotherapy planning [14], as well as to monitor the tumor recurrence and tumor response to treatment [15]. While these new applications have not yet been proven or even widely studied in multicenter clinical trials, many individual practioners are, nonetheless, actively using DTI. The reasoning behind such use typically begins with the logic that the aim of a surgical brain tumor treatment is as complete lesion resection as possible without harming vital brain functions [16,17]. Therefore, a preoperative approach which maps, even imperfectly, the tumor and its relationship to nearby functional structures may facilitate patient's outcome [18]. Because of this logic and related rationales, considerable interest in DTI has arisen. In this chapter, therefore, we will review the methodology and uses of DTI as it has been studied in these anecdotal and early studies.

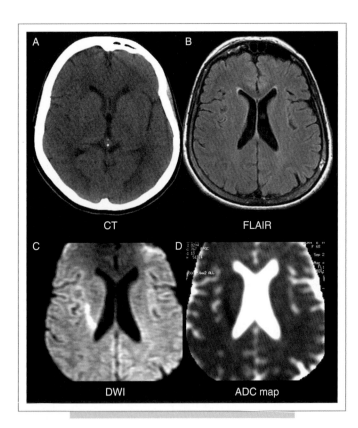

FIG. 25.1. A 67-year-old woman presenting with new onset of left hemiparesis at 2 hours. **(A)** Non-enhancing CT and **(B)** T2-weighted image do not demonstrate any abnormality. **(C)** Diffusion-weighted image demonstrates a hyperintense area in the posterior limb of the right internal capsule and corona radiata. An area of hyperacute infarct could be diagnosed, based on the findings of restricted diffusion, reflecting cytotoxic edema, confirmed by the apparent diffusion coefficient map – black area on the ADC map **(D)**.

PHYSICAL BASIS

Diffusion-Weighted MR Imaging

Diffusion-weighted imaging is based on the random or Brownian motion of water molecules, typically in relation to their thermal energy, although influenced by the surrounding environment. In the brain, the presence of some tissue structures restricts free water motion [19], for example, the diffusion of water molecules is higher in the ventricles than in the parenchyma, as there are few barriers. MRI makes it possible to estimate the diffusivity of water molecules.

Most diffusion measurements today are made using a variant of the diffusion-weighted sequence first described by Stejskal and Tanner. Their initial approach described a spin-echo (SE) sequence together with two equal and opposite extra gradient pulses [20]; the amount of signal loss can be related to the magnitude of diffusion. For practical purposes, an echoplanar imaging (EPI) SE T2-weighted

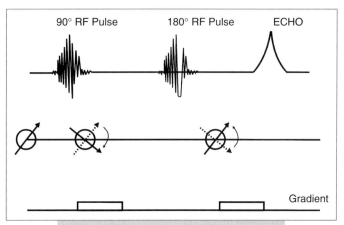

FIG. 25.2. A schematic figure demonstrating the diffusion-weighted sequence described by Stejskal and Tanner.

sequence is utilized, because of its ability to reduce motion artifacts and its rapid acquisition time [21]. Stejskal and Tanner's approach uses two magnetic pulses or gradients to label the spins: the application of the first diffusion gradient causes a dephasing of water protons; because they move randomly, not all water protons are in place for rephasing from the application of the second diffusion gradient. Thus, there is a signal decrease that depends on how far the water molecules move (Figure 25.2) [22]. The net signal on the final diffusion-weighted image is therefore influenced by both the T2 tissue effect and by the tissue diffusion characteristics. By acquiring an image with little diffusion-weighting and another image with substantial diffusion-weighting, the apparent diffusion coefficient (ADC) can be calculated on a voxel-by-voxel basis, allowing the generation of a map that reflects solely the diffusion influence excluding the T2 effects, which prevents misinterpretation from the so-called 'T2 shine-through' effect [21,22].

Some pathologic processes appear to change the characteristic of the brain microenvironment, which in turn results in alteration of the apparent diffusion coefficient [22]. In clinical practice, diffusion imaging has been used to assess acute cerebral ischemia [23–25], as diffusion water motion decreases after the onset of ischemia. The mechanism underlying the observed decrease in ADC in acute ischemia is still poorly understood. Failure of the Na^+/K^+ adenosine triphosphatase pump is believed to play an important role in this pathophysiological process, however, and this state of ischemia is often referred to as cytotoxic edema [26,27]. While ischemia was the first application of DWI, diffusion imaging has also been applied to the evaluation of many other neurological conditions. The appearance of lesions in diseases such as multiple sclerosis [28–31] (Figure 25.3), encephalitis [32] and Creutzfeldt-Jakob disease [33] (Figure 25.4) can be contrasted with the appearance of diffusion in neoplastic brain lesions (see later Figures).

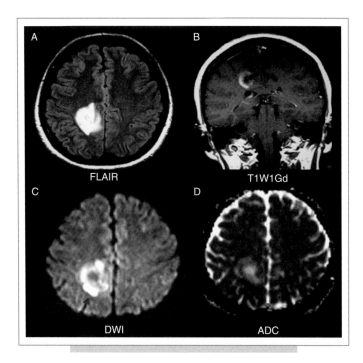

FIG. 25.3. (A) Axial FLAIR image showing the presence of demyelinating plaques on the brainstem and in the cerebellum peduncles that enhance in the post-contrast T1-weighted image **(B,C)**. These lesions may consist of acute demyelinating plaques. Most of acute multiple sclerosis plaques may have restricted diffusion, characterized by hyperintensity on DWI **(D)** and low ADC values on ADC map.

Diffusion Tensor MR Imaging

The mathematical description of the movement of water can be complex. While water motion occurs in all three cardinal directions, the description of this movement may be highly variable depending on local restrictions. In an attempt to simplify the analysis of diffusion imaging, often water mobility is assumed to behave in a manner that can be described using a Gaussian approximation. As we will see below, this is a powerful approach and is widely popular; however, in more recent years the limitations of this technique have begun to become apparent, as will also be discussed below. Nevertheless, the Gaussian approximation has definite utility, and therefore we will use this approximation as well for much of this chapter.

When water molecules diffuse equally in all directions, this is termed isotropic diffusion. This phenomenon is typical in the ventricles and, at the resolution of typical clinical MRI, such isotropic behavior seems to be the case in the gray matter. In the white matter, however, free water molecules diffuse anisotropically, that is, the water diffusion is not equal in all three orthogonal directions [34,35]. This is likely because tissue structures cause impediment of the water motion; these structures likely include the cell membranes but, more importantly, the myelin sheath surrounding myelinated white matter [36]. Put another

way, isotropic diffusion can be graphically represented as a sphere, whereas anisotropic diffusion can be graphically expressed as an ellipsoid [37], with water molecules moving farther along the long axis of a fiber bundle and moving less perpendicularly (Figure 25.5) [38].

In a tensor model, water mobility is described by a 3×3 matrix. To estimate the nine-tensor matrix elements, the diffusion gradients must be applied to at least six non-collinear directions (it is six rather than nine because only some degree of symmetry is assumed, and so only six of the nine elements are unique under this assumption) [38]. The eigenvalues represent the three principal diffusion coefficients measured along the three coordinate directions of the ellipsoid [38]. The eigenvectors represent the directions of the tensor [39]. Because interpreting a tensor representation can be non-intuitive, scalar metrics have been proposed to simplify DTI data [36]. For example, fractional anisotropy (FA) measures the fraction of the total magnitude of diffusion anisotropy. FA values vary from completely isotropic diffusion (graded as 0) up to completely anisotropic diffusion (graded as 1) [36,37].

In addition to assessment of the diffusion in a single voxel, DTI has been used to attempt to map the white matter fiber tracts. This is typically done by connecting each voxel eigenvector to its adjacent in accordance with the direction the fibers are pointing at [40,41]. DTI can also determine a color-coded FA map of fiber orientation [42]. A different color has been attributed to represent a different fiber orientation along the three orthogonal spatial axes: in the standard convention, red stands for the left-right direction of x-oriented fibers, the blue color stands for the superior-inferior direction of y-oriented fibers, and green for the anterior-posterior direction of z-oriented fibers (Figure 25.6) [42,43].

CLINICAL USE

As mentioned above, there are no multicenter clinical trials demonstrating the benefit of DTI (or DWI) in brain tumors. Therefore, formally speaking, there is no scientific evidence for the utility of DTI in tumor imaging. Nevertheless, in the following sections we will summarize current opinion based on single-center studies and clinical anecdotes; clearly, more well-controlled studies are needed to state definitively the benefit of these techniques to patients.

Assessment of Extracranial Tumors

The diagnosis of epidermoid tumor can be challenging in clinical practice, as its signal intensity is frequently quite similar to cerebrospinal fluid on conventional T1- and T2-weighted images. Under special conditions, arachnoid cysts may have an atypical appearance on conventional MRI, because of blood and high protein contents, complicating the differential diagnosis between these two extra-axial lesions [22]. DWI can also be used as an effective

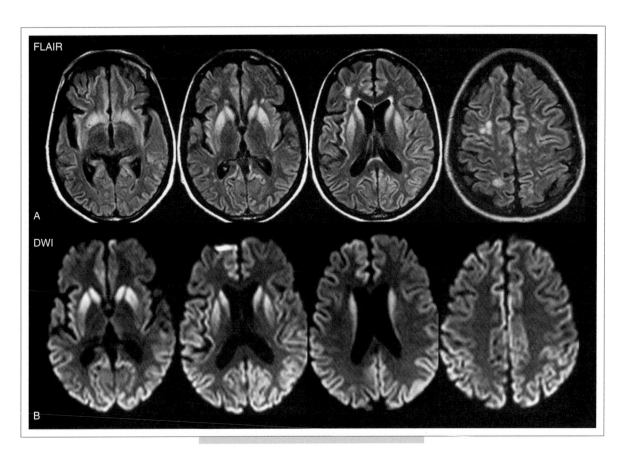

FIG. 25.4. A 45-year-old woman complaining of mental disturbance with the diagnosis of Creutzfeldt-Jakob disease. Axial FLAIR images **(A)** and diffusion-weighted images **(B)** demonstrate gyriform increased signal intensity predominantly in the cortex of the occipito-parietal lobes and also in the basal ganglia, consistent with restricted diffusion.

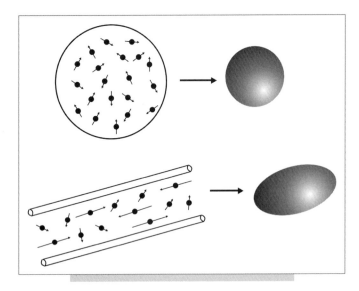

FIG. 25.5. Free water molecules diffuse equally in all different directions. This is termed isotropic diffusion and can be graphically represented as a sphere. When there is a preferred direction for the free water molecules to diffuse, the diffusion is termed anisotropic and can be graphically represented as an ellipsoid.

way of differentiating an arachnoid cyst from epidermoid tumors [44]. Both lesions present the same T1 and T2 signal intensity characteristic of cerebrospinal fluid. On DWI, epidermoid tumors are hyperintense, because they are solidly composed, whereas arachnoid cysts are hypointense, demonstrating high diffusivity [44]. The ADC values of epidermoid tumors are similar to those of the brain parenchyma, whereas the ADC values of arachnoid cysts are similar to those of CSF [45]. Water molecules in arachnoid cysts demonstrate a high diffusibility as there is no structure impeding their motion (Figure 25.7). On the other hand, the water molecules are impeded by tumor contents, like keratinous debris (Figure 25.8) [46]. As a result, DWI can be used better to assess follow-up of surgically resected epidermoid tumors, proving efficacious in the detection of residual lesions (Figure 25.9) [47].

Most meningiomas, especially malignant and atypical samples, also have a restricted diffusion, displaying low ADC values, when compared with typical meningiomas (Figure 25.10) [48,49]. However, meningiomas rarely pose difficulty to diagnosis on conventional MR imaging. Some aspects have been postulated to contribute to

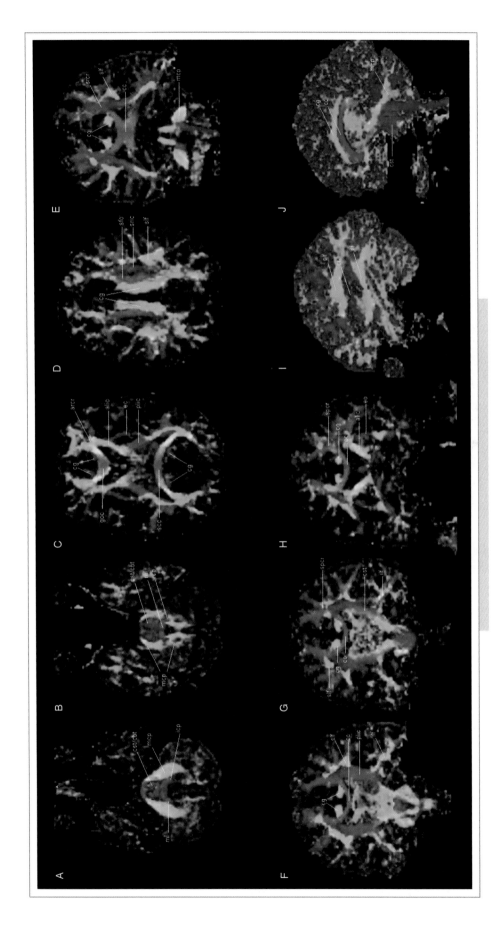

FIG. 25.6. Diffusion tensor imaging (DTI) color-coded map of a healthy volunteer. Locations of white matter tracts are assigned on color maps. The direction of the main fiber tracts is represented by red (right-left), green-yellow (anterior-posterior) and blue (superior-inferior). Several main fiber tracts visible on color maps are annotated on the basis of anatomic knowledge. **(A–D)** Axial fractional anisotropic (FA) color maps. **(E–H)** Coronal FA color maps. **(I,J)** Sagittal FA color maps. Mcp, middle cerebral peduncle; cst, corticospinal tract; cbt, corticobulbar tract; ml, medial lemniscus; icp, inferior cerebellar peduncle; cg, cingulum; cc, corpus callosum; gcc, genu of corpus callosum; scc, splenium of corpus callosum; arcr, anterior region of corona radiata; alic, anterior limb of internal capsule; plic, posterior limb of internal capsule; ec, external capsule; sric, superior region of internal capsule; sfof, superior fronto-occipital fasciculus; ifof, inferior fronto-occipital fasciculus; slf, superior longitudinal fasciculus; ilf, inferior longitudinal fasciculus.

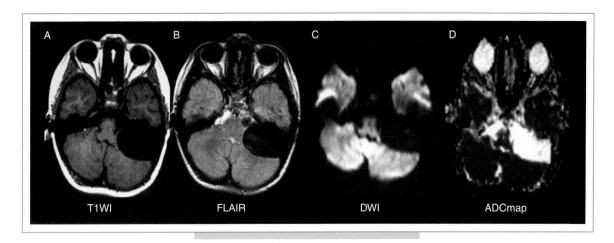

FIG. 25.7. A 3-year-old girl with arachnoid cyst in the left cerebellopontine angle cistern. **(A)** Axial T1-weighted image and **(B)** FLAIR image show a hypointense round lesion on the left cerebellopontine angle cistern that causes compression to the left cerebellum hemisphere. **(C)** Diffusion-weighted image demonstrates a hypointense signal, with high diffusibility seen on the apparent diffusion coefficient map **(D)**, which is isointense relative to cerebrospinal fluid.

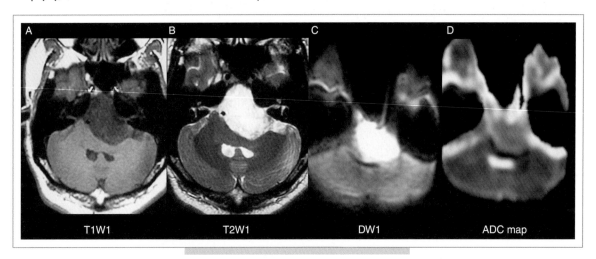

FIG. 25.8. A 27-year-old man with epidermoid tumor in the left cerebellopontine angle cistern. An expansive rounded lesion in the left cerebellopontine angle cistern presenting with hypointense on axial T1-weighted image **(A)** and hyperintense on axial T2-weighed image **(B)** is demonstrated. **(C)** Axial diffusion-weighted image shows a hyperintense lesion, of which an apparent diffusion coefficient map **(D)** shows a nearly isointense relative to normal brain parenchyma.

decrease ADC values in malignant and atypical meningiomas – hypercellularity, a high nuclear-to-cytoplasmic ratio, prominent nucleoli and decreased extracellular water [49]. The preoperative measurement of ADC values may predict the malignancy of meningiomas, which may be useful to decide on the surgical strategy or to prescribe the adjunctive therapy. As a consequence, a more accurate prognosis may be obtained.

Characterization of Intracranial Cystic Masses

A number of groups have suggested that DWI can aid in the distinction between brain abscesses and necrotic and cystic neoplasms on MRI. This differentiation is still a challenge in both clinical and radiological settings. Early

studies suggest that DWI can provide a sensitive and specific method for differentiating tumor from abscess in certain settings [50–53]. The abscesses have a high signal on DWI and a reduced ADC within the cavity (Figure 25.11). This restricted diffusion is thought to be related to the characteristic of the pus in the cavity. Because pus is a viscous fluid that consists of inflammatory cells, debris and macromolecules like fibrinogen [54], this may, in turn, lead to reduced water mobility, lower ADC and bright signal on DWI. On the other hand, necrotic and cystic tumors display a low signal on DWI (similar to the CSF in the ventricles), with an increased ADC as well as isointense or hypointense DWI signal intensity in the lesion margins (Figure 25.12) [52]. Although these findings can be helpful, they are of course

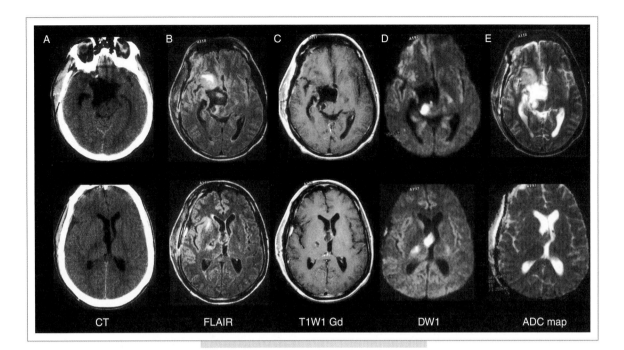

FIG. 25.9. Follow-up MR exam performed 3 months after surgical resection of an epidermoid tumor in a 38-year-old man. **(A)** CT images show an abnormal hypodense area at the surgical site. **(B)** Axial FLAIR images and **(C)** axial contrast-enhanced T1-weighted images demonstrate post-surgical changes in the brain parenchyma, as well as an area of slightly hyperintensity on FLAIR images and isointensity on T1-weighted image that may represent residual tumor. **(D)** Axial diffusion-weighted images can easily show a hyperintense lesion, which represents residual tumor, surrounded by a hypointense area, corresponding to post-surgical changes. **(E)** ADC maps obtained also demonstrate the difference between the residual tumor, with isointensity relative to normal brain tissue, and the hyperintense area corresponding to post-surgical changes, filled with cerebrospinal fluid.

not absolute: under certain conditions, restricted diffusion has been documented in hemorrhagic metastases, radiation necrosis and cystic astrocytoma [55].

Tumor Grading

Tumor grading is very important in treatment decision and evaluation of prognosis. While tissue samples are obtained as part of most therapeutic approaches, heterogeneity in tissue sampling, as well as monitoring malignant dedifferentiation and other needs, have led to a desire to use imaging better to ascertain tumor grade. In certain settings, diffusion imaging appears to increase both the sensitivity and specificity of MR imaging in the evaluation of brain tumors by providing information about tumor cellularity, which may, in turn, improve the prediction of tumor grade [12,56–64]. The mechanism in which DWI may help in tumor grading is not clear, but speculation is that the free water molecules' diffusivity is restricted by cellularity increases present in high-grade lesions [38,65]. The reduction in extracellular space, as well as the high nuclear-to-cytoplasmic ratios of some cancer cells, causes a relative reduction in the ADC values (Figure 25.13) [66]. Some studies suggest a correlation between the ADC values and tumor cellularity [48,65], with lower ADC values suggesting high-grade lesions (Figure 25.14) [48,67]. In some studies,

however, ADC values found in high- and low-grade gliomas have overlapped somewhat [48]. Evaluation of tumor grade with diffusion image remains uncertain and still cannot be considered a reliable tool for this purpose, since ADC values overlap considerably in different brain neoplasms. Indeed, many factors besides tumor cellularity might be responsible for the observed reduction in diffusibility. One such factor that might contribute to the reduction of ADC values in high-grade gliomas is the presence of hydrophilic glycosaminoglycans, such as hyaluronan, in the extracellular space, which may decrease water content [68].

DWI has also been shown to assist in assessing high cellularity of other brain neoplasms. In some studies, lymphoma, a highly cellular tumor, has been found to present hyperintense signal intense on DWI and reduced ADC values [69], and it may be in differentiating lymphoma from other CNS lesions that DWI has its greatest value (Figure 25.15). In another study, the ratio of ADC values in the center of rim-enhancing intracranial lesions relative to normal appearing white matter was significantly higher in patients with toxoplasmosis than in patients with lymphoma [70].

DWI may also be helpful in distinguishing medulloblastoma, a primitive neuroectodermal tumor, from other pediatric brain tumors. Some investigators report that medulloblastoma has displayed restricted diffusion, again

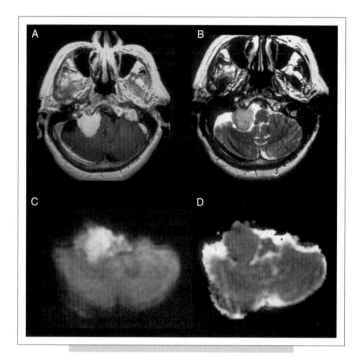

FIG. 25.10. An 84-year-old woman with meningioma. **(A)** Axial post-contrast-enhanced T1-weighted image **(B)** and axial T2-weighted image **(C)** show an extra-axial enhancing lesion in the right cerebello-ponto-medullary angle cistern. Axial DWI **(D)** demonstrates increased signal intensity due to restricted diffusion, caused by increase in cellularity.

presumably because of the densely packed tumor cells and high nuclear-to-cytoplasmic ratio (Figure 25.16) [71,72]. In one report, the solid enhancing portion of cerebellar hemangioblastomas on post-contrast T1-weighted images was, together with other posterior fossa neoplasms, the only one to demonstrate hypointensity on DWI. The main explanation for the high ADC values in these tumors was suggested to be due to the rich vascular spaces present in hemangioblastomas [73].

Diffusion tensor imaging has also been used to attempt grading brain neoplasms and seems to provide some utility by assessment of tumor cellularity. To date, the additional information provided by DTI has not yet been shown to correlate with tumor cellularity [74], since a high degree of fiber tract disorganization in the tumor core is thought to be present [75]. Nevertheless, one report has argued that fractional anisotropy can distinguish high-grade gliomas from low-grade gliomas [76]. This study found a significant difference between the fractional anisotropy when only the solid portion of the lesion was analyzed, avoiding the necrotic and cystic portions. The FA values in high-grade gliomas were higher than those in the low-grade gliomas, which was taken to suggest higher symmetry of histologic organization. However, these results are somewhat contradictory to the usual understanding of the microstructure of high-grade gliomas. The histological characteristics of high-grade

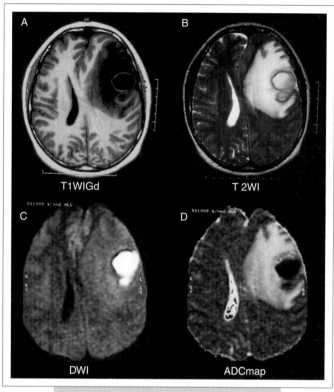

FIG. 25.11. A 23-year-old man presenting with early onset of headache and speech difficulties. **(A)** Axial post-contrast T1-weighted image and **(B)** T2-weighted image show an expansive, necrotic rim-enhancing lesion, surrounded by vasogenic edema, causing mass effect to the adjacent brain parenchyma compressing the ipsilateral ventricle. The lesion has restricted diffusion, due to its high viscosity content, presenting with high signal intensity on DWI **(C)** and low ADC values **(D)**.

gliomas compared with those of low-grade gliomas typically reveal pleomorphologic structures and a regressive organization rather than an increase in parallel histological organization [76]. A separate report found no differences between low- and high-grade gliomas with regards to FA values in the tumor center. This may be consistent with the disorganization of fiber tracts in the center of both entities, resulting in a loss of structural organization [75]. FA values within epidermoid tumors were higher than in the normal white matter, perhaps because of the high packing density of the cells and their solid-state cholesterol [77].

The hope for mean diffusibility in brain tumors to be a parameter for tumor classification has yet to be fulfilled. Further complicating this situation is that mean diffusibility in brain tumors is likely influenced by tumor cellularity, intra- or extracellular edema and tumor necrosis, at least at the typical b value of $1000 \, s/mm^2$. This multifactorial influence likely makes mean diffusibility too non-specific a parameter for tumor grade or infiltration and does not give a specific perception of tumor classification [75].

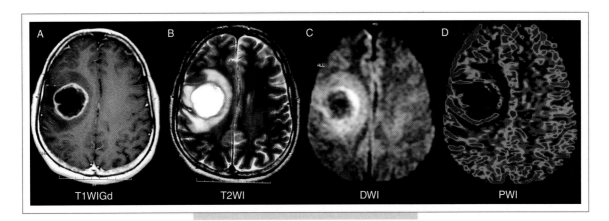

FIG. 25.12. A 48-year-old man with glioblastoma multiforme. **(A)** Contrast-enhanced, axial T1-weighted image shows an enhancing necrotic mass, surrounded by abnormal hyperintense area on the axial T2WI **(B)**. This abnormal hyperintense T2WI area can represent peritumoral edema and/or infiltrating tumor. The tumor is iso-hypointense on the T2WI, indicating high cellularity, which is also demonstrated as a restricted diffusion on the DWI **(C)** and on the ADC map **(D)**.

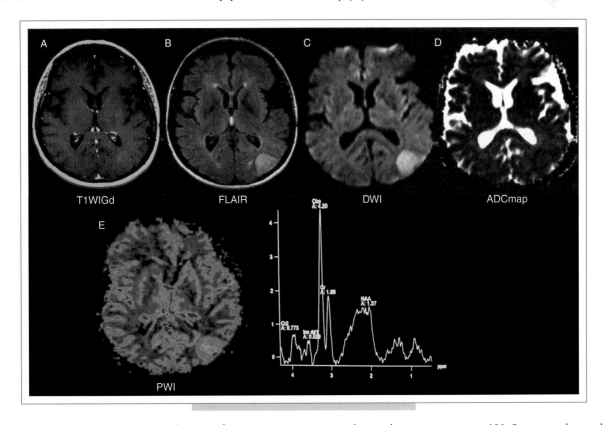

FIG. 25.13. A 65-year-old man complaining of cognitive impairment with anaplastic astrocytoma. **(A)** Contrast-enhanced, axial, T1-weighted image shows a round cortical lesion with a slight contrast enhancing in the left occipito-parietal lobe, with hyperintensity on FLAIR **(B)** and diffusion-weighted images **(C)** and restricted diffusion on ADC maps **(D)**. **(E)** The lesion also has hyperperfusion representing neoangiogenesis.

PREOPERATIVE PLANNING AND PERITUMORAL MARGINS

The precise determination of the margins of the tumor is considered by many investigators to be of the utmost importance to the management of brain tumors. The goal of a surgical approach to the brain neoplasm is the complete resection of the tumor, coupled with minimum neurological deficit [78]. While a variety of approaches has been used, diffusion imaging has also been enlisted in the attempt to determine the margins of tumors in the brain. Some suggest DWI can provide information about

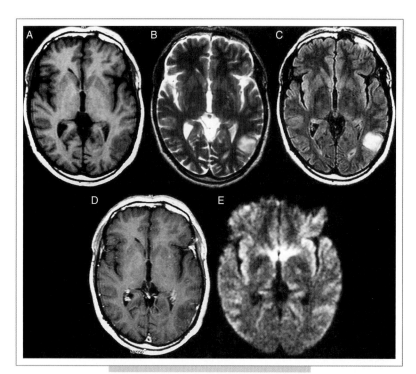

FIG. 25.14. A biopsy-proven low-grade glioma in a 43-year-old man, presenting with focal partial seizures in the right arm for 15 days. An expansive lesion with hypointensity on T1-weighted image **(A)** and hyperintensity on T2-weighted image **(B)** and FLAIR image **(C)**, which does not enhance in the post-contrast enhanced T1-weighted image **(D)**. Axial diffusion-weighted image does not show restricted diffusion **(E)**.

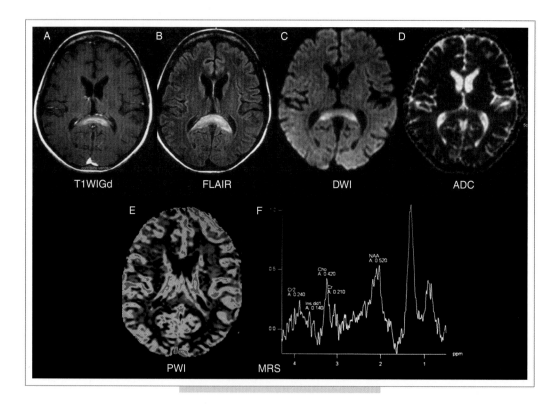

FIG. 25.15. A 72-year-old female, presenting with cognitive impairment and mental and language disturbance for 20 days, with a biopsy proven diagnosis of lymphoma. **(A)** Axial T1-weighted image shows a hypointense lesion in the corona radiata, which enhances after intravenous injection of contrast **(B)** and has hyperintensity on FLAIR image **(C)**. The lesion does not have hyperperfusion on relative cerebral blood volume maps **(D)**. **(E)** Magnetic resonance spectroscopy demonstrates markedly elevated lactate/lipids peaks and a low N-acetyl-aspartate peak. There is also noted a slight increase on choline peak. The lesion has restricted diffusion demonstrated by hyperintensity on diffusion-weighted image **(F)**, confirmed by low apparent diffusion coefficient values.

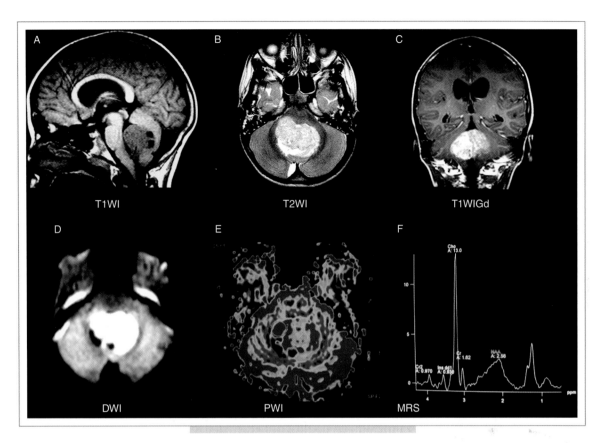

FIG. 25.16. A 7-year-old boy with medulloblastoma, presenting with headache, vomiting, gait disturbance and behavior impairment. An expansive heterogeneous lesion, localized in the fourth ventricle, has isointense signal intensity on T1-weighted axial image **(A)**, hyperintense on T2-weighted axial image **(B)** and enhances after contrast intravenous injection **(C)**. The lesion has restricted diffusion on diffusion-weighted image **(D)**, due to the hypercellularity. The lesion also presents hyperperfusion **(E)** and on the MR-spectroscopy **(F)** a high peak of choline, low peak of NAA and an elevation on lactate/lipids levels are also noted.

peritumoral neoplastic cell infiltration [12,56–61], perhaps even help discriminate the boundaries between tumor, infiltrating tumor, peritumoral edema and normal brain parenchyma [12,23,56,79]. It is well known that high-grade tumors tend to spread diffusely across the brain, moving along the fiber tracts [80,81]. However, not all studies have found DWI to be helpful in the evaluation of tumor extensions [48,64,82], most likely because of the difficulty in finding the borders of some tumors, even on histopathology. Given these conflicting findings, it remains unclear whether diffusion-weighted imaging or diffusion tensor imaging is a useful tool to predict the true extent of a brain tumor. Due to the challenges of obtaining a histologic standard of reference ('gold standard'), it is reasonable to question DWI's ability to distinguish neoplastic cell infiltration beyond the enhancing portion of the lesion on the abnormal hyperintense T2-weighted image, which traditionally is thought to represent vasogenic edema with or without tumor invasion.

Some investigators believe that metastases with perilesional edema have higher ADC values than a primary brain tumor with peritumoral edema and these investigators have suggested that higher ADC values may allow better

differentiation between metastases and primary lesions [58]. In high-grade gliomas, the abnormal hyperintense T2-weighted image is thought to represent not only vasogenic edema, but also infiltrative neoplasm cells. The normal white matter extracellular space contains a complex matrix, which restricts the free motion of water molecules [83]. Increased permeability of the blood–brain barrier produces a vasogenic edema that surrounds the brain tumor [84]. Invasive neoplasm cells have been found to destroy the extracellular matrix [85]. Following this logic, cell infiltration alters the integrity of extracellular matrix, reducing the obstruction for free water molecules to diffuse. However, this logic is somewhat speculative and vasogenic edema with neoplasm cell infiltration may be found in high-grade gliomas, while in cases of meningiomas, metastases and low-grade gliomas vasogenic edema without neoplasm infiltration is seen – both with similar ADC values.

While the specific biophysics of perilesional edema remains unclear, and ADC alone may not differentiate peritumoral edema with infiltration from peritumoral edema without infiltration, a number of investigators have reported that DTI can detect variations in FA values around lesions. This has, in turn, led some to speculate that these variations may be due to the presence of infiltrative tumor. However, it is not clear if infiltrating tumor will destroy fibers and therefore result in lower FA, or if the tumor infiltrating along the fibers actually increases FA by having more ultrastructure aligned with the fibers. One report did not find any significant difference in the FA values of either the enhancing or non-enhancing portion of the two neoplastic lesions [86], while another report has verified that the periphery of low-grade gliomas, without abnormal hyperintense signal on the T2-weighted images, contains a considerable amount of preserved fiber tracts (high FA values), whereas most tracts are disarranged in high-grade gliomas (low FA values) [75]. Diffusion tensor imaging seems to demonstrate that the periphery of low-grade gliomas contains preserved fiber tracts, whereas most tracts are disarranged in grade III gliomas. However, the presence of edema can obscure fiber tracts that are present.

The involvement of the white matter tracts can often be identified in brain tumor patients by using either anisotropy maps (FA maps are the most widely used) or tractography or both. Based on DTI findings resulting from studies of brain tumor patients, the white matter involvement by a tumor can be arranged into five different categories [62,63]:

1 Displaced: maintained normal or slightly decreased anisotropy relative to the contralateral tract in the corresponding location, but situated in an abnormal T2WI signal intensity area or presented in an abnormal orientation (Figure 25.17)
2 Invaded: reduced anisotropy; the main fiber tracts remained identifiable on orientated color-coded FA maps (Figure 25.18)
3 Disrupted: marked reduced anisotropy; the main fiber tracts are unidentifiable on oriented color-coded FA maps (Figure 25.19)
4 Infiltrated: slightly reduced anisotropy without displacement of white matter architecture; the fibers remain identifiable on orientation maps (Figure 25.20)
5 Edematous: marked reduced anisotropy with normal anisotropy and normal orientated on color-coded FA maps, but located in an abnormal T2WI signal intensity area (Figure 25.21) [74].

Displacement rather than destruction of white matter fibers (as suggested by FA) around low-grade gliomas has been described [14,77] and this is consistent with structural imaging approaches. Low-grade neoplasms (see Figure 25.5) are typically well-circumscribed lesions that do not cause invasion or destruction of fiber tracts. Such lesions tend to produce a displacement or deviation of surrounding white matter fibers (Figure 25.22). This is not absolute: one study described a case in which the corticospinal tract had been infiltrated by an oligodendroglioma that, nevertheless, had spared the motor strip and the posterior limb of the internal capsule [87]. Furthermore, displacement rather than

FIG. 25.17. Pattern of main fiber tract involvement: displaced. Maintained normal or slightly decreased anisotropy, situated in an abnormal location or presenting in an abnormal orientation. It confirmed an intact tract.

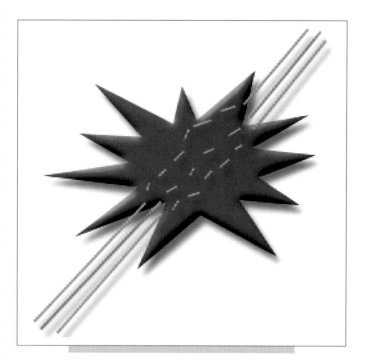

FIG. 25.18. Pattern of main fiber tract involvement: disrupted. Marked reduced anisotropy and unidentifiable orientation color-coded fractional anisotropy maps.

FIG. 25.19. Pattern of main fiber tract involvement: invaded. Reduced anisotropy, but remaining identifiable on orientation color-coded fractional anisotropy maps.

FIG. 25.21. Pattern of main fiber tract involvement: edematous. Marked reduced anisotropy with normal orientation, but located in an abnormal T2-weighted signal intensity area.

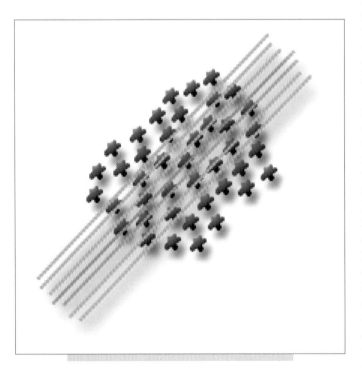

FIG. 25.20. Pattern of main fiber tract involvement: infiltrated. Slightly decreased anisotropy without displacement of white matter architecture, remaining identifiable on orientation color-coded fractional anisotropy maps and tractography.

infiltration of the adjacent white matter tracts has also been described in cerebral metastases [14] and meningiomas [88]. Finally, in some cases of high-grade gliomas, the tumor could cause not only invasion and disruption of the main fiber tracts, but also associated displacement (Figure 25.23). So while DTI can identify tracts and help determine the presence or absence of invasion versus displacement, the presence or absence of displacement does not appear to be highly diagnostic for high- or low-grade malignancy.

Both vasogenic edema and tumor tissue appear to be more isotropic than normal white matter. Thus, the anisotropy in the T2WI hyperintense area that surrounds the tumor is typically reduced, either because of invasion of neoplastic cells or because of edema. One report has noted that, in patients with high-grade gliomas, but not with low-grade gliomas or cerebral metastases, the anisotropy is also low in the white matter areas adjacent to tumors that look normal on T2WI (as compared with the contralateral hemisphere) [14]. The same situation has been reported in lymphomas. When compared with the abnormal white matter adjacent to metastases, decreased anisotropy in the abnormal white matter that surrounds the gliomas was demonstrated [87]. FA values decrease in the abnormal area that surrounds high-grade tumors on T2-weighted imaging. Again, this presumably happens because of increased water content and also tumor invasion [15]. This is not uniform: a certain number of studies have not found any difference in the analyses of abnormal white matter adjacent to high-grade gliomas and metastases [15].

One important finding, the presence of tract disruption, is mostly found in high-grade tumors and is thought

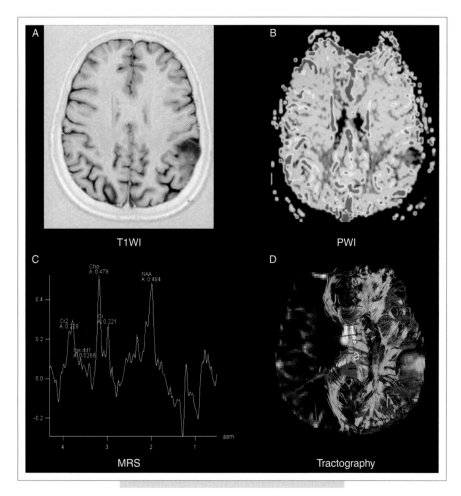

FIG. 25.22. A 42-year-old man with a diagnosis of a low-grade astrocytoma presented with early onset of focal seizures. **(A)** The MRI examination demonstrates an expansive lesion in the perirolandic area, which does not have hyperperfusion. **(B)** The mass lesion causes displacement of the main fiber tracts adjacent to the tumor, which is well demonstrated on tractography. There seems to be invasion or disruption of these tracts.

to be caused by a combination of peritumoral edema, tumor mass effect and tumor infiltration effect (Figure 25.24) [14,88]. While anatomic disruption and functional disruption are not necessarily equivalent, the ability to visualize tracts has led to a number of investigations in this area. The main fiber tracts are infiltrated in cases of gliomatosis cerebri, which has a specific histopathologic behavior. In this lesion, the neoplastic cells form parallel rows among nerve fibers, preserving them. However, there is destruction of the myelin sheaths. Thus, the anisotropy is slightly reduced when compared to normal subjects, but increased when compared to high-grade gliomas [89]. The main fiber tracts can remain identifiable on orientation maps and also on the tractography (Figure 25.25).

Metastatic lesions are surrounded by abnormal T2WI that most likely consist of vasogenic edema. The edematous areas have reduced FA values. This fact is typically explained by an increase in water content rather than by destruction or invasion of nerve fibers to be consistent with neuropathologic findings (Figure 25.26). DTI has not been reported to help in differentiation of apparently normal white matter from edematous brain and enhancing peritumoral margins [90]. The drop in FA values of the area

infiltrated by cell tumors is lower than that in the peritumoral edema [2,13,74]. DTI can distinguish edematous areas with intact fibers – mostly found in metastases – from disrupted fibers – mostly found in high-grade gliomas [91].

Intracranial neoplasms may involve both the functional cortex and the corresponding white matter tracts. The preoperative identification of eloquent areas through non-invasive methods, such as blood oxygen level dependent (BOLD) functional MR imaging (fMRI) and DTI tractography, offers some advantages; not only can it reduce the time of surgery in some instances, but it may also minimize some intraoperative cortical stimulation methods, such as the identification of language cortex [91]. Until recently, preoperative and perioperative methods to evaluate brain function of patients with brain tumors were restricted to cortical activation studies. Increasingly, however, investigators are beginning to combine fMRI with DTI. The attraction is that fMRI can be an accurate and non-invasive method used to map functional cerebral cortex, identifying eloquent areas in the cortex and displaying their relationship to the lesion [92], whereas DTI may be able to identify accurately the main fiber tracts to be avoided during surgery so as to safely guide the tumor resection [2]. Consequently,

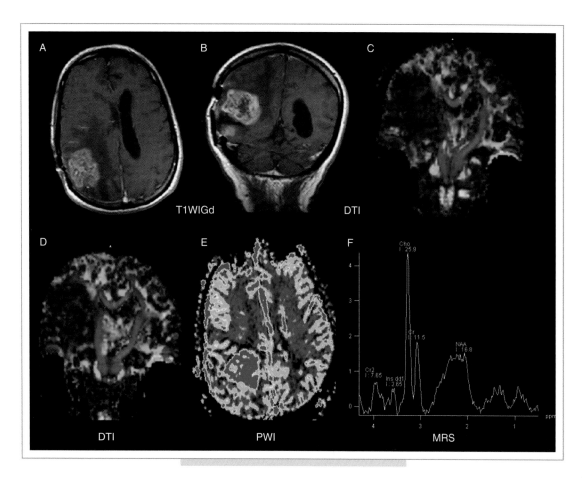

FIG. 25.23. **(A,B)** An expansive and infiltrating lesion in a 73-year-old man with left hemiparesis and seizures, with the diagnosis of glioblastoma multiforme. The lesion has hyperperfusion **(C)**, markedly elevated choline and lactate/lipids peaks and a low N-acetyl-aspartate peak **(D)**. **(E,F)** Coronal diffusion tensor imaging-fractional anisotropy color-coded maps show that the lesion dislocates and infiltrates the corticospinal tract and the superior longitudinal fasciculus. There is also distortion of the corpus callosum.

the combination of DTI tractography and fMRI might allow us to map precisely an entire functional circuit (Figure 25.27) [93]. Even though fMRI locates eloquent cortical areas, determination of the course and integrity of the fiber tracts remains essential to the surgical planning [91,94]. This identification of the fiber tracts can facilitate reaching a decision regarding the likelihood of an operation [2]. As a result, neurosurgeons may have more information to inform the choice of surgical approach to be taken. This better evaluation of risks by neurosurgeons is possible if they can know the spatial relation between the tumor and major fiber tracts [41] and thereby avoid postoperative neurological deficit [3,94]. However, this remains to be proven in randomized trials.

Many investigators hope that the combined use of fMRI and DTI tractography might define the structural basis of functional connectivity in normal and pathological brains [95] and allow for improved intraoperative guidance. Since fMRI is, in many instances, able to depict the exact location of the motor cortex, it should be possible to delineate the corticospinal tract (CST) by DTI tractography.

Modern neurosurgical navigation systems typically provide a probe-guided intraoperative MR display of the brain [96]. Such systems are already widely used and, in some cases, able to combine the information of fMRI [91,96] with that of DTI tractography [93,97] or even of both together [88,92].

INTRAOPERATIVE DIFFUSION IMAGING

The extent of resection of a brain tumor appears to be a prognostic factor for survival and time to recurrence [98,99]. Therefore, the ability to determine intraoperatively whether the resection is incomplete should be highly advantageous [100]. Intraoperative MR imaging (iMRI) has been used to guide brain tumor resection [101]. With the advent of DTI, iMRI can be used to assist in the preservation of the main fiber tracts; this is desirable to avoid postoperative neurological deficits.

The rationale for iMRI is straightforward: surgical manipulations and maneuvers alter the anatomic position of brain structures and the tumor [102], and morphological changes of the brain may also occur between the time of

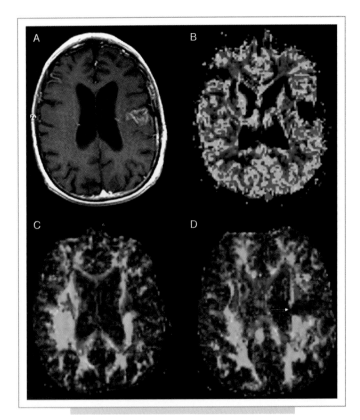

FIG. 25.24. A 56-year-old man with an anaplastic astrocytoma, presented with right hemiparesis. The contrast-enhanced T1-weighted image **(A)** shows a left frontal lesion that has hyperperfusion on the relative cerebral blood volume map **(B)** – note the black signal caused by excessive enhancement with resulting T1 effect. **(C,D)** The axial diffusion-tensor imaging-fractional anisotropy color-coded maps demonstrate disruption of the left corona radiata (arrow).

the preoperative MRI exams and the time of the surgery [103]. Therefore, the exact location of brain tumors based on preoperative exams may not be the same (i.e. brain shift) [104]. Intraoperative MRI has been proposed as a possible way to enable neurosurgeons to optimize their surgical approaches by avoiding critical structures and the adjacent normal brain parenchyma [102]. Some reports suggest that in 65–92% of the cases in which neurosurgeons are believed to have performed a complete and thorough tumor resection, iMRI still demonstrates tissue to be resected [101,105], because such lesions can be difficult to differentiate from the normal brain parenchyma visually.

In like manner, many investigators have suggested that DTI performed intraoperatively could add information regarding the integrity of main fiber tracts, which again may have shifted during the procedure. Findings at iMRI led to modification of surgical procedure in almost 30% of patients in a recent report [104], with investigators reporting that intraoperative DTI is able to depict changes in fiber orientation secondary to craniotomy or burr hole procedures [100].

Intraoperative MRI can have the full range of MRI capabilities, including functional sequences such as fMRI [101] and diffusion-weighted imaging (e.g. screening for ischemia) [93]. For example, one report described how intraoperative diffusion imaging was performed during neurosurgery for the resection of a tumor using an interventional MRI system [93]. Intraoperative development of hyperacute cerebral ischemia had been previously detected in two patients and this was confirmed later by a follow-up MR exam. DTI, together with a neuronavigation system, was performed in a third patient as an integral part of an image-guided tumor resection. After processing the DTI data, DTI tractography was performed. The relation of the tumor to the anatomy of the white matter fiber tracts adjacent to it was clearly and plainly demonstrated in a case of oligodendroglioma. The fiber tracts were displaced, without being infiltrated or disrupted by the tumor. The complete tumor resection was performed without any postoperative neurological deficit.

Registration of the relation between brain tumor lesion and main fiber tracts in the neuronavigator system facilitates the preservation of these fibers. For this purpose, to compensate the brain shift, updating the neuronavigating information with intraoperative data should be necessary [100].

Although anecdotal, such reports suggest that intraoperative diffusion imaging may provide important clinical information, adding substantially to the intraoperative information available about the pathologic state of the brain parenchyma and the structure of white matter.

POST-TREATMENT EVALUATION

Assessment of treatment response typically relies on the assessment of contrast enhancement on subsequent imaging studies within 24 hours after the surgical procedure [106] and then later by the evaluation of tumor size weeks to months after conclusion of therapy [107]. The appearance of a new enhancing area often results in management alterations, frequently leading to adjuvant therapy. A recent report has observed the benefits of performing DWI in immediate postoperative MR examinations [106]. Areas of restricted diffusion were described adjacent to low- or high-grade glioma tumor resection cavity. Follow-up MR examinations revealed that these areas of restricted diffusion resolved and that contrast enhancing appears in the corresponding location. This enhancement subsequently regressed to form an area of encephalomalacia. Because conventional MR examination at the time of the enhancement period is easily misdiagnosed as tumor recurrence or tumor progression, such findings can lead to erroneous interpretation of treatment failure and the initiation of a new adjuvant therapy. Investigators have concluded that a corresponding area of restricted diffusion almost always precedes the delayed contrast enhancement described, which invariably evolves into encephalomalacia or gliotic

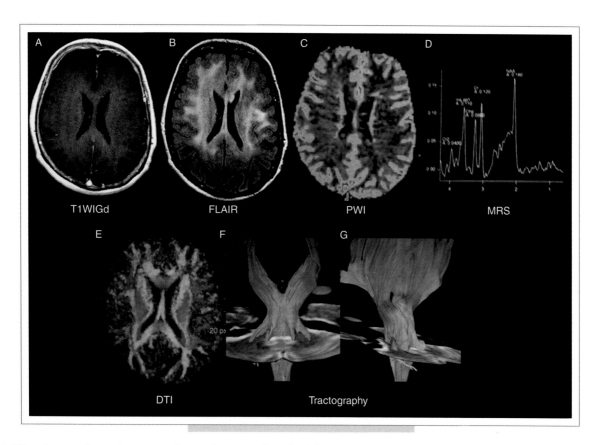

FIG. 25.25. An invading white matter lesion that extends to both hemispheres, involving the fronto-parietal lobes through the corpus callosum in a 68-year-old woman who presented with cognitive impairment. The diagnosis of gliomatosis cerebri was made after a biopsy. The lesion does not enhance on the post-contrast T1-weighted image **(A)**, has hyperintense signal on the FLAIR image **(B)** and does not have hyperperfusion on the relative cerebral blood volume map **(C)**. Magnetic resonance spectroscopy shows a high myoinositol peak, a moderately high choline peak and a subtle reduction on the N-acetyl-aspartate peak **(D)**. The diffusion-tensor images-fractional anisotropy color-coded map **(E)** and tractography **(F,G)** demonstrate that the main fiber tracts are preserved. This is probably explained by the fact that gliomatosis cerebri is a diffusely invading lesion that preserves the normal underlying cytoarchitectural pattern because it does not destroy the nerve fibers.

cavity on long-term follow-up studies. While this needs to be reproduced in other studies, one can speculate that reduced diffusibility in and around the surgery bed may represent areas of infarct, ischemia or even venous congestion, secondarily to acute cellular damage, such as surgical trauma, retraction and vascularity damage, and tumor devascularization [66]. Such findings are unlikely to represent early recurrence (Figure 25.28).

Diffusion-weighted imaging might also be a marker for the response to therapy because changes in tumor water diffusibility may occur secondarily to changes in cell density, a method referred to as 'functional diffusion imaging'. In this methodology, ADC values are measured over time and compared on a coregistered voxel-by-voxel basis. In this setting, ADC appears to be a sensitive and early predictor of therapeutic efficacy [108]. Specifically, investigators prospectively compared tumor diffusion values at 3 weeks after initiation of therapy with pretreatment images so as to measure ADC changes induced by therapy [109]. If

this methodology holds up in multicenter trials, it would allow a lack of change in ADC values in the tumor to indicate a failure in therapy. This would in turn provide an opportunity to switch to a more beneficial therapy, minimizing the morbidity associated with a prolonged and inefficient treatment. The logic behind this methodology is that successful treatment will result in extensive cell damage, leading to a reduction in cell density. The neoplasm cell loss results in an increase in extracellular space that can raise free water molecule diffusibility. As increases in brain tumor ADC correspond to decreases in tumor volume in long-term follow-up studies, DWI may, therefore, be an important surrogate marker for quantification of treatment response (Figure 25.29).

DTI may also play a role in the management of patients undergoing radiation therapy and chemotherapy. By adding information about the location of white matter tracts, DTI tractography might be successfully utilized alongside fMRI for radiosurgery planning. In theory, this should allow a reduction of the dose applied, as well as of the volume of normal brain irradiated with a high dose, hopefully reducing necrosis [14]. DTI may also help in the early detection of white matter injuries caused by chemotherapy and radiation therapy. A report showed a correlation between the reductions of FA values, young age at treatment, an increased interval since the beginning of treatment and poor intellectual outcome in patients with medulloblastoma [110]. The possibility of using FA or other DTI changes as a biomarker for neurotoxicity is enticing.

POTENTIAL FUTURE APPLICATIONS

While diffusion-weighted imaging has been recently used to assess brain tumor diagnoses and treatments, to help in the preoperative planning and to guide surgery intraoperatively, there are still a number of new areas and possibly future applications. These include therapy monitoring (such as the functional diffusion method described above), monitoring radiation or other treatments via quantitative assessment of white matter changes as a marker for toxicity, plasticity after surgery, radiation, or other treatment, and so on. However, many of these techniques will require improved image quality. Specifically, diffusion-weighted imaging is an echoplanar sequence that suffers from a low signal-to-noise ratio (SNR) due to the need for rapid acquisition approaches. The types of quantification of diffusibility or anisotropy that might allow these advanced applications are very sensitive to inadequate SNR. Mean diffusibility and fractional anisotropy values may vary

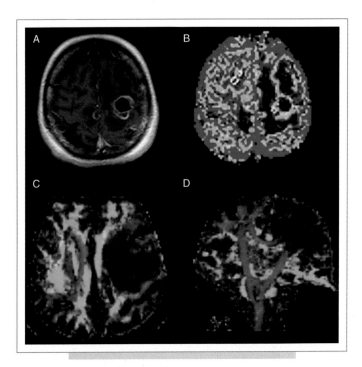

FIG. 25.26. A 50-year-old woman with new onset of seizures and history of breast cancer. A round rim-enhancing with a necrotic center **(A)** and hyperperfusion **(B)** lesion is surrounded by peritumoral edema, consisting of breast cancer metastasis. Axial **(C)** and coronal **(D)** diffusion-tensor images-fractional anisotropy maps show the edematous changes in the FA values. Thus, it is difficult to identify the main fiber tracts within the vasogenic edema. This does not necessarily mean that these fibers are infiltrated or disrupted, however.

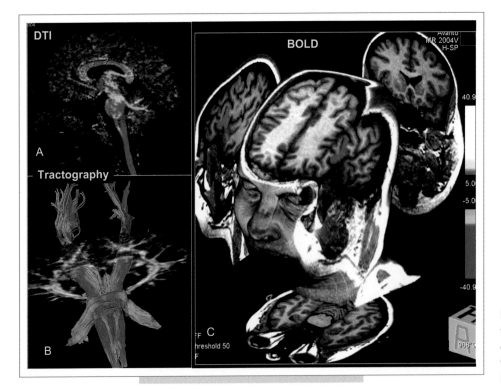

FIG. 25.27. The whole brain functional circuit demonstrated with the diffusion-tensor MR-image **(A)** and tractography **(B)** together with the BOLD sequence **(C)**.

with different SNR. With the appropriate SNR, and if more powerful gradients could be used, a smaller voxel could be obtained in a clinically acceptable measurement time. The voxel size typically used in diffusion images is big enough to cause partial-volume effect and this makes the correct identification of the intersection fibers difficult within the same voxel. A smaller voxel size may be helpful in solving

the problem of crossing fibers and could be useful to deline-ate small white matter structures. Some have demonstrated a better SNR in a 3.0T scanner. However, some challenges remain including geometric distortions [111]. More impor-tantly for the future, however, will be the move beyond the tensor representation of diffusion. This model fails to account for cases of intersection and dispersion of fibers

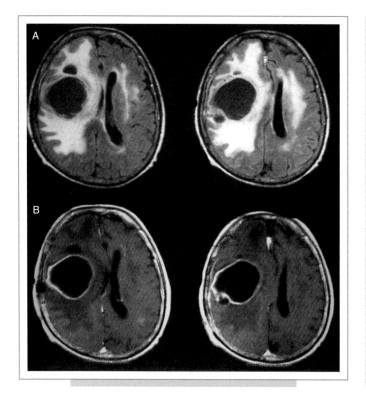

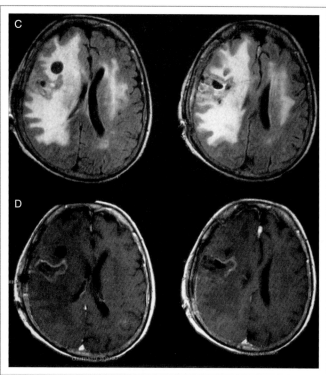

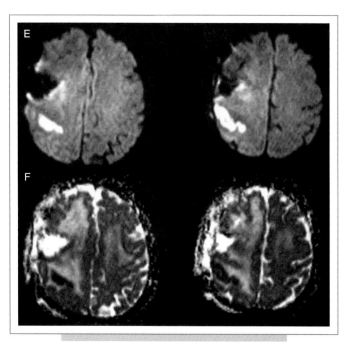

FIG. 25.28. A patient presenting with an expansive round rim-enhancing lesion on the post-contrast T1-weighted images **(A)**, with necrotic center and surrounded by hyperintense area on FLAIR images **(B)**, that may consist of neoplasic infiltration and/or vasogenic edema. The diagnosis of glioblastoma mul-tiforme was done after brain biopsy. The follow-up exam, after surgery, shows a partial resection of the lesion **(C,D)**. The diffu-sion-weighted images **(E)** and the apparent diffusion coefficient maps **(F)** show an area of restricted diffusion just adjacent to the surgery bed, consisting of acute infarct.

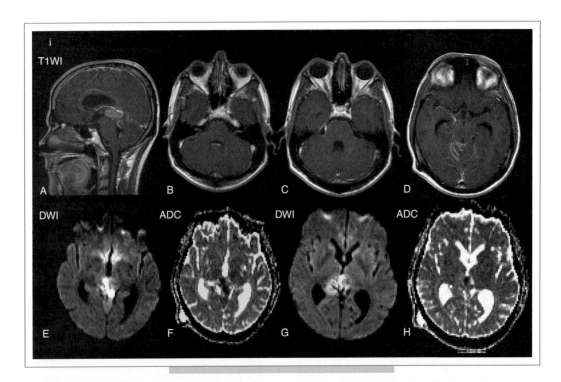

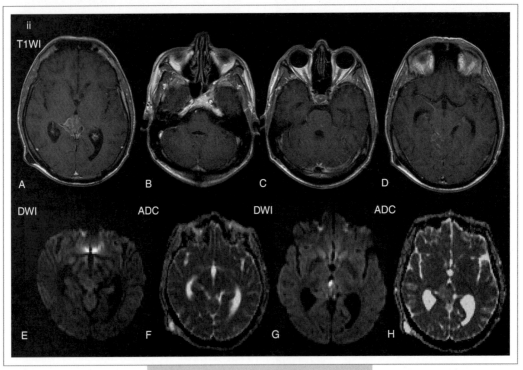

FIG. 25.29. A 43-year-old man with a biopsy-proven diagnosis of pinealoblastoma. The post-contrast T1-weighted images **(iA–D)** show an expansive enhancing round lesion in the pineal. There is also leptomeningeal dissemination to the cranial nerves and the cerebellum sulcus. The lesion also has areas of restricted diffusion **(iE–H)**. The follow-up exam after chemotherapy and radiotherapy shows that the lesion decreased in size, there is no more leptomeningeal enhancement on the post-contrast T1-weighted images **(iiA–D)**. In the axial diffusion-weighted images there are no more restricted diffusion areas within the lesion. The apparent diffusion coefficient values are higher, this might suggest a positive response to therapy **(iiE–H)**.

in a same voxel [112]. Because insertion of fibers is the key event in linking white matter to gray matter, and because so much of the brain is composed of regions of intersecting fibers, it seems highly likely that a method that goes beyond the tensor, to so-called 'supertensor' representations will become useful once the technology allows these techniques to be practical. One such approach is called diffusion spectrum imaging (DSI); other variants of this same approach go by other acronyms (e.g. high angular resolution diffusion imaging, or HARDI) but the concept is the same: to move beyond the limitations of the tensor model. With the increasing popularity of DTI, its limitations will become familiar to more investigators and these more advanced methods will likely become the dominant approach for investigating the brain. The better understanding and assessment of both large and small fibers will almost certainly help in the evaluation of the relationship between brain neoplasm lesions and the underlying normal anatomy.

LIMITATIONS

Although initial reports suggest advantages of DWI and DTI in the evaluation of patients with brain tumors, these reports are largely single-center, uncontrolled, preliminary findings. Therefore, these results must be cautiously interpreted. Multicenter studies are sorely needed. There are difficulties in designing prospective studies that assess certain methods such as intraoperative MRI, where blinding is difficult and bias inevitably is present.

As suggested above, there remain substantial technical hurdles, even though the rapid evolution of MRI systems is making ever more powerful approaches possible. Improvements are particularly welcome given the limited signal-to-noise ratio of diffusion overall. For example, the limited spatial resolution of EPI approaches may lead to reduced sensitivity. The method herein assessed is only capable of depicting the prominent fiber tracts [13,112] and more advanced approaches (e.g. DSI) may be much more useful in the future. Susceptibility artifacts can cause image distortion that prevents DTI data from being accurately analyzed [13] and numerous other technical challenges remain.

SUMMARY

Diffusion imaging seems to offer the possibility to add important information to presurgical planning. Although experience is limited, DTI seems to provide useful local information about the structures near the tumor and this appears to be useful in planning. In the future, DTI may provide an improved way to monitor intraoperative surgical procedures as well as their complications. Furthermore, evaluation of the response to treatment with chemotherapy and radiation therapy might also be possible. Although diffusion imaging has some limitations, it is an area of active investigation and implementation in clinical practice.

REFERENCES

1. Knisely JPS, Rockwell S (2002). Importance of hypoxia in the biology and treatment of brain tumors. Neuroimag Clin N Am 12:525–536.
2. Landis SH, Murray T, Bolden S (1998). Cancer statistics, 1998. CA Cancer J Clin 48:6–29.
3. Berens ME, Rutka JT, Rosenblum ML (1990). Brain tumor epidemiology, growth, and invasion. Neurosurg Clin North Am 1:1–18.
4. Jemal A, Thomas A, Murray T et al (2002). Cancer statistics 2002. CA 52(1):23–47.
5. Koeller KK, Henry JM (2001). Superficial gliomas: radiologic-pathologic correlation. Radiographics 21:1533–1556.
6. Hunt D, Treasure P (1998). Year of life lost due to cancer in East Anglia 1990–1994. East Anglian Cancer Intelligence Unit, Institute of Public Health, Cambridge.
7. Brain MRC (1990). Tumor Working Party. Prognostic factor for high-grade malignant glioma: development of a prognostic index. A report of the Medical Research Council Brain Tumor Working Party. J Neuro-Oncol 9:47–55.
8. Felix R, Schorner W, Laniado M et al (1985). Brain tumors: MR imaging with gadolin-DPTA. Radiology 156:681–688.
9. Tovi M (1993). MR imaging in cerebral gliomas analysis of tumor tissue components. Acta Radiol Suppl 384:1–24.
10. Knopp EA, Cha S, Johnson G et al (1999). Glial neoplams: dynamic contrast-enhanced T2*-weighted MR imaging. Radiology 211:791–798.

11. Sorensen AG (2006). Magnetic resonance as a cancer imaging biomarker. J Clin Oncol 24:3274–3281.
12. Brunberg JA, Chenevert TL, McKeever PE et al (1995). In vivo MR determination of water diffusion coefficients and diffusion anisotropy: correlation with structural alteration in gliomas of the cerebral hemispheres. Am J Neuroradiol 16:361–371.
13. Mori S, Frederiksen K, Van Zijl PCM et al (2002). Brain white matter anatomy of tumor patients evaluated with diffusion tensor imaging. Ann Neurol 51:377–380.
14. Price SJ, Burnet NG, Donovan T et al (2003). Diffusion tensor imaging of brain tumors at 3T: a potential tool for assessing white matter tract invasion. Clin Radiol 58:455–462.
15. Lu S, Ahn D, Johnson G, Cha S (2003). Peritumoral diffusion tensor imaging of high-grade gliomas and metastatic brain tumors. Am J Neuroradiol 24:937–941.
16. Maldjian JA, Schulder M, Liu WC et al (1997). Intraoperative functional MRI using a real-time neurosurgical navigation system. J Comput Assist Tomogr 21:910–912.
17. Schulder M, Maldjian JA, Liu WC et al (1998). Functional image-guided surgery of intracranial tumors located in or near the sensorimotor cortex. J Neurosurg 89:412–418.
18. Laundre BJ, Jellinson BJ, Badie B et al (2005). Diffusion tensor imaging of the corticospinal tract before and after mass resection as correlated with clinical motor findings: preliminary data. Am J Neuroradiol 26:791–796.

19. Holodny AI, Ollenschlager M (2002). Diffusion imaging in brain tumor. Neuroimag Clin N Am 12:107–124.

20. Stejskal E, Tanner J (1965). Spin diffusion measurements: spin echos in the presence of time-dependent field gradient. J Chem Phys 42:288–292.

21. Romero JM, Schaefer PW, Grant PE, Becerra L, Gonzáles RG (2002). Diffusion MR imaging of acute ischemic stroke. Neuroimag Clin N Am 12:35–53.

22. Barboriak DP (2003). Imaging of brain tumors with diffusion-weighted and diffusion tensor MR imaging. Magn Reson Imaging Clin N Am 11:379–401.

23. Maier SE, Gudbjartsson H, Patz SL et al (1998). Line scan diffusion imaging: characterization in healthy subjects and stroke patients. Am J Roentgenol 171:85–93.

24. Warach S, Chien D, Li W, Ronthal M, Edelman RR (1992). Fast magnetic resonance diffusion-weighted imaging of acute human stroke. Neurology 42:1717–1723.

25. Sunshine JL, Tarr RW, Lanzieri CF, Landis DM, Selman WR, Lewin JS (1999). Hyperacute stroke: ultrafast MR imaging to triage patients prior to therapy. Radiology 212:325–332.

26. Beauchamp NJ, Uluğ AM, Passe TJ, van Zijl PC (1998). MR diffusion imaging in stroke: review and controversies. RadioGraphics 18:1269–1283.

27. Schaefer PW (2001). Applications of DWI in clinical neurology. J Neurol Sci 186 (Suppl 1):S25–35.

28. Larsson HBW, Thomsen C, Frederiksen J, Stubgaard M, Henriksen O (1993). In vivo magnetic resonance measurement in the brain of patients with multiple sclerosis. Magn Reson Imaging 10:712.

29. Horsfield MA, Lai M, Webb SL et al (1996). Apparent diffusion coefficients in benign and secondary progressive multiple sclerosis by nuclear magnetic resonance. Magn Reson Med 36:393–400.

30. Castriota-Scanderberg A, Tomaiuolo F, Sabatini U, Nocentini U, Grasso MG, Caltagirone C (2000). Demyelinating plaques in relapsing-remitting and secondary proggressive multiple sclerosis: assessment with diffusion MR imaging. Am J Neuroradiol 21:862–868.

31. Tsuchiya K, Hachiya J, Maehara T (1999). Diffusion-weighted MR imaging in multiple sclerosis. Eur J Radiol 31(3):165–169.

32. Tsuchiya K, Katase S, Yoshino A, Hachiya J (1999). Diffusion-weighted MR imaging of encephalitis. Am J Roentegenol 173:1097–1099.

33. Na DL, Suh CK, Choi SH et al (1999). Diffusion-weighted magnetic resonance imaging in probable Creutzfeldt-Jakob disease. Arch Neurol 56:951–957.

34. Chenevert TL, Brunberg JA, Pipe JG (1990). Anisotropic diffusion in human white matter: demonstration with MR techniques in vivo. Radiology 177:401–405.

35. Moseley ME, Cohen Y, Kucharczyk W, Mintorovitch J, Asgari HS, Wendland MF (1990). Diffusion-weighted MR imaging of anisotropic water diffusion in cat central nervous system. Radiology 176:439–445.

36. Melhem ER, Mori S, Mukundan G, Kraut MA, Pomper MG, Van Zijl PCM (2002). Diffusion tensor MR imaging of the brain and white matter tractography. Am J Roentgenol 178(1):3–16.

37. Ito R, Mori S, Melhem ER (2002). Diffusion tensor MR imaging and tractography. Neuroimag Clin N Am 12(1):1–19.

38. Pierpaoli C, Jezzard P, Basser P, Barnett A, Di Chiro G (1996). Diffusion tensor MR imaging of the human brain. Radiology 201:637–648.

39. Basser PJ, Mattiello J, LeBihan D (1994). Estimation of the effective self-diffusion tensor from the NMR spin echo. J Magn Reson B 103:247–254.

40. Basser PJ, Pajevic S, Pierpaoli C, Duda J, Aldroubi A (2000). In vivo fiber tractography using DT-MRI data. Magn Reson Med 44:625–632.

41. Mamata H, Mamata Y, Westin CF et al (2002). High-resolution line scan diffusion tensor MR imaging of white matter fiber tract anatomy. Am J Neuroradiol 23:67–75.

42. Douek P, Turner R, Pekar J, Patronas N, Le Bihan D (1991). MR color mapping of myelin fiber orientation. J Comput Assist Tomogr 15(6):923–929.

43. Pajevic S, Pierpaoli C (1999). Color schemes to represent the orientation of anisotropic tissues from diffusion tensor data: application to white matter fiber tract mapping in the human brain. Magn Reson Med 42:526–540.

44. Tsuruda JS, Chew WM, Moseley ME, Norman D (1990). Diffusion-weighted MR imaging of the brain: value of differentiating between extra-axial cysts and epidermoid tumors. Am J Neuroradiol 155:1049–1065.

45. Chen S, Ikawa F, Kurisu K, Arita K, Takaba J, Kanou Y (2001). Quantitative MR evaluation of intracranial epidermoid tumors by fast fluid-attenuated inversion recovery imaging and echo-planar diffusion-weighted imaging. Am J Neuroradiol 22(6):1089–1096.

46. Tsuruda JS, Chew WM, Moseley ME et al (1999). Diffusion-weighted MR imaging of extraaxial tumors. Magn Reson Med 9:352–361.

47. Laing AD, Mitchell PJ, Wallace D (1999). Diffusion-weighted magnetic resonance imaging of intracranial epidermoid tumors. Aust Radiol 43:16–19.

48. Kono K, Inoue Y, Nakayama K et al (2001). The role of diffusion-weighted imaging in patients with brain tumors. Am J Neuroradiol 22:1081–1088.

49. Filippi CG, Edgar MA, Ulug AM et al (2001). Appearance of meningiomas on diffusion-weighted images: correlation diffusion constants with histopathologic findings. Am J Neuroradiol 22:65–72.

50. Guo AC, Provenzale JM, Cruz Jr LCH, Petrella JR (2001). Cerebral abscesses: investigation using apparent diffusion coefficient maps. Neuroradiology 43:370–374, 2001.

51. Bergui M, Zhong J, Bradac GB, Sales S (2001). Diffusion-weighted images of intracranial cyst-like lesions. Neuroradiology 439(10):824–829.

52. Chang SC, Lai PH, Chen WL, Weng HH, Ho JT, Wang JS (2002). Diffusion-weighted MRI features of brain abscess and cystic or necrotic brain tumors: comparison with conventional MRI. Clin Imaging 26(4):227–236.

53. Chan JH, Tsui EY, Chau LF et al (2002). Discrimination of an infected brain tumor from a cerebral abscess by combined MR perfusion and diffusion imaging. Comput Med Imaging Graph 26(1):19–23.

54. Ebisu T, Tanaka C, Umeda M et al (1996). Discrimination of brain abscess from necrotic or cystic tumors by diffusion-weighted echo planar imaging. Reson Imaging 14(9):1113–1116.

55. Hartmann M, Jansen O, Heiland S, Sommer C, Munkel K, Sartor K (2001). Restricted diffusion within ring enhancement is not pathognomonic for brain abscess. Am J Neuroradiol 22(9):1738–1742.

56. Tien RD, Feldesberg GJ, Friedman H, Brown M, MacFall J (1994). MR imaging of high-grade cerebral gliomas: value of diffusion-weighted echo-planar pulse sequence. Am J Roentgenol 162:671–677.

57. Eis M, Els T, Hoehn-Berlage M, Hossman KA (1994). Quantitative diffusion MR imaging of cerebral tumor and edema. Acta Neuro-chir Suppl (Wien) 60:344–346.

58. Krabbe K, Gideon P, Wang P, Hansen U, Thomsen C, Madsen F (1997). MR diffusion imaging of human intracranial tumours. Neuroradiology 39:483–489.

59. Le Bihan D, Douek P, Argyropoulou M, Turner R, Patronas N, Fulham M (1993). Diffusion and perfusion magnetic resonance imaging in brain tumors. Top Magn Reson Imaging 5:25–31.

60. Tsuruda JS, Chew WM, Moseley ME, Norman D (1991). Diffusion-weighted MR imaging of extraaxial tumors. Magn Reson Med 19:316–320.

61. Yanaka K, Shirai S, Kimura H, Kamezaki T, Matsumura A, Nose T (1995). Clinical application of diffusion-weighted magnetic resonance imaging to intracranial disorders. Neurol Med Chir 16:361–371.

62. Cruz, Jr LCH, Sorensen GS (2005). Diffusion tensor magnetic resonance imaging of brain tumors. Neurosurg Clin N Am 16:115–134.

63. Cruz, Jr LCH, Sorensen GS (2006). Diffusion tensor magnetic resonance imaging of brain tumors. Magn Reson Imaging Clin N Am 14(2):183–202.

64. Stadnik TW, Chaskis C, Michotte A et al (2001). Diffusion-weighted MR imaging of intracerebral masses: comparison with conventional MR imaging and histologic findings. Am J Neuroradiol 22:969–976.

65. Sugahara T, Korogi Y, Kochi M et al (1999). Usefulness of diffusion-weighted MRI with echo-planar technique in the evaluation of cellularity in gliomas. J Magn Reson Imaging 9:53–60.

66. Cha S (2006). Update on brain tumor imaging: from anatomy to physiology. Am J Neuroradiol 27:475–487.

67. Noguchi K, Watanabe N, Nagayoshi T et al (1999). Role of diffusion-weighted echo-planar MRI in distinguishing between brain abscess and tumor: a preliminary report. Neuroradiology 41:171–174.

68. Sadeghi N, Camby I, Goldman S et al (2003). Effect of hydrophilic components of the extracellular matrix on quantifiable diffusion-weighted imaging of human glioma: preliminary results of correlating apparent coefficient diffusion values and hyaluronan expression level. Am J Roentgenol 181(1):235–241.

69. Guo AC, Cummings TJ, Dash RC, Provenzale JM (2002). Lymphomas and high-grade astrocytomas: comparison of water diffusibility and histologic characteristics. Radiology 224(1):177–183.

70. Camacho DLA, Smith JK, Castillo M (2003). Differentiation of toxoplasmosis and lymphoma in AIDS patients by using apparent diffusion coefficients. Am J Neuroradiol 24(4):633–637.

71. Gauvain KM, McKinstry RC, Mukherjee P et al (2001). Evaluating pediatric brain tumor cellularity with diffusion-tensor imaging. Am J Roentgenol 177:449–454.

72. Koetsenas AL, Roth TC, Manness WK, Faeber EN (1999). Abnormal diffusion-weighted MRI in medulloblastoma: Does it reflect small cell histology? Pediatr Radiol 29:524–526.

73. Quadrery FA, Okamoto K (2003). Diffusion-weighted MRI of haemangioblastomas and other cerebellar tumours. Neuroradiology 45(4):212–219.

74. Witwer BP, Moftakhar R, Hasan KM et al (2002). Diffusion-tensor imaging of white matter tracts in patients with cerebral neoplasm. J Neurosurg 97:568–575.

75. Goebell E, Paustenbach S, Vaeterlein O et al (2006). Low-grade and anaplastic gliomas: differences in architecture evaluated with diffusion-tensor MR imaging. Radiology 239:217–222.

76. Inoue T, Ogasawara K, Beppu T, Ogawa A, Kabasawa H (2005). Diffusion tensor imaging for preoperative evaluation of tumor grade in gliomas. Clin Neurol Neurosurg 107:174–180.

77. Weishmann UC, Symms MR, Parker GJM et al (2000). Diffusion tensor imaging demonstrates deviation of fibers in normal appearing white matter adjacent to a brain tumour. J Neurol Neurosurg Psychiatr 68:501–503.

78. Jellinson BJ, Field AS, Medow J et al (2004). Diffusion tensor imaging of cerebral white matter: a pictorial review of physics, fiber tract anatomy, and tumor imaging patterns. Am J Neuroradiol 23:356–369.

79. Yoshiura T, Wu O, Zaheer A, Reese TG, Sorensen AG (2001). Highly diffusion-sensitized MRI of brain: dissociation of gray and white matter. Magn Reson Med 45:734–740.

80. Burger PC, Heinz ER, Shibata T, Kleihues P (1988). Topographic anatomy and CT correlations in the untreated glioblastoma multiforme. J Neurosurg 68:698–704.

81. Johnson PC, Hunt SJ, Drayer BP (1989). Human cerebral gliomas: correlation of postmortem MR imaging and neuropathologic findings. Radiology 170:211–217.

82. Castillo M, Smith JK, Kwock L, Wilber K (2001). Apparent diffusion coefficients in the evaluation of high-grade cerebral gliomas. J Neuroradiol 22(1):60–64.

83. Giese A, Westphal M (1996). Glioma invasion in the central nervous system. Neurosurgery 39:235–252.

84. Papadopoulos MC, Saadoun S, Davies DC, Bell BA (2001). Emerging molecular mechanisms of brain tumour oedema. Br J Neurosurg 15:101–108.

85. Yamamoto M, Ueno Y, Hayashi S, Fukushima T (2002). The role of proteolysis in tumor invasiveness in glioblastoma multiforme and metastatic brain tumors. Anticancer Res 22:4265–4268.

86. Tsuchiya K, Fujikawa A, Nakajima M, Honya K (2005). Differentiation between solitary metastasis and high-grade gliomas by diffusion tensor imaging. Br J Radiol 78:533–537.

87. Holodny AI, Schwartz TH, Ollenschleger M, Liu WC, Schulder M (2001). Tumor involvement of the corticospinal tract: diffusion magnetic resonance tractography with intraoperative correlation. J Neurosurg 95(6):1082.

88. Holodny AI, Ollenschleger M, Liu WC, Schulder M, Kalnin AJ (2001). Identification of the corticospinal tracts achieved using blood-oxygen-level-dependent and diffusion functional MR imaging in patients with brain tumors. Am J Neuroradiol 22:83–88.

89. Akai H, Mori H, Aoki S et al (2005). Diffusion tensor tractography of gliomatosis cerebri: fiber tracking through the tumor. J Comput Assist Tomogr 29(1):127–129.

90. Sha S, Bastin ME, Whittle IR, Wardlaw JM (2002). Diffusion tensor MR imaging of high-grade cerebral gliomas. Am J Neuroradiol 23:520–527.

91. Tummala RP, Chu RM, Liu H, Truwit C, Hall WA (2003). Application of diffusion-tensor imaging to magnetic-resonance-guided brain tumor resection. Pediatr Neurosurg 39:39–43.

92. Schulder M, Maldjian JA, Liu W-C et al (1998). Functional image-guided survey of intracranial tumors located in or near the sensorimotor cortex. J Neurosurg 89:412–418.

93. Krings T, Reiges MH, Thiex R, Gilsbach JM, Thron A (2001). Functional and diffusion-weighted magnetic resonance images of space-occupying lesions affecting the motor system: imaging the motor cortex and pyramidal tracts. J Neurosurg 95(5):816–824.

94. Mamata Y, Mamata H, Nabavi A et al (2001). Intraoperative diffusion imaging on a 0.5 Tesla interventional scanner. J Magn Reson Imaging 13:115–119.

95. Guye M, Parker GJM, Symms M, Boulby P, Wheeler-Kingshott CAM, Salek-Haddadi A et al (2003). Combined functional MRI and tractography to demonstrate the connectivity of the human primary motor cortex in vivo. Neuroimage 19:1349–1360.

96. Maldjian JA, Schulder M, Liu W-C et al (1997). Intraoperative functional MRI using a real-time neurosurgical navigation system. J Comput Assist Tomogr 21(6):910–912.

97. Krings T, Coenen VA, Axer H et al (2001). In vivo 3D visualization of normal pyramidal tracts in human subjects using diffusion-weighted magnetic resonance imaging and a neuronavigator system. Neurosci Lett 307:192–196.

98. Keles GE, Chang EF, Lamborn KR et al (2006). Volumetric extent of resection and residual contrast enhancement on initial surgery as predictors of outcome in adult patients with hemispheric anaplastic astrocytoma. J Neurosurg 105(1):34–40.

99. Keles GE, Anderson B, Berger MS (1999). The effect of extent of resection on time to tumor progression and survival in patients with glioblastoma multiforme of the cerebral hemisphere. Surg Neurol 52(4):371–379.

100. Nimsky C, Ganslandt O, Hastreiter P et al (2005). Intraoperative diffusion-tensor MR imaging: shifting of white matter tracts during neurosurgical procedures – initial experience. Radiology 234:218–225.

101. Bradley WG (2002). Achieving gross total resection of brain tumors: intraoperative MR imaging can make a big difference (editorial). Am J Neuroradiol 23:348–349.

102. Jolesz FA, Talos I-F, Schwartz RB et al (2002). Intraoperative magnetic resonance imaging and magnetic resonance imaging-guided therapy of brain tumors. Neuroimag Clin N Am 12:665–683.

103. Martin AJ, Hall WA, Liu H et al (2000). Brain tumor resection: intraoperative monitoring with high-field-strength MR imaging-initial results. Radiology 215:222–228.

104. Nimsky C, Ganslandt O, Keller BV, Romstock J, Fahlbush R (2004). Intraoperative high-field-strength MR imaging: implement and experience in 200 patients. Radiology 233:67–78.

105. Albert FK, Forsting M, Sartor K, Adams HP, Kunze S (1994). Early post-operative magnetic resonance imaging after resection of malignant glioma: objective evaluation of residual tumor growth and its influence on regrowth and prognosis. Neurosurgery 34:45–61.

106. Smith JS, Cha S, Catherine M et al (2005). Serial diffusion-weighted magnetic resonance imaging in case of glioma: distinguishing tumor recurrence from postresection injury. J Neurosurg 103:428–438.

107. James K, Eisenhauer E, Christian M et al (1999). J Natl Cancer Inst 91:523–528.

108. Ross BD, Chenevert TL, Kim B, Ben-Joseph O (1994). Magnetic resonance imaging and spectroscopy: application to experimental neuro-oncology. Q Magn Reson Biol Med 1:89–106.

109. Moffat BA, Chenevert TL, Lawrence TS et al (2005). Functional diffusion map: a noninvasive MRI biomarker for early stratification of clinical brain tumor response. Proc Natl Acad Sci USA 12;102(15):5524–5529.

110. Khong P-L, Kwong DL, Chan GCF, Sham JST, Cham F-L, Ooi G-C (2003). Diffusion-tensor imaging for the detection and qualification of treatment-induced white matter injury in children with medulloblastoma: a pilot study. Am J Neuroradiol 24:734–740.

111. Hunsche S, Mosely ME, Stoeter P, Hedehus M (2001). Diffusion-tensor MR imaging at 1.5T and 3.0T: initial observations. Radiology 221:550–556.

112. Wiegell MR, Larsson HBW, Wedeen VJ (2000). Fiber crossing in human brain depicted with diffusion tensor MR imaging. Radiology 217:897–903.

CHAPTER 26

Diffusion Imaging of Brain Tumors

Stephan E. Maier and Hatsuho Mamata

INTRODUCTION

The capability of MR to measure and image molecular diffusion has provided a new source of image contrast that gives information clearly different from that provided by T1- and T2-weighted images. Stejskal and Tanner [1] introduced a spin echo sequence with strong gradient pulses on both sides of the inverting 180° pulse. In clinical scanners, each gradient is typically applied for a duration of several ten-milliseconds, during which time the average water molecule in brain tissues may migrate 10 or more micrometers in a random direction. In pure water, with a diffusion constant D and a diffusion-weighting b exerted by the gradients, the irregularity of molecular motion causes an attenuation of the MR signal by the factor $\exp(-Db)$. The diffusion coefficient of water in tissues is much lower than the diffusion coefficient of pure water, since molecular displacement of water is impeded by the many intracellular or extracellular structures, such as macromolecules, organelles and membranes. An obvious application of this difference in diffusion is the differentiation between fluid-filled cysts and tissue. Diffusion differences among diverse tissues and their pathologic alterations are less striking, but may still aid in the diagnostic image interpretation. In clinical diffusion imaging and many research studies, images are sampled only for two or at most a few b-factors under $1000\,s/mm^2$ and results obtained in tissues are analyzed assuming the monoexponential signal decay observed in pure water. Theoretical monoexponential model extension of data collected in the conventional b-factor range would predict the absence of any interpretable tissue signal at very high diffusion-weighting. Experimental extension of the b-factor range well beyond the typical $1000\,s/mm^2$ limit, however, reveals a more complex behavior of the diffusion-related tissue water signal loss. The correct and meaningful physical and physiological interpretation of the observed deviation from a monoexponential signal loss in tissues is the subject of an ongoing and intense discussion among leading experts. Nevertheless, it appears that analysis of the non-monoexponential signal loss can contribute to the tissue differentiation.

In many tissues, when averaged over the macroscopic scale of image voxels, the diffusion restriction exerted by the intra- or extracellular structures is identical in every direction, i.e. the observed diffusion is isotropic. Gray matter exhibits such diffusion that is largely isotropic. In contrast, cerebral white matter is a very structured tissue, where the axon arrangement and myelin sheaths ensure anisotropic diffusion, such that diffusion parallel to the white matter is less restricted than the water diffusion perpendicular to it. Diffusion in tumors, except for areas of white matter invasion, shows little anisotropy. Therefore, the primary value of imaging the preferred diffusion direction and diffusion anisotropy in brain tumor patients lies rather in the evaluation of white matter fiber tract integrity than tumor tissue characterization. Based on geometry and the degree of anisotropy loss, white matter tract alterations caused by tumor growth, such as dislocation, swelling, infiltration and disruption, can be documented in great detail.

Image data presented here were collected with the so-called line scan diffusion imaging (LSDI) technique [2]. While this technique is inherently slower than the more commonly used single-shot echoplanar diffusion imaging technique [3], image distortions due to susceptibility variations, commonly present in echoplanar images of the orbita, maxillary cavities and the inferior fossa, are greatly reduced [4,5]. Moreover, since chemical shift artifacts tend to be negligible with LSDI, fat signal suppression is not required. Unless otherwise noted, LSDI images were obtained with a rectangular 128×96 imaging matrix interpolated to a 256×192 matrix, $220\,mm \times 165\,mm$ field of view and 4mm slice thickness, 2592ms repetition time (TR) and 69ms echo time (TE).

CONVENTIONAL DIFFUSION IMAGING OF BRAIN TUMORS

Most conventional diffusion imaging protocols acquire diffusion-weighted image data with a b-factor of around $1000\,s/mm^2$. Except for the characterization of fluid-filled cysts, such diffusion-weighted images do not permit any meaningful interpretation without further post-processing. To eliminate signal variations due to T2 decay and non-uniform proton density, usually at least two measurements, one with diffusion gradients and one without, are obtained and maps of the diffusion constant are then calculated. To avoid any ambiguities in the image interpretation, due to the directionally dependent restricted diffusion in white matter tracts, image data are collected sequentially with diffusion-weighting along three or more orthogonal directions and the directionally independent trace or mean apparent diffusion constant (ADC) is calculated. An example of a trace diffusion-weighted image and trace diffusion map with the corresponding T1-weighted and T2-weighted images is presented in Figure 26.1.

In general, diffusion constants of tumor tissue are higher than the diffusion constants of normal white and gray matter. The highest diffusion values are found in cystic and necrotic parts of lesions. One of the first reported diagnostic applications of diffusion imaging in brain tumors was the differentiation between epidermoid tumors and extra-axial cysts [6,7]. Although fluid attenuated inversion recovery (FLAIR) imaging is also capable of this differentiation, diffusion-weighted imaging (DWI) provides the best lesion conspicuity [8]. The water molecules in the fluid of cystic lesions experience no restrictions by cell structures, and the diffusion constant is therefore very high and can easily be distinguished from the diffusion constant of any other tissue, including tumor tissue (Figure 26.2). Similarly, diffusion values are also helpful to achieve a higher confidence in differentiating cystic or necrotic tumor lesions from an abscess than with conventional MRI alone [9–12]. Abscesses are characterized by a diffusion constant that is slightly lower than the diffusion constant of white and gray matter. Thus, with the exception of the early stage of tumor necrosis [11,13] that may also appear hyperintense on diffusion-weighted images, a cerebral abscess should be suspected in all cases of cystic or necrotic masses with high signals on diffusion-weighted images and low values on the diffusion maps.

Different tumors exhibit a wide range of consistent and dissimilar diffusion values. Some pertinent literature values

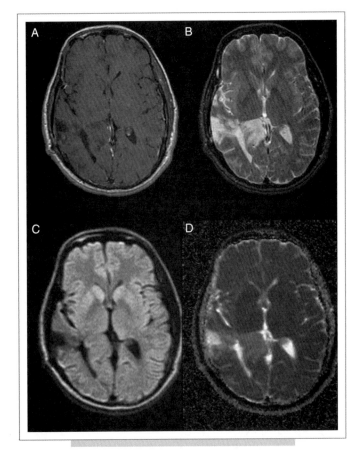

FIG. 26.1. Axial brain images of a 43-year-old male patient with a recurrent low-grade astrocytoma. **(A)** T1-weighted post-contrast spin-echo image (TR 700 ms/TE 14 ms) shows a non-enhancing lesion. **(B)** T2-weighted spin-echo image (TR 300 ms/TE 80 ms) shows the lesion and minor oedema. **(C)** Trace diffusion-weighted image (b-factor 1000 s/mm^2) reveals a distorted geometry of the right ventricle and brain surface. **(D)** On the trace diffusion map, in accordance with a high diffusion, the cerebrospinal fluid space appears bright (D=3.21 μm/ms^2). The diffusion within the tumor lesion is elevated (D=1.36 μm/ms^2). Gray and white matter show no obvious difference and appear dark gray (D = 0.80 μm/ms^2).

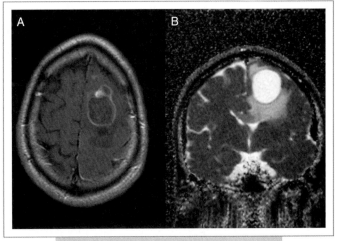

FIG. 26.2. Brain images of a giant cell astrocytoma with cyst formation in a 51-year-old male patient. **(A)** Axial T1-weighted post-contrast spin-echo image (TR 600 ms/TE 19 ms) shows a round lesion with an enhancing rim. **(B)** The same lesion is shown in a coronal trace diffusion map. The non-enhancing central portion discerned on the T1-weighted image is distinguished by a high diffusion constant (D = 2.82 μm/ms^2), which is comparable to the diffusion constant of water at body temperature. The diffusion within the enhancing rim is much lower, but partial volume effects do not permit an exact determination of the diffusion constant. The diffusion within the corpus callosum on the contralateral side is elevated due to edema formation (D=1.18 μm/ms^2) (see also Figure 26.5).

and their standard deviations are presented in Table 26.1. The most comprehensive list of brain tumor values can be found in a study by Yamasaki et al [14]. The values published by Yamasaki et al agree well with values of several other studies that were smaller in scope [15–19], and also with values derived from the image data presented here. A consistent finding is that diffusion values of astrocytic tumors seem to be negatively correlated with their respective World Health Organization (WHO) grade. In agreement with the observed correlation, it has been confirmed both qualitatively [20–22] and quantitatively [18,23–27] that the diffusion value in tumors is inversely correlated with cell density and/or nucleus to cytoplasm ratio. Unfortunately, there is a considerable overlap between the tumor specific ranges of diffusion values. Therefore, the observed diffusion value by itself, despite the very intriguing correlation with cell architecture, does not generally permit the diagnosis of a particular brain tumor. Another confounding factor is that the range of diffusion values representative for peritumoral edema overlaps with the diffusion value range of most tumors. Although Morita et al [28] found that the higher diffusion values of the peritumoral edema observed in high-grade gliomas (see Table 26.1) may reflect the destruction of the extracellular matrix ultrastructure by malignant cell infiltration, the usefulness of diffusion values to demarcate tumor lesion boundaries is very limited. Nevertheless, DWI does certainly aid the diagnostic decision based on anatomic features and signal intensities manifest on the T1- and T2-weighted images. Yamasaki et al [14] point out that once more confidence has been gained with the diffusion values of certain tumor types, DWI may play an important role to differentiate between meningiomas and schwannomas, as well as between craniopharyngiomas and pituitary adenomas or meningiomas (see Table 26.1).

A unique ability of DWI is to follow drug [29] or radiation therapy [30] induced changes in brain tumors. Different tissue reactions can be documented: tumor areas resistant to therapy exhibit unaltered diffusion values; areas of transient cell swelling or ischemia are characterized by decreased diffusion values; meanwhile, areas experiencing cell lysis or apoptosis present increased diffusion values. Analysis of such diffusion changes during different stages of therapy, preferably with the image data coregistered to pretreatment scans, permits quantitative documentation of the response to therapy. It has also been demonstrated that changes observed during the early stages of therapy are an early indicator of the final treatment response [31,32] and the correct interpretation, should a tumor be unresponsive to therapy, may facilitate a timely decision for alternative therapy.

HIGHLY DIFFUSION-WEIGHTED IMAGING OF BRAIN TUMORS

Diffusion-weighted brain images with very high b-factors of $5000\,s/mm^2$ or more, which clearly depict tissue structures above the noise threshold, can readily be obtained on clinical scanners. Examples of such images obtained in a brain tumor patient are presented in Figure 26.3. The additional information gained by high-b DWI without further post-processing is fairly limited [33,34]. Several analysis methods have been proposed to describe the non-monoexponential diffusion-related signal decay. The obvious biexponential or multiexponential analysis yields two or more diffusion constants and respective volume fractions [35,36]. Interpretation of these diffusion constants and volume fractions as an expression of intra- and extracellular diffusion has been discussed [37], but has not gained unequivocal acceptance. The lack of physiological interpretation notwithstanding, added parameters of a biexponential analysis do permit a better tissue characterization than a monoexponential analysis alone. Yoshiura et al [38] and Maier et al [39] found that biexponential diffusion signal analysis, unlike monoexponential analysis, permits a clear separation of gray and white matter. Maier et al [40] also found that biexponential fit parameters derived from wide-range diffusion-weighted image scans are useful for tumor tissue characterization. They demonstrated that tumor tissue can be differentiated both from normal and edematous white matter: from normal white matter by the diffusion constant of the fast diffusing component, and from edematous white matter by the volume fraction of the slow

Tissue	ADC $\mu m/ms^2$
TABLE 26-1 Diffusion values of normal and pathologic tissues in the brain	
Normal white matter	0.705 ± 0.014 [4]
Deep gray matter	0.75 ± 0.03 [62]
Cystic/necrotic tumor areas	2.70 ± 0.31 [11]
Vasogenic (peritumoral) edema	1.30 ± 0.11 [16]
Peritumoral edema (high-grade glial tumors)	1.825 ± 0.115 [28]
Cytotoxic (ischemic) edema	1.04 ± 0.05 [16]
Abscess	0.65 ± 0.16 [11]
WHO grade I Pilocytic astrocytoma	1.659 ± 0.260 [14]
WHO grade II Diffuse astrocytoma	1.530 ± 0.148 [14]
WHO grade II ependymoma	1.230 ± 0.119 [14]
WHO grade III Anaplastic astrocytoma	1.245 ± 0.153 [14]
WHO grade IV Glioblastoma	1.079 ± 0.154 [14]
WHO grade IV PNET	0.835 ± 0.122 [14]
WHO grade IV Medulloblastoma	0.66 ± 0.15 [15]
Craniopharyngioma	1.572 ± 0.210 [14]
Schwannoma	1.384 ± 0.140 [14]
Epidermoid	1.263 ± 0.174 [14]
Germ cell tumor	1.189 ± 0.175 [14]
Pituitary adenoma	1.121 ± 0.202 [14]
Meningioma	1.036 ± 0.270 [14]
Malignant lymphoma	0.725 ± 0.192 [14]
Metastatic tumor	1.149 ± 0.192 [14]

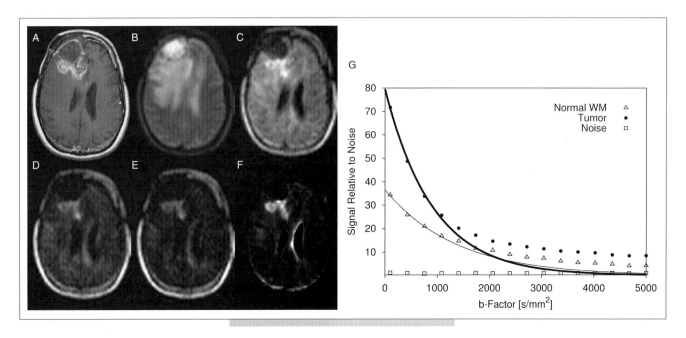

FIG. 26.3. Axial brain images and diffusion-signal plot of a frontal glioblastoma with postoperative cyst formation in a 38-year-old male patient. **(A)** Axial T1-weighted post-contrast spin-echo image (TR 600 ms/TE 25 ms) predominantly shows the margins and not the solid part of the tumor. **(B)** LSDI image (64×48 imaging matrix interpolated to a 256×192 matrix, 220 mm×165 mm field of view, 7.3 mm slice thickness, 2040 ms TR, and 94 ms TE) with a b-factor of 5 mm/s^2 appears basically T2-weighted and visualizes the extent of edema. **(C)** LSDI image with a b-factor of 1000 mm/s^2 demonstrates the appearance with conventional diffusion weighting. **(D)** Highly diffusion-weighted LSDI image with a b-factor of 3000 mm/s^2. **(E)** Very high diffusion-weighted LSDI image with a b-factor of 5000 mm/s^2 exhibits signal above the noise threshold for all tissues. Extraordinarily high residual signal, despite high diffusion weighting, is observed in the solid part of the tumor. LSDI images were all obtained with a single diffusion encoding direction. **(F)** Computed χ^2 error map of the monoexponential fit to the diffusion related signal decay reveals enhancement of the solid part of the tumor. Other areas of enhancement are due to partial volume effects in voxels with non-uniform tissues. **(G)** Plots of diffusion-attenuated signal versus b-factor for normal white matter (WM), tumor, and noise. The solid lines show the signal decay predicted by a monoexponential fit through the first four signal values up to a b-factor of 1000 mm/s^2. Evidently the signal decays much slower than a monoexponential fit of the signal measured over the conventional b-factor range would predict. The diffusion values measured with conventional diffusion imaging (b=1000 mm/s^2) were 1.10 μm/ms^2 for tumor and 0.74 μm/ms^2 for normal-appearing white matter.

diffusing component. The separation is possible in principle, but the computed maps are plagued by low signal-to-noise ratio and limited spatial resolution. Moreover, the simultaneous interpretation of multiple computed parameter maps is cumbersome. In a totally different approach, it was demonstrated that the χ^2 (chisquare) error parameter associated with monoexponential fits of the tissue water signals, measured at typically between four to sixteen different b-factors, can also be used to characterize the deviation from a simple monoexponential signal decay on a pixel-by-pixel basis (see Figure 26.3F) [41]. Indeed, such χ^2-maps are convenient to interpret and permit good visualization of spatial details, and the resulting lesion contrast by far exceeds the contrast of biexponential parameter maps. In peritumoral edema, χ^2 values were on average 68% higher than in normal white matter. A remarkable departure from a simple monoexponential signal decay, with average χ^2 values almost 400% higher than in normal white matter, occurred in highly

malignant primary brain tumors, such as glioblastomas (see Figure 26.3) or anaplastic astrocytomas. Unlike the enhancement of T1-weighted post-contrast images, such enhancement was consistently observed in the solid part of tumors. On the other hand, low-grade astrocytomas, one case of ganglioglioma (Figure 26.4), and metastases, demonstrated χ^2 values that were not profoundly different from the χ^2 value of white matter.

Presently, clinical routine MRI characterizes the extent of brain tumors by the appearance of the margins on T1 contrast-enhanced images. In brain tumors, the mechanism of T1-weighted contrast after paramagnetic contrast agent injection is known to be a local breakdown of the blood–brain barrier. However, tumor growth is not always associated with a leaking blood–brain barrier and the contrast agent enhancement is not specific. It could be due to tumor tissue itself, inflammation, increased permeability of blood vessels or other abnormal changes. High-b

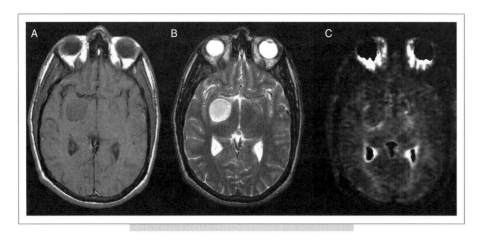

FIG. 26.4. Axial brain images of a 24-year-old male patient with a ganglioglioma (WHO grade I). **(A)** T1-weighted post-contrast spin-echo image (TR 500 ms/TE 14 ms) exhibits no enhancement at the location of the tumor. **(B)** The T2-weighted spin-echo image (TR 5600 ms/TE 96 ms) shows the tumor lesion. **(C)** The χ^2 error parameter map computed from trace LSDI data shows no enhancement, except along the rim of the tumor. The enhancement in the orbital cavity visible on the χ^2 map is artifactual and caused by eye motion. The tumor diffusion value measured with conventional diffusion imaging (b=1000 mm/s^2) was 1.37 μm/ms^2. LSDI image parameters were: 128×96 imaging matrix interpolated to a 256×192 matrix, 220 mm×165 mm field of view, 6 mm slice thickness, 2682 ms TR and 92 ms TE.

DWI, although currently purely experimental, seems to provide tumor contrast clearly different from the contrast observed on T1-weighted images. It should be added that diffusion-based tissue differentiation does not depend on the effect of contrast agents, since with the common imaging protocols the diffusion-weighted images are only minimally T1-weighted. A remarkable finding of the analysis of the non-monoexponential diffusion-related signal decay in tumor patients is the observation of strong enhancement outside the tumor boundaries confirmed by T1 contrast-enhanced imaging. Although in some instances this can be attributed to partial volume effects, such enhancement seems to indicate malignant cell infiltration [41]. Indeed, average cell size (cell density), cell-size distribution (cell-size irregularity) and nucleus to cytoplasm size ratio, are contrast agent independent parameters that appear to have profound effects on diffusion and, in particular, the observed deviation from a monoexponential signal decay. The very same parameters permit definitive differentiation between pathologic and normal tissue, when histological analysis is applied.

DIFFUSION TENSOR IMAGING OF NERVE FIBER TRACT INTEGRITY IN BRAIN TUMOR PATIENTS

A general model for the restricted diffusion is well represented by the work of Basser and colleagues who have incorporated the diffusion tensor formalism into diffusion data analyses [42]. With diffusion tensor imaging (DTI) a minimum of six diffusion-weighted images with different diffusion encoding directions is acquired. Subsequent computation of the diffusion tensor yields for each voxel three diffusion coefficients (diffusion eigenvalues) along three orthogonal principal directions (diffusion eigenvectors). In white matter, the eigenvector associated with the largest eigenvalue defines the tissue's fiber tract axis [43]. The directional variability of restricted diffusion is usually quantified by the rotationally invariant fractional anisotropy (FA) index [44], which ranges from 0 (isotropic diffusion) to 1.0 (totally restricted diffusion). The different white matter tracts are well characterized by their different anisotropy values [45,46] and range from 0.81 in the corpus callosum, over 0.62 in the internal capsule, to 0.09 in cortical gray matter [46].

Only very few studies have explored diffusion anisotropy inside brain tumors and its potential value for tissue characterization. Inoue et al [47] measured the lowest anisotropy in grade I gliomas (FA = 0.150) and the highest anisotropy in grade III and IV gliomas (FA = 0.23 and 0.229). The same group of investigators found that FA in glioblastomas is an indicator of cell density and proliferation activity [48]. The authors discuss the possibility of residual white matter as a factor contributing to the observed diffusion anisotropy in glioblastomas, but have no explanation why cell density in the solid core would correlate with the measured FA. In an earlier study [49], that concluded diffusion anisotropy added no benefit to tissue differentiation, FA values in the non-enhancing core of high-grade glioma tumors were 0.09, in the enhancing tumor core 0.13, and in the enhancing tumor margin 0.16. In a more recent study [50], FA values differed significantly between low-grade and high-grade gliomas in the tumor margin, but not

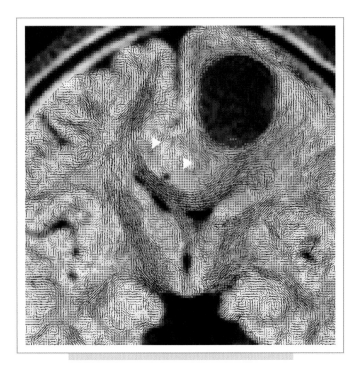

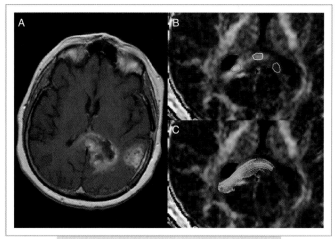

FIG. 26.5. Coronal diffusion tensor map of giant cell astrocytoma with cyst (same patient as shown in Figure 26.2). In-plane component of the primary diffusion eigenvector is depicted with lines overlaid on the diffusion-weighted image. The out-of-plane component of the eigenvector is visualized with red (strong component) and blue (intermediate size component) dots. The diffusion tensor map reveals a marked lateral shift of the corpus callosum and the cingulum tracts (arrows). Moreover, a widening of the left half of the corpus callosum can be discerned. In the immediate surrounding of the cyst, fiber structures appear to be intact, but severely distorted.

FIG. 26.6. Axial brain images of a 57-year-old female patient with a multifocal glioblastoma. **(A)** The T1-weighted post-contrast spin-echo image (TR 750 ms/TE 14 ms) shows two lesions, one of them with invasion into the splenium of the corpus callosum. **(B)** Detail of diffusion anisotropy map with three regions of interest (ROI) drawn in the splenium of the corpus callosum (fractional anisotropy: red ROI 0.65, yellow ROI 0.46, green ROI 0.20). **(C)** Same anisotropy map as shown in (B) with overlaid in-plane fiber tracking path, documenting the presence of intact fibers in the tumor-invaded splenium.

in the tumor center. This significant difference was believed to be the result of a more extensive fiber tract destruction in high grade gliomas than in low grade gliomas, whereas in the center of both low grade and high grade gliomas virtually no fibers remain intact. A modifying or cautionary detail to be considered when evaluating areas of low anisotropy is the difficulty in obtaining reliable anisotropy measures under unfavorable signal-to-noise ratio conditions [44,46]. Indeed, the observed glioma grade dependency of the FA value may also be the effect of different enhancement and consequently different signal-to-noise ratios, rather than directionally dependent diffusion restrictions.

The phenomenon of diffusion anisotropy is of particular interest to studies that evaluate the integrity of white matter fiber tracts. Based on geometry and the degree of anisotropy loss, white matter tract alterations, such as dislocation, swelling, infiltration and disruption, can be documented. In the case of brain tumors, such changes are characteristically present in various degree and forms. Figure 26.5 shows a case where DTI adds valuable information

regarding fiber tract dislocation and swelling. A case, where DTI reveals partial destruction of a fiber tract by tumor infiltration, can be seen in Figure 26.6. An example of complete fiber tract disruption visualized with diffusion anisotropy imaging is documented in Figure 26.7. Several studies have explored the potential of DTI to detect white matter infiltration [51–53]. In high-grade gliomas and meningiomas, both Tropine et al [53] and Provenzale et al [52] observed for the peritumoral region with increased T2-signal a trend towards greater FA reduction in high-grade gliomas. Within the normal appearing peritumoral white matter, FA reduction in high-grade gliomas was significantly larger than the FA reduction in meningiomas [52] or metastases [51]. Reduced FA values in high-grade gliomas seem to indicate changes that are not only related to vasogenic edema, but also related to tumor infiltration. These findings suggest a potential role for diffusion imaging in the detection of tumoral infiltration, particularly in areas that appear normal on conventional MR images.

More recently, diffusion tensor tractography [54] has emerged as a diagnostically helpful application. Diffusion tensor tractography uses the principal diffusion direction, measured with DTI, to compute the pathways of complete nerve fiber tracts. The tracing is performed by first defining positions of interest in a white matter tract. By following repetitively and in small steps along spatially interpolated directions of maximum diffusion, a contigu-

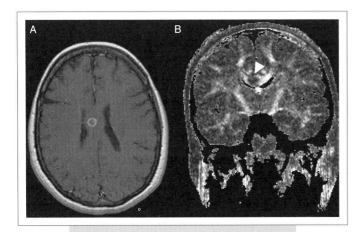

FIG. 26.7. Brain images of a 52-year-old female patient with a melanoma metastasis. **(A)** The axial T1-weighted post-contrast spin-echo image (TR 750 ms/TE 14 ms) shows a small lesion with an enhancing rim within the body of the corpus callosum. **(B)** On the coronal anisotropy map, the same lesion is clearly discerned (arrow) as an area with low anisotropy (fractional anisotropy: 0.11; $D = 0.99 \, \mu m/ms^2$) within the high anisotropy region of the corpus callosum.

ous path that passes through the initial seed positions is defined. Visualized with suitable software in two or three dimensions, these tracts depict the involved anatomy of white matter fibers (see Figure 26.6C). Thus the departure from the basic assessment of the anisotropy of each diffusion tensor to the more elaborate analysis of neighbor relations among diffusion tensors opens the possibility for assessing in vivo axonal fiber connectivity and functional links between brain regions. Diffusion tensor tractography, especially fused together with conventional and functional image information, provides a powerful tool to avoid or minimize the injury of displaced but still intact fiber tracts during neurosurgery or radiation therapy [55–57].

SUMMARY

Diffusion imaging clearly adds a new dimension to the diagnostic imaging of brain tumors. Of all the parameters that can be obtained with diffusion imaging, mean diffusion appears to be the most ubiquitously useful and most readily interpreted parameter. The value of this parameter lies not in the demarcation of tumor boundaries, but rather in the definitive tissue characterization in some cases where T1- and T2-weighted images alone do not provide enough diagnostic information. The ability of diffusion imaging to document cellular response during drug or radiation therapy is also unique, but requires image registration for reliable detection of changes. The application of an extended b-factor range and analysis of the non-monoexponential diffusion signal decay shows some interesting potential for diagnostic imaging in brain tumors. At this point, however, this approach should be considered experimental and the capability of detecting lesions may only bring to bear when faster sequences with improved spatial resolution and better signal-to-noise ratio become available. The diffusion anisotropy parameter, which can be obtained with DTI, seems to provide useful local information about white matter integrity. DTI is also valuable in the precise localization of white matter tracts. Tractography gives a quick and intuitive overview of the displaced course of white matter tracts in two or three-dimensional space. High-spatial resolution DTI with complete brain coverage requires relatively long scan times, but the information that can be extracted is extremely valuable for brain surgery and radiation therapy. The ability to obtain such information during surgical procedures, when brain structures are likely to shift position, has been demonstrated [57,58].

One major limitation of the magnetic resonance diffusion images in comparison to the standard T1- and T2-weighted images is the inherent low signal-to-noise ratio and coarse spatial resolution. Partial volume effects prevent an equally good delineation of lesion boundaries and the images obtained with the commonly used single-shot echoplanar diffusion imaging method are plagued by geometric distortions. Such distortions may severely limit the use of such images for surgical planning or tractography [59]. Image post-processing or alternative acquisition techniques, such as LSDI [2], parallel single-shot echoplanar diffusion imaging [60], or PROPELLER diffusion imaging [61] can overcome these limitations.

ACKNOWLEDGMENT

The research data presented in this chapter were obtained with grant support from the National Institute of Neurological Disorders and Stroke (R01 NS39335 to SEM) and the National Institute of Biomedical Imaging (R01 EB006867 as a continuation R01 NS39335, also to SEM).

REFERENCES

1. Stejskal EO Tanner JE (1965). Spin diffusion measurements: spin echoes in the presence of a time-dependent field gradient. J Chem Phys 42:288–292.

2. Gudbjartsson H, Maier SE, Mulkern RV, M'orocz IA, Patz S, Jolesz FA (1996). Line scan diffusion imaging. Magn Reson Med 36(4):509–519.

3. Turner R, Le Bihan D, Maier J, Vavrek R, Hedges LK, Pekar J (1990). Echo-planar imaging of intravoxel incoherent motions. Radiology 177:407–414.

4. Maier SE, Gudbjartsson H, Patz S et al (1998). Line scan diffusion imaging: characterization in healthy subjects and stroke patients. Am J Roentgenol 17(1):85–93.

5. Maeda M, Maier SE, Sakuma H, Ishida M, Takeda K (2006). Apparent diffusion coefficient in malignant lymphoma and carcinoma involving cavernous sinus evaluated by line scan diffusion-weighted imaging. J Magn Reson Imaging 24(3):543–548.

6. Tsuruda JS, Chew WM, Moseley ME, Norman D (1990). Diffusion-weighted MR imaging of the brain: value of differentiating between extraaxial cysts and epidermoid tumors. Am J Neuroradiol 11(5):925–931.

7. Maeda M, Kawamura Y, Tamagawa Y et al (1992). Intravoxel incoherent motion (IVIM) MRI in intracranial, extraaxial tumors and cysts. J Comput Assist Tomogr 16(4):514–518.

8. Chen S, Ikawa F, Kurisu K, Arita K, Takaba J, Kanou Y (2001). Quantitative MR evaluation of intracranial epidermoid tumors by fast fluid-attenuated inversion recovery imaging and echo-planar diffusion-weighted imaging. Am J Neuroradiol 22(6):1089–1096.

9. Ebisu T, Tanaka C, Umeda M et al (1996). Discrimination of brain abscess from necrotic or cystic tumors by diffusion-weighted echo planar imaging. Magn Reson Imaging 14(9):1113–1116.

10. Noguchi K, Watanabe N, Nagayoshi T et al (1999). Role of diffusion-weighted echo-planar MRI in distinguishing between brain abscess and tumour: a preliminary report. Neuroradiology 41(3):171–174.

11. Chang SC, Lai PH, Chen WL et al (2002). Diffusion-weighted MRI features of brain abscess and cystic or necrotic brain tumors: comparison with conventional MRI. Clin Imaging 26(4):227–236.

12. Reddy JS, Mishra AM, Behari S et al (2006). The role of diffusion-weighted imaging in the differential diagnosis of intracranial cystic mass lesions: a report of 147 lesions. Surg Neurol 66(3):246–250.

13. Holtas S, Geijer B, Stromblad LG, Maly-Sundgren P, Burtscher IM (2000). A ring-enhancing metastasis with central high signal on diffusion-weighted imaging and low apparent diffusion coefficients. Neuroradiology 42(11): 824–827.

14. Yamasaki F, Kurisu K, Satoh K et al (2005). Apparent diffusion coefficient of human brain tumors at MR imaging. Radiology 235(3):985–991.

15. Rumboldt Z, Camacho DL, Lake D, Welsh CT, Castillo M (2006). Apparent diffusion coefficients for differentiation of cerebellar tumors in children. Am J Neuroradiol 27(6):1362–1369.

16. Pronin IN, Kornienko VN, Fadeeva LM, Rodionov PV, Golanov AV (2000). Diffusion-weighted image in the study of brain tumors and peritumoral edema (article in Russian). Zh Vopr Neirokhir Im N N Burdenko Jul-Sep(3):14–16.

17. Dorenbeck U, Grunwald IQ, Schlaier J, Feuerbach S (2005). Diffusion-weighted imaging with calculated apparent diffusion coefficient of enhancing extra-axial masses. J Neuroimaging 15(4):314–317.

18. Guo AC, Cummings TJ, Dash RC, Provenzale JM (2002). Lymphomas and high-grade astrocytomas: comparison of water diffusibility and histologic characteristics. Radiology 224(1):177–183.

19. Kono K, Inoue Y, Nakayama K et al (2001). The role of diffusion-weighted imaging in patients with brain tumors. Am J Neuroradiol 22(6):1081–1088.

20. Uhl M, Altehoefer C, Kontny U, Il'yasov K, Buchert M (2002). MRI-diffusion imaging of neuroblastomas: first results and correlation to histology. Eur Radiol 12(9):2335–2338.

21. Rodallec M, Colombat M, Krainik A, Kalamarides M, Redondo A, Feydy A (2004). Diffusion-weighted MR imaging and pathologic findings in adult cerebellar medulloblastoma. J Neuroradiol 31(3):234–237.

22. Herneth A, Guccione S, Bednarski M (2003). Apparent diffusion coefficient: a quantitative parameter for in vivo tumor characterization. Eur J Radiol 45(3):208–213.

23. Sugahara T, Korogi Y, Kochi M et al (1999). Usefulness of diffusion-weighted MRI with echo-planar technique in the evaluation of cellularity in gliomas. J Magn Reson Imaging 9(1):53–60.

24. Lyng H, Haraldseth O, Rofstad E (2000). Measurement of cell density and necrotic fraction in human melanoma xenografts by diffusion-weighted magnetic resonance imaging. Magn Reson Med 43(6):828–836.

25. Gupta RK, Cloughesy TF, Sinha U et al (2000). Relationships between choline magnetic resonance spectroscopy, apparent diffusion coefficient and quantitative histopathology in human glioma. J Neuro-Oncol 50(3):215–226.

26. Gauvain KM, McKinstry RC, Mukherjee P et al (2001). Evaluating pediatric brain tumor cellularity with diffusion-tensor imaging. Am J Roentgenol 177(2):449–454.

27. Hayashida Y, Hirai T, Morishita S et al (2006). Diffusion-weighted imaging of metastatic brain tumors: comparison with histologic type and tumor cellularity. Am J Neuroradiol 27(7):1419–1425.

28. Morita K, Matsuzawa H, Fujii Y, Tanaka R, Kwee IL, Nakada T (2005). Diffusion tensor analysis of peritumoral edema using lambda chart analysis indicative of the heterogeneity of the microstructure within edema. J Neurosurg 102(2):336–341.

29. Mardor Y, Roth Y, Lidar Z et al (2001). Monitoring response to convection-enhanced Taxol delivery in brain tumor patients using diffusion-weighted magnetic resonance imaging. Cancer Res 61(13):4971–4973.

30. Mardor Y, Pfeffer R, Spiegelmann R et al (2003). Early detection of response to radiation therapy in patients with brain malignancies using conventional and high b-value diffusion-weighted MRI. J Clin Oncol 21(6):1094–2000.

31. Mardor Y, Roth Y, Ochershvilli A et al (2004). Pre-treatment prediction of brain tumors response to radiation therapy using high b-value diffusion-weighted MRI. Neoplasia 6(2): 136–142.

32. Moffat BA, Chenevert TL, Lawrence TS et al (2005). Functional diffusion map: a noninvasive MRI biomarker for early stratification of clinical brain tumor response. Proc Natl Acad Sci USA 102(15):5524–5529.

33. DeLano MC, Cooper TG, Siebert JE, Potchen MJ, Kuppusamy K (2000). High-b-value diffusion-weighted MR imaging of adult brain: image contrast and apparent diffusion coefficient map features. Am J Neuroradiol 21(10):1830–1836.

34. Burdette JH, Durden DD, Elster AD, Yen YF (2001). High b-value diffusion-weighted MRI of normal brain. J Comput Assist Tomogr 25(4):515–519.

35. Mulkern RV, Gudbjartsson H, Westin CF et al (1999). Multicomponent apparent diffusion coefficients in human brain. NMR Biomed 12:51–62.

36. Maier SE, Vajapeyam S, Mamata H, Westin CF, Jolesz FA, Mulkern RV (2004). Biexponential diffusion tensor analysis of human brain diffusion data. Magn Reson Med 51(2): 321–330.

37. Niendorf T, Dijkhuizen RM, Norris DG, van Lookeren Campagne M, Nicolay K (1996). Biexponential diffusion attenuation in various states of brain tissue: implications for diffusion-weighted imaging. Magn Reson Med 36(6):847–857.

38. Yoshiura T, Wu O, Zaheer A, Reese TG, Sorensen AG (2001). Highly diffusion-sensitized MRI of brain: dissociation of gray and white matter. Magn Reson Med 45(5):734–740.

39. Maier SE, Mamata H, Mulkern RV (2003). Biexponential analysis of diffusion related signal decay in human cortical grey matter. In Book of Abstracts, Eleventh Annual Meeting, Toronto, Canada, Society of Magnetic Resonance, Berkeley California. 595.

40. Maier SE, Bogner P, Bajzik G et al (2001). Normal brain and brain tumor: multicomponent apparent diffusion coefficient line scan imaging. Radiology 219:842–849.

41. Maier SE, Mamata H, Mulkern RV (2003). Characterization of normal brain and brain tumor pathology by chisquares parameter maps of diffusion-weighted image data. Eur J Radiol 45(3):199–207.

42. Basser P, Mattiello J, Le Bihan D (1994). MR diffusion tensor spectroscopy and imaging. Biophys J 66(1):259–267.

43. Lin CP, Tseng WY, Cheng HC, Chen JH (2001). Validation of diffusion tensor magnetic resonance axonal fiber imaging with registered manganese-enhanced optic tracts. Neuroimage 14(5):1035–1047.

44. Papadakis NG, Xing D, Houston GC et al (1999). A study of rotationally invariant and symmetric indices of diffusion anisotropy. Magn Reson Imaging 17(6):881–892.

45. Pierpaoli C, Jezzard P, Basser P, Barnett A, Di Chiro G (1996). Diffusion tensor MR imaging of the human brain. Radiology 201(3):637–648.

46. Mamata H, Jolesz FA, Maier SE (2004). Characterization of central nervous system structures by magnetic resonance diffusion anisotropy. Neurochem Int 45(4):553–560.

47. Inoue T, Ogasawara K, Beppu T, Ogawa A, Kabasawa H (2005). Diffusion tensor imaging for preoperative evaluation of tumor grade in gliomas. Clin Neurol Neurosurg 107(3):174–180.

48. Beppu T, Inoue T, Shibata Y et al (2005). Fractional anisotropy value by diffusion tensor magnetic resonance imaging as a predictor of cell density and proliferation activity of glioblastomas. Surg Neurol 63(1):56–61.

49. Sinha S, Bastin ME, Whittle IR, Wardlaw JM (2002). Diffusion tensor MR imaging of high-grade cerebral gliomas. Am J Neuroradiol 23(4):520–527.

50. Goebell E, Paustenbach S, Vaeterlein O et al (2006). Low-grade and anaplastic gliomas: differences in architecture evaluated with diffusion-tensor MR imaging. Radiology 239(1): 217–222.

51. Price S, Burnet N, Donovan T et al (2003). Diffusion tensor imaging of brain tumours at 3T: a potential tool for assessing white matter tract invasion? Clin Radiol 58(6):455–462.

52. Provenzale JM, McGraw P, Mhatre P, Guo AC, Delong D (2004). Peritumoral brain regions in gliomas and meningiomas: investigation with isotropic diffusion-weighted MR imaging and diffusion-tensor MR imaging. Radiology 232(2): 451–460.

53. Tropine A, Vucurevic G, Delani P et al (2004). Contribution of diffusion tensor imaging to delineation of gliomas and glioblastomas. J Magn Reson Imaging 20(6):905–912.

54. Basser PJ, Pajevic S, Pierpaoli C, Duda J, Aldroubi A (2000). In vivo fiber tractography using DT-MRI data. Magn Reson Med 44(4):625–632.

55. Holodny AI, Schwartz TH, Ollenschleger M, Liu WC, Schulder M (2001). Tumor involvement of the corticospinal tract: diffusion magnetic resonance tractography with intraoperative correlation. J Neurosurg 95(6):1082.

56. Yu CS, Li KC, Xuan Y, Ji XM, Qin W (2005). Diffusion tensor tractography in patients with cerebral tumors: a helpful technique for neurosurgical planning and postoperative assessment. Eur J Radiol 56(2):197–204.

57. Nimsky C, Ganslandt O, Hastreiter P et al (2005). Preoperative and intraoperative diffusion tensor imaging-based fiber tracking in glioma surgery. Neurosurgery 56(1):130–137.

58. Mamata Y, Mamata H, Nabavi A et al (2001). Intraoperative diffusion imaging on a 0.5 Tesla interventional scanner. J Magn Reson Imaging 13(1):115–119.

59. Hahn HK, Klein J, Nimsky C, Rexilius J, Peitgen HO (2006). Uncertainty in diffusion tensor based fibre tracking. Acta Neurochir Suppl 98:33–41.

60. Bammer R, Auer M, Keeling SL et al (2002). Diffusion tensor imaging using single-shot SENSE-EPI. Magn Reson Med 48(1):128–136.

61. Pipe JG, Farthing VG, Forbes KP (2002). Multishot diffusion-weighted FSE using PROPELLER MRI. Magn Reson Med 47(1):42–52.

62. Helenius J, Soinne L, Perkio J et al (2002). Diffusion-weighted MR imaging in normal human brains in various age groups. Am J Neuroradiol 23(2):194–199.

CHAPTER 27

Functional MRI

Nicole Petrovich Brennan and Andrei Holodny

INTRODUCTION

In the brain tumor patient, quality of life is an important and often predominant goal of clinical management. Commonly, this entails maintaining speech and motor function. This is often achieved by treating or offering a resection to a patient whose lesion has not yet invaded eloquent cortical areas, but may be affecting function by mass effect, edema or seizure genesis. Knowing the location of eloquent brain regions can help plan the surgical approach and sway decisions on whether to perform intraoperative mapping or awake craniotomy. Additionally, it can identify those patients who may have anomalous functional organization, potentially changing the clinical management of the patient.

There are many techniques used to localize brain function in the brain tumor patient including: blood oxygen level dependent functional MRI (BOLD fMRI); positron emission tomography (PET); near infrared spectroscopy (NIRS); direct cortical stimulation; magnetoencephalography (MEG) and transcranial magnetic stimulation (TMS). Each of these techniques has distinct advantages and disadvantages in terms of spatial resolution, temporal resolution, invasiveness, sensitivity and interpretation. BOLD fMRI is currently the most frequently utilized for the following reasons: BOLD fMRI has excellent spatial resolution; is entirely non-invasive (and therefore repeatable) and has a sufficient signal-to-noise ratio on a standard clinical 1.5T magnet to yield reproducible maps of function in most cases.

Here we will discuss the use of BOLD functional MRI in the management of brain tumor patients. We will focus particular attention on presurgical planning using fMRI as this tends to be one of its most common uses in the clinical setting. We will detail the procedures for mapping motor and language areas, as well as the patient considerations when attempting such. We will then talk about specific considerations for BOLD fMRI in the presence of a brain tumor. Finally, we will outline some common pitfalls in the interpretation of fMRI maps.

GENERAL FMRI CONSIDERATIONS

fMRI takes advantage of the fact that oxyhemoglobin is diamagnetic and deoxyhemoglobin is paramagnetic.

For example, this means that the gray scale values of the voxels covering the motor gyrus in an MR image when the patient is resting are different than the gray scale values in those voxels when the patient's fingers are moving. As a result, because of the need for a statistical comparison to generate a functional map, most fMRI paradigms alternate between a rest and an active task state. The task is then repeated multiple times for statistical power. There are many statistical tests and many statistical packages used to analyze the data, but most use either cross correlation, t-test or the general linear model.

With this in mind, there are two main types of fMRI stimulus presentation paradigms. Figure 27.1 illustrates the main differences between event-related and block design. Event-related paradigms display stimuli in short spurts (one or two stimuli per image acquisition) and then return to baseline. This type of paradigm can be 'fast' event-related or 'slow' event-related depending on the interstimulus interval. Fast event-related presents the stimuli in quick succession whereas slow event-related designs may give a stimulus or two and then allow the hemodynamic response to recover over the course of a 10 or 12 second baseline, for example (the hemodynamic response can take up to 30 seconds to recover completely) about the exact time course of the response to a single or near single event.

The second standard stimulus presentation design is termed block design. Block designed paradigms are typically organized such that alternating blocks of the same type of stimuli are presented together. For example, the patient might rest or fixate on a crosshair for 20 seconds and then name pictures for 20 seconds. The block design averages many images of the same type presented together in time, and as a result tends to yield a stronger signal overall than you would get from single events in an event-related paradigm (when comparing the same number of acquisitions). Birn et al describe the block and event-related designs as the difference between detecting the signal and estimating the signal respectively [1,2]. This is to say that the event-related design is better at estimating the details of the hemodynamic response to a single event. In our experience, block designs are unequivocally better for brain tumor patients.

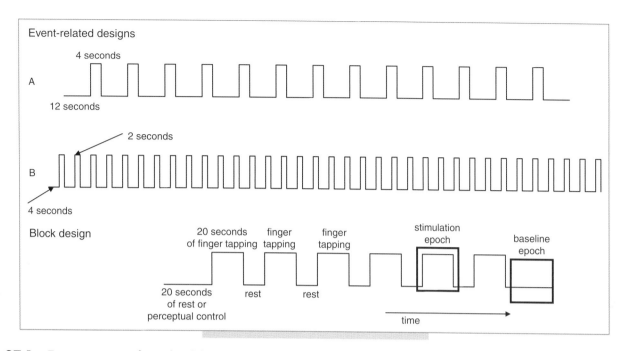

FIG. 27.1. Two main types of stimulus delivery protocols. Block design groups stimuli of the same type with multiple repetitions per epoch. The example shown here is for finger tapping, but the stimulus is easily replaced with pictures or some other linguistic task for language assessment. Slow **(A)** and fast **(B)** event-related designs allow the estimation of the hemodynamic response to a single stimulus (or minimal number of stimuli). Block design alternates blocks of stimuli of the same type and is generally preferred for use with brain tumor patients when there is no need for temporal information and the goal is to assess basic function.

This may be because their performance is variable and benefits from signal averaging. Regardless, using event-related designs with patients in our institution is only driven by necessity. These paradigms are helpful to researchers interested in the effects of a brain tumor on brain functioning, as the specific brain based responses can be studied in detail. Additionally, event-related designs have proven useful with patients performing vocalized speech paradigms where head motion is an issue and the contaminated images can be easily extracted when they are in isolation.

SENSORY/MOTOR fMRI

The sensory/motor system is among the most straightforward systems to map using fMRI. Common paradigms include bilateral finger tapping where the patient is instructed to move their fingers sequentially at specified times. Sensory stimulation is sometimes achieved by scrubbing the patient's hand with a textured object. Importantly though, the selection of task performance should be based on the location of the lesion in relation to the motor homunculus where each area of the body is spatially represented in distinct cortical areas. While hand clenching or finger tapping are common approaches for confirming the location of the motor gyrus, it can be hard to extrapolate the location of the motor gyrus inferiorly and medially given this information alone [3–5]. As a result, a more

lateral/inferior lesion might benefit from the additional inclusion of a tongue movement paradigm. Figure 27.2 shows how tongue and hand motor tasks can effectively map the different cortical representations of the motor homunculus. Similarly, a dorsal, medial lesion may require that foot/leg stimulation be performed. Localizing foot at the most medial aspect of the motor strip can nicely define the location of foot (and by extrapolation leg function) relative to the supplementary motor area and the more lateral (often slightly anterior) portion of the motor gyrus. It goes without saying that a smaller matrix size will enable more specific, segregated maps of function (Figure 27.2 = 128 × 128). In some cases, the sacrifice in the signal-to-noise ratio that one makes going from 64 × 64 to 128 × 128 may be advantageous in those cases. In our experience, the motor signal tends to be strong and resilient to head motion even in brain tumor patients, allowing us routinely to use the higher spatial resolution of 128 × 128.

Another important consideration is the reciprocity of neural connectivity between the pre-central and postcentral gyri. This has implications in mapping brain tumor patients who may be paretic or completely paralyzed. As we will see with language fMRI, it is important to choose a task that the patient can do, otherwise, the interpretation of the map is difficult. This is not the case with localizing the motor cortex. If a patient cannot move due to the presence of a lesion or mass effect, it is reasonable to do a task

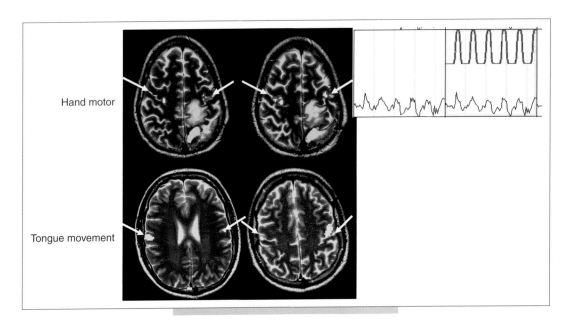

FIG. 27.2. Hand motor fMRI and tongue movement fMRI in two brain tumor patients illustrating the ability of fMRI to distinguish between the more dorsal/superior and the lateral/inferior portions of the motor homunculus. Arrows indicate the location of the central sulcus. The boxes represent the actual signal in two voxels in the motor cortex. The black is the measured signal and the red is the ideal or model for the stimulus presentation. In this case, the patient was instructed to move fingers six times each separated by rest.

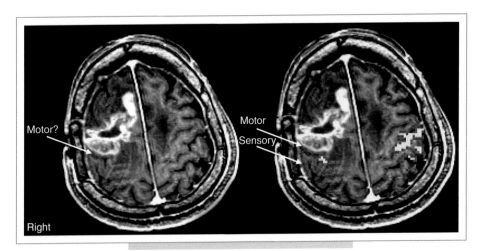

FIG. 27.3. T1-post-contrast image of a patient with a glioblastoma multiforme in the right hemisphere. This extensive lesion makes it difficult to determine whether the lesion is within the motor gyrus. fMRI of bilateral finger tapping makes this determination clearer. Note both the more anterior motor activity and the posterior sensory activity.

involving passive sensory stimulation. Granted, in many cases the resultant fMRI map will be biased toward the estimation of the location of the sensory gyrus but, in our experience, the pre-central gyrus will also yield reliable signal. There are two main patterns of activity one should expect from a finger-tapping paradigm. In one case, the motor gyrus activates nearly to the exclusion of the sensory gyrus. Commonly, however, both the motor and sensory gyri activate.

One should consider how the neurosurgeon will use the motor fMRI map to plan or guide oncologic surgery. First, the map serves to confirm or refute the surgeon's impression of the location of the lesion relative to the motor gyrus. It is useful to know whether the lesion is in the motor strip itself or anterior or posterior to it based on the fMRI measurement. These judgments can be difficult when a patient's clinical presentation is misleading. This is particularly true in patients with brain tumors where low-grade lesions or diffuse high-grade lesions may not affect the patient's function, leading the clinician to wonder whether the lesion is actually in the eloquent cortex.

Additionally, a tumor may have mass effect that effaces sulci and renders them difficult to discern anatomically via traditional imaging. Figure 27.3 shows a T1-post-contrast image of a patient with an extensive glioblastoma multiforme in the right hemisphere. Without fMRI, it was difficult to know whether this lesion extended into the motor

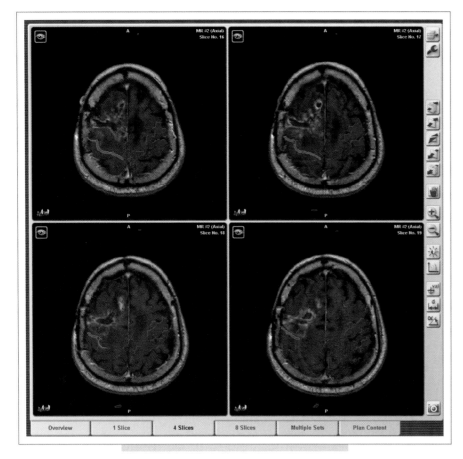

FIG. 27.4. Integration of foot (pink) and hand (purple) fMRI on the patient from Figure 27.3 into a neurosurgical navigation system. The green line indicates the location of central sulcus.

gyrus itself. Following fMRI mapping, it was determined that the lesion was within the motor gyrus and the patient had an awake intraoperative procedure as a result.

Intraoperative direct cortical bipolar stimulation is considered the 'gold standard' in neurosurgery to localize function. fMRI can help direct this stimulation. The fMRI map can be fused to the intraoperative guidance system allowing the surgeon to begin stimulation based not only on anatomical landmarks but the fMRI localization. Figure 27.4 shows the integration of the hand and foot fMRI on the patient in Figure 27.3 into a neurosurgical navigation system. The surgeon was able to direct the intraoperative direct cortical stimulation before resecting this lesion. The hand fMRI was concordant with the direct cortical stimulation (foot/leg was not attempted).

LANGUAGE fMRI

Language lateralization and language localization are the two main types of language mapping for oncologic neurosurgery.

Lateralization

The intracarotid sodium amobarbital procedure, better known as the WADA test [6], is currently the gold standard for determining language and memory dominance in patients. It entails anesthetizing one hemisphere of the brain at a time while a clinician monitors the patient's performance on language and memory tasks. However, the WADA test is invasive and can cause stroke or other complications. Consequently, there is a significant effort to replace the WADA test. For the purpose of language lateralization, there are a number of studies confirming the concordance of fMRI with WADA and direct cortical stimulation [3,7–9].

There are a variety of language tasks to choose from and there is currently no consensus in the field as to which is the most specific/sensitive. Some common tasks are verb generation (where the patients generates verbs to visually presented nouns), semantic fluency (where the patient generates words that fit a particular category like 'fruits'), phonemic fluency (where the patient is asked to generate words to a given letter), sentence completion (where the patient reads a portion of a sentence and chooses an appropriate last word) and narrative listening (where the patient answers questions about aurally presented narratives).

Typically, the most useful way to measure essential language function for presurgical planning in brain tumor patients is to do a combined task analysis as described by Ramsey et al [10]. The output of this analysis shows what is activated in common during different language tasks.

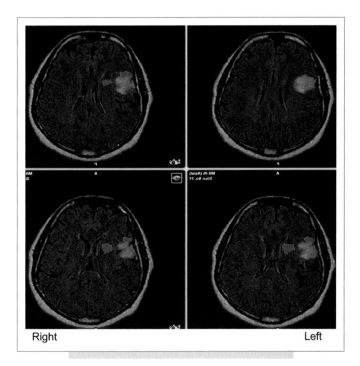

Right   Left

FIG. 27.5. Flair image of a right-handed man with dysarthria and seizures. MRI revealed a low-grade minimally enhancing lesion in the expected location of Broca's area. fMRI revealed both insular and inferior frontal activity. The patient was operated on awake with direct cortical stimulation and the location of Broca's area (speech arrest) was concordant with intraoperative cortical stimulation.

Arguably, this helps to decrease variability in the fMRI map and presumably de-emphasizes the supportive language areas and highlights the essential language areas. This, in turn, may provide a more accurate measure of language laterality in the brain tumor patient.

Localization

Localization of speech areas using fMRI for neurosurgical planning in brain tumor patients is nuanced. fMRI can localize the gyrus or set of gyri where Broca's area and Wernicke's area are generally located. Additionally, as with the motor maps, fMRI of language is also a useful guide for intraoperative mapping. Ideally, fMRI would localize the essential areas of the language networks to such a degree that the surgeon could plan margins during oncologic surgery. Unfortunately, this is not the case. The fMRI activations are subject to variations in size depending on the type of statistical analysis and statistical thresholding. If, for example, one were to change the correlation coefficient on a language map from 0.76 to 0.50, the spatial extent of the fMRI activity would grow. This may or may not reflect the actual extent of neuronal activity and is based on a complicated array of factors. As a result, surgeons will not plan surgical margins during brain tumor surgery based on fMRI maps of language function.

In addition to the variability of the spatial extent of the fMRI activations, fMRI language maps reveal both essential and supportive function. This means that some activations may be essential for the task (and therefore not resectable) and some may be supportive in such a way that they are resectable without risk of deficit to the patient. Currently, there is no way to segregate the activations seen on an fMRI language map into essential and supportive function. This makes the use of direct cortical stimulation or other complementary techniques necessary in many cases for surgical decision-making in brain tumor patients.

Figure 27.5 illustrates how language localization can guide the surgeon and aid in risk assessment and patient counseling. A 54-year-old man presented with seizures and slight dysarthria. MRI revealed a low-grade slightly enhancing lesion on T1. fMRI (during phonemic fluency) placed the lesion within Broca's area and lateral to insular language activity. The patient was counseled as to the risk of operating given the fMRI localization (and lateralization). The patient was operated on awake and direct cortical bipolar stimulation confirmed the location of Broca's area as measured by fMRI. In this case, only a biopsy was attempted due to the location of the tumor within Broca's area and flanked by fMRI activity. The patient had no new deficits following surgery.

SPECIAL CONSIDERATIONS FOR fMRI MEASUREMENTS USING BRAIN TUMOR PATIENTS

Patients with brain tumors present unique challenges to the fMRI researcher/clinician. It is important the patients are scanned using tasks that they can perform [11], otherwise, the interpretation of these scans is difficult. As a result, standardizing paradigms for use with patients will be difficult. In many institutions, patients receive neuropsychological exams (these can be brief) ahead of the scanning session in order to tailor the fMRI paradigm to the patient's strengths and weaknesses. A less time-consuming option is to design variations of the stimuli. For example, because the rate of stimulus presentation has an effect on the fMRI map, during a picture naming language task for example (where the patient is asked to name visually presented pictures), perfectly intact patients will receive nearly the same stimulus presentation that a normal subject might [12]. However, if the patient is aphasic, the rate and difficulty of the pictures can easily be adjusted to account for the patient's disabilities.

Thus, being able to monitor the patient's responses is important. Some have included vocalized speech paradigms for this purpose. However, there are many issues and a broad spectrum of literature discussing the difficulties with the use of vocalized speech during fMRI. Most of these are technical considerations and deal with the head motion inherent in vocalized paradigms. Head motion is one of the biggest issues associated with fMRI imaging in general, but is particularly an issue when scanning patients. Head motion can cause both false positive and false negative predictions of function. When it is correlated with the timing of the stimulus presentation it can be impossible to extract from real signal. With this said, some researchers have recently taken advantage of the hemodynamic lag to extract head motion [13,14]. By presenting the patient with an event-related design, the image where the patient actually spoke can be removed and the subsequent images (where the hemodynamic response occurred) can be measured relatively free of motion artifact.

This technique however, necessitates short stimulus durations and typically means that the experiment will be designed as an event-related paradigm. As aforementioned, when using event-related paradigms, many repetitions of the stimuli need to be presented in order to make up for the loss in statistical power associated with single (or near single) events. These types of paradigms, while they may be appropriate in measuring a more complete picture of the speech network, are generally not optimal for patients.

The Effects of Brain Tumors and Prior Surgery on the BOLD fMRI Signal

The BOLD fMRI signal is generally dependent upon a normally working neurovascular system. It is thought that in some types of tumors (particularly high-grade tumors with associated hypervascularity and abnormal vasculature) the vascular reactivity normally associated with moving one's hand for example, will no longer be as tightly coupled with the vascular response. It only follows that this may affect the reliability/accuracy of the BOLD signal. Holodny et al showed this effect in patients with brain tumors in the motor strip [15]. They showed a significant difference in fMRI signal volume during finger tapping when comparing the unaffected and affected hemispheres. Hou et al characterized this difference further and showed an inverse relationship between perfusion (rCBV) and the fMRI activation volume in grade IV gliomas [16]. That is, as perfusion measures increased, fMRI volume decreased.

Prior surgery is common when brain tumor patients come for treatment. This is an important consideration for fMRI because blood products and surgical hardware can distort the magnetic field and affect the fMRI results. This is pronounced in T2*-based fMRI images where dropout can cause a misinterpretation of fMRI results. Kim et al compared fMRI volumes of activity during finger tapping in the motor strip in the unaffected hemisphere and the hemisphere with prior surgery [17]. They showed a reduction in the volume of fMRI activity in the motor strip in the hemisphere with prior surgery. Figure 27.6 demonstrates the type of signal void that can affect T2* scans. Of course, such a dropout will adversely affect the accuracy of the fMRI measurement. As a result, the authors suggest that the raw T2* images be inspected before data analysis in order to avoid misleading interpretations of the fMRI data.

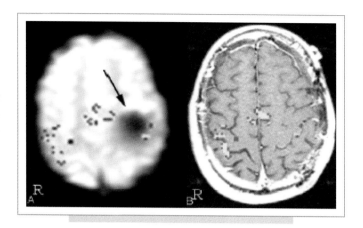

FIG. 27.6. The signal dropout in the T2* signal due to the effects of prior surgery. Also note that the volume of fMRI activity is decreased relative to the unaffected hemisphere. (From MJ Kim et al. The effect of prior surgery on blood oxygen level-dependent functional MR imaging in the preoperative assessment of brain tumors, Am J Neuroradiol 26(8):1980-1985, 2005, © by American Society of Neuroradiology.)

Peck et al did a similar study in patients with brain tumors, but asked whether the fMRI-determined laterality index (a measurement of the degree of language laterality; see pitfalls section for more information) is affected in patients with prior surgery when using direct cortical stimulation as a ground truth [18]. There were no statistically significant differences in the hemispheric determination of language laterality between patients with and those without prior surgery. However, there were discrepancies in the region of interest (Broca's area), laterality indices and the results of intraoperative mapping. In conclusion, cases where the patient has an extensive malignant lesion and/or the patient has had prior surgery, the fMRI should be interpreted with caution.

INTERPRETATION PITFALLS

It goes without saying that the value of the fMRI map in clinical practice is directly related to the end-stage interpretation of the fMRI data. Consequently, it is worth outlining some of the more common mistakes in fMRI interpretation.

Motor

Most post-processing protocols include some degree of spatial smoothing. This is widely accepted and not only boosts the signal-to-noise ratio, but eliminates spurious, pixilated activations and enhances the 'blob' appearance of the fMRI activations. Depending on how much smoothing is applied (twice the voxel size is typical), the motor activations can appear to cover gyri, cross sulci or even be located in white matter. The posterior bank of the motor gyrus and the posterior bank of the sensory gyrus are the portions of gray matter that typically activate in an fMRI scan. Using a 3T scanner where minimal smoothing is often required,

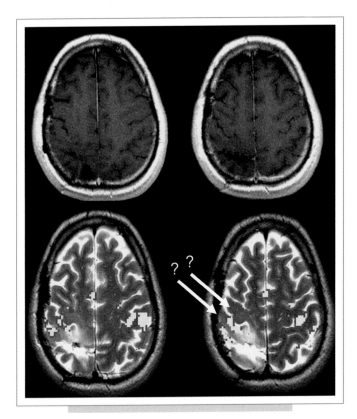

FIG. 27.7. A common interpretation pitfall. While the distribution of fMRI activity on this 1.5 T finger tapping fMRI (spatially smoothed at 4 mm) might lead one to believe that the posterior gyrus is the motor gyrus, the correct answer is the anterior gyrus as it is the posterior bank of the motor and sensory gyri that activate during an fMRI task. This T2* scan was not misregistered with the anatomical scan and is a common spatial distribution.

one can easily visualize the spatial extent of the gray matter participating. Most institutions use 1.5 T scanners, however, where one measures about one fourth of the signal. When smoothing is applied, these activations can be misleading. Without knowledge of how smoothing extends the spatial extent of the activations and some information about the exact anatomy expected to activate, one might assume that the motor gyrus is the more posterior gyrus in Figure 27.7. However, with a little information about the physiology and the spatial smoothing, the correct interpretation places the motor gyrus as the more anterior gyrus. (In this case, there was no misregistration between the anatomical and the functional scans, but that should be considered carefully.) This example points to a common distribution of fMRI activity and a potential for misinterpretation.

Language

In most completely right-handed, normal control subjects, despite clear left lateralization for language, there is often some right hemispheric activity while performing a language task. Because the laterality index (number of pixels in left hemisphere − number of pixels in the right hemisphere/number of pixels in left hemisphere + number of pixels in the right hemisphere) is a measure of ratios, this activity may decrease the degree of laterality but, in our experience, tends not to render a proven left-dominant individual bilateral or right dominant for language. Within the homologue of Broca's area in the right hemisphere, this is commonly attributed to the prosodic component of speech (broadly defined as intonation). It is important to expect right hemispheric activity, particularly in brain tumor patients where a degree of subclinical compensation may have occurred. In Figure 27.8, a patient with

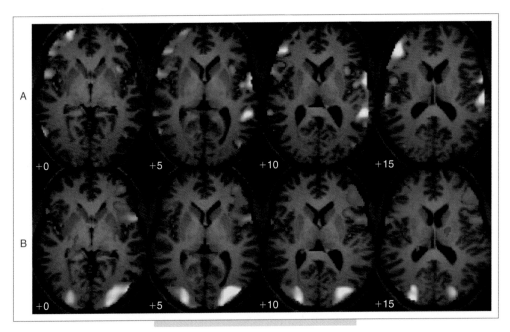

FIG. 27.8. Further potential for an error in interpretation. fMRI map of a right-handed patient with a left-hemispheric lesion in the supplementary motor area. **(A)** Patient silently responded to aural autobiographical questions. **(B)** Silent word generation to visually presented letters. The patient was left-hemisphere dominant on direct cortical stimulation. The examples illustrates the need to examine more than one task when determining dominance for language.

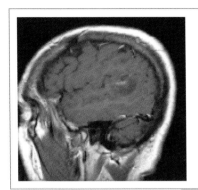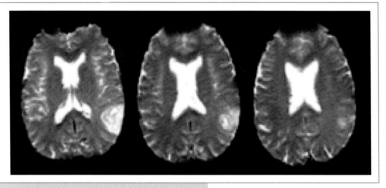

FIG. 27.9. A 62-year-old man with a minimally enhancing low-grade glioma in the expected location of Wernicke's area (left-superior temporal gyrus). fMRI suggested translocation of the patient's Wernicke's function that was confirmed both with direct cortical stimulation and gross total resection.

a low-grade glioma in the left superior frontal and frontal lobes performed both auditory responsive questions and phonemic fluency (where the patient silently generated words to visually presented letters). In the responsive questions task, there is a significant amount of bilateral activity that is not unexpected given the aural nature of the task. However, during the phonemic fluency task, the patient demonstrates the expected strong left dominance for language. The figure demonstrates the variability of the fMRI measurement and the need to scan patients on multiple language trials before final interpretation.

While it is likely that there is an abundance of brain-based compensation in brain tumor patients that is measurable on fMRI but does not change the patient's management, there are instances of plasticity that are clinically relevant. In our institution, we saw a 100% right-handed 62-year-old man with a low-grade neoplasm in the superior temporal gyrus (the expected location of Wernicke's area) [19]. He was completely intact on aphasia testing and fMRI suggested that while his Broca's function was nicely lateralized to the left-hemisphere as expected, his Wernicke's area only activated in the right hemisphere. Direct cortical stimulation confirmed the translocation and the patient was completely intact after a gross total resection of the lesion (Figure 27.9). This case nicely evidences the fact that while fMRI has its limitations in brain tumor patients, it can be a powerful first step to further testing when atypical patterns of activity are found.

REFERENCES

1. Birn RM, Cox RW, Bandettini PA (2002). Detection versus estimation in event-related fMRI: choosing the optimal stimulus timing. Neuroimage 15(1):252–264.
2. Mechelli A, Henson RN, Price CJ, Friston KJ (2003). Comparing event-related and epoch analysis in block design fMRI. Neuroimage 18(3):806–810.
3. Hirsch J, Ruge MI, Kim KH et al (2000). An integrated functional magnetic resonance imaging procedure for preoperative mapping of cortical areas associated with tactile, motor, language, and visual functions. Neurosurgery 47(3):711–721; discussion 721–722.
4. Mueller WM, Yetkin FZ, Hammeke TA et al (1996). Functional magnetic resonance imaging mapping of the motor cortex in patients with cerebral tumors. Neurosurgery 39(3):515–520.
5. Nimsky C, Ganslandt O, Kober H et al (1999). Integration of functional magnetic resonance imaging supported by magnetoencephalography in functional neuronavigation. Neurosurgery 44(6):1249–1255; discussion 1255–1256.
6. Wada J (1949). A new method for determination of the side of cerebral speech dominance: a preliminary report on the intracarotid injection of sodium Amytal in man. Iqakaa te Seibutzuquki 14:221–222.
7. Medina L, Aguirre E, Bernal B, Altman N (2004). Functional MR imaging versus Wada test for evaluation of language lateralization: cost analysis. Radiology 230:49–54.
8. Gaillard W, Balsamo M, Xu B et al (2002). Language dominance in partial epilepsy patients identified with an fMRI reading task. Neurology 59:256–265.
9. Binder J, Swanson S, Hammke T et al (1996). Determination of language dominance using functional MRI: a comparison with the Wada test. Neurology 46(4):978–984.
10. Ramsey NF, Sommer IE, Rutten GJ, Kahn RS (2001). Combined analysis of language tasks in fMRI improves assessment of hemispheric dominance for language functions in individual subjects. Neuroimage 13(4):719–733.
11. Price CJ, Crinion J, Friston KJ (2006). Design and analysis of fMRI studies with neurologically impaired patients. J Magn Reson Imaging 23(6):816–826.
12. Mechelli A, Friston KJ, Price CJ (2000). The effects of presentation rate during word and pseudoword reading: a comparison of PET and fMRI. J Cogn Neurosci 12 Suppl 2:145–156.
13. Huang J, Carr TH, Cao Y (2002). Comparing cortical activations for silent and overt speech using event-related fMRI. Hum Brain Mapp 15(1):39–53.

14. Abrahams S, Goldstein LH, Simmons A et al (2003). Functional magnetic resonance imaging of verbal fluency and confrontation naming using compressed image acquisition to permit overt responses. Hum Brain Mapp 20(1):29–40.

15. Holodny AI, Schulder M, Liu WC, Wolko J, Maldjian JA, Kalnin AJ (2002). The effect of brain tumors on BOLD functional MR imaging activation in the adjacent motor cortex: implications for image-guided neurosurgery. Am J Neuroradiol 21(8):1415–1422.

16. Hou BL, Bradbury M. Peck KK, Petrovich NM, Gutin PH, Holodny AI (2006). Effect of brain tumor neovasculature defined by rCBV on BOLD fMRI activation volume in the primary motor cortex. Neuroimage 32(2): 489–497.

17. Kim MJ, Holodnya AI, Hou BL et al (2005). The effect of prior surgery on blood oxygen level-dependent functional MR imaging in the preoperative assessment of brain tumors. Am J Neuroradiol 26(8):1980–1985.

18. Peck K, Brennan N, Hou B, Gutin P, Tabar V, Holodny A (2006). The reliability of presurgical fMRI language lateralization in patients with primary brain tumors and prior surgery. 14th ISMRM, Seattle, Washington, USA, May, 2006. 604.

19. Petrovich NM, Holodny AI, Brennan CW, Gutin PH (2004). Isolated translocation of Wernicke's area to the right hemisphere in a 62-year-old-man with a temporo-parietal glioma. Am J Neuroradiol 25(1):130–133.

CHAPTER 28

Magnetic Resonance Spectroscopy

T. Mikkelsen and D. Hearshen

INTRODUCTION

Proton magnetic resonance spectroscopy (^{1}H MRS) has been used in a number of applications for the diagnosis and management of patients with glioma. The technique uses a method which suppresses the water proton signal and allows the resolution of spectral signals for other molecules biochemically related to the biology of the tumor. These are identifiable by their resonance frequencies and typically include choline (Cho), which is thought to reflect membrane turnover, creatine (Cr) and N-acetylaspartate (NAA), a neuronal marker, and a frequency range containing overlapping signals which consists of resonances from lipid and lactate, associated with necrosis. Since these molecules are many times less abundant than water, larger voxels of acquisition are required. In addition, there are significant issues related to partial volume averaging from mixed tissues and those related to tissue/fluid boundaries, which may confound signal acquisition. MRS images (also called chemical shift imaging) can be generated by integrating the signals beneath these metabolite peaks, but the large volumes of acquisition required practically limit image resolution. There are also technical issues with regard to normalization and a number of different metabolite ratios are typically generated. Few validation studies have been carried out to verify the utility of these methods in glioma management, but several studies have correlated signal characteristics with tissue histology [1–4].

CLINICAL APPLICATIONS

The application of so-called metabolic imaging in glioma patients has been recently reviewed [5,6] and placed into context with other physiologic imaging techniques, including dynamic susceptibility contrast-enhanced magnetic resonance imaging (MRI).

Neoplastic Versus Non-Neoplastic

While MRS has been used to distinguish neoplastic from non-neoplastic lesions, the detection of gliomas is typically not a major issue [7,8].

Grading and Differential Diagnosis

A major issue in the assessment of brain tumors is grading and differential diagnosis [9]. Several investigators have used statistical correlation with primary diagnosis and predictive modeling to suggest that MRS is able to define lesion type and grade [10]. The number of independent variables available from the analysis of MRS data is large and there are many analytic approaches, including qualitative, quantitative and pattern recognition techniques. Most of these studies employ quantitative analysis, using measures combining multiple parameters to increase sensitivity and decrease the number of dimensions in the statistical analysis [1,10–12]. These measures usually involve ratios of metabolic indices with the numerator and denominator from the same volume (ipsilateral) or denominator from normal appearing tissue, usually white matter (contralateral, NAWM). Cho/NAA$_{ips}$, Cho/Cr$_{ips}$, Cho/Cr$_{con}$ and Cho/Cho$_{con}$ are widely used. Measures involving lactate and lipid signals have also been used [13]. Lactate, reflecting the presence of anaerobic metabolism, is visible in various stages of both treatment-naïve and treated tumor metabolism and also in the response of normal tissue to treatment. Signal from lipid is not seen in MRS of normal brain tissue, but is observed in both tumor and radiation necrosis. Measures involving lipid and lactate are subject to greater variability due to overlap between lipid and lactate resonant frequencies, and the range and overlap of resonant frequencies between various types of lipids, which are present in necrotic tissue in variable amounts. Lactate editing sequences have been used to separate lactate and lipid signals and may be useful in characterizing stages in treated tumor metabolism [14].

An analogous but distinct issue is the definition of lesion boundaries. This, of course, is related directly to prognosis, the measurement of the extent of resection, the definition of radiation ports and efficacy measurement. Gliomas in general, and more anaplastic gliomas in particular, infiltrate and spread great distances in the brain [15]. The regional infiltration during tumor progression has been most strikingly shown in the whole-mount studies of

Scherer and Burger, where glioblastomas have a central area of necrosis, a highly vascularized cellular rim of tumor and a peripheral zone of infiltrating cells [16,17]. Studies have shown that tumor cells have migrated from the primary site of malignant gliomas, resulting in the almost inevitable local recurrence and tumor progression seen clinically [17]. Recurrence of human gliomas following surgery and radiation is most commonly seen in the margin adjacent to the initial tumor where leaking tumor neovasculature is permeable to imaging contrast agents, but may also be remote [18].

Biopsy studies have shown that the conventional contrast-enhancing margin in MRI does not accurately reflect the lesion boundary. As a result, we have carried out a number of studies employing the techniques of multi-voxel MRS and directed surgical biopsies in order to validate alternative measures of the burden of disease [1]. Since conventional MRI does not adequately reflect the intratumoral heterogeneity of gliomas, we sought to determine a correlation between different ^{1}H MRSI metabolic ratios and the degree of tumor infiltration in diffusely infiltrating gliomas. Image-guided biopsies with semi-quantitative and qualitative histopathological analyses from a series of 31 untreated patients with low- and high-grade gliomas were correlated with multivoxel ^{1}H MRSI referenced to the same spatial coordinates. With 247 tissue samples and 307 observations, choline-containing compounds using contralateral creatine and choline for normalization or ipsilateral N-acetylaspartate, appeared to correlate best with the degree of tumor infiltration, regardless of tumor histologic grade. Our results show that MRSI is more accurate than conventional MRI in defining indistinct tumor boundaries and quantifying the degree of tumor infiltration. These data suggest that MRSI could play a role in the selection of stereotactic biopsy sites, especially in non-enhancing tumors. We would caution, though, that MRS should not be used as a substitute for histopathological diagnosis. A corollary would be that it may be possible to assist in the planning of surgical resection using this technique. It follows also that the assessment of residual disease after surgery is another important potential application. This would be particularly true for low-grade glioma or non-enhancing tumor components in high-grade lesions. This would follow from data in high-grade glioma, suggesting improved prognosis in patients who have gross total resections or imaging-complete resections, employing contrast enhancement [19,20]. Measurement of the contrast-enhancing compartment may also not be the optimal parameter for tumor response assessment. This holds true especially in this era of cytostatic agents which interfere with the proliferation, invasion, angiogenesis and differentiation processes. Several correlative histopathological studies have shown that the contrast-enhancing margin is not representative of the lesion boundary [21,22].

Prognosis and Survival

Several investigators have suggested that MRS features, including maximal choline and lactate levels, correlate with survival at least as well as standard clinicopathological parameters [23]. Prospective validation studies, however, have not been done [23–26].

MRS Guidance for Therapy

Surgical Planning

Determination of the extent of resection and biopsy site choice are two perisurgical applications for MRS. There are problems, typically, in performing MRS studies immediately postoperatively. Postoperative studies are often confounded by postoperative features, including intracavitary air and tissue/fluid boundaries, each of which can introduce significant artifacts [20].

Radiation Planning [27–29]

Another use for MRS is in image-guided therapy, including definition of radiation ports. Enhancement on MRI of gliomas acquired after the administration of contrast agents containing gadolinium (Gd) has traditionally been used to delineate tumor margin for surgical and radiation treatment planning. However, increased Cho and/or decreased NAA have been found well outside these margins [28], implying that images of Cho, NAA, or Cho/NAA$_{ips}$ (MRSI) may better reflect the distribution of infiltrating tumor. An example is shown in Figure 28.1. Though Gd enhancement is restricted to a small volume near the lateral ventricles, significantly increased Cho is found in a much larger volume, extending out to the cortical surface. This has prompted the development of strategies to combine MRI and MRS in radiation treatment planning [29]. Since MRSI has inherently lower resolution, these images are fused with higher resolution MRI, acquired during the same examination. Many treatment plans are generated using computed tomography (CT) images. This requires an intermediate fusion of the CT and MR coordinate systems [30]. An example of MRSI and MRI retrospectively fused with a treatment plan is shown in Figure 28.2. In this case, increased Cho suggestive of tumor is found outside the 80% isodose line. Subsequent MRI and tumor resection 9 months later indicated tumor in this region. In a retrospective review of radiation treatment plans for 10 patients, MRSI acquired prior to the treatment plan (but not used for planning) demonstrated increased Cho outside the 80% isodose line in three cases. Radiation treatment plans which rely on Gd enhancement may underestimate areas of infiltrative disease, while T2-weighted MR images probably overestimate its extent. MRSI adds additional information which may help to refine the treatment plans, however, MRSI has invariably lower resolution than MRI and the margins delineated by features on either image should be interpreted with caution when superimposed on the other.

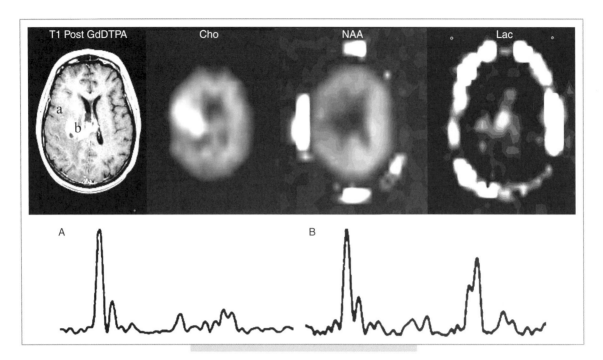

FIG. 28.1. Images and spectra depicting spectroscopic abnormalities associated with a high-grade glioblastoma multiforme (GBM). Upper left shows Gd enhancement of a mass within the periventricular white matter with invasion of the lateral ventricles. Spectroscopic images show increased choline (Cho) (top middle left) and decreased N-acetylaspartate (NAA) (top middle right) over a much larger area, while the image of lactate (top right) shows increased anaerobic metabolism in the medial aspect of the lesion. Decreased NAA and increased Cho indicate infiltrating disease at the lateral aspect of the abnormality.

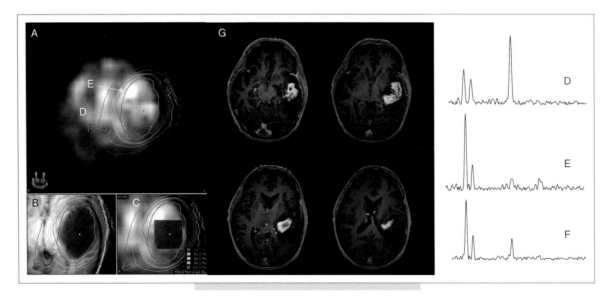

FIG. 28.2. The presenting diagnosis for this patient was grade 2 astrocytoma with atypical features. Only subtle contrast enhancement is shown **(B)**. The radiation treatment plan is shown fused to SI and Gd MRI in **(A)**, **(B)** and **(C)**. The isodose curve representing 80% of the 2.22 Gy delivered to the target is shown in green. Normal appearing contralateral white matter is shown in **(D)**. The heterogeneous lesion has a solid component (very high choline (Cho) and very low N-acetylaspartate (NAA), spectrum, **(E)**) within the 80% isodose curve and a component suggesting infiltrating disease (Cho not as high as **(E)** and NAA higher, spectrum **(F)**) outside the 80% isodose line. Follow-up MRI at 9 months after initial radiation treatment showed a new area of contrast enhancement **(G)** confirmed as recurrent tumor.

These preliminary data suggest that maximal efficacy from local field radiation treatment (RT) will not be achieved unless the spectroscopic abnormality is included within the treatment ports [31]. MRSI would theoretically be a more appropriate tool for target delineation for glioma radiation treatment planning; however, several series using modern radiation techniques with 3D conformal planning with or without dose escalation have shown that the majority of recurrences still occur in-field and not remotely or at the margin [18]. In this setting, MRSI appears unlikely to improve local control or change the pattern of recurrence of fractionated radiation therapy unless it proves to be more specific than conventional MRI. This technique would be particularly useful if higher doses appeared to be better tolerated by minimizing inclusion of normal brain parenchyma. Unfortunately, even with aggressive dose escalation for the treatment of glioblastoma, a high in-field failure rate is still observed [32].

Assessment of Treatment Efficacy/Response

The role of MRS in efficacy evaluation has been the subject of a recent NCI workshop report [33].

Radiation

The effect of therapeutic radiation on the brain, including decreases in NAA and subacute increases in choline have been described and can be difficult to discriminate from disease progression [34–36].

Chemotherapy

Decreases in choline signal in response to chemotherapy have been seen and monitored using MRS [34,37–40]. The technical issues of which volumes or features should be used in such monitoring, such as the peak heights, median or mean values and which normalization corrections should be done, remain to be determined.

Serial

Serial monitoring using MRSI appears to be a promising tool for tumor response assessment with novel non-cytotoxic agents. It will need to be compared with conventional contrast enhancement measurements [13,28]. Validation studies, including prospective use of these imaging modalities will be necessary before these can be used reliably as effective surrogates for therapy efficacy.

Differential Diagnosis

Recurrent Disease

Given the impact of specific treatment modalities on the MRS parameters described above in newly diagnosed patients, it has been noted recently that the spectroscopic features of recurrent disease are rather different than treatment-naïve tumors. This obviously must lead to caution regarding the interpretation of MRS in suspected recurrent disease using values derived from studies on newly diagnosed patients [2,41,42].

Malignant Transformation of Low-grade Tumors

Malignant degeneration in low-grade neoplasms of the brain has also been detected using MRS [43]. Discordant conventional and spectroscopic findings may also be used to define a more aggressive clinical course with earlier disease progression [44].

Differentiation of Radiation Necrosis and Recurrent Tumor

Early studies using single voxel techniques could not adequately sample heterogeneous tissues found at various stages in the treatment of gliomas. Since recurrent tumor and radiation necrosis can both result in enhancement with Gd, studies utilizing MRSI have examined spectral patterns from tissues exhibiting different features on MRI, with and without administration of Gd, in an attempt to differentiate recurrent tumor from radiation necrosis. Multivoxel MRSI allows simultaneous sampling of necrotic core, lesion margin defined by enhancement with Gd, and abnormal tissue outside lesion margins defined by hyperintensity on T2-weighted MRI, which may include remote foci and/or infiltrative disease.

Figure 28.3 shows an MRS study comparison of an image-guided-biopsy proven recurrence at the inferior location (seen on the left, site B) and biopsy-proven radiation necrosis at the superior location (seen on the right, site C). It is of note that at the radiation necrosis biopsy site, the ipsilateral Cho/NAA ratio is clearly abnormal, which would suggest tumor recurrence, but the Cho/nCho ratio is less than one. This situation might arise if treatment-naïve tumor and either treated tumor or radiation necrosis had either a different spectral pattern, or different admixtures of tissue types within the MRSI voxel itself.

The spectral characteristics of radiation-induced changes in normal and tumor tissue depend on the time interval between the application of radiation and the acquisition of MRSI. Reports of early changes to normal tissue include an increase in Cho and a moderate decrease in NAA, while spectral features of delayed radiation-induced necrosis have been reported to include an overall decrease in Cho, Cr and NAA and the presence of lipid and/or lactate. An example of early-induced radiation change is shown in Figure 28.4. The MRSI was acquired approximately one month after treatment with 2.22 Gy targeted to the Gd-enhanced volume. Note that the increased Cho in white matter contralateral to the site of the original lesion coincides with the 30% isodose line, a relatively low radiation dose. This is in contrast to the MRSI from the superior location (see Figure 28.4C), where the time interval between treatment and acquisition of MRSI was 9 months.

Our group has used the intraoperative biopsy correlation method to attempt to define diagnostic imaging parameters

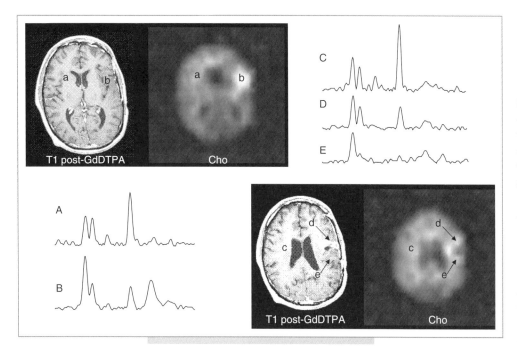

FIG. 28.3. Images and spectra of a patient with initial glioblastoma multiforme (GBM) who was evaluated for recurrent disease versus radiation necrosis. Image-guided stereotactic biopsy was performed within 24 hours of the MR examination. Biopsy confirmed both recurrent tumor **(B)** in the inferior location and radiation necrosis on the superior slice (D and E). Note that the Cho/NAA$_{ips}$ on **(D)** and **(E)** is clearly abnormal, but the Cho/Cho$_{con}$ is <1 while the Cho/Cho$_{con}$ is >1 at the recurrent tumor location (B).

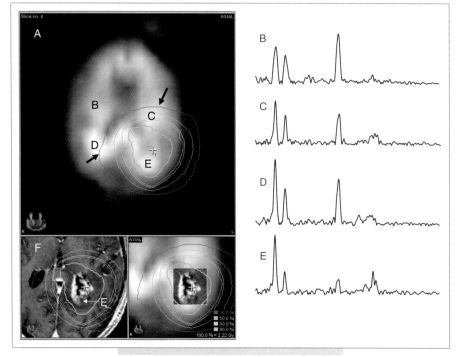

FIG. 28.4. Radiation treatment plan fused to SI and MRI **(A)**; Cho image, top and Gd MRI, lower left). MRSI was collected one month after 2.22 Gy was delivered to the target (80% isodose curve depicted in light green). Site **(B)** was normal contralateral white matter; site **(C)** showed increased Cho, decreased Cr and NAA along with the presence of Lac/Lip indicating tumor (spectra acquired from rim of contrast enhancement); at site **(D)** Cho is increased through the white matter tracts of the corpus callosum, but normal amounts of Cr and NAA suggest early radiation injury or infiltrative disease; site **(E)** shows increased Cho, decreased Cr and NAA along with the presence of Lac/Lip indicating tumor (spectra acquired from center of lesion). At the right parietal area of increased Cho, NAA occurs at normal levels and Lac/Lip is not present. Both of these findings are consistent with early radiation effect.

capable of discriminating recurrent tumor from radiation necrosis [36]. In that study, we looked at 27 patients who had been treated previously with surgery, radiotherapy and chemotherapy and reoperated for clinical and/or radiographic signs that caused suspicion for recurrent disease. Tissues were categorized into four groups:

1 spectroscopically normal
2 pure tumor
3 mixed tumor with necrosis
4 pure radiation necrosis.

Analysis was performed on 99 ^{1}H MRSI observations to determine whether the ^{1}H MRSI ratios varied according to tissue category. ^{1}H MRSI ratios were found to distinguish pure tumor from pure necrosis. No values suggested that mixed specimens could be distinguished in a statistically significant way from either pure tumor or pure necrosis.

The data suggest that metabolite ratios derived on the basis of ^{1}H MRSI spectral patterns do allow reliable differential diagnostic statements to be made when the tissues are composed of either pure tumor or pure necrosis, but the spectral patterns are less definitive when tissues composed of varying degrees of mixed tumor and necrosis are examined. This work is ongoing in an attempt to improve the discrimination of mixed cases, which is a condition commonly seen clinically.

MRSI nominal voxel volume, $\sim{<}1\,$ml, can still include admixtures of different tissue types in heterogeneous lesions as well as normal tissue, which potentially exist, even on high resolution MRI used for reference. In a preliminary study comparing MRSI with image-guided stereotactic biopsies contained within the MRSI voxel and characterized as greater than 80% pure tumor, lesions prior to and following therapy had significantly different spectroscopic features, probably reflecting the presence of micro-necrosis in the treated voxel [45]. This suggests that given the current resolution constraints in MRSI, it may not be possible to isolate the spectral characteristics of 'pure' treated/recurrent tumor. It is also possible that MRSI voxels contain admixtures of tumor, radiation necrosis and normal tissues.

There are several ways to apply MRS in the management of patients with malignant gliomas. While a number of proof-of-concept studies have been done as demonstrations, no prospective studies validating MRS methods as a surrogate for treatment or differential diagnosis have been completed. However, promising data suggest that these methods may find a role in the assessment of therapy efficacy and differential diagnosis.

REFERENCES

1. Croteau D et al (2001). Correlation between magnetic resonance spectroscopy imaging and image-guided biopsies: semiquantitative and qualitative histopathological analyses of patients with untreated glioma. Neurosurgery 49(4):823–829.

2. Rock JP et al (2004). Associations among magnetic resonance spectroscopy, apparent diffusion coefficients, and image-guided histopathology with special attention to radiation necrosis. Neurosurgery 54(5):1111–1117; discussion 1117–1119.

3. Cheng LL et al (1998). Correlation of high-resolution magic angle spinning proton magnetic resonance spectroscopy with histopathology of intact human brain tumor specimens. Cancer Res 58(9):1825–1832.

4. Dowling C et al (2001). Preoperative proton MR spectroscopic imaging of brain tumors: correlation with histopathologic analysis of resection specimens. Am J Neuroradiol 22(4):604–612.

5. Nelson SJ (2003). Multivoxel magnetic resonance spectroscopy of brain tumors. Mol Cancer Ther 2(5):497–507.

6. Cao Y et al (2006). Physiologic and metabolic magnetic resonance imaging in gliomas. J Clin Oncol 24(8):1228–1235.

7. Vuori K et al (2004). Low-grade gliomas and focal cortical developmental malformations: differentiation with proton MR spectroscopy. Radiology 230(3):703–708.

8. Hourani R, Horska A, Albayram S (2006). Proton magnetic resonance spectroscopic imaging to differentiate between non-neoplastic lesions and brain tumors in children. J Magn Reson Imaging 23(2):99–107.

9. Gruber S et al (2005). Proton magnetic resonance spectroscopic imaging in brain tumor diagnosis. Neurosurg Clin N Am 16(1):101–114, vi.

10. Preul MC et al (1996). Accurate, noninvasive diagnosis of human brain tumors by using proton magnetic resonance spectroscopy. Nat Med 2(3):323–325.

11. Cohen BA et al (2005). Assessing global invasion of newly diagnosed glial tumors with whole-brain proton MR spectroscopy. Am J Neuroradiol 26(9):2170–2177.

12. Stadlbauer A, Gruber S, Nimsky C et al (2006). Preoperative grading of gliomas by using metabolite quantification with high-spatial-resolution proton MR spectroscopic imaging. Radiology 238(3):958–969.

13. Li X et al (2005). Relationship of MR-derived lactate, mobile lipids, and relative blood volume for gliomas in vivo. Am J Neuroradiol 26(4):760–769.

14. Star-Lack J, Spielman D, Adalsteinsson E, Kurhanewicz J, Terris DJ, Vigneron DB (1998). In vivo lactate editing with simultaneous detection of choline, creatine, NAA, and lipid singlets at 1.5 T using PRESS excitation with applications to the study of brain and head and neck tumors. J Magn Reson 133(2):243–254.

15. Mikkelsen T et al (1998). *Brain Tumor Invasion: Clinical, Biological and Therapeutic Considerations*. Wiley-Liss, New York.

16. Scherer HJ (1940). The forms of growth in gliomas and their practical significance. Brain 63(3):1–35.

17. Burger PC, Kleihues P (1989). Cytologic composition of the untreated glioblastoma with implications for evaluation of needle biopsies. Cancer 63(10):2014–2023.

18. Hochberg FH, Pruitt A (1980). Assumptions in the radiotherapy of glioblastoma. Neurology 30(9):907–911.

19. Keles GE, Lamborn KR, Berger MS (2001). Low-grade hemispheric gliomas in adults: a critical review of extent of resection as a factor influencing outcome. J Neurosurg 95(5):735–745.

20. Fountas KN, Kapsalaki EZ (2005). Volumetric assessment of glioma removal by intraoperative high-field magnetic resonance imaging. Neurosurgery 56(5):E1166; author reply E1166.

21. Burger PC et al (1983). Computerized tomographic and pathologic studies of the untreated, quiescent, and recurrent glioblastoma multiforme. J Neurosurg 58(2):159–169.

22. Daumas-Duport C et al (1987). Serial stereotactic biopsies and CT scan in gliomas: correlative study in 100 astrocytomas, oligo-astrocytomas and oligodendrocytomas. J Neuro-Oncol 4(4):317–328.

23. Kuznetsov YE et al (2003). Proton magnetic resonance spectroscopic imaging can predict length of survival in patients with supratentorial gliomas. Neurosurgery 53(3):565–574; discussion 574–576.

24. Oh J et al (2004). Survival analysis in patients with glioblastoma multiforme: predictive value of choline-to-n-acetyl-aspartate index, apparent diffusion coefficient, and relative cerebral blood volume. J Magn Reson Imaging 19(5):546–554.

25. Li X et al (2004). Identification of MRI and 1H MRSI parameters that may predict survival for patients with malignant gliomas. NMR Biomed 17(1):10–20.

26. Fountas KN, Karampelas I (2004). Proton magnetic resonance spectroscopic imaging can predict length of survival in patients with supratentorial gliomas. Neurosurgery 55(1):257–258; author reply 258.

27. Pirzkall A et al (2004). 3D MRSI for resected high-grade gliomas before RT: tumor extent according to metabolic activity in relation to MRI. Int J Radiat Oncol Biol Phys 59(1):126–137.

28. Nelson SJ, Vigneron DB, Dillon WP (1999). Serial evaluation of patients with brain tumors using volume MRI and 3D 1H MRSI. NMR Biomed 12(3):123–138.

29. Nelson S, Graves E, Pirzkall A et al (2002). In vivo molecular imaging for planning radiation therapy of gliomas: An application of 1H MRSI. J Magn Reson Imaging 16(4):464–476.

30. Chang J et al (2006). Image-fusion of MR spectroscopic images for treatment planning of gliomas. Med Phys 33(1):32–40.

31. Chan AA et al (2004). Proton magnetic resonance spectroscopy imaging in the evaluation of patients undergoing gamma knife surgery for Grade IV glioma. J Neurosurg 101(3):467–475.

32. Souhami L et al (2004). Randomized comparison of stereotactic radiosurgery followed by conventional radiotherapy with carmustine to conventional radiotherapy with carmustine for patients with glioblastoma multiforme: report of Radiation Therapy Oncology Group 93–05 protocol. Int J Radiat Oncol Biol Phys 60(3):853–860.

33. Evelhoch J et al (2005). Expanding the use of magnetic resonance in the assessment of tumor response to therapy: workshop report. Cancer Res 65(16):7041–7044.

34. Lichy MP et al (2004). Monitoring individual response to brain-tumour chemotherapy: proton MR spectroscopy in a patient with recurrent glioma after stereotactic radiotherapy. Neuroradiology 46(2):126–129.

35. Kaminaga T, Shirai K (2005). Radiation-induced brain metabolic changes in the acute and early delayed phase detected with quantitative proton magnetic resonance spectroscopy. J Comput Assist Tomogr 29(3):293–297.

36. Rock JP et al (2002). Correlations between magnetic resonance spectroscopy and image-guided histopathology, with special attention to radiation necrosis. Neurosurgery 51(4):912–919; discussion 919–920.

37. Balmaceda C et al (2006). Magnetic resonance spectroscopic imaging assessment of glioma response to chemotherapy. J Neuro-Oncol 76(2):185–191.

38. Murphy PS et al (2004). Monitoring temozolomide treatment of low-grade glioma with proton magnetic resonance spectroscopy. Br J Cancer 90(4):781–786.

39. Preul MC et al (2000). Using proton magnetic resonance spectroscopic imaging to predict in vivo the response of recurrent malignant gliomas to tamoxifen chemotherapy. Neurosurgery 46(2):306–318.

40. Lodi R et al (2003). Gliomatosis cerebri: clinical, neurochemical and neuroradiological response to temozolomide administration. Magn Reson Imaging 21(9):1003–1007.

41. Lehnhardt FG, Bock C, Rohn G, Ernestus RI, Hoehn M (2005). Metabolic differences between primary and recurrent human brain tumors: a (1)H NMR spectroscopic investigation. NMR Biomed 18(6):371–382.

42. Preul MC et al (1998). Magnetic resonance spectroscopy guided brain tumor resection: differentiation between recurrent glioma and radiation change in two diagnostically difficult cases. Can J Neurol Sci 25(1):13–22.

43. Tedeschi G et al (1997). Increased choline signal coinciding with malignant degeneration of cerebral gliomas: a serial proton magnetic resonance spectroscopy imaging study. J Neurosurg 87(4):516–524.

44. Wu W-C et al (2002). Discrepant MR spectroscopic and perfusion imaging results in a case of malignant transformation of cerebral glioma. Am J Neuroradiol 23(10):1775–1778.

45. Hearshen DS et al (2004). Comparison of MRSI and image guided histopathology in treatment naive and treated patients with malignant glioma. Proc Intl Soc Mag Reson Med 11:389.

CHAPTER 29

Perfusion Imaging for Brain Tumor Characterization and Assessment of Treatment Response

James M. Provenzale and Kathleen Schmainda

INTRODUCTION

This chapter elucidates the more common types of magnetic resonance imaging (MRI) and computed tomography (CT) perfusion imaging methods to assess brain tumors. These methods require dynamic imaging such that images are acquired quickly to capture the fast physiologic (i.e. perfusion) information. Dynamic MRI methods for which an exogenous contrast agent is administered are referred to as dynamic contrast enhanced (DCE) or dynamic susceptibility contrast (DSC) studies depending on the primary sensitivity of the examination to either relaxivity (T1) or susceptibility (T2, T2*) contrast agent characteristics. While most perfusion techniques (both MRI and CT) rely on the administration of an exogenous contrast agent, newer dynamic MRI perfusion techniques, referred to as arterial spin labeling (ASL) techniques, do not. Instead they are capable of non-invasively 'labeling' blood water with a radiofrequency (RF) pulse resulting in MRI signal changes proportional to tissue perfusion.

Because dynamic contrast-agent enhanced MRI perfusion methods represent the majority of perfusion studies performed in brain tumors, the primary focus of this chapter will be on those methods. The general methodology for performing these studies will be discussed along with a review of existing data and challenges to applying these methods for diagnosis and assessment of treatment response. That discussion will be followed by a more concise summary of dynamic CT and ASL-MRI techniques for the study of brain tumors.

DYNAMIC CONTRAST-AGENT MRI PERFUSION TECHNIQUES

The term perfusion, as employed in the MRI literature, is typically used as an all-encompassing term describing both morphologic and functional parameters that can be derived from dynamic contrast-agent studies. This more generalized definition of perfusion differs from other non-MRI approaches that define perfusion specifically as the delivery of an agent to the tissue in ml/s per gram of tissue. Instead, while MRI perfusion studies can include measures of cerebral blood flow (CBF), which is measured in ml/s per ml of tissue, they also include measures of cerebral blood volume (CBV) and other derived parameters such as the tissue mean transit time (MTT), mean vessel diameter (mVD), plasma-tissue contrast agent transfer constant (K_{trans}) and compartmental volumes, such as v_e or EES, i.e. the extravascular–extracellular space. These derived parameters are of great interest as indicators of tumor angiogenesis, which is the new vessel formation required for tumor growth. As such, the parameters have potential applications in diagnosis, determining patient prognosis, predicting which patients will respond to a particular treatment and evaluating treatment efficacy.

With MRI there exist two general approaches to the dynamic imaging of contrast agent for the purposes of measuring perfusion parameters. The first approach relies on collecting T1-weighted images, over many minutes, during the uptake and accumulation of a low dose of a Gd (gadolinium)-chelated contrast agent. This dynamic T1-weighted approach is referred to as dynamic contrast enhanced (DCE) imaging. The second approach, termed dynamic susceptibility contrast (DSC) imaging, relies on collecting T2- or T2*-weighted images over a period of seconds during which a high concentration (bolus-dose) of Gd contrast agent passes through the tissue. Each approach measures a different subset of perfusion parameters and has a different set of challenges for the evaluation of tumors, as described below.

Dynamic Contrast Enhanced (DCE) Imaging
Basic Principles

Dynamic contrast-enhanced MRI (DCE-MRI) is the acquisition of a time-series of T1-weighted MR images acquired during the administration of a Gd-chelated contrast agent. In most studies, the Gd-chelate used is a low molecular-weight agent because, for many years, only these agents have been approved by the US Food and Drug Administration (FDA) for clinical use. Outside of the brain these agents distribute as extracellular agents, meaning that they distribute throughout the tissue, but are excluded from intact cells. However, in brain when the blood–brain barrier (BBB) is intact, low molecular weight Gd chelates remain intravascular. These organ-dependent distributions dictate in part the utility of DCE methods for deriving tissue specific information. For some animal DCE studies, large molecular weight Gd agents are also used. Because these agents are not cleared as readily from the vascular space, they offer some distinct advantages for more specifically measuring vascular permeability, as described in more detail below. Here, we will first describe DCE methods assuming an extracellular distribution, which reflects how the models were initially derived.

Unlike conventional contrast agent enhanced MRI, which simply provides a snapshot of enhancement at one time point, DCE-MRI provides a fuller description of the inflow ('wash-in') and outflow ('wash-out') contrast kinetics within tissue. Often the analysis of such data has taken the form of simple descriptors such as time to peak (i.e. enhancement slope) or peak enhancement. These parameters have been used in an attempt to differentiate tumor types and predict tumor grade, with mixed results [1–6]. In an effort to provide more quantitative data directly related to underlying tissue parameters, a variety of pharmacokinetic models have been applied to fit the data. Of the various different models used to analyze DCE-MRI data, the most commonly applied has been the Tofts and Kermode model first described in 1991 [7]. This model is described by the following equation:

$$dC_t/dt = K^{trans} (C_p - C_t/v_e) \qquad (1)$$

where C_p and C_t are the plasma and tissue contrast agent concentrations, K^{trans} is the volume transfer constant and v_e is the fractional extravascular–extracellular volume. The parameter labels and definitions used in Equation **1** are those recently recommended by an international group of scientists which reached a consensus regarding the DCE pharmacokinetic model and the underlying assumptions, definitions and labeling of the fitted parameters [8]. Therefore, this model is now often referred to as the consensus DCE model.

The transfer constant K^{trans}, which is equivalent to the product of extraction-fraction (E) and flow (F), is a function of flow and the permeability-surface (PS) area product $(EF = F (1-\exp^{(-PS/F)}))$, which has been extensively described in the earlier tracer kinetic literature. The EF product has several physiologic interpretations, depending on the balance between capillary permeability and blood flow. This balance must be well understood for the proper interpretation of studies measuring this index. Under high permeability conditions where PS is much greater than blood flow (F), the flux of contrast agent across the capillary membrane is flow-limited, meaning that K^{trans} is primarily a measure of flow ($K^{trans} \approx F$). Conversely, when the permeability is low such that PS << F, K^{trans} is more accurately used as an index of vascular permeability ($K^{trans} \approx PS$). Therefore, whether K^{trans} is more appropriately interpreted as an indicator of flow or vascular permeability, or some combination thereof, depends on the relative condition of the capillary membrane, flow rates and contrast agent size. So, especially for low molecular weight extracellular Gd contrast agents, K^{trans} is not a pure index and must not be interpreted as such. Alternatively, distributed parameter models have been proposed that enable the measurement of both PS and F separately rather than indirectly through the measure of K^{trans} [9,10]. These approaches have shown some promise for the characterization of gliomas [11,12] and information about the tumor microenvironment that enables discrimination of tumor types such as gliomas, meningiomas and brain metastases [13]. However, as with any multiparameter fitting approach, such approaches can be very sensitive to initial conditions; estimating an increasing number of parameters leads to increasing instability of a model, especially when the signal-to-noise (SNR) is low. Accordingly, a low contrast-to-noise ratio can prevent a fit on a per-voxel basis. Furthermore, the contrast material dosage may not be increased in a sufficient manner both because of FDA limits on total dose of contrast agent that can be administered and because of the need to avoid saturating the arterial input function (AIF; C_p in Equation **1**) [14].

Another approach to the measure of vascular permeability is to use large molecular weight agents. For these agents, PS << F, even under many pathologic conditions, making it more likely that K^{trans} is a more direct measure of vascular PS. Several recent animal studies have demonstrated that large molecular weight contrast agents, such as albumin-Gd-DTPA [15,16] or Gadomer-17 [14], may be more effective in detecting small permeability differences than Gd-DTPA [14,17]. The disadvantage to this approach is that large molecular weight Gd contrast agents, which have been used in animal models, have not yet been approved for human use.

In the original consensus model (see Equation **1**), the assumption is made that the effective vascular space is negligible. While this may be valid in some normal tissues such as brain and muscle, whose vascular fractions are only a few percent [8], it most assuredly is not true in highly vascular tumors. The assumption has been charged with resulting in erroneously high K^{trans} values. For these reasons, several investigators have modified the original consensus model to include the fractional vascular volume, so that Equation **1** is recast to define the transfer and rate constants explicitly in terms of the plasma and EES tracer:

$$v_e \, dC_e/dt = K^{trans} (C_p - C_e) \qquad (2)$$

where C_e is the contrast agent concentration in the EES and the contribution of the intravascular tracer is then added to give the total tissue contrast agent concentration:

$$C_t = v_p C_p + v_e C_e \tag{3}$$

where v_p is the fractional volume of the intravascular (plasma) space. To determine the effect of this modification, Harrer et al [12] compared the original model [7] (which also used an AIF previously determined from a sample of the normal population) with DCE for which v_p was not assumed negligible and the AIF was measured for each glioma patient. Their results suggest that more accurate values of K^{trans} were, in fact, obtained. Areas of high K^{trans}, as determined with the original model, were visible as areas of high v_p on the maps that included AIF measurements and v_p.

Alternatively, the multiple time-graphic method or Patlak plot [18] has been used to analyze the dynamic T1-weighted data, thereby also providing measures of the vascular transfer constants (K^{trans}) and plasma volume fraction (v_p) [19,20]. The theory underlying this approach is the same as that described above. The difference is that the Patlak model assumes that the measurement is made during the period where the plasma concentration is linear and no back-flux of contrast agent from the extravascular space occurs. Under these conditions the accumulation of tissue Gd is described by:

$$C_t(t)/C_p(t) = K^{trans} \int C_p(t')dt' + v_p \tag{4}$$

Consequently, plotting the ratio of $C_t(t)/C_p(t)$ as a function of $\int (C_p(t')dt'$ gives a plot where the slope is equal to K_{trans} and the intercept equal to V_p. The transfer constants determined with this approach were found to agree with those determined with ^{14}C-sucrose measurements [20].

Imaging Methodology

The general approach to collection of DCE data is to use a fast gradient-echo T1-weighted sequence. Typically a three-dimensional spoiled gradient-echo image with short repetition time (TR) (TR <7 ms), minimum echo time (TE) (TE <1.5 ms) and low flip angle ($\approx 30°$) is used [21]. The images are acquired every few minutes. Improved quantification of estimated gadolinium concentration requires the acquisition of a pre-contrast T1 map. For the contrast agent injection it is recommended that a power injector be used to minimize inter-study variation [22]. The total contrast dose should be administered over 15–30 seconds, followed immediately by a saline flush. For these pharmacokinetic models, a measure of the AIF is also required. To obtain the AIF, the measurement protocol must have sufficient temporal resolution [23], especially when the contrast agent is administered as a bolus. Often the AIF cannot be measured directly and a common AIF is used. However, determination of individualized AIFs has been shown to improve study reproducibility [24].

Use of DCE to Characterize Non-CNS Tumors

Some of the fundamental data validating DCE techniques in the assessment of tumors has come from investigations of non-CNS tumors [21,25–27]. Of the parameters determined, K^{trans}, i.e. the plasma-tissue contrast agent transfer constant, has thus far proven to be the most promising by showing changes in response to anti-angiogenesis therapy [28] and a correlation with tumor vascular endothelial growth factor (VEGF) expression and microvascular density in breast tumors [29,30]. In another study, investigators attempted to increase the specificity for predicting breast tumor malignancy based on permeability measurements by employing a macromolecular contrast agent [17]. This study used dynamic data to yield estimates of the endothelial transfer coefficient, K^{PS} [=K^{trans}] and the fractional plasma volume of the tumors. The authors found a moderate yet statistically significant correlation for K^{PS} with presence of malignant histological features, but did not find such a correlation for fractional plasma volume. Using a criterion that a K^{PS} value greater than 0 indicated a malignant lesion, the investigators found both a sensitivity and specificity for detecting malignant tumor of 0.73 and a positive predictive value of 0.79.

Distinction of High-grade Gliomas from Other Neoplasms

Studies correlating DCE measurements in brain tumors [12,24,31] with histological features are fewer in number compared with studies performing the same comparison in non-CNS tumors. However, in such brain tumor studies, investigators have found good correlation between the major forms of hemodynamic measures available from DCE-based studies (e.g. various forms of permeability measurements and fractional blood volume) and tumor grade. In one study, good correlation between fractional vascular volume measurements and grade of brain tumors was seen [32]. One study measured both fractional blood volume and microvascular permeability in high-grade brain tumors along with immunohistochemical labeling (anti-Ki-67 monoclonal antibody, or MIB-1 index) indicative of mitotic activity [33]. Strong correlation between microvascular permeability and MIB-1 index was seen; however, the correlation between fractional blood volume and MIB-1 index was not statistically significant. Example DCE results obtained from a patient with residual high-grade glioma are shown in Figure 29.1.

The primary issue with regard to the use of DCE techniques in brain tumors is that its utility and information depends on whether the blood–brain barrier is intact. Fractional vascular volume measurements are more applicable when the blood–brain barrier is intact and rCBV measurements are more applicable when the blood–brain barrier is interrupted [11].

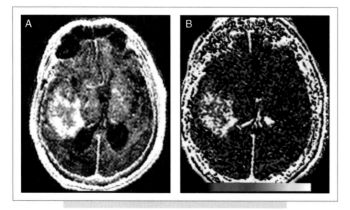

FIG. 29.1. Use of T1-weighted dynamic contrast-enhanced technique to evaluate permeability in a patient with residual high-grade glioma after previous resection. Technique uses dynamic contrast-enhanced 3D spoiled gradient imaging sequence obtained every 6.45 s for 58 s after IV infusion of 0.1 mmol/kg of gadopentetate dimeglumine. Before contrast material infusion, five unenhanced T1-weighted spoiled gradient images are obtained, each using different flip angle to derive T1 values of tissue needed for analysis of permeability. **(A)** Anatomic image obtained using contrast-enhanced T1-weighted spoiled gradient sequence through tumor shows lesion to have inhomogeneous contrast enhancement, with periphery of tumor having more marked contrast enhancement than central portion. **(B)** Quantitative color-coded K^{trans} (volume transfer constant between blood plasma and extravascular extracellular space, min^{-1}) map obtained at same location as **(A)** showing range of K^{trans} values from 0.0 min^{-1} (dark end of scale) to 0.1 min^{-1} (bright end of scale) shows marked increase in permeability within much of periphery of tumor. (Map of K^{trans} generated using TOPPCAT Software, courtesy of Daniel Barboriak MD). (Reprinted from Provenzale JM, Mukundan S, Dewhirst MW. The role of blood–brain barrier permeability in brain tumor imaging and therapeutics. Am J Roentgenol 2005; 185:763–767 by permission of the American Roentgen Ray Society.)

Dynamic Susceptibility Contrast (DSC) Imaging

Basic Principles

In 1990, Rosen et al [34] demonstrated that if imaging was performed during administration of a bolus of a Gd (gadolinium)-chelated contrast agent, a transient decrease in signal intensity could be seen (Figure 29.2). This approach is based on the principle of susceptibility contrast. Specifically, if a high concentration of a Gd contrast agent is confined to the vasculature, as is the case under bolus conditions, a difference in susceptibility or 'magnetizability' between the contrast-containing vessel and tissue occurs. The signal change can be converted into a relaxation rate change, which is proportional to the fraction of blood volume within each image pixel. Because the arterial contrast agent concentration is typically not measured, the resulting CBV image maps show solely relative CBV values and are therefore termed rCBV maps.

In addition to measuring CBV, cerebral blood flow (CBF) can also be determined from the same dynamic data. The quantification of CBF from DSC data requires the measurement of the arterial input function (AIF) and the deconvolution of the tissue concentration time curve. While there are many ways to perform the deconvolution, the model-independent singular value decomposition (SVD) approach is emerging as the most reliable [35].

DSC Imaging Methodology

A typical rCBV image protocol lasts only about 1–3 minutes, with images collected approximately every 1 second throughout this time period (see Figure 29.2). For the first 30–60 seconds baseline images are acquired. Next, a bolus of contrast agent is administered while the collection of images continues both during and after contrast-agent administration. To create the rCBV maps, the signal versus time data are converted into T2* (or T2) relaxation rate (ΔR2* or ΔR2) versus time data, which are proportional to

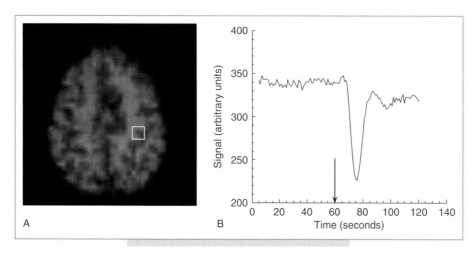

FIG. 29.2. Dynamic susceptibility contrast (DSC) data. **(A)** An echoplanar image (EPI) collected during the bolus administration of a gadolinium contrast agent. **(B)** A signal time course from a voxel within the echoplanar image in (A). Note the transient decrease in the signal intensity as the bolus of contrast agent passes through the brain tissue. This decrease is due to the susceptibility effect of the contrast agent.

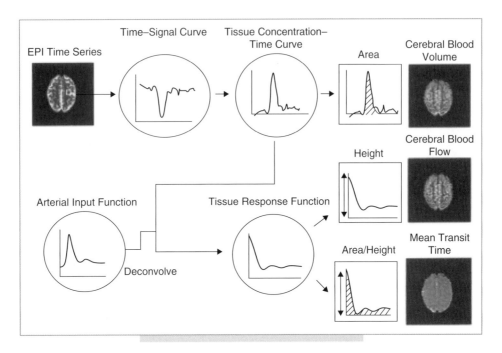

FIG. 29.3. Diagram explaining calculation of relative cerebral blood volume (rCBV), cerebral blood flow (CBF) and mean transit time (MTT) using dynamic susceptibility contrast technique. Signal-time course data for each voxel is converted to T2* relaxation rate changes (DR2*) which are assumed proportional to the tissue contrast agent concentration. Maps of rCBV are obtained by determining area below the trace concentration-time curve. Maps of relative cerebral blood flow are obtained by determining height of the tissue response function. To obtain the tissue response function, the arterial input function (AIF) must be deconvolved from the measured tissue concentration-time curve. (Adapted by permission from Petrella JR, Provenzale JM. MR perfusion imaging of the brain: techniques and applications. Am J Roentgenol 2000; 175:207–220 by permission of the American Roentgen Ray Society.)

tissue contrast agent concentration. These data can then be integrated (i.e. added up over time) to give rCBV for each image pixel (Figure 29.3). This description provides only the most basic implementation of a DSC rCBV protocol and the data analysis. Though seemingly simple, there are several issues regarding implementation and analysis that must be considered in order appropriately to use these methods. Several of these issues are briefly discussed here.

Choice of DSC Imaging Sequence

To acquire DSC image data, fast imaging techniques are required. In most cases, echoplanar imaging (EPI) methods are used. Though some have used fast gradient echo methods such as FLASH [36–39], the time resolution is not as good compared to EPI methods and, therefore, the quality of FLASH data is less than that of EPI. Most, if not all, newer clinical MRI systems have the necessary high-speed EPI capabilities.

There also exists a choice in the type of EPI sequence to use. Investigators use either gradient-echo (GE)-EPI techniques or spin-echo (SE)-EPI techniques, or an image sequence that acquires both GE and SE data [40,41]. GE and SE methods differ in their sensitivity to vessel diameter. While GE techniques are equally sensitive to all vessel diameters,

SE methods have a maximal sensitivity to microvessels [42]. In recent publications, we reported results using an image sequence that simultaneously collected both GE and SE data [40,41]. Besides providing maps of total and microvascular blood volume we were able to compute a GE:SE ratio related to the mean vessel diameter, as had been similarly done in a rat tumor model [43,44]. Example results using this simultaneous GE/SE method are given in Figure 29.4.

The image sequences used in rCBV mapping are T2- or T2*-weighted to enable them to be sensitive to the susceptibility effect described above. To maximize this sensitivity, attention must be paid to the choice of imaging parameters. Detailed analyses of the signal-to-noise ratio of rCBV maps and its dependence on the choice of experimental parameters and processing strategies, has been undertaken [45,46]. To maximize the signal-to-noise ratio of rCBV maps, it was recommended that the input bolus duration be as short as possible, that a sufficient number of baseline images are collected and that the imaging parameters (TE, TR) be optimized for the dose of contrast agent.

Post-Processing of DSC Data

In the study just described [45], it was also suggested that the straightforward integration of the signal time course,

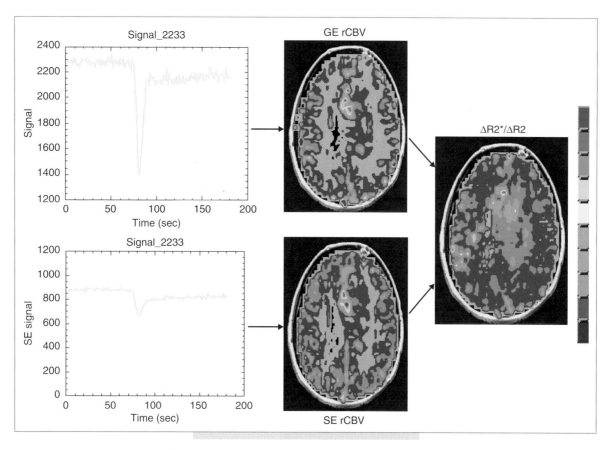

FIG. 29.4. Example of data and image information that can be collected when simultaneously collecting GE (gradient-echo) and SE (spin-echo) DSC data. Example GE and SE time courses are shown at left. The resulting GE and SE rCBV maps from a patient with a grade III astrocytoma are displayed in the center. The GE rCBV provides information about total blood volume, while the SE rCBV map gives information about the microvascular blood volume. The GE to SE relaxation rate ratio map, shown at right, provides a measure of the mean vessel diameter.

rather than a gamma variate fit of the time course, is the most reliable and practical method of determining rCBV, at least in normal brain. In another study, four different post-processing methods were compared for the determination of rCBV and CBF using Monte-Carlo simulations and in patients with cerebral ischemia [47]. The rCBV was determined by two different methods that numerically integrate the tracer concentration-time curves, with one approach integrating over the whole time series and the other limiting the integration to the first pass. In addition, two other methods were applied that take recirculation into account. One method used a gamma-variate fit, while the other method determined CBV by deconvolving the tissue time curve with the arterial input function and then integrating the resulting residue function. The study showed clear differences in precision and accuracy among these methods for the determination of relative as well as absolute CBV. For relative CBV measurements, simple numerical integration over the whole image range had clear advantages in terms of high SNR and modest computational requirements. Although gamma-variate fitting and deconvolution both demonstrated

accurate relative CBV measurements, these numerical manipulations introduced additional noise in the maps and increased computational complexity. For the determination of absolute CBV, the two methods that minimize recirculation effects were shown to be most accurate. Of the two approaches, the deconvolution approach gave maps with higher SNR compared to gamma-variate fitting and, because the deconvolution approach is necessary for the determination of CBF, this choice seems the most logical when absolute values are desired.

Contrast Agent Leakage Effects

Additional issues arise when using DSC methods to determine rCBV in brain tumors in particular. Specifically, determination of brain tumor rCBV from the contrast-enhanced MRI signal can be complicated by a leaky blood–brain barrier, as is often the case with tumors. Under these conditions, contrast agent leaks out of the vasculature into the brain or tumor tissue, thereby diminishing the susceptibility effect (i.e. producing a smaller signal decrease) of the contrast agent passage through the vasculature.

This feature may lead to an underestimation of rCBV. Various levels of consideration have been given to this issue. Many investigators perform rCBV imaging without leakage correction [48–51], but do mention it as a possible confounding factor. Others authors include data in their studies only if leakage is not demonstrated [51]. Some investigators use low flip angle GRE [52–54] or dual-TE [55] methods to minimize the T1 effects that occur with agent extravasation. Yet another approach is to manage this effect by both minimizing its impact during data collection and applying a correction during post-processing [40,41]. The approach is first to administer a preload of contrast agent to diminish any T1 changes that might subsequently occur during the bolus administration. Then, after collecting the dynamic data, leakage is detected and fitted on a pixel-wise basis [56,57]. With this approach, it was demonstrated that GE rCBV significantly correlated with tumor grade while the uncorrected maps did not [40,41,57]. Contrast agent leakage can also confound the determination of CBF. Though the effect is less profound compared to CBV, it does contribute inaccuracies. Consequently, a post-processing correction algorithm was also recently developed for purposes of measuring CBF in brain tumors [58].

As just described, most attention has been paid to minimizing the T1 effects resulting from agent extravasation. Although the T1 minimization sequences and post-processing algorithms correct for T1 enhancement effects, they may not correct for possible T2- and T2*-weighted signal changes that result from loss of compartmentalization or increased sensitivity to dipolar T2 effects [59]. These latter effects can be significant and result in *overestimations* of blood volume. Therefore, we need a better understanding of the various DSC methods, their respective accuracies and sensitivities to T1, T2 and T2* effects in the presence of leakage.

Distinction of Glioma Tumor Grade

One of the characteristic features of high-grade tumors relative to low-grade tumors is greater capillary density. Because rCBV is generally considered to reflect capillary density, it is reasonable to suppose that high-grade tumors have higher rCBV than low-grade tumors. Therefore, by using this dynamic susceptibility contrast (DSC) MRI method, several laboratories have demonstrated the feasibility of determining rCBV in patients with brain tumors [37–40,45,46, 48–50,52,60–62]. Several studies suggest that MRI-derived rCBV may better differentiate histologic tumor types than conventional MRI and provide information to predict glial tumor grade [38,40,49,50,52,63]. Example post-contrast and rCBV image maps of low- and high-grade gliomas are given in Figure 29.5.

Correspondingly, rCBV has been correlated with angiographic features of tumors [64]. In such studies, rCBV is typically measured within the tumor region and then expressed as a percentage of rCBV in a region of normal brain white matter. In most studies, no attempt has been

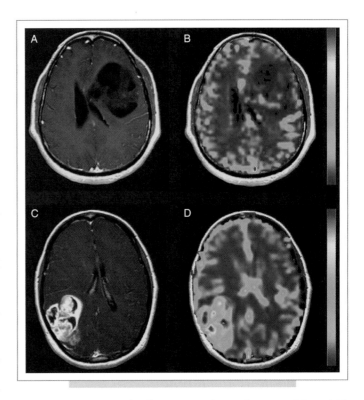

FIG. 29.5. T1-weighted contrast enhanced images (**A** and **C**) and GE rCBV maps (**B** and **D**) from a 46-year-old male diagnosed with a low-grade astrocytoma (**A,B**) and (**C,D**) a 45-year-old female diagnosed with a glioblastoma multiforme.

made to measure the same control region in all patients, but instead the control region typically differs in location from patient to patient [65]. On rCBV maps, normal gray matter regions are typically seen to have higher rCBV than normal white matter regions. The major reason cortical gray matter regions have not standardly been used as control regions is the fact that rCBV values in various cortical gray matter regions can vary widely in the same individual according to whether cortical vessels, which would elevate rCBV measures, are included in the region of interest used for rCBV measurement.

In most studies, the greatest rCBV values measured in a high-grade brain neoplasm are generally found to be much higher than in normal white matter and approximately the same as, or higher than, rCBV values in gray matter [66]. In these studies, investigators have typically measured the region of highest rCBV within a tumor, taking into account the fact that tumors are characteristically heterogeneous with regard to vascular supply and, thus, rCBV. Many studies have been performed to attempt to distinguish high-grade gliomas from other types of primary brain tumors (e.g. low-grade gliomas, central nervous system lymphoma and meningiomas) and metastases. The rCBV values in various series are often expressed as a ratio of highest rCBV in tumor compared to a representative rCBV value in a normal

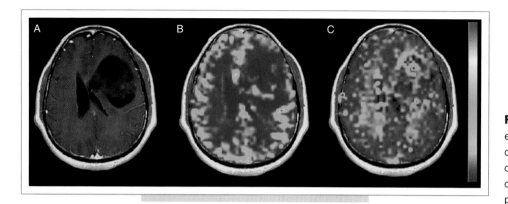

FIG. 29.6. T1-weighted contrast-enhanced images **(A)** and **(B)** GE rCBF and **(C)** GE MTT maps from a 46-year-old male diagnosed with a low-grade astrocytoma. These are from the same patient as shown in Figure 29.3.

brain region (typically normal appearing white matter in either the same hemisphere or contralateral hemisphere). The resultant ratio is referred to using various terms, of which the most common is the rCBV ratio.

A recent study using the combined GE/SE sequence [41], described above, emphasizes the importance of the choice of the tumor region of interest (ROI) on the study results. When extracting mean data from ROI of the entire tumor, the GE rCBV and ratio information showed statistically significant differences between low-grade and high-grade tumors, while the SE rCBV information did not [40,41]. However, when extracting the rCBV information from blood volume 'hot spots' or areas with the highest microvessel blood volume, a significant correlation was also found between SE rCBV and glioma grade [41]. This finding is consistent with the histopathologist's approach of assessing tumor angiogenesis, which is performed by evaluating areas of highest microvessel density.

Distinction of High-grade Gliomas from other Neoplasms

DSC-based studies have been generally successful in distinguishing high-grade glial neoplasms from other cerebral neoplasms. A number of studies have indeed verified the hypothesis that the mean values of regions of highest rCBV in high-grade gliomas are substantially higher than the mean values of regions of highest rCBV in low-grade gliomas [67,68]. One exception that has been noted is that of low-grade oligodendrogliomas. In one series that used a threshold rCBV ratio of 1.5 to distinguish high-grade tumors from low-grade tumors, the investigators found that this threshold had 100% sensitivity for detecting high-grade gliomas but that two of four low-grade oligodendrogliomas had rCBV ratios above this threshold [68]. In the same study, two of nine low-grade astrocytomas had rCBV ratios that exceeded the threshold ratio.

Distinction of high-grade gliomas from central nervous system lymphoma is another difficulty that neuroradiologists face with some frequency based on the two entities, on similar imaging features of the two types of tumors on conventional MR imaging studies. In one study, lymphoma lesions could be distinguished from high-grade glial neoplasms on the basis of lower mean rCBV values [69]. The authors also found that primary CNS lymphoma lesions typically had a time-signal intensity curve characterized by an increase in signal intensity above baseline following the initial decrease in signal intensity. Such an increase has been shown to be associated with marked leakage of contrast material across a very leaky blood–brain barrier; if not taken into account when analyzing DSC images, such an increase in signal intensity can produce lower than expected rCBV measurements [70].

In another such example, dural metastases from non-CNS sources were found to have significantly lower rCBV values than lesions with which they could be confused on radiological grounds, i.e. meningiomas [71].

Potential Role of CBF to Characterize Brain Tumors

Measurement of CBF provides information not derived from CBV measures. For example, it has been shown that areas of high vessel density may not necessarily correspond to well-perfused (oxygenated) areas [72]. Blood flow within solid tumors is spatially and temporally heterogeneous [72–75]. The importance of tumor heterogeneity as a factor influencing the response to treatment has been well described [15,72,76]. Blood flow is necessary for the hematogenous delivery of anti-cancer agents. Decreased perfusion can result in tissue hypoxia, which diminishes the tumor's sensitivity to radiation treatment [77]. Thus, measurement of CBF should aid in monitoring the progress of therapies [78]. This may be of particular importance for antivascular agents that have been shown to induce acute and massive tumor hemostasis [79,80]. Therefore, measurement of CBF provides functional information about the tumor vasculature, which is complementary to the morphologic measurement provided by CBV images. Example CBF images, also derived from DSC data, are given in Figure 29.6. These images were determined using the singular value decomposition method after correcting the data for contrast agent leakage effects [58].

Assessment of Tumor Vascular Architecture

As mentioned above, use of combined GE/SE DSC techniques has also shown that a measure of a voxel's mean vessel diameter (mVD) can also be determined. In particular, due to the different sensitivities of GE and SE to vessel diameter, theory has shown that the ratio of GE and SE relaxation rates ($\Delta R2^*/\Delta R2$) is proportional to the mVD [42]. A strong correlation between this ratio and an independent measure of vessel diameter has been demonstrated in animal tumor models [43,44]. We have also demonstrated, for the first time in patients, that the ratio was significantly correlated with tumor grade [40,41].

Assessment of tumor vessel architecture, with measurements of such parameters as mVD, may be important to our understanding of treatment effects. Several studies have demonstrated changes in vascular architecture with tumor growth [81], in response to treatment and under various conditions such as hypoxia. It has been suggested that administration of individual antiangiogenic agents may block specific angiogenic pathways and thus result in different patterns of tumor vessel formation [82]. In one study using DSC methods, rats treated with the corticosteroid dexamethasone showed that the mean vessel diameter and overall architecture was much more like that in normal brain [83]. This result may be explained by the fact that dexamethasone inhibits VEGF expression [84,85] and vessels become dilated when both VEGF and angiopoietin-2 are present [81].

MR Perfusion Imaging Assessment of Peritumoral Regions

Definition of tumors margins on cross-sectional imaging studies is one of the more difficult tasks the neuroradiologist encounters in assessing imaging studies of patients with contrast-enhancing brain tumors. The problem would be relatively simple if one could assume that the margins of a contrast-enhancing lesion coexisted with the margins of the tumor. However, numerous reasons exist for believing this assumption is generally invalid. First, histological analysis of high-grade brain neoplasms frequently shows gross or microscopic evidence of tumor extension beyond the regions that contrast enhance on imaging studies. Second, many tumors do not contrast enhance (or weakly enhance) on imaging studies. Therefore, use of enhancing regions to demarcate tumor extent is fruitless. For these reasons, neuroradiologists have turned to imaging techniques other than solely conventional contrast-enhanced CT or MR imaging to attempt to define actual tumor extent. The major techniques for this application are advanced MR imaging techniques (e.g. MR spectroscopy, diffusion imaging and perfusion imaging) and nuclear medicine techniques (e.g. positron emission tomography). This review will focus solely on the role of MR perfusion imaging to define tumor margins.

In one study, investigators using DSC methods to study tumor margins found that rCBV values in the peritumoral regions of high-grade gliomas were significantly higher than in metastases [86]. However, in the same study, rCBV in metastases was substantially higher than in gliomas, which seems at variance with the finding of higher rCBV in peritumoral regions of metastases. Another study also found significantly higher rCBV values in the peritumoral regions of high-grade gliomas than in metastases; in that study, rCBV values in the two types of tumors being studied were very similar [87].

MR Perfusion Imaging for Assessment of Tumor Therapy

Whereas a number of well-designed studies to monitor response to tumor therapy have been performed in animal models, relatively few studies have used MR perfusion imaging to assess tumor response in humans. Animal studies will be discussed first.

In one study using a murine breast tumor model and DCE technique, investigators used a heavily T1-weighted three-dimensional spoiled gradient-refocused acquisition in a steady-state pulse (SPGR) sequence and a macromolecular contrast medium to assess response following treatment with anti-VEGF antibody [88,89]. The study found both decreased tumor growth rates and decreases in microvascular permeability following treatment. In another study that also used DCE technique and rats harboring a tumor raised from human glioblastoma multiforme cell lines, investigators showed a significant decrease in microvascular permeability after therapy with a monoclonal antibody raised against VEGF [90].

With regard to therapeutic trials in humans, in one of the few trials of an antiangiogenesis therapy for human brain tumors, early decreases in rCBV (measured on DSC MR imaging) and degree of contrast enhancement (measured on DCE imaging) were seen [91]. In another study, investigators sought to determine whether the rCBV measured in tumors prior to therapy would allow prediction of treatment outcome [92]. The pretreatment rCBV did not allow such prediction but, instead, the change in rCBV after the first 6 weeks of therapy did allow prediction of treatment outcome at 6 months [92]. Change in rCBV was substantially more sensitive in prediction of outcome than change in tumor size over the same period.

Distinction of Tumor Recurrence from Radiation Necrosis

Another problem that neuroradiologists encounter on a fairly regular basis is distinction of tumor recurrence from radiation necrosis. The problem lies in the fact that both processes can produce marked interruption of the blood–brain barrier (and subsequent contrast enhancement after intravenous infusion of contrast material) and mass effect. Thus, the distinction of radiation necrosis from tumor recurrence using conventional contrast-enhanced MRI and CT imaging is often difficult. Alternative imaging techniques, such as FDG-PET and MR spectroscopy, are increasingly used in an attempt to distinguish radiation necrosis from residual tumor [93]. MR imaging perfusion imaging has also been brought to bear on distinguishing these two entities. The theoretical basis for the use of perfusion MR imaging lies in the fact that recurrent high-grade tumor is marked by angiogenesis (which would typically be associated with high rCBV) and tumor necrosis is generally associated with lack of rCBV elevation [94]. In one study, investigators used dynamic MR imaging to use rate of contrast enhancement to distinguish high-grade tumor recurrence from radiation necrosis [95]. In that study, signal enhancement-time curves were analyzed by fitting to a sigmoidal-exponential function and the resultant maximal enhancement rates were calculated as the first derivative of the fitted curve. The mean maximal enhancement rate in tumor recurrence was significantly higher than in radiation necrosis; cases of mixed tumor recurrence and radiation necrosis had values that were intermediate between solely tumor recurrence and solely radiation necrosis. Similarly, another study used DSC imaging in an attempt to distinguish recurrent tumor from radiation necrosis in 20 cases [62]. Recurrent tumor was consistently found to have high rCBV ratio whereas radiation necrosis was seen to have low rCBV relative to contralateral normal brain tissue. Finally, another report has shown that it is possible to depict a difference by DSC technique between low-grade astrocytomas following radiation treatment and normal brain [96].

Arterial Spin Labeling Techniques

Recently, MR perfusion imaging using arterial spin labeling has begun to show promise as a technique for hemodynamic assessment of brain tumors. At the time of writing, the technique is not widespread and only a small number of clinical studies have been performed in humans. Nonetheless, because of its promise, the technique will be reviewed here.

Arterial spin labeling is a completely non-invasive MR imaging technique because it does not require infusion of exogenous tracer material. Instead, a subject's own blood is used as an intrinsic tracer. Arterial blood outside the imaging plane is labeled using an inversion pulse and, upon entering the imaging section, spins within blood exchange with tissue water within capillaries [65]. By subtracting an image of the target tissue obtained prior to blood labeling from an image obtained after magnetized spins have entered the target slice, a hemodynamic image is obtained. Preliminary studies using animals harboring tumors have shown promising results for measuring various hemodynamic parameters within tumors [97]. Work in humans with brain tumors has shown ability to distinguish high-grade gliomas from low-grade gliomas, as well as very good correlation of the ratio of tumor blood flow to cerebral blood flow as determined by arterial spin labeling technique and

dynamic susceptibility contrast technique [65]. Example ASL images, obtained from a healthy volunteer are given in Figure 29.7.

Although investigators have been heavily reliant on DCE techniques to determine degree of permeability within tumors, permeability can also be assessed from DSC images within certain limitations. Such measurements are possible because two dominant features contribute to the time signal-intensity curve generated from DSC-based data after rapid passage of a bolus on gadolinium-based contrast material. The predominant feature is the so-called K_1 effect, which represents the effect of intravascular contrast material; this effect is used to derive rCBV maps. However, another feature contributes to the time-signal intensity curve, the so-called K_2 effects generated by the T1-shortening effects of the extravascular component of the infused contrast material. The K_2 effects can be analyzed separately from the K_1 effects to produce maps that depict relative degrees of leakage of contrast material. In one study, investigators found a correlation between increasing degree of permeability derived using DSC technique and increasing tumor grade [98]. Although some degree of overlap was seen between permeability values for low-grade tumors and high-grade tumors, in general, the two tumor types could be distinguished from one another.

CT Perfusion Imaging

In recent years, CT perfusion imaging has gained acceptance as a valuable imaging technique for assessment of some central nervous system diseases, such as cerebral ischemia. However, because MR imaging remains the preferred technique for assessment of brain tumors, reports of CT perfusion studies of rCBV for evaluation of brain tumors are much fewer in number than reports of MR imaging for the same purpose. However, CT perfusion imaging for assessment of brain tumors has been well-validated in animal studies [99]. In humans, CT perfusion imaging has been reported to be of value in assessment of head and neck tumors (possibly reflecting the fact that CT imaging remains the preferred technique for evaluation of neck tumors at many institutions). For instance, using a technique that assesses not only the initial phase of contrast enhancement, but also the time-signal intensity curve over the first few minutes after infusion of contrast material, investigators have examined regional blood flow, capillary permeability and relative compartmental volumes in human neck tumors [100]. Those investigators found higher regional blood flow in neck neoplasms compared to normal muscle, but attributed that higher blood flow to a higher regional blood volume rather than to higher flow per unit plasma volume. Published examples of CT perfusion imaging for assessment of brain tumors remain confined at this point to small series or case reports. In a report of five cases of brain tumors assessed using CT perfusion imaging, high-grade gliomas were found to produce heterogeneous

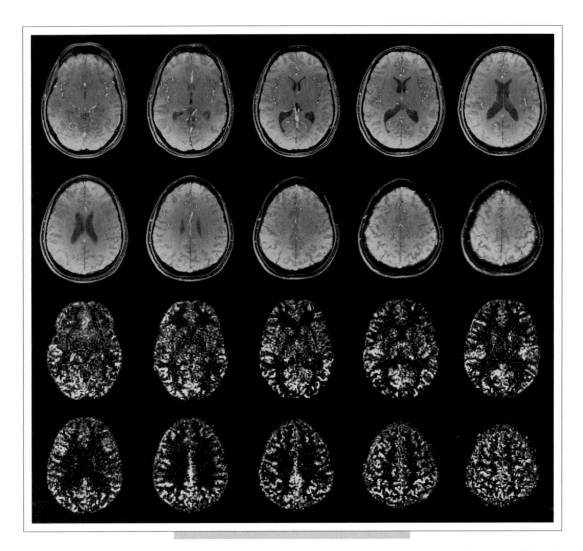

FIG. 29.7. T1-weighted anatomic (upper images) and ASL images (lower images) obtained from a healthy volunteer on a 3T MRI System. The ASL images were obtained using a background suppressed PICORE/QUIPSSII tagging scheme implemented with a multi-shot, spin-echo spiral-out sequence, TI2=1600 ms, TI1=700 ms, tag thickness=20 cm, tag gap=1 cm, TE=11 ms, TR=3000 ms. (Images are courtesy of Eric Paulson, Medical College of Wisconsin.)

rCBV maps with regions of high rCBV, whereas low-grade gliomas solely had only low rCBV [101]. These findings are similar to those reported using MR imaging. In one series of patients with various diagnoses studied by CT perfusion, the investigators reported markedly elevated rCBV and rCBF within contrast-enhancing tumor in a patient with a malignant brain neoplasm [102]. Another case report reported CT perfusion imaging findings in a patient with a high-grade glioma who could not undergo MR imaging because of the presence of a cardiac pacemaker [103]. CT perfusion imaging showed moderately elevated rCBV and markedly elevated rCBF and mean permeability surface flow within the tumor compared to normal tissue. Finally, CT perfusion findings have been reported in two patients with brain metastases developing from a non-CNS origin, of which one also underwent MR perfusion imaging using DCE technique [104]. In both patients, the tumors were seen to contrast-enhance on conventional CT imaging and both had markedly increased permeability-surface area measurements on CT perfusion. One tumor was seen to have normal rCBF and only mildly elevated rCBV, while the other tumor had both markedly increased rCBF and rCBV. In the latter patient, regions of highest permeability slightly differed on CT perfusion imaging compared to regions of highest permeability on MR imaging. However, the regions of high rCBV on CT perfusion imaging conformed relatively well to regions of tumor having high rCBV on MR perfusion imaging.

REFERENCES

1. Kaiser WA, Zeitler E (1989). MR imaging of the breast: fast imaging sequences with and without Gd-DTPA. Radiology 17:681–686.

2. Stack JP et al (1990). Breast disease: tissue characterization with Gd-DTPA enhancment profiles. Radiology 174:491–494.

3. Kelcz F et al (1993). Reducing the false positive gadolinium-enhanced breast MRI results through parameters analysis of the enhancement profile. In *12th Annual Meeting of the Society of Magnetic Resonance in Medicine*. New York.

4. Noseworthy MD, Morton G, Wright GA (1999). A comparison of normal and cancerous prostate using dynamic T1 and T2* weighted MRI. In *7th Annual Meeting of the International Society of Magnetic Resonance in Medicine*. Philadelphia.

5. Hawihorst, H. et al (1998). Uterine cervical carcinoma: comparison of standard and pharmacokinetic analysis of time-intensity curves for assessment of tumor angiogenesis and patient survival. Cancer Res 58:3598–3602.

6. Buckley DL et al (1997). Microvessel disunity in invasive breast cancer assessed by dynamic Gd-DTPA enhanced MRI. J Magn Reson Imaging 7:461–464.

7. Tofts PS, Kermode AG (1991). Measurement of the blood–brain barrier permeability and leakage space using dynamic MR imaging: 1. Fundamental concepts. Magn Reson Med 17:357–367.

8. Tofts PS et al (1999). Estimating kinetic parameters from dynamic contrast-enhanced T1-weighted MRI of a diffusable tracer: standardized quantities and symbols. J Magn Reson Imaging 10:223–232.

9. StLawrence KS, Lee TY (1998). An adiabatic approximation to the tissue homogeneity model for water exchange in the brain: I. Theoretical derivation. J Cereb Blood Flow Metab 18:1365–1377.

10. Johnson JA, Wilson TA (1966). A model for capillary exchange. Am J Physiol 210:1299–1303.

11. Ludemann L, Hamm B, Zimmer C (2000). Pharmacokinetic analysis of glioma compartments with dynamic Gd-DTPA-enhanced magnetic resonance imaging. Magn Reson Imaging 18:1201–1214.

12. Harrer JU et al (2004). Comparative study of methods for determing vascular permeability and blood volume in human gliomas. J Magn Reson Imaging 20(5):748–757.

13. Ludemann L et al (2000). Quantitative measurement of leakage volume and permeability in gliomas, meningiomas and brain metastases with dynamic contrast-enhanced MRI. Magn Reson Imaging 23:833–841.

14. Henderson E et al (2000). Simultaneous MRI measurement of blood flow, blood volume, and capillary permeability in mammary tumors using two different contrast agents. J Magn Reson Imaging 12:991–1003.

15. Vajkoczy P, Menger MD (2000). Vascular microenvironment in gliomas. J Neuro-Oncol 50(1–2):99–108.

16. vanDijke CF et al (1996). Mammary carcinoma model: correlation of macromolecular contrast-enhanced MR imaging characterizations of tumor microvascultature and histologic capillary density. Radiology 198(3):813–818.

17. Daldrup H et al (1998). Correlation of dynamic contrast-enhanced MR imaging with histologic tumor grade: comparison of macromolecular and small-molecular contrast media. Am J Roentgenol 171(4):941–949.

18. Patlak CS, Blasberg RG, Fenstermacher JD (1983). Graphical evaluation of blood-to-brain transfer constants from multiple-time uptake data. J Cerebr Blood Flow Metab 3:1–7.

19. Kenney J et al (1992). Measurement of blood–brain barrier permeability in a tumor model using magnetic resonance imaging with gadolinium-DTPA. Magn Reson Med 27:68–75.

20. Ewing JR et al (2003). Patlak plots of Gd-DTPA MRI data yield blood–brain transfer constants concordant with those of 14C-sucrose in areas of blood–brain opening. Magn Reson Med 50(2):283–292.

21. Choyke PL, Dwyer AJ, Knopp MV (2003). Functional tumor imaging with dynamic contrast enhanced magnetic resonance imaging. J Magn Reson Imaging 17:509–520.

22. Evelhoch J et al (2005). Expanding the use of magnetic resonance in the assessment of tumor response to therapy: workshop report. Cancer Res 65(16):7041–7044.

23. Henderson E, Rutt BK, Lee TY (1998). Temporal sampling requirements for the tracer kinetics modeling of breast disease. Magn Reson Imag 16:1057–1073.

24. Rijpkema M et al (2001). Method for quantitative mapping of dynamic MRI contrast agent uptake in human tumors. J Magn Reson Imaging 14(4):457–463.

25. Degani H et al (1997). Mapping of patholophysiologic features of breast tumors by MRI at high spatial resolution. Nat Med 3:780–782.

26. Mayr NA et al (1999). MR microcirculation assessment in cervical cancer: correlations with histomorphological tumor markers and clinical outcome. J Magn Reson Imaging 10:267–276.

27. Padhani AR (2002). Dynamic contrast-enhanced MRI in clinical oncology: current status and future directions. J Magn Reson Imaging 16: 407–422.

28. Gossmann A et al (2002). Dynamic contrast-enhanced magnetic resonance imaging as a surrogate marker of tumor response to anti-angiogenic therapy in a xenograft model of glioblastoma multiforme. J Magn Reson Imaging 15:223–240.

29. Knopp MV et al (1999). Pathophysiologic basis of contrast enhancement in breast tumors. J Magn Reson Imaging 10:260–266.

30. Hulka CA et al (1997). Dynamic echo-planar imaging of the breast: experience in diagnosing breast carcinoma and correlation with tumor angiogenesis. Radiology 205(3): 837–842.

31. Yang S et al (2003). Dynamic contrast-enhanced perfusion MR imaging measurements of endothelial permeability: differentiation between atypical and typical meningiomas. Am J Neuroradiol 24:1554–1559.

32. Lund R et al (2005). Using rCBV to Distinguish Radiation Necrosis from Tumor Recurrence in Malignant Gliomas. In *47th Annual Meeting of American Society for Therapeutic Radiology and Oncology*. Denver.

33. Roberts HC et al (2001). Correlation of microvascular permeability derived from dynamic contrast-enhanced MR imaging with histologic grade and tumor labeling index: a study in human brain tumors. Acad Radiol 8:384–391.

34. Rosen BR et al (1990). Perfusion imaging with NMR contrast agents. Magn Reson Med 14:249–265.

35. Ostergaard L et al (1996). High resolution measurement of cerebral blood flow using intravascular tracer bolus passages. Part I: Mathematical approach and statistical analysis. Magn Reson Med 36:715–725.

36. Hacklander T et al (1996). Cerebral blood volume maps with dynamic contrast-enhanced T1-weighted FLASH imaging: normal values and preliminary clinical results. J Comput Assist Tomogr 20(4):532–539.

37. Hacklander T, Reichenbach JR, Modder U (1997). Comparison of cerebral blood volume measurements using the T1 and T2* methods in normal human brains and brain tumors. J Comput Assist Tomogr 21(6):857–866.

38. Bruening R et al (1996). Echo-planar MR determination of relative cerebral blood volume in human brain tumors: T1 versus T2 weighting. Am J Neuroradiol 17:831–840.

39. Siegal T et al (1997). Utility of relative cerebral blood volume mapping derived from perfusion magnetic resonance imaging in the routine follow up of brain tumors. J Neurosurg 86:22–27.

40. Donahue KM et al (2000). Utility of simultaneously acquired gradient-echo and spin-echo cerebral blood volume and morphology maps in brain tumor patients. Magn Reson Med 43:845–853.

41. Schmainda KM et al (2004). Characterization of a first-pass gradient-echo spin-echo method to predict brain tumor grade and angiogenesis. Am J Neuroradiol 25:1524–1532.

42. Boxerman J et al (1995). The intravascular contribution to fMRI signal change: Monte Carlo modeling and diffusion-weighted studies in vivo. Magn Reson Med 34:4–10.

43. Dennie J et al (1998). NMR imaging of changes in vascular morphology due to tumor angiogenesis. Magn Reson Med 40:793–799.

44. Tropres I et al (2001). Vessel size imaging. Magn Reson Med 45:397–408.

45. Boxerman JL, Rosen BR, Weisskoff RM (1997). Signal-to-noise analysis of cerebral blood volume maps from dynamic NMR imaging studies. J Magn Reson Imaging 7(3):528–537.

46. Hou L et al (1999). Optimization of fast acquisition methods for whole-brain relative cerebral blood volume (rCBV) mapping with susceptibility contrast agents. J Magn Reson Imaging 9:233–239.

47. Perkio J et al (2002). Evaluation of four postprocessing methods for determination of cerebral blood volume and mean transit time by dynamic susceptibility contrast imaging. Magn Reson Med 47:973–981.

48. Aronen HJ et al (1993). Ultrafast imaging of brain tumors. Top Magn Reson Imaging 5:14–24.

49. Aronen H et al (1994). Cerebral blood volume maps of gliomas: comparison with tumor grade and histologic findings. Radiology 191:41–51.

50. Aronen HJ et al (2000). High microvascular blood volume is associated with high glucose uptake and tumor angiogenesis in human gliomas. Clin Cancer Res 6(6):2189–2200.

51. Fuss M et al (2001). Tumor angiogenesis of low-grade astrocytomas measured by dynamic susceptibility contrast-enhanced MRI (DSC-MRI) is predictive of local tumor control after radiation therapy. Int J Radiat Oncol Biol Phys 51(2):478–482.

52. Maeda M et al (1993). Tumor vascularity in the brain: evaluation with dynamic susceptibility-contrast MR imaging. Radiology 189:233–238.

53. Cha S (2004). Perfusion imaging of brain tumors. Top Magn Reson Imaging 15(5):279–289.

54. Knopp EA et al (1999). Glial neoplasms: dynamic contrast-enhanced T2*-weighted MR imaging. Radiology 211(3):791–798.

55. Heiland S et al (1999). Simultaneous assessment of cerebral hemodynamics and contrast agent uptake in lesions with disrupted blood–brain barrier. Magn Reson Imaging 17(1):21–27.

56. Weisskoff RM et al (1994).Simultaneous blood volume and permeability mapping using a single Gd-based contrast injection. In *2nd Annual Meeting of the Society of Magnetic Resonance in Medicine.* San Francisco.

57. Boxerman J, Schmainda KM, Weisskoff RM (2006). Relative cerebral blood volume maps corrected for contrast agent extravasation significantly correlate with glioma tumor grade whereas uncorrected maps do not. Am J Neuroradiol 27:859–867.

58. Quarles CC, Ward BD, Schmainda KM (2005). Improving the reliability of obtaining tumor hemodynamic parameters in the presence of contrast agent extravasation. Magn Reson Med 53:1307–1316.

59. Quarles CC et al (20050. Dexamethasone normalizes brain tumor hemodynamics as indicated by dynamic susceptibility contrast MRI perfusion parameters. Technol Cancer Res Treat 4(3):245–249.

60. Guckel F et al (1994). Assessment of cerebral blood volume with dynamic susceptibility contrast enhanced gradient-echo imaging. J Comput Assist Tomogr 18(3):344–351.

61. Zhu XP et al (2000). Quantification of endothelial permeability, leakage space, and blood volume in brain tumors using combined T1 and T2* contrast-enhanced dynamic MR imaging. J Magn Reson Imaging 11:575–585.

62. Sugahara T et al (2000). Posttherapeutic intraaxial brain tumor: the value of perfusion-sensitive contrast-enhanced MR imaging for differentiating tumor recurrence from nonneoplastic contrast-enhancing tissue. Am J Neuroradiol 21:901–909.

63. Sugahara T et al (2001). Perfusion-sensitive MR imaging of gliomas: comparison between gradient-echo and spin-echo echo-planar imaging techniques. Am J Neuroradiol 22(7):1306–1315.

64. Sugahara T, Korogi Y, Kochi M (1998). Correlation of MR imaging-determined cerebral blood volume maps with histologic and angiographic determination of vascularity in gliomas. Am J Roentgenol 171:1479–1486.

65. Warmuth C, Gunther M, Zimmer C (2003). Quantification of blood flow in brain tumors: comparison of arterial spin labeling and dynamic susceptibility-weighted contrast-enhanced MR imaging. Radiology 228:523–532.

66. Aronen HJ et al (1995). Echo-planar MR cerebral blood volume mapping of gliomas. Clinical utility. Acta Radiol 36(5):520–528.

67. Bulakbasi N et al (2005). Assessment of diagnostic accuracy of perfusion MR imaging in primary and metastatic solitary malignant brain tumors. Am J Neuroradiol 26:2187–2199.

68. Lev MH et al (2004). Glial tumor grading and outcome prediction using dynamic spin-echo MR susceptibility mapping

compared with conventional contrast-enhanced MR: confounding effect of elevated rCBV of oligodendrogliomas. Am J Neuroradiol 25:214–221.

69. Hartmann M et al (2003). Distinguishing of primary cerebral lymphoma from high-grade glioma with perfusion-weighted magnetic resonance imaging. Neurosci Lett 338(2):119–122.

70. Wong JC, Provenzale JM, Petrella JR (2000). Perfusion MR imaging of brain neoplasms. Am J Neuroradiol 174:1147–1157.

71. Kremer S et al (2004). Contribution of dynamic contrast MR imaging to the differentiation between dural metastasis and meningioma. Neuroradiology 46(8):642–648.

72. Gillies RJ et al (1999). Causes and effects of heterogeneous perfusion in tumors. Neoplasia 1(3):197–207.

73. Jain RK (1988). Determinants of tumor blood flow: a review. Cancer Res 48:2641–2658.

74. Hamberg LM et al (1994). Spatial heterogeneity in tumor perfusion measured with functional computed tomography at 0.05 μl resolution. Cancer Res 54:6032–6036.

75. Eskey CJ et al (1992). 2H-Nuclear magnetic resonance imaging of tumor blood flow: spatial and temporal heterogeneity in a tissue-isolated mammary adenocarcinoma. Cancer Res 52: 6010–6019.

76. Jain RK (1993). Physiological resistance to the treatment of solid tumors. In *Drug Resistance in Oncology*. Teicher BA (ed.). Marcel Dekker, Inc., New York. 87–105.

77. Durand RE, LePard NE (1994). Modulaton of tumor hypoxia by conventional chemotherapeutic agents. Int J Radiat Oncol 29(3):481–486.

78. Gee MS et al (2001). Doppler ultrasound imaging detects changes in tumor perfusion during antivascular therapy associated with vascular anatomic alterations. Cancer Res 61:2974–2982.

79. Tozer GM et al (1999). Combretastatin A-4 phosphate as a tumor vascular-targeting agent: early effects in tumors and normal tissues. Cancer Res 59:1626–1634.

80. Dark GG et al (1997). Combretastatin A-4, an agent that displays potent and selective toxicity toward tumor vasculature. Cancer Res 57:1829–1834.

81. Holash J et al (1999). Vessel cooption, regression, and growth in tumors mediated by angiopoietins and VEGF. Science 284:1994–1998.

82. Bernsen HJ et al (1999). Suramin treatment of human glioma xenografts; effects on tumor vasculature and oxygenation status. J Neuro-Oncol 44:129–136.

83. Badruddoja MA et al (2003). Antiangiogenic effects of dexamethasone in 9L gliosarcoma assessed by MRI cerebral blood volume maps. Neuro-Oncology 5(4):235–243.

84. Heiss JD et al (1996). Mechanism of dexamethasone suppression of brain tumor-associated vascular permeability in rats. J Clin Invest 98:1400–1408.

85. Machein MR et al (1999). Differential downregulation of vascular endothelial growth factor by dexamethasone in normoxic and hypoxic rat glioma cells. Neuropathol Appl Neurobiol 25:104–112.

86. Chiang IC, Kuo YT, Lu CY (2004). Distinction between high-grade gliomas and solitary metastases using peritumoral 3T magnetic resonance spectroscopy, diffusion, and perfusion imaging. Neuroradiology 46:619–627.

87. Law M et al (2002). High-grade gliomas and solitary metastases: differentiation by using perfusion and proton spectroscopic MR imaging. Radiology 222:715–721.

88. Pham CD et al (1998). Magnetic resonance imaging detects suppression of tumor vascular permeability after administration of antibody to vascular endothelial growth factor. Cancer Invest 16(4):225–230.

89. Brasch R et al (1997). Assessing tumor angiogenesis using macromolecular MR imaging contrast agents. J Magn Reson Imaging 7(1):68–74.

90. Gossmann A et al (2002). Dynamic contrast-enhanced magnetic resonance imaging as a surrogate marker of tumor response to anti-angiogenic therapy in xenograft model of glioblastoma multiforme. J Magn Reson Imaging 15:233–240.

91. Provenzale JM, Galvez Moya M, Leeds NE et al (2002). MR imaging evaluation of a novel anti-angiogenesis agent, PTK787. Meeting of the Radiological Society of North America.

92. Essig M et al (2003). Assessment of brain metastases with dynamic susceptibility-weighted contrast-enhanced MR imaging: inital results. Radiology 228:193–199.

93. Rock JP et al (2004). Associations among magnetic resonance spectroscopy, apparent diffusion coefficients, and image-guided histopathology with special attention to radiation necrosis. Neurosurgery 54(5):1111–1117.

94. Cha S et al (2003). Dynamic, contrast-enhanced perfusion MRI in mouse gliomas: correlation with histopathology. Magn Reson Med 49:848–855.

95. Hazle JD, Jackson EF, Schomer DF, Leeds NE (1997). Dynamic imaging of intracranial lesions using fast spain-echo imaging: differentiation of brain tumors and treatment effects. J Magn Reson Imaging 7:1084–1093.

96. Wenz F et al (1996). Effect of radiation on blood volume in low-grade astrocytomas and normal brain tissue: quantification with dynamic susceptibility contrast MR imaging. Am J Roentgenol 166(1):187–193.

97. Silva AC, Kim SG, Garwood M (2000). Imaging blood flow in brain tumors using arterial spin labeling. Magn Reson Med 44:169–173.

98. Provenzale JM et al (2002). Comparison of permeability in high-grade and low-grade brain tumors using dynamic susceptibililty contrast MR imaging. Am J Neuroradiol 178:711–716.

99. Cenic A et al (2000). A CT method to measure hemodynamics in brain tumors: validation and application to cerebral blood flow maps. Am J Neuroradiol 21:462–470.

100. Brix G et al (1999). Regional blood flow, capillary permeability, and compartmental volumes: measurement with dynamic CT-initial experience. Radiology 210:269–276.

101. Leggett D, Miles KA, Kelly BB (1999). Blood–brain barrier and blood volume imaging of cerebral glioma using functional CT: a pictorial review. Eur J Radiol 30:185–190.

102. Nabavi DG et al (1999). CT assessment of cerebral perfusion: experimental validation and initial clinical experience. Radiology 213:141–149.

103. Eastwood JD, Provenzale JM (2003). Cerebral blood flow, blood voume, and vascular permeability of cerebral glioma assessed with dynamic CT perfusion imaging. Neuroradiology 45:373–376.

104. Roberts HC et al (2002). Dyanmic, contrast-enhanced CT of human brain tumors: quantitative asessment of blood volume, blood flow, and microvascular permeability: report of two cases. Am J Neuroradiol 23:828–832.

CHAPTER 30

Positron Emission Tomography (PET) and Single Photon Emission Computed Tomography (SPECT) Imaging

Masanori Ichise and Ronald L. Van Heertum

OVERVIEW

Incidence and Prevalence

Brain tumors are generally divided into primary and secondary or metastatic brain tumors. There are approximately 14 000 new primary brain tumors diagnosed each year with a slightly greater incidence in men (57%) as compared to women (43%), with the peak incidence, in adults, occurring between the ages of 60 and 80 years [1]. Overall, brain tumors account for approximately 2% of all malignant tumors. The most common form of adult primary brain tumor is the glioma, which arises from glial cells, with diffuse fibrillary astrocytomas the most common type in adults. Gliomas account for greater than 90% of primary brain tumors in patients over age 20 years and 60% of all primary brain tumors [2]. Typically, gliomas are classified into increasing grades of malignancy based on histopathology. The World Health Organization (WHO) grading system, which is based on cellularity, is the most commonly used classification system [3]. In children, brain tumors are most frequently infratentorial in location. The most common tumor type is the medulloblastoma, which arises in the posterior fossa. Pediatric astrocytomas, the majority (80%) of which are of low-grade histology and infratentorial in location, are also seen relatively frequently [4].

Brain Tumor Classification

In general, brain tumors are classified as primary or secondary, based on whether the tumor originated in the brain or spread hematogenously from another tumor site in the body. The WHO brain tumor classification system is based on typing the cell of origin in the brain such as the glial cell. The celluarity of astrogliomas is further characterized as to the degree of malignancy ranging from well differentiated low-grade tumors (grade I), high-grade (grade II) to anaplastic (grade III) and the least well differentiated glioblastoma multiforme (grade IV) tumors [5]. As would be expected, the survival rate diminishes significantly as the tumor grade progresses from grade I to IV. Early PET studies [6] also reported that the degree of glial cell malignancy, based on histopathology, directly correlated with the degree of tumor glycolytic activity observed on [18F]fluoro-deoxyglucose (FDG) PET imaging. Furthermore, the degree of tumor metabolic activity was inversely correlated to prognosis as patients with the highest metabolic activity tumors had the shortest long-term survival [7].

NEUROIMAGING

CT and MRI

There are a wide variety of neuroimaging procedures, both invasive and non-invasive, that have been employed in the evaluation of the brain. In the evaluation of brain tumors, however, the most frequently employed neuroimaging procedures are CT and MR imaging. Each of these techniques has a number of very significant strengths and weaknesses. As a result, both CT and MRI are often required in the assessment of individual brain tumors. CT and MR, as anatomic techniques, demonstrate a number of direct and indirect findings that are useful in characterizing brain tumor type, extent and response to treatment. Among the direct findings are tumor density, which may be secondary to such factors as water content, regressive changes, vascularity, calcifications and contrast enhancement [2]. Other observed changes, which are more indicative of the presence of a mass and less specific for the presence of a tumor, are referred to as indirect findings or signs [2]. The most common indirect signs include presence of a mass effect, edema and changes in the adjacent bone. Overall, the accuracy of CT and MRI for detecting the presence of brain tumors is typically >95% [8]. Unfortunately, even though CT and MRI are superb imaging techniques, they are continuing to be refined with new imaging sequences with and without contrast media. These techniques are somewhat less useful in characterizing the tumor as to cell type and degree of malignancy. As a result, other techniques, including magnetic resonance spectroscopy and nuclear medicine techniques, are being employed to further the understanding of the underlying biochemical processes occurring at the tumor cellular level [9].

Magnetic Resonance Spectroscopy

The field of magnetic resonance spectroscopy (MRS) is continuing to evolve. Specifically, the technique is progressing from single voxel to multivoxel and two-dimensional techniques have resulted in improved spatial resolution, which in turn is supplying additional information regarding tumor heterogeneity. This approach supplies unique information regarding specific intracellular metabolites such as choline, creatine, phosphocreatine, N-acetylaspartate (NAA), lactic acid and various mobile lipids and triglycerides. Measurements of these metabolites, in particular, and the comparison of metabolite ratios have been of value in tumor grading, analysis of tumor heterogeneity, degeneration to more malignant tumor grade, response to therapy and differentiation of radiation necrosis from viable tumor [10]. Unfortunately, although MRS is a very sensitive technique, it still lacks the specificity to be used solely as a stand-alone technique clinically.

SPECT Imaging

Nuclear medicine techniques employed for the evaluation of brain tumors consist of SPECT and PET imaging utilizing a variety of radiopharmaceuticals. The radiopharmaceutical employed with SPECT can be categorized into a number of broad categories as follows: regional cerebral blood flow compounds ([^{99m}Tc]HMPAO and [^{99m}Tc]ECD); cationic compounds (thallium-201(^{201}Tl), [^{99m}Tc]MIBI, [^{99m}Tc]tetrofosmin); labeled amino acids ([^{123}I] iodo-α-methyltyrosine ([^{123}I]IMT)); labeled antibodies; labeled somatostatin analogs ([^{111}In]octreotide) and apoptosis compounds ([^{123}I]anexcin). SPECT has been widely available for the past two decades, although PET and particularly PET/CT are also becoming increasingly available for clinical use. The disadvantage of SPECT is its slightly inferior spatial resolution to that of PET. However, this disadvantage of SPECT could be offset by some of the major advantages of SPECT that are related to the more cost-effective availability of SPECT tracers compared with PET tracers. For SPECT or PET to be a viable clinical or research tool for imaging of brain tumors, however, it must provide information that cannot be obtained by other imaging modalities such as CT,

MRI and MRS [1,11]. The pertinent information expected to be obtained from SPECT or PET imaging of brain tumors relates to whether: (1) tracer uptake correlates with tumor histological grade and hence tracer uptake may be used as a diagnostic or prognostic indicator; and (2) tracer uptake can be used to indicate postoperative tumor recurrence or it can distinguish between tumor recurrence and radiation necrosis.

Regional Cerebral Blood Flow

The two most popular brain SPECT compounds for assessment of cerebral blood flow (rCBF) are [^{99m}Tc] HMPAO and [^{99m}Tc] ECD. Both of these compounds are lipophilic and readily cross the intact blood–brain barrier (BBB) where they convert to hydrophilic compounds that are not able to easily cross back across the BBB. In general, rCBF compounds are not routinely used in the assessment of brain tumors, although in the past these compounds were used in combination with ^{201}Tl for differentiation of malignant brain lesions from more benign tumors [12]. At present, much of the interest in rCBF SPECT for evaluation of brain tumors has switched to functional magnetic resonance imaging (fMRI).

Labeled Cations: Thallium-201 (^{201}Tl) and [^{99m}Tc]MIBI

Thallium-201 (^{201}Tl) is chemically a potassium analogue and its tumor uptake appears to be related to multiple factors including alterations in the BBB, Na$^+$/K$^+$ ATPase pump activity and blood flow [13,14]. Although low energy x-rays (68–80 KeV) from ^{201}Tl decays are not optimal for imaging, the long physical half-life of 73 h is a convenient feature of ^{201}Tl. Typically, ^{201}Tl SPECT brain imaging is performed at 15 min after bolus intravenous injection of 2–5 mCi of ^{201}Tl and, additionally, delayed SPECT imaging may be performed a few hours after the ^{201}Tl administration.

To date, many studies have been performed to investigate the potential utility of ^{201}Tl SPECT imaging for gliomas and other brain tumors [13,15–23]. The results of these studies suggest that ^{201}Tl uptake generally correlates with histological tumor grades. However, there is a considerable overlap between ^{201}Tl uptake and tumor grades, probably because of the heterogeneity of tissue within the tumor (necrosis or edema) and the partial volume effects (e.g. see [16,20]) as well as the fact that some low-grade tumors such as a pilocytic astrocytoma may show high ^{201}Tl uptake [17]. In a large retrospective study of 90 patients, in the differential diagnosis of various brain tumors [19], the sensitivity and specificity of ^{201}Tl brain SPECT were 72% and 82%, respectively. The false positives and negatives in this study were related to the limited resolution of the ^{201}Tl brain SPECT, as well as the areas of normal physiological ^{201}Tl activity (false positives). In this respect, the inferior spatial resolution of SPECT confounded by the inferior physical property of ^{201}Tl (low x-ray energies) may be a major limitation

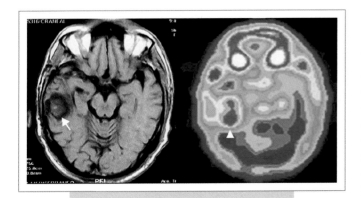

FIG. 30.1. [201]Tl SPECT images of tumor recurrence versus radiation necrosis. Inconclusive MRI scan (left) for tumor recurrence in the right temporal lobe (arrow), alongside a ^{201}Tl-SPECT image (right) of local radiotracer uptake at the margins of the primary lesion (arrowhead). (Modified and reproduced with permission from Gomez-Rio et al (201)Tl-SPECT in low-grade gliomas: diagnostic accuracy in differential diagnosis between tumour recurrence and radionecrosis. Eur J Nucl Med Mol Imaging 31:1237–1243. Copyright © 2004 Springer-Verlag Gmbh, Heidelberg Germany. All rights reserved.)

of ^{201}Tl SPECT for brain tumor imaging. Thus, ^{201}Tl SPECT may not be the study of choice to be used to diagnose or predict the prognosis of brain tumors.

However, evidence suggests ^{201}Tl SPECT allows differentiation of a high-grade tumor recurrence from radiation necrosis (Figure 30.1) and the results of a more recent study [23] suggest that ^{201}Tl-SPECT has adequate diagnostic accuracy to be part of routine algorithms in the follow-up of patients even with low-grade glioma suspected of tumor recurrence. Gomez-Rio et al [23] studied 84 patients with resected low-grade gliomas who had suspicion of tumor recurrence during their follow-up. ^{201}Tl-SPECT images were assessed by visual analysis and by estimation of the uptake index (ratio of mean counts in the lesion to those in the contralateral mirror area). The global visual analysis diagnostic accuracy for ^{201}Tl-SPECT was 83%, with a sensitivity of 88% and a specificity of 76%. An uptake index cut-off value of 1.25 showed a sensitivity of 90% and specificity of 80% for detection of tumor activity. Diagnostic pitfalls were observed in regions with physiological ^{201}Tl uptake, i.e. the posterior cranial fossa, diencephalon, lateral ventricles and cavernous and longitudinal venous sinuses. Because of more favorable prognosis of patients with low-grade gliomas, ^{201}Tl SPECT may thus play an important role as a clinical tool to detect tumor recurrence.

[99m Tc]MIBI is also a myocardial-imaging agent like [201]Tl. [99m Tc]MIBI is a cationic compound and accumulates in cytoplasm and mitochondria of the myocardium as a result of passive diffusion across the negative cellular/organelle membrane. However, [99m Tc]MIBI is normally excluded from the brain by the BBB and, unlike ^{201}Tl,

[99m Tc]MIBI accumulates in the choroid plexus, which cannot be prevented by the administration of potassium perchlorate. Thus, [99m Tc]MIBI uptake in brain tumors appears to relate mainly to the breakdown of BBB in the tumor. Although [99m Tc]MIBI has better imaging characteristics (140 keV and higher allowable injection dose of up to 30 mCi), studies designed to evaluate the usefulness of [99m Tc]MIBI for diagnosis, prognosis and detection of tumor recurrence have not yet convincingly shown its superiority to [201]Tl [24–28] and the clinical utility of [99m Tc]MIBI SPECT needs yet to be defined. [99m Tc]MIBI is eliminated from cells by P-glycoprotein, which also acts as energy-driven efflux pump for several antineoplastic agents. However, multiple drug resistance (MDR)-1 gene expression, as demonstrated by [99m Tc]MIBI, does not appear to correlate with chemoresistance in gliomas [27].

Labeled Amino Acids: [123 I] Iodo-α-Methyltyrosine ([123 I]IMT)

[123 I]IMT is a SPECT tracer developed as an alternative to methyl-[11 C]-L-methionine ([11 C]MET). [123 I]IMT is a radiolabeled amino acid (tyrosine) analogue, which is taken up in the brain by carrier mediated, stereo-selective active transport systems across the BBB and cell membranes. [123 I]IMT is not incorporated into cellular proteins. However, cellular uptake of [123 I]IMT in glioma cells appears to be a biomarker of cellular proliferation [29]. [123 I] possesses favorable SPECT imaging characteristics for imaging (γ rays of 159 keV and a physical half life of 13 h) and well-known and characterized synthetic procedures to incorporate into biologically relevant molecules such as an amino acid analogue, methyltyrosine. In North America, the commercial availability of [123 I] has been limited to one supplier/manufacturer, although this situation may improve in the near future with efforts to increase [123 I] availability from additional sources.

Despite [123 I]IMT being theoretically a marker for cellular proliferation, it may be limited in predicting histological tumor grades. However, the results of several studies suggest that [123 I]IMT SPECT can detect postoperative tumor recurrence and identify residual tumor tissue [30–35]. Weber et al [35], for example, performed [123 I]IMT SPECT imaging for 114 consecutive patients with newly diagnosed gliomas, including 71 patients who had undergone tumor resection 4–6 weeks before SPECT imaging and 43 patients with unresectable tumors who all underwent conformal radiotherapy. In the former group, median survival was only 13 months in 52/71 patients with focal [123 I]IMT uptake, while median survival was reached in patients without focal [123 I]IMT uptake (19/71) (Figure 30.2). In contrast, in the latter 43 patient group with unresectable high-grade gliomas, [123 I]IMT uptake did not correlate with survival. Thus, [123 I]IMT SPECT may be a viable alternative to more expensive [11 C]MET PET, which requires on-site cyclotron facility to produce short-lived [11]C.

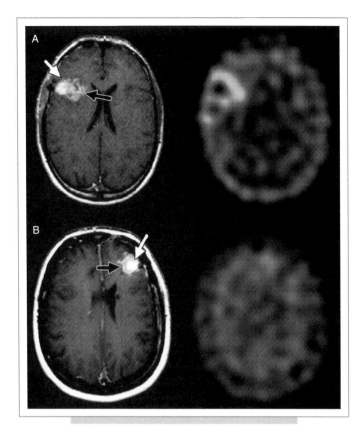

FIG. 30.2. In patients with anaplastic astrocytoma, examples of [^{123}I]IMT SPECT studies with **(A)** and without **(B)** focal IMT uptake at resection site. **(A)** On left, MR image (T1-weighted after administration of gadolinium-DTPA) shows resection cavity (white arrow) and rim of contrast enhancement (black arrow). Resection cavity contains blood and is, therefore, hyperintense in T1-weighted image. On right, [^{123}I]IMT SPECT image shows intense uptake of [^{123}I]IMT at resection margin. This patient died 6 months after SPECT study. **(B)** On left, MR image shows resection cavity (white arrow) and surrounding contrast enhancement (black arrow). However, on right, [^{123}I]IMT SPECT image shows no clear focal IMT uptake. This patient was still alive after 38 months of follow-up. (Reproduced with permission from Weber et al Correlation between postoperative 3-[(123)I]iodo-L-alpha-methyltyrosine uptake and survival in patients with gliomas. J Nucl Med 42:1144–1150. Copyright © 2001, Society of Nuclear Medicine. All rights reserved.)

PET Imaging
[^{18}F]FDG (FDG) PET

The role of PET brain imaging in the evaluation of brain tumors is continuing to evolve and expand as new radiopharmaceuticals become available. Unfortunately, at the present time, the only compound approved by the Food and Drug Administration (FDA) is FDG. The role of FDG PET imaging is essentially as an adjunct technique in

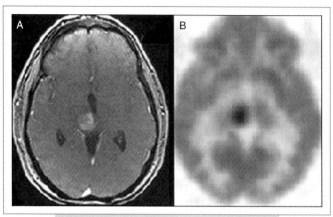

FIG. 30.3. Right para-thalamic astrocytoma in a 44-year-old man complaining of headaches. **(A)** Post-contrast T-1 axial MR reveals a focal enhancing lesion medial to the right thalamus. **(B)** Axial FDG-PET study at the same level reveals a corresponding hypermetabolic lesion confirming a malignant tumor (grade III astrocytoma).

cases where CT and MRI are unable to establish a diagnosis or answer the clinical question that needs to be addressed. The areas where FDG-PET has been shown to be of value include tumor detection, defining tumor extent and degree of malignancy, post-treatment assessment of tumor response and identification of tumor recurrence versus radiation necrosis [36].

Tumor Detection

Since the first introduction of FDG PET imaging, it has been recognized that the technique is very sensitive for detecting high-grade gliomas (Figure 30.3) and less accurate for the identification of lower grade lesions [6]. Additionally, certain extra-axial tumors such as pituitary tumors may show FDG avidity (Figure 30.4). Furthermore, in the detection of brain tumors, FDG-PET is somewhat hampered by the high uptake in cortical gray matter and somewhat confounded by uptake in areas of inflammation and/or infection. It has recently been reported [37] that serial delayed scanning, post-FDG administration, can help to enhance visualization of some brain tumor since washout of the tumor is delayed relative to the surrounding brain parenchyma. In addition, it is helpful to delay scanning follow-up post-chemotherapy, 1–2 weeks, and post-radiation treatment, 6–8 weeks, so as to reduce uptake from accompanying inflammation post-treatment.

Other radiopharmaceuticals, in particular labeled amino acids, are very useful and often complementary to the role of FDG imaging. The most commonly utilized PET labeled amino is [^{11}C]methionine (MET) (see below). This compound, which is incorporated into the tumor protein synthesis pathway, has a much higher accuracy rate than FDG for detecting low-grade brain tumors. In addition, MET

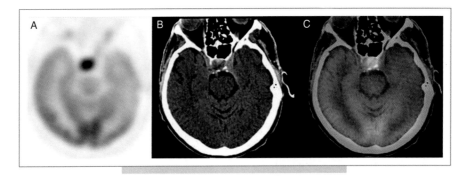

FIG. 30.4. Unsuspected pituitary tumor in an elderly male referred for a PET/CT study to assess progressive cognitive dysfunction. **(A)** Axial FDG-PET scan reveals a focal, hypermetabolic, midline lesion at the level of the temporal lobes. **(B)** Axial non-contrast CT scan reveals a soft tissue mass with bone erosion of the surrounding sella turcica. **(C)** Fused PET/CT images confirming a pituitary tumor.

does not demonstrate the same degree of confounding uptake in surrounding brain [38].

Staging the Degree of Malignancy

Since FDG crosses the intact BBB and becomes incorporated into the glycolytic pathway of neurons and tumor cells, where it becomes trapped at the level of phosphorylation, it has the unique ability to map regional metabolic brain activity. In addition, the degree of uptake in tumor cells has been shown to correlate directly with the degree of malignancy noted on histopathology. Some investigators [39] have reported that the level of FDG uptake is a better measure of the malignancy potential of individual brain tumors, than actual histopathology. Specifically, patients with histological low-grade brain tumors and markedly elevated glucose metabolism have a shorter survival time than patients with low-grade tumors and lower glucose metabolism [7]. The ability of FDG to assess tumor glucose metabolism is a very important aspect of brain tumor evaluation, particularly in the follow-up assessment of patients post-treatment.

Follow-up Assessment and Monitoring Treatment

There are a number of areas where PET imaging has proven to be useful in the follow-up assessment of brain tumors after initial assessment (Figures 30.5, 30.6 and 30.7). There are also a number of significant drawbacks to the use of PET imaging, in particular FDG-PET, in the follow-up assessment of brain tumors.

The major issue issues to be answered following treatment can generally be addressed using MRI. This technique is usually quite adequate for assessing the response to treatment by monitoring the reduction in tumor size, edema and lesion contrast enhancement. MRI may, however, be less useful, at times, for the determination of the degree of malignancy in a patient with a tumor recurrence and also for differentiating radiation necrosis from viable tumor residual or recurrence [2]. In such cases, PET imaging and magnetic resonance spectroscopy (MRS) can be useful adjuvant imaging techniques.

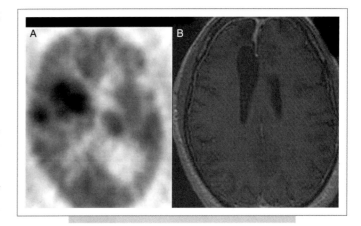

FIG. 30.5. Recurrent grade IV astrocytoma post-treatment in a 41-year-old man, who is currently referred for evaluation of new onset seizures post-treatment (surgery and radiation). **(A)** Axial, **(C)** coronal and **(D)** sagittal plane FDG-PET reveals evidence of focal hypermetabolic lesions representing recurrent disease at the posterior and inferior surgical margins. **(B)** Axial MR revealing post-surgical changes along with probable tumor recurrence at the periphery of the surgical site.

First, caution should employed when using FDG-PET imaging in the follow-up of brain tumors after treatment with radiation as there is a significant degree of inflammatory response that can significantly confound the uptake within the tumor, particularly when the study is performed too soon after the completion of radiation treatment. As a general rule, the first FDG-PET study should not be performed until a minimum of 8 weeks following treatment has elapsed. In addition, since FDG is very avidly taken up in the highly metabolic cortical gray matter, lower metabolically active residual or recurrent lesions may be easily masked by surrounding normal brain activity. In spite of these drawbacks, FDG imaging can be very useful for detecting lesions that have transformed from low-grade histopathology lesions to higher-grade tumors with a resultant greater degree of malignancy and aggressiveness [36].

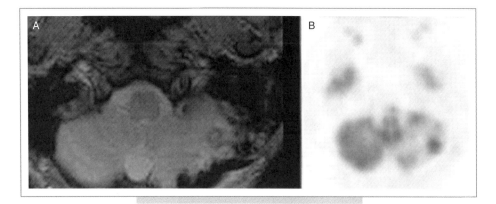

FIG. 30.6. Recurrent cerebellar tumor in a 52-year-old man who is post-treatment (surgery and radiation) for an astrocytoma grade III involving the left cerebellum hemisphere. **(A)** Axial MR, at the level of the cerebellum, reveals post-surgical changes and a non-enhancing focus involving the lateral aspect of the left cerebellum. **(B)** Axial FDG-PET, at the comparable level, reveals diffuse hypometabolism of the left cerebellar hemisphere secondary to prior radiation treatment to this region and a hypermetabolic focus corresponding to the lesion seen on MR that is consistent with recurrent tumor.

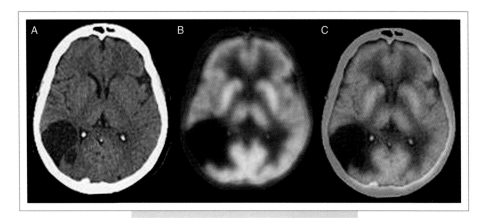

FIG. 30.7. PET/CT postoperative assessment of a grade III/IV right parietal temporal asrtocytoma. **(A)** Axial non-contrast CT, **(B)** FDG-PET and **(C)** fused FDGPET/CT scan reveals postoperative changes with no evidence of tumor recurrence.

Such a determination is very important, as the follow-up treatment management of such a tumor would likely be changed significantly. PET imaging is also useful for the differentiation of radiation necrosis versus viable tumor in cases where MRI is unable to make the distinction [40]. Unfortunately, FDG-PET may be misleading in these cases because of concomitant post-radiation changes that could result in an equivocal or even false positive interpretation. At present, there is not very much literature to support the concept of using labeled amino acids for the differentiation of radiation necrosis versus viable tumor. On the other hand, MRS is a technique that is useful for detecting viable tumor and distinguishing tumor from areas of radiation necrosis [41].

[^{11}C]MET and [^{18}F]FLT

[^{11}C]methyl-methionine ([^{11}C]MET) is one of several amino acid analogues that have been developed as tumor imaging agents [42]. [^{11}C]MET uptake in tumors generally reflects the protein synthesis rate because [^{11}C]MET, like other amino acids, is transported via amino acid transport across the cell membrane and is incorporated into proteins. However, [^{11}C]MET also seems to be incorporated into phospholipids through the S-adenylmethionine pathway [43]. Other radiolabeled amino acids that have been developed include [^{11}C]tyrosine, [^{18}F]fluoro-tyrosine and O-(2-[^{18}F]-fluoroethyl)-L-tyrosine. For tumor detection and tumor delineation, these tracers have the advantage of their low uptake in normal brain as compared with [^{18}F]FDG. Disadvantages of [^{11}C]MET are its uptake in acutely ischemic and inflammatory brain tissue [44], its short half-life of 20 minutes and the need for an on-site cyclotron facility, although the [^{18}F] versions of [^{11}C]MET may overcome the latter disadvantages. [^{11}C]MET, for example, performs better in defining the extent of tumor involvement and recurrence of brain tumors (Figure 30.8) including low-grade gliomas than does [^{18}F]FDG. Amino-acid tracers, such as [^{11}C]MET, thus appear to play a complementary role to [^{18}F]FDG [34,45–50]. In addition, [^{11}C]MET may be useful for monitoring chemotherapy in gliomas (Figure 30.9) [51,52]. Because of the current efforts to develop effective treatments for gliomas, [^{11}C]MET

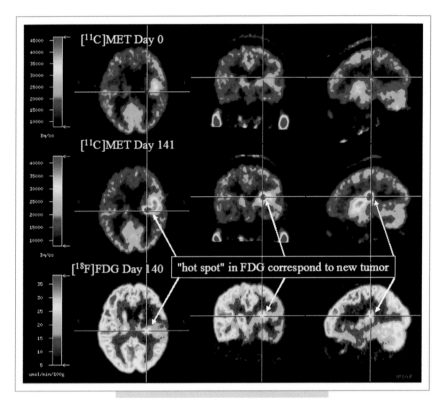

FIG. 30.8. Growth of glioblastoma. [^{11}C]methyl-methionine ([^{11}C]MET) PET brain images of a patient with glioblastoma at day 0 (top row) and day 141 (middle row) and [^{18}F]FDG images of the brain of same patient at day 140. Due to lower background brain activity, the delineation of tumor extent and visualization of new tumor growth is much better for ([^{11}C]MET than for [^{18}F]FDG. (Reproduced with permission from Karl Herholtz, MD, University of Cologne, Germany. All rights reserved.)

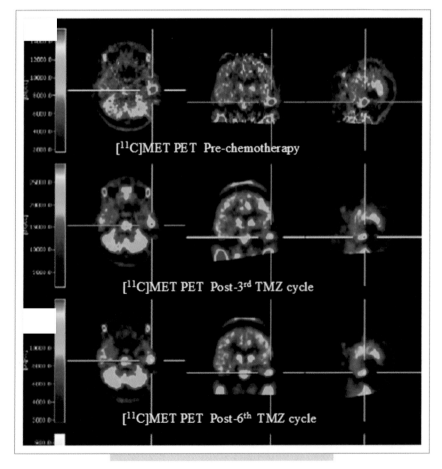

FIG. 30.9. [^{11}C]methyl-methionine ([^{11}C] MET) PET in monitoring chemotherapy. Patient with oligoastrocytoma (WHO grade III) was treated with 6 cycles of temozolomide (TMZ). Reduced [^{11}C]MET uptake in the left temporal lobe tumor as a metabolic response to TMZ chemotherapy is evident after third TMZ cycle (middle and bottom rows). (Modified and reproduced with permission from Galldiks et al Use of (11)C-methionine PET to monitor the effects of temozolomide chemotherapy in malignant gliomas. Eur J Nucl Med Mol Imaging. Feb 1;:1–9 [Epub ahead of print]. Copyright © 2006 Springer-Verlag Gmbh, Heidelberg Germany. All rights reserved.)

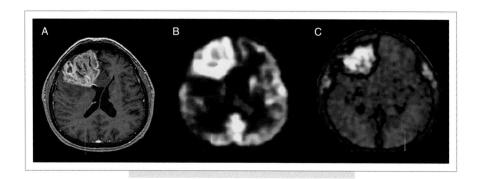

FIG. 30.10. Newly diagnosed glioblastoma. **(A)** MRI (contrast-enhanced T1-weighted image) shows large area of contrast enhancement in right frontal lobe. Both [^{18}F]FDG PET **(B)** and [^{18}F]FLT PET **(C)** show increased uptake in same area. (Reproduced with permission from Chen et al Imaging proliferation in brain tumors with ^{18}F-FLT PET: comparison with ^{18}F-FDG. J Nucl Med 46: 945–952. Copyright © 2005, Society of Nuclear Medicine. All rights reserved.)

and other amino acid tracers are expected to play an important role for the evaluation of treatment response in experimental protocols designed to improve the patients' prognosis.

Another index of tumor cellular proliferation that can be non-invasively assessed by PET is the incorporation of nucleosides into DNA in proliferating cells. Radiolabeled thymidine ([3H]TdR) is the gold standard for determination of cell proliferation in cell culture and, to date, ^{11}C- and ^{18}F-labeled thymidine compounds have been developed to allow a non-invasive assessment of tumor proliferation, as well as early response to chemotherapy by PET. A recently developed thymidine analogue, 3'-deoxy-3'-[^{18}F]fluorothymidine ([^{18}F]FLT), has been shown to be a very promising tracer for in vivo imaging of cellular proliferation [53]. [^{18}F]FLT uptake reflects the activity of thymidine kinase-1 (TK1), an enzyme expressed during the DNA synthesis phase of the cell cycle [54]. Significant correlations of quantitative [^{18}F]FLT uptake with the immunohistochemistry marker of cell proliferation Ki-67 have been shown in several extracranial tumors [55–58]. For the evaluation of gliomas, relative [^{18}F]FLT uptake within gliomas is greater than relative [^{18}F]FDG uptake (Figure 30.10) or [^{11}C]MET uptake and [^{18}F]FLT uptake seems to correlate more strongly with Ki-67 index than does [^{18}F]FDG [11,53,59–62]. Thus, [^{18}F]FLT PET is expected to be an improved alternative or complementary tool to [^{18}F]FDG and [^{11}C]MET PET in evaluating brain tumors, particularly chemotherapeutic tumor response.

CONCLUSIONS

SPECT and PET imaging of brain tumors provides information that cannot be obtained by other imaging modalities such as CT, MRI and MRS. These functional brain tumor imaging techniques are evolving with the continuing development of new radiopharmaceuticals that ultimately aim to be a biomarker of the tumor histological grade and hence a diagnostic or prognostic marker, as well as a biomarker of postoperative tumor recurrence or a marker that can distinguish between tumor recurrence and radiation necrosis.

REFERENCES

1. Benard F, Romsa J, Hustinx R (2003). Imaging gliomas with positron emission tomography and single-photon emission computed tomography. Semin Nucl Med 33:148–162.
2. Del Sole A, Falini A, Ravasi L et al (2001). Anatomical and biochemical investigation of primary brain tumours. Eur J Nucl Med 28:1851–1872.
3. Smirniotopoulos JG (1999). The new WHO classification of brain tumors. Neuroimaging Clin N Am 9:595–603.
4. Farinotti M, Ferrarini M, Solari A, Filippini G (1998). Incidence and survival of childhood CNS tumours in the Region of Lombardy, Italy. Brain 121:1429–1436.
5. Kleihues P, Louis DN, Scheithauer BW et al (2002). The WHO classification of tumors of the nervous system. J Neuropathol Exp Neurol 61:215–225.
6. Dichiro G, Patronas NJ, Delapaz RL et al (1982). Glucose-utilization changes in cerebral-cortex and cerebellum associated with brain tumors. Ann Neurol 12:75.
7. Dichiro G (1987). Positron emission tomography using [F-18] fluorodeoxyglucose in brain tumors – a powerful diagnostic and prognostic tool. Invest Radiol 22:360–371.
8. Ricci PE (1999). Imaging of adult brain tumors. Neuroimaging Clin N Am 9:651–661.

9. Tovi M, Thuomas KA, Bergstrom K et al (1986). Tumour delineation with magnetic resonance imaging in gliomas. A comparison with positron emission tomography and computed tomography. Acta Radiol Suppl 369:161–163.

10. Magalhaes A, Godfrey W, Shen YM, Hu JN, Smith W (2005). Proton magnetic resonance spectroscopy of brain tumors correlated with pathology. Acad Radiol 12:51–57.

11. Jacobs AH, Kracht LW, Gossmann A et al (2005). Imaging in neuro-oncology. NeuroRx 2:333–347.

12. Schwartz RB, Holman BL, Polak JF et al (1998). Dual-isotope single-proton emission computerized tomography scanning in patients with glioblastoma multiforme: association with patient survival and histopathological characteristics of tumour after high-dose radiotherapy. J Neurosurg 89:60–68.

13. Kaplan WD, Takvorian T, Morris JH, Rumbaugh CL, Connolly BT, Atkins HL (1987). Tl-201 brain-tumor imaging – a comparative-study with pathological correlation. J Nucl Med 28:47–52.

14. Sehweil AM, Mckillop JH, Milroy R, Wilson R, Abdeldayem HM, Omar YT (1989). Mechanism of Tl-201 uptake in tumors. Eur J Nucl Med 15:376–379.

15. Ancri D, Basset JY, Lonchampt MF, Etavard C (1978). Diagnosis of cerebral lesions by thallium 201. Radiology 128:417–422.

16. Kim KT, Black KL, Marciano D et al (1990). Tl-201 SPECT imaging of brain-tumors – methods and results. J Nucl Med 31:965–969.

17. Oriuchi N, Tamura M, Shibazaki T et al (1993). Clinical-evaluation of Tl-201 SPECT in supratentorial gliomas – relationship to histologic grade, prognosis and proliferative activities. J Nucl Med 34:2085–2089.

18. Ishibashi M, Taguchi A, Sugita Y et al (1995). Tl-201 in brain tumors – relationship between tumor-cell activity in astrocytic tumor and proliferating cell nuclear antigen. J Nucl Med 36:2201–2206.

19. Dierckx RA, Martin JJ, Dobbeleir A, Crols R, Neetens I, Dedeyn PP (1994). Sensitivity and specificity of Tl-201 single-photon emission tomography in the functional detection and differential-diagnosis of brain tumors. Eur J Nucl Med 21:621–633.

20. Ricci M, Pantano P, Pierallini A et al (1996). Relationship between thallium-201 uptake by supratentorial glioblastomas and their morphological characteristics on magnetic resonance imaging. Eur J Nucl Med 23:524–529.

21. Sun D, Liu QC, Liu WG, Hu WW (2000). Clinical application of Tl-201 SPECT imaging of brain tumors. J Nucl Med 41:5–10.

22. Staffen W, Hondl N, Trinka E, Iglseder B, Unterrainer J, Ladurner G (1998). Clinical relevance of Tl-201-chloride SPECT in the differential diagnosis of brain tumours. Nucl Med Commun 19:335–340.

23. Gomez-Rio M, Torres DMD, Rodriguez-Fernandez A et al (2004). Tl-201-SPECT in low-grade gliomas: diagnostic accuracy in differential diagnosis between tumour recurrence and radionecrosis. Eur J Nucl Med Mol Imaging 31:1237–1243.

24. Otuama LA, Treves ST, Larar JN et al (1993). Tl-201 versus technetium-99M-MIBI SPECT in evaluation of childhood brain tumors – a within-subject comparison. J Nucl Med 34:1045–1051.

25. Bagni B, Pinna L, Tamarozzi R et al (1995). SPECT imaging of intracranial tumors with Tc-99(M)-Sestamibi. Nucl Med Commun 16:258–264.

26. Soler C, Beauchesne P, Maatougui K et al (1998). Technetium-99 m sestamibi brain single-photon emission tomography for detection of recurrent gliomas after radiation therapy. Eur J Nucl Med 25:1649–1657.

27. Yokogami K, Kawano H, Moriyama T et al (1998). Application of SPECT using technetium-99 m sestamibi in brain tumours and comparison with expression of the MDR-1 gene: is it possible to predict the response to chemotherapy in patients with gliomas by means of Tc-99 m-sestamibi SPECT? Eur J Nucl Med 25:401–409.

28. Nishiyama Y, Yamamoto Y, Fukunaga K, Satoh K, Kunishio K, Ohkawa M (2001). Comparison of Tc-99(m)-MIBI with Tl-201 chloride SPECT in patients with malignant brain tumours. Nucl Med Commun 22:631–639.

29. Langen KJ, Muhlensiepen H, Holschbach M, Hautzel H, Jansen P, Coenen HH (2000). Transport mechanisms of 3-[I-123]Iodo-alpha-methyl-L-tyrosine in a human glioma cell line: comparison with [H-3-methyl]-L-methionine. J Nucl Med 41:1250–1255.

30. Biersack HJ, Coenen HH, Stocklin G et al (1989). Imaging of brain-tumors with L-3-[I-123]iodo-alpha-methyl tyrosine and Spect. J Nucl Med 30:110–112.

31. Kuwert T, Woesler B, Morgenroth C et al (1998). Diagnosis of recurrent glioma with SPECT and iodine-123-alpha-methyl tyrosine. J Nucl Med 39:23–27.

32. Schmidt D, Gottwald U, Langen KJ et al (2001). 3-[I-123] Iodo-alpha-methyl-L-tyrosine uptake in cerebral gliomas: relationship to histological grading and prognosis. Eur J Nucl Med 28:855–861.

33. Bader JB, Samnick S, Moringlane JR et al (1999). Evaluation of L-3-[I-123]iodo-alpha-methyltyrosine SPECT and [F-18]fluorodeoxyglucose PET in the detection and grading of recurrences in patients pretreated for gliomas at follow-up: a comparative study with stereotactic biopsy. Eur J Nucl Med 26:144–151.

34. Sasaki M, Kuwabara Y, Yoshida T et al (1998). A comparative study of thallium-201 SPECT, carbon-11 methionine PET and fluorine-18 fluorodeoxyglucose PET for the differentiation of astrocytic tumours. Eur J Nucl Med 25:1261–1269.

35. Weber WA, Dick S, Reidl G et al (2001). Correlation between postoperative 3-[I-123]iodo-L-alpha-methyltyrosine uptake and survival in patients with gliomas. J Nucl Med 42:1144–1150.

36. Van Heertum RL, Greenstein EA, Tikofsky RS (2004). 2-deoxy-fluorglucose-positron emission tomography imaging of the brain: current clinical applications with emphasis on the dementias. Semin Nucl Med 34:300–312.

37. Spence AM, Muzi M, Mankoff DA et al (2004). F-18-FDG PET of gliomas at delayed intervals: improved distinction between tumor and normal gray matter. J Nucl Med 45:1653–1659.

38. Pirotte B, Goldman S, Massager N et al (2004). Comparison of F-18-FDG and C-11-methionine for PET-guided stereotactic brain biopsy of gliomas. J Nucl Med 45:1293–1298.

39. Alavi JB, Alavi A, Chawluk J et al (1988). Positron emission tomography in patients with glioma – a predictor of prognosis. Cancer 62:1074–1078.

40. Kim EE, Wong WH, Haynie TP, Leeds N, Podoloff DA, Tilbury RS (1992). Differentiation of recurrent brain tumors from posttreatment changes by means of PET with C-11 methionine. Radiology 185:298.

41. Lichy MP, Bachert P, Henze M, Lichy CM, Debus J, Schlemmer HP (2004). Monitoring individual response to brain-tumour

chemotherapy: proton MR spectroscopy in a patient with recurrent glioma after stereotactic radiotherapy. Neuroradiology 46:126–129.

42. Kubota K, Yamada K, Fukada H et al (1984). Tumor-detection with carbon-11-labeled amino-acids. Eur J Nucl Med 9:136–140.

43. Ishiwata K, Kubota K, Murakami M, Kubota R, Senda M (1993). A comparative-study on protein incorporation of L-[methyl-H-3]methionine, L-[1-C-14]leucine and L-2-[F-18]fluorotyrosine in tumor-bearing mice. Nucl Med Biol 20:895–899.

44. Jacobs A (1995). Amino acid uptake in ischemically compromised brain tissue. Stroke 26:1859–1866.

45. Herholz K, Holzer T, Bauer B et al (1998). C-11-methionine PET for differential diagnosis of low-grade gliomas. Neurology 50:1316–1322.

46. Kaschten B, Stevenaert A, Sadzot B et al (1998). Preoperative evaluation of 54 gliomas by PET with fluorine-18-fluorodeoxyglucose and/or carbon-II-methionine. J Nucl Med 39:778–785.

47. Chung JK, Kim YK, Kim SK et al (2002). Usefulness of C-11-methionine PET in the evaluation of brain lesions that are hypo- or isometabolic on F-18-FDG PET. Eur J Nucl Med Mol Imaging 29:176–182.

48. Jager PL, Vaalburg W, Pruim J, de Vries EGE, Langen KJ, Piers DA (2001). Radiolabeled amino acids: basic aspects and clinical applications in oncology. J Nucl Med 42:432–445.

49. Ogawa T, Inugami A, Hatazawa J et al (1996). Clinical positron emission tomography for brain tumors: comparison of fludeoxyglucose F 18 and L-Methyl-C-11-methionine. Am J Neuroradiol 17:345–353.

50. Pauleit D, Floeth F, Tellmann L et al (2004). Comparison of O-(2-F-18-fluoroethyl)-L-tyrosine PET and 3-I-123-iodo-alpha-methyl-L-tyrosine SPECT in brain tumors. J Nucl Med 45:374–381.

51. Herholz K, Kracht LW, Heiss WD (2003). Monitoring the effect of chemotherapy in a mixed glioma by C-11-methionine PET. J Neuroimaging 13:269–271.

52. Galldiks N, Kracht LW, Burghaus L et al (2006). Use of C-11-methionine PET to monitor the effects of temozolomide chemotherapy in malignant gliomas. Eur J Nucl Med Mol Imaging 33:516–524.

53. Shields AF, Grierson JR, Dohmen BM et al (1998). Imaging proliferation in vivo with [F-18]FLT and positron emission tomography. Nat Med 4:1334–1336.

54. Rasey JS, Grierson JR, Wiens LW, Kolb PD, Schwartz JL (2002). Validation of FLT uptake as a measure of thymidine kinase-1 activity in A549 carcinoma cells. J Nucl Med 43:1210–1217.

55. Vesselle H, Grierson J, Muzi M et al (2002). In vivo validation of 3' deoxy-3'-[F-18]fluorothymidine ([F-18]FLT) as a proliferation imaging tracer in humans: correlation of [F-18]FLT uptake by positron emission tomography with Ki-67 immunohistochemistry and flow cytometry in human lung tumors. Clin Cancer Res 8:3315–3323.

56. Francis DL, Freeman A, Visvikis D et al (2003). In vivo imaging of cellular proliferation in colorectal cancer using positron emission tomography. Gut 52:1602–1606.

57. Cobben DCP, Jager PL, Elsinga PH, Maas B, Suurmeijer AJH, Hoekstra HJ (2003). 3'-F-18-fluoro-3'-deoxy-L-thymidine: a new tracer for staging metastatic melanoma? J Nucl Med 44:1927–1932.

58. Wagner M, Seitz U, Buck A et al (2003). 3'-[F-18]fluoro-3'-deoxythymidine ([F-18]-FLT) as positron emission tomography tracer for imaging proliferation in a murine B-cell lymphoma model and in the human disease. Cancer Res 63:2681–2687.

59. Krohn KA (2001). Evaluation of alternative approaches for imaging cellular growth. Q J Nucl Med 45:174–178.

60. Mier W, Haberkorn U, Eisenhut M (2002). [F-18]FLT; portrait of a proliferation marker. Eur J Nucl Med Mol Imaging 29:165–169.

61. Chen W, Cloughesy T, Kamdar N et al (2005). Imaging proliferation in brain tumors with F-18-FLT PET: comparison with F-18-FDG. J Nucl Med 46:945–952.

62. Choi SJ, Kim JS, Kim JH et al (2005). [F-18]3'-deoxy-3'-fluorothymidine PET for the diagnosis and grading of brain tumors. Eur J Nucl Med Mol Imaging 32:653–659.

CHAPTER **31**

Positron Emission Tomography (PET) and Single Photon Emission Computed Tomography (SPECT) Physics

Marie Foley Kijewski

INTRODUCTION

The purpose of emission tomography is to estimate the distribution of a radiotracer, in this context a tracer of tumor activity or blood distribution, from external measurements of the pattern of photons emerging from the brain. Some of these photons are detected, and certain information about them recorded, by the scanner. These external measurements are termed 'projections' and each measurement in a projection represents, ideally, the sum of radioactivity concentration along a line through the brain. From these measured projection datasets and knowledge of certain aspects of the SPECT or PET instrument, estimated images of the distribution of radioactivity are mathematically reconstructed. All modern SPECT and PET scanners image the three-dimensional (3D) distribution of radioactivity, either as a stack of two-dimensional (2D) transaxial images or directly as a 3D volume.

Image reconstruction techniques fall into two categories: analytic and iterative. Reconstruction by analytic methods is rapid and straightforward, as it involves solution of the inverse reconstruction problem (i.e. determining the original distribution from knowledge of its projections) by an algorithm. The underlying mathematics were published in 1917 by Radon [1]. Iterative reconstruction consists of a repeated updating of an initial image estimate (often obtained by analytic reconstruction) based on comparison of calculated projections of the current image estimate to the measured projections. Many iterative approaches have been developed; they differ in the criterion to be minimized and the algorithm for updating the image estimates, as well as in the statistical and physical models they incorporate. This approach has the potential to be more accurate, because it can incorporate more information about the physics of the imaging process, and have better noise properties, as it uses a more realistic statistical model.

Iterative reconstruction has been, until recently, used only in research applications because of the long reconstruction times. These times have been reduced significantly because of improvements in computer processing speed and advances in efficiency of the reconstruction algorithms, and iterative algorithms are now coming into clinical use.

Brain SPECT or PET images are interpreted by visual inspection by a physician or analyzed for measurements of physical quantities, such as activity concentration, size of tumor, or parameters of dynamic acquisitions.

SPECT IMAGING

Introduction to SPECT Physics

SPECT imaging is based on detection of gamma radiation from radioisotopes that emit one photon of a particular energy per nuclear decay, characterized by a particular half-life. (Some radioisotopes used in SPECT emit significant numbers of photons at each of several energies.) For optimal imaging, photon energy should be in the hundred-keV range; the energy must be high enough that significant numbers of photons escape the body, even from deep structures, but not so high that they have little probability of being stopped by the detector (see Anger camera below). The energy (140 keV) of the photon emitted by the most commonly used SPECT radioisotope, ^{99m}Tc, is ideal. The injected tracer amounts are limited by dosimetry considerations because, in contrast to x-ray imaging, the radioactivity remains in the patient for an extended period of time, in declining concentrations as it is excreted and decays. Therefore, the emission rate is low and imaging normally takes a long time (20–30 minutes for SPECT versus a few seconds for x-ray CT). Furthermore, the statistical quality of emission CT images is low, because they are formed with relatively few photons.

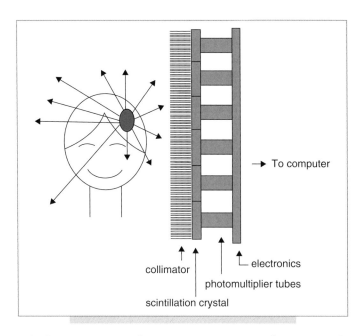

FIG. 31.1. Diagram of Anger camera. Courtesy of Dr Mi-Ae Park.

SPECT Instrumentation

Anger Camera

The most commonly used SPECT instrument is the Anger camera [2] (Figure 31.1). A photon which traverses a collimator hole strikes a scintillating crystal, most commonly made of sodium iodide (NaI). This material provides good detection efficiency and relatively good energy resolution (9% FWHM at 140 keV) at reasonable cost. Each photon is converted in the crystal to many lower-energy photons in the visible light range. These, in turn, are detected by an array, usually hexagonal, of photomultiplier tubes, which convert the light photons to electrical current. Each scintillation in the crystal generates a signal in several photomultiplier tubes; those closest to the interaction will receive the greatest amount of light. Positioning electronics yield an estimate of the location of the interaction based on the relative signal from each tube. The energy of each detected photon is also estimated by a pulse height analyzer.

Collimator

A collimator is a device used to restrict the photon acceptance angle in order to provide positional information for detected photons. A SPECT collimator is a thick sheet of metal of high atomic number, usually lead or tungsten, pierced by an array of holes. In theory, only photons which traverse these holes without striking the septa between them can be detected. In practice, some photons penetrate the septa and are detected, degrading the resolution and contrast of the reconstructed image. This is more likely for isotopes which emit higher energy photons, such as ^{111}In, frequently used for tumor imaging [3], with primary gamma-ray emissions at 171 and 245 keV.

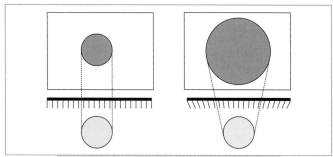

FIG. 31.2. Diagrams of parallel-hole (left) and cone-beam (right) collimators. Courtesy of Dr Mi-Ae Park.

In parallel-hole collimators (Figure 31.2), all holes are parallel to one another and perpendicular to the detector face. The fundamental design parameter for such a collimator is the hole size; smaller holes provide better information on the location of the emission and, consequently, better spatial resolution, but accept fewer photons, thereby increasing noise. Sensitivity is a crucial consideration for SPECT imaging, as the fraction of photons striking a collimator which traverse it without being absorbed is on the order of 1/1000. In an effort to increase sensitivity, a number of focusing collimator designs have been proposed [4–11]; fan-beam collimators focus only in the transaxial (within-plane) direction, while cone-beam collimators focus in both transaxial and axial directions. A recent design by Park et al [12], for example, which consists of a very highly focused cone-beam collimator paired with a long-focal-length fan-beam collimator, is expected to yield sensitivity greater than that of dual-parallel beam collimators by a factor of 14 at the center of the brain and an average factor of 10 throughout the brain. Focusing collimators lead to improved spatial resolution, as well as improved sensitivity, due to magnification (see Figure 31.2).

SPECT Image Reconstruction and Correction

The major physical effects which cause the image measurements to deviate from their assumed line projections are limited spatial resolution, photon attenuation and photon scatter. Spatial resolution is determined by characteristics of the collimator and of the crystal. Because each collimator hole accepts photons originating within an extended volume rather than along a line, the pattern of photons striking the detector is a blurred version of the true projection through the radioactivity distribution. This blurring increases with distance from the collimator, leading to variable spatial resolution within the reconstructed image. Further blurring takes place within the crystal. Because the intrinsic resolution of the crystal and the geometry of the collimators are well characterized, it is, in theory, possible to mathematically remove the blurring from the projections. This can be accomplished prior to analytical reconstruction by inverse filtering, or within an iterative reconstruction

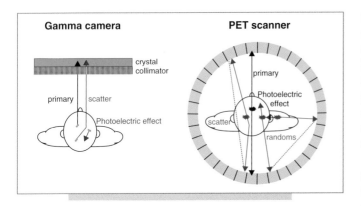

FIG. 31.3. *Illustration of types of events in SPECT (left) and PET (right) scanners. Courtesy of Dr Georges El Fakhri.*

algorithm by incorporating the resolution characteristics into the system model. In practice, because the measured projections are noisy versions of the true projections, resolution recovery is accompanied by amplification of image noise, no matter how it is implemented. The desirability of compensating for limited spatial resolution depends on the purpose for which the image is to be used. For example, resolution recovery leads to more accurate but less precise estimates of activity concentration within a tumor.

Some emitted photons do not escape the brain, but undergo interactions with the tissue (Figure 31.3). The likelihood of an emitted photon escaping the brain depends on the location of the emission and the density and atomic number of the anatomical structures along the photon path. The probability of an emitted photon being lost to attenuation is greatest for photons originating near the center of the brain, because these photons must traverse the greatest amount of tissue (averaged over all projection angles) in order to escape the brain and be detected. The effects of attenuation on reconstructed images are a negative bias in reconstructed activity concentration and an increase in image noise; both of these are greatest at the center of the brain. Several exact analytical correction methods [13–16] have been proposed; however, the most widely used remains the approximate technique of Chang [17]. These methods do not take into account the heterogeneous density of brain structures, most importantly the higher density of the skull. Attenuation correction can also be incorporated into an iterative reconstruction algorithm, either assuming that the brain is a uniform attenuator or incorporating patient-specific attenuation maps derived either by adding transmission capability to the SPECT instrument [18,19] or from CT images [20]. It should be noted that, although correction methods compensate for the negative bias due to attenuation, thereby improving the accuracy of the reconstructed voxel values, they do not affect the increase in image noise [21].

Although patient-specific attenuation maps are crucial for accurate attenuation correction in whole-body imaging,

they are less important for brain SPECT because of the smaller attenuating volume and more homogeneous density. The value of non-uniform attenuation correction in the brain depends on the imaging task; for example, Rajeevan et al [22] reported that non-uniform attenuation correction yielded more accurate estimates of absolute activity concentration within brain regions of interest (ROI) than did uniform attenuation correction, while the accuracy of ROI-to-background ratio estimates was not improved.

Attenuated photons are either fully absorbed by patient tissue in a photoelectric event, or partially absorbed, with the residual energy carried off as a lower-energy photon in a different direction (Compton scatter) (see Figure 31.3). Those scattered photons which retain enough energy to be accepted within the energy window may be detected; because the emission event is assumed to have occurred along the path of the scattered photon, rather than along the path of the original photon, scattered photons contribute to the reconstructed image at the wrong location. The effects of detection of scattered photons on the reconstructed image are reduced contrast and a positive bias in reconstructed voxel values. The acceptance of scattered photons is controlled by the energy window settings; narrow energy windows reduce the amount of scatter in the dataset. Because the energy resolution of the detectors is not perfect, however, narrow energy windows also lead to rejection of some primary photons. There are two basic approaches to scatter compensation. The first approach estimates the amount of scatter at each point in the projection dataset, so that it can be subtracted from the projections, and the second incorporates scatter modeling into the reconstruction algorithm. The first approach, subtracting an estimate of the scatter from the projections, leads to a more accurate image but increases noise, because fewer photons contribute to the image. A number of methods based on this approach, and differing in the method used to estimate the contribution of scatter, have been proposed; the most commonly used are the dual-energy-window method of Jaszczak [23] and the triple-energy-window method of Ogawa [24]. Several multiple-energy-window methods have also been proposed [25–28]; this more complex approach can, for brain imaging, lead to performance in estimation tasks approaching that of perfect scatter-rejection [25]. Few SPECT instruments, however, have the capability to collect photons in multiple energy windows. The second approach, incorporating scatter modeling into the reconstruction algorithm, essentially uses the scattered photons by restoring them to their correct location, rather than removing them. This approach should, in theory, lead to more accurate images without increasing noise as much as the scatter-rejection techniques. The major disadvantage of this approach is that accurate modeling of scatter is computationally demanding. A comprehensive review of scatter correction techniques is given by Zaidi and Koral [29].

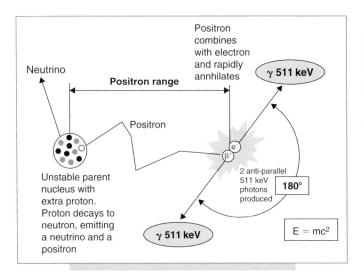

FIG. 31.4. Illustration of positron emission and decay. Courtesy of Dr Georges El Fakhri.

PET IMAGING

Introduction to PET Physics

PET radioisotopes have unstable nuclei in which a proton decays to a neutron, emitting a positron, the antiparticle of the electron, with the same rest mass (511 keV) but opposite charge, and a neutrino, a chargeless and (almost) massless particle which does not interact with matter (Figure 31.4). The positron travels some distance through tissue depending on its initial kinetic energy, before interacting with an electron; this annihilation yields two 511 keV photons traveling in (nearly) opposite directions. The positron range in tissue, on the order of a few milli-meters, places a fundamental limit on the spatial resolution of PET imaging. Although the annihilation photon energy is much higher than optimal for imaging purposes, the potential for coincidence imaging more than compensates for this shortcoming. When two photons are detected in opposed detectors within the coincidence timing window, a 'true' coincidence event is recorded, and the annihilation is assumed to have occurred somewhere along the line of response (LOR) connecting the two detectors (see Figure 31.3). This electronic collimation has advantages over the physical collimation which is necessary for SPECT. The absence of the physical collimator, which absorbs a large fraction of the incident photons, is the major reason for the higher (by one or two orders of magnitude) sensitivity of PET compared to SPECT.

Some annihilation events do not generate a coincidence event, because only one of the photons strikes a detector; the 'singles' count rate is the total of all detected counts. There are also some erroneous coincidences, termed 'randoms,' which occur when single photons from each of two annihilations are detected in coincidence

(see Figure 31.3). Although in whole-body PET imaging the randoms rate sometimes exceeds the true coincidence rate, in brain imaging the randoms rate seldom exceeds 25% of the true rate. The randoms rate can be minimized by using fast scintillators and fast detector electronics.

Annihilation photons may undergo Compton scatter, just as single photons may, and strike a detector in coincidence with the other photon in the pair (see Figure 31.3). If the energy of the scattered photon falls within the detector energy window, then a spurious true coincidence will be recorded. This yields inaccurate positioning, analogous to inaccurate positioning due to detection of scattered radiation in SPECT. Furthermore, some positron emitters generate additional gamma rays, which can also result in false coincidences.

PET Instrumentation
Detector Materials

The choice of radiation detectors for PET instruments requires tradeoffs among several detector characteristics, including physical density, atomic number, scintillation speed and light output. The stopping power is determined by physical density and effective atomic number; high stopping power is desirable, as it leads to increased sensitivity. Atomic number also determines the ratio of photoelectric events to Compton scattering events; because photoelectric events are preferred for imaging, high atomic number is desirable. Faster crystals make possible narrower coincidence timing windows and, therefore, fewer random coincidences.

The original PET instruments used thallium-doped sodium iodide (NaI(Tl)), but modern PET scanners use other crystals with higher stopping power, such as bismuth germinate (BGO), cerium-doped lutetium oxyorthosilicate (LSO) and gadolinium oxyorthosilicate (GSO). LSO and GSO are faster than the other materials and GSO has the best energy resolution, making possible better scatter rejection.

Detector Configurations

Most modern PET scanners use multiple rings of detectors, with each detector in coincidence with a number of detectors on the opposite side of the ring (and, in some cases, with detectors in other rings; see below). A variant of the ring configuration is a polygonal array, in which each detector is in coincidence with detectors on the opposing face. Less expensive options include partial rings, which require rotation of the detector array for acquisition of a complete set of data, and single-crystal detectors, similar to those in SPECT systems, but with the capability for coincidence imaging [30]. One such system, the PENNPET, consists of a hexagonal array of large-area detectors; it has been designed for brain imaging and is in use at a number of centers.

The spatial resolution of a typical PET scanner is determined by the size of the individual crystals. Block detectors were recently introduced to avoid the cost of more, smaller crystals and their associated individual PMT and electronics [31]. A block detector consists of a large crystal that is divided into an array of subelements by partial cuts, normal to the face, that are filled with reflective material. Modern systems using rings of block detectors can have up to 20 000 detector elements. This design improves spatial resolution, but at the cost of some loss of sensitivity due to the reduction in active crystal surface. Another recent innovation, used in several commercial systems, is the pixelated detector matrix, which consists of a set of small crystals coupled to a single light guide, viewed by an array of photomultiplier tubes.

The original PET scanners used septa between slices to define image planes, preventing cross-slice coincidences (2D PET). This configuration leads to the most accurate imaging, because it minimizes random coincidences and detection of scattered photons; it also, however, reduces sensitivity. In some systems, coincidences between adjacent slices are also acquired. In 3D PET, septa are not used; coincidences are accepted from the entire field of view, leading to greatly increased sensitivity compared to 2D. There is significant axial variation in sensitivity, with the highest sensitivity at the center and lowest sensitivity at the most extreme axial positions. The disadvantages of the 3D geometry, increased scatter and randoms, are less significant for brain than for whole-body PET.

PET Image Reconstruction and Correction

Before PET data can be reconstructed, they must be corrected for several physical effects. Corrections for dead time, a count-rate dependent phenomenon by which the detector is paralyzed and stops recording counts, and for varying sensitivity among detectors, are relatively straightforward for brain imaging. A certain number of detected coincidences are random, i.e. single photons from each of two different annihilations are detected in coincidence (see Figure 31.3). The most commonly used approach to correction for randoms is to subtract from the coincidence data for each LOR an estimate of the number of random coincidences, obtained by using a delayed timing window [32].

Attenuation correction of PET data is simpler than correction of SPECT data because, in contrast to SPECT, the probability of detection is uniform everywhere along an LOR. The correction factor depends only on the total thickness and density of tissue along the LOR. All that is necessary, then, for attenuation correction in PET is an estimate of the total attenuation along the LOR. Patient-specific measurements can be obtained by a transmission scan, using either an external radioisotope source or an x-ray tube mounted on the PET scanner or, more recently, an integrated PET/CT system.

Detection of scattered photons emerging from the patient that retain enough energy to be accepted by the detection system causes the coincidence event to be assigned to the wrong LOR; as in SPECT, this leads to reduced contrast and a positive bias in reconstructed voxel activity concentration values. The first line of defense against scattered radiation is discrimination on the basis of energy; this is not very effective in PET, because of the relatively poor energy resolution of PET detector materials. The simplest approach to scatter correction is to estimate the amount of scatter at each point in the measured projection data and mathematically remove it from the projection dataset. The estimates of scatter can be based on the measurements outside the projection of the body contour (presumably consisting only of scatter) [33] or on data acquisition in two [34], three [35], or many [36] energy windows. A more accurate, but much more complex, approach is to use Monte Carlo simulations to determine scatter estimates [37,38]. Finally, scatter modeling can be incorporated into an iterative reconstruction algorithm [39–42]. A review of scatter correction for brain PET imaging is given by Zaidi [43].

REFERENCES

1. Radon J (1917). Über die bestimmung von funktionen durch ihre integralwerte längs gewisser mannigfaltigkeiten. Ber Verh Sachs Ak Mn 69:262–277.
2. Anger HO (1958). Scintillation camera. Rev Sci Instrum 29:27–33.
3. Bakker WH, Albert R, Bruns C et al (1999). [In-111-DTPA-D-Phe1]-octreotide, a potential radiopharmaceutical for imaging of somatostatin receptor-positive tumors – synthesis, radiolabeling and in vitro validation. Life Sci 49:1583–1591.
4. Lalush DS (1998). Dual-planar circular-orbit cone-beam SPECT. J Nucl Med 39:22P.
5. Kamphuis C, Beekman FJ (1998). The use of offset cone-beam collimators in a dual head system for combined emission transmission brain SPECT: a feasibility study. IEEE Trans Nucl Sci 45:1250–1254.
6. Li JY, Jaszczak RJ, VanMullekom A, Scarfone C, Greer KL, Coleman RE (1996). Half-cone beam collimation for triple-camera SPECT systems. J Nucl Med 37:498–502.
7. Gullberg GT, Zeng GL, Datz FL, Christian PE, Tung CH, Morgan HT (1992). Review of convergent beam tomography in single photon emission computed tomography. Phys Med Biol 37:507–534.
8. Jaszczak RJ, Greer KL, Coleman RE (1988). SPECT using a specially designed cone beam collimator. J Nucl Med 29:1398–1405.
9. Jaszczak RJ, Floyd CE, Manglos SH, Greer KL, Coleman RE (1986). Cone beam collimation for single photon emission computed tomography – analysis, simulation, and image reconstruction using filtered backprojection. Med Phys 13:484–489.

10. Hawman EG, Hsieh J (1986). An astigmatic collimator for high-sensitivity SPECT of the brain. J Nucl Med 27:930.

11. Jaszczak RJ, Chang LT, Murphy PH (1979). Single photon emission computed tomography using multi-slice fan beam collimators. IEEE Trans Nucl Sci 26:610–618.

12. Park MA, Moore SC, Kijewski MF (2005). Brain SPECT with short focal-length cone-beam collimation. Med Phys 32:2236–2244.

13. Bellini S, Piacentini M, Cafforio C, Rocca F (1979). Compensation of tissue absorption in emission tomography. IEEE Trans ASSP 27:213–218.

14. Gullberg GT, Budinger TF (1981). The use of filtering methods to compensate for constant attenuation in single photon emission computed tomography. IEEE Trans Biomed Eng 28:142–157.

15. Metz CE, Pan XC (1995). A unified analysis of exact methods of inverting the 2D exponential radon transform, with implications for noise control in SPECT. IEEE Trans Med Imaging 14:643–658.

16. Tretiak O, Metz C (1980). The exponential Radon transform. SIAM J Appl Math 39:341–354.

17. Chang LT (1978). Method for attenuation correction in radionuclide computed tomography. IEEE Trans Nucl Sci 25:638–643.

18. Murase K, Tanada S, Inoue T, Sugawara Y, Hamamoto K (1993). Improvement of brain single photon emission tomography (SPET) using transmission data acquisition in a 4-head SPET scanner. Eur J Nucl Med 20:32–38.

19. Bailey DL, Hutton BF, Walker PJ (1987). Improved SPECT using simultaneous emission and transmission tomography. J Nucl Med 28:844–851.

20. Tang HR, Da Silva AJ, Matthay KK et al (2001). Neuroblastoma imaging using a combined CT scanner-scintillation camera and I-131-MIBG. J Nucl Med 42:237–247.

21. Kijewski MF, Müller SP, Moore SC (1997). Nonuniform collimator sensitivity: improved precision for quantitative SPECT. J Nucl Med 38:151–156.

22. Rajeevan N, Zubal IG, Ramsby SQ, Zoghbi SS, Seibyl J, Innis RB (1998). Significance of nonuniform attenuation correction in quantitative brain SPECT imaging. J Nucl Med 39:1719–1726.

23. Jaszczak RJ, Chang LT, Stein NA, Moore FE (1979). Whole-body single-photon emission computed tomography using dual, large-field-of-view scintillation cameras. Phys Med Biol 24:1123–1143.

24. Ogawa K, Harata Y, Ichihara T, Kubo A, Hashimoto S (1991). A practical method for position-dependent Compton scatter correction in single photon emission CT. IEEE Trans Med Imaging 10:408–412.

25. Moore SC, Kijewski MF, Müller SP, Rybicki F, Zimmerman RE (2001). Evaluation of scatter compensation methods by their effects on parameter estimation from SPECT projections. Med Phys 28:278–287.

26. Haynor DR, Kaplan MS, Miyaoka RS, Lewellen TK (1995). Multiwindow scatter correction techniques in single-photon imaging. Med Phys 22:2015–2024.

27. Gagnon D, Todd-Pokropek A, Arsenault A, Dupras G (1989). Introduction to holospectral imaging in nuclear medicine for scatter subtraction. IEEE Trans Med Imaging 8:245–250.

28. Koral KF, Wang XQ, Rogers WL, Clinthorne NH, Wang XH (1988). SPECT Compton scattering correction by analysis of energy spectra. J Nucl Med 29:195–202.

29. Zaidi H, Koral KF (2004). Scatter modelling and compensation in emission tomography. Eur J Nucl Med Mol Imaging 31:761–782.

30. Karp JS, Muehllehner G, Mankoff DA et al (1990). Continuous-slice Penn-PET – a positron tomograph with volume imaging capability. J Nucl Med 31:617–627.

31. Casey ME, Nutt R (1986). A multicrystal 2-dimensional BGO detector system for positron emission tomography. IEEE Trans Nucl Sci 33:460–463.

32. Hoffman EJ, Huang SC, Phelps ME, Kuhl DE (1981). Quantitation in positron emission computed tomography. 4. Effect of accidental coincidences. J Comput Assist Tomogr 5:391–400.

33. Bailey DL, Meikle SR (1994). A convolution-subtraction scatter correction method for 3D PET. Phys Med Biol 39:411–424.

34. Grootoonk S, Spinks TJ, Sashin D, Spyrou NM, Jones T (1996). Correction for scatter in 3D brain PET using a dual energy window method. Phys Med Biol 41:2757–2774.

35. Shao LX, Freifelder R, Karp JS (1994). Triple energy window scatter correction technique in PET. IEEE Trans Med Imaging 13:641–648.

36. Bentourkia M, Msaki P, Cadorette J, Lecomte R (1995). Energy-dependence of scatter components in multispectral PET imaging. IEEE Trans Med Imaging 14:138–145.

37. Levin CS, Dahlbom M, Hoffman EJ (1995). A Monte-Carlo correction for the effect of Compton scattering in 3D PET brain imaging. IEEE Trans Nucl Sci 42:1181–1185.

38. Zaidi H (1999). Relevance of accurate Monte Carlo modeling in nuclear medical imaging. Med Phys 26:574–608.

39. Floyd CE, Jaszczak RJ, Greer KL, Coleman RE (1986). Inverse Monte-Carlo as a unified reconstruction algorithm for ECT. J Nucl Med 27:1577–1585.

40. Beekman FJ, Kamphuis C, Frey EC (1997). Scatter compensation methods in 3D iterative SPECT reconstruction: a simulation study. Phys Med Biol 42:1619–1632.

41. Hutton BF, Baccarne V (1998). Efficient scatter modelling for incorporation in maximum likelihood reconstruction. Eur J Nucl Med 25:1658–1665.

42. Kadrmas DJ, Frey EC, Karimi SS, Tsui BMW (1998). Fast implementations of reconstruction-based scatter compensation in fully 3D SPECT image reconstruction. Phys Med Biol 43:857–873.

43. Zaidi H (2000). Comparative evaluation of scatter correction techniques in 3D positron emission tomography. Eur J Nucl Med 27:1813–1826.

CHAPTER 32

Ultra-high Field Magnetic Resonance Imaging of Brain Tumors

Gregory A. Christoforidis

INTRODUCTION

The Food and Drug Administration has approved human imaging up to 8.0 Tesla (T). Currently, imaging up to 4.0T is available for clinical practice, whereas higher field imaging up to 8.0T is only available on research systems. Ultra-high field (UHF) magnetic resonance imaging (MRI) generally refers to imaging at field strengths of 7T or more. This chapter reviews the current experience with the imaging of brain tumors at ultra-high field strengths and potential clinical applications currently in transitional research.

Imaging of brain tumors attempts to define the tumor's anatomic location, its extent, the neurologic function that may be affected and intrinsic characteristics which may predict tumor behavior. Magnetic resonance imaging at higher field strengths has the potential to improve all of these goals. The higher signal-to-noise ratio (SNR) obtained at higher field strengths allows greater spatial resolution. Greater spatial resolution can provide a detailed assessment of the relationship of the tumor to adjacent structures such as gray matter, suabarachnoid space or vascular structures. Although higher field strength has resulted in greater spatial resolution, the visualized extent of brain tumors imaged at 8T and 1.5T has not differed. Finally, because greater susceptibility effects occur at higher field strengths, elements present within the tumor bed with greater magnetic susceptibility are expected to contrast against the tumor background. Such components may include foci of hemorrhage, iron-containing structures or components labeled with susceptibility based contrast agents, and vessels containing deoxyhemoglobin. Technical and clinical challenges exist which need to be overcome in order to attain the full potential of brain tumor imaging at ultra-high field strengths.

SPATIAL RESOLUTION

Higher SNR is the major advantage with ultra-high field MRI. Images comparing 8T and 1.5T images are shown in Figure 32.1. SNR was computed as the ratio of these signals and the noise in an air filled region outside the head.

CNR is the signal difference divided by the noise. These measurements and histogram analysis showed that there is more than a factor of 2 signal variation due to radiofrequency (RF) inhomogeneity (see below) across the 8T image, thus 8T signal measurements are only approximate. For some regions, up to a 15-fold SNR improvement is achieved with 8T. Note in Figure 32.2 that with similar pulse sequences, differences in SNR are clearly visualized. However, practical limitations, most notably RF inhomogeneity, susceptibility artifacts, the tissue relaxation times and suboptimal pulse sequences still hamper ultra-high field MRI. The RF coils used for ultra-high field (>7T) are transverse electromagnetic (TEM) resonators [1–3]. At this field strength, these volume coils are unfortunately quite inhomogeneous in terms of the B1 fields, not only for the volume, but for various locations on the same image. These coils tend to demonstrate the best image quality (highest SNR) within the central portion of the brain.

Although increased spatial resolution is possible at lower field strengths with surface coil imaging and smaller field of view, the signal is limited by radiofrequency penetration. The higher SNR possible at higher field strengths

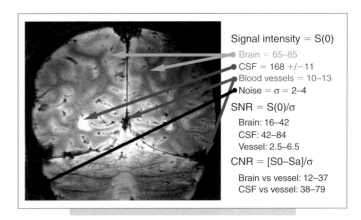

FIG. 32.1. Calculation of signal-to-noise ratio (SNR) and contrast-to-noise ratio (CNR) in a normal subject imaged at 8 T.

allows for deeper radiofrequency penetration. Imaging at lower field strengths using surface coils and small field of view has resulted in improved SNR derived from the surface of the brain [4]. Such techniques have been applied toward the detections of cortical dysplasia in the setting of childhood epilepsy [5] and smaller voxel size for MR spectroscopy along the surface of the brain, but not the deeper structures. UHFMRI allows for greater SNR through both superficial and deeper structures of the brain. This has allowed for imaging with larger matrix size.

Comparison of 1.5T and 8.0T images derived from patients with brain tumors has not been shown to demonstrate greater tumor extent based on signal changes associated with tumor infiltration. However, the relationship of involved tumor to adjacent anatomic structures is more clearly depicted with higher resolution imaging. In addition, areas of questionable infiltration at lower field strengths are more clearly depicted with higher field strength imaging. Higher resolution images allow for a more detailed characterization of tumoral architecture and relationship to adjacent anatomic structures (Figure 32.3). Septations within the tumor bed stand out more readily when present. Using a 9.4 Tesla MRI unit, Fatterpekar et al were able to demonstrate lamination of the cerebral cortex within fixed specimens using acquisition times which lasted albeit 14 hours [6]. Although such image acquisitions are unrealistic in a clinical setting, the imaging is useful in the understanding of expected signal intensity profiles from the human cortex.

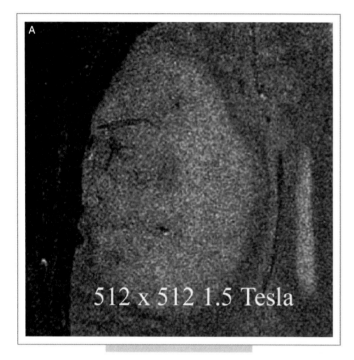

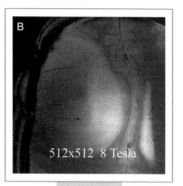

FIG. 32.2. Improved signal-to-noise ratio at 8T versus 1.5T. The 4–5 fold signal to noise advantage of 8T is clearly demonstrated when comparing gradient echo MR images using similar parameters acquired at 1.5T **(A)** and 8T **(B)** in a subject with anaplastic astrocytoma.

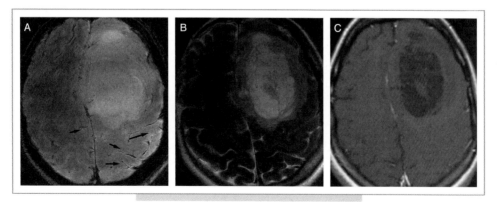

FIG. 32.3. 8T MRI in a patient with a low-grade oligodendroglioma. **(A)** 8T GE, **(B)** RARE and **(C)** 1.5T T1W gadolinium images of a WHO grade II oligodendroglioma. Note the ability to resolve fine septations within the tumor on the RARE image. Even though cortical penetrating veins are visualized within the tumor bed (arrows), no foci of irregular vessels suggestive of neovascularity were resolved within the tumor bed of 8T GE imaging.

High resolution image acquisitions acquired in vivo have been obtained at matrix sizes as large as 2048 × 2048 [7]. In our experience, in vivo imaging of the cortex of the human brain often displays a laminar appearance on gradient echo UHF MRI using a 1024 × 1024 matrix size in which the middle layers have slightly lower signal intensity relative to the cortex. This can be explained by the higher concentration of normal microvascular structures in the middle layers of the cortex. It is suspected that higher resolution imaging will lead to the observation of findings that would otherwise not be resolved and allow the depiction of smaller normal anatomic structures and the person interpreting the examination better to localize an abnormality. Thus, imaging at ultra-high field strengths has the potential for image interpretation with greater sophistication than possible at lower field strengths.

MAGNETIC SUSCEPTIBILITY

Magnetic susceptibility is the magnetic response of a substance to a magnetic field and can result in local magnetic field inhomogeneities and signal loss. These effects are proportional to field strength and the differences in susceptibility of two regions. Magnetic susceptibility increases with field strength. It can result from macroscopic effects such as the interface between substances (i.e. around air cavities or at the interface between bone and soft tissues) or it can result from microscopic effects such as paramagnetic materials (i.e. unpaired electrons in blood products) or ferromagnetic materials (i.e. iron). The effects can vary depending on the size, shape and direction of the object relative to the external field. Magnetic susceptibility can result in the ability to visualize microscopic structures with susceptibility effects. Because gradient echo images do not have a refocusing pulse, they are especially sensitive to effects from magnetic susceptibility. Such structures include microhemorrhage, deposition of paramagnetic substances within the tissues, or deoxyhemoglobin within blood vessels

The most unique results with brain tumor imaging thus far have been found using high-resolution (<200 μm) proton density/T2*-weighted gradient echo axial images. On these images, hundreds of small linear low signal regions correspond to the microvasculature of the brain (Figures 32.4–32.6).

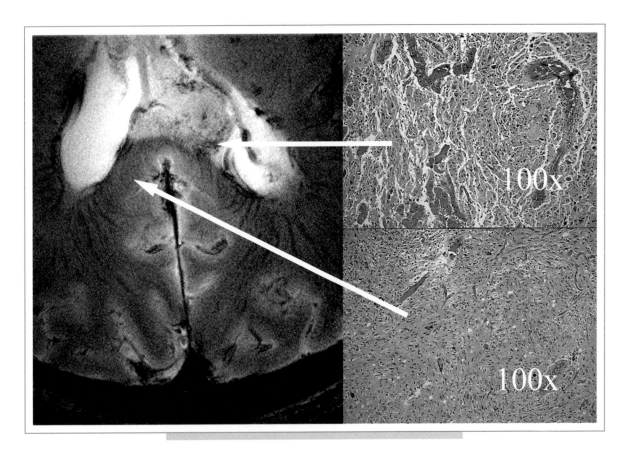

FIG. 32.4. Histopathologic correlation to high resolution 8T gradient echo (GE) MRI. Glioblastoma was imaged within an intact fresh cadaver using 8T GE MRI using a 1024 × 1024 matrix giving an in-plane resolution of 196 μm. Directed specimens were obtained at autopsy. Areas of serpiginous signal voids seen on 8T MRI (in this case probably related to deoxy- or methemoglobin in vessels) corresponded to areas of enlarged microvessels and increased microvascular density on hematoxylin and eosin stains. Areas of the tumor with smaller microvessels and low microvascular density had few signal voids on 8T MRI.

As mentioned earlier, the image demonstrates a susceptibility based signal void for cerebral vasculature with visualization of the cortical penetrating vessels, transcallosal veins and transmedullary veins. The expected diameter of some of these vessels is less than 100 μm (8). This susceptibility-based image contrast is attributed to the presence of deoxyhemoglobin, primarily within the venous structures, and results in a blood oxygen level dependent (BOLD) contrast. T2* in venous blood is very short. The vessels are seen at low signal regions. Because of their sensitivity to subtle local field variations, gradient echo imaging is most suitable for this purpose. Similar BOLD contrast images have been acquired at lower magnetic field strength [3,9,10]. Susceptibility effect from small lesions containing paramagnetic substances may become more conspicuous using high-resolution UHFMRI (Figure 32.7).

Tumor growth is believed to depend on angiogenesis. Folkman proposed that without angiogenesis, tumor growth is limited to several millimeters in thickness [11]. Furthermore, tumor cells are believed to induce angiogenesis by secreting a number of growth factors (angiogenesis factors), which stimulate endothelial cells [11–14]. Glioblastoma has been shown to be one of the most vascularized tumors for humans [15]. One therapeutic strategy for patients with brain tumors involves the development of agents that modulate angiogenesis [11,14,15]. Validation of techniques which monitor angiogenesis is key to the development of such strategies. It therefore follows that the imaging identification of angiogenic vessels may serve as a marker for response to antiangiogenesis treatment. Histopathologic methods grading angiogenesis have been

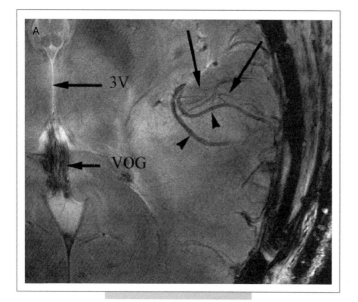

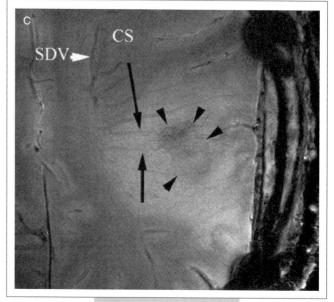

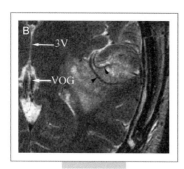

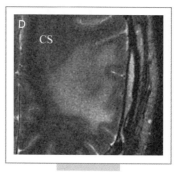

FIG. 32.5. Microvascularity within a glioblastoma. Gradient echo (GE) 8 T MRI with a 1024 × 1024 matrix and in-plane resolution of 196 μm (**A** and **C**) are compared to fast spine echo (FSE) T2 images with a 512 × 512 matrix (**B** and **D**) in a patient with glioblastoma. Images were cropped to demonstrate microvascular structures. Arrowheads (**A** and **B**) demonstrate arterial structures within the sylvian fissure (identified in both exams), the smaller venous structures draining the tumor (arrows) are only observed on the 8 T images (arrowheads). Signal loss adjacent to a focus of distorted microvasculature (**C**, arrowheads) is thought to represent microvasculature beyond the resolution of this scan (VOG = vein of Galen; 3V = third ventricle, CS = centrum semiovale).

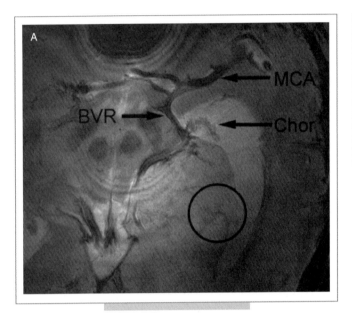

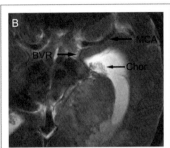

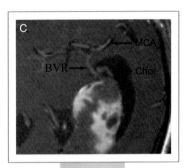

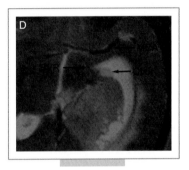

FIG. 32.6. Microvascularity within a high-grade pleomorphic xanthoastrocytoma. 8T gradient echo (GE) **(A)**, 8T RARE **(B)**, 1.5T T1 with gadolinium **(C)** and 1.5T FSE T2 **(D)** of a high grade pleomorphic xantoastrocytoma. Note the identification of foci of microvascularity within the tumor bed (circle). Venous structures such as the basal vein of Rosenthal (BVR) appear substantially larger relative to arterial structures such as the middle cerebral artery (MCA) of equivalent size on the GE images relative to the 8T RARE and the 1.5T images due to susceptibility. Finally, 8T images display detail within anatomic structures such as the choroid plexus (Chor) better than 1.5T images.

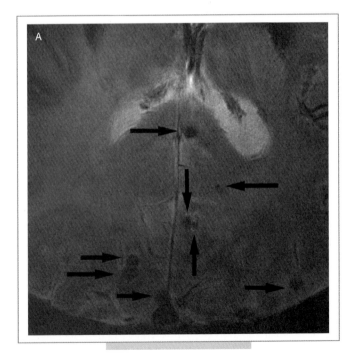

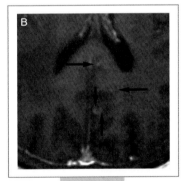

FIG. 32.7. Identification of metastases on high resolution 8T gradient echo imaging. Comparison of high resolution 8T gradient echo MRI **(A)** to post-gadolinium 1.5T 3D FSPGR **(B)** images in a patient with metastases along the occipital lobes and splenium of the corpus callosum. Note the difference in the conspicuity of the metastatic foci (arrows) between the two studies. Some of the metastases identified on the 1.5T SPGR image are not seen on the 8T image. This is likely related to a combination of the higher signal-to-noise ratio attained at 8T imaging and susceptibility effects resulting from the metastatic foci.

based on the number of vessels, degree of glomeruloid vascular structure formation and endothelial cytology [15]. More recently, microvascular density (MVD) counting using panendothelial staining techniques (i.e. factor VIII, CD34) have been used to quantify angiogenesis and have been found to act as independent prognosticators [16–18]. It should therefore follow that an imaging assessment of vascular density within a tumor bed may act as an indicator for tumoral angiogenesis.

High field magnetic resonance imaging at 8 Tesla using 1024×1024 matrix size has been able to visualize vasculature to approximately 100 μm in size [8]. This type of high-resolution imaging has been able to identify directly vascular features within a tumor bed [19]. The MRI findings of distorted vasculature included enlargement of transmedullary veins, increased tortuosity and increased apparent vessel density [19,20]. Previous studies have demonstrated that apparent vessel density in a nude mice glioma model as identified by 4.7 Tesla gradient echo magnetic resonance images significantly correlates with the histopathologically identified density of blood containing vessels [21]. High-resolution MRI of apparent vascular density has been successfully used as an assay for angiogenesis in this same animal model [22,23]. Using gradient echo 8T imaging, it has been possible to identify microvascularity within brain tumors in humans (see Figures 32.5 and 32.6). The tumor bed can be analyzed for foci with vascular distortion. These foci of microvascularity can then be assessed for microvessel size and concentration compared with that found in gray matter and white matter. Semiquantitative determination of the number of foci with areas of abnormal vasculature, as well as the size and density of the abnormal microvascular structures relative to gray matter and white matter, has been noted to correlate with tumoral grade [24]. In addition, foci of vascular distortion within the tumor bed on gradient echo 8T MRI correspond to areas of increased cerebral blood volume on dynamic contrast enhanced 1.5T MRI [25]. Finally, in our experience, directed brain tumor biopsies derived from foci corresponding to microvascular distortion on gradient echo 8T MRI have corresponded to areas of increased microvascularity and neovascularity on histopathology [26]. Further improvement in spatial resolution would be possible with larger matrix size, however, the acquisition times may be prohibitively long for routine in vivo work.

Conventional MRIs display images in terms of signal amplitude (magnitude images). Phase imaging (also known as susceptibility weighted imaging (SWI)) displays images in terms of phase shifts. As such they highlight anatomic structures with susceptibility effects including iron-containing structures, venous vasculature and cortex, and filter out susceptibility effects due to field inhomogeneities. RF inhomogeneity is reflected as a variation in signal-to-noise [27]. As a result, flow voids from arteries (see arrows in Figure 32.8) are suppressed and venous structures (Figure

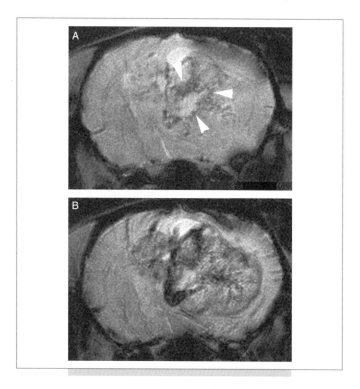

FIG. 32.8. USPIO enhancement of microvascularity in the F98 glioma. Fischer rats bearing the F98 glioma were initially imaged while ventilated on 100% O_2 **(A)** and following intravenous administration of 2 mgFe/kg ultra-small superparamagnetic iron oxide (USPIO) intravascular contrast agent. These rats were imaged at 8T using high resolution gradient echo imaging. In-plane resolution here is 78 μm. Compare microvascularity identified on the pre-contrast image with microvascularity identified on the post-contrast image. Note the increase in conspicuity in hypointense serpiginous vessels within the tumor bed. Vessel conspicuity centrally within the tumor bed prior to contrast injection in these rats, which were ventilated with 100% oxygen, may be related to hypoxia (arrowheads). Their visibility on gradient echo T2 images results from deoxyhemoglobin within those vessels. This method can increase the conspicuity of microvascularity within tumors and potentially identify regions of hypoxia within the tumor bed.

32.8, arrowheads) are enhanced. Phase imaging therefore has the potential to allow the differentiation of arteries from veins.

More recently, the use of ultra-small particle iron oxide (USPIO) contrast agents has been shown to increase the visibility of these microvascular structures in a rodent model. The total volume of increased microvascularity within the tumor bed identified on 8T MRI following the intravenous administration of USPIO (Figure 32.9) is substantial [28]. It is interesting that foci exist within the tumor bed which show microvascularity prior to USPIO injection and other foci which do not. Since visualization of microvasculature prior to USPIO injection depends on BOLD effect, these differences may relate to differences in deoxyhemoglobin

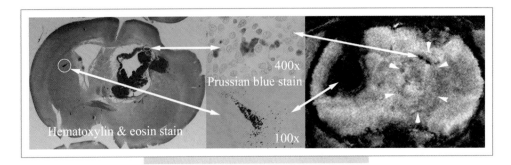

FIG. 32.9. Detection of microglia migration in the F98 rat glioma model. 200 μl ultra-small superparamagnetic iron oxide (USPIO) intravascular contrast agent (0.1 mmol/ml) was injected into right hemisphere of Fischer rat brain to allow endocytosis by microglia. Two weeks later, 10^5 F98 glioma cells were injected into contralateral caudate nucleus. Serial 8T gradient echo high resolution MR scans with in-plane resolution of 78 μm were performed to monitor the microglia migration. Ten days after implantation of F98 glioma cells, the rat was sacrificed. On 8T GE images immediately prior to sacrifice display profound signal loss at the USPIO injection site. Note the relative size of this signal loss relative to the iron particles on histopathology. A few microglia cells were identified in the tumor region containing the USPIO particles on histology. It is impressive that 8T GE MRI is sensitive enough to visualize only a few USPIO labeled cells.

concentration between different regions within the tumor bed. As such it would lend support to the hypothesis that this type of imaging may be useful in identifying foci of hypoxia within the tumor bed (Figure 32.9).

IMAGING TUMOR OXYGENATION

Tumor hypoxia is known to protect against radiation treatment. Fractionated doses of radiation delivered during periods of higher oxygenation result in more effective tumoral suppression. The identification of tumor tissues which are hypoxic by nature can result in improved treatment planning [29–33]. The identification of deoxyhemoglobin is thought to assess indirectly tumoral oxygenation via non-invasive means. Deoxyhemoglobin can be assessed via direct measurements of T2* relaxation or blood oxygen level dependent (BOLD) signal behavior (Figure 32.9). Studies using electron paramagnetic resonance (EPR) oximetry have been able to demonstrate that BOLD imaging can be used to assess and quantify tumoral oxygenation within tumor models [29]. Similarly, near infrared spectroscopy (NIRS) measurements of deoxyhemoglobin have been shown to correlate well with tissue deoxyhemoglobin concentration [34].

It has been shown that tumor perfusion and vessel reactivity can vary between tumor types [35,36]. A decrease in vessel reactivity can result in inefficient oxygen delivery. On the other hand, this difference may result from vessel tortuosity which, in turn, may limit erythrocyte access in certain parts of the tumor bed, resulting in lower deoxyhemoglobin concentration and thereby less BOLD effect with lesser visualization of these vascular structures. Exogenous contrast agents would, however, have the opportunity to visualize the total perfused vascular volume [36]. Because tumoral oxygenation is critical in influencing tumor response to radiation and some chemotherapeutic agents,

imaging of hypoxic regions could provide an in vivo marker for potential treatment response [37]. Preliminary data indicate foci of microvascularity identified within the F98 glioma on 8T gradient echo MR images without contrast agent correspond to microvascular foci in hypoxic regions on immunohistochemical stains using a nitroimidazole agent. This indicates that BOLD effect identified within abnormal microvasculature of brain tumors may be able indirectly to identify hypoxic regions associated with microvascularity [28].

CELLULAR MAGNETIC RESONANCE IMAGING

Currently approved MRI contrast agents are water-soluble gadolinium-based conjugates which depend on breakdown of the blood–brain barrier and case enhancement either by increased vascularization of the tumor bed or accumulation in the interstitial tissues. As such they are not necessarily an accurate representation of tumor extent.

Iron oxide agents can become incorporated by cells and tag them. This leads to the ability to image them using susceptibility based sequences on MRI. Via the process of endocytosis, cell types which have been incubated and successfully labeled with ferumoxides have included mesenchymal stem cells, astrocytes, oligodendrocyte progenitor cells, mononuclear cells and microglia cells [38,39]. These labeled cells have then been directly implanted into the brain and followed over time. In the setting of CNS pathology, microglia cells appear to take up iron oxides most efficiently. Alternatively, USPIO particles directly injected into the brain have been shown to be taken up predominantly by microglia cells via phagocytosis. Microglial cells can then be identified to migrate toward glial tumors and can be used to outline the tumor on MRI. Higher resolution

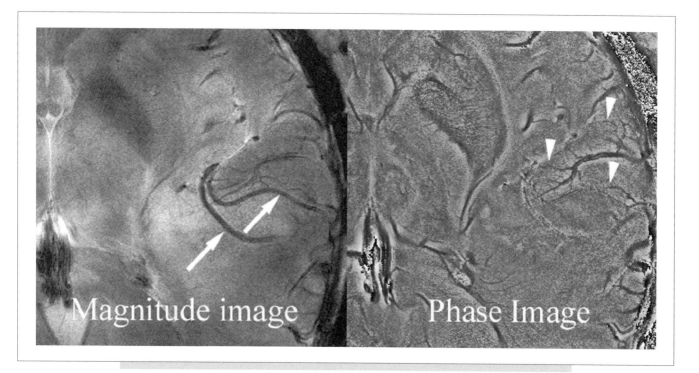

FIG. 32.10. Conventional MRIs display images in terms of signal amplitude (magnitude images). Phase images (susceptibility weighted imaging (SWI)) display images in terms of phase shifts. As such they highlight anatomic structures with susceptibility effects, including iron-containing structures, paramagnetic effect from deoxyhemoglobin within venous vasculature and cortex and filter out susceptibility effects due to field inhomogeneities. RF inhomogeneity is reflected as a variation in signal-to-noise. As a result flow voids from arteries (arrows) are suppressed and venous structures containing paramagnetic deoxyhemoglobin (arrowheads) are enhanced. This can serve as a method to distinguish arterial from venous structures.

and higher susceptibility effects are expected to favor the use of UHF MRI over lower field strengths in detecting cells labeled with USPIO. An example is provided in Figure 32.10.

In tumor models, macrophage and microglia cells have been shown to surround and infiltrate the tumor in amounts ranging from 20 to 40% of the tumor cell mass [38,40–42]. It is not clear whether these cells act as immune defense or promote glioma invasion and proliferations [43–45]. Glioma cells have the ability to secrete cytokines and attract microglia cells [38,46]. Thus microglia cells labeled with iron oxide cells can be used to localize gliomas. Even solitary invading tumor cells have been detected with USPIO-labeled microglia cells [38]. Potential pitfalls in imaging the trafficking of mononuclear cells and stem cells within the central nervous system contrast agents include potential inability to distinguish readily from other pathologies, as well as similar signal loss from tumor components with susceptibility effects such as hemorrhage.

MR SPECTROSCOPY

MR spectroscopy (MRS) has the potential to provide insight into the metabolite make-up within brain tissue and neoplasms. Higher field strengths are expected to afford proton MRS of cerebral metabolites with increased sensitivity and spectral dispersion. In addition, MRS spectra are expected to allow for the differentiation of more peaks with higher field strengths [47]. There are limitations from field inhomogeneities generated by greater susceptibility effects field strengths up to 7T. In recent investigations, the sensitivity threshold to detect concentration changes was $0.2\,\mu$mol/g for most of the quantified metabolites, in a screen of 17 different metabolites [48]. Furthermore, changes of the concentration of metabolites have been observed during physiologic stimulation of the visual cortex [49]. In addition, proton phosphorus MRS has been demonstrated with a spatial resolution of 7.5 ml using surface coil imaging [50–53]. These preliminary studies indicate that higher field strength magnets show some promise, however, their application to brain neoplasms has yet to be assessed.

The capabilities of UHF MRI as they apply to brain tumor imaging have yet to be fully explored. It is expected that, with time, the higher signal-to-noise ratios derived from higher field strengths will result in improved spatial resolution, not only of microvascularity but also functional imaging, dynamic contrast enhancement, spectroscopic imaging and MRI derived from protons other than water such as phosphorus, sodium or fluorine. These methods have yet to be applied to brain tumor imaging at ultra-high field strengths.

ACKNOWLEDGMENTS

This research was supported by a grant from the National Cancer Institute, NCI 1R21CA/NS92846-01A1.

REFERENCES

1. Haacke EM, Tkach JA, Parrish TB (1989). Reducing T*, dephasing in gradientfield-echo imaging. Radiology 170:457–462.

2. Ibrahim TS, Abduljalil AM, Lee R, Baertlein BA, Robitaille PML (2001). Analysis of B1 field profiles and SAR values for multistrut transverse electromagnetic RF coils in high field MRI application. Phys Med Biol 46:2545–2455.

3. Ibrahim TS, Lee R, Abduljalil AM, Baertlein BA, Robitaille PML (2001). Effect of RF coil excitation on field inhomogeneity at ultra high fields: a field optimized tem resonator. Magn Reson Imaging 19:1339–1347.

4. Wald LL, Crvajal L, Moyer SE et al (1995). Phased array detectors and an automated intensity correction algorithm for high resolution MR imaging of the human brain. Magn Reson Med 34:433–439.

5. Grant PE, Barkovich AJ, Wald LL, Dillon WP, Laxe KD, Vigneron DB (1997). High-resolution surface coil MR of cortical lesions in medically refractory epilepsy: a prospective study. Am J Neuradiol 18:29–31.

6. Fatterpekar GM, Naidich TP, Delman BN et al (2002). Cytoarchitecture of the human cerebral cortex: MR microscopy of excised specimens at 9.4 Tesla. Am J Neuroradiol 23:1313–1321.

7. Robitaille PM, Abduljalil AM, Kangarlu A (2000). Ultra high resolution imaging of the human head at 8 Tesla: 2K × 2K for Y2K. J Comput Assist Tomogr 24:2–8.

8. Christoforidis GA, Bourekas EC, Baujan M et al (1999). High resolution MRI of the deep brain vascular anatomy at 8 Tesla: susceptibility-based enhancement of the venous structures. J Comput Assist Tomogr 23:857–866.

9. Reichenbach JR, Essig M, Haacke EM et al (1998). High-resolution venography of the brain using magnetic resonance imaging. Magn Reson Mater Phys Biol Med 6:62–69.

10. Reichenbach JR, Barth M, Haacke EM, Klarhofer M, Kaiser WA, Moser E (2000). High-resolution MR venography at 3.0 Tesla. J Comput Assist Tomogr 24:949–957.

11. Folkman J (1995). Angiogenesis in cancer, vascular, rheumatoid and other disease. Nat Med 1:27–31.

12. Plate KH, Risau W (1995). Angiogenesis in malignant gliomas. Glia 15:339–347.

13. Weller RO, Foy M, Cox S (1977). The development and ultrastructure of the microvasculature in malignant gliomas. Neuropathol Appl Neurobiol 3:307–322.

14. Haddad SF, Moore SA, Schlefer RL, Goeken JA (1992). Vascular smooth muscle hyperplasia underlies the formation of glomeruloid vascular structures in glioblastoma multiforme. J Neuopath Exp Neurol 51:488–492.

15. Brem S, Cotran R, Folkman J (1972). Tumor angiogenesis: a quantitative method for histologic grading. J Nat Cancer Inst 48:347–356.

16. Weidner N (1998). Tumoral vascularity as a prognostic factor in cancer patients: the evidence continues to grow. J Pathol 184:119–122.

17. Fox SB (1997). Tumour angiogenesis and prognosis. Histopathology 30:294–301.

18. Eberhard A, Kahlert S, Goede V, Hemmerlein B, Plate KH, Augustin HG (2000). Heterogeneity of angiogenesis and blood vessel maturation in human tumors: implications for antiangiogenic tumor therapies. Cancer Res 60:1388–1393.

19. Christoforidis GA, Grecula JC, Newton HB et al (2002). Visualization of microvascularity within glioblastoma multiforme utilizing 8 Tesla high-resolution magnetic resonance imaging. Am J Neuroradiol 23:1553–1556.

20. Christoforidis GA, Kangarlu A, Abduljalil AM et al (2004). Susceptibility based imaging of glioblastoma microvascularity at 8T: correlation of MR imaging and postmortem pathology. Am J Neuroradiol 25:756–760.

21. Abramovich R, Frenkiel D, Neeman M (1998). Analysis of subcutaneous angiogenesis by gradient echo magnetic resonance imaging. Mag Reson Med 39:813–824.

22. Abramovich R, Marikovsky M, Meir G, Neeman M (1998). Stimulation of tumour angiogenesis by proximal wound: spatial and temporal analysis by MRI. Br J Cancer 77:440–447.

23. Abramovich R, Meir G, Neeman M (1995). Neovascularization induced growth of implanted C6 glioma multicellular spheroids: magnetic resonance microimaging. Cancer Res 5:1956–1962.

24. Christoforidis GA, Yang M, Figueredo T et al (2004). Microvascularity within human gliomas identified at 8T high resolution MRI and histopathology. Proceedings ASNR 2004.

25. Christoforidis GA, Varakis K, Newton HB (2002). Comparison of brain tumor microvascularity identified on gradient echo 8T MRI with perfusion MRI at 1.5 T. Comparison of brain tumor microvascularity identified on gradient echo 8T MRI with perfusion MRI at 1.5 T. International Society for Magnetic Resonance in Medicine Tenth Scientific Meeting and Exhibition Program 2002. 317.

26. Christoforidis GA, Yang M, Abduljalil A et al (2007). Correlation of microvascularity identified within human gliomas between 8T gradient echo MRI. Joint Annual Meeting International Society for Magnetic Resonance in Medicine – European Society for Magnetic Resonance in Medicine and Biology 2007 Program 551.

27. Abduljalil AM, Schmalbrock P, Novak V, Chakeres DW (2003). Enhanced gray and white matter contrast of phase susceptibility-weighted images in ultra-high field magnetic resonance imaging. J Magn Reson Imaging 18:284–290.

28. Novak V, Christoforidis G (2006). Clinical promise: clinical imaging at ultra high field. In *Ultrahigh Field Magnetic Resonance Imaging (Biological Magnetic Resonance)*. Robitaille, PM, Berliner L (eds). Springer, Berlin.

29. Dunn JF, O'Hara JA, Zaim-Wadhiri Y et al (2002). Changes in oxygenation of intracranial tumors with carbogen: a bold MRI and EPR oximetry study. J Magn Resonan Imaging 16:511–521.

30. Veninga T, Langendijk HA, Slotman BJ et al (2001). Reirradiation of primary brain tumors: survival, clinical response and prognosic factors. Radiother Oncol 59:127–137.

31. Hockel M, Schlenger K, Aral B (1996). Association between tumor hypoxia and malignant progression in advanced cancer of the uterine cervix. Cancer Res 56:4509–4515.

32. Overgaard J, Horsman MR (1996). Modification of hypoxia-induced radioresistance in tumors by the use of oxygen and sensitizers. Semin Radiat Oncol 6:10–21.

33. Okunieff P, Hoeckel MK, Dunphy EP et al (1993). Oxygen tension distributions are sufficient to explain the local response of human breast tumors treated with radiation alone. Int J Radiat Oncol Biol Phys 26:631–636.

34. Punwani S, Ordidge RJ, Cooper CE, Amess P, Clemence M (1998). MRI measurements of cerebral deoxyhemoglobin concentraton – correlation with near infrared spectroscopy (NIRS). NMR Biomed 11:281–289.

35. Neeman M, Dafni H, Bukhari O, Braun RD, Dewirst MW (2002). In vivo BOLD contrast MRI mapping of subcutaneous vascular function and maturation: validation by intravital microscopy. Magn Reson Med 45:887–898.

36. Robinson SP, Rijken PFJW, Howe FA et al (2003). Tumor vascular architecture and function evaluated by non-invasive susceptibility MRI methods and immunohistochemistry. J Magn Res Imaging 17:445–454.

37. Yetkin FZ, Mendelsohn D (2002). Hypoxia imaging in brain tumors. Neuroimaging Clin N Am 12:537–552.

38. Fleige G, Nolte C, Synowitz, F et al (2001). Magnetic labeling of activated microglia in experimental gliomas. Neoplasia 3:489–499.

39. Moore A, Marecos E, Bogdanove A Jr, Weissleder R (2000). Tumoral distribution of long-circulating dextran coated iron oxide nanoparticles in a rodent model. Radiology 214: 568–574.

40. Roggendorf W, Strupp S, Paulus W (1996). Distribution and characterization of microglia/macrophages in human brain tumors. Acta Neuropathol (Berlin) 92:288–293.

41. Morimura T, Neuchrist C, Kitz K et al (1990). Monocyte subpopulations in human gliomas: expression of Fc and complement receptors and correlation with tumor proliferation. Acta Neuropathol (Berlin) 80:287–294.

42. Wood GW, Morantz RA (1979). Immunohistologic evaluation of the lymphoreticular infiltrate of human central nervous sytem tumors. J Natl Cancer Inst 62:485–491.

43. Huettner C, Czub S, Kerkau S, Roggendorf W, Tonn JC (1997). Interleukin 10 is expressed in human gliomas in vivo and increases glioma cell proliferation and motility in vitro. Anticancer Res 17:3217–3224.

44. Taniguchi Y, Ono K, Yoshida S, Tanaka R (2000). Antigen-presenting capability of glial cells under glioma-harboring conditions and the effect of glioma derived factors on antigen presentation. J Neuroimmunol 111:177–185.

45. Wagner S, Czub S, Greif M et al (1999). Microglial/macrophage expression of interleukin 10 in human glioblastomas. Int J Cancer 82:12–16.

46. Badie B, Schartner J, Klaver J, Vorpahl J (1999). In vitro modulation of microglia motility by glioma cells is mediaed by hepatocyte growth factor/scatter factor. Neurosurgery 44:1077–1082.

47. Juchem C, Muller-Bierl B, Schick F, Logothetis NK, Pfeuffer J (2006). Combined passive and active shimming for in vivo MR spectroscopy at high magnetic fields. J Magn Reson 183: 278–89. Epub 2006 Sep 29.

48. Mangia S, Tkac I, Gruetter R et al (2006). Sensitivity of single-voxel H-1-MRS in investigating the metabolism of the activated human visual cortex at 7T. Magn Reson Imaging 24:343–348.

49. Mangia S, Tkac I, Gruetter R et al (2006). Sustained neuronal activation raises oxidative metabolism to a new steady-state level: evidence from (1)H NMR spectroscopy in the human visual cortex. J Cereb Blood Flow Metab 27:1055–1063.

50. Lei H, Zhu XH, Zhang, XL, Ugurbil K, Chen W (2003). In vivo ^{31}P magnetic resonance spectroscopy of human brain at 7T: an initial experience. Magn Reson Med 49:199–205.

51. Maudsley AA, Hilal SK, Perman WH, Simon HE (1983). Spatially resolved high-resolution spectroscopy by 4-dimensional NMR. J Magn Reson 51:147–152.

52. Sammet S, Koch RM, Murdoch JB, Schmalbrock P, Duraj J, Knopp MV (2006). Chemical shift imaging of the human brain at 7T. Radiological Society of North America (RSNA) 92nd Annual Meeting, Chicago, USA.

53. Koch RM, Sammet S, Schmalbrock P, Duraj J, Thompson M, Knopp MV (2006). Diffusion tensor imaging and fiber tracking of the optic nerve at 7T. Society of Molecular Imaging (SMI), 5th Annual Meeting, Hawaii USA.

CHAPTER 33

Neuroimaging of Brain Tumors in Animal Models of CNS Cancer

Yanping Sun and Karl F. Schmidt

INTRODUCTION

For several decades, animal models of human cancers of the central nervous system (CNS) have played essential roles as tools for the development of novel anticancer therapies. Orthotopically implanting xenografts of human tumors into immunodeficient rodents has been a valuable preclinical research methodology for more than 30 years [1–5], and recent advances in genetically engineered mouse models (GEMs) that spontaneously develop CNS tumors hold promise as a new class of models that may more closely mimic the human pathology [6–9]. Increasingly, neuroimaging is being used in combination with these animal models to gather preclinical data that are unavailable with endpoint histological analysis alone, and that are readily translated to the observations in clinical settings, where similar diagnostic imaging protocols are used.

Due to inter-animal variability, histological evaluations of the effects of therapeutic interventions on GEMs or xenograft models of brain cancer typically require large numbers of animals in order to attain statistical significance for survival curves and other measurements of disease parameters, such as tumor mass or proliferation. Non-invasive imaging methods can be used to monitor disease progression in vivo within the same animal and, in contrast to the clinic, where patients typically undergo neuroimaging evaluation after presenting with symptoms that result from advanced disease, longitudinal experiments in GEMs or xenograft animals can provide detailed information regarding disease progression, as well as the effects of treatment. Imaging modalities such as MRI, microPET, SPECT and optical imaging give the research investigator a powerful set of tools with which to investigate the biological mechanisms of disease and the effects of novel treatments.

MAGNETIC RESONANCE IMAGING OF SMALL ANIMAL MODELS OF CNS CANCER

Magnetic resonance imaging of small animals is a versatile tool for investigating tumor extent and physiology, allowing the investigator to probe a wide variety of anatomical and physiological parameters non-invasively and in a single experimental setting. Dedicated MRI systems for animal research typically operate at much higher magnetic field strengths, ranging from 4.7 to 11.7 Tesla (T), as opposed to the 1.5 to 3.0T range of most clinical scanners [10]. The higher field strength of these systems allows investigators to obtain images at higher spatial resolution than is possible using most clinical systems and enables a variety of measurements that can provide insights into tumor physiology and vasculature [11]. Dedicated animal scanners have a smaller bore (inner diameter of the superconducting magnet) making higher magnetic field strengths easier to achieve; nonetheless, the large bore, lower field strength systems used clinically can also be used for the investigation of small animals [10].

Distinguishing Malignant Tissue from Normal Brain Parenchyma

Strategies that are used in clinical MRI for investigating cancer are also applicable to imaging experiments using animal models [12]. Spin-spin or transverse relaxation (T2) and spin-lattice or longitudinal relaxation (T1) rates are intrinsic tissue parameters whose differences produce the exceptional soft tissue contrast that is visible in magnetic resonance images [11,13]. Factors affecting T1 and T2 include water mobility, membrane integrity, apoptosis and other physiological features at the cellular and subcellular level. Prolonged T1 and T2 values are common features of malignant brain tumors and a change in T1

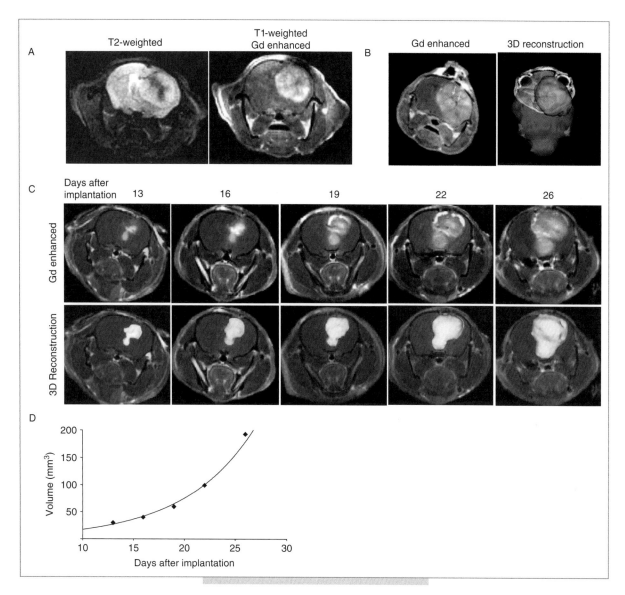

FIG. 33.1. Identification of malignant tumor tissue in animal models of glioblastoma multiforme by MRI. **(A)** MRI images that distinguish tumor tissue from surrounding brain parenchyma by using T2 contrast (T2-weighted) and T1 contrast following the administration of an exogenous contrast agent, gadolinium-DPTA (Gd enhanced). **(B)** 3D reconstruction of tumor, produced by segmentation of Gd-enhanced images. **(C)** Time course, Gd-enhanced images and corresponding 3D reconstructions of tumors from a single animal in a longitudinal investigation. **(D)** Quantitative measurements of tumor volume as a function of time from the subject in C. A, C, D: Courtesy of Dr Y Sun, Small Animal Imaging, Brigham and Women's Hospital, Harvard Medical School. B: Reprinted from Ref [20a] with kind permission of the American Association for Cancer Research.

and T2 can be indicative of a response to treatment [14]. T1 and T2 relaxation times of human U87 brain tumors have been reported in a mouse model of glioblastoma multiforme (GBM) [15] and more detailed measurements of T2-related relaxation have revealed biexponential signal decay characteristics that may reflect intra- and extracellular water volumes [16,17]. Figure 33.1A illustrates the contrast produced by a T2-weighted image to distinguish malignant tumor tissue from the surrounding brain parenchyma in a xenograft model of GBM in a mouse bearing a large tumor. Viable tumor tissue near the border of the proliferating tumor appears hyperintense, while necrotic tissue in the tumor core appears darker than tissue in the surrounding tumor and brain. While T1 and T2 contrast are valuable parameters to differentiate tumor tissue from normal brain neuropil, they often provide suboptimal contrast for segmentation and volume reconstruction, due to the heterogeneity of contrast throughout the tumor and due to the frequent presence of poorly defined transitional zones between core tumor tissue and the normal brain.

To provide enhanced differentiation of tumor tissue from normal brain tissue, contrast agents such as gadolinium-DTPA (Gd-DTPA) [18,19], that perturb the local magnetic field to create contrast detectable by MRI, can be introduced into the blood of the animal. Ordinarily, these contrast agents are restricted to the vascular compartment in the brain, but in the vicinity of tumor tissue, the compromised blood–brain barrier (BBB) causes leakage of the contrast agent into the surrounding tissue where it remains for a period of time after the agent has been cleared from the blood [20]. This residual contrast agent causes local changes in T1 and T2 relaxation that can be accentuated by MR imaging sequences. Figure 33.1A illustrates the comparison of a Gd-DPTA contrast agent enhanced image of an orthotopic GBM xenograft tumor to a T2-weighted image acquired in the absence of the contrast agent. The clearly delineated regions of compromised BBB integrity (presumed to reflect the presence of tumor tissue) provide the investigator with a clear picture of tumor extent that can be used for reliable segmentation and volume assessment (see Figure 33.1B) [20a].

Following tumor growth within the same animal over time using quantitative volume measurements (see Figures 33.1B,C and D) provides valuable information describing disease progression that can be used to assess the impact of treatments. High resolution multislice MRI that differentiates tumor tissue from the surrounding neuropil of the brain enables the investigator to construct three-dimensional models of the tumor (see Figure 33.1B). These models enable a more accurate calculation of tumor volume than estimation using geometric primitives (i.e. spheres or ellipsoids), as is often done following histological measurements of maximum tumor length, depth and height [21]. This improved accuracy can be especially relevant during advanced stages of disease, when tumor infiltration into surrounding areas of the brain is often asymmetrical.

In addition to producing T1- and T2-weighted images with tissue contrast that is a complex function of both of these parameters, it is possible to quantify explicitly T1 and T2, as well as other measures such as proton density (ρ_1) at each pixel in the image. When combined with anatomical images of the same slices, the T1 and T2 of various anatomical aspects of the tumor are easily evaluated. Figure 33.2A illustrates the changes in calculated T1 and calculated T2 between normal brain tissue and GBM tissue, providing quantitative information that can be more easily compared with results from other experiments.

Measuring Vascular Parameters by MRI

Cerebral blood volume (CBV), cerebral blood flow (CBF) and macrovascular architecture can also be assessed by MRI to provide information about tumor physiology and the effects of treatment. Tumor-related angiogenesis often causes a disorganized proliferation of vessels that deleteriously affects blood flow within the tumor with respect to healthy tissue [22] and alters the volume of blood in the vascular compartment within the tumor in a manner that can be specific to tumor type [23]. The perfusion and altered CBV of the local tissue can be detected by MRI with the use of blood pool contrast agents in dynamic contrast enhanced (DCE) imaging protocols [24], or through the use of water in arterial blood as an endogenous tracer in a variety of techniques referred to as arterial spin labeling (ASL) [25]. Additionally, macrovascular changes in the cerebral vasculature can result from the angiogenic processes of advanced tumors and can be directly visualized using magnetic resonance angiography (MRA).

Blood pool contrast agents such as Gd-DPTA alter the magnetic susceptibility of the blood and induce magnetic field gradients within and around blood vessels, causing a detectable drop in the MR signal that can be large (30–50%) in T2*-weighted images and that is dependent on the volume of blood in the voxel from which the signal is detected [26]. In DCE experiments, rapid image acquisitions record this transient drop in signal as a bolus of injected contrast agent arrives at the tissue of interest. The dynamics and extent of signal change in response to the intravenously administered contrast agent permit a calculation of local CBV or CBF. In animal studies, it has been reported that MRI-derived cerebral blood volume (CBV) in 9L rat brain tumors correlated with histological measures of microvessel density [27], observations that complement clinical work in glioma patients where rCBV maps permit the differentiation of glioma vascularization types and show a good correlation with the histological grading of the tumors [28].

The ASL technique offers particular advantages over contrast enhanced imaging by eliminating the need for exogenous contrast agents [29–32]. In ASL imaging methods, spatially selective radiofrequency pulses are used to perturb the nuclear spins of the water in arterial blood that is in, or flowing to, the brain. After a period of time has elapsed to allow these perturbed spins to flow into or wash out from the imaged tissue, exchange of perturbed spins with unperturbed spins produces a measurable change in signal intensity reflective of perfusion. When the appropriate pulse sequences are used, this signal change can be converted into a quantitative value for blood flow, as shown in Figure 33.2A (CBF) [33]. Figure 33.2B illustrates the labeled and unlabeled images acquired in a typical ASL experiment and the subtraction process yielding a perfusion weighted image (PWI) that reveals a deficit of perfusion (hypointense area) within the tumor.

Blood flowing in macrovessels within the brain can be visualized by MRA using a variety of pulse sequences that are sensitive to signal from blood that flows into the field of view during the course of acquisition (i.e. time-of-flight MRA protocols) or by using exogenous contrast agents. This measurement can be valuable to assess vascular changes that result from pathogenic angiogenesis. Figure 33.3 illustrates pathological changes in the vasculature in the vicinity of a spontaneously occurring choroid plexus tumor in genetically engineered mice [34] and illustrates the high level of detail that can be obtained with MRA to investigate the architecture of cerebral vessels in small animals.

Probing Cellular Structure and Organization by Diffusion MRI

As water molecules diffuse through space and through the magnetic field gradients that are applied during an MRI scan, the protons within acquire shifts in phase that erode the detected magnetic resonance signal. Imaging protocols specifically designed to exploit this phenomenon are able to detect the ability of water in the extracellular spaces of the brain to diffuse microscopic distances in one or several directions [35–37]. Diffusion-weighted images (DWI) are sensitive to this molecular movement of water and can be used to probe the structure of biologic tissues at a

microscopic level. Differences in cellular organization and differences in the integrity of cellular membranes within tumors affect the movement of water molecules and can distinguish tumor tissue from normal brain in DWI images [38].

The diffusion characteristics of extracellular water can be quantified by MRI in experiments that directly assess the apparent diffusion coefficient (ADC) by acquiring images with varying amounts of diffusion weighting, referred to as the b factor. At high diffusion weighting (high b value), tissue with lower ADC appear hyperintense relative to regions of higher ADC. Figure 33.2C illustrates a T2-weighted image without diffusion weighting ($b=0$); the tumor appears hyperintense relative to normal brain tissue due to the longer T2 value of this tissue. The adjacent panel shows a more heavily diffusion weighted image

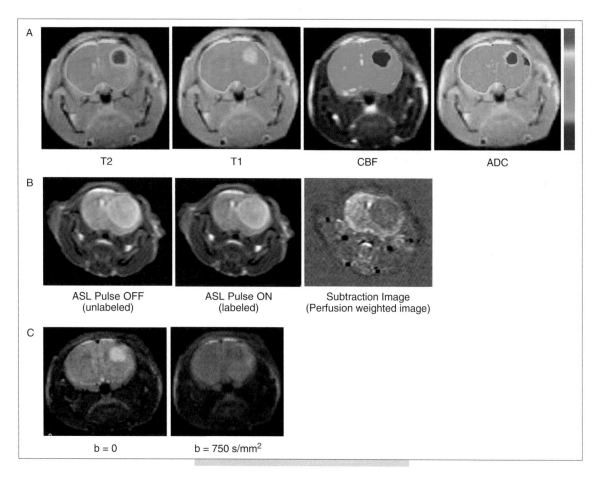

FIG. 33.2. Quantitative measurements of tumor physiology and perfusion by MRI. **(A)** False color images illustrating quantitative values of T2, T1, CBF and apparent diffusion coefficient (ADC) obtained in vivo from a mouse bearing a GBM tumor. The color bar to the right indicates the color legend for values between the maximum (red) and minimum (blue) values of the reported parameters with ranges: T2: 0 to 72 ms, T1: 500 to 3000 ms, ADC: 0 to 1.0 mm²/s, CBF: 0 to 400 ml/100g/min. **(B)** Images illustrating the arterial spin labeling (ASL) acquisitions that reveal perfusion-related contrast, apparent as a hypointense region (reflecting reduced perfusion) within the tissue of the tumor. **(C)** Signal attenuation in the tumor region was observed on high diffusion weighted image (b=750 s/mm²) due to the higher ADC value of tumor tissue relative to normal brain tissue. A: T2, T1, ADC and C: Reprinted from Ref [15] with kind permission of John Wiley and Sons, Inc. A: CBF: Reprinted from Ref [33] with kind permission of John Wiley and Sons, Inc. B: Courtesy of Dr Y Sun, Small Animal Imaging, Brigham and Women's Hospital, Harvard Medical School.

($b = 750\,\mathrm{s/mm}^2$), where signal within the tumor is attenuated due to the higher ADC value in this region. Quantitative maps of ADC can be calculated on a pixel-by-pixel basis as shown in Figure 33.2A (ADC). Tumor tissue (red) is readily differentiated from normal brain tissue by the significantly higher ADC value than that of normal brain (green). Analyses of changes in ADC from the tumor core, through the tumor edge, to the surrounding tissue have been reported previously [15].

Quantitative diffusion measurements have been shown to be sensitive to tissue cellular size, extracellular volume and membrane permeability [39]. DWI has been used in the clinical management of primary brain tumors [40,41] and has been used to distinguish tumor types and viability [42] in animal models. Changes in ADC have been used as

a biomarker for changes in the cellular structure of tumor tissue during tumor progression and to distinguish tissue compartments in early, intermediate and advanced stages of disease [38]. Consistently high diffusion in necrotic tissue, relative to solid tumor, provides a rationale for the use of diffusion imaging to monitor cellular changes following anticancer therapies [14] making ADC quantification a potentially valuable means of assessing treatment response in animal experiments.

Assessment of Tissue Protein/Peptide Fragments Using Amide Proton Transfer (APT)

Recently, a novel MRI method has been developed in which contrast is generated based on the tissue content of mobile protein and peptide fragments. These fragments contain amide protons in each of the peptide bonds that are in direct exchange with water protons. Recently, it was shown that this interaction can be used to detect the presence of these fragments using the approach of chemical exchange saturation transfer (CEST) MRI [43,44]. When exchangeable protons are selectively saturated by radiofrequency irradiation, the saturation is transferred to the water protons through proton exchange. The solvent (water) pool is considerably larger than the solute pool and proton exchange quickly returns non-saturated protons to the peptide and protein fragments. When an amide proton saturating pulse is applied for a period of several seconds, an enhanced saturation of water occurs at a concentration that is much higher than that of the solute and this enhanced saturation can be detected by MRI. This CEST principle was first demonstrated on small solutes [45] and later on proteins in solution [46,47]. Using this technique, a study of 9L gliomas in rat brain revealed that tumor protein contrast can be valuable for distinguishing tumor tissue from normal brain [48]. Figure 33.4 illustrates that APT weighted imaging can be used to delineate malignant tumor tissue in certain cases where T2-weighted contrast is ambiguous. In the 9L tumor illustrated in the figure, both edema and malignant tumor tissue appear as

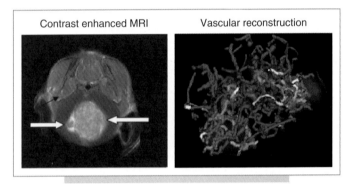

FIG. 33.3. Magnetic resonance angiography (MRA) investigation of a spontaneously occurring choroid plexus tumor in a genetically engineered mouse. (Left) Gadolinium-enhanced T_1 image showed the location of the large tumor. (Right) Three-dimensional visualization of the magnified segmented tumor (*gray*) drawn at partial opacity and with vessels color-coded relative to the tumor surface. Blue, outside; red, inside; gold, traverse; cyan, entering or exiting; white, traversing vessels with less than five vessel points within the tumor. There is marked neoangiogenesis, producing an increase in vessel number both within and surrounding the tumor. Reprinted from Ref [34] with kind permission of the American Association for Cancer Research.

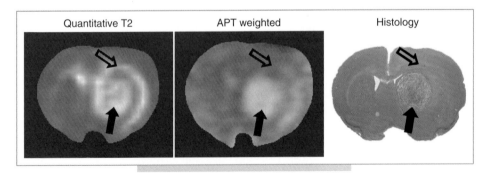

FIG. 33.4. Quantitative T2- and amide proton transfer- (APT) weighted images of a 9L brain tumor with corresponding histology. Comparison of T2 map, APT-weighted image and corresponding histology for a rat 9L brain tumor. The hyperintense peritumoral tissue (open arrow) in the T2 map is normal in the APT-weighted image. Courtesy of Drs J Zhou and P van Zijl, Johns Hopkins University Medical School.

hyperintense areas in the T2-weighted image, but only the tumor tissue appears hyperintense on the APT weighted image, as confirmed by histological analysis. Recently, it was demonstrated that this contrast can also be detected in brain tumor patients [49]. The relatively uniform intensity of the tumor in the APT-weighted images may further benefit investigators by reducing the confusion of tumors that present with highly heterogeneous contrast using traditional imaging modalities, such as T2 contrast.

Probing Tumor Metabolism by Magnetic Resonance Spectroscopy (MRS)

Nuclear magnetic resonance (NMR) spectroscopy has been used by physicists and chemists to study molecular structure and mobility since the 1950s. Spatially selective NMR spectroscopy, or MRS, can complement other MRI methods to obtain detailed biochemical information from a specific anatomical location within the living animal that can be used as a prognostic or diagnostic marker in studies of cancer. Several NMR-active isotopes are available for MRS study: ^{31}P and ^{1}H are the most common nuclei used in vivo; ^{1}H is an effective nucleus for investigating metabolite concentrations, while ^{31}P can be used in bioenergetics and in the investigation of local lipid content. ^{23}Na can be applied in investigations of cellular energetics and ^{13}C and ^{19}F have been used to investigate several other aspects of metabolism. These are just a few applications of this active field of research and the reader is directed to several reviews that discuss the use of MRS in detail [50–53].

Among investigations of cancer, numerous studies have demonstrated that MRS can characterize the chemical fingerprint for different tumor types, as well as discriminate tumor from non-neoplastic lesions and from the effects of radiation [54–57]. In general, decreased N-acetylaspartate (NAA) and creatine (Cr) concentrations and increased choline (Cho) concentrations correlate with tumor grade. Increases in lipid and lactate concentrations are also observed in some gliomas. Reduction of NAA is likely due to neuronal death or damage, while the decrease in Cr may arise from changes in cell energetics. Increased Cho (and in some cases lipid) is believed to reflect changes in increased membrane synthesis. Lactate is a by-product of anaerobic metabolism and is believed to accumulate when tumor growth outpaces neovascularization. More recently, it has been demonstrated that MRS is useful in predicting response to therapy and in assessing response to therapy [58–60]. It has also been shown that changes in the Cho resonance can predict response to therapy weeks to months before changes are observed by anatomical imaging. ^{1}H MRS has been used to study the correlation between the ^{1}H-MRS lipid signal, necrosis and lipid droplets during C6 rat glioma development [61]. ^{31}P MRS spectra was used to evaluate the effects of the Ionidamine in the 9L tumor model [62] and ^{13}C and ^{31}P MRS was used to investigate the regulation of phosphocholine and phosphoethanolamine level in rat glioma [63].

Figure 33.5 illustrates ^{1}H MRS spectra obtained from an intracranial 9L tumor in rat which demonstrated a decrease in choline/creatine ratio and increase in the lipid/lactate resonance peak (1.3 ppm) intensity for treated 9L tumor compared with untreated tumor spectra [64]. The sensitivity of serial MRS measurement has revealed that MRS can deliver important data for the evaluation of tumor responses to treatment. The integration of spectroscopic studies and imaging techniques into preclinical animal model investigations provides a valuable avenue in drug discovery.

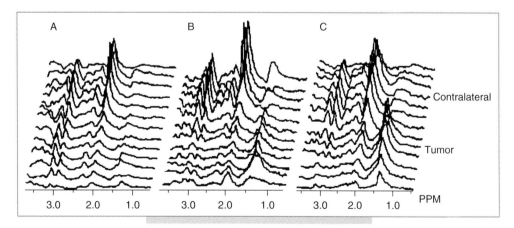

FIG. 33.5. Spatially localized ^{1}H magnetic resonance spectra obtained from a rat brain harboring an **(A)** untreated intracranial 9L and a treated glioma on the **(B)** sixth and **(C)** ninth day following initiation of daily treatments with FMdC (15 mg/kg, i.p.). Each series of spectra were obtained from 25 µl voxels along a column through the rat brain. In the contralateral side, resonances were observed corresponding to NAA, creatine and choline at 2.0, 3.0 and 3.2 ppm, respectively. Untreated 9L tumors showed absence of NAA, increased choline and decreased creatine as compared with the contralateral brain spectra. FMdC treated tumors showed a progressive increase in the lipid+lactate resonance intensity (1.3 ppm) and decreased choline intensity on days 6 and 9 of treatment. Reprinted from Ref [64] with kind permission of John Wiley and Sons, Inc.

IN VIVO BIOLUMINESCENT IMAGING

Bioluminescent imaging (BLI) is an experimental technique that uses light emitting proteins (luciferases) found naturally in insects, bacteria and plants as optical reporters ideal for cell labeling in vivo. Luciferases are proteins with enzymatic activity that, in the presence of ATP, oxygen and the appropriate substrate (typically luciferin), catalyze the oxidation of the substrate in a reaction that results in the emission of a photon. The photons emitted have broad spectral distributions with peaks between 538 and 582 nm for various firefly luciferases, but considerable emission occurs at longer wavelengths (>600 nm) and this radiation propagates more easily through tissue for detection by sensitive external cameras. Photons readily traverse several millimeters of tissue and thin bones, such as the mouse cranium, and can be detected non-invasively by cooled, charge-coupled device (CCD) cameras housed inside compartments sealed against external light [65–67].

In orthographic xenograft models using transplanted cells that have been genetically engineered to express luciferase, tumor proliferation can be monitored using straightforward techniques and a modest detection system [68–70]. Tumor cells harboring a gene that encodes luciferase under the control of a constitutively activated promoter accumulate luciferase in the cytoplasm. Because luciferase coding DNA is propagated as tumor cells divide, this optical label endures in future generations of cells and can be used to monitor tumor growth within the same animal over time. To assay the quantity of cells containing luciferase, luciferin is administered systemically to anesthetized animals before a period of photo-acquisition. Luciferin is a small molecule that readily crosses the BBB and cellular membranes and can be easily administered systemically at doses on the order of 150 mg/kg [68,71,72]. Luminescence reaches its peak level within 10–20 minutes and the image acquisition times are relatively short (from several seconds to a few minutes) [73,74]. Several studies have documented a high degree of correlation between bioluminescence and MRI volumetric measurements [68,70,75]. Rapid image acquisition and ease of the imaging procedure allow multiple mice to be imaged simultaneously, making this a very high throughput technique; it is not unreasonable for one system to process more than 200 mice per day.

The dependence of the light emitting reaction on the presence of luciferase from sustained expression of the luciferase gene, ATP, oxygen and the presence of unmetabolized substrate makes BLI subject to cellular viability. This sensitivity has been examined in conjunction with diffusion MRI to suggest that BLI signal loss may be a surrogate marker of cell death, for example, in response to a chemotherapeutic being investigated [76]. Other factors, such as translocation to deeper tissue with increased absorption, or loss of luciferase expression in tumor cell progeny may also affect the detected signal in longitudinal studies. It is also probable that volume measurement, as a function of photon emission, is less sensitive to tumor morphology than volume measurements using 3D reconstruction techniques, which may be sensitive to partial volume effects in MRI. However, factors such as substrate concentration, reaction rates, rate of substrate clearance and internal absorption and scattering of photons complicate precise quantitative analyses. The scattering of emitted light presents an additional challenge for the spatial resolution of luciferase harboring cells, but work to develop reliable bioluminescent tomography is promising [77,78].

Figure 33.6 illustrates the application of BLI to track intracranial tumor growth in a xenograft mouse model of GBM. Bioluminescence increases exponentially over time (Figure 33.6B), consistent with the expected change in tumor volume [70,74,75]. Quantitative volume measurements determined by MRI are highly correlated with BLI measurements (Figure 33.6C) and have been shown to exhibit a log-linear correlation for up to 4 weeks following implantation [70,75].

MOLECULAR IMAGING OF SMALL ANIMALS

MicroPET

Positron emission tomography (PET) is a medical imaging technique that is used clinically to assess local tissue metabolism and the local concentration of molecules of interest, such as cell surface receptors, often at picomolar concentrations [79]. MicroPET technology was developed as a scaled-down version of clinical PET systems to bring these imaging capabilities to preclinical research using animals [80]. Technologically, microPET systems have delivered a solution to the problems of increased demand for sensitivity and spatial resolution as a result of the small size of the animals being investigated, with resolution that surpasses 2 mm in each direction (<8 mm^3). In both clinical and animal imaging, radioactive isotopes of oxygen (^{15}O), nitrogen (^{13}N), carbon (^{11}C), or fluorine (^{18}F), are added to molecules that either participate in a metabolic pathway of interest or bind selectively with a molecular target of interest, and are administered to the patient or animal and allowed to accumulate in the target tissue prior to scanning.

Image formation is made possible through the acquisition of many detected events that occur as these radioactive isotopes decay. The isotopes used in imaging experiments decay with varying half-lives and, in all cases, the decay results in the production of a positron that quickly annihilates with a nearby electron resulting in the emission of two gamma rays. The gamma rays travel away from the locus of annihilation on opposed and nearly collinear trajectories. The near-collinearity of these trajectories permits the calculation of a line of response (LOR) when arrayed sensors in the scanner register simultaneous gamma ray detections. By collecting multiple LORs over a period of minutes, a 3D reconstruction of the emission volume can be produced using traditional tomographic methods [80,81].

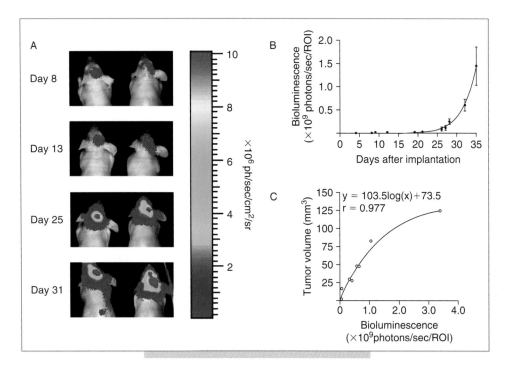

FIG. 33.6. Bioluminescence imaging of GBM xenografts reveal increases in luminescence that are correlated with tumor growth and quantitative volume measurements determined by MRI. Luciferase-expressing U87 cells were stereotactically implanted into the right hemisphere of NCr nude mice. **(A)** A dim light photograph of two representative animals is overlaid with a false-color representation of photonic flux (photons/s/cm²/sr) at the indicated time after intracranial tumor cell implantation. **(B)** The bioluminescence through standardized regions of interest (photons/s/ROI) was found to increase exponentially and **(C)** was highly correlated with tumor volume measurements determined by MR imaging. A: Courtesy of Dr. A. Kung, Dana-Farber Cancer Institute. B, C: Reprinted from Ref [70] with permission of the National Academy of Sciences.

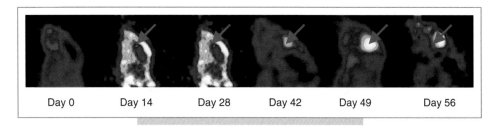

FIG. 33.7 MicroPET images of GBM tumor growth within a single rat using ¹⁸F integrated into an RGD peptide. Serial microPET scans of the same mouse before (Day 0) and after inoculation of U87 human glioblastoma cells to the forebrain. Sagittal images were obtained as 10-min static scans at 30-min post-administration of 200 mCi of ¹⁸F FB-RGD and reconstructed using filtered back-projection (FBP) algorithm. Reprinted from Ref [86] with kind permission of Springer-Verlag GmbH.

In oncological investigations, tissue metabolism is often measured by PET with the administration of radiolabeled fluorodeoxyglucose (FDG) which accumulates in cells as a function of cellular demand for glucose. FDG-PET is commonly employed as a diagnostic tool in clinical oncology, owing to the higher-than-normal metabolism of most malignant neoplasms including many tumors of the brain [82–84]. However, the ability of PET and microPET to reveal the presence and distribution of specific cell surface receptors has shifted much of the focus of current animal research toward the investigation of molecules that may play pathogenic roles in the cancers of the brain. Integrins represent one

example of molecular targets being investigated by microPET, on the basis that tumor expression of integrins has been shown to play an essential role in tumor angiogenesis and antagonists of integrin receptors are currently being evaluated as potential therapies for tumorogenic cancers [85].

Figure 33.7 illustrates the results of a longitudinal microPET study following tumor growth within the same animal in an orthotopic U87 xenograft model of GBM. ¹⁸F was incorporated into an RGD peptide, a potent alphav-integrin antagonist, and administered to localize brain regions containing high concentrations of alphav-integrin receptors, areas co-localized with tumor tissue and

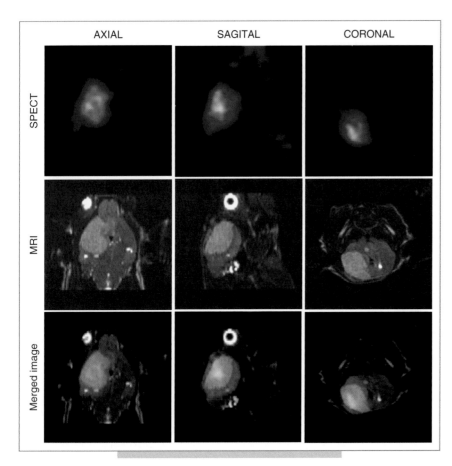

FIG. 33.8. MicroSPECT images of GBM tumor in mice using a vitronectin binding radioligand (99mTc-NC100692). Multimodality SPECT and MRI images reveal a large tumor in axial, coronal and sagittal planes in a mouse bearing a U87 cell xenograft. The coregistered data confirm that both modalities reveal tumor tissue in the same loci. Courtesy of JW Barnes and Dr M W Kieran, Department of Pediatric Oncology, Dana-Farber Cancer Institute, Harvard Medical School.

vascular endothelium resulting from pathological angiogenesis. The technique used by Chen et al is applicable to microPET investigations of other cell surface targets with the availability of appropriate radiolabeled ligands [86].

MicroSPECT

Single photon emission computed tomography (SPECT) operates on similar principles as PET with the important distinction that the radionuclides used in SPECT experiments exhibit longer half-lives and emit gamma photons of varying energies that are generally lower than those emitted by isotopes used in PET experiments. For small-animal imaging, the energy of the photons is in the range of 25–511 keV [87]. The formation of SPECT images requires the use of a collimator that determines the direction of the photon as it travels between the source and the detector. In clinical SPECT systems, the collimator has parallel holes that are oriented to image a large field of view appropriate for a human subject. This resolution is not good enough for imaging small animals, however, so most SPECT experiments investigating animals typically require the higher spatial resolution afforded by pinhole collimators. Several dedicated small animal pinhole SPECT systems have been developed in recent years [88–92] and it has been reported that volumetric resolution on the order of 0.1 μml can be achieved [91], making this modality well suited for studies of

molecular binding or uptake with submillimeter resolution in the living animal. Even though the detection efficiency of microSPECT systems is much lower than that of microPET, dynamic studies are still often possible because of the longer half-life of SPECT tracers.

High resolution SPECT provides a valuable tool in development and evaluation of radiopharmaceutical agents and in molecular research. In combination with small animal MRI, multipinhole microSPECT with submillimeter resolution has been used to follow the development of disease in mouse models over time [93,94] and the use of multiple energy windows to perform dual-isotope imaging has also been reported [93]. Figure 33.8 illustrates concomitant SPECT and MRI imaging of a U87 GBM xenograft in a mouse model and reflects the ability of SPECT to distinguish malignant tumor tissue by means of a vitronectin binding radioligand (99mTc-NC100692, 1 mCi) [95]. The development of highly sensitive SPECT systems with exceptional spatial resolution has created a broad range of new applications of molecular imaging in animal research.

SUMMARY

Animal models will continue to play an indispensable role in the investigation of CNS cancer and in the search for novel therapies. In studies using these models, non-invasive neuroimaging techniques afford the investigator important capabilities that include and exceed the capabilities of clinical diagnostic imaging. Common among all non-invasive imaging modalities is the benefit of performing measurements in vivo, typically in longitudinal evaluations within the same animal that can provide a more sensitive assay of therapeutic response. A variety of measurements of disease status and drug action can be obtained, whether using MRI and MR spectroscopy, optical imaging technologies, molecular techniques such as PET and SPECT, or other technologies that are being continually improved for application to the investigation of brain tumors in small animals such as microCT [96–99], contrast enhanced ultrasounography [100] and photoacoustic tomography [101]. Each imaging technique has specific limitations and the combination of multiple techniques has become an increasingly popular strategy to leverage the strengths of different modalities within a single experiment, as in Figure 33.8. The combination of molecular imaging techniques with high resolution anatomical or physiological imaging can be used to determine the pharmacodynamic and pharmacokinetic actions of drugs and to determine whether novel therapeutics modulate their intended targets in vivo. These capabilities, and the increasing frequency of use of small animal imaging bodes well for the translation of in vitro biology to in vivo animal models and, ultimately, to observations expected in the clinic.

REFERENCES

1. Rygaard J, Povlsen CO (1969). Heterotransplantation of a human malignant tumour to 'Nude' mice. Acta Pathol Microbiol Scand 77(4):758–760.

2. Giovanella BC et al (1972). Development of invasive tumors in the 'nude' mouse after injection of cultured human melanoma cells. J Natl Cancer Inst 48(5):1531–1533.

3. Shapiro WR et al (1979). Human brain tumor transplantation into nude mice. J Natl Cancer Inst 62(3):447–453.

4. Bosma GC, Custer RP, Bosma MJ (1983). A severe combined immunodeficiency mutation in the mouse. Nature 301(5900):527–530.

5. Alley M et al (2004). Human tumor xenograft models in NCI drug development. In *Anticancer drug development guide: preclinical screening, clinical trials, and approval.* Teicher B, Andrews P (eds). Humana Press, Inc., Totowa. 125–152.

6. Goodrich LV et al (1997). Altered neural cell fates and medulloblastoma in mouse patched mutants. Science 277(5329): 1109–1113.

7. Lu X et al (2001). Selective inactivation of p53 facilitates mouse epithelial tumor progression without chromosomal instability. Mol Cell Biol 21(17):6017–6030.

8. Hallahan AR et al (2004). The SmoA1 mouse model reveals that notch signaling is critical for the growth and survival of sonic hedgehog-induced medulloblastomas. Cancer Res 64(21):7794–800.

9. Lin W et al (2004). Interferon-gamma induced medulloblastoma in the developing cerebellum. J Neurosci 24(45): 10074–1083.

10. Pirko I et al (2005). Magnetic resonance imaging, microscopy, and spectroscopy of the central nervous system in experimental animals. NeuroRx 2(2):250–264.

11. Westbrook C (1999). *Handbook of MRI technique,* 2nd edn. Blackwell Publishing, Oxford.

12. Sorensen AG (2006). Magnetic resonance as a cancer imaging biomarker. J Clin Oncol 24(20):3274–3281.

13. Cameron IL, Ord VA, Fullerton GD (1984). Characterization of proton NMR relaxation times in normal and pathological tissues by correlation with other tissue parameters. Magn Reson Imaging 2(2):97–106.

14. Chenevert TL, McKeever PE, Ross BD (1997). Monitoring early response of experimental brain tumors to therapy using diffusion magnetic resonance imaging. Clin Cancer Res 3(9):1457–1466.

15. Sun Y et al (2004). Quantification of water diffusion and relaxation times of human U87 tumors in a mouse model. NMR Biomed 17(6):399–404.

16. Schad LR et al (1989). Multiexponential proton spin-spin relaxation in MR imaging of human brain tumors. J Comput Assist Tomogr 13(4):577–587.

17. Macri MA et al (1992). Spin-lattice relaxation in murine tumors after in vivo treatment with interferon alpha/beta or tumor necrosis factor alpha. Magn Reson Med 23(1): 12–20.

18. Villringer A et al (1988). Dynamic imaging with lanthanide chelates in normal brain: contrast due to magnetic susceptibility effects. Magn Reson Med 6(2):164–174.

19. Rosen BR, Belliveau JW, Chien D (1989). Perfusion imaging by nuclear magnetic resonance. Magn Reson Q 5(4):263–281.

20. Runge VM et al (1994). Visualization of blood–brain barrier disruption on MR images of cats with acute cerebral infarction: value of administering a high dose of contrast material. Am J Roentgenol 162(2):431–435.

20a. Schmidt NO et al (2004). Antiangiogenic therapy by local intracerebral microinfusion improves treatment efficiency and survival in an orthotopic human glioblastoma model. Clinical Cancer Research 10:1255–1262.

21. Schmidt KF et al (2004). Volume reconstruction techniques improve the correlation between histological and in vivo tumor volume measurements in mouse models of human gliomas. J Neuro-Oncol 68(3):207–215.

22. Lim M, Cheshier S, Steinberg GK (2006). New vessel formation in the central nervous system during tumor growth, vascular malformations, and Moyamoya. Curr Neurovasc Res 3(3):237–245.

23. Cha S et al (2005). Differentiation of low-grade oligodendrogliomas from low-grade astrocytomas by using quantitative blood-volume measurements derived from dynamic susceptibility contrast-enhanced MR imaging. Am J Neuroradiol 26(2):266–273.

24. Hylton N (2006). Dynamic contrast-enhanced magnetic resonance imaging as an imaging biomarker. J Clin Oncol 24(20):3293–3298.

25. Barbier EL, Lamalle L, Decorps M (2001). Methodology of brain perfusion imaging. J Magn Reson Imaging 13(4): 496–520.

26. Buxton RB (2002). *Introduction to functional magnetic resonance imaging: principles and techniques.* Cambridge University Press, New York. 523.

27. Pathak AP et al (2001). MR-derived cerebral blood volume maps: issues regarding histological validation and assessment of tumor angiogenesis. Magn Reson Med 46(4):735–747.

28. Ludemann L, Hamm B, Zimmer C (2000). Pharmacokinetic analysis of glioma compartments with dynamic Gd-DTPA-enhanced magnetic resonance imaging. Magn Reson Imaging 18(10):1201–1214.

29. Williams DS et al (1992). Magnetic resonance imaging of perfusion using spin inversion of arterial water. Proc Natl Acad Sci USA 89(1):212–216.

30. Calamante F et al (1999). Measuring cerebral blood flow using magnetic resonance imaging techniques. J Cereb Blood Flow Metab 19(7):701–735.

31. Detre JA et al (1998). Noninvasive MRI evaluation of cerebral blood flow in cerebrovascular disease. Neurology 50(3):633–641.

32. Silva AC, Kim SG, Garwood M (2000). Imaging blood flow in brain tumors using arterial spin labeling. Magn Reson Med 44(2):169–173.

33. Sun Y et al (2004). Perfusion MRI of U87 brain tumors in a mouse model. Magn Reson Med 51(5):893–899.

34. Brubaker LM et al (2005). Magnetic resonance angiography visualization of abnormal tumor vasculature in genetically engineered mice. Cancer Res 65(18):8218–8223.

35. Westbrook C, Kaut C (1998). *MRI In Practice*, 2nd edn. Blackwell Science, Inc., Oxford. 326.

36. Bammer R (2003). Basic principles of diffusion-weighted imaging. Eur J Radiol 45(3):169–184.

37. Mitchell D (1999). *MRI Principles*. W.B. Saunders Company, New York.

38. Herneth AM, Guccione S, Bednarski M (2003). Apparent diffusion coefficient: a quantitative parameter for in vivo tumor characterization. Eur J Radiol 45(3):208–213.

39. Latour LL et al (1994). Time-dependent diffusion of water in a biological model system. Proc Natl Acad Sci USA 91(4):1229–1233.

40. Krabbe K et al (1997). MR diffusion imaging of human intracranial tumours. Neuroradiology 39(7):483–489.

41. Chenevert TL et al (2000). Diffusion magnetic resonance imaging: an early surrogate marker of therapeutic efficacy in brain tumors. J Natl Cancer Inst 92(24):2029–2036.

42. Eis M, Els T, Hoehn-Berlage M (1995). High resolution quantitative relaxation and diffusion MRI of three different experimental brain tumors in rat. Magn Reson Med 34(6): 835–844.

43. Zhou J et al (2003). Using the amide proton signals of intracellular proteins and peptides to detect pH effects in MRI. Nat Med 9(8):1085–1090.

44. van Zijl PC et al (2003). Mechanism of magnetization transfer during on-resonance water saturation. A new approach to detect mobile proteins, peptides, and lipids. Magn Reson Med 49(3):440–449.

45. Ward KM, Aletras AH, Balaban RS (2000). A new class of contrast agents for MRI based on proton chemical exchange dependent saturation transfer (CEST). J Magn Reson 143(1): 79–87.

46. Goffeney N et al (2001). Sensitive NMR detection of cationic-polymer-based gene delivery systems using saturation transfer via proton exchange. J Am Chem Soc 123(35):8628–8629.

47. McMahon MT et al (2006). Quantifying exchange rates in chemical exchange saturation transfer agents using the saturation time and saturation power dependencies of the magnetization transfer effect on the magnetic resonance imaging signal (QUEST and QUESP): Ph calibration for poly-L-lysine and a starburst dendrimer. Magn Reson Med 55(4):836–847.

48. Zhou J et al (2003). Amide proton transfer (APT) contrast for imaging of brain tumors. Magn Reson Med 50(6):1120–1126.

49. Jones CK et al (2006). Amide proton transfer imaging of human brain tumors at 3T. Magn Reson Med 56(3):585–592.

50. Gillies RJ, Morse DL (2005). In vivo magnetic resonance spectroscopy in cancer. Annu Rev Biomed Eng 7:287–326.

51. Nelson SJ (2003). Multivoxel magnetic resonance spectroscopy of brain tumors. Mol Cancer Ther 2(5):497–507.

52. Bachelard H (1998). Landmarks in the application of 13C-magnetic resonance spectroscopy to studies of neuronal/glial relationships. Dev Neurosci 20(4–5):277–288.

53. Howe FA et al (1993). Proton spectroscopy in vivo. Magn Reson Q 9(1):31–59.

54. Preul MC et al (1998). Using pattern analysis of in vivo proton MRSI data to improve the diagnosis and surgical management of patients with brain tumors. NMR Biomed 11(4–5):192–200.

55. Castillo M, Kwock L (1998). Proton MR spectroscopy of common brain tumors. Neuroimaging Clin N Am 8(4):733–752.

56. Wald LL et al (1997). Serial proton magnetic resonance spectroscopy imaging of glioblastoma multiforme after brachytherapy. J Neurosurg 87(4):525–534.

57. Olsen KI et al (2005). Advanced magnetic resonance imaging techniques to evaluate CNS glioma. Expert Rev Neurother 5(6 Suppl):S3–11.

58. Lazareff JA, Gupta RK, Alger J (1999). Variation of post-treatment H-MRSI choline intensity in pediatric gliomas. J Neuro-Oncol 41(3):291–298.

59. Graves EE et al (2001). Serial proton MR spectroscopic imaging of recurrent malignant gliomas after gamma knife radiosurgery. Am J Neuroradiol 22(4):613–624.

60. Tzika AA et al (2001). Proton magnetic spectroscopic imaging of the child's brain: the response of tumors to treatment. Neuroradiology 43(2):169–177.

61. Zoula S et al (2003). Correlation between the occurrence of 1H-MRS lipid signal, necrosis and lipid droplets during C6 rat glioma development. NMR Biomed 16(4):199–212.

62. Ben-Yoseph O et al (1998). Mechanism of action of lonidamine in the 9L brain tumor model involves inhibition of lactate efflux and intracellular acidification. J Neuro-Oncol 36(2):149–157.

63. Gillies RJ, Barry JA, Ross BD (1994). In vitro and in vivo 13C and 31P NMR analyses of phosphocholine metabolism in rat glioma cells. Magn Reson Med 32(3):310–318.

64. Ross BD et al (2003). Evaluation of (E)-2-deoxy-2'-(fluoromethylene)cytidine on the 9L rat brain tumor model using MRI. NMR Biomed 16(2):67–76.

65. Viviani VR (2002). The origin, diversity, and structure function relationships of insect luciferases. Cell Mol Life Sci 59(11): 1833–1850.

66. Shah K, Weissleder R (2005). Molecular optical imaging: applications leading to the development of present day therapeutics. NeuroRx 2(2):215–225.

67. Choy G, Choyke P, Libutti SK (2003). Current advances in molecular imaging: noninvasive in vivo bioluminescent and fluorescent optical imaging in cancer research. Mol Imaging 2(4):303–312.

68. Rehemtulla A et al (2000). Rapid and quantitative assessment of cancer treatment response using in vivo bioluminescence imaging. Neoplasia 2(6):491–495.

69. Vorechovsky I et al (1997). Somatic mutations in the human homologue of Drosophila patched in primitive neuroectodermal tumours. Oncogene 15(3):361–366.

70. Rubin JB et al (2003). A small-molecule antagonist of CXCR4 inhibits intracranial growth of primary brain tumors. Proc Natl Acad Sci USA 100(23):13513–13518.

71. Contag CH et al (1997). Visualizing gene expression in living mammals using a bioluminescent reporter. Photochem Photobiol 66(4):523–531.

72. Honigman A et al (2001). Imaging transgene expression in live animals. Mol Ther 4(3):239–249.

73. Edinger M et al (1999). Noninvasive assessment of tumor cell proliferation in animal models. Neoplasia 1(4):303–310.

74. Kung AL (2005). Harnessing the power of fireflies and mice for assessing cancer mechanisms. Drug Discov Today 2:153–158.

75. Szentirmai O et al (2006). Noninvasive bioluminescence imaging of luciferase expressing intracranial U87 xenografts: correlation with magnetic resonance imaging determined tumor volume and longitudinal use in assessing tumor growth and antiangiogenic treatment effect. Neurosurgery 58(2):365–372; discussion 365–372.

76. Rehemtulla A et al (2002). Molecular imaging of gene expression and efficacy following adenoviral-mediated brain tumor gene therapy. Mol Imaging 1(1):43–55.

77. Chaudhari AJ et al (2005). Hyperspectral and multispectral bioluminescence optical tomography for small animal imaging. Phys Med Biol 50(23):5421–5441.

78. Wang G et al (2006). In vivo mouse studies with bioluminescence tomography. Optics Express 14(17):7801–7809.

79. Bergstrom M, Grahnen A, Langstrom B (2003). Positron emission tomography microdosing: a new concept with application in tracer and early clinical drug development. Eur J Clin Pharmacol 59(5–6):357–366.

80. Chatziioannou AF (2002). Molecular imaging of small animals with dedicated PET tomographs. Eur J Nucl Med Mol Imaging 29(1):98–114.

81. Turkington TG, Coleman RE (2002). Clinical oncologic positron emission tomography: an introduction. Semin Roentgenol 37(2):102–109.

82. Weber WA, Avril N, Schwaiger M (1999). Relevance of positron emission tomography (PET) in oncology. Strahlenther Onkol 175(8):356–373.

83. Di Chiro G (1987). Positron emission tomography using [18F] fluorodeoxyglucose in brain tumors. A powerful diagnostic and prognostic tool. Invest Radiol 22(5):360–371.

84. Del Sole A et al (2001). Anatomical and biochemical investigation of primary brain tumours. Eur J Nucl Med 28(12):1851–1872.

85. Tucker GC (2006). Integrins: molecular targets in cancer therapy. Curr Oncol Rep 8(2):96–103.

86. Chen X et al (2006). Longitudinal microPET imaging of brain tumor growth with F-18-labeled RGD peptide. Mol Imaging Biol 8(1):9–15.

87. King MA et al (2002). Introduction to the physics of molecular imaging with radioactive tracers in small animals. J Cell Biochem Suppl 39:221–230.

88. Jaszczak RJ et al (1994). Pinhole collimation for ultra-high-resolution, small-field-of-view SPECT. Phys Med Biol 39(3):425–437.

89. Ishizu K et al (1995). Ultra-high resolution SPECT system using four pinhole collimators for small animal studies. J Nucl Med 36(12):2282–2287.

90. Weber DA, Ivanovic M (1999). Ultra-high-resolution imaging of small animals: implications for preclinical and research studies. J Nucl Cardiol 6(3):332–344.

91. Beekman FJ et al (2005). U-SPECT-I: a novel system for sub-millimeter-resolution tomography with radiolabeled molecules in mice. J Nucl Med 46(7):1194–1200.

92. Moore S et al (2004). A triple-detector, multiple-pinhole system for SPECT imaging of rodents. J Nucl Med 45(5):97P.

93. Mahmood A et al (2006). Following disease progression in small animals with μ-SPECT. In *Technetium, Rhenium and other metals in Chemistry and Nuclear Medicine 7*. SGEditoriali, Padova. 465–470.

94. Hua J et al (2005). Noninvasive imaging of angiogenesis with a 99mTc-labeled peptide targeted at alphavbeta3 integrin after murine hindlimb ischemia. Circulation 111(24):3255–3260.

95. Barnes JW et al (2006). EMD121974 targeting of αvβ3 integrin inhibits orthotopic glioblastoma multiforme. In *American Association for Cancer Research*. Washington, DC.

96. Ritman EL (2003). Molecular imaging in small animals – roles for micro-CT. J Cell Biochem Suppl 39:116–124.

97. Psarros TG, Mickey B, Giller C (2005). Detection of experimentally induced brain tumors in rats using high resolution computed tomography. Neurol Res 27(1):57–59.

98. Seemann MD, Beck R, Ziegler S (2006). In vivo tumor imaging in mice using a state-of-the-art clinical PET/CT in comparison with a small animal PET and a small animal CT. Technol Cancer Res Treat 5(5):537–542.

99. Winkelmann CT et al (2006). Microimaging characterization of a B16-F10 melanoma metastasis mouse model. Mol Imaging 5(2):105–114.

100. Ellegala DB et al (2003). Imaging tumor angiogenesis with contrast ultrasound and microbubbles targeted to alpha(v)beta3. Circulation 108(3):336–341.

101. Ku G et al (2005). Imaging of tumor angiogenesis in rat brains in vivo by photoacoustic tomography. Appl Opt 44(5):770–775.

CHAPTER 34

Advances in Contrast Agent Development for CT/MRI

Michael V. Knopp

INTRODUCTION

Neuroimaging began utilizing contrast agents for cross-sectional imaging immediately after their availability, due to the uniqueness of their impermeability through the blood–brain barrier (BBB) and ready leakage when a breakdown of the BBB occurs. Additionally, contrast agents can enhance the delineation of the intracranial vasculature for high resolution non-invasive computed tomography (CT) and magnetic resonance (MR) angiography. While x-ray contrast agents lead to an attenuation of the transmitted x-rays, MR agents exhibit their properties by locally changing the relaxivity. All contrast agents for CT utilize iodine as their central ion while MR contrast agents can be classified into the paramagnetic (mostly gadolinium-based) and superparamagnetic (mostly iron oxide-based). For both CT and MR, neuroimaging consists of the majority of procedures performed on those modalities in our environment. In CT, no distinctive differences in the visualization of contrast enhancement is recognized between the different agents currently available for clinical use while in MR imaging, the distinctive characteristics of the contrast agents can lead to substantially different visualization and enhancement patterns that have to be recognized.

Contrast agents are always dosed at the lowest effective dose to enable the diagnostically appropriate visualization of the intracranial vasculature, disturbances in the BBB or other microcirculatory changes. From a safety perspective, the rapid elimination from the body, no or limited drug–drug interactions, and no or limited toxicity are the key desirable aspects. CT does have its advantage in terms of rapid acquisition, especially in emergency situations, while MR allows not only a morphological, but also functional and molecular pathway oriented assessment. CT and MR enable the perfusion/microcirculatory assessment during the rapid bolus passage of the contrast agent, whereas only MR allows more differentiated contrast approaches. While no reliable data are available on the global use of x-ray contrast agents for CT neuro-applications, it can be estimated that we currently have about six million dose utilizations of MR contrast agents per year for neuroimaging [1]. It is further anticipated that the utilization, especially in neuro-oncology, will further shift toward MR and, therefore, the understanding of MR contrast agents for these applications is of major importance.

CLASSIFICATION OF MR CONTRAST AGENTS FOR NEURO-ONCOLOGIC IMAGING

MR contrast agents currently fall into two broad categories; those based on gadolinium, which are predominately paramagnetic in nature, and those based upon iron oxide particles of different coating and size that are superparamagnetic in nature (Figure 34.1). Currently, the overwhelming utilization for neuro-oncologic imaging is based on gadolinium chelates that can be further subclassified into agents that show no interaction with proteins or those that have varying temporary interaction with proteins that lead to increases in relaxivity and different elimination pathways. Table 34.1 summarizes the contrast agents that are currently available or in clinical trials at varying stages relevant for neuro-oncologic MR imaging [2]. Contrast agents are locally approved by the regulatory authority, such as the Food and Drug Administration (FDA) in the USA, similar to therapeutic drugs. The clinical development of contrast agents is currently an overall 10-year plus process that starts with the preclinical assessment and toxicology, followed by a Phase I feasibility assessment. Phase II studies are designed to assess the dose imaging characteristics, with the goal to identify the lowest diagnostically sufficient dose. Subsequent Phase III studies focus on diagnostic efficacy and safety to support the desired indication and regulatory label.

Non-protein Interacting Standard Gadolinium Chelates

Currently, six gadolinium chelate contrast agents are approved in one or more countries and typically always have neuro-imaging as a labeled indication. This group of

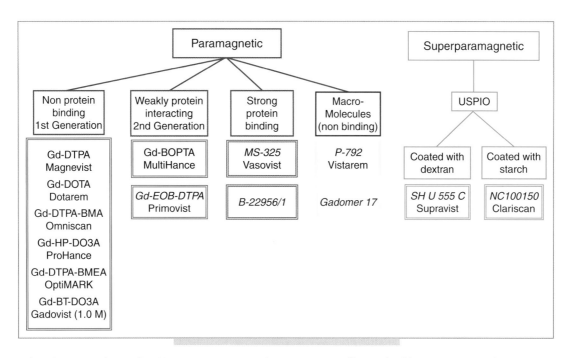

FIG. 34.1. Classification scheme for MR contrast agents that are potentially applicable to neuro-oncologic imaging. The paramagnetic gadolinium chelates can be classified according to their degree of protein interaction. The ultra small iron oxide particles are 'blood pool' agents which demonstrate long intravascular enhancement. (n.b. The products in italics are still in the developmental phase at the time of writing.)

TABLE 34-1	Physicochemical characteristics of commercially-available first pass gadolinium-based MR contrast agents						
Characteristic	Gd-DTPA Magnevist (0.5 mol/l)	Gd-DOTA Dotarem (0.5 mol/l)	Gd-HP-DO3A ProHance (0.5 mol/l)	Gd-DTPA-BMA Omniscan (0.5 mol/l)	Gd-BT-DO3A Gadovist (1.0 mol/l)	Gd-DTPA-BMEA OptiMARK (0.5 mol/l)	MultiHance (0.5 mol/l)
Molecular structure	Linear, ionic	Cyclic, ionic	Cyclic, non-ionic	Linear, non-ionic	Cyclic, non-ionic	Linear, non-ionic	Linear, ionic
Thermodynamic stability constant (log K_{eq})	22.1	25.8	23.8	16.9	21.8	16.6	22.6
Osmolality (Osm/kg)	1.96	1.35	0.63	0.65	1.6	1.11	1.97
Viscosity (mPa · s at 37°C)	2.9	2.0	1.3	1.4	4.96	2.0	5.3
T1 relaxivity (l/mmol/s), plasma	4.9	4.3	4.6	4.8	5.6	N/A	9.7

N/A = not available

'conventional' gadolinium chelate agents was introduced more than 15 years ago with nearly simultaneous approval of gadopentetate dimeglumine (Gd-DTPA, Magnevist, Bayer Schering AG) in all three key markets – Europe, USA and Japan. Five of these agents are available as 0.5 molar formulations and one, gadobutrol (Gd-BT-DO3A, Gadovist, Bayer Schering AG) is being marketed at a 1.0 molar formulation. Although differences exist between these agents in terms of their molecular structure and chemical and physical properties (Figure 34.2; see Table 34.1), all agents are non-specific and are eliminated unchanged via the renal pathway by glomerular filtration. The T1 relaxation rates of these agents are comparable and fall in the range between 4.3 and 5.6 l/mmol/s. These similarities therefore lead to equivalent imaging characteristics at the same dose and injection rate. A differentiating factor of substantial relevance has been the binding strength of the gadolinium by its surrounding chelating complex. Two agents have substantially lower binding that has led to the inclusion of excess chelate in the formulation in order to trap any dissociated gadolinium ion. The two agents that exhibit these characteristics and therefore have different

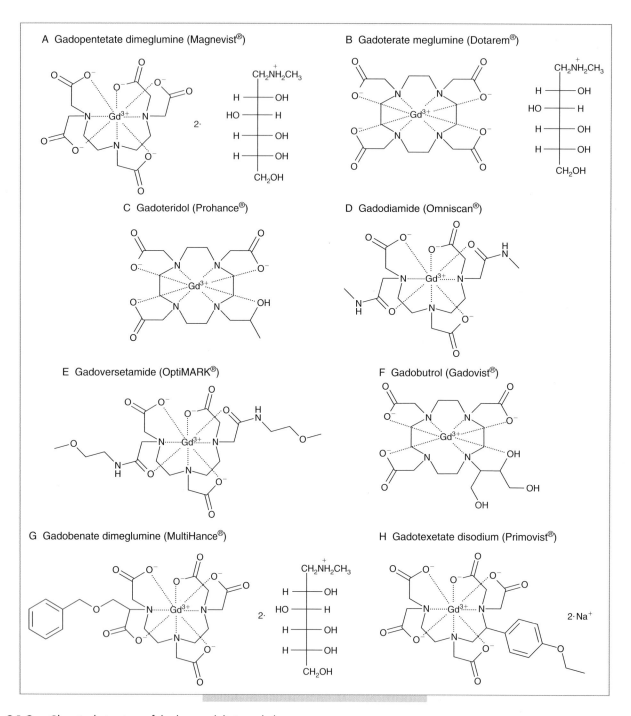

FIG. 34.2. Chemical structure of the key gadolinium chelates.

formulations are gadodiamide (Gd-DTPA-BMA, Ominscan, GE Healthcare/Amersham) and gadoversetamide (Gd-DTPA-BMEA, Optimark, Mallinckrodt). The excess chelate in these agents has been identified as a causative interference with the colorimetric method for serum calcium that have resulted in reported spurious hypocalcemia in clinical laboratory tests using the colorimetric test for serum calcium [3,4]. Another safety aspect that has become of relevance is the association of the use of gadolinium chelates with the development of nephrogenic systemic fibrosis (NSF), which has led to a recent public health advisory by the FDA and the European regulatory authorities. Nephrogenic systemic fibrosis or nephrogenic fibrosing dermopathy (NFD) is a new, rare, disease that affects patients with renal failure [5,6]. Cases have only been identified after 1997 and it appears to be a systemic disorder with most prominent and visible effects in the skin. It is currently postulated that there is a causative relationship with a deposition of gadolinium in the tissue, with the majority of cases currently reported with gadodiamide. It is therefore currently recommended that should patients with a higher risk for this rare disease entity, such as patients with renal impairment, renal dialysis, after transplantation or surgery, or thrombosis, have a strong medical need for use of a gadolinium contrast agent, the goal is to use the lowest possible dose and to avoid those agents that have the lowest binding of gadolinium, gadodiamide or gadoversetamide.

Other than these issues, gadolinium chelates have shown on pharmacovigilance analysis an outstanding safety profile with spontaneous event reporting analysis on Gd-DTPA identifying an adverse event reporting rate of 7.9 cases per 100 000 administrations [1]. Gadobutrol is the only agent that is available at 1.0 molar formulation which enables twice the concentration of gadolinium to be delivered into the vasculature per unit volume, thereby enabling a stronger vasculature signal for perfusion and vasculature imaging which has led to the initial preferred utilization of this type of agent for susceptibility weighted perfusion imaging of the brain.

GADOLINIUM CHELATES WITH WEAK PROTEIN INTERACTION

This class represents a second generation of gadolinium chelates that possess a higher T1 relaxivity in blood such as for gadobenate (Gd-BOPTA, MultiHance, Bracco) $(9.7\,mM^{-1}s^{-1})$ due to the weak transient interactions between the agent and serum proteins, particularly albumin. This higher T1 relaxivity manifests as a significantly greater intravascular signal intensity enhancement compared to that achieved with conventional gadolinium chelates at equivalent doses, with the benefits of a more pronounced effect in smaller vessels, as well as in the margins of tumors. In order to assess objectively if differences in the visualization of intracranial lesions exist between the first group of standard gadolinium chelates and the new group, an intra-individual crossover study was performed that revealed that gadobenate dimeglumine shows a significantly stronger contrast enhancement and improved delineation of brain tumors compared to the representative of standard agents gadopentetate dimeglumine [7]. This finding was confirmed in a subsequent larger study [8]. The practical impact is that intracranial lesions can be significantly better delineated and detected using the standard dosage (Figure 34.3). It has to be considered for follow-up examinations that the improved enhancement and delineation has to be differentiated from disease progression.

The clinical advantages of the increased relaxivity have also been demonstrated for all vascular territories from the carotid vasculature to the distant run-off vessels [2]. Like the conventional non-protein interacting gadolinium chelates, gadobenate dimeglumine has an excellent safety

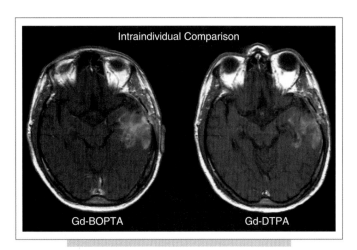

FIG. 34.3. The intraindividual comparison of a representative of the standard gadolinium chelates (Gd-DTPA, Magnevist) with the Gd-BOPTA (MultiHance) reveals a more intense and extensive contrast enhancement for the representative of the class of weakly interacting, higher relaxivity agents. The margins of the glioma are better visualized.

profile with a very low incidence of adverse events noted for the clinical development program as a whole [2].

A second agent with elevated T1 relaxivity (8.2 l/mmol/s) in human plasma is gadotexetate disodium (Primovist, Bayer Schering AG) (see Figure 34.2H) [2]. This agent has been developed and is being marketed in some European countries for liver imaging and is currently not being used nor clinically evaluated for neuroimaging.

GADOLINIUM CHELATES WITH STRONG PROTEIN INTERACTION

Currently, two agents with strong affinity for serum proteins are in clinical development or approved in some countries. The agent approved in Europe for MR angiography and overall furthest along the developmental process is gadofosveset trisodium (MS-325, Vasovist, Bayer Schering AG) which is being dosed for contrast enhanced MRA at a dose of 0.03 mmol/kg bodyweight [9]. This agent has been reported to be 88–96% non-covalently bound to albumin in human plasma and to exhibit a relaxivity at 0.5T that is 6 to 10 times that of gadopentetate dimeglumine [2]. Studies have shown that this agent can be utilized both for first pass contrast enhanced MRA and for steady-state imaging in a number of vascular territories, including the carotid arteries [2]. The agent does also exhibit an extravasation in the case of BBB breakdown.

The second agent with a strong affinity for serum proteins is gadocoletic acid (B22956, Bracco). This agent is currently undergoing Phase II trials for enhanced coronary MRA and has been shown to have even stronger affinity for serum albumin than MS-325 (approximately 94% bound non-covalently), with a similarly long intravascular residence time [2]. Both these agents appear to have potential neuroimaging applications and also have excellent safety profiles.

Both of these so-called paramagnetic 'blood pool' contrast agents are agents for which the intravascular residence time is considerably extended compared to the conventional 'first pass' gadolinium agents. With these agents, the intravascular signal remains high for an extended period of time, thereby permitting MR imaging during a more prolonged 'steady-state' timeframe, in addition to conventional first pass contrast enhanced-MRA. For potential neuroimaging, the visualization of the vasculature in addition to the BBB breakdown extravasation can be either a potential advantage or a potential disadvantage. As no formal clinical trials on neuro-oncologic patient populations have been performed up to now, no confirmative data are currently available. There are two principal types of paramagnetic 'blood pool' contrast agents: those whose intravascular residence time is prolonged due to a capacity of the gadolinium chelate for strong interaction with serum proteins, and those that have a macromolecular structure whose large size limits the extent of extravasation compared to the first pass gadolinium agents. Another important aspect to classify blood pool agents is regarding their capability and efficacy to be used, both in first pass as well as for steady state vascular imaging.

GADOLINIUM CONTRAST AGENTS WITH MACROMOLECULAR STRUCTURES

Examples of gadolinium-based blood pool agents with macromolecular structures are P792 (Vistarem, Guebert) and Gadomer-17 (Bayer Schering AG) [2]. These agents differ from the currently available low molecular weight gadolinium agents in possessing large molecular structures that prevent extravasation of the molecules from the intravascular space following injection, but do have slow, reduced leakage in case of BBB breakdown. The molecular weights of P792 and gadomer-17 are 6.5 kDa and 35 kDa, respectively, which compare with the weights of between approximately 0.56 kDa and 1.0 kDa for the purely first pass gadolinium agents [2]. Whereas the structure of P792 is based on that of gadoterate, substituted with four large hydrophilic spacer arms, gadomer-17 is a much larger polymer of 24 gadolinium cascades [2]. In addition to differences in molecular weight and structure, these two agents appear to differ in terms of their rates of vascular clearance, with P792 considered a rapid clearance blood pool agent [2].

Despite these differences, both agents are currently under investigation for possible applications in contrast enhanced MRA of coronary arteries [2].

SUPERPARAMAGNETIC IRON OXIDE AGENTS

The second major category of potential contrast agents for neuroimaging consists of the superparamagnetic group, which is based on particles of iron oxide (PIO) that are differentiated by the size and by the coating. Those with a diameter larger than 50 nanometers are referred to as small (SPIO) and those smaller as ultrasmall (USPIO). Most iron oxide particles have either a starch or dextran coating and biologic characteristics are predominately dependent on coating, while its imaging characteristics as T1-w or T2*-w agent depend on its size. The only agent that is current approved for clinical use in the USA is AMI 25, also known as Endorem (Guebert) or Feridex (Berlex). This SPIO has been also used for cell-tracking and has therefore the potential for molecular-based neuroimaging applications, however its labeled indication is actually liver imaging, as it is taken up by the reticuloendothelial system (RES) [10]. As with several of the gadolinium-based agents, none of the USPIO agents are yet approved for clinical use. The clinical development of USPIOs in the recent past had focused on vascular imaging, such as NC100150, a USPIO molecule coated with starch and SHU 555 C, a USPIO molecule coated with dextran. Their potential to visualize the intracranial microvasculature at ultra-high field has been recently demonstrated in rodents and is being further evaluated [11,12]. AMI 227 (Ferumoxtran, Guerbert), another USPIO, has been evaluated for lymphatic MR imaging [13,14]. As with the gadolinium blood pool agents, the potential advantages of this agent include a long intravascular half-life with minimal leakage into the interstitial space, thereby permitting the steady-state vascular imaging. The iron oxide particles do have the potential to be carrier molecules of choice, especially for high and ultra-high field MR imaging with the potential to create targeted nano-molecules for MR based in vivo molecular imaging.

While contrast agents for both CT and MR did not reveal distinctively different imaging characteristics in the past, now new agents provide truly distinctive characteristics that advance the capabilities in non-invasive disease detection and characterization. The advent of molecular targeted agents is on the horizon for neuro-oncologic cross-sectional imaging that will enable us further to improve imaging capabilities.

REFERENCES

1. Knopp M, Balzer T, Esser M, Kashanian F, Paul P, Niendorf H (2006). Assessment of utilization and pharmacovigilance based on spontaneous adverse event reporting of gadopentetate dimeglumine as a magnetic resonance contrast agent after 45 million administrations and 15 years of clinical use. Investig Radiol 41:491–499.

2. Knopp M, Kirchin M (2006). Contrast agents for magnetic resonance angiography: Current status and future perspectives.

In *Magnetic Resonance Angiography*, 2nd edn. Arlart I, Bongartz G, Marchal G (eds). Springer, Berlin.

3. Prince M, Erel H, Lent R et al (2003). Gadodiamide administration causes spurious hypocalcemia. Radiology 227:639–646.

4. Choyke P, Knopp M (2003). Pseudohypocalcemia with MR imaging contrast agents: a cautionary tale. Radiology 227:627–628.

5. Marckmann P, Skov L, Rossen K et al (2006). Nephrogenic systemic fibrosis: suspected causative role of gadodiamide used for contrast-enhanced magnetic resonance imaging. J Am Soc Nephrol 17:2359–2362.

6. Marckmann P, Skov L, Rossen K, Heaf J, Thomsen H (2007). Case-control study of gadodiamide-related nephrogenic systemic fibrosis. Nephrol.Dial.Transplant. Epub ahead of print.

7. Knopp M, Runge V, Essig M et al (2004). Primary and secondary brain tumors at MR imaging: bicentric intraindividual crossover comparison of gadobenate dimeglumine and gadopentetate dimeglumine. Radiology 230:55–64.

8. Maravilla K, Maldjian J, Schmalfuss I et al (2006). Contrast enhancement of central nervous system lesions: multicenter intra-individual crossover comparative study of two MR contrast agents. Radiology 240:389–400.

9. Hartman M, Wiethoff A, Hentrich H, Rohrer M (2006). Initial imaging recommendations for vasovist angiography. Eur Radiol Suppl 2:B15–B23.

10. Anderson S, Glod J. Arbab A et al (2005). Noninvasive MR imaging of magnetically labeled stem cells to identify neovasculature in a glioma model. Blood 105:420–425.

11. Yang M, Christoforidis G, Figueredo T, Heverhagen J, Abduljalil A, Knopp M (2005). Dosage determination of ultrasmall particles of iron oxide for the delineation of microvasculature in the Wistar rat brain. Investigat Radiol 40:655–660.

12. Muldoon L, Tratnyek P, Jacobs P et al (2006). Imaging and nanomedicine for diagnosis and therapy in the central nervous system: report of the eleventh annual Blood–Brain Barrier Disruption Consortium Meeting. Am J Neuroradiol 27:715–721.

13. Bellin M, Roy C, Kinkel K et al (1998). Lymph node metastases: safety and effectiveness of MR imaging with ultrasmall superparamagnetic iron oxide particles – initial clinical experience. Radiology 207:799–808.

14. Harisinghani M, Barentsz J, Hahn P et al (2003). Noninvasive detection of clinically occult lymph-node metastases in prostate cancer. New Engl J Med 348:2491–2499.

SECTION IV

Neuroimaging of Brain Tumors

CHAPTER 35

Malignant Astrocytomas

A. Drevelegas, D. Chourmouzi and N. Papanicolaou

INTRODUCTION

Gliomas are the most common primary brain tumors in adults and account for 40–50% of all intracranial tumors. They may manifest at any age, but preferentially affect adults. Their peak incidence is in the fifth and sixth decades of life. They are slightly more common in men than women (1.5/1 ratio) and significantly more common in white than black people. Gliomas can affect any part of the CNS, but they usually occur more supratentorially in adults and infratentorially in children [1].

The clinical symptoms of the tumors depend on the anatomic location of the neoplasm in the brain. Headache, seizures, hemiparesis, personality changes, visual loss, gait disturbances and signs of increased intracranial pressure are among the most common clinical manifestations [2].

More than half of all glial tumors are astrocytic tumors. The pathologic classification and grading of astrocytomas is controversial but, on the other hand, it is also critical for assessment of their prognosis and treatment. A simple grading system for gliomas relies upon recognition of four parameters: nuclear atypia, mitoses, endothelial proliferation and necrosis. The presence of two or more of the above-described features in a glioma places the tumor in the high-grade category [3]. High-grade or malignant gliomas are tumors with both expansive and infiltrative growth. They show some degree of anaplasia, without any cleavage plane and, in microscopic examination, tumor cells are known to extend beyond the tumor margins. Anaplastic astrocytoma (AA; WHO grade III) and glioblastoma multiforme (GBM; WHO grade IV), the most common primary malignant brain tumors, are classified as high-grade gliomas or malignant astrocytomas. This histologic behavior is a consequence of multiple genetic changes that accumulate during stepwise progression [4]. During the past 10 years, tremendous progress has been made in our understanding of the critical events that accompany astroglial transformation and malignant progression [5] (Table 35.1). Gliosarcoma, a rare tumor (WHO grade IV) composed both of neoplastic glial and mesenchymal cells, will be also reviewed with the high-grade neoplasms.

TABLE 35-1	Genetic abnormalities in astrocytic tumors		
Tumor type	Chromosome deviation	Gene alteration	Growth factors amplification/ over expression
Astrocytoma	+7q, −22q	−p53	PDGFR-α
Anaplastic astrocytoma	1, +7q, −9p, −10, v13q, +19, +20, −22q	−p53 −p16 −Rb	
Primary glioblastoma	1, −4q, −6q, +7q, 8q, −9p, −10, −13q, −17p, 19, +20, −22q	+MDM2 −PTEN −p16 −Rb	EGFR
Secondary glioblastoma	−4q, −9p, −10, −13q, −17p	−p53 −p16 −Rb −DCC	PDGFR-α

+, chromosomic gains or gene amplification/overexpression; −, chromosomic losses or gene inactivation. PDGFR-α; platelet-derived growth factor receptor alpha; EGFR, epidermal growth derived factor receptor. (From *Imaging of brain tumors with pathological correlations.* A.Drevelegas (ed.) With kind permission of Springer Science and Business Media.)

ANAPLASTIC ASTROCYTOMA

Anaplastic astrocytoma (WHO grade III) is an infiltrating lesion with biology and average age of diagnosis intermediate between simple astrocytomas and GBM. Although some AA arise as new primaries, 65–75% result from dedifferentiation of low-grade gliomas [6]. The course of progression of low-grade gliomas to AA varies considerably, with time intervals ranging from less than 1 year to more than 10 years, the mean interval being 4–5 years. The designation of a tumor as AA reflects a distinct histologic classification of malignant glioma characterized by an abundance of pleomorphic astrocytes with evidence of mitosis [7]. On the other hand, they lack the necrosis and/or significant vascular proliferation that characterize GBM.

Anaplastic astrocytomas appear in a slightly higher age group than low-grade astrocytomas. They can occur at any age, but most commonly in the fourth to sixth decades of life. Cerebral hemispheres are the most common location of these tumors. Seizures and focal neurologic deficits represent the most common clinical symptoms.

Although AA are malignant tumors, they have a better prognosis and a higher likelihood of response to treatment than GBM. The median survival rate for the patients with these tumors is 2 to 3 years.

On CT scans, AA present as ill-defined inhomogeneous lesions. Calcification is rarely encountered and only in cases of pre-existing low-grade gliomas with malignant transformation. Peritumoral edema may be present as a hypodense area. After the administration of contrast material, the tumors either do not enhance or they show focal, nodular or heterogeneous enhancement (Figure 35.1A,B) [8,9]. On MRI, the appearance of AA may be quite variable. On T1-weighted images they appear heterogeneously iso- to hypointense. T2-weighted images reflect better the heterogeneous composition of the AA with areas of high signal intensity representing the tumor itself and the peritumoral vasogenic edema. Tumor cells can be identified either in most distant aspect of peritumoral edema or occasionally in areas depicted as normal on T2-weighted images outside the margins of peritumoral edema. On T1-weighted images after the administration of contrast medium they show heterogeneous or patchy enhancement (Figure 35.1C–E) [10].

In terms of imaging characteristics, AAs may be difficult to differentiate from GBMs. However, the margins of AA are less defined and exhibit a moderate amount of mass effect, vasogenic edema and heterogeneity. They show a minimal amount of hemorrhage, as opposed to findings in GBM. Necrosis, the imaging hallmark of GBM, is absent. AAs may also mimic the appearance of low-grade astrocytomas and can present as a well-demarcated, homogeneous non-enhancing mass [11,12]. A non-enhancing supratentorial neoplasm does not always equate with low-grade malignancy. In one study, 40% of non-enhancing lesions proved to be anaplastic astrocytomas (Figure 35.2) [13]. MRI is the modality of choice for tumor surveillance and potential malignant transformation over time (Figure 35.3).

GLIOBLASTOMA MULTIFORME

Glioblastoma multiforme (WHO grade IV) is the most common brain tumor, accounting for more than half of all gliomas and 15–20% of all intracranial tumors. About 50–60% of all astrocytic tumors are classified as GBM. Although they represent only 1–2% of all malignancies, the incidence of GBMs is in the range of two to three new cases per 100 000 population per year [14].

GBMs may occur at any age, but most commonly affect adults with a peak incidence between 45 and 70 years. In childhood, GBM are relatively uncommon compared to adults and account for 7–9% of all pediatric intracranial tumors.

GBMs occur most commonly in the subcortical white matter of the temporal, parietal, frontal and occipital lobes. In a series of 987 GBMs from University Hospital Zurich, the most frequently affected sites were the temporal (31%), parietal (24%), frontal (23%) and occipital (16%) lobes [15]. The basal ganglia may also be involved. Brainstem GBMs are infrequent and occur most commonly in children, as a result of transformation of pre-existing low-grade glioma, while the cerebellum and spinal cord are rare sites for this neoplasm.

The most common clinical symptoms in GBMs include seizures, headaches, personality changes, focal neurologic deficits, speech impairment, cranial nerve involvement and increased intracranial pressure. Glioblastoma represent an aggressive tumor with an ominous prognosis. Despite the progress in surgery, radiation therapy or chemotherapy, the mean survival time ranges between 6 and 12 months. Only exceptionally patients survive beyond 2 years.

Glioblastomas can arise 'de novo' (primary GBM), or after progression of an AA (secondary GBM). Primary GBMs account for the vast majority of cases (60%) in adults older than 50 years. After a short clinical history, usually less than 3 months, they manifest de novo (i.e. without clinical or histopathological evidence of a pre-existing, less malignant precursor lesion). Secondary GBMs (40%) typically develop in younger patients (<45 years) through malignant progression from a low-grade astrocytoma (WHO grade II) or AA (WHO grade III) [16].

On unenhanced CT, GBMs appear as central low-density lesions that reflect necrosis, located usually in the centrum semiovale (see Figures 35.5A and 35.6A). Necrosis is the imaging hallmark of GBMs [17,18]. Calcification is rare in GBMs and, when present, is thought to be related to a pre-existing low-grade astrocytoma. They are heterogeneous due to the reflecting sites of necrosis, hemorrhage and increased cellularity (Figure 35.4). Hemorrhage was reported in approximately 19% of patients with GBM and in approximately 12% of patients with low-grade lesions [19]. Besides necrosis and hemorrhage, another characteristic radiologic feature is the presence of edema, which surrounds the tumor, extends along the adjacent white matter tracts and usually produces significant mass effect. After the administration of contrast material they usually show marked heterogeneous ring-like enhancement with thick, shaggy, irregular and nodular wall (Figures 35.5B and 35.6B).

The multiplanar capability of MRI and its higher contrast resolution make it the modality of choice in the evaluation of brain tumors.

On T1-weighted images, GBMs appear as heterogeneous masses with central necrosis, thick irregular margin and peritumoral edema (Figures 35.7A and 35.8A). After the administration of contrast medium they show inhomogeneous

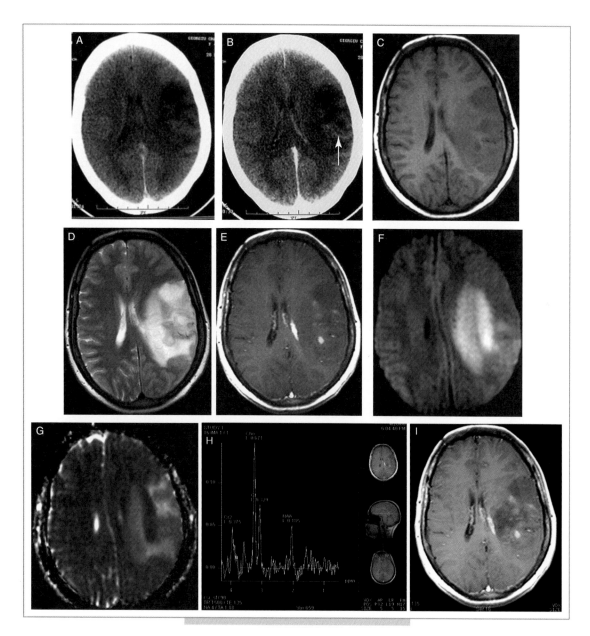

FIG. 35.1. Anaplastic astrocytoma. **(A)** NCCT shows a heterogeneous, hypodense left parietal lobe mass. **(B)** After the administration of contrast medium, the mass remains predominantly hypodense and only a small area of enhancement is present (arrow). **(C)** Axial T1-weighted image shows a slightly hypointense lesion with thickened left parietal cortex. **(D)** On axial T2-weighted image, the lesion is heterogeneously hyperintense with peritumoral edema. **(E)** On axial post-contrast T1WI the hypointense lesion shows patchy enhancement that exhibits high signal intensity on diffusion weighted image **(F)**, and low signal intensity on ADC map **(G)** due to restricted diffusion. **(H)** Long TE (135 ms) spectra demonstrate elevated choline and reduced NAA at the area of tumor plus edema. **(I)** Cho/Cr spectral map shows abnormal Cho/Cr with red color.

or irregular ring-like enhancement (Figures 35.7B and 35.8B). The thick and irregular enhanced rim surrounds a hypointense central area consistent with necrosis within the tumor. The peripheral non-enhancing low attenuation zone represents vasogenic edema. The enhanced ring-like structure does not represent the outer tumor border

because infiltrating glioma cells can be identified easily within, and occasionally beyond, a 2-cm margin.

On T2-weighted images, GBM appears as a heterogeneous mass with high signal intensity. The heterogeneous appearance may be due to central necrosis, hemorrhage and tumor vascularity. Prominent peritumoral vasogenic

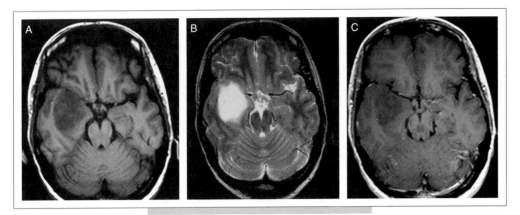

FIG. 35.2. Anaplastic astrocytoma of the right temporal lobe. **(A)** Axial T1-weighted image shows a space occupying hypointense mass. **(B)** On T2-weighted image the mass shows homogeneous high signal intensity. **(C)** On post-contrast T1-weighted image the mass is unenhanced.

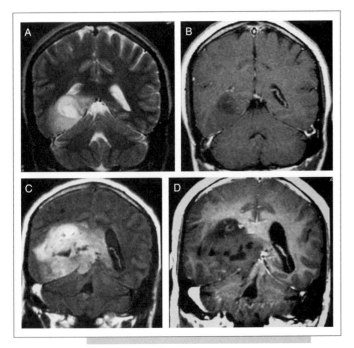

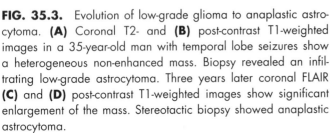

FIG. 35.3. Evolution of low-grade glioma to anaplastic astrocytoma. **(A)** Coronal T2- and **(B)** post-contrast T1-weighted images in a 35-year-old man with temporal lobe seizures show a heterogeneous non-enhanced mass. Biopsy revealed an infiltrating low-grade astrocytoma. Three years later coronal FLAIR **(C)** and **(D)** post-contrast T1-weighted images show significant enlargement of the mass. Stereotactic biopsy showed anaplastic astrocytoma.

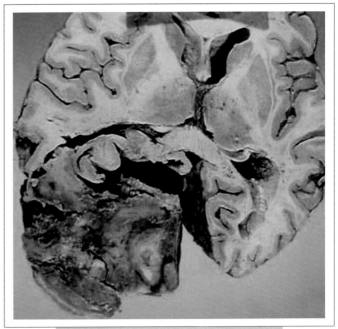

FIG. 35.4. Glioblastoma multiforme. Gross specimen shows a large heterogeneous tumor occupying the right occipital lobe with necrotic and hemorrhagic areas.

edema is always present and is better appreciated in T2- as compared with T1-weighted images. The vasogenic edema is produced by abnormal neoplastic vessels, which lack the normal blood–brain barrier, resulting in transudation of fluids and proteins into the extracellular space. Although MR imaging is the most sensitive method for depicting abnormal amounts of tissue water, discrimination of tumor tissue from edema in terms of signal characteristics has proved unreliable. The white matter edema produced by

GBMs is very extensive and actually represents tumor plus edema (see Figures 35.7C and 35.8C)[20–23].

GBMs may disseminate along various pathways but most commonly along the white matter tracts to involve the contralateral hemisphere. Spread across the corpus callosum, internal capsule, optic radiations and the anterior and posterior commisure is typical. Symmetric extension through the corpus callosum gives rise to butterfly appearance (Figure 35.9). Subependymal spread of GBM can occur and is correlated with a poor prognosis. Spread along the neuraxis via the CSF is also well documented and, incidentally, has been found to have a 6–20% incidence at autopsy series. Extracranial metastases of GBMs are very rare and occur

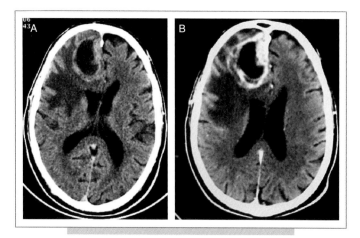

FIG. 35.5. Axial non- **(A)** and post-contrast **(B)** CT in a patient with glioblastoma multiforme show a thick and nodular ring-like enhancing mass with central necrosis. Note also the mass effect and the surrounding white matter edema.

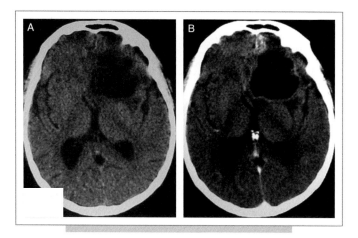

FIG. 35.6. Glioblastoma multiforme. **(A)** Non-contrast CT shows a hypodense mass in the left frontal lobe. **(B)** Post-contrast CT shows a centrally necrotic mass with thin ring-like enhancement. Note also the surrounding mass effect and the midline shift.

most commonly in patients after surgery. However, surgery is not a prerequisite for extracranial metastatic spread of primary CNS malignant disease; extraneural metastases have been described in patients who had not undergone previous surgery. The incidence of symptomatic metastases is certainly lower than the incidence seen at post-mortem and is due to the short survival of the affected patients [24–28].

The vast majority of GBMs are solitary lesions. Multifocal or multicentric tumors occur rarely, in only 0.5–1% of cases [29]. Multicentric tumors are described as widespread lesions in different lobes or hemispheres, without the evidence of possible intracranial spread or microscopic connection. On the other hand, those with either gross or microscopic continuity are defined as multifocal (Figure 35.10). The most frequent dissemination route in the latter group is the meningeal-subarachnoid space, followed by the subependymal, intraventricular route and direct brain penetration through white matter tracts [30–32].

Studies of human neoplasms have demonstrated that increased malignancy is associated with increased neovascularity. Microvascular proliferation (i.e. angiogenesis) is one of the most important criteria for determining the malignancy of gliomas. Tumor capillaries not only provide nutrients for growing tumor cells, but also a route for spread and infiltration far beyond the site of the parent tumor [33]. Angiogenesis is an important parameter in assessing the grading of gliomas.

Although MRI is the modality of choice in the evaluation of brain tumors, conventional T1 and T2 sequences, as well as post-contrast MRI, failed to discriminate neoplastic cells from peritumoral edema and to determine tumor margins accurately. On the other hand, radiologic grading of tumors with conventional MR imaging is not always accurate, with sensitivity in identifying high-grade gliomas ranging from 55.1 to 83.3% [34–37].

The development of techniques capable of accurately depicting tumor margins and grades in vivo is important for determination of the most appropriate treatment for gliomas. The implementation of echoplanar imaging allowed

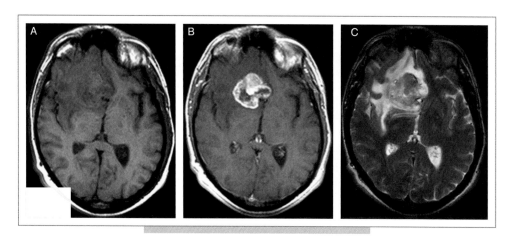

FIG. 35.7. Glioblastoma multiforme. **(A)** Non-contrast CT shows a heterogeneous mass in the right frontal lobe. **(B)** After the administration of contrast medium, thick and irregular ring-like enhancement is seen with central necrosis. **(C)** Axial T2-weighted image shows a heterogeneous mass with peritumoral vasogenic edema extending along the adjacent white matter.

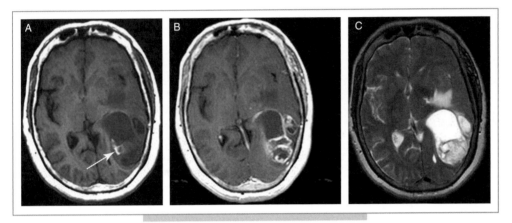

FIG. 35.8. Glioblastoma multiforme. **(A)** Non-contrast T1WI shows a large cystic left parieto-occipital tumor with intratumoral hemorrhage (arrow). **(B)** Post-contrast T1WI shows irregular ring-like enhancement. **(C)** On T2-weighted image the cavitary necrosis shows high signal intensity and the solid component of the tumor appears heterogeneous. Note also the high signal peritumoral edema.

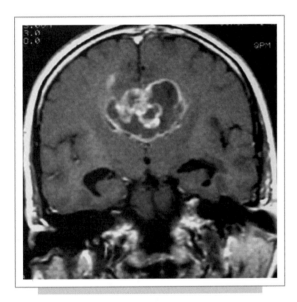

FIG. 35.9. Coronal post-contrast T1-weighted MR image shows an enhancing mass extending through the corpus callosum to both frontal and parietal lobes (butterfly GBM).

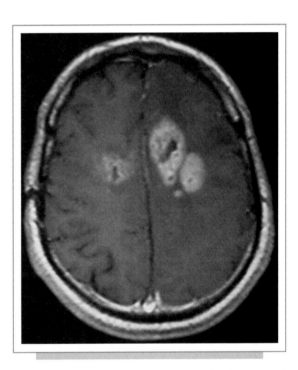

FIG. 35.10. Axial post-contrast T1-weighted image demonstrates multiple, nodular enhanced lesions in both hemispheres. Biopsy proved multifocal GBM.

the development of advanced imaging techniques such as diffusion-weighted imaging (DWI) and dynamic contrast enhanced perfusion, providing physiologic information that complements the anatomic information available with conventional MR imaging. Today, advanced MR imaging is a key modality not only for lesion diagnosis, but also to evaluate the extension, type and grade of the tumor. Additionally, it provides some new insights for neuroradiology practice, such as showing tumor areas before stereotactic biopsy, distinguishing radiation necrosis from tumor infiltration and assessing tumor response to therapy [38–39].

Diffusion-weighted (DW) imaging uses a pair of strong magnetic gradient pulses to dephase and subsequently rephase protons. Protons with higher diffusion rates show a loss of phase coherence and a low MR signal, while protons with slow or restricted diffusion will largely rephase and appear as a high MR signal. The information provided

reflects the viability and structure of tissue on a cellular level. Quantitative information on restricted diffusion of water molecules can be obtained by calculating the apparent diffusion coefficient (ADC). DW imaging has proved clinically useful in the evaluation of cerebral ischemia, infection and tumors [40–42].

The role of diffusion-weighted imaging in patients with brain tumors has been investigated. ADCs could provide useful information in the diagnosis of patients with brain tumors, such as tumor malignancy and peritumoral infiltration. Tumor cellularity and tumor grade have been correlated with ADC values. Brain neoplasms with higher cellularity or higher grades show a significant reduction

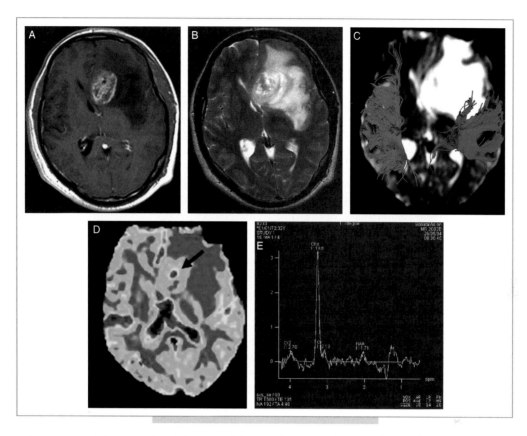

FIG. 35.11. Glioblastoma multiforme. **(A)** Axial post-gadolinium T1-weighted image shows inhomogeneous contrast enhancement of the tumor, located in the left basal ganglia area. **(B)** On T2-weighted image the tumor appears inhomogeneous surrounded by extensive edema. **(C)** Diffusion tensor based tractography reveals extensive infiltration and displacement of internal capsule fibers nearby the tumor. **(D)** Perfusion map demonstrates increased CBV at the center of the tumor (arrow). **(E)** Significantly elevated choline peak (Cho/Cr = 8.51) is shown on a long TE spectra (TE = 135 ms) obtained from the enhancing part of the tumor.

in the value of the apparent diffusion coefficient and a marked increase in the signal of diffusion-weighted images. The low ADC values in tumor probably reflect a decreased volume of extracellular space due to increased tumor cellularity and increased intracellular viscosity, with subsequent restriction of water motion. Lower ADCs indicate mostly malignant glioma, whereas higher ADCs indicate low-grade gliomas (see Figure 35.1F,G) [43–45]. Other investigators failed to find a significant difference between the ADC values of high-grade gliomas and low-grade tumors.

Recently, there has been an effort to define glioma margins using diffusion tensor imaging. MR diffusion tensor imaging (DTI) allows identification and characterization of white matter tracts according to the direction and degree of their anisotropic water diffusion. Quantifying the degree of anisotropy in terms of metrics, such as the fractional anisotropy, offers insight into white matter development and degradation. There are fractional anisotropy changes in the white matter of glioblastomas that might indicate cellular infiltration beyond the area of the tumor enhancement. Within the tumor center, white matter fibers are displaced by cellular infiltration and fractional anisotropy is reduced,

whereas in the periphery and in a narrow rim of white matter surrounding the tumor, this parameter could preserve or even be increased by fiber compression due to the space occupying effect of the tumor (Figure 35.11A–C) [46–48].

Several studies suggest that contrast enhancement alone is not sufficient to predict tumor grade, since 20% of low-grade gliomas demonstrate contrast enhancement while one third of high-grade tumors do not [49]. Although contrast enhancement indicates disruption of the blood–brain barrier, it is not particularly effective at revealing the underlying regional vascularity. Precise determination of the regional vascularity is important, given that the degree of vascular proliferation is an important parameter for the histopathologic grading of gliomas.

The introduction into clinical practice of echoplanar MR imaging allowed the estimation of parameters that reflect tissue vascularization. Perfusion-weighted imaging (PWI) provides useful information about the microcirculation in brain tissue. This technique requires the dynamic (bolus) intravenous administration of MR contrast agent. As the paramagnetic contrast agent passes through the intravascular compartment, it causes local field inhomogeneities that result in magnetic susceptibility effects with a decrease in signal on multiple repeated T2* images that can be measured. This signal drop depends on both the vascular concentration of contrast agent and the concentration of small vessels per voxel of tissue [40–51]. Changes in signal intensity may be used to calculate an image of the relative cerebral blood volume (rCBV).

In brain tumors, cerebral blood volume (CBV) maps are particularly sensitive for depicting the microvasculature of a tumor and therefore its aggressiveness and proliferative potential. Several studies have found statistically significant correlations between tumor rCBV and glioma grade. High-grade gliomas demonstrate higher rCBV values than low-grade tumors and the low-grade tumors with contrast enhancement present with low rCBV (see Figures 35.11D and 35.12) [52–55]. Several studies have shown that low-grade gliomas (LGGs) have maximal rCBV values of between 1.11 and 2.14, while high-grade glioma (HGGs) maximal rCBV values range between 3.54 and 7.32 [56]. In clinical practice, 95–100% sensitivity has been reported for differentiating high-grade from low-grade gliomas using thresholds of 1.75 and 1.5 for rCBV respectively. CBV measurements can also be used to reduce the sampling error in the histopathologic diagnosis of gliomas, improving the selection of targets for stereotactic biopsy. High CBV areas represent the best stereotactic biopsy site for accurate grading of astrocytomas [57,58].

MR spectroscopy (MRS) is a non-invasive technique that allows in vivo measurements of certain tissue metabolites. By suppressing water signal, the relative concentrations of non-water, proton-containing metabolites from discrete tissue regions can be quantified. There is certainly compelling evidence that MR spectroscopy provides important supplementary information to that of conventional MR imaging. In normal brain, the principal metabolite signals that can be measured by MRS are choline (Cho), creatine (Cr), N-acetylaspartate (NAA) and lactate (Lac) [59,60].

Each metabolite resonates at a specific frequency (ppm) and each one reflects specific cellular and biochemical processes.

Choline is a cell membrane marker. The increase in Cho in simple terms is attributed to cell membrane turnover and proliferation. An elevation in Cho may be due to either cell membrane synthesis, destruction or both. Cho is most prominent in regions with high neoplastic cellular density and is progressively lower in moderate and low-grade tumors. Some highly malignant tumors and some GBMs may show low Cho because of extensive necrosis [61–63].

Cr is a marker of 'energy metabolism' and is reduced in tumors due to the increased metabolic activity of tumors consuming energy. NAA is a marker of neuronal density and viability and it is generally felt to be neuron specific. The decrease in NAA represents the replacement of normal functioning neurons and axons by any disease that adversely affects neuronal integrity [60,61].

Lactate is a product of anaerobic glycolysis and is present only in minute amounts in normal brain. It appears when tumors outgrow their blood supply and start utilizing anaerobic glycolysis. Lactate levels increase significantly in necrotic and cystic tumors.

Lipids are products of brain destruction and are found in necrotic portions of tumors. Myo-inositol is a glial cell marker and osmolyte hormone receptor mechanism. The basic metabolite changes common to brain tumors include elevation in choline (Cho), lactate (Lac), lipids (L), decrease in N-acetylaspartate (NAA) and decrease in creatine (Cr) (see Figures 35.1H,I and 35.11E). As a general rule, as malignancy increases, NAA and creatine decrease and choline, lactate and lipids increase. Myo-inositol can also be used to differentiate LGGs and HGGs. LGGs express higher levels of myo-inositol compared with HGGs. This may be due to the lack of activation of phosphatidylinositol metabolism resulting in accumulation of myo-inositol in LGGs [64–66].

There is extensive literature demonstrating the metabolite ratios of Cho/Cr, NAA/Cr, NAA/Cho, Cho/NAA and the presence of lipids and lactate to be useful in grading tumors and predicting tumor malignancy.

A significant difference is noted in several investigations in Cho/Cr, Cho/NAA and NAA/Cr ratios for differentiating between low- and high-grade gliomas. Law et al demonstrated a threshold value of 1.56 for Cho/Cr to provide sensitivity, specificity, and positive and negative predictive values of 75.8%, 47.5%, 81.2% and 39.6%, respectively,

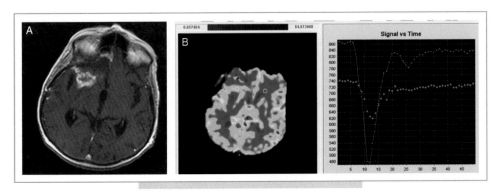

FIG. 35.12. High-grade astrocytoma presents with rim enhancement pattern on post-gadolinium T1-weighted axial image **(A)**, while on perfusion color map demonstrates increased rCBV **(B**-left**)**. Time-intensity curve shows significant signal drop in the area of the tumor as opposed to normal contralateral white matter, indicative of high-grade tumor.

for the determination of high-grade gliomas versus a low-grade glioma. In the same study, a threshold value of 1.6 for Cho/NAA provided 74.2%, 62.5%, 85.6% and 44.6% for the sensitivity, specificity and positive and negative predictive values, respectively, for determination of high-grade gliomas [67].

The differential diagnosis of GBM includes abscess, metastasis, lymphoma and tumefactive multiple sclerosis (MS). Brain abscess remains a diagnostic challenge because the presenting clinical manifestations and neuroradiologic appearances are often non-specific. Imaging findings helpful in differential diagnosis include a thin wall with ring-like enhancement that is often thinner along the medial margin, daughter rings and hypointense rim on T2-weighted images (Figure 35.13A,B) [68,69].

Most studies have reported increased ADC values in necrotic component of tumors as opposed to abscesses. Noguchi et al found that the necrotic components of tumors had ADC values in the range of $2.2–3.2 \times 10^{-3}\,mm^2/s$, in contrast to abscesses that had ADC values of less than $0.7 \times 10^{-3}\,mm^2/s$ [70]. In another study, the calculated ADC values of the enhancing component of GBM were in the range of 0.78×10^{-3} to 1.79×10^{-3} (mean 1.14×10^{-3}) (see Figure 35.13C,D) [71]. However, there are reported cases of tumors demonstrating marked hyperintensity on DW-MRI with extremely low ADC values in the necrotic centre, similar to what has been reported for abscesses. Restricted diffusion within these lesions could be due to a variety of causes, including intratumoral hemorrhage, cytotoxic edema in the early phase of cell death, thick sterile liquefaction or pyogenic superinfection [72–75].

MR spectroscopy can also non-invasively contribute to the establishment of the differential diagnosis between tumors and abscesses. The spectrum of the abscess cavity shows elevation of lactate, acetate, succinate, alanine and some amino acids and this spectrum appears significantly different from that of necrotic or cystic brain tumor [76].

Intracranial metastases and primary malignant gliomas are two common brain tumors encountered in adults. The management of these two tumors is different and can potentially affect the clinical outcome. When the clinical findings are non-specific, conventional MR imaging characteristics may be similar and unable to differentiate the two entities.

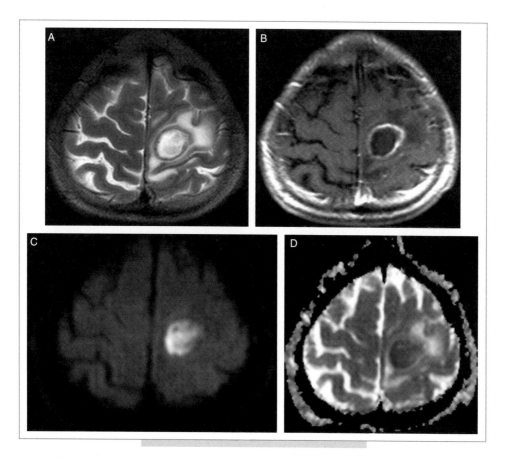

FIG. 35.13. Diffusion study in a patient with brain abscess. **(A)** Axial T2-weighted image shows a well defined lesion with hypointense rim and perilesional edema. **(B)** Post-gadolinium T1-weighted image demonstrates ring like enhancement. **(C)** Diffusion weighted image (b = 1000 mm²/s) shows high signal intensity within the lesion, while the same area in the corresponding **(D)** ADC map presents with low signal intensity due to restricted diffusion.

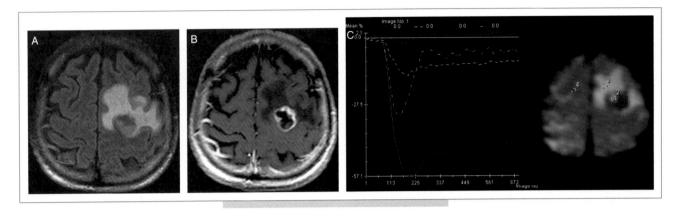

FIG. 35.14. Perfusion study in a patient with brain metastasis. **(A)** Axial FLAIR and **(B)** post-gadolinium T1-weighted images show an enhancing lesion surrounded by extensive edema. **(C)** Time-intensity curves demonstrate significantly higher signal drop at the site of the enhancing rim (red) as opposed to normal contralateral brain tissue (cyan). Note that the central area (yellow) of the lesion presents with lower signal drop, while in the area of perilesional edema (green) there is no signal drop due to capillaries compression by vasogenic edema.

Diffusion-weighted imaging techniques can be helpful in distinguishing solitary intra-axial metastatic lesions from high-grade gliomas. Studies showed that both contrast enhancing portions and peritumoral edema of metastasis have higher ADC than high-grade gliomas. However, the distinction between metastases and high-grade gliomas is often difficult to make based on ADC values, as some high-grade gliomas also exhibit high ADC values [73,77–80].

Perfusion MR imaging and spectroscopy may also be used to differentiate high-grade primary gliomas and solitary metastases. Law et al found significant difference between the mean rCBV value within the peritumoral region in high-grade gliomas, which was 1.31 ± 0.97, suggesting increased peritumoral perfusion due to tumor infiltration, and the mean rCBV surrounding metastatic tumors which was 0.39 ± 0.19, consistent with compression of capillaries by vasogenic edema (Figure 35.14) [81]. No difference has been demonstrated between the intratumoral rCBV values of high-grade gliomas, and metastases. In the same study, they have found elevated Cho/Cr (2.28 ± 1.24) in the peritumoral regions of high-grade gliomas in keeping with tumoral infiltration. No increase in the Cho/Cr (0.76 ± 0.23) was found in the peritumoral region of metastases, which again suggests vasogenic edema (Figure 35.15). In the same study, there was also no significant difference in the peritumoral NAA/Cr between the two groups because there is no neuronal replacement or destruction in the peritumoral region of either pathologic condition.

The differential diagnosis of primary CNS lymphoma (PCNSL) from malignant astrocytoma can be based on the periventricular and deep gray matter location of lymphoma and on hypo- or isointensity of PCNSL on T2-weighted images while high-grade glioma shows high signal intensity. On post-contrast T1-weighted images PCNSL show intense homogeneous enhancement (Figure 35.16A,B). However, less commonly PCNSL may be hyperintense on T2-weighted images and may show ring-like enhancement, especially in immunocompromised patients. In these cases, it can be difficult to distinguish PCNSL from high-grade glioma on the basis of conventional MR imaging features.

Diffusion weighted imaging techniques showed that the mean ADC value in lymphomas is lower than that of high-grade gliomas reflecting the increased cellularity of lymphomas (see Figure 35.16C,D) [82–83]. Neovascularization is absent in PCNSLs, in contrast to high-grade gliomas, therefore perfusion weighted images can also be used to distinguish the two entities. The lower rCBV values of PCNSLs may be helpful in differentiating from other brain tumors such as high-grade gliomas (see Figure 35.16E)[84].

The MR spectrum of PCNSL is characterized by increased Cho, lactate and lipids and is associated with decreases in NAA, Cr and myo-inositol levels. The latter metabolite pattern can also be found in malignant astrocytomas. However, if the analysis of the solid portion of the tumor shows a significant increase in lipids, it is probably lymphoma [85].

The conventional MRI features of tumefactive MS may mimic that of high-grade gliomas. Both lesions may show variable contrast enhancement, perilesional edema, mass effect and central necrosis. The pathologic difference between tumefactive MS and high-grade gliomas is the absence of angiogenesis in MS, which demonstrates low rCBV values, as opposed to gliomas that are characterized by neovascularization and angiogenesis resulting in a significant elevation of rCBV [86].

In spectroscopy, there is some overlap between tumefactive MS and high-grade gliomas. There is Cho elevation in MS from astrogliosis, demyelination and inflammation, lactate is found from anaerobic glycolysis and decrease of NAA from neuronal and axonal damage. However, NAA/Cr ratio is higher in both the enhancing and central regions

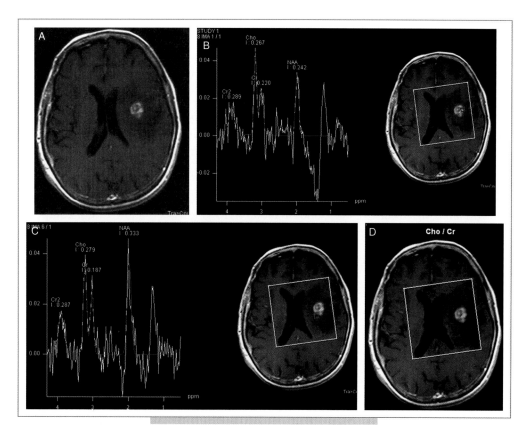

FIG. 35.15. Brain metastasis. **(A)** Axial post-gadolinium T1-weighted image shows focal areas with intense enhancement and a low signal intensity peritumoral edema. **(B)** Long TE (135 ms) spectra obtained from the enhancing lesion shows elevation of Cho and decrease of NAA, while in **(C)** peritumoral edema Cho is within the normal range. **(D)** Cho/Cr map demonstrates focal increase at the site of enhancing lesion (red color) while normal Cho/Cr is noted in the peritumoral areas.

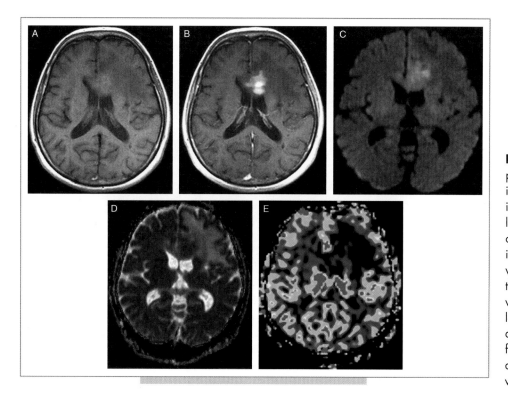

FIG. 35.16. Primary cerebral lymphoma. **(A)** Non-contrast T1-weighted image shows an isointense mass involving the genu of the corpus callosum. **(B)** After the administration of contrast medium the mass shows intense enhancement. **(C)** On diffusion weighted image (b = 1000 mm²/s) the tumor shows high signal intensity while on the ADC map **(D)** exhibits low signal intensity due to restricted diffusion(arrowhead). **(E)** On perfusion color map the tumor and the adjacent edema demonstrate low CBV values.

in MS. Hence, the combination of reduced perfusion and moderate NAA reduction can help differentiate tumefactive MS from high-grade gliomas [87,88].

Post-therapeutic MR examinations are used routinely to differentiate therapy-induced necrosis from residual or recurrent tumor. On conventional MRI, abnormal contrast enhancement on post-contrast T1-weighted images and the pattern of hyperintensity on T2-weighted images are used to discriminate these entities. Features that are inspected for assessment of tumor progression are regions of abnormal contrast enhancement on post-contrast T1-weighted images and the volume of hyperintensity on T2-weighted images. Other certain conventional MR imaging findings suggesting predominant glioma progression are corpus callosum involvement, conjunction with multiple enhancing lesions without crossing of the midline and subependymal spread [89]. Although such morphologic changes are indicative of the existence of the disease, it is often difficult to determine whether an enhanced lesion represents tumor recurrence or not, especially when the enhancement is initially observed.

Perfusion MR imaging may be helpful for differentiating recurrent tumors from post-therapeutic necrosis. Elevated rCBV values most likely represent recurrent tumor while decreased rCBV values are likely to represent radionecrosis due to vascular injury (Figure 35.17). Studies found that if the rCBV ratio of enhancing lesion is more than 2.6 or less than 0.6 tumor recurrences or radiation necrosis should be strongly suspected, respectively [90].

MRS has a significant clinical impact on post-treatment evaluation of intracranial neoplasms and it is useful to identify recurrent tumor earlier than conventional MRI and differentiate residual or recurrent tumor from post-treatment abnormality. Evidence of radiation necrosis is typically observed within 3–6 months after therapy and is characterized by decreases in Cho, Cr, NAA and increased lipid and lactate levels. On the other hand, elevated Cho levels and greater Cho/NAA ratio suggest tumor recurrence (Figure 35.18).

Increases in lipid and lactate levels are less specific than elevated Cho levels because they may result from both tissue necrosis and radiation-induced necrosis. Follow up of these lesions is crucial for accurate diagnosis [91].

GLIOSARCOMA

Gliosarcoma (WHO grade IV) is a rare tumor composed of glial and sarcomatous components. The glial component usually demonstrates the typical features of GBM. Although oligodendroglial cells are occasionally recognized, in most cases, the glial cells are astrocytic, with a tissue pattern of a GBM. As expected, heterogeneity is a feature of the glial component. Additional variation is recognized in the mesenchymal component. Malignant fibrous histiocytoma is the most frequent sarcomatous component. Fibrosarcoma, smooth and striated muscle sarcoma, as well as bone or cartilaginous sarcoma may also be found [92–94]. Gliosarcomas are peripherally located and involve the temporal, parietal and occipital lobes [95–97]. Posterior fossa gliosarcomas have also been reported. Radiation induced gliosarcoma may appear at the site of a treated intracranial neoplasm [95,98]. Most patients with gliosarcoma are in their fifth to seventh decades. Extracranial metastasis of the sarcomatous component is common, occurring in 15–30% of all gliosarcomas [93,99].

On plain CT, most tumors appear as slightly hyperdense lesions because of their high vascularity and cellularity. After the contrast administration, gliosarcomas show

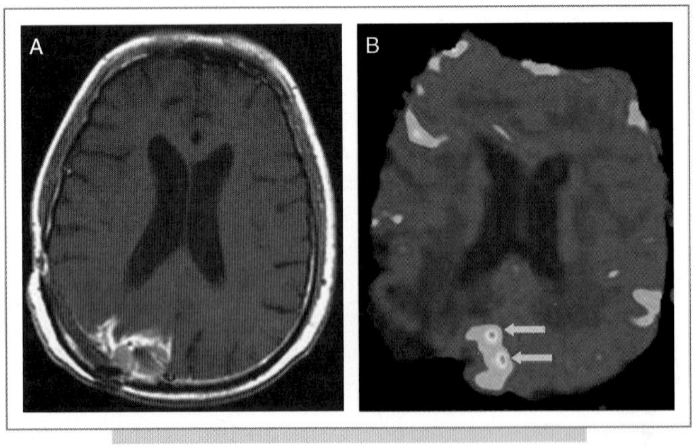

FIG. 35.17. High grade astrocytoma reccurence. **(A)** Postoperative axial post-gadolinium T1-weighted image shows inhomogeneous peripheral enhancement. **(B)** Perfusion map demonstrates two foci (arrows) with increased CBV indicating tumor recurrence.

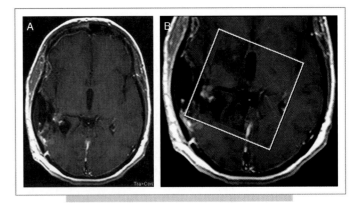

FIG. 35.18. Glioblastoma multiforme reccurence. **(A)** Postoperative axial post-gadolinium T1-weighted image demonstrates enhanced areas in a patient who underwent radiation treatment. **(B)** Cho/Cr color metabolite map reveals a focal area (red spot) with elevated Cho/Cr value, indicating tumor recurrence. The remaining enhancing tissue presents with low Cho/Cr values reflecting radiation necrosis.

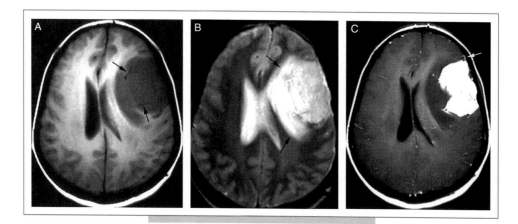

FIG. 35.19. Gliosarcoma. **(A)** Axial T1WI shows a hypointense lesion with intratumoral vessels (arrows). **(B)** On axial T2WI the mass as well as the adjacent edema (arrows) shows high-signal intensity. **(C)** On axial post-contrast T1WI the tumor shows intense homogeneous enhancement. Note the close relation of the mass with the dura that is enhanced in a way mimicking meningioma (arrow). (From *Imaging of brain tumors with pathological correlations.* A.Drevelegas (ed.) With kind permission of Springer Science and Business Media.)

marked enhancement and may mimic a meningioma, when the tumor is located near the skull or falx. Gliosarcomas, however, are less homogeneously hyperdense than meningiomas, do not have a large base in contact with the skull and are virtually always associated with peritumoral edema. In other cases, the CT appearance is that of an intracerebral mass with irregular and peripheral enhancement and large necrotic areas, similar to malignant astrocytomas or glioblastomas [100,101].

On MR they have inhomogeneous or cystic appearance with surrounded vasogenic edema. They are intra-axial but abutting a dural surface with intense heterogeneous tumor enhancement. On T2-weighted images they have intermediate signal intensity with peripheral high-signal intensity due to the surrounding edema (Figure 35.19). Hemorrhage and necrosis are common [102,103].

Gliosarcoma should be included in the differential diagnosis of any tumor that appears to be intra-axial but abuts a dural surface and shows imaging characteristics similar to gray matter on T2-weighted images. Gliosarcomas are considered to have an aggressive course, with prognosis rather similar to that of GBMs [93,103].

REFERENCES

1. Davis FG, McCarthy B, Jukich P (1999). The descriptive epidemiology of brain tumors. Neuroimaging Clin N Am 9: 581–594.
2. Karkavelas G, Taskos N (2002). Epdemiology, histological classification and clinical course of brain tumors. In *Imaging of Brain Tumors with Histological Correlations.* Drevelegas A (ed.). Springer-Verlag, Berlin.1–10.
3. Daumas-Duport C, Scheihauer B, O'Fallon J, Kelly P (1988). Grading of astrocytomas. A simple and reproducible method. Cancer 62:2152–2165.
4. Louis DN, Cavenee WK (1997). Molecular biology of central nervous system neoplasms. In *Cancer: Principles and Practice of Oncology* De Vita VT, Hellman S, Rosenberg SA (eds). Lippincott Raven Publishers, Philadelphia. 2013–2022.
5. Goussia AC, Polyzoidis K, Kyritsis AP (2002) Molecular abnormalities in gliomas. In *Imaging of Brain Tumors with Histological Correlations.* Drevelegas A (ed.). Springer-Verlag, Berlin. 27–35.
6. Paris JE, Scheithauer BW (1993). Glial tumors. In *Principles and Practice of Neuropathology.* Nelson JS, Parisi JE, Scheithauer BW (eds). Mosby, St Louis. 123–183.
7. See SJ, Gilbert MR (2001). Anaplastic astrocytoma: diagnosis, prognosis, and management. Semin Oncol 31: 618–634.
8. Philippon JH, Clemenceau SH, Fauchon FH, Foncin JF (1993). Supratentorial low-grade astrocytomas in adults. Neurosurgery 32:554–559.
9. Castillo M, Scatliff JH, Bouldin TW, Suzuki K (1992). Radiologic-pathologic correlation: intracranial astrocytoma. Am J Neuroradiol 13:1609–1616.
10. Watanabe M, Tanaka R, Takeda N (1992). Magnetic resonance imaging and histopathology of cerebral gliomas. Neuroradiology 35:463–469.
11. Barker FG 2nd, Chang SM, Huhn SL et al (1997). Age and the risk of anaplasia in magnetic resonance-nonenhancing supratentorial cerebral tumors. Cancer 80:936–941.
12. Kondziolka D, Lunsford LD, Martinez AJ (1993). Unreliability of contemporary neurodiagnostic imaging in evaluation of suspected adult supratentorial (low-grade) astrocytoma. J Neurosurg 79:533–536.
13. Ginsberg LE, Fuller GN, Hashmi M, Leeds NE, Schomer DF (1998). The significance of lack of MR contrast enhancement

of supratentorial brain tumors in adults: histopathological evaluation of a series. Surg Neurol 49:436–440.

14. Boring CC, Squires TS, Tong T (1993). Cancer statistics. CA 43:7–26.

15. Kleihues P, Burger PC, Collins VP, Newcomb EW, Ohgaki H, Cavenee WK (2000). Glioblastoma. In *Pathology and Genetics of Tumours of the Nervous System*. Kleihues P, Cavenee WK (eds).1993; FIARC Press, Lyon. 29–39.

16. Becker LE (1995). Central neuronal tumors in childhood: relationship to dysplasia. J Neuro-Oncol 24:13.

17. Burger PC, Heinz ER, Shibata T, Kleihues P (1988). Topographic anatomy and CT correlations in the untreated glioblastomas multiforme. J Neurosurg 68:698–704.

18. Lilja A, Bergstrom K, Spannare B, Olsson Y (1981). Reliability of CT in assessing histopathological features of malignant supratentorial gliomas. J Comput Assist Tomogr 5:625–636.

19. Kondziolka D, Bernstein M, Resch L et al (1987). Significance of hemorrhage into brain tumors: clinicopathological study. J Neurosurg 67:852–857.

20. Earnest F, Kelly PJ, Scheithauer BW et al (1988). Cerebral astrocytomas: histopathologic correlation of MR and CT contrast enhancement with stereotactic biopsy. Radiology 166:823–827.

21. Atlas SW (1990). Adult supratentorial tumors. Sem Roentgenol 25:130–154.

22. Tovi M, Lilja A, Erickson A (1994). MR imaging in cerebral gliomas: tissue component analysis in correlation with histopathology of whole-brain specimens. Acta Radiol 35:495–505.

23. Drevelegas A, Karkavelas G (2002). High-grade gliomas. In *Imaging of Brain Tumors with Histological Correlations*. Drevelegas A (ed.). Springer-Verlag, Berlin. 109–135.

24. Boleslaw H, Liwinics L, Rubinstein J (1979). The pathways of extraneural spread in metastasizing gliomas. Hum Pathol 10:453–467.

25. Dolman CL (1974). Lymph node metastases as a first manifestation of glioblastoma multiforme. J Neurosurg 41:607–609.

26. Vestosick FT, Selker RG (1990). Brain stem and spinal metastases of supratentorial glioblastomas multiforme: a clinical series. Neurosurgery 27:516–522.

27. Taha M, Ahmad A, Wharton S, Jellinek D (2005). Extra-cranial metastasis of glioblastoma multiforme presenting as acute parotitis. Br J Neurosurg 193:48–51.

28. Rees JH, Smirniotopoulos JG, Jones RV, Wong K (1996). Glioblastoma mutliforme: radiologic-pathologic correlation. Radiographics 16:1413–1438.

29. Barnard RO, Geddes JF (1987). The incidence of multifocal gliomas: a histologic study of large hemisphere sections. Cancer 60:1519–1531.

30. Prather JL, Long JM, van Heertum R, Hardman J (1975). Multicentric and isolated multifocal glioblastomas multiforme simulating metastatic disease. Br J Radiol 48:10–15.

31. Lafitte F, Morel-Precetti S, Martin-Duverneuil N et al (2001). Multiple glioblastomas: CT and MR features. Eur Radiol 11: 131–136.

32. Jawahar A, Weilbaecher C, Shorter C, Stout N, Nanda A (2003). Multicentric glioblastoma multiforme determined by positron emission tomography: a case report. Clin Neurol Neurosurg 106:38–40.

33. Scherer HJ (1940). The role of vascular proliferation in the growth of brain tumors. Brain 63:1–35.

34. Dean BL, Drayer BP, Bird CR et al (1990). Gliomas: classification with MR imaging. Radiology 174:411–415.

35. Moller-Hartmann W, Herminghaus S, Krings T et al (2002). Clinical application of proton magnetic resonance spectroscopy in the diagnosis of intracranial mass lesions. Neuroradiology 44:371–381.

36. Knopp EA, Cha S, Johnson G et al (1999). Glial neoplasms: dynamic contrast-enhanced T2*-weighted MR imaging. Radiology 211:791–798.

37. Law M, Yang S, Wang H et al (2003). Glioma grading: sensitivity, specificity, and predictive values of perfusion MR imaging and proton MR spectroscopic imaging compared with conventional MR imaging. Am J Neuroradiol 241: 1989–1998.

38. Chang YW, Yoon HK, Shin HJ, Roh HG, Cho JM (2003). MR imaging of glioblastoma in children: usefulness of diffusion/perfusion-weighted MRI and MR spectroscopy. Pediatr Radiol 33:836–842.

39. Moffat BA, Chenevert TL, Lawrence TS et al (2005). Functional diffusion map: a noninvasive MRI biomarker for early stratification of clinical brain tumor response. Proc Natl Acad Sci 102:5524–5529.

40. Drevelegas A (2002). Imaging modalities in brain tumors. In *Imaging of Brain Tumors with Pathologic Correlations*. Drevelegas A (ed.). Springer-Verlag, Berlin. 11–25.

41. Mullins ME, Schaefer PW, Sorensen AG et al (2002). CT and conventional and diffusion-weighted MR imaging in acute stroke: study in 691 patients at presentation to the emergency department. Radiology 224:353–360.

42. Schaefer PW, Ozsunar Y, He J et al (2003). Assessing tissue viability with MR diffusion and perfusion imaging. Am J Neuroradiol 24:436–443.

43. Lam WW, Poon WS, Metreweli C (2002). Diffusion MR imaging in glioma: does it have any role in the preoperation determination of grading of glioma? Clin Radiol 57:219–225.

44. Kono K, Inoue Y, Nakayama K et al (2001). The role of diffusion-weighted imaging in patients with brain tumors. Am J Neuroradiol 22:1081–1088.

45. Krabbe K, Gideon P, Wagn P, Hansen U, Thomsen C, Madsen F (1997). MR diffusion imaging of human intracranial tumors. Neuroradiology 39:483–489.

46. Tropine A, Vucurevic G, Delani P et al (2004). Contribution of diffusion tensor imaging to delineation of gliomas and glioblastomas. J Magn Reson Imaging 20:905–912.

47. Roberts TP, Liu F, Kassner A, Mori S, Guha A (2005). Fiber density index correlates with reduced fractional anisotropy in white matter of patients with glioblastoma. Am J Neuroradiol 26:2183–2186.

48. Lu S, Ahn D, Johnson G, Cha S (2003). Peritumoral diffusion tensor imaging of high-grade gliomas and metastatic brain tumors. Am J Neuroradiol 24:937–941.

49. Knopp EA, Cha S, Johnson G et al (1999). Glial neoplasms: dynamic contrast-enhanced T2*-weighted MR imaging. Radiology 211:791–798.

50. Baird AE, Benfield A, Schlaug G et al (1997). Enlargement of human cerebral ischemic lesions volumes measured by diffusion-weighted magnetic resonance imaging. Ann Neurol 41:581–589.

51. Wilms G, Sunaert S, Flamen P (2001). Recent developments in brain tumour diagnosis. In *Recent Advances in Diagnostic*

Neuroradiology. Demaerel P, Baert AL, Demaerel Ph (eds). Springer, Berlin. 119–135.

52. Cha S, Knopp EA, Johnson G, Wetzel SG, Litt AW, Zagzag D (2002). Intracranial mass lesions: dynamic contrast enhanced susceptibility-weighted echo-planar perfusion MR imaging. Radiology 223:11–29.

53. Maia AC Jr, Malheiros SM, da Rocha AJ et al (2005). MR cerebral blood volume maps correlated with vascular endothelial growth factor expression and tumor grade in nonenhancing gliomas. Am J Neuroradiol 26:777–783.

54. Yang D, Korogi Y, Sugahara T et al (2002). Cerebral gliomas: prospective comparison of multivoxel 2D chemical-shift imaging proton MR spectroscopy, echoplanar perfusion and diffusion-weighted MRI. Neuroradiology 44:656–666.

55. Rollin N, Guyotat J, Streichenberger N et al (2006). Clinical relevance of diffusion and perfusion magnetic resonance imaging in assessing intra-axial brain tumors. Neuroradiology 48:150–159.

56. Law M, Yang S, Wang H et al (2003). Glioma grading: sensitivity, specificity and predictive value of perfusion MRI and proton spectroscopic imaging compared with conventional MR imaging. Am J Neuroradiol 24:1989–1998.

57. Lev MH, Rosen BR (1999). Clinical applications of intracranial perfusion MR imaging. Neuroimaging Clin N Am 9:309–331.

58. Yang D, Korogi Y, Sugahara T et al (202). Cerebral gliomas: prospective comparison of multivoxel 2D chemical-shift imaging proton MR spectroscopy, echoplanar perfusion and diffusion-weighted MRI. Neuroradiology 44:656–666.

59. Law M (2005). Perfusion and MRS for brain tumor diagnosis. In *Clinical Magnetic Resonance Imaging.* Edelman RR (ed.). Saunders, Philadelphia. 1215–1247.

60. Pomper MG, Port JD (2000). New techniques in MR imaging of brain tumors. Magn Reson Imaging Clin N Am 8:691–713.

61. Castillo M, Kwock L (1998). Proton MR spectroscopy of common brain tumors. Neuroimaging Clin N Am 8:733–752.

62. Brandao LA (ed.) (2004). Introduction and technique. In *MR Spectroscopy of the Brain.* Lippincott Williams&Wilkins, Philadelphia. 1–15.

63. Gupta RK, Cloughesy TF, Sinha U et al (2000). Relationships between choline magnetic resonance spectroscopy, apparent diffusion coefficient and quantitative histopathology in human glioma. J Neuro-Oncol 50:215–226.

64. Tedeschi G, Lundbom N, Raman R et al (1997). Increased choline signal coinciding with malignant degeneration of cere-bral gliomas: a serial proton magnetic resonance spectroscopy imaging study. J Neurosurg 87:516–524.

65. Urenjak J, Williams SR, Gadian DG, Noble M (1992). Specific expression of N-acetylaspartate in neurons, oligodendrocyte-type-2 astrocyte progenitors, and immature oligodendrocytes in vitro. J Neurochem 59:55–61.

66. Negendank WG, Sauter R, Brown TR et al (1996). Proton magnetic resonance spectroscopy in patients with glial tumors: a multicenter study. J Neurosurg 84:449–458.

67. Law M, Yang S, Wang H et al (2003).Glioma grading: sensitivity, specificity and predictive value of perfusion MRI and proton spectroscopic imaging compared with conventional MR imaging. Am J Neuroradiol 24:1989–1998.

68. Haimes AB, Zimmerman RD, Morgello S et al (1989). MR imaging of brain abscesses. Am J Roentgenol 152:1073–1085.

69. Kim YJ, Chang KH, Song IC et al (1998). Brain abscess and necrotic or cystic brain tumor: discrimination with signal intensity on diffusion-weighted MR imaging Am. J Roentgenol 171:1487–1490.

70. Noguchi K, Watanabe N, Nagayoshi T et al (1999). Role of diffusion-weighted echo-planar MRI in distinguishing between brain abscess and tumor: a preliminary report. Neuroradiology 41:171–174.

71. Stadnik TW, Chaskis C, Michotte A et al (2001). Diffusion-weighted MR imaging of intracerebral masses: comparison with conventional MR imaging and histologic findings. Am J Neuroradiol 22:969–976.

72. Batra A, Tripathi RP (2004). Atypical diffusion-weighted magnetic resonance findings in glioblastoma multiforme. Australas Radiol 48:388–391.

73. Tung GA, Evangelista P, Rogg JM, Duncan JA 3rd (2001). Diffusion-weighted MR imaging of rim-enhancing brain masses: is markedly decreased water diffusion specific for brain abscess? Am J Roentgenol 177:709–712.

74. Hakyemez B, Erdoga C, Yildirim N, Parlak M (2005). Glioblastoma multiforme with atypical diffusion-weighted MR findings. Br J Radiol 78:989–992.

75. Holtas S, Geijer B, Stromblad LG, Maly-Sundgren P, Burtscher IM (2000). A ring-enhancing metastasis with central high signal on diffusion-weighted imaging and low apparent diffusion coefficients. Neuroradiology 42:824–827.

76. Lai PH, Li KT, Hsu SS et al (2005). Pyogenic brain abscess: findings from in vivo 1.5-T and 11.7-T in vitro proton MR spectroscopy. Am J Neuroradiol 26:279–288.

77. Krabbe K, Gideon P, Wagn P, Hansen U, Thomsen C, Madsen F (1997). MR diffusion imaging of human intracranial tumors. Neuroradiology 39:483–489.

78. Stadnik TW, Demaerel P, Luypaert RR et al (2003). Imaging tutorial: differential diagnosis of bright lesions on diffusion-weighted MR images. Radiographics 23:e7.

79. Chiang IC, Kuo YT, Lu CY et al (2004). Distinction between high-grade gliomas and solitary metastases using peritumoral 3-T magnetic resonance spectroscopy, diffusion, and perfusion imagings. Neuroradiology 46:619–627.

80. Lu S, Ahn D, Johnson G, Law M, Zagzag D, Grossman RI (2004). Diffusion-tensor MR imaging of intracranial neoplasia and associated peritumoral edema: introduction of the tumor infiltration index. Radiology 232:221–228.

81. Law M, Cha S, Knopp EA, Johnson G, Arnett J, Litt AW (2002). High-grade gliomas and solitary metastases: differentiation by using perfusion and proton spectroscopic MR imaging. Radiology 222:715–721.

82. Guo AC, Cummings TJ, Dash RC, Provenzale JM (2002). Lymphomas and high-grade astrocytomas: comparison of water diffusibility and histologic characteristics. Radiology 224:177–183.

83. Hiwatashi A (2003). Brain neoplasms. In *Diffusion-Weighted MR Imaging of the Brain.* Moritani T, Ekholm S, Westesson P-L (eds). Springer, Berlin. 161–179.

84. Hartmann M, Heiland S, Harting I et al (2003). Distinguishing of primary cerebral lymphoma from high-grade glioma with perfusion-weighted magnetic resonance imaging. Neurosci Lett 338:119–122.

85. Harting I, Hartmann M, Jost G et al (2003). Differentiating primary central nervous system lymphoma from glioma in

humans using localised proton magnetic resonance spectroscopy. Neurosci Lett 342:163–166.

86. Cha S, Pierce S, Knopp EA et al (2001). Dynamic contrast-enhanced T2*-weighted MR imaging of tumefactive demyelinating lesions. Am J Neuroradiol 22:1109–1116.

87. Arnold DL, Matthews PM, Francis GS, O'Connor J, Antel JP (1992). Proton magnetic resonance spectroscopic imaging for metabolic characterization of demyelinating plaques. Ann Neurol 31:235–241.

88. Davie CA, Hawkins CP, Barker GJ (1994). Serial proton magnetic resonance spectroscopy in acute multiple sclerosis lesions. Brain 117:49–58.

89. Mullins ME, Bares GD, Schaefer PW, Hochberg FH, Gonzalez RG, Lev MH (2005). Radiation necrosis versus glioma recurrence: conventional MR imaging clues to diagnosis. Am J Neuroradiol 26:1967–1972.

90. Sugahara T, Korogi Y, Tomiguchi S et al (2000). Post-therapeutic intraaxial brain tumor: the value of perfusion sensitive contrast-enhanced MR imaging for differentiating tumor recurrence from nonneoplastic contrast-enhancing tissue. Am J Neuroradiol 21:901–919.

91. Law M, Hamburger M, Johnson G (2004). Differentiating surgical from non-surgical lesions using perfusion MR imaging and proton MR spectroscopic imaging. Technol Cancer Res Treat 3:557–565.

92. Banerjee AK, Sharma BS, Kak VK, Ghatak NR (1989). Gliosarcoma with cartilage formation. Cancer 63:518–523.

93. Morantz RA, Feigin I, Ransohoff J 3rd (1976). Clinical and pathological study of 24 cases of gliosarcoma. Neurosurgery 45:398–408.

94. Coons SW, Ashby LS (1999). Pathology of intracranial neoplasms. Neuroimaging Clin N Am 9:615–649.

95. Beute BJ, Fobben ES, Hubschmann O, Zablow A, Eanelli T, Solitare GB (1991). Cerebellar gliosarcoma: report of a probable radiation-induced neoplasm. Am J Neuroradiol 12:554–556.

96. Kim DS, Kang SK, Chi JG (1999). Gliosarcoma: a case with unusual epithelial feature. J Korean Med Sci 14:345–350.

97. Meis JM, Martz KL, Nelson JS (1991). Mixed glioblastoma multiforme and sarcoma. A clinicopathologic study of 26 radiation therapy oncology group cases. Cancer 67:2342–2349.

98. Lach M, Wallace CJ, Krcek J, Curry B (1996). Radiation-associated gliosarcoma. Can Assoc Radiol J 47:209–212.

99. Cerame MA, Guthikonda M, Kohli CM (1985). Extraneural metastases in gliosarcoma: a case report and review of the literature. Neurosurgery 17:413–418.

100. Maiuri F, Stella L, Benvenuti D, Giamundo A, Pettinato G (1990). Cerebral gliosarcomas: correlation of computed tomographic findings, surgical aspect, pathological features, and prognosis. Neurosurgery 26:261–267.

101. Sakurai T, Abe J, Hayashi T, Sekino H, Tadokoro M (1993). A case of gliosarcoma associated with large cyst. No Shinkei Geka 21:637–640.

102. Dwyer KW, Naul LG, Hise JH (1996). Gliosarcoma: MR features. J Comput Assist Tomogr 20:719–723.

103. Galanis E, Buckner JC, Dinapoli RP (1998). Clinical outcome of gliosarcoma compared with glioblastoma multiforme: North Central Cancer Treatment Group results. J Neurosurg 89:425–430.

CHAPTER 36

Low-Grade Astrocytomas

Gregory A. Christoforidis

INTRODUCTION

Astrocytic neoplasms include a group of central nervous system tumors derived from astrocytic cells which vary in degree of malignancy, location and degree of infiltration. A variety of tumor types comprise this set of tumors which affect different locations, age groups and gender distributions [1,2]. Astrocytomas can be divided into infiltrative types (75%) and non-infiltrative types (25%). Non-infiltrative astrocytomas include pilocytic astrocytomas, pilomyxoid astrocytomas, pleomorphic xanthoastrocytomas, subependymal giant cell astrocytomas and desmoplastic cerebral astrocytoma of infancy [3–6]. Infiltrative astrocytomas include anaplastic and well-differentiated types [1,2].

The differentiation of low-grade from high-grade astrocytomas relies on certain histopathologic features. These include cellularity, nuclear atypia, mitotic activity, necrosis and vascular endothelial proliferation. According to the Mayo-St. Anne's system of astrocytoma grading, the presence of any mitotic figures, necrosis or vascular proliferation place a tumor in the high-grade category [7]. Low-grade astrocytomas are differentiated from reactive astrocytes or normal brain tissue based on cellularity and size of abnormal astrocytes [1,2,7]. Furthermore, a diverse pattern of genomic lesions can lead to the transformation of low-grade astrocytomas into higher tumor grades [8]. After reviewing imaging methods which can potentially distinguish low- from high-grade astrocytomas, this chapter reviews the imaging characteristics of individual types of low-grade astrocytomas.

PHYSIOLOGIC IMAGING

Many imaging modalities have been employed in recent years in an attempt to differentiate non-invasively high- or low-grade gliomas. Hydrogen proton magnetic resonance spectroscopy (MRS), positron emission tomography (PET) imaging, cerebral blood volume (CBV) mapping and single-photon emission computed tomography (SPECT) are all modalities which have been employed towards this end. Contrast enhancement is often associated with more aggressive tumors, however, this is not always the case (9).

MRS can be used to quantify various metablolites within a sample of tumor tissue. These metabolites include N-acetylaspartate (NAA), a neuronal marker, choline, a cell membrane component, lactate, a marker for glycolysis and necrosis, creatine, a marker for energy metabolism and myo-inositol, a metabolite that is primarily found in astrocytes and has been shown to be more abundant in low-grade gliomas relative to higher-grade gliomas [10]. Hydrogen MRS of brain tumors has been studied for over a decade now [11]. In general, primary brain tumors demonstrate reduced levels of NAA and increased levels of choline relative to normal brain tissue. Elevated choline levels are thought to represent areas of increased tumor cellularity and proliferative activity [12–16], whereas decreased NAA is thought to represent decreased neuronal density and viability found in gliomas [13,14]. Elevated levels of lactate have been identified in higher-grade tumors, whereas no elevation of lactate peak has been associated with low grade tumors [17,18]. There is confounding evidence to support that lactate may not be as accurate an indicator of malignancy, however [15,19].

Contrast reagents have been used to characterize brain tumors. Enhancement of tumors has been shown to depend on the degree of disruption of the blood–brain barrier as well as tumoral microvascularity. Histologic studies demonstrate that astrocytomas can infiltrate well past the margins of enhancement [20]. The extent and pattern of enhancement can vary depending on the type of contrast agent used, the use of corticosteroid, pulse sequence used, dosage of contrast reagent used and field strength [21–23]. As field strength increases both T1 and T1 shortening increase. As a result the degree of contrast enhancement increases as well. Tumor to brain contrast after gadolinium administration at 3T imaging compared to 1.5T has been shown to be significantly higher [24]. Contrast-to-noise ratio (CNR) has been shown to be 2.8 times higher than at 1.5 T the same contrast dosing, and 1.3 times when using half the contrast dose at 3.0T and full dose at 1.5T. Extent of contrast enhancement increases with contrast dose [25]. The time course of MR imaging following contrast administration affects signal intensity. Peak signal intensity occurs at

25–35 minutes, however, there is no significant change that occurs after 5 minutes. It is generally accepted that contrast enhanced MR imaging brain tumors should not begin until 2–5 minutes after contrast administration [26]. Comparison of gadolinium dimeglumine versus gadopentate dimeglumine contrast reagents in a multisite study has shown that gadopentate was superior in both qualitative and quantitative assessments of brain to tumor contrast [27]. Contrast enhancement has been shown to occur in 20% of low-grade gliomas, whereas lack of contrast enhancement gives a two in three chance of low tumor grade [22,23]. As a result alternate imaging methods predictive of tumor grade have been studied.

Dynamic MR imaging studies attempt to characterize tumor vascular dynamics by imaging differences in the signal intensity of a tumor as contrast reagents course through the vascular bed using T1-weighted, T2-weighted and T2*-weighted techniques. Contrast leakage across a vascular bed is influenced by vascular permeability, contrast binding within the vascular bed, hematocrit and contrast reagent concentration within the blood. The choice of post-processing methods influences the signal-to-noise ratio and the overall precision and accuracy obtained with dynamic contrast imaging [28]. This information can then be used to study the cerebral blood volume, blood flow, blood–brain barrier permeability and water exchange kinetics of the neoplasm relative to normal brain tissue. Of these parameters, relative cerebral blood volume mapping has received the most attention in characterizing tumor grade. These methods have been used as surrogate markers for microvascularity. Relative cerebral blood volume (rCBV) mapping uses MRI perfusion imaging techniques in order to construct a map of the relative perfusion of blood within the brain. Animal studies have indicated that rCBV corresponds to vascular concentration of contrast agent [29,30]. Areas of low rCBV within a tumor, compared to normal white matter or gray matter, have been suggested to correlate with lower tumor grade whereas areas of higher perfusion correlate with higher grade [29,30]. As a result surgical biopsies may be directed toward more aggressive areas of the tumor using this information. MR rCBV maps have also been shown to correlate with vascular endothelial growth factor (VEGF) expression in non-enhancing gliomas and tumor grade [21]. Finally, rCBV measurements have been shown to assist in identifying low-grade gliomas that are either high-grade gliomas misdiagnosed as a result of sampling error, or low-grade gliomas undergoing transformation to a higher grade [31]. Vessel permeability within the vascular bed of an astrocytoma has been suggested to relate to the presence of vascular endothelial growth factor and angiogenesis [32]. Indeed, dynamic contrast enhancement methods using T1 and T2* techniques have been used to assess vascular permeability within a tumor bed and have been shown to correlate to tumor grade as well [33]. It is noteworthy, that although use of corticosteroids is known to affect peritumoral edema, blood flow within edematous tissue and contrast enhancement, the influence of dexamethasone use on blood flow within enhancing tumor bed may not be significant [34].

Diffusion-weighted imaging (DWI) techniques have allowed insight in characterizing tumor behavior. Isotropic DWI characterizes diffusion in three directions. Diffusion tensor imaging is able to display and quantify the relationship between intra-axial tumors and adjacent white matter fiber tracts in multiple directions. This method allows the post-processed formation of diffusion tractography, which has been shown to display tumor white matter infiltration patterns. Preoperative identification of the anatomic relationship between the neoplastic tissue and eloquent cortical and white matter regions may assist the surgeon when planning the extent to which a brain tumor can be removed and preserve functionality. It has been hypothesized that identification of these patterns may be able to differentiate tumor types [35]. DWI provides information on diffusion of water within the tumor bed and can be used to calculate the apparent diffusion coefficients within regions of interest. The DWI appearance of a tumor has been suggested to be influenced by tumor cellularity and total nuclear area and can thus have the potential to assist in the characterization and grading of brain tumors [36]. High-grade gliomas have been reported to be associated with decreased apparent diffusion coefficient (ADC) values. Although in theory ADC values may assist the differentiation of high- from low-grade astrocytomas or help distinguish non-gliomatous tumors from metastases or lymphoma, this has not always been shown to be consistently reliable [36]. Some authors suggest that ADC maps can help determine the extent of tumor invasion, however, this remains controversial. Some preliminary evidence indicates that ADC mapping may help in the assessment of brain tumor response to therapy and differentiate radiation necrosis from tumor recurrence [36]. Fractional anisotropy (FA) is a measure of water diffusion favoring one direction over others. Because FA values differ depending on location, they are compared to similar regions of interest in the normal appearing brain in the contralateral hemisphere. Unlike ADC values, FA values have been shown to differ in peritumoral signal abnormality associated with infiltrating tumors such as gliomas when compared to non-infiltrating tumors such as meningiomas [37]. Although FA values have been found to be low within the tumor beds of both high- and low-grade gliomas, interrogation of the central portion of the tumor compared with the peripheral portion of the tumor appear to differ slightly in low- but not high-grade gliomas [38]. This suggests greater disorganization within the peripheral portion of tumor bed of high-grade gliomas than low-grade gliomas. It is critical to point out that measurements obtained via diffusion-weighted imaging methods could potentially be influenced by therapeutic effects such as radiation, chemotherapy and steroid use.

PET imaging has also been used to separate high- from low-grade tumors by distingushing hypometabolic from hypermetabolic tumors. Metabolic activity within brain tumors using PET has been studied extensively with a large number of markers. Applications include measurement of blood flow, blood volume, oxygen use, glucose use, glucose transport, amino acid uptake, protein synthesis, blood–brain barrier integrity, cerebral pH, membrane metabolism and nucleic acid synthesis [39]. Tumor grade has been studied by either measuring glucose uptake using 18F-fluoro-deoxyglucose (FDG) or amino acid uptake by using 11C-methylmethionine. Low-grade gliomas demonstrate normal or lower glucose or amino acid uptake relative to normal brain tissue [11,13,39]. PET imaging of astrocytomas with FDG has been shown to be predictive of tumor progression and able to differentiate tumor recurrence from radiation necrosis. More recently PET imaging with VEGF has been shown to be a promising agent which may be able to demonstrate angiogenesis in a manner which differs from relative cerebral blood volume mapping which correlates with microvascularity [40].

SPECT imaging of brain tumors has been studied predominantly with thallium-201 (201-Tl) and technetium-99m hexamethyl propyleneamine oxime (99mTc- HMPAO). Because thallium behaves biologically like potassium, imaging of 201-thalium can act as a marker for the function of the sodium-potassium pump. Thallium can thus act as a measure of metabolic activity. Using the ratio of 201-thallium uptake in the tumor in question relative to normal brain has been used to predict which tumors are low-grade [41,42]. Furthermore, thallium can be used to differentiate tumor necrosis from recurrent tumor [41]. 99mTC-HMPAO can act as a measure of tissue perfusion. Much like perfusion MRI, tumor recurrence shows an increased uptake of 99mTc-HMPAO and can help increase the accuracy of 201-Tl-SPECT differentiation of recurrent tumor from radiation necrosis.

WELL DIFFERENTIATED DIFFUSE INFILTRATIVE ASTROCYTOMA

Well differentiated diffuse infiltrative astrocytomas represent WHO classification grade II astrocytomas [1,2] and comprise approximately 25% of all gliomas. They are characterized by a high degree of cellular differentiation located in an environment of neuroglial fibrils and are often accompanied by degenerative microscopic cysts. Mean age at presentation is 34 years with a slight male predilection. Approximately 10% occur under the age of 20 and 30% over the age of 45 [2]. The most frequently encountered subtype is the fibrillary astrocytoma. Other subtypes include the gemistocytic astrocytoma which requires a minimum 20% fraction of gemistocytic astrocytes and may be associated with large cysts and particularly prone to transformation to higher grade. The protoplasmic astrocytoma is a rare variant with low cellular activity, with frequent mucoid degeneration and microcyst formation.

On macroscopic examination these tumors have a firm rubbery consistency, which expands the involved brain, sometimes containing calcifications. Macroscopic cysts and microcysts can occur with these tumors and, if extensive, may give them a gelatinous appearance [2]. They infiltrate adjacent and distant brain irrespective of histologic grade and often have a tendency to progress to a more malignant phenotype. Progression to higher grade (Figure 36.1) can result from a variety of pathways, including the overexpression of proteins involved with cellular proliferation, the inactivation of genes responsible for apoptosis and the deregulation of the normal cell cycle [8]. Markers for proliferative activity such as MIB-1 index and molecular markers such as p53 overexpression have been found useful in predicting which low-grade diffuse astrocytomas will transform into higher grades [2,8,43].

They most frequently arise from the supratentorial brain and the brainstem. Supratentorial location is the most frequent location in all age groups and they have a special predilection for the frontal and temporal lobes [2]. Brainstem involvement, is the second most frequent site of involvement but more frequently seen with children. Cerebellar location is rare. Brainstem location is often associated with poorer prognosis (Figure 36.2), in part due to limited ability to completely resect tumors in such locations. Tectal plate lesions often produce symptoms earlier due to the potential to obstruct the aqueduct of Sylvius and produce hydrocephalus. In this location, low-grade astrocytomas can be distinguished from other lesions, such as hamartomas, based on size, with most lesions under 4 ml in size tending to be hamartomas, whereas most lesions over 10 ml volume tend to be neoplastic [44].

On imaging these tumors appear homogeneous. They can be difficult to detect on CT and may manifest with only a slight density difference and usually do not enhance (see Figure 36.1). MRI is more sensitive than CT in detecting these tumors [11,45–47]. It has been well documented that these tumors may extend well beyond the margins of the tumor as identified on MR [48–50]. Indeed, a threshold cellular density has been calculated in determining the detectability of these tumors on MRI [51]. They display a low signal on T1WI and a higher signal on T2WI with little or no enhancement following gadolinium administration, however, because contrast enhancement has been reported in up to 40% of cases, it is not thought to be a reliable marker for high histopathologic grade [43,48,52]. CT reveals tumoral calcifications in 20% of cases whereas none are identified on MRI [43,53]. Well differentiated diffuse astrocytomas may spread to gray matter and may have cystic foci on imaging, but typically lack peritumoral edema and have nearly no mass effect [47,48]. When the tumor infiltrates the cortex it may be confused with infarction or cortical based tumors such as oligodendroglioma or ganglion cell tumor.

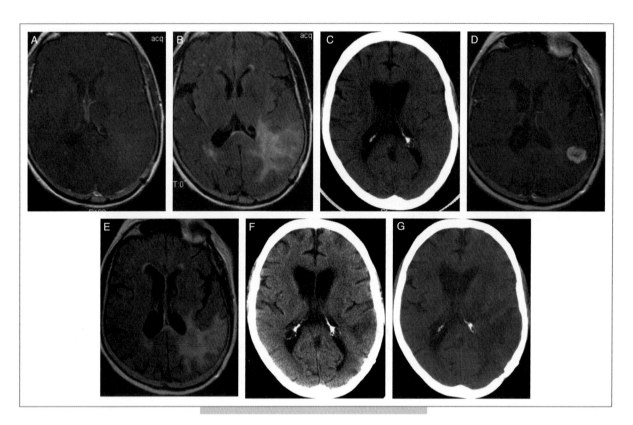

FIG. 36.1. Anaplastic transformation of low grade astrocytoma. A 61-year-old female patient who presented with aphasia was discovered to have a WHO grade II astrocytoma involving left temporal and parietal lobes. The lesion did not enhance on gadolinium MRI **(A)** but demonstrates extensive signal abnormality along on FLAIR MRI **(B)**. The tumor is much less conspicuous on a contrast enhanced CT of the head **(C)**. She underwent conformal radiation and chemotherapy. One year later the imaging characteristics of the tumor changed. The tumor enhanced **(D)**, but the extent of the signal abnormality on FLAIR MRI did not change significantly **(E)**. There was some hypodensity noted on CT within the tumor bed **(F)**. Three months later the tumor showed significant progression on CT **(G)** and the patient died shortly thereafter.

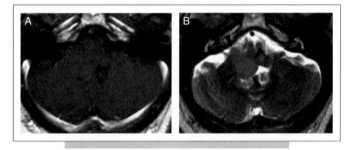

FIG. 36.2. Brainstem well differentiated diffuse astrocytoma. A 43-year-old male patient presenting with difficult swallowing, weakness and tremors was found to have a WHO grade II astrocytoma which involved the right middle cerebellar peduncle and the medulla oblongata. The tumor did not enhance following gadolinium administration **(A)** and is better depicted on the T2 weighted MRI **(B)** (arrowheads).

Imaging has been shown to have the potential to predict progression of WHO grade II astrocytomas to higher grades or poor outcome. Larger tumor volumes and involvement of more than one lobe as depicted on T2 MRI have been shown to be associated with worse outcomes and transformation to higher grades [54]. In a study of 35 patients, relative cerebral blood volume maps were shown to predict progression as an adjunct to histopathologic findings [55]. Higher relative cerebral blood volumes are thought to be associated with microvascularity and predictive of angiogenesis. Inconsistencies between histopathology and dynamic susceptibility contrast enhanced MRI may be attributable to misdiagnosis or sampling error on biopsy. In a study of 25 patients, fibrillary WHO grade II astrocytomas associated with higher regional cerebral blood volumes were shown to be associated with earlier recurrence following fractionated stereotactic radiotherapy relative to those with lower regional cerebral blood volumes (Figure 36.3) [56]. Sudden change in contrast enhancement may be an indication

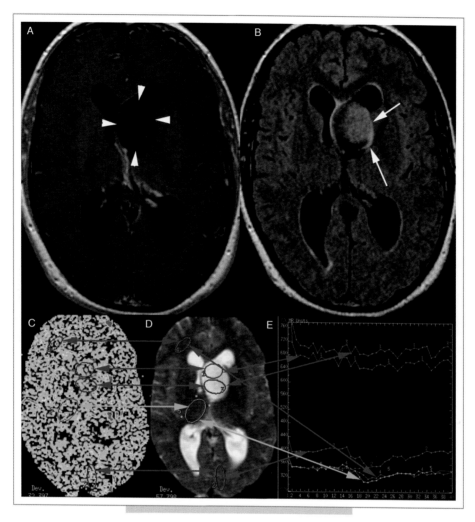

FIG. 36.3. Exophytic low grade astrocytoma near the foramen of Monro. A 35-year-old male who presented with headaches and nausea was found to have a mass on MRI (arrowheads A) associated with ventriculomegaly. It did not enhance **(A)** and appeared to arise from the region of the foramen of Monro. Its lack of enhancement and heterogeneity suggested against most tumors that typically occur in this location such as subependymal giant cell astrocytoma and central neurocytoma. Subependymomas, although they do not enhance and can have a similar appearance, are unusual in this age group. T2 echo planar perfusion studies with cerebral blood volume maps **(C)** and regions of interest **(D)** depicted on time activity curves **(E)** indicate low microvascularity suggestive of low-grade neoplasm. Biopsy revealed WHO grade II diffuse astrocytoma on MRI.

of conversion to higher grade (Figure 36.1) or radiation necrosis.

NON-INFILTRATIVE DIFFUSE ASTROCYTOMAS

These tumors are relatively well circumscribed when compared to diffuse astrocytomas. In general they have a good prognosis. Several subtypes have been defined.

Pilocytic Astrocytoma

Pilocytic astrocytomas usually present in the first two decades of life. They often arise from the hypothalamus, optic nerve, optic chiasm, brainstem, cerebellum or, less commonly, in the cerebral hemispheres. They are included in the few tumors that characteristically arise from the corpus callosum. These tumors are classified as WHO type I neoplasms, classically composed of compact and spongy tissue. The name 'pilocytic' (directly translated as 'hair cell') is derived from the long hair-like projections originating from the neoplastic astrocytes. Spongiform tissue contains microcysts. Pilocytic astrocytomas are slow growing and have been found to stabilize spontaneously or regress. Microvascular proliferation is a frequent feature of pilocytic astrocytomas and accounts for the contrast enhancement accompanying these tumors on cross-sectional imaging.

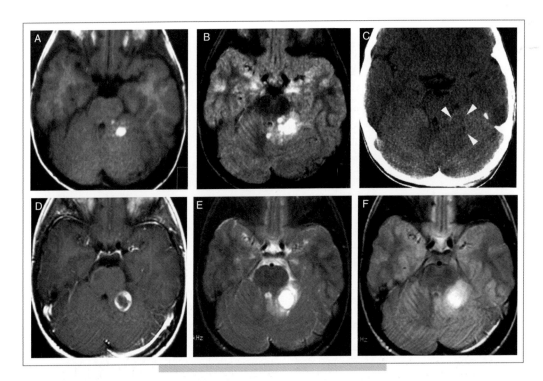

FIG. 36.4. Pilocytic astrocytoma with cystic generation. A 9-year-old boy presenting with headache due to a left cerebellar neoplasm. Gadolinium-enhanced T1 MRI demonstrated an enhancing focus within the tumor as well as T2 signal abnormality beyond the enhancing borders of the tumor (**A, B**). The tumor is less conspicuous on CT (**C**, arrowheads). One year later, the tumor developed a cystic component with ring enhancement on gadolinium enhanced T1 imaging (**D**), which follows cerebrospinal fluid intensity on T2 (**E**) but not proton density (**F**) imaging.

Other histopathologic features which may accompany this tumor include Rosenthal fibers, eosinophilic granular bodies and ganglion cells. Rosenthal fibers represent filaments found in the cell processes of these tumor cells, but can also be seen with reactive astrogliosis. The presence of esosinophilic granular bodies in these tumors is considered an important marker for low-grade neoplasms such as this one. These tumors exhibit a paucity of consistent genetic abnormalities predictive of outcome [8]. It is, however, interesting that these tumors may display abnormalities on chromosome 17 on the same region as neurofibromatosis type 1 (NF1) gene and patients with NF1 have a higher incidence of pilocytic astrocytomas [57]. Pilocytic astrocytomas may spread locally but are occasionally found to spread via CSF [3,43,58]. Malignant transformation has been reported with pilocytic astrocytomas. The acquisition of histopathologic features suggestive of malignant transformation, such as multiple mitoses per high power field or palisading necrosis, has been reported to occur, however, the outcome is not always grim in such cases [3]. Indeed, proliferative indices such as MIB-1, ranging up to 4%, and TP53 genetic mutations which have been shown to have poor a prognosis in diffuse astrocytomas, do not have the same implication for pilocytic astrocytomas [3].

On cross-sectional imaging they are identified as well circumscribed tumors (Figures 36.4–36.8). Classically, they present as a cystic tumor with an enhancing mural nodule (Figures 36.4 and 36.5). Cystic components are identified in approximatley 68% of cases and may develop on follow-up exam (Figure 36.4). In general, chiasmatic location is associated with a lesser incidence of cyst formation (Figure 36.7). On CT, the solid component of the tumor typically appears hypodense (43%) or isodense (51%) or, less commonly, hyperdense (6%). Calcifications have been reported in approximately 11% of CT cases (Figure 36.6). On MRI, they present as well circumscribed tumors (96%) with benign morphologic features and rare evidence for vasogenic edema pattern (5%). On MRI, they display contrast enhancement in 94% of cases, which is thought to be related to the vascular nature of these tumors (Figure 36.7). Despite the presence of increased vascularity in these tumors, this is known not to be a sign for higher grade in this tumor type. On T1-weighted imaging they tend to be lower signal intensity relative to gray matter. On T2-weighted imaging they are of higher signal relative to gray matter. The T2 signal intensity of the solid component is similar to that of cerebrospinal fluid in 50% of cases. This differs from medulloblastomas, for example, which are

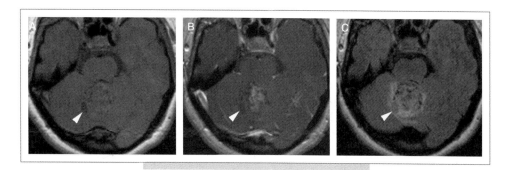

FIG. 36.5. Cerebellar pilocytic astrocytoma. A 36-year-old female presenting with headache. T1 **(A)**, post-gadolinium T1 **(B)** and FLAIR **(C)** MR images demonstrate a well-circumscribed heterogeneously enhancing pilocytic astrocytoma. A cyst-like component is identified with an arrowhead.

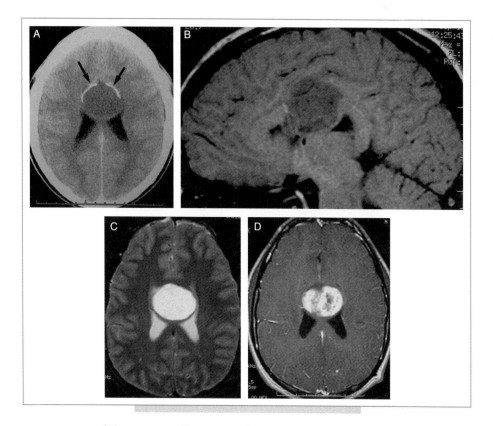

FIG. 36.6. Pilocytic astrocytoma of the corpus callosum. Axial CT **(A)**, sagittal T1 **(B)**, axial T2 **(C)** and axial post-gadolinium T1 **(D)** MR images of a pilocytic astrocytoma centered in the corpus callosum. Note the calcific focus which occasionally accompanies this tumor type (arrows). Pilocytic astrocytomas differ from other tumors involving the corpus callosum because they are usually well circumscribed and not typically associated with vasogenic edema.

isointense to gray matter and can thus help distinguish medulloblastomas from pilocytic astrocytomas [59]. Varying imaging presentations may include solid tumor as well as multicystic tumors. In most locations they have a round or oval shape, however, in the chiasmatic location, they have been found to have a multilobulated shape.

MR spectroscopy of these tumors can reveal an elevated lactate peak in these tumors, however, this is not a sign of malignancy for pilocytic astrocytomas [19]. In general, they are distinguished from other cystic tumors on the basis of location and patient age [3,11,43,60,61]. Although involvement of the subarachnoid space is frequent with pilocytic

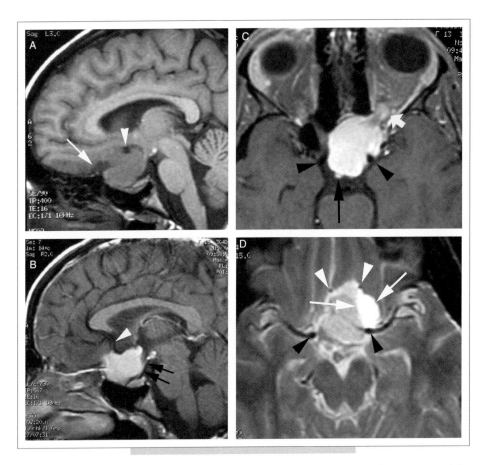

FIG. 36.7. Pilocytic astrocytoma involving the hypothalamus and optic nerves. T1 parasagittal **(A)**, post-gadolinium T1 midline sagittal **(B)**, post-gadolinium T1 axial image through the orbits **(C)** and axial T2 image through the suprasellar region **(D)** demonstrate a pilocytic astrocytoma that has a cystic component (white arrows) and is isointense to gray matter on T1 and slightly hyperintense relative to gray matter on T2-weighted images. Note the tumor's relationship to the anterior cerebral arteries (white arrowheads), the distal internal carotid arteries (black arrowheads) and the infundibular stalk of the pituitary as well as growth of the tumor along the optic nerve (wide white arrow).

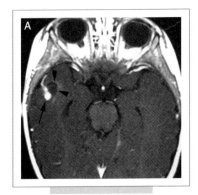

FIG. 36.8. Pilocytic astrocytoma involving the temporal lobe. Pilocytic astrocytomas can occasionally arise from the supratentorial brain as depicted on this T1 post-gadolinium MRI image. This cyst **(A, arrowheads)** and enhancing mural nodule (arrow) appearance in this location can also be seen with gangliogliomas and pleomorphic xanthoastrocytomas.

astrocytoma, leptomeningeal dissemination is rare [3,62]. Pilocytic astrocytomas with atypical features on histopathology may display aggressive features such as invasion of adjacent structures and rapid growth on imaging (Figure 36.9).

Pilomyxoid Astrocytoma

Pilomyxoid astrocytoma (PMA) is a low-grade astrocytoma previously considered to be classified as a pilocytic astrocytoma. This group of astrocytomas has a monophasic pilomyxoid angiocentric histologic pattern, lacks the Rosenthal fibers that are characteristic of pilocytic astrocytomas and rarely have eosinophilic granular bodies. They tend to occur in early childhood, but have been identified in adults [63]. After resection, progression-free survival has been reported in 38.7% of cases [64,65]. These tumors tend to occur in the hypothalamic/chiasmatic region (Figure 36.10) and are likely to be solid enhancing tumors

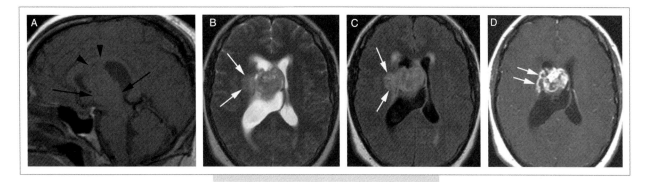

FIG. 36.9. Atypical pilocytic astrocytoma. Sagittal T1 **(A)**, axial T2 **(B)**, axial FLAIR **(C)** and axial post-gadolinium T1 **(D)** images of a pilocytic astrocytoma which displayed aggressive features on histopathology including an increase in proliferative activity. The neoplasm spans between the corpus callosum (black arrowheads) and the thalamus (black arrows). There is apparent invasion of adjacent centrum semiovale (white arrows) suggestive of a more aggressive neoplasm.

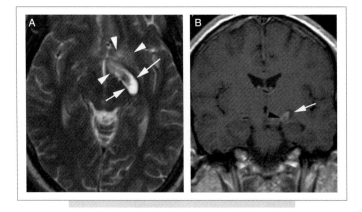

FIG. 36.10. Pilomyxoid astrocytoma. T2 axial **(A)** and post-gadolinium coronal T1 **(B)** images through the brain of a 36-year-old male with a hypothalamic cystic lesion. Note the cystic component (arrows) and the infiltration within adjacent brain **(B)**.

with homogenously high signal on T2-weighted images with signal extending into the adjacent white matter and cerebrospinal fluid dissemination. The larger and homogenously enhancing solid component, T2 signal abnormality extending into the gray matter and white matter, CSF dissemination and attendant hydrocephalus help distinguish PMA from pilocytic astrocytomas. Proton MRS of PMA in two cases reported in the literature, have shown a decreased concentration of choline, creatinine and NAA which differs from pilocytic astrocytomas which tend to display elevated choline and decreased creatine and NAA. Because it is only recently described, imaging findings are not frequently reported in the literature [66].

Pleomorphic Xanthoastrocytoma

Pleomorphic xanthoastrocytomas (PXAs) are tumors with a varied histologic appearance. They are generally classified as WHO grade II neoplasms, but have been known to undergo malignant transformation. In general, the extent of nuclear pleomorphism associated with these tumors places them in the St Anne-Mayo grade II category. Presence of mitotic activity elevates these tumors to an anaplastic grade. As the name implies, histopathologic features distinguishing pleomorphic xanthoastrocytomas include spindly cells, as well as mononucleated or multinucleated giant cells with intracellular accumulation of lipid droplets often surrounded by a thick reticulen laden basement membrane. Gross evaluation indicates that they involve the leptomeninges as well as the underlying brain. Their peripheral location suggests that they arise from subpial astrocytes [67]. These tumors have been rarely known to transform into gangliogliomas [4,7,68,69]. If only a small amount of tumor is provided to the pathologist for analysis, PXA may be confused with a glioblastoma. MRI and CT appearance of the tumor may assist pathologic interpretation of this tumor [70].

Like pilocytic astrocytomas, PXAs occur more frequently in the first two decades of life. Unlike pilocytic astrocytomas they occur more commonly in the cerebral hemispheres with a predilection in the temporal lobes. On imaging, PXAs have a cyst and mural nodule appearance and are difficult to distinguish from gangliogliomas (Figure 36.11). Imaging features of low-grade such as lack of peritumoral edema and calvarial scalloping frequently accompany these tumors. On CT, these tumors have a cystic appearance in half the cases and rarely contain calcifications. Angiography reveals that these tumors are hypervascular and receive supply from the meningeal arteries [71]. MRI reveals that relative to gray matter the solid component of these tumors is of similar signal intensity on T1-weighted images and increased signal on T2-weighted sequences [71,72]. Differentiation of high-grade from low-grade PXA is confounding. Perfusion MRI may show some promise, however (Figure 36.12).

Because of their peripheral location, they can be confused with meningioma on imaging [11,71]. Other differential considerations when identifying a tumor with characteristics

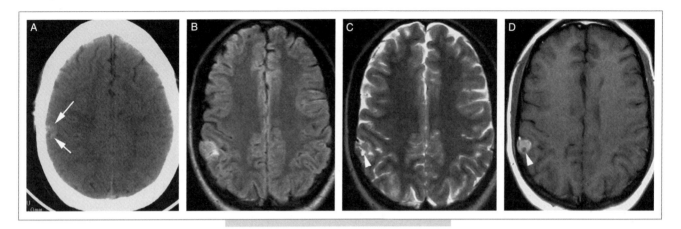

FIG. 36.11. Pleomorphic xanthoastrocytoma. A 25-year-old male presenting with seizure. Axial CT **(A)**, axial FLAIR MRI **(B)**, axial T2 MRI **(C)**, and axial post-gadolinium T1 MRI **(D)** of a low-grade pleomorphic xanthoastrocytoma (PXA) located in the right parietal lobe cortex. Note the hyperdense appearance on CT (arrow). Distinguishing features for this tumor type include a well-circumscribed appearance, cyst formation (arrowheads), cortical location and contrast enhancement. It would be difficult to distinguish this tumor from other cortically based tumors such as metastasis or ganglioglioma. Because of frequent association with a dural tail they may appear similar to cystic meningiomas.

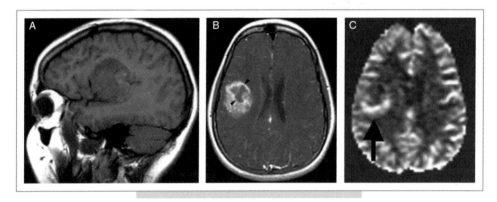

FIG. 36.12. Pleomorphic xanthoastrocytoma. Sagittal T1 **(A)**, axial post-gadolinium T1 **(B)** and axial relative cerebral blood volume (rCBV) MRI map **(C)** of a WHO grade III pleomorphic xanthoastrocytoma centered in the inferior right frontal gyrus. Note the cortical location, the presence of a cystic-appearing component (arrowheads), and an enhancing component as well as isointense signal relative to gray matter on T1, all features of PXA. Because the tumor is well circumscribed and not associated with vasogenic edema, it is not readily distinguished from a low-grade pleomorphic xanthoasrocytomas on conventional MRI. Although, higher cerebral blood volume in a portion of the tumor (arrow) is suggestive of higher grade in astrocytomas, in a PXA it may just be an indication of increased vascularity unrelated to grade.

of PXA include pilocytic astrocytoma, ganglioglioma, and oligodendroglioma.

Subependymasl Giant Cell Astrocytoma

Subependymal giant cell astrocytoma (SGA) is a low-grade primary brain tumor assigned a WHO grade I classification. As the name implies, these tumors are composed of large ganglioid astrocytes which are located along the wall of the lateral ventricle. They invariably occur in the setting of tuberous sclerosis and affect the region near the foramen of Monro eventually obstructing this structure and causing hydrocephalus. In general, more aggressive histopathologic features such as mitosis and cellular pleomorphism in these neoplasms have not been associated with shorter survival times as they would in other tumors [5,7,43,68,69].

Subependymal giant cell astrocytomas occur in 6–16% of patients with tuberous sclerosis. In addition to identifying this tumor in its typical location near the foramen of Monro, the identification of stigmata related to tuberous sclerosis in the same patient confirms the diagnosis of SGA [55,73–77]. Neonatal ultrasound imaging has identified this tumor as one of the few brain tumors which may be identifed at birth and should therefore be included in the differential diagnosis of neonatal tumors when appropriate. The mass tends to be isoechoic with hyperechoic foci representing calcification or hemorrhage. Intrinsic CT features

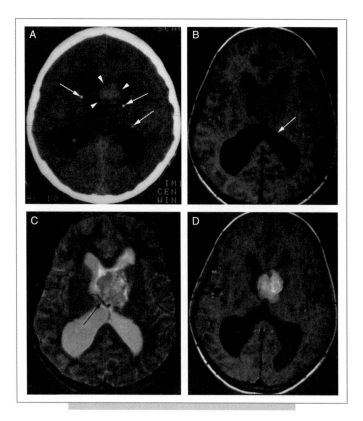

FIG. 36.13. Subependymal giant cell astrocytoma. Axial CT **(A)**, axial T1 MRI **(B)**, axial T2 MRI **(C)** and axial post-gadolinium T1 MRI **(D)** of a subependymal giant cell astrocytoma (SGA) located in the right foramen of Monro (arrowheads) in a patient with tuberous sclerosis. Note that subependymal nodules (white arrows) are more conspicuous on the CT examination than the MRI. Note the relationship of the tumor to the internal cerebral vein (black arrow) which may affect surgical planning. The tumor is hyperdense on CT, has a heterogeneous appearance on T2 and T1 MRI and enhances following contrast on MRI. These features although typical of SGA are also features found in most lateral ventricular tumors. The diagnosis is made on the basis of stigmata of tuberous sclerosis such as the subependymal nodules.

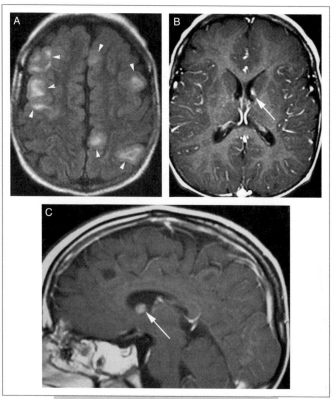

FIG. 36.14. Cortical tubers with supependymal giant cell astrocytoma. FLAIR **(A)**, post-gadolinium axial **(B)** and sagittal **(C)** T1 MRI of a cognitively impaired patient with seizures and adenomata sebaceum. Among pulse sequences FLAIR MRI appears to display cortical tubers (arrowheads) most effectively. CT is less sensitive to detecting cortical tubers. A 5 mm enhancing lesion near the foramen of Monro (arrows) most likely represents a subependymal giant cell astrocytoma. Due to their propensity to enlarge, aggressive follow-up MRI or surgical removal is often suggested.

include the presence of calcifications and a hyperdense appearance relative to cortex (Figure 36.13). Tumoral calcifications are thought to relate to small areas of hemorrhage. Contrast enhancement is common with these tumors on both CT and MRI. Associated stigmata of tuberous sclerosis include the presence of cortical tubers and subependymal nodules. In general, cortical tubers are more readily apparent on MRI (Figure 36.14), whereas calcified subependymal nodules are more readily identified on CT (Figure 36.13) [55,74,75]. The extent of brain involvement with cortical tubers has been shown to correlate with the severity of disease in these patients [55, 75,76]. Patients with tuberous sclerosis likely benefit from annual surveillance for these tumors during childhood [75]. Patients with biopsy-proven asymptomatic subependymal giant cell astrocytoma who

do not undergo resection should be imaged annually. If the tumor grows or changes enhancement pattern, this may serve as an indication to increase the frequency of imaging surveillance or to surgical removal of the tumor [78]. It should be underscored that larger tumors near the foramen of Monro and symptomatic presentation is associated with higher morbidity. It has been suggested that any lesion near the foramen of Monro, greater than 5 mm in size, with incomplete calcification and increasing in size should be removed as soon as there is clear evidence of growth on subsequent scans [79]. In this manner, early resection of these tumors when they arise results in improved overall outcome. Differential considerations for this tumor on imaging include other intraventricular tumors such as central neurocytoma, metastasis, oligodendroglioma, pilocytic astrocytoma and meningioma. SGA can be distinguished from these tumors on imaging by the identification of features of tuberous sclerosis as described above [77].

REFERENCES

1. Cavanee WK, Furnari FB, Nagane M et al (2000). Diffusely infiltrating astrocytomas. In *Pathology and Genetics of Tumours of the Nervous System*. Kleihues P, Cavanee WK (eds). International Agency for Research on Cancer Press, Lyon. 9–21.

2. Kleihues P, Davis RL, Ohgaki H, Burger PC, Westphal MM, Cavanee WK (2000). Diffuse astrocytoma. In *Pathology and Genetics of Tumours of the Nervous System*. Kleihues P, Cavanee WK (eds). International Agency for Research on Cancer Press, Lyon. 22–26.

3. Burger PC, Sheithauer BW, Paulus W et al (2000). *Pathology and Genetics of Tumours of the Nervous System*. Kleihues P, Cavanee WK (eds). International Agency for Research on Cancer Press, Lyon. 45–51.

4. Kepes JJ, Louis DN, Giannini C, Paulus W (2000). Pleomorphic xanthoastrocytoma. In *Pathology and Genetics of Tumours of the Nervous System*. Kleihues P, Cavanee WK (eds). International Agency for Research on Cancer Press, Lyon. 52–54.

5. Wiestler OD, Lopes BS, Green AJ, Vinters HV (2000). Tuberous sclerosis complex and subependymal giant cell astrocytoma. In *Pathology and Genetics of Tumours of the Nervous System*. Kleihues P, Cavanee WK (eds). International Agency for Research on Cancer Press, Lyon. 227–230.

6. Taruto AL, VandenBerg SR, Rorke LB (2000). Despoplastic infantile astrocytoma and ganglioglioma. In *Pathology and Genetics of Tumours of the Nervous System*. Kleihues P, Cavanee WK (eds). International Agency for Research on Cancer Press, Lyon. 99–102.

7. Daumas-Duport C, Scheihauer B, O'Fallon J, Kelly P (1988). Grading of astrocytomas. A simple and reproducible method. Cancer 62:2152–2165.

8. Pollack IF, Hamilton RL, Finkelstein SD, Lieberman F (2002). Molecular abnormalities and correlations with tumor response and outcome in glioma patients. Neuroimaging Clin N Am 12:627–639.

9. Chamberlain MC, Murovic JA, Levin VA (1988). Absence of contrast enhancement on CT brain scans of patients with supratentorial malignant gliomas. Neurology 38:1371–1374.

10. Castillo M, Smith JK, Kwock L (2000). Correlation of myo-inositol levels and grading of cerebral astrocytomas. Am J Neuroradiol 21:1645–1649.

11. Sanders WP, Christoforidis GA (1999). Imaging of low-grade primary brain tumors. In *The Practical Management of Low-Grade Primary Brain Tumors*. Rock JP, Rosenblum ML, Shaw EG, Caincross JG (eds). Lippincott Williams & Wilkins, Philadelphia. 5–32.

12. Gupta RK, Sinha U, Cloughesy TF et al (1999). Inverse correlation between choline magnetic resonance spectroscopy signal intensity and the apparent diffusion coefficient in human glioma. Magn Reson Med 41:2–7.

13. Pomper MG, Port JD (2000). New techniques in MR imaging of brain tumors. MRI Clin N Am 8:691–713.

14. Castillo M, Kwock L (1998). Proton MR spectroscopy of common brain tumors. Neuroimaging Clin N Am 8:733–752.

15. Ott D, Hening J, Ernst T (1993). Human brain tumors: assessment with in vivo proton MR spectroscopy. Radiology 186:745–752.

16. Shimizu H, Kumabe T, Shirane R, Yoshimoto T (2000). Correlation between level measured by proton MR spectroscopy and Ki-67 labeling index in gliomas. Am J Neuroradiol 21:659–665.

17. Fulham MJ, Bizzi A, Dietz MJ et al (1992). Mapping of brain tumor metabolites with proton spectroscopic imaging: clinical relevance. Radiology 185:675–686.

18. Gotsis ED, Fountas K, Kapsalaki E, Toulas P, Peristeris G, Papadakis N (1996). In vivo proton MR spectroscopy: the diagnostic possibilities of lipid resonances in brain tumors. Anticancer Res 16:1565–1568.

19. Hwang JH, Egnaczyk GF, Ballard E, Dunn RS, Holland SK, Ball WS (1998). Proton MR spectroscopic characteristics of pediatric pilocytic astrocytomas. Am J Neuroradiol 19:535–540.

20. Earnest F, Kelly PJ, Sheithauer BW (1988). Cerebral astrocytomas: histopathologic correlation of MR and CT contrast enhancement with stereotactic biopsy. Radiology 166:823–827.

21. Maia ACM, Malheiros SMF, da Rocha AJ et al (2005). MR cerebral blood volume maps correlated with vascular endothelial growth factor expression and tumor grade in nonenhancing gliomas. Am J Neuroradiol 26:777–783.

22. Barker FG, Chang SM, Huhn SL et al (1997). Age and the risk of anaplasia in magnetic resonance-nonenhancing supratentorial cerebral tumors. Cancer 80:936–941.

23. Scott JN, Bracher PM, Sevick RJ, Rewcastle NB, Forsyth PA (2002). How often are nonenhancing supratentorial gliomas malignant? A population study. Neurology 59:947–949.

24. Trattnig S, Ba-Ssalamah A, Noebauer-Huhmann IM et al (2003). MR contrast agent at high-field MRI (3 Tesla). Top Magn Reson Imaging 14(5):365–375.

25. Yuh WT, Fisher DJ, Engelken JD (1991). MR evaluation of CNS tumors: dose comparison study with gadopentate dimeculmine and gadoteridol. Radiology 180:485–491.

26. Abdullah ND, Mathews VP (1999). Contrast issues in brain tumor imaging. Neuroimaging Clin N Am 9:733–749.

27. Knopp MV, Runge VM, Essig M et al (2004). Primary and secondary brain tumors at MR imaging: bicentric intraindividual crossover comparison of gadobenate dimeglumine and gadopentetate dimeglumine. Radiology 230(1):55–64.

28. Kotys MS, Akbudak E, Markham J, Conturo TE (2007). Precision, signal-to-noise ratio, and dose optimization of magnitude and phase arterial input functions in dynamic susceptibility contrast MRI. J Magn Reson Imaging 25:598–611.

29. Aronen HJ, Gazit IE, Louis DN et al (1994). Cerebral blood volume maps of gliomas: comparison with tumor grade and histologic findings. Radiology 191:41–51.

30. Lev MH, Rosen R (1999). Clinical applications of intracranial perfusion MRI imaging. Neuroimaging Clin N Am 9:309–331.

31. Law M, Oh S, Johnson G et al (2006). Perfusion magnetic resonance imaging predicts patient outcome as an adjunct to histopathology: a second reference standard in the surgical and nonsurgical treatment of low-grade gliomas. Neurosurgery 58:1099–1107.

32. Dvorak HF, Brown LF, Detmar M, Dvorak AM (1995). Vascular permeability factor/vascular enhdothelial growth factor, microvascular hyperpermeability and angiogenesis. Am J Pathol 146:1029–1039.

33. Provenzale JM, Wang GR, Brenner T, Petrella JR, Sorensen AG (2002). Comparison of permeability in high-grade and low grade brain tumors using dynamic susceptibility contrast MR imaging. Am J Roentgenol 178:711–716.
34. Bastin ME, Carpenter TK, Armitage PA, Sinha S, Wardlaw JM, Whittle IR (2006). Effects of dexamethasone on cerebral perfusion and water diffusion in patients with high-grade glioma. Am J Neuroradiol 27:402–408.
35. Talos IF, Zou KH, Kikinis R, Jolesz FA (2007). Volumetric assessment of tumor infiltration of adjacent white matter based on anatomic MRI and diffusion tensor tractography. Acad Radiol 14: 431–436.
36. Moritani T, Ekholm S, Westessib PL (2004). *Diffusion-weighted MR Imaging of the Brain.* Springer Verlag, Heidelberg. 161–179.
37. Provenzale JM, McGraw P, Mhatre P, Alexander CG, Delong D (2004). Peritumoral brain regions in gliomas and meningiomas: investigation with isotropic diffusion weighted imaging and diffusion-tensor MR imaging. Radiology 232:451–460
38. Goebell E, Paustenbach S, Vaeterlein P et al (2006). Low-grade and anaplastic gliomas: differences in architecture evaluated with diffusion-tensor MR imaging. Radiology 239:217–222.
39. Budinger TF, Brennan KM (2000). Metabolic imaging. In *Neuro-Oncology: The Essentials.* Berenstein M, Berger MS (eds). Thieme Medical Publishers, New York. 79–93.
40. Cai W, Chen K, Mohamedali KA et al (2006). PET of vascular endothelial growth factor receptor expression. J Nucl Med 47(12):2048–2056.
41. Schwartz RB, Varvalho PA, Alexander ED (1991). Radiation necrosis vs high grade recurrent glioma: differentiation by using dual-isotope SPECT with 201Tl and 99mTc-HMPAO. Am J Neuroradiol 12:1187–1192.
42. Hustinix R, Alavi A (1999). SPECT and PET imaging of brain tumors. Neuroimaging Clin N Am 9:751–766.
43. Atals SW, Lavi E (1996). Intraaxial brain tumors. In *Magnetic Resonance Imaging of the Brain and Spine*, 2nd edn. Atlas SW (ed.). Lippincott Raven, Philadelphia. 423–488.
44. Ternier J, Wray A, Puget S, Bodaert N, Zerah M, Sainte-Rose C (2006). Tectal plate lesions in children. J Neurosurg 104 (6 Suppl):369–376.
45. McGinnis BD, Brady TJ, New PF et al (1983). Nuclear magnetic resonance (NMR) imaging of tumors in the posterior fossa. J Comput Assist Tomogr 7:575–584.
46. Garin von Eisiedel H, Loffler W (1982). Nuclear magnetic resonance imaging of brain tumors unrevealed by CT. Eur J Radiol 2:226–234.
47. Holland BA, Brandt-Zawadzki M, Norman D, Newton TH (1985). Magnetic resonance imaging of primary intracranial tumors: a review. Int J Radiat Oncol Biol Phys 11:315–321.
48. Earnest F, Kelly PJ, Scheithauer BW et al (1988). Cerebral astrocytomas: histopathologic correlation of MR and CT contrast enhancement with stereotactic biopsy. Radiology 166:823–827.
49. Kelly P, Caumas-Duport C, Kispert D, Kall B, Sheihauer B, Illig J (1987). Imaging-based stereotactic serial biopsies in untreated intracranial glial neoplasms. J Neurosurg 66:865–874.
50. Greene G, Hitchon P, Schelper R, Yuh W, Dyste G (1989). Diagnostic yield in CT-guided stereotactic biopsies of gliomas. J Neurosurg 71:494–497.
51. Swanson KR, Alvord EC, Murray JD (2000). A quantatative model for differential motility of gliomas in grey and white matter. Cell Prolif 33:317–329.
52. Marks JE, Gado M (1977). Serial computed tomography of primary brain tumors following surgery, irradiation, and chemotherapy. Radiology 125:119–125.
53. Holland BA, Kucharcyzk A, Brandt-Zawadzki M, Norman D, Haas DK, Harper PS (1985). MR imaging of calcified intracranial lesions. Radiology 157:353–356.
54. Mariani L, Siegenthaler P, Guzman R et al (2004). The impact of tumour volume and surgery on the outcome of adults with supratentorial WHO grade II astrocytomas and oligoastrocytomas. Acta Neurochir (Wien) 146(5):441–448.
55. Altman NR, Purser RK, Post MJD (1988). Tuberous sclerosis: characteristics at CT and MR imaging. Radiology 167:527–532.
56. Fuss M, Wenz F, Essig M et al (2001). Tumor angiogenesis of low-grade astrocytomas measured by dynamic susceptibility contrast-enhanced MRI (DSC-MRI) is predictive of local tumor control after radiation therapy. Int J Radiat Oncol Biol Phys 51(2):478–482.
57. Raffel C (1996). Molecular biology of pediatric gliomas. J Neuro-Oncol 28:121–128.
58. Guitierrez JA (1999). Classification and pathobiology of low-grade glial and glioneuronal neoplasms. In *The Practical Management of Low-Grade Primary Brain Tumors.* Rock JP, Rosenblum ML, Shaw EG, Caincross JG (eds). Lippincott Williams & Wilkins, Philadelphia. 33–67.
59. Arai K, Sato N, Aoki J et al (2006). MR signal of the solid portion of pilocytic astrocytoma on T2-weighted images: is it useful for differentiation from medulloblastoma? Neuroradiology 48(4):233–237.
60. Lee YY, Van Tassel P, Bruner JM, Moser RP, Share JC Juvenile pilocytic astrocytomas: CT and MR characteristics. Am J Roentgenol 152:1263–1270.
61. Coakley KJ, Huston J, Sheithauer BW, Forbes G, Kelly PJ (1995). Pilocytic astrocytomas: well-demarcated magnetic resonance appearance despite frequent infiltration histologically. Mayo Clinic Proc 70:747–751.
62. Abel TJ, Chowdhary A, Thapa M et al (2006). Spinal cord pilocytic astrocytoma with leptomeningeal dissemination to the brain. Case report and review of the literature. J Neurosurg 105(6 Suppl):508–514.
63. Komotar RJ, Mocco J, Zacharia BE (2006). Astrocytoma with pilomyxoid features presenting in an adult. Neuropathology 26(1):89–93.
64. Tihan T, Fisher PG, Kepner JL (1999). Pediatric astrocytomas with monomorphous pilomyxoid features and a less favorable outcome. J Neuropathol Exp Neurol 58(10):1061–1068.
65. Arslanoglu A, Cirak B, Horska A (2003). MR imaging characteristics of pilomyxoid astrocytomas. Am J Neuroradiol 24(9):1906–1908.
66. Cirak B, Horska A, Barker PB, Burger PC, Carson BS, Avellino AM (2005). Proton magnetic resonance spectroscopic imaging in pediatric pilomyxoid astrocytoma. Childs Nerv Syst 21(5):404–409.
67. Weldon-Linne CM, Victor TA, Groothuis DR, Vick NA (1983). Pleomorphic xanthoastrocytoma: ultrastructural and immunohistochemical study of a case with a rapidly fatal outcome following surgery. Cancer 52:2055–2063.

68. Zulch KJ (1986). *Brain Tumors: Their Biology and Pathology*, 3rd edn. Springer-Verlag, Berlin.15:210–341.

69. Russel DS, Rubinstein LJ (1989). *Pathology of Tumors of the Nervous System*, 5th edn. Williams & Wilkins, Baltimore. 3:83–350.

70. Rippe DJ, Boyko OB, Radi M, Worth R, Fuller GN (1992). MRI of temporal lobe pleomorphic xanthoastrocytoma. J Comput Assist Tomogr 16:856–859.

71. Yoshino MT, Lucio R (1992). Pleomorphic xanthoastrocytoma. Am J Neuroradiol 13:1330–1332.

72. Petropoulou K, Whiteman ML, Altman NR, Bruce J, Morrison G (1995). CT and MRI of pleomorphic xanthoastrocytoma: unusual biologic behavior. J Comput Assist Tomogr 19:860–865.

73. Turgut M, Akalan N, Ozgen T, Ruacan S, Erbengi A (1996). Subependymal giant cell astrocytoma associated with tuberous sclerosis: diagnostic and surgical characteristics of five cases with unusual features. Clin Neurol Neurosurg 98:217–221.

74. Hahn JS, Bejar R, Gladson CL (1991). Neonatal subependymal giant cell astrocytoma with tuberous sclerosis: MRI, CT, and ultrasound correlation. Neurology 41:124–128.

75. Nabbout R, Santos M, Rolland Y, Delalande O, Dulac O, Chiron C (1999). Early diagnosis of subependymal giant cell astrocytoma in children with tuberous sclerosis. J Neurol Neurosurg Psychiatr 66:370–375.

76. Jeong MG, Chung TS, Coe CJ, Jeon TJ, Kin DI, Joo AY (1997). Application of magnetization transfer imaging for intracranial lesions of tuberous sclerosis. J Comput Assist Tomogr 21:8–14.

77. Jelenik J, Smirniatopoulos JG, Parisi JE, Kanzer M (1990). Lateral ventricular neoplasms of the brain: differential diagnosis based on clinical, CT, and MR findings. Am J Neuroradiol 11:567–574.

78. Clarke MJ, Foy AB, Wetjen N, Raffel C (2006). Imaging characteristics and growth of subependymal giant cell astrocytomas. Neurosurg Focus 20:1–4.

79. de Ribaupierre S, Dorfmuller G, Bulteau C et al (2007). Subependymal giant-cell astrocytomas in pediatric tuberous sclerosis disease: when should we operate? Neurosurgery 60(1):83–89; discussion 89–90.

CHAPTER 37

Imaging of Oligodendrogliomas

William Ankenbrandt and Nina Paleologos

INTRODUCTION

The role of clinical imaging of brain tumors is to provide sufficient data for diagnosis and appropriate care of the patient. After initial tumor detection, additional imaging techniques may help, by narrowing the differential diagnosis, biopsy or resection planning, prognosis, radiation therapy planning and follow-up evaluation. We will review standard CT and MR brain imaging useful in the evaluation of oligodendrogliomas and review what is known about the characteristics of oligodendroglioma on PET and the newer MR methods of imaging, such as MRS and dynamic contrast-enhanced MRI (DCE).

CHARACTERISTICS OF OLIGODENDROGLIOMAS ON ROUTINE CT AND MR IMAGES

In most cases, standard neuroimaging with either CT or MRI allows primary brain tumors to be distinguished from other intra-axial lesions, such as cerebritis, abscess, infarct and vascular malformation. Specific cellular histopathology may be suspected due to imaging characteristics, however, it is not specifically distinguishable by neuroimaging. Astrocytomas and oligodendrogliomas in particular, cannot be differentiated from one another based on imaging alone. However, there are some characteristics on routine imaging that may favor oligodendroglioma over astrocytoma. These include predominance of cortical involvement; presence of calcification on CT, heterogeneous attenuation/signal including in some cases presence of intratumoral cysts and presence of subtle, patchy or 'lacy' enhancement [1]. (See Figures 37.1–37.3 for variable appearances of grade II oligodendrogliomas and Figures 37.4 and 37.5 for two different appearing images showing what proved to be grade III (anaplastic) oligodendrogliomas in both.) MRI is better than CT at demonstrating enhancement in malignant brain tumors [2]. Contrast enhancement and endothelial hyperplasia have been shown to correlate negatively with survival in 'classical' oligodendrogliomas [3].

On CT, low-grade oligodendrogliomas typically appear as non-enhancing low density masses. High-grade or

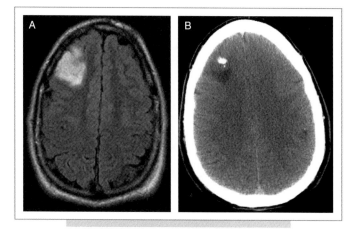

FIG. 37.1. WHO grade II oligodendroglioma. The lesion is a small frontal cortically-based tumor with high signal on FLAIR **(A)**. Note coarse calcification within the lesion on CT **(B)**. The lesion did not enhance with gadolinium.

anaplastic tumors may enhance to a variable degree. Calcification is found microscopically in 90% of oligodendrogliomas [4]. Calcifications may also be seen on CT and, when detectable, are often coarse (see Figures 37.1B and 37.4A). Calcification is not specific to oligodendroglioma, but is quite common. The reported prevalence of calcification detected on CT varies widely but is at least 20% [5,6].

MRI is superior to CT in delineating these tumors, which appear as an increased signal abnormality on T2-weighted and fluid attenuated inversion-recovery (FLAIR) images. They frequently involve the cortex which likely explains the high incidence of seizures in these patients and they may be large at the time of presentation. Oligodendrogliomas are frequently heterogeneous on T1-weighted images and cysts and areas of hemorrhage may be present. Low-grade tumors characteristically do not show enhancement. Enhancement is important in identifying high-grade or anaplastic tumors [2]. Assessment of tumor grade by stereotactic biopsy or from tissue obtained at subtotal resection is prone to sampling error because oligodendrogliomas, especially large

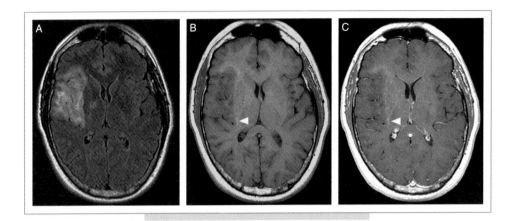

FIG. 37.2. WHO grade II oligodendroglioma. This right insular/subinsular tumor has little heterogeneity, the sort of mass that could be misinterpreted as an infarct. Some infarcts in this location are misinterpreted as tumor. Note cortical predominance of the lesion. There are areas of subtle T1 shortening at the margin of the tumor, but careful comparison between pre- and post-gadolinium T1-weighted images shows no enhancement separable from baseline increased T1 signal **(B** and **C**, arrowheads**)**.

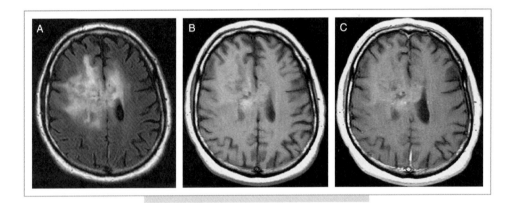

FIG. 37.3. WHO grade II oligodendroglioma. This lesion is markedly heterogeneous and presented at very large size. At presentation it had already crossed the body of the corpus callosum. There is no demonstrable enhancement separable from baseline T1-shortening **(B** and **C)**.

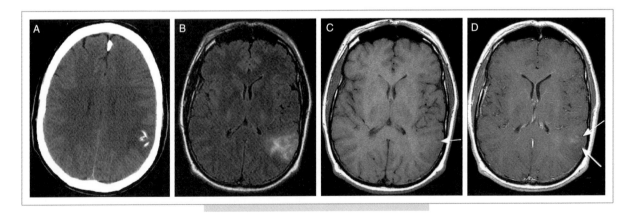

FIG. 37.4. Anaplastic (WHO grade III) oligodendroglioma. CT shows coarse calcification within the lesion **(A)**. Note the lack of heterogeneity and minimal enhancement. By carefully comparing pre- and post-gadolinium images it can be established that there is faint enhancement **(C** and **D**, arrows**)**. Interestingly, both this tumor and the much more 'aggressive'-appearing tumor in Figure 37.5 were found to have chromosome 1p and 19q deletions, predicting favorable response to therapy.

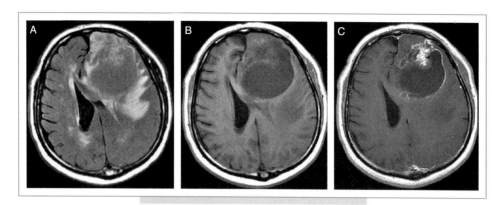

FIG. 37.5. Anaplastic (WHO grade III) oligodendroglioma. Partially-cystic, partially-solid, heterogeneous mass with prominent gadolinium enhancement **(C)**. There is marked edema **(A)** and mass effect, with subfalcine herniation. 1p and 19q deletions were confirmed, just as in the more 'benign'-appearing anaplastic oligodendroglioma in Figure 37.4.

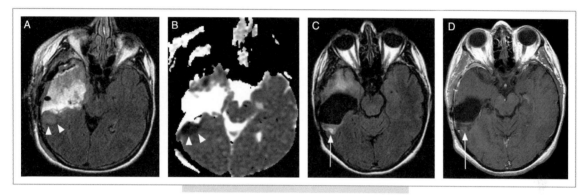

FIG. 37.6. Postoperative day 1 status post resection of right temporal oligodendroglioma. There is very subtle increased signal posterior to the right temporal surgical resection cavity on FLAIR images, with corresponding low signal on ADC maps indicating restricted diffusion and acute ischemia **(A** and **B,** arrowheads). Follow-up MRI shows a more distinct, focal area of increased signal on FLAIR **(C,** arrow), with focal enhancement **(D,** arrow). Because DWI was performed on the immediate postoperative scan, it can be confidently predicted that the enhancement that subsequently developed in the location where there was restricted diffusion represents blood–brain barrier breakdown secondary to infarction and *not* recurrent tumor.

ones, contain regions of variable histology. Small areas of enhancement in part of a large predominantly non-enhancing tumor likely represents high-grade pathology in that area. Non-enhancing areas may reflect areas of lower grade pathology. Histology obtained from non-enhancing areas of the same tumor may be of lower grade, thus resulting in 'sampling error' or undergrading of the tumor histologically. This reflects the variable histology seen in different areas of tumor. The absence of contrast enhancement does not rule out high-grade histology. White and colleagues, in a study of conventional MRI features of oligodendroglioma, found that the presence of contrast enhancement had a sensitivity of only 63% and a specificity of 50% in differentiating anaplastic from low-grade tumors [7].

Diffusion-weighted Imaging

Typically part of a routine MR examination of the brain, diffusion-weighted imaging (DWI) has revolutionized imaging of brain infarction and is classically thought of as a stroke imaging tool, but is helpful in tumor imaging as well.

The physical assumption behind the technique is that water in tissue is constantly in motion, flowing from intracellular to extracellular space and from one volume box of imaging data (voxel) to another. Energy, in the form of rapid magnetic field gradient changes, is applied to hydrogen atoms of water, which precess or 'spin'. Hydrogen atoms in water molecules free to move within a gradient field will have random phase. They will lose phase coherence and the resultant image will lose signal intensity. Water trapped, or whose free diffusion is 'restricted', will have hydrogen atoms spinning in greater phase coherence with one another, with resultant greater signal intensity [8,9].

Restricted diffusion, with a low apparent diffusion coefficient (ADC), in the postoperative setting adjacent to a resection cavity, may indicate postoperative infarct. This is very helpful to recognize, since the infarcted brain will enhance with gadolinium in the subacute phase, which may complicate interpretation of follow-up scans since it could be confused with recurrent, enhancing tumor (Figure 37.6).

Low ADC within a tumor has been correlated with increased tumor cellularity and higher grade neoplasia [10]. Other processes, such as abscess following biopsy, or intra-tumoral infarction, can also have restricted diffusion. Restricted diffusion related to infarction typically reverses within about 5 days, so short-interval follow-up MRI may be helpful. Persistently low ADC in part of a tumor probably indicates higher cellularity within that part. Low ADC has been found to correlate with elevated choline on MR spectroscopy of gliomas [11].

MR Spectroscopy

Malignancy is characterized by accelerated cell division (high 'mitotic index'), which is detectable on proton magnetic resonance spectroscopy (1H-MRS, hereinafter 'MRS') as elevated choline. Choline elevation occurs when there is rapid cell membrane synthesis.

The N-CH3 group of choline (Cho), the first major peak from the left on MRS, has a resonance at 3.2 parts per million (ppm). This is positioned immediately to the left of creatine (Cr), which has a peak at 3.0 ppm. N-acetylaspartate (NAA) is the third major peak from the left, with a resonance at 2.0 ppm. These three peaks represent the most common metabolites evaluated, with Cr typically used as the putative reference, since it changes little with most disease processes. NAA, a marker of intact neurons, will be depressed in many disease processes including in some oligodendrogliomas and other primary brain tumors.

Long echo time (TE) multivoxel MRS may be used to evaluate the three principal metabolites (Cho, Cr, NAA) in each box of a grid placed over a larger part of the tumor and usually also including a portion of presumably normal brain. Long TE single voxel MRS in abnormal brain is compared to normal as well; 3-Tesla (T) MRS may be preferable, however, 1.5T scanners are more readily available for clinical scanning. The signal-to-noise ratio (SNR) is acceptable on a 1.5T magnet, especially if a phased array head coil is used [12].

MRS as an initial diagnostic tool for oligodendrogliomas can be misleading if considered in isolation from other imaging findings, since low-grade gliomas may only have modest depression of NAA and little or no demonstrable elevation of choline (Figure 37.7). The finding of faint or lacy enhancement in a heterogeneous, cortically-based primary tumor with less than 2:1 elevation of Cho:Cr ratio may favor an oligodendroglioma. A Cho:Cr ratio of greater than 2:1 is suggestive of high-grade neoplasm, as is a Cho:NAA ratio greater than 0.8 [13].

One MRS finding that may help distinguish oligodendroglioma from astrocytoma preoperatively is myo-inositol (MI), an astroglial marker which is particularly elevated in low-grade astrocytomas [14]. A caveat is that glycine (resonance 3.4–3.7 ppm) overlaps the MI peak and has been found in high-grade primary tumors, and has been demonstrated to be elevated on MRS in a rare case of

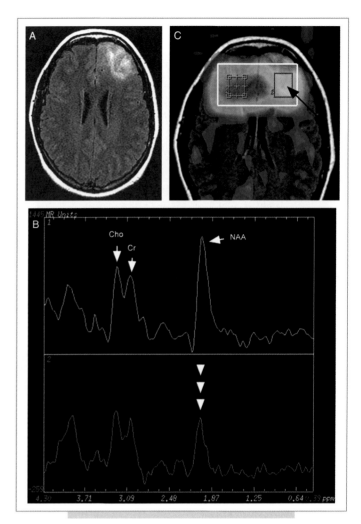

FIG. 37.7. Grade II mixed glioma (oligoastrocytoma). Many low-grade tumors will not show significant elevation of choline on hydrogen-MRS. This left frontal infiltrating tumor shows moderate depression of NAA as the only abnormality of the three major peaks (**B**, arrowheads). (**C**) is a spectroscopic image map showing distribution of NAA within the sampled area. Note decrease in NAA levels (in this image, higher concentrations are red, lower are blue or green: the arrow points to a predominantly blue and green region in the left frontal lobe) approximately corresponding with the area of increased signal on FLAIR (**A**).

oligodendroglial gliomatosis cerebri [15]. Glycine elevation has also been reported in astrocytoma [16]. The glutamine plus glutamate peak may be higher in oligodendrogliomas than in astrocytomas and one of the resonances is immediately adjacent to MI, further muddying the waters. Elevation of MI is also found in MS plaques, HIV infection and metachromatic leukodystrophy, and it is normally much higher in infants than in adults [17]. Absence of myo-inositol elevation in a suspected low-grade glioma may favor oligodendroglioma. An elevated lipid-lactate

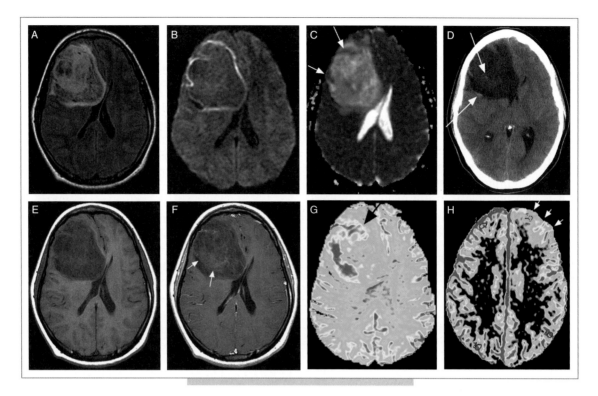

FIG. 37.8. Grade II mixed glioma (oligoastrocytoma). **(A)** FLAIR. **(B** and **C)** DWI, B = 1000, and ADC map. Areas of predominantly T2 shine-through are present at the periphery of the tumor **(B)**. The ADC map helps determine which areas have restricted diffusion **(C**, arrows). **(D)** CT shows faint calcifications (arrows). **(E** and **F)** Pre- and post-gadolinium T1W images suggest faint, patchy enhancement **(F**, arrows). **(G)** Negative enhancement integral image (approximating rCBV) shows areas of markedly increased relative cerebral blood volume in this low grade tumor. Note that not all of the areas of increased blood volume correspond with areas of gadolinium enhancement (black arrow). **(H)** One caveat: some low-grade oligodendrogliomas and mixed gliomas have low rCBV. **(H)** is a different grade II mixed glioma and is the same tumor as in Figure 37.7. There is predominantly low rCBV in the region of tumor infiltration (arrows).

combined peak suggests higher grade oligodendroglioma, but the combined peak is also elevated in infections [18].

MRS is perhaps most helpful in following oligodendrogliomas and other primary gliomas that have been treated. After surgery, with or without radiation therapy, there is often a margin of increased T2 signal intensity (T2 prolongation) adjacent to the resection cavity. If routine follow-up MRI shows growth of the area of abnormal signal, MRS can be used to attempt to distinguish incidental postoperative gliosis or signal changes related to radiation therapy from those changes related to tumor progression. Sometimes, however, MRS is non-diagnostic due to the proximity of the cavity and magnetic field heterogeneity caused by blood degradation products and other contents of the cavity. In cases where MRS is equivocal, MR perfusion may add diagnostic confidence.

MR Perfusion and Permeability Imaging

Dynamic Susceptibility-weighted Contrast-enhanced Imaging (DSC)

High-grade oligodendrogliomas, as well as many low-grade oligodendrogliomas, have elevated relative cerebral blood volume (rCBV) compared to normal white matter

(Figure 37.8). The conclusion that high-grade tumors can be distinguished from low-grade tumors based on elevated rCBV may not be true for oligodendrogliomas but, nonetheless, DSC may be helpful to distinguish tumor from other etiologies and to distinguish recurrent tumor from radiation necrosis [19]. Herpes encephalitis and toxoplasmosis tend to have low rCBV; most infarcts have low rCBV; bacterial abscesses are variable; and tumefactive demyelination has slightly lower rCBV than high-grade tumor, with presence of vessels running through the lesion on the echoplanar source images possibly favoring demyelination over tumor [20].

Relative cerebral blood volume is essentially represented as the area under the curve in a perfusion study performed with DSC (dynamic susceptibility-weighted contrast-enhanced) technique. Each curve represents the signal loss due to the T2* effect of concentrated gadolinium as the contrast material passes through each image slice over time. Normally, the curve has a steep downslope and relatively steep upslope as well. In tumors that have elevated rCBV, the peak is deeper, with a larger area under the curve representing higher rCBV. The area under the curve can be depicted for each voxel as a particular color in a scale typically progressing through a rainbow spectrum from blue to red. With a color map of rCBV for the entire slice being evaluated, it is straightforward for non-color-blind individuals to determine whether the rCBV for the area of signal abnormality in question is greater than that of adjacent presumably normal white matter.

Unfortunately for color-blind individuals, the color scale most commonly employed includes a red-green scale. Also, many standard PACS systems are black-and-white systems. A web-based program is needed for most to view color maps. It is also possible to use a region of interest (ROI) over the area in question, another over presumably normal brain tissue in an analogous 'mirror' location in the contralateral hemisphere, and superimpose the baselines visually to inspect the size of the area under the curve. We have found it most helpful in clinical cases visually to inspect color maps. If relying on comparison of dynamic T2* signal drop-out curves, in a region where there is increased T2 signal intensity either due to tumor or edema, the baseline of the curve will be higher than areas that have normal T2 signal.

Although most oligodendrogliomas will not be confused with metastases, high-grade lesions with ring enhancement could have a similar appearance to solitary metastasis. Evaluation of the perfusion curve may help distinguish metastasis from primary tumor. Primary tumor is more likely to have a curve that rapidly returns to near baseline, whereas metastasis is more likely to return to less than 50% of baseline. Also, it has been shown that rCBV in the peritumoral region of primary enhancing brain tumors is higher (as is Cho:Cr ratio and possibly also ADC) within this region of so-called 'infiltrative edema' compared to the analogous peritumor region, representing vasogenic edema surrounding a solitary metastasis [21,22]. DSC perfusion MR has also been shown to be helpful in distinguishing high-grade brain neoplasm from brain abscess [23].

MR Permeability Imaging – Dynamic Contrast-enhanced Imaging (DCE)

For trials of new therapies for oligodendroglioma, it may be helpful to attempt to quantify the degree of vascular permeability in a tumor, since it has been proposed that tumors with greater vascular permeability may have improved response to chemotherapy due to more efficient drug delivery. Permeability as measured by dynamic MR perfusion techniques may correlate with tumor grade [24,25]. Dynamic contrast-enhanced imaging (DCE) has been proposed as a quantitative method of measuring vascular permeability by assessing a time-intensity curve after administration of contrast [26]. Unlike DSC, which relies on a first-pass, T2* effect, DCE is a dynamic T1-weighted imaging method that measures the extent of contrast enhancement over time. Thus far it has been employed predominantly as a research tool. It has been shown that

mapping permeability can be accomplished concurrently with DSC imaging using a single gadolinium bolus [27].

Functional MRI and Fractional Anisotropy
Functional MRI

Blood oxygen level dependent (BOLD) functional MRI (fMRI) of the brain is helpful in the preoperative evaluation of tumors, especially those that are thought to encroach on measurably 'eloquent' anatomic areas, such as those functional regions responsible for motor activity and for generation and comprehension of language. The technique relies on detection of very small differences in blood flow due to upregulation of flow in metabolically-active functional areas. Paradoxically, there will be a measurable *increase* in blood oxygen in metabolically active areas because the upregulation of flow due to regional vasodilatation overcomes the increased oxygen extraction related to metabolic activity. This is measurable on echoplanar T2*-weighted imaging because deoxyhemoglobin has a strong paramagnetic effect that reduces signal by increasing spin phase-incoherence. Areas that have a higher ratio of oxyhemoglobin to deoxyhemoglobin will therefore have slightly higher signal intensity on T2*-weighted images [28,29].

Differences in signal are very small and interpretation requires post-processing of source image maps that are created during a task (e.g. finger tapping) and at rest. Essentially, the block of slices obtained during the task is compared to the block of slices at rest, and areas that have statistically significantly higher signal during the task are considered to be functionally active areas. In clinical functional MRI, it is helpful to have a reference of activation in grouped normal volunteers so that the patient's fMRI can be compared to the expected pattern of activation with each language and motor task.

Oligodendrogliomas are often cortically based and may infiltrate functionally-active, eloquent brain areas. fMRI is most helpful for preoperative planning when complete resection is contemplated in tumors which are thought, based either on clinical data (e.g. aphasia or hemiparesis) or routine imaging findings (e.g. tumor encroaching on or infiltrating precentral gyrus, frontal operculum or superior temporal gyrus), to involve highly eloquent brain regions.

Functional data sets are acquired with 5 mm thick echoplanar (EPI) sections. Since the EPI images have poor tissue contrast relative to standard diagnostic imaging sequences, the processed data showing color maps of areas of activation are superimposed not on the original EPI images, but on a separately acquired set of images. Software is used to merge the image locations of the original functional data set with diagnostic quality brain images, typically FLAIR and spoiled gradient echo (SPGR) images.

Diffusion Tensor Imaging – Fractional Anisotropy

Diffusion tensor imaging is sometimes employed preoperatively to attempt to establish the location of important fiber tracts, such as the corticospinal tract, and to determine whether there is displacement, infiltration or destruction of such tracts. The diffusion-weighted images displayed as part of routine MRI are 'isotropic' – that is the differences in directional water diffusivity (greater ease of diffusion along axonal tracts, for example) are essentially nullified by averaging of directional anisotropy. The goal of diffusion tensor imaging is essentially to discover for each voxel which way water diffuses most readily and this has been shown, in an experimental cat model, to be along the orientation of major white matter fiber tracts [30]. A map of directional anisotropy (most commonly calculated as 'fractional anisotropy') can approximate white matter tract configurations by displaying diffusivity along three principal axes in three different colors [31,32].

Cerebral Blood Volume

In one study, the role of rCBV was examined in the characterization of low-grade versus high-grade oligodendroglial tumors and no statistical difference was found in rCBV between the low- and high-grade tumors [33]. Sample size in this study was small and only three anaplastic tumors were analyzed with large ROIs (measuring 20–40 mm^2) which are known to affect sensitivity negatively in the detection of focal areas of increased vascularization in anaplastic neoplasms. In another study, using smaller ROIs, rCBV was significantly different between low-grade (1.61 ± 1.20) and high-grade tumors (5.45 ± 1.96) with an optimal rCBV ratio cutoff value in identification of anaplastic tumors of 2.14 [34]. The authors found that the sensitivity of rCBV in grading oligodendroglial tumors was 100%, indicating that all high-grade tumors were correctly classified, but that the specificity of rCBV measurement was 86% with a cutoff value of 2.14, indicating that a few low-grade oligodendroglial tumors may be incorrectly classified as high-grade. They suggest that rCBV ratios can be used more confidently to exclude the presence of anaplasia within oligodendroglial tumors than vice versa. Whether low-grade oligodendrogliomas with high rCBV behave more aggressively than those with lower rCBV or whether those tumors are being undergraded due to sampling error needs further investigation.

Positron Emission Tomography (PET)

Positron emission tomography gives metabolic information on tumor in vivo and is discussed in detail in regards to its usage in differentiating recurrent brain tumor from radiation injury, and in guiding stereotactic biopsy and radiation target volumes, elsewhere in this book. Oligodendrogliomas show a higher uptake on ^{11}C-methionine PET than astrocytomas, and anaplastic oligodendrogliomas have an increased uptake when compared to low-grade tumors [35,36]. In one study of 27 low-grade tumors which had undergone diagnostic surgery, but no other treatment, low uptake of ^{11}C-methionine was found to be a predictor of long time to tumor progression in oligodendrogliomas, but not in astrocytomas or oligoastrocytomas [37]. The activity volume index (AVI), a metabolic index generated from an automated semi-quantification of PET with methionine, has been shown to decrease in patients with low-grade oligodendrogliomas responding to procarbazine (PCV) chemotherapy [38].

CONCLUSION

Physiologic imaging techniques developed over the past two decades can add helpful information to aid in diagnosis, treatment planning and follow-up evaluation of oligodendrogliomas and mixed gliomas. While MRI is no substitute for histology, diagnostic confidence can be increased by selective application of these newer imaging techniques. The goal of imaging is to assess, relatively non-invasively, oligodendrogliomas, hopefully to monitor effective therapy (Figure 37.9), but also to identify potential tumor recurrence in a reliable and timely fashion.

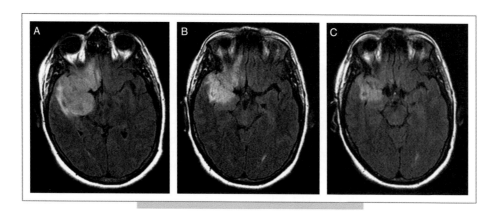

FIG. 37.9. **(A)** Grade II oligodendroglioma, baseline after biopsy. **(B)** Following 10 months of antineoplastic therapy with temodar. **(C)** Following 20 months of temodar.

REFERENCES

1. Daumas-Duport C, Varlet P, Tucker ML et al (1997). Oligodendrogliomas. Part 1: patterns of growth, histological diagnosis, clinical and imaging correlations: a study of 153 cases. J Neuro-Oncol 34:37–59.

2. Graif M, Bydder GM, Steiner RE et al (1985). Contrast-enhanced MR imaging of malignant brain tumors. Am J Neuroradiol 6:855–862.

3. Daumas-Duport C, Tucker ML, Kolles H et al (1997). Oligodendrogliomas. Part II: a new grading system based on morphological and imaging criteria. J Neuro-Oncol 34:61–78.

4. Burger PC, Scheithauer BW, Vogel FS (2002). *Surgical Pathology of the Nervous System and its Coverings*. Churchill Livingstone, New York.

5. Lee YY, Van Tassel P (1989). Intracranial oligodendrogliomas: imaging findings in 35 treated cases. Am J Roentgenol 152:361–369.

6. Vonofakos D, Marcu H, Hacker H (1979). Oligodendrogliomas: CT patterns with emphasis on features indicating malignancy. J Comput Assist Tomogr 3:783–788.

7. White ML, Zhang Y, Kirby P et al (2005). Can tumor contrast enhancement be used as a criterion for differentiating tumor grades of oligodendrogliomas? Am J Neuroradiol 26:784–790.

8. Mosely ME, Kucharczyk J, Mintorovitch J et al (1990). Diffusion-weighted MR imaging of acute stroke: correlation with T2-weighted and magnetic susceptibility-enhanced MR imaging in cats. Am J Neuroradiol 11:423–429.

9. Atlas SW (1998). The basis of MR contrast from flow and diffusion. In *Magnetic Resonance Imaging of the Brain and Spine*. Lippincott-Raven Publishers, New York.

10. Sugahara T, Korogi Y, Kochi M et al (1999). Usefulness of diffusion-weighted MRI with echo-planar technique in the evaluation of cellularity in gliomas. J Magnet Res Imaging 9:53–60.

11. Gupta RK, Sinha U, Cloughesy TF, Alger JR (1999). Inverse correlation between choline magnetic resonance spectroscopy signal intensity and the apparent diffusion coefficient in human glioma. Magnet Reson Med 41:2–7.

12. Inglese M, Spindler M, Babb JS et al (2006). Field, coil, and echo-time influence on sensitivity and reproducibility of brain proton MR spectroscopy. Am J Neuroradiol 27:684–688.

13. Stadlbauer A, Gruber S, Nimsky C et al (2006). Preoperative grading of gliomas by using metabolite quantification with high-spatial-resolution proton MR spectroscopic imaging. Radiology 238:958–969.

14. Castillo M, Smith JK, Kwock L (2000). Correlation of myo-inositol levels and grading of cerebral astrocytomas. Am J Neuroradiol 21:1645–1649.

15. Gutowski NJ, Gómez-Ansón B, Torpey N et al (1999). Oligodendroglial gliomatosis cerebri: ^{1}H-MRS suggests elevated glycine/inositol levels. Neuroradiology 41:650–653.

16. Londono A, Castillo M, Armao D et al (2003). Unusual MR spectroscopic imaging pattern in an astrocytoma: lack of elevated choline and high myo-inositol and glycine levels. Am J Neuroradiol 24:942–945.

17. Ross BD, Colletti P, Lin A (2006). Magnetic resonance spectroscopy of the brain: neurospectroscopy. In *Clinical Magnetic Resonance Imaging*, 3rd edn. Edelman RR, Hellelink JR, Zlatkin MB, Crues JV (eds). Saunders/Elsevier, New York. 1850–1851.

18. Rijpkema M, Schuuring J, van der Meulin Y et al (2003). Characterization of oligodendrogliomas using short echo time ^{1}H MR spectroscopic imaging. NMR Biomed 16:12–18.

19. Law M, Yang S, Wang H et al (2003). Glioma grading: sensitivity, specificity, and predictive values of perfusion MR imaging and proton MR spectroscopic imaging compared to conventional MR imaging. Am J Neuroradiol 24:1989–1998.

20. Cha S, Knopp EA, Johnson G et al (2002). Intracranial mass lesions: dynamic contrast-enhanced susceptibility-weighted echo-planar perfusion MR imaging. Radiology 223:11–29.

21. Law M, Cha S, Knopp EA et al (2002). High-grade gliomas and solitary metastases: differentiation by using perfusion and proton spectroscopic MR imaging. Radiology 222:715–721.

22. Cha S (2006). Update on brain tumor imaging: from anatomy to physiology. Am J Neuroradiol 27:475–487.

23. Holmes TM, Petrella JR, Provenzale JM (2004). Distinction between cerebral abscesses and high-grade neoplasms by dynamic susceptibility contrast perfusion MRI. Am J Radiol 183:1247–1252.

24. Provenzale JM, Wang GR, Brenner T et al (2002). Comparison of permeability in high-grade and low-grade brain tumors using dynamic susceptibility contrast MR imaging. Am J Radiol 178:711–716.

25. Roberts HC, Roberts TP, Brasch RC et al (2000). Quantitative measurement of microvascular permeability in human brain tumors achieved using dynamic contrast-enhanced MR imaging: correlation with histologic grade. Am J Neuroradiol 21:891–899.

26. Tofts PS, Kermode AG (1991). Measurement of the blood–brain barrier permeability and leakage space using dynamic MR imaging. Fundamental concepts. Magn Reson Med 17:357–367.

27. Donahue KM, Krouwer HG, Rand SD et al (2000). Utility of simultaneously acquired gradient-echo and spin-echo cerebral blood volume and morphology maps in brain tumor patients. Magn Reson Med 43:845–853.

28. Brown GG, Jernigan TL, Cato M (2006). Functional magnetic resonance imaging in neuropsychiatric disorders. In *Clinical Magnetic Resonance Imaging*, 3rd edn. Edelman RR, Hellelink JR, Zlatkin MB, Crues JV (eds). Saunders/Elsevier, New York. 1807.

29. Ogawa S, Lee TM, Kay AR, Tank DW (1990). Brain magnetic resonance imaging with contrast dependent on blood oxygenation. Proc Natl Acad Sci USA 87:9868–9872.

30. Moseley ME, Cohen Y, Kucharczyk J et al (1990). Diffusion-weighted MR imaging of anisotropic water diffusion in cat central nervous system. Radiology 176:439–446.

31. Witwer BP, Moftakhar R, Hasan KM et al (2002). Diffusion-tensor imaging of white matter tracts in patients with cerebral neoplasm. J Neurosurg 97:568–575.

32. Jellison BJ, Field AS, Medow J et al (2004). Diffusion tensor imaging of cerebral white matter: a pictorial review of physics, fiber tract anatomy, and tumor imaging patterns. Am J Neuroradiol 25:356–369.

33. Xu M, See SJ, Ng WH et al (2005). Comparison of magnetic resonance spectroscopy and perfusion-weighted imaging in

presurgical grading of oligodendroglial tumors. Neurosurgery 56:919–926.

34. Spampinato MV, Smith JK, Kwock L et al (2007). Cerebral blood volume measurements and proton MR spectroscopy in grading of oligodendroglial tumors. Am J Roentgenol 188:204–212.

35. Van den Bent MJ, Kros JM (2005). Oligodendrogliomas and mixed gliomas. In *Principles of Neuro-Oncology*. Schiff D, O'Neill BP (eds). McGraw-Hill, New York. 15:311–332.

36. Derlon JM, Chapon F, Noel MH et al (2000). Non-invasive grading of oligodendrogliomas: correlation between in vivo metabolic pattern and histopathology. Eur J Nucl Med 27:778–787.

37. Ribum D, Smits A (2005). Baseline ^{11}C-methionine PET reflects the natural course of grade 2 oligodendrogliomas. Neurol Res 27:516–521.

38. Tang BN, Sadeghi N, Branle F et al (2005). Semi-quantification of methionine uptake and flair signal for the evaluation of chemotherapy in low-grade oligodendroglioma. J Neuro-Oncol 71:161–168.

CHAPTER 38

Primary CNS Lymphoma

Thinesh Sivapatham, Herbert B. Newton, Eric C. Bourekas and H. Wayne Slone

INTRODUCTION

Primary central nervous system lymphoma (PCNSL) is defined as the presentation of extranodal lymphoma confined to the CNS. Although the vast majority of CNS lymphomas (80–90%) are primary lesions, they are relatively uncommon, accounting for only about 1% of all lymphomas. The incidence of PCNSL has increased over the past three decades in both the immunocompromised and immunocompetent populations, placing it behind only meningiomas and low-grade astrocytomas in prevalence. Previously accounting for only about 1% of all CNS neoplasms, they currently represent up to 15% of all primary brain tumors [1].

The prevalence of PCNSL continues to be significantly higher in immunocompromised than immunocompetent patients. A dramatic increase in the incidence of PCNSL in the immunocompromised population occurred in the early 1980s, coinciding with the first reported cases of acquired immunodeficiency syndrome (AIDS). Patients with AIDS continue to account for the largest group of immunocompromised patients with CNS lymphoma, and PCNSL in an HIV-seropositive patient is an AIDS-defining condition. Other immunocompromised states associated with CNS lymphoma include organ transplantation, Wiskott-Aldrich syndrome, congenital immune deficiency syndromes and prolonged immunosuppressive therapy [1]. Additionally, there have been reports of CNS lymphoma related to autoimmune diseases such as Sjogren's syndrome and systemic lupus erythematosus [2,3]. Clinical data suggest that up to 6% of AIDS patients will develop PCNSL [2]. In transplant patients, the risk is about 1–5% (1–2% for renal transplant patients [4] and 2–7% for cardiac, lung and liver transplant patients [4–6]). Patients with congenital immune deficiency syndromes carry a risk of 4% [2]. A growing body of evidence has linked Epstein-Barr virus with the pathogenesis of CNS lymphoma in AIDS patients [7].

Although CNS lymphomas are responsive to chemotherapy and radiation, prognosis remains poor, with high rates of recurrence. Chemotherapy can be delivered systemically or intrathecally. More recently, some patients have been treated with intra-arterial delivery of chemotherapeutic agents following disruption of the blood–brain barrier [8–10]. Although surgical intervention is often necessary for diagnostic purposes, surgical resection provides no therapeutic benefit and is generally reserved for cases in which the tumor might cause brain herniation [11].

Lymphomas can be divided into the Hodgkin's and non-Hodgkin's types. Non-Hodgkin's lymphoma can be further subdivided by cell type and grade and these classifications have long been a source of controversy. The vast majority of PCNSLs are high-grade tumors of the non-Hodgkin's B-cell type [1]. Primary T-cell lymphoma of the CNS is rare [12,13] and CNS involvement by Hodgkin's lymphoma is even rarer [14]. The imaging features of CNS lymphoma are variable and attempts at correlating imaging features with histologic subtypes have yielded inconsistent results [15,16]. Therefore, the diagnosis of PCNSL hinges upon stereotactic biopsy and histologic assessment. Nevertheless, neuroimaging can suggest the diagnosis, leading to timely initiation of appropriate therapy. In this chapter, we will discuss the characteristic imaging features of this aggressive neoplasm.

GENERAL IMAGING FEATURES

Anatomical Distribution

An overwhelming majority of PCNSL lesions occur in the brain, but they can also occur in the spinal cord, meninges, globe and CSF. Of those primary CNS lesions occurring in the brain, supratentorial lesions occur much more commonly than those in the posterior fossa, accounting for 80–90% of all lesions. One-half to two-thirds of lesions have been reported to occur in the cerebral hemispheres, with the white matter of the frontal lobes being the most common location, followed by the temporal, parietal and occipital lobes. The central deep gray matter of the basal ganglia, thalamus and hypothalamus, often reported as a classic location for CNS lymphoma, accounts for about one-third of lesions. Approximately 10% of lesions occur in the posterior fossa and only 1% are found in the spinal cord. [1,17–19]

A characteristic feature of PCNSL is its tendency to contact ependymal or meningeal surfaces, with leptomeningeal involvement seen in about 12% of cases [1]. Corpus callosum involvement is also relatively common, reported in some series to be nearly as common as involvement of the deep gray matter [19]. Frontal lobe lesions that cross the genu of the corpus callosum create the classic 'butterfly' pattern. Dural involvement is uncommon and often mimics meningioma in its presentation. In untreated CNS lymphoma, calcification, hemorrhage and cysts are only rarely seen [20]. Another characteristic finding of CNS lymphoma is the relatively little mass effect and edema relative to lesion size [17].

CT and MRI

On computed tomography (CT), CNS lymphoma usually presents as an iso- or hyperdense mass relative to brain parenchyma on unenhanced scans, and almost all lesions demonstrate contrast enhancement (Figure 38.1) [1,17,21–24]. However, a negative CT examination does not exclude the diagnosis, with false negative rates of up to 38% having been reported on initial diagnostic scans [25](although these rates are currently likely to be lower given advances in CT technology). Magnetic resonance imaging (MRI) is more sensitive [23] and is currently the imaging modality of choice in the evaluation of CNS lymphoma. On MRI, the lesions usually present as well-demarcated masses that are isointense to hypointense relative to gray matter on T1-weighted sequences. The T2-weighted imaging characteristics can be more variable. Initial reports described these lesions to be predominantly hyperintense to gray matter on T2-weighted sequences. More recent studies indicate that more than half the lesions are hypointense to isointense relative to gray matter [1,17,21]. The hypointensity on T2-weighted imaging, as well as the hyperattenuation on CT, is attributable to the dense cellularity of the tumor and high nucleus-to-cytoplasmic ratio [1]. Higher T2 signal intensity within CNS lymphoma lesions has been correlated with higher degrees of necrosis [15]. On fluid attenuated inversion recovery (FLAIR) sequences which are T2-weighted, the imaging characteristics are similar to traditional T2 sequences. A variable pattern can be seen on diffusion-weighted imaging (DWI). While most tumors demonstrate elevated water diffusibility, the dense cellularity of CNS lymphoma often results in restricted diffusion (high signal) on DWI [26]. As with CT, most lesions enhance with contrast administration (Figures 38.2 and 38.3); treatment with steroids may result in decreased enhancement or non-enhancement [15,17]. Imaging findings of PCNSL in children are variable, but often similar to those seen in adults [27].

Angiography

The angiographic appearance of CNS lymphoma is highly variable. A common finding is that of an avascular mass with little vessel displacement or involvement [28]. Jack et al described a homogeneous vascular stain that

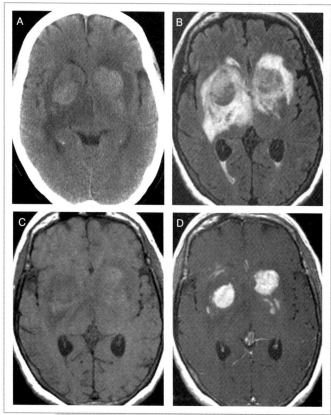

FIG. 38.2. A 73-year-old female with progressive motor weakness and biopsy-proven PCNSL of the B-cell type. **(A)** Hyperdense masses are seen involving the basal ganglia bilaterally on non-contrast CT, with surrounding edema. **(B)** On MRI, FLAIR sequence shows the masses to be isointense to gray matter, with surrounding high-T2 signal related to edema. On pre- **(C)** and post-gadolinium **(D)** sequences, the lesions demonstrate fairly homogeneous contrast enhancement.

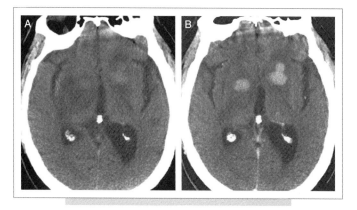

FIG. 38.1. A 61-year-old male who presented with gait disturbance and slurred speech. Biopsy yielded B-cell PCNSL. **(A)** Non-contrast CT demonstrates bilateral hyperdense lesions involving the deep gray and white matter. Surrounding low density is consistent with edema. **(B)** The lesions demonstrate homogeneous enhancement with the administration of contrast.

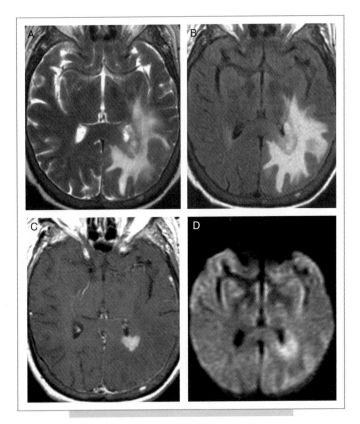

FIG. 38.3. In this 62-year-old female with PCNSL, a periventricular lesion in the deep white matter of the left cerebral hemisphere is isointense to gray matter on the T2 **(A)** and FLAIR **(B)** sequences, with surrounding hyperintensity related to edema. The lesion demonstrates homogeneous enhancement on the postgadolinium T1 sequence **(C)** and restricted diffusion (high signal) on DWI **(D)**.

appeared in the late arterial or early venous phase in 12 of 29 patients with PCNSL who underwent cerebral angiography [29]. Vascular encasement may rarely be seen [18]. Lesions involving the meninges may demonstrate a dural supply, also a rare finding [30]. Angiography plays little role in the diagnosis of CNS lymphoma. There has been an increasing role for endovascular techniques in the treatment of CNS lymphoma, with the use of blood–brain barrier disruption to deliver intra-arterial chemotherapy [9,10]. Osmotic disruption of the blood–brain barrier is achieved by the infusion of intra-arterial mannitol into the desired cerebrovascular distribution, followed by the intra-arterial administration of the chemotherapeutic agent (Figure 38.4) [8].

Other Imaging Modalities

Cross-sectional imaging studies (CT and MRI) are the most commonly utilized imaging modalities in the diagnosis and management of CNS lymphoma. However, several studies have reported on the utility of thallium-201 single photon emission computed tomography (SPECT) and F-18

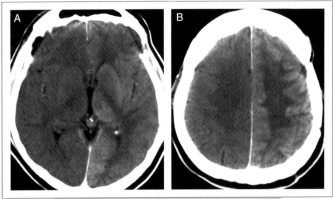

FIG. 38.4. In this 46-year-old male with PCNSL, osmotic disruption of the blood–brain barrier (BBB) was achieved with the intra-arterial infusion of mannitol into the left internal carotid artery (ICA), immediately followed by infusion of the chemotherapeutic regimen. Contrast-enhanced CT was then performed to evaluate the degree of BBB disruption. The study shows asymmetrically increased enhancement of the left cerebral hemisphere in the distribution of the left ICA (A,B) consistent with adequate BBB disruption.

fluoro-deoxyglucose (FDG) positron emission tomography (PET) in CNS lymphoma [31–35]. Generally, CNS lymphoma lesions demonstrate hypermetabolic activity on PET imaging and increased radiopharmaceutical uptake with SPECT imaging. While most brain neoplasms will demonstrate similar findings, PET and SPECT may help differentiate neoplastic processes from benign ones, where the activity will be relatively less intense. These imaging modalities may be particularly useful in differentiating CNS lymphoma from infectious or inflammatory lesions in AIDS patients (see below), and in monitoring response to therapy.

Imaging Patterns in Immunocompetent Versus Immunocompromised Patients

In immunocompetent patients, PCNSL most commonly presents as a solitary supratentorial lesion in the cerebral hemispheres [17,19]. The vast majority of lesions in immunocompetent patients demonstrate homogeneous contrast enhancement [15,17,24]. Lesions in the immunocompromised host are more likely to be multiple, with a more variable enhancement pattern [15,36]. Central necrosis is also more common in immunocompromised patients [15]. This often results in a ring pattern on imaging: the central area of necrosis results in T2 hyperintensity, with a densely cellular, T2 hypointense rim, surrounded by T2 hyperintensity peripherally related to edema [1]. The pattern is reversed on T1-weighted imaging, with a T1 hyperintense ring surrounded by T1 hypointense edema. There is usually intense ring enhancement with contrast, which may be smooth or irregular (Figure 38.5) [15,25]. Periventricular enhancement, although relatively uncommon, is very specific for CNS lymphoma in AIDS patients [37].

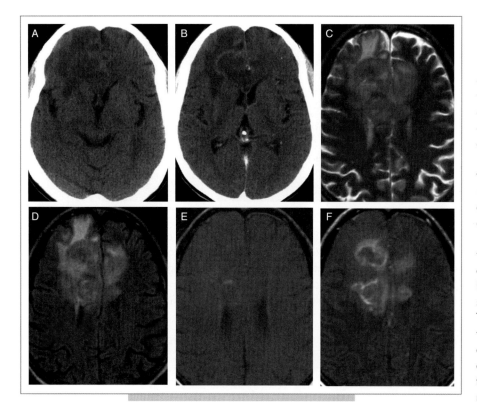

FIG. 38.5. A 44-year-old female with a history of HIV who presented with high fevers and headache. Differential diagnosis included toxoplasmosis and PCNSL. She was started empirically on anti-toxoplasma therapy, but did not respond. An alternative diagnosis of PCNSL was then made, and the patient improved with steroids. CT without (**A**) and with (**B**) contrast demonstrates a lesion in the right frontal lobe, extending into the corpus callosum, with a ring pattern of enhancement and surrounding edema. On MRI, T2 (**C**) and FLAIR (**D**) sequences show lesions involving the frontal lobes bilaterally, extending into the corpus callosum. Increased signal within these lesions is likely related to necrosis, with the surrounding high signal secondary to edema. The areas of the lesion that are isointense to normal brain parenchyma indicate the densely cellular areas. These are also the areas that enhance with gadolinium, seen on the pre- (**E**) and post-contrast (**F**) sequences, resulting in the ring enhancement pattern.

Neurological complications are common in patients infected with the human immunodeficiency virus (HIV). These patients often present with neurological symptoms and multifocal brain lesions on CT or MRI, many of which demonstrate ring enhancement. The most likely etiology in this scenario is cerebral toxoplasmosis, but the differential diagnosis includes PCNSL, as well as other (less common) infectious processes including neurosyphilis and progressive multifocal leukoencephalopathy (PML). Definitive diagnosis requires brain biopsy, which is associated with the morbidity and mortality of an invasive procedure. Standard practice has been to place the patient empirically on anti-toxoplasmosis therapy for a 2-week period. If the lesions respond, the diagnosis is confirmed. If the lesions fail to respond or worsen, an alternative diagnosis such as PCNSL is entertained and the patient may go on to brain biopsy if indicated (see Figure 38.5). The obvious downfall to this approach is that if the presumptive diagnosis of toxoplasmosis is incorrect, the underlying disease process may progress and the patient's chances for prolonged survival may be decreased. In this scenario, thallium-201 SPECT or F-18 FDG PET, particularly when combined with toxoplasma serology, may be helpful in arriving at the correct diagnosis in a more timely fashion [31–35].

Differential Diagnosis of Primary CNS Lymphoma

Differential considerations in the diagnosis of PCNSL based on imaging findings include gliomas (usually higher-grade lesions like glioblastoma multiforme) (Figure 38.6), metastases (from extracranial sources or from other brain primaries), abscess and other types of infection, and granulomatous processes such as sarcoidosis (Figure 38.7) and tuberculosis [1]. Active demyelinating disease can mimic PCNSL [38] and primitive neuroectodermal tumor should be considered in the pediatric population. T2 hypointensity, when present, can help differentiate CNS lymphoma from other brain neoplasms and demyelinating disease, which are both more likely to be hyperintense on T2-weighted sequences. Distinguishing brain abscess from CNS lymphoma can be more difficult, as both can demonstrate ring-like enhancement and increased signal on DWI. Brain abscesses are typically associated with a moderate to large amount of surrounding vasogenic edema, while PCNSL typically generates only mild edema relative to lesion size, and this feature can sometimes be useful in making the diagnosis. The differentiation of PCNSL from toxoplasmosis can be virtually impossible in the immunocompromised population based on conventional imaging techniques. As discussed in the previous section, both can demonstrate ring enhancement patterns and multicentricity and alternative imaging modalities may help make the diagnosis more rapidly (see above). Secondary CNS lymphoma usually involves the leptomeninges and patients typically present with cranial nerve palsies (Figure 38.8).

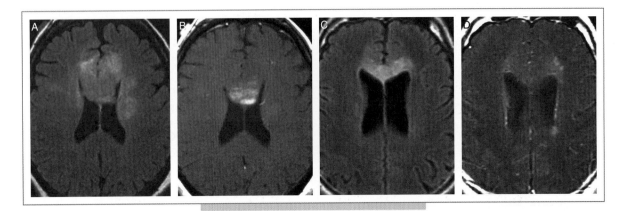

FIG. 38.6. Two patients with similar imaging findings. The first is a 68-year-old female with new onset seizure. On MRI, FLAIR sequence **(A)** shows abnormal high signal in the corpus callosum and periventricular regions. The post-contrast T1 sequence **(B)** shows enhancement in the corpus callosum. The second patient is a 51-year-old male with progressive neurologic deterioration. Again, there is abnormal signal in the periventricular region on FLAIR **(C)** and subtle enhancement in the corpus callosum and periventricular region on post-contrast T1 **(D)**. The first patient had a high-grade astrocytoma, while the second patient had PCNSL.

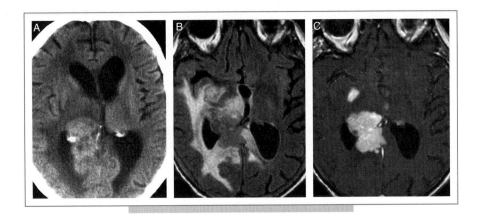

FIG. 38.7. A 74-year-old male with changes in mental status and difficulty with ambulation. Non-contrast CT showed a hyperdense mass involving the deep gray matter on the right, extending into the splenium of the corpus callosum, with surrounding edema **(A)**. MRI showed an isointense lesion in the same distribution with surrounding high signal edema on the FLAIR sequence **(B)**. On the post-contrast T1 sequence **(C)**, the lesion demonstrated homogeneous enhancement, with an enhancing satellite lesion more anteriorly. The patient had a known history of sarcoidosis, and the initial explanation for the neurological symptoms and imaging findings was neurosarcoidosis. He went on to biopsy and the diagnosis was B-cell PCNSL.

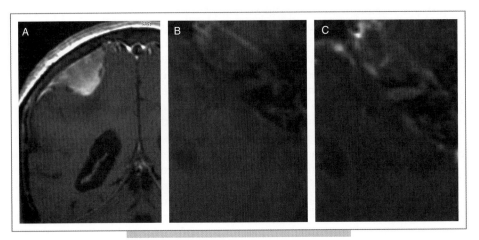

FIG. 38.8. A 71-year-old female with systemic follicular lymphoma who presented with ataxia, slurred speech and diplopia. Coronal post-gadolinium T1-weighted sequence shows an enhancing dural-based mass overlying the right parietal convexity **(A)**. Pre- **(B)** and post-contrast **(C)** axial images through the region of the internal auditory canal (IAC) on the left demonstrates abnormal and asymmetric enhancement in the left IAC. The diagnosis was secondary leptomeningeal lymphoma.

UNCOMMON MANIFESTATIONS OF PRIMARY CNS LYMPHOMA

Primary Leptomeningeal Lymphoma

Primary leptomeningeal lymphoma is a rare presentation of PCNSL characterized by lymphomatous involvement of the leptomeninges in the absence of parenchymal CNS or systemic involvement. It accounts for fewer than 8% of all PCNSLs [39]. The most common clinical symptoms are those of raised intracranial pressure [40]. Meningismus, cranial neuropathies or spinal radiculopathies have also been described [39–41]. Cerebrospinal fluid analysis has been inconsistent in showing the presence of malignant cells. Histologic analyses have revealed both B- and T-cell primary lymphoma of the leptomeninges [42,43].

Neuroimaging can often be unremarkable, or show non-specific findings such as hydrocephalus. Less common findings include meningeal calcification [43] and meningeal or cranial nerve enhancement. Proton density or FLAIR MRI sequences may demonstrate high signal intensity in the subarachnoid space, although this is a non-specific finding and can also be seen with other processes such as subarachnoid hemorrhage and meningeal inflammation of any cause. A very rare presentation of primary leptomeningeal lymphoma is that of a cerebellopontine angle mass [44]. Only a few cases have been reported, and a lesion in this location may show effacement of the basal cisterns and enlargement of the ventricles on imaging. Due to its non-specific clinical and imaging findings, the diagnosis of primary leptomeningeal lymphoma can be difficult and requires a high index of suspicion (Figure 38.9). Primary parenchymal CNS lymphoma with meningeal extension and systemic lymphoma with meningeal spread are both more common entities (see Figure 38.8) and must be excluded.

Intravascular Lymphomatosis

Intravascular lymphomatosis is a rare type of B-cell non-Hodgkin's lymphoma characterized by the aggressive intravascular proliferation of lymphoid cells. The process is systemic and any organ can be involved, with or without clinical manifestation. The skin and central nervous system are the most frequent sites of involvement, with a neurological presentation in over 80% of cases having been reported [45]. Isolated CNS involvement is frequent and is difficult to distinguish from other cerebral microangiopathies [46]. Presenting symptoms include mental status changes, dementia and focal or non-localizing neurologic deficits as a result of ischemia caused by vessel occlusion [46–48]. The disease affects predominantly small blood vessels, more specifically arterioles, capillaries and post-capillary venules [46]. A few reports have described involvement of the cortical veins and dural sinuses, although this is rare [49].

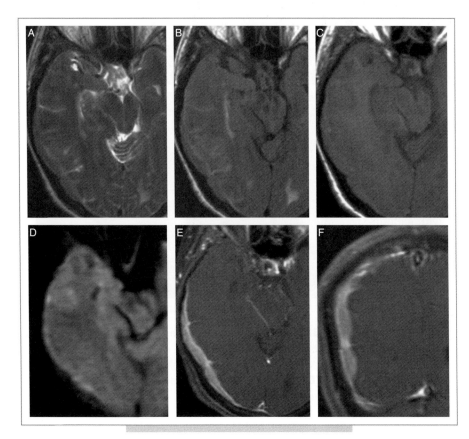

FIG. 38.9. A 64-year-old female with recurrent falls, headache and vertigo. Biopsy confirmed primary leptomeningeal lymphoma. On MRI, axial T2 **(A)**, FLAIR **(B)** and pre-contrast T1 **(C)** sequences demonstrate a dural based lesion that is nearly isointense to brain parenchyma on all pulse sequences. There is also signal abnormality involving the subarachnoid space along the right cerebral convexity on the FLAIR sequence. On DWI **(D)**, the lesion demonstrates slightly increased signal (restricted diffusion). Axial **(E)** and coronal **(F)** post-gadolinium sequences show homogeneous enhancement of the lesion.

Diagnosis can be elusive, as hematologic abnormalities are commonly absent and CSF analysis is non-specific. Prognosis remains poor, with a mortality rate over 80%. The success of therapy hinges on an early diagnosis prior to irreversible ischemic damage. Diagnosis is based upon biopsy of suspected sites of involvement. However, as the lesions are small and often located deep within the brain parenchyma, they are prone to sampling error; it is not uncommon for biopsy results to be non-diagnostic or fail to show the presence of intravascular tumor cells. Additionally, lesions found in sensitive areas of the brain may preclude biopsy. For these reasons it is estimated that, in up to 80% of cases, diagnosis is not made until histologic evaluation at autopsy [46].

Imaging findings are variable, but intravascular lymphoma most commonly presents as multiple lesions scattered throughout the cerebral hemispheres, with the brainstem and cerebellum less frequently involved [46,47]. Of the hemispheric lesions, the subcortical white matter is more commonly affected than the cortex and subcortical nuclei. These lesions typically demonstrate hyperintensity on T2-weighted MR sequences, which appears to correlate with edema and gliosis on biopsy [47]. This can be a non-specific finding, often difficult to distinguish from age-related white matter changes. Another common presentation of intravascular lymphomatosis is that of multifocal areas of ischemia and infarction [50], not surprising given the underlying pathology of small vessel occlusion. Lesions in various stages of evolution can be seen at one time. In the acute phase, ischemic lesions demonstrate high signal intensity on diffusion-weighted imaging (DWI). Hyperintensity can also initially be seen on FLAIR sequences (Figure 38.10). As these lesions evolve, they follow the expected signal characteristics of an ischemic small vessel stroke; the DWI hyperintensity will resolve, with persistence of abnormal FLAIR signal [46]. The chronic stage of these lesions is characterized by tissue loss and T2 hyperintensity, likely representing gliosis as a result of chronic ischemia and tissue damage. Subsequently, the T2 hyperintense lesions described earlier in this paragraph may be along the spectrum of ischemic changes related to vessel occlusion. Enhancement patterns are also variable, with less than one-third of lesions demonstrating enhancement with gadolinium on initial imaging. When enhancement is present, it can manifest as parenchymal mass-like enhancement or meningeal enhancement [46,47]. Spinal cord involvement is fairly common, but the sensitivity of MRI in detecting these lesions is low; less than 40% of patients with autopsy proven cord lesions had an abnormal MRI. When present, the most common signal abnormality was T2 hyperintensity. DWI and FLAIR signal abnormalities can be partially reversible and may correlate with response to chemotherapy [46].

Lymphomatosis Cerebri

In contrast to the usual presentation of PCNSL as solitary or multiple mass lesions within the CNS, lymphomatosis cerebri is an extremely rare form of CNS lymphoma that manifests as a diffuse, infiltrative process without formation of a distinct mass. The entity was first described in 1999 [51], and only about five cases have been reported in the literature since that time. In all of the reported cases, the clinical presentation was that of a rapidly progressive dementia. MRI demonstrated a diffuse white matter process with relative sparing of gray matter and a diffusely infiltrating B-cell lymphoma was confirmed in those cases that went to autopsy [52]. Contrast enhancement may be absent, suggesting that the blood–brain barrier remains intact [53]. Because the clinical and imaging features are similar to those of other diffuse white matter disorders, such as demyelinating disease, leukodystrophy, small vessel ischemic disease, viral disease, Binswanger's disease, or infiltrating glioma (gliomatosis cerebri), biopsy is necessary for accurate diagnosis.

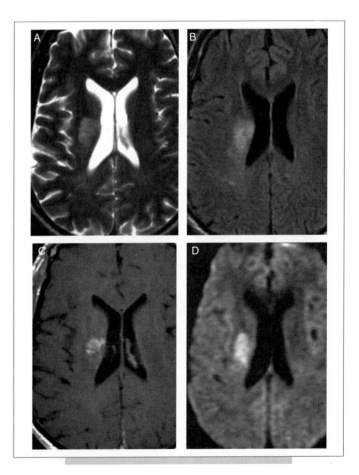

FIG. 38.10. A 48-year-old male with diffuse large B-cell lymphoma, who presented with an acute basal ganglia infarct. Biopsy showed intravascular lymphomatosis. Axial T2 **(A)** and FLAIR **(B)** MRI sequences demonstrate abnormal high signal in the basal ganglia on the right. The post-contrast axial T1 sequence shows the lesion to enhance **(C)**. On DWI **(D)**, the lesion demonstrates restricted diffusion, consistent with history of recent infarct.

Primary Lymphoma of the Spinal Cord

Accounting for about 1% of all PCNSLs [1], primary intramedullary spinal lymphoma is extremely rare, with only a handful of reported cases. Patients commonly present with focal neurologic deficits, including pain, paresthesias and paresis. MRI typically reveals an intramedullary lesion with T2 hyperintensity and intense enhancement with the administration of gadolinium [54–56]. Surrounding edema may be present as well. Lymphomas of both B- and T-cell origin have been described in this location. The imaging findings are non-specific and differential considerations include malignant glioma, metastatic disease and inflammatory processes such as multiple sclerosis and acute transverse myelitis. CSF analysis can often be non-specific and histopathological evaluation may be necessary for diagnosis.

Primary CNS Hodgkin's Lymphoma

CNS involvement in Hodgkin's lymphoma is uncommon and is seen in less than 1% of patients with the disease [57]. When present, CNS involvement is usually secondary to disseminated disease elsewhere in the body and occurs at the time of relapse in most patients. Hodgkin's lymphoma as a primary intracranial lesion is exceedingly rare, with only a few case reports in the literature [57–59]. Clinical symptoms at presentation are variable, ranging from memory loss and hemiparesis to cranial nerve palsies. Although meningeal involvement is common, neuroimaging may also reveal a parenchymal lesion without dural attachment. In cases where the meninges are involved, lesions may be indistinguishable from meningioma in clinical presentation and imaging features [60].

Primary Intraocular Lymphoma

Intraocular lymphoma is a subset of PCNSL that may occur as a solitary lesion, or in association with lymphoma elsewhere in the CNS. Primary lymphoma of the globe often progresses to the CNS parenchyma and meninges. Patients often present with symptoms of chronic vitreitis, including complaints of blurred vision and floaters [61]. The diagnosis is contingent upon histologic analysis and is often delayed until a prolonged course of treatment for idiopathic vitreitis has failed [61,62]. Vitrectomy specimen or vitreous aspirate reveals diffuse large B-cell lymphoma in most cases (although T-cell lymphomas have also been reported). Neuroimaging plays little role in the diagnosis, but is routinely performed to assess for concurrent CNS disease.

Low-grade PCNSL

Low-grade lymphoma of the CNS is relatively rare, although its true incidence is currently unknown. One series of 40 patients with low-grade PCNSL described differences in clinical, pathological and radiological features compared to the more common high-grade subtype [63]. In this series, seizures were the most common symptom at presentation, while mental status changes were rare. The reverse is true in high-grade PCNSL, where mental status changes are common at presentation and seizures are less frequent. Due to the indolent course of the low-grade subtype, mean time to diagnosis from symptom onset is often longer relative to the high-grade subtype (approximately 15 months in this series compared to approximately 3 months for high-grade PCNSL). Prognosis is typically better for the low-grade subtype. Both B-cell and T-cell histologic types were identified. While B-cell cases were much more common, the incidence of T-cell cases in this series (20%) was much higher than that reported for PCNSL of the high-grade subtype. On imaging, lesions of the low-grade subtype were more likely to demonstrate hyperintensity on T2-weighted MRI sequences and moderate or heterogeneous contrast enhancement, and less likely to be periventricular in location compared to the high-grade subtype.

MALT Lymphoma of the CNS

Mucosa-associated lymphoid tissue (MALT) lymphomas are characterized as low-grade B-cell lymphomas that occur in extranodal tissue. They were initially described in the GI tract, but have since been reported to develop in a variety of extranodal tissues. More recently, there have been reports of MALT lymphomas arising from the meninges. Now classified as extranodal marginal zone B-cell lymphoma of MALT-type, these lesions often present in the CNS as well-defined, dural-based masses that mimic meningiomas both in clinical presentation and imaging features [64–66]. There have also been a few reports of MALT lymphoma presenting as a cerebellopontine angle mass [67]. Due to its localized nature and slow growth, surgical resection with or without adjuvant therapies is often curative.

Primary T-cell CNS Lymphoma

PCNSL of the T-cell type (TPCNSL) is rare, with the majority of cases having been reported in the immunocompetent population. There have been a few reports of AIDS-related TPCNSL and, in these patients, an association with the human T-cell lymphotropic virus type I (HTLV-I) has been found [68]. In the largest published series of TPCNSL (45 patients), the presentation and outcome were similar to the B-cell subtype [69]. Imaging may reveal solitary or multiple lesions, which demonstrate variable enhancement patterns. The following imaging characteristics were described on MRI in a series of seven patients: tendency to be subcortical in location; relatively high incidence of hemorrhage; ring pattern of enhancement; and cystic areas consistent with necrosis (Figure 38.11) [70]. Primary spinal and leptomeningeal lymphomas of the T-cell subtype have also been reported [71].

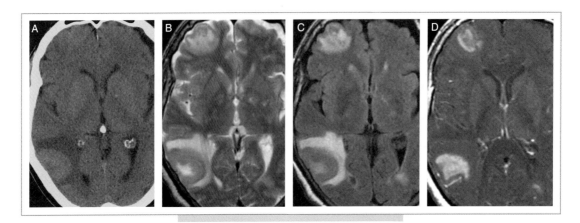

FIG. 38.11. A 79-year-old male with seizures and biopsy-proven PCNSL of the T-cell subtype. Non-contrast CT **(A)** shows a sub-cortical hyperdense mass in the right parieto-occipital region with surrounding edema. On MRI, T2 **(B)** and FLAIR **(C)** sequences show additional lesions in the right frontal lobe and left periventricular region. All three lesions enhance with gadolinium on the post-contrast sequence **(D)**.

REFERENCES

1. Koeller KK, Smirniotopoulos JG, Jones RV (1997). Primary central nervous system lymphoma: radiologic-pathologic correlation. RadioGraphics 17:1497–1526.
2. Schabet M (1999). Epidemiology of primary CNS lymphoma. J Neuro-Oncol 43:199–210.
3. Pisoni CN, Grinberg AR, Plana JL, Freue RD, Manni JA, Paz L (2003). Primary central nervous system lymphoma in a patient with systemic lupus erythematosus (article in Spanish). Medicina (B. Aires) 63(3):221–223.
4. Boubenider S, Hiesse C, Goupy C, Kriaa F, Marchand S, Charpentier B (1997). Incidence and consequences of post-transplantation lymphoproliferative disorders. J Nephrol 10:136–145.
5. Penn I (1993). Incidence and treatment of neoplasia after transplantation. J Heart Lung Transpl 12(6 Pt 2):S328–336.
6. Penn I (1996). Posttransplantation de novo tumors in liver allograft recipients. Liver Transpl Surg 2(1):52–59.
7. Jellinger KA, Paulus W (1995). Primary central nervous system lymphomas – new pathological developments. J Neuro-Oncol 24(1):33–36.
8. Doolittle ND, Petrillo A, Bell S, Cummings P, Eriksen S (1998). Blood–brain barrier disruption for the treatment of malignant brain tumors: The National Program. J Neurosci Nurs 30(2):81–90.
9. Fortin D, Desjardins A, Benko A, Niyonsega T, Boudrias M (2005). Enhanced chemotherapy delivery by intraarterial infusion and blood–brain barrier disruption in malignant brain tumors: the Sherbrooke experience. Cancer 103(12):2606–2615.
10. Siegal T, Zylber-Katz E (2002). Strategies for increasing drug delivery to the brain: focus on brain lymphoma. Clin Pharmacokinet 41(3):171–186.
11. Batchelor T, Loeffler JS (2006). Primary CNS lymphoma. J Clin Oncol 24(8):1281–1288.
12. Gijtenbeek JM, Rosenblum MK, DeAngelis LM (2001). Primary central nervous system T-cell lymphoma. Neurology 57(4):716–718.
13. Lee DK, Chung CK, Kim HJ et al (2002). Multifocal primary CNS T-cell lymphoma of the spinal cord. Clin. Neuropathol 21(4):149–155.
14. Hirmiz K, Foyle A, Wilke D et al (2004). Intracranial presentation of systemic Hodgkin's disease. Leuk Lymphoma 45(8):1667–1671.
15. Johnson BA, Fram EK, Johnson PC, Jacobowitz R (1997). The variable MR appearance of primary lymphoma of the central nervous system: comparison with histopathologic features. Am J Neuroradiol 18:563–572.
16. Jack CR Jr, O'Neill BP, Banks PM, Reese DF (1988). Central nervous system lymphoma: histologic types and CT appearance. Radiology 167(1):211–215.
17. Coulon A, Lafitte F, Hoang-Xuan K et al (2002). Radiographic findings in 37 cases of primary CNS lymphoma in immuno-competent patients. Eur Radiol 12:329–340.
18. Jiddane M, Nicoli F, Diaz P et al (1986). Intracranial malignant lymphoma. Report of 30 cases and review of the literature. J Neurosurg 65(5):592–599.
19. Küker W, Nágele T, Korfel A et al (2005). Primary central nervous system lymphomas (PCNSL): MRI features at presentation in 100 patients. J Neuro-Oncol 72:169–177.
20. Jenkins CN, Colquhoun IR (1998). Characterization of primary intracranial lymphoma by computed tomography: an analysis of 36 cases and a review of the literature with particular reference to calcification haemorrhage and cyst formation. Clin Radiol 53(6):428–434.
21. Slone HW, Blake JJ, Shah R, Guttikonda S, Bourekas EC (2005). CT and MRI findings of intracranial lymphoma. Am J Roentgenol 184(5):1679–1685.
22. Gliemroth J, Kehler U, Gaebel C, Arnold H, Missler U (2003). Neuroradiological findings in primary cerebral lymphomas of non-AIDS patients. Clin Neurol Neurosurg 105(2):78–86.
23. Suwanwela N, Tantanatrakool B, Suwanwela NC (2001). Intracranial lymphoma: CT and MR findings. J Med Assoc Thai 84(Suppl 1):S228–243.

24. Herrlinger U, Schabet M, Bitzer M, Petersen D, Krauseneck P (1999). Primary central nervous system lymphoma: from clinical presentation to diagnosis. J Neuro-Oncol 43(3):219–226.

25. Remick SC, Diamond C, Migliozzi JA et al (1990). Primary central nervous system lymphoma in patients with and without the acquired immune deficiency syndrome. Medicine 69(6):345–360.

26. Cotton F, Ongolo-Zogo P, Louis-Tisserand G et al (2006). Diffusion and perfusion MR imaging in cerebral lymphomas (French). J Neuroradiol 33(4):220–228.

27. Porto L, Kieslich M, Schwabe D, Yan B, Zanella FE, Lanfermann H (2005). Central nervous system lymphoma in children. Pediatr Hematol Oncol 22(3):235–246.

28. Tallroth K, Katevuo K, Holsti L, Andersson U (1981). Angiography and computed tomography in the diagnosis of primary lymphoma of the brain. Clin Radiol 32(4):383–388.

29. Jack CR, Reese DF, Scheithauer BW (1986). Radiographic findings in 32 cases of primary CNS lymphoma. Am J Roentgenol 146(2):271–276.

30. Spillane JA, Kendall BE, Moseley IF (1982). Cerebral lymphoma: clinical radiological correlation. J Neurol Neurosurg Psychiatr 45(3):199–208.

31. Heald AE, Hoffman JM, Bartlett JA, Waskin HE (1996). Differentiation of central nervous system lesions in AIDS patients using positron emission tomography (PET). Int J STD AIDS 7:337–346.

32. Lorberboym M, Estok L, Machac J et al (1996). Rapid differential diagnosis of cerebral toxoplasmosis and primary central nervous system lymphoma by thallium-201 SPECT. J Nucl Med 37(7):1150–1154.

33. Pierce MA, Johnson MD, Maciunas RJ et al (1995). Evaluating contrast-enhancing brain lesions in patients with AIDS by using positron emission tomography. Ann Intern Med 123(8):594–598.

34. Pomper MG, Constantinides CD, Barker PB et al (2002). Quantitative MR spectroscopic imaging of brain lesions in patients with AIDS: correlation with [11C-methyl]thymidine PET and thallium-201 SPECT. Acad Radiol 9(4):398–409.

35. Skiest DJ, Erdman W, Chang WE, Oz OK, Ware A, Fleckenstein J (2000). SPECT thallium-201 combined with Toxoplasma serology for the presumptive diagnosis of focal central nervous system mass lesions in patients with AIDS. J Infect 40(3):274–281.

36. Thurnher MM, Rieger A, Kleibl-Popov C et al (2001). Primary central nervous system lymphoma in AIDS: a wider spectrum of CT and MRI findings. Neuroradiology 43(1):29–35.

37. Dina TS (1991). Primary central nervous system lymphoma versus toxoplasmosis in AIDS. Radiology 179(3):823–828.

38. DeAngelis LM (1990). Primary central nervous system lymphoma imitates multiple sclerosis. J Neuro-Oncol 9(2):177–181.

39. Lachance DH, O'Neill BP, Macdonald DR et al (1991). Primary leptomeningeal lymphoma: report of 9 cases, diagnosis with immunocytochemical analysis, and review of the literature. Neurology 41(1):95–100.

40. Kim HJ, Ha CK, Jeon BS (2000). Primary leptomeningeal lymphoma with long-term survival: a case report. J Neuro-Oncol 48(1):47–49.

41. Shenkier TN (2005). Unusual variants of primary central nervous system lymphoma. Hematol Oncol Clin N Am 19(4): 651–664, vi.

42. Grove A, Vyberg M (1993). Primary leptomeningeal T-cell lymphoma: a case and a review of primary T-cell lymphoma of the central nervous system. Clin Neuropathol 12(1):7–12.

43. King A, Wilson H, Penney C, Michael W (1998). An unusual case of primary leptomeningeal marginal zone B-cell lymphoma. Clin Neuropathol 17(6):326–329.

44. Berciano J, Jimenez C, Figols J et al (1994). Primary leptomeningeal lymphoma presenting as cerebellopontine angle lesion. Neuroradiology 36(5):369–371.

45. Treves TA, Gadoth N, Blumen S, Korczyn AD (1995). Intravascular malignant lymphomatosis: a cause of subacute dementia. Dementia 6(5):286–293.

46. Baehring JM, Henchcliffe C, Ledezma CJ, Fulbright R, Hochberg FH (2005). Intravascular lymphoma: magnetic resonance imaging correlates of disease dynamics within the central nervous system. J Neurol Neurosurg Psychiatr 76(4):540–544.

47. Williams RL, Meltzer CC, Smirniotopoulos JG, Fukui MB, Inman M (1998). Cerebral MR imaging in intravascular lymphomatosis. Am J Neuroradiol 19(3):427–431.

48. Albrecht R, Krebs B, Reusche E, Nagel M, Lencer R, Kretzschmar HA (2005). Signs of rapidly progressive dementia in a case of intravascular lymphomatosis. Eur Arch Psychiatr Clin Neurosci 255(4):232–235.

49. Kenez J, Barsi P, Majtenyi K et al (2000). Can intravascular lymphomatosis mimic sinus thrombosis? A case report with 8 months' follow-up and fatal outcome. Neuroradiology 42(6):436–440.

50. Kinoshita T, Sugihara S, Matusue E et al (2005). Intravascular malignant lymphomatosis: diffusion-weighted magnetic resonance imaging characteristics. Acta. Radiol 46(3):246–249.

51. Bakshi R, Mazziotta JC, Mischel PS, Jahan R, Seligson DB, Vinters HV (1999). Lymphomatosis cerebri presenting as a rapidly progressive dementia: clinical, neuroimaging and pathologic findings. Dement Geriatr Cogn Disord 10(2):152–157.

52. Rollins KE, Kleinschmidt-DeMasters BK, Corboy JR, Damek DM, Filley CM (2005). Lymphomatosis cerebri as a cause of white matter dementia. Hum Pathol 36(3):282–290.

53. Lai R, Rosenblum MK, DeAngelis LM (2002). Primary CNS lymphoma: a whole-brain disease? Neurology 59(10):1557–1562.

54. Herrlinger U, Weller M, Küker W (2002). Primary CNS lymphoma in the spinal cord: clinical manifestations may precede MRI detectability. Neuroradiology 44(3):239–244.

55. Schild SE, Wharen RE Jr, Menke DM, Folger WN, Colon-Otero G (1995). Primary lymphoma of the spinal cord. Mayo Clin Proc 70(3):256–260.

56. Bekar A, Cordan T, Evrensel T, Tolunay S (2001). A case of primary spinal intramedullary lymphoma. Surg Neurol 55(5):261–264.

57. Deckert-Schlüter M, Marek J, Šetlík M et al (1998). Primary manifestation of Hodgkin's disease in the central nervous system. Virchows Arch 432(5):477–481.

58. Ashby MA, Barber PC, Holmes AE, Freer CE, Collins RD (1988). Primary intracranial Hodgkin's disease. A case report and discussion. Am J Surg Pathol 12(4):294–299.

59. Clark WC, Callihan T, Schwartzberg L, Fontanesi J (1992). Primary intracranial Hodgkin's lymphoma without dural attachment. Case report. J Neurosurg 76(4):692–695.

60. Johnson MD, Kinney MC, Scheithauer BW et al (2000). Primary intracerebral Hodgkin's disease mimicking meningioma: case report. Neurosurgery 47(2):454–456; discussion 456–457.

61. Hormigo A, Abrey L, Heinemann MH, DeAngelis LM (2004). Ocular presentation of primary central nervous system lymphoma: diagnosis and treatment. Br J Haematol 126(2):202–208.

62. Levy-Clarke GA, Chan CC, Nussenblatt RB (2005). Diagnosis and management of primary intraocular lymphoma. Hematol Oncol Clin N Am 19(4):739–749, viii.

63. Jahnke K, Korfel A, O'Neill BP et al (2006). International study on low-grade primary central nervous system lymphoma. Ann Neurol 59(5):755–762.

64. Pavlou G, Pal D, Bucur S, Chakrabarty A, van Hille PT (2006). Intracranial non-Hodgkin's MALT lymphoma mimicking a large convexity meningioma. Acta Neurochir (Wien) 148(7):791–793; discussion 793.

65. Rottnek M, Strauchen J, Moore F, Morgello S (2004). Primary dural mucosa-associated lymphoid tissue-type lymphoma: case report and review of the literature. J Neuro-Oncol 68(1):19–23.

66. Tu PH, Giannini C, Judkins AR et al (2005). Clinicopathologic and genetic profile of intracranial marginal zone lymphoma: a primary low-grade CNS lymphoma that mimics meningioma. J Clin Oncol 23(24):5718–5727.

67. Itoh T, Shimizu M, Kitami K et al (2001). Primary extranodal marginal zone B-cell lymphoma of the mucosa-associated lymphoid tissue type in the CNS. Neuropathology 21(3): 174–180.

68. Calderon EJ, Japon MA, Chinchon I, Soriano V, Capote FJ (2002). Primary lymphoma of the central nervous system and HTLV-I infection. Haematologia (Budap) 31(4):365–367.

69. Shenkier TN, Blay JY, O'Neill BP et al (2005). Primary CNS lymphoma of T-cell origin: a descriptive analysis from the international primary CNS lymphoma collaborative group. J Clin Oncol 23(10):2233–2239.

70. Kim EY, Kim SS (2005). Magnetic resonance findings of primary central nervous system T-cell lymphoma in immunocompetent patients. Acta Radiol 46(2):187–192.

71. Lee DK, Chung CK, Kim HJ et al (2002). Multifocal primary CNS T-cell lymphoma of the spinal cord. Clin Neuropathol 21(4):149–155.

CHAPTER 39

Pituitary and Sellar Region Lesions

Eric C. Bourekas, Lilja Solnes and H. Wayne Slone

INTRODUCTION

Sellar and juxtasellar lesions are common, with up to 27% of people having incidental pituitary tumors at autopsy [1]. Clinically, tumors in this region frequently present with endocrine abnormalities or symptoms related to mass effect on subjacent structures. Mass lesions in this region are not uncommon, accounting for 15–20% of all intracranial neoplasms. The most common lesions in adults are pituitary adenomas and meningiomas, while the most common lesions in children are craniopharyngiomas and hypothalamic/chiasmatic gliomas. In a review of 131 cases, Johnsen et al found that 54.2% of all lesions in this region were adenomas, 10.7% meningiomas, 9.6% craniopharyngiomas, 4.6% chiasmatic or hypothalamic gliomas [2]. The radiologic evaluation of suspected sellar and suprasellar lesions should include a contrast enhanced multiplanar MRI with dedicated small field-of-view images through the region of interest. The exact protocol varies by institution and various protocols will be discussed later. In those patients in whom MRI is contraindicated, a contrast enhanced CT scan may be done, although the soft tissue resolution is poor and significant artifact renders this examination suboptimal. Often the imaging characteristics of a lesion, along with clinical history including patient's age, will be helpful in formulating a differential diagnosis and aid in the clinical decision-making tree.

ANATOMY

The sella turcica is a bony depression in the sphenoid bone. The sella is bordered laterally by the cavernous sinuses, superiorly by the diaphragma sella (dural fold), anteroinferiorly by the sphenoid sinus and posteriorly by the pontine cistern. The pituitary gland normally sits within the sella. The gland itself is composed of an anterior lobe, intermediate (vestigial) and posterior lobe. The anterior lobe, adenohypophysis, forms about 75–80% of the gland and is a center for hormone synthesis, including thyroid stimulating hormone (TSH), adrenocorticotrophic hormone (ACTH) and prolactin. The posterior lobe, neurohypophysis, is connected to the hypothalamus by the pituitary stalk and receives hormones from the hypothalamus via the pituitary stalk (i.e. antidiuretic hormone (ADH)). The two can be differentiated on imaging because the posterior lobe characteristically demonstrates increased T1 signal on unenhanced images while the anterior lobe is isointense to gray matter on T1-weighted images. However, in the first six weeks of life the anterior pituitary is bright as well on T1-weighted images in 82% [3].

The size and configuration of the pituitary gland is thought to vary by age and sex. In adults, the gland is slightly larger in females when compared to males. The height of a normal gland can be up to 9 mm [4]. The gland tends to enlarge during puberty and pregnancy. Therefore, caution must be exercised in suggesting pituitary gland pathology in young adolescent patients and women of child-bearing age.

IMAGING

Although CT is used at some institutions for the evaluation of the sellar and juxtasellar regions, evaluation of this region is hampered by beam hardening artifacts related to the dense bone and by metallic artifacts from dental fillings. Soft tissue resolution is poor when compared with MRI. However, CT is excellent for the evaluation of adjacent bony structures and the presence or absence of calcifications, both of which can be important in the differential diagnosis. CT should only be used as the primary mode of evaluation in patients in which there is a contraindication to MR. In these patients, 1 mm axial and coronal sections are obtained post-contrast, with pre-contrast images in one plane being useful.

MRI, because of its multiplanar capability and superior soft tissue resolution, is the examination of choice for evaluation of sellar and juxtasellar lesions. Involvement of the subjacent structures such as the optic chiasm, cavernous sinus, sphenoid sinus, orbit, temporal lobes and carotid arteries can be better evaluated with MRI. A small field of view of 16–20, with thin sections of 3 mm or less and high resolution, with a matrix of 256 × 256 are essential. T1- and T2-weighted images are generally obtained, with T1-weighted images after contrast administration.

Sagittal and coronal imaging is most useful for imaging the pituitary and cavernous sinus region. At our institution we obtain a high-resolution three-dimensional sequence in the coronal plane with contrast for optimal evaluation of the pituitary. Because the pituitary lacks a blood–brain barrier, it enhances intensely, early and more so than tumors, so that tumors generally appear as areas of non-enhancement. Dynamic T1-weighted fast spin-echo imaging after a bolus of contrast can detect lesions not seen on standard imaging in 10–14% of cases [5,6]. However, in 12.5–17% of cases, lesions are better seen with standard post-contrast images and in 8–9% of cases lesions are seen with standard imaging and not with dynamic imaging [6]. For these reasons, standard imaging is used in most cases, with dynamic imaging performed in problem cases.

Incidental lesions are frequently identified on MR in the sellar region due to increased soft tissue resolution. Autopsy studies reveal incidental pituitary pathology in up to 27% of cases, most being incidental microadenomas or pars intermedia cysts [1]. Focal 2–3 mm lesions on imaging will prove to be incidental as often as they are endocrinologically significant [1]. In patients with incidental lesions, with normal laboratory values and no clinical symptoms, stability over a 2-year period on follow-up imaging suggests a benign etiology and no further studies are necessary [7].

Venous sampling of the inferior petrosal or cavernous sinus is an invasive study that can be valuable in the diagnosis of a suspected adenoma and, in particular, Cushing's disease, in cases where imaging has failed to identify a lesion. Catheters are placed from each of the femoral veins into the internal jugular veins and then advanced into each of the inferior petrosal sinuses [8]. Blood samples are obtained from each side both before and after stimulation with corticotropin releasing hormone (CRH) and analyzed. This technique can reliably distinguish pituitary Cushing's disease due to a microadenoma from ectopic ACTH syndrome, in patients with negative imaging and can frequently lateralize the lesion [9,10]. Although inferior petrosal sinus sampling is usually performed, it has been shown that bilateral, simultaneous cavernous sinus sampling, using corticotropin-releasing hormone, is as accurate as inferior petrosal sinus sampling in detecting Cushing's disease and perhaps more accurate in lateralizing the abnormality within the pituitary gland [11]. However, this procedure is not without risk. Miller et al reported that 0.2–0.5% of patients undergoing the procedure suffered a major neurologic complication thought to be related to venous hypertension [12].

The discussion that follows is based on whether a lesion is solid, mixed solid and cystic or cystic, although this is a somewhat oversimplified approach. Clearly, lesions that tend to be solid such as adenomas can be cystic or mixed and lesions that tend to be mixed solid and cystic such as craniopharyngiomas can be purely solid or purely cystic.

SOLID LESIONS

Pituitary Adenoma

Adenomas arise from the adrenohypophyseal cells and are the most common tumors of the sella turcica in adults. Although many different classification schemes exist, pituitary adenomas are most commonly classified according to size and function (functioning versus non-functioning adenomas) [13–15]. Lesions smaller than 1 cm are classified as *microadenomas* and those larger than 1 cm are classified as *macroadenomas*. Microadenomas tend to be functioning lesions, whereas macroadenomas are usually non-functioning lesions.

Johnsen et al reported that 54% of sellar and juxtasellar lesions in their series were pituitary adenomas [2]. Pituitary adenomas are usually seen in adults and are uncommon in children, representing less than 3% of all pediatric intracranial tumors [16]. However, pituitary adenomas occurring in the pediatric population are thought to be more aggressive lesions and warrant more aggressive therapy than their adult counterparts [17]. The clinical presentation may be related to size of the lesion, presence or absence of hormone secretion or degree of local invasion.

Approximately 60–70% of adenomas present with symptoms related to hormone secretion while the remainder present with symptoms related to mass effect [18]. Microadenomas most commonly present with a clinical picture reflecting the hormone excess. The most common functioning adenomas are prolactinomas. Other functioning adenomas include adrenocorticotropic hormone-secreting tumors, thyroid-stimulating hormone secreting tumors and growth hormone secreting tumors. Non-secreting adenomas go unrecognized until they produce symptoms such as visual compromise, due to compression of the optic chiasm, nerve or tract, or cranial nerve pathology due to local invasion of the cavernous sinus.

Although adenomas are overwhelmingly benign, they have been known to metastasize, with seeding of the CSF [19], and are associated with an increased morbidity and mortality due to an increased incidence of cardiovascular disease and cerebrovascular disease [20,21]. Patients with non-functioning adenomas and acromegaly have also been shown to have a significantly higher incidence of malignancy than the general population [22].

The imaging appearance of pituitary adenomas is non-specific, and no inference to histology can be made from the sellar patterns. On MR, microadenomas appear as areas of non-enhancement (Figure 39.1). This reflects the difference in enhancement pattern of normal pituitary tissue when compared with enhancement patterns of adenomas. The gland is not generally enlarged, although they may result in a change in the contour of the gland, with the upper margin becoming convex, the floor of the sella demonstrating a down-sloping, or with deviation of the infundibulum. It has been reported that more

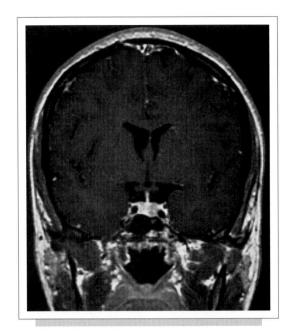

FIG. 39.1. Microadenoma. Conventional coronal T1-weighted image with contrast through the pituitary. The patient is a 21-year-old female with headaches. She was not found to have significant endocrine abnormalities. The image demonstrates a small hypointense (non-enhancing relative to the pituitary) lesion of the left pituitary consistent with a microadenoma.

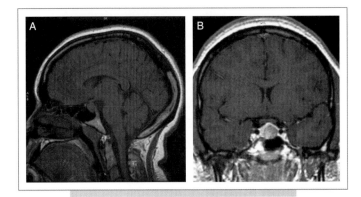

FIG. 39.2. Macroadenoma. **(A)** Sagittal T1-weighted image pre-contrast. **(B)** Coronal T1-weighted image post-contrast. Images demonstrate a mass lesion of the sella and suprasellar region. The unenhanced lesion has signal similar to gray matter on T1-weighted images and demonstrates mild to moderate contrast enhancement.

than 90% of microadenomas can be detected by contrast enhanced MRI [1].

The MR appearance of macroadenomas can be variable, although typically they follow gray matter signal on all sequences (Figure 39.2). Cystic changes can be seen in up to 18% of cases (Figure 39.3) [23]. Enhancement is also variable although generally intense and somewhat heterogeneous (Figures 39.4 and 39.5). A typical figure-of-eight

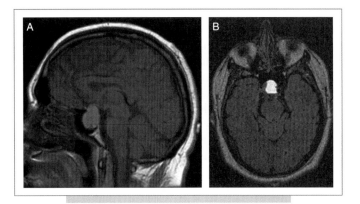

FIG. 39.3. Cystic macroadenoma. **(A)** Sagittal T1-weighted image shows a hyperintense lesion involving the sella and suprasellar region. **(B)** Axial FLAIR image shows a bright focus within the lesion thought to represent cystic change.

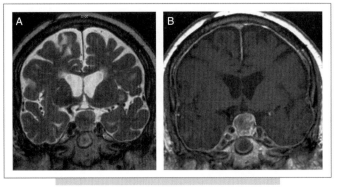

FIG. 39.4. Macroadenoma. **(A)** Coronal T2-weighted image. **(B)** Coronal T1-weighted image post-contrast. Images reveal a predominantly peripherally enhancing macroadenoma in a 78-year-old male, invading the right cavernous sinus and compressing the optic chiasm.

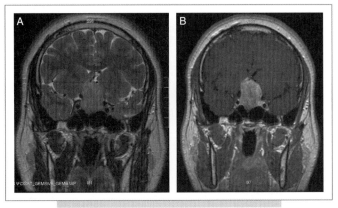

FIG. 39.5. Macroadenoma. **(A)** Coronal T2-weighted image. **(B)** Coronal T1-weighted post-contrast. Sellar and suprasellar pituitary adenoma which displaces adjacent vasculature and optic chiasm. Patient's initial complaint was of visual disturbances.

configuration can be seen due to compression of the tumor at the diaphragma sella (Figure 39.6). Calcifications can be seen although they are not commonly associated with adenomas [24]. Adenomas may also hemorrhage especially in the setting of treatment with bromocriptine (Figure 39.7). Although more commonly seen in the setting of craniopharyngiomas, edema can be seen along the optic tracts in the setting of any pituitary lesion compressing the optic tracts (Figure 39.8) [25]. Saeki et al proposed that this edema-like pattern is seen because of dilated Virchow-Robinson spaces due to drainage obstruction related to the presence of a pituitary lesion [25].

Invasion of adjacent structures is seen in up to 35% of cases and is not indicative of malignancy (Figures 39.8, 38.9 and 38.10) [26]. Approximately 10% of pituitary adenomas involve the cavernous sinus (see Figures 39.4, 39.7 and 39.10) [27]. Knowledge of involvement of the cavernous sinus is important in surgical planning because of increased morbidity/mortality related to the surgery regardless of histologic grade of the tumor. It has been reported that

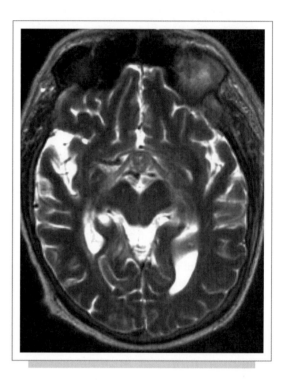

FIG. 39.8. Macroadenoma with edematous changes involving the optic tracts bilaterally. Axial T2-weighted image shows abnormal high signal in the optic tracts bilaterally compatible with edema related to a pituitary adenoma occupying the suprasellar region.

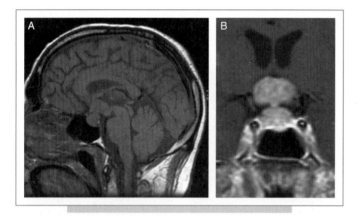

FIG. 39.6. Macroadenoma. **(A)** Sagittal T1-weighted image pre-contrast. **(B)** Coronal T1-weighted post contrast. The patient is a 46-year-old male with lung cancer metastatic to brain. However, given the stability of this lesion and its configuration 'classic figure-of-eight', it was thought that the pituitary lesion represented an incidental macroadenoma. The waist of the lesion is caused by the diaphragma sella through which the lesion has passed.

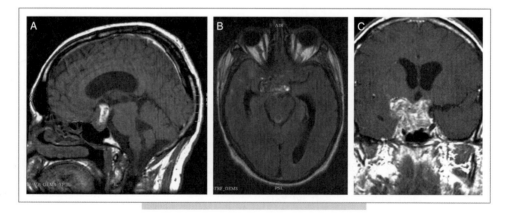

FIG. 39.7 Invasive macroadenoma with possible calcifications versus hemorrhage. **(A)** Sagittal T1-weighted image. **(B)** Axial FLAIR. **(C)** Post-contrast enhanced coronal T1-weighted image. This is an invasive macroadenoma involving the cavernous sinus which demonstrates areas of increased signal on T1-weighted images before the administration of contrast, as well as areas of decreased signal on FLAIR and T2-weighted images. These foci may represent calcifications or hemorrhage.

encasement of the carotid artery is the most specific sign of cavernous sinus invasion [28]. Cottier et al suggest that if the tumor encases the internal carotid artery by 25% or less, or the margin drawn between the medial aspect of the supra and intracavernous internal carotid artery was not invaded by tumor, the negative predictive value was 100% [29]. Asymmetric tentorial enhancement has also been described as a useful sign for suggesting invasion or severe compression by a sellar tumor on the cavernous sinus [30].

Prolactinomas and growth-hormone secreting tumors can be treated medically [31]. Bromocriptine is commonly used in prolactinomas, with MRI used to evaluate the patient's response to therapy [32]. A decrease in tumor size can be seen as early as 1 week after the start of therapy. Additionally, MRI can detect post-therapy hemorrhage into macroadenomas and mass effect or inferior herniation of the chiasm as a result of a decrease in the tumor size [33]. Bromocriptine has been associated with an increased incidence of intratumoral hemorrhage, also known as pituitary apoplexy [34]. The clinical syndrome is characterized by sudden headache, vomiting, visual impairment and meningismus, caused by rapid enlargement of an adenoma due to hemorrhagic infarction [35]. Subarachnoid hemorrhage and vasospasm have been reported [36]. Octreotide is commonly used in patients with acromegaly due to excess secretion of growth hormone [37]. Trans-sphenoidal surgery is the preferred approach to the resection of pituitary adenomas because it is associated with lower morbidity and mortality than the transcranial approach, which is generally the preferred approach for large tumors [38].

Meningioma

Meningiomas are the second most common primary brain tumors of the sellar and juxtasellar region representing 11% of tumors in this region and second only to pituitary adenomas which represent 54% of all lesions [2,23]. Ten percent of all meningiomas occur in the parasellar region. Meningiomas can arise from the diaphragma sella (Figure 39.11), tuberculum sella, medial sphenoid ridge, optic nerve sheaths, clinoids (Figure 39.12), cavernous sinus (Figure 39.13) or clivus, as well as the planum sphenoidale (Figure 39.14). Rarely, meningiomas can be intrasellar, probably arising from the diaphragma or tuberculum sella and growing downward. Clinically, patients usually present with visual disturbances, because of involvement of the cavernous sinus, and without endocrine dysfunction.

On CT, calcification is common, as is hyperostosis. Contrast enhancement is typically intense. MR is the examination of choice for meningiomas because of its multiplanar capability. On MR, meningiomas are typically isointense to gray matter on T1- and isointense or mildly to moderately hyperintense on T2-weighted images and demonstrate homogeneous and relatively intense contrast enhancement (see Figure 39.14), in contrast to

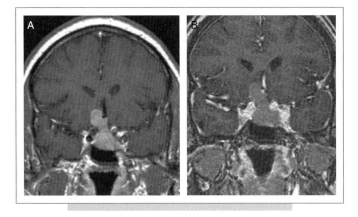

FIG. 39.10. Invasive macroadenoma. **(A)** Coronal T1-weighted post contrast. **(B)** Thin section T1-weighted post contrast. Images show a macroadenoma extending into the inferior right frontal lobe. The lesion invades the left cavernous sinus. Patient presented due to visual disturbances in the left eye.

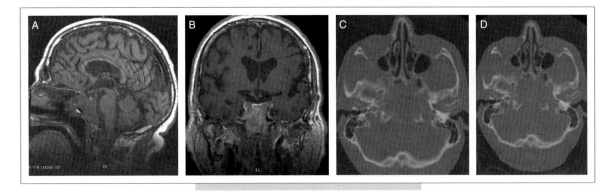

FIG. 39.9. Invasive macroadenoma. **(A)** Sagittal T1-weighted image without contrast. **(B)** Coronal T1-weighted post-contrast enhanced image. Images demonstrate a mass in the pituitary region which appears to invade the clivus. **(C)** and **(D)** Axial CT image at the level of the clivus (bone windows) shows destruction of much of the clivus related to an invasive adenoma. This is an 80-year-old male who had a CT scan of the head following a syncopal episode.

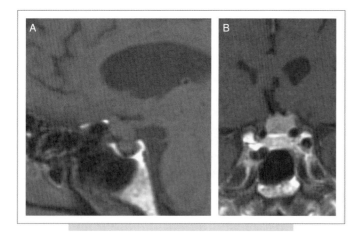

FIG. 39.11. Meningioma of the diaphragma sella. **(A)** Pre-contrast T1-weighted sagittal image. **(B)** Coronal post-contrast enhanced image shows a uniformly enhancing lesion arising from the diaphragmatic sella. The normally enhancing pituitary can be seen as a separate structure inferior to the lesion.

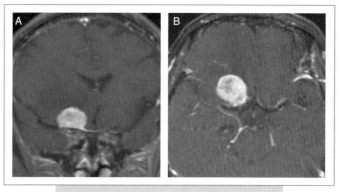

FIG. 39.12. Clinoid meningioma. **(A)** Coronal T1-weighted post-contrast enhanced image. **(B)** Axial T1-weighted post-contrast enhanced image. Images show a homogeneously enhancing lesion which abuts the left anterior clinoid process. The coronal image also shows a classic dural tail (enhancement of the subjacent dura) extending across the planum sphenoidale suggesting the diagnosis of meningioma.

macroadenomas where enhancement is not as intense and somewhat heterogeneous [39]. A dural tail is helpful although not diagnostic, being present in up to 57% of cases (see Figure 39.12A) [2]. The epicenter of the lesion is usually not in the sella, with the sella almost always being normal and the pituitary gland easily identified. This is the major distinguishing factor from macroadenomas [39]. Another distinguishing factor is the fact that meningiomas typically encase and constrict the carotids (see Figure 39.13), whereas macroadenomas may encase but typically do not constrict vessels. Invasion of the carotid artery is frequently seen, even when there is no evidence of narrowing of the artery [40]. Making the diagnosis preoperatively is important since meningiomas are treated via craniotomy rather than by a trans-sphenoidal approach. Surgical resection is

considered to be the treatment of choice for meningiomas, whenever this can be accomplished with acceptable morbidity. However, radiosurgery can also be an effective management strategy for many patients with meningiomas [41].

Chiasmatic and Hypothalamic Gliomas

Gliomas of the hypothalamus and optic pathways represent 5% of pediatric intracranial tumors [42]. The vast majority represent slow growing pilocytic astrocytomas, although malignant gliomas and, in particular, glioblastoma multiforme may occur, especially in adults [43]. The most common locations for pilocytic astrocytomas are the cerebellum, optic nerve and chiasm and hypothalamic region [44]. Pilocytic astrocytomas occur most commonly

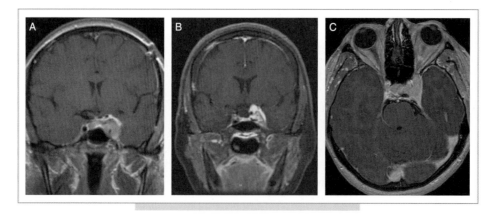

FIG. 39.13. Meningioma of the cavernous sinus. **(A** and **B)** Coronal post-contrast enhanced T1-weighted image. **(C)** Axial T1-weighted post-contrast enhanced image. The patient has transient vision loss. Images show a relatively homogeneously enhancing mass involving the left cavernous sinus, encasing and narrowing the internal carotid artery which is characteristic of meningiomas. Patient was a 51-year-old woman with a history of left-sided visual disturbances.

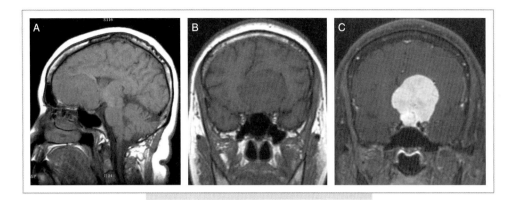

FIG. 39.14. Meningioma of the planum sphenoidale. **(A)** Sagittal T1-weighted image demonstrates a mass lesion arising from the region of the planum sphenoidale along the midline, which is essentially isointense to gray matter. **(B)** Coronal T1-weighted image without contrast shows an isointense lesion in the anterior cranial fossa. **(C)** Coronal T1-weighted image post-contrast shows intense and uniform enhancement. The imaging findings are classic for a meningioma of the planum sphenoidale.

in the first two decades of life, in 75% of cases [45]. Mean age of presentation is 5 years of age [46]. An association with neurofibromatosis 1 (NF1) has been reported with up to 15–21% of NF1 patients developing pilocytic astrocytomas, usually involving the optic nerve or chiasm [47, 48].

It is difficult to discern the site of origin of chiasmatic/hypothalamic gliomas by imaging in most cases, since both the optic chiasm and the hypothalamus are involved regardless of the site of origin [49]. For this reason, both are discussed as one entity. Chiasmatic gliomas may occasionally demonstrate extension along the optic tracts or optic nerves thus indicating their site of origin. Patients usually present with visual symptoms, headaches and endocrine abnormalities. Endocrine abnormalities occur in 42%, with growth hormone deficiency being the most common [46,50].

MR is the examination of choice. The lesions tend to be solid with microcyst formation. They are usually iso- or hypointense on T1-weighted images, hyperintense on T2-weighted images and demonstrate variable enhancement with contrast. These lesions do not typically hemorrhage or calcify. The sella and its contents are usually normal [46,50].

Generally, the treatment of choice for pilocytic astrocytomas is resection, except for lesions involving the optic pathway and hypothalamic region where the treatment of choice is chemotherapy and radiation [44]. Chemotherapy may be used to postpone radiation treatment until after the age of 5, which may reduce neurological morbidity [42]. Chiasmatic/hypothalamic gliomas are more aggressive in the very young and in adults [42,51,52]. In NF1I the course is more indolent [42]. Survival is 93% at 5 years and 74% at 10 years with chiasmatic lesions having a 19-year 44% survival [53]. The prognosis of chiasmatic gliomas is worse than with optic nerve gliomas [53].

Germinoma

The sellar and suprasellar region is the second most common location of CNS germ cell tumors. The most common location is the pineal gland. The most commonly occurring CNS germ cell tumor is a germinoma. Other germ cell tumors such as teratomas, embryonal carcinomas, choriocarcinomas and mixed tumors are much less common intracranially, particularly in the sellar and suprasellar region. However, these lesions are not as common as other CNS tumors, representing approximately 0.5–3% of all CNS tumors [54]. These lesions occur in children and young adults. Synchronous pineal and suprasellar lesions occur in 6–12% of germinomas and are considered diagnostic [55]. Primary suprasellar germinomas have no sex predilection, in contrast to pineal germinomas which show a male predominance [50]. The clinical presentation often includes hypopituitarism, diabetes insipidus and visual disturbances.

On MR, the mass appears as a well marginated, round or lobulated, homogeneous tumor with prolonged T1 and T2 relaxation times, which strongly enhances after gadolinium administration (Figure 39.15). The presence of these imaging findings in a young patient, along with a clinical history of diabetes insipidus, is a strong clue to the diagnosis of germinoma [56]. Also, germinomas may secrete beta-human chorionic gonadotropin or alpha-fetoprotein, which can be detected in the CSF or blood and can aid in the preoperative diagnosis. Germinoma cells frequently spread through the CSF. Therefore, imaging the entire neural axis may be helpful.

Biopsy is necessary prior to treatment except in the case of synchronous suprasellar and pineal lesions. The lesions are extremely radiosensitive, with over 90% of patients being effectively treated with radiation therapy alone. Germinomas are, however, also very chemosensitive, with reports suggesting that the dose and volume of radiation required can be lessened with the addition of adjuvant chemotherapy [57]. At our institution, intra-arterial chemotherapy after disruption of the blood–brain barrier has been used with good results. Radical resection offers no benefit over biopsy with subsequent radiation and or chemotherapy, making the preoperative diagnosis very important in decision-making [58]. Prognosis is good with 10- and 20-year survival rates of 92.7% and 80.6% respectively [59].

Metastasis

Metastases to the pituitary gland represent 1–4% of pituitary tumors, with 50% being of breast origin and 20% from lung [60]. Metastases to the sellar, suprasellar or parasellar regions may arise in the sphenoid bone or sinus, cavernous sinus, pituitary gland, hypothalamus or surrounding soft tissues. It may be difficult to distinguish a metastasis from a primary pituitary abnormality on the basis of imaging alone; however, the presence of bony destruction or the history of a known primary tumor may be helpful. The incidence of pituitary metastases ranges from 0.14 to 28.1% of

all brain metastases, the incidence being higher in autopsy series [61]. Sellar metastases appear to favor the posterior lobe of the pituitary and the pituitary stalk which may be related to the difference in blood supply when compared with the anterior lobe [62]. The posterior lobe of the pituitary derives its blood supply primarily from arterial blood flow, whereas the anterior lobe receives most of its blood supply from portal vessels. These patients frequently present with diabetes insipidus which occurs in 33–70% of patients with symptomatic pituitary metastases, in contrast to only 1–2% of patients with pituitary adenomas (Figures 39.16 and 38.17). Primary carcinomas of the pituitary gland are rare [63]. Metastases to the gland are more common.

Schwannoma

Schwannomas, which are tumors derived from the myelin sheath of peripheral nerves, can be found involving

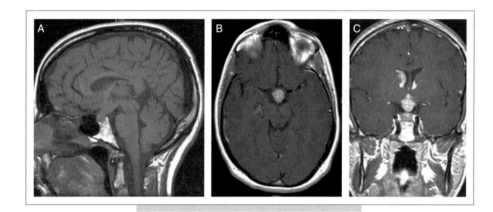

FIG. 39.15. Germinoma. **(A)** T1-weighted sagittal. **(B)** Axial T1-weighted post contrast. **(C)** Coronal T1-weighted post contrast. Images show a lobulated mass which appears to enhance uniformly in the sellar/suprasellar region. Patient was a 28-year-old female complaining of abnormal menses. Abnormal enhancement is also seen in the region of the lateral ventricles bilaterally compatible with tumor involvement.

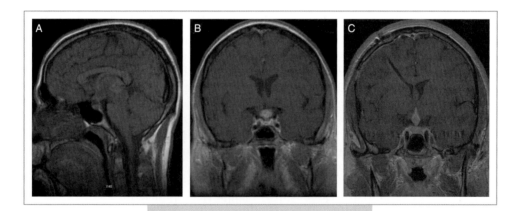

FIG. 39.16. Metastatic anaplastic oligodendroglioma involving the pituitary stalk. **(A)** T1-weighted sagittal image. **(B** and **C)** Post-contrast enhanced coronal T1-weighted images show an irregular suprasellar lesion involving the infundibulum. The patient is a 19-year-old male with a history of anaplastic oligodendroglioma in the right frontal lobe. Multiple lesions seen on brain MRI were thought to represent multifocal metastases due to CSF seeding.

the cranial nerves within the cavernous sinus and parasellar regions. In general, pituitary function is not affected; however, often the cranial nerves III, IV, V and VI are affected within the cavernous sinuses or in the suprasellar and prepontine cisterns. Schwannomas most commonly arise from the trigeminal nerve, but may involve other cranial nerves as well [64]. Remodeling of the skull base foramina where individual nerves exit may occur. When multiple lesions are seen, neurofibromatosis should be considered. On CT, schwannomas are usually hyperdense lesions with homogeneous enhancement and they may be hard to differentiate from meningiomas. On MRI, they may be isointense or hyperintense to gray matter on T1-weighted images and they enhance homogeneously [64].

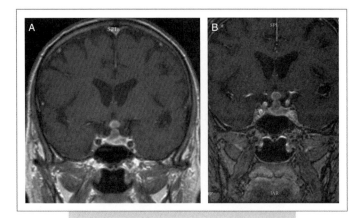

FIG. 39.17. Infundibular metastases. **(A)** Coronal T1-weighted image through the sellar region. **(B)** Coronal SPGR post-contrast enhanced images through the infundibulum demonstrate a small enhancing lesion of the infundibulum in a patient with renal cell carcinoma representing metastatic disease.

Hypothalamic Hamartoma

A hamartoma of the tuber cinereum usually presents as precocious puberty in a young child [65]. It is important to differentiate this lesion from a hypothalamic glioma, because the prognosis for hamartoma is much more favorable. Imaging is best with thin-section coronal and sagittal MRI. The findings are usually characteristic: the mass arises from the undersurface of the hypothalamus and is exophytic. The nodular mass (<1 cm) hangs into the suprasellar cistern adjacent to the mammillary bodies. On T1-weighted images, the signal is isointense with normal brain and on T2-weighted images there is mild hyperintensity or isointensity. The distinguishing feature is that these lesions do not usually enhance with contrast administration.

Pituitary Hyperplasia

The pituitary gland commonly enlarges during puberty and pregnancy. The gland may also increase in size in the setting of endocrine dysequilibrium. For example, in the setting of primary hypothyroidism, the pituitary gland will grow in size. This hyperplasia is thought to be related to lack of negative feedback on the hypothalamus by circulating thyroid hormone [66]. On imaging, the pituitary gland will appear enlarged, but will not show any non-enhancing lesions or non-uniformity on post-contrast enhanced images (Figure 39.18). Following therapy for hypothyroidism, the gland will decrease in size [67].

MIXED SOLID AND CYSTIC LESIONS

Craniopharyngioma

Craniopharyngiomas are the most common intracranial tumor of non-glial origin in children, comprising up to 10% of pediatric brain tumors [31,68,69]. They are formed from ectodermal remnants of Rathke pouch and are

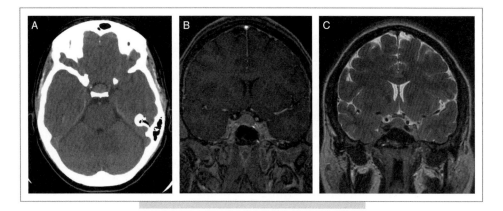

FIG. 39.18. Pituitary hyperplasia. **(A)** Axial CT scan through the region of the pituitary shows a homogeneously enhancing mass. **(B)** Coronal post-contrast enhanced SPGR images demonstrated a prominent pituitary gland which enhances homogeneously. **(C)** Coronal T2-weighted image shows a prominent pituitary with uniform signal. Patient was a young female who had undergone a total thyroidectomy followed by I–131 therapy for thyroid cancer.

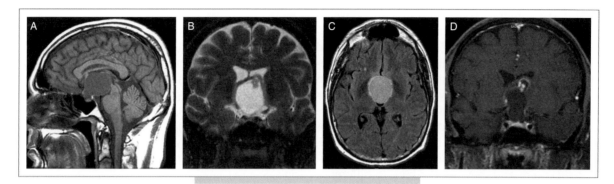

FIG. 39.19. Craniopharyngioma-cystic. **(A)** Sagittal T1-weighted image. **(B)** Axial T2-weighted image. **(C)** Axial FLAIR. **(D)** Coronal T1-weighted post-contrast enhanced image. The images demonstrate a predominantly cystic mass lesion in the suprasellar region. Post-contrast enhanced images, as well as the T2-weighted images, show an enhancing solid nodular density associated with this cystic mass suggesting the diagnosis of craniopharyngioma.

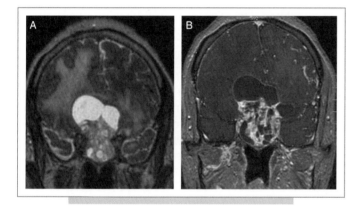

FIG. 39.20. Mixed solid and cystic craniopharyngioma. **(A)** Coronal T2-weighted image. **(B)** Coronal T1-weighted post-contrast enhanced image. The images show a mixed solid/cystic lesion centered in the suprasellar cistern with marked enhancement of the solid component on post-contrast enhanced images. The mass compresses and may invade the third ventricle resulting in hydrocephalus.

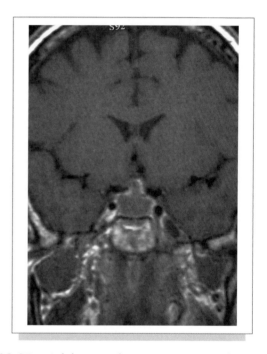

FIG. 39.21. Solid craniopharyngioma. Coronal T1-weighted post-contrast enhanced image shows a solid lesion in the suprasellar region which is a non-specific finding. This lesion was resected and found to represent a solid craniopharyngioma.

composed of a squamous epithelium. Craniopharyngiomas can occur anywhere from the floor of the third ventricle (hypothalamus) to the pharyngeal tonsils, with 67% being found in the suprasellar region. These tumors have a bimodal incidence, with peaks in the first and fifth decades. Although they are histologically benign, they are behaviorly aggressive, invading adjacent structures and thus making resection difficult. Recurrence is local, although meningeal seeding has been described [70].

Two distinct subtypes are recognized: the adamantinous, which tend to occur in children, and the squamous-papillary variants, which tend to occur in adults [71]. Craniopharyngiomas are thought to be part of a continuum of ectodermally derived cystic epithelial lesions which includes arachnoid cysts, Rathke's cleft cysts, epidermoids and dermoids, with craniopharyngiomas being at the more

aggressive end of the spectrum [72]. They typically present because of mass effect on the chiasm and hydrocephalus, with visual disturbances, headaches, pituitary and hypothalamic dysfunction. Endocrine deficiency is seen in 80%, with growth hormone deficiency noted in 75% [73].

Craniopharyngiomas can be predominantly cystic (Figure 39.19), mixed cystic/solid (Figure 39.20), or entirely solid (Figure 39.21), although the mixed solid and cystic appearance is rather characteristic. Hemorrhage is not an uncommon finding, particularly within cystic portions of the tumors. Craniopharyngiomas may grow to compress the optic chiasm superiorly, to displace the normal pituitary gland and stalk, invade the cavernous sinuses, and even to encase or occlude the carotid arteries. CT is the examination of choice for evaluation of calcification which is seen in 87% of cases (see Figure 39.23) [74]. On MRI, high signal intensity on T1- and T2-weighted images is seen in cysts with high cholesterol content or with subacute hemorrhage. Craniopharyngiomas can also be of low signal intensity on T1-weighted images if the cyst contains a large amount of keratin [75]. Fluid levels can be seen in cystic regions. Adamantinous craniopharyngiomas tend to be primarily cystic or mixed cystic-solid lesions that occur in children, whereas squamous-papillary subtypes tend to be predominately solid or mixed solid-cystic and occur in adults [71]. Distinguishing between the two has a prognostic significance since adamantinous tumors tend to recur. MRI can be helpful in distinguishing between the two, with encasement of vessels, a lobulated shape and the presence of hyperintense cysts favoring adamantinous tumors and a round shape, presence of hypointense cysts and a predominately solid appearance being seen with squamous-papillary tumors [71]. Calcification, encasement of vessels and recurrence favor adamantinous craniopharyngiomas [71]. Enhancement tends to be heterogeneous with the solid portions of the lesion enhancing after the administration of contrast (Figure 39.20). Preoperative differentiation from arachnoid cysts, Rathke cleft cysts, epidermoids and dermoids can be difficult. Calcification and solid components are features more commonly seen with craniopharyngiomas [74]. In adults, calcified aneurysms must be part of the differential of calcified lesions in the sellar, parasellar and suprasellar region.

Treatment is primarily surgical, with the efficacy of radiotherapy being well documented. Gamma knife radiosurgery may also play a role, although there are no specific guidelines [76]. Recurrence-free survival after total resection is 86.9% at 5 years, but falls to 48.8% with subtotal resection [77].

Epidermoid and Dermoid

Epidermoids and dermoids are uncommon, slow growing masses that account for 1% of all intracranial neoplasms [78]. These lesions are similar in their development, histology, behavior and imaging and for this reason are discussed together. Both lesions are generally considered developmental/congenital masses rather than neoplastic, arising from ectodermal heterotopia. Both cysts are lined with stratified squamous epithelium, with dermoids adding mesodermal elements such as hair, sebaceous and sweat glands.

Epidermoids are slightly more common than dermoids intracranially. They typically spread along the basal surfaces, with the cerebellopontine angles being the most common location, followed by the parasellar region [79,80,81]. They are extra-axial lesions with only 1.5% being intracerebral

[82]. They are overwhelmingly benign, although rarely they can be malignant. Average age of presentation is 37.3, with a male to female ratio of 3:2 [78]. The symptomatic onset is generally slow, lasting 2 years or more, although for suprasellar lesions it is much shorter [82]. Presenting symptoms may include headaches, visual problems, cranial nerve symptoms and seizures, which typically indicate rupture. Rupture can produce aseptic meningitis, which can be lethal, although not necessarily so.

Epidermoids on CT appear as hypodense masses, with irregular borders and rare contrast enhancement. Dense lesions have been reported and calcification is occasionally seen [81,83]. On MR, they typically are of low signal on T1- and of increased signal on T2-weighted images, following that of CSF on all pulsing sequences [78]. Classically, they can be differentiated from arachnoid cysts on diffusion-weighted images where these lesions show restricted diffusion and are bright, whereas arachnoid cysts are not bright.

Dermoids are midline lesions, occurring in the parasellar, frontobasal region or posterior fossa [84]. Average age of presentation is 36.2 with a male to female ratio of 3:1 [78]. The complications of dermoids are similar to epidermoids. They can present with headaches, seizures, meningeal signs and transient ischemic attacks (TIAs) [84,85]. Most of these symptoms are indicative of rupture, which produces a chemical or aseptic meningitis and which can be lethal [84,86]. The CT appearance of dermoids is similar to epidermoids. Their MR appearance depends on the amount of fat present, although generally they are of increased signal on both T1- and T2-weighted images [78]. CT or MR can both make the diagnosis of rupture, although MR is the preferred preoperative study [85].

Treatment is surgery, with 86% being in good or excellent condition postoperatively. The 20-year survival of epidermoids is 92.8%, with good survival even with recurrence [78]. Epidermoids have a classic mother-of-pearl appearance at surgery.

Rathke Cleft Cyst

Rathke cleft cysts are epithelial cysts that arise from Rathke's pouch remnants. Small Rathke cleft cysts are commonly found at autopsy in 12–33% of normal pituitary glands with cysts noted by imaging being much less common [87]. They are usually found in the pars intermedia between the anterior and posterior lobes of the pituitary [88]. Histologically, they have a single row of cuboidal or columnar epithelial lining in contrast to craniopharyngiomas, which have a squamous epithelial lining.

Rathke cleft cysts are usually intrasellar cysts (Figure 39.22) but may be intrasellar and suprasellar, rarely purely suprasellar and have been described even in the sphenoid sinus [89]. They may be asymptomatic, although they frequently present with endocrine abnormalities, headaches and visual field defects, especially when there is suprasellar extension [74]. Nishioka et al found that the presence of anterior pituitary dysfunction or headaches was not related to cyst size, but was more likely to be present in patients whose cyst demonstrated a high/isointense T1-weighted signal as opposed to those demonstrating low signal intensity [90]. Mean age of presentation is 38 years of age, with a female predominance of 2:1 [31,91]. Teramoto et al found that 87% of incidental lesions found in the medial aspect of the pituitary gland were Rathke cleft cysts, whereas 74% of incidental lesions centered in the lateral gland represented adenomas [92].

On CT, the sella may be normal or slightly expanded. The cysts can usually be seen as an area of low attenuation that is similar to CSF. Calcification is seen in approximately 13% of cases usually peripherally, in contrast to craniopharyngiomas in which calcification is seen in 87% of cases [74]. Peripheral rim enhancement may be seen [23]. On MRI, the

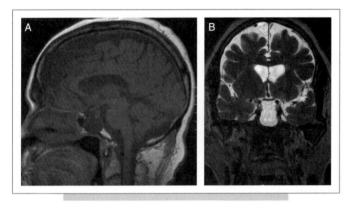

FIG. 39.22. Rathke's cleft cyst. **(A)** Sagittal T1-weighted image without contrast demonstrates a cystic intrasellar lesion. **(B)** Coronal T2-weighted images show that this cystic structure follows CSF signal. The sella is expanded. This lesion was thought to represent a Rathke's cleft cyst. Similar findings could be seen in the setting of empty sella or arachnoid cyst.

cysts are usually isointense with CSF on all pulse sequences (see Figure 39.22). Occasionally, they may have a more unusual signal due to varying cyst fluid composition. Cyst wall biopsy and aspiration, usually trans-sphenoidal, is considered to be curative with a low recurrence rate of 5% [72,91].

Arachnoid Cyst

These non-neoplastic cysts are a rare cause of cystic lesions in the sellar region. As noted previously, it has been suggested that they are a part of a continuum of epithelial cysts at the more benign end of the spectrum behaviorally. They are thought to be developmental lesions arising from an imperforate membrane of Liliequist, which develops into a diverticulum and subsequently a cyst when it loses its communication with the subarachnoid space [93]. They tend to present at an older age, usually the fifth decade, with headaches, visual field defects and impotence [74]. On MR, these sellar and/or suprasellar cysts follow CSF signal on all pulsing sequences (Figure 39.23) as do epidermoids, but with no restricted diffusion which helps distinguish them from epidermoids. They are well defined, with no calcification and no enhancement [50,74].

Empty Sella

The term empty sella refers to a condition in which the sella is filled with CSF with significant flattening of the pituitary gland with possible enlargement of the sella turcica (Figure 39.24). The etiology for this finding is thought to be a defect in the diaphragma sella which exposes the sella and its contents to CSF pulsations which, over time, enlarges the sella and flattens the gland. It is generally considered an incidental finding with most patients with this finding demonstrating no clinical sequelae. However, some patients may have endocrine deficiencies, headaches or visual disturbances. A strong statistical correlation between

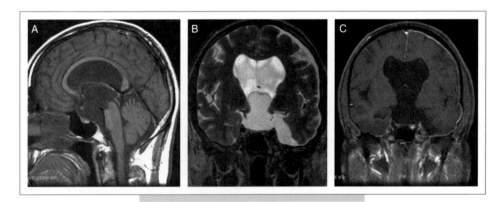

FIG. 39.23. Arachnoid cyst. **(A)** Sagittal T1-weighted image. **(B)** Coronal T2-weighted image. **(C)** Coronal T1-weighted post-contrast enhanced image. Images show a non-enhancing lesion centered in the sellar/suprasellar region which follows CSF signal on all sequences and was thought to represent an arachnoid cyst. Patient was scanned following a traumatic event. The arachnoid cyst was an incidental finding.

an empty sella and idiopathic intracranial hypertension has been established [94].

PITUITARY STALK LESIONS

The thickness of the normal pituitary stalk is approximately 2 mm. The normal stalk enhances markedly on CT and MRI with contrast. The most common clinical problem associated with disease of the pituitary stalk is diabetes insipidus. When this is present, there usually is absence of the normal hyperintensity of the posterior pituitary noted on T1-weighted MRI. Diabetes insipidus may also occur as a result of transection of the pituitary stalk.

The differential diagnosis of a thickened stalk includes sarcoidosis, tuberculosis, histiocytosis X, germinoma, lymphoma (Figure 39.25), leukemia, metastasis (see Figures

39.17 and 39.18) and ectopic posterior pituitary. A thickened stalk can also be due to an extension of a glioma from the hypothalamus.

Clinical findings may help in sorting out the etiology for pituitary stalk thickening. For example, in patients with neurosarcoidosis or tuberculous infiltration of the stalk, the chest radiograph is generally abnormal and may be helpful in the differentiation from histiocytosis X. Clinically, patients with histiocytosis X may have skin lesions, otitis media, or bone lesions, in addition to interstitial lung disease [95].

Neurosarcoidosis causing clinical symptoms is an uncommon manifestation of sarcoidosis. It has been reported that symptomatic neurosarcoidosis occurs in up to 5% of patients [96]. The most common clinical symptoms include cranial nerve dysfunction, aseptic meningitis related to leptomeningeal spread, hydrocephalus and dysfunction of the hypothalamic-pituitary axis such as diabetes insipidus [96,97]. Neurosarcoidosis commonly affects the leptomeninges with findings of basilar leptomeningitis affecting the pituitary gland region and subjacent structures [97]. Mass lesions as a result of sarcoid are less common (Figure 39.26).

MISCELLANEOUS LESIONS

Lymphocytic hypophysitis is a rare entity affecting the pituitary gland and stalk. It is thought to represent an inflammatory disorder, although the exact cause is unknown [98]. Clinically, patients may complain of headache, visual disturbances, symptoms related to diabetes insipidus or abnormal menses. The pituitary hormone levels may be depressed. On imaging, the lesion may present as a mass involving the adenohypophysis and mimicking pituitary adenomas (Figure 39.27) and/or present as a mass

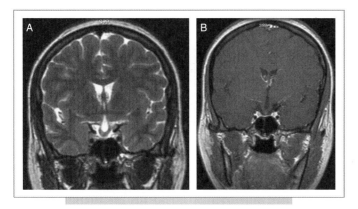

FIG. 39.24. Empty sella. **(A)** Coronal T2-weighted image. **(B)** Coronal T1-weighted post contrast. Images show an empty sella with downward herniation of the optic chiasm. Patient was a young female with panhypopituitarism.

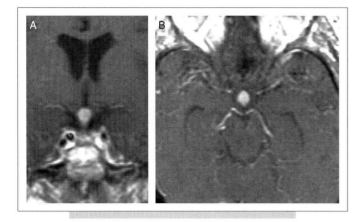

FIG. 39.25. Lymphoma. **(A)** Coronal and **(B)** axial T1-weighted images following the administration of contrast demonstrate an enhancing lesion involving the pituitary stalk in a patient with a history of systemic lymphoma. Findings resolved following chemotherapy for lymphoma.

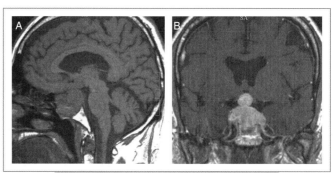

FIG. 39.26. Neurosarcoid. **(A)** Sagittal T1-weighted image. **(B)** Coronal T1-weighted image after contrast administration. Images show a lobulated mass involving the sellar and suprasellar regions which enhances on post-contrast images. The patient was a 62-year-old man with known sarcoidosis complaining of visual disturbances. His symptoms improved following a steroid regimen.

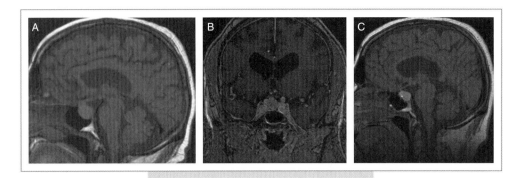

FIG. 39.27. Lymphocytic hypophysitis. **(A)** T1-weighted sagittal image showing a pituitary mass. **(B)** Coronal thin section T1-weighted image following the administration of contrast shows enhancement of this lesion. Patient was a 74-year-old male who presented with headaches. **(C)** Follow-up T1-weighted sagittal image showing increased T1-signal within the lesion thought to represent hemorrhage.

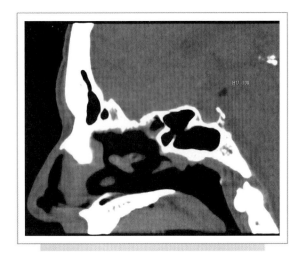

FIG. 39.28. Suprasellar lipoma. Midline sagittal CT image without contrast reveals a low density lesion, measuring [m]108 HU, of the suprasellar cistern representing an incidental lipoma. This was an incidental finding in a patient complaining of headache.

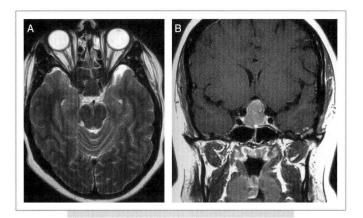

FIG. 39.29. Choristoma. **(A)** Axial T2-weighted image and **(B)** axial T1-weighted post-contrast enhanced image show a well-circumscribed sellar/juxtasellar mass which enhances uniformly. Its appearance is suggestive of an adenoma. Histology after resection revealed a choristoma.

primarily involving the infundibulum and the neurohypophysis. Lymphocytic hypophysitis may regress following steroid therapy.

Other entities that should be considered in the differential diagnosis of lesions centered in the sellar and juxtasellar regions include aneurysms, lymphoma, leukemia, teratoma, chordoma, chondrosarcoma, melanoma, nasopharyngeal carcinoma, mucoceles, lipoma (Figure 39.28), hemangioma [99], pituitary astrocytomas [100], xanthogranuloma and choristoma (Figure 39.29) [101].

REFERENCES

1. Elster AD (1993). Modern imaging of the pituitary. Radiology 187:1–14.
2. Johnsen DE, Woodruff WW, Allen IS et al (1991). MR imaging of the sellar and juxtasellar regions. Radiographics 11:727–758.
3. Dietrich RB, Lis LE, Greensite FS, Pitt D (1995). Normal MR appearance of the pituitary gland in the first 2 years of life. Am J Neuroradiol 16:1413–1419.
4. Wolpert SM, Molitch ME, Goldman JA, Wood JB (1984). Size, shape and appearance of the normal female pituitary gland. Am J Roentgenol 143:377–381.
5. Miki Y, Matsuo M, Nishizawa S et al (1990). Pituitary adenomas and normal pituitary tissue: enhancement patterns on Gadopentetate-enhanced MR imaging. Radiology 177:36.

6. Bartynski WS, Lin L (1997). Dynamic and conventional spin-echo MR of pituitary microlesions. Am J Neuroradiol 18(5):965–972.

7. Naidich MJ, Russell EJ (1999). Current approaches to imaging of the sellar region and pituitary. Endocrinol Metab Clin N Am 28(1):45–79.

8. Miller DL, Doppman JL, Nieman LK et al (1990). Petrosal sinus sampling: discordant lateralization of ACTH-secreting pituitary microadenomas before and after stimulation with corticotropin-releasing hormone. Radiology 176:429.

9. Graham KE, Samuel MH, Nesbit GM et al (1999). Cavernous sinus sampling is highly accurate in distinguishing Cushing's disease from the ectopic adrenocorticotropin syndrome and in predicting intrapituitary tumor location. J Clin Endocrinol Metab 84(5):1602–1610.

10. Mamelak AN, Dowd CF, Tyrrell JB, McDonald JF, Wilson CB (1996). Venous angiography is needed to interpret inferior petrosal sinus and cavernous sinus sampling data for lateralizing adrenocorticotropin-secreting adenomas. J Clin Endocrinol Metab 81(2):475–481.

11. Oliverio PJ, Monsein LH, Wand GS, Debrun GM (1996). Bilateral simultaneous cavernous sinus sampling using corticotropin-releasing hormone in the evaluation of Cushing's disease. Am J Neuroradiol 17:1669–1674.

12. Miller DL, Doppman JL, Peterman SB, Nieman LK, Oldfield EH, Chang R (1992). Neurologic complication of petrosal sinus sampling. Radiology 185:143–147.

13. Asa SL (1999). The pathology of pituitary tumors. Clin N Am 28(1):13–43.

14. Kovacs K, Scheithauer B, Horvath E, Lloyd R (1996). The world health organization classification of adenohypophysial neoplasms. Cancer 78(3):502–510.

15. Mindermann T (1997). Letter to the Editor: classification of pituitary adenomas. Acta Neurochir 139:267–270.

16. Poussaint TY, Barnes PD, Anthony DC, Spack N, Scott RM, Tarbell NJ (1996). Hemorrhagic pituitary adenomas of adolescence. Am J Neuroradiol 17(10): 1907–1912.

17. Fisher B, Gasper LE, Stitt LW, Noone BE (1994). Pituitary adenomas in adolescents: a biologically more aggressive disease? Radiology 192:869–872.

18. Evanson EJ (2002). Imaging of the pituitary gland. Imaging 14:93–102.

19. De Boucaud L, Dousset V, Caillaud P, Viaud B, Guerin J, Caille JM (1999). Metastases from a pituitary adenoma: MRI. Neuroradiology 41(10):785–787.

20. Brada M, Burchell L, Ashley S, Traish D (1999). The incidence of cerebrovascular accidents in patients with pituitary adenoma. Int J Radiat Oncol Biol Phys 45(3):693–698.

21. Nilsson B, Gustavsson-kadaka E, Bengtsson B, Jonsson B (2000). Pituitary adenomas in Sweden between 1958 and 1991: incidence, survival and mortality. J Clin Endocrinol Metab 85(4):1420–1425.

22. Popovic V, Damjanovic S, Micic D et al (1998). Increased incidence of neoplasia in patients with pituitary adenomas. Clin Endocrinol 49:441–445.

23. Simonetta AB (1999). Imaging of suprasellar and parasellar tumors. Neuroimaging Clin N Am 9(4):717–732.

24. Majos C, Coll S, Aguilera C, Acebes JJ, Pons LC (1998). Imaging of giant pituitary adenomas. Neuroradiology 40:651–655.

25. Saeki N, Uchino Y, Murai H et al (2003). MR imaging study of edema-like change along the optic tract in patients with pituitary region tumors. Am J Neuroradiol 24:336–342.

26. Levy RA, Quint DJ (1998). Giant pituitary adenoma with unusual orbital and skull base extension. Am J Roentgenol 170(1):194–196.

27. Ahmadi J, North CM, Segall HD, Zee CS, Weiss MH (1986). Cavernous invasion by pituitary adenomas. Am J Roentgenol 146:257–262.

28. Scotti G, Yu C, Dillon WP et al (1988). MR imaging of cavernous sinus involvement by pituitary adenomas. Am J Roentgenol 151:799–806.

29. Cottier JP, Destrieux C, Vinikoff-Sonier C, Jan M, Herbreteau D (2000). MRI diagnosis of cavernous sinus invasion by pituitary adenomas. Ann Endocrinol 61(3)269–274.

30. Nakasu Y, Nakasu S, Ito R, Mitsuya K, Fujimoto O (2001). Tentorial enhancement on MR image is a sign of cavernous sinus involvement in patients with sellar tumors. Am J Neuroradiol 22:1528–1533.

31. Freda PU, Post KD (1999). Differential diagnosis of sellar masses. Endocrinol Metab Clin N Am 28(1): 81–117.

32. Pinzone JJ, Katznelson L, Danila DC, Pauler DK, Miller CS, Klibanski A (2000). Primary medical therapy of micro- and macroprolactinomas in men. J Clin Endocrinol Metab 85(9):3053–3057.

33. Lundin P, Bergstrom K, Nyman R et al (1992). Macroprolactinomas: serial MR imaging in long-term bromocriptine therapy. Am J Neuroradiol 13:1287.

34. Yousem DM, Arrington JA, Zinreich JS, Kumar AJ, Bryan RN (1989). Pituitary adenomas: possible role of bromocriptine in intratumoral hemorrhage. Radiology 170 (1):239–243.

35. Randeva H, Schoebel J, Byrne J, Esiri M, Adams CBT, Wass JAH (1999). Classical pituitary apoplexy: clinical features, management and outcome. Clin Endocrinol 51:181–188.

36. Sanno N, Ishii Y, Sugiyama M et al (1999). Subarachnoid hemorrhage and vasospasm due to pituitary apoplexy after pituitary function tests. Acta Neurochir 141:1009–1010.

37. Melmed S (1996). Acromegaly. Metabolism 45(8):51–52.

38. Giovanelli M, Losa M, Mortini P (1996). Surgical therapy of pituitary adenomas. Metabolism 45(8):115–116.

39. Cappabianca P, Cirillo S, Alfieri A et al (1999). Pituitary macroadenoma and diaphragma sellae meningioma: differential diagnosis on MRI. Neuroradiology 41:22–26.

40. Sen C, Hague K (1997). Meningiomas involving the cavernous sinus: histological factors affecting the degree of resection. J Neurosurg 87:535–543.

41. Stafford SL, Pollock BE, Foote, RL et al (2001). Meningioma radiosurgery: tumor control, outcomes, and complications among 190 consecutive patients. Neurosurgery 49(5): 1029–1038.

42. Janss AJ, Grundy R, Cnaan A et al (1995). Optic pathway and hypothalamic/chiasmatic gliomas in children younger than age 5 years with a 6-year follow-up. Cancer 75(4):1051–1059.

43. Barbaro NM, Rosenblum ML, Maitland CG, Hoyt WF, Davis RL (1982). Malignant optic glioma presenting radiologically as a 'cystic' suprasellar mass: case report and review of the literature. Neurosurgery 11(6):787–789.

44. Koeller KK, Rushing EJ (2004). Pilocytic astrocytoma: radiology-pathologic correlation. Radiographics 24:1693–1708.

45. Wallner KE, Gonzales MF, Edwards MSB, Wara WM, Sheline GE (1988). Treatment of juvenile pilocytic astrocytomas. J Neurosurg 69:171–176.

46. Collet-Solberg PF, Sernyak H, Satin-Smith M et al (1997). Endocrine outcome in long-term survivors of low-grade hypothalamic/chiasmatic glioma. Clin Endocrinol 47(1):79–85.

47. Deliganis AV, Geyer JR, Berger MS (1996). Prognostic significance of type 1 neurofibromatosis (von Recklinghausen disease) in childhood optic glioma. Neurosurgery 38:1114–1119.

48. Lewis RA, Gerson LP, Axelson KA, Riccardi VM, Whitford RP (1984). Von Recklinghausen neurofibromatosis. II. Incidence of optic nerve gliomata. Ophthalmology 91:929–935.

49. Albert A, Lee BC, Saint-Louis L, Deck MD (1986). MRI of optic chiasm and optic pathways. Am J Neuroradiol 7(2):255–258.

50. Freda PU, Wardlaw SL (1999). Clinical review 110: diagnosis and treatment of pituitary tumors. J Clin Endocrinol Metab 84(11):3859–3866.

51. Alshail E, Rutka JT, Becker LE, Hoffman HJ (1997). Optic chiasmatic-hypothalamic glioma. Brain Pathol 7(2):799–806.

52. Nishio S, Takeshita I, Fujiwara S, Fukui M (1993). Optico-hypothalamic glioma: an analysis of 16 cases. Childs Nerv Syst 9(6):334–338.

53. Rodriguez LA, Edwards MS, Levin VA (1990). Management of hypothalamic gliomas in children: an analysis of 33 cases. Neurosurgery 26(2):242–246.

54. Fischbein NJ, Dillon Wp, Barkovich AJ (2000). *Teaching Atlas of Brain Imaging*. Thieme, New York. 62–64.

55. Sugiyama K, Uozumi T, Kiya K et al (1992). Intracranial germ cell tumor with synchronous lesions in the pineal and suprasellar regions: six cases and review of the literature. Surg Neurol 38:114–120.

56. Kollias SS, Barkovich AJ, Edwards MS (1991). Magnetic resonance analysis of suprasellar tumors of childhood. Pediatr Neurosurg 17(6):284–303.

57. Packer RJ, Cohen BH, Consy K (2000). Intracranial germ cell tumors. Oncologist 5(4): 312–20.

58. Sawamura Y, de Tribolet N, Ishii N, Abe H (1997). Management of primary intracranial germinomas: diagnostic surgery or radical resection. J Neurosurg 87(2):262–6.

59. Matsutani M, Sano K, Takakura K et al (1997). Primary intracranial germ cell tumors: a clinical analysis of 153 histologically verified cases. J Neurosurg 86(3):446–455.

60. McCutcheon IE, Kitagawa RH, Sherman SI, Bruner JM (2001). Adenocarcinoma of the salivary gland metastatic to the pituitary gland: case report. Neurosurgery 48(5):1161–1166.

61. Chiang MF, Brock M, Patt S (1990). Pituitary metastases. Neurochirurgia 33:127–131.

62. Buonaguidi R, Ferdeghini M, Faggionato F, Tusini G (1983). Intrasellar metastases mimicking a pituitary adenoma. Surg Neurol 20:373–378.

63. Pernicone PJ, Scheithauer BW, Sebo TJ (1997). Pituitary carcinoma. Cancer 79:804–812.

64. Eisenberg MD, Al-Mefty O, DeMonte F, Burson GT (1999). Benign nonmeningeal tumors of the cavernous sinus. Neurosurgery 44 (5): 949–954.

65. Hahn FJ, Leinbrock LG, Huseman CA, Makos MM (1988). The MR appearance of hypothalamic hamartoma. Neuroradiology 30:67.

66. Shimoto T, Hiroto H, Kanji K et al (1999). Rapid progression of pituitary hyperplasia in humans with primary hypothyroidism: demonstration with MR imaging. Radiology 213:383–388.

67. Hutchins WW, Crues JV, Miya P, Pojunas KW (1990). MR demonstration of pituitary hyperplasia and regression after therapy for hypothyroidism. Am J Neuroradiol 11:410.

68. Young SC, Zimmerman REA, Nowell MA et al (1987). Giant cystic craniopharyngiomas. Neuroradiology 29:468.

69. Khafaga Y, Jenkin D, Kanaan I, Hassounah M, Shabanah MA, Gray A (1998). Craniopharyngioma in children. Int J Radiat Oncol Biol Phys 42(3): 601–606.

70. Gupta k, Kuhn MJ, Shevlin DW, Wacaser LE (1999). Metastatic craniopharyngioma. Am J Neuroradiol 20:1059–1060.

71. Sartoretti-Schefer S, Wichmann W, Aguzzi A, Valavanis A (1997). MR differentiation of adamantinous and squamous-papillary craniopharyngiomas. Am J Neuroradiol 18(1):77–87.

72. Harrison MJ, Morgello S, Post KD (1994). Epithelial cystic lesions of the sellar and parasellar region: a continuum of ectodermal derivatives? J Neurosurg 80:1018–1025.

73. Lafferty AR, Chrousos GP (1999). Pituitary tumors in children and adolescents. J Clin Endocrinol Metab 84(12):4317–4323.

74. Shin JL, Asa SL, Woodhouse LJ, Smyth HS, Ezzat S (1999). Cystic lesions of the pituitary: clinicopathological features distinguishing craniopharyngioma, Rathke's cleft cyst, and arachnoid cyst. J Clin Endocrinol Metab 84(11):3972–3982.

75. Pusey E, Kortman KE, Flannigan BD et al (1987). MR of craniopharyngiomas: tumor delineation and characterization. Am J Neuroradiol 8:443.

76. Laws ER, Vance ML (1999). Radiosurgery for pituitary tumors and craniopharyngiomas. Neurosurg Clin N Am 10(2): 327–336.

77. Fahlbusch R, Honegger J, Paulus W, Huk W, Buchfelder M (1999). Surgical treatment of craniopharyngiomas: experience with 168 patients. J Neurosurg 90:237–250.

78. Yaşargil MG, Abernathey CD, Sarioglu A (1989). Microsurgical treatment of intracranial dermoid and epidermoid tumors. Neurosugery 23(4):561–567.

79. Mori K, Handa H, Moritake K, Takeuchi J, Nakano Y (1982). Suprasellar epidermoid. Neurochirurgia 25(4):138–142.

80. Yamakawa K, Shitara N, Genka S, Manaka S, Takakura K (1989). Clinical course and surgical prognosis of 33 cases of intracranial epidermoid tumors. Neurosurgery 24(4): 568–573.

81. Horowitz BL, Chari MV, James R, Bryan RN (1990). MR of intracranial epidermoid tumors: correlation of in vivo imaging with in vitro 13C spectroscopy. Am J Neuroradiol 11(2):299–302.

82. Netsky MG (1988). Epidermoid tumors. Review of the literature. Surg Neurol 29(6):477–483.

83. Braun IF, Naidich TP, Leeds NE, Koslow M, Zimmerman HM, Chase NE (1977). Dense intracranial epidermoid tumors. Computed tomographic observations. Radiology 122(3):717–719.

84. Wilms G, Casselman J, Demaerel P, Plets C, De Haene I, Baert AL (1991). CT and MRI of ruptured intracranial dermoids. Neuroradiology 33(2):149–151.

85. Smith AS, Benson JE, Blaser SI, Mizushima A, Tarr RW, Bellon EM (1991). Diagnosis of ruptured intracranial dermoid cyst: value MR over CT. Am J Neuroradiol 12(1):175–180.

86. Cohen JE, Abdallah JA, Garrote M (1997). Massive rupture of suprasellar dermoid cyst into ventricles. Case illustration. J Neurosurg 87(6):963.

87. El-Mahdy W, Powall M (1998). Transsphenoidal management of 28 symptomatic Rathke's cleft cysts, with special references to visual and hormonal recovery. Neurosurgery 42:7–16.

88. Takanashi J, Tada H, Barkovich AJ, Saeki N, Kohno Y (2005). Pituitary cysts in childhood evaluated by MR imaging. Am J Neuroradiol 26:2144–2147.

89. Megdiche-Bazarbacha H, Hammouda KB, Aicha AB et al (2006). Intrasphenoidal Rathke cleft cyst. Am J Neuroradiol 27:1098–1100.

90. Nishioka H, Haraoka J, Izawa H, Ikeda Y (2006). Magnetic resonance imaging, clinical manifestations, and management of Rathke's cleft cyst. Clin Endocrinol 64(2):184–188.

91. Voelker JL, Campbell RL, Muller J (1991). Clinical, radiographic and pathologic features of symptomatic Rathke's cleft cysts. J Neurosurg 74(4):535–544.

92. Teramoto A, Hirakawa K, Sanno N, Osamura Y (1994). Incidental pituitary lesions in 1000 unselected autopsy specimens. Radiology 193:161–164.

93. Fox JL, Al-Mefty O (1980). Suprasellar arachnoid cysts: an extension of the membrane of Liliequist. Neurosurgery 7:615–618.

94. Foley KM, Posner JB (1975). Does pseudotumor cerebri cause the empty sella syndrome and benign intracranial hypertension? Neurology 25:565–569.

95. Tien RD, Newton TH, McDermott MW et al (1990). Thickened pituitary stalk on MR images in patients with diabetes insipidus and Langerhans cell histiocytosis. Am J Neuroradiol 11:707.

96. Smith JK, Matheus MG, Castillo M (2004). Imaging manifestations of neurosarcoidosis. Am J Roentgenol 182:289–295.

97. Hollander MD, Friedman DP (1998). Neuroradiology case of the day. Radiographics 18:1608–1611.

98. Ahmadi J, Meyers GS, Segall HD, Sharma OP, Hinton DR (1995). Lymphocytica adenohypophysitis: contrast-enhanced MR imaging in five cases. Radiology 195:30–34.

99. Bourekas EC, Tzalonikou M, Christoforidis GA (2000). Cavernous hemangioma of the optic chiasm. Am J Roentgenol 175(3):888–891.

100. Nishizawa S, Yokoyama T, Hinokuma K et al (1997). Pituitary astrocytoma: magnetic resonance and hormonal characteristics. J Neurosurg 87:131.

101. Paulus W, Honegger J, Keyvani K, Fahlbusch R (1999). Xanthogranuloma of the sellar region: a clinicopathological entity different from adamantinomatous craniopharyngioma. Acta Neuropathol 97:377–382.

CHAPTER 40

Meningeal Tumors

Gregory A. Christoforidis

INTRODUCTION

Neoplasms arising from the meninges predominantly have meningothelial cell origin. These include meningiomas, hemangiopericytoma and melanocytic tumors. In addition, mesenchymal non-meningothelial tumors and metastases can arise from meninges. These tumors are distinguished from other intracranial tumors by their extra-axial location and their rich vascular supply. Their extra-axial location often makes them amenable to surgical resection and cure. Imaging and image interpretation for these lesions aims to establish their extra-axial location, identify any invasion of adjacent structures such as the calvarium, the brain or the dural venous sinuses, confirm their vascular nature and identify associations with congenital diseases such as neurofibromatosis. After reviewing a basic approach to imaging of these tumors, this chapter goes on to take a closer look at individual meningeal tumor types.

IMAGING OF EXTRA-AXIAL TUMORS

Features common to extra-axial meningeal tumors that can distinguish them from intra-axial tumors frequently include a dural attachment, displacement of the cortex away from the calvarium and often times a thin cerebrospinal fluid space adjacent to the tumor (Figures 40.1 and 40.2). Because of their tendency to be vascular, these tumors often have high cerebral blood volumes when applying perfusion magnetic resonance imaging (MRI) techniques and lack an N-acetylaspartate (NAA) peak on MRI spectroscopy. Preoperative planning for meningeal tumor surgery and embolization includes imaging evaluation. This can help identify the location and extent of the tumour, the vascularity of the tumor, the potential invasion of large vessels, such as the internal carotid artery, and the potential involvement of draining veins such as the dural venous sinuses. The location of the tumor should give the neurointerventionalist insight to the vascular supply of the tumor [1,2–6]. Imaging features, such as pattern of enhancement or presence of calcifications, may help assess tumoral vascularity and tumor type. Imaging may help assess whether sacrifice of the internal carotid artery may be necessary or whether the tumor has occluded the dural sinuses. If there is suspicion for invasion of adjacent dural venous sinuses, angiographic methods which evaluate these structures such as computed tomography angiography (CTA), MR venography or conventional angiography often assist the surgeon for surgical planning.

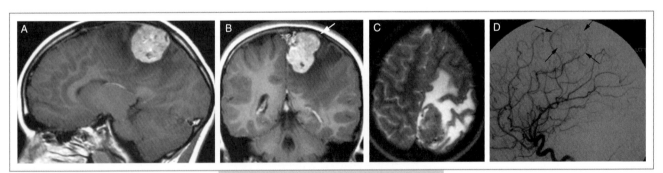

FIG. 40.1. Intra-axial glioma mimicking a meningioma. **(A)** Sagittal **(B)** coronal post-gadolinium T1 and **(C)** axial T2 images of a heterogeneously enhancing mass closely applied to the dura. On coronal imaging there appears to be cortex partially surrounding the mass (arrow). There is no cleft of cerebrospinal fluid separating the tumor from adjacent brain to confirm extra-axial location. **(D)** Lateral view of left internal carotid angiogram demonstrates microvascularity within the tumor bed (arrows) but no tumor blush. This helps confirm intra-axial location.

MENINGIOMAS

Meningiomas were first described in 1614 by Felix Paster [1,7,8]. Meningiomas are tumours which are thought to arise from the arachnoid cap cells found in the arachnoid granulations [7,9]. Meningiomas represent the most common benign intracranial tumor comprising 13–26% of all intracranial tumors [10]. It is noteworthy that, although the annual incidence is approximately 6 per 100 000 population, a 1.4% incidence has been reported in autopsy series. Many meningiomas are small and go unnoticed during life. The peak incidence is in the sixth and seventh decades of life and they have a 2:1 female predilection. They often occur in multiples, especially when associated with neurofibromatosis type 2 (Figure 40.3) or when hereditary (Figure 40.4) [11,12]. Meningiomas have also been associated with radiation exposure and head trauma [13,14].

Meningiomas have historically been divided into many subtypes of which meningothelial, fibrous and transitional are the most common, however, the World Health Organization (WHO) classification scheme is independent of subtypes [7–9]. The WHO classification of meningiomas includes three grades. Grade I meningiomas are considered benign. Grade II and grade III meningiomas are termed atypical, comprising 4.7–7.2% of all meningiomas, and anaplastic, comprising 1.0–2.8% of all meningiomas [11,12]. WHO grade II and III meningiomas occur more frequently with children and males [12]. Anaplastic meningiomas have been known to metastasize to the lung, pleura, bone and liver [12].

Surgical removal of meningiomas continues to be the most effective method of treating these tumors. Preoperative embolization has become an accepted procedure which serves to devascularize highly vascular meningiomas without disturbing the blood supply to the adjacent normal structures. Embolization of a meningioma softens the tumour, reduces intraoperative blood loss and surgical time, and assists in more complete tumor removal. Advances in microcatheter technology and embolic agents and advances in the understanding of the functional anatomy of the head and neck have further contributed to the refinement and extent of endovascular techniques [1,6,10].

Imaging Evaluation

Meningiomas are typically intracranial in origin, but can be found within the orbit and spinal canal [11]. They rarely have been reported in almost every organ. Intracranially, meningiomas occur most frequently along the cerebral convexities, especially along the falx cerebri. Intracranial origin in order of decreasing frequency of occurrence includes the parafalcine location, the falx, the sphenoid ridges and the posterior fossa (including petroclival location and foramen magnum). In addition, they are found at the tuberculum sella, olfactory grooves, the tentorium, optic nerves, petrous ridges and, rarely, intraventicular or even extracranial origin [15]. Symptoms and signs of meningioma vary according to size and location. Patients present with headache 36% of the time (Figures 40.4–40.6)

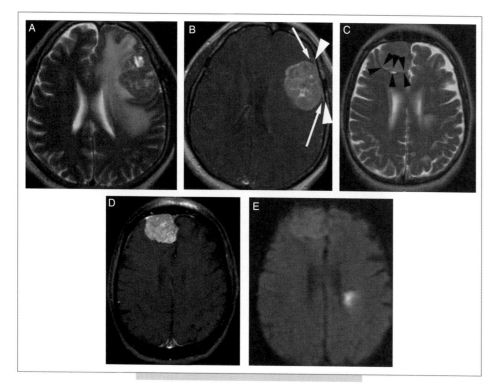

FIG. 40.2. Intra-axial metastasis compared to an extra-axial meningioma. **(A)** Axial T2 and **(B)** axial post-gadolinium T1 images of a peripherally located mass with dural tail sign (white arrowheads), raises the question of extra-axial tumor with extensive vasogenic edema. Normal tissue begins to surround the tumor (white arrows) and no cerebrospinal fluid cleft is identified adjacent to the tumor. Meningioma, however, is still a differential consideration. Compare this to **(C)** T2 and **(D)** post-gadolinium T1 images from a different patient with a homogeneously enhancing mass extra-axial meningioma. Note the cerebrospinal fluid cleft surrounding the mass (black arrowheads). **(E)** Diffusion MRI of the meningioma demonstrates no significant diffusion restriction (high signal focus in the left hemisphere is a subacute infarction).

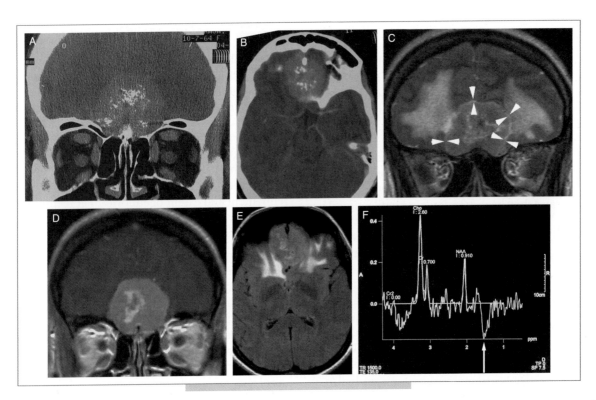

FIG. 40.3. Olfactory groove meningioma with MR spectroscopy. Patient presenting with anosmia. **(A)** Non-contrast coronal and **(B)** contrast enhanced axial CT scans demonstrate a peripherally located mass similar in density to gray matter with a broad base to the dura associated with calcifications within the tumor bed. Further investigation was performed with **(C)** T2 coronal, **(D)** post-gadolinium coronal, **(E)** axial FLAIR and **(F)** MR spectroscopy. Extra-axial location is confirmed by a narrow cerebrospinal fluid cleft (arrows) on T2 imaging (C). The mass is associated with vasogenic edema and is heterogeneous in signal characteristics and contrast enhancement. Heterogeneity can be in part attributed to calcification. MR spectroscopy reveals low NAA, high choline and inverse lipid lactate (arrow) peaks.

[16,17], whereas other symptoms may be due to compression of adjacent structures such as cranial nerves (Figures 40.7–40.9) or obstruction of the ventricles (Figure 40.10). Atypical imaging features do not imply atypical histology (Figures 40.11 and 40.12) [18].

CT imaging of meningiomas demonstrates isodense or slightly hyperdense tumours with a 15–20% incidence of tumoral calcifications (see Figure 40.8). Calcifications are more common in fibroblastic or transitional subtypes with dense calcifications occurring in psammomatous subtypes. On unenhanced CT scans they are sharply circumscribed and tend to be hyperdense or isodense relative to gray matter. Meningiomas which are isodense relative to gray matter may be difficult to identify on non-contrast CT (see Figure 40.4). Meningiomas enhance uniformly following contrast administration. They demonstrate bony changes, usually in the form of hyperostosis in 20% of cases, (see Figures 40.3, 40.13 and 40.14) with more benign varieties [18]. Hyperostosis associated with meningiomas has not been shown to be associated with either tumor size or grade [17,19]. Meningiomas associated with hyperostosis not invading the skull often display higher levels of alkaline phosphatase [20]. More aggressive varieties may demonstrate destructive changes (Figure 40.15). Lytic changes have been associated with intraosseous meningiomas (Figure 40.16) [21]. Abnormal enlargement of the paranasal sinuses has been associated with meningiomas as well [22]. Hypodensity within meningiomas has been attributed to the presence of cysts. These cysts may be a result of arachnoid cyst formation, cystic degeneration or cerebrospinal fluid trapped within the tumor [17]. Fatty transformation may also result in low density within the tumor, as with the rare lipomatous meningiomas [23,24]. Meningiomas which lack contrast enhancement or have heterogeneous or ring-like contrast enhancement patterns may mimic other tumor types. Hemorrhage and necrosis within the tumor bed may also result in a heterogeneous appearance on both non-contrast and contrast-enhanced CT [25,26].

Edema within the white matter adjacent to a meningioma occurs in up to 75% of cases [27] and is best displayed on MRI (see Figures 40.4, 40.5, 40.8, 40.10, 40.11, 40.14, 40.15, 40.17). The source of this edema is controversial, but thought to arise from either repeated mechanical injury to the adjacent brain, secretions from the tumor, or induction

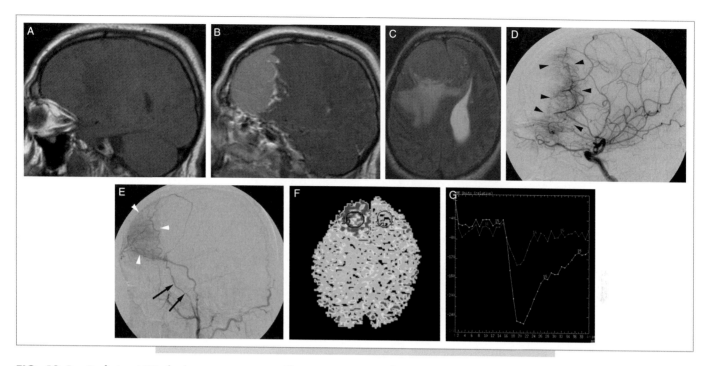

FIG. 40.4. Perfusion MRI of a large meningioma. This patient presented with headaches, with no other symptoms. He was found to have a sizeable meningioma. **(A)** T1 sagittal, **(B)** T1 post-gadolinium sagittal, **(C)** T2 axial, **(D)** lateral internal carotid artery injection, **(E)** lateral external carotid artery injection, **(F)** axial perfusion cerebral blood volume map and **(G)** time intensity curves are shown here. Note the pial supply to the meningioma (black arrows) on the internal carotid artery injection (D), as well as the middle meningeal artery (arrow) supply on the external carotid artery injection. Notice that on perfusion imaging the meningioma typically demonstrates a hyperperfusing mass with delayed contrast washout on time-intensity curves (arrows) relative to normal brain tissue.

of edema from a factor secreted by the tumor [28]. It is not possible to predict histologic subtype, venous occlusion or tumoral vascularity based on the association of edema on imaging [17,29–31]. Some studies have suggested that there may be an association between edema, brain invasion and recurrence [32]. Secretory meningiomas have been shown to produce a greater amount of edema [33].

MAGNETIC RESONANCE IMAGING

MR signal intensity of meningiomas varies on T1, T2 and proton density (PD) sequences. In general, meningiomas are hyperintense on T2 and more so on PD. MRI better differentiates enhancing tumor from dura, however, one should be aware of dural reaction to the tumor in the form of an enhancing dural tail which does not contain tumor cells (see Figures 40.4, 40.6 and 40.7) [7–9]. The dural tail, however, is not specific for meningiomas and has been shown also to occur with lymphoma, peripheral gliomas and schwannomas [34]. The presence of adjacent osseous changes may help differentiate dural reaction from neoplastic dural thickening.

Dynamic contrast MRI may be useful in separating typical from atypical meningiomas (see Figure 40.5); of the parameters tested, volume transfer constant, a measure of capillary permeability, appears to be able to distinguish typical from atypical meningiomas [35]. On diffusion-weighted imaging, malignant meningiomas tend to display higher signal intensities (see Figures 40.2 and 40.16). Lower diffusion constants on diffusion-weighted MRI have been shown to be predictive of higher tumor grades [36]. MR spectra are dominated by choline signal, reduced signal from creatinine and phosphocreatinine and lack N-acetyl-aspartate (see Figure 40.8) [37,38].

ANGIOGRAPHIC EVALUATION

Angiographic assessment of meningiomas is useful for preoperative planning and preoperative embolization. Angiographic evaluation of these tumors depends on their location and the vascular territory that they occupy as depicted on multiplanar contrast enhanced MRI. Information derived from angiography should include [5]: dural arteries supplying the tumor; pial arteries supplying

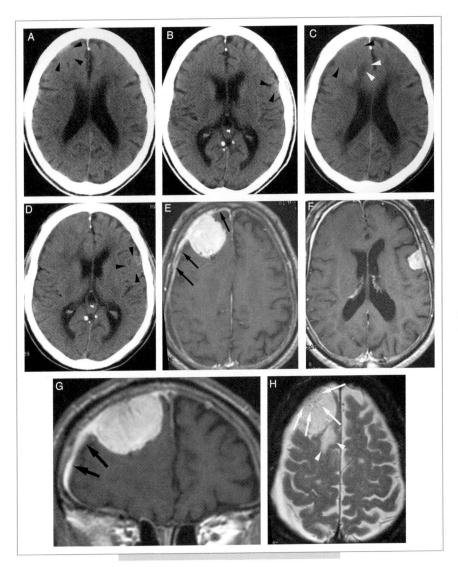

FIG. 40.5. Subtle meningioma on non-contrast CT. A patient who presented with headache underwent a head CT. Two meningiomas are not readily identified on non-contrast CT images of the brain (**A** and **B**, arrowheads) due to small size and isodensity relative to adjacent brain. Follow-up CT of the head after a second presentation 4 years later displays a small amount of vasogenic edema (**C**, white arrowheads) adjacent to one of the two neoplasms (**C** and **D**). Post-gadolinium axial (**E** and **F**) and coronal (**G**) and T2-weighted (**H**) MRI of the brain suggest that this patient has multiple meningiomas. Note the dural reaction adjacent to the tumor (black arrows) and the flow voids within the tumor (white arrows).

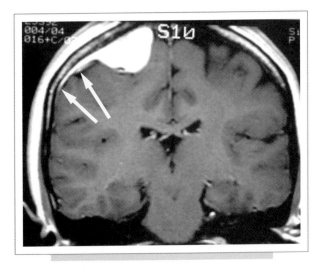

FIG. 40.6. Low-grade meningioma with dural tail. Post-gadolinium coronal T1 MRI demonstrating a dural tail (arrow) and no evidence for vasogenic edema.

the tumor; the relationship of dural/pial arterial supply to the tumor; the detection of arterio-arterial anastomoses; the detection of abnormal origin of dural supplying arteries; the displacement of pial arteries; the transosseous supply of the tumor; the compression and/or encasement of the internal carotid, vertebral, or basilar arteries with skull meningiomas; the collateral circulation to the brain; the intensity of tumor staining and flow characteristics of the tumor; the venous drainage of the tumor; cortical venous changes and the status and patency of adjacent major dural venous sinuses (see Figures 40.5, 40.14 and 40.16).

Characteristic angiographic findings for meningiomas include dilation of meningeal feeding arteries, radiating intratumoral dural arteries and a fairly homogeneous tumoral blush of varying intensity which persists into the late venous phase [15,39]. Degree of vascularity can vary depending on histologic subtype. Angioblastic and transitional types tend to be more vascular than fibroblastic or syncytial varieties. Psammomatous subtypes are sometimes

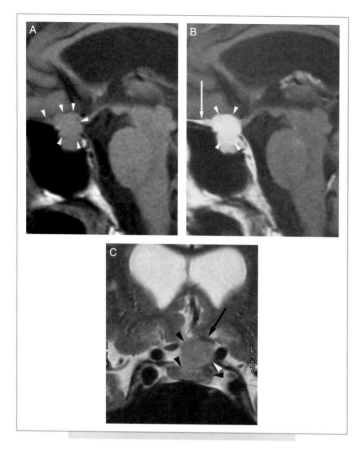

FIG. 40.7. Meningioma of the diaphragma sella. **(A)** Sagittal T1, **(B)** post-gadolinium sagittal T1 and **(C)** coronal T2 demonstrate a neoplasm along the sella. The most common tumor in this location is pituitary adenoma, however, certain features may suggest a different diagnosis. The dural attachment and dural tail suggest the diagnosis of meningioma. A dural tail is less commonly seen with pituitary adenoma. Note a clear separation between the meningioma (white arrows) and the pituitary gland black arrows especially on the T2 images. Note the displacement of the optic nerve which can account for the patient's symptoms of loss of visual acuity.

avascular [11]. The blood supply to a meningioma is typically from meningeal vasculature (see Figures 40.5 and 40.14), however, additional supply or exclusive supply can be derived from transosseous vessels (see Figure 40.16), pial vessels (see Figure 40.5) and even choroidal vessels with intraventricular meningiomas (see Figure 40.10) [6].

At the current state of development of endovascular techniques, safe and efficient embolization of pial supply to a meningioma is rarely possible [40]. As a result, embolization of meningiomas is most effective when an exclusive meningeal supply is present. In the case of meningioma with dominant pial supply, embolization is indicated if the dural supply to the tumor is difficult to access at surgery.

This is often the case with basal meningiomas and posterior fossa meningomas.

In general, meningiomas of the convexity, parasagittal area and the falx cerebri receive dural supply from the branches of the ipsilateral middle meningeal artey. Those located closer to the midline may also receive supply from the contralateral middle meningeal artery. When meningiomas involving the convexities extend through the calvaria into the subgaleal space, they may receive transosseous supply from scalp arteries. Depending on their location, these meningiomas may receive supply from the superficial temporal, posterior auricular or occipital arteries [5,6].

Meningiomas involving the anterior portion of the falx may also receive supply from the artery of the falx cerebri which arises from the anterior ethmoidal artery of the ophthalmic artery. Furthermore, meningiomas involving the free margin of the tentorium and the wall of the inferior sagittal sinus often receive supply from the dural branches of the pericallosal and callosomarginal arteries. These arteries are usually not accessible to superselective embolization due to the size and distal origin of these branches [5,6]. Because the territories involved are more dangerous, special care should be taken when embolization via the ophthalmic artery or cerebral vasculature is considered as described earlier. In a similar manner, meningiomas arising from the olfactory groove and those arising from the planum sphenoidale may receive dural supply from the anterior and posterior ethmoidal arterial branches of the ophthalmic artery.

Meningiomas involving the cavernous sinus and perisellar region typically receive arterial supply from branches of the cavernous internal carotid artery such as the inferolateral trunk (see Figure 40.14) and the meningohypophyseal trunk, the accessory meningeal artery, the cavernous branch of the middle meningeal artery and the recurrent meningeal branches of the ophthalmic artery (see Figures 40.14 and 40.16) [4–6]. Special considerations for embolization via arterial feeders in this region have been described earlier in this chapter. In addition, in cases where there is encasement of the internal carotid artery or suspicion for invasion of the internal carotid artery as depicted on MRI or angiography, permanent balloon occlusion preoperatively should be considered.

When meningiomas are located in the middle cranial fossa or on the greater wing of the sphenoid, they typically receive supply from the sphenoidal branch of the middle meningeal artery. Meningiomas arising from the tentorial leaf can be divided into three types on the basis of location. Those arising from the medial tentorial leaf, those arising from the anterolateral tentorial leaf and those arising from the posterior tentorial leaf. Meningiomas arising from the medial tentorial leaf have a tendency to recruit supply from the marginal tentorial artery which usually arises from the posterior bend of the cavernous carotid artery, however, it has also been found to arise from the

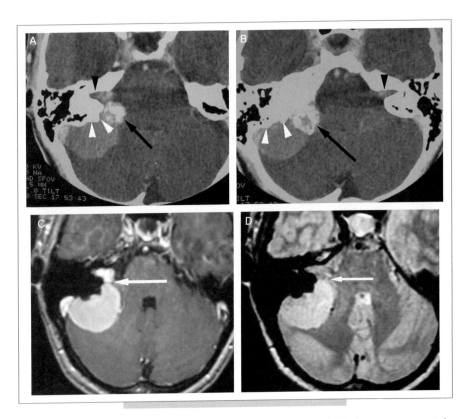

FIG. 40.8. Meningioma in neurofibromatosis type II. Axial enhanced CTs **(A** and **B)** demonstrate a right cerebellopontine angle (CPA) cistern extra-axial mass associated with calcifications (black arrows) and hyperostosis (white arrowheads) which are features suggestive of meningioma. Enlargement of the internal auditory canals associated with enhancing lesions (black arrowheads) suggests acoustic schwannomas, which are the most common CPA cistern lesions. **(C)** Post-gadolinium MRI identifies bilateral acoustic schwannomas and a cleft separating the CPA cistern lesion from the internal auditory canal lesion. Furthermore, the internal auditory canal lesion and the CPA cistern lesion differ in signal intensity on proton density MRI **(D)**. This suggests the diagnosis of neurofibromatosis II with bilateral acoustic schwanomas with a right CPA cistern meningioma.

horizontal portion of the cavernous carotid, the ophthalmic artery, the lacrimal artery and the accessory meningeal artery [5]. Meningiomas arising from the anterolateral portion of the tentorial leaf typically receive arterial supply from the lateral clival branch of the posterior bend of the cavernous carotid artery, and the basal tentorial branch of the petrosquamosal branch of the middle meningeal artery. Arterial supply to meningiomas arising from the posterior portion of the tentorium can be divided into supratentorial supply and infratentorial supply. The supratentorial supply usually includes middle meningeal supply via petrosquamosal branches and parieto-occipital tentorial branches. Infratentorial supply includes branches of the mastoid artery, the posterior meningeal artery and the posterior falx artery. Origins of the posterior meningeal artery and the posterior falx artery are variable and may arise from the vertebral artery, the posterior inferior cerebellar artery (PICA) or the occipital artery. In addition, the posterior meningeal artery has been found to arise from the neuromeningeal trunk of the ascending pharyngeal artery [6].

Petroclival meningiomas and meningiomas along the posterior surface of the petrous bone and of the cerebellopontine angle are supplied by: clival branches of the internal carotid artery (ICA) anteriorly; branches of the petrosal branch of the middle meningeal artery superiorly; the ascending branch of the mastoid artery posteriorly and the jugular and hypoglossal branches of the ascending pharyngeal artery inferiorly and centrally. The anterior ICA (AICA) often provides additional supply via the subarcuate artery [6]. Meningiomas adjacent to the cerebellar convexity typically receive supply from the posterior meningeal artery and the artery of the falx cerebelli. In addition, posterior fossa meningiomas frequently receive supplemetary pial supply via the superior cerebellar artery (SCA), the AICA and/or the PICA [6,41].

IMAGING MENINGIOMAS POST-EMBOLIZATION

Maps of relative regional cerebral blood volume have been also shown to provide hemodynamic information

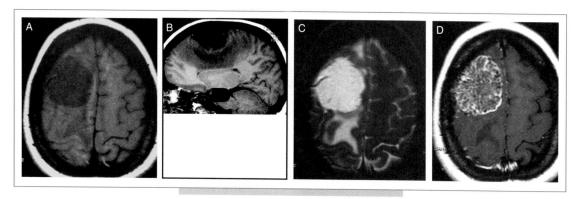

FIG. 40.9. Atypical meningioma. **(A)** Axial T1, **(B)** sagittal T1, **(C)** axial T2 and post-gadolinium T1 **(D)** MR images demonstrate an extra-axial mass adjacent to the right frontal lobe associated with vasogenic edema and mass effect. Histopathology revealed atypical meningioma.

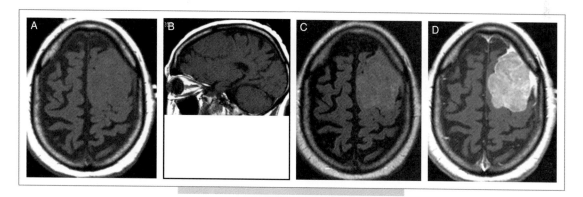

FIG. 40.10. Atypical meningioma. **(A)** Axial T1 **(B)** sagittal T1 **(C)** axial T2 and post-gadolinium T1 **(D)** MR images demonstrate an extra-axial mass adjacent to the right frontal lobe associated with vasogenic edema and mass effect. Histopathology revealed atypical meningioma.

regarding tumor perfusion and, as such, can monitor the treatment effect of embolization in meningiomas [42]. Comparing T1 post-gadolinium imaging to dynamic MR using a contrast bolus indicates that, following embolization, some areas of embolized tumor may enhance in a delayed fashion due to slow collateral flow (see Figure 40.16) [42]. Tumor necrosis has also been demonstrated on the basis of MR spectroscopy. The spectroscopic data indicate high concentrations of lactate 24 hours following embolization followed by broad aliphatic signal several days later [37,38,43–46]. Lipid signal appears to be associated with softer tumor at surgery [46]. Persistence of a lactate signal is felt to indicate incomplete tumor embolization [37]. Spectroscopic experiments have suggested that the process of tumoral necrosis is complete four days following embolization. Based on this information, surgery should be planned following completion of the process of necrosis [37,38]. Sequential MRIs following embolization have shown that areas, which in the immediate post-embolization period did not enhance, variably did so over the next few days, especially along the periphery of the tumor bed. Histopathologically, this area of 're-enhancement' corresponded to vital tissue [46]. On the basis of MR findings, it has been suggested that MRI and MRS following embolization may serve a purpose to identify revascularization and establish the presence of fatty degeneration of the meningioma [46].

MESENCHYMAL, NON-MENINGOTHELIAL TUMORS

As the name implies, mesenchymal tumors originate from non-meningothelial cells arising from the meninges. The most common of these tumors is the hemangiopericytoma (see Figures 40.18 and 40.19). They also include tumors arising from adipose tumors, fibrous tumors, fibrohistiocytic tumors, muscle tumors, osteocartilaginous tumors, neoplasms arising from elements of the vessel wall and meningeal sarcomas. Their imaging varies depending on the cell of origin. For example, lipomas

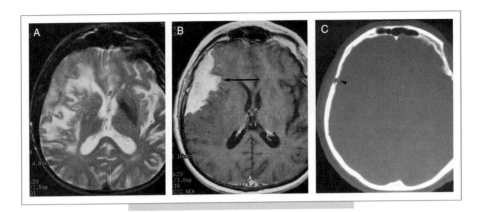

FIG. 40.11. Malignant meningioma. **(A)** Axial T2 and **(B)** axial post-gadolinium T1 MRI demonstrate a mass with a broad base towards the dura with irregular margins with suggestion of tumor invasion into the brain (arrow) and adjacent vasogenic edema. **(C)** An axial bone window CT demonstrated bone lysis (arrowhead). On histopathology this was found to be a malignant meningioma.

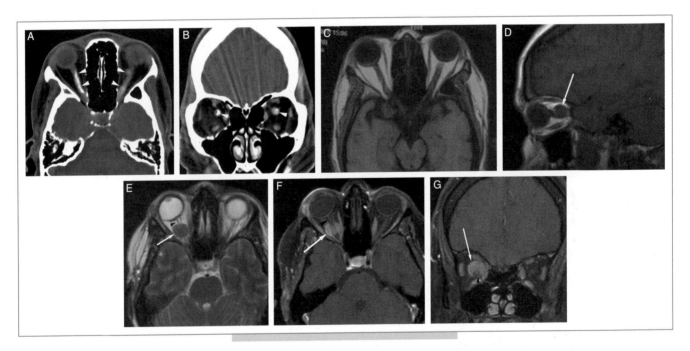

FIG. 40.12. Optic nerve meningiomas. **(A)** Axial and **(B)** coronal CT scans of a patient with 'tram track calcifications' along the optic nerves (white arrowheads) which are not visible on T1 MRI **(C)**. They appear as dark outlines but are less conspicuous than on CT. 'Tram track calcifications' favor the diagnosis of meningioma. **(D)** Sagittal T1, **(E)** axial T2 and **(F)** post-gadolinium fat suppressed axial and **(G)** coronal T1 images of the orbits in a different patient demonstrate a more mass-like meningioma (arrows) arising from the nerve sheath and surrounding the optic nerve (black arrowhead).

tend to be hypodense on CT and display high signal on T1-weighted imaging and suppress with pulse sequence in which a fat saturation pulse is applied and chemical shift artifact on T2 images (Figure 40.20). Fibrous tumors tend to be very dense and on MRI display low signal due to lack of water. Although lipomas represent 0.4% of intracranial tumors, and hemangiopericytomas represent 0.4–1% of all intracranial tumors, sarcomas represent less than 0.1% of all intracranial tumors and have been predominantly

reported in single case reports [47,48]. With the exception of rhabdomyosarcomas, which occur predominantly in children, there is no documented predilection for age or sex and no documented preferential location.

HEMANGIOPERICYTOMA

Intracranial hemangiopericytomas are rare tumors of mesenchymal origin arising from vascular pericytes within the meninges which are indistinguishable on histopathologic

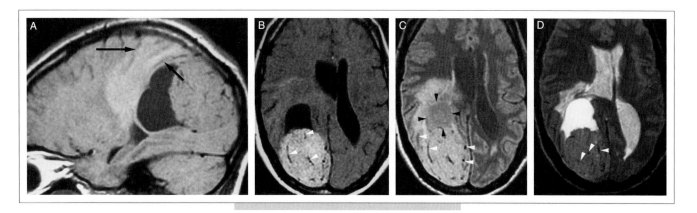

FIG. 40.13. Cystic meningioma. **(A)** Sagittal T1, **(B)** axial post-gadolinium T1, **(C)** axial proton density and **(D)** axial T2 MR images of a cystic meningioma. Note that the cystic component does not follow cerebrospinal fluid on the proton density image (black arrowheads). Note the vascular flow voids within the tumor bed (white arrowheads). Note the accordion-like effect of this extra-axial mass on the adjacent cortical layers.

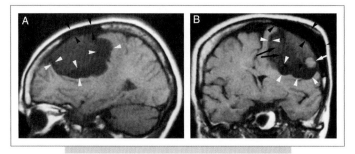

FIG. 40.14. Cystic meningioma. **(A)** Sagittal and **(B)** coronal T1 MR images of a cystic meningioma. Compared to Figure 40.13, the cystic component (white arrowheads) surrounds the solid component of the tumor. Note the hyperostosis (black arrowheads) depicted by loss of narrow and thickening of clavarium, as well as a small hemorrhage (white arrow) within the tumor and vascular flow voids (black arrows); all features which can be seen with meningiomas.

analysis from hemangiopericytomas arising from elsewhere in the body. Although hemangiopericytomas were described elsewhere in the body, intracranial hemangiopericytomas were first described in 1954 by Begg and Garret, identifying one case of dural based hemangiopericytoma and reviewing six cases of dural based tumors previously labeled as an angioblastic subtype of meningioma [49–51]. They constitute approximately 0.4–1% of all CNS neoplasms and tend to occur in younger age groups relative to meningiomas, with a mean presenting age of 37–44 and an M:F ratio of 1.4. Males tend to present at a younger age than females [52]. Hemangiopericytomas are highly vascular neoplasms with a propensity to invade adjacent structures, a tendency to recur following surgical resection and an ability to metastasize. Aggressive management with complete resection, radiation therapy and frequent and prolonged follow-up is advocated [53].

Their extra-axial location may confuse them with more aggressive meningiomas, however, more aggressive features may help suggest the diagnosis on imaging. They tend to be large (greater than 4 cm), lobular, extra-axial, hyperdense tumors, which enhance on CT with no evidence of calcification [50,51,53]. They often have a broad based attachment to the dura, but often (11/34 cases) have a narrow dural attachment (see Figure 40.19). A narrow dural attachment is unusual for meningiomas. On CT, unlike meningiomas, they are more frequently associated with lysis rather than hyperostosis of adjacent bone (19/32 cases). There is some evidence to suggest that bone lysis is more frequently associated with anaplastic hemangiopericytoma, whereas lack of bone lysis on CT is more frequently associated with well differentiated forms. Hyperostosis has not been reported with hemangiopericytomas. These tumors tend to be heterogeneously hyperdense on CT, but can be homogeneously hyperdense. Similarly, contrast enhancement tends to be heterogeneous but can be homogeneous.

On MRI studies they are usually (13/17 cases) either homogeneously or heterogeneously isointense to gray matter on T1WI, but can be hypointense or hyperintense. They are usually (10/17 cases) either homogeneously or heterogeneously isointense relative to gray matter on T2WI [50,53]. As a rule they tend to have prominent serpentine vascular signal voids. A significant amount of adjacent brain edema is almost always seen (30/34 cases) with these tumors [50]. Contrast MRI usually displays heterogeneous enhancement and may be associated with a dural tail.

Angiographic findings include supply similar to meningiomas with numerous corkscrew vessels. Early venous drainage typically seen with hemangiopericytomas is not typically seen with meningiomas. MRS demonstrates increased

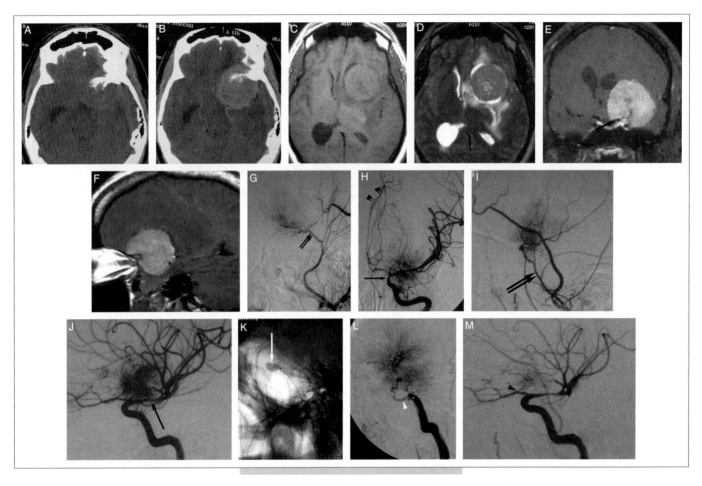

FIG. 40.15. Paracavernous meningioma with angiography and embolization. Unlike pituitary adenomas, meningiomas often narrow the intracranial carotid artery. Axial CT scans without (**A**) and with (**B**) contrast demonstrate an enhancing mass isodense to gray matter associated with adjacent hypersostosis of the sphenoid bone. (**C**) Axial T1, (**D**) T2 and (**E, F**) post-gadolinium T1 coronal MRI images demonstrate homogeneous enhancement of this mass which surrounds the internal carotid artery (black arrow). AP and lateral projections of external carotid artery (**G** and **I**) and internal carotid artery (**H** and **J**) arteriograms demonstrate the vascularity of the meningioma predominantly supplied via the internal carotid artery branches. Unlike pituitary adenomas, meningiomas tend to narrow the internal carotid artery as in this case (black arrow). There is also proximal round shift of the anterior cerebral artery (double arrowheads) due to this subfrontal mass located adjacent to the anterior cerebral artery. Embolization was performed via the inferior frontal branch of the middle meningeal artery (double arrowheads) and via the inferolateral trunk (**L** white arrowhead) of the internal carotid artery using balloon assistance (**K** white arrow). At the end of the embolization, the only supply left was via the recurrent meningeal artery from the ophthalmic artery (**M**). Embolization assists the surgeon in removing meningiomas more completely and with less blood loss.

choline and myo-inositol peaks and mildly increased lipid peaks [53].

MELANOCYTIC LESIONS

Melanocytic lesions are rare (0.06–0.1% of brain tumors) neoplasms arising from leptomeningeal melanocytes. These can be either diffuse or circumscribed and include: malignant melanoma, melanocytoma and neurocutaneous melanosis [54]. Melanocytoma is considered the most common of these and has an annual incidence of 1 per 10 million population [54].

Melanocytomas are considered benign lesions with a tendency to invade and recur locally. Melanocytomas are suspected to arise from melanocytes which are of neural crest origin frequently found scattered along the surface of the brain. Case reports in the literature indicate that they tend to be isodense or hyperdense relative to gray matter on CT with homogeneous enhancement and display high signal on T1- and low signal on T2-weighted imaging. Hemorrhage within the tumor may influence its imaging appearance. To a large degree, the signal characteristics depend on the degree of melanin concentration within the

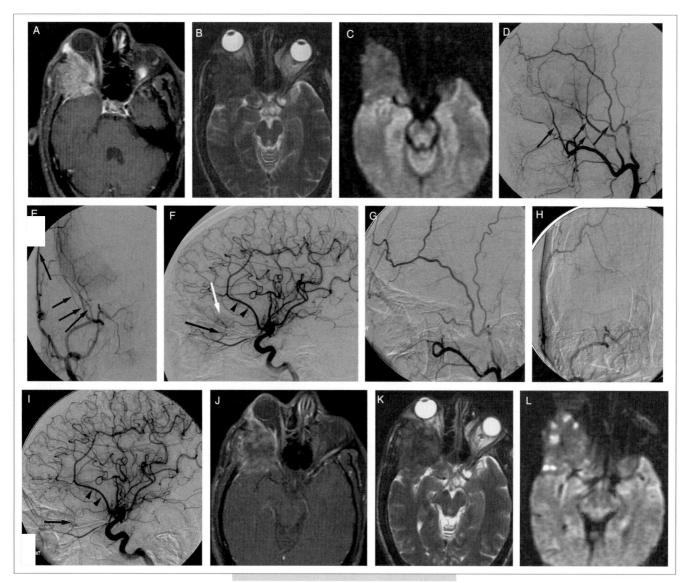

FIG. 40.16. Intraosseous meningioma before and after embolization. **(A)** Axial post-gadolinium T1, **(B)** axial T2 and **(C)** axial diffusion MR images before embolization of an intraosseous meningioma involving the right sphenoid bone, orbit and middle cranial fossa. The signal on diffusion imaging suggest lower grade. Pre- and post-embolization lateral **(D** and **G)** and anteroposterior (AP) **(E** and **H)** external carotid artery (ECA) and lateral internal carotid artery (ICA) **(F** and **I)** arteriograms demonstrate arterial supply (black arrows) via ECA branches including superficial temporal, deep temporal, accessory meningeal and middle meningeal as well as ICA supply via the recurrent meningeal artery from the ophthalmic artery. Post-embolization reveals minimal tumor blush on the ICA injection (white arrow). Note the mass effect on the anterior cerebral artery (black arrowheads). Post-embolization, there is substantially reduced enhancement **(J)**. High signal foci within the tumor represent restricted diffusion **(K** and **L)** due to embolization rather than higher grade.

tumor. The paramagnetic effect of melanin resulting in both T1 and T2 shortening influence the imaging appearance of such tumors [55–58]. Differential considerations for melanocytoma include pigmented meningioma, melanotic schwannoma and melanoma. Progression of melanocytoma to malignant melanoma has been described [59].

Diffuse melanocytosis represents leptomeningeal proliferation of melanocytic cells. On imaging, diffuse melanocytosis displays diffuse leptomeningeal thickening with high signal on T1-weighted MRI [57]. Diffuse melanocytosis carries a poor prognosis [54]. Malignant melanomas display more aggressive features such as pleomorphism, high mitotic rates, necrosis, hemorrhage, invasion of the brain, radioresistance and a poor prognosis. Meningeal melanomatosis refers to malignant melanoma with diffuse meningeal spread [54].

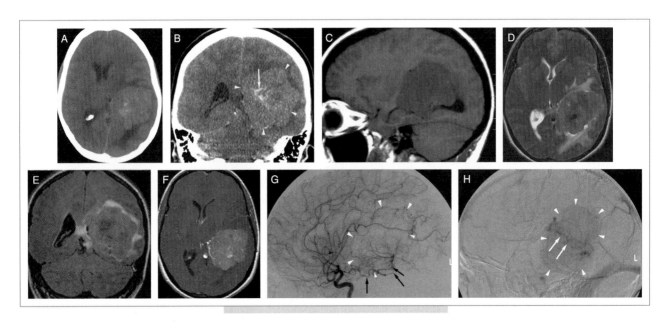

FIG. 40.17. Intraventricular meningioma. **(A)** Axial and **(B)** coronal CT demonstrate an intraventricular mass (white arrowheads) associated with calcifications (white arrow). **(C)** Sagittal T1, **(D)** axial T2, **(E)** coronal FLAIR and **(F)** post-gadolinium T1 images confirm an intraventricular mass. Lateral left internal carotid angiogram demonstrates anterior choroidal artery supply (black arrow) on arterial phase **(G)** and sustained vascular blush with enlarged draining veins (white arrows) on late venous phase **(H)**.

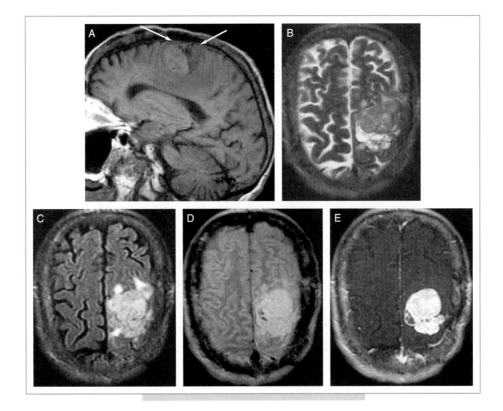

FIG. 40.18. Recurrent hemangiopericytoma. **(A)** Sagittal T1, **(B)** axial T2, **(C)** axial FLAIR, **(D)** axial proton density and **(E)** post-gadolinium T1 MR images of an extra-axial mass heterogeneous in signal characteristics on T1 and T2 but similar to gray matter. This mass has a narrow dural attachment (arrows) and associated vasogenic edema.

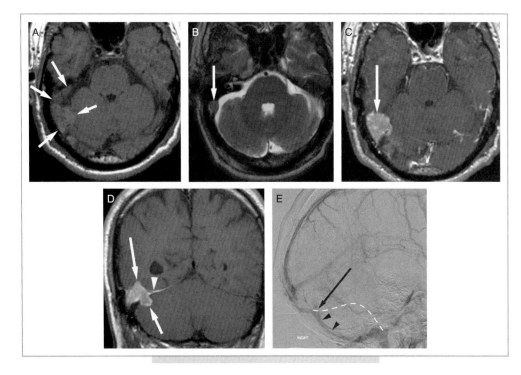

FIG. 40.19. Recurrent hemangio-pericytoma with dural sinus invasion. **(A)** Axial T1, **(B)** axial T2, **(C)** post-gadolinium axial and **(D)** coronal T1 MR images of a recurrent hemangiopericytoma (white arrows) invading the transverse and sigmoid sinuses (white arrowheads). **(E)** Venous phase of a lateral right internal carotid arteriogram demonstrates occlusion of the right transverse sinus (black arrow) with collateral venous drainage (black arrowheads). The normal course of the sigmoid sinus is indicated by the dashed white line.

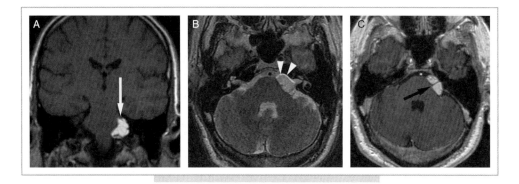

FIG. 40.20. Cerebellopontine angle cistern lipoma. **(A)** Coronal T1, **(B)** axial T2 and **(C)** post-gadolinium axial SPGR T1 MR images of a right cerebellopontine angle cistern lipoma (white arrow). Chemical shift artifact (white arrowheads) suggests lipid content within the tumor. The seventh and eighth cranial nerves are noted to course through the tumor (black arrow).

REFERENCES

1. Valavanis A, Christoforidis GA (2002). Tumors of the head and neck. In *Interventional Neuroradiology – Theory and Practice*. Byrne J (ed.). Oxford University Press, Oxford. 213–234.
2. Valavanis A (1998). Interventional radiology in the neck. In *Head and Neck Surgery, Vol. 3: Neck*. Panje WR, Herbehold C (eds). Georg Thieme, Stuttgart. 487–493.
3. Valavanis A (1990). Interventional neuroradiology for head and neck surgery. In *Otolaryngology – head and neck surgery*. Update II. Cummings CW, Fredrickson JM, Harker LA, Krause CJ, Schuller DE (eds). Mosby Year Book, St Louis. 206–223.
4. Lasjaunias P, Berenstein A (1987). *Surgical neuroangiography, vol I: Functional anatomy of craniofacial arteries*. Springer, Berlin.
5. Valavanis A (1993). Embolization of intracranial and skull base tumors. In *Interventional Neuroradiology*. Valavanis A (ed.). Springer, Berlin. 63–91.
6. Christoforidis GA, Valavanis A (2003). Embolization of meningiomas. Neurointerventionist 4:43–53.
7. Black PMcL (1993). Meningiomas. Neurosurgery 32:643–657.
8. Bondy M, Ligon BL (1996). Epidemiology and etiology of intracranial meningiomas: a review. J Neuro-Oncol 29:197–206.
9. Langford LA (1996). Pathology of meningiomas. J Neuro-Oncol 29:217–221.
10. Engelhard HH (2001). Progress in the diagnosis and treatment of patients with meningiomas: diagnostic imaging, preoperative embolization. Surg Neurol 55:89–101.

11. Louis DN, Budka H, von Deimling A (1997). Meningiomas. In *Pathology and Genetics of Tumours of the Nervous System*. Kleihues P, Cavenee WK (eds). International Agency for Research on Cancer, Lyon. 134–141.

12. Louis DN, Sheithauer BW, Budka H, von Deimling A, Kepes JJ (2000). Meningiomas. In *Pathology and Genetics of Tumours of the Nervous System*. Kleihues P, Cavanee WK (eds). International Agency for Research on Cancer Press, Lyon. 176–184.

13. Sadetzki S, Flint-Richter P, Ben-Tal T et al (2002). Radiation induced meningiomas: a descriptive study of 253 cases. J Neurosurg 97:1078–1082.

14. Phillips LE, Koepsell TD, Van Belle G et al (2002). History of head trauma and risk of intracranial meningioma: population-based case-control study. Neurology 58: 1849–1852.

15. Hasso AN, Bell SA, Tadmor R (1994). Intracranial vascular tumors. Neuroimaging Clin N Am 4: 849–870.

16. Drevelengas A, Karkavelas G, Hourmouzi D, Boulogianni G, Petridis A, Dimitriadis A (2002). *Meningeal Tumors in Imaging of Brain Tumors with Histological Correlations*. Drevelengas A (ed.). Springer, Berlin. 177–214.

17. Buetow MP, Buetow PC, Smirniotopoulos JG (1991). Typical, atypical, and misleading features in meningioma. Radio-Graphics 11:1087–1106.

18. O'Leary S, Adams WM, Parrish RW, Mukonoweshuro W (2007). Atypical imaging appearances of intracranial meningiomas. Clin Radiol 62:10–17.

19. Pieper DR, Al-Mefty O, Hanada Y et al (1999). Hyperostosis associated with meningioma of the cranial base: secondary changes or tumor invasion. Neurosurgery 44:742–746.

20. Heick A, Mosdal C, Jorgensen K, et al (1993). Localized cranial hyperostosis of meningiomas: a result of neoplastic enzymatic activity? Acta Neurol Scand 87:243–247.

21. Nil TN, Oner YA, Kaymaz M et al (2005). Primary intraosseous meningioma: CT and MRI appearance. Am J Neuroradiol 26:2053–2056.

22. Miller NR, Golnik KC, Zeidman SM et al (1996). Pneumosinus dilatans: a sign of intracranial meningioma. Surg Neurol 46: 471–474.

23. Bolat F, Kayaselcuk F, Aydin MV et al (2003). Lipidized or lipomatous meningioma, which is more appropriate? A case report. Neurol Res 25:764–766.

24. Kubota Y, Ueda T, Kagawa Y et al (1997). Microcystic meningioma without enhancement on neuroimaging. Neurol Med Chir (Tokyo) 37:407–410.

25. Roosen N, Deckert M, Lumenta CB et al (1989). Endotheliomatous meningioma with coalescing microcysts, presenting as a hypodense lesion with ring enhancement on computed tomography. Neurochirurgia 32:160–163.

26. Masahiro Y, Masahiro S, Tsuyoshi O (2003). Ring-enhanced malignant meningioma mimicking a brain metastasis from a renal cell carcinoma. Urol Int 70:80–82.

27. Bradac GB, Ferszt R, Bender A, Schorner W (1986). Peritumoral edema in meningioma: a radiological and histological study. Neuroradiology 28:304–312.

28. Buetow MP, Buetow PC, Smirniatopoulos JG (1991). From the archives of the AFIP: typical, atypical, and misleading features in meningioma. Radiographics 11:1087–1106.

29. Kizanna E, Lee R, Toung N et al (1996). A review of the radiological features of intracranial meningiomas. Australas Radiol 40:454–462.

30. Domingo Z, Rowe G, Blamire AM et al (1998). Role of ischaemia in the genesis of oedema surrounding meningiomas assessed using magnetic resonance imaging and spectroscopy. Br J Neurosurg 12:414–418.

31. Pistolesi S, Fontanini G, Camacci T et al (2002). Meningioma-associated brain oedema: the role of angiogenic factors and pial blood supply. J Neuro-Oncol 60:159–164.

32. Mantle RE, Lach B, Delgado MR et al (1999). Predicting the probability of meningioma recurrence based on the quantity of peritumoral brain edema on computerized tomography scanning. J Neur-Osurg 91:375–383.

33. Probst-Cousin S, Villagran-Lillo R, Lahl R et al (1997). Secretory meningioma: clinical, histologic, and immunohistochemical findings in 31 cases. Cancer 79:2003–2015.

34. Nagele T, Peterson D, Klose U, et al (1994). The 'dural tail' adjacent to meningiomas studied by dynamic contrast-enhanced MRI: a comparison with histopathology. Neuroradiology 36: 303–307.

35. Yang S, Law M, Zagzag D et al (2003). Dynamic contrast-enhanced perfusion MR imaging measurements of endothelial permeability: differentiation between atypical and typical meningiomas. Am J Neuroradiol 24:1554–1559.

36. Filippi CG, Edgar MA, Ulug AM, Prowda JC, Heier LA, Zimmerman RD (2001). Appearance of meningiomas on diffusion-weighted images: correlating diffusion constants with histopathologic findings. Am J Neuroradiol 22:65–72.

37. Tymianski M, Willinsky RA, Tator CH, Mikulis D, TerBrugge KG, Markson L (1994). Embolization with temporary balloon occlusion of the internal carotid artery and in vivo proton spectroscopy improves radical removal of petrous-tentorial meningioma. Neurosurgery 35:974–977.

38. Jungling FD, Wakhloo AK, Henning J (1993). In vivo proton spectroscopy of meningioma after preoperative embolization. Magn Reson Med 30:155–160.

39. Lasjaunias P, Berenstein A (1987). *Surgical Neuroangiography, vol II: Endovascular Treatment of Craniofacial Lesions*. Springer, Berlin.

40. Kaji T, Hama Y, Iwasaki Y, Kyoto Y, Kusano S (1999). Preoperative embolization of meningiomas with pial supply: successful treatment of two cases. Surg Neurol 52:270–273.

41. Masters LT, Nelson PK (1998). Pre-operative angiography and embolisation of petroclival meningiomas. Intervention Neuroradiol 4:209–221.

42. Bruening R, Wu RH, Yousry TA et al (1998). Regional relative blood volume MR maps of meningiomas before and after partial embolization. J Comput Assist Tomogr 22:104–110.

43. Wakhloo AK, Juengling FD, Van Velhoven V, Schumacher M, Henning J, Schwechheimer K (1993). Extended preoperative polyvinyl-alcohol microembolization of intracranial meningiomas: assessment of two embolization techniques. Am J Neuroradiol 14:571–582.

44. Kallmes DE, Evans AJ, Kaptain GJ et al (1997). Hemorrhagic complications in embolizaion of a meningioma: case report and review of the literature. Neuroradiology 39:877–880.

45. Houkin K, Kamada K, Sawamura Y, Iwasaki Y, Abe H, Kashiwaba T (1995). Proton magnetic resonance spectroscopy (1H-MRS) for the evaluation of treatment of brain tumours. Neuroradiology 37:99–103.

46. Bendszus M, Warmuth-Metz M, Klein R et al (2002). Sequential MRI and MR spectroscopy in embolized meningiomas:

correlation with surgical and histopathologic findings. Neuroradiology 44:77–82.

47. Paulus W, Scheithauer BW (2000). Mesenchymal, non-meningothelial tumors. In *Pathology and Genetics of Tumours of the Nervous System*. Kleihues P, Cavanee WK (eds). International Agency for Research on Cancer Press, Lyon. 185–189.

48. Jellinger K, Paulus W (1991). Mesenchymal, non-meningothelial tumors of the central nervous system. Brain Pathol 1(2):79–87.

49. Begg CF, Garret R (1954). Hemangiopericytoma occurring in the meninges. Cancer 7:602–606.

50. Chiechi MV, Smirniotopoulos JG, Mena H (1996). Intracranial hemangiopericytomas: MR and CT features. Am J Neuroradiol 17(7):1365–1371.

51. Alen JF, Lobato RD, Gomez PA et al (2001). Intracranial hemangiopericytoma: study of 12 cases. Acta Neurochir (Wien) 143(6):575–586.

52. Jaaskelainen J, Louis DN, Paulus W, Haltia MJ (2000). Haemangiopericytoma. In *Pathology and Genetics of Tumours of the Nervous System*. Kleihues P, Cavanee WK (eds). International Agency for Research on Cancer Press, Lyon. 190–192.

53. Fountas KN, Kapsalaki E, Kassam M et al (2006). Management of intracranial meningeal hemangiopericytomas: outcome and experience. Neurosurg Rev 29(2):145–153.

54. Jellinger K, Chou P, Paulus W (2000). Melanocytic lesions. In *Pathology and Genetics of Tumours of the Nervous System*. Kleihues P, Cavanee WK (eds). International Agency for Research on Cancer Press, Lyon. 193–195.

55. Hamasaki O, Nakahara T, Sakamoto S, Kutsuna M, Sakoda K (2002). Intracranial meningeal melanocytoma. Neurol Med Chir (Tokyo) 42(11):504–509.

56. Chow M, Clarke DB, Maloney WJ, Sangalang V (2001). Meningeal melanocytoma of the planum sphenoidale. Case report and review of the literature. J Neurosurg 94(5):841–845.

57. Offiah CJ, Laitt RD (20060. Case report: intracranial meningeal melanocytoma: a cause of high signal on T1- and low signal on T2-weighted MRI. Clin Radiol 61(3):294–298.

58. Kiecker F, Hofmann MA, Audring H et al (2007). Large primary meningeal melanoma in an adult patient with neurocutaneous melanosis. Clin Neurol Neurosurg 109:448–451.

59. Roser F, Nakamura M, Brandis A, Hans V, Vorkapic P, Samii M (2004). Transition from meningeal melanocytoma to primary cerebral melanoma. Case report. J Neurosurg 101(3): 528–531.

CHAPTER **41**

Intracranial Schwannomas

Liangge Hsu

INTRODUCTION

Intracranial schwannomas (also known as neurinomas, neurilemmomas, neuromas) are often solitary encapsulated tumors that grow along the sensory distribution of cranial and spinal nerves or nerve roots. They were first described 200 years ago, comprise approximately 5–10% of all intracranial tumors and are mostly located at the cerebellar pontine angle (CPA) [1,2]. Incidence is around 0.92–1.9 per 100 000 and they often present between the ages of 40 and 60 years [2,3]. Among these, 80–90% arise from the vestibular branch or ganglion of cranial nerve VIII [4]. There is a female to male predilection ratio of approximately 1.5–2:1 [3]. Prevalence of incidental or asymptomatic acoustic schwannoma is about 2 in 10 000 people, slightly more in males [5]. These are rare tumors in children.

About 90–95% of vestibular schwannomas are solitary and 5–10% bilateral as seen in neurofibromatosis 2 (NF2)

patients. Conversely, 95% of NF2 patients develop bilateral vestibular schwannomas [4]. While almost all isolated vestibular schwannoma patients have somatic involvement, about 6% are mosaic for NF2 gene mutation [6]. In these instances, the age of the individual at presentation is the most important determinant of risk of NF2, with patients less than 30 years of age at high risk for the disease and those older than 55 years of age with negligible risk.

After CN VIII, the next most common location of intracranial schwannoma is CN V, representing approximately 0.2% of intracranial tumors (Figure 41.1) [7]. These arise either from nerve root at the cerebellopontine angle or Meckel's cave for intradural location, or extradurally from Gasserian ganglion. They tend to be more varied in appearance and are more often associated with cystic components (Figure 41.2). There are also sporadic reports of trochlear, abducens, facial, vagal, glossopharyngeal and hypoglossal schwannomas in the literature (Figures 41.3–41.5) [8–11]. Facial schwannomas tend to arise at the geniculate ganglion (Figures 41.6 and 41.7) presenting with sensory hearing loss rather than facial nerve palsy as the acoustic nerve

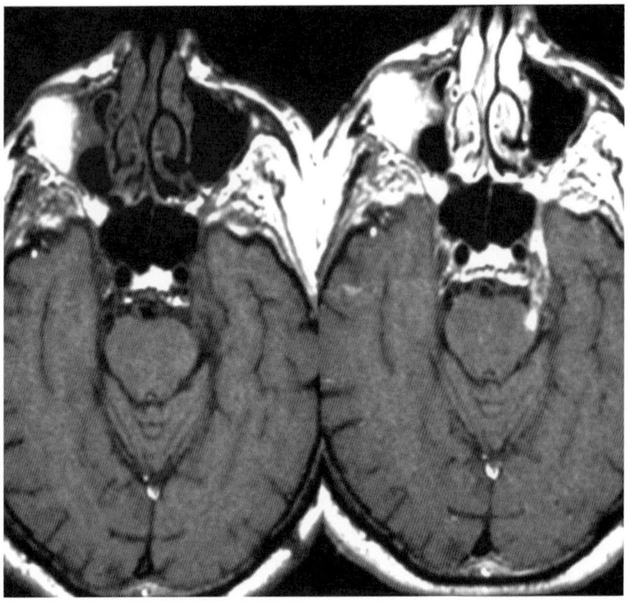

FIG. 41.1. Axial pre- (left) and post-gadolinium (right) images of a left fifth nerve schwannoma extending into Meckel's cave.

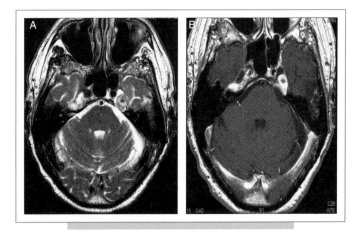

FIG. 41.2. Axial T2 **(A)** and post-gadolinium images **(B)** of left fifth nerve schwannoma within Meckel's cave with cystic component.

408

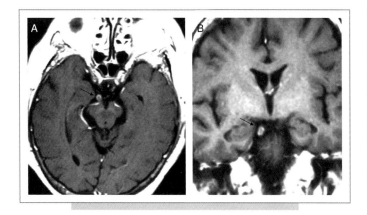

FIG. 41.3. Axial **(A)** and coronal **(B)** post-gadolinium T1 images of small right third nerve schwannoma (arrows).

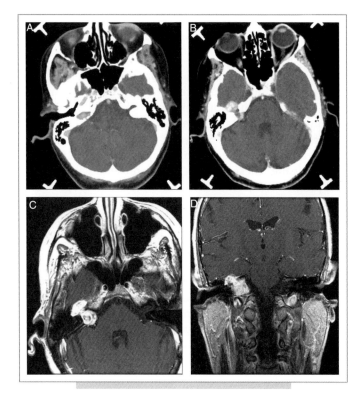

FIG. 41.4. Axial post-contrast CT **(A and B)**, axial **(C)** and coronal **(D)** post-gadolinium T1 images of right dumb-bell shaped facial nerve schwannoma.

is more thinly myelinated. Intraparenchymal schwannomas are exceedingly rare with a slight male predilection of 1.6:1 ratio. They are seen in a younger age group (<30 years) and have been reported in periventricular, brainstem and sellar locations [12–14].

Schwann cells are thought to originate from neural crest cells that migrate with the growing neurites during nerve development. All myelinated fibers are ensheathed by Schwann cells, except at the nodes of Ranvier, to facilitate

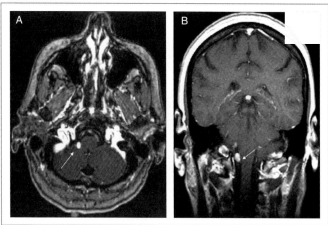

FIG. 41.5. Small homogeneously enhancing glossopharyngeal schwannoma (arrows) on axial **(A)** and coronal **(B)** T1 post-gadolinium images.

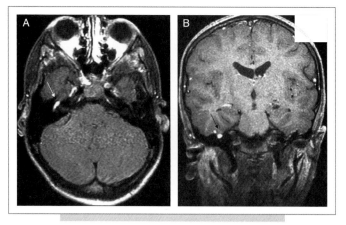

FIG. 41.6. Axial **(A)** and coronal **(B)** post-gadolinium T1 images of a small right facial nerve schwannoma (arrows) arising from the geniculate ganglion.

rapid transmission of electric impulses via saltatory conduction. Unlike neurofibromas that contain all elements of nerve including Schwann cells, nerve fibers and fibroblast causing splaying of the axon, schwannomas do not consist of nerve tissue [15].

PATHOLOGY

Histologically, vestibular schwannomas are composed of a variable percentage of two components. The usually more abundant Antoni type A (Figure 41.8) component is more compact consisting of interwoven bundles of fusiform cells and collagen whorls and reticulin fibers. They may arrange in a palisading pattern around a nuclear free zone, named a Verocay body. The more loosely packed Antoni type B (Figure 41.8) is made up of spindle and polygonal cells randomly scattered within a loose matrix, often with

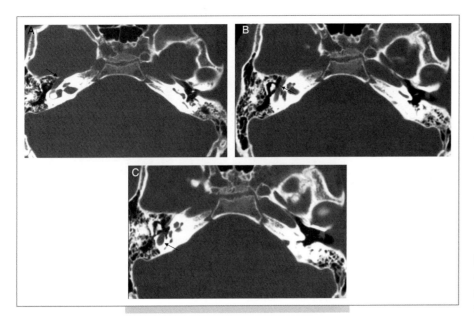

FIG. 41.7. Same patient as in Figure 41.6 demonstrating (clockwise top left) bony enlargement of the course of right facial nerve secondary to schwannoma (arrows).

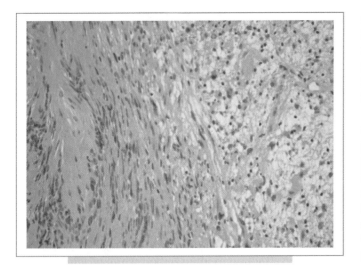

FIG. 41.8. Histology showing both Antoni type A (more compact left) and type B (loosely packed right) cells in the same schwannoma.

microcystic or mucinous areas that may coalesce to form larger cysts. Occasionally, other variants are seen consisting of fibroblastic changes with dense collagen bands, calcification, fatty or xanthomatous changes and intratumoral vascular thrombosis (hyaline thickening of vascular wall) or sometimes even hemorrhage. Ancient change in schwannoma describes a histologic variant with biphasic features that result from degenerative changes from a long-standing tumor [16]. The spindle shaped cells stain positively for S-100 protein and vimentin.

Malignant schwannomas (1%) are very rare, but are often seen in NF patients, characterized by focal areas of extremely high cellularity, pleomorphism and mitosis in an otherwise typical benign tumor. They can either arise from pre-existing tumors (latency 10–20 years) or can be de novo, often seen in a younger age group (15–39 years). The rare intraparenchymal schwannomas are thought to originate from perivascular Schwann cells of innervated arteries, dural cranial nerves site or ectopic cell crests [15].

Clinically, these tumors often present with gradual and progressive asymmetric and unilateral (except in NF2 patients) hearing loss, especially of the higher frequency pattern. About 10% have sudden hearing loss thought to be secondary to acute vascular compromise of auditory or cochlear nerve [4,14]. Tinnitis and dysequilibrium result from pressure exerted on the cochlear and vestibular divisions of the acoustic nerve. When involving cranial nerve V, they have weakness in the masticator muscles and less commonly present with facial pain. Diplopia suggests cavernous sinus extension and vertigo and lower cranial nerve palsies signify lower or skull base involvement [17].

IMAGING

Acoustic neuroma is somewhat of a misnomer as it mostly arises from the vestibular division of CN VIII, originates from Schwann cells and does not contain nerve tissue. A better and more accurate term would perhaps be vestibular schwannoma. The development of MR imaging, especially with contrast, has markedly facilitated the diagnosis of eighth nerve schwannomas, especially the small intracannalicular ones that do not cause widening of the internal auditory canal (IAC). These tumors enhance intensely due to a lack of blood–brain barrier. Contrast enhancement tends to have a rapid initial decrease (15 min after injection) followed by a slower decline [18]. Maximum enhancement can be reasonably anticipated within half an hour after injection. Contrast gives an overall sensitivity of

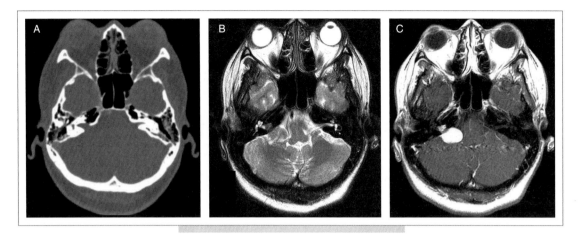

FIG. 41.9. Axial bone window CT **(A)**, axial T2 **(B)** and post-gadolinium T1 **(C)** images show widened right IAC from a right vestibular schwannoma.

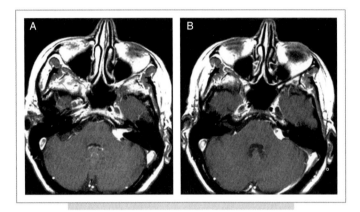

FIG. 41.10. Left vestibular schwannoma with both intra- and extracanalicular components on axial T1 post-gadolinium images.

almost 100% with the ability to detect a volume as small as 0.06 mm^3 [19,20]. On CT, schwannomas are iso- or low density relative to brain and sometimes contain cystic component or rarely calcification. Rarely, a lipid rich and larger Antoni A component tumor can impart a lower attenuation on CT [21].

About 70% of the time there is widening of the IAC, best seen with bone windows on thin section CT images (Figure 41. 9). The majority of eighth nerve schwannomas have both intra- and extracanalicular components. The reason is that the transition zone between the oligodendrocytes and Schwann cells lies in close proximity to the vestibular ganglion at the opening of the porus acousticus and schwannomas tend to grow bi-directionally from this junction (Figure 41.10) [22,23]. The next most common pattern is purely intracanalicular (Figure 41.11) followed rarely by a pure extracanalicular lesion (Figure 41.12).

On T1 MR images they tend to be iso- or hypointense to brain while heterogeneously hyperintense on T2 sequence.

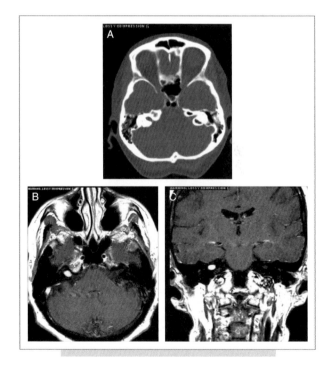

FIG. 41.11. Axial bone window CT **(A)**, axial **(B)** and coronal **(C)** post-gadolinium T1 images show widened right IAC due to an intracanalicular vestibular schwannoma.

It is thought that mucopolysaccharide contributes to lower T1 signal, while collagen imparts a low T2 signal [24,25]. In fact there has been description of 'target' sign on T2 sequence whereby centrally a low signal is seen secondary to fibrocollagen surrounded by peripheral high signal from myxomatous matrix. It was thought that this so-called 'target' sign is not seen in malignant schwannomas [26]. Long-standing schwannomas are also more likely to have cystic degenerative changes and hemorrhage (Figure 41.13).

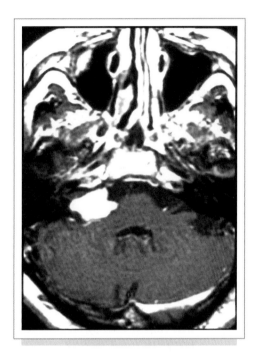

FIG. 41.12. Axial T1 post-gadolinium image with a purely extracanalicular right vestibular schwannoma.

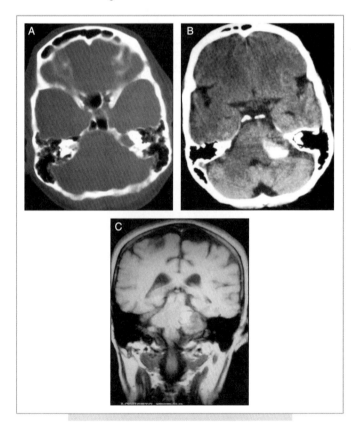

FIG. 41.13. Bony **(A)**, non-contrast **(B)** CT and coronal non-contrast T1 **(C)** images show a large left vestibular schwannoma with hemorrhagic component seen as high attenuation on CT and bright signal on T1.

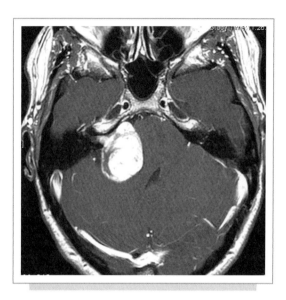

FIG. 41.14. Giant inhomogeneously enhancing right vestibular schwannoma on axial post-gadolinium T1 image.

Schwannomas tend to be slow growing with most at a rate of about 0.02 cm/year, some 0.2 cm/year and rarely at 1 cm/year [27]. Most present as medium sized lesions of about 2 cm with 2–4 cm defined as large lesions and >4 cm as giant schwannomas (Figure 41.14). The latter tend not to contain any canalicular component. Pregnancy may accelerate the growth of schwannomas with worsening of symptoms [28].

A standard MR protocol at our institution for IAC lesions includes axial T1, T2 (5–6 mm slice), FLAIR (fluid attenuated inversion recovery) and diffusion sequences of the brain with additional thin section (1–3 mm) FIESTA (fast imaging employing steady-state acquisition) images acquired; the latter is best for identifying facial, cochlear, and vestibular nerves, as well as the anterior inferior cerebellar artery (AICA) (Figure 41.15). Thin section axial and coronal (1–3 mm) pre- and post-gadolinium T1 images through the IAC are also performed. Some have also found that hearing preservation is better if labyrinth signal is normal on 3D T2-weighted volume images [29]. Main differentials for schwannoma are from meningioma (Figure 41.16), metastasis (Figure 41.17) and other extra-axial processes such as epidermoid (Figure 41.18) and non-acoustic schwannoma.

As most schwannomas occur at the cerebellopontine angle, one often needs to differentiate this from a meningioma. There are a few observations that can help in such instances. Schwannomas tend to be isodense and calcify less commonly on CT. They are globular, centered at the porus acousticus with acute margins, while meningiomas often arise along the petrous bone and are more sessile in shape with obtuse margins. Schwannomas are also more heterogeneous in appearance with cystic changes and tend

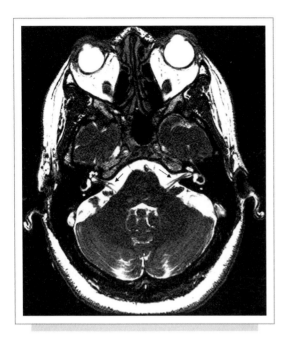

FIG. 41.15. FIESTA image demonstrates high-resolution visualization of bilateral seventh-eighth nerve complex (arrows).

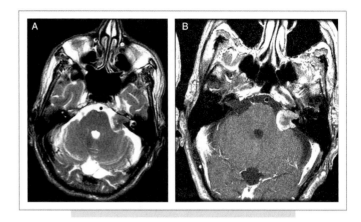

FIG. 41.17. A rare mesothelioma metastasis to the left IAC on axial T2 **(A)** and post-gadolinium T1 **(B)** images.

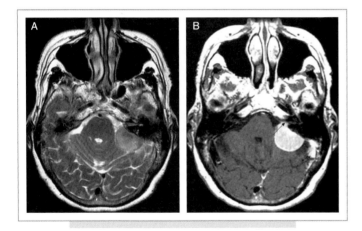

FIG. 41.16. Axial T2 **(A)** and post-gadolinium T1 **(B)** images show a broad based homogeneously enhancing left meningioma centered at the left petrous bone with dural tail (arrow) extending into the left IAC.

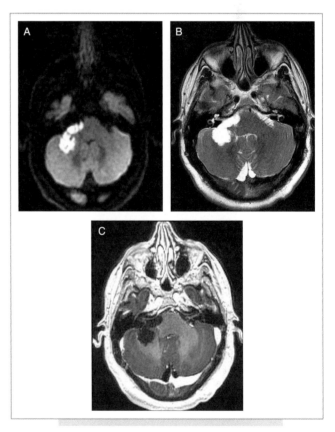

FIG. 41.18. Non-enhancing amorphous right CP angle epidermoid on axial diffusion **(A)**, T2 **(B)** and post-gadolinium T1 **(C)** images.

to widen the internal auditory canal. Meningiomas, on the other hand, are more commonly associated with a dural tail (Figures 41.16 and 41.19).

After radiotherapy, more than 80% (most centrally) show loss of enhancement (Figure 41.20), 5% show increase and about 11% show no changes. There is, however, no correlation between the volume of tumor to enhancement [30]. Adjacent brain, such as the pons, may also show mild enhancement after radiation [31,32].

NEUROFIBROMATOSIS II

Neurofibromatosis II (NF2) is an autosomal dominant highly penetrant phakomatosis that does not have any sex or racial predilection [33]. Half of affected patients are due to new gene mutations while the other 50% are inherited

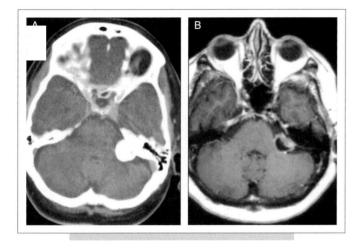

FIG. 41.19. Axial non-contrast CT **(A)** and post-gadolinium T1 images **(B)** show a calcified left CP angle meningioma.

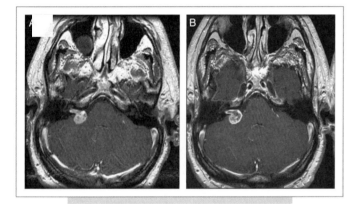

FIG. 41.20. Axial post-gadolinium T1 images show heterogeneous enhancing right vestibular schwannoma before **(A)** and decreased enhancement and necrosis after radiation **(B)**.

[34,35]. They often present with multiple cranial nerves especially bilateral vestibular schwannomas (VS) by the age of 30. About 18% of patients manifest before 15 years of age and often develop spinal tumors and skin lesions. The Wishart type has an early onset (early 20s), multiple tumors including bilateral vestibular schwannomas and a severe clinical and rapid course. A second or Gardner type has later presentation, bilateral VS and a more benign course with fewer tumors. The third or segmental NF2, which constitutes about 25% of cases, has somatic mosaicism with mutation occurring during embryogenesis rather than germ line DNA, resulting in altered genetic material in only a portion of the patient's cells [4,36].

It is thought that the disorder arises due to the alteration or loss of genetic material and occasional translocation of the NF2 gene locus [37] on chromosome 22q12.2 that codes for a protein named 'merlin' (*moezin-ezrin-radixin-like-protein*). It has high homology to a family of cytoskeletal-associated proteins and possesses tumor suppression

activity [38]. This gene consists of 110 kilobases with 16 constitutive exons and alternatively spliced exon [38,39] whose function is thought to include regulation of the signaling of Rho GTPase. There are at least 150 different mutations described where most are point mutations [40]. Missense mutations [41] are thought to give a milder form of the disease while nonsense, frame shifting [42,43] and truncating mutations [44] are associated with more severe phenotypes. The deletion or mutation results in increased risk of development of schwannomas, meningiomas, ependymomas, gliomas and juvenile posterior subcapsular lenticular opacity (pediatric population) [45,46]. Variability in expression results in differing size, location and number of tumors. Larger deletions can also cause mental retardation and congenital abnormalities [47]. Milder phenotypic form of the disease is often seen in individuals with somatic mosaicism with unilateral vestibular schwannoma and other ipsilateral tumors [48]. Genetic testing of these patients may give misleading information because their lymphocytes can be normal, thus necessitating direct tumor tissue analysis for proper diagnosis [6]. Recent reports indicate that 25–30% of NF2 patients without a family history are mosaic for the mutation [49,50]. Prevalence of NF2 is approximately 1 in 210 000 with incidence of 1 in 30–40 000 [51].

Over the years there have been various modifications of the NIH consensus criteria for the diagnosis of NF2 currently listed as [4]:

- Bilateral vestibular schwannomas
- First-degree relative with NF2 *and*
    - Unilateral vestibular schwanoma *or*
    - Any two of: meningioma, schwannoma, glioma, neurofibroma, cataract
- Unilateral vestibular schwannoma *and* any two of meningioma, schwannoma, glioma, neurofibroma, posterior subcapsular lenticular opacities
- Multiple menigiomas *and*
    - Unilateral vestibular schwannoma *or*
    - Any two of: schwannoma, glioma, neurofibroma, Cataract.

Besides using for diagnostic purposes, genetic testing is also employed for early detection of patients (primarily children) of classically affected parents, for better management of the disease over time.

Unlike sporadic schwannomas, the tumors associated with NF2 tend to be more invasive [52] and encase the entire vestibulocochlear and facial nerves, are bilateral and occur at a younger age (20–30s) [45]. Those with concomitant meningiomas have a ten-fold increase in growth rate. Patients often present with hearing loss, facial weakness, tinnitus, balance problems and seizures [53]. Malignant transformation is rare and can be iatrogenic from prior radiation. There have been reports of malignant transformation of about 5% in radiated compared to 1% in nonradiated patients. Other symptoms are mass effect from

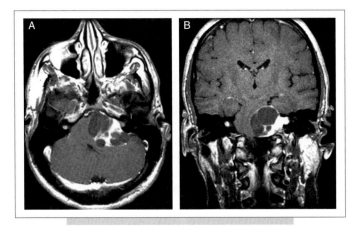

FIG. 41.21. Bilateral vestibular schwannomas with a large cystic left-sided lesion in an NF2 patient on axial (**A**) and coronal (**B**) post-gadolinium T1 images.

meningiomas, non-acoustic schwannomas, skin tumors and visual loss, the latter two most often being the initial presentation in children.

Spinal tumors are also commonly found in two-thirds of patients and are debilitating and difficult to manage [54]. Although the tumors are rarely malignant, they are, however, situated at locations and in proximity to vital structures that result in high morbidity and mortality. Usual time of survival from initial diagnosis is about 15 years with average age of death at 36 years old [51]. The age at presentation, presence of meningioma and whether treatment was conducted at a specialty center are all predictors of mortality risk. In recent years, mononeuropathy in childhood and polyneuropathy in adults not directly associated with tumors are increasingly recognized [55].

Schwannomatosis is a genetically and clinically distinct disease from NF2 characterized by multiple schwannomas including intracranial, spinal root and peripheral nerves except the vestibular location. There are familial cases with an autosomal dominant inheritance but with variable gene expression and penetrance. The genetic locus of this disease has been found to be near but separate from the NF2 gene [49,56,57].

As in the sporadic form, contrast MR imaging is the gold standard for detecting vestibular schwannomas and other intracranial lesions (Figure 41.21). This is due to the multiplanar, non-invasive, radiation free and numerous pulse sequence capabilities of MR imaging. CT may be better in visualizing choroid plexus and cortical calcifications and is also performed in patients who cannot undergo MR imaging.

MANAGEMENT AND TREATMENT

The average growth rate for vestibular schwannoma is approximately 1 mm^3 per year. The tumor size and location, age, NF2 and degree of hearing loss are important factors in the management of these patients. There is no correlation between initial tumor size and growth rate. In the sporadic population, most can be followed with MR imaging every 6–12 months under the so-called expectant management [58].

For microsurgery, there are primarily three basic surgical approaches, each with their unique advantages and risks. In 1960, Dr William House introduced the translabyrinthine approach (TLA) for patients with poor preoperative hearing. It is not used in patients where the hearing needs to be preserved (30 dB and speech discrimination score [SDS] of 70%) or if it is the site of the only functional hearing as this procedure sacrifices whatever preoperative hearing function there was. In NF2 patients, this surgical approach is often used in conjunction with auditory brainstem implant (ABI) [58,59].

Another technique is the retrosigmoid/suboccipital approach (RSA) indicated for patients with reasonable hearing and greater than 0.5 cm extension of tumor into the cerebellopontine angle. It is limited for lesions that extend laterally into the internal auditory canal where 18% have residual tumor within the fundus after surgery. This approach also has a slightly higher risk of axonal and vascular injury [58,59].

The third approach is via the middle cranial fossa (MCF) and is best for small intracanalicular (<0.5 cm) tumors with good preoperative hearing. The risks of this technique include manipulation of the facial nerve, higher incidence of seizure due to retraction of the temporal lobe and increased postoperative CSF leak due to thinner dura. In other words, the MCF approach tends to preserve hearing but comes with higher risks [58,59].

In NF2 patients a multidisciplinary approach to diagnosis, evaluation and treatment at an experienced center is the key for the optimal management of their disease. Surgical resection remains the cornerstone for treatment of vestibular schwannomas with the best results for small (<1.5 cm), medially positioned intracanalicular tumors that preserve both hearing and facial nerve function [60]. Debulking or subtotal resection of larger tumor is reserved for patients with worsening symptoms or brainstem compression [61]. For large bilateral tumors (>2 cm) and good hearing (SDS > 50%), patients may be observed. If SDS is less than 50%, surgery can be contemplated. For small bilateral tumors (<2 cm) and good preoperative hearing (SDS > 50%), the smaller one is removed while the larger lesion may be resected if there is brainstem compression. If the tumor size is equal bilaterally, the side with worse hearing will undergo resection. In the case of one side with small tumor and good hearing while the other with large tumor and bad hearing (SDS< 50%), the latter tumor will be removed with option of cochlear implant. When the converse is true, i.e. a small tumor with bad hearing on one side and large tumor with good hearing on the other,

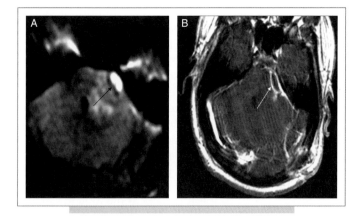

FIG. 41.22. Postoperative complication with small abscess (arrows) at the surgical site after resection of left vestibular schwannoma with restricted diffusion **(A)** and ring enhancement **(B)** on axial DWI and post-gadolinium T1 images respectively.

then the smaller side will be removed. Then, finally, if both tumors are either large or small but both with poor hearing then the larger one is first removed with consideration for auditory brainstem implant [36].

In deciding between the different surgical options, factors that need to be considered are long-term facial nerve function and postoperative rate of preserved hearing and balance. Postoperative complications are also important including CSF leak, infection and postoperative headache. Good hearing is defined as having four frequency pure tone with an average of better than 50 decibel (dB) and SDS greater or equal to 50%. The overall rate of hearing preservation is about 44% best in MCF (55%) and less in RSA (30%). Facial nerve function is graded from I to VI (good to poor) at 6–12 months after surgery. It may be splayed during surgery, especially for >1.5 cm lesions that are cystic and thus more adherent. Overall rate is 82% with good follow-up (grade I/II), breaking down as 92% in RSA, 89% in MCF and 73% for TLA [58,59].

CSF leak is often seen during the first 2–3 or 10–14 days after surgery. It can occur at various locations including the incision site, tympanic membrane, Eustachian tube and nasopharynx. Overall rate is 8%, with a breakdown of 6% for MCF, 9% TLA and 11% RSA [62,63]. Postoperative infections (Figure 41.22) are often aseptic and thought to be due to irritation from bone dust or hemorrhage. It occurs in about 2% MCF, and 3% in both RSA and TLA. Steroid treatment in this population has been helpful [62,64–66]. Finally, headache may persist up to 3 months after surgery with overall rate of 10% seen most in RSA (21%) and less in MCF and TLA [66–69].

In general, tumors that measure greater than 1.5 cm in axial plane at the cerebellopontine angle have higher risk for facial nerve injury from surgery than radiation therapy. Solid tumors also fare better than cystic ones. Outpatient stereotactic radiation treatment (single dose or fractionated), primarily gamma knife, is also an alternative to surgery with better long-term control in non-NF2 patients (95% versus 60%) [70]. In the late 1980s and early 1990s, about 79% had normal facial nerve function after 5 years from radiation with 51% without hearing loss. With lower doses of about 13 Gy in recent years there is further decrease in post-treatment dysfunction, approaching 1.1% in facial and 2.6% in trigeminal nerves; 71% hearing preservation and with overall 97% control rate. On imaging, tumors may initially enlarge during the first 6 months to a year after radiation due to edema. Very infrequently a large tumor undergoes initial surgical debulking followed by radiation [59].

The risks and benefits of radiation therapy should also be carefully weighed to prevent untoward transformation of existing tumors, especially in individuals with a copy of inactive tumor suppression gene. There are sporadic reports of squamous cell carcinoma and malignant transformation secondary to radiation [71,72]. It should be noted that the experience with radiation is in the relatively early stages, therefore, the surgical option is still the first consideration. The upper limit for tumor size for radiation therapy is about 3 cm maximum in the axial plane.

Training with an audiologist is important to maintain functional hearing and speech production. Sudden hearing loss from vascular compromise may benefit from cochlear implants where the nerves are relatively intact. Sometimes, hearing evaluation can detect subtle dysfunction of auditory nerves before the visualization of any abnormality on MRI.

Imaging provides information on size, site, the presence or absence of and preoperative planning and postoperative follow-up of tumors. In addition, they also serve as a non-invasive way to monitor tumor growth, as well as recurrent or residual lesions over time with or without prior treatment.

In summary, sporadic and NF2 vestibular schwannomas are typically benign tumors with morbidity and mortality mostly attributed to lesion size and location, and presenting age of the patient. Thin section MRI, especially with contrast, remains the imaging modality of choice for the diagnosis of these tumors. It is also a non-invasive technique for pre- and postoperative planning and follow-up before and after treatment. Microsurgery remains the corner stone for treatment with radiation gradually assuming a larger role as the technique continues to improve over time. TLA approach is best for large tumors and poor preoperative hearing, RSA for small and medial lesions, and MCF for lateral tumors. For NF2 patients, a multidisciplinary involvement appears to be the optimal approach in the management of this population.

REFERENCES

1. Chandler CL, Ramsden RT (1993). Acoustic schwannoma. Br J Hosp Med 49:335–343.
2. D'Alessandro G, Giovnni M, Iannizzi L et al (1995). Epidemiology of primary intracranial tumors in the Valle d'Aosta during the 6-yr period 1986–1991. Neuroepidemiology 14:139–146.
3. Barker DJP, Weller RO, Garfield JS (1976). Epidemiology of primary tumors of the brain and spinal cord: a regional survey in southern England. J Neurol Neurosurg Psychiatr 39: 290–296.
4. Consensus Development panel (1994). National Institute of Health Consensus Development Conference statement on Acoustic Neuroma Dec 11–13, 1991. Arch Neurol 51:201–207.
5. Lin D, Hegarty JL, Fischbein NJ et al (2005). The prevalence of 'incidental' acoustic neuroma. Arch Otolaryngol Head Neck Surg 131:241–244.
6. Mohyuddin A, Neary WJ, Wallace A et al (2002). Molecular genetic analysis of the NF2 gene in young patients with unilateral vestibular schwannomas. J Med Genet 39:315–322.
7. Shisano G, Olivecrona H (1960). Neurinomas of the Gasserian ganglion and trigeminal root. J Neurosurg 17:306–322.
8. Celli P, Ferrante L, Acqui M et al (1992). Neurinoma of the third, fourth and sixth cranial nerves and report of a new fourth nerve case. Surg Neurol 38:216–224.
9. Morey RA, Halliday GC (1965). Neurinomas of facial nerve. Report of a case. J Neurosurg 23:539–541.
10. Lanotte M, Giordana MT, Forni C et al (1992). Schwannoma of the cavernous sinus. J Neurosurg Sci 36:233–238.
11. Fugiwara S, Hachisuga S, Numaguchi Y (1980). Intracranial hypoglossal neurinoma. Report of a case. Neuroradiology 20:87–90.
12. Ezura M, Ikeda H, Ogawa A et al (1992). Intracerebral Schwannoma, case report. Neurosurgery 30:97–100.
13. Frim DM, Ogilvy CS, Vonsattal JP et al (1992). Is intracerebral schwannoma a developmental tumor of children and young adults? Pediatr Neurosurg 1:190–194.
14. Sharma PR, Gurusinghe NT, Lynch PG et al (1993). Parenchymatous schwannoma of the cerebellum. Br J Neurosurg 7:83–90.
15. Lantos PL, Louis DN, Rosenblum MK, Kleihues P (2002). Tumors of the nervous system. In Greenfields Neuropathology, 7th edn Vol 2. Graham DI, Lantos PL (eds). Arnold, London. 11:897–909.
16. Ugokwe K, Nathoo N, Prayson R et al (2005). Trigeminal nerve schwannoma with ancient change. Case report and review of literature. J Neurosurg 102:1163–1165.
17. Chandler CL, Ramsden RT (1993). Acoustic schwannoma. Br J Hosp Med 49:335–343.
18. Hatam A, Bergstrom M, Moller A et al (1978). Early contrast enhancement of acoustic neuroma. Neuroradiology 17:31–33.
19. Curtin HD (1997). Rule out eighth nerve tumor: contrast enhanced T1-weighted or high resolution T2-weighted MR? Am J Neuroradiol 18:1834–1838.
20. Schmalbrock P, Chakeres DW, Monroe JW et al (1999). Assessment of internal auditory canal tumors: a comparison of contrast-enhanced T1-weighted and steady-state T2-weighted gradient-echo MR imaging. Am J Neuroradiol 20:1207–1213.
21. Wu E, Tang Y, Zhang Y et al (1986). CT in diagnosis of acoustic neuromas. Am J Neuroradiol 7:745–750.
22. Schuknecht H (1974). Pathology of the Ear. Harvard University Press, Cambridge.
23. Bebin J (1979). Pathophysiology of Acoustic Tumors. Vol I. Diagnosis. University Park Press, Baltimore.
24. Press GA, Hesselink JR (1988). MR imaging of cerebellopontine angle and internal auditory canal lesions at 1.5T. Am J Neuroradiol 9:241–251.
25. Daniels DL, Millen SJ, Meyer GA et al (1987). MR detection of tumor in the internal auditory canal. Am J Neuroradiol 8:249–252.
26. Varma DG, Moulopoulos A, Sara AS et al (1992). MR imaging of extracranial nerve sheath tumors. J Comput Assist Tomogr 16:448–453.
27. Lanser MJ, Sussman SA, Frazer K (1992). Epidemiology, pathogenesis and genetics of acoustic tumors. Otolaryngol Clin N Am 25:499–520.
28. Kasantikul V, Brown WJ (1981). Estrogen receptors in acoustic neurilemmomas. Surg Neurol 15:105–109.
29. Somers T, Casselman J, de Ceulaer G et al (2001). Prognostic value of magnetic resonance imaging findings in hearing preservation surgery for vestibular schwannoma. Otol Neurotol 22:87–94.
30. Nakamura H, Jokura H, Takahashi K et al (2000). Serial follow-up MR imaging after gamma knife radiosurgery for vestibular schwannoma. Am J Neuroradiol 21:1540–1546.
31. Noren G, Arndt J, Hindmarsh T (1983). Stereotactic radiosurgery in cases of acoustic neurinoma: further experiences. Neurosurgery 13:12–22.
32. Thomsen J, Tos M, Borgesen SE (1990). Gamma knife: hydrocephalus as a complication of stereotactic radiosurgical treatment of an acoustic neuroma. Am J Otol 11:330–333.
33. Martuza RL, Eldridge R, Wertelecki W et al (1988). Neurofibromatosis 2: clinical and DNA linkage studies of a large kindred. N Engl J Med 318:684–688.
34. MacCollin M, Ramesh V, Jacoby LB et al (1994). Mutational analysis of patients with neurofibromatosis 2. Am J Hum Genet 55:314–320.
35. Parry DM, MacCollin MM, Kaiser-Kupfer MI et al (1996). Germ-line mutations in the neurofibromatosis 2 gene: correlations with disease severity and retinal abnormalities. Am J Hum Genet 59:529–539.
36. Neff BA, Welling DB (2005). Current concepts in the evaluation and treatment of NF Type II. Otolaryngol Clinics N Am 38:671–684.
37. Tsilchorozidou T, Menko F, Lalloo F et al (2004). Constitutional rearrangements of chromosome 22 as a cause of neurofibromatosis. J Med Genet 41:529–534.
38. Trofatter JA, MacCollin MM, Rutter JL et al (1993). A novel moesin-, ezrin, radixin-like gene is a candidate for the neurofibromatosis 2 tumor suppressor. Cell 72:791–800.
39. Rouleau GA, Merel P, Lutchman M et al (1993). Alteration in a new gene encoding a putative membrane-organizing protein causes neuro-fibromatosis type 2. Nature 363:515–521.
40. Legoix P, Sarkissian HD, Cazes L et al (2000). Molecular characterization of germline NF2 gene rearrangements. Genomics 65:62–66.

41. Evans DG, Trueman L, Wallace A et al (1998). Genotype/phenotype correlations in type 2 neurofibromatosis (NF2): evidence for more severe disease associated with truncating mutations. J Med Genet 35:450–455.

42. Ruttledge MH, Andermann AA, Phelan CM et al (1996). Type of mutation in the neurofibromatosis 2 gene (NF2) frequently determines severity of disease. Am J Hum Genet 59:331–342.

43. Sainz J, Figueroa K, Baser ME et al (1995). High frequency of nonsense mutations in the NF2 gene caused by C to t transitions in five CGA codons. Hum Mol Genet 4:137–139.

44. Baser ME, Friedman JM, Aeschliman D et al (2002). Predictors of the risk of mortality in neurofibromatosis 2. Am J Hum Genet 71:715–723.

45. Evans DG, Huson SM, Donnai D et al (1992). A clinical study of type 2 neurofibromatosis. Q J Med 84:603–618.

46. Mautner VF, Tatagiba M, Guthoff R et al (1993). Neurofibromatosis 2 in the pediatric age group. Neurosurgery 33:92–96.

47. Barbi G, Rossier E, Vossbeck S et al (2002). Constitutional de novo interstitial deletion of 8 Mb on chromosome 22q12.1–12.3 encompassing the neurofibromatosis type 2 (NF2) locus in a dysmorphic girl with severe malformations. J Med Genet 39:E6.

48. MacCollin M, Jacoby L, Jones D et al (1997). Somatic mosaicism of the neurofibromatosis 2 tumor suppressor gene. Neurology 48A:429.

49. Moyhuddin A, Baser ME, Watson C et al (2003). Somatic mosaicism in neurofibromatosis 2: prevalence and risk of disease transmission to offspring. J Med Genet 40:459–463.

50. Kluwe L, Mautner V, Heinrich B et al (2003). Molecular study of frequency of mosaicism in neurofibromatosis 2 patients with bilateral vestibular schwannomas. J Med Genet 40:109–114.

51. Evans DG, Huson SM, Donnai D et al (1992). A genetic study of type 2 neurofibromatosis in the United Kingdom. I. Prevalence, mutation rate, fitness, and confirmation of maternal transmission effect on severity. J Med Genet 29:841–846.

52. Jaaskelainen J, Paetau A, Pyykko I et al (1994). Interface between the facial nerve and large acoustic neurinomas. Immunohistochemical study of the cleavage plane in NF2 and non-NF2 cases. J Neurosurg 80:541–547.

53. Selesnick SH, Jackler RK (1992). Clinical manifestations and audiologic diagnosis of acoustic neuromas. Otolaryngol Clin N Am 25:521–551.

54. Parry DM, Eldridge R, Kaiser-Kupfer MI et al (1994). Neurofibromatosis 2 (NF2): clinical characteristics of 63 affected individuals and clinical evidence for heterogeneity. Am J Med Genet 52:450–461.

55. Sperfeld AD, Hein C, Schroder JM et al (2002). Occurrence and characterization of peripheral nerve involvement in neurofibromatosis type 2. Brain 125:996–1004.

56. MacCollin M, Woodfin W, Kronn D et al (1996). Schwannmatosis: a clinical and pathologic study. Neurology 46:1072–1079.

57. MacCollin M, Willet C, Heinrich B et al (2003). Familial schwannomatosis: exclusion of the NF2 locus as the germline event. Neurology 60:1968–1974.

58. Harsha WJ, Backous DP (2005). Counseling patients on surgical options for treating acoustic neuroma. Otolaryngol Clin N Am 38:643–652.

59. Wackym PA (2005). Stereotactic radiosurgery, microsurgery and expectant management of acoustic neuroma. Basis for informed consent. Otolaryngol Clin N Am 38:653–670.

60. Briggs RJ, Brackmann DE, Baser ME et al (1994). Comprehensive management of bilateral acoustic neuromas. Current perspectives. Arch Otolaryngol Head Neck Surg 120:1307–1314.

61. Ojemann RG (1993). Management of acoustic neuromas (vestibular schwannomas). Clin Neurosurg 40:498–535.

62. Sanna M, Taibah A, Russo A et al (2004). Perioperative complications in acoustic neuroma (vestibular schwannoma) surgery. Otol Neurotol 25:379–386.

63. Becker S, Jackler R, Pitts L (2003). CSF leak after acoustic neuroma surgery, a comparison of translabyrinthine, middle fossa and retrosigmoid approaches. Otol Neurotol 24:107–112.

64. Mamikoglu B, Wiet R, Esquivel C (2002). Translabyrinthine approach for the management of large and giant vestibular schwannoma. Otol Neurotol 23:224–227.

65. Lanman T, Brackmann D, Hitselberger W et al (1999). Report of 190 consecutive cases of large acoustic tumor (vestibular schwannoma) removed via the translabyrinthine approach. J Neurosurg 90:617–623.

66. Mass S, Wiet R, Dinces E (1999). Complications of the translabyrinthine approach for the removal of acoustic neuroma. Arch Otolaryngol Head Neck Surg 125:801–804.

67. Weber P, Gantz B (1996). Results and complications from acoustic neuroma excision via middle fossa approach. Am J Otol 17:669–675.

68. Andersson G, Ekvall L, Kinnefors A et al (1997). Evaluation of quality of life and symptoms after translabyrinthine acoustic neuroma surgery. Am J Otol 18:421–426.

69. Schaller B, Baumann A (2003). Headache after removal of vestibular schwannoma via the retrosigmoid approach: a long-term follow-up study. Otolaryngol Head Neck Surg 128:387–395.

70. Rowe JG, Radatz MW, Walton L et al (2003). Clinical experience with gamma knife stereotactic radiosurgery in the management of vestibular schwannomas secondary to type 2 neurofibromatosis. J Neurol Neurosurg Psychiatr 74:1288–1293.

71. Lustig LR, Jackler RK, Lanser MJ (1997). Radiation-induced tumors of the temporal bone. Am J Otol 18:230–235.

72. Hanabusa K, Morikawa A, Murata T et al (2001). Acoustic neuroma with malignant transformation. Case report. J Neurosurg 95:518–521.

CHAPTER 42

Diencephalic and Other Deep Brain Tumors

A. Drevelegas and E. Xinou

INTRODUCTION

The diencephalon, one of the most highly developed structures of the human central nervous system, consists of two major components: the thalamus, a key structure for transmitting information to the cerebral hemispheres; and the hypothalamus, which integrates the functions of the autonomic nervous system and endocrine hormone release from the pituitary gland [1].

Tumors that affect the thalamus and the basal ganglia can develop as primary tumors or secondarily from neighbouring tumors which invade the deep portions of the cerebrum [2].

Therefore, we will classify deep brain tumors either as primary neoplasms or as neighboring tumors that invade the diencephalic or the deep telecenphalic structures.

PRIMARY DIENCEPHALIC TUMORS

Basal Ganglia–Thalami

Thalamic tumors are generally considered together with basal ganglionic tumors, although the thalami are part of the diencephalon. Basal ganglia and thalamic tumors account for 10% of supratentorial tumors in children and nearly two-thirds of these neoplasms involve the thalami [3]. Thalamic tumors account for approximately 1–1.5% of all brain tumors [4–7]; they are slightly more common in children than in adults and involve significant diagnostic, prognostic and therapeutic problems [5,8,9].

Their deep location makes them difficult to treat, mainly because surgery is rarely feasible [4,10–15]. Radiation therapy with or without adjuvant chemotherapy [8,16] is still considered to be the basic treatment.

They are predominantly low-grade astrocytomas [11]; but oligodendrogliomas, ganglion cell tumors, germinal neoplasms, primitive neuroectodermal tumors, lymphomas and metastases can also grow up from the basal ganglia and thalamic region [2].

Gliomas

Gliomas can be divided into two broad categories: well-circumscribed gliomas like pilocytic astrocytomas, and infiltrating or diffuse gliomas, including fibrillary astrocytomas, oligodendrogliomas and gliomatosis cerebri. In the first category, the extent of excision is a powerful prognostic factor, whereas the prognosis in the latter is related more closely to the histologic grade [17]. According to Wald, the number of astrocytic tumors was 147 (77%) among 191 cases of thalamic tumors assessed by histological examination [3].

Pilocytic Astrocytoma (PA)

PAs represent the prototype of benign circumscribed astrocytomas, usually without a tendency to infiltrate and progress toward anaplasia [5]. Macroscopically, PAs can be seen as homogeneous solid, heterogeneous multicystic or predominantly cystic tumors, occasionally with calcifications developing in the solid portions. Hemorrhages are very unusual [18,19]. PAs are tumors with a characteristic, heterogeneous 'biphasic' histological appearance, consisting of alternating spongy areas with other more cellular regions [18]. In the thalami, PAs are much less frequent than in the cerebellum, optic-chiasmatic region and cerebral hemispheres [5].

Computed tomography (CT)/magnetic resonance (MRI) features are characterized by the well-circumscribed appearance, with sharp margins and variable size, mass effect and degree of perilesional edema. The typical cerebellar presentation of a cystic tumor with a mural nodule (Figure 42.1) or a thick solid rim is rarely encountered in the thalami, where solid tumors predominate, with or without small cystic areas [5].

CT may demonstrate a well-circumscribed mass that is homogeneous in cases of solid or predominantly solid tumors and heterogeneous in multicystic neoplasms. Cystic areas should be barely hypodense and similar to cerebrospinal fluid. Solid portions may be iso- to hypodense to brain parenchyma [20]. Sometimes hyperdense foci of calcification can be detected [2].

On MRI, cysts may be iso- or hyperintense to CSF in all pulse sequences. Nevertheless, the degree of hyperintensity of the cysts depends on the biochemical characteristics of the cystic fluid. Solid areas are usually isointense or discretely hypointense relative to normal brain on T1-weighted images. On intermediate and T2-weighted images, solid areas are usually hyperintense [20–22]. Usually, the solid portion of the tumor enhances after contrast administration (Figure 42.2) [21,23,24]. If MRI depicts calcifications, they normally induce a decrease in signal in all sequences. Rarely, they are depicted as areas of hyperintensity on T1-weighted images. Surrounding edema should be minimal or undetectable [25].

The inhomogeneity of PA may lead to the suspicion of malignant astrocytoma, but the well-defined margins, bright contrast enhancement and the lack of necrosis usually lead to a correct preoperative diagnosis [5].

Diffuse or Infiltrating Astrocytoma

Diffuse astrocytomas are less common in children than in adults. They can be classified as low-grade diffuse astrocytomas (LGDA), anaplastic astrocytomas (AA) and glioblastoma multiforme (GBM) [20,24].

LGDA usually enlarge but do not destroy brain structures [25]. Cyst and hemorrhage are uncommon and calcification may be seen in up to 20% of cases [24,26]. CT may demonstrate a homogeneous, poorly circumscribed hypodense lesion with occasional contrast enhancement [20,24]. Sometimes, the tumor is relatively isodense and mass effect may be the only detectable imaging sign.

LGDA are usually hypointense on T1-weighted images, hyperintense or intermediate on T2-weighted images and demonstrate no or variable enhancement [20,22,24]. Surrounding vasogenic edema is an uncommon radiological finding (Figure 42.3) [20,22,24]. On FLAIR (fluid attenuated inversion-recovery) and IR images, ill-defined hyperintense regions are noted in the subthalamic areas, the cerebral peduncles or the internal capsule, revealing marginal infiltration as perilesional edema is usually lacking or extremely limited [5]. AA are biologically more aggressive neoplasms with imaging findings similar to those of LGDA, except for the more frequent contrast enhancement and vasogenic edema [24].

On the other hand, the typical features of AA are the inhomogeneity of MR signal on T1- and T2-weighted images, due to variable cellularity and vascularity, and the presence of heterogeneous contrast enhancement and extensive edema. Thus, inhomogeneity, irregular contrast enhancement and extensive edema should be considered as characteristic of anaplasia, but also the extensive infiltrative growth is to be maintained as an indicator of malignancy and of dismal prognosis (Figure 42.4) [5]. GBM is far more common in adults than in children, as is progression from astrocytoma to GBM. Thalamic GBM appears as a fairly localized mass lesion with extensive vasogenic edema [5]. Necrosis and microvascular proliferation are histological characteristics that induce typical imaging manifestations [24,26]. CT and MRI appearance of thalamic GBM is markedly inhomogeneous, usually with central necrosis, possible cavitation and fluid/fluid levels and frequent hemorrhage at all stages of evolution (Figure 42.5A,B). On T2-weighted images, the

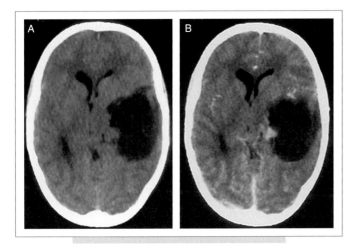

FIG. 42.1. Pilocytic astrocytoma. **(A)** Axial non-contrast CT scan shows an isodense left thalamic mural nodule with a hypodense cyst extending to the adjacent structures. **(B)** On axial post-contrast CT scan the mural nodule shows intense and homogeneous enhancement. The wall of the cyst remains unenhanced.

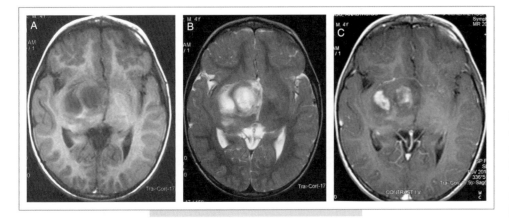

FIG. 42.2. Right thalamic pilocytic astrocytoma. The tumor appears hypointense on T1-weighted image **(A)**, hyperintense on T2-weighted image **(B)** and shows marked enhancement after contrast injection **(C)**.

recognition of central necrosis is more distinct, showing heterogeneous intensity and degraded blood products, while peripheral areas exhibit relatively low signal intensity (high cellularity) and strong enhancement after contrast medium administration (Figure 42.5C,D,E; also see Figure 35.12A–C) [5]. Flow voids related to vascular proliferation may also be observed [24]. Despite its circumscribed appearance, thalamic GBM may have small finger-like marginal extensions that represent their tendency to infiltrate adjacent tissues or spread more diffusely, producing intraventricular and craniospinal leptomeningeal CSF seeding [5,27]. Little or no vasogenic edema is typical with all grades of thalamic astrocytoma. This may be due to a paucity of white matter pathways in this area [28].

MR spectroscopy (MRS) is useful in distinguishing between low and high aggressive neoplasms. In general, brain tumors are characterized by a decrease in creatine (Cr) and in N-acetylaspartate (NAA), which reflects loss or dysfunction of the neurons, with simultaneous increase in choline (Cho), which is related to the tumor malignancy [25,29,30]. Necrotic tumors, such as GBM, may show low Cho levels, which paradoxically indicate lower grade neoplasm but, in such cases, the levels of lactic/lipids can increase, allowing the correct grading of these tumors [31].

Bilateral Thalamic Glioma

Among the primary thalamic gliomas, a type known as bilateral thalamic glioma (BTG) has been identified, in which a large tumor appears symmetrically in both thalami and is accompanied by behavioral impairments varying from personality changes to dementia [6,32–34]. BTG is rare among the thalamic tumors and has a very poor prognosis despite therapy [33].

These tumors are generally low-grade at presentation and produce remarkably symmetric areas of radiographic abnormality in some cases, with extension, which is equally symmetric, to the adjacent basal ganglia and midbrain [9,35,36]. They are usually undetectable on CT, whereas on FLAIR and T2-weighted images they appear as diffuse, homogeneous high-intensity areas (Figure 42.6A–C) [37–39]. The differential diagnosis of BTG includes brainstem encephalitis, acute necrotizing encephalopathy, mitochondrial encephalomyopathy and deep cerebral venous thrombosis [9]. The acute onset of signs and symptoms

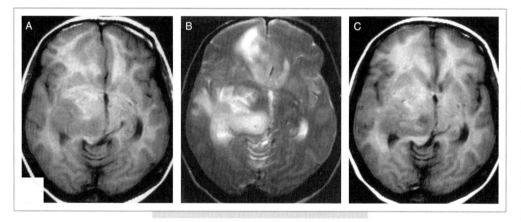

FIG. 42.3. Right thalamic glioma. The tumor appears hypointense on non-contrast T1-weighted image **(A)** and hyperintense on T2-weighted image **(B)** with infiltration of the adjacent structures (internal capsule, parietal and frontal lobe). Post-gadolinium T1-weighted image **(C)** shows subtle contrast enhancement.

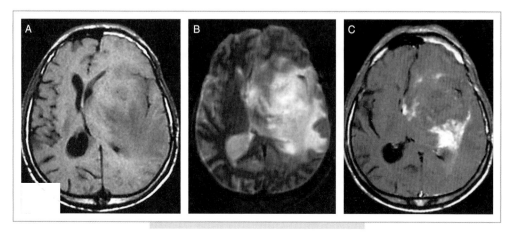

FIG. 42.4. Anaplastic astrocytoma. **(A)** Axial T1-weighted image shows a low signal mass in the left thalamus extending to adjacent structures and displacing the midline structures. **(B)** On axial T2-weighted image the mass shows high signal intensity extending to the ipsilateral parietal lobe. **(C)** After contrast administration the mass shows inhomogeneous enhancement.

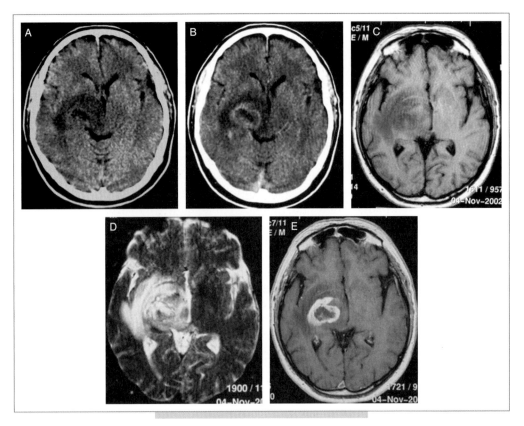

FIG. 42.5. Glioblastoma multiforme. **(A)** Non-contrast CT scan shows a hypodense left thalamic mass, which shows ring-like enhancement after the contrast medium administration **(B)**. On non-contrast T1-weighted image **(C)** the mass appears inhomogeneously hypointense and on T2-weighted image **(D)** inhomogeneously hyperintense surrounded by high signal intensity edema. **(E)** Axial post-contrast T1-weighted image shows a rim-enhancing mass.

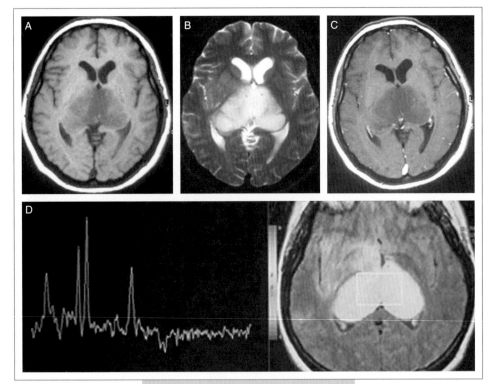

FIG. 42.6. Bilateral thalamic glioma. Axial T1- **(A)** and T2-weighted images **(B)** show enlargement of both thalami with low signal intensity on T1, high signal intensity on T2 and infiltration of the right basal ganglia and adjacent parietal cortex. **(C)** On post-contrast MR image the tumor remains unenhanced. **(D)** MR spectroscopy demonstrates higher peak of creatine than that of choline and reduced NAA. Biopsy proved a fibrillary astrocytoma.

favors infarction or inflammatory process over thalamic neoplasm [40].

BTG may originate in the subependymal glia, and hence have primary relation to the medial areas of the thalamus [41]. The tumor can spread across the midline via the interthalamic adhesion and the roof or floor of the third ventricle [6,41].

Metabolic analysis of BTGs by MR spectroscopy shows a higher peak of creatine than choline, in contrast to the increased peak of choline usually observed in low-grade gliomas. The MR NAA signal is decreased in the BTGs, as is usually observed in low-grade gliomas (see Figure 42.6D) [35].

Oligodendrogliomas and Mixed Gliomas

The pathological features of oligodendrogliomas (ODG) and mixed gliomas (MG) provide the basis for imaging findings: they are circumscribed, heterogeneous lesions that frequently contain calcification, cysts and hemorrhage or blood degradation products; calcification is the most typical finding of ODG. Thus, a thalamic ODG or MG resembles a malignant astrocytoma, but appears more circumscribed and more frequently calcified and without necrosis. The signal intensity is heterogeneous on both T1- and T2-weighted images, because calcification and proteinaceous cystic content can also induce hyperintensity counterparts in T1-weighted images (Figure 42.7). The extent of edema and rapid increase in size support the hypothesis of the tumor being anaplastic [5].

Ganglion Cell Tumors

Gangliogliomas and gangliocytomas (GG-GC) are more common in children than in adults [42]. The thalamic location is exceedingly rare and a definite preoperative neuroradiological diagnosis is virtually impossible because their imaging features are similar to their astrocytic counterparts (pilocytic, fibrillary, anaplastic) and because in the deep location they lack the calvarial scalloping that supports correct recognition in the typically superficial temporal location [5]. They are usually well-circumscribed, benign and non-progressive tumors that are frequently cystic and calcify [5,19]. In the rarely encountered thalamic GG-GC, the cystic appearance predominates, with sharp margins and focal areas of contrast enhancement (Figure 42.8B) [5,43]. Calcification results in areas of T1-hyperintensity, and signal intensity is highly variable on T2-weighted images (Figure 42.8A) [5].

MRS analysis indicates that GG-GC has higher NAA/Cho and Cr/Cho ratios relative to other tumors [44]. These ratios can be related to both higher levels of NAA and a discrete drop in Cho reflecting the low aggressiveness of these tumors. As ganglion cell tumors are frequently cystic, they may show a peak of lactic acid, like many other cystic processes [44,45]. Because of their deep location and benign clinical behavior, long-term imaging follow-up is often carried out and no progression confirmed [5].

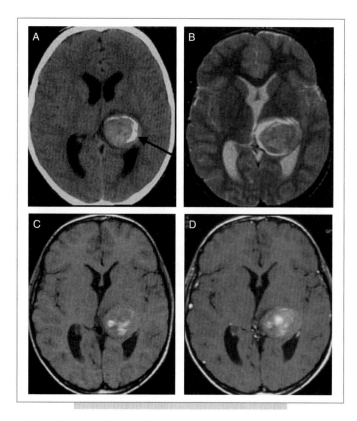

FIG. 42.7. Oligodendroglioma of the left thalamus in a 7-year-old female. **(A)** Axial CT scan demonstrates a rounded, well-circumscribed and slightly isodense mass with peripheral calcification (arrow) located in the left thalamus. The lesion appears inhomogeneously hypointense on T2-weighted image **(B)**, with little central hyperintense hemorrhagic foci on T1-weighted image **(C)**, and mild enhancement after contrast injection **(D)**. (Reprinted from the article Neuroimaging of thalamic tumors in children, Colosimo C et al Childs Nerv Syst 2002;18:426–439, with kind permission of Springer Science and Business Media.)

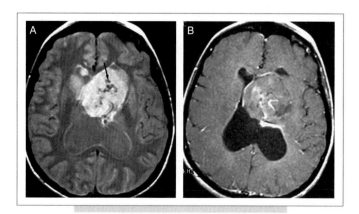

FIG. 42.8. Gangliocytoma in the left basal ganglia. **(A)** Axial proton density image shows a mass with high signal intensity and central calcifications (arrow). **(B)** Post-gadolinium T1-weighted image shows internal focal and linear enhancement.

Germ Cell Tumors

Germ cell tumors (GCT) usually grow up from the pineal gland or in the suprasellar area, but up to 15% occur in the thalamus and the basal ganglia, and are usually unilateral [42,46]. Most thalamic GCTs actually originate outside the thalami and infiltrate the thalami from the posterior or anterior walls of the third ventricle [5]. The most common presenting symptom is hemiparesis, caused by tumor invasion of the internal capsule, followed by mental retardation [47].

Germinoma is the most common germ cell tumor, usually appears in the second or early third decade of life and shows a striking male predominance [21,42,46,48,49]. A high prevalence of intracranial germinoma in the Far East is well documented, but the reason is unknown [49,50]. Histologically, germinomas range from benign processes to highly malignant neoplasms with a tendency to metastasize throughout the CSF spaces [51].

The prognosis of germinoma is significantly better than with the other GCTs, with the 5-year survival rate approaching 80% [51–53]. Because of the good response to chemotherapy and radiotherapy, early detection and treatment is very important [54]. However, diagnosis of germinoma in the basal ganglia and thalamus is usually delayed because of minimal or non-specific clinical findings in the relatively early stage [55].

In the early stage of basal ganglionic germinomas, CT may not show any abnormality, despite neurological symptoms [55,56]. An irregularly defined, slightly hyperdense area without mass effect is the early CT sign of germinoma in the basal ganglia (Figure 42.9) [56–59]. After the administration of the contrast medium, they show intense enhancement. On the other hand, demonstration of an ill-defined hyperintense area on both T2- and T1-weighted images without contrast enhancement and, especially, atrophy of the ipsilateral basal ganglia, are early diagnostic MR imaging features [60–62].

In advanced stages, germinomas show a typical infiltrative pattern and CT density/MR signal intensity that make them difficult to differentiate from neuroepithelial tumors [5]. In fact, the rich cellularity of germinomas is represented by moderate hyperintensity on unenhanced CT and quite intense contrast enhancement (Figure 42.10A) [5,63]. The infiltrating tumor appears heterogeneous on T1- and T2-weighted images, with cystic areas and hemorrhage, especially in large lesions, and shows heterogeneous intense enhancement and moderate perilesional edema (Figure 42.10B–D) [5,55]. Intratumoral calcification is frequent and presents as nodular or spotty hyperdensity on CT and on T1-weighted images. Unlike typical germinomas in pineal or suprasellar regions, thalamic and basal ganglionic germinomas show a higher tendency to cystic formation, hemorrhage, calcification and progressive infiltration into the internal capsule, which may in turn cause cerebral hemiatrophy [57]. Cystic changes and hemorrhage seem to

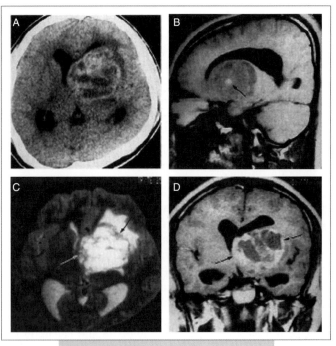

FIG. 42.10. Basal ganglia germinoma in a 17-year-old male with two months history of vomiting and seizures. **(A)** Non-contrast CT scan shows a multicystic mass with hyperdense solid areas located in the left basal ganglia. **(B)** Left parasagittal T1-weighted image shows a slightly hypointense mass with central focal hemorrhage (arrow). **(C)** Axial T2-weighted image shows a hyperintense multiseptated mass with moderate peritumoral edema (arrows). **(D)** Post-contrast T1-weighted image shows intense enhancement of the solid portion of the mass (arrows). (Reprinted from the article Germinomas of the basal ganglia and thalamus: MR findings and a comparison between MR and CT, Moon WK et al Am J Roentgenol 1994;162:1413-1417, with permission from the American Journal of Roentgenology.)

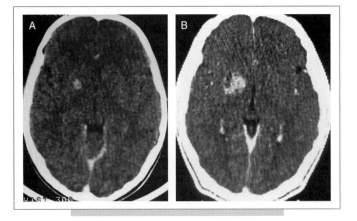

FIG. 42.9. Basal ganglia germinoma. Post-contrast CT scan **(A)** shows a small strongly enhanced lesion. On post-contrast CT scan **(B)** one year later, the lesion shows increase in size and cystic degeneration. Histologic examination after stereotactic biopsy revealed a germinoma.

be attributed to the more rapid enlargement of the germinomas in the basal ganglia and thalamus compared to those in the pineal body and suprasellar region [54,55].

Choriocarcinoma, yolk-sac tumor, embryonic cell carcinoma and endodermal sinus tumors are highly malignant subtypes of GCT [42]. These high-grade tumors are heterogeneous on CT and MRI and, except for the occurrence of hemorrhage in choriocarcinoma, there is no imaging peculiarity that could help in distinguishing them from other tumors [21,42,46]. For this purpose, laboratory examinations of tumor markers are useful: choriocarcinoma releases human chorionic gonadotropin, yolk sac tumor α-fetoprotein and embryonic cell carcinoma releases both human chorionic gonadotropin and α-fetoprotein [57].

When GCT is suspected, contrast material is very useful in detecting CSF spread in the brain and spine, a relatively common and early finding [20–42]. The differential diagnosis should include glioma, primary CNS lymphoma and primitive neuroectodermal tumors [5].

Primitive Neuroectodermal Tumors

Primitive neuroectodermal tumors (PNET) account for 15–25% of pediatric brain tumors, typically appear during the first 10 years of life, but only rarely arise in the thalamus [20,26,42,64]. The vast majority are medulloblastomas, developing characteristically in the posterior fossa. Those PNET arising inside the supratentorial compartment are rare neoplasms, comprising 15% of all central nervous system PNET [42,65]. Pineoblastoma, which is the most frequent neoplasm of this group of supratentorial PNET, grows up from the pineal gland, and we will not consider it as a primary basal ganglia–thalamic tumor [2,65]. Generally, PNET growing up from the supratentorial brain usually arise from the deep cerebral white matter [19,42].

While PNET are dense cellular well-circumscribed neoplasms, they are microscopically invasive, with extension of tumor cells beyond a reactive desmoplastic glial pseudocapsule [2,51]. Supratentorial PNET are remarkable for their inhomogeneity [2]. CT and MRI show a well-defined heterogeneous expansive lesion, with cystic and necrotic areas, and calcifications in most of the cases [20–22,24,42]. The solid portions of the tumor are often hyperdense on CT and relatively hypointense on T2-weighted images, due to their high cellularity and nuclear-to-cytoplasmic ratio [21]. Despite their large size, PNET typically induce only minimal surrounding edema, which may be the best clue to correct preoperative diagnosis [19,20,66]. Enhancement is usually diffuse and heterogeneous (Figure 42.11) [28]. Contrast material administration is also useful in demonstrating the presence of leptomeningeal tumor implants, which are common in PNET [20,21,23,42].

MRS may be used to define the malignant behavior of PNET. As in astrocytomas, the rise of the choline peak is an indicator of their aggressiveness [45]. The NAA/Cho and Cr/Cho ratios are lower in PNET than in astrocytomas and ependymomas, which is related to the more significant drop of choline levels in PNET [2,31].

Atypical Rhabdoid-teratoid Tumors

Atypical rhabdoid-teratoid tumors (ARTT) are very rare neoplasms, which are included within the group of embryonal tumors, together with the PNET. The microscopic distinctive feature of ARTT is the presence of rhabdoid cells [2,5]. These highly malignant tumors affect almost exclusively infants and children younger than 3 years and, in this age group, they develop characteristically in the posterior fossa [2,5]. In older patients, the incidence of supratentorial ARTT is strikingly higher [67]. They can rarely arise from the basal ganglia and thalamus, but those that appear in the cerebral hemispheres or the pineal gland can also invade through to the deep brain [68]. Imaging cannot distinguish ARTT from classic PNET [2]. Consequently, these malignant tumors appear at diagnosis as a large heterogeneous mass that contains necrosis, areas of high cellularity (hypointense on T2-weighted images) and hemorrhage. They also demonstrate extensive surrounding edema and mass effect and intense heterogeneous enhancement (Figure 42.12). The prognosis is poor and CSF seeding is common [5].

Lymphomas

Primary cerebral lymphomas (PCL) are infrequent non-glial intracranial neoplasms that constitute 1–6% of malignant CNS tumors and are predominantly of B-cell origin [2,69,70]. During the last two decades, their incidence has increased, since acquired immunodeficiency syndrome (AIDS) was first recognized [71,72]. PCL can occur anywhere in the brain, but the most common locations are the periventricular white matter and the corpus callosum

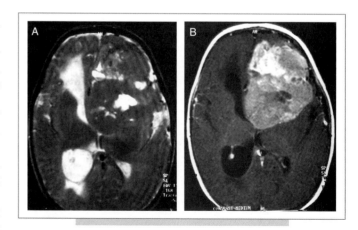

FIG. 42.11. PNET in a one-year-old male. **(A)** Axial T2-weighted image demonstrates a slightly hypointense lesion with focal hyperintensities located in the left basal ganglia. **(B)** Post-gadolinium T1-weighted image shows diffuse intense inhomogeneous enhancement of the mass extending to the adjacent frontal lobe.

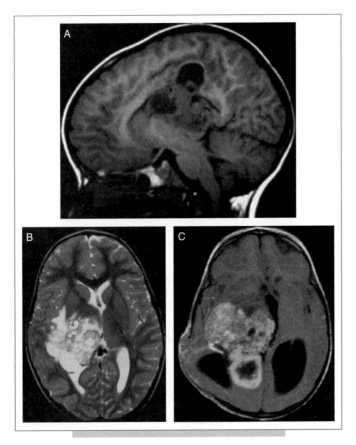

FIG. 42.12. Atypical teratoid-rhabdoid tumor in a 2-year-old female. **(A)** Sagittal T1-weighted image, **(B)** axial T2-weighted image and **(C)** axial T1-weighted image after contrast injection demonstrate a huge inhomogeneous deep located mass which has its epicenter in the right thalamus. The lesion shows multiple macro- and micro-cystic components, as well as necrosis. The signal of the solid component appears mildly hypointense on T1-weighted, and mildly hyperintense on T2-weighted images (suggesting high cellularity), with intense contrast enhancement. (Reprinted from the article Neuroimaging of thalamic tumors in children, Colosimo C et al Childs Nerv Syst 2002;18:426–439, with kind permission of Springer Science and Business Media.)

[73,74]. Up to 17% of PCL arise in the basal ganglia, thalamus and hypothalamus [75,76]. It has also been reported that up to 75% of lymphomatous masses are seen to be in contact with ependyma, meninges, or both [21]. PCL has a distinct tendency for perivascular extension, whereas leptomeningeal or ependymal involvement occurs in about 12% of the cases [77].

A 5-year survival rate of more than 30% can be achieved when combining chemotherapy and radiotherapy treatment [78]. Nevertheless, because of the rapid course of PCL, a delay in whole brain irradiation and chemotherapy markedly decreases the effectiveness of the treatment and survival. Therefore, early diagnosis is critical [79].

In most cases, PCL in the basal ganglia and thalamus show distinctive neuroimaging features that allow differentiation from the more common glial tumors [5]. PCL classically appears as a predominantly hyperdense mass on CT and relatively hypointense on T2-weighted images, features that are attributed to dense cellularity and high nuclear-to-cytoplasmic ratio of neoplastic cells [24,76,80]. Surrounding edema is mild and much less extensive than is seen with primary glial tumors or metastases [81]. Calcifications and hemorrhage are uncommon before treatment [24,82]. Both in post-contrast CT and MRI, PCL usually shows strong homogeneous enhancement in immunocompetent patients and ring enhancement in AIDS-patients, due to central necrosis (Figures 42.13 and 42.14) [21,25,72,76]. The detection of enhancement along perivascular spaces on MRI is also characteristic of PCL, allowing differential diagnosis from other malignant processes [21].

MR spectroscopy of PCL shows almost complete loss of NAA, decrease of creatine, massive increase of choline, lactate and lipids, whereas the myo-inositol peak appears normal or slightly decreased [31,75].

The differential diagnosis of PCL in immunocompetent patients includes glioma, metastasis, PNET, multiple sclerosis and meningioma [83,84]. A hyperdense lesion on CT and a periventricular T2 hypointense mass with mild surrounding edema and ependymal seeding on MRI,

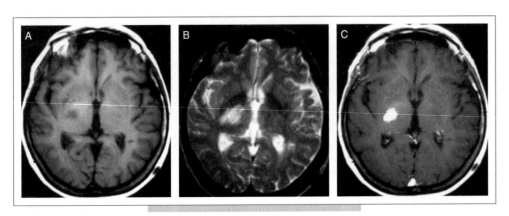

FIG. 42.13. Lymphoma of the right thalamus. **(A)** Axial T1-weighted image shows a low signal lesion, which shows high signal intensity on T2-weighted image **(B)**. Axial post-contrast T1-weighted image **(C)** demonstrates intense and homogeneous enhancement.

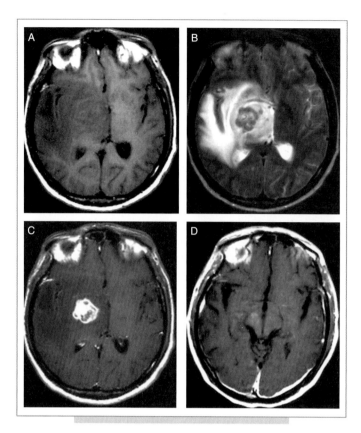

FIG. 42.14. Thalamic lymphoma in a patient with AIDS. **(A)** Axial T1-weighted image shows an extensive low signal area located in the right parietal and thalamic region. **(B)** On axial T2-weighted image a low signal right thalamic lesion with high signal central foci is seen surrounded by high signal intensity peritumoral edema. **(C)** Post-gadolinium axial T1-weighted image shows intense and inhomogeneous enhancement with central necrosis. **(D)** A few months after therapy the tumor is almost completely resolved, but simultaneous meningeal enhancement is seen.

favors the diagnosis of PCL [72]. In immunocompromised patients, PCL should be differentiated from toxoplasmosis abscesses [73,85].

Metastases

Metastatic CNS tumors now account for up to 50% of all brain tumors and affect primarily the brain parenchyma [86]. It has been estimated that 80–85% of brain metastases are located supratentorialy. The frontal and parietal lobes of the cerebral hemispheres are most commonly affected, with the corticomedullary junction being the earliest site of involvement [87]. Basal ganglia and thalamus are rare locations and are involved only in 3% of the cases [86].

Metastases may result from hematogenous spread of distant tumors or from CSF seeding of CNS tumors [2]. In adults, hematogenous metastases mostly come from tumors of lung, breast, gastrointestinal tract, genitourinary tract and skin (melanoma), whereas in children, they mostly come

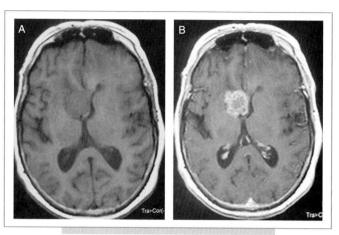

FIG. 42.15. Basal ganglia metastasis. Pre- **(A)** and post-contrast **(B)** axial T1-weighted images show an isointense lesion involving the right caudate nucleus, which enhances markedly and heterogeneously.

from sarcomas, Wilms tumor and hepatoblastoma [2,21]. PNET, astrocytomas, ependymomas, germinomas, pineoblastomas and choroids plexus carcinomas in children and glioblastomas, lymphomas in adults often show CSF spread when the diagnosis is made.

Most intracerebral metastases are multiple, regardless of the site of origin. However, there is a high incidence of solitary metastasis, estimated to range from 30% to 50%, which is especially common in melanoma, lung and breast carcinoma [21]. The imaging findings of hematogenous metastases are non-specific. They are usually surrounded by prominent edema which follows white matter boundaries and does not usually cross the corpus callosum. Contrast enhancement can be solid and nodular, as well as ring-like (Figure 42.15).

The differential diagnosis includes abscess, demyelinating disease, thromboembolic stroke and multifocal malignant glioma [86].

Hypothalamus–Optic Chiasm

In midsagittal sections, the human hypothalamus is bound anteriorly by the lamina terminalis, posteriorly by a plane drawn between the posterior commissure and the caudal limit of the mammillary body and superiorly by the hypothalamic sulcus. Ventrally, the hypothalamus encompasses the floor of the third ventricle, the inferior surface of which presents to the subarachnoid space below and is termed the tuber cinereum (grey swelling). The lateral boundaries include the internal capsule, cerebral peduncle and subthalamus on each side [88].

Hypothalamic and Chiasmatic Gliomas

Gliomas of the hypothalamus and optic pathways represent 5% of pediatric intracranial tumors and 25–30% of

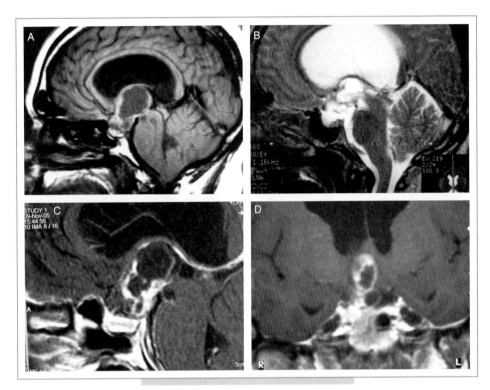

FIG. 42.16. Chiasmatic/hypothalamic pilocytic astrocytoma. **(A)** Sagittal T1-weighted image shows a cystic tumor involving the hypothalamus and optic chiasm. **(B)** On T2-weighted image the tumor appears heterogeneously hyperintense. Post-contrast sagittal **(C)** and coronal **(D)** images show intense heterogeneous enhancement. Note the enlargement of the lateral ventricles due to the invagination of the tumor within the third ventricle.

pediatric suprasellar tumors with 60% involving the optic chiasm and hypothalamus [89–91]. These tumors occur mostly in children, especially in those under 10 years of age. There is an equal male to female distribution [92]. A definite association of hypothalamic/chiasmatic gliomas with neurofibromatosis type I (NF1) is noted in 33% of patients [89,90]. The vast majority are slow-growing pilocytic astrocytomas, although malignant histology and, in particular, glioblastoma multiforme may occur, especially in adults over 50 years of age [93,94].

The distinction between hypothalamic and chiasmatic gliomas often depends on the predominant position of the tumor. Chiasmatic gliomas may occasionally demonstrate extension along the optic tracts or optic nerves, whereas hypothalamic tumors grow into the third ventricle, often leading to hydrocephalus [51,95]. In many cases, especially in larger gliomas, it is difficult to determine the exact site of origin, as the hypothalamus and chiasm are inseparable. For this reason, they are discussed as a single entity based on their similar histology, clinical presentation and management [92].

Patients with hypothalamic/chiasmatic gliomas usually present with visual symptoms, headaches and endocrine abnormalities, due to hypothalamic dysfunction [90,95]. A rare but characteristic presentation of these tumors is the diencephalic syndrome, which is characterized by profound emaciation with normal caloric intake, absence of cutaneous adipose tissue, locomotor hyperactivity, euphoria and alertness [96]. Typically, gliomas associated with this syndrome are larger, occur at a younger age and are often more aggressive than other gliomas arising in this region [97].

Hypothalamic and chiasmatic gliomas are predominantly solid tumors, with cystic changes occurring in larger hypothalamic tumors [90]. They are usually isointense or hypointense relative to gray matter on T1-weighted images and moderately to markedly hyperintense on T2-weighted images. After contrast administration, chiasmatic tumors usually show intense, heterogeneous enhancement, whereas pre- and post-chiasmatic tumors show lesser degrees of enhancement (Figure 42.16) [51,90,97,98]. Hyperintense signal may be seen extending into the optic radiations and lateral geniculate bodies on T2-weighted images, due either to tumor infiltration or to tumor-related edema (Figure 42.17) [51].

Only subtotal tumor resection is possible because of the involvement of visual pathways and the pituitary-hypothalamic axis [99]. While the rate of recurrence or progression of untreated lesions is similar regardless of location, mortality from those involving both the chiasm and hypothalamus is much higher. However, 20–50% of untreated lesions will not progress [94].

Hamartoma

Hypothalamic hamartomas are rare congenital non-neoplastic heterotopias with prevalence estimated as high as 1 in 50000 to 100000 [100]. They typically present in young males but are not uncommon in females [51]. Although they do not invade surrounding tissues, they do have the potential for slow growth and may be quite large [92]. The hypothalamic hamartoma syndrome traditionally comprises the clinical triad of central precocious puberty, epilepsy (gelastic seizures) and developmental retardation [101].

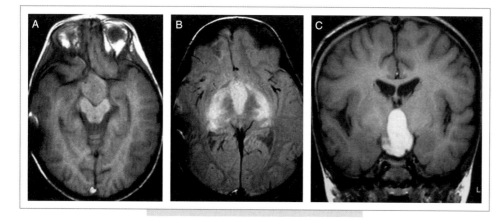

FIG. 42.17. Chiasmatic/hypothalamic pilocytic astrocytoma. **(A)** T1-weighted image shows an isointense suprasellar mass. **(B)** On proton-density image the mass is hyperintense and the optic radiations and the lateral geniculate bodies also show high signal intensity. **(C)** Coronal post-contrast T1-weighted image shows intense and homogeneous enhancement of the mass.

Hypothalamic hamartomas have been classified radiologically on the basis of the breadth of attachment to the tuber cinereum (sessile versus pedunculated), the presence of more than minimal distortion of the outline of the third ventricle (intrahypothalamic versus parahypothalamic), or using a combination of size, breadth of attachment, distortion of the hypothalamus and location of attachment (types Ia, Ib, IIa and IIb) [102–104]. These classifications have attempted to correlate the clinical features observed with the physical properties of the hamartomas [103,105]. Hypothalamic hamartomas associated with epilepsy usually have sessile attachment to the hypothalamus and displace normal hypothalamic structures, whereas those associated with precocious puberty are usually pedunculated [101–104,106–108].

Most patients with hypothalamic hamartoma have a sporadic form of the disease, without family history or risk of recurrence and without associated congenital anomalies. However, roughly 5% of the population with the intrahypothalamic subtype of hypothalamic hamartoma have Pallister-Hall syndrome, which can include anomalies such as postradial polydactyly, bifid epiglottis and imperforate anus [101,109].

On MRI, hypothalamic hamartomas are uniformly isointense to gray matter on T1-weighted images and slightly hyperintense or isointense on T2-weighted images with no enhancement after gadolinium administration, thereby demonstrating the integrity of the blood–brain barrier (Figure 42.18) [102,108,110–117]. Cyst formation and calcifications have been rarely reported [93,108,118]. Arachnoid cysts occurring in association with hypothalamic hamartomas have been also rarely reported, suggesting either a common antecedent or causal relationship [108,119–121].

MR spectroscopy has shown lower NAA and higher myo-inositol concentration than normal thalamic gray matter or frontal lobe white and gray matter [108,122]. These findings suggest reduced neuronal density and relative gliosis compared with normal gray matter [108].

Differential diagnosis includes craniopharyngioma, hypothalamic and chiasmatic glioma, germ cell tumor and metastasis.

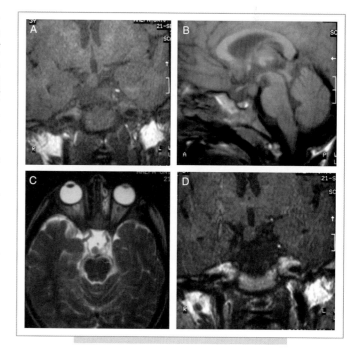

FIG. 42.18. Tuber cinereum hamartoma. Coronal **(A)** and sagittal **(B)** T1-weighted images show an isointense mass located between the infundibulum and the mammillary bodies. On axial T2-weighted image **(C)** the mass remains isointense and does not enhance after contrast injection on T1-weighted image **(D)**.

Chordoid Glioma

Chordoid glioma is a new clinicopathological entity that occurs in the region of the hypothalamus–anterior third ventricle [123]. This tumor was named chordoid glioma because of its distinctive histologic appearance, reminiscent of chordoma, and its avid staining with glial fibrillary acidic protein [124]. Although the incidence of chordoid glioma cannot yet be evaluated because of its recent description, it seems to be an uncommon tumor. The clinical presentation is related to the local mass effect of the tumor, and includes headaches, obstructive hydrocephalus, homonymous hemianopsia and hypothalamic dysfunction [123]. Distinctive imaging characteristics are its consistent location, ovoid shape with the greater dimension in the superoinferior orientation, hyperdensity on CT scans, relative isointensity on T2-weighted images and uniform intense contrast enhancement. Sagittal MR images clearly depict the infundibulum to be displaced posteriorly, whereas Rathke cleft cysts and tuber cinereum hamartomas generally displace the infundibulum anteriorly (Figure 42.19) [123,125]. The differential diagnosis also includes lymphomas, meningiomas and hypothalamic/chiasmatic gliomas. The current treatment of choice for chordoid glioma is surgical resection [123].

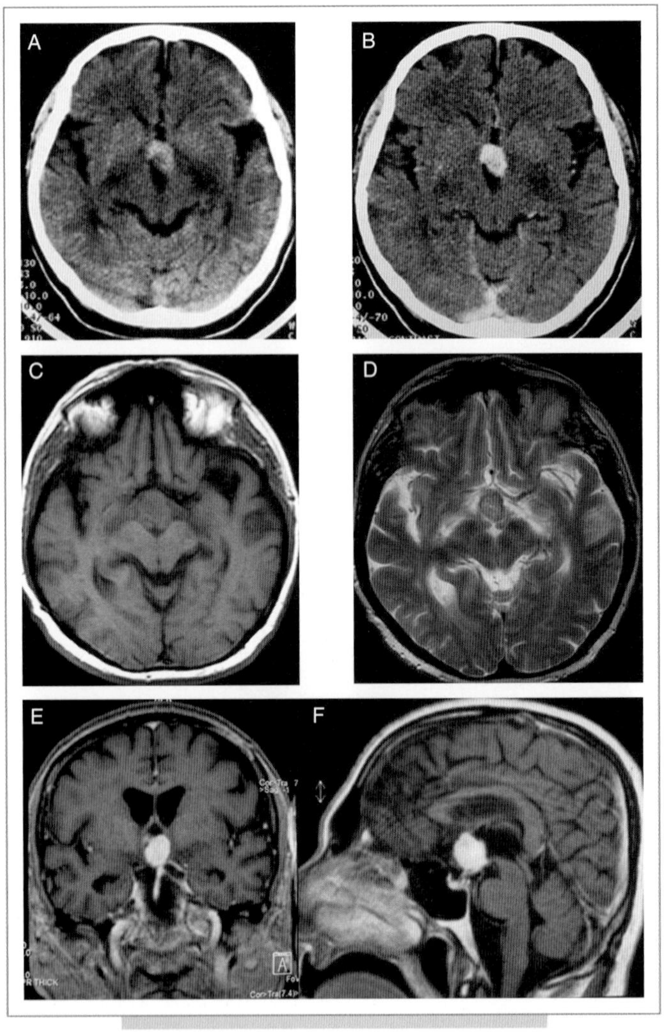

Lymphoma

Primary CNS lymphomas are infrequent tumors that may arise from different parts of the brain, with deep hemispheric periventricular white matter being the most common. Lymphomatous involvement of the hypothalamic–third ventricular area is extremely rare and appears usually in patients with AIDS [79,126,127]. The usual clinical manifestations are hypopituitarism, psychiatric and memory disturbances [126,127]. As a diffuse, intrinsic tumor that tends to infiltrate or replace the brain tissue rather than displace or compress it, a hypothalamic–third ventricular lymphoma should more accurately be considered a hypothalamic lymphoma that has secondarily invaded the third ventricle cavity [128,129]. Unfortunately, in many cases, neither CT nor MRI images can provide an exact delimitation between the tumor and the third ventricle margins and this may lead to topographical misdiagnosis [130]. On CT and MRI, hypothalamic lymphomas appear as homogeneous isointense masses, which show marked homogeneous enhancement after contrast administration (Figure 42.20) [127].

FIG. 42.19. Chordoid glioma in the hypothalamic-third ventricle region in a 59-year-old male. Non-contrast CT scan **(A)** demonstrates a hyperdense suprasellar lesion, which on post-contrast CT scan **(B)** shows intense and homogeneous enhancement. On axial T1- **(C)** and T2- **(D)** weighted images the lesion in the hypothalamic area is relatively isointense. Contrast-enhanced coronal **(E)** and sagittal **(F)** T1-weighted images show dense and homogeneous enhancement. Note the ovoid shape of the lesion with the greater dimension in the superoinferior orientation and the posterior displacement of the infundibulum.

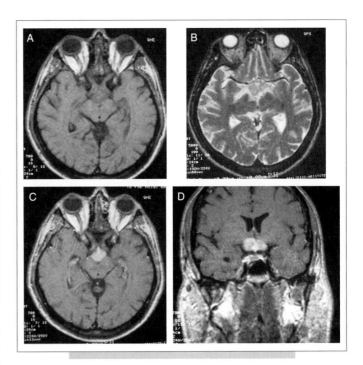

FIG. 42.20. Hypothalamic non-Hodgkin's lymphoma. **(A)** Axial T1-weighted image shows a hypointense lesion in the suprasellar region. **(B)** On axial T2-weighted image the lesion appears slightly hyperintense. Axial **(C)** and coronal **(D)** T1-weighted images after contrast injection show intense and homogenous enhancement of the lesion. Biopsy proved a hypothalamic non-Hodgkin's B-cell lymphoma.

Miscellaneous Masses

Other entities that should be considered in the differential diagnosis of a hypothalamic lesion are neurosarcoidosis, lipoma [131] and osteolipoma [132–134].

TUMORS THAT SECONDARILY INVADE THE DEEP BRAIN

Tumors that arise from the cerebral hemispheres, the cerebral ventricles and the pineal gland can invade the diencephalon, but they are not actually deep brain tumors.

Peripheral Brain Tumors

The same tumors that arise from the deep brain can be diagnosed peripherally in the cerebral hemispheres.

Low- and high-grade astrocytomas, PNET and supratentorial ependymomas can invade the deep brain after developing from the white matter of the cerebral hemispheres.

Ventricular Tumors

Choroid plexus tumors, subependymal giant cell tumors, ependymomas, subependymomas and central neurocytomas are ventricular tumours that can occasionally spread to the deep brain.

Pineal Gland Tumors

Pineal gland tumors include germ cell tumors, pineocytomas and pineoblastomas, which are discussed in detail in Chapter 44.

REFERENCES

1. Martin JH (ed.) (1989). Development of CNS. In *Neuroanatomy: Text and Atlas*. Elsevier, Amsterdam. 32–53.
2. Garcia-Santos JM, Torres del Rio S, Sanchez A, Martinez-Lage JF (2002). Basal ganglia and thalamic tumours: an imaging approximation. Childs Nerv Syst 18:412–425.
3. Wald SL, Fogelson H, McLaurin RL (1982). Cystic thalamic gliomas. Childs Brain 9:381–393.
4. Cheek WR, Taveras JM (1966). Thalamic tumors. J Neurosurg 24:505–513.
5. Colosimo C, di Lella GM, Tartaglione T, Riccardi R (2002). Neuroimaging of thalamic tumors in children. Childs Nerv Syst 18:426–439.
6. McKissock W, Paine KW (1958). Primary tumours of the thalamus. Brain 81:41–63.
7. Tovi D, Schisano G, Liljeqvist B (1961). Primary tumors of the region of the thalamus. J Neurosurg 18:730–740.
8. Allen JC (2000). Initial management of children with hypothalamic and thalamic tumors and the modifying role of neurofibromatosis-1. Pediatr Neurosurg 32:154–162.
9. Yoshida M, Fushiki S, Takeuchi Y et al (1998). Diffuse bilateral thalamic astrocytomas as examined serially by MRI. Childs Nerv Syst 14:384–388.
10. Arseni C (1958). Tumors of the basal ganglia; their surgical treatment. Arch Neurol Psychiatr 80:18–24.
11. Bernstein M, Hoffman HJ, Halliday WC, Hendrick EB, Humphreys RP (1984). Thalamic tumors in children. Long-term follow-up and treatment guidelines. J Neurosurg 61:649–656.
12. Drake JM, Joy M, Goldenberg A, Kreindler D (1991). Computer- and robot-assisted resection of thalamic astrocytomas in children. Neurosurgery 29:27–33.
13. Lyons MK, Kelly PJ (1992). Computer-assisted stereotactic biopsy and volumetric resection of thalamic pilocytic astrocytomas. Report of 23 cases. Stereotact Funct Neurosurg 59:100–104.
14. Matsumoto K, Higashi H, Tomita S, Furuta T, Ohmoto T (1995). Resection of deep-seated gliomas using neuroimaging for stereotactic placement of guidance catheters. Neurol Med Chir (Tokyo) 35:148–155.
15. Steiger HJ, Gotz C, Schmid-Elsaesser R, Stummer W (2000). Thalamic astrocytomas: surgical anatomy and results of a pilot series using maximum microsurgical removal. Acta Neurochir (Wien) 142:1327–1336; discussion 1336–1337.
16. Packer RJ, Ater J, Allen J et al (1997). Carboplatin and vincristine chemotherapy for children with newly diagnosed progressive low-grade gliomas. J Neurosurg 86:747–754.
17. Burger PC, Cohen KJ, Rosenblum MK, Tihan T (2000). Pathology of diencephalic astrocytomas. Pediatr Neurosurg 32:214–219.
18. Becker LE (1999). Pathology of pediatric brain tumors. Neuroimaging Clin N Am 671–690.
19. Naidich TP, Zimmerman RA (1984). Primary brain tumors in children. Semin Roentgenol 19:100–114.
20. Luh GY, Bird CR (1999). Imaging of brain tumors in the pediatric population. Neuroimaging Clin N Am 9:691–716.
21. Atlas SW, Lavi E (1996). Intra-axial brain tumors. In *Magnetic Resonance Imaging of the Brain and Spine*, 2nd edn. Atlas SW (ed.). Lippincott Williams & Wilkins, Philadelphia. 315–422.
22. Edwards-Brown MK (1994). Supratentorial brain tumors. Neuroimaging Clin N Am 4:437–455.
23. Castillo M (1994). Contrast enhancement in primary tumors of the brain and spinal cord. Neuroimaging Clin N Am 4:63–80.
24. Koeller KK (2000). Central nervous system neoplasms: intraaxial. RSNA Categorical Course in Diagnostic Radiology. Neuroradiology 105–121.
25. Gore JC, Kennan RP (1996). Contrast agents and relaxation effects. In *Magnetic Resonance Imaging of the Brain and Spine*, 2nd edn. Atlas SW (ed.). Lippincott Williams & Wilkins, Philadelphia. 89–107.
26. Bleyer WA (1999). Epidemiologic impact of children with brain tumors. Childs Nerv Syst 15:758–763.
27. Kumar R, Jain R, Tandon V (1999). Thalamic glioblastoma with cerebrospinal fluid dissemination in the peritoneal cavity. Pediatr Neurosurg 31:242–245.

28. Lefton DR, Pinto RS, Silvera VM, DeLara FA, Schwartz JB, Haller JO (2000). Radiologic features of pediatric thalamic and hypothalamic tumors. Crit Rev Diagn Imaging 41:237–278.

29. Fujimoto T, Nakano T (2000). Understanding spectra of 31P and 1H MRS. In *MRS of the Brain and Neurological Disorders*. Igata A, Asakura T, Fujimoto T (eds). CRC, New York. 40–58.

30. Sasahira M (2000). Brain tumors. In *MRS of the Brain and Neurological Disorders*. Igata A, Asakura T, Fujimoto T (eds). CRC, New York. 89–158.

31. Castillo M, Kwock L (1998). Proton MR spectroscopy of common brain tumors. Neuroimaging Clin N Am 8:733–752.

32. Hirano H, Yokoyama S, Nakayama M, Nagata S, Kuratsu J (2000). Bilateral thalamic glioma: case report. Neuroradiology 42:732–734.

33. Partlow GD, del Carpio-O'Donovan R, Melanson D, Peters TM (1992). Bilateral thalamic glioma: review of eight cases with personality change and mental deterioration. Am J Neuroradiol 13:1225–1230.

34. Uchino M, Kitajima S, Miyazaki C, Shibata I, Miura M (2002). Bilateral thalamic glioma – case report. Neurol Med Chir (Tokyo) 42:443–446.

35. Esteve F, Grand S, Rubin C et al (1999). MR spectroscopy of bilateral thalamic gliomas. Am J Neuroradiol 20:876–881.

36. Reardon DA, Gajjar A, Sanford RA et al (1998). Bithalamic involvement predicts poor outcome among children with thalamic glial tumors. Pediatr Neurosurg 29:29–35.

37. Carter DJ, Wiedmeyer DA, Antuono PG, Ho KC (1989). Correlation of computed tomography and postmortem findings of a diffuse astrocytoma: a case report. Comput Med Imaging Graph 13:491–494.

38. Rogers LR, Weinstein MA, Estes ML, Cairncross JG, Strachan T (1994). Diffuse bilateral cerebral astrocytomas with atypical neuroimaging studies. J Neurosurg 81:817–821.

39. Yanaka K, Kamezaki T, Kobayashi E et al (1992). MR imaging of diffuse glioma. Am J Neuroradiol 13:349–351.

40. Levy RA, Quint DJ (1994). Bilateral thalamic venous infarctions mimicking tumor. Am J Roentgenol 163:226–227.

41. Smyth GE, Stern K (1938). Tumors of the thalamus, a clinicopathological study. Brain 61:339–374.

42. Barkovich AJ (ed.) (1990). Brain tumors in childhood. In *Paediatric Neuroimaging*. Raven, New York. 149–204.

43. Fujimura M, Kayama T, Kumabe T, Yoshimoto T (1995). Ganglioglioma in the basal ganglia totally resected by a transdistal Sylvian approach. Tohoku J Exp Med 175:211–218.

44. Wang Z, Sutton LN, Cnaan A et al (19950. Proton MR spectroscopy of pediatric cerebellar tumors. Am J Neuroradiol 16:1821–1833.

45. Taylor JS, Ogg RJ, Langston JW (1998). Proton MR spectroscopy of pediatric brain tumors. Neuroimaging Clin N Am 8:753–779.

46. Sumida M, Uozumi T, Kiya K et al (1995). MRI of intracranial germ cell tumours. Neuroradiology 37:32–37.

47. Tamaki N, Lin T, Shirataki K et al (1990). Germ cell tumors of the thalamus and the basal ganglia. Childs Nerv Syst 6:3–7.

48. Huh SJ, Kim IH, Ha SW, Park CI (1992). Radiotherapy of germinomas involving the basal ganglia and thalamus. Radiother Oncol 25:213–215.

49. Kobayashi T, Kageyama N, Kida Y, Yoshida J, Shibuya N, Okamura K (1981). Unilateral germinomas involving the basal ganglia and thalamus. J Neurosurg 55:55–62.

50. Shih CJ (1977). Intracranial tumors in Taiwan. A cooperative survey of 1,200 cases with special reference to intracranial tumors in children. Taiwan Yi Xue Hui Za Zhi 76:515–527.

51. Jones BV, Patterson RJ (1997). Supratentorial tumors. In *Pediatric Neuroradiology*. Ball WS Jr (ed.). Lippincott-Raven, Philadelphia. 369–442.

52. Allen JC (1991). Controversies in the management of intracranial germ cell tumors. Neurol Clin 9:441–452.

53. Jennings MT, Gelman R, Hochberg F (1985). Intracranial germ-cell tumors: natural history and pathogenesis. J Neurosurg 63:155–167.

54. Kim DI, Yoon PH, Ryu YH, Jeon P, Hwang GJ (1998). MRI of germinomas arising from the basal ganglia and thalamus. Neuroradiology 40:507–511.

55. Moon WK, Chang KH, Kim IO et al (1994). Germinomas of the basal ganglia and thalamus: MR findings and a comparison between MR and CT. Am J Roentgenol 162:1413–1417.

56. Komatsu Y, Narushima K, Kobayashi E et al (1989). CT and MR of germinoma in the basal ganglia. Am J Neuroradiol 10:S9–11.

57. Higano S, Takahashi S, Ishii K, Matsumoto K, Ikeda H, Sakamoto K (1994). Germinoma originating in the basal ganglia and thalamus: MR and CT evaluation. Am J Neuroradiol 15:1435–1441.

58. Mutoh K, Okuno T, Ito M et al (1988). Ipsilateral atrophy in children with hemispheric cerebral tumors: CT findings. J Comput Assist Tomogr 12:740–743.

59. Soejima T, Takeshita I, Yamamoto H, Tsukamoto Y, Fukui M, Matsuoka S (1987). Computed tomography of germinomas in basal ganglia and thalamus. Neuroradiology 29:366–370.

60. Okamoto K, Ito J, Ishikawa K et al (2002). Atrophy of the basal ganglia as the initial diagnostic sign of germinoma in the basal ganglia. Neuroradiology 44:389–394.

61. Takano T, Matsui E, Yamano T, Shimada M, Nakasu Y, Handa J (1993). Sequential MRI findings in a patient with a germ cell tumor in the basal ganglia. Brain Dev 15:283–287.

62. Takeda N, Fujita K, Katayama S (2004). Germinoma of the basal ganglia. An 8-year asymptomatic history after detection of abnormality on CT. Pediatr Neurosurg 40:306–311.

63. Wong LW, Jayakumar CR (1997). Germinoma of the basal ganglia and thalamus – CT and MRI findings. Singapore Med J 38:444–446.

64. Rorke LB, Trojanowski JQ, Lee VM (1997). Primitive neuroectodermal tumors of the central nervous system. Brain Pathol 7:765–784.

65. Becker LE, Halliday WC (1987). Central nervous system tumors of childhood. Perspect Pediatr Pathol 10:86–134.

66. Figeroa RE, el Gammal T, Brooks BS, Holgate R, Miller W (1989). MR findings on primitive neuroectodermal tumors. J Comput Assist Tomogr 13:773–778.

67. Rorke LB, Packer RJ, Biegel JA (1996). Central nervous system atypical teratoid/rhabdoid tumors of infancy and childhood: definition of an entity. J Neurosurg 85:56–65.

68. Martinez-Lage JF, Nieto A, Sola J et al (1997). Primary malignant rhabdoid tumor of the cerebellum. Childs Nerv Syst 13:418–421.

69. Jellinger KA, Paulus W (1992). Primary central nervous system lymphomas – an update. J Cancer Res Clin Oncol 119:7–27.

70. Ling SM, Roach M 3rd, Larson DA, Wara WM (1994). Radiotherapy of primary central nervous system lymphoma

in patients with and without human immunodeficiency virus. Ten years of treatment experience at the University of California San Francisco. Cancer 73:2570–2582.

71. Belman AL (1997). Pediatric neuro-AIDS. Update. Neuroimaging Clin N Am 7:593–613.

72. Ruiz A, Post MJ, Bundschu C, Ganz WI, Georgiou M (1997). Primary central nervous system lymphoma in patients with AIDS. Neuroimaging Clin N Am 7:281–296.

73. Drevelegas A (ed.) (2002). Lymphomas and hematopoietic neoplasms. In *Imaging of Brain Tumors with Histological Correlation*. Springer-Verlag, Berlin. 215–225.

74. Lee YY, Bruner JM, Van Tassel P, Libshitz HI (1986). Primary central nervous system lymphoma: CT and pathologic correlation. Am J Roentgenol 147:747–752.

75. Kuker W, Nagele T, Korfel A (2005). Primary central nervous system lymphomas (POCNSL): MRI features at presentation in 100 patients. J Neuro-Oncol 72:169–177.

76. Zimmerman RA (1990). Central nervous system lymphoma. Radiol Clin N Am 28:697–721.

77. Rosenthal MA, Green MD (1995). Cerebral lymphoma. In *Brain Tumors*. Kaye AH, Laws ER Jr (eds). Churchill-Livingstone, Edinburgh. 861–869.

78. Gliemroth J, Kehler U, Gaebel C, Arnold H, Missler U (2003). Neuroradiological findings in primary cerebral lymphomas of non-AIDS patients. Clin Neurol Neurosurg 105:78–86.

79. Erdag N, Bhorade RM, Alberico RA, Yousuf N, Patel MR (2001). Primary lymphoma of the central nervous system: typical and atypical CT and MR imaging appearances. Am J Roentgenol 176:1319–1326.

80. Buhring U, Herrlinger U, Krings T, Thiex R, Weller M, Kuker W (2001). MRI features of primary central nervous system lymphomas at presentation. Neurology 57:393–396.

81. Vandermarcq P, Drapeau C, Ferrie JC (19970. Imaging aspects of primary cerebral lymphoma. Neurochirurgie 43: 363–368.

82. Ricci PE (1999). Imaging of adult brain tumors. Neuroimaging Clin N Am 9:651–669.

83. Kelly WM, Brant-Zawadzki M (1983). Acquired immunodeficiency syndrome: neuroradiologic findings. Radiology 149:485–491.

84. Tubman DE, Frick MP, Hanto DW (1983). Lymphoma after organ transplantation: radiologic manifestations in the central nervous system, thorax, and abdomen. Radiology 149:625–631.

85. Cordoliani YS, Derosier C, Pharaboz C, Jeanbourquin D, Schill H, Cosnard G (1992). Primary brain lymphoma in AIDS. 17 cases studied by MRI before stereotaxic biopsies. J Radiol 73:367–376.

86. Osborn AG (2004). Neoplasms and tumorlike lesions. In *Diagnostic Imaging: Brain*. Osborn AG et al (eds). Amirsys, Salt Lake City. 140–143.

87. Patronas NJ (2002) Brain metastasis. In *Imaging of Brain Tumors with Histological Correlation*. Drevelegas A (ed.). Springer-Verlag, Berlin. 253–276.

88. Nauta WH, Haymaker W (1969). Hypothalamic nuclei and fiber connections. In *The Hypothalamus*. Haymaker W, Anderson E, Nauta WJH (eds). Thomas, Springfield. 136–209.

89. Janss AJ, Grundy R, Cnaan A et al (1995). Optic pathway and hypothalamic/chiasmatic gliomas in children younger than age 5 years with a 6-year follow-up. Cancer 75:1051–1059.

90. Collet-Solberg PF, Sernyak H, Satin-Smith M et al (1997). Endocrine outcome in long-term survivors of low-grade hypothalamic/chiasmatic glioma. Clin Endocrinol 47:79–85.

91. Rodriguez LA, Edwards MS, Levin VA (1990). Management of hypothalamic gliomas in children: an analysis of 33 cases. Neurosurgery 26:242–246.

92. Kucharczyk W, Montanera WJ, Becker LE (1996). The sella turcica and parasellar region. In *Magnetic Resonance Imaging of the Brain and Spine*, 2nd edn. Atlas WS (ed.). Lippincott Williams & Wilkins, Philadelphia. 871–930.

93. Rauschning W (1994). Brain tumors and tumorlike masses: classification and differential diagnosis. In *Diagnostic Neuroradiology*. Osborn AG (ed.). Mosby, St Louis. 401–528.

94. Alvord EC Jr, Lofton S (1988). Gliomas of the optic nerve or chiasm. Outcome by patients' age, tumor site, and treatment. J Neurosurg 68:85–98.

95. Bourekas EC, Miller JW, Christoforidis GA (2002). Masses of the sellar and juxtasellar region. In *Imaging of Brain Tumors with Histological Correlation*. Drevelegas A (ed.). Springer-Verlag, Berlin. 227–252.

96. Russell A (1951). A diencephalic syndrome of emaciation in infancy and childhood. Arch Dis Child 26:274.

97. Poussaint TY, Barnes PD, Nichols K (1997). Diencephalic syndrome: clinical features and imaging findings. Am J Neuroradiol 18:1499–1505.

98. Osborn AG (2004). Sella and pituitary. In *Diagnostic Imaging: Brain*. Osborn AG et al (eds). Amirsys, Salt Lake City. 12–15.

99. Namba S, Nishimoto A, Yagyu Y (1985). Diencephalic syndrome of emaciation (Russell's syndrome). Long-term survival. Surg Neurol 23:581–588.

100. Weissenberger AA, Dell ML, Liow K et al (2001). Aggression and psychiatric comorbidity in children with hypothalamic hamartomas and their unaffected siblings. J Am Acad Child Adolesc Psychiatr 40:696–703.

101. Kerrigan JF, Ng YT, Chung S, Rekate HL (2005). The hypothalamic hamartoma: a model of subcortical epileptogenesis and encephalopathy. Semin Pediatr Neurol 12:119–131.

102. Boyko OB, Curnes JT, Oakes WJ, Burger PC (1991). Hamartomas of the tuber cinereum: CT, MR, and pathologic findings. Am J Neuroradiol 12:309–314.

103. Arita K, Ikawa F, Kurisu K et al (1999). The relationship between magnetic resonance imaging findings and clinical manifestations of hypothalamic hamartoma. J Neurosurg 91:212–220.

104. Valdueza JM, Cristante L, Dammann O et al (1994). Hypothalamic hamartomas: with special reference to gelastic epilepsy and surgery. Neurosurgery 34:949–958.

105. Freeman JL (2003). The anatomy and embryology of the hypothalamus in relation to hypothalamic hamartomas. Epileptic Disord 5:177–186.

106. Debeneix C, Bourgeois M, Trivin C, Sainte-Rose C, Brauner R (2001). Hypothalamic hamartoma: comparison of clinical presentation and magnetic resonance images. Horm Res 56: 12–18.

107. Jung H, Neumaier Probst E, Hauffa BP, Partsch CJ, Dammann O (2003). Association of morphological characteristics with precocious puberty and/or gelastic seizures in hypothalamic hamartoma. J Clin Endocrinol Metab 88:4590–4595.

108. Freeman JL, Coleman LT, Wellard RM et al (2004). MR imaging and spectroscopic study of epileptogenic hypothalamic

hamartomas: analysis of 72 cases. Am J Neuroradiol 25: 450–462.

109. Hall JG, Pallister PD, Clarren SK et al (1980). Congenital hypothalamic hamartoblastoma, hypopituitarism, imperforate anus and postaxial polydactyly – a new syndrome? Part I: clinical, causal, and pathogenetic considerations. Am J Med Genet 7:47–74.

110. Barral V, Brunelle F, Brauner R, Rappaport R, Lallemand D (1988). MRI of hypothalamic hamartomas in children. Pediatr Radiol 18:449–452.

111. Burton EM, Ball WS Jr, Crone K, Dolan L (1989). Hamartoma of the tuber cinereum: a comparison of MR and CT findings in four cases. Am J Neuroradiol 10:497–501.

112. Chong BW, Newton TH (1993). Hypothalamic and pituitary pathology. Radiol Clin N Am 31:1147–1153.

113. Hahn FJ, Leibrock LG, Huseman CA, Makos MM (19880. The MR appearance of hypothalamic hamartoma. Neuroradiology 30:65–68.

114. Lona Soto A, Takahashi M, Yamashita Y, Sakamoto Y, Shinzato J, Yoshizumi K (1991). MRI findings of hypothalamic hamartoma: report of five cases and review of the literature. Comput Med Imaging Graph 15:415–421.

115. Marliani AF, Tampieri D, Melancon D, Ethier R, Berkovic SF, Andermann F (1991). Magnetic resonance imaging of hypothalamic hamartomas causing gelastic epilepsy. Can Assoc Radiol J 42:335–339.

116. Striano S, Striano P, Sarappa C, Boccella P (2005). The clinical spectrum and natural history of gelastic epilepsy-hypothalamic hamartoma syndrome. Seizure 14:232–239.

117. Turjman F, Xavier JL, Froment JC, Tran-Minh VA, David L, Lapras C (1996). Late MR follow-up of hypothalamic hamartomas. Childs Nerv Syst 12:63–68.

118. Prasad S, Shah J, Patkar D, Gala B, Patankar T (2000). Giant hypothalamic hamartoma with cystic change: report of two cases and review of the literature. Neuroradiology 42:648–650.

119. Booth TN, Timmons C, Shapiro K, Rollins NK (2004). Pre- and postnatal MR imaging of hypothalamic hamartomas associated with arachnoid cysts. Am J Neuroradiol 25:1283–1285.

120. Kuzniecky R, Guthrie B, Mountz J et al (1997). Intrinsic epileptogenesis of hypothalamic hamartomas in gelastic epilepsy. Ann Neurol 42:60–67.

121. Nishio S, Morioka T, Hamada Y, Kuromaru R, Fukui M (2001). Hypothalamic hamartoma associated with an arachnoid cyst. J Clin Neurosci 8:46–48.

122. Martin DD, Seeger U, Ranke MB, Grodd W (2003). MR imaging and spectroscopy of a tuber cinereum hamartoma in a patient with growth hormone deficiency and hypogonadotropic hypogonadism. Am J Neuroradiol 24:1177–1180.

123. Pomper MG, Passe TJ, Burger PC, Scheithauer BW, Brat DJ (2001). Chordoid glioma: a neoplasm unique to the hypothalamus and anterior third ventricle. Am J Neuroradiol 22:464–469.

124. Brat DJ, Scheithauer BW, Staugaitis SM, Cortez SC, Brecher K, Burger PC (1998). Third ventricular chordoid glioma: a distinct clinicopathologic entity. J Neuropathol Exp Neurol 57:283–290.

125. Simonetta AB (1999). Imaging of suprasellar and parasellar tumors. Neuroimaging Clin N Am 9:717–732.

126. Lee MT, Lee TI, Won JG (2004). Primary hypothalamic lymphoma with panhypopituitarism presenting as stiff-man syndrome. Am J Med Sci 328:124–128.

127. Pascual JM, Gonzalez-Llanos F, Roda JM (2002). Primary hypothalamic-third ventricle lymphoma. Case report and review of the literature. Neurocirugia (Astur) 13:305–310.

128. Lanzieri CF, Sabato U, Sacher M (1984). Third ventricular lymphoma: CT findings. J Comput Assist Tomogr 8:645–647.

129. Namasivayam J, Teasdale E (1992). The prognostic importance of CT features in primary intracranial lymphoma. Br J Radiol 65:761–765.

130. Shelton CH 3rd, Phillips CD, Laws ER, Larner JM (1999). Third ventricular lesion masquerading as suprasellar disease. Surg Neurol 51:177–180.

131. Kurt G, Dogulu F, Kaymaz M, Emmez H, Onk A, Baykaner MK (2002). Hypothalamic lipoma adjacent to mamillary bodies. Childs Nerv Syst 18:732–734.

132. Bognar L, Balint K, Bardoczy Z (2002). Symptomatic osteolipoma of the tuber cinereum. Case report. J Neurosurg 96:361–363.

133. Mackenzie IR, Girvin JP, Lee D (1996). Symptomatic osteolipoma of the tuber cinereum. Clin Neuropathol 15:60–62.

134. Wittig H, Kasper U, Warich-Kirches M, Dietzmann K, Roessner A (1997). Hypothalamic osteolipoma: a case report. Gen Diagn Pathol 142:361–364.

CHAPTER 43

Neuronal Tumors

Jeffrey J. Raizer and Michelle J. Naidich

INTRODUCTION

Neuronal and mixed neuronal-glial tumors are a rare but important group of primary brain tumors. With the exception of gangliogliomas and dysplastic gangliocytomas of the cerebellum, the remaining tumors have been described in the past three decades [1–7]. The benign nature of these tumors makes histologic diagnosis important as most of these neoplasms can be cured surgically, thereby avoiding radiation therapy (RT) or chemotherapy (CTX), which may have long-term consequences. The imaging characteristics of each of these tumors will be reviewed including modern techniques; these findings in conjunction with the clinical history may suggest the diagnoses.

CENTRAL NEUROCYTOMA

Clinical

Central neurocytoma was first described in 1982 and further characterized in 1993 [3,8]. They make up 0.25–0.5% of intracranial tumors [8]. The majority of these tumors (75%) present in patients between 20 and 40 years of age, without a gender predilection [9,10]. The intraventricular location of the central neurocytoma makes them unique among glioneuronal tumors, as the rest are primarily confined to the brain parenchyma. They usually present with symptoms of increased intracranial pressure, due to their intraventricular location within the lateral and third ventricles [8,10].

Patients complain of headache and visual changes and on exam may demonstrate papilledema and ataxia [10,11]. The clinical course is typically less than 6 months. Rare cases of sudden death due to acute ventricular obstruction have been reported [9]. Their intraventricular location lends neurocytomas to disseminate within the neuroaxis and even intraperitoneally, via ventriculoperitoneal (VP) shunt [12–16]. Extraventricular neurocytomas have been described less commonly, with symptoms consistent with the location of the tumor [17,18].

Pathology

Central neurocytomas are thought to arise from germinal matrix cells in the septum pellucidum or the periventricular region. Germinal matrix cells are progenitor cells which can differentiate into both neuronal and glial elements. Consequently, central neurocytomas may have both cell types present, although extraventricular tumors are predominantly glial [9,12,19].

Central neurocytomas consist of small uniform cells with round or oval nuclei. There is scant surrounding cytoplasm creating a perinuclear halo or 'fried egg' appearance, which can superficially resemble an oligodendroglioma [9,10,20–22]. Focal calcifications may be present but mitoses, vascular proliferation and necrosis are infrequent. Central neurocytomas stain positive for markers of neuronal differentiation, thereby differentiating them from oligodendrogliomas. Neuronal characteristics seen on electron microscopy define central neurocytomas [9,10,20–22].

Central neurocytoma is a relatively benign tumor; however, several cases of poor outcome secondary to aggressive behavior, recurrence and CSF dissemination led to the reclassification of this tumor in 1999 to a WHO grade II [12]. The best indicator of the behavior of this tumor is the labeling index (LI), which is usually low in most cases. In one study, if the LI was <2% there was a relapse rate of 22%, while an LI > 2% was associated with a relapse rate of 63% [16]. Atypical neurocytomas have an LI > 5%, focal necrosis, vascular proliferation and increased mitotic activity [17].

Imaging (Figures 43.1 and 43.2)

A central neurocytoma is typically a supratentorial tumor arising from the septum pellucidum or ventricular wall. Koeller et al report that 50% are located in the lateral ventricle near the foramen of Monro, 15% are both in the lateral and third ventricle and 3% are isolated to the third ventricle [9]. Bilateral involvement of the ventricles is present in 13% of cases. There is also a single report of a central neurocytoma in the fourth ventricle [19]. These tumors are occasionally reported in extraventricular locations such as the cerebrum, cerebellum and spinal cord; CSF dissemination may be noted in some cases [9,19,23].

Computerized tomography shows a well-circumscribed intraventricular mass. The tumor abuts and may bow the

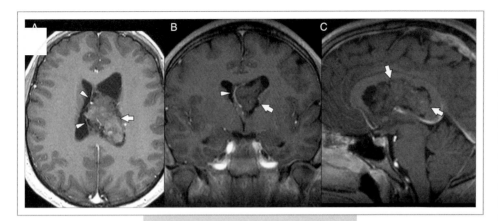

FIG. 43.1. Central neurocytoma. Axial **(A)**, coronal **(B)** and sagittal **(C)** post-contrast T1 images show the tumor (arrows) filling and expanding the left lateral ventricle. Note the close association with the septum pellucidum (arrowheads) on the axial and coronal views. The mass is predominately iso-intense on T1-weighted imaging with only minimal enhancement.

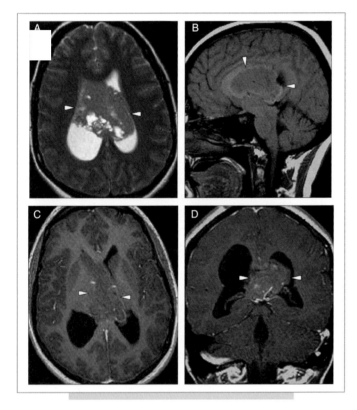

FIG. 43.2. Central neurocytoma. Axial T2-weighted images **(A)** of a different patient show a large tumor involving both lateral ventricles (arrowheads). It is isointense to gray matter with some irregular areas of high T2 signal correlating to cystic regions. There are a few very small foci of very low signal present which may represent calcifications, hemorrhage or vessels. **(B)** A non-contrast sagittal T1 image shows the isointense mass (arrowheads) filling the lateral ventricles. Axial and coronal **(C, D)** post-contrast images demonstrate the lesion to have only minimal enhancement. Courtesy of Dr Eric Bartlett.

septum pellucidum. The tumor is of equal or higher attenuation relative to normal brain parenchyma. Low attenuation cystic areas are often present. Clumped, amorphous,

globular calcifications are seen in 50% of tumors. Following administration of contrast there is mild to moderate enhancement. Associated findings of hydrocephalus and monoventricular dilatation may be present [3,8,9,11,21].

On MRI, the solid portions of the tumor are isointense to gray matter on T1-weighted images. The T2 characteristics are variable with areas that are isointense to mildly hyperintense. Areas of high T2 signal with corresponding low T1 signal correlate to cystic regions. Areas of very low T1 and T2 signal or flow voids may represent calcifications, hemorrhagic regions or vessels. There is variable enhancement [3,8,9,11,21]. Peritumoral edema is infrequent and, if present, may be an indicator of invasion through the ventricular wall into the brain parenchyma or of a highly aggressive tumor [24,25].

Angiographically, some central neurocytomas show moderate vascularity, but most are relatively non-vascular demonstrating displacement of the normal vascular structures.

The list of intraventricular tumors for differential consideration based on the classic imaging modalities includes intraventricular oligodendrogliomas, astrocytomas, or meningiomas, subependymomas, ependymomas, choroids plexus papillomas, colloid cysts, intraventricular teratomas and dermoids. However, these tumors have some variable clinical or imaging characteristics that may differentiate them from a central neurocytoma. In the absence of such distinguishing features, advanced imaging techniques may help make the diagnosis.

Magnetic resonance spectroscopy (MRS) is a method whereby metabolites of a sampled tissue are measured (see Chapter 28). Based on the various ratios of the metabolites detected, conclusions can be made regarding the nature of the tissue present. N-acetylaspartate (NAA), a neuronal marker, is produced by normally functioning mature neurons. In conditions of neuronal cell loss or dysfunction, the amount of NAA decreases. Choline and its derivatives (phosphocholine, phosphatidylcholine and glycerophosphocholine) participate in cell membrane synthesis and breakdown. Elevation of choline corresponds to increased cell proliferation. Higher choline levels suggest a more aggressive tumor type [20,26]. Creatine is a measure of energy metabolism. Lactate occurs when there is evidence of anaerobic metabolism and lipid correlates with necrosis.

When central neurocytomas are evaluated by MRS, the amount of NAA present is decreased relative to normal values. It might be expected that NAA would be increased in these tumors, but the immature neuronal cells likely account for this discrepancy. MRS of central neurocytomas also shows increased choline. In most tumors, there is a linear correlation between the increased choline and the MIB-1 labeling index; in central neurocytomas this is not the case [26]. Elevated choline has been reported in other 'benign' tumors besides central neurocytomas; the mechanism for this elevation is unclear. Several MRS studies also demonstrate the presence of a glycine peak, which is not present in normal brain or in most other brain tumors [20,26,27]. This metabolite, so far, appears unique to the central neurocytomas, although the absence of glycine does not exclude the diagnosis [20,27,28]. MRS of extraventricular central neurocytomas is equivalent to the intraventricular version [28].

Positron emission tomography (PET) is an imaging technique that measures various metabolic indices. PET uses various labeled tracers to assess homodynamic and metabolic information about various tumors. In tumors, the cerebral circulation, as measured by regional cerebral blood flow and volume, varies depending on the tumor's histology. In central neurocytomas, these perfusion parameters are elevated. Oxygen metabolism is always reduced in tumors due to anaerobic metabolism. However, the reduction of the cerebral metabolic rate of oxygen is less in central neurocytomas than other brain tumors. This may help distinguish this tumor from other more malignant histology. ^{18}F-fluorodeoxyglucose (FDG) can be used to measure the metabolic rate of glucose within the tumor. Increased metabolic rate of glucose is a characteristic of malignant brain tumors [12,28,29]. In a study performed by Mineura et al, most of the central neurocytomas had low metabolic rates of glucose utilization [24]. One of the cases had a relatively increased glucose metabolism; this particular tumor showed rapid re-growth after partial resection. Ohtani et al present a case report where a central neurocytoma had an extremely high uptake of FDG, even greater than the normal brain cortex [29]. Upon resection, this tumor was noted to have atypical histological features and an elevated LI of 7%. Consequently, the cerebral metabolic rate of glucose maybe a useful predictor of the proliferative potential of the tumor, not only at the time of diagnosis, but also as a tool to assess residual tumor during treatment.

Thallium-201 single photon emission computed tomography (^{201}Tl SPECT) is a technique that reflects changes in blood–brain barrier permeability, regional blood flow and activity of sodium-potassium adenosine triphosphatase pumps on cell membranes. Elevation of ^{201}Tl typically reflects malignancy in brain tumors. However, studies have shown increased uptake of ^{201}Tl in central neurocytomas that does not correspond with the LI [26]. The underlying reason for this discrepancy is unknown.

Treatment

The primary treatment of these tumors is surgical, irrespective of location [10,17,18,30,31]. Gross total resection (GTR) has resulted in 5-year local control rates of >95% and 5-year survival rates of >90% [30]. Radiation therapy has been advocated for lesions that are subtotally resected (STR) to increase local control and survival [15,30]. However, the exact role of radiation is ill defined as some patients with an STR do not recur and have similar survival to those with GTR. Radiosurgery has been used in cases of STR with good local control rates [32,33]. The use of CTX in these tumors is limited [14,30,31,34–36].

GANGLIOGLIOMA AND GANGLIOCYTOMA

Clinical

Gangliogliomas were first named by Perkins in 1926 [5]. Gangliogliomas account for 1–11% of tumors in children and less than 1% in adults (1.3% of all brain tumors). Gangliogliomas (WHO grade I or II) are composed of neuronal and glial elements, while gangliocytomas (WHO grade I) are composed predominantly of neoplastic neurons [38]. They most commonly present with seizures, but other symptoms do occur depending on the location of the tumor. They usually present before the age of 30 and have a female predominance [39].

Pathology

Gangliogliomas have a neoplastic glial component, usually astrocytes [38]. Gangliocytomas are composed of irregular groups of large multipolar neurons often with dysplastic features; non-neoplastic glial elements and reticulin fibers make up the stroma [38]. Labeling indices are often less than 3%, but an increased growth fraction and the presence of p53 may be associated with tumor recurrence. Despite these tumors being benign, reports of malignant degeneration, anaplastic gangliogliomas, drop metastases and dissemination via a VP shunt are published [39–45]. Features of anaplasia appear to be a negative prognostic factor [44,46,47]. Malignant change is usually of the glial elements, which may lead to features similar to a glioblastoma multiforme.

Imaging (Figures 43.3 and 43.4)

These tumors can be solid, cystic, or both with little or no surrounding edema or mass effect [9]. The location of gangliogliomas is most common in the cerebral hemispheres at the cortical or corticomedullary junction, with involvement of the temporal, parietal, occipital and frontal

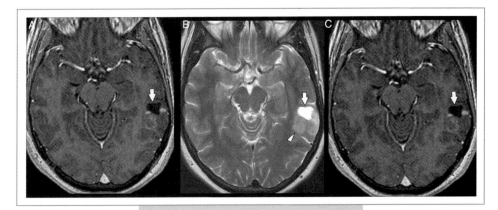

FIG. 43.3. Ganglioglioma. **(A)** An axial T1-weighted image shows an irregularly shaped hypointense lesion within the left temporal lobe. **(B)** On the T2-weighted sequence, the lesion is high signal (arrow) with surrounding less intense edema (arrowhead). **(C)** There is peripheral nodular enhancement on the post-contrast axial T1-weighted image.

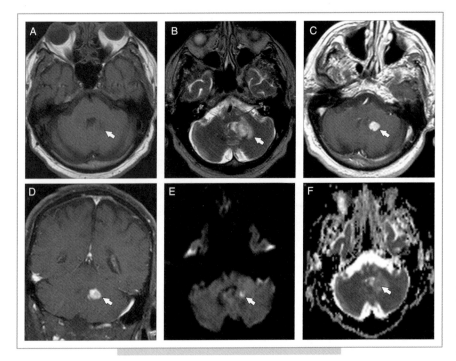

FIG. 43.4. Ganglioglioma. **(A)** An axial T1-weighted image faintly shows an area of diminished signal (arrow) in the medial aspect of the left cerebellum. **(B)** The T2-weighted image shows a larger area of abnormal hyperintensity (arrow) corresponding to the lesion and the surrounding edema. **(C)** Post-contrast axial and **(D)** coronal show avid enhancement of the lesion (arrow). **(E)** The B1000 diffusion-weighted sequence demonstrates high signal with faint corresponding low ADC signal **(F)** within a portion of the tumor (arrows) consistent with an area of restricted diffusion.

lobes and basal ganglia [48–53]. The temporal lobe is affected more in older patients [54]. Much less common is the posterior fossa where involvement of the cerebellum and brainstem is reported [48,53,55,56]. Other intracranial sites of involvement include the trigeminal nerve [57], choroid plexus [58], cerebellopontine angle [49,59], dura mimicking a meningioma [60], pineal gland [61,62], optic chiasm [63,64], intraventricular [65] and hypothalamus [55]. Haddad [55] noted that midline lesions tended to occur in younger patients [55]. Size can range from 1 to 10.5 cm and there appears to be an increase in tumor diameter and volume in patients <10 years of age compared to those who are older [48,49,51,53,54].

Imaging on CT scan with contrast often shows a well circumscribed solid mass or cyst with a mural nodule; the latter is seen in about 40–50% of cases [11,48,52,53,55].

The cystic lesion is often hypodense or isodense, while the solid can be any density [49,52]. Most tumors enhance after the administration of contrast which can be homogeneous, heterogeneous, or have a ring pattern [66]. Solid tumors are more likely to enhance homogeneously while cystic lesions enhance inhomogeneously [48,55,66]; cystic lesions can have laminar or nodular enhancement [9,48,49,52,53,55]. CT scan may be normal in some cases even with the administration of contrast [9,48,49,52,53,55]. Calcifications are seen in 25–30% of cases, but less so in solid appearing lesions [9,39,49,52,53]. Intratumoral hemorrhage is a rare event [49,55]. For peripherally placed tumors, remodeling of the skull may be seen [11,48].

MRI appearance is hypointense, isointense, hyperintense or mixed relative to brain on T1 (in decreasing order) and almost always hyperintense on T2, but can be isointense in some cases [49,52,66]. A well-circumscribed cyst with a mural nodule is typically seen on MRI. On MRI, the cystic component is heterogeneous and mainly hypointense on T1 compared to CSF. On T2 they are usually hyperintense. Solid lesions on MRI are hypo-, iso-, or hyperintense on T1 but display increased intensity on T2 [49,66]. They are often well circumscribed. Enhancement is more common than not, but is variable and may be solid, rim, or nodular [38,39,49]. Cystic type lesions are more common in children [54]. Peritumoral edema may be seen but is infrequent; when present it may correlate with an increased grade [49–51,53]. In one report of a patient with leptomeningeal (LM) spread there was enhancement of the meninges in several areas of the brain [67].

On MRS, when compared to the contralateral normal hemisphere, there are reduced Cho/Cr and NAA/Cr ratios and an increased Cho/NAA ratio; the NAA/Cr ratio is relatively higher than gliomas, likely from the neuronal component of the tumor [49]. A lactate peak was seen in two separate studies evaluating single gangliogliomas [68,69]. Wang et al found that the NAA/Cho and Cr/Cho ratios were higher than in other tumors evaluated in the cerebellum [69]. MR perfusion of these tumors demonstrates a relative cerebral blood volume greater than that of other low-grade gliomas; there was a non-significant trend towards correlation with enhancement, but not with increased grade [51].

On angiography, these lesions are essentially avascular or hypovascular with abnormal vascularity, but a vascular blush can be seen [48,53,55,70].

PET imaging with FDG shows these tumors to be hypometabolic relative to normal brain, but may be normometabolic, corresponding to a low-grade histology; one exception is a patient with a grade I/II ganglioglioma that was hypermetabolic [49,50]. Selch et al reported one patient with normal glucose metabolism while the rest were less active than normal brain [53]. Provenzale et al coregistered PET with MRI, thereby allowing a better delineation of tumor borders, but found that lesions that were hypometabolic before coregistration actually had a heterogeneous metabolism and hypermetabolism compared to white matter (all tumors were low grade) [71]. The dimensions on PET correlated well with size on MRI [71]. PET with C-11 methionine may be hypometabolic or moderately hypermetabolic [49,72]. SPECT imaging also correlates with grade and is increased if the tumor is high grade; in most cases it demonstrates reduced or normal perfusion [49,50,53].

Gangliogliomas may also occur in the spinal cord where they have an average length of eight vertebral bodies compared to four for spinal cord astrocytomas or ependymomas [73]. Most occur in the cervical spine but can involve the whole cord and are eccentrically located [73]. Tumoral cysts are more common than in astrocytomas or ependymomas, as is boney erosion. On T1, they most often have mixed signal intensity and less often are hypointense; the mixed density differs from astrocytomas or ependymomas.

On T2, they more often have a homogeneous than heterogeneous signal [73]. Edema is not a common finding. The enhancement pattern is variable, but most often is patchy and often reaches the surface of the cord.

Gangliocytomas occur most commonly in the cerebrum – usually the temporal lobe, followed by the frontal and parietal lobes [11,66,74]. They may also be seen in the cerebellum, hypothalamus adjacent to the floor of the third ventricle, pineal region, cerebellum and pituitary [75–78]. They have been reported to mimic extra-axial lesions like a meningioma [74,79]. Gangliocytomas may be associated with dysplastic or malformed brains [66].

On CT, they are hypointense and calcification and cysts may be seen [74,79]. On T1 MRI, they are usually hypointense and appear hyperintense on T2-weighted images; after the administration of contrast there is strong homogeneous enhancement [74]. MRI will reveal low signal intensity on T1, high signal on T2, and enhancement in many cases [74]. A peripherally located lesion may have a dural tail and appear like a meningioma.

Gangliocytomas have been reported in the spinal cord, usually at the cervicothoracic junction [80]. On T1, there is a heterogeneous mass with cord expansion. On T2, there was heterogeneous increased signal intensity throughout the mass and a moderate amount of heterogeneous contrast enhancement. Jacob reported a cord lesion without enhancement but increased signal on T2 [81].

The radiographic differential diagnoses for gangliogliomas or gangliocytomas includes low-grade gliomas, hamartomas and pleomorphic xanthoastrocytoma.

Treatment

Many large retrospective case series highlight the important role of surgery in these tumors, which can lead to long-term survival rates [39,49,53,82]. Several series have found no increased survival with adjuvant RT but, for anaplastic lesions, this may be of some benefit, especially when incompletely resected [44,46,47,53,55, 83,84,85]. There are a handful of case reports of patients being treated with various CTX regimens for gangliogliomas, some of which where anaplastic or malignant [43,46,53,82,84,85–93]. In most cases, patients with CTX have also received RT, so the true benefit is difficult to assess; in several series there was no benefit, patients did poorly, or there was no mention of outcome.

DESMOPLASTIC INFANTILE GANGLIOGLIOMA (DIG) AND DESMOPLASTIC INFANTILE ASTROCYTOMA (DIA)

Desmoplastic infantile ganglioglioma (DIG) was first recognized by Vandenberg et al in 1987 [7]. A few years before, Taratuto et al described the desmoplastic infantile astrocytoma (DIA) [6]. These are rare tumors, designated as grade I by the WHO. Both tumors have identical clinical presentations, imaging and histology with one exception. The DIA lacks the neuronal component seen in DIG.

Clinical

Presentation is usually within the first 4 months of life, although older infants have been described, but almost all cases occur in children <2 years with a male predominance [7,94]. There is a short time course of symptoms. Infants present with a rapid increase in head circumference, bulging fontanelles, downward ocular deviation ('sunset sign'); other symptoms include seizures, paresis, increased muscle tone and reflexes [6,94,95].

Pathology

These tumors present as a large bulky mass, usually involving the parietal and frontal lobes and, occasionally, the temporal lobe. Over 60% involve more than one lobe, although they never cross the midline. The tumor has two distinct components. There is a large cystic component that may be a part of the tumor or a reaction to the second, solid component. The solid component contains neuronal elements in variable stages of differentiation. There may be very cellular areas of undifferentiated cells with mitotic figures and necrosis; however, the LI is typically low. There is an intense desmoplastic reaction at the periphery in combination with fibroblastic elements resulting in attachment to the dura. As stated above, DIA differs from DIG in that it lacks the later neuronal component [6,7,65,96,97].

Although this is a WHO grade 1 tumor, the typical benign nature of the neoplasm is variable. It is suggested that the amount or degree of anaplasia seen within the solid component will predict the tumor's behavior [96]. There are occasional reports of recurrence and CSF spread [95,97,98].

Imaging (Figure 43.5)

There are several key features of the DIG. It is a bulky tumor within the frontal and parietal lobes. There is a solid component that abuts the meningeal surface with contrast enhancement along this side of the mass; there is an associated large cystic component [99]. Plain films may demonstrate diastasis of the sutures. Ultrasound shows a multicystic mass. On CT, the tumor should have well-defined cystic components with attenuation equal to CSF. The solid component is superficial in location at the cortical margin and is isodense to slightly hyperdense. The solid component shows avid enhancement. There are no calcifications or blood products. MRI shows a hypointense cystic mass on T1 with an isointense peripheral solid component that enhances; on T2, the cyst is hyperintense and the solid portion is heterogeneous but will intensely enhance [6]. Occasionally, there is mild surrounding vasogenic edema.

Common differential considerations will include primitive neuroectodermal tumor (PNET), typical gangliogliomas, supratentorial ependymomas and pleomorphic

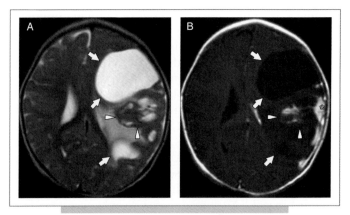

FIG. 43.5. DIG. **(A)** An axial T2-weighted image demonstrates a tumor involving predominately the left frontal lobe extending into the parietal lobe. There is a large, high signal, cystic component, as well as some additional smaller cysts (arrows). The solid components (arrowheads) are isointense to brain parenchyma. There is surrounding high T2 edema and mass effect on the left lateral ventricle. **(B)** A post-contrast axial T1-weighted image demonstrates the low signal cystic components (arrows). There is enhancement of the solid component (arrowheads) that is avid at the site of attachment to the dura (asterisk). Courtesy of Dr Francine Kim.

xanthroastrocytomas. However, these tumors have various characteristics that distinguish them from the DIG [95,97,99,100]. Imaging demonstrating a large supratentorial, cystic and solid mass with intense enhancement at its dural base of attachment in a patient less than 18 months of age suggests a DIG in the majority of cases.

Treatment

The superficial location of the tumor and excellent discernible border between the solid tumor and the adjacent normal brain tissue should allow for gross total resection, except if there is secondary dural attachment [101].

Although the tumors appear histologically malignant, their course is usually benign [7,101,102]. At least one report of CNS dissemination exists [95,97,98]. The primary mode of treatment is surgical with the aim of gross total resection [7,94,101–103]. In fact, some reports have demonstrated stable follow-up and even spontaneous regression after partial resection [103,104]. One author even recommends repeat surgery before resorting to RT [105]. There are only a few patients who have received CTX for DIG/DIA [7,68,96,106,107,108].

DYSEMBRYOPLASTIC NEUROEPITHELIAL TUMOR

Clinical

Dysembryoplastic neuroepithelial tumors (DNET) are low-grade, cortically based tumors that were described as a distinct entity by Daumas-Duport et al in 1988 [2]. DNET are located primarily in the temporal lobes and often lead to drug resistant epilepsy. They occur most frequently in the second and third decades with a male predominance.

Pathology

These are nodular appearing lesions with glioneuronal elements that have some glial fibrillary acidic protein (GFAP) positivity, with 'floating neurons' that are synaptophysin positive. The LI is low and cortical dysplasia may be seen. These are usually benign tumors but there are two reports of malignant transformation [109,110].

Imaging (Figure 43.6)

These lesions are most commonly found in the medial or lateral temporal lobes, but can also occur in the frontal, parietal and occipital lobes, with size ranging from 1 to 5 cm [110–112]. Other intracranial locations include the pons, thalamus, basal ganglia, cerebellum and third ventricle [113–118]. Baisden et al reported on 10 cases of DNET that arose from the anterior septum pellucidum [119]. They were lobulated lesions, with the larger ones extending into the third ventricle. Varying degrees of hydrocephalus were present, with the tumors ranging from 1 to 3 cm. They were non-enhancing in all but one case, which had minimal peripheral enhancement. The lesions were hypointense on T1 and hyperintense on T2 or proton density; mild edema was noted.

On CT scan, the tumor may be hypodense with a pseudocystic appearance, isodense, or hyperdense with calcifications in about 20% of cases [11,112]. They are primarily well demarcated hypodense lesions [2,66,112,120]. Osternun et al found the solid portion to be moderately hypodense with marked hypodensity of the cystic lesion [112]. Calcification is seen in about 20% of cases. Most are inhomogeneous in density without mass effect or edema. About one-third enhance after administration of contrast.

On T1 MRI, the tumors are hypointense and, on T2, hyperintense; they are well demarcated, multilobulated or gyriform in appearance [111,120,121]. Most lesions are cystic or multicystic, have a shorter T1 than CSF and appeared hypointense, isointense or hyperintense on proton density imaging [111,112]. This signal pattern is due to the abundant myxoid interstitial matrix and accounts for the differences with CSF. The solid portion is isointense and approximately one-third to one-half have contrast enhancement (ring-like features or a faint margin of enhancement) especially with a nodular pattern [11,111,112]. The margins on MRI may not be as well defined as on CT. Calvarial changes can be seen, especially for lesions extending to the cortical margin.

On T2 images, the lesions are hyperintense, with a well demarcated margin [110,111,122]. Despite the cystic appearance on MRI, cysts are infrequently seen histologically [111,112]. Edema and mass effect are not seen in most

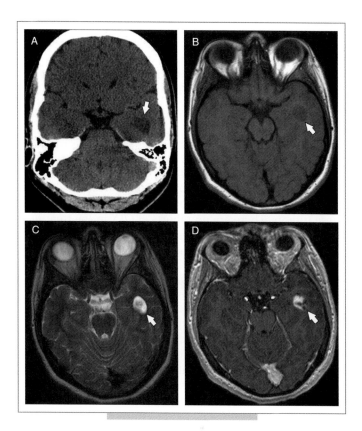

FIG. 43.6. DNET. **(A)** Axial CT shows a small rounded area of low attenuation in the left temporal lobe. **(B)** Axial T1- and **(C)** T2- weighted images show the same rounded lesion with low T1 and high T2 signal. There is no surrounding edema. **(D)** A post-contrast axial T1 image shows heterogeneous enhancement.

cases, but bone remodeling may be noted [110–112,122]. Rare cases may demonstrate regions of hemorrhage. With high-resolution imaging, septations can be seen on MRI within the lesion [110]. Fernandez et al also described a 'spatial' distribution of DNETs as mostly triangular, or as rectangular or round [110]. On MRI, these lesions can change over time with the development of enhancement, without necessarily indicating a change in grade [123]. They often have a gyriform configuration [122].

MRS in one patient showed a reduced NAA and the Cho/Cr and Cho/NAA ratios are lower than other gliomas [111]. Bulakbasi et al evaluated three patients with DNET and found they had normal spectra, which was one way to differentiate them from other lesions; they also had the highest apparent diffusion coefficient (ADC) values among low-grade tumors [124]. Vuori et al also report essentially normal spectra in DNET with better preservation of the NAA peak than in low-grade gliomas [125]. Increased Cho and decreased NAA has been reported in a DNET that developed contrast enhancement, but remained benign on pathology. These MRS findings may differentiate from other primary brain tumors.

On FDG and C-11 methionine PET, these lesions are hypometabolic, especially if in the mesial temporal lobe, or have mild uptake; uptake has no bearing on the level of enhancement [72,126]. Increased activity may suggest another diagnosis [72]. SPECT with Tc99m-HMPAO or IMP can have normal perfusion or hypoperfusion and no uptake on thallium-201 [111,127].

Treatment

Surgical resection is the main treatment with excellent long-term outcomes even for subtotally resected tumors [2,120,121,128]. There does not appear to be a role for RT for these tumors. There are no reports of CTX in the treatment of DNET, except in cases where the diagnosis was incorrect, but there does not appear to be any indication for its use.

DYSPLASTIC GANGLIOCYTOMA OF THE CEREBELLUM

Dysplastic cerebellar gangliocytoma (WHO grade I) or Lhermitte-Duclos (LD) is a lesion of the posterior fossa that causes unilateral expansion of the cerebellar hemisphere. This rare condition is named after the first clinicians to report it in 1920 [4]. Historically, this lesion has been known by at least ten different names, reflecting the confusion regarding its pathogenesis. Some features of this lesion suggest a neoplastic origin while other features support this lesion as hamartomatous. Despite the historically murky classification, the laminated imaging pattern of the affected cerebellar hemisphere is so characteristic that a diagnosis can be made based on imaging alone.

Clinical

Although LD has been reported in a wide range of ages, it classically presents in the third to fourth decades of life; there is no gender predominance [129,130]. Patients experience headaches, nausea, vomiting, papilledema, cerebellar symptoms and cranial nerve palsies, and visual disturbances due to the progressive mass effect in the posterior fossa leading to increased intracranial pressure and hydrocephalus [129–131]. LD has been associated with several other developmental abnormalities. These include megalencephaly, microgyria, syringomyelia, heterotopia, various skeletal abnormalities (polydactly) and cutaneous hemangiomas [129,132,133]. There is also a strong association with Cowden's syndrome (CS), an autosomal dominant phakomatosis characterized by multiple hamartomas of ectodermal, endodermal and mesodermal origin. There are facial tricholemmomas, acral keratosis, oral papillamatosis, intestinal polyposis and fibrocystic disease of the breast. There is also an increased incidence of malignant tumors involving the breast, thyroid, GI, GU and CNS systems [129,130]. The association between CS and LD was first proposed in the early 1990s [134]. It is believed that LD and CS represent different ends of the spectrum of the same phakomatosis.

Lok et al investigated 20 patients with CS without neurologic disease, 35% of these had an abnormality identified on brain MR, including three patients with unsuspected LD [135]. In fact, 40% of patients diagnosed with LD have been found to have CS [136]. The gene responsible for both Cowden's syndrome and LD has been mapped to chromosome 10. The gene is known as PTEN. This is a very powerful tumor suppressor gene because it plays a role in cell division, survival and migration [134,137,138]. Alterations of the PTEN tumor suppressor gene lead to development of the Cowden's syndrome–LD complex.

Pathology

On histopathology, the lesion is characterized by large neuronal cells that expand the granular and molecular layers [129,130]. In LD, there are diffuse or focal areas of cerebellar cortical hypertrophy. This typically involves one cerebellar hemisphere, left greater than right. There are occasional reports of involvement of the vermis or both hemispheres. Macroscopically, the folia become thickened and expanded, lose their secondary folding and cause expansion of the involved tissue. There is replacement of the inner granular layer and Purkinje layer with large neuronal cells. These neuronal cells send their myelinated axons into the overlying molecular layer causing the molecular layer to increase in thickness. There is thinning of the central white matter. The contrast between the hypermyelinated superficial layer and the atrophic white matter gives the impression of an 'inverted cortex' [139]. Calcifications are frequently present in the molecular layer associated with a variable degree of microvascular proliferation [131]. The borders between the lesion and normal cerebellar tissue are poorly demarcated. The absence of necrosis, the rare mitosis and lack of significant proliferative activity favors a benign, hamartomatous nature of this lesion. This is, therefore, classified as a WHO grade I lesion. However, there are reports of recurrence following resection even after relatively long disease-free intervals, raising the suspicion of perhaps a more ominous nature of this lesion [129,132,138–142].

Imaging (Figure 43.7)

The imaging appearance of this lesion reflects the altered cytoarchitecture within the mass. The central portion of the enlarged folia consisting of the diminished white matter core and the abnormal granular layer of the cortex will appear as a low attenuation layer on CT and low T1 signal with corresponding high T2 signal on MR. The cortical surface (outer molecular layer) of the expanded folia and the leptomeningeal surfaces in the effaced sulci are indistinguishable. These regions are isodense on CT and isointense of both T1- and T2-weighted imaging. Calcifications (high signal on T1-weighted sequences) and enhancement may occur. This corresponds either to the proliferation of blood vessels or anomalous veins seen along the

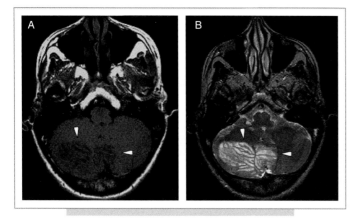

FIG. 43.7. Lhermitte-Duclos. **(A)** Axial T1 and **(B)** T2 images show the classic striated cerebellum of Lhermitte-Duclos. This lesion involves both a portion of the right cerebellum and the vermis. There are stripes of low T1 signal with corresponding high T2 signal representing the abnormal granular layer and the diminished white matter. These stripes alternate with isointense regions that are made up of the expanded outer molecular layer and the adjacent leptomeningeal surfaces. Courtesy of Dr Eric Bartlett.

cortical surface and calcareous deposits within the vessel walls [129,131,134,140–142]. Restricted diffusion has been reported in a few cases [142,143]. This is secondary to the increased cellularity of the granular layer and the increased density of the myelinated axons in the molecular layer. MRI imaging shows a hypointense lesion on T1 with little or no enhancement and hyperintense on T2 [129,130].

Various advanced imaging techniques have looked at LD not as a diagnostic tool, but rather to further investigate the underlying nature of the lesion. MRS shows decreased NAA. This likely reflects a lack of normal neuronal architecture or immature embryonic tissue unable to express NAA rather than a neoplastic behavior. Furthermore, the ratio of choline to creatine is normal, confirming the absence of cell turnover and proliferation as would be expected in neoplastic processes [136]. Imaging techniques employed to evaluate vascular perfusion document elevated cerebral blood volume and flow in these lesions [132,143–145].

This is felt to correspond with the vascular proliferation seen histologically. Ogasawara et al also demonstrated that the cerebral metabolic rate of oxygen within an LD lesion was equivalent to the contralateral cerebellum [144]. As stated earlier, oxygen metabolism should be decreased in tumors because of their anaerobic metabolism. An additional SPECT study employed technetium-99m-ethyl cysteinate dimer (^{99m}Tc-ECD) as a tracer [146]. Brain tumors lack the enzymatic process to retain this tracer so there is a defect on the SPECT images corresponding to the tumor. However, there is uptake within LD indicating enzymatically normally functioning tissue. These advanced imaging modalities lend support to the theory that LD is not a malignancy, but rather a hamartoma.

Treatment

There have been no reports of malignant transformation of this tumor and its clinical course is benign. The optimal treatment is surgical excision and decompression to relieve symptoms of increased intracranial pressure [128,129]. There are no reports of CTX or RT for this lesion and there does not appear to be any indication to use these, even when the lesion is incompletely resected.

CEREBELLAR LIPONEUROCYTOMA

First recognized by Bechtel in 1978, this is a well-differentiated neurocytic tumor (WHO grade I or II) that occurs in the fifth and six decades, with a slight male predominance [1,147,148].

Clinical

They typically occur in the cerebellar lobes or vermis or cerebellopontine angle and present with symptoms referable to the posterior fossa – dizziness, headache, vomiting and gait disturbance, similar to dysplastic gangliogliocytoma of the cerebellum [149]. They have been reported to occur within the lateral ventricle [150]. Patients can have recurrence of their tumor [147].

Pathology

They contain round neoplastic cells with a consistent neuronal and focal lipomatous differentiation and a low rate of proliferation. There is focal GFAP expression and also neuron specific enolase and synaptophysin staining; the LI is typically low [147–149].

Imaging

CT scan imaging shows a well circumscribed mass in the posterior fossa, often with associated hydrocephalus. The tumor is iso- to hypodense to brain with moderate patchy contrast enhancement [147]. They may have a cystic component. One case report of a presumed liponeurocytoma had evidence of hemorrhage on CT scan without contrast [147].

On MR, T1 the mass is iso- to hypointense, with areas of hyperintensity correlating with the fat density seen on CT scans; they may have a cystic component [147,149]. There is irregular enhancement post-contrast. On T2 and proton density images, the tumor is heterogeneous and has variable hyperintensity compared to normal cortex [147]. The cyst rim is non-contrast enhancing and is isointense to CSF on T2 and proton density. Edema is minimal when present [147,149]. Hemorrhage may be seen on MRI that is subacute to chronic [149]. The lipomatous component may be hyperintense to a variable degree on T1 MRI and enhancement is minimal [148,151]. The differential diagnosis on imaging for this tumor only includes medulloblastoma and ependymoma.

Treatment

The treatment in general is surgical but, for supratentorial lesions, radiation may prevent relapse; however, the limited number of cases makes any definite recommendations difficult [149]. There are no reports of CTX for these tumors, so its role is unknown. One report for this tumor acting more aggressively exists; this tumor recurred despite radiation therapy; the patient underwent a second resection and was treated with chemotherapy per a medulloblastoma protocol, but outcome or response was not noted [152]. Relapses of this tumor have been reported.

REFERENCES

1. Bechtel JT, Patton JM, Takei Y (1978). Mixed mesenchymal and neuroectodermal tumor of the cerebellum. Acta Neuropathol (Berl) 41:261–263.

2. Dumas-Duport C, Scheithauer BW, Chodkiewicz JP, Laws ER Jr, Vedrenne C (1988). Dysembryoplastic neuroepithelial tumor: a surgically curable tumor of young patients with intractable partial seizures. Report of thirty-nine cases. Neurosurgery 23:545–556.

3. Hassoun J, Gambarelli D, Grisoli F et al (1982). Central neurocytoma. An electron-microscopic study of two cases. Acta Neuropathol (Berl) 56:151–156.

4. Lhermitte J, Duclos P (1920). Sur un ganglioneurome diffus du coertex du cervelet. Bull Assoc Fran Etude Cancer 9:99–107.

5. Perkins OC (1926). Ganglioglioma. Arch Pathol Lab Med 2:11–17.

6. Taratuto AL, VandenBerg SR, Rorke LB (2000). Desmoplastic infantile astrocytoma and ganglioglioma. In *Pathology & Genetics: Tumours of the Nervous System*. Kleihues P, Cavenee WK (eds). IARC Press, Lyon. 99–102.

7. VandenBerg SR, May EE, Rubinstein LJ et al (1987). Desmoplastic supratentorial neuroepithelial tumors of infancy with divergent differentiation potential ('desmoplastic infantile gangliogliomas'). Report on 11 cases of a distinctive embryonal tumor with favorable prognosis. J Neurosurg 66: 58–71.

8. Hassoun J, Soylemezoglu F, Gambarelli D et al (1993). Central neurocytoma: a synopsis of clinical and histological features. Brain Pathol 3:297–306.

9. Koeller KK. Sandberg GD (2002). From the archives of the AFIP. Cerebral intraventricular neoplasms: radiologic-pathologic correlation. Radiographics 22:1473–1505.

10. Schmidt MH, Gottfried ON, von Koch CS, Chang SM, McDermott MW (2004). Central neurocytoma: a review. J Neuro-Oncol 66:377–384.

11. Shin JH, Lee HK, Khang SK et al (2002). Neuronal tumors of the central nervous system: radiologic findings and pathologic correlation. Radiographics 22:1177–1189.

12. Coelho Neto M, Ramina R, de Meneses MS, Arruda WO, Milano JB (2003). Peritoneal dissemination from central neurocytoma: case report. Arq Neuropsiquiatr 61:1030–1034.

13. Elek G, Slowik F, Eross L, Toth S, Szabo Z, Balint K (1999). Central neurocytoma with malignant course. Neuronal and glial differentiation and craniospinal dissemination. Pathol Oncol Res 5:155–159.

14. Eng DY, DeMonte F, Ginsberg L, Fuller GN, Jaeckle K (1997). Craniospinal dissemination of central neurocytoma. Report of two cases. J Neurosurg 86:547–552.

15. Rades D, Fehlauer F (2002). Treatment options for central neurocytoma. Neurology 59:1268–1270.

16. Soylemezoglu F, Scheithauer BW, Esteve J, Kleihues P (1997). Atypical central neurocytoma. J Neuropathol Exp Neurol 56:551–556.

17. Brat DJ, Scheithauer BW, Eberhart CG, Burger PC (2001). Extraventricular neurocytomas: pathologic features and clinical outcome. Am J Surg Pathol 25:1252–1260.

18. Giangaspero F, Cenacchi G, Losi L, Cerasoli S, Bisceglia M, Burger PC (1997). Extraventricular neoplasms with neurocytoma features. A clinicopathological study of 11 cases. Am J Surg Pathol 21:206–212.

19. Hsu PW, Hsieh TC, Chang CN, Lin TK (2002). Fourth ventricle central neurocytoma: case report. Neurosurgery 50:1365–1367.

20. Chuang MT, Lin WC, Tsai HY, Liu GC, Hu SW, Chiang IC (2005). 3-T proton magnetic resonance spectroscopy of central neurocytoma: 3 case reports and review of the literature. J Comput Assist Tomogr 29:683–688.

21. Goergen SK, Gonzales MF, McLean CA (1992). Interventricular neurocytoma: radiologic features and review of the literature. Radiology 182:787–792.

22. Kiehl TR, Kalkanis SN, Louis DN (2004). Pigmented central neurocytoma. Acta Neuropathol (Berl) 107:571–574.

23. Martin AJ, Sharr MM, Teddy PJ, Gardner BP, Robinson SF (2002). Neurocytoma of the thoracic spinal cord. Acta Neurochir (Wien) 144:823–828.

24. Mineura K, Sasajima T, Itoh Y et al (1995). Blood flow and metabolism of central neurocytoma: a positron emission tomography study. Cancer 76:1224–1232.

25. Wichmann W, Schubiger O, von, DA, Schenker C, Valavanis A (1991). Neuroradiology of central neurocytoma. Neuroradiology 33:143–148.

26. Kanamori M, Kumabe T, Shimizu H, Yoshimoto T (2002). (201)Tl-SPECT, (1)H-MRS, and MIB-1 labeling index of central neurocytomas: three case reports. Acta Neurochir (Wien) 144:157–163.

27. Kim DG, Choe WJ, Chang KH et al (2000). In vivo proton magnetic resonance spectroscopy of central neurocytomas. Neurosurgery 46:329–333.

28. Moller-Hartmann W, Krings T, Brunn A, Korinth M, Thron A (2002). Proton magnetic resonance spectroscopy of neurocytoma outside the ventricular region – case report and review of the literature. Neuroradiology 44:230–234.

29. Ohtani T, Takahashi A, Honda F et al (2001). Central neurocytoma with unusually intense FDG uptake: case report. Ann Nucl Med 15:161–165.

30. Schild SE, Scheithauer BW, Haddock MG et al (1997). Central neurocytomas. Cancer 79:790–795.

31. Sgouros S, Carey M, Aluwihare N, Barber P, Jackowski A (1998). Central neurocytoma: a correlative clinicopathologic and radiologic analysis. Surg Neurol 49:197–204.

32. Anderson RC, Elder JB, Parsa AT, Issacson SR, Sisti MB (2001). Radiosurgery for the treatment of recurrent central neurocytomas. Neurosurgery 48:1231–1237.

33. Tyler-Kabara E, Kondziolka D, Flickinger JC, Lunsford LD (2001). Stereotactic radiosurgery for residual neurocytoma. Report of four cases. J Neurosurg 95:879–882.

34. Brandes AA, Gardiman M, Volpin L et al (2000). Chemotherapy in patients with recurrent and progressive central neurocytoma. Cancer 88:169–174.

35. Dodds D, Nonis J, Mehta M, Rampling R (1997). Central neurocytoma: a clinical study of response to chemotherapy. J Neuro-Oncol 34:279–283.

36. Louis DN, Swearingen B, Linggood RM et al (1990). Central nervous system neurocytoma and neuroblastoma in adults – report of eight cases. J Neuro-Oncol 9:231–238.

37. von Koch CS, Schmidt MH, Uyehara-Lock JH, Berger MS, Chang SM (2003). The role of PCV chemotherapy in the treatment of central neurocytoma: illustration of a case and review of the literature. Surg Neurol 60:560–565.

38. Nelson JS, Bruner JM, Wiestler OD, VandenBerg SR (2000). Ganglioglioma and gangliocytoma. In *Pathology & Genetics: Tumours of the Nervous System*. Kleihues P, Cavenee WK (eds). IARC Press, Lyon. 96–98.

39. Zentner J, Wolf HK, Ostertun B et al (1994). Gangliogliomas: clinical, radiological, and histopathological findings in 51 patients. J Neurol Neurosurg Psychiatr 57:1497–1502.

40. Hukin J, Siffert J, Velasquez L, Zagzag D, Allen J (2002). Leptomeningeal dissemination in children with progressive low-grade neuroepithelial tumors. Neuro-Oncol 4:253–260.

41. Jay V, Squire J, Blaser S, Hoffman HJ, Hwang P (1997). Intracranial and spinal metastases from a ganglioglioma with unusual cytogenetic abnormalities in a patient with complex partial seizures. Childs Nerv Syst 13:550–555.

42. Kurian NI, Nair S, Radhakrishnan VV (1998). Anaplastic ganglioglioma: case report and review of the literature. Br J Neurosurg 12:277–280.

43. Nakajima M, Kidooka M, Nakasu S (1998). Anaplastic ganglioglioma with dissemination to the spinal cord: a case report. Surg Neurol 49:445–448.

44. Rumana CS, Valadka AB (1998). Radiation therapy and malignant degeneration of benign supratentorial gangliogliomas. Neurosurgery 42:1038–1043.

45. Wacker MR, Cogen PH, Etzell JE, Daneshvar L, Davis RL, Prados MD (1992). Diffuse leptomeningeal involvement by a ganglioglioma in a child. Case report. J Neurosurg 77:302–306.

46. Krouwer HG, Davis RL, McDermott MW, Hoshino T, Prados MD (1993). Gangliogliomas: a clinicopathological study of 25 cases and review of the literature. J Neuro-Oncol 1:139–154.

47. Lang FF, Epstein FJ, Ransohoff J et al (1993). Central nervous system gangliogliomas. Part 2: Clinical outcome. J Neurosurg 79:867–873.

48. Castillo M, Davis PC, Takei Y, Hoffman JC Jr (1990). Intracranial ganglioglioma: MR, CT, and clinical findings in 18 patients. Am J Neuroradiol 11:109–114.

49. Im SH, Chung CK, Cho BK et al (2002). Intracranial ganglioglioma: preoperative characteristics and oncologic outcome after surgery. J Neuro-Oncol 59:173–183.

50. Kincaid PK, El-Saden SM, Park SH, Goy BW (1998). Cerebral gangliogliomas: preoperative grading using FDG-PET and 201Tl-SPECT. Am J Neuroradiol 19:801–806.

51. Law M, Meltzer DE, Wetzel SG et al (2004). Conventional MR imaging with simultaneous measurements of cerebral blood volume and vascular permeability in ganglioglioma. Magn Reson Imaging 22:599–606.

52. Matsumoto K, Tamiya T, Ono Y, Furuta T, Asari S, Ohmoto T (1999). Cerebral gangliogliomas: clinical characteristics, CT and MRI. Acta Neurochir (Wien) 141:135–141.

53. Selch MT, Goy BW, Lee SP et al (1998). Gangliogliomas: experience with 34 patients and review of the literature. Am J Clin Oncol 21:557–564.

54. Provenzale JM, Ali U, Barboriak DP, Kallmes DF, Delong DM, McLendon RE (2000). Comparison of patient age with MR imaging features of gangliogliomas. Am J Roentgenol 174:859–862.

55. Haddad SF, Moore SA, Menezes AH, VanGilder JC (1992). Ganglioglioma: 13 years of experience. Neurosurgery 31:171–178.

56. Lagares A, Gomez PA, Lobato RD, Ricoy JR, Ramos A, de la LA (2001). Ganglioglioma of the brainstem: report of three cases and review of the literature. Surg Neurol 56:315–322.

57. Athale S, Hallet KK, Jinkins JR (1999). Ganglioglioma of the trigeminal nerve: MRI. Neuroradiology 41:576–578.

58. Jaeger M, Hussein S, Schuhmann MU, Brandis A, Samii M, Blomer U (2001). Intraventricular trigonal ganglioglioma arising from the choroid plexus. Acta Neurochir (Wien) 143:953–955.

59. Kwon JW, Kim IO, Cheon JE et al (2001). Cerebellopontine angle ganglioglioma: MR findings. Am J Neuroradiol 22:1377–1379.

60. Siddique K, Zagardo M, Gujrati M, Olivero W (2002). Ganglioglioma presenting as a meningioma: case report and review of the literature. Neurosurgery 50:1133–1135.

61. Faillot T, Sichez JP, Capelle L, Kujas M, Fohanno D (1998). Ganglioglioma of the pineal region: case report and review of the literature. Surg Neurol 49:104–107.

62. Tender GC, Smith RD (2004). Pineal ganglioglioma in a young girl: a case report and review of the literature. J La State Med Soc 156:316–318.

63. Liu GT, Galetta SL, Rorke LB et al (1996). Gangliogliomas involving the optic chiasm. Neurology 46:1669–1673.

64. Pant I, Suri V, Chaturvedi S, Dua R, Kanodia AK (2006). Ganglioglioma of optic chiasma: case report and review of literature. Childs Nerv Syst 22:1–4.

65. Majos C, Aguilera C, Ferrer I, Lopez L, Pons LC (1998). Intraventricular ganglioglioma: case report. Neuroradiology 40:377–379.

66. Koeller KK, Henry JM (2001). From the archives of the AFIP: superficial gliomas: radiologic-pathologic correlation. Armed Forces Institute of Pathology. Radiographics 21:1533–1556.

67. Tien RD, Tuori SL, Pulkingham N, Burger PC (1992). Ganglioglioma with leptomeningeal and subarachnoid spread: results of CT, MR, and PET imaging. Am J Roentgenol 159:391–393.

68. Bock D, Rummele P, Friedrich M, Wolff JE (2002). Multifocal desmoplastic astrocytoma, frontal lobe dysplasia, and simian crease. J Pediatr 141:445.

69. Wang Z, Sutton LN, Cnaan A et al (1995). Proton MR spectroscopy of pediatric cerebellar tumors. Am J Neuroradiol 16:1821–1833.

70. Kalyan-Raman UP, Olivero WC (1987). Ganglioglioma: a correlative clinicopathological and radiological study of ten surgically treated cases with follow-up. Neurosurgery 20: 428–433.

71. Provenzale JM, Arata MA, Turkington TG, McLendon RE, Coleman RE (1999). Gangliogliomas: characterization by registered positron emission tomography-MR images. Am J Roentgenol 172:1103–1107.

72. Rosenberg DS, Demarquay G, Jouvet A et al (2005). {11C}-Methionine PET: dysembryoplastic neuroepithelial tumours compared with other epileptogenic brain neoplasms. J Neurol Neurosurg Psychiatr 76:1686–1692.

73. Patel U, Pinto RS, Miller DC et al (1998). MR of spinal cord ganglioglioma. Am J Neuroradiol 19:879–887.

74. Kim HS, Lee HK, Jeong AK, Shin JH, Choi CG, Khang SK (2001). Supratentorial gangliocytoma mimicking extra-axial tumor: a report of two cases. Korean J Radiol 2:108–112.

75. Beal MF, Kleinman GM, Ojemann RG, Hochberg FH (1981). Gangliocytoma of third ventricle: hyperphagia, somnolence, and dementia. Neurology 31:1224–1228.

76. Ebina K, Suzuki S, Takahashi T, Iwabuchi T, Takei Y (1985). Gangliocytoma of the pineal body. A case report and review of the literature. Acta Neurochir (Wien) 74:134–140.

77. Furie DM, Felsberg GJ, Tien RD, Friedman HS, Fuchs H, McLendon R (1993). MRI of gangliocytoma of cerebellum and spinal cord. J Comput Assist Tomogr 17:488–491.

78. Geddes JF, Jansen GH, Robinson SF et al (2000). 'Gangliocytomas' of the pituitary: a heterogeneous group of lesions with differing histogenesis. Am J Surg Pathol 24:607–613.

79. Peretti-Viton P, Perez-Castillo AM, Raybaud C et al (1991). Magnetic resonance imaging in gangliogliomas and gangliocytomas of the nervous system. J Neuroradiol 18:189–199.

80. Russo CP, Katz DS, Corona RJ Jr, Winfield JA (1995). Gangliocytoma of the cervicothoracic spinal cord. Am J Neuroradiol 16:889–891.

81. Jacob JT, Cohen-Gadol AA, Scheithauer BW, Krauss WE (2005). Intramedullary spinal cord gangliocytoma: case report and a review of the literature. Neurosurg Rev 28:326–329.

82. Luyken C, Blumcke I, Fimmers R, Urbach H, Wiestler OD, Schramm J (2004). Supratentorial gangliogliomas: histopathologic grading and tumor recurrence in 184 patients with a median follow-up of 8 years. Cancer 101:146–155.

83. Celli P, Scarpinati M, Nardacci B, Cervoni L, Cantore GP (1993). Gangliogliomas of the cerebral hemispheres. Report of 14 cases with long-term follow-up and review of the literature. Acta Neurochir (Wien) 125:52–57.

84. Silver JM, Rawlings CE III, Rossitch E Jr, Zeidman SM, Friedman AH (1991). Ganglioglioma: a clinical study with long-term follow-up. Surg Neurol 35:261–266.

85. Dash RC, Provenzale JM, McComb RD, Perry DA, Longee DC, McLendon RE (1999). Malignant supratentorial ganglioglioma (ganglion cell-giant cell glioblastoma): a case report and review of the literature. Arch Pathol Lab Med 123:342–345.

86. Hakim R, Loeffler JS, Anthony DC, Black PM (1997). Gangliogliomas in adults. Cancer 79:127–131.

87. Hassall TE, Mitchell AE, Ashley DM (2001). Carboplatin chemotherapy for progressive intramedullary spinal cord low-grade gliomas in children: three case studies and a review of the literature. Neuro-oncology 3:251–257.

88. Hirose T, Schneithauer BW, Lopes MB, Gerber HA, Altermatt HJ, VandenBerg SR (1997). Ganglioglioma: an ultrastructural and immunohistochemical study. Cancer 79: 989–1003.

89. Jay V, Squire J, Becker LE, Humphreys R (1994). Malignant transformation in a ganglioglioma with anaplastic neuronal and astrocytic components. Report of a case with flow cytometric and cytogenetic analysis. Cancer 73:2862–2868.

90. Johnson JH Jr, Hariharan S, Berman J et al (1997). Clinical outcome of pediatric gangliogliomas: ninety-nine cases over 20 years. Pediatr Neurosurg 27:203–207.

91. Kaba SE, Langford LA, Yung WK, Kyritsis AP (1996). Resolution of recurrent malignant ganglioglioma after treatment with cis-retinoic acid. J Neuro-Oncol 30:55–60.

92. Prados MD, Edwards MS, Rabbitt J, Lamborn K, Davis RL, Levin VA (1997). Treatment of pediatric low-grade gliomas with a nitrosourea-based multiagent chemotherapy regimen. J Neuro-Oncol 32:235–241.

93. Sasaki A, Hirato J, Nakazato Y, Tamura M, Kadowaki H (1996). Recurrent anaplastic ganglioglioma: pathological characterization of tumor cells. Case report. J Neurosurg 84:1055–1059.

94. Tamburrini G, Colosimo C Jr, Giangaspero F, Riccardi R, Di RC (2003). Desmoplastic infantile ganglioglioma. Childs Nerv Syst 19:292–297.

95. Taranath A, Lam A, Wong CK (2005). Desmoplastic infantile ganglioglioma: a questionably benign tumour. Australas Radiol 49:433–437.

96. De MK, Van GS, Van CF et al (2002). Desmoplastic infantile ganglioglioma: a potentially malignant tumor? Am J Surg Pathol 26:1515–1522.

97. Nikas I, Anagnostara A, Theophanopoulou M, Stefanaki K, Michail A, Hadjigeorgi C (2004). Desmoplastic infantile ganglioglioma: MRI and histological findings case report. Neuroradiology 46:1039–1043.

98. Setty SN, Miller DC, Camras L, Charbel F, Schmidt ML (1997). Desmoplastic infantile astrocytoma with metastases at presentation. Mod Pathol 10:945–951.

99. Martin DS, Levy B, Awwad EE, Pittman T (1991). Desmoplastic infantile ganglioglioma: CT and MR features. Am J Neuroradiol 12:1195–1197.

100. Serra A, Strain J, Ruyle S (1996). Desmoplastic cerebral astrocytoma of infancy: report and review of the imaging characteristics. Am J Roentgenol 166:1459–1461.

101. Trehan G, Bruge H, Vinchon M et al (2004). MR imaging in the diagnosis of desmoplastic infantile tumor: retrospective study of six cases. Am J Neuroradiol 25:1028–1033.

102. Sugiyama K, Arita K, Shima T et al (2002). Good clinical course in infants with desmoplastic cerebral neuroepithelial tumor treated by surgery alone. J Neuro-Oncol 59:63–69.

103. Sperner J, Gottschalk J, Neumann K, Schorner W, Lanksch WR, Scheffner D (1994). Clinical, radiological and histological findings in desmoplastic infantile ganglioglioma. Childs Nerv Syst 10:458–462.

104. Mallucci C, Lellouch-Tubiana A, Salazar C et al (2000). The management of desmoplastic neuroepithelial tumours in childhood. Childs Nerv Syst 16:8–14.

105. Takeshima H, Kawahara Y, Hirano H, Obara S, Niiro M, Kuratsu J (2003). Postoperative regression of desmoplastic

infantile gangliogliomas: report of two cases. Neurosurgery 53:979–983.

106. Bachli H, Avoledo P, Gratzl O, Tolnay M (2003). Therapeutic strategies and management of desmoplastic infantile ganglioglioma: two case reports and literature overview. Childs Nerv Syst 19:359–366.

107. Fan X, Larson TC, Jennings MT, Tulipan NB, Toms SA, Johnson MD (2001). December 2000: 6 month old boy with 2 week history of progressive lethargy. Brain Pathol 11: 265–266.

108. Hammond RR, Duggal N, Woulfe JM, Girvin JP (2000). Malignant transformation of a dysembryoplastic neuroepithelial tumor. Case report. J Neurosurg 92:722–725.

109. Rushing EJ, Thompson LD, Mena H (2003). Malignant transformation of a dysembryoplastic neuroepithelial tumor after radiation and chemotherapy. Ann Diagn Pathol 7:240–244.

110. Fernandez C, Girard N, Paz PA, Bouvier-Labit C, Lena G, Figarella-Branger D (2003). The usefulness of MR imaging in the diagnosis of dysembryoplastic neuroepithelial tumor in children: a study of 14 cases. Am J Neuroradiol. 24: 829–834.

111. Lee DY, Chung CK, Hwang YS et al (2000). Dysembryoplastic neuroepithelial tumor: radiological findings (including PET, SPECT, and MRS) and surgical strategy. J Neuro-Oncol 47:167–174.

112. Ostertun B, Wolf HK, Campos MG et al (1996). Dysembryoplastic neuroepithelial tumors: MR and CT evaluation. Am J Neuroradiol 17:419–430.

113. Leung SY, Gwi E, Ng HK, Fung CF, Yam KY (1994). Dysembryoplastic neuroepithelial tumor. A tumor with small neuronal cells resembling oligodendroglioma. Am J Surg Pathol 18:604–614.

114. Whittle IR, Dow GR, Lammie GA, Wardlaw J (1999). Dsyembryoplastic neuroepithelial tumour with discrete bilateral multifocality: further evidence for a germinal origin. Br J Neurosurg 13:508–511.

115. Kuchelmeister K, Demirel T, Schlorer E, Bergmann M, Gullotta F (1995). Dysembryoplastic neuroepithelial tumour of the cerebellum. Acta Neuropathol (Berl) 89:385–390.

116. Cervera-Pierot P, Varlet P, Chodkiewicz JP, Dumas-Duport C (1997). Dysembryoplastic neuroepithelial tumors located in the caudate nucleus area: report of four cases. Neurosurgery 40:1065–1069.

117. Guesmi H, Houtteville JP, Courtheoux P, Derlon JM, Chapon F (1999). Dysembryoplastic neuroepithelial tumors. Report of 8 cases including two with unusual localization. Neurochirurgie 45:190–200.

118. Yasha TC, Mohanty A, Radhesh S, Santosh V, Das S, Shankar SK (1998). Infratentorial dysembryoplastic neuroepithelial tumor (DNT) associated with Arnold-Chiari malformation. Clin Neuropathol 17:305–310.

119. Baisden BL, Brat DJ, Melhem ER, Rosenblum MK, King AP, Burger PC (2001). Dysembryoplastic neuroepithelial tumor-like neoplasm of the septum pellucidum: a lesion often misdiagnosed as glioma: report of 10 cases. Am J Surg Pathol 25:494–499.

120. Dumas-Duport C, Varlet P, Bacha S, Beuvon F, Cervera-Pierot P, Chodkiewicz JP (1999). Dysembryoplastic neuroepithelial tumors: nonspecific histological forms – a study of 40 cases. J Neuro-Oncol 41:267–280.

121. Dumas-Duport C (1993). Dysembryoplastic neuroepithelial tumours. Brain Pathol 3:283–295.

122. Kuroiwa T, Bergey GK, Rothman MI et al (1995). Radiologic appearance of the dysembryoplastic neuroepithelial tumor. Radiology 197:233–238.

123. Jensen RL, Caamano E, Jensen EM, Couldwell WT (2005). Development of contrast enhancement after long-term observation of a dysembryoplastic neuroepithelial tumor. J Neuro-Oncol 78:1–4.

124. Bulakbasi N, Kocaoglu M, Ors F, Tayfun C, Ucoz T (2003). Combination of single-voxel proton MR spectroscopy and apparent diffusion coefficient calculation in the evaluation of common brain tumors. Am J Neuroradiol 24:225–233.

125. Vuori K, Kankaanranta L, Hakkinen AM et al (2004). Low-grade gliomas and focal cortical developmental malformations: differentiation with proton MR spectroscopy. Radiology 230:703–708.

126. Kaplan AM, Lawson MA, Spataro J et al (1999). Positron emission tomography using [18F] fluorodeoxyglucose and [11C] l-methionine to metabolically characterize dysembryoplastic neuroepithelial tumors. J Child Neurol 14:673–677.

127. Abe M, Tabuchi K, Tsuji T, Shiraishi T, Koga H, Takagi M (1995). Dysembryoplastic neuroepithelial tumor: report of three cases. Surg Neurol 43:240–245.

128. Raymond AA, Halpin SF, Alsanjari N et al (1994). Dysembryoplastic neuroepithelial tumor. Features in 16 patients. Brain 117 (Pt 3):461–475.

129. Nowak DA, Trost HA (2002). Lhermitte-Duclos disease (dysplastic cerebellar gangliocytoma): a malformation, hamartoma or neoplasm? Acta Neurol Scand 105:137–145.

130. Wiestler OD, Padberg GW, Steck PA (2000). Cowden disease and dysplastic gangliocytoma of the cerebellum/Lhermitte-Duclos disease. In *Pathology & Genetics: Tumours of the Nervous System*. Kleihues P, Cavenee WK (eds). IARC Press, Lyon. 235–237.

131. Kulkantrakorn K, Awwad EE, Levy B et al (1997). MRI in Lhermitte-Duclos disease. Neurology 48:725–731.

132. Spaargaren L, Cras P, Bomhof MA et al (2003). Contrast enhancement in Lhermitte-Duclos disease of the cerebellum: correlation of imaging with neuropathology in two cases. Neuroradiology 45:381–385.

133. Williams DW III, Elster AD, Ginsberg LE, Stanton C (1992). Recurrent Lhermitte-Duclos disease: report of two cases and association with Cowden's disease. Am J Neuroradiol 13:287–290.

134. Murray C, Shipman P, Khangure M et al (2001). Lhermitte-Duclos disease associated with Cowden's syndrome: case report and literature review. Australas Radiol 45:343–346.

135. Lok C, Viseux V, Avril MF et al (2005). Brain magnetic resonance imaging in patients with Cowden syndrome. Medicine (Baltimore) 84:129–136.

136. Nagaraja S, Powell T, Griffiths PD, Wilkinson ID (2004). MR imaging and spectroscopy in Lhermitte-Duclos disease. Neuroradiology 46:355–358.

137. Robinson S, Cohen AR (2000). Cowden disease and Lhermitte-Duclos disease: characterization of a new phakomatosis. Neurosurgery 46:371–383.

138. Vantomme N, Van CF, Goffin J, Sciot R, Demaerel P, Plets C (2001). Lhermitte-Duclos disease is a clinical manifestation of Cowden's syndrome. Surg Neurol 56:201–204.

139. Derrey S, Proust F, Debono B et al (2004). Association between Cowden syndrome and Lhermitte-Duclos disease: report of two cases and review of the literature. Surg Neurol 61:447–454.

140. Awwad EE, Levy E, Martin DS, Merenda GO (1995). Atypical MR appearance of Lhermitte-Duclos disease with contrast enhancement. Am J Neuroradiol 16:1719–1720.

141. Meltzer CC, Smirniotopoulos JG, Jones RV (1995). The striated cerebellum: an MR imaging sign in Lhermitte-Duclos disease (dysplastic gangliocytoma). Radiology 194:699–703.

142. Moonis G, Ibrahim M, Melhem ER (2004). Diffusion-weighted MRI in Lhermitte-Duclos disease: report of two cases. Neuroradiology 46:351–354.

143. Klisch J, Juengling F, Spreer J et al (2001). Lhermitte-Duclos disease: assessment with MR imaging, positron emission tomography, single-photon emission CT, and MR spectroscopy. Am J Neuroradiol 22:824–830.

144. Ogasawara K, Beppu T, Yasuda S, Kobayashi M, Yukawa H, Ogawa A (2001). Blood flow and oxygen metabolism in a case of Lhermitte-Duclos disease: results of positron emission tomography. J Neuro-Oncol 55:59–61.

145. Pirotte B, Goldman S, Baleriaux D, Brotchi J (2002). Fluorodeoxyglucose and methionine uptake in Lhermitte-Duclos disease: case report. Neurosurgery 50:404–407.

146. Ogasawara K, Yasuda S, Beppu T et al (2001). Brain PET and technetium-99m-ECD SPECT imaging in Lhermitte-Duclos disease. Neuroradiology 43:993–996.

147. Alkadhi H, Keller M, Brandner S, Yonekawa Y, Kollias SS (2001). Neuroimaging of cerebellar liponeurocytoma. Case report. J Neurosurg 95:324–331.

148. Kleihues P, Chimelli L, Giangaspero F Cerebellar liponeurocytoma. In *Pathology & Genetics: Tumours of the Nervous System*. Kleihues P, Cavenee WK (eds). IARC Press, Lyon. 110–111.

149. Jackson TR, Regine WF, Wilson D, Davis DG (2001). Cerebellar liponeurocytoma. Case report and review of the literature. J Neurosurg 95:700–703.

150. Kuchelmeister K, Nestler U, Siekmann R, Schachenmayr W (2006). Liponeurocytoma of the left lateral ventricle – case report and review of the literature. Clin Neuropathol 25:86–94.

151. Cacciola F, Conti R, Taddei GL, Buccoliero AM, Di LN (2002). Cerebellar liponeurocytoma. Case report with considerations on prognosis and management. Acta Neurochir (Wien) 144:829–833.

152. Jenkinson MD, Bosma JJ, Du PD et al (2003). Cerebellar liponeurocytoma with an unusually aggressive clinical course: case report. Neurosurgery 53:1425–1427.

CHAPTER 44

Pineal Region Tumors

V. Michelle Silvera

INTRODUCTION

Preoperative imaging of pineal region tumors may suggest a histologic diagnosis based on signal characteristics, enhancement pattern and morphologic features, but only a few of these tumors have specific imaging findings. This lack of pathognomonic imaging features limits the ability of the radiologist to arrive at a reliable preoperative diagnosis. The additional challenge of accessing pineal region tumors for the purpose of histologic sampling adds to the difficulty of diagnosing and managing pathology in the pineal region.

Most neoplasms encountered in the pineal region are somewhat unique to this area and, as such, the clinician should be able to produce a short list of diagnostic possibilities aided by the knowledge of the clinical presentation, age and gender of the patient, radiological imaging characteristics and the presence or absence of biological markers (α-fetoprotein (AFP), (β-human chorionic gonadotropin (β-HCG) and placental alkaline phosphatase (PLAP)). Ultimately however, as the aforementioned features demonstrate variable overlap, histologic sampling is usually necessary for obtaining a definitive diagnosis.

ANATOMY AND FUNCTION

The pineal gland is a midline, ovoid structure that measures approximately $8 \times 4\,mm$ in size and is located within the quadrigeminal plate cistern, where it lies between the two superior colliculi of the tectum. The gland arises from the posterior wall of the third ventricle, to which it is attached by the pineal stalk. The pineal stalk consists of a superior lamina, which contains the habenular commissure, and an inferior lamina, which contains the posterior commissure. The two laminae are separated by the pineal recess of the third ventricle. The internal cerebral veins, the tela choroidea of the third ventricle and the splenium of the corpus callosum are located superior to the pineal gland.

The pineal gland is predominantly made up of pineolocytes, which are specialized neuroepithelial cells with cytoplasmic processes. A small portion of the pineal gland consists of astrocyte-like cells.

The gland plays a role in the regulation of circadian biologic rhythms and cyclic biologic responses to environmental changes and regulation of certain circulating hormone levels, the most significant of which are serotonin, melatonin and pineal peptides.

Diverse processes such as development and growth, gonadal function, sleep, body temperature, motor activity and aging are influenced by the pineal gland [1]. Despite the multitude of biological systems in which the pineal gland participates, the destruction of the gland by tumor, treatment or resection, rarely produces measurable effect.

EPIDEMIOLOGY AND PATHOLOGIC CLASSIFICATION

Tumors of the pineal region are relatively uncommon and account for approximately 3–8% of brain tumors in children and 0.4–1% of brain tumors in patients of all ages [2,3]. Interestingly, the occurrence of pineal masses is notably higher in Asia, where the incidence is 10–12.5% of all pediatric brain tumors and 2.2–8% of brain tumors overall [4].

Pineal region tumors can be divided into three categories: germ cell tumors; pineal parenchymal tumors and tumors of supporting tissues of the pineal gland or adjacent structures.

The most common tumors in the pineal region are the germ cell tumors, of which 65% are comprised of pure germinomas. The remainder of germ cell tumors is represented by either non-germinomatous germ cell tumors (NGGCT) such as embryonal carcinoma, endodermal sinus tumor (yolk sac tumor), choriocarcinoma and teratoma or mixed germ cell tumors.

Tumors arising from pineal parenchymal cells account for 5–20% of all pineal region tumors. These tumors can be divided into pineoblastomas, pineocytomas and mixed pineocytoma/pineoblastomas.

The remainder of the pineal region tumors arise from the supporting cells within the pineal gland or from the surrounding structures. These include astrocytomas, ependymomas, choroid plexus papillomas, primitive neuroectodermal tumors (PNETs), lymphomas, gangliogliomas, dermoid/epidermoid cysts, meningiomas and metastases.

CLINICAL PRESENTATION

The most common clinical findings in patients with pineal region tumors are related to increased intracranial pressure secondary to compression of the sylvian aqueduct and hydrocephalus. These symptoms include headaches, nausea and vomiting, drowsiness, blurred and double vision, sixth cranial nerve palsy and papilledema. Ophthalmologic signs such as impairment of upward gaze, also known as Parinaud's syndrome, as well as dilated pupils unreactive to light but responsive to accommodation, convergence retractory nystagmus (rhythmic convergence of the eyes and retraction of the eyes into the orbits, usually elicited by attempting to look upward) are seen with compression or invasion of the adjacent midbrain. Less common symptoms include hemiparesis, third nerve palsy, ataxia, precocious puberty, hypogonadism and seizures. Diabetes insipidus may occur if the floor of the third ventricle or the hypothalamus is involved by tumor.

IMAGING

Pineal Calcification

Calcification of the pineal gland is rare under the age of 6 and occurs in fewer than 11% of children between the ages of 11 and 14. Forty percent of the normal population has pineal calcifications by the age of 20 [5]. Normal pineal calcification has been described as compact, sand-like or granular in appearance and less than 1 cm in size (Figure 44.1).

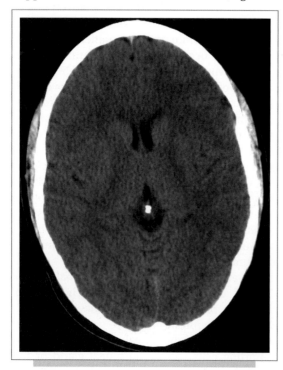

FIG. 44.1. Normal pineal gland. CT scan shows calcification in a normal pineal gland. The calcification is 3 mm in size and compact.

Pineal Region Germ Cell Tumors

The most common tumors of the pineal region are germ cell tumors. These are non-neuroectodermal tumors. The origin of these tumors is unknown, but the tumor cells are believed to arise from fetal germinal cells, similar to gonadal germ cells, which migrated to the fetal central nervous system early in development and persisted [6]. Germ cell tumors are classified by cell type and may be benign or malignant.

Germinoma

Germinomas are the most common tumor of the pineal region. These account for two-thirds of germ cell tumors and 40% of all pineal region tumors [7]. Germinomas are histologically identical to testicular seminomas and ovarian dysgerminomas and are histologically characterized by large multipotential primitive germ cells and cells that resemble lymphocytes. Males are far more commonly affected than females and the peak age of presentation is within the second and third decades of life. Germinomas per se are not specific to the pineal region and can be found as primary tumors within the suprasellar region, periventricular white matter and within the basal ganglia.

On computed tomography (CT), germinomas are typically well defined, hyperdense and demonstrate homogeneous enhancement. Tumoral calcification is variably seen (Figure 44.2). A calcified pineal gland engulfed by

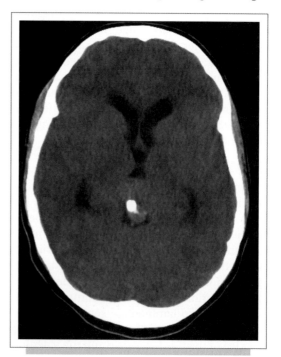

FIG. 44.2. Pineal germinoma. Non-enhanced CT scan demonstrates a mass in the region of the pineal gland, extending into the posterior aspect of the third ventricle. The calcified pineal gland is displaced to the right by tumor. The tumor substance is homogeneous in appearance and of slightly higher attenuation than brain. Hydrocephalus is present.

non-calcified tumor has been reported to be a specific imaging appearance of germinomas (Figure 44.3) [8]. On magnetic resonance (MR) images, germinomas tend to be homogeneous in signal characteristics with usually few if any small tumoral cysts. The enhancement pattern is generally homogeneous and intense. The tumor substance typically approximates that of gray matter on both short and long time to recovery/echo time (TR/TE) images (Figure 44.4) [9].

The growth vector of tumor is usually in a ventral direction, with encroachment or extension into the posterior aspect of the third ventricle (Figure 44.5). Invasion into the adjacent thalami or midbrain is not infrequently seen (Figure 44.6) [10]. Leptomeningeal dissemination of tumor into the intracranial or spinal compartment is frequently seen [8]. In the series by Maity et al, 36% of patients had spinal dissemination of tumor at the time of diagnosis [11].

Non-germinomatous Germ Cell Tumors

These tumors, which include embryonal carcinomas, yolk sac tumors, choriocarcinomas, teratomas and mixed germ cell tumors, tend to be more heterogeneous in signal characteristics and enhancement pattern than pure germinomas [12,13]. Small intratumoral cysts and tumoral calcifications are more commonly seen (Figures 44.7, 44.8 and 44.9). Overall, however, the appearance of these tumors is non-specific.

Teratomas are the second most common pineal region masses and account for 15% of all pineal region tumors [7]. These tumors are much more common in males. The five teratomas reported by Ganti et al were all male [8]. In their series, all teratomas were heterogeneous in appearance with cysts, calcifications and variable enhancement. Eighty percent demonstrated calcification and the presence of a fatty component was found to be specific (Figure 44.10).

Choriocarcinomas are rare and their imaging appearance is non-specific. However, the presence of intratumoral hemorrhage has been reported to be suggestive of a choriocarcinoma [9].

No specific radiologic features have been described for yolk sac tumors or mixed germ cell tumors.

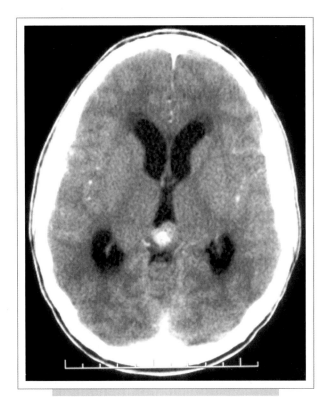

FIG. 44.3. Pineal germinoma. Enhanced CT scan shows a homogeneously enhancing, non-calcified mass which surrounds the anterior aspect of the calcified pineal gland.

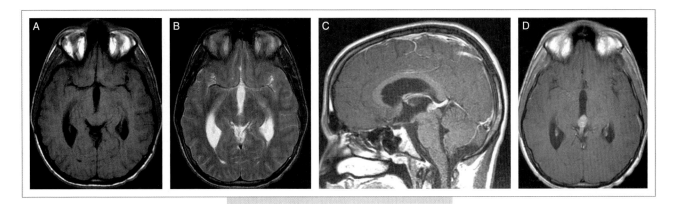

FIG. 44.4. Pineal germinoma. **(A)** Axial T1-weighted MR image demonstrates an isointense pineal region mass. **(B)** Axial T2-weighted MR image shows the mass to be homogeneously hypointense. **(C)** Sagittal T1-weighted post-contrast MR image shows homogeneous enhancement of the pineal mass, which compresses the tectum. **(D)** Axial T1-weighted post-contrast image shows homogeneous enhancement of the pineal region mass.

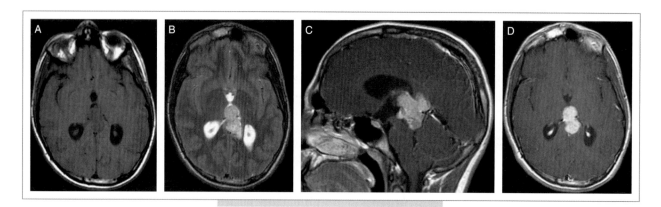

FIG. 44.5. Pineal germinoma. **(A)** Axial T1-weighted MR image demonstrates a homogeneous, hypointense pineal region tumor which extends into the posterior aspect of the third ventricle. **(B)** Axial T2-weighted MR image shows a homogeneous appearing hypointense mass. **(C)** Sagittal T1-weighted post-contrast MR image shows a lobulated, mildly heterogeneous enhancing mass that compresses the midbrain. **(D)** Axial T1-weighted post-contrast MR image shows mildly heterogeneous enhancement of the pineal germinoma.

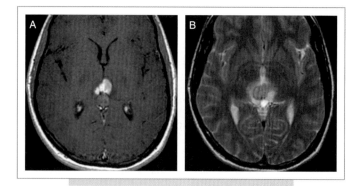

FIG. 44.6. Pineal germinoma. **(A)** Axial T1-weighted post-contrast MR image shows a circumscribed, intensely enhancing pineal mass with edema within the adjacent thalami, suggestive of parenchymal invasion. **(B)** Axial T2-weighted MR image shows the tumor to be hypointense relative to brain. Edema is seen within the adjacent thalami.

Pineal Parenchymal Tumors

Pineal parenchymal tumors account for less than 15% of pineal region masses. These tumors are subdivided into pineocytomas, pineoblastomas and mixed pineocytoma/pineoblastomas. There is no gender predilection [14].

Pineocytomas

Pineocytomas are well differentiated tumors that retain the morphological and immunohistochemical features of pineal parenchymal cells. Lobular architecture and pineocytomatous rosettes are typical features. The pineocytoma is a circumscribed though unencapsulated, slow growing tumor, which does not usually seed the CSF pathways.

In the series of pineal parenchymal tumors by Chiechi et al, 10 patients were found to have pineocytomas, eight

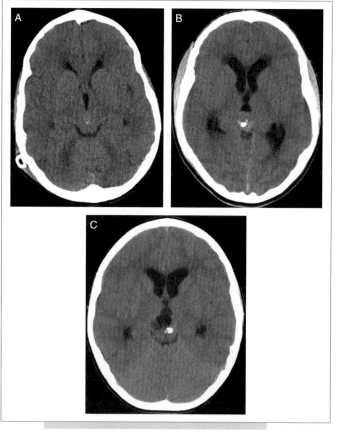

FIG. 44.7. Non-germinomatous germ cell tumor in three different patients. **(A)** Non-enhanced CT scan shows a mildly hyperdense pineal tumor with punctuate tumoral calcifications. **(B)** Non-enhanced CT scan shows a heterogeneous pineal tumor with heterogeneous tumoral calcification. **(C)** Non-enhanced CT scan shows a heterogeneous, partially cystic tumor with calcification.

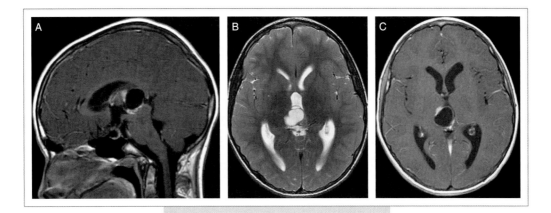

FIG. 44.8. Non-germinomatous germ cell tumor. **(A)** Sagittal T1-weighted MR image shows a partially cystic and partially enhancing pineal tumor. **(B)** Axial T2-weighted MR image shows a partially cystic, heterogeneous appearing tumor. The solid component of tumor is hypointense. **(C)** Axial T1-weighted post-contrast MR image shows the partially cystic and partially solid heterogeneously enhancing tumor.

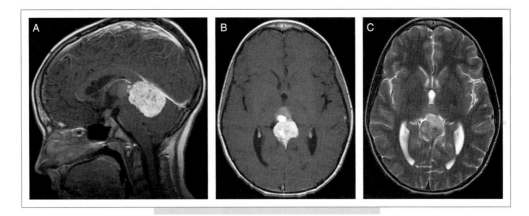

FIG. 44.9. Non-germinomatous germ cell tumor with seeding. **(A)** Sagittal T1-weighted post-contrast MR image shows a very large, very heterogeneous pineal region tumor. The different components of the tumor demonstrate variable enhancement. This is suggestive of an NGGCT with several different tumoral elements. Subarachnoid seeding is observed within the suprasellar cistern. **(B)** Axial T1-weighted post-contrast MR image again demonstrates the variable enhancement of the solid components of tumor within the mass. **(C)** Axial T2-weighted image shows variable hypointensity of the solid tumor components.

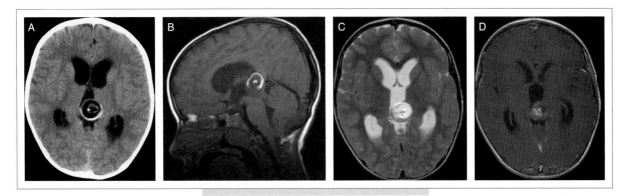

FIG. 44.10. Teratoma. **(A)** Non-enhanced CT scan shows a heterogeneous mass with tumoral calcification and a fatty component. **(B)** Sagittal T1-weighted MR image without contrast shows the fatty component in an arc-like configuration. The calcific component is less well appreciated. **(C)** Axial T2-weighted MR image shows the fatty component to be associated with chemical shift artifact. **(D)** Axial T1-weighted post-contrast MR image demonstrates mild tumoral enhancement.

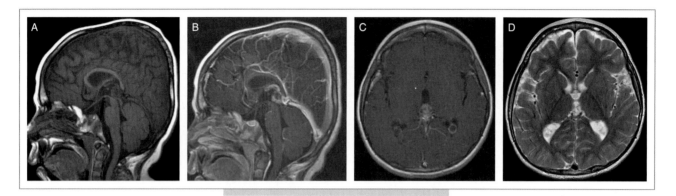

FIG. 44.11. Pineocytoma. **(A)** Sagittal T1-weighted MR image without contrast shows a heterogeneous, hypointense pineal region mass. **(B)** Sagittal T1-weighted post-contrast MR image shows heterogeneous enhancement. **(C)** Axial T1-weighted post-contrast MR image shows heterogeneous enhancement. **(D)** Axial T2-weighted MR image shows a relatively hyperintense pineal mass.

of which were in adult patients and two of which were in children [15]. All pineocytomas in their series were solid in appearance with either homogeneous or heterogeneous signal characteristics with the exception of one pineocytoma, which simulated a pineal cyst. In contrast, in a series of 13 cystic pineal region masses with histologic correlation by Engel et al, six of these cystic pineal masses were proven to be pineocytomas [16]. All patients in this series were under the age of 10.

Solid pineocytomas are typically hypodense on CT and demonstrate various enhancement patterns, either homogeneous or heterogeneous [15].

On magnetic resonance imaging (MRI), solid pineocytomas show low, iso- or mixed low and isointense signal relative to gray matter on short TR/TE images and more commonly high signal on long TR/TE images. The enhancement pattern is either homogeneous or heterogeneous and the tumor margins are usually well defined (Figure 44.11). Tumoral calcification is common [9]. In Engel's series of cystic pineocytomas, all pineocytomas were well defined, measured 8–20 mm in diameter, demonstrated enhancing walls and were indistinguishable from pineal cysts.

Pineoblastomas

Pineoblastomas are embryonal tumors that resemble primitive neuroectodermal tumors (PNET). Pineoblastomas are distinct from PNETs in other sites due to their photosensory differentiation. Pineoblastomas are malignant unencapsulated tumors, which frequently invade the adjacent brain parenchyma and seed the CSF pathway.

On CT, pineoblastomas are typically hyperdense, show variable calcification and are either homogeneous or heterogeous in appearance (Figure 44.12). In the series of pineoblastomas reported by Chiechi et al, four of the eight patients with pineoblastomas had CTs available and none of these demonstrated tumoral calcification [15].

The 'exploded pineal pattern' of calcification has been described in both pineocytomas and pineoblastomas

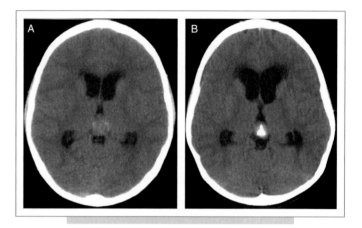

FIG. 44.12. Pineoblastoma, two different patients. **(A)** Axial non-contrast CT image demonstrates punctuate calcification within the tumor substance. **(B)** Axial non-contrast CT image demonstrates calcification with hyperdense tumor.

(Figure 44.13). The pattern describes the peripheral displacement of native pineal gland calcifications within tumor that is more typical of pineal parenchymal tumors than of germ cell tumors [9].

MRI generally shows low to isointense signal to gray matter on short TR/TE images and variable signal with low, high and mixed signal seen on long TR/TE images. When compared to pineocytomas, pineoblastomas are usually slightly larger, their contours more frequently lobulated and their enhancement pattern more homogeneous (Figures 44.14 and 44.15). These tumors are less frequently calcified than pineocytomas [15].

Patients with mixed pineocytoma/pineoblastoma tumors show a mixture of the imaging features mentioned above.

Pineal Astrocytomas

These tumors are thought to arise from the pineal interstitial cells and may be difficult to discern from astrocytomas

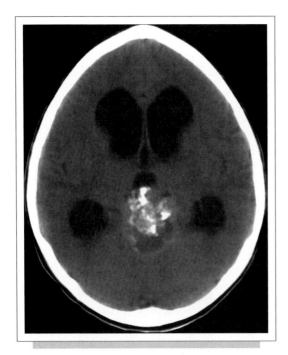

FIG. 44.13. Pineoblastoma. Non-contrast CT demonstrates the 'exploded pineal pattern' with peripheral displacement of native pineal calcifications within the tumor substance.

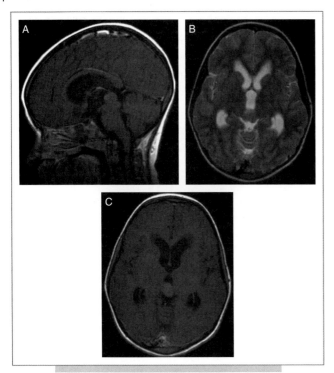

FIG. 44.14. Pineoblastoma. **(A)** Sagittal T1-weighted post-contrast MR image shows mildly heterogeneous enhancement of the pineal region tumor. **(B)** Axial T2-weighted MR image shows markedly hypointense tumor substance. **(C)** Axial T1-weighted post-contrast MR image shows a circumscribed tumor with mild enhancement.

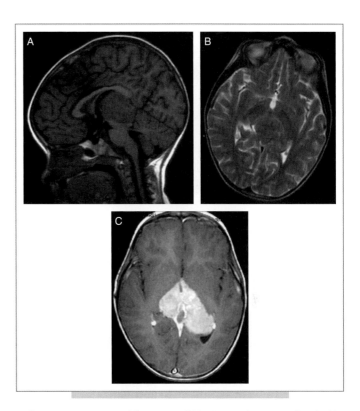

FIG. 44.15. Pineoblastoma. **(A)** Sagittal T1-weighted MR image shows a very large pineal region tumor with marked mass effect. **(B)** Axial T2-weighted MR image shows markedly hypointense tumor substance. **(C)** Axial T1-weighted post-contrast MR image shows a large, lobulated intensely enhancing mass.

that arise from the parapineal structures, such as the midbrain, thalami or splenium, which then extend secondarily into the pineal region.

Other Tumors

Other tumors may occur in the pineal region, such as ependymomas (Figure 44.16), PNETs (Figure 44.17), meningiomas, dermoid/epidermoid cysts and metastases. Dermoid cysts are circumscribed, non-enhancing masses, which may contain fat. Epidermoid cysts typically demonstrate striking restricted diffusion and meningiomas are characterized by a dural attachment to the free edge of the tentorium. Non-neoplastic masses include vascular malformations such as cavernous malformations and vein of Galen aneurysms (Figure 44.18), arachnoid cysts (Figure 44.19) and congenital lesions such as lipomas (Figure 44.20).

Pineal Cyst

Pineal cysts are common incidental findings in the normal population. Cysts can be seen in up to 5% of the normal population [17], or more frequently in up to 40% of unselected patients [18]. Pineal cysts may be unilocular or multilocular and are filled with proteinaceous material. Histologically, the fluid is surrounded by cyst wall and

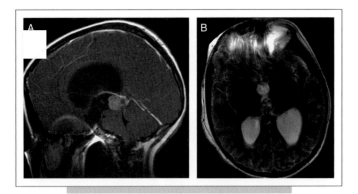

FIG. 44.16. Ependymoma. **(A)** Sagittal T1-weighted post-contrast MR image shows a lobulated enhancing pineal region mass. **(B)** Axial T2-weighted images are degraded by brace artifact. Tumor substance is markedly hypointense relative to brain.

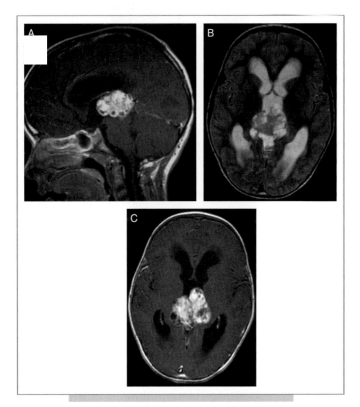

FIG. 44.17. PNET. **(A)** Sagittal T1-weighted post-contrast MR image shows a large heterogeneous lobulated mass with small tumoral cysts. **(B)** Axial T2-weighted MR image shows a heterogeneous mass with multiple small tumoral cysts. **(C)** Axial T1-weighted post-contrast MR image shows a large tumor with intense enhancement and multiple small tumoral cysts.

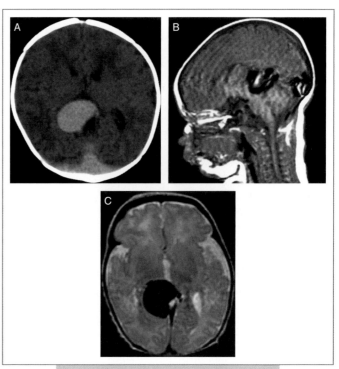

FIG. 44.18. Vein of Galen malformation. **(A)** Axial post-contrast CT scan of the brain demonstrates enhancement within the large Vein of Galen malformation. **(B)** Sagittal T1 non-contrast T1 weighted image demonstrates the large flow void of the Vein of Galen malformation with pulsation artifact, denoting its vascular nature. **(C)** T2-weighted image again shows the large flow void with pulsation artifact.

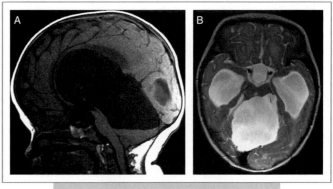

FIG. 44.19. Arachnoid cyst. **(A)** Sagittal T1-weighted MR image shows a very large cyst with mass effect. **(B)** Axial T2-weighted MR image shows the cyst with marked mass effect upon the tectum and hydrocephalus.

pineal parenchyma. The cyst wall is typically composed of three layers: an inner gliotic layer with Rosenthal fibers, a middle layer with columns of pineal parenchyma and a thin fibrous external layer. As the pineal gland is located outside the blood–brain barrier, the periphery of the cyst will typically enhance (Figure 44.21). Size varies and when pineal cysts are large enough, symptoms may occur due to local mass effect.

The majority of pineal region tumors grow as solid masses occasionally with areas of small cystic change.

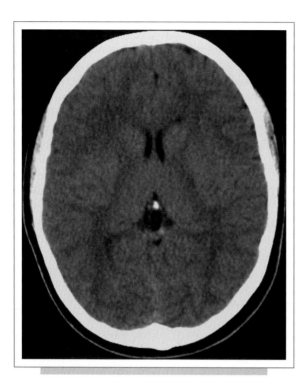

FIG. 44.20. Pineal lipoma. Non-contrast CT scan demonstrates a pineal region lipoma. The calcified pineal gland lies anterior to the fatty mass.

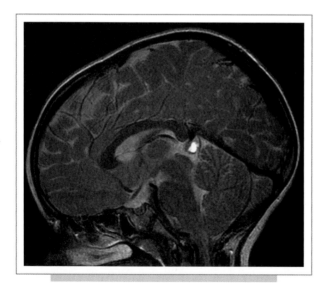

FIG. 44.21. Pineal cyst. Sagittal T2-weighted image demonstrates a simple pineal cyst.

A cystic appearance of a pineal neoplasm is unusual [7]. However, in the aforementioned series of 13 cystic pineal region masses by Engel et al, six of these cystic pineal masses were proven to be pineocytomas [16]. The remainder of the cystic lesions in the series was represented by four glial cysts, one arachnoid cyst, a low-grade glioma and a teratoma.

TUMOR MARKERS

Tumors of germ cell origin may produce specific embryonic markers such as β-HCG, AFP and PLAP. The marker profile in CSF and serum is helpful in the differential diagnosis of germ cell tumors.

Choriocarcinoma produces β-HCG. Endodermal sinus tumor produces AFP. Embryonal cell carcinoma produces both AFP and β-HCG. Mature teratomas do not express any tumor markers, whereas malignant teratomas may produce low levels of β-HCG or AFP. Mixed germ cell tumors may produce a combination of tumor markers.

Germinomas are either marker negative or may demonstrate elevations of PLAP [19]. Low levels of β-HCG can be seen in germinomas with syncytiotrophoblastic giant cells.

CSF is obtained from a lumbar puncture or from the lateral ventricles at the time of surgery and sent for analysis. The concentration of tumor markers found in the CSF is usually higher than the levels obtained from serum. The concentration of tumor markers found in the lumbar CSF is generally higher than the levels found in CSF obtained from the ventricles [20].

MAKING A DIAGNOSIS

In the absence of detectable tumor markers within the CSF or serum, a tissue diagnosis will be necessary. Diagnostic options include ventriculoscopic biopsy at the time of third ventriculostomy. As patients with pineal region tumors usually present with obstructive hydrocephalus, this technique is probably the simplest and safest option for obtaining tissue. If this approach is not feasible, either stereotactic biopsy or an open biopsy can be performed.

If tumor markers are positive, the diagnosis of an NGGCT, namely embryonal carcinoma, endodermal sinus tumor, choriocarcinoma, immature or malignant teratoma or mixed germ cell tumor can be presumed. As treatment is directed towards the most malignant element of tumor and considering that the treatment protocols for NGGCTs are the same, the precise histologic diagnosis is not necessary as the result would not alter the treatment plan and, as such, a tissue diagnosis can be foregone.

If the serum and CSF are negative for tumor markers, the differential considerations include germinoma, pineoblastoma, pineocytoma as well as non-germ cell origin tumors and non-pineal parenchymal tumors, and tissue sampling is necessary.

TREATMENT AND PROGNOSIS

Germinomas are highly radiosensitive and the majority of germinomas are curable by radiation therapy alone. In the absence of subarachnoid seeding, patients receive cranial radiation to the whole ventricular field. There is some debate over the extent of this field of radiation. If there is evidence for tumor dissemination, patients receive craniospinal radiation. Five-year survival rates are generally over

85% [21]. Supplemental platinum-based chemotherapy may allow reduced dose radiation therapy for similar tumor control rates [22].

NGGCTs are treated with multiagent platinum-based chemotherapy. Surgery may be performed prior to chemotherapy or after reduction of tumor with chemotherapy. A surgical gross total resection is the goal if there is residual tumor. Radiation therapy to the craniospinal axis is then recommended only if the patient is rendered disease-free post-chemotherapy and surgery. Five-year survival rates between 40 and 70% have been reported [23].

Pineocytomas are generally treated with surgical resection. Supplemental radiation therapy may be considered up-front or postponed until recurrence is determined [24]. Five-year survival rates over 80% are reported [24].

Pineoblastoma is treated with surgical resection followed by adjuvant craniospinal radiation therapy and multiagent chemotherapy. The prognosis is generally poor as with other PNETs, with a 5-year survival rate of 30% in childhood [25].

Benign pineal region tumors such as mature teratomas, meningiomas, dermoid and epidermoid cysts and choroid plexus papillomas are curable by surgery alone. Low-grade astrocytomas may benefit from surgery, radiation therapy or chemotherapy.

SUMMARY

Tumor pathology found in the pineal region is somewhat unique to this area of the brain. The imaging features of pineal region tumors are unfortunately varied, with considerable overlap. Therefore, making an accurate preoperative diagnosis in these patients is usually not possible based on imaging alone.

Certain observations discussed in this chapter may help in limiting the differential considerations. For instance, pineal parenchymal tumors have no gender predilection, whereas germinomas are more common in males. Therefore, a homogeneously enhancing pineal mass in a young male is most likely going to be a germinoma. CSF dissemination suggests either a germ cell tumor or a pineoblastoma. The presence of intratumoral fat is seen in teratomas and dermoid cysts. Strong restricted diffusion is seen in epidermoid cysts. A pineal region mass with a dural attachment to the free edge of the tentorium is seen in meningiomas.

In the absence of any of the aforementioned imaging characteristics or demonstrable tumor markers, biopsy is necessary for a diagnosis. The prognosis is variable as described, dependent on the tumor histology.

REFERENCES

1. Apuzzo M, Gruen P (1996). Pineal physiology. In *Neurosurgery*. Wilkins R, Rengachary S (eds). Mcgraw-Hill, New York. 977–994.
2. Hoffman HJ, Yoshida M, Becker LE (1983). Pineal region tumors in childhood. Experience at the Hospital for Sick Children. In Humphreys RP (ed.). *Concepts in Pediatric Neurosurgery* 4. Karger, Basel. 360–386.
3. Russel DS, Rubinstein LJ (1977). *Pathology of Tumors of the Nervous System*, 4th edn. Williams & Wilkins, Baltimore. 284–295.
4. Koide O, Watanabe Y, Sato K (1980). Pathological survey of intracranial germinoma and pinealoma in Japan. Cancer 45:2119–2130.
5. Zimmerman RA, Bilaniuk LT (1982). Age-related incidence of pineal calcification detected by computed tomography. Radiology 142:659–662.
6. Jennings MT, Gelman R, Hochberg F (1985). Intracranial germ-cell tumors: natural history and pathogenesis. J Neurosurg 63:155–167.
7. Smirniotopoulos JG, Rushing EJ, Mena H (1992). Pineal region masses: differential diagnosis. Radiographics 12:577–596.
8. Ganti SR, Hilal SK, Stein BM, Silver AJ, Mawad M, Sane P (1986). CT of pineal region tumors. Am J Roentgenol 146:451–458.
9. Zee CS, Segall H, Apuzzo M et al (1991). MR imaging of pineal region neoplasms. J Comput Assist Tomogr 15:56–63.
10. Tien RD, Barkovich AJ, Edwards MS (1990). MR imaging of pineal tumors. Am J Roentgenol 155:143–151.
11. Maity A, Shu HK, Janss A et al (2004). Craniospinal radiation in the treatment of biopsy-proven intracranial germinomas: twenty-five years' experience in a single center. Int J Radiat Oncol Biol Phys 58:1165–1170.
12. Kilgore DP, Strother CM, Starshak RJ, Haughton VM (1986). Pineal germinoma: MR imaging. Radiology 158:435–438.
13. Muller-Forell W, Schroth G, Egan PJ (1988). MR imaging in tumors of the pineal region. Neuroradiology 30:224–231.
14. Rubenstein I (1970). Tumors of the central nervous system. In *Atlas of Tumor Pathology*, 2nd series. Armed Forces Institute of Pathology, Washington, DC. 269–284.
15. Chiechi MV, Smirniotopoulos JG, Mena H (1995). Pineal parenchymal tumors: CT and MR features. J Comput Assist Tomogr 19:509–517.
16. Engel U, Gottschalk S, Niehaus L et al (2000). Cystic lesions of the pineal region – MRI and pathology. Neuroradiology 42:399–402.
17. Mamourian AC, Towfighi J (1986). Pineal cysts: MR imaging. Am J Neuroradiol 7:1081–1086.
18. Tapp E, Huxley M (1972). The histological appearance of the human pineal gland from puberty to old age. J Pathol 108:137–144.
19. Shinoda J, Yamada H, Sakai N, Ando T, Hirata T, Miwa Y (1988). Placental alkaline phosphatase as a tumor marker for primary intracranial germinoma. J Neurosurg 68:710–720.
20. Allen JC, Nisselbaum J, Epstein F, Rosen G, Schwartz MK (1979). Alphafetoprotein and human chorionic gonadotropin determination in cerebrospinal fluid. An aid to the

diagnosis and management of intracranial germ-cell tumors. J Neurosurg 51:368–374.

21. Chao CK, Lee ST, Lin FJ, Tang SG, Leung WM (1993). A multivariate analysis of prognostic factors in management of pineal tumor. Int J Radiat Oncol Biol Phys 27:1185–1191.

22. Choi JU, Kim DS, Chung SS, Kim TS (1998). Treatment of germ cell tumors in the pineal region. Childs Nerv Syst 14:41–48.

23. Robertson PL, DaRosso RC, Allen JC (1997). Improved prognosis of intracranial non-germinoma germ cell tumors with multimodality therapy. J Neuro-Oncol 32:71–80.

24. Schild SE, Scheithauer BW, Schomberg PJ et al (1993). Pineal parenchymal tumors. Clinical, pathologic, and therapeutic aspects. Cancer 72:870–880.

25. Dirks PB, Harris L, Hoffman HJ, Humphreys RP, Drake JM, Rutka JT (1996). Supratentorial primitive neuroectodermal tumors in children. J Neuro-Oncol 29:75–84.

CHAPTER 45

Chordomas and Other Skull Base Tumors

Eric C. Bourekas and H. Wayne Slone

CHORDOMA

Chordomas are rare, locally aggressive, low-grade to intermediate-grade malignant tumors, usually of bone, that arise from embryonic notochord remnants or ectopic notochordal tissue. They represent 4% of all primary bone tumors, but only 0.2–1% of primary intracranial tumors [1,2]. Since first being described in 1894 by Ribbert, not many more than 1000 cases have been reported [2]. They usually occur along the midline, anywhere from the spheno-occipital synchondrosis of the clivus to the sacro-coccygeal region (Figures 45.1–45.4). Cranial chordomas represent approximately one-third of all chordomas, with spinal chordomas being 15–33% and sacral chordomas 29–49% [2,3]. Cranial chordomas typically occur at the spheno-occipital synchondrosis of the clivus and less commonly at the basisphenoid and basiocciptal synchondroses, but may occur anywhere along the midline, having been described in the nose [4], nasal septum [5], oropharynx [6], ethmoid sinus [7], sphenoid sinus, maxilla, retropharyngeal extraosseous [8], retroclival extradural [9], intradural [10], sella tursica (see Figure 45.4) [11–13], suboccipital region [14] and nasopharynx [15]. Cranial chordomas can occur off midline in 14% of cases and have been described in the cavernous sinus (Figure 45.5), jugular foramen (Figure 45.6) [16–18] cerebellopontine angle [19], intradural cerebellar [20] and temporal bone at the petrous apex [21].

The overall incidence of chordomas is 0.08 per 100 000. They are more common in whites than blacks (4:1) [2] and more common in males than females, as much as 2:1 in one series [3]. Median age of diagnosis is 58.5 years, although in blacks the median age of diagnosis is much younger at 27. The median age for cranial tumors is 49, which is much younger than the median age of presentation for sacral tumors – 69 years. The younger age of cranial presentation may in part relate to the earlier manifestation of symptoms. Females also have a greater likelihood of cranial presentation, as do blacks. Cranial chordomas are more common in females and typically occur at a younger age [2].

Pathology

Grossly, chordomas are soft, gelatinous, semitranslucent blue-gray, multilobulated tumors. Histologically, although there are no absolute or specific histologic features, there are four criteria used for the diagnosis of chordoma that are fairly constant:

1 a lobular arrangement of cells
2 a tendency for the cells to grow in cords, irregular bands, or pseudoacinar forms
3 production of abundant intercellular mucinous matrix and, most characteristically
4 the presence of large physaliphorous and vacuolated cells [15].

There are two different histologic types of chordomas: typical or classic chordomas and chondroid chordomas. Typical chordomas are primarily made up of cords of physaliphorous cells. The chondroid variant of chordomas, first described by Heffelfinger et al in 1973, has a striking histologic resemblance to chondrosarcomas and chondromas, with the stroma resembling hyaline cartilage with neoplastic cells in lacunae [3]. The chondroid variant represents approximately 14% of all chordomas, although almost all chondroid chordomas occur in the cranial region and represent a third of all chordomas in the spheno-occipital region. They are more common in females, present at a younger age and are believed to be associated with a better prognosis [3]. Typical chordomas are almost always positive for epithelial markers such as cytokeratin (CK) and epithelial membrane antigen (EMA) and positive for S-100 protein in 44–76.5% of cases [22,23]. Chondroid chordoma stain positive for S-100 protein in 85–100% of cases, but in only one-third of cases for CK and EMA [22,23]. Chordomas also stain strongly with vimentin.

Clinical

The symptoms generally depend on the location of the tumor. Diplopia is the most common symptom, being

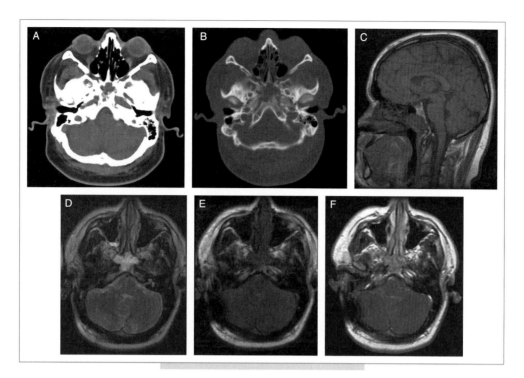

FIG. 45.1. Typical chordoma. A 30-year-old male with a history of a cochlear implant presents with right facial pain, headache and fever. Immunostains were positive for cytokeratin AE1/AE3 and S-100 protein. **(A)** CT head without contrast and **(B)** CT head with bone windowing demonstrate a lytic lesion of the clivus. **(C)** MRI sagittal T1-weighted image without contrast demonstrates the mass of the clivus, isointense to muscle and projecting into the nasopharynx. **(D)** MRI axial T2-weighted image without contrast shows that the mass is lobulated and of increased signal. The signal is relatively uniform which occurs in only 21%. **(E)** MRI axial T1-weighted image without contrast. The lesion is of low signal. **(F)** MRI axial T1-weighted image with contrast demonstrates relatively homogeneous enhancement.

noted in 64% of cases in one series [24–29]. Diplopia is due to cranial nerve palsies of which cranial nerve VI is the most commonly involved [25,28,30]. Headaches are the second most common symptom in cranial skull base chordomas and are noted in 32% [24–29]. They are usually occipital or retro-orbital. Tinnitus and hearing loss may also be symptoms [31] as can neck pain with clival chordomas [32]. Other symptoms may include facial pain and dysesthesia, decreased vision, gait disturbance and vertigo [24].

Imaging

Magnetic resonance imaging (MRI) and computed tomography (CT) are both useful modalities used in the evaluation of chordomas and generally both are obtained. Both modalities are equivalent in the detection of the tumor, but MRI is better in evaluating the extent of the tumor and delineating its relationship to the vasculature, which is critical in surgical planning [33]. CT is better in demonstrating bone destruction and the presence of calcifications.

On CT, chordomas are typically homogeneous and isodense to muscle, expansile, destructive and lytic lesions of the skull base. CT is excellent in assessing the degree of osteolysis [34]. Thin section CT is most useful in defining the anatomy of bone destruction at the skull base and the extent of calcifications. Foci of calcification may also be seen in less than 50% of cases and are difficult to distinguish from sequestered bone fragments [34]. In typical chordomas, calcifications within the lesion are more likely to represent sequestered bone fragments, while in chondroid chordomas these are more likely to be tumoral dystrophic calcifications. After contrast, the tumors demonstrate heterogeneous enhancement (see Figures 45.1A,B, 45.2A,B, 46.6A,B and C).

MRI, because of its superior soft tissue contrast and multiplanar capability, is the exam of choice in delineating the anatomic extension of chordomas and defining involvement of adjacent and critical structures. Chordomas involve the cavernous sinus in 60% of cases, the petrous apex in 48%, intradural 48%, upper clivus 48%, sphenoid sinus 44%, middle clivus 32%, lower clivus 24%, occipital condyle 24% with less common involvement of the nasopharynx, infratemporal fossa, jugular foramen and C2–3 to name a few [24]. Sagittal images are ideal in defining the posterior margin of the tumor, its relationship to the brainstem and nasopharynx and intradural extension. Coronal images are most useful in showing involvement of the cavernous sinus,

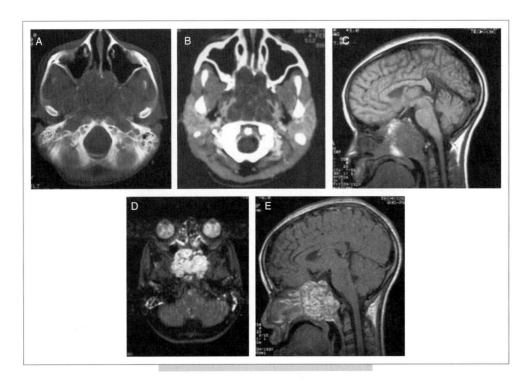

FIG. 45.2. Typical chordoma. **(A)** CT of the head with bone window demonstrates a large, destructive, multilobulated mass of the clivus and nasopharynx. **(B)** CT of the head with contrast: after contrast the mass shows enhancing septations. **(C)** MRI sagittal T1-weighted image without contrast shows that the mass has destroyed the clivus posteriorly but maintains a CSF interface between it and the brainstem. **(D)** MRI axial T2-weighted image without contrast: on this image the mass is hyperintense with septations of low signal separating lobulated areas of increased signal, rather classic for a chordoma. **(E)** MRI sagittal T1-weighted image with contrast; the mass enhances rather intensely as most chordomas do with a lobulated, honeycomb appearance which is characteristic for chordomas.

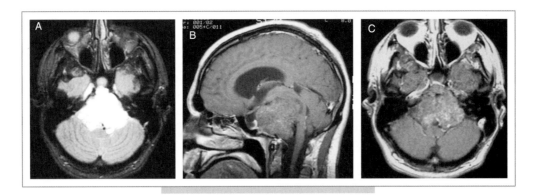

FIG. 45.3. Typical chordoma. **(A)** MRI axial T2-weighted image reveals a very large, lobulated mass of the clivus that is hyperintense. **(B)** MRI sagittal and **(C)** axial T1-weighted images with contrast demonstrate only mild enhancement of the large clival lesion with severe compression and posterior displacement of the brainstem. Courtesy of Steven P. Meyers MD, University of Rochester, Rochester, New York.

which is most common, and of the optic chiasm [33,35,36]. Chordomas are generally isointense to hypointense on T1-weighted images (T1WI) with 75% being isointense [33,37]. The T1WI images are especially useful in defining tumor–CSF interfaces [22]. Foci of increased signal on T1WI may be seen reflecting hemorrhage and mucinous collections [37]. On T2-weighted images (T2WI), they are typically of increased signal that is heterogeneous in 79% [37], with 70% demonstrating septations of low signal, separating lobulated areas of increased signal (see Figures 45.2D and

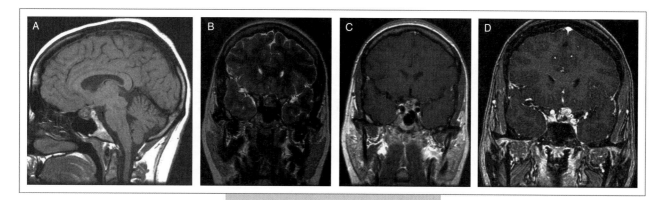

FIG. 45.4. Chondroid chordoma of the sella. A 22-year-old female with irregular menses. The imaging was initially interpreted as an adenoma. **(A)** MRI sagittal T1-weighted image demonstrates a heterogeneous mass of the sella with foci of increased signal which may represent hemorrhage or calcification. **(B)** MRI coronal T2-weighted image: the sellar and slightly suprasellar mass is heterogeneous but primarily isointense to cortex. **(C)** MRI coronal T1-weighted image with contrast. **(D)** Coronal T1-weighted 3D-SPGR (spoiled gradient echo) image post-contrast demonstrates heterogeneous but intense enhancement of the sellar lesion which extends to the suprasellar region anterior to the optic chiasm and involves the intracranial segment of the left optic nerve.

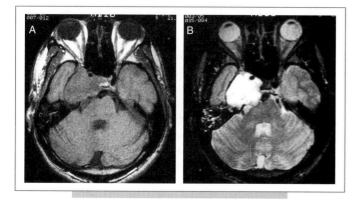

FIG. 45.5. Chondroid chordoma. **(A)** MRI axial T1-weighted image without contrast demonstrates the large mass in the region of the cavernous sinus, isointense to brain with partial encasement of the carotid artery. **(B)** MRI axial T2-weighted image shows the hyperintense lesion. Courtesy of Steven P. Meyers MD, University of Rochester, Rochester, New York.

45.6E) [33,37,38]. Chondroid chordomas tend to be brighter on T1WI and less intense than typical chordomas on T2WI, secondary to lower water content in the cartilage-like tissue [33]. Almost all chordomas enhance after intravenous contrast administration, with 81% enhancing heterogeneously (see Figures 45.1–45.6) [37]. A lobulated, honeycomb appearance after gadolinium contrast is rather characteristic (see Figure 45.2E) [39].

Overall, CT and MRI are complementary exams in the imaging of chordomas and both are usually needed for treatment planning. The imaging appearance of chordoma, however, is not pathognomonic. Imaging studies cannot reliably distinguish chordoma from chondrosarcoma or other skull base lesions. Histological studies are necessary for confirmation of the specific diagnosis [38].

Chordomas have been detected on positron emission tomography (PET) using F-18 fluoro-deoxyglucose (FDG) and C-11 methionine (MET) [40,41]. In 12 of 15 (80%) patients imaged prior to treatment with carbon-ion radiotherapy (CIRT), the chordomas were clearly visible on the baseline MET-PET study with a mean tumor to non-tumor ratio of 3.3+/−1.7. The methionine uptake decreased significantly to 2.3+/−1.4 after carbon-ion radiotherapy. A significant reduction in tumor MET uptake of 24% was observed after carbon-ion radiotherapy; 93% of patients showed no local recurrence after CIRT with a median follow-up time of 20 months. This indicates that MET-PET could potentially provide important tumor metabolic information for the therapeutic monitoring of chordomas after treatment [41]. Much work remains prior to PET being routinely used.

Treatment

Surgery is the mainstay of treatment of chordomas. Radical resection is the key factor in longer survival and improved quality of life with longer survival associated with more extensive tumor removal [29,42–46]. The location of the tumor, its invasive nature, its proximity to and involvement of critical neural (cranial nerves, optic chiasm, brainstem) and vascular structures (carotid, vertebral, basilar arteries) frequently prevents complete resection with radical removal achieved in only 44% in one series [24]. Radical surgical removal and high dose radiation therapy, particularly proton beam, are effective in tumor control. High-dose radiation therapy, especially fractionated proton beam or combined proton-photon beam therapy, has been shown to

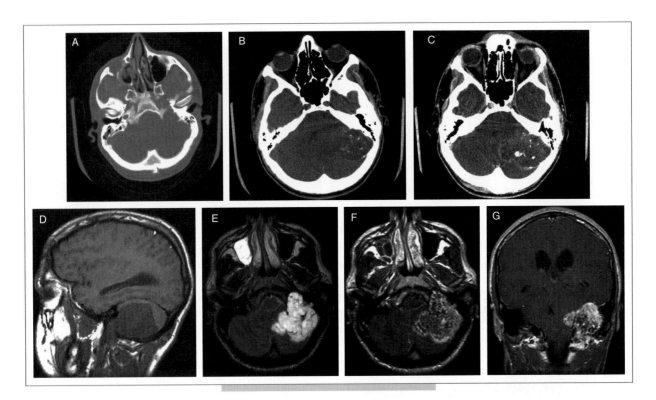

FIG. 45.6. Jugular fossa chondroid chordoma. A 22-year-old white male who presents to the emergency department complaining of a 3–4-day history of emesis, vertigo, headache and right eye central vision loss. **(A)** Axial temporal bone CT demonstrates erosion of the bony margins of the jugular foramen on the left. **(B)** Unenhanced axial CT image demonstrates a large, slightly hypodense, left posterior fossa mass, with scattered calcifications, that results in compression of the fourth ventricle and displacement of the fourth ventricle to the right. **(C)** CT after contrast, there is mild heterogeneous and primarily peripheral enhancement. **(D)** MRI sagittal unenhanced T1-weighted image shows the large, hypointense posterior fossa mass extending through the jugular foramen. **(E)** MRI axial fast spin-echo T2-weighted image shows the large, multilobulated, heterogeneous high T2 signal lesion in the posterior fossa with septations of low signal. **(F)** Post-contrast axial T1-weighted image and **(G)** sagittal post-contrast T1-weighted image demonstrate the heterogeneous, honeycomb enhancement of the lesion, typical of chordomas.

provide good tumor control and acceptable rates of complication [47–49]. Radiation is more effective when given after surgery rather than after tumor recurrence and its effectiveness has been related to the volume of residual tumor [44]. Gamma knife has been used as adjunctive therapy, but limited experience does not permit meaningful conclusions [50,51]. The same is true of brachytherapy with implantation of I-125 seeds [52,53]. Chemotherapy generally plays no role in the initial treatment of chordomas. The role of chemotherapy is in patients with metastatic disease and locally advanced recurrent disease, where systemic palliative therapy that includes thalidomide retards disease progression [54]. Imatinib mesylate has been found to have antitumor activity in patients with chordomas [55]. Local relapse is the predominant type of treatment failure seen in up to 95% of treatment failures, with surgical pathway recurrence and metastatic disease also possible [56]. Metastatic disease has been reported in 7–14% of cases with metastases to bone and lungs being most common and lymph node, brain and abdominal visceral involvement reported [56–58]. In the

largest series of chordomas by McMaster et al, the median survival for all races, genders and locations was 6.29 years. Median survival was longer for females (7.25 years) than for males (5.93 years), but this was not statistically significant [2]. Other studies, however, have suggested that female gender, as well as tumor necrosis of greater than 10% in biopsy specimens and tumor volume of greater than 70 ml, are independent predictors of shortened survival after radiation therapy [59]. In the McMaster series, median survival was higher for cranial chordomas (6.94 years) compared to spinal (5.88 years) and sacral (6.48 years), with median survival for females higher than males (7.70 years versus 6.42 years) [2]. The overall 5-, 10- and 20-year survival rates were 67.7%, 39.9% and 13.1% respectively [2]. Following local relapse 5-year survival has been reported as 5% [56].

CHONDROSARCOMA

Chondrosarcomas of the skull base are uncommon neoplasms that can resemble chordomas and, indeed, are

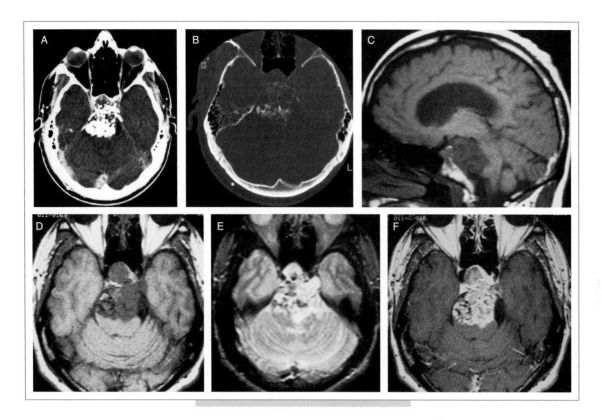

FIG. 45.7. Conventional grade I chondrosarcoma. **(A)** Axial CT of the head with contrast with soft tissue window and **(B)** bone window demonstrates a heavily calcified mass of the clivus with arcuate and ring-shaped mineralization classic for a chondrosarcoma. **(C)** Sagittal T1-weighted and **(D)** axial T1-weighted images, both without contrast, demonstrate a large hypodense and slightly heterogeneous mass of the clivus causing compression of the brainstem. **(E)** Axial T2-weighted image demonstrates the mass to be hyperintense but heterogeneous with areas of low signal caused by matrix mineralization, fibrocartilaginous elements or both. **(F)** Axial T1-weighted image post-contrast demonstrates intense but heterogeneous enhancement. Courtesy of Robert Tarr MD, Case Western Reserve University, Cleveland, Ohio.

frequently misdiagnosed as such. In the largest review of 200 skull base chondrosarcomas by Rosenberg et al, in some 37% of lesions diagnosed as chordomas at other institutions, the diagnosis was changed to chondrosarcoma [60]. They are malignant tumors that produce cartilage matrix. Primary chondrosarcomas arise de novo, whereas secondary chondrosarcomas arise from pre-existing lesions such as Paget's, enchondromas, osteochondromas, etc. Overall, they are the third most common primary malignant bone tumor after multiple myeloma and osteosarcoma, and represent 20–27% of all primary malignant bone tumors [61]. Craniofacial chondrosarcomas represent 2% of all chondrosarcomas and have a predilection for skull base [61]. Only 6% of skull base tumors are chondrosarcomas, with chordomas being more common skull base tumors [61]. Unlike chordomas, which tend to be midline lesions, the majority of chondrosarcomas occur off midline with 6% of the tumors arising in the sphenoethmoid complex, 28% originating in the clivus (Figure 45.7) and 66% occurring in the temporo-occipital junction (Figure 45.8) [60,62]. The mean age of presentation is 39 years of age, which is

younger than for chordomas. They are slightly more common in females with a female to male ratio of 1.3:1 [60].

Pathology

Histologically, the majority of skull base chondrosarcomas (63%) are mixed hyaline and myxoid tumors, with 29.5% being pure myxoid and 7.5% classified as hyaline chondrosarcomas [60]. The myxoid cartilage is the component that causes diagnostic confusion with chordoma. Myxoid chondrosarcomas can be distinguished from chordoma on morphologic grounds because the neoplastic chondrocytes are relatively small, contain diminished amounts of cytoplasm and do not grow in cohesive groups, nests or sheets. Most tumors (50.5%) are grade I, 28.5% mixed grades I and II, 21% pure grade II with grade III being uncommon, in fact 0% in Rosenberg's study [60]. Chondrosarcomas are subclassified into conventional or classic (hyaline/myxoid), dedifferentiated, clear cell, and mesenchymal subtypes. The conventional type of chondrosarcoma is the most common cartilage tumor to develop in the skull base and the least aggressive [60,63]. Typically,

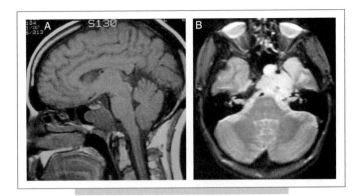

FIG. 45.8. Conventional grade I chondrosarcoma. **(A)** MRI sagittal T1-weighted image demonstrates a lobulated hypointense mass of the clivus, extending into the sphenoid sinus. **(B)** Axial T2-weighted image shows the mass is heterogeneous, hyperintense and encasing the left carotid artery. Courtesy of Steven P. Meyers MD, University of Rochester, Rochester, New York.

low-grade chondrosarcomas may undergo dedifferentiation and become highly malignant tumors with a poor prognosis [64]. They almost always (98.9%) stain for S-100 protein and do not stain for keratin. Faint staining for epithelial membrane antigen can be seen in a minority of cases [60].

Clinical

The presenting symptoms of chondrosarcomas are similar to those of chordomas previously described and include hearing loss, pulsatile tinnitus, vertigo, headache, diplopia due to cranial nerve palsies of cranial nerves III, IV and VI, facial pain and paresthesias due to cranial nerve V palsy and facial spasm or facial paresis due to cranial nerve VII involvement to name a few [65].

Imaging

Imaging, unfortunately, is non-specific and generally unable to distinguish between chordomas and chondrosarcomas [66]. On CT, chondrosarcomas frequently contain matrix calcifications or sequestered bony fragments (see Figures 45.7A,B). Matrix mineralization is seen in 44–80% of cases [62,67]. Matrix calcification may demonstrate the typical chondroid pattern of ring, arc or snowflake-like foci. Peripheral eggshell-like calcification has also been

described. Extensive chondroid calcification tends to occur in lower grade classic chondrosarcomas [63,68]. On unenhanced CT, the masses may be of increased or decreased density relative to brain parenchyma and may mildly or moderately enhance [66]. On MRI, chondrosarcomas are of low to intermediate signal on T1 (see Figures 45.7C,D and 45.8A) and high signal on T2 with heterogeneous signal noted in 59% caused by matrix mineralization, fibrocartilaginous elements or both (see Figures 45.7E and 45.8B). The margins tend to be multilobulated and well defined. Contrast enhancement is noted in 100% and is heterogeneous in 73% (see Figure 45.7F) [62].

Treatment

The goal of surgery is complete resection [69]; however, gross total removal is only achieved in a small percentage due to tumor location and frequent involvement of adjacent critical structures such as the cavernous sinus, carotid and basilar arteries, optic chiasm, etc [60]. Surgery combined with proton beam irradiation can yield excellent results with 5- and 10-year progression-free survival rates of 70% and 45% respectively compared to 5- and 10-year survival rates of chordoma, which have been reported to be approximately 51% and 35% respectively, despite similar aggressive management. Rosenberg et al reported using high-dose postoperative fractionated precision radiation therapy, combining conformal megavoltage and charged particle irradiation using the 160-MeV proton beam, with a median dose of 72.1 Cobalt-Gray-equivalent, given in 38 fractions [60].

OTHER SKULL BASE TUMORS

Malignant tumors outnumber benign ones at the skull base and metastases are more common than primary tumors [60]. Of the primary tumors, an overwhelming majority are chordomas and chondrosarcomas [60]. Besides chordoma and chondrosarcoma, there are many other tumors that have been reported to involve the skull base and that should be considered. These include meningioma, metastasis, invasive adenoma, craniopharyngioma, glomus tumors, plasmacytoma, lymphoma, giant cell tumor, nasopharyngeal carcinoma, eosinophilic granuloma, rhabdomyosarcoma, adenoid cystic carcinoma, benign fibrous histiocytoma, chondroblastoma, ameloblastoma, chondroma, osteogenic sarcoma and enchondroma [70].

REFERENCES

1. Dahlin DC, MacCarty CS (1952). Chordoma. Cancer 5:1170–1178.
2. McMaster ML, Goldstein AM, Bromley CM, Ishibe N, Parry DM (2001). Chordoma: incidence and survival patterns in the United States, 1973–1995. Cancer Causes Control 12:1–11.
3. Heffelfinger MJ, Dahlin DC, MacCarty CS, Beabout JW (1973). Chordomas and cartilaginous tumors at the skull base. Cancer 32:410–420.
4. Mou Z, Liu Z (2003). Primary chordoma of the nose. Chin Med J 116:154–156.

5. Scartozzi R, Couch M, Sciubba J (2003). Chondroid chordoma of the nasal septum. Arch Otolaryngol Head Neck Surg 129:244–246.

6. Gladstone HB, Bailet JW, Rowland JP (1998). Chordoma of the oropharynx: an unusual presentation and review of the literature. Otolaryngol Head Neck Surg 118:104–107.

7. Loughran S, Badia L, Lund V (2000). Primary chordoma of the ethmoid sinus. J Laryngol Otol 114:627–629.

8. Wang YP, Lee KS, Chen YJ, Huang JK (2004). Extraosseous chordoma of the retropharyngeal space. Otolaryngol Head Neck Surg 130:383–385.

9. Warakaulle DR, Anslow P (2002). A unique presentation of retroclival chordoma. J Postgrad Med 48:285–287.

10. Mapstone TB, Kaufman B, Ratcheson RA (1983). Intradural chordoma without bone involvement: nuclear magnetic resonance (NMR) appearance. Case report. J Neurosurg 59:535–537.

11. Kakuno Y, Yamada T, Hirano H, Mori H, Narabayashi I (2002). Chordoma in the sella turcica. Neurol Med Chir (Tokyo) 42:305–308.

12. Haridas A, Ansari S, Afshar F (2003). Chordoma presenting as pseudoprolactinoma. Br J Neurosurg 17:260–262.

13. Thodou E, Kontogeorgos G, Scheithauer BW et al (2000). Intrasellar chordomas mimicking pituitary adenoma. J Neurosurg 92:976–982.

14. Carpentier A, Blanquet A, George B (2001). Suboccipital and cervical chordomas: radical resection with vertebral artery control. Neurosurg Focus 10:E4.

15. Batsakis JG, Kittleson AC (1963). Chordomas. Otorhinolaryngologic presentation and diagnosis. Arch Otolaryngol 78:168–175.

16. Ramina R, Maniglia JJ, Fernandes YB et al (2004). Jugular foramen tumors: diagnosis and treatment. Neurosurg Focus 17:E5.

17. Iwasaki S, Ito K, Takai Y, Morita A, Murofushi T (2004). Chondroid chordoma at the jugular foramen causing retro-labyrinthine lesions in both the cochlear and vestibular branches of the eighth cranial nerve. Ann Otol Rhinol Laryngol 113:82–86.

18. Itoh T, Harada M, Ichikawa T, Shimoyamada K, Katayama N, Tsukune Y (2000). A case of jugular foramen chordoma with extension to the neck: CT and MR findings. Radiat Med 18:63–65.

19. Okamoto K, Furusawa T, Ishikawa K, Sasai K, Tokiguchi S (2006). Focal T2 hyperintensity in the dorsal brain stem in patients with vestibular schwannoma. Am J Neuroradiol 27:1307–1311.

20. Dow GR, Robson DK, Jaspan T, Punt JA (2003). Intradural cerebellar chordoma in a child: a case report and review of the literature. Childs Nerv Syst 19:188–191.

21. Khmel'nitskaia NM, Zavarzin BA (2000). Chordoma of the temporal bone. Vestn Otorinolaringol 58–59.

22. Mitchell A, Scheithauer BW, Unni KK, Forsyth PJ, Wold LE, McGivney DJ (1993). Chordoma and chondroid neoplasms of the spheno-occiput. An immunohistochemical study of 41 cases with prognostic and nosologic implications. Cancer 72:2943–2949.

23. Meis JM, Giraldo AA (1988). Chordoma. An immuno-histochemical study of 20 cases. Arch Pathol Lab Med 112: 553–556.

24. Al Mefty O, Borba L (1997). Skull base chordomas: a management challenge. J Neurosurg 86:182–189.

25. Forsyth PA, Cascino TL, Shaw EG et al (1993). Intracranial chordomas: a clinicopathological and prognostic study of 51 cases. J Neurosurg 78:741–747.

26. Raffel C, Wright DC, Gutin PH, Wilson CB (1985). Cranial chordomas: clinical presentation and results of operative and radiation therapy in twenty-six patients. Neurosurgery 17:703–710.

27. Rich TA, Schiller A, Suit HD, Mankin HJ (1985). Clinical and pathologic review of 48 cases of chordoma. Cancer 56:182–187.

28. Volpe R, Mazabraud A (1983). A clinicopathologic review of 25 cases of chordoma (a pleomorphic and metastasizing neoplasm). Am J Surg Pathol 7:161–170.

29. Watkins L, Khudados ES, Kaleoglu M, Revesz T, Sacares P, Crockard HA (1993). Skull base chordomas: a review of 38 patients, 1958–88. Br J Neurosurg 7:241–248.

30. Chetiyawardana AD (1984). Chordoma: results of treatment. Clin Radiol 35:159–161.

31. Zhang H, Zhang D, Sun Y et al (2006). The study of 11 cases of chordoma in the skull base or neck. Lin Chuang Er Bi Yan Hou Ke Za Zhi 20:251–253.

32. Alvarado R, Gomez J, Morale SG, Davis CP (2004). Neck pain: common complaint, uncommon diagnosis – symptomatic clival chordoma. South Med J 97:83–86.

33. Sze G, Uichanco LS III, Brant-Zawadzki MN et al (1988). Chordomas: MR imaging. Radiology 166:187–191.

34. Meyer JE, Oot RF, Lindfors KK (1986). CT appearance of clival chordomas. J Comput Assist Tomogr 10:34–38.

35. Oot RF, Melville GE, New PF et al (1988). The role of MR and CT in evaluating clival chordomas and chondrosarcomas. Am J Roentgenol 151:567–575.

36. Erdem E, Angtuaco EC, Van Hemert R, Park JS, Al Mefty O (2003). Comprehensive review of intracranial chordoma. Radiographics 23:995–1009.

37. Meyers SP, Hirsch WL Jr, Curtin HD, Barnes L, Sekhar LN, Sen C (1992). Chordomas of the skull base: MR features. Am J Neuroradiol 13:1627–1636.

38. Doucet V, Peretti-Viton P, Figarella-Branger D, Manera L, Salamon G (1997). MRI of intracranial chordomas. Extent of tumour and contrast enhancement: criteria for differential diagnosis. Neuroradiology 39:571–576.

39. Drummond DS, Wenzel DM, Kudryk W (1997). Spheno-occipital chondroid chordoma. J Otolaryngol 26:197–200.

40. Lin CY, Kao CH, Liang JA, Hsieh TC, Yen KY, Sun SS (2006). Chordoma detected on F-18 FDG PET. Clin Nucl Med 31:506–507.

41. Zhang H, Yoshikawa K, Tamura K et al (2004). Carbon-11-methionine positron emission tomography imaging of chordoma. Skeletal Radiol 33:524–530.

42. Hug EB, Loredo LN, Slater JD et al (1999). Proton radiation therapy for chordomas and chondrosarcomas of the skull base. J Neurosurg 91:432–439.

43. Arnold H, Herrmann HD (1986). Skull base chordoma with cavernous sinus involvement. Partial or radical tumour-removal? Acta Neurochir (Wien) 83:31–37.

44. Berson AM, Castro JR, Petti P et al (1988). Charged particle irradiation of chordoma and chondrosarcoma of the base of skull and cervical spine: the Lawrence Berkeley

Laboratory experience. Int J Radiat Oncol Biol Phys 15: 559–565.

45. Gay E, Sekhar LN, Rubinstein E et al (1995). Chordomas and chondrosarcomas of the cranial base: results and follow-up of 60 patients. Neurosurgery 36:887–896.

46. Sen CN, Sekhar LN, Schramm VL, Janecka IP (1989). Chordoma and chondrosarcoma of the cranial base: an 8-year experience. Neurosurgery 25:931–940.

47. Austin JP, Urie MM, Cardenosa G, Munzenrider JE (1993). Probable causes of recurrence in patients with chordoma and chondrosarcoma of the base of skull and cervical spine. Int J Radiat Oncol Biol Phys 25:439–444.

48. Hug EB, Fitzek MM, Liebsch NJ, Munzenrider JE (1995). Locally challenging osteo- and chondrogenic tumors of the axial skeleton: results of combined proton and photon radiation therapy using three-dimensional treatment planning. Int J Radiat Oncol Biol Phys 31:467–476.

49. Austin-Seymour M, Munzenrider J, Goitein M et al (1989). Fractionated proton radiation therapy of chordoma and low-grade chondrosarcoma of the base of the skull. J Neurosurg 70:13–17.

50. Lanzino G, Dumont AS, Lopes MB, Laws ER Jr (2001). Skull base chordomas: overview of disease, management options, and outcome. Neurosurg Focus 10:E12.

51. Muthukumar N, Kondziolka D, Lunsford LD, Flickinger JC (1998). Stereotactic radiosurgery for chordoma and chondrosarcoma: further experiences. Int J Radiat Oncol Biol Phys 41:387–392.

52. Gutin PH, Leibel SA, Hosobuchi Y et al (1987). Brachytherapy of recurrent tumors of the skull base and spine with iodine-125 sources. Neurosurgery 20:938–945.

53. Kumar PP, Good RR, Skultety FM, Leibrock LG (1988). Local control of recurrent clival and sacral chordoma after interstitial irradiation with iodine-125: new techniques for treatment of recurrent or unresectable chordomas. Neurosurgery 22:479–483.

54. Schonegger K, Gelpi E, Prayer D et al (2005). Recurrent and metastatic clivus chordoma: systemic palliative therapy retards disease progression. Anticancer Drugs 16:1139–1143.

55. Casali PG, Messina A, Stacchiotti S et al (2004). Imatinib mesylate in chordoma. Cancer 101:2086–2097.

56. Fagundes MA, Hug EB, Liebsch NJ, Daly W, Efird J, Munzenrider JE (1995). Radiation therapy for chordomas of the base of skull and cervical spine: patterns of failure and outcome after relapse. Int J Radiat Oncol Biol Phys 33:579–584.

57. Singh W, Kaur A (1987). Nasopharyngeal chordoma presenting with metastases. Case report and review of literature. J Laryngol Otol 101:1198–1202.

58. Yuh WT, Flickinger FW, Barloon TJ, Montgomery WJ (1988). MR imaging of unusual chordomas. J Comput Assist Tomogr 12:30–35.

59. O'Connell JX, Renard LG, Liebsch NJ, Efird JT, Munzenrider JE, Rosenberg AE (1994). Base of skull chordoma. A correlative study of histologic and clinical features of 62 cases. Cancer 74:2261–2267.

60. Rosenberg AE, Nielsen GP, Keel SB et al (1999). Chondrosarcoma of the base of the skull: a clinicopathologic study of 200 cases with emphasis on its distinction from chordoma. Am J Surg Pathol 23:1370–1378.

61. Murphey MD, Walker EA, Wilson AJ, Kransdorf MJ, Temple HT, Gannon FH (2003). From the archives of the AFIP: imaging of primary chondrosarcoma: radiologic-pathologic correlation. Radiographics 23:1245–1278.

62. Meyers SP, Hirsch WL Jr, Curtin HD, Barnes L, Sekhar LN, Sen C (1992). Chondrosarcomas of the skull base: MR imaging features. Radiology 184:103–108.

63. Cure JK, Bhatia M, Richardson MS, Conway WF (1995). General case of the day. Myxoid chondrosarcoma of the cavernous sinus. Radiographics 15(5):1231–1234.

64. Littrell LA, Wenger DE, Wold LE et al (2004). Radiographic, CT, and MR imaging features of dedifferentiated chondrosarcomas: a retrospective review of 174 de novo cases. Radiographics 24(5):1397–1409.

65. Neff B, Sataloff RT, Storey L, Hawkshaw M, Spiegel JR (2002). Chondrosarcoma of the skull base. Laryngoscope 112(1):134–139.

66. Pamir MN, Ozduman K (2006). Analysis of radiological features relative to histopathology in 42 skull-base chordomas and chondrosarcomas. Eur J Radiol 58(3):461–470.

67. Grossman RI, Davis KR (1981). Cranial computed tomographic appearance of chondrosarcoma of the base of the skull. Radiology 141(2):403–408.

68. Lee YY, Van Tassel P (1989). Craniofacial chondrosarcomas: imaging findings in 15 untreated cases. Am J Neuroradiol 10:165–170.

69. Stapleton SR, Wilkins PR, Archer DJ, Uttley D (1993). Chondrosarcoma of the skull base: a series of eight cases. Neurosurgery 32(3):348–355.

70. Laine FJ, Nadel L, Braun IF (1990). CT and MR imaging of the central skull base. Part 2. Pathologic spectrum. Radiographics 10:797–821.

CHAPTER 46

Pediatric Brain Tumors

Tina Young Poussaint

INTRODUCTION

The most common type of solid tumor among children is the pediatric brain tumor, which is the second most frequent childhood malignancy after leukemia, and the leading cause of death from solid tumors in this population [1]. About 9% of the brain tumors reported to the Central Brain Tumor Registry of the USA (CBTRUS) between 1998 and 2002 occurred in persons younger than 20 [2]. Among children aged 0–19, the prevalence rate for all primary brain and central nervous system tumors was roughly 9.5 per 100 000, with an estimated 26 000 children living with this diagnosis in the USA in 2000; the incidence rate was 29.1 cases per 1 000 000 from 1996 to 2003 [2,3].

Our ability to reach an accurate preoperative diagnosis of the pediatric brain tumor hinges largely on magnetic resonance imaging (MRI), which is the ideal modality for defining tumor extent, developing an effective treatment plan and implementing image-guided therapies involving surgery, radiotherapy and chemotherapy. Indeed, because of MRI's unique multiplanar capability, its ability to provide detailed structural and anatomical information and its ability to produce images of superior resolution, MRI is unrivalled as a superior diagnostic tool [4]. It is also the central modality for tumor follow-up including tumor response, progression and treatment effects. Craniospinal MR with gadolinium is the modality of choice for detection of tumor dissemination in the CSF pathways [5]. CT is used for the assessment of calcification and cortical bony involvement, especially in lesions such as craniopharyngioma and orbital, nasosinus, petrous temporal, skull base and cranial calvarial lesions. Other imaging procedures that aid in localization and in guiding surgical and radiation therapy include stereotactic MR and CT, the open magnet and intraoperative sonography. Image fusion with reconstructions is used for stereotaxis, radiosurgery, stereotactic radiotherapy and image-guided surgery.

Functional assessment of the brain can be performed using modalities such as single-photon emission computed tomography (SPECT) and positron emission tomography (PET). Depending on the type of tracer used, PET and SPECT are also used to assess metabolic activity in brain tumors and are useful in distinguishing tumor from radiation necrosis or scar [6].

Using advanced MR techniques, such as magnetic resonance perfusion imaging, MR diffusion imaging and magnetic resonance spectroscopy (MRS), further characterizes the hemodynamics, cellularity and metabolism of any suspected malignancies.

The ability of these advanced MR techniques, separately and in combination, to elucidate the physiologic characteristics of pediatric tumors is proving essential in reaching accurate and timely diagnoses, developing effective postsurgical treatment plans and in implementing regular and responsive follow-up. MR perfusion imaging, MR diffusion imaging and MR spectroscopy are described briefly below. They were also reviewed in a recent article [7] and are covered in depth by other contributors to this book.

MR perfusion imaging is used to evaluate cerebral perfusion dynamics by analysis of the hemodynamic parameters of relative cerebral blood volume, relative cerebral blood flow and mean transit time. The techniques used to perform perfusion imaging include T2*-weighted dynamic susceptibility, arterial spin labeling techniques and T1-weighted dynamic contrast-enhanced perfusion techniques [7]. MR perfusion can be used for preoperative tumor grading, evaluating tumor margins and distinguishing radiation necrosis versus tumor recurrence.

MR diffusion imaging using predominately echoplanar techniques has also been useful in characterizing tissue and tumor cellularity, in gauging tumor response to treatment [8], in grading tumor and in distinguishing tissue types [9–12]. Diffusion tensor imaging (DTI) provides visualization of fiber bundle direction and integrity with in vivo characterization of the rate and direction of white matter diffusion useful for presurgical planning or coregistration of tractography data with radiosurgical planning and functional MR data [13]. Moreover, fractional anisotropy using DTI may prove helpful in assessing treatment-induced white matter changes in children [14,15].

Proton nuclear magnetic resonance spectroscopy – 1H-NMRS, a non-invasive in vivo technique that provides additional metabolic diagnostic indices beyond anatomic information, has been extensively used to evaluate brain tumors. Specifically, it enables the radiologist to identify tumor tissue, to grade tumors, to differentiate tumor types, to distinguish active tumor from radiation necrosis or scar tissue, to guide stereotactic biopsy sites and to determine early response to treatment [7].

SPECIFIC PEDIATRIC BRAIN TUMORS

Tumors such as meningiomas, schwann-cell tumors, metastases from outside the brain and pituitary tumors rarely present in children, but are commonly found in adults [16]. In children, the vast majority of brain tumors are primary CNS lesions [17]. Pediatric brain tumors are typically classified pathologically or by location; classification by location helps limit the differential diagnosis. In this chapter, tumors of the cerebral hemispheres, posterior fossa tumors, sellar and suprasellar tumors and parameningeal tumors will be classified by location. Pineal region tumors will be covered elsewhere in this book (see Chapter 44).

Tumors of the Cerebral Hemispheres

Astrocytomas

Gliomas account for 56% of all tumors and 67% of malignant tumors in children aged 0–14 [2]. The World Health Organization (WHO) has classified four basic glial tumors: pilocytic astrocytoma; diffuse astrocytoma; anaplastic astrocytoma and glioblastoma multiforme for which there is a parallel grading system (I–IV) that defines the microscopic appearance of each tumor [18].

Pilocytic astrocytomas are typically well-demarcated with the T2 signal abnormality matching the amount of gadolinium enhancement, occasionally presenting with an associated cyst. Higher grades of malignancy are usually heterogeneous in signal intensity with ill-defined margins, edema, hemorrhage, necrosis, mass effect and irregular enhancement (Figure 46.1A–C).

Perfusion studies in adults and children have measured the relative cerebral blood volume (rCBV) in the assessment of glioma grade and have demonstrated that low-grade astrocytomas have a significantly lower average rCBV than high-grade astrocytomas, such as anaplastic astrocytomas or glioblastoma multiforme [19–25]. One of the pitfalls of perfusion imaging in children is that pilocytic astrocytomas can have a high blood volume that mimics high-grade tumors on perfusion imaging alone [26].

Diffusion imaging studies in astrocytomas have demonstrated a significant negative correlation between apparent diffusion coefficient (ADC) values and WHO astrocytic tumor grades for tumor grades II through IV [27]. Moreover, ADC values have been found to correlate inversely with tumor cellularity (see Figure 46.1D) [9].

MR spectroscopy is uniquely able to measure the higher choline (Cho) to creatine (Cr) ratio that is characteristic of high-grade gliomas; MRS has thus shown great

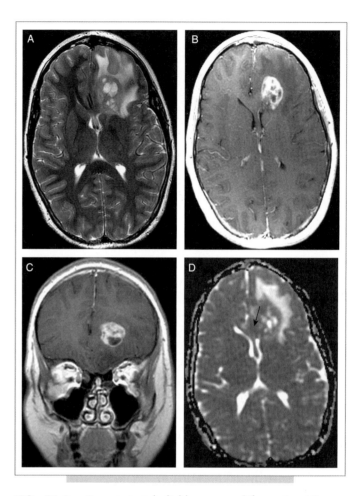

FIG. 46.1. Supratentorial glioblastoma multiforme in a 15-year-old boy with seizures. **(A)** Axial T2 MR image demonstrates heterogeneous mass within left frontal lobe with extensive surrounding edema. **(B** and **C)** Axial and coronal T1 image with gadolinium demonstrates heterogeneous enhancing tumor. **(D)** Axial diffusion ADC (apparent diffusion coefficient) map demonstrates restricted diffusion (arrow) in the solid component of the mass.

utility in distinguishing high-grade gliomas from low-grade gliomas [28] and has enhanced overall diagnostic accuracy. Along these same lines, Tzika et al have examined normalized choline and lipids as the physiologic characteristic that distinguishes high-grade from low-grade tumors in children [29].

Supratentorial PNET Tumors

Supratentorial primitive neuroectodermal tumors (PNET) are found in younger children with a mean age of 5 years [18]. These tumors are often large and sharply demarcated, may be solid and homogeneous, or can be heterogeneous with cyst, calcification or hemorrhage on CT and MRI [30]. Frequently, calcification is seen on CT. Following contrast, there is heterogeneous enhancement. These tumors can occur in either the cerebral hemispheres or lateral

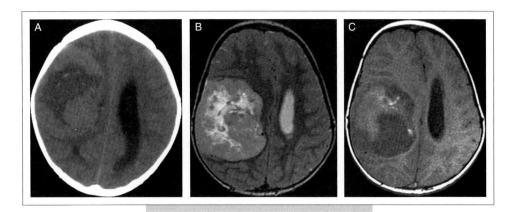

FIG. 46.2. Supratentorial PNET in a 3-year-old boy with headache, nausea and vomiting. **(A)** CT image demonstrates large right parietal heterogeneous dense mass. **(B)** Axial T2 MR image demonstrates heterogeneous T2 signal within the mass. **(C)** Axial T1 MR image before gadolinium shows small areas of bright T1 signal consistent with foci of hemorrhage.

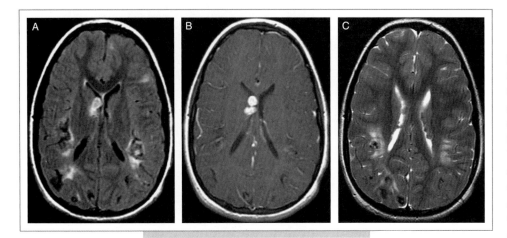

FIG. 46.3. Tuberous sclerosis and subependymal giant cell astrocytoma. **(A)** Axial FLAIR MR image demonstrates multiple calcified cortical tubers and hyperintense mass at the right foramen of Monro. **(B)** Axial T1 MR image with gadolinium demonstrates enhancing subependymal giant cell astrocytoma at the right foramen of Monro. **(C)** Axial T2 MR image shows hypointense calcified cortical tubers and bilateral subependymal nodules.

ventricles (Figure 46.2). The differential diagnosis of supratentorial tumors similar to PNET includes other large heterogeneous tumors such as atypical teratoid/rhabdoid tumors and ependymoma.

Supratentorial Ependymoma

Supratentorial ependymomas represent 40% of all ependymomas. These tumors typically occur in children younger than 5 years. On CT, these tumors are heterogeneous with calcification and cyst formation. On MR imaging, these tumors are heterogeneous containing cysts, calcification and occasional hemorrhage [31]; they usually demonstrate irregular, heterogeneous enhancement with gadolinium.

Subependymal Giant Cell Tumor (SEGA)

Tuberous sclerosis (TS), an autosomal dominant multisystem disease linked to chromosomes 9 and 16, is caused by a mutation of the TSC1 and TSC2 genes that encode for hamartin and tuberin proteins, respectively [32,33]. Neuroimaging findings include subependymal and cortical tubers, white matter lesions and subependymal giant cell astrocytomas [34,35]. Subependymal giant cell tumors are of mixed glioneuronal lineage associated with subependymal nodules that have symptoms such as hydrocephalus, interval growth or papilledema. These tumors are seen in approximately 10% of patients with TS [34]. They arise from the lateral ventricle near the foramen of Monro and present with obstructive hydrocephalus between the ages of 10 and 30 [36]. These tumors are often calcified on CT; on MR, they are hypointense or isointense on T1-weighted images and are isointense to hyperintense on T2-weighted sequences. Following contrast, these tumors uniformly enhance, and are often associated with, subependymal nodules and cortical tubers (Figure 46.3).

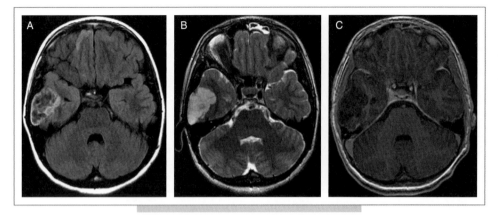

FIG. 46.4. Dysembryoplastic neuro-epithelial tumor in an 11-year-old boy with seizures. **(A and B)** Axial FLAIR and T2 MR image demonstrates sharply circumscribed, non-enhancing, multi-septated hyperintense mass extending to the cortex of the right temporal lobe. **(C)** Axial T1 image with gadolinium shows the tumor does not enhance.

Neuronal and Mixed Neuronal-glial Tumors

These tumors are characterized by neuronal and glial differentiation. Tumors in this category include ganglio-cytoma, ganglioglioma, dysembryoplastic neuroepithelial tumor (DNET), central neurocytoma, paraganglioma and cerebellar liponeurocytoma [18].

Dysembryoplastic Neuroepithelial Tumor (DNET)

DNET is a benign tumor associated with intractable, partial complex seizures occurring in children and young adults. These tumors are located either cortically or supratentorially and are often associated with cortical dys-plasias occurring most commonly in the temporal and fron-tal lobes [37]. On CT, the tumor is typically hypodense with minimal or absent enhancement. On MRI, the lesion is well-circumscribed; it is hypointense on T1-weighted images and hyperintense on T2-weighted images without surrounding edema and with minimal enhancement (Figure 46.4). There may be remodeling of the calvarium.

Choroid Plexus Tumors

The most common types of choroid plexus tumors in childhood are choroid plexus papilloma and choroid plexus carcinoma. Choroid plexus tumors represent 2–4% of brain tumors in children [18].

Choroid Plexus Papilloma

Choroid plexus papillomas are intraventricular tumors that are well-circumscribed. In children, these tumors arise in the lateral ventricle, often the trigone, whereas, in adults, they are found in the fourth ventricle. Choroid plexus pap-illomas are lobulated in appearance and are associated with hydrocephalus often containing punctate regions of calcification, hemorrhage and vascular flow voids. The hydrocephalus may arise from CSF overproduction and/or obstruction. Imaging features on CT include a lobulated intraventricular mass with calcification and homogeneously intense enhancement. MRI features include T1 isointense to hypointense signal with T2 hypointensity and intense

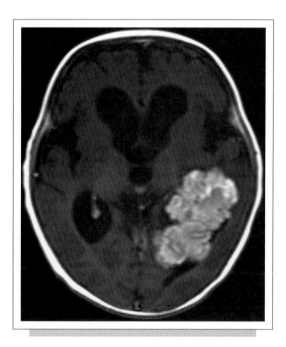

FIG. 46.5. Choroid plexus papilloma in a 7-month-old boy with irritability, nausea and vomiting. Axial MR image with gado-linium shows large intraventricular lobulated mass in left lateral ventricle with intense enhancement. There is moderate enlarge-ment of the lateral and third ventricles.

contrast enhancement (Figure 46.5). On MR spectroscopy, these tumors exhibited a significantly higher level of the metabolite, myo-inositol, low creatine and decreased choline when compared to choroid plexus carcinomas [38,39].

Choroid Plexus Carcinoma

Choroid plexus carcinomas represent 20–40% of choroid plexus tumors. They are heterogeneous on MRI, extend beyond the margin of the ventricle and are associated with edema and mass effect (Figure 46.6). Both choroid plexus papilloma and carcinoma can seed the CSF pathways.

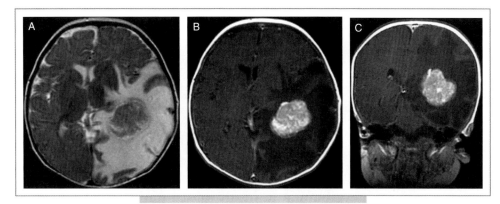

FIG. 46.6. Choroid plexus carcinoma in a 5-month-old with nausea, vomiting, lethargy and apnea. **(A)** Axial T2 MR image demonstrates hypointense mass centered in the trigone of the left lateral ventricle with extensive surrounding edema in the adjacent white matter. **(B** and **C)** Axial and coronal T1 MR images with gadolinium demonstrate enhancing mass in the left trigone of the lateral ventricle.

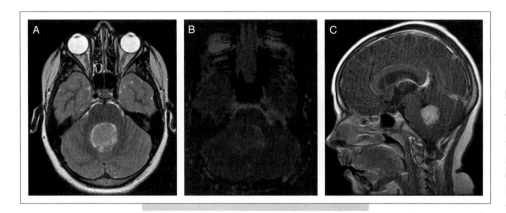

FIG. 46.7. Medulloblastoma. **(A)** Axial T2 MR image shows hypointense mass in fourth ventricle. **(B)** Axial diffusion MR images of medulloblastoma shows restricted diffusion in mass. **(C)** Sagittal T1 MR image with gadolinium shows enhancing fourth ventricular mass.

On MR spectroscopy, choroid plexus carcinomas have elevated choline levels compared to choroid plexus papillomas and lower creatine and creatine/total choline ratios than other pediatric brain tumors [38,39].

Posterior Fossa Tumors

The common posterior fossa tumors of childhood include medulloblastoma, cerebellar astrocytoma, brainstem glioma and ependymoma. Less frequent posterior fossa tumors include atypical teratoid/rhabdoid tumors, hemangioblastoma, dermoid-epidermoid tumors, acoustic schwannomas, meningioma, teratoma and skull base tumors. Skull base tumors such as carcinoma, metastasis, sarcomas, histiocytosis, chordoma and paraganglioma may invade the posterior fossa [40]. Presenting symptoms include head tilt, cranial nerve palsies, ataxia, headache, nausea and vomiting.

Medulloblastoma

Medulloblastoma is the most common tumor of the posterior fossa in childhood, representing 15–20% of brain tumors in children and 30–40% of posterior fossa neoplasms [40]. Medulloblastomas usually arise in the midline within the vermis with growth into the fourth ventricle [41–46]. This differs from adults and older children in which the tumor is found laterally in the cerebellar hemispheres [42,47]. On CT, the tumor is typically characterized as hyperdense and on MRI, the tumor is viewed on T1-weighted images as hypointense and on T2-weighted images as hypointense to gray matter [43]. Homogeneous enhancement is usually seen following gadolinium administration (Figure 46.7). To identify leptomeningeal enhancement and seeding, the entire craniospinal axis must be screened (Figure 46.8). On diffusion MR imaging, these tumors demonstrate restricted diffusion due to high cellularity within the tumor (Figure 46.7C) [48]. In addition, MR spectroscopy of untreated pediatric brain tumors has demonstrated a significantly elevated taurine concentration when compared to other tumors [39,49,50]. Creatine/choline and N-acetylaspartate to choline ratios have been used to differentiate between normal cerebellar tissue and posterior fossa tumors such as PNET, ependymoma and juvenile pilocytic astrocytomas [51].

Cerebellar Astrocytomas

Cerebellar astrocytoma, one of the most common posterior fossa tumors (second only to medulloblastoma), is often diagnosed as a juvenile pilocytic astrocytoma with excellent survival after gross total surgical resection [52–54]. Cerebellar astrocytomas can occur in the midline or in the cerebellar hemispheres. These tumors are often associated

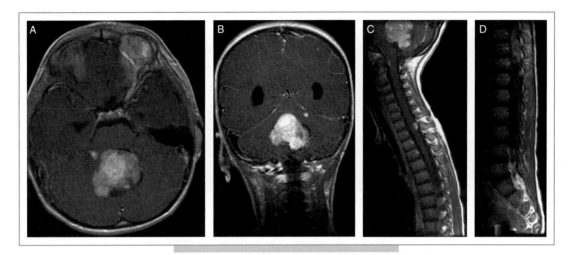

FIG. 46.8. Medulloblastoma with spinal leptomeningeal seeding. **(A** and **B)** Axial and coronal T1 MR images with gadolinium show enhancing posterior fossa mass. **(C** and **D)** Sagittal T1 MR images with gadolinium show spinal leptomeningeal seeding.

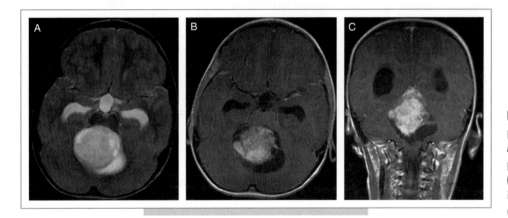

FIG. 46.9. Posterior fossa juvenile pilocytic astrocytoma. **(A)** Axial T2 MR image of hyperintense mass in posterior fossa with hydrocephalus. **(B** and **C)** Axial and coronal T1 MR images show enhancing tumor nodule with cysts within the tumor.

with hydrocephalus due to compression of the aqueduct or fourth ventricle. The classic appearance is a cerebellar mass consisting of a large cyst with a solid tumor nodule. However, the imaging spectrum of appearances may include cystic, solid, or a mix of cystic and solid presentations. Macrocysts or microcysts can form and the tumor can be grossly cystic or solid. On CT and MR, a cyst with enhancement of the solid nodular component is typically found (Figure 46.9).

The solid, enhancing component of cerebellar pilocytic astrocytomas has greater ADC values than other pediatric cerebellar tumors such as ependymoma, rhabdoid tumor and medulloblastoma [48].

MR spectrosopy of these tumors has demonstrated high lactate concentrations and consistently high choline content and choline/NAA and choline/creatine ratios despite the benign clinical course of this tumor type [55].

Brainstem Tumors

Brainstem tumors represent 10–20% of all CNS tumors in childhood. These tumors can be categorized as diffuse and focal.

Pontine Tumors

Patients with diffuse pontine gliomas commonly present with multiple cranial nerve palsies, long tract signs and ataxia [56,57]. Typically, these tumors can cause the pons to expand by more than 50% and can infiltrate into the medulla or midbrain. Diagnosis is based on the characteristic changes on MR imaging of diffuse, T2-weighted, hyperintense expansion of the brainstem without biopsy. Despite multiple treatment approaches, the prognosis is poor with long-term survival <10% [58]. From a pathological standpoint, these tumors represent high-grade lesions. On CT, these tumors are of low density or isodense. On MRI, they are typically isointense to hypointense on T1-weighted images and are hyperintense on T2 images. Enhancement is usually absent or minimal (Figure 46.10). Calcification and hemorrhage within the tumor is rare.

Using multivoxel MR spectroscopy, Laprie et al followed eight patients with diffuse pontine glioma after radiotherapy (RT) [59]. Spectroscopy in these tumors was evaluated before RT, at response and at recurrence. Cho:NAA and Cho:Cr values within the imaging abnormalities were

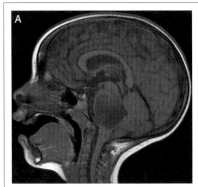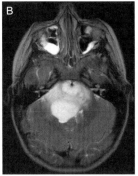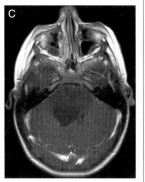

FIG. 46.10. Diffuse pontine glioma. **(A)** Sagittal T1 MR image shows expansile hypointense mass in the pons with mass effect on the fourth ventricle. **(B)** Axial T2 MR image shows encasement of the basilar artery and large hyperintense pontine mass with mass effect on the fourth ventricle. **(C)** Axial T1 MR image with gadolinium shows no enhancement within the mass.

significantly higher than the mean values in normal-appearing regions. Cho:NAA values decreased in studies performed from diagnosis to the time of response to RT, followed by an increase at the time of relapse. In another study, two pediatric patients with diffuse pontine tumors improved clinically following radiotherapy; however, MR spectroscopy showed an overall increase in Cho/Cr and Cho/NAA ratios indicating tumor progression [50]. Unfortunately, the disease progressed in both patients. Each study demonstrates the potential utility of MR spectroscopy in determining tumor response or failure to therapy.

MR diffusion tensor imaging of brainstem gliomas has been reported by Helton et al [51]. This technique can provide visualization and quantification of tumor involvement in the brainstem by measuring the degree to which the tumor has invaded white matter tracts. They found that: (1) the corticospinal tracts and transverse pontine fibers were more affected than the medial lemnisci; and (2) fractional anisotropy and the apparent diffusion coefficient was significantly ($P < .05$) altered in the tracts of all patients with pontine tumors compared to control groups.

FDG-PET, along with MR findings in a study of 20 pediatric patients with brainstem glioma, helped to differentiate between anaplastic astrocytomas and glioblastomas among high-grade tumors. Glioblastomas demonstrated hypermetabolic or hypometabolic lesions and anaplastic astrocytomas showed no metabolic change or hypometabolic lesions [52].

Focal pontine gliomas are uncommon, representing 5–10% of brainstem gliomas [58]. They may have an exophytic component [63]. These tumors also demonstrate marked enhancement and have a better prognosis than diffuse pontine gliomas [64].

Medullary Tumors

Cervicomedullary astrocytomas are a unique group of brainstem tumors with a good prognosis when compared to the diffuse pontine group of brainstem gliomas [65,66]. On CT, these tumors are isodense; on MRI, these tumors are isointense on T1-weighted images and are hyperintense on

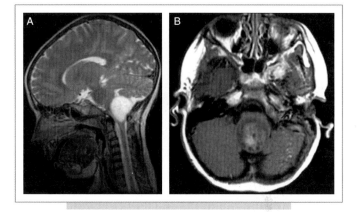

FIG. 46.11. Cervicomedullary astrocytoma. **(A)** Sagittal T2 MR image demonstrates hyperintense cervicomedullary mass. **(B)** Axial T1 MR image with gadolinium shows mild enhancement and tiny cysts posteriorly.

T2-weighted images. They are typically well circumscribed, often demonstrate enhancement and are described pathologically as pilocytic astrocytomas (Figure 46.11) [67]. These tumors likely originate in the upper cervical cord and grow upward toward the obex into the medulla and are dorsally exophytic[68].

Diffuse medullary tumors are poorly defined lesions centered in the medulla that can extend into the pons or upper cervical cord and have a poorer prognosis than the dorsally exophytic type of tumor.

Midbrain Tumors

Tumors of the midbrain include focal and diffuse midbrain tumors and tectal tumors. The focal midbrain tumor has a better prognosis than the diffuse. Tectal tumors are a unique subset of tumors of the brainstem. These tumors are often low-grade pilocytic astrocytomas, present with obstructive hydrocephalus and usually do not require treatment beyond CSF diversion such as third ventriculostomy. Before the advent of MRI, patients with this tumor were frequently misdiagnosed as having aqueductal stenosis on CT. On MRI, the lesions are usually isointense on T1-weighted images and hyperintense on T2-weighted images without enhancement (Figure 46.12). Tumors larger than 2.5 cm in diameter with enhancement are significant radiologic predictors of those patients who will need further treatment beyond CSF diversion [69]. In a study of 40 children with tectal tumors, the only factor predictive of tumor enlargement was lesion volume at presentation ($P = 0.002$). Lesions with a volume less than 4 cm^3 were likely to follow a benign course. All large lesions, defined as a volume greater than 10 cm^3 at presentation, eventually required treatment [70].

Ependymomas

Ependymomas constitute 6–12% of all intracranial tumors in children and 8–15% of posterior fossa tumors in children [18]. They are the fourth most common posterior fossa tumor in children following medulloblastoma, cerebellar astrocytoma

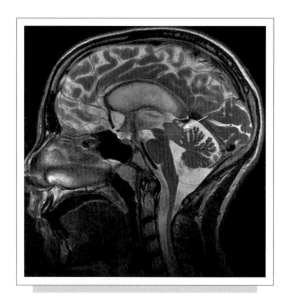

FIG. 46.12. Tectal glioma. Sagittal T2 image shows expansile tectal mass (arrow) obstructing the aqueduct of Sylvius.

and brainstem glioma. Ependymomas arise from ependymal cells lining the ventricles; they grow out of the fourth ventricle via the foramina of Luschka and Magendie into the cisterna magna, basilar cisterns, cerebellopontine angles and through the foramen magnum into the upper cervical canal around the spinal cord. On CT, the tumor reveals mixed density with punctate calcification in 50% of cases with variable enhancement. These tumors are heterogeneous on MRI, reflecting a combination of solid component, cyst, calcification, necrosis, edema or hemorrhage [71]. On T1-weighted images, ependymomas are usually hypointense; and on T2-weighted images, the mass is often isointense to gray matter with foci of dark T2 signal related to calcification or blood, and foci of bright T2 signal related to cyst or necrosis within the tumor. Following contrast administration, there is heterogeneous enhancement in the tumor (Figure 46.13).

Atypical Teratoid/rhabdoid Tumors

Of all atypical teratoid/rhabdoid tumors, the majority (94%) are intra-axial, occurring in the infratentorial component in 47% of cases [72]. The imaging features are identical to medulloblastoma, however, the prognosis is worse with a median survival rate with disseminated leptomeningeal tumor of 16 months (Figure 46.14) [73]. These tumors occur at a younger age, are predominately seen in females and have a propensity to seed [73]. At the molecular level, they are distinguished by mutation or deletion of both copies of the hSNF5/INI1 gene that maps to chromosome band 22q11.2, and is observed in approximately 70% of primary tumors [74].

Sellar and Suprasellar Tumors

Sellar and suprasellar tumors in children include craniopharyngioma, chiasmatic/hypothalamic glioma, hypothalamic hamartoma, pituitary adenoma, germ cell tumors, Langerhans cell histiocytosis, Rathke's cleft cysts, arachnoid cysts and dermoid/epidermoid cysts. The most common sellar and suprasellar tumors are craniopharyngioma and chiasmatic/hypothalamic glioma.

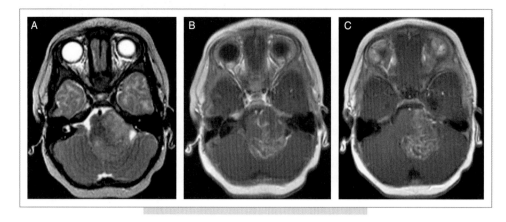

FIG. 46.13. Posterior fossa ependymoma. **(A)** Axial T2 MR image demonstrates mass within fourth ventricle extending through the left foramen of Luschka. **(B** and **C)** Axial T1 MR images with gadolinium demonstrate heterogeneously enhancing fourth ventricular tumor extending through the left foramen of Luschka.

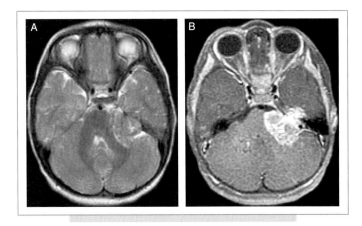

FIG. 46.14. Posterior fossa rhabdoid tumor in a 20-month-old infant with left facial nerve paralysis. **(A)** Axial T2 MR image shows hypointense mass in left cerebellopontine angle cistern. **(B)** Axial T1 MR image with gadolinium shows enhancing left cerebellopontine angle mass.

Craniopharygioma

Craniopharyngioma is a benign (WHO grade I) neoplasm arising in the suprasellar or intrasellar region. These tumors arise from remnants of Rathke's pouch (i.e. the craniopharyngeal duct) and are composed of characteristic squamous epithelium [75]. They represent 1–5% of all intracranial tumors and constitute 50% of the suprasellar tumors seen in childhood, and they have a bimodal distribution occurring in childhood, adolescence and older adults in their fifties [76]. In children, the clinical presentation includes headache, visual field defects, diplopia and short stature with occasional hydrocephalus and papilledema present. Classically, these are cystic tumors filled with a cholesterol-rich fluid grossly resembling motor oil. The tumors are usually intrasellar, suprasellar or both.

Craniopharyngiomas can compress, envelop or infiltrate adjacent structures and produce a reactive gliosis, although they are histologically benign. Hetelekidis et al have demonstrated that if the tumor is greater than 5 cm, there is a greater likelihood of recurrence [77]. In many cases, surgical extirpation is difficult, and with a high rate of recurrence, adjuvant radiotherapy is often necessary. On CT, 90% of craniopharyngiomas are cystic and calcified [78,79]. On MRI, these tumors have variable signal characteristics depending on the contents of the cyst and the presence of calcium. On CT, the solid component of the tumor may be isodense or hypodense and on MRI, T1 isointense to hypointense with T2 isointense to hyperintense. The calcified component is of increased attenuation on CT and often T2 hypointense on MRI. The cystic component may be of high or low T1 intensity and T2 hyperintense. Following gadolinium administration, peripheral enhancement of the cyst, as well as heterogeneous enhancement of the solid component, generally occurs (Figure 46.15).

Chiasmatic/Hypothalamic Gliomas

These tumors constitute 5% of pediatric intracranial tumors [80]. Chiasmatic/hypothalamic tumors are often low-grade fibrillary and pilocytic astrocytomas. Between 20 and 50% of patients with chiasmatic/hypothalamic tumors have neurofibromatosis type 1 (NF1). NF1 is the most common of the phakomatoses, inherited as an autosomal dominant disorder localized on chromosome 17 with variable penetrance. Optic gliomas are the most common tumor of the CNS in NF1 [81]. These gliomas may involve any portion of the optic pathway including one or both optic nerves, the chiasm (Figure 46.16), tracts, the lateral geniculate bodies, or the optic radiations. Other intracranial lesions that may be seen in NF1 include vacuolization in the myelin (i.e. NF spots), plexiform neurofibromas and other astrocytomas (Figure 46.17) [82]. The clinical presentation of optic glioma is most often characterized by decreased visual acuity; though other symptoms including visual field defects, optic atrophy, hydrocephalus, or hypothalamic dysfunction and

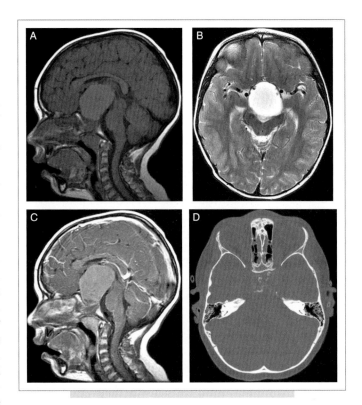

FIG. 46.15. Craniopharyngioma. **(A)** Sagittal T1 MR image shows large sellar and suprasellar mass with bright T2 signal superiorly and small amount of hypointense signal inferiorly in the sella. **(B)** Axial T2 MR image shows predominantly hyperintense mass in the suprasellar region. **(C)** Sagittal T1 MR image with gadolinium shows mild enhancement of the inferior solid component in the sella and peripheral enhancement of the superior cystic component. **(D)** Axial CT image shows stippled calcification within the mass.

papilledema may be seen [83]. These lesions are usually T1-hypointense and T2-hyperintense with typically homogeneous gadolinium enhancement and, in large tumors, heterogeneous enhancement. MRI with fat suppression leads to optimal visualization of the optic pathways [84]. Spontaneous regression of optic gliomas in patients with and without NF1 has been reported [85].

Diencephalic syndrome is also commonly associated with hypothalamic/chiasmatic astrocytomas and may be the underlying cause of failure to thrive in infancy in a rare number of cases. Clinical characteristics include severe emaciation in the setting of normal linear growth. The tumors associated with this unusual syndrome are often larger in size, occur at a younger age and are more aggressive than those associated with other presentations

(Figure 46.18). Despite low-grade histologic findings, these tumors may seed throughout the CSF pathways [86]. This syndrome serves as a model for studying growth hormone resistance and metabolic regulation of adiposity [87].

Hypothalamic Hamartoma

Hypothalamic hamartomas are masses of mature ganglionic tissue, located between the pituitary stalk and mamillary bodies involving the region of the tuber cinereum. They do not demonstrate invasion or growth [88]. Often, the patients are male presenting with symptoms including precocious puberty, gelastic seizures, hyperactivity and developmental delay. On CT, there is a suprasellar mass, isodense with gray matter, non-calcified, and rarely cystic. On MRI, a well-demarcated mass is present within or adjacent to, the tuber cinereum or mamillary bodies. The hamartomas are T1-isointense and T2-isointense to slightly hyperintense to gray matter without enhancement (Figure 46.19). In a study using short-echo-time point-resolved spectroscopy (PRESS) sequences, Amstutz et al sampled hypothalamic hamartomas [89]. Sequences were used to compare choline (Cho), N-acetylaspartate (NAA) and myo-inositol (mI) resonances by using a creatine (Cr) reference. Spectral ratios and T2 signal intensity ratios of the hamartomas were then compared with histopathologic findings. In 14 hypothalamic hamartomas, a spectrum of increased mI/Cr ratios was seen. Those tumors with markedly elevated

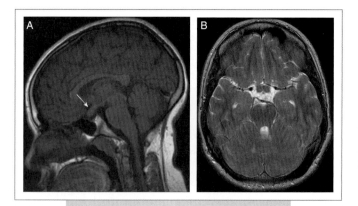

FIG. 46.16. Optic chiasm glioma in a 14 year old with NF1. **(A)** Sagittal T1 MR shows a thickened optic chiasm (arrow). **(B)** Axial T2 MR image shows thickened optic chiasm and NF spots in cerebellum.

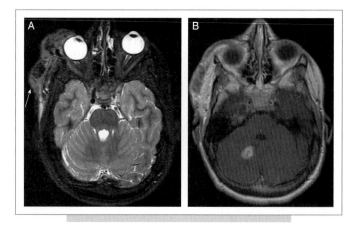

FIG. 46.17. An 18-year-old girl with NF-1, right facial plexiform neurofibroma and left cerebellar astrocytoma. **(A)** Axial FSEIR MR image demonstrates right facial plexiform neurofibroma (arrow). **(B)** Axial T1 MR image with gadolinium shows right cerebellar pilocytic astrocytoma.

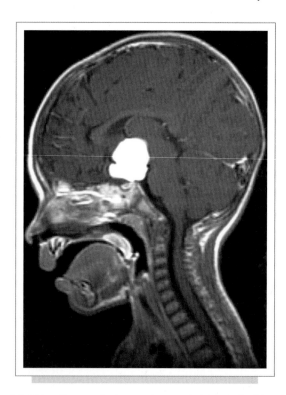

FIG. 46.18. Diencephalic syndrome. A 13-month-old girl with failure to thrive. Sagittal T1 MR image with gadolinium shows enhancing suprasellar mass.

mI/Cr demonstrated an increased glial component when compared with the remaining tumors. Increased glial component was also found to have a positive correlation with hyperintensity of lesions on T2-weighted images [89].

Other studies have identified a subgroup of patients with hypothalamic hamartoma and associated congenital anomalies including hypoplasia of the olfactory bulbs, absence of the pituitary gland, cardiac and renal anomalies, imperforate anus, craniofacial anomalies, syndactyly and a short metacarpal – clinical features characterizing the autosomal dominant Pallister-Hall syndrome [90,91].

Germ Cell Tumors

CNS germ cell tumors in childhood occur commonly in the suprasellar region and pineal region and may be germinomas or non-germinomatous germ cell tumors. These patients may clinically present with central diabetes insipidus (DI), wasting, precocious puberty or growth failure. On imaging, a mass may be seen involving the suprasellar region or there may be thickening of the infundibulum (Figure 46.20). On CT, these lesions are well-defined and slightly hyperdense. On MRI, they are T1 hypointense, T2 hypointense and markedly enhancing. CSF seeding is not uncommon. On occasion, there may be an associated pineal region germinoma, which may be metastatic or synchronous.

Suprasellar germinomas often present with central DI. On MRI, the normally seen posterior pituitary bright spot is absent [92]. The mechanism of central DI is thought to be the interruption of transport of the vasopressin neurosecretory granules along the hypothalamic-neurohypophyseal pathway. Central DI and the absence of the posterior bright spot may precede other clinical and imaging features of hypothalmic tumor by months or years. Therefore, follow-up MRI with gadolinium is recommended in all of these patients [93]. Other causes of pituitary stalk thickening include Langerhans histiocytosis, lymphocytic hypophysitis and rare entities such as sarcoidosis and lymphoma.

Teratomas occur more often in the pineal region than in the suprasellar region. They are classified as mature or immature depending on whether the tissue components resemble mature adult tissues or immature embryonic tissues. On CT and MRI, the mass often contains fat, bone or cartilage with heterogeneous density and intensity characteristics. These patients may present with precocious puberty.

Pituitary Adenomas

Pituitary adenomas are uncommon in childhood and represent less than 3% of all intracranial tumors [94]. The clinical presentation depends on tumor size, hormonal activity and extrasellar extent. Pituitary adenomas are divided into hormonally active and inactive types, with the majority being hormonally active and most commonly prolactin-secreting. These lesions tend to occur in adolescence; they are often microadenomas (<1 cm) and are most often prolactinomas (Figure 46.21). Macroadenomas are greater than 1 cm in diameter and are often prolactinomas in patients presenting with neuroendocrine symptoms, visual field deficits and headache. The majority of macroadenomas in adolescence are hemorrhagic and often occur in males [95].

On MRI, pituitary adenomas are T1 hypointense in the majority of cases. Less frequently, adenomas are T1 isointense or hyperintense [96]. On T2 images, adenomas are often hyperintense or isointense and macroadenomas are

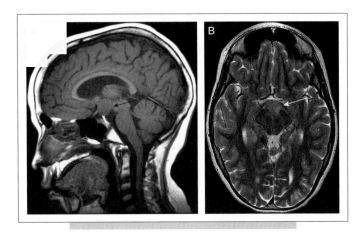

FIG. 46.19. Hypothalamic hamartoma. **(A)** Sagittal T1 MR image shows hypointense hypothalamic mass (arrow). **(B)** Axial T2 MR image shows isointense hypothalamic mass (arrow).

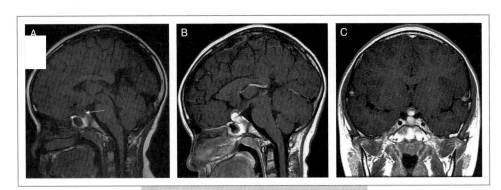

FIG. 46.20. Non-germinomatous germ cell tumor of suprasellar region in a 10-year-old girl with central diabetes insipidus. **(A)** Sagittal T1 MR image shows loss of posterior pituitary bright spot (arrow). **(B and C)** Sagittal and coronal T1 image shows thickened enhancing infundibulum and sellar mass.

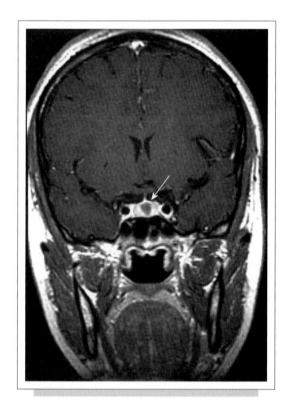

FIG. 46.21. Prolactinoma. Coronal T1 MR image with gadolinium shows non-enhancing pituitary microadenoma (arrow).

usually hyperintense [96]. Macroadenomas may invade the cavernous sinuses, extend superiorly into the suprasellar cistern and can compress the optic chiasm. Immediately following gadolinium administration, the lesion will be hypointense compared to the normal gland.

Langerhans' Cell Histiocytosis

Systemic Langerhans' cell histiocytosis may include involvement of the pituitary stalk and hypothalamus. These patients may present with diabetes insipidus; MRI shows loss of the normal posterior pituitary bright spot [97]. MRI may also demonstrate a solitary mass in the region of the median eminence of the pituitary stalk or thickening of the infundibulum. On CT, the mass is isodense and on MRI, the lesion is T1 isointense and T2 hyperintense. Following gadolinium administration, there is marked enhancement.

Parameningeal and Metastatic Tumors

Parameningeal tumors are extradural but contiguous with the CNS. They may arise from or involve the scalp, cranial vault, cranial base, orbits, sinuses or pharynx, petrous temporal structures or soft tissues of the face or neck. The parameningeal neoplastic processes encountered in childhood include those of osseous, chondroid or myeloid (reticuloendothelial) origin, other mesenchymal origin tumors and those arising from notochordal elements including neuroblastoma, rhabdomyosarcoma, histiocytosis, plexiform neurofibroma, and angiofibroma (Figure 46.22) [98].

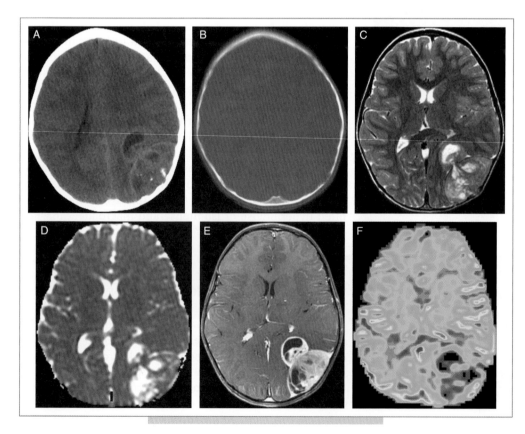

FIG. 46.22. Supratentorial extra-axial Ewing's sarcoma. **(A)** Axial CT image demonstrates heterogeneously dense mass lesion in the left parietal lobe with scattered calcification. **(B)** Axial CT bone window image demonstrates lytic destruction of the left parietal bone. **(C)** Axial T2 MR image demonstrates large heterogeneous extra-axial tumor compressing the left occipital lobe with mass effect on the left parietal lobe. **(D)** Axial MR diffusion ADC map demonstrates restricted diffusion within the mass. **(E)** Axial T1 MR image with gadolinium demonstrates nodular heterogeneous enhancement of the mass. **(F)** Axial T2* MR perfusion image demonstrates increased perfusion in central areas of the mass.

Metastatic disease to the brain is rare and, when present, is commonly leptomeningeal disease from a primary brain tumor. However, when brain metastases in childhood occur, they generally arise from sarcomas such as Ewing's sarcoma and rhabdomyosarcoma [99].

CONCLUSION

With the introduction of MRI as a diagnostic tool in the 1980s, radiological science in general, and neuroimaging in particular, has undergone spectacular change. As a non-invasive modality that provides incredibly detailed structural, anatomic, metabolic and functional data without the potentially harmful effects of ionizing radiation, MRI has emerged as the imaging 'gold standard' for a host of diseases and anomalies, including pediatric brain tumors. Indeed, without the precise pre-, intra- and postoperative images produced by MRI, the physician's ability accurately to detect, diagnose, treat and follow a wide range of malignancies affecting the CNS would still be at a most rudimentary stage.

Most recently, MRI has proven an indispensible tool in implementing extremely sophisticated, technologically advanced surgeries that are executed under computer-assisted image guidance. Because of MRI's ability to achieve precise localization and targeting, brain tumor excision is now accomplished with exceptional accuracy and without causing damage to surrounding healthy tissue, which is of particular benefit to the developing pediatric brain.

As ever faster and more efficient image processing techniques are deployed, conventional MRI and its imaging 'offspring', including MR perfusion imaging, MRS, diffusion imaging and DTI, are effecting rapid, ground-breaking change on the quality and kind of data that are generated. These techniques, when combined with other modalities such as CT, SPECT and PET, are resulting in: (1) more accurate and timely diagnoses; (2) interventions tailored to the individual needs of the patient; and (3) improved assessment of response to treatment.

Thus equipped with extremely detailed disease profiles, the entire medical team – from the diagnosing neuroradiologist to the treating neurosurgeon and neuro-oncologist – are able to implement the most responsive and effective therapeutic interventions from the moment the tumor is detected to complete eradication or remission. In the near future, computer-assisted robotics should also play an important role in maximizing the utility of MRI and other imaging modalities. Of immediate benefit is the practice of combining modalities to generate the most comprehensive 'picture' of the tumor including accurate staging which, in turn, leads to successful surgical and postoperative outcomes. In adopting the most effective treatment approach to pediatric brain tumors, the physician increasingly chooses MRI as the imaging foundation upon which suspected malignancies are diagnosed and established malignancies are evaluated.

In summary, this chapter aims to convey: (1) the overwhelming advantages of MRI in treating pediatric brain tumors (especially when used in tandem with other modalities); and (2) the MR imaging features that are unique to specific lesions affecting the CNS. An understanding of these two points should have an immediate and positive impact on 'best practice' in treating pediatric brain tumors and result in vastly improved patient outcomes.

REFERENCES

1. Gurney J, Smith M, Bunin G (1999). CNS and miscellaneous intracranial and intraspinal neoplasms. In *Cancer Incidence and Survival among Children and Adolescents: United States SEER Program 1975–1995 NIH Publication No. 99–4649*. Ries L, Smith M, Gurney J (eds). National Cancer Institute SEER Program, Bethesda. 51–63.
2. CBTRUS (2005). *Statistical Report: Primary Brain Tumors in the United States, 1998–2002*. Published by the Central Brain Tumor Registry of the United States.
3. Ries L, Harkins D, Krapcho M et al (2006). *SEER Cancer Statistics Review, 1975–2003*. National Cancer Institute, Bethesda.
4. Poussaint TY (2001). Magnetic resonance imaging of pediatric brain tumors: state of the art. Top Magn Reson Imaging 12:411–433.
5. Dunbar SF, Barnes PD, Tarbell NJ (1993). Radiologic determination of the caudal border of the spinal field in cranial spinal irradiation. Int J Radiat Oncol Biol Phys 26:669–673.
6. Hustinx R, Alavi A (1999). SPECT and PET imaging of brain tumors. Neuroimaging Clin N Am 9:751–766.
7. Poussaint TY, Rodriguez D (2006). Advanced neuroimaging of pediatric brain tumors: MR diffusion, MR perfusion, and MR spectroscopy. Neuroimaging Clin N Am 16:169–192, ix.
8. Provenzale JM, Mukundan S, Barboriak DP (2006). Diffusion-weighted and perfusion MR imaging for brain tumor characterization and assessment of treatment response. Radiology 239:632–649.
9. Gauvain KM, McKinstry RC, Mukherjee P et al (2001). Evaluating pediatric brain tumor cellularity with diffusion-tensor imaging. Am J Roentgenol 177:449–454.
10. Kono K, Inoue Y, Nakayama K et al (2001). The role of diffusion-weighted imaging in patients with brain tumors. Am J Neuroradiol 22:1081–1088.
11. Mardor Y, Roth Y, Lidar Z et al (2001). Monitoring response to convection-enhanced taxol delivery in brain tumor patients using diffusion-weighted magnetic resonance imaging. Cancer Res 61:4971–4973.
12. Ross BD, Chenevert TL, Rehemtulla A (2002). Magnetic resonance imaging in cancer research. Eur J Cancer 38:2147–2156.

13. Witwer BP, Moftakhar R, Hasan KM et al (2002). Diffusion-tensor imaging of white matter tracts in patients with cerebral neoplasm. J Neurosurg 97:568–575.

14. Khong PL, Kwong DL, Chan GC et al (2003). Diffusion-tensor imaging for the detection and quantification of treatment-induced white matter injury in children with medulloblastoma: a pilot study. Am J Neuroradiol 24:734–740.

15. Khong PL, Leung LH, Chan GC et al (2005). White matter anisotropy in childhood medulloblastoma survivors: association with neurotoxicity risk factors. Radiology 236:647–652.

16. Pollack IF (1994). Brain tumors in children. New Engl J Med 331:1500–1507.

17. Packer R, Pollack I (2000). *Pediatric Brain Tumors, Brain Tumor Progress Review Group*. National Cancer Institute, Bethesda.

18. Kleihues P, Cavenee WK (2000). *World Health Organization Classification of Tumours: Pathology and Genetics of Tumours of the Nervous System*. IARC Press, Lyon.

19. Aronen HJ, Gazit IE, Louis DN et al (1994). Cerebral blood volume maps of gliomas: comparison with tumor grade and histologic findings. Radiology 191:41–51.

20. Fayed N, Modrego PJ (2005). The contribution of magnetic resonance spectroscopy and echoplanar perfusion-weighted MRI in the initial assessment of brain tumours. J Neuro-Oncol 72:261–265.

21. Jackson A, Kassner A, Annesley-Williams D et al (2002). Abnormalities in the recirculation phase of contrast agent bolus passage in cerebral gliomas: comparison with relative blood volume and tumor grade. Am J Neuroradiol 23:7–14.

22. Knopp EA, Cha S, Johnson G et al (1999). Glial neoplasms: dynamic contrast-enhanced T2*-weighted MR imaging. Radiology 211:791–798.

23. Law M, Yang S, Wang H et al (2003). Glioma grading: sensitivity, specificity, and predictive values of perfusion MR imaging and proton MR spectroscopic imaging compared with conventional MR imaging. Am J Neuroradiol 24:1989–1998.

24. Sugahara T, Korogi Y, Kochi M et al (1998). Correlation of MR imaging-determined cerebral blood volume maps with histologic and angiographic determination of vascularity of gliomas. Am J Roentgenol 171:1479–1486.

25. Sugahara T, Korogi Y, Kochi M et al (2001). Perfusion-sensitive MR imaging of gliomas: comparison between gradient-echo and spin-echo echo-planar imaging techniques. Am J Neuroradiol 22:1306–1315.

26. Ball WS Jr, Holland SK (2001). Perfusion imaging in the pediatric patient. Magn Reson Imaging Clin N Am 9:207–230.

27. Yamasaki F, Kurisu K, Satoh K et al (2005). Apparent diffusion coefficient of human brain tumors at MR imaging. Radiology 235:985–991.

28. Magalhaes A, Godfrey W, Shen Y et al (2005). Proton magnetic resonance spectroscopy of brain tumors correlated with pathology. Acad Radiol 12:51–57.

29. Tzika AA, Astrakas LG, Zarifi MK et al (2003). Multiparametric MR assessment of pediatric brain tumors. Neuroradiology 45:1–10.

30. Figeroa RE, el Gammal T, Brooks BS et al (1989). MR findings on primitive neuroectodermal tumors. J Comput Assist Tomogr 13:773–778.

31. Furie DM, Provenzale JM (1995). Supratentorial ependymomas and subependymomas: CT and MR appearance. J Comput Assist Tomogr 19:518–526.

32. Haines JL, Short MP, Kwiatkowski DJ et al (1991). Localization of one gene for tuberous sclerosis within 9q32–9q34, and further evidence for heterogeneity. Am J Hum Genet 49:764–772.

33. Kandt RS, Haines JL, Smith M et al (1992). Linkage of an important gene locus for tuberous sclerosis to a chromosome 16 marker for polycystic kidney disease. Nat Genet 2:37–41.

34. Goh S, Butler W, Thiele EA (2004). Subependymal giant cell tumors in tuberous sclerosis complex. Neurology 63:1457–1461.

35. Mizuguchi M, Takashima S (2001). Neuropathology of tuberous sclerosis. Brain Dev 23:508–515.

36. Houser OW, Gomez MR (1992). CT and MR imaging of intracranial tuberous sclerosis. J Dermatol 19:904–908.

37. Koeller KK, Dillon WP (1992). Dysembryoplastic neuroepithelial tumors: MR appearance. Am J Neuroradiol 13:1319–1325.

38. Krieger MD, Panigrahy A, McComb JG et al (2005). Differentiation of choroid plexus tumors by advanced magnetic resonance spectroscopy. Neurosurg Focus 18:E4.

39. Panigrahy A, Krieger MD, Gonzalez-Gomez I et al (2006). Quantitative short echo time 1H-MR spectroscopy of untreated pediatric brain tumors: preoperative diagnosis and characterization. Am J Neuroradiol 27:560–572.

40. Barkovich A (2005) *Pediatric Neuroimaging*, 4th edn. Lippincott Williams & Wilkins, Philadelphia.

41. Bourgouin PM, Tampieri D, Grahovac SZ et al (1992). CT and MR imaging findings in adults with cerebellar medulloblastoma: comparison with findings in children. Am J Roentgenol 159:609–612.

42. Koci TM, Chiang F, Mehringer CM et al (1993). Adult cerebellar medulloblastoma: imaging features with emphasis on MR findings. Am J Neuroradiol 14:929–939.

43. Meyers SP, Kemp SS, Tarr RW (1992). MR imaging features of medulloblastomas. Am J Roentgenol 158:859–865.

44. Rollins N, Mendelsohn D, Mulne A et al (1990). Recurrent medulloblastoma: frequency of tumor enhancement on Gd-DTPA MR imaging. Am J Neuroradiol 11:583–587.

45. Zerbini C, Gelber RD, Weinberg D et al (1993). Prognostic factors in medulloblastoma, including DNA ploidy. J Clin Oncol 11:616–622.

46. Kuhl J (1998). Modern treatment strategies in medulloblastoma. Childs Nerv Syst 14:2–5.

47. Maleci A, Cervoni L, Delfini R (1992). Medulloblastoma in children and in adults: a comparative study. Acta Neurochirurg 119:62–67.

48. Rumboldt Z, Camacho DL, Lake D et al (2006). Apparent diffusion coefficients for differentiation of cerebellar tumors in children. Am J Neuroradiol 27:1362–1369.

49. Kovanlikaya A, Panigrahy A, Krieger MD et al (2005). Untreated pediatric primitive neuroectodermal tumor in vivo: quantitation of taurine with MR spectroscopy. Radiology 236:1020–1025.

50. Tong Z, Yamaki T, Harada K et al (2004). In vivo quantification of the metabolites in normal brain and brain tumors by proton MR spectroscopy using water as an internal standard. Magn Reson Imaging 22:1017–1024.

51. Wang Z, Sutton LN, Cnaan A et al (1995). Proton MR spectroscopy of pediatric cerebellar tumors. Am J Neuroradiol 16:1821–1833.

52. Gjerris F, Klinken L (1978). Long-term prognosis in children with benign cerebellar astrocytoma. J Neurosurg 49:179–184.

53. Pencalet P, Maixner W, Sainte-Rose C et al (1999). Benign cerebellar astrocytomas in children. J Neurosurg 90: 265–273.

54. Campbell JW, Pollack IF (1996). Cerebellar astrocytomas in children. J Neuro-Oncol 28:223–231.

55. Hwang JH, Egnaczyk GF, Ballard E et al (1998). Proton MR spectroscopic characteristics of pediatric pilocytic astrocytomas. Am J Neuroradiol 19:535–540.

56. Epstein F, Wisoff JH (1988). Intrinsic brainstem tumors in childhood: surgical indications. J Neuro-Oncol 6:309–317.

57. Littman P, Jarret P, Bilaniuk L et al (1980). Pediatric brainstem gliomas. Cancer 45: 2787–2792.

58. Freeman CR, Farmer JP (1998). Pediatric brain stem gliomas: a review. Int J Radiat Oncol Biol Phys 40:265–271.

59. Laprie A, Pirzkall A, Haas-Kogan DA et al (2005). Longitudinal multivoxel MR spectroscopy study of pediatric diffuse brainstem gliomas treated with radiotherapy. Int J Radiat Oncol Biol Phys 62:20–31.

60. Thakur SB, Karimi S, Dunkel IJ et al (2006). Longitudinal MR spectroscopic imaging of pediatric diffuse pontine tumors to assess tumor aggression and progression. Am J Neuroradiol 27:806–809.

61. Helton KJ, Phillips NS, Khan RB et al (2006). Diffusion tensor imaging of tract involvement in children with pontine tumors. Am J Neuroradiol 27:786–793.

62. Kwon JW, Kim IO, Cheon JE et al (2006). Paediatric brain-stem gliomas: MRI, FDG-PET and histological grading correlation. Pediatr Radiol Online, 18 July 2006.

63. Hoffman HJ (1996). Dorsally exophytic brain stem tumors and midbrain tumors. Pediatr Neurosurg 24:256–262.

64. Farmer JP, Montes JL, Freeman CR et al (2001). Brainstem gliomas. A 10-year institutional review. Pediatr Neurosurg 34:206–214.

65. Robertson PL, Allen JC, Abbott IR et al (1994). Cervicomedullary tumors in children: a distinct subset of brainstem gliomas. Neurology 44:1798–1803.

66. Epstein F, Wisoff J (1987). Intra-axial tumors of the cervicomedullary junction. J Neurosurg 67:483–487.

67. Poussaint TY, Yousuf N, Barnes PD et al (1999). Cervicomedullary astrocytomas of childhood: clinical and imaging follow-up. Pediatr Radiol 29:662–668.

68. Epstein F, Farmer J (1993). Brain-stem glioma growth patterns. J Neurosurg 78:408–412.

69. Poussaint TY, Kowal JR, Barnes PD et al (1998). Tectal tumors of childhood: clinical and imaging follow-up. Am J Neuroradiol 19:977–983.

70. Ternier J, Wray A, Puget S et al (2006). Tectal plate lesions in children. J Neurosurg 104:369–376.

71. Chen CJ, Tseng YC, Hsu HL et al (2004). Imaging predictors of intracranial ependymomas. J Comput Assist Tomogr 28:407–413.

72. Meyers SP, Khademian ZP, Biegel JA et al (2006). Primary intracranial atypical teratoid/rhabdoid tumors of infancy and childhood: MRI features and patient outcomes. Am J Neuroradiol 27:962–971.

73. Rorke LB, Packer RJ, Biegel JA (1996). Central nervous system atypical teratoid/rhabdoid tumors of infancy and childhood: definition of an entity. J Neurosurg 85:56–65.

74. Biegel JA (2006). Molecular genetics of atypical teratoid/rhabdoid tumor. Neurosurg Focus 20:E11.

75. Petito CK, DeGirolami U, Earle KM (1976). Craniopharyngiomas: a clinical and pathological review. Cancer 37:1944–1952.

76. Bunin GR, Surawicz TS, Witman PA et al (1998). The descriptive epidemiology of craniopharyngioma. J Neurosurg 89:547–551.

77. Hetelekidis S, Barnes PD, Tao ML et al (1993). 20-year experience in childhood craniopharyngioma. Int J Radiat Oncol Biol Phys 27:189–195.

78. Pusey E, Kortman KE, Flannigan BD et al (1987). MR of craniopharyngiomas: tumor delineation and characterization. Am J Roentgenol 149:383–388.

79. Kollias S, Barkovich A, Edwards M (1991–1992). Magnetic resonance analysis of suprasellar tumors of childhood. Pediatr Neurosurg 17:284–303.

80. Janss AJ, Grundy R, Cnaan A et al (1995). Optic pathway and hypothalamic/chiasmatic gliomas in children younger than age 5 years with a 6-year follow-up. Cancer 75:1051–1059.

81. Aoki S, Barkovich AJ, Nishimura K et al (1989). Neurofibromatosis types 1 and 2: cranial MR findings. Radiology 172:527–534.

82. Rodriguez D, Young Poussaint T (2004). Neuroimaging findings in neurofibromatosis type 1 and 2. Neuroimaging Clin N Am 14:149–170, vii.

83. Barnes PD, Robson CD, Robertson RL et al (1996). Pediatric orbital and visual pathway lesions. Neuroimaging Clin N Am 6:179–198.

84. Simon J, Szumowski J, Totterman S et al (1988). Fat suppression MR imaging of the orbit. Am J Neuroradiol 9:961–968.

85. Parsa CF, Hoyt CS, Lesser RL et al (2001). Spontaneous regression of optic gliomas: thirteen cases documented by serial neuroimaging. Arch Ophthalmol 119:516–529.

86. Poussaint TY, Barnes PD, Nichols K et al (1997). Diencephalic syndrome: clinical features and imaging findings. Am J Neuroradiol 18:1499–1505.

87. Fleischman A, Brue C, Poussaint TY et al (2005). Diencephalic syndrome: a cause of failure to thrive and a model of partial growth hormone resistance. Pediatrics 115:e742–748.

88. Poussaint TY, Gudas T, Barnes PD (1999). Imaging of neuroendocrine disorders of childhood. Neuroimaging Clin N Am 9:157–175.

89. Amstutz DR, Coons SW, Kerrigan JF et al (2006). Hypothalamic hamartomas: correlation of MR imaging and spectroscopic findings with tumor glial content. Am J Neuroradiol 27:794–798.

90. Clarren SK, Alvord EC Jr, Hall JG (1980). Congenital hypothalamic hamartoblastoma, hypopituitarism, imperforate anus, and postaxial polydactyly – a new syndrome? Part II: Neuropathological considerations. Am J Med Genet 7:75–83.

91. Hall JG, Pallister PD, Clarren SK et al (1980). Congenital hypothalamic hamartoblastoma, hypopituitarism, imperforate anus and postaxial polydactyly – a new syndrome? Part I: clinical, causal, and pathogenetic considerations. Am J Med Genet 7:47–74.

92. Tien R, Kucharczyk J, Kucharczyk W (1991) MR imaging of the brain in patients with diabetes insipidus. Am J Neuroradiol 12:533–542.

93. Appignani B, Landy H, Barnes P (1993). MR in idiopathic central diabetes insipidus of childhood. Am J Neuroradiol 14:1407–1410.

94. Haddad S, VanGilder J, Menezes A (1991). Pediatric pituitary tumors. Neurosurgery 29:509–514.

95. Poussaint TY, Barnes PD, Anthony DA et al (1996). Hemorrhagic pituitary adenomas of adolescence. Am J Neuroradiol 17:1907–1912.

96. Kucharczyk W, Davis D, Kelly W et al (1986). Pituitary adenoma: high resolution MRI at 1.5 T. Radiology 161:761–765.

97. Maghnie M, Arico M, Villa A et al (1992). MR of the hypothalamic-pituitary axis in Langerhans cell histiocytosis. Am J Neuroradiol 13:1365–1371.

98. Barnes PD, Robertson RL, Young Poussaint T (1997). Structural imaging of CNS tumors. In *Cancer of the Nervous System*. Black PM, Loeffler JS (eds). Blackwell Science, Cambridge. 54–97.

99. Bouffet E, Doumi N, Thiesse P et al (1997). Brain metastases in children with solid tumors. Cancer 79:403–410.

CHAPTER 47

Intracranial Metastases

Arastoo Vossough, R. Gilberto Gonzalez, John W. Henson and Pamela W. Schaefer

INTRODUCTION

Brain metastases are the most common intracranial tumors in adults. Approximately 97 800–170 000 new cases are diagnosed in the USA each year [1,2]. There is an equal male to female ratio. The most common causes of brain metastases are cancers of the lung, breast, unknown primary, melanoma and colon. Melanoma is the most likely tumor to produce brain metastases. The majority of metastases from extracranial primary tumors to the central nervous system are intraparenchymal. Approximately 80% of metastases occur supratentorially, 15% in the cerebellum and 5% in the brainstem [3,4]. Gastrointestinal, prostate and uterine cancers disproportionately metastasize to the posterior fossa [3]. Multiple metastatic lesions are seen in 60–75% of all cases as determined by gadolinium-enhanced MR imaging [2,5]. Melanoma and small cell lung carcinoma metastases are most likely to be multiple. Metastases are often located at gray–white matter junctions. It is thought that tumor emboli lodge in arterioles where changes in vessel caliber trap metastatic emboli. Breakdown of the blood–brain barrier and increased vascular permeability lead to characteristic surrounding vasogenic edema. Metastases typically have a large amount of surrounding vasogenic edema, in comparison to the size of the metastatic nidus. Despite the above typical characteristics, in a prospective study of 48 adult patients who were initially thought to have a single brain metastasis by MRI, 11% were found on biopsy to have a different diagnosis [6].

Modern neuroimaging techniques have the potential of increasing the sensitivity and specificity of cerebral metastasis detection.

COMPUTED TOMOGRAPHY (CT)

On non-contrast-enhanced head CT scans, metastases are typically hypodense to isodense, well-circumscribed solid lesions (Figure 47.1). Some may have necrotic low attenuation centers. Low attenuation vasogenic edema surrounds most metastases and is often disproportionately large compared to the size of the mass. Less commonly, metastases are hyperdense, which include hemorrhagic metastases, such as melanoma, choriocarcinoma, renal cell carcinoma, thyroid carcinoma, lung and breast. Hyperdensity is also seen in tumors with high cellularity and high nuclear-to-cytoplasmic ratio, such as lymphoma and small cell lung cancer. Metastases are infrequently hyperdense secondary to calcification, though a variety of metastases can calcify.

Most metastases densely enhance with iodinated contrast secondary to breakdown of the blood–brain barrier and angiogenesis (Figure 47.1B). Enhancement can be homogenous or a ring configuration. High-dose contrast with delayed imaging at 1 hour may detect additional lesions in a significant number of patients [7]. CT may fail to identify posterior fossa lesions and isodense hemorrhagic metastases. The ability to detect a lesion on any imaging modality is dependent on both lesion contrast, with respect to background, and lesion

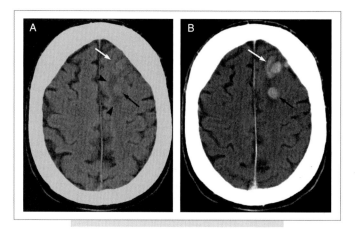

FIG. 47.1. Computed tomography in a 51-year-old patient with brain metastases from breast cancer. **(A)** Non-contrast CT scan showing hypodense edema (arrowheads) surrounding an isodense round brain metastatic focus in the subcortical left frontal white matter (black arrow). A second isodense focus within the cortex is barely seen given the similar attenuation to the cortical gray-matter (white arrow). **(B)** CT after the administration of iodinated contrast greatly increases the conspicuity of the metastatic foci and now the cortical lesion (white arrow) is clearly visualized.

size. Large lesions and lesions with high contrast are detected more easily than lesions with small size and low contrast. A small lesion may be detected if it possesses enough inherent contrast to become conspicuous. Lesions missed on CT are almost always less than 10 mm in size. Immediate imaging after injection of contrast, such as when a CT angiography (CTA) is performed, is not suitable for metastasis detection, since more time is needed for contrast to accumulate within the tumor. Therefore, CTA is not an adequate post-contrast study for evaluation of metastases. Although not as sensitive as MR imaging, CT remains very useful for the detection of parenchymal metastases.

MAGNETIC RESONANCE IMAGING (MRI)

Conventional MRI

On MRI, most metastases are slightly hypointense to isointense with respect to gray matter on T1-weighted images with surrounding hypointense edema (Figure 47.2). Marked central hypointensity is characteristic of necrotic or cystic tumors. On T2-weighted images, most metastases are isointense to hyperintense when compared to gray matter with the surrounding edema being even more hyperintense (Figure 47.2). Marked central hyperintensity is again seen in necrotic or cystic tumors. Hemorrhagic metastases are an exception to this pattern and demonstrate variable signal intensities depending on the stage of hemorrhage. Adenocarcinomas demonstrate T2 hypointensity for unknown reasons and without correlation to the presence of mucin, blood products, iron or calcium [8]. Cellular metastases with high nuclear-to-cytoplasmic ratios, such as lymphomas, and calcified metastases may also demonstrate T2 hypointensity. Peripheral gray matter lesions and lesions without significant edema can be missed on unenhanced MR imaging. Unenhanced MR imaging can also fail to demonstrate small nodules hidden by edema from other more dominant lesions. Lesions missed on CT and unenhanced MR imaging are almost always less than 10 mm in size.

Almost all metastases enhance following intravenous contrast injection on both CT and MR imaging secondary to blood–brain barrier breakdown and tumor angiogenesis (Figure 47.2D). Davis et al have shown that gadolinium-enhanced MR imaging is superior to double-dose delayed CT for lesion detection, anatomic localization and differentiation of solitary versus multiple metastases [7]. Metastases may enhance in a solid or ring-like configuration; the wall of the ring is usually thick and irregular. There may be a stage when metastases grow avascularly, therefore, they do not enhance on routine MR imaging, but are large enough to be detected on T2-weighted MR sequences [7]. Contrast enhancement may be strikingly decreased by steroid therapy. In addition, contrast enhancement may be altered by radiation or chemotherapy.

At times, single or multiple metastases may be differentiated from non-neoplastic entities on the basis of their

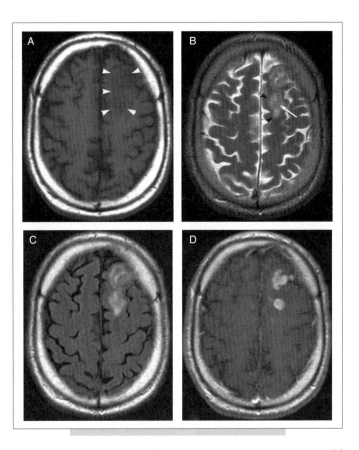

FIG. 47.2. Conventional MRI sequences in a 51-year-old patient with brain metastases from breast cancer. **(A)** Axial T1-weighted image without contrast shows a slightly hypointense area in the left frontal lobe (arrowheads). **(B)** Axial T2-weighted image showing hyperintense vasogenic edema (black arrowheads), surrounding a slightly less bright metastatic lesion in the left frontal lobe white matter (white arrow). The metastatic lesion itself is relatively hypointense to the edematous surrounding white matter, presumably due to denser cell packing within the tumor. **(C)** Axial FLAIR image showing hyperintense vasogenic edema. Often, the FLAIR image more conspicuously shows the edema surrounding metastatic lesions. **(D)** Axial T1-weighted image after the intravenous administration of gadolinium contrast shows avid, solid enhancement of the metastatic foci.

pattern of rim enhancement. Abscesses, demyelinating lesions and resolving hematomas frequently have smooth, thin walls and metastases and primary tumors usually have thick irregular walls. Primary tumors and metastases frequently have similar enhancement patterns and it is not possible to differentiate between the two on conventional MRI. Extension of edema into the corpus callosum or gray matter favors a high-grade glioma and the presence of multiple lesions favors metastases. Differentiating hemorrhagic metastases from non-neoplastic hemorrhagic lesions is frequently difficult. Findings suggesting tumor are non-hemorrhagic areas, which enhance, an incomplete

T2 hypointense hemosiderin ring, persistent edema and delayed evolution of hemorrhage. Carcinomatous encephalitis or miliary metastasis is a rare form of metastatic disease to the brain, usually from primary lung carcinoma. It is characterized by diffuse spread of small tumor nodules to the brain and meninges in a perivascular distribution.

There are several strategies to improve further the detection of metastases by MRI.

Double- and Triple-dose Contrast

A useful technique for increasing lesion detection is increasing the contrast dose beyond the standard 0.1 mmol/kg of body weight. There is a relatively linear relationship between lesion contrast ratio and intravenous contrast dose over the range of 0.05 mmol/kg to 0.3 mmol/kg [9]. In a larger series specifically evaluating brain metastasis, it was shown that triple-dose gadolinium MRI increases mean lesion contrast, increases qualitative conspicuity of lesions, detects additional small lesions and, in a subset of patients, would have changed choice of therapy [10]. Increased contrast-to-noise ratio and visual assessment ratings were also shown to be superior in a randomized trial [11]. Lesions greater than 10 mm will be seen at all dose levels. The utility of higher dose gadolinium mainly lies in the ability to reveal smaller lesions, especially less than 5 mm [10,12].

Delayed Imaging

Several studies of standard-dose gadolinium enhanced MR imaging also looked at delayed imaging and found marginal benefit. The findings show improvement in lesion detection for small lesions with delayed imaging. Although delayed imaging improves detection of small metastases, high dose (0.3 mol/kg) imaging shows the most lesions [10]. One study of the effective time window for scanning concluded that the post-contrast scan should be started with a 2–5 minute delay after injection of contrast medium and there is no major advantage in waiting longer [13].

Magnetization Transfer (MT)

Magnetization transfer MR imaging is a technique that increases the contrast of enhancing and non-enhancing lesions by suppressing background signal and is achieved by selectively saturating the signal from the immobile water protons. MT can be used to improve lesion contrast of gadolinium-enhanced brain lesions on T1-weighted spin-echo images. In one study, using MT resulted in a 108% improvement in the contrast-to-noise compared with conventional T1-weighted gadolinium-enhanced images [14]. Some studies have suggested that use of standard gadolinium dose with MT are equivalent to those reported for triple-dose gadolinium-enhanced MR imaging with conventional spin-echo techniques [14,15]. While use of triple-dose with MT further increases contrast, it does not result in detection of additional tumors [15].

Newer Contrast Agents

MRI contrast agents with higher T1 relativity have been shown to improve the sensitivity of brain metastasis detection. Gadobenate dimeglumine significantly increased the sensitivity from gadopentetate dimeglumine, gadodiamide and gadoterate meglumine at similar doses [16]. Gadobenate dimeglumine has also been shown to have similar sensitivity at reduced doses when compared to standard dose gadodiamide [17].

Higher Field MRI

Higher field MR magnets above 1.5 Tesla now have widespread clinical availability. It has been shown that 3.0T imaging increases both the subjective assessment of brain metastases and also produces significantly higher signal-to-noise and contrast-to-noise ratios when compared to 1.5T [18]. Gadolinium administration produces higher contrast between tumor and normal brain on 3.0T than on 1.5T [19], resulting in better detection of brain metastases and leptomeningeal involvement [18]. The relatively improved detection rate by using higher contrast doses is maintained at high field.

Gradient Echo T2* Sequences and Melanoma Imaging

Despite its relative rarity as a systemic neoplasm, melanoma is the third most common primary to produce brain metastasis. Approximately one-half of melanoma metastases are hyperintense on T1-weighted images prior to the administration of gadolinium, while other cerebral metastases rarely demonstrate T1 hyperintensity [20]. Melanin itself may lead to T1 shortening and melanoma metastases have a propensity for hemorrhage, with methemoglobin also producing T1 shortening. Additionally, melanin and blood products may also produce susceptibility effect on T2* images due to the presence of metal ions including iron, copper, manganese and zinc (Figure 47.3). T2*-weighted signal loss and T1 shortening are both five times more common in melanoma metastases than in lung cancer metastases. Three-quarters of melanoma metastases had either susceptibility effect or intrinsic T1 hyperintensity and 7% of melanoma lesions were detected principally on T2*-weighted sequences [20]. These findings underline the value of the T2*-weighted sequence in patients with known melanoma who are undergoing CNS staging.

Diffusion-Weighted Imaging (DWI)

Technical aspects of DWI are discussed in Chapter 25. Metastatic tumors with dense cell packing or high nuclear-to-cytoplasmic ratio, such as lymphoma and small cell lung cancer can show restricted diffusion on DWI (Figure 47.4). Diffusion imaging aids in the differentiation of cystic neoplasms and abscesses and can play a role in delineating cystic from solid components in metastases [21].

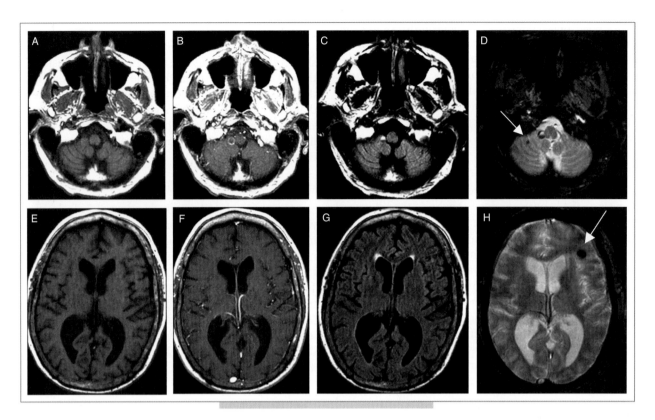

FIG. 47.3. Melanoma brain metastases detected on T2* imaging sequences. T1-weighted sequences before **(A, E)** and after **(B, F)** administration of gadolinium, FLAIR sequences **(C, G)** and T2*-weighted sequences **(D, H)** are shown. T1 isointense lesions are seen in the right cerebellar hemisphere lesion (top row) and left frontal lobe (bottom row). An additional small lesion is seen on T2* image in the right cerebellar hemisphere. A subtle abnormality can be seen in the left frontal lobe on the FLAIR image **(G)**, but the lesion is markedly more conspicuous on T2*-weighted image **(H)**.

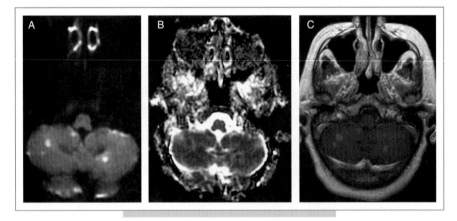

FIG. 47.4. Diffusion-weighted imaging in metastatic small cell lung cancer to the brain. **(A)** DWI imaging shows multiple foci of increased signal in the cerebellum bilaterally, initially raising the question of embolic infarcts. **(B)** ADC maps confirm low ADC and restricted diffusion in these lesions, along with surrounding hyperintensity and vasogenic edema. **(C)** Post-contrast T1-weighted image demonstrates solid and punctate enhancement in these lesions.

There is no significant difference in the fractional anisotropy (FA) values of either the enhancing or non-enhancing peritumoral portions of high-grade gliomas and solitary brain metastases [22,23]. However, when peritumoral FA values were expressed as a differential from values 'expected' in bland edema, there was a significant difference between metastases and gliomas. 'Expected' FA values were predicted from a linear regression of FA onto apparent diffusion coefficient (ADC) for a series of bland edema cases; the difference between 'expected' and observed FA was named the 'tumor infiltration index' [24]. Also, the peritumoral mean diffusivity (MD) of metastatic lesions measured significantly greater than that of gliomas [23,24]. Displacement of subcortical white matter fibers by subjective visual assessment of diffusion tensor tractography images (as opposed to tract disruption or invasion) is also more commonly seen in metastases compared to high-grade gliomas [22,25]. Diffusion anisotropy is highly sensitive to microstructural changes that do not appear on conventional imaging. Unfortunately, this high sensitivity is accompanied by relatively low pathologic specificity, such that useful DTI based tissue characterization may ultimately require the use of more sophisticated approaches. For example, the major eigenvalue of the diffusion tensor (reflecting diffusivity in the longitudinal direction) was found to be significantly lower in the peritumoral white matter surrounding high-grade gliomas than metastases, even when the anisotropy showed no difference [26].

Perfusion-Weighted Imaging (PWI)

Technical aspects of PWI are discussed in Chapter 29. The importance of tumor angiogenesis in regulation of tumor growth and development of metastatic disease has been demonstrated. Metastatic tumors to the brain induce neovascularization as they grow and expand. These neovessels resemble those of the primary systemic tumor with fenestrated membranes and open endothelial junctions, in contrast to normal brain capillaries with an intact blood–brain barrier with its tight junctions and continuous basement membrane [27]. It has been shown that PWI can be useful in differentiating solitary cerebral metastases from high-grade gliomas. PWI could not reliably differentiate metastatic tumors from high-grade gliomas when relative cerebral blood volume (rCBV) was measured in the enhancing portion of the tumors, since both are highly vascular tumors and demonstrate increased rCBV (Figure 47.5) [28]. However, rCBV was significantly higher within the peritumoral region of primary gliomas compared to metastatic lesions. The reason is thought to be in the different histopathologic milieu of the peritumoral region. In primary gliomas, the peritumoral region represents a variable combination of vasogenic edema and tumor cell infiltration, whereas in metastatic lesions, the peritumoral region represents pure vasogenic edema due to leaky capillaries [29,30]. Differentiation of metastases from low-grade gliomas is usually not a diagnostic dilemma, due to the general lack of contrast enhancement and minimal peritumoral edema in low-grade gliomas. In any case, low-grade gliomas often have low rCBV when compared to metastases and high-grade gliomas [30].

Magnetic Resonance Spectroscopy (MRS)

Technical aspects of MRS are discussed in Chapter 28. N-acetylaspartate (NAA) is low or absent in metastases, consistent with the lack of neuroglial elements [31–33]. Metastatic lesions also show an elevated choline/creatine

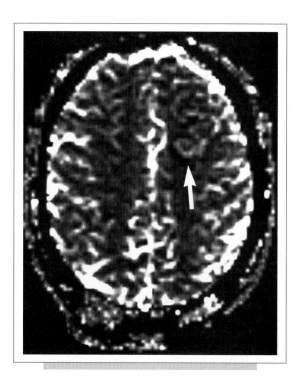

FIG. 47.5. Perfusion-weighted imaging (PWI) in a patient with breast cancer brain metastasis. This is a relative cerebral blood volume (rCBV) weighted image of the perfusion study, demonstrating increased rCBV in the round metastatic focus in the left frontal white matter (arrow).

ratio, as do gliomas [33–35]. Metastatic lesions were initially thought to be distinguishable from high-grade gliomas by the presence of lipid and lactate peaks, lack of creatine and increased lipid/creatine ratio [36,37]. However, in areas of central necrosis in high-grade gliomas, there is often presence of lactate and lipid, as well as a lack of creatine, therefore decreasing the specificity of the above findings [38]. Nevertheless, the finding of high choline in the *peritumoral* region of a lesion is more likely to represent a glioma rather than a metastasis, since as noted earlier, the peritumoral region of metastases is pure vasogenic edema, whereas in gliomas it represents a combination of tumor cell infiltration and vasogenic edema [38]. Increased choline/creatine ratio can sometimes distinguish areas of tumor recurrence from radiation necrosis.

NUCLEAR MEDICINE TECHNIQUES

Single-photon emission computed tomography (SPECT) is a method in which gamma photons emitted from a patient intravenously injected with a radiotracer are detected with a rotating gamma camera. Thallium-201 is a potassium analogue tracer that localizes in tumors. SPECT imaging with thallium-201 has been used chiefly in the evaluation of primary tumors, but also shows accumulation in metastatic tumors [39,40]. One group has shown that delayed thallium imaging may have some utility in differentiation of high-grade glioma and metastases [40]. One group reported a sensitivity of 70% for both thallium-201 and technetium-99m MIBI SPECT in the detection of cerebral metastases. SPECT thallium scanning is not useful for following tumors that do not demonstrate thallium uptake and cannot reliably be used for screening.

Positron emission tomography (PET) has been used extensively in the evaluation and follow-up of primary brain tumors, especially for distinguishing between necrosis from radiation and chemotherapy and recurrent tumor; residual tumor usually demonstrates increased FDG uptake, and necrosis from radiation or chemotherapy usually demonstrates decreased FDG uptake. In one study, FDG-PET detected only 50–68% of metastases from non-CNS neoplasms [41] and in another study, only 61% of metastatic lesions in the brain were identified by PET [42]. Coregistration of PET with MRI improved the sensitivity for detecting metastatic recurrence from 65% to 86% [43]. Unsuspected brain metastases were detected in only 0.7% undergoing PET scanning for evaluation of extracranial neoplasms [44]. In another large study, unsuspected cerebral or skull metastases were detected in only 0.4% of patients and the authors concluded that FDG-PET screening for cerebral lesions in patients with body malignancy has little clinical impact [45]. One reason for the low sensitivity of PET may be the fact that metastases are often located close to the high FDG uptake gray matter at the gray–white junction, and hence are more difficult to identify.

There is variability in FDG accumulation in metastases of different pathologies, as well as among lesions of similar pathology. When a metastasis demonstrates increased FDG uptake similar to a high-grade primary tumor, FDG PET could be useful in following response to treatment and in distinguishing recurrent tumor from necrosis secondary to radiation or chemotherapy. FDG PET should not be used as a screening method or for following lesions that are isometabolic or hypometabolic. Newer tracers are continuously being developed for PET imaging, which may further increase the utility of this modality in evaluation of brain metastases.

MENINGEAL METASTASES

Approximately 8% of patients with systemic cancer have evidence of meningeal metastasis at autopsy [7]. The most common clinical presentations are headache, mental status change, cranial and spinal nerve dysfunction, gait disturbance and leg and lower back pain. Cerebrospinal fluid (CSF) cytology is specific, but remains negative in 5–10% of patients [7]. The most common primary tumors to spread to the meninges are breast carcinoma, lung carcinoma, melanoma, lymphoma and leukemia. Most frequently, metastatic disease spreads to the meninges hematogenously via small meningeal vessels. Superficial parenchymal lesions and skull metastases may, however, directly invade the meninges. Dissemination to the CSF via perivascular and perineural lymphatics may also occur. Leptomeningeal metastases are more common than pachymeningeal metastases. Pachymeningeal metastases are frequently associated with calvarial metastases, but isolated dural metastases can be seen in breast cancer, lymphoma, leukemia, prostate cancer and neuroblastoma. Metastases involving the dura may result in subdural effusions and can also lead to venous sinus thrombosis, either by direct invasion or by compression.

Non-contrast-enhanced CT is quite insensitive in detecting meningeal carcinomatosis. Contrast-enhanced CT is more sensitive than non-enhanced T1- and T2-weighted MR images, but inferior to contrast MR imaging. MR imaging with gadolinium is most commonly used to detect pachymeningeal and leptomeningeal metastases, but FLAIR (fluid attenuated inversion-recovery) imaging can be quite sensitive in detection of diffuse leptomeningeal disease (Figure 47.6). Uniform diffuse meningeal enhancement is the most common MR finding. Other findings include focal meningeal enhancement, nodular meningeal enhancement and hydrocephalus. Infrequently, meningeal metastases invade the underlying parenchyma with parenchymal T2 hyperintensity, swelling and contrast enhancement.

Although MR imaging with gadolinium is sensitive for detecting meningeal disease, it is not specific. Diffuse linear or focal linear can be seen in both neoplastic and inflammatory lesions (see Figure 47.6E). Postoperatively, most patients have brain, pial and dural enhancement. Brain or pial enhancement longer than 1 year may suggest recurrent tumor. In patients with mass lesions causing obstructive hydrocephalus, meningeal enhancement may be secondary to vascular stasis rather than to carcinomatosis. In addition, dural enhancement adjacent to a parenchymal tumor does not always indicate invasion. Discontinuous enhancing dura adjacent to a mass correlates with dural invasion, but uniformly enhancing dura near a parenchymal tumor is often not due to dural invasion.

IMAGING APPROACH IN BRAIN METASTASES

New developments in neuroimaging are increasing the sensitivity for the detection of brain metastases. Given these advances, one question raised is whether or not the additional information has clinical usefulness, especially in terms of patient outcome and in relation to new therapeutic choices. Given a choice of multiple treatment options, the second question that must be answered is which imaging modality is appropriate for which group of patients. The answer cannot easily be derived from the literature, as there are inconsistencies in the literature, rapid advances in neuroimaging increasing the sensitivity and specificity of metastasis detection, continuous development of new treatments and regional variation in the availability of these various diagnostic and

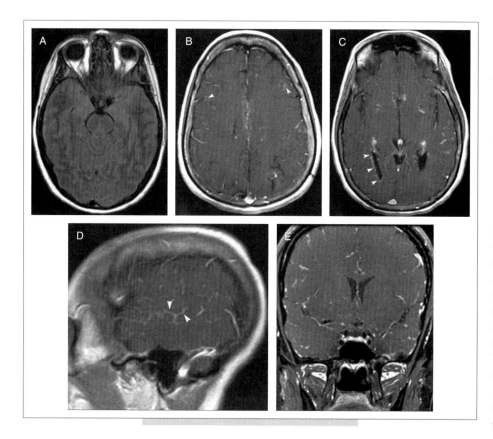

FIG. 47.6. MRI of leptomeningeal metastases and potential mimickers. **(A)** Axial FLAIR MRI image demonstrating abnormal diffuse sulcal hyperintensity in a 58-year-old patient with non-small cell lung carcinoma. Normally, the sulci should be dark on FLAIR. **(B)** Axial post-contrast T1-weighted MRI in same patient, showing abnormal enhancement of the sulci and leptomeninges, some of which have a nodular appearance (arrowheads). **(C)** Axial post-contrast T1-weighted MRI in same patient, showing abnormal enhancement of the ependymal surface of the lateral ventricles (arrowheads). **(D)** Sagittal post-contrast T1-weighted MRI with better demonstration of sulcal leptomeningeal enhancement (arrowheads). **(E)** Abnormal diffuse sulcal enhancement on coronal post-contrast T1-weighted MRI in a 32-year-old patient with neurosarcoidosis. Inflammatory and infectious processes of the meninges may at times be indistinguishable from meningeal metastases.

therapeutic modalities. A reasonable approach would base screening on neurologic symptoms, primary tumor staging and cell type. Neuroimaging screening for brain metastases is generally indicated in patients with any stage of disease or cell type who present with focal or non-focal neurologic symptoms or unexplained systemic symptoms. Screening of asymptomatic patients is controversial and dependent on the primary tumor. A large number of malignant melanoma patients harbor brain metastases. Therefore, most physicians would agree that routine brain screening for asymptomatic melanoma metastases is indicated.

Another issue is the imaging evaluation of patients presenting with a brain mass. Patients present to their physician or to the emergency ward with neurological symptoms and a brain mass is discovered on neuroimaging. Since the first step in the management of these patients is to obtain a histopathologic diagnosis, the search for a systemic tumor can be seen as an approach to selecting a biopsy site. In the absence of a history of systemic cancer, a series of radiological studies usually follows, which may include CT imaging of the chest, abdomen and pelvis, radionuclide bone scan and whole-body FDG-PET imaging. An extended search for the extent of systemic tumor prior to establishing a histopathologic diagnosis could lead to an inappropriate expenditure of resources, including unnecessary imaging studies and longer interval to diagnostic

biopsy [46]. Staging examinations are appropriate only for patients with a systemic malignancy and are inappropriate for most patients with primary brain tumors. Mavrakis et al performed a retrospective analysis of the presenting features and diagnostic workup of patients with neurologic symptoms as the presenting manifestation of cancer [46]. Clinical features were not useful in predicting primary versus metastatic tumor. Brain MRI and chest CT identified the ultimate site of diagnostic biopsy in 97% of patients who present with a newly detected brain mass. The clinical and imaging features of these patients can be used to optimize the diagnostic workup of patients with newly detected brain masses.

CONCLUSION

High-resolution MR imaging with intravenous gadolinium has greatly enhanced the ability to detect and characterize metastases as well as to reasonably differentiate them from other intracranial processes. Combination of conventional MRI with more advanced techniques of DWI, PWI, MRS and PET can be useful in the evaluation and management of patients with intracranial metastases. Nevertheless, additional research is needed to determine further which imaging modality is most appropriate for specific patient populations and to determine whether or how the newer imaging techniques are useful in terms of patient outcome.

REFERENCES

1. Johnson JD, Young B (1996). Demographics of brain metastasis. Neurosurg Clin N Am 7(3):337–344.

2. Hutter A et al (2003). Brain neoplasms: epidemiology, diagnosis, and prospects for cost-effective imaging. Neuroimaging Clin N Am 13(2):237–250, x–xi.

3. Posner JB (1992). Management of brain metastases. Rev Neurol (Paris) 148(6–7):477–487.

4. Delattre JY et al (1988). Distribution of brain metastases. Arch Neurol 45(7):741–744.

5. Sze G et al (1990). Detection of brain metastases: comparison of contrast-enhanced MR with unenhanced MR and enhanced CT. Am J Neuroradiol 11(4):785–791.

6. Patchell RA et al (1990). A randomized trial of surgery in the treatment of single metastases to the brain. N Engl J Med 322(8):494–500.

7. Schaefer PW, Budzik RF Jr, Gonzalez RG (1996). Imaging of cerebral metastases. Neurosurg Clin N Am 7(3):393–423.

8. Carrier DA et al (1994). Metastatic adenocarcinoma to the brain: MR with pathologic correlation. Am J Neuroradiol 15(1):155–159.

9. Yuh WT et al (1992). Review of the use of high-dose gadoteridol in the magnetic resonance evaluation of central nervous system tumors. Invest Radiol 27 Suppl 1:S39–44.

10. Yuh WT et al (1995). The effect of contrast dose, imaging time, and lesion size in the MR detection of intracerebral metastasis. Am J Neuroradiol 16(2):373–380.

11. Haustein J et al (1993). Triple-dose versus standard-dose gadopentetate dimeglumine: a randomized study in 199 patients. Radiology 186(3):855–860.

12. Akeson P et al (1995). Brain metastases – comparison of gadodiamide injection-enhanced MR imaging at standard and high dose, contrast-enhanced CT and non-contrast-enhanced MR imaging. Acta Radiol 36(3):300–306.

13. Akeson P, Nordstrom CH, Holtas S (1997). Time-dependency in brain lesion enhancement with gadodiamide injection. Acta Radiol 38(1):19–24.

14. Finelli DA et al (1994). Improved contrast of enhancing brain lesions on postgadolinium, T1-weighted spin-echo images with use of magnetization transfer. Radiology 190(2):553–559.

15. Knauth M et al (1996). MR enhancement of brain lesions: increased contrast dose compared with magnetization transfer. Am J Neuroradiol 17(10):1853–1859.

16. Colosimo C et al (2001). Detection of intracranial metastases: a multicenter, intrapatient comparison of gadobenate dimeglumine-enhanced MRI with routinely used contrast agents at equal dosage. Invest Radiol 36(2):72–81.

17. Runge VM, Parker JR, Donovan M (2002). Double-blind, efficacy evaluation of gadobenate dimeglumine, a gadolinium chelate with enhanced relaxivity, in malignant lesions of the brain. Invest Radiol 37(5):269–280.

18. Ba-Ssalamah A et al (2003). Effect of contrast dose and field strength in the magnetic resonance detection of brain metastases. Invest Radiol 38(7):415–422.

19. Nobauer-Huhmann IM et al (2002). Magnetic resonance imaging contrast enhancement of brain tumors at 3 tesla versus 1.5 tesla. Invest Radiol 37(3):114–119.

20. Gaviani P et al (2006). Improved detection of metastatic melanoma by T2*-weighted imaging. Am J Neuroradiol 27(3):605–608.

21. Bukte Y et al (2005). Role of diffusion-weighted MR in differential diagnosis of intracranial cystic lesions. Clin Radiol 60(3):375–383.

22. Tsuchiya K et al (2005). Differentiation between solitary brain metastasis and high-grade glioma by diffusion tensor imaging. Br J Radiol 78(930):533–537.

23. Lu S et al (2003). Peritumoral diffusion tensor imaging of high-grade gliomas and metastatic brain tumors. Am J Neuroradiol 24(5):937–941.

24. Lu S et al (2004). Diffusion-tensor MR imaging of intracranial neoplasia and associated peritumoral edema: introduction of the tumor infiltration index. Radiology 232(1):221–228.

25. De La Cruz L et al (2005). Assessment of brain tumor with diffusion tensor imaging and tractography. In *RSNA Annual Meeting*. Radiological Society of North America, Chicago.

26. Wiegell MR, Henson JW, Tuch DS (2003). Diffusion tensor imaging shows potential to differentiate infiltrating from non-infiltrating tumors. In *Proceedings of the (ISMRM) Eleventh Scientific Meeting*. International Society for Magnetic Resonance in Medicine, Toronto.

27. Groothuis DR (2000). The blood-brain and blood-tumor barriers: a review of strategies for increasing drug delivery. Neuro-oncology 2(1):45–59.

28. Law M et al (2002). High-grade gliomas and solitary metastases: differentiation by using perfusion and proton spectroscopic MR imaging. Radiology 222(3):715–721.

29. Bertossi M et al (1997). Ultrastructural and morphometric investigation of human brain capillaries in normal and peritumoral tissues. Ultrastruct Pathol 21(1):41–49.

30. Cha S (2004). Perfusion MR imaging of brain tumors. Top Magn Reson Imaging 15(5):279–289.

31. Kwock L et al (2002). Clinical applications of proton MR spectroscopy in oncology. Technol Cancer Res Treat 1(1):17–28.

32. Poptani H et al (1995). Characterization of intracranial mass lesions with in vivo proton MR spectroscopy. Am J Neuroradiol 16(8):1593–1603.

33. Bruhn H et al (1989). Noninvasive differentiation of tumors with use of localized H-1 MR spectroscopy in vivo: initial experience in patients with cerebral tumors. Radiology 172(2):541–548.

34. Castillo M, Kwock L, Mukherji SK (1996). Clinical applications of proton MR spectroscopy. Am J Neuroradiol 17(1):1–15.

35. Sijens PE et al (1995). 1H MR spectroscopy in patients with metastatic brain tumors: a multicenter study. Magn Reson Med 33(6):818–826.

36. Ishimaru H et al (2001). Differentiation between high-grade glioma and metastatic brain tumor using single-voxel proton MR spectroscopy. Eur Radiol 11(9):1784–1791.

37. Bulakbasi N et al (2003). Combination of single-voxel proton MR spectroscopy and apparent diffusion coefficient calculation in the evaluation of common brain tumors. Am J Neuroradiol 24(2):225–233.

38. Law M (2004). MR spectroscopy of brain tumors. Top Magn Reson Imaging 15(5):291–313.

39. Dierckx RA et al (1994). Sensitivity and specificity of thallium-201 single-photon emission tomography in the functional detection and differential diagnosis of brain tumours. Eur J Nucl Med 21(7):621–633.

40. Kojima Y et al (1994). Differentiation of malignant glioma and metastatic brain tumor by thallium-201 single photon emission computed tomography. Neurol Med Chir (Tokyo) 34(9):588–592.

41. Griffeth LK et al (1993). Brain metastases from non-central nervous system tumors: evaluation with PET. Radiology 186(1):37–44.

42. Rohren EM et al (2003). Screening for cerebral metastases with FDG PET in patients undergoing whole-body staging of non-central nervous system malignancy. Radiology 226(1):181–187.

43. Chao ST et al (2001). The sensitivity and specificity of FDG PET in distinguishing recurrent brain tumor from radionecrosis in patients treated with stereotactic radiosurgery. Int J Cancer 96(3):191–197.

44. Larcos G, Maisey MN (1996). FDG-PET screening for cerebral metastases in patients with suspected malignancy. Nucl Med Commun 17(3):197–198.

45. Ludwig V et al (2002). Cerebral lesions incidentally detected on 2-deoxy-2-[18F]fluoro-D-glucose positron emission tomography images of patients evaluated for body malignancies. Mol Imaging Biol 4(5):359–362.

46. Mavrakis AN et al (2005). Diagnostic evaluation of patients with a brain mass as the presenting manifestation of cancer. Neurology 65(6):908–911.

CHAPTER 48

Non-Neoplastic Mass Lesions of the CNS

Amir A. Zamani

INTRODUCTION

Effective treatment of neurological diseases can begin only after a positive diagnosis has been made. Many neurological conditions have characteristic clinical histories and physical findings. Based on these, a positive diagnosis can be made in the majority of cases. Laboratory tests and radiological studies help to confirm the clinical diagnosis or narrow the list of differential diagnoses. At times, imaging studies are diagnostic enough to help clinicians arrive at a definite diagnosis; at other times, the imaging is confusing and produces a differential diagnosis of its own. This chapter deals with non-neoplastic conditions and other entities that simulate tumors. Before starting to discuss some of the more common entities, it is noteworthy to mention that a great deal of effort can be spared if proper clinical information is provided to the radiologist at the time of interpretation. Also, it is essential that the neuroradiologists have a thorough knowledge of the neurological diseases they deal with.

CYSTS AND CYST-LIKE LESIONS

Perivascular Spaces (PVS)

With the advent of magnetic resonance imaging (MRI), observation of perivascular spaces has become routine. With computed tomography (CT) only the most conspicuous spaces (in the inferolateral basal ganglia) were seen. Before the innocuous and ubiquitous nature of these spaces was recognized, erroneous diagnosis of lacunar infarction was frequently made. These spaces are fairly well defined, cystic looking regions with density/intensity characteristics of CSF [1]. The adjacent brain is normal in intensity, a fact that differentiates these from old lacunar infarcts. Perivascular spaces seen in high convexity white matter and those seen in periatrial regions commonly appear as punctuate or short linear spaces. Occasionally, bizarre shapes and distributions are seen, challenging the diagnostic acumen of the radiologist (Figure 48.1).

Widened perivascular spaces do not communicate with CSF spaces; they are merely spaces between vessels and wrapping pia containing interstitial fluid. They are seen where perforating vessels enter the brain. They are seen in inferolateral basal ganglia at a level that passes through the anterior commisure, in the brainstem and thalami, in the dentate nuclei, in the hippocampi and in white matter of high convexities, etc. Naturally, the great majority of these spaces are asymptomatic and remain so. It is claimed that, under certain conditions, these spaces enlarge or become more prominent. For example, in patients with suprasellar lesions, including craniopharyngiomas, the PVS surrounding the optic tracts are seen to enlarge. Enlargement

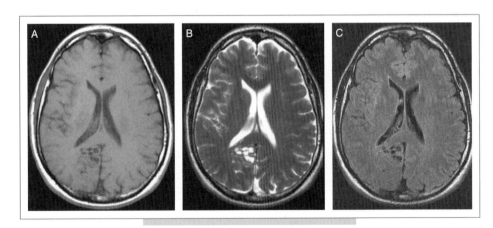

FIG. 48.1. Prominent perivascular spaces can occur at unusual sites. **(A)** and **(B)** T1 and T2WI of a cluster of these spaces posterior to the splenium. Note there is no hyperintensity around these lesions on FLAIR image **(C)**.

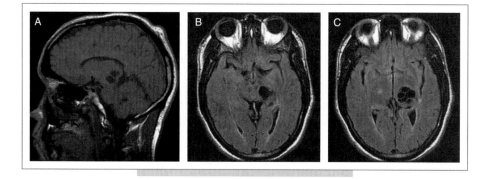

FIG. 48.2. Shows these spaces at their thalamomesencephalic site. Enlargement of these has been reported to cause hydrocephalus.

of high convexity spaces has been reported in patients with mild traumatic brain injury [2] and multiple sclerosis [3] and has been attributed to the associated inflammatory response. Very large spaces are called tumefactive spaces [4]. They appear as clusters of CSF intensity cysts and do not enhance. Widened perivascular spaces in the brainstem and thalami deserve special mention (Figure 48.2).

According to Saeki et al [5], widened perivascular spaces in the brainstem were first described by Elster and Richardson [6]. In a recent study of 115 patients with unrelated ocular symptoms, Saeki et al found widened PV spaces in the pontomesencephalic junction in 87% and in the thalamo-mesencephalic region in 63% of patients [5]. The great majority of these spaces were less than 4 mm in size. The authors mention that disturbance of drainage in these spaces can cause enlargement and lead to hydrocephalus, as a result of aqueductal stenosis. In a review of the literature, they found that Parkinson's syndrome and motor weakness were the other symptoms.

Other normal spaces containing CSF include cavum septum pellucidum, cavum Verage and cavum velum interpositum. These are rarely associated with symptoms and, on rare occasions, can be mistaken for a neoplasm; they usually do not require treatment and will not be discussed further.

Pineal Cysts

These cysts are fairly common. The incidence of pineal cysts at autopsy is about 20–40%. The great majority of these are quite small, only a few millimeters in size. At imaging, the incidence varies, but about 10% may show cysts in this region. Eighty percent of these are quite small (Figure 48.3) measuring less than 1 cm [7]. Larger lesions, with hemorrhage, may be seen. They are more common in females (F:M ratio 3:1) and they usually grow very slowly. On histology, they show an inner gliotic layer, a middle pineal parenchymal layer and an outer fibrovascular capsule [8].

On imaging, these cysts are usually sharply defined round or oval masses in the pineal region. The fluid content varies, hence the variability of signal intensity on MR images. They enhance with contrast and ring-like and nodular enhancements are most common. Larger lesions may cause hydrocephalus. Smaller lesions are usually asymptomatic.

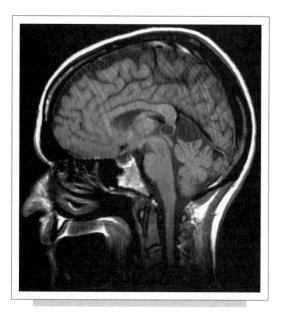

FIG. 48.3. Pineal cyst: typical pineal cyst is less than 1 cm in size and has a smooth thin wall. (This figure also shows an arachnoid cyst behind the pineal cyst and causing an indentation of the upper vermis.)

Parinaud's syndrome may be seen as a result of tectal compression.

Although the clinical course of these lesions is usually benign, the therapeutic decision-making is, at times, difficult. A cyst larger than 1.5 cm, compressing the tectum and causing symptoms, probably deserves histological diagnosis by biopsy and likely, surgical removal. A cyst measuring only a few millimeters found incidentally needs no intervention. Between these two extremes, there are many situations in which a definitive management strategy cannot be formulated. A developing consensus, however, is that these need to be followed clinically, rather than by repeated MRI, as these lesions grow very slowly [9].

Arachnoid Cysts

These are benign well-defined extra-axial cystic lesions containing CSF and enclosed by pia-arachnoid. The most

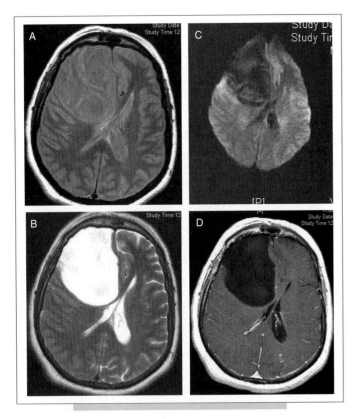

FIG. 48.4. Arachnoid cysts can be quite large and cause remodeling of adjacent inner table (**A** and **B**). They are not associated with diffusion restriction (**C**) and do not enhance (**D**).

common site for arachnoid cysts is within the middle cranial fossa [7]. A theory for formation of this type of arachnoid cyst is that the temporal lobe meninges fail to fuse as the sylvian fissure develops, thus forming an arachnoid cyst. Besides developmental causes, trauma and inflammation have been implicated in the formation of arachnoid cysts. Barkovich points out that these secondary cysts are different pathologically from congenital arachnoid cysts [10]. Other important sites include suprasellar, CP angles, sylvian fissures, quadrigeminal plate, interhemispheric, and convexity regions. Arachnoid cysts account for 1% of all intracranial non-traumatic masses. They are three times more common in males.

Many arachnoid cysts are asymptomatic. Depending on location however, headaches, seizures, hearing loss, tic and hydrocephalus can be seen. Middle fossa arachnoid cysts are associated with some degree of temporal lobe hypogenesis. Arachnoid cysts usually do not enlarge. Hemorrhage and sudden enlargement have been described rarely.

Radiologically, these lesions are sharply defined CSF density/intensity extra-axial lesions measuring from less than 1 cm to many centimeters. Large masses cause remodeling of the adjacent inner table and compress brain (Figure 48.4). On MRI, they have the same intensity characteristics as CSF. They tend to displace adjacent vascular structures

and cranial nerves rather than engulf them, a point helpful in differentiating them from epidermoids. These cysts do not have the serrated margins of an epidermoid and do not cause restricted diffusion. In addition, they do not enhance with contrast material.

Treatment of arachnoid cysts is controversial. A small asymptomatic arachnoid cyst is best left alone. For symptomatic ones, resection, shunting and fenestration have been recommended.

Epidermoid and Dermoid Cysts

The great majority of these are ectodermal inclusion cysts [11]. Intraspinal epidermoids may be acquired, resulting from repeated lumbar puncture. Epidermoids are usually intradural lesions. Extradural lesions are usually within the bone (intradiploic). Epidermoids tend to be off midline. The most common sites are the CP angle and suprasellar locations. Occasionally, they can be intraventricular. Dermoids tend to be close to midline: vermian and parasellar sites are most common. Both enlarge as a result of epithelial desquamation. Dermoids contain ectodermal elements not seen with epidermoids. These include hair, sebaceous and sweat glands and, occasionally, dental enamel. Rupture of a dermoid releases its fatty content into the ventricles and subarachnoid spaces and can cause chemical meningitis.

Epidermoids are lobulated, cauliflower-like, cystic lesions. On CT, they may appear hypodense. Occasionally, a hyperdense 'white' epidermoid is seen [12]. They tend to surround and engulf vascular structures and cranial nerves. They can insinuate themselves in the crevices of the brain making their complete resection impossible. On MRI, they tend to be hypointense on T1W images. They usually have irregular margins against the brain. On T2W images, these lesions are usually hyperintense and occasionally inhomogeneous (Figure 48.5). On FLAIR (fluid attenuated inversion-recovery images), the signal does not suppress completely giving rise to an inhomogeneous lesion. Arachnoid cysts on FLAIR images are completely hypointense. On diffusion-weighted images, epidermoids are hyperintense in contradistinction from arachnoid cysts that are completely hypointense [13]. The epidermoids usually do not enhance; occasionally peripheral incomplete enhancement is seen. They grow very slowly in a linear fashion similar to skin. Malignant degeneration is extremely rare.

Dermoids are much rarer than epidermoids and are usually well-defined hypodense masses on CT. Calcification is common (Figure 48.6). On MRI, they are hyperintense on T1W images but variable on T2WI. In a ruptured dermoid, fat/CSF levels are seen on both CT and MRI images in the ventricles and subarachnoid spaces. Again, potential for growth is limited. Malignant degeneration is extremely rare.

Treatment of these lesions is surgical. Incomplete resection leads to recurrence many years later as these lesions tend to grow slowly.

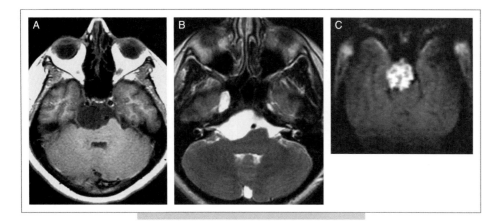

FIG. 48.5. Epidermoid cysts are often hypointense on T1- and hyperintense on T2-weighted images. They are associated with restricted diffusion.

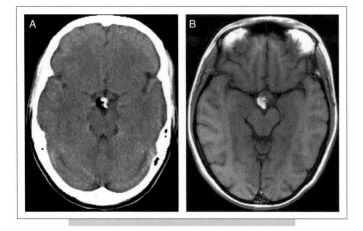

FIG. 48.6. Dermoid cysts tend to be midline in location. They contain fatty tissue and calcify frequently. **(A)** non-contrast CT. The fatty content is easy to see on MR **(B)**.

Intracranial Lipoma

Although not really cystic, we include intracranial lipomas here because of some similarities with the lesions described above.

These lesions result from maldevelopment of the meninx primitive that forms the subarachnoid cisterns. They are usually asymptomatic and found incidentally during cranial CT/MRI exams. Occasionally, headaches, seizures and cranial nerve findings are seen, some caused by the lipoma and some related to the associated other abnormalities. The incidence of lipoma at autopsy is up to 0.21%. Lipomas can be large lesions usually related to the posterior aspect of corpus callosum. These are usually associated with some degree of callosal dysgenesis (Figure 48.7). Other lipomas are small lesions seen in the chiasmatic cistern, quadrigeminal cistern (Figure 48.8) and CP angle cistern [14]. Cranial nerves and vessels usually course through lipomas, a fact that prevents complete resection of these lesions if surgery is deemed indicated. The great majority of

lipomas, with the exception of CP angle lipomas, are at or near the midline.

On CT, lipomas are well-defined lesions with very low density values characteristic of adipose tissue. They may show peripheral calcification, common in callosal lipomas but rare in others. MRI shows a mass that is hyperintense on T1W images and hyperintense on fast spin-echo T2W images. Typical chemical shift artifacts may be seen and aid in diagnosis. On conventional SE sequence and on IR sequences these lesions are hypointense. Fat suppression techniques can be used to confirm the adipose nature of these lesions. They do not enhance with contrast administration.

Treatment by surgical means is rarely indicated, as these lesions are usually incidental findings. Complete surgical excision may not be possible because of vascular and cranial nerve structures coursing through these lesions.

Neuroglial (Neuroepithelial) Cysts

This is an inhomogeneous group of cysts that includes ependymal cysts, choroid plexus cysts and choroidal fissure cysts. We will discuss choroidal fissure cysts. These lesions are easy to identify because of their characteristic location along the choroidal fissure between the medial temporal lobe and diencephalon. They have the same density/intensity as CSF. The size varies from only a few millimeters to a few centimeters. They are not surrounded by glial tissue and do not enhance. They can have an ependymal or arachnoid lining [7].

Colloid Cysts

The great majority of these cysts are located at the foramen of Monro. The cell of origin is the subject of considerable controversy. They are encapsulated cystic lesions with a lining epithelium that may be flat, cuboidal or pseudostratified [15]. Their size varies from a few millimeters to a few centimeters; most measure about 1 cm in size. Very often they are found incidentally during imaging. Headache is the most common presenting symptom. They usually do not enlarge, but sudden enlargement causing obstructive

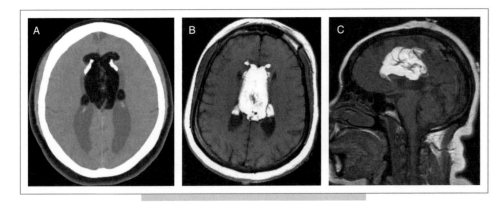

FIG. 48.7. Large intracranial lipomas are midline structures associated with hypogenesis of corpus callosum and are very often calcified (non-contrast CT) **(A)**. MRI **(B** and **C)** shows the extent of lipoma and absence of corpus callosum.

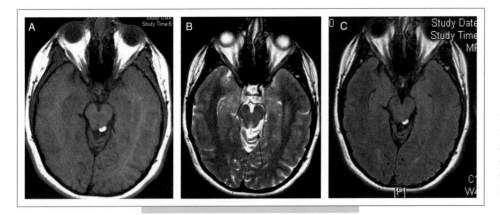

FIG. 48.8. Small quadrigeminal plate lipoma is easy to see on T1WI **(A)** because of distinct hyperintensity of fat on these sequences. On fast spin-echo images they are harder to see, as signal intensity of fat remains high. Arrow on **(B)** shows the typical chemical shift artifact. On FLAIR image **(C)**, these lipomas are again well seen because of attenuation of signal arising from CSF.

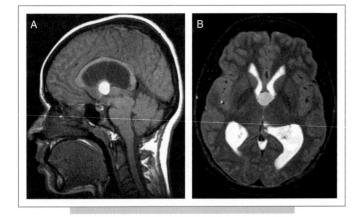

FIG. 48.9. Colloid cysts are easily identified thanks to their characteristic location near the foramen of Monro. Signal intensity is variable. In this case, the cyst is hyperintense on both T1- **(A)** and T2- **(B)** weighted images. Other patterns of signal intensity may be seen. Note also hydrocephalus with dilatation of lateral ventricles.

varies (Figure 48.9). They usually do not enhance with gadolinium. Positive identification of these lesions is fairly simple because of typical location and typical imaging findings. Treatment is by surgical resection. A transcallosal approach into a lateral ventricle and foramen of Monro is the most commonly used technique. Proximity of these lesions to the fornix explains the relatively high frequency of memory problems as presenting symptoms and postoperative complications.

Neuroenteric Cysts

These are usually intradural spinal canal lesions. Here, they are more common in the cervical region if there is no associated osseous malformation, and more common in the lumbar region when there are associated osseous/visceral/cardiac abnormalities. Occasionally, they can be seen in the posterior cranial fossa in the CP angles, anterior to the brainstem and at the foramen magnum. Intraventricular and intramedullary lesions have been reported. The lining epithelium could be GI type, respiratory type, or mixed. The shape of the cyst varies and tubular shapes may be seen. The MR appearance varies with the cyst contents. T1 hyperintense lesions are not uncommon. On T2WIs, they are usually hyperintense relative to CSF. Spinal neuroenteric cysts may be associated with vertebral anomalies such as spina bifida and butterfly vertebrae [16,17].

hydrocephalus, even death, has been reported. On CT, many are seen as well-defined hyperdense lesions that enhance mildly. With MRI, many of these lesions are hyperintense on T1W images. The appearance on T2W images

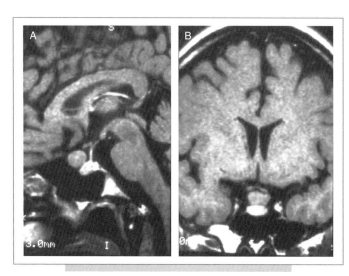

FIG. 48.10. Hamartomas are most common in suprasellar region. They are similar in intensity to the cortex on T1–weighted images (**A** and **B**).

HAMARTOMAS

These are lesions composed of near normal brain tissues but with abnormal proportions for the regions where they are located. They are most common at the floor of the third ventricle. Microscopic asymptomatic hamartomas in the region of hypothalamus are fairly common; larger lesions seen by the naked eye are far less common. They are seen most commonly inferior to tuber cinereum and can fill the chiasmatic cistern or interpeduncular cistern. They may be asymptomatic or present with headaches, visual disturbances and gelastic seizures (spells of uncontrollable laughing). Precocious puberty as a result of production of gonadotropin-releasing hormone (GnRH) is the most common endocrine abnormality. Other hormonal disturbances (acromegaly, for example) are very rare but reported. As these lesions are composed of normal brain tissue, they have the same intensity as cortex on T1WI (Figure 48.10). Their intensity may be slightly higher on T2WI compared with cortex. The lesions do not enhance and do not have calcifications. They have very limited growth potential. Pedunculated lesions may be able to be resected completely, if symptomatic. If not pedunculated, resection may be incomplete but, as mentioned earlier, these lesions grow very slowly, if at all [18].

OTHER NON-CYSTIC LESIONS THAT MAY MIMIC NEOPLASTIC CONDITIONS

Infarcts may mimic neoplasms. Infarcts, however, are usually in a vascular territory and present with an acute onset. They do not keep their mass effect for long. Usually mass effect is maximal about the fourth or fifth day and decreases afterwards. If mass effect persists beyond the first month or two, the diagnosis of infarction is probably incorrect. The enhancement pattern is different in infarction – infarcts do not enhance in the first few days. Gyral enhancement begins after the third or fourth day and continues for 2 or 3 weeks. A round, ring-like enhancement is very uncommon. The single most important imaging clue, however, is provided by diffusion-weighted images. Acute infarcts are associated with restricted diffusion. They are hyperintense on diffusion-weighted imaging (DWI) and hypointense on apparent diffusion coefficient (ADC) maps [19]. Inclusion of DWIs in every brain imaging protocol has become routine and will help reduce the incidence of erroneous diagnosis of an infarct as a tumor.

Venous infarcts, because they do not occur in an arterial territory, can cause more significant problems in differential diagnosis. The clinical history is important, with altered sensorium, headaches and seizures and a more acute onset than neoplastic conditions. Lesions tend to be outside arterial territories and may be bilateral. These lesions tend to become hemorrhagic. Diffusion-weighted images are less frequently positive here than with arterial infarcts (Figure 48.11) and imaging abnormalities tend to reverse frequently [20]. In patients with a proper clinical history and these imaging findings, thrombosed vessels (superior sagittal sinus thrombosis, lateral sinus thrombosis, deep vein thrombosis, etc.) should be suspected. MR venography will be diagnostic.

A subacute intracranial hemorrhage can mimic a neoplasm on CT: there is mass effect, surrounding edema and ring-like enhancement. With MRI, differentiating subacute hemorrhage from neoplasia is much easier.

Inflammatory Conditions Simulating a Neoplasm

Brain abscesses present with symptoms related to increased intracranial pressure and focal symptoms. Symptoms related to the site of lesion include seizures, hemiparesis, disorder of language function and ataxia. Progression of symptoms over days to a few weeks may suggest a diagnosis of neoplasia. Constitutional symptoms such as fever may be absent. On imaging, usually there are extensive signal abnormalities, mass effect, edema and, in due course, ring enhancement (Figures 48.12 and 48.13). All these imaging findings may also be seen in neoplasia (primary tumors and metastatic disease). The anatomy of the enhancement is different, however. Nodularity of the ring suggests a neoplastic condition. The ring is usually thinner on the side closer to the ventricle. Diffusion-weighted images show restricted diffusion (Figure 48.12) in a large percentage of cases of brain abscess. MR spectroscopy of the fluid part of the lesion will show lysosomal amino acids at 0.9 ppm. These are not seen in neoplastic conditions [21].

Encephalitis may be so focal that it may simulate a neoplastic condition (Figure 48.14). This confusing picture may be seen in a variety of encephalitides and has been

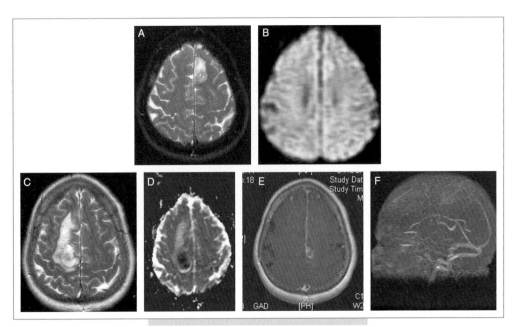

FIG. 48.11. Venous infarcts may present with confusing pictures leading to erroneous diagnoses. The infarct seen on **(A)** was diagnosed as a possible tumor. Note lack of restricted diffusion on **(B)**. A month later, the patient returned. The original lesion seen on **(A)** was no longer seen. A hemorrhagic lesion of medial left frontal lobe was present **(C** and **D)**. The lesion itself showed almost no enhancement. Nodular enhancement was present on the medial aspect of left hemisphere and multiple linear vascular-appearing enhancements were seen in both frontal regions. MRV showed occlusion of anterior two-thirds of the superior sagittal sinus.

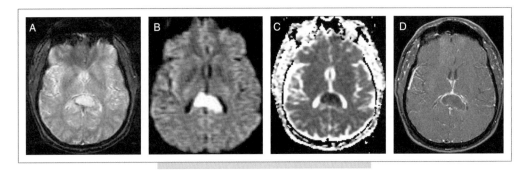

FIG. 48.12. *Stretococcus epidermidis* abscess in an unusual location, splenium of corpus callosum **(A)**. DWI and ADC map **(B** and **C)** show restricted diffusion. There is ring enhancement **(D)**. There were no constitutional symptoms and, although abscess was considered, other possibilities were also entertained in the differential diagnosis and included neoplastic conditions and demyelinating lesion.

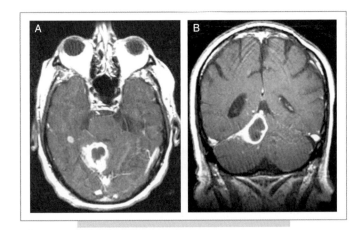

FIG. 48.13. *Nocardia* abscesses. Note multiplicity and solid enhancement of smaller lesions **(A)**. Coronal image **(B)** shows the largest lesion is contiguous with the undersurface of tentorium. Tuberculosis can produce lesions similar to **(B)**.

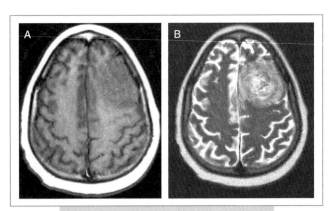

FIG. 48.14. Encephalitis may present with a radiological picture mimicking a neoplasm. This patient's lesion, due to *Aspergillus* infection, presents with a round inhomogeneous left frontal mass. The lesion is slightly hemorrhagic as expected from an angiophilic fungal infection. There was no contrast enhancement. There was also a temporal lobe lesion (not shown) and both showed restricted diffusion.

seen in herpes simplex virus, toxoplasmosis and cryptococcal infections. Altered intensity, mass effect and enhancement may be seen. History is usually quite helpful in this differentiation. Also, CSF analysis, if performed, will show a meningitic picture quite different from that seen with neoplasia. On imaging, encephalitis is not usually confined to a small region; often, the involvement is bilateral and asymmetric. Again, diffusion-weighted images may show restricted diffusion. Enhancement pattern may also be different. Occasionally, when diagnostic uncertainty persists, medical treatment for encephalitis is instituted (e.g. when toxoplasmosis is suspected); in the meanwhile diagnostic workup continues. One of the more common diagnostic dilemmas is differentiation of toxoplasmosis from CNS lymphoma in a patient with AIDS. Both can cause multiple lesions. Periventricular/subependymal location is far more common in lymphoma and positron emission tomography/single-photon emission computed tomography (PET/SPECT) may show increased uptake with lymphoma. MR perfusion may show increased cerebral blood volume in lymphomas versus diminished blood volume in toxoplasmosis, although some recent studies have shown lymphoma is very often associated with decreased relative cerebral blood volume [22]. Spectroscopy of lymphoma shows increased choline and lipids and decreased N-acetyl-aspartate (NAA), in toxoplasmosis the lactate peak is the dominant peak.

Primary CNS vasculitis can present with symptoms suggestive of a mass lesion. Headaches are the most common presenting symptom. Depending on the location of the lesions, weakness, language disorders, seizures, etc., may occur. If not treated, the disease is usually fatal. The sedimentation rate is usually increased, but not as high as seen in temporal arteritis. Angiography is positive in a large percentage of patients. On MRI, brain lesions are usually seen in the white matter; extension to cortex can occur (Figure 48.15). Some of these lesions are undoubtedly infarcts. Small hemorrhages may be seen. Other lesions result from inflammation with lymphocytic and giant cell infiltrates. The lesions may be single or multiple and in one or both hemispheres. Mass effect may be associated with the lesion. Enhancement is variable and ranges from little enhancement to linear to nodular. Response to corticosteroids and immunosuppressants can be dramatic. Brain lesions mimicking neoplasm have been described recently in patients with lymphocytic vasculitis [23] and neuro-Behcet [24].

Sarcoidosis and other granulomatous diseases may present with a protean picture. CNS sarcoidosis may be the first site of involvement. In systemic sarcoidosis, clinical CNS involvement may be present in 5% of patients [25]. Meningeal involvement is the most common CNS pathology and often causes leptomeningeal enhancement. Diffuse thickening of the falx and tentorium may be seen. Masses arising from the meninges may simulate a menigioma. Mass effect may be present. These lesions, however, tend

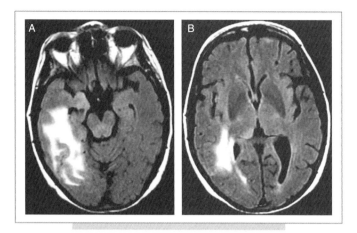

FIG. 48.15. Primary angiitis of CNS can present with lesion mimicking neoplasia. **(A)** and **(B)** are FLAIR images showing a right temporo-occipital lesion with involvement of white matter on **(A)** and right thalamus on **(B)**, where compression of atrium is also evident. There was no significant enhancement. Patient had presented with worsening headaches and a diagnosis of diffuse infiltrative glioma was considered. Biopsy showed primary angiitis of central nervous system (PACNS).

to be hypointense on T2W MR images. The margins tend to be less sharp and more infiltrative-looking than a meningioma. Meningeal lesions at the floor of the third ventricle may be present and cause thickening of infundibulum (Figure 48.16) and optic nerves. Infundibular thickening is a good sign and should be sought in every case. Bilateral cavernous sinus enlargement may be seen. Sarcoid granuloma is another form of CNS involvement and is easier to detect with MRI than CT [26]. Chest x-ray may show bilateral hilar adenopathy and diffuse increased reticular markings. These helpful signs are not seen in every case, however. Not infrequently, the final differentiation is made by pathological examination of biopsy specimens. Spinal involvement may also be seen.

Tuberculosis can cause chronic basal meningitis. Occasionally, masses are produced (i.e. tuberculomas) that are usually at the gray–white junction and tend to be small, ring-enhancing and multiple. Masses attached to the undersurface of tentorium are seen (see Figure 48.13) and should raise the suspicion of tuberculosis.

Tolossa-Hunt syndrome is a granulomatous disease characterized by painful ophthalmoplegia which responds to corticosteroid treatment. It causes unilateral cavernous sinus enlargement with moderate enhancement. It is one of the causes of cavernous sinus lesions, with meningiomas, aneurysms, lymphoma, metastases, etc., being some of the others. Administration of steroids is both diagnostic and therapeutic.

Demyelinating disease may present with intracranial masses [27]. Tumefactive multiple sclerosis (MS) is the term used for these types of large, focal lesions. Presentation may

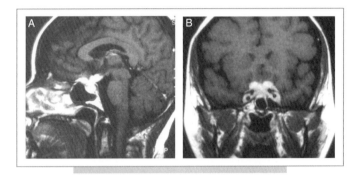

FIG. 48.16. Sarcoidosis may cause meningeal thickening. This is common at the suprasellar cistern. Thickening of infundibulum and cranial end of optic nerves and chiasm may be seen **(A)**. There may be bilateral enlargement of the cavernous sinuses **(B)**.

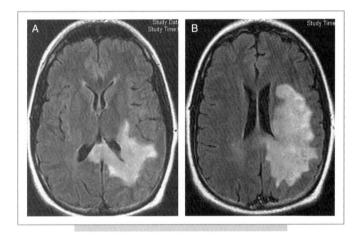

FIG. 48.17. Demyelinating lesions can cause extensive white matter involvement **(A)**. Mild mass effect may be associated **(B)**. Margins are not sharply defined.

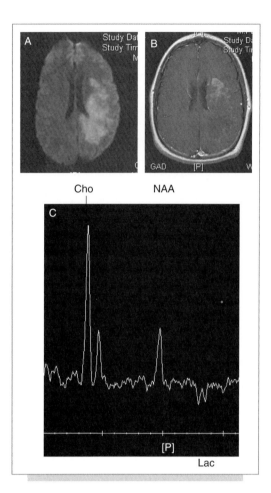

FIG. 48.18. Same patient depicted on Figure 48.17. Enhancement may be mild and at the margin of the lesion **(B)**. Concentric layers of enhancement may be seen. **(C)** MR spectroscopy shows elevated choline (Cho), depressed NAA and a lactate (Lac) doublet.

be acute or subacute. Hemiparesis, language disorders, hemianesthesia, etc., may be seen. These patients usually do not progress to clinical MS. Imaging shows a region of white matter signal abnormality with mild mass effect. The borders may be poorly defined. Enhancement varies, but peripheral enhancement or concentric enhancement may be seen. There may be associated signal abnormality on DWIs. Occasionally, a brainstem mass is found with expansion of the brainstem simulating a neoplasm. In these cases, finding other white matter lesions in typical locations (periventricular white matter, corpus callosum, middle cerebellar peduncles, etc.) may help, but usually the lesion is the only lesion. Multiple complete or incomplete rings of enhancement are in favor of a demyelinating disorder. Patients are usually young. MR spectroscopy shows decreased NAA and increased choline. A large choline peak may be seen (Figures 48.17 and 48.18) making differentiation from a high-grade tumor even more difficult [28]. CSF analysis may help by showing oligoclonal bands. The

lesions may regress with time or respond dramatically to steroids. Biopsy may be the only way to make a definitive diagnosis.

Extramedullary hematopoiesis can cause intracranial mass lesions. The history may be helpful in these cases (i.e. severe, prolonged anemia). Biopsy may be necessary to make a definitive diagnosis.

Lhermitte-Duclos disease is discussed here briefly because, although not decidedly neoplastic, it can present with mass-like lesions of the cerebellum. It is not clear if this entity is a dysplasia, a hamartoma or a neoplasm. WHO classification of CNS tumors regards them as dysplastic gangliocytomas. There is significant thickening of cerebellar folia [29]. The lesion presents as a sharply defined region of cerebellum with a characteristic laminated appearance on T2W images, and mild mass effect on the fourth ventricle. It presents in young patients with signs of increased intracranial pressure and cerebellar signs. The lesions do not enhance with gadolinium.

REFERENCES

1. Osborn AG (2004). Enlarged perivascular spaces. In *Diagnostic Imaging: Brain*. Amirsys, Salt Lake City. I-7–22.
2. Ingles M, Bomsztyk E, Gonen O et al (2005). Dilated perivascular spaces: hallmark of mild traumatic brain injury. Am J Neuroradiol 26:719–724.
3. Ge Y, Law M, Herbert J et al (2005). Prominent perivenular spaces in multiple sclerosis as a sign of perivascular inflammation in primary demyelination. Am J Neuroradiol 26:2316–2319.
4. Salzman KL, Osborn AG, House P et al (2005). Giant tumefactive perivascular spaces. Am J Neuroradiol 26:298–305.
5. Saeki N, Sato M, Kobuta M et al (2005). MR imaging of normal perivascular space expansion at midbrain. Am J Neuroradiol 26:566–571.
6. Elster AD, Richardson DN (1991). Focal high signal on MR scan of the midbrain caused by enlarged perivascular spaces: MR-pathologic correlation. Am J Neuroradiol 11:1119–1122.
7. Osborn AG, Preece MT (2006). Intracranial cysts: radiologic-pathologic correlation and imaging approach. Radiology 239(3):650–663.
8. Lantos PL, Louis DN, Rosenblum MK et al (2002). Tumors of the nervous system. In *Greenfield's Neuropathology*, 7th edn, Vol 2. Graham DI, Lantos PL (eds). Arnold Publishers, London. 873.
9. Barboriak DP, Lee L, Provenzale JM (2001) Serial MR imaging of pineal cysts: implications for natural history and follow-up. Am J Roentgenol 176:737–743.
10. Barkovich AJ (1995). In *Pediatric Neuroimaging*, 2nd edn. Raven Press, New York. 448–453.
11. Lantos PL, Louis DN, Rosenblum MK et al (2002). Tumors of the nervous system. In *Greenfield's Neuropathology*, 7th edn, Vol 2. Graham DI, Lantos PL (eds). Arnold Publishers, London. 964–966.
12. Ikushima I, Korogi Y, Hirai T et al (1997). MR of epidermoids with a variety of pulse sequences. Am J Neuroradiol 18:1359–1363.
13. Tsuruda JS, Chew WM, Moseley ME et al (1990). Diffusion weighted MR imaging of the brain: value of differentiating between extra-axial cyst and epidermoid tumor. Am J Neuroradiol 11:925–931.
14. Lalwani AK (1992). Meningiomas, epidermoids and other non-acoustic tumors of the cerebellopontine angle. Otolaryngol Clin N Am 25:707–728.
15. Adams JH, Graham DI (1994). In *An Introduction to Neuropathology*, 2nd edn. Churchill Livingstone, London. 298.
16. Khosla A, Wippold FJ (2002). Pictorial assay. CT myelography and MR imaging of extramedullary cysts of the spinal canal in adults and pediatric patients. Am J Roentgenol 178:202–207.
17. Gao PY, Osborn AG, Smirniotopoulis JG (1995). Neuroenteric cysts: pathology, imaging spectrum and differential diagnosis. Int J Neuroradiol 1:17–27.
18. Burton EM, Ball WS, Crone K et al (1989). Hamartomas of tuber cinereum: a comparison of MR and CT findings in four cases. Am J Neuroradiol 10:497–501.
19. Eastwood JD, Engelter ST, MacFall JF et al (2003). Quantitative assessment of the time course of infarct signal intensity on diffusion-weighted images. Am J Neuroradiol 24:680–687.
20. Rottger C, Trittmacher S, Gerriets T et al (2005). Reversible MR imaging abnormalities following cerebral venous thrombosis. Am J Neuroradiol 26:607–613.
21. Kim SH, Chang KH, Song IC et al (1997). Brain abscess and brain tumor: discrimination with in vivo H-1 MR spectroscopy. Radiology 204:239–245.
22. Cha S, Knopp EA, Johnson G et al (2002). Intracranial mass lesions: dynamic contrast-enhanced susceptibility-weighted echo planar perfusion MR imaging. Radiology 223:11–29.
23. Panchal NJ, Niku S, Imbesi SG (2005). Lymphocytic vasculitis mimicking aggressive multifocal cerebral neoplasm: MR imaging and MR spectroscopic appearance. Am J Neuroradiol 26:642–645.
24. Matsuo K, Yanada K, Nakajima K et al (2005). Neuro-Behcet disease mimicking brain tumor. Am J Neuroradiol 26:650–653.
25. Christoforidis GA, Spickler EM, Recio MV et al (1999). MR of CNS sarcoidosis: correlation of imaging features to clinical symptoms and response to treatment. Am J Neuroradiol 20:655–669.
26. Miller DH, Kendall BE, Barter S et al (1988). Magnetic resonance imaging in central nervous system sarcoidosis. Neurology 38:378–383.
27. Giang DW (1992). Multiple sclerosis masquerading as a mass lesion. Neuroradiology 34:150–154.
28. Saindee AM, Cha S, Law M et al (2002). Proton MR spectroscopy of tumefactive demyelinating lesions. Am J Neuroradiol 23:1378–1386.
29. Adams JH, Graham DI (1994). In *An Introduction to Neuropathology*, 2nd edn. Churchill Livingstone London. 13:234–235.

CHAPTER 49

Neuroimaging Issues in Assessing Response to Brain Tumor Therapy

Stephan Ulmer, Gordon J. Harris and John W. Henson

IMAGING APPEARANCE OF GLIOMAS OF THE ADULT CEREBRAL HEMISPHERES

Gliomas of the adult cerebral hemispheres are remarkable for the degree to which their imaging characteristics reflect their histopathologic features (Figure 49.1).

In low-grade diffuse fibrillary astrocytomas (LGA; WHO grade II), there is mild expansion of the affected brain without evidence of abnormal gadolinium enhancement or significant surrounding vasogenic edema. These imaging findings reflect the histopathological findings of mildly increased cellularity without significant angiogenesis.

Anaplastic astrocytomas (AA; WHO grade III) typically have nodular areas of homogeneous (i.e. without ring-enhancement) gadolinium enhancement, with the enhancement reflecting the presence of newly formed blood vessels which have an abnormal blood–brain barrier and are thus permeable to gadolinium. These tumors tend to produce moderate expansion of the involved brain regions due to

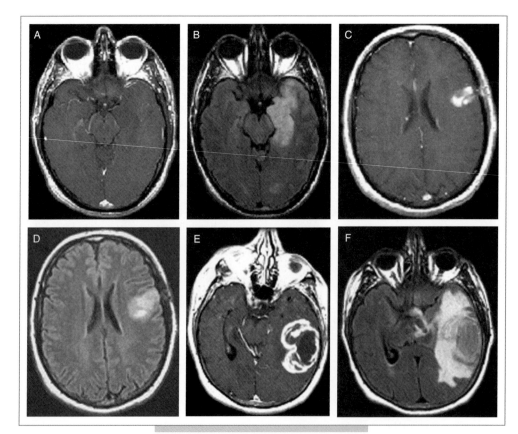

FIG. 49.1. Imaging findings of astrocytomas reflect the histopathological features of these tumors. Comparison of the imaging findings in low-grade astrocytoma **(A,B)**, anaplastic astrocytoma **(C,D)** and glioblastoma multiforme **(E,F)**. The left column shows axial gadolinium-enhanced T1-weighted images and the right column shows FLAIR images. See the text for discussion of the correlation between imaging features of these tumor grades and histological findings.

greater cellular proliferation and may or may not show evidence of significant surrounding vasogenic edema.

In glioblastoma multiforme (GBM; WHO grade IV), there is usually more mass effect and extensive surrounding T2-weighted hyperintensity due to a combination of vasogenic edema and infiltrating tumor. Centrally non-enhancing regions that correlate with histopathologic findings of necrosis are surrounded by areas of peripheral enhancement.

While the above features are typical for astrocytomas of the adult cerebral hemispheres, there are several important exceptions and additional features that must be noted. First, diffuse infiltrating gliomas of the cerebral hemispheres are commonly disseminated throughout the brain at the time of diagnosis and this feature is not well delineated on imaging. Indeed, precisely because these tumors are infiltrating, intensification of local treatments, such as with brachytherapy, does not significantly improve overall survival [1]. Secondly, the presence of abnormal enhancement in a tumor shown to be a diffuse astrocytoma implies the presence of high-grade tumor, even if the biopsy specimen shows only LGA. Therefore, enhancing 'low-grade' astrocytomas should be followed radiologically with the same frequency as AA, since there is a significant likelihood of biopsy sampling error. Low-grade oligodendrogliomas often demonstrate enhancement and elevated relative cerebral blood volume (rCBV) and, in distinction to astrocytomas, these findings do not imply higher grade, a factor which will be discussed in more detail below under cerebral blood volume measurements [2].

On the other hand, the absence of enhancement does not reliably indicate the absence of high-grade histopathology, since one-third of non-enhancing diffuse gliomas in adults will be high-grade. The risk of high-grade histopathology in a non-enhancing astrocytoma rises with age, such that by 45 years of age, the likelihood of finding high-grade tumor is 50% [3]. These facts not only indicate that all non-enhancing masses suspected to represent a glioma should be biopsied, but also that imaging approaches other than contrast enhancement are needed to target the most anaplastic portion of the tumor.

EVOLUTION OF DIFFUSE GLIOMAS IN ADULTS

Adult gliomas have a well-known tendency to become more histologically anaplastic and clinically aggressive over time, evolving towards GBM [4–6]. This evolution is called anaplastic progression and results from a stepwise accumulation of molecular genetic alterations within tumor cells. Although the precise numbers are difficult to determine, approximately 50% of LGA undergo radiographically detectable anaplastic progression, with the remainder of LGA showing gradual growth as a low-grade lesion [5].

GBM arising within a known LGA is termed 'secondary GBM'. On the other hand, patients whose GBM are diagnosed after a short interval of symptoms where there is no known prior LGA are said to have de novo, or primary GBM. There are a number of important distinctions between these two types of GBM. De novo tumors typically arise in older adults. Conversely, patients with secondary GBM tend to be younger at the time of diagnosis. The two types of tumors have distinguishing genetic alterations, for instance TP53 is usually unaltered and EGFR is activated in patients with de novo GBM, whereas TP53 tends to be inactivated and EGFR is unaltered in patients with secondary GBM. The dichotomies in the age, rate of anaplastic progression and genetic changes in GBM give strong evidence that distinct genetic pathways produce subsets of malignant gliomas. According to this hypothesis, the tumors of younger patients tend to show anaplastic progression over an interval of years, whereas the tumors in patients of older age undergo progression from initial formation to GBM over months. This is directly supported by data showing that LGA progress more rapidly to high-grade in older patients [7]. Although other explanations for tumor behavior have been proposed [8], there are few existing data or theoretical support for these hypotheses.

IMAGING DIFFUSE GLIOMAS FOLLOWING DIAGNOSIS

Low-Grade Gliomas

Patients with low-grade gliomas usually undergo maximal possible tumor resection, with or without adjuvant therapy. Thus, the rationale behind serial imaging surveillance in these patients is the early detection of tumor progression.

Low-grade gliomas often show progression as gradual enlargement of the region of non-enhancing hyperintense T2-weighted signal surrounding the resection cavity. Mandonnet, Delattre and co-workers found that in untreated low-grade oligodendrogliomas there is approximately 4 mm annual increase in cross-sectional diameter of the tumor each year [9]. Figure 49.2 illustrates this pattern of growth in a patient with a left frontal low-grade oligoastrocytoma who reported increasing severity of the seizures affecting language function. When compared to the immediate prior study, serial MRI examinations did not provide an explanation for the clinical deterioration. However, comparison to an earlier baseline study revealed a clear, gradual increase in tumor size. This observation provided a reasonable radiographic basis for understanding the clinical progression and the patient was begun on oral chemotherapy with subsequent improvement in her symptoms. Thus, it is of utmost importance to compare each new MRI study with prior studies over as long a period as possible in order to detect the presence of gradual interval growth which could escape detection if comparison is limited to the most recent prior examination.

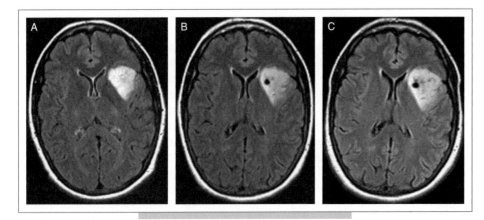

FIG. 49.2. Gradual growth of an untreated low-grade oligoastrocytoma over a 4-year interval ((**A**) in 1999 at diagnosis, (**B**) in 2002, (**C**) in 2003). Comparison of axial FLAIR images reveals slow enlargement of the tumor. The patient's seizures began to worsen in 2003. Mandonnet and coworkers (2003) showed that untreated low-grade oligodendrogliomas enlarge about 4 mm in cross-sectional diameter each year.

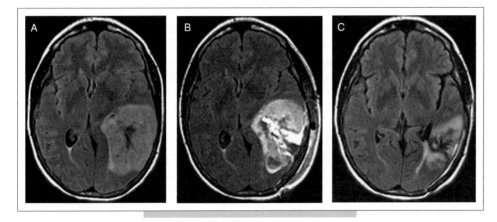

FIG. 49.3. Marked shrinkage of brain tumors is commonly seen with 1p- anaplastic oligodendrogliomas and primary central nervous system lymphomas. In this example, a 1p- anaplastic oligodendroglioma involuted following three cycles of oral chemotherapy without irradiation. (**A**) Preoperative, (**B**) following a subtotal resection, (**C**) following 3 monthly cycles of oral chemotherapy with temozolomide.

Patients with low-grade gliomas may receive treatment when their tumors show growth or when there are symptoms, such as poorly controlled seizures, that might benefit from therapy. Treatment may consist of involved field irradiation or chemotherapy without irradiation. In these patients, 'response' of the tumor may consist of shrinkage of the lesion or stabilization of size, with or without a decrease in target symptoms (e.g. decrease in seizure frequency). Clinical experience indicates that stabilization is the most common form of response. When shrinkage does occur in LGA or low-grade oligodendrogliomas, it is usually minor in degree (i.e. insufficient to qualify for a partial response as defined below) and is often apparent only after 1 year or more following initiation of therapy. However, responses are frequently seen in anaplastic oligodendrogliomas which have loss of heterozygosity of the short arm of chromosome 1 (1p-tumors), where best response is often seen after only 3 months of therapy.

In the case of anaplastic progression within low-grade gliomas, a new focus of rapidly growing, contrast-enhancing tumor appears within an existing, non-enhancing mass. The interval from initial diagnosis to anaplastic progression is strongly age-dependent [7]. These patients are treated aggressively with irradiation and chemotherapy at the time progression is recognized. The imaging considerations regarding assessment of treatment responses in these patients are the same as those described below for high-grade gliomas.

High-Grade Gliomas

Patients who are diagnosed with high-grade gliomas (i.e. WHO grade III and IV) are usually treated in the postoperative setting with a combination of involved field irradiation and chemotherapy. Increasingly, 1p- anaplastic oligodendrogliomas are treated initially with chemotherapy alone, reserving irradiation until there is evidence of tumor progression.

High-grade tumors may show shrinkage, stability or progression during these treatments. Shrinkage is common only with 1p- anaplastic oligodendrogliomas, in which case it is often seen within 3 months of initiation of therapy (Figure 49.3). Shrinkage may occur in the size of nodular enhancing disease and in the area of hyperintense T2-weighted signal. Tumor cysts often do not change in size even when the surrounding tumor is shrinking.

Stability in the size of the tumor is a more common outcome with treatment, especially with astrocytomas. As discussed below, this fact has led to the use of 'progression-free survival' as a common radiographic endpoint in clinical trials for high-grade gliomas.

As practitioners of neuro-oncology are acutely aware, the vast majority of high-grade gliomas demonstrate progressive enlargement within months of diagnosis. Progression most often is declared when there is enlargement of one or more enhancing nodules within the tumor or appearance of a new nodular area of enhancement either locally at the tumor margin or distantly elsewhere in the brain. Increases in size of a tumor cyst, in the extent of hyperintense T2-weighted signal, and the appearance of hemorrhage when there is stability of the remainder of the lesion, do not qualify as tumor progression.

USE OF IMAGING IN CLINICAL TRIALS

Enormous effort has been put into clinical trials of new agents for malignant gliomas over the past 30 years. In the testing of new treatments, the standard 3-phase design is employed [10]. Phase I and phase III studies, with their respective goals of maximal tolerated dose determination and overall survival, do not rely heavily upon a neuroimaging endpoint. Imaging is a most important tool in the phase II study, however, since the major goal of this stage is to look for evidence of effectiveness of the new agent as determined by radiographic response rate and clinical status.

Thus, whereas survival is considered the gold standard of efficacy at the phase III stage, a surrogate measure of survival (e.g. radiographic response rate) is employed in phase II studies. Phase II studies usually are performed in patients with progressive tumors and radiographic response rates are used because they can be measured much more quickly than survival and because any survival measurement would be complicated by the administration of prior therapy. The use of radiographic response as a surrogate endpoint for survival at the phase II stage makes a crucial assumption, however, and that is that radiographic response is a valid surrogate measure for a meaningful response as defined by survival.

The terms complete response (CR), partial response (PR), stable disease (SD) and progressive disease (PG) are used in neuro-oncology trials and these terms are defined below under 'Response criteria'. Complete response and partial response are seen in approximately 10% of patients in clinical trials for newly diagnosed malignant gliomas [11] and in a smaller fraction of patients with progressive disease. For that reason, it is typical to employ endpoints that measure the duration of SD, such as 'progression-free survival' (PFS). Thus, a common end point for phase II clinical trials is PFS after a 6-month or 12-month interval. Phase II studies compare the percentage of patients with 6-month

or 12-month PFS to that of a historical control group. PFS is used because it is a more concrete endpoint than is 'time to progression' (TTP), in which the interval to SD is measured in each study subject based on imaging data despite variation in the intervals between studies in individual patients [12]. There are significant issues with the use of PFS in a phase II study. Not only does the use of SD measure a weaker biological effect than does tumor shrinkage, but there is great difficulty with the choice of a valid comparison group. Together, these problems render the validity of current methods for measurement of efficacy in phase II trials questionable.

Another major issue relates to whether radiographic response is a valid surrogate measure for overall survival. Studies have provided strong evidence that radiographic response predicts survival in anaplastic oligodendroglioma (AO), where response correlates with longer survival and with the loss of heterozygosity of chromosome 1p [13]. For newly diagnosed AA, a study by the Radiation Therapy Oncology Group (RTOG) randomized patients to receive irradiation plus procarbazine, CCNU and vincristine (PCV) with or without 5-bromodeoxyuridine (BudR) [12]. Patients with no progression at 6 months had a median survival of 67 months versus a median survival of 19 months for patients who had shown progression by 6 months. Thus, for newly diagnosed anaplastic gliomas, there is evidence that response correlates with survival. A study by Galanis et al of the relationship between response and survival in patients with newly diagnosed 'enhancing gliomas' (n = 36 with low-grade glioma 25%, AA 14%, and GBM 61%) found no relationship between response and survival [11]. It is important to note, however, that this was a study of adjuvant therapy, not salvage therapy, whereas most phase II trials are conducted in patients with progressive disease.

For the most common type of high-grade glioma, GBM, there is even less evidence to document the validity of radiographic response as a surrogate measure of efficacy (i.e. overall survival) in the phase II setting.

Grant et al reviewed the imaging studies of 136 patients with progressive malignant gliomas following two cycles of nitrosourea-based therapy and found no correlation between response as determined by the Macdonald criteria and time to progression or survival [14]. Shah et al were unable to find a relationship between 6-month progression-free survival and overall survival in 53 patients entered into a number of different clinical trials for progressive malignant gliomas [15]. Thus it appears unlikely that response and duration of stable disease are valid measures of a survival-associated response. This does not mean that these two measures are not valid indicators of efficacy of a new agent, however. In fact, a measurement approach that is highly sensitive to response would likely provide the best measure of efficacy, as discussed below under 'Comparison of diameter and volumetric approaches'.

TECHNICAL ASPECTS OF MEASURING RESPONSE AND PROGRESSION

General Considerations

Although the problems discussed above represent serious limitations for clinical trials of new brain tumor therapies, neuroimaging is still the best available tool for evaluating response in the phase II setting. There are two basic approaches in common use for the measurement of tumor activity on MRI scans. These are diameter-based methods and computer-assisted volumetric methods.

With both of these approaches a baseline study is required, ideally within 14 days prior to initiation of treatment. Follow-up studies are then performed according to treatment protocol, but most commonly after every second cycle or every two months. The same imaging technique must be used at every time point (i.e. MRI and CT measurements cannot be mixed). Other considerations for these studies include whether there is a requirement for specific sequences (e.g. a 3 mm/skip 0 mm slice thickness T1-weighted gadolinium-enhanced axial image for volumetric analysis, see below) and whether anonymized data must be recorded onto a compact disc for transmission to a central neuroradiology review site. Central review is usually employed to confirm response status, with the primary clinical interpretation being the responsibility of the patient's individual neuro-oncologist. It may be important to record whether there are lesions that have been treated with radiosurgery.

An issue that often arises is the choice of the portion of the lesion to measure or, when there are multiple or multifocal lesions, the choice of the specific lesion to measure. Cystic or centrally necrotic lesions are problematical, since a large portion of the lesion cannot reasonably be expected to respond to therapy. In this case, it is best to use a peripheral, nodular component of the lesion for response assessment when using diameter-based methods. This problem is more easily navigated with computer-assisted volumetric assessment (see below). Another problem is distinguishing contrast enhancement from intrinsic T1-weighted hyperintensity due to blood products. This latter issue is especially difficult on the immediate postoperative MRI scan (i.e. the study performed within 72 hours of surgery), since extensive signal changes from blood products are routinely present. In this circumstance, a detailed, side-by-side comparison of the pre- and post-contrast enhanced images is required. Many trials have a minimum tumor size requirement for inclusion: if a 10 mm diameter enhancing nodule cannot be distinguished, then the patient will be ineligible for some protocols. Occasionally, the intrinsically T1-weighted hyperintense postoperative blood products along the surgical margin can obscure the presence of residual enhancing disease and these foci only become apparent on further follow-up studies when the blood product related signal has resolved.

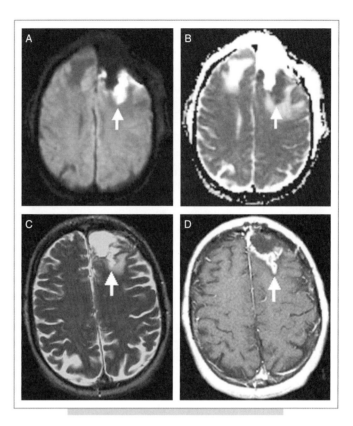

FIG. 49.4. Two-thirds of patients undergoing surgery for malignant gliomas have infarcts in their tumors on the immediate postoperative MRI scan. These infarcts often demonstrate nodular gadolinium enhancement on subsequent studies, a finding which could be easily confused with tumor. Immediate postoperative DWI **(A)** and ADC **(B)** show restricted diffusion (arrows), followed by a 3-month postoperative T2-weighted image **(C)** and gadolinium-enhanced T1-weighted imaging showing enhancement within the infarct **(D)**.

Areas of restricted diffusion occur in two-thirds of patients following surgery for malignant gliomas (Figure 49.4). These foci evolve into encephalomalacia, consistent with the presence of infarction. Follow-up studies reveal enhancement in about one-half of these infarcts, a finding that could easily be misinterpreted as persistent or progressive tumor if the presence of the perioperative infarct is not realized. Thus, it is important to review the immediate postoperative diffusion-weighted image (DWI) when evaluating an area of focal nodular enhancement for serial measurement.

Gadolinium enhancement is a time dependent phenomenon and this fact can result in significant changes in apparent lesion size between the initial and final enhanced sequences (Figure 49.5). Typically, the images are acquired in the sequence of axial, followed by sagittal, and then coronal planes, with the initial axial image being acquired within 3–4 minutes after gadolinium injection. The images

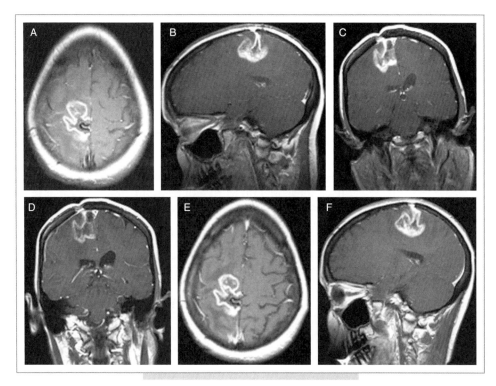

FIG. 49.5. Contrast enhancement within brain tumors is a dynamic process. A standardized imaging protocol is important for the purposes of direct comparison. In this case, the usual acquisition order **(A,B,C)** of the triplanar images was reversed in **(D,E,F)**, producing marked differences between the coronal images.

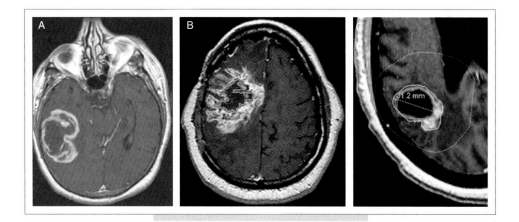

FIG. 49.6. **(A)** Illustration of the common, orthogonal diameters approach to brain tumor measurement. The difficulties with this approach are described in detail in the text. **(B)** Computer-aided volumetric approach for measurement of an irregularly shaped glioblastoma containing a significant region of non-enhancing, necrotic tumor. The program calculates enhancing volume, central non-enhancing volume and total lesion volume. **(C)** Another case of glioblastoma comparing RECIST and volumetric measurements.

should always be obtained in the same sequence at about the same time interval in order to avoid misinterpretation of response or progression. Evaluation of the degree of enhancement within the deep cerebral veins (e.g. internal cerebral veins) and within the nasal mucosa can help to confirm the presence of an adequate dose of gadolinium.

It is important to exclude resection cavities from measurements. In many cases, the presence of a cavity is best demonstrated on the T2-weighted images.

Response Criteria and Diameter-Based Measurement Methods

In the standard approach employed in most neuro-oncology clinical trials, two orthogonal measurements are made of the gadolinium-enhancing portion of the tumor on the axial slice containing the most extensive portion of the lesion (Figure 49.6). The same measurement, using the scan slice that is as closely matched as possible from each study, is measured on each exam obtained during the clinical trial. The two diameters are multiplied to give an estimate of tumor area on that image. While this approach has the advantage of simplicity, there are also substantial disadvantages to this approach compared to the computer-assisted volumetrics. These relative merits are discussed in more detail below.

Once the baseline and serial follow-up exams are available on a patient in a clinical trial, the so-called Macdonald criteria are applied to the serial imaging measurements and clinical status to determine response [16]. These criteria are:

- Complete response (CR): complete disappearance of all enhancing tumor on consecutive scans performed at least 1 month apart. No new lesions. No administration of glucocorticoids and neurologically stable or improved.
- Partial response (PR): greater than or equal to 50% decrease from baseline in the enhancing tumor measurement on consecutive scans performed at least 1 month apart. No new lesions. Stable or decreasing glucocorticoid dose, neurologically stable or improved.
- Stable disease (SD): other than CR, PR, or PD.
- Progressive disease (PD): greater than or equal to 25% increase in the tumor measurements or appearance of new lesions on consecutive scans performed at least 1 month apart or decline in neurological status with stable or increased dose of steroids. ('New lesion' must be defined by each protocol and often is designated as the appearance of a new enhancing lesion of more than 5 mm diameter.)

In 2000, an international committee published new imaging response guidelines known as Response Evaluation Criteria in Solid Tumors (RECIST; see http://ctep.cancer.gov/guidelines/recist.html) [17]. The criteria were designed as a uniform, simplified and conservative criterion to determine response to therapy for all types of solid tumors. However, RECIST has not been widely adopted for clinical trials involving malignant gliomas, in part for historical reasons and in part because RECIST specifies solid, non-cystic tumors, while GBM frequently presents with a non-enhancing necrotic core.

The basic measurement in RECIST is a single linear dimension across the largest diameter of the tumor on a selected axial image. This measurement is repeated with each study, always using the largest cross-sectional diameter, even if it varies from the original orientation.

For patients with multiple lesions within the brain, RECIST categorizes the lesions within one organ as measurable (i.e. target lesions) when the lesion is at least 10 mm in longest diameter as seen on a single axial slice, or non-measurable (i.e. non-target lesions) if smaller than 10 mm in longest diameter. The diameter of 10 mm is based on the RECIST rule that the smallest measurable lesion should be at least double the size of the commonly used slice thickness of 5 mm. Predominantly cystic and leptomeningeal lesions are considered non-measurable. Lesions that are not suitable for accurate repeated measurements include those where two or more lesions are abutting each other, producing difficulty in establishing the margin of the tumors and those near rounded or bony anatomic surfaces (e.g. skull base or vertex where volume averaging might lead to inaccuracy of measurement).

Baseline documentation includes all measurable lesions up to a maximum of 10 lesions throughout the body and no more than 5 lesions per organ. When the brain is the sole organ site to be assessed some brain metastasis trials have measured up to 10 lesions in the brain. All other non-target lesions are to be identified and recorded. The recorded measure for each scan is the sum of all of the 'measurable' single lesion diameters.

The evaluation of target lesions is defined by RECIST in four response categories:

- Complete response (CR): complete disappearance of all target lesions.
- Partial response (PR): 30% decrease in the sum of the longest diameter of target lesions.
- Stable disease (SD): changes that do not meet criteria for CR, PR, or PD.
- Progressive disease (PD): 20% increase in the sum of the longest diameter of target lesions.

Note that the RECIST criteria, unlike the Macdonald criteria, do not include a measure of clinical status or steroid dosage.

Non-target lesions are defined by three categories. Complete response refers to disappearance of all non-target lesions, incomplete response/stable disease to persistence of one or more non-target lesions, progressive disease to appearance of one or more new lesions. A new lesion is defined as a focus of tumor not previously seen that measures at least 5 mm in diameter.

Based on all possible combinations of the above mentioned response categories of target and non-target lesions, the overall response represents the best response of the tumor to treatment, and these combinations are detailed in the criteria available in publication [17] and at the world wide website noted above.

Difficulties with the cross-sectional diameter approach include the fact that malignant gliomas are typically irregular in shape such that the standard cross-sectional diameter measurements do not truly estimate the size of the lesion. There often are large areas of cystic necrosis which, although included in the measurements, are in fact unlikely to respond to treatment (see Figure 49.6) and RECIST specifically denotes such tumors as non-measurable.

Finally, progression often occurs within a focal region of the tumor and, therefore, serial linear measurements across the identical portion of the tumor from study to study may not detect evidence of early progression.

Computer-Aided Volumetric Methods

In the computer-aided volumetric, or perimeter approach, we use semi-automated tumor segmentation software developed by one of the authors (GJH); semi-automated tumor volumetric analysis software is available in the FDA-approved 3D imaging workstation Vitrea™ (Vital Images, Minetonka, MN). To segment the tumor volume, first a region of interest is drawn around a gadolinium-enhancing mass lesion on a series of T1-weighted axial images to define the general location. The computer then draws a border along the margin between the enhancing and non-enhancing regions on all adjacent images, using a combination of image histogram statistics and morphological filtering. Likewise, any non-enhancing cystic areas are segmented by the software. These images are reviewed and adjusted, if necessary, by a neuroradiologist and the program then calculates an enhancing, non-enhancing (e.g. the centrally necrotic or cystic area) and a total, or combined, volume in cubic centimeters (see Figure 49.6). Computer-aided volumetric analysis has been shown to be more sensitive in the early detection of progression, especially with smaller lesions [18,19]. The computer-aided volumetric approach showed excellent intra-observer and inter-observer reliability.

Technical issues with the perimeter method include the choice of lesion or enhancing area within the lesion, confirmation of the margins of enhancement and exclusion of blood products that are hyperintense on T1-weighted images, enhancing perioperative infarctions and resection cavity. Malignant gliomas often show extreme geographic complexity (see Figure 49.6) and the final decision regarding the correct perimeter may involve subjective judgments based on experience on the part of the reviewing neuroradiologist. It is valuable to have one neuroradiologist make all of the measurements during central review in order to reduce variability. When follow-up studies are obtained, it is important to review the volumetric analysis of the new study in close comparison to the prior study in order to maintain consistency. When volumetric analysis is used to follow the size of smaller lesions with a minimum lesion diameter of 10 mm (e.g. brain metastasis), a slice thickness of 3 mm skip 0 mm is an ideal compromise between high resolution and increasing time required for image acquisition. Ideally, a tumor should be visible and measured on at least three slices for volumetric assessment to limit variability due to inadequate sampling.

Comparison of Diameter and Volumetric Approaches

At least four studies have compared the value of diameter and volumetric approaches to the detection of response in clinical trials [15,18–20]. These studies have shown that three perpendicular diameters to estimate volume had the lowest degree of concordance with 1D (single linear measurement as in RECIST), 2D (two orthogonal measurements as in the Macdonald criteria) and volumetric approaches. The volumetric approach gave a very low degree of intra- and inter-reader variability. Of significant interest, the computer-assisted volumetric approach yielded a higher response rate than did the single linear (1D) measurements. This is likely secondary to the higher sensitivity of the computer assisted measurements to tumor size, as the volumetric approach excludes the non-enhancing portion of the tumor (which is unlikely to respond to treatment) and also covers the entire enhancing tumor volume rather than measuring only a single diameter [15]. For instance, PR would have been declared in 8% of 284 scans using the 1D measurement compared to 17% using the volumetric approach.

Correlations of response criteria for 1D (RECIST) and 2D (Macdonald) diameter approaches versus the computer-assisted volumetric approach have not been closely examined. Table 49.1 provides a comparison based on diameters of idealized spherical lesions. Further research will be necessary to determine the correct values for most accurate response rate comparisons.

TABLE 49-1 Comparison of response criteria for different measurement schemes. Percent change refers to change in measurement compared to baseline (CR, PR and SD) or best response/nadir (SD, PD)					
	CR	PR	SD	PD	Comment
RECIST (1D)	100%↓	≥30%↓	All others	≥20%↑	Sum of single linear measures (see text)
Macdonald (2D) [16]	100%↓	≥50%↓	All others	≥25%↑	Product of two orthogonal diameters
Volumetric extrapolated from RECIST* (volume versus 1D)	100%↓	≥66%↓	All others	≥73%↑	Computer-assisted volumetrics
Volumetric extrapolated from Macdonald (volume versus 2D)	100%↓	≥65%↓	All others	≥40%↑	Computer-assisted volumetrics
Volumetric (as per [21])	100%↓	≥50%↓	All others	≥25%↑	Computer-assisted volumetrics

* Extrapolation refers to taking the single largest diameter of a lesion (i.e. RECIST) and converting that to a volume ($V = 4/3\pi r^3$) assuming a spherical lesion. CR: complete response; PR: partial response; SD: stable disease; PD: progressive disease; RECIST: response evaluation criteria in solid tumors.

NEW IMAGING TECHNIQUES

A number of exciting new approaches in brain tumor imaging have become possible as a result of advances in MRI technology. While the value of these techniques is still being defined, it is likely that they will find their place alongside routine anatomical imaging in both clinical trials and routine clinical patient care, particularly when highly specific antitumor treatment strategies, such as antiangiogenesis factors, are administered.

Proton Magnetic Resonance Spectroscopy (1H-MRS)

Specific metabolic changes within tumor tissue can be determined with 1H-MRS in a semiquantitative manner, thus providing complementary information to routine anatomical imaging [22]. A number of changes in proton spectra are seen within glial tumors. Elevation of the 'choline' resonance is thought to be a consequence of increased synthesis and turnover of membrane phospholipids that correlates with cell density in gliomas [23]. The choline resonance is compared to the creatine resonance from contralateral normal brain [24] or, in some studies, to NAA (N-acetylaspartate). Concentrations of the lipids and lactate can also be determined.

Intrinsic limitations of 1H-MRS at 1.5T include insufficient spectral resolution and signal-to-noise to measure additional specific metabolites of interest, poor spatial resolution with resultant volume averaging and technical challenges involved in obtaining adequate spectra. Careful choice of the site of voxel placement is crucial, since a number of artifacts can degrade the signal. Voxels taken along the rim of peripheral enhancement yield more informative spectra than those taken from a centrally necrotic region of a malignant glioma. The heterogeneity of tumors argues for the use of multivoxel spectroscopic imaging over the use of single voxel MRS.

To date, studies have shown decreases in levels of choline during or following therapy in brainstem gliomas in children treated with radiotherapy [25], low-grade gliomas treated with temozolomide [26] and brain metastases treated with radiosurgery [27]. A caveat is that the presentation of averaged values from a study group may obscure a substantial degree of variation that can lessen the value of the findings in individual cases.

Dynamic Vascular Imaging Techniques

Dynamic MRI techniques can be used to measure key aspects of tumor vascularity in vivo. Two important features of tumor vasculature are the density of capillaries and the permeability of these vessels. Increased vessel numbers lead to elevated cerebral (i.e. tumor) blood volume (rCBV), which can be measured relative to normal brain by perfusion MRI techniques. Thus, rCBV provides regional analysis of capillary density in gliomas. In adult astrocytomas,

rCBV values correlate well with histopathologic grade and survival [2]. An rCBV value of ≥ 1.5 correlates well with the presence of high-grade tumor and with decreased survival in patients with diffuse astrocytomas. Low-grade oligodendrogliomas may have elevated rCBV and enhancement which is likely secondary to the increased capillary density in all grades of this histopathologic subset. Treatment related decreases in rCBV have been demonstrated in patients with progressive malignant gliomas undergoing therapy with the antiangiogenesis agent thalidomide [28].

Capillary permeability can be determined by rate and degree of tumor tissue enhancement following bolus administration of gadolinium. Early data have suggested that permeability analysis is more sensitive and specific than blood volume maps and routine gadolinium enhanced MRI in identifying high-grade tumors and in distinguishing progressive tumor from radiation necrosis [29]. Relationships between capillary permeability and response to treatment have not been reported.

Diffusion Imaging

Diffusion-weighted imaging (DWI) is sensitive to the Brownian movement of water molecules within and surrounding tumor cells [30]. This movement can be measured in multiple planes, providing not only a diffusion rate, but also a direction of movement (i.e. a diffusion tensor). Diffusion tensor imaging (DTI) has been shown to be of value in differentiating primary and metastatic tumors [31]. In the vasogenic edema surrounding brain tumors, there is elevated diffusion of water along compact white matter tracts such as those found in the corpus callosum and the optic radiations. A lower rate of diffusion and more isotropic movement is seen in the white matter surrounding primary tumors, presumably due to the presence of infiltrating tumor, compared to the diffusion features of vasogenic edema surrounding a brain metastasis [32].

In densely cellular tumors, there is relative restriction in the diffusion of water, whereas areas of necrosis or less densely cellular tumor have elevated levels of diffusion. For this reason, the value of diffusion imaging in assessment of therapeutic response is under investigation [22,33]. If tumors or areas within tumors with the highest cell density are more likely respond to treatment, then DWI may be able to predict response or provide a tool to measure early response.

Molecular Imaging

This is a rapidly emerging field of research that seeks to achieve in vivo imaging of molecular processes within brain tumor cells [34]. In one example of the use of MRI to image 'smart probes', tumor cell specific uptake of monocrystalline iron-oxide nanoparticles (MION) by tumor cells can be achieved due to the high concentration of transferrin receptors (ETR) on the cells of some tumors. These cells thus have a high rate of MION endocytosis and the MION particles can be imaged with MRI. In another example that has been tested in experimental models, cells can be induced to overexpress ETR and these cells can then be selectively detected in vivo by MRI after administration of MION particles. Stem cells labeled with superparamagnetic ferumoxide-poly-L-lysine complexes have permitted imaging analysis of the localization of these cells to tumor in glioma-bearing mice [35].

Paramagnetic chelates that have the capacity to change their magnetic properties upon enzymatic hydrolysis are under investigation. For instance, a gadolinium galactopyranose substrate demonstrates increased relaxivity following β-galactosidase mediated hydrolysis. Thus, the presence of ectopically expressed β-galactosidase can be indirectly detected by MRI. 'Magnetic nanosensors' are being developed that may be able to detect specific DNA or RNA sequences. Enzymes such as tyrosinase, which have a high metal binding capacity, might also be imaged with MRI when overexpressed. The value of these new approaches has yet to be determined [34].

SUMMARY

MRI is an important tool in the day-to-day clinical management of patients with brain tumors and in the assessment of new therapeutic agents in clinical trials. In this chapter, we have pointed out the special issues relating to the use of imaging in these trials and discussed the exciting technological advances that are already in use, as well as those that are in development.

ACKNOWLEDGMENT

This work was supported by the Stephen E. and Catherine Pappas Brain Tumor Imaging Research Program, Massachusetts General Hospital.

REFERENCES

1. Selker RG et al (2002). The Brain Tumor Cooperative Group NIH Trial 87-01: a randomized comparison of surgery, external radiotherapy, and carmustine versus surgery, interstitial radiotherapy boost, external radiation therapy, and carmustine. Neurosurgery 51(2):343–355; discussion 355–357.
2. Lev ML et al (2004). Glial tumor grading and outcome prediction using dynamic spin-echo MR susceptibility mapping compared to conventional contrast enhanced MRI: confounding effect of elevated relative cerebral blood volume of oligodendrogliomas. Am J Neuroradiol 25:214–221.
3. Barker FG et al (1997). Age and the risk of anaplasia in magnetic resonance-nonenhancing supratentorial cerebral tumors. Cancer 80:936–941.
4. Recht LD, Lew R, Smith TW (1992). Suspected low-grade glioma: is deferring treatment safe? Ann Neurol 31:431–436.
5. Recht LD, Bernstein M (1995). Low-grade gliomas. Neurol Clin 13:847–859.
6. Kleihues P et al (1995). Histopathology, classification, and grading of gliomas. Glia 15:211–221.
7. Shafqat S, Hedley-Whyte ET, Henson JW (1999). Age-dependent rate of anaplastic transformation in low-grade astrocytoma. Neurology 52:867–869.
8. James CD et al (1988). Clonal genomic alterations in glioma malignancy stages. Cancer Res 48:5546–5551.
9. Mandonnet E et al (2003). Continuous growth of mean tumor diameter in a subset of grade II gliomas. Ann Neurol 53(4): 524–528.
10. Batchelor T, Stanley K, Andersen J (2001). Clinical trials in neuro-oncology. Curr Opin Neurol 14(6):689–694.
11. Galanis E et al (2006). Validation of neuroradiologic response assessment in gliomas: measurement by RECIST, two-dimensional, computer-assisted tumor area, and computer-assisted tumor volume methods. Neuro-oncology 8(2):156–165.
12. Prados MD et al (2004). Phase III randomized study of radiotherapy plus procarbazine, lomustine, and vincristine with or without BUdR for treatment of anaplastic astrocytoma: final report of RTOG 9404. Int J Radiat Oncol Biol Phys 58(4):1147–1152.
13. Cairncross JG et al (1998). Specific genetic predictors of chemotherapeutic response and survival in patients with anaplastic oligodendrogliomas. J Natl Cancer Inst 90:1473–1479.
14. Grant R et al (1997). Chemotherapy response criteria in malignant glioma. Neurology 48(5):1336–1340.
15. Shah GD et al (2006). Comparison of linear and volumetric criteria in assessing tumor response in adult high-grade gliomas. Neuro-oncology 8(1):38–46.
16. Macdonald DR et al (1990). Response criteria for phase II studies of malignant glioma. J Clin Oncol 8:1277–1280.
17. Therasse P et al (2000). New guidelines to evaluate the response to treatment in solid tumors. European Organization for Research and Treatment of Cancer, National Cancer Institute of the United States, National Cancer Institute of Canada. J Natl Cancer Inst 92(3):205–216.
18. Sorensen AG et al (2001). Comparison of diameter and perimeter methods for tumor volume calculation. J Clin Oncol 19(2):551–557.
19. Warren KE et al (2001). Comparison of one-, two-, and three-dimensional measurements of childhood brain tumors. J Natl Cancer Inst 93(18):1401–1405.
20. Poussaint TY et al (2003). Interobserver reproducibility of volumetric MR imaging measurements of plexiform neurofibromas. Am J Roentgenol 180(2):419–423.
21. Batchelor TT et al (2004). Phase 2 study of weekly irinotecan in adults with recurrent malignant glioma: final report of NABTT 97-11. Neuro-oncology 6(1):21–27.
22. Lee KC et al (2006). Dynamic imaging of emerging resistance during cancer therapy. Cancer Res 66(9):4687–4692.

23. Croteau D et al (2001). Correlation between magnetic resonance spectroscopy imaging and image-guided biopsies: semiquantitative and qualitative histopathological analyses of patients with untreated glioma. Neurosurgery 49(4):823–829.

24. Rabinov JD et al (2002). In vivo MRS at 3 Tesla predicts recurrent glioma vs. radiation effects – initial experience. Radiology 225:871–879.

25. Laprie A et al (2005). Longitudinal multivoxel MR spectroscopy study of pediatric diffuse brainstem gliomas treated with radiotherapy. Int J Radiat Oncol Biol Phys 62(1):20–31.

26. Murphy PS et al (2004). Monitoring temozolomide treatment of low-grade glioma with proton magnetic resonance spectroscopy. Br J Cancer 90(4):781–786.

27. Kimura T et al (2004). Evaluation of the response of metastatic brain tumors to stereotactic radiosurgery by proton magnetic resonance spectroscopy, 201TlCl single-photon emission computerized tomography, and gadolinium-enhanced magnetic resonance imaging. J Neurosurg 100(5):835–841.

28. Cha S et al (2000). Dynamic contrast-enhanced T2-weighted MR imaging of recurrent malignant gliomas treated with thalidomide and carboplatin. Am J Neuroradiol 21(5):881–890.

29. Roberts HC et al (2000). Quantitative measurement of microvascular permeability in human brain tumors achieved using dynamic contrast-enhanced MR imaging: correlation with histologic grade. Am J Neuroradiol 21(5):891–899.

30. Schaefer PW, Grant PE, Gonzalez RG (2000). Diffusion-weighted MR imaging of the brain. Radiology 217:331–345.

31. Lu S et al (2003). Peritumoral diffusion tensor imaging of high-grade gliomas and metastatic brain tumors. Am J Neuroradiol 24(5):937–941.

32. Provenzale JM et al (2004). Peritumoral brain regions in gliomas and meningiomas: investigation with isotropic diffusion-weighted MR imaging and diffusion-tensor MR imaging. Radiology 232(2):451–460.

33. Roth Y et al (2004). High-b-value diffusion-weighted MR imaging for pretreatment prediction and early monitoring of tumor response to therapy in mice. Radiology 232(3):685–692.

34. Shah K et al (2004). Molecular imaging of gene therapy for cancer. Gene Ther 11(15):1175–1187.

35. Shah K (2005). Current advances in molecular imaging of gene and cell therapy for cancer. Cancer Biol Ther 4(5):518–523.

CHAPTER 50

Pitfalls in the Neuroimaging of Brain Tumors

Erik B. Nine, Haricharan Reddy, Eric C. Bourekas and H. Wayne Slone

INTRODUCTION

Each year, more than 200 000 people are newly diagnosed with a primary or metastatic brain tumor. Accurate diagnosis and grading of brain tumors are critical to determining prognosis and therapy. Imaging plays a critical role in these processes. Certain imaging characteristics, such as the pattern of enhancement and alteration of signal on magnetic resonance imaging (MRI), or density alterations on computed tomography (CT), suggest a brain tumor. However, imaging of brain tumors is frequently not so straightforward. Many entities can mimic a brain tumor. These mimics of tumors include inflammatory, infectious, demyelination and cerebrovascular diseases, among others. Once the diagnosis has been made and treatment initiated, it is important to determine response to therapy and differentiate tumor recurrence from treatment-related complications. In the following chapter, we discuss pitfalls in the neurological imaging of brain tumors categorized by location: extra-axial, the meninges and intra-axial. We will focus on the mimics of brain tumors in each location.

IMAGING FEATURES

Extra-axial and Meningeal Tumors

Meningioma and lymphoma are common extra-axial brain tumors. A common mimic of these lesions is sarcoidosis, which may be difficult to distinguish from a brain tumor [1]. Low signal intensity on diffusion-weighted images (DWI) helps distinguish sarcoidosis from lymphoma or meningioma, which is often mildly hyperintense on DWI.

Infectious granulomas such as tuberculosis may mimic meningiomas. Pulmonary involvement, as well as clinical parameters, can help differentiate these entities. Other infectious etiologies which can present with ring enhancing lesions simulating brain tumor include aspergillosis and toxoplasmosis. These patients are often immunocompromised. Paranasal sinus disease is usually the origin for the intracranial aspergillus.

Intracranial foreign body granulomas can simulate meningiomas. Gel foam, shunt tubing, suture, cotton or acrylate monomer can cause similar MRI imaging characteristics. Low signal on T1-weighted images, heterogeneous high signal on T2-weighted images and contrast enhancement are the typical MRI features [1]. Close correlation with history regarding type of surgery and length of time from surgery can assist in the diagnosis.

Whenever meningioma is a diagnostic consideration, other neoplastic processes which should be considered include lymphoma, dural-based metastasis and plasmacytoma. Of note, the dural tail sign, which is often associated with a meningioma, is a pitfall, as it can be seen with any lesion involving the dura.

Meningeal enhancement is seen in leptomeningeal metastases or even primary tumors such as lymphoma. It is important to differentiate leptomeningeal from pachymeningeal or dural enhancement. Other causes of leptomeningeal enhancement include infectious meningitis (bacterial, tuberculosis, fungal, syphilis and viral), granulomatous meningitis (sarcoidosis), subacute subarachnoid hemorrhage, subacute cerebral infarction and Sturge-Weber syndrome. Differential considerations for dural or pachymeningeal enhancement include infection (skull base osteomyelitis), intracranial hypotension, iatrogenic etiologies (post-craniotomy or shunt insertion), venous thrombosis and granulomatous meningitis (sarcoidosis). Correlation with the history usually narrows the differential diagnosis. Idiopathic intracranial hypotension shows diffuse pachymeningeal enhancement, as well as other findings including the 'sagging' brain.

In addition to enhancement, hyperintensity of the cerebrospinal fluid (CSF) on fluid attenuated inversion-recovery (FLAIR) imaging is often noted with leptomeningeal metastases. Other processes, such as subarachnoid hemorrhage and infection, can result in similar hyperintensity. Bright CSF on FLAIR imaging is not always indicative of subarachnoid disease. If the patient has been administered 100% O_2, the CSF may be bright. Bright CSF can also be a

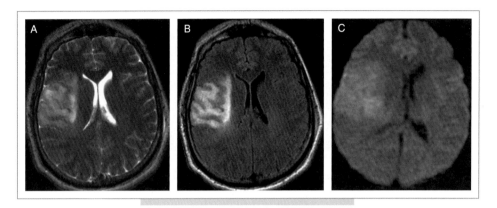

FIG. 50.1. A 49-year-old male with seizure activity and loss of consciousness. **(A)** T2-weighted MR image shows infiltrative area of increased signal involving both gray and white matter in the right MCA distribution mimicking infarct. **(B)** FLAIR sequence reveals mild gyral thickening and surrounding edema and mild mass effect. **(C)** Of note, there is mild restricted diffusion on the DWI sequences, not to the degree expected with infarction. Findings suggest infiltrative tumor. Subsequent biopsy yielded oligodendroglioma.

result of diffusion of gadolinium contrast into the CSF in the setting of renal insufficiency.

Intra-Axial Lesions

Cerebral infarction is one of the most common mimics of a brain tumor. It is usually straightforward to differentiate brain tumor from infarct, especially if the clinical history is known and the characteristic imaging findings are seen in a vascular distribution. When there is an atypical clinical course or unreliable history, differentiating cerebral infarction from brain tumor may be difficult (Figure 50.1).

An acute ischemic stroke may show an ill-defined cerebral lesion with mass effect, with or without parenchymal enhancement. These findings are similar to a glioma. DWI is crucial in this setting and usually will demonstrate a greater degree of hyperintensity (i.e. restricted diffusion) in the setting of an infarct, as compared to neoplasm. However, this may not be conclusive and short-term follow-up is necessary in these patients.

Subacute infarction is even more likely to mimic a brain tumor. Subacute infarction often demonstrates contrast enhancement. Moreover, DWI may be falsely negative as pseudonormalization of the apparent diffusion coefficient can mask the infarction on imaging. Variable contrast enhancement can also mimic a neoplasm. However, the vascular distribution, gyral enhancement and minimal to no mass effect are more characteristic of subacute infarction.

Venous infarction may simulate brain tumor as it does not occur in a vascular territory, has more extensive white matter lesions, can show hemorrhagic change, has irregular contrast enhancement, and can appear similar to peritumoral edema. If a lesion is not in a typical vascular distribution or there are multifocal high signal changes on T2-weighted imaging, further inspection for evidence of venous infarction should be made via MR venography (Figure 50.2).

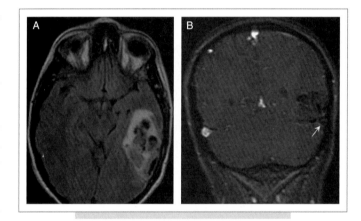

FIG. 50.2. This is a 36-year-old HIV-positive female with a 3-day history of global headache. **(A)** On MRI, FLAIR sequence shows a heterogeneous signal mass involving white and gray matter, with surrounding edema which could suggest neoplasm, however, this is non-specific. **(B)** Coronal MR venogram shows left transverse dural venous sinus thrombosis (arrow) and indicates the parenchymal lesion is actually a hemorrhagic venous infarct.

Hematoma may be indistinguishable from brain tumor. History and follow-up imaging can help make this differentiation.

A large aneurysm with surrounding edema may mimic a brain tumor. If the aneurysm is partially or completely thrombosed, it may exhibit ring enhancement or peripheral calcifications as imaging features. Laminated calcification morphology may be characteristic.

Tumefactive multiple sclerosis (MS) can mimic brain tumors such as gliomas (Figure 50.3). The lack of mass effect and edema may help in the differential diagnosis. Also, incomplete rim enhancement and mildly restricted diffusion are typical of MS lesions.

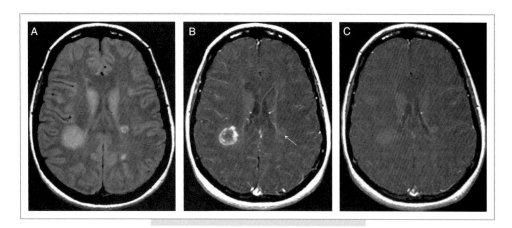

FIG. 50.3. In this 48-year-old female with weakness and suspected multiple sclerosis, there are several periventricular lesions in the deep white matter of both cerebral hemispheres present on the initial FLAIR sequences **(A)**. One of the lesions shows thick rim enhancement in the posterior right frontal lobe following gadolinium administration raising the possibility of neoplasm **(B)**. However, the lack of enhancement of several other periventricular lesions (arrow), as well as partial interval decrease in size and enhancement on the 7-month follow-up examination, suggests demyelination **(C)**.

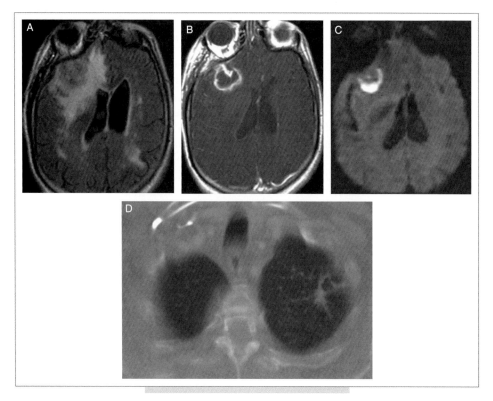

FIG. 50.4. In this 69-year-old male with 3-day history of weakness and dizziness a contrast-enhanced MRI was performed which demonstrated a heterogeneous mass in the right frontal lobe with surrounding increased FLAIR signal consistent with vasogenic edema **(A)**. Administration of intravenous gadolinium confirms circumferential rim enhancement suggesting primary or metastatic brain malignancy **(B)**. Diffusion-weighted imaging (DWI) confirms marked increased signal **(C)**, suggesting parenchymal abscess. Further work-up later revealed a cavitary left upper lobe lung lesion **(D)**. Biopsy revealed Nocardia infection.

Intracranial parenchymal sarcoidosis can also mimic a brain tumor. The findings of neurosarcoidosis are non-specific. Parenchymal lesions may be either solitary or multiple and can be infratentorial or supratentorial. They usually exhibit moderate edema. It is difficult to distinguish sarcoid from brain tumors such as glioma, metastases or lymphoma. Associated findings may include involvement of the basal leptomeninges and cranial nerves.

Neurosyphilis is becoming more common, due to the spread of AIDS worldwide. Focal meningovascular neurosyphilis (cerebral gumma) can present as solitary or multiple ring enhancing subcortical masses.

Differentiating cerebral abscesses from necrotic tumor is difficult on conventional MRI. DWI helps in distinguishing these entities. Increased signal in diffusion-weighted imaging and decreased apparent diffusion coefficient (ADC) values suggest brain abscesses (Figure 50.4). The converse

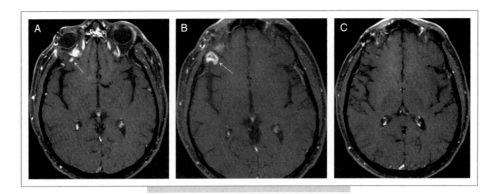

FIG. 50.5. A 51-year-old male with history of right lacrimal adenocystic carcinoma following intraoperative high dose rate brachytherapy and external beam radiation therapy now presents with persistent headache. Initial T1-weighted gadolinium enhanced MRI images reveal a rim-enhancing lesion in the frontal lobe adjacent to the orbital apex which enlarged over a 7-month period (arrow) **(A, B)**. Differential diagnosis of a recurrent rim-enhancing lesion included tumor extension versus radiation necrosis. Initial PET evaluation of the brain was normal. Two year delayed follow-up T1-weighted contrast enhanced images reveal resolution of the rim-enhancing signal abnormality **(C)**, confirming radiation necrosis.

is true for brain tumors. In addition, the ring enhancement of an abscess is typically smooth and thinner adjacent to the ventricular system. The capsule of an abscess often shows low T2 and high T1 signal. In contrast, an irregular and thickened rim of enhancement is typical of brain tumors.

In the HIV patient, differentiating toxoplasmosis from lymphoma may be difficult, as both may demonstrate ring enhancement. Further differentiation may be made utilizing positron emission tomography (PET) imaging and MR spectroscopy.

RECURRENCE VERSUS TREATMENT-RELATED COMPLICATIONS

The diagnosis of radiation necrosis on imaging remains challenging, in part because the pattern of abnormal enhancement closely mimics that of recurrent brain tumor (Figure 50.5). Tumor recurrence is suggested by new, nodular enhancement. In contrast, treatment-related complications, such as post-radiation therapy changes, necrosis and abscess formation, usually will not demonstrate new nodular enhancement.

MRI manifestations of radiation necrosis are cystic or finger-like foci of abnormally increased signal intensity within white matter on T2WI images. Rim or possibly nodular enhancement may be present [2–4], as there is often diffuse damage to the blood–brain barrier. The degree of enhancement can vary with time. The enhancement pattern may be heterogeneous and lesions may appear discontinuous with surrounding areas of non-enhancing necrosis. Vasogenic edema and regional mass effect may be minimal or extensive. These findings are often non-specific and may not permit differentiation from tumor recurrence.

Assessment of ADC ratios in the enhancing regions of a brain tumor, during the follow-up evaluation after therapy, may be useful in differentiating radiation effects from tumor recurrence or progression [5]. Additionally, numerous metabolic imaging techniques have been evaluated to help distinguish recurrent neoplasm from radiation related post-treatment changes. Previous studies have demonstrated diminished levels of choline-containing compounds in brain parenchyma affected by chronic radiation necrosis as compared with tumor, which typically shows an elevated choline to creatine ratio [6]. Positron emission tomography utilizing ^{18}F-FDG will also assist in differentiating neoplasm from radiation necrosis by determining relative glucose metabolism [7,8]. Despite advances in metabolic imaging, differentiating recurrent neoplasm from radiation necrosis remains difficult.

REFERENCES

1. Report, Primary Brain Tumors in the United States, 1992–1997.
2. Lee AW, Cheung LO, Ng SH et al (1990). Magnetic resonance imaging in the clinical diagnosis of late temporal lobe necrosis following radiotherapy for nasopharyngeal carcinoma. Clin Radiol 12:256–270.
3. Chan YL, Leung SF, King AD, Choi PH, Metreweli C (1999). Late radiation injury to the temporal lobes: morphologic evaluation at MR imaging. Radiology 213:800–807.
4. Kumar AJ, Leeds NE, Fuller GN et al (2000). Malignant gliomas: MR imaging spectrum of radiation therapy- and

chemotherapy-induced necrosis of the brain after treatment. Radiology 217:377–384.

5. Hein PA, Eskey CJ, Dunn JF, Hug EB (2004). Diffusion-weighted imaging in the follow-up of treated high-grade gliomas: tumor recurrence versus radiation injury. Am J Neuroradiol 25:201–209.

6. Fulham MJ, Bizzi A, Dietz MJ et al (1992). Mapping of brain tumor metabolites with proton MR spectroscopic imaging: clinical relevance. Radiology 185:675–686.

7. Chao ST, Suh JH, Raja S, Lee SY, Barnett G (2001). The sensitivity and specificity of FDG PET in distinguishing recurrent brain tumor from radionecrosis in patients treated with stereotactic radiosurgery. Int J Cancer 96:191–197.

8. Langleben DD, Segall GM (2000). PET in differentiation of recurrent brain tumor from radiation injury. J Nucl Med 41:1861–1867.

SECTION V

Neuroimaging of Other Neuro-Oncological Syndromes

CHAPTER 51

Neuroradiology of Leptomeningeal Metastases

Marc C. Chamberlain

INTRODUCTION

Leptomeningeal meningitis (LM) is the third most common CNS metastatic complication of systemic cancer and, additionally, is seen in patients with primary brain cancers. Due to the fact that the entire neuraxis is contained by a cerebrospinal fluid (CSF) compartment which circulates, all regions of the CNS may be involved by LM which can result in multifocal neurological signs and symptoms. Diagnosis is problematic, as the disease is often not considered due to the pleomorphic manifestations of LM and occurrence in patients with extra-neural sites of metastases. Useful tests to establish diagnosis and guide treatment include magnetic resonance imaging (MRI) of the brain and spine, CSF cytology and radioisotope CSF flow studies. Assessing the extent of CNS disease radiographically in patients with LM is of clinical value, as documenting large volume subarachnoid bulky disease or CSF flow obstruction is prognostically significant and may change treatment. This review summarizes the three most frequently used neuroradiographic imaging modalities to evaluate patients with LM, including cranial and spine magnetic resonance imaging (MRI) and radioisotope CSF flow studies (Table 51.1).

COMPARATIVE CRANIAL COMPUTED TOMOGRAPHY AND MR IMAGING

Leptomeningeal metastasis is found at autopsy in 5% of patients with systemic cancers [2]. Premorbid diagnosis is much less frequent, reflecting failure to consider the diagnosis of LM or difficulty in making the diagnosis [2–4].

TABLE 51-1 Leptomeningeal metastases disease staging [1]

CSF analysis for pathologic confirmation
Contrast-enhanced MR or CT brain scans
Contrast MR spine imaging or CT myelography in patients with spinal cord dysfunction
CSF flow study using ventricular reservoir
Spine imaging in patients with spinal cord CSF flow block if not previously performed

LM is diagnosed clinically when malignant cells are found by cytologic examination of the CSF. Multiple examinations of the CSF may be required to make the diagnosis [2–4]. CSF cytology is negative in 40–50% of patients in whom the diagnosis is confirmed at the time of autopsy. Both cranial (CT) and MRI following IV contrast administration (CE-CT/MR-Gd) have demonstrated abnormalities consistent with LM and may be used to make the diagnosis of LM regardless of CSF cytology [5].

Cranial CE-CT is abnormal in approximately 26–56% of patients with LM [5–9]. The CE-CT abnormalities include:

1 sulcal-cisternal enhancement
2 ependymal-subependymal enhancement
3 irregular tentorial enhancement
4 cisternal or sulcal obliteration
5 subarachnoid enhancing nodules
6 intraventricular enhancing nodules
7 communicating hydrocephalus [5–9]. In addition, up to 60% of patients with cranial CE-CT findings suggestive of LM have coexistent parenchymal metastases.

Two prior reports discuss results of MR in patients with LM; in both, MR performed without contrast enhancement was compared to CE-CT [10,11]. Of the total of 30 patients reported, MR without gadolinium was abnormal in 35% compared with 53% evaluated with CE-CT. CE-CT was, therefore, considered superior to non-contrast MR in the evaluation of patients with LM. The regions of abnormal sulcal, cisternal, ependymal and tentorial enhancement seen with CE-CT had normal signal intensity on non-contrast MR.

Mathews compared CE-CT with Gd-MR in experimental bacterial meningitis in a canine model [12]. Unenhanced MR was not helpful in identifying meningitis in that study. However, T1-weighted (T1W) Gd-MR was superior to CE-CT in detecting abnormal leptomeningeal enhancement and complications of meningitis including cerebritis and ventriculitis. Similar findings were demonstrated also by Frank in a rabbit model of meningeal carcinomatosis [13]. Sze compared spinal Gd-MR with CT-myelography in 12 patients with LM [14]. They concluded that spinal Gd-MR was competitive with, and sometimes superior to, CT-myelography in demonstrating intradural extramedullary spinal diseases in LM.

Data from Chamberlain indicate that Gd-MR is the preferred imaging modality for diagnosing LM and suggest that CE-CT may be unnecessary [15]. These authors found T1W post-contrast axial and coronal images especially useful for demonstrating the abnormalities seen in LM. Gd-MR had a higher specificity and sensitivity than CE-CT in the evaluation of LM (Table 51.2). All abnormalities detected by CE-CT were detected also by Gd-MR. Despite the superiority of Gd-MR to CE-CT in the evaluation of LM, both modalities had a high incidence of false negative studies (30% by Gd-MR, 58% by CE-CT) in this study of 14 patients with CSF positive LM. Normal findings by either or both examinations do not exclude a diagnosis of LM. LM is a histologic diagnosis and often is diagnosed by the demonstration of neoplastic cells in the CSF. However, in instances of a non-diagnostic CSF examination, Gd-MR may be diagnostic of LM (Figures 51.1–51.14).

TABLE 51-2 Comparison of MR imaging and cranial CT in leptomeningeal metastases [15]		
Findings	Imaging techniques	
Abnormal enhancement	71%	29%
Nodules	43%	36%
Subarachnoid	36%	29%
Parenchymal	43%	29%
Hydrocephalus	7%	7%

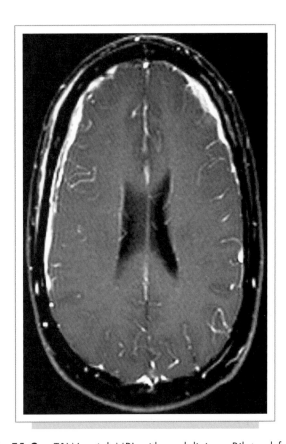

FIG. 51.2. T1W axial MRI with gadolinium. Bilateral frontal lateral convexities subarachnoid tumor (breast cancer).

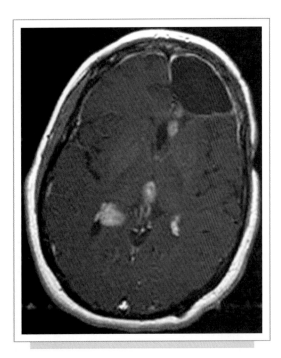

FIG. 51.1. T1W axial MRI with gadolinium. Intraventricular nodules in the left ventricular atrium, third and left lateral ventricles (breast cancer).

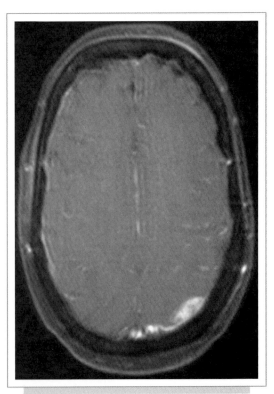

FIG. 51.3. T1W axial MRI with gadolinium. Focal subarachnoid tumor left lateral parietal convexity (breast cancer).

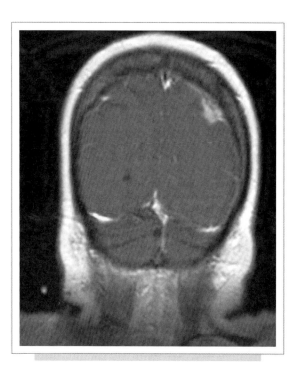

FIG. 51.4. T1W coronal MRI with gadolinium. Focal subarachnoid tumor left lateral parietal convexity (breast cancer).

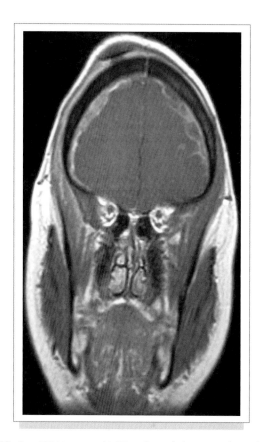

FIG. 51.6. T1W coronal MRI with gadolinium. Bilateral frontal lateral convexities subarachnoid tumor (breast cancer).

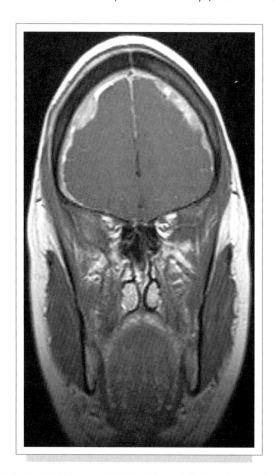

FIG. 51.5. T1W coronal MRI with gadolinium. Bilateral frontal lateral convexities subarachnoid tumor (breast cancer).

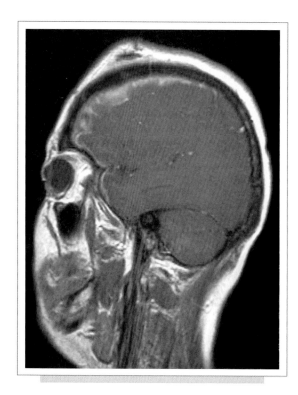

FIG. 51.7. T1W sagittal MRI with gadolinium. Subarachnoid tumor over superior convexity frontal lobe (breast cancer).

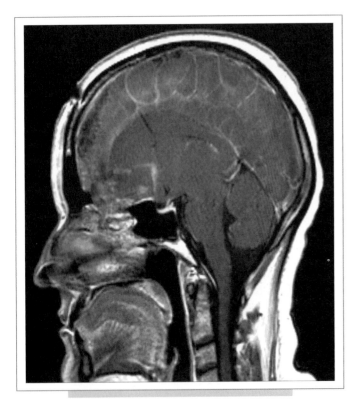

FIG. 51.8. T1W sagittal MRI with gadolinium. Enhancement of falx (melanoma).

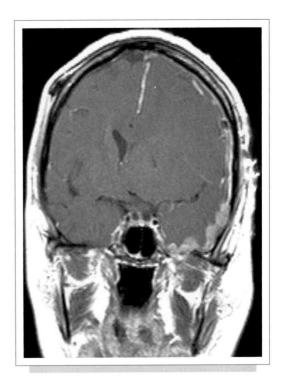

FIG. 51.10. T1W coronal MRI with gadolinium. Subarachnoid tumor with lateral and inferior subarachnoid space of left temporal lobe (melanoma).

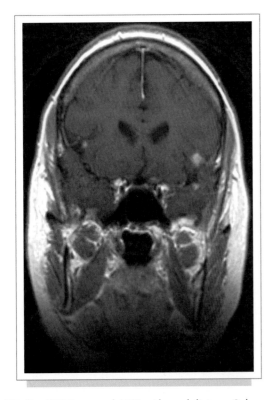

FIG. 51.9. T1W coronal MRI with gadolinium. Subarachnoid nodules in right and left sylvian cisterns (melanoma).

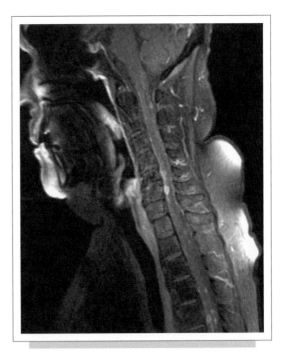

FIG. 51.11. T1W sagittal MRI with gadolinium. Multiple subarachnoid nodules both ventral and dorsal to cervical and upper thoracic spinal cord (non-small cell lung cancer).

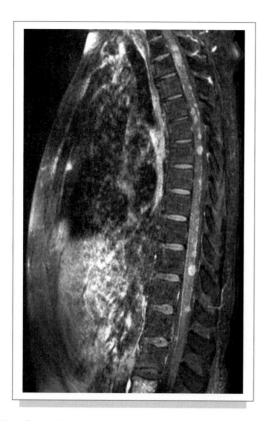

FIG. 51.12. T1W sagittal MRI with gadolinium. Multiple subarachnoid nodules both ventral and dorsal (dorsal > ventral) to thoracic spinal cord (non-small cell lung cancer).

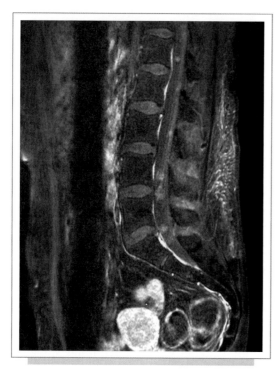

FIG. 51.13. T1W sagittal MRI with gadolinium. Multiple subarachnoid nodules involving the lumbar sac and cauda equina (non-small cell lung cancer).

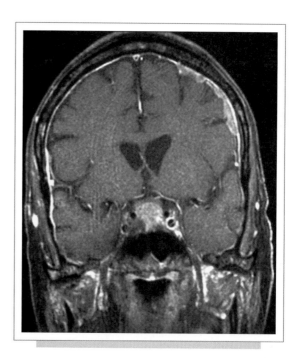

FIG. 51.14. T1W coronal MRI with gadolinium. Thick subarachnoid tumor overlying the left frontal lateral convexity (prostate cancer).

In the majority of patients with LM, Gd-MR is most useful in demonstrating bulky disease, a pattern of disease most responsive to radiotherapy.

COMPARATIVE SPINAL IMAGING

Only a limited number of studies have evaluated spine imaging in patients with LM [4,7,15–25]. However, a significant percentage of patients with LM will require spine imaging for one of several indications, including: supportive evidence in arriving at a diagnosis of LM; defining regions of bulky disease (e.g. subarachnoid nodules); evaluating possible coexistent epidural spinal cord compression or intraparenchymal spinal metastases; and defining regions of CSF flow interruption. Three methods of spine imaging are available including computerized tomographic myelography (CT-M), contrast-enhanced spine magnetic resonance imaging (S-MR) and ^{111}indium-DTPA CSF flow studies (FS). A study by Chamberlain compared the three methods of spine imaging in a series of 61 patients with LM [27] (Table 51.3).

Leptomeningeal metastasis occurs in approximately 5% of patients with cancer, of whom in 10–15% the initial presentation of cancer is with LM [15]. The diagnosis of LM is made by sampling CSF for malignant cells, however, approximately 40% of patients proven to have LM at postmortem will have negative CSF cytology ante-mortem [27]. This suggests a high incidence of LM which is cytologically negative thereby requiring corroborative tests to assist in diagnosing LM. These tests include a clinical syndrome

TABLE 51-3 Abnormalities of spine imaging as a function of clinical presentation

	Clinical presentation				
	Normal neurological examination	Cerebral dysfunction	Cranial nerve palsies	Spinal cord dysfunction	Multilevel dysfunction
Number of patients (% Total)	14 (23%)	3 (5%)	21 (34%)	30 (49%)	6 (10%)
Imaging modality	(# of patients with abnormal study)				
CT-M	0	1	7	16	5
S-MR	1	3	5	15	4
FS	1	2	6	18	6

CT-M: computerized tomographic myelography; S-MR: contrast enhanced spine magnetic resonance imaging; FS: [111]indium-DTPA CSF flow study.

consistent with LM, CSF biochemical markers of LM (e.g. chorio-embryonic antigen) and neuroradiographic studies documenting findings compatible with LM [4,15–18]. In the study by Chamberlain cited above, 15% of patients (9/61) were felt to have LM notwithstanding negative CSF cytology [26]. The diagnosis of LM in this group of patients was confirmed by either a compatible clinical syndrome or neuroradiographic findings (cranial or spinal) consistent with LM. In two-thirds of these patients (6/9), post-mortem examination of the central nervous system was performed confirming the ante-mortem diagnosis of LM and substantiating the utility of both neuroradiographic and clinical examination in diagnosing patients with LM and negative CSF cytology. From the post-mortem study of LM by Glass referenced above, a discrepancy exists regarding the incidence of patients with negative CSF cytology and autopsy proven LM. It is probable that a number of patients with LM and negative CSF cytology were not diagnosed, either because they were not referred for evaluation or these patients had extensive systemic disease and were managed with supportive care only.

A number of studies have evaluated patients with LM by spine imaging, usually employing either CT-M or S-MR [4,16–25]. However, relatively few studies have compared both CT-M and S-MR in patients with LM [7,20]. Myelographic features characteristic of LM include: parallel longitudinal striations due to thickened nerved roots in the cauda equine; nerve root sleeve filling defects; widening of the spinal cord; varying degrees of CSF flow block; subarachnoid nodules and scalping of the subarachnoid membranes [4,7,16,19,20,22]. Similar findings are seen with contrast-enhanced spine-MR which, in addition, demonstrates pial-enhancement of the spinal cord, termed by Kramer 'sugar coating' [7,20–25].

The study of 61 patients with LM cited above sequentially studied by FS, CT-M and S-MR demonstrated similar findings to those previously described [4,7,15–25]. However, in contrast to children, the incidence of spinal imaging abnormalities in patients with asymptomatic LM appears lower in adults (approximately 20% for any given spinal imaging study). An increased rate of spinal imaging abnormalities is seen in adult patients with symptomatic

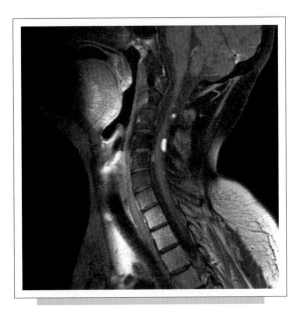

FIG. 51.15. T1W sagittal MRI with gadolinium. Multiple subarachnoid nodules both ventral and dorsal to cervical spinal cord (ependymoma).

LM defined as either cerebral, cranial nerve or spinal cord dysfunction with percentages approximating those found in children (approximately 50–70% for any given spinal imaging study). Very few discordant studies (<10%) were observed when comparing CT-M and S-MR in the above mentioned study and, in general, CT-M and S-MR were complementary in evaluating patients with LM. The differences were usually quantitative and not qualitative, notwithstanding that S-MR better demonstrated coexisting epidural spinal cord compression and intradural intramedullary spinal cord metastases, two CNS complications of metastatic systemic cancer requiring emergent radiotherapy. Therefore, S-MR would appear to be the preferable spine imaging modality for patients with LM requiring spine imaging, not only because of its comparability to CT-M but, additionally, because S-MR is non-invasive, has a very low rate of procedure-related morbidity and is associated with greater patient acceptability (Figures 51.15–51.20).

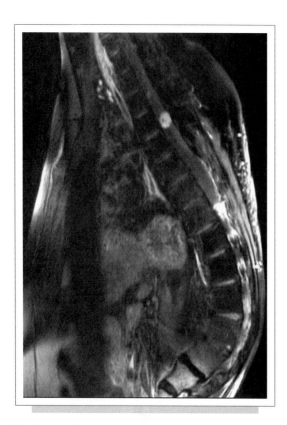

FIG. 51.16. T1W sagittal MRI with gadolinium. A large subarachnoid nodule ventral to the thoracic spinal cord immediately above tumor affecting the lumbar space (ependymoma).

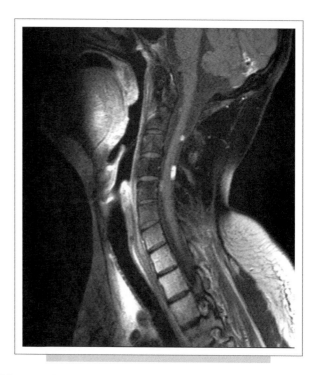

FIG. 51.18. T1W sagittal MRI with gadolinium. Subarachnoid tumor nodules ventral and dorsal to cervical spinal cord (ependymoma).

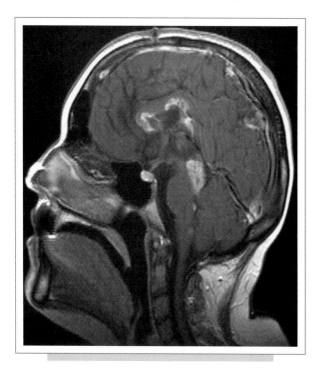

FIG. 51.17. T1W sagittal MRI with gadolinium. A corpus callosum based tumor with dissemination to the collicular cistern and floor of the fourth ventricle (glioblastoma).

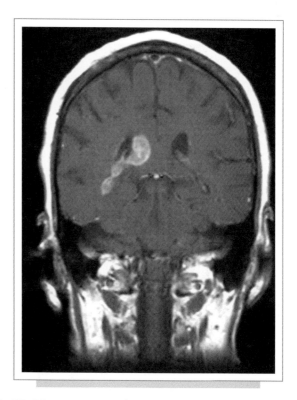

FIG. 51.19. T1W coronal MRI with gadolinium. Intraventricular tumor extending from the right lateral ventricle subsaendyma (primary CNS lymphoma).

FS have in prior reports demonstrated their superiority in detecting interruption of CSF flow in patients with LM when compared with CT-M and S-MR [15,18]. The above mentioned study corroborates and expands upon this prior experience [26]. Interruption of CSF flow with loculation or compartmentalization of CSF as documented by FS pathologically appears to be a consequence of minimal tumor volume resulting in tumor adhesions of the subarachnoid space. These subarachnoid adhesions are below the radiographic sensitivity of either CT or MR imaging. However, FS are informative only with respect to compartmentalization of CSF and provide no information regarding bulky leptomeningeal disease, an aspect of LM best addressed by CT-M or, preferably S-MR. Failure of radionuclide to appear in a given CSF compartment is operationally defined as CSF flow block [18]. In vivo CSF flow is assessable by MR techniques and, in the future, detection and quantification of CSF flow by MR may obviate the need for radionuclide CSF flow studies [29,30]. Finally, although infrequently present in patients with LM, both CT-M and S-MR are superior to FS in detecting epidural spinal cord metastases. Abnormalities of CSF flow, as demonstrated by FS, are therapeutically addressed by administering involved-field radiotherapy to the obstructed site of CSF flow. Approximately 50% of patients with an intracranial disturbance of CSF flow and 30% of patients with a disturbance of CSF flow intraspinally will have restored normal CSF flow as assessed by post-radiotherapy FS after

irradiation [18]. Radionuclide CSF flow studies, or so-called radionuclide ventriculography, provide a safe physiological assessment of the functional anatomy of the CSF spaces [15,18]. Compartmentalization of CSF as documented by FS is an important consideration in patients with LM considered for intra-CSF chemotherapy, as homogeneous distribution of administered chemotherapy within CSF is critical for effective treatment [15,18]. In patients with LM not considered candidates for intra-CSF therapy, for example in a patient with leukemic meningitis being treated with craniospinal irradiation, CSF flow studies would not be required as part of staging the extent of LM.

In conclusion, patients with LM frequently require spine imaging and, based on the discussion above, it is suggested both S-MR and FS provide the best neuroradiographic assessment. S-MR is superior for detecting bulky disease, whereas FS best demonstrates interruption of CSF flow. Patients with LM require an extent of leptomeningeal disease assessment before institution of radiotherapy or intra-CSF chemotherapy. Spine imaging is an important component of this neuroradiographic assessment best served by utilizing both S-MR and FS. In patients with a normal neurological examination and LM, S-MR evaluation may be deleted. FS, however, should be performed in this patient population with LM and normal neurological examination. If interruption of CSF flow is documented by FS, S-MR should then be performed to assess for bulky leptomeningeal disease. In patients with abnormal neurological examinations, either bulky disease (best detected by S-MR) or CSF compartmentalization (best detected by FS) occurs sufficiently often to warrant combined study with both S-MR and FS.

RADIOISOTOPE CSF FLOW STUDIES

Radionuclide cerebrospinal fluid (CSF) flow studies or so-called radionuclide ventriculography provide a safe physiologic assessment of the functional anatomy of the CSF spaces [31–33]. CSF circulates through the ventricular system and subarachnoid space that surrounds both the brain and spinal cord (Figure 51.21). Normally, CSF flows anteriorly in the lateral ventricles through the foramen of Monro and into the third ventricle. CSF then flows caudally from the third ventricle through the aqueduct of Sylvius into the fourth ventricle. Exit of CSF from the fourth ventricle is directed into the dorsal spinal subarachnoid space through the foramen of Magendie and into the basal cisterns through the lateral foramina of Luschka. Passage of CSF through the foramen of Magendie into the vallecula and the beginning of downward flow into the dorsal cervical subarachnoid space precedes the exit of CSF from the foramina of Luschka. CSF then flows caudally through the dorsal spinal subarachnoid space, followed by ascent of CSF in the ventral spinal subarachnoid space. Completion of the normal pattern of CSF circulation is by ascent from

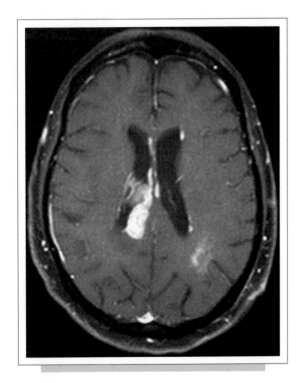

FIG. 51.20. T1W axial MRI with gadolinium. Intraventricular tumor dissemination with ependymal implants in the right and left lateral ventricles (primary CNS lymphoma).

the basal cisterns toward the superior sagittal sinus by way of migration over the cerebral convexities and along medial routes through the suprasellar and quadrigeminal cisterns [31–33]. Normal times of radionuclide appearance in anatomically separable CSF compartments have been established by radionuclide ventriculography [18,34–36].

CSF flow studies permit detection of CSF flow abnormalities which are often not apparent by conventional neuroradiographic imaging [26,34–36]. Approximately 50% of patients with LM will manifest CSF flow abnormalities, most commonly at the base of brain, spinal subarachnoid space (especially at the level of the conus medullaris and cauda equina), and over the cerebral convexities [26,34–36]. CSF flow studies are more sensitive in demonstrating compartmentalization of CSF pathways than are MR spine, CT myelography and cranial MR or CT imaging [18,26,34]. This review summarizes five studies highlighting the importance of radionuclide CSF flow studies in both the treatment and outcome of patients with leptomeningeal metastases [18,26,34,37,38].

Table 51.4 demonstrates the normal times of appearance of [111]indium-DTPA in various CSF compartments following intraventricular injection of radionuclide in 30 adult patients with leptomeningeal metastases [18]. These studies were performed in a consecutive series of 60 adult patients (Figures 51.22 and 51.23). In a similar study, performed in children, the normal time to appearance of radioisotope in various CSF compartments following ventricular injection is seen in Table 51.5 [34]. In a separate study, 20 consecutive adult patients with leptomeningeal metastases were evaluated by intralumbar administered [111]indium-DTPA (Table 51.6). Results from patients in this study with normal CSF flow studies can be seen in Table 51.3 [36].

In a study of 30 consecutive patients with cytologically documented leptomeningeal metastases, all patients underwent intraventricular [111]indium-DTPA CSF flow studies [18]. Sixteen patients (53%) had normal [111]indium-DTPA CSF flow studies. Fourteen patients (47%) had documented compartmentalization of the CSF system. In four patients (13%), unsuspected CSF block was confined to the base of the brain with absent spinal descent and poor cerebral convexity ascent of [111]indium-DTPA. All patients received whole-brain irradiation and two (50%) were subsequently shown to have restored normal CSF flow. In 10 other patients (33%), suspected (8) and unsuspected (2) spinal blocks were seen. Patients with spinal blocks underwent CT myelography or spine MR and were subsequently treated with involved-field irradiation. Following radiation therapy, normal CSF flow studies were achieved in only four patients (40%) with documented spinal cord blocks.

In a study of 61 consecutive adult and pediatric patients with leptomeningeal metastases as defined by positive CSF cytology or a clinical syndrome and compatible neurographic findings (cited above), all patients at presentation underwent CT myelography (CT-M), spine MR (S-MR) and ventricular [111]indium-DTPA CSF flow studies [26]. In 57% of patients, all three spine imaging modalities were normal. Forty-three percent of patients demonstrated abnormalities on spine imaging; 33% had abnormal radioisotope CSF flow studies, 34% showed abnormalities on S-MR and 33% had abnormalities by CT-M. Radioisotope

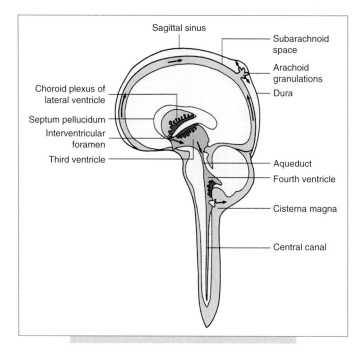

FIG. 51.21. Diagram of normal CSF flow.

	mCi[111]indium-DTPA	Ventricular system	CSF compartment: time to appearance (minutes)				
			Cisterna magna/basal cisterns	Spinal subarachnoid			Sylvian cisterns
				Cervical cord	Thoracic cord	Lumbar cord	
Median	700	1	5	15	20	30	50
Range	500–900	0–5	5–15	5–20	10–20	25–50	35–90

TABLE 51-4 Adult leptomeningeal metastases: normal compartmental appearance of [111]indium-DTPA following intraventricular administration [41]

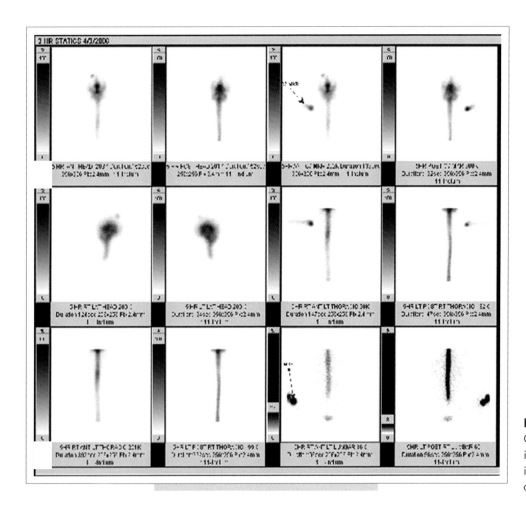

FIG. 51.22. [111]-indium DTPA CSF flow study. Radioisotope injected into an Ommaya device implanted in the right lateral ventricle showing normal CSF flow.

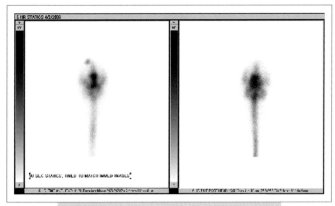

FIG. 51.23. [111]-indium DTPA CSF flow study. 24-hour delayed study showing CSF obstruction over lateral cerebral convexities.

TABLE 51-5 Pediatric leptomeningeal metastases: normal compartmental appearance of [111]indium-DTPA CSF flow studies following intraventricular administration [41]

	CSF compartment: time to appearance (minutes)					
	Ventricular system	Cisterna magna/basal cisterns	Spinal subarachnoid			
			Cervical cord	Thoracic cord	Lumbar cord	Sylvian cisterns
Median	1	5	8	15	35	80
Range	0–3	3–5	5–10	10–30	20–45	60–90

	mCi ^{111}indium-DTPA	CSF compartment: time to appearance (minutes)						
		Spinal subarachnoid			Cisterna magna/basal cisterns	Sylvian cisterns	Ventricular system	Cerebral convexities
		Lumbar Cord	Thoracic cord	Cervical cord				
Median	500	1	22.5	32.5	37.5	65	1440	1440
Range	450–600	0–1	20–25	30–35	35–40	60–70	1440	1440

TABLE 51-6 Normal compartmental appearances of ^{111}indium-DTPA following intra-lumbar administration to adult patients with leptomeningeal metastases [40]

flow studies were most sensitive for detecting interruption of CSF flow, whereas CT-M and S-MR better demonstrated nerve root thickening, cord enlargement, subarachnoid nodules, intraparenchymal cord tumor and epidural spinal cord compression.

In a study of 40 adult patients with cytologically documented leptomeningeal metastases, all were demonstrated to have interruption of CSF flow by radionuclide ventriculography [38]. All patients were treated with radiotherapy (30 Gy and 10 fractions) to the site of CSF obstruction after which intra-CSF chemotherapy (methotrexate or cytarabine followed by cytarabine or thio-TEPA if clinically indicated) was administered. Twenty patients (Group I) after radiotherapy to the site of CSF flow abnormality, demonstrated re-establishment of normal CSF flow. By contrast, 20 patients (Group II) treated in a similar manner had persistency of CSF flow obstruction. All patients were treated with intraventricular chemotherapy. Median survival was 6 months in Group I (range 3–15 months) compared with 1.75 months in Group II (range 1–4 months). Cause of death differed between groups with 20% of Group I patients dying of progressive leptomeningeal disease compared with 70% of Group II patients.

Radionuclide CSF flow studies provide a physiologic assessment of the functional anatomy of the various CSF compartments [18,31–36]. Normal times of radionuclide appearance in CSF compartments in these various studies are similar to those reported in the literature. The pattern of CSF circulation time to appearance of regionally administered chemotherapy is similar whether given antegrade by Ommaya reservoir administration or by intralumbar administration. The results of these various studies suggest that ^{111}indium-DTPA or ^{99}Tc macro-aggregated albumin CSF flows studies are more sensitive in demonstrating compartmentalization of CSF pathways than are enhanced spine MR imaging, CT myelography or enhanced cranial MR/CT imaging [18,26,34,37]. The reasons for this increased sensitivity may reflect the dynamic nature of radioisotope CSF flow studies, whereby radionuclide is passively carried by CSF bulk flow in a physiologic antegrade manner. Therefore, radioisotope CSF flow studies in both adult and children with LM demonstrate CSF

compartmentalization, a critical issue in intra-CSF drug distribution. Furthermore, both intracranial compartmentalization and spinal subarachnoid block are better visualized by radioisotope CSF flow studies than comparative neuro-radiographic studies [18]. Treatment of CSF flow blocks by involved-field radiotherapy and re-establishment of normal CSF flow documented by post-treatment radioisotope CSF flow studies ensure both homogeneous distribution of chemotherapy administered via CSF and treatment of the entire neuraxis. A mechanism of leptomeningeal metastatic relapse appears to involve CSF compartmentalization, thereby interrupting intra-CSF drug distribution [38]. Recurrent leptomeningeal metastases with CSF compartmentalization is therapeutically difficult to treat, notwithstanding the application of intra-CSF chemotherapy, often with a change in drug therapy and radiotherapy to involved regions.

CSF flows as a consequence of both transmitted choroidal arterial pulsations and due to bulk flow resulting from continuous CSF production; therefore, when radioisotope is administered intraventricularly, flow is expected to the level of CSF obstruction [31,33]. It is less clear how CSF flows below a CSF obstruction, as for example when documented by an intralumbar radionuclide CSF flow study, perhaps spinal dural venous plexus pulsations contribute to cephalad flow.

In general, CSF compartmentalization, demonstrated by radionuclide CSF flow studies, often proves recalcitrant to medical therapy [18,26,34,37]. Similar to prior reports, following CSF block directed radiotherapy, only 50% of base of brain and 40% of spinal subarachnoid block respond with restoration of normal CSF flow. No patient failing radiotherapy in a variety of studies has demonstrated re-establishment of normal CSF flow when treated with intra-CSF chemotherapy [18,26,34,37,38]. Interruption of CSF flow with compartmentalization of CSF, as documented by ^{111}indium-DTPA CSF flow studies, pathologically appears to be a consequence of minimal tumor volume resulting in tumor adhesions of the subarachnoid space. These subarachnoid space adhesions are below the radiographic sensitivity of either CT or MR imaging. However, CSF flow studies are informative only with respect to compartmentalization of

TABLE 51-7 Comparison of spine imaging in leptomeningeal metastases [27]

Patient population (n = 63)	Image modality		
	Spine MR	CT-myelography	CSF flow study
57% normal			
43% abnormal	34%	33%	33%

the CSF and provide no information regarding bulky leptomeningeal disease, an aspect of leptomeningeal metastases best addressed by conventional cranial or spinal neuroradiographic imaging (Table 51.7)[15,18,26,34,40,41].

The rationale for radioisotope CSF flow studies is straightforward; intra-CSF chemotherapy, to be effective, assumes homogeneous distribution of chemotherapy within CSF compartments, thereby ensuring exposure of tumor cells to chemotherapy regardless of location in the CSF or leptomeninges. Therefore, patients studied by CSF flow studies before administration of intra-CSF chemotherapy who demonstrate obstruction of CSF flow, would appear to be poor candidates for regional chemotherapy unless normal communications of CSF can be re-established. Two recent studies suggest that radioisotope CSF flow studies have prognostic significance in patients with leptomeningeal metastases and independently predict patient's survival in patients with leptomeningeal metastases who are considered candidates for intra-CSF therapy [15,38,41].

In conclusion, radioisotope CSF flow studies in patients with leptomeningeal metastases appear to have at least two practical uses. First, radioisotope CSF flow studies, by documenting CSF flow, predict for homogeneous distribution of intra-CSF chemotherapy. Notwithstanding this observation, no study has documented differences in CSF drug levels above and below sites of CSF flow interruption as demonstrated by radioisotope CSF flow study. Secondly, in patients with CSF flow obstruction refractory to site of obstruction therapy including involved-field radiotherapy and intra-CSF chemotherapy, limited survival, rapid leptomeningeal disease progression, and death due to progressive CSF disease is predicted. CSF flow obstruction documented by radioisotope flow study is another criterion by which to assess patients for applicability of intra-CSF chemotherapy, independent of the patient's Karnofsky performance status, neurologic disability and extent and activity of systemic disease.

CONCLUSIONS

Magnetic resonance imaging with gadolinium enhancement is the technique of choice to evaluate patients with

suspected leptomeningeal metastasis [15]. Because LM involves the entire neuraxis, imaging of the entire CNS is required in patients considered for further treatment. T1-weighted sequences, with and without contrast, combined with fat suppression T2-weighted sequences, constitute the standard examination [42]. MRI has been shown to have a higher sensitivity than CE-CT in several series [15,21] and is similar to CT-M for the evaluation of the spine, but significantly better tolerated [26,43].

Any irritation of the leptomeninges (i.e. due to blood, infection or cancer) will result in their enhancement on MRI, which is seen as a fine signal-intense layer that follows the gyri and superficial sulci. Subependymal involvement of the ventricles often results in ventricular enhancement. Some changes, such as cranial nerve enhancement on cranial imaging and intradural extramedullary enhancing nodules on spinal MR (most frequently seen in the cauda equina) can be considered diagnostic of LM in patients with cancer [44]. Lumbar puncture itself can rarely cause a meningeal reaction leading to dural-arachnoidal enhancement, so imaging should be obtained preferably prior to the procedure [45]. MR-Gd still has a 30% incidence of false negative results so that a normal study does not exclude the diagnosis of LM. On the other hand, in cases with a typical clinical presentation, abnormal MR-Gd alone is adequate to establish the diagnosis of LM [15,33,44].

Radionuclide studies using either [111]indium-diethyl-enetriamine pentaacetic acid or ^{99}Tc macro-aggregated albumin, constitute the technique of choice to evaluate CSF flow dynamics [46,47]. Abnormal CSF circulation has been demonstrated in 30–70% of patients with LM, with blocks commonly occurring at the skull base, the spinal canal and over the cerebral convexities [26,47,48]. Patients with interruption of CSF flow demonstrated by radionuclide ventriculography have been shown in three clinical series to have decreased survival when compared to those with normal CSF flow [38,47,49]. Involved-field radiotherapy to the site of CSF flow obstruction restores flow in 30% of patients with spinal disease and 50% of patients with intracranial disease [50]. Re-establishment of CSF flow with involved-field radiotherapy followed by intrathecal chemotherapy led to longer survival, lower rates of treatment-related morbidity and lower rate of death from progressive LM, compared to the group that had persistent CSF blocks [26,47]. These findings may reflect that CSF flow abnormalities prevent homogeneous distribution of intrathecal chemotherapy, resulting in: (1) protected sites where tumor can progress; and (2) accumulation of drug at other sites leading to neurotoxicity and systemic toxicity. Based on this, many authors recommend that intrathecal chemotherapy be preceded by a radionuclide flow study and, if a block is found, that radiotherapy be administered in an attempt to re-establish normal flow [50,51].

REFERENCES

1. Chamberlain MC (1992). Leptomeningeal metastasis: review of current concepts. Curr Opin Oncol 4(3):533–539.
2. Theodore WH, Gendelman S (1981). Meningeal carcinomatosis. Arch Neurol 38:696–699.
3. Olso ME, Dhernik NL, Posner JB (1974). Infiltration of the leptomeninges by systemic cancer: a clinical and pathologic study. Arch Neurol 30:122–137.
4. Wasserstrom WR, Glass JP, Postner JB (1985). Diagnosis and treatment of leptomeningeal metastases from solid tumors: experience with 90 patients. Cancer 49:759–772.
5. Lee Y, Glass JP, Geoffrey A, Wallace S (1994). Cranial-computed tomographic abnormalities in leptomeningeal metastasis. Am J Roentgenol 143:1035–1039.
6. Jaeckle KA, Krol G, Posner JB (1985). Evolution of computed tomographic abnormalities in leptomeningeal metastasis. Ann Neurol 17:85–89.
7. Krol G, Sze G, Malkin M, Walker R (1988). MR of cranial and spinal meningeal carcinomatosis: comparison with CT and myelography. Am J Neuroradiol 9:709–714.
8. Enzmann DR, Krikorin CY, Hayward R (1978). Computed tomography in leptomeningeal spread in tumor. J Comput Assist Tomogr 2:445–448.
9. Ascherl GF, Hilal SK, Brisman R (1981). Computed tomography of disseminated meningeal and ependymal malignant neoplasms. Neurology 31:567–574.
10. Davis PC, Freedman NC, Fry SM (1987). Leptomeningeal metastasis: MR imaging. Radiology 163:449–454.
11. Barloon TJ, Uyuh WT, Yang CJ (1987). Spinal subarachnoid tumor seeding from intracranial metastasis: MR findings. J Comput Assist Tomogr 11:242–244.
12. Mathews VP, Kuharik MA, Edwards MK, D'Amour PG, Azzarelli B, Dreesen RG (1988). Gd-DTPA-enhanced MR imaging of experimental bacterial meningitis: evaluation and comparison with CT. Am J Neuroradiol 9:1045–1050.
13. Frank JA, Girton M, Dwyer AJ, Wright DC, Cohen PJ, Doppman JL (1988). Meningeal carcinomatosis in the VX2 rabbit tumor model: detection with Gd-DTPA-enhanced MR imaging. Radiology 167:825–829.
14. Sze G, Abramson A, Krol G et al (1988). Gadolinium-DTPA in the evaluation of intradural extramedullary spinal disease. Am J Neuroradiol 9:153–163.
15. Chamberlain MC, Sandy AD, Press GA (1990). Leptomeningeal metastasis: a comparison of gadolinium-enhanced MR and contrast-enhanced CT of the brain. Neurology 40:435–438.
16. Little JR, Dale AJD, Okazaki, H (1974). Meningeal carcinomatosis: clinical manifestations. Arch Neurol 30:138–143.
17. Olson ME, Chernik NL, Posner JB (1974). Infiltration of the leptomeninges by systemic cancer. A clinical and pathologic study. Arch Neurol 30:122–137.
18. Chamberlain MC, Corey-Bloom J (1991). Leptomeningeal metastases: [111]Indium-DTPA CSF flow studies. Neurology 41:1765–1769.
19. Pedersen AG, Paulson OB, Gyldensted C (1985). Metrizamide myelography in patients with small cell carcinoma of the lung suspected of meningeal carcinomatosis. J Neuro Oncol 3:85–89.
20. Kramer ED, Rafto S, Packer RJ, Zimmerman RA (1991). Comparison of myelography with CT follow-up versus gadolinium MRI for subarachnoid metastatic disease in children. Neurology 41:46–50.
21. Sze G, Abramson A, Krol G et al (1988). Gadolinium-DTPA in the evaluation of intradural extramedullary spinal disease. Am J Roentgenol 9:153–163.
22. Kim KS, Ho SU, Weinberg PE, Lee C (1982). Spinal leptomeningeal infiltration by systemic cancer: myelographic features. Am J Roentgenol 139:361–365.
23. Wiener MD, Boyko OB, Friedman HS, Hockenberger B, Oakes WJ (1990). False-positive spinal MR findings for subarachnoid spread of primary CNS tumor in postoperative pediatric patients. Am J Neuroradiol 11:1100–1103.
24. Rippe DF, Boyko OB, Friedman HS et al (1990). Gd—DTPA-enhanced MR imaging of leptomeningeal spread of primary CNS tumor in children. Am J Neuroradiol 11:329–332.
25. Lim V, Sobel DF, Zyroff J (1990). Spinal cord pial metastases: MR imaging with gadopentetate dimeglumine. Am J Neuroradiol 11:975–982.
26. Glass JP, Melamed M, Chernik NL, Posner JB (1979). Malignant cells in cerebrospinal fluid (CSF): the meaning of a positive CSF cytology. Neurology 28:1369–1375.
27. Chamberlain MC (1995). Comparative spine imaging in leptomeningeal metastases. J Neuro-Oncol 23:233–238.
28. Enzmann DR, Pelc NJ (1992). Brain motion: measurement with phase-contrast MR imaging. Radiology 185:653–660.
29. Feinberg DA (1992). Modern concepts of brain motion and cerebrospinal fluid flow. Radiology 185:630–632.
30. Schellinger D, LeBihan D, Sunder SR et al (1992). MR of slow CSF flow in the spine. Am J Neuroradiol 13:1393–1403.
31. Larson SM, Johnson GS, Ommaya AK, Jones AE, DiChiro G (1973). The radionuclide ventriculogram. J Am Med Assoc 224:853–857.
32. Di Chiro G, Hammock MK, Bleyer WA (1976). Spinal descent of cerebrospinal fluid in man. Neurology 25:1–8.
33. Lyons MK, Meyer FB (1990). Cerebrospinal fluid physiology and the management of increased intracranial pressure. Mayo Clin Proc 75:684–707.
34. Chamberlain MC (1994). Pediatric leptomeningeal metastasis: [111]Indium-DTPA CSF flow studies. J Child Neurol 9:150–154.
35. Haaxma-Reiche H, Piers DA, Beekhuis H (1989). Normal cerebrospinal fluid dynamics: a study with intraventricular injection of ^{111}In-DTPA in leukemia and lymphoma without meningeal involvement. Arch Neurol 46:997–999.
36. Grossman SA, Trump CL, Chen DCP, Thompson G, Camargo E (1982). Cerebrospinal fluid flow abnormalities in patients with neoplastic meningitis. Am J Med 73:641–647.
37. Chamberlain MC (1995). Spinal [111]Indium-DTPA CSF flow studies in leptomeningeal metastasis. J Neuro Oncol 25:135–141.
38. Chamberlain MC, Kormanik P (1996). Prognostic significance of [111]Indium-DTPA CSF flow studies. Neurology 46(6):1674–1677.
39. Chamberlain MC (1994). New approaches to and current treatments of leptomeningeal metastases. Curr Opin Neurol 7:492–500.

40. Glantz M, Hall WA, Cole BF et al (1995). Diagnosis, management, and survival of patients with leptomeningeal cancer based on cerebrospinal fluid-flow studies. Cancer 75:2919–2931.

41. Chamberlain MC (1998). Radioisotope CSF flow studies in leptomeningeal metastases. J Neuro-Oncol 38:135–140.

42. Grossman SA, Krabak MJ (1999). Leptomeningeal carcinomatosis. Cancer Treat Rev 25(2):103–119.

43. Schumacher M, Orszagh M (1998). Imaging techniques in neoplastic meningiosis. J Neuro Oncol 38(2–3):111–120.

44. Schuknecht B, Huber P, Buller B et al (1992). Spinal leptomeningeal neoplastic disease. Evaluation by MR, myelography and CT myelography. Eur Neurol 32(1):11–16.

45. Freilich RJ, Krol G, DeAngelis LM (1995). Neuroimaging and cerebrospinal fluid cytology in the diagnosis of leptomeningeal metastasis. Ann Neurol 38(1):51–57.

46. Mittl RL Jr, Yousem DM (1994). Frequency of unexplained meningeal enhancement in the brain after lumbar puncture. Am J Neuroradiol 15(4):633–638.

47. Glantz MJ, Hall WA, Cole BF et al (1995). Diagnosis, management, and survival of patients with leptomeningeal cancer based on cerebrospinal fluid-flow status. Cancer 75(12):2919–2931.

48. Trump DL, Grossman SA, Thompson G et al (1982). CSF infections complicating the management of neoplastic meningitis. Clinical features and results of therapy. Arch Intern Med 142(3):583–586.

49. Mason WP, Yeh SD, DeAngelis LM (1998). [111]Indium-diethylenetriamine pentaacetic acid cerebrospinal fluid flow studies predict distribution of intrathecally administered chemotherapy and outcome in patients with leptomeningeal metastases. Neurology 50(2):438–444.

50. Chamberlain MC, Kormanik P, Jaeckle KA et al (1999). [111]Indium-diethylenetriamine pentaacetic acid CSF flow studies predict distribution of intrathecally administered chemotherapy and outcome in patients with leptomeningeal metastases. Neurology 52(1):216–217

51. Gleissner B, Chamberlain MC (2006). Clinical presentation and therapy of neoplastic meningitis. Lancet Neurology 5:443–452.

CHAPTER 52

Imaging of Epidural Spinal Cord Compression

Lubdha M. Shah, Jonathan P. Gordon, C. Douglas Phillips and David Schiff

CLINICAL BACKGROUND

Spinal cord compression is a dreaded complication of metastatic cancer that can lead to paraplegia or quadriplegia making early diagnosis imperative. It has been estimated that 5% of patients with systemic cancer at autopsy were found to have pathologic evidence of tumor invading the extradural space [1,2]. Although nearly all malignancies can involve the spine and/or epidural space, myeloma, breast carcinoma, prostate carcinoma, lung carcinoma and lymphoma are the most common because of the frequency of these tumors and their propensity to metastasize [3].

Multiple epidural metastases have been reported in approximately 9–49% of patients with epidural spinal cord compression (ESCC) [4–8] and the site of epidural metastatic disease varies from 68% in the thoracic spine, 16% in the lumbosacral spine and 15% in the cervical spine [9]. Of those patients with spinal metastatic disease, close to 20% will have at least two sites of epidural involvement at some time during the course of the disease [10]. Though multiplicity has been associated with specific tumor type [6], this has not been confirmed by other investigators [11]. The incidence of ESCC is expected to increase due to improved survival of cancer patients. Additionally, in those patients who survive long enough, recurrent spinal epidural metastasis is not uncommon [12,13]. In a study of 103 patients, van der Sande et al showed that recurrent spinal epidural metastasis occurred in 19% of the patients with prior spinal epidural metastases, and slightly more often in the patients with breast cancer (21%) than in the patients with other primary tumors (17%) [12]. Because of this potential for multiple epidural metastases and their association with ESCC, many authorities advocate diagnostic imaging of the entire spine [5,11].

Clinically, metastatic epidural spinal cord compression is nearly always preceded by back pain or neck pain, which typically is weeks to months in duration [14]. In fact, back pain is the initial symptom in 80–96% of patients [3] and may be local or radicular. Adams et al illustrated the importance of ensuring congruity of the localizing sensory level and the radiographic findings as they showed false-localization of thoracic sensory level in the setting of cervical cord compression [15]. Weakness (76%), bowel and bladder function (57%) and sensory loss (35–51%) may also be seen at the time of diagnosis [3].

Many variables influence the functional outcome of ESCC with the pre-treatment neurologic status being the most important determinant of functional prognosis [16–18]. Bowel and bladder dysfunction is an unfavorable prognostic sign [3] and the extent of subarachnoid block is related to degree of motor and sphincter dysfunction [18,19]. Furthermore, the degree of thecal sac/cord compression correlates positively with the degree of neurologic impairment [20,21] and is directly related to functional outcome [22]. The presence of vertebral body collapse is also an ominous finding [23]. The duration of sensory and motor changes has been considered an important prognostic factor [24]. However, other studies have shown that the duration of the worst degree of neurologic dysfunction may be a more important prognostic factor than the entire duration of weakness [19].

The radiosensitivity of the primary tumor is also an important prognostic factor for the clinical response [3,18,25–27]. Breast, prostate, small cell lung cancer, lymphoma and multiple myeloma are most radiosensitive, while renal cell and melanoma are extremely radioresistant [25,27]. Studies have found that 74% of paraparetic patients with radiosensitive tumors became ambulatory after radiation therapy; however, only 34% of those harboring less radiosensitive tumors were able regain ambulation [3,28].

PATHOLOPHYSIOLOGY

Spinal tumors can be organized according to location as extradural, intradural, extramedullary or intramedullary. However, this is a simplification as a single tumor can reside in more than one compartment. Alternatively, lesions with the same pathology may occur in different compartments. In general, metastatic tumors are far more common in the extradural space [29].

Several anatomic metastatic patterns of spinal epidural metastases have been proposed [1,19,25–27,30]. The most commonly proposed mechanism is by arterial spread to bone marrow, which results in vertebral body collapse and formation of an anterior epidural mass [1,26]. In a prospective study of 59 patients with ESCC, Kim et al found hematogenous spread to the spine in 51 cases (86%) [19]. Experimental models have supported this theory of arterial seeding by demonstrating medullary expansion of tumor cells upon systemic injection of tumor cells in mice [31]. The vertebral body is involved more commonly than the posterior elements, presumably because of its large volume and its highly vascular marrow, which promotes proliferation of tumor cells [30]. In a study based on computed tomography (CT) of the vertebral column, pedicle destruction was identified only in the presence of vertebral body involvement [32]. The initial location of metastases is within the posterior vertebral body correlating to the basivertebral arterial plexus [32]. Subsequently, the basivertebral veins or other penetrating vessels provide entry into the valveless epidural venous sinuses [31,33], which can lead to circumferential soft tissue compression of the spinal cord in 22% of cases [30]. Such hematogenous dissemination from vertebral bone metastases to the dura is a frequent pathway in patients with breast carcinoma [34]. Alternatively, there may be direct infiltration of the dura from vertebral lesions [30,35].

Another suggested mechanism of spinal epidural metastases is via the intervertebral foramina from a paravertebral source [19,25,30]. Kim et al found direct extension from a paravertebral mass through bone or intervertebral foramina in eight of 59 cases (14%) [19]. This transforaminal pattern comprises a significantly smaller proportion, approximately 10%, of spinal epidural metastases cases [30] and is frequently associated with lung cancer and lymphoma [19,36]. Epidural extension of lymphoma from a vertebral body is also common [37]. Spread of epidural tumor along nerve roots or their lymphatics is a controversial route of spread [33,38].

The pathophysiology of neurologic symptoms may be due to a vascular phenomenon. It is hypothesized that the soft tissue material may impinge upon the epidural venous plexus resulting in venous hypertension and vasogenic edema [1,25,39]. This venous hypertension may be complicated by hematomyelia at the site of vascular compression [31]. Epidural tumor itself [25,30] or vertebral body collapse with extradural mass [27] may also have direct mass effect on the spinal cord leading to neurologic deterioration.

EPIDURAL ANATOMY

Given the frequent involvement of the extradural compartment in neoplastic disease, it is important to

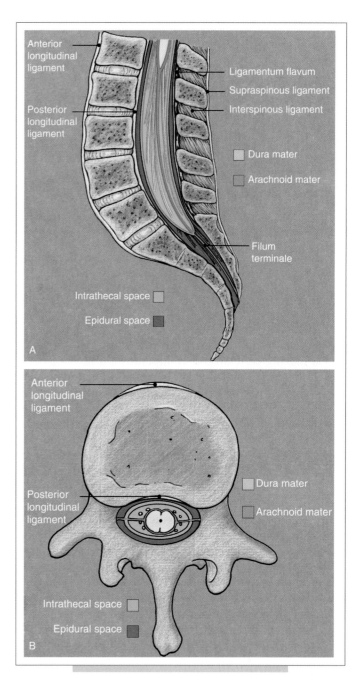

FIG. 52.1. **(A)** Illustration in the sagittal plane of the individual components of the lumbar spine including intrathecal and extradural/epidural spaces. Illustration by Laurie Persson. University of Virginia Health System. **(B)** Illustration in the axial plane through the lumbar spine demonstrates the soft tissue components of the spinal canal, including the leptomeningeal and dural layers. Illustration by Laurie Persson. University of Virginia Health System.

understand the anatomy of the epidural space (Figure 52.1). The epidural space surrounds the dural sac and is bounded by the posterior longitudinal ligament anteriorly and the ligamenta flava and the periosteum of the laminae posteriorly. The lateral boundaries are the pedicles of the spinal column and the intervertebral foramina containing their neural elements. The epidural space communicates freely with the paravertebral space through the interverte-bral foramina. At its superior extent, the space is anatomi-cally closed at the foramen magnum where the spinal dura attaches with the endosteal dura of the cranium. Caudally, the epidural space ends at the sacral hiatus, which is closed by the sacrococcygeal ligament [40,41]. As the size of the dural sac relative to the epidural space decreases at the L4–L5 level, the posterior longitudinal ligament falls away from the anterior dura and fat fills the anterior epidural space [41].

The epidural space contains loose areolar connective tissue, fat, lymphatics, arteries and an extensive plexus of veins, as well as the spinal nerve roots as they exit the dural sac. The epidural venous plexus is a valveless system that communicates with the basivertebral vein, the intra-cranial sigmoid, occipital and basilar venous sinuses and the azygous system [40,41]. In the thoracolumbar region (T10–L2), the basivertebral vein originates from this venous plexus and extends into the vertebral bodies.

In the cervicothoracic region, there is minimal epidural fat and the dura contacts the lamina. The posterior epidural space is continuous with a thin layer of epidural fat between the lamina and the dura (Figure 52.2). In comparison, in the lumbar region, the anteroposterior dimension of the posterior epidural space is greater, averaging 5.0–6.0 mm in adult males (Figure 52.3) [40]. Areas of epidural fat under the ligamentum flavum of the lumbar spine extend under the laminae, but are separated by areas where the posterior dura contacts, but does not adhere to, the periosteum of the lamina. Contact with the pedicles also divides the posterior epidural space from the lateral epidural space [42].

The anterior dura adheres tightly to the posterior lon-gitudinal ligament, which stretches across the interverte-bral disks to form the anterior epidural space between the posterior longitudinal ligament and the periosteum of the vertebral body [40]. The dura and posterior longitudinal liga-ment blend with the annular ligament, dividing the anterior epidural space into vertical compartments at each vertebral level. In areas immediately next to the intervertebral disks, dense connective tissue extensions extend superiorly and inferiorly, further dividing the anterior epidural space into lateral halves. In the lumbar region, a membranous extension of the posterior longitudinal ligament joins with the neural elements laterally and isolates the anterior epidural space from the posterior and lateral epidural space [42]. In case of lumbar vertebral metastasis associated with anterior epi-dural carcinomatous infiltration, Hutzelmann et al observed that infiltrations tend to respect the midline [43]. This phe-nomenon was observed to be uni- or bilateral in 88.3% of all cases with intraspinal anterior epidural carcinomatous infiltration, especially in that part of the vertebral body where the basal vertebral venous plexus was located [43].

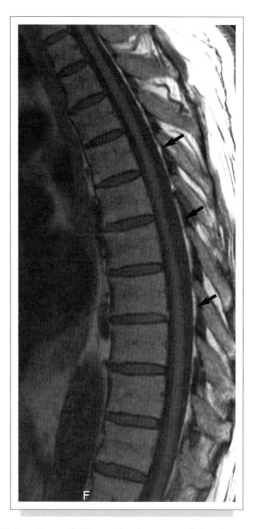

FIG. 52.2. Sagittal T1-weighted image of the thoracic spine demonstrates the relatively diminutive amount of epidural fat (arrows) in comparison to the lumbar spine in Figure 52.3.

It is likely due to the anatomy of the vertebral body and espe-cially its stabilization by the posterior longitudinal ligament. This imaging feature may be helpful in delineating vertebral body metastases with epidural infiltration from intraspinal processes, which proceed with the destruction of the verte-bral body.

IMAGING

The diagnosis of epidural spinal cord compression requires demonstration of extrinsic compression of the the-cal sac by tumor. Currently, magnetic resonance imaging (MRI) is the most specific and sensitive diagnostic tool for evaluating spinal lesions due to neoplasm [31]. However, it is useful to consider the appearance of spinal tumor with other modalities. Specifically, this may be of importance in patients in whom MRI is contraindicated (e.g. pacemaker, aneurysm clip, certain cochlear implants).

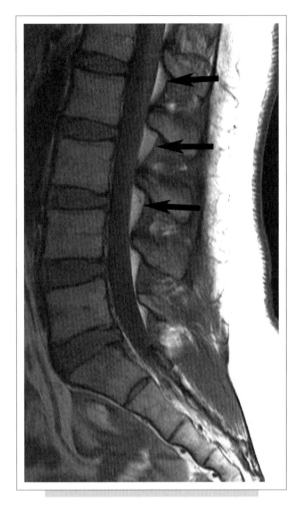

FIG. 52.3. Sagittal T1-weighted image of the lumbar spine demonstrates relatively abundant amount of epidural fat (arrows) in comparison to the thoracic spine.

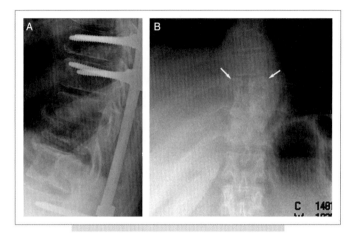

FIG. 52.4. **(A)** Lateral radiograph of the thoracic spine demonstrating pathologic compression deformities of two mid-thoracic vertebral bodies at the level of known ESCC. **(B)** Posterior anterior radiograph of the thoracic spine in a different patient demonstrates loss of height and pedicular erosion (arrows) of a lower thoracic vertebral body secondary to metastatic disease.

Radiography

Plain radiographs are widely utilized and inexpensive, but a have high false negative rate of 10–17% [2]. At least 50% of bone must be destroyed before it is radiographically detected [26,27] and pedicular erosion predicts epidural disease in only 31% of cases (Figure 52.4) [26]. Moreover, radiographs cannot evaluate paraspinal tumor invasion through the neural foramina or soft tissue impingement of the thecal sac [27]. If there are multiple lesions, it may be difficult to assess which is the clinically relevant lesion with radiographs alone [25].

Despite these limitations, bone destruction is often seen with ESCC; the series by Kim et al demonstrated 86% of patients had vertebral body destruction at the level of ESCC [19]. In another study [44], epidural tumor occurred in 87% of patients with major vertebral body collapse, in 31% of those with pedicle erosion and no major collapse and in only 7% of those with metastases restricted to the vertebral body without collapse. Radiographic abnormalities

with symptoms have an even greater predictive value; Rodichok et al found that 91% of patients with back pain and myelopathy/radiculopathy and evidence of bone destruction on radiographs had ESCC [45,46]. However, in cases of lymphoma, plain radiographs are not as useful, as bony destruction is seen only in 30–42% of patients [36]. Preservation of the disk space suggests infectious disease over metastases [26].

Myelograpghy and CT Myelography

Before the widespread availability of MRI, myelography was the gold standard for the evaluation of ESCC, with the advantage of obtaining a CSF sample at the time of the procedure. In myelography, intrathecal contrast opacifies the intradural/subarachnoid space and demonstrates epidural metastases as extradural defects. Typically, conventional myelography and post-myelographic CT are equally sensitive to MR imaging for epidural metastases that cause cord compression. However, Carmody et al found MRI to be more sensitive than conventional myelography when cord compression had not yet developed, enabling detection of spinal metastases that are subtly encroaching upon the subarachnoid space due to improved cross-sectional resolution [47]. The addition of cross-sectional anatomy of CT to myelography increases detection of bony and paravertebral metastases. CT myelography may be helpful in cases where magnetic susceptibility artifact from orthopedic hardware on MRI obscures the spinal canal [48].

Myelography is relatively safe and accurate for the evaluation of ESCC. However, numerous disadvantages make it a less attractive diagnostic procedure. Although not a significant concern in this population of patients, exposure to ionizing radiation should be considered seriously.

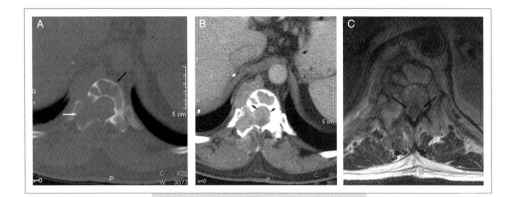

FIG. 52.5. **(A)** Unenhanced axial CT image (bone algorithm) through thoracic vertebra demonstrates a destructive lesion involving the body (black arrow) and right pedicle (white arrow) from metastatic renal cell carcinoma. **(B)** Unenhanced axial CT image (soft tissue algorithm) through thoracic vertebra demonstrates soft tissue component eroding the vertebral body and pedicle, extending into the epidural space and compressing the spinal cord (arrows). **(C)** Axial T2-weighted MRI image through the same level as (A)and (B) illustrates better soft tissue delineation, particularly the mass effect on the thecal sac (arrows), as compared to the CT.

Myelography requires an invasive procedure to introduce intrathecal contrast agents and both the puncture and the contrast agent can produce side effects and, rarely, significant adverse reactions. The side effects of water-soluble contrast media are well known. Certain medications are contraindicated as they lower the seizure threshold and have to be withheld for a period of time. Some patients are unable to tolerate the procedure because of discomfort. There is also a risk of further neurologic deterioration after lumbar puncture [49], nerve root avulsion [50] and puncture site hematomas that can occasionally lead to death [51–53]. When there are two areas of myelographic block, the study may not be able to demonstrate the extent of either lesion or detect an intervening lesion in the unopacified segment. Although CT myelography will allow assessment of the spinal canal in cases with apparent block, in some cases, a second puncture with contrast injection rostral to the block may be required to delineate the superior extent of the epidural disease [11]. Because of these limitations, myelography should be reserved for those patients who cannot undergo a technically adequate MRI. For example, this would include those patients with MRI incompatible hardware including pacemakers or with metallic foreign bodies close to vital structures.

Computed Tomography

Unenhanced CT is relatively insensitive for identifying extradural metastatic involvement. However, as bone destruction is often seen with ESCC, O'Rourke et al found the presence of cortical disruption around the neural canal to be highly associated with epidural compression [54] and CT, in particular, provides better delineation of cortical bone than MRI. Current multidetector spiral CT with reformations may be able to provide a screening tool; however, the soft tissue contrast is poor as compared to MRI (Figure 52.5). Other

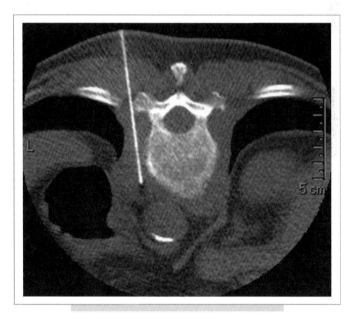

FIG. 52.6. Axial CT image with the patient in the prone position demonstrates biopsy needle traversing percutaneouly into the paravertebral mass, which was proven to be lymphoma.

applications of CT, which are useful to obtain histologic diagnosis, include CT-guided or CT fluoroscopically guided biopsy of paraspinal and epidural masses (Figure 52.6) [9].

Nuclear Medicine

Nuclear medicine modalities (such as bone scan and [18F]fluoro-deoxyglucose positron emission tomography (PET FDG)) are often used in the assessment of metastatic disease; however, they are of limited utility in the evaluation of spinal epidural disease. Radionuclide bone scanning is a relatively sensitive tool for diagnosing vertebral body metastases. However, there is a high frequency of false positive results with a specificity of only 53% [26], and it is limited in its ability to define the anatomic extent of destruction which requires further imaging for further delineation. Furthermore, it is not helpful in the detection of spinal cord compression. Some studies have found that MRI is more sensitive than single-photon emission computed tomography (SPECT) bone scintigraphy for the detection of metastatic disease [47,48,55,56], but may not be as sensitive for the detection of small metastases in the posterior elements [57]. Very active lesions that might not be visible on a bone scan are generally detectable on MRI [29]. Metser et al studied the role of ^{18}F-FDG positron emission computed tomography (PET)/CT in the assessment of secondary malignant involvement of the spinal column. Of 242 lesions, ^{18}F-FDG PET alone detected significantly more malignant lesions than did CT alone (96% versus 68%, respectively); however, this modality only detected epidural

extension of tumor, neural foramen involvement of tumor, or a combination of both in 33% of cases [58]. Fusion of metabolic PET data with cross-sectional CT anatomy can help localize metastatic disease to the epidural space (Figure 52.7) [59].

Magnetic Resonance Imaging

With advancements in imaging techniques, MRI has become the fundamental diagnostic imaging modality in the evaluation of spinal epidural metastasis [14,47,60–63]. MRI not only demonstrates the presence and extent of bony involvement but also, most importantly, delineates the presence and location of paravertebral and epidural extension and the degree of neural compromise and thecal sac impingement. It is non-invasive and has increased sensitivity relative to CT or myelography. In contrast to plain radiographs, CT, myelography and CT myelography, MRI does not use ionizing radiation, which is particularly

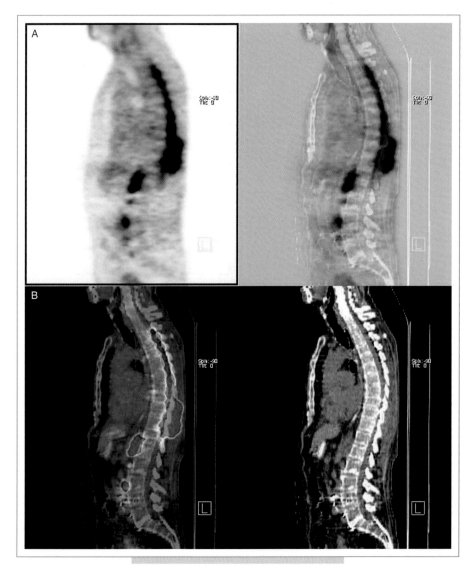

FIG. 52.7. PET imaging was performed 45 minutes after the injection of 15 mCi F-18 FDG. **(A)** Maximum intensity projection image shows the overall pattern of extensive hypermetabolic metastatic disease. **(B)** Sagittal PET-CT images demonstrate intense hypermetabolism along the distribution of the epidural space from T1 to T10 (upper panel: FDG-PET alone on the left, combined more PET and less CT on the right; lower panel: combined less PET and more CT on the left, CT alone on the right). Courtesy of H. Jadvar, MD PhD. (From Fusion positron emission tomography-computed tomography demonstration of epidural metastases. Clinical Nuclear Medicine. 2004;29(1): 39–40).

advantageous in the lumbar area where gonadal exposure may occur. Additional advantages of MRI include the ability to evaluate the spinal cord, nerve roots and disks, while their location and morphology can only be inferred on plain radiography and less completely evaluated on myelography. Furthermore, it is also the only modality able to evaluate the internal structure of the cord.

The multiplanar capabilities, as well as the superior soft tissue discrimination and contrast resolution of MRI, enable a highly accurate evaluation of the nature and extent of epidural lesions [61]. MRI can detect widely separated lesions, which may not be identified by CT myelography if they cause minimal subarachnoid compression or if they are located between two areas of myelographic block. The MR protocol in some institutions includes a sagittal screen of the entire spine in cases of ESCC. However, this may not identify far lateral epidural disease or may be an insufficient imaging sequence in severely scoliotic spines. In symptomatic epidural metastases, identifying multiple areas of metastatic disease is critical, particularly when deciding between surgery or radiotherapy or in the delineation of radiation ports. For asymptomatic epidural metastases, Schiff et al showed that it may be safe to omit MRI of the cervical spine when there is radiographically verified thoracic or lumbar ESCC [11]. However, since the likelihood of asymptomatic or unsuspected second site of epidural disease is higher in the thoracic and lumbosacral spine, thorough imaging of these regions is advisable [11].

The fundamentals of MR physics are detailed in other sections of this text. Briefly, an MR image is determined by the spatial allocation of individual MR signals that represent the corresponding anatomical structure. The magnetic field is spatially varied such that the nuclear spins demonstrate different precessional frequencies at different positions. The frequency and phase of each nuclear spin are acquired in the presence of radiofrequency pulsations and magnetic gradients and stored in a matrix of voxels (k-space). The MR image is a mathematical reconstruction (Fourier transformation) of these data.

Imaging of the cervical and thoracic spine requires higher spatial resolution in comparison to the lumbar spine. Also, motion artifact related to swallowing and respiration and pulsation from the thoracoabdominal vessels is more problematic in the cervical and thoracic spine, respectively. To address some of these imaging issues, advanced acquisition techniques have been developed. In conventional MR imaging, the phase-encoding steps are performed in sequential order by switching a magnetic field gradient and this determines the measurement time. With recent techniques such as parallel imaging, the raw data are simultaneously acquired at an accelerated pace via two or more receiver coils with varied spatial sensitivity [64,65]. Parallel imaging exploits the difference in sensitivities between individual coil elements in a receive array to reduce the number of gradient encoding steps required for imaging.

The shortened measurement time is of particular benefit in examinations with breathhold technique. Although there may be lower signal-to-noise ratio as compared to 'complete' measurement results, higher resolution may be achieved with similar measurement times through phase oversampling and averaging. Phased array surface coils, which cover a large volume with several smaller coils, are used in spine parallel imaging and can be used to achieve an optimized signal-to-noise ratio (SNR) over a large field of view (FOV) [66]. Phased array coils have an increased number of elements to enable faster dynamic scans with the consequent reduction of motion artifacts.

Sagittal and axial T1-weighted (short TR [time to recovery]/short TE [echo time]) and T2-weighted (long TR/long TE) are the typical sequences used in the evaluation of spinal metastatic disease. Fast spin-echo (FSE) imaging, also known as turbo spin-echo, has replaced conventional spin-echo (SE) imaging of the spine due to improved time efficiency and image quality [67]. FSE sequences collect small segments of k-space after each radiofrequency (RF) pulse for excitation [67]. This enables decreased overall imaging time, thereby lessening the potential for patient motion. Moreover, the time saved can be used for obtaining additional signal averages to improve signal.

The FSE strategy is particularly helpful in T2-weighted imaging (WI) to take advantage of the long TR intervals to accommodate a relatively long echo train. These heavily T2-weighted studies produce marked contrast between the CSF, the conus medullaris and the extradural tissues. However, the fat remains relatively bright in spite of its short T2 on FSE sequences and may decrease detection of tumor [29,68]. The degree of compromise of the thecal sac, spinal cord and nerve roots can be easily assessed on heavily T2-weighted images on which the cerebrospinal fluid (CSF) is particularly bright (Figure 52.8) [69]. However, CSF flow can create problems including signal loss, spurious signal and blurred interfaces, particularly on those sequences that produce myelographic contrast and for those that suppress CSF signal intensity (Figure 52.9) [67]. However, in the lumbar spine, CSF pulsation is dramatically dampened [67]. While lymphoid malignancies are often isointense with marrow on T2-weighted images, most metastatic carcinomas and sarcomas give higher signal intensity than fat, suggesting that MRI findings may be used to distinguish between these pathologies [70]. Sze et al found long TR sequences without contrast material showed equally good delineation [71]. Additionally, the presence of apparently bright disks may be a subtle sign of diffuse replacement of normal fatty bone marrow in the vertebrae [72].

T1-weighted FSE imaging is helpful to identify marrow replacement processes, such as metastases. Normal vertebra in young adults is composed primarily of hematopoietic bone marrow containing significant amounts of fat (from 25 to 50%) [73]. With aging, the marrow is converted to even larger amounts of fat and,

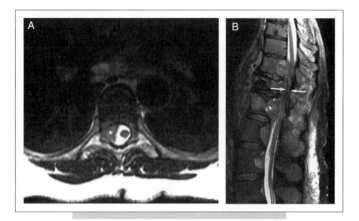

FIG. 52.8. **(A)** Axial T2-weighted image demonstrates a hypointense soft tissue mass in the right lateral epidural space (asterisk *) causing mass effect on the lateral aspect of the thecal sac. **(B)** Sagittal T2-weighted image in a different patient with proven metastatic renal cell carcinoma demonstrates a large hypointense, heterogenous mass causing mass effect on the thoracic spinal cord (arrows).

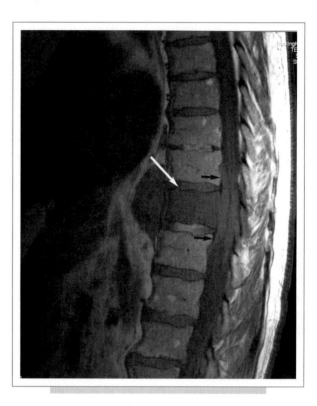

FIG. 52.10. Sagittal T1-weighted image demonstrates abnormal hypointense marrow signal throughout a mid-thoracic vertebral body (white arrow). A T1 hypointense epidural component insinuates anteriorly and posteriorly in the epidural space (black arrows). This was biopsy proven lymphoma.

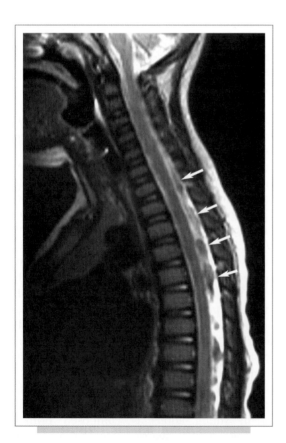

FIG. 52.9. Sagittal T2-weighted image demonstrates CSF pulsation artifact in the posterior aspect of the thecal sac (arrows).

correspondingly, the mean percent volume of hematopoietic marrow decreases progressively, with the result that in the eighth decade of life it is about half of that present in the first decade (29.2% versus 57.9%) [74,75]. This is useful in detection of pathology, as an abnormal hypointense focus is sharply contrasted to the normal hyperintense fatty marrow in the adult spine. Tumors tend to be hypointense on non-contrast T1-weighted images, reflecting replacement of fatty bone marrow, increased water content and hypercellularity (Figure 52.10) [76]. However, in acute compression fractures, the distinction between non-neoplastic involvement and tumor replacement of marrow is difficult because both may display low T1 signal. The former is due to marrow edema and the latter due to the tumor itself (Figure 52.11) [29]. Post-gadolinium imaging may not be helpful because lesions may be obscured as they enhance to the isointensity of the adjacent marrow on the T1-weighted sequences (Figure 52.12) [60]. Also, acute compression fractures can enhance [29].

Epidural space lesions, however, may be difficult to detect on unenhanced T1-weighted sequences and further evaluation with contrast enhanced images or long TR images may be necessary [29,60]. Sze et al demonstrated the utility of gadolinium in characterizing possible epidural

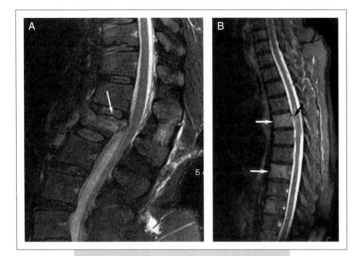

FIG. 52.11. **(A)** Sagittal T2-weighted image demonstrates diffusely increased marrow signal due to acute compression fracture (arrow). **(B)** Sagittal STIR image demonstrates diffusely increased marrow signal in multiple vertebral bodies (white arrows). There is a ventral epidural component associated with the more superior lesion (black arrow). Patient has history of metastatic well-differentiated neuroendocrine carcinoma. Courtesy of Alford Bennet, MD, University of Virginia Health System.

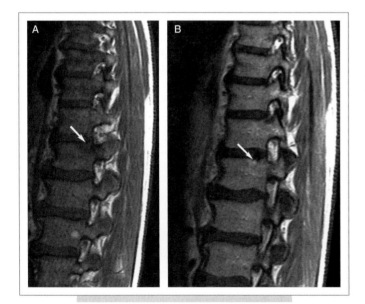

FIG. 52.12. **(A)** Unenhanced sagittal T1-weighted image demonstrates an ill-defined lesion involving the posterior body and pedicle (arrow) of a lumbar vertebral body. The abnormal hypointensity is conspicuous in a background of normal marrow hyperintensity. **(B)** Gadolinium-enhanced T1-weighted image demonstrates diffuse enhancement of the lesion (arrow), which obscures it in the background of hyperintense marrow.

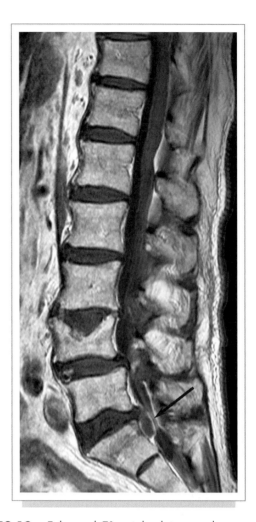

FIG. 52.13. Enhanced T1-weighted image shows a peripherally enhancing lesion in the ventral epidural space (arrow) extending inferiorly from the L5–S1 disk intervertebral disk space which represents herniated disk material.

tumor, in delineating the extent of tumor and in outlining regions of spinal cord compression [60,70]. This is helpful in the cervical and thoracic spine where there is a relative lack of prominent ligaments and epidural fat, which typically increase the conspicuity of lesions projecting from the spine. The timing of post-contrast imaging can be used to differentiate tumor from a herniated intervertebral disk; the latter demonstrates delayed enhancement (Figure 52.13) [77]. Early contrast images may help localize tumor for biopsy. Enhanced MRI may be used for post-treatment assessment of epidural tumor. After radiation therapy, decreased vascularity can explain relatively diminished enhancement on MR. It is prudent to compare serial images with the baseline as untreated epidural neoplasms can vary in the degree of enhancement [60]. In some patients, there may be a suspicion of spinal cord compression and post-contrast imaging is essential for demonstrating that there is no compressive lesion, but rather an intradural and even intramedullary lesion, which can mimic extradural compressive lesions [29]. Thus, gadolinium-enhanced MRI is also helpful to evaluate for subarachnoid seeding (Figure 52.14). Sze et al have reported the efficacy with which Gd-DTPA can detect even small lesions in the intradural extramedullary space [71].

Gradient-echo (GRE) imaging is more often used in the cervical and thoracic spine to produce myelogram-like images [78,79]. In GRE imaging, the RF pulse is applied with a partial flip angle (less than 90 degrees) and the gradients are then inverted to refocus the excited spins without the application of a 180 degree pulse. Although the absent 180 degree pulse enables faster imaging, it makes the images more susceptible to local field inhomogeneities (Figure 52.15A). The relatively long TEs and large voxel size in routine two-dimensional GRE studies result in considerable susceptibility artifact at the bone–soft issue interface, which exaggerates the true degree of bony canal and neural foraminal stenosis [67]. However, by producing a myelographic

effect with bright CSF, GRE images can delineate regions of impingement (Figure 52.15B) [29].

Diffusion-weighted imaging (DWI) is an established method for studying cerebral disease processes, but previous use of diffusion-weighted imaging in the spine has been limited for technical reasons [67]. Bauer et al were able to use DWI with a steady state free procession sequence to distinguish benign vertebral compression fractures from pathologic compression fractures [80]. However, Castillo et al showed that DWI of the spine had no advantage in the detection and characterization of vertebral metastases as compared with non-contrast T1-weighted imaging, but was considered superior to T2-weighted imaging [81]. Though epidural metastatic disease was not specifically addressed in their study, Eastwood et al demonstrated the utility of DWI in diagnosing epidural abscess [82]. There is the theoretical possibility that hypercellular epidural metastasis may be detected by DWI similar to the utility of apparent diffusion coefficient (ADC) values in determining high-grade intracranial tumors [83,84].

Fat saturation techniques can be used to suppress the signal from normal adipose tissue to reduce chemical shift artifact or improve visualization of uptake of contrast material. A frequency-selective saturation radiofrequency pulse with the same resonance frequency as that of lipids is applied to each slice-selection RF pulse and a homogeneity spoiling gradient pulse is applied immediately after the saturation pulse to diphase the lipid signal. Thus, the signal excited by the subsequent slice selection pulse contains no contribution from lipid. This method is lipid-specific, reliable for tissue characterization and allows good demarcation of small anatomic details [85]. Because subtle low-intensity lesions can be obscured by the high signal intensity of marrow fat on routine T1-weighted spin-echo imaging, suppression of marrow fat can be helpful to detect

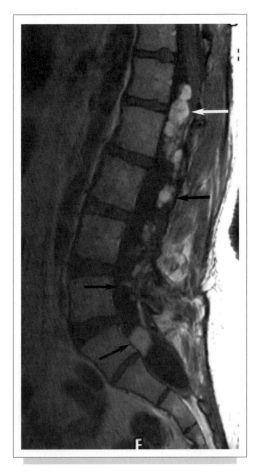

FIG. 52.14. Sagittal post-gadolinium T1-weighted image demonstrating an enhancing lobulated mass (myxopapillary ependymoma) involving the conus medullaris and filum terminale (white arrow) with additional nodular enhancing foci along the cauda equine (black arrows).

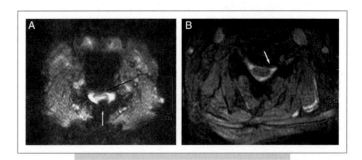

FIG. 52.15. **(A)** Axial gradient echo (GRE) image illustrates overestimation of spinal canal narrowing due to magnetic susceptibility artifact caused by ligamentum flavum calcification (white arrow) and posterior osteophytes (black arrow). **(B)** Axial GRE image shows effacement of the ventral thecal sac and narrowing of the left neural foramen due to a disk-osteophyte complex (arrow).

metastatic lesions in the spine. This is particularly important in the elderly population because of the heterogeneous fatty vertebral marrow [86]. Low intensity lesions enhance after the administration of gadolinium and can be obscured; therefore, fat suppression of post-contrast images can be useful to differentiate normally enhancing marrow from a pathologic lesion (Figure 52.16). The disadvantages of this technique are its susceptibility to static field inhomogeneities, misregistration artifact, particularly due to foreign bodies like metal or air collections, and unreliability with low-field-strength magnets.

Inversion recovery imaging is another method for suppression of fat signal based on differences in the T1 of tissues. In a short T1(tau) inversion-recovery (STIR) sequence, a 180 degree pulse is applied and the longitudinal magnetization of adipose tissue recovers faster than that of water (shorter T1). Subsequently, a 90 degree pulse is applied at the null point of adipose tissue, which produces no signal whereas water will still produce a signal. This technique is less sensitive to magnetic field inhomogeneities and has a high sensitivity for the detection of neoplasia because of its ability to show the combined effects of prolonged T1 and

T2 relaxation times of these pathologic tissues in a background of hypointense fat suppressed marrow (Figure 52.17) [67,87,88]. However, this imaging sequence has a low signal-to-noise ratio and is limited because of the motion artifact caused by CSF pulsation. The drawback of STIR and IRFSE (inversion recovery fast spin echo sequences) is their failure to show epidural metastatic disease because CSF and epidural tumor both have high signal intensity making CSF indistinguishable from metastatic disease [86]. Mehta et al demonstrated that T1-weighted, FSE and fat-saturated FSE sequences were superior to STIR and IRFSE in the detection of epidural metastatic disease [86].

Fluid attenuated inversion-recovery (FLAIR) is an inversion recovery sequence in which the inversion time is chosen to null the CSF signal and the TR and TE are chosen to provide heavy T2-weighting. This is useful in the spine because pulsation artifact from CSF flow, which is troublesome on SE and FSE T2-weighted sequences, is negated with FLAIR. T1-FLAIR provides thinner slices and higher

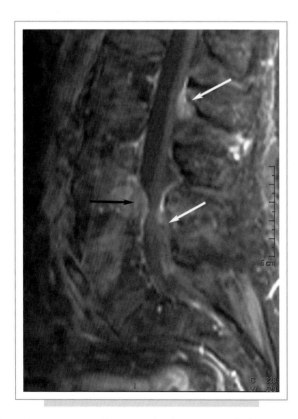

FIG. 52.16. Sagittal T1-weighted image with fat saturation shows a heterogeneously enhancing lesion (lymphoma) extending from the L4 vertebral body into the ventral epidural space (black arrow). Enhancing foci in the posterior epidural space at the L4–L5 and the L2–L3 levels (white arrows) have increased conspicuity due to the fat saturated technique.

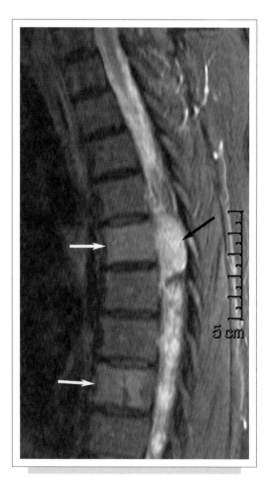

FIG. 52.17. Sagittal STIR sequence demonstrates hyperintense foci in two mid-thoracic spine vertebral bodies (white arrows). The more superior lesion has a hyperintense epidural component (black arrow), which is difficult to delineate from the intrathecal CSF. This is metastatic neuroendocrine tumor.

spatial resolution in less time than the conventional 2D SE imaging for routine spine studies. In addition, it provides greater CSF suppression and improved cord-to-CSF contrast by using the inherent magnetization transfer effect to suppress CSF in post-contrast studies [89]. This sequence also allows increased conspicuity of lesions of the spinal cord and bone marrow and reduced hardware-related artifacts as compared with conventional T1-weighted spin-echo sequences [89].

CSF flow imaging based on the phase contrast technique provides information about the phase (or direction) of flow and the velocity (or magnitude) of CSF flow such that no signal is detected from stationary tissue. Prominent epidural venous plexus can be differentiated from solid tumor, metastatic disease, inflammatory disease or herniated disk with MR CSF flow images [90]. Flow images offer additional specificity by demonstrating the epidural venous plexus contribution to an enhancing mass lesion. When there is prominence of epidural plexus due to spondylosis, the extent of plexus can be seen on flow images. However, epidural involvement by tumor or scar component will not show flow [91].

Alteration of the spinal cord structural integrity due to various diseases such as metastases, abscess or spondylosis is a major cause of motor dysfunction and can be assessed by using diffusion tensor imaging (DTI) methods. In a study of 15 symptomatic and 11 normal volunteers, Facon et al evaluated the diagnostic accuracy of apparent diffusion coefficient (ADC), fractional anisotropy (FA) and fiber tracking in both acute and slowly progressive spinal cord compressions [92]. They found that FA has the highest sensitivity and specificity in the detection of acute spinal cord abnormalities. In this way, DTI may be helpful to assess integrity of the spinal cord fiber tracts with acute or chronic spinal cord compression from epidural metastases.

With higher magnetic strength (i.e. 3.0T), there are advantages of higher signal-to-noise ratio (SNR), which would be double at 3.0T in comparison to 1.5T, enabling improved contrast or spatial resolution [93–95]. There is also the possibility of decreased scan times without reduction in image quality, which would be helpful in reducing patient motion artifacts. The improved SNR may reveal subtle structural and pathologic details involving the cervical and lumbar spinal elements, such as nerve roots, neural foramina and intracanalicular lesions [96]. Increased magnetic strength brings with it issues of increased energy deposition (specific absorption rate), increased effects of magnetic susceptibility differences of tissues, particularly at tissue–bone and tissue–air interfaces, increased chemical shift artifact which degrades standard spine echo images, and alterations in the T1 and T2 relaxation times [96]. At this time, 3.0T MRI still requires optimizing imaging sequences and improving coil technology to utilize fully the advantages of higher magnetic field strength.

While MRI has revolutionized the detection of spinal epidural metastases, it has inherent disadvantages. Some patients who have contraindications to MRI, such as pacemakers or cochlear implants, will require other modalities for primary evaluation. Although not a contraindication to spine MR, metallic hardware in the area of scanning may, in some cases, limit the delineation of anatomic detail. Sedation and adequate analgesia can facilitate obtaining an MRI in those patients with pain and/or claustrophobia. Nevertheless, patient motion cannot always be controlled by sedation or analgesia and continues to be the most vexing problem during sequences requiring long acquisition times. Marrow heterogeneity due to osteopenia, myeloinfiltrative processes and anemia can complicate interpretation of MR imaging. Hematopoetic marrow of younger patients is also difficult to assess in the setting of a neoplastic process involving the extradural spine. As mild to moderate degenerative changes evolve, the marrow may demonstrate T1 hypointensity and T2 hyperintensity and may enhance to some extent after administration of Gd-DTPA, which may confound the imaging diagnosis. Mild hyperemia and marrow edema associated with fractures can also enhance [60]. In selected cases, more than one of these modalities, such as CT for better delineation of the osseous structures, will be needed for a complete evaluation.

Interventional MRI is a newly evolving field which has been used guiding tumor ablation, aspiration cytology and surgical biopsy [97]. Preoperative endovascular embolization of highly vascular metastases such as those from melanoma, renal cell carcinoma and thyroid carcinoma can be performed to minimize intraoperative blood loss (Figure 52.18). Angiographic embolization within 48 hours of the operative procedure may preclude the recruitment of adjacent non-embolized vessels [98].

IMAGING DIFFERENTIAL DIAGNOSIS

The radiological differential diagnosis of an epidural mass is extensive and includes metastases, lymphoma, hemorrhage, radiation-induced tumor and inflammatory/infectious process as the following examples will illustrate. History is critical in establishing the neuro-oncologic clinical setting to aid in obtaining the appropriate imaging study and in accurate interpretation of the diagnostic images. In addition to evaluating the imaging characteristics of the epidural mass, it is also important to assess any additional abnormalities further to narrow the differential diagnosis.

The patholophysiology and examples of metastases to the epidural space have been discussed earlier in the chapter. It is important to emphasize lymphoma as it can commonly extend into the epidural space directly from a vertebral body, which is usually the cause of cord compression with lymphoma [37]. The soft tissue component demonstrates iso- to hypointensity relative to muscle on the T1WI and variable iso- to hyperintensity on T2WI and enhances uniformly and intensely. It is often multisegmental

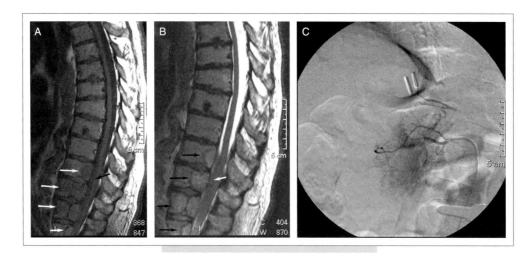

FIG. 52.18. **(A)** Sagittal T1-weighted image of metastatic renal cell carcinoma illustrates hypointense lesions in multiple vertebral bodies (white arrows) and extending to the epidural space (black arrow). **(B)** Sagittal T2-weighted image of metastatic renal cell carcinoma demonstrates multiple hyperintense lesions in lower thoracic vertebral bodies (black arrows) and extending to the epidural space (white arrow). **(C)** Selective digital substraction spinal angiographic image shows a vascular tumor blush in the paravertebral/epidural region, which was subsequently treated by endovascular embolization prior to operative decompression.

and can infiltrate through the neural foramina, which is typical of lymphoma. Isolated hematogenous involvement of only the epidural space occurs rarely in metastatic lesions and more frequently in lymphoma [36]. Lymphoma has a predilection for the thoracic spine and has a tendency to spread over several vertebral levels [99]. The marrow signal of the affected osseous structures is T1 hypointense. STIR sequences may show osseous T2 characteristics better [100]. There may be thickened enhancing nerve roots with or without focal nodules. Intramedullary tumors may be present, which show variable patchy and/or confluent enhancement and may be infiltrative or discrete (Figure 52.19). Other hematologic malignancies, such as multiple myeloma, have also been associated with spinal epidural disease [101]. In multiple myeloma, the vertebral body marrow may demonstrate an inhomogeneous pattern with foci of T1 hypointensity in a background of diffusely low signal [102]. Spinal cord compression due to amyloid deposits has been reported in multiple myeloma [103,104].

Differentiating features of infection versus malignancy were described in a study by Hovi et al [105]. Infectious processes involving the spine usually involve more than one vertebra. While intervertebral disk involvement is seen more often with infection and very rarely with malignant disease, neural arch involvement is seen more often with malignant disease. In cases of direct spread of infection into the epidural space from diskitis/osteomyelitis, the abscesses are often located in the anterior aspect of the spinal canal. Epidural abscesses are often multisegmental and fusiform, centered on and contiguous with the diseased disk and adjacent vertebral bodies. Spinal epidural abcesses can also result from hematogenous spread of bacteria usually from a cutaneous or mucosal source. They can be circumferential,

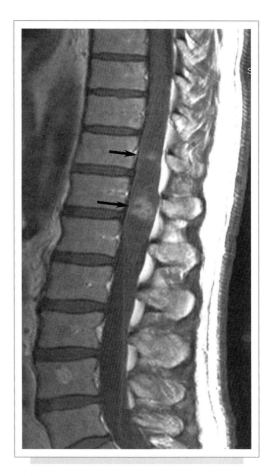

FIG. 52.19. Gadolinium enhanced sagittal T1-weighted image demonstrates an expansile low intensity intramedullary lesion with irregular heterogeneous foci of enhancement (arrows). This was shown to be an ependymoma.

and the posterior epidural space is involved in 80% of the cases from a hematogenous source [106]. Radiography may demonstrate end-plate erosion and vertebral body height loss. Myelographic findings of extradural mass impeding upon the thecal sac are non-specific. CT shows an enhancing epidural mass narrowing the canal along with the osseous changes. The lesion is iso- to hypointense on T1WI and hyperintense on T2WI and STIR. There may be homogeneous or heterogeneous phlegmon, peripherally enhancing necrotic lesion and prominent enhancing anterior epidural veins or basivertebral plexus above or below the abscess (Figure 52.20). In cases of extensive spinal epidural abscess, there may be diffuse dural enhancement. Signal alterations in the cord may be due to direct infection, cord compression and/or cord ischemia.

The differential diagnosis of an epidural mass in a patient with treated cancer includes radiation-induced secondary tumor and inflammatory/infectious process. However,

osteoradionecrosis has been reported to present as a metastatic epidural spinal lesion within a previously irradiated port [107] and is a serious complication in the treatment of head and neck malignancies. The lesion demonstrates hypointensity on T1WI and variable hyperintensity on T2WI (Figure 52.21). Unlike tumor, osteoradionecrosis does not typically enhance and is usually hypometabolic on PET imaging [108].

Inflammatory processes including granulomatous disease can also involve the epidural space [109–111]. Granulomatous infections caused by tuberculosis (TB), brucellosis, fungi and parasites, including hydatid disease (Echinococcus), have some imaging findings different from those seen with non-specific bacterial infection [112]. For example, TB spreads via the anterior longitudinal ligament with relatively limited disk involvement in comparison to degree of vertebral body and paraspinal infection and with frequent thoracic segment involvement [112]. In a study by Colmenero et al, epidural and paravertebral masses were present in 68.3% and 78% of cases of TB osteomyelitis, respectively [107]. This was significantly greater than in cases of pyogenic and brucella osteomyelitis [112]. Brucella infection also commonly involves the lumbar spine and the epidural space (Figure 52.22) [110].

Rarely, inflammatory arthropathy (e.g. tophaceous gout, renal osteoarthropathy, calcium pyrophosphate dihydrate crystal deposition) can have a soft tissue component in

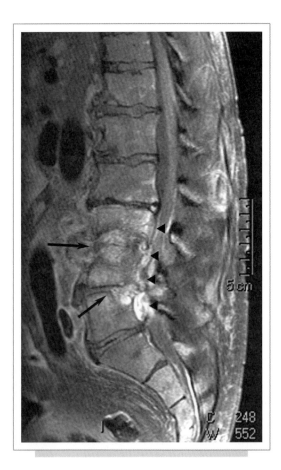

FIG. 52.20. Enhanced T1-weighted fat saturated image demonstrates a heterogeneously enhancing mass (epidural abscess secondary to bacterial infection) in the ventral epidural space extending from the L3–L4 level to the L5–S1 level (arrowheads). End-plate irregularity and abnormal disk space enhancement due to diskitis/osteomyelitis is noted at the L3–L4 and L4–L5 levels (black arrows).

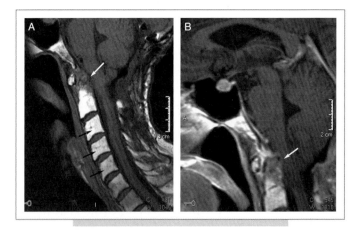

FIG. 52.21. **(A)** Sagittal unenhanced T1-weighted image demonstrates marrow replacement process involving the clivus and odontoid process with adjacent dural thickening (white arrow). The remainder of the cervical spine displays radiation treatment related fatty replacement as homogeneous T1 hyperintensity of the marrow (black arrows). Courtesy of D. Schiff, MD. (Osteoradionecrosis mimicking metastatic epidural spinal cord compression. Neurology; Jan; 64:396–397.) **(B)** Sagittal gadolinium-enhanced T1-weighted image illustrates marrow replacement process involving the clivus and odontoid process with adjacent mildly enhancing dural based mass (arrow), which compresses the medulla oblongata.

addition to the osseous disease. Gout of the axial skeleton is unusual, presenting with a range of symptoms from neck or back pain to various neurologic syndromes, including radiculopathy, myelopathy and cauda equine syndrome [113]. Radiographs may be normal or may show non-specific degenerative changes. CT is useful to demonstrate focal facet joint erosion, which is a more characteristic imaging feature of gout than with osteoarthritis (Figure 52.23A) [114]. Tophaceous lesions produce abnormal

T1 and T2 signal intensities and enhance with gadolinium (Figure 52.23B,C). Epidural tophus can extend beneath the posterior longitudinal ligament over two vertebral levels, as well as involve the intervening intervertebral disks and portions of the adjacent endplates [115]. Similar MR imaging findings have been reported in cases of cord compression secondary to gout of the cervical spine [116,117].

Epidural/subdural hematoma should be considered in the differential diagnosis when a long segmental epidural mass displays high signal on both T1- and T2-weighted images [101]. There is usually no significant post-gadolinium enhancement; however, there are case reports of patchy enhancement of hyperacute/acute epidural hematoma [118,119]. In the absence of trauma, as in spontaneous epidural hemorrhage, the vertebrae are intact [120]. Depending on the age of the blood products, the hemorrhage may be hypo-, iso- or hyperintense. It can be eccentric, multilocular or multisegmental (Figure 52.24). The cord or cauda equina may be displaced or encased. Uncommonly, an epidural CSF collection can be seen after trauma or after surgery. It demonstrates CSF signal intensities on all sequences (Figure 52.25) as opposed to a soft tissue epidural mass.

When a focal epidural mass at the level of the intervertebral disk space is identified, a disk extrusion should be considered. An intervertebral disk extrusion has a narrow contiguous segment with the parent disk while a sequestered disk has no continuity with the parent disk. The disk material may migrate away from the site of herniation regardless of continuity. On T1WI the disk material is isointense to the parent disk and iso- to hyperintense on T2WI depending on the degree of hydration (Figure 52.26A,B). The disk demonstrates peripheral enhancement (Figure 52.26C), but can demonstrate diffuse enhancement if imaged greater than 30 minutes after contrast injection. These are most commonly seen at the L4–L5 or L5–S1 levels.

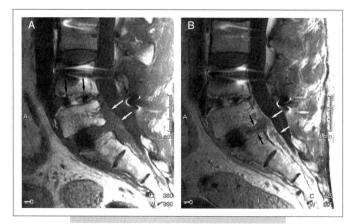

FIG. 52.22. **(A)** Unenhanced sagittal T1-weighted image of the lumbar spine shows an abnormal amorphous mass (brucellosis abscess) in the ventral epidural space from the L4–L5 level to the S2 level (white arrows). Heterogeneous signal in the L4–L5 intervertebral disk may be due to calcification and/or vacuum phenomenon (black arrows). **(B)** Enhanced sagittal T1-weighted image of the lumbar spine shows heterogeneous enhancement of the abnormal amorphous mass (brucellosis abscess) in the ventral epidural space from the L4–L5 level to the S2 level (white arrows). There is also abnormal enhancement along the margins of the L5–S1 disk space suggesting involvement by the infectious process (black arrows).

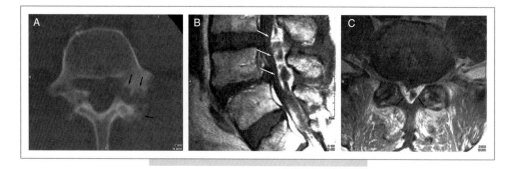

FIG. 52.23. **(A)** Unenhanced axial CT with bone algorithm through the lumbar spine demonstrates characteristic findings of gout, including well-defined erosions with sclerotic borders and overhanging edges (arrows). **(B)** Enhanced sagittal T1-weighted image demonstrates a heterogeneously enhancing lesion in the posterior epidural space extending from the L3–L4 to the L4–L5 levels and anteriorly displacing the thecal sac (arrows). This is an epidural soft tissue manifestation of gout. **(C)** Enhanced axial T1-weighted image demonstrates a heterogeneously enhancing tophus (arrows) in the posterior epidural space anteriorly displacing the thecal sac.

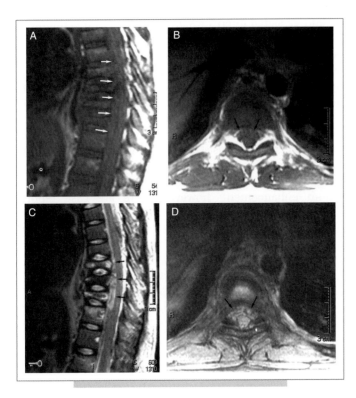

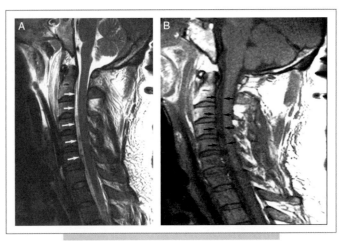

FIG. 52.24. (A) Sagittal T1-weighted image through the thoracic spine demonstrates isointense amorphous mass (epidural hematoma) in the ventral epidural space spanning multiple levels (arrows). **(B)** Axial T1-weighted image through the thoracic spine demonstrates isointense epidural hematoma (arrows) in the ventral epidural space displacing the cord posteriorly. **(C)** Sagittal T2-weighted image through the thoracic spine demonstrates heterogeneously hyperintense epidural hematoma in the ventral epidural space spanning multiple levels (arrows). There is hyperintensity in multiple vertebral bodies compatible with marrow edema from acute compression fractures. **(D)** Axial T2-weighted image through the thoracic spine demonstrates heterogeneously hyperintense epidural hematoma (arrows) in the ventral epidural space displacing the cord posteriorly.

Sometimes extensive soft tissue material, which is hyperintense on T1WI and low density on CT insinuates in the epidural space. This may represent epidural lipomatosis, which is an aberrant excess of epidural fat, often related to exogenous or endogenous steroids. The majority of cases are idiopathic and the thoracic and lumbar spinal canal are equally involved. There can be mass effect on the thecal sac and nerve roots. Bone et al quantified normal and pathological amounts of epidural fat grading it from I through III [121]. They found that all grade III patients exhibited radiculopathy [121]. Radiographs may demonstrate osteopenia or vertebral compression fractures due to the steroids, however, no bone erosion is detected on CT. The lesion is homogeneously hyperintense on T1WI, hypointense on T2WI and does not enhance (Figure 52.27A). Fat-saturation

FIG. 52.25. (A) Sagittal unenhanced T2-weighted image demonstrates an anterior epidural collection (arrows), which is isointense to CSF, and does not cause mass effect on the cord. **(B)** Sagittal unenhanced T1-weighted image shows an anterior epidural collection representing epidural hygroma (arrows), which is isointense to CSF.

techniques (such as STIR) can be used to confirm adipose tissue (Figure 52.27B).

The rare entity of ossification of the posterior longitudinal ligament (OPLL) can have a confounding T2 hypointense appearance if first encountered on MRI. OPLL is ventrally located in the cervical spine and CT is best at demonstrating the bone. There is no enhancement of this ossified ligament (Figure 52.28).

Extramedullary hematopoiesis (EH) has also been described to be a cause of epidural spinal cord compression [122–124]. EH is a compensatory mechanism of coping with several hematological disorders, but is always of long duration. It is hypothesized that such hematopoietic tissue in the spinal cord vicinity arises from embryonal rests in the extradural areolar tissue of mesodermal origin [124]. These patients are expected to have good recovery despite long-standing neurological deficits [123]. On MR imaging, it can manifest as paravertebral (Figure 52.29) and epidural masses, which are isointense on T1- and T2-weighted sequences and demonstrate intermediate enhancement. An important finding to recognize is the diffuse low to intermediate T1 signal of the vertebral body marrow as a result of displacement of fatty marrow by hematopoetic elements. Technetium colloid scanning can also be used to demonstrate ectopic marrow formation [123].

Primary tumors of the dura, such as fibromas, fibromatosis and sarcomas, are rare [34], but given their extradural location, they may cause spinal cord compression. Meningiomas can rarely be intraosseous, extradural (7%) or paraspinous in location. Extradural meningiomas are thought to arise from arachnoid rests [125]. An extradural meningioma may easily be mistaken for a spinal

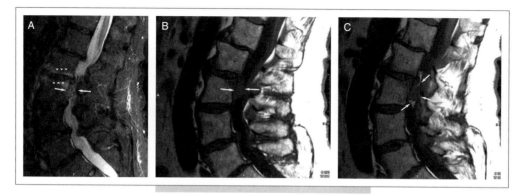

FIG. 52.26. **(A)** Sagittal STIR sequence shows degenerative disk disease and diskogenic end-plate changes at the L2–L3 level (small arrowheads). There is a heterogeneously hyperintense lesion in the ventral epidural space along the posterior margin of the L3 vertebral body (arrows) representing disk sequestration. Its hydration state is different to that of the parent disk and it appears discontinuous with the disk space. **(B)** Sagittal unenhanced T1-weighted sequence demonstrates sequestered disk in the ventral epidural space along the posterior margin of the L3 vertebral body (arrows) which appears isointense to the intervertebral disk and seems to cascade inferiorly from the L2–L3 intervertebral disk space. **(C)** Enhanced sagittal T1-weighted sequence demonstrates peripheral enhancement of the sequestered disk (arrow) in the ventral epidural space along the posterior margin of the L3 vertebral body.

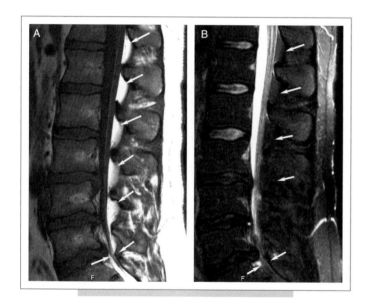

FIG. 52.27. **(A)** Sagittal T1-weighted image demonstrates prominent amount of fat in the posterior epidural space throughout the lumbar spine and circumferentially at the L5–S1 level (arrows). **(B)** Sagittal STIR sequence shows nulling of posterior and anterior epidural fat (arrows).

metastasis even at surgery, therefore it requires intraoperative histology [126]. Most often these lesions are isointense to the cord on T1 and T2WI, but some may be hyperintense on T2WI. Hypointensity within the lesion is likely due to calcification. Very vascular meningiomas can have prominent flow-voids. Bony changes are more frequent when a meningioma is extradural, including widened intravertebral

foramen, pedicle erosion, or rarely, invasion of the vertebral body [125].

Schwannomas are an additional extradural neoplasm to consider, as 15% of schwannomas are purely extradural involving the epidural space [100]. Schwannomas are well-marginated lesions, which can be very hyperintense on T2WI due to intratumoral cystic formation or necrosis. Areas of T2 hypointensity may be due to dense cellularity, hemorrhage or collagen deposition (Figures 52.30A,B) [126]. Intense enhancement may be uniform, heterogeneous or peripheral (Figure 52.30C). They typically do not have a dural attachment and very rarely occur posterior to the cord. CT can demonstrate adjacent bone erosion and/or remodeling, such as enlarged neural foramina, expansion of the central canal, or posterior body of the vertebrae.

CONCLUSION

In summary, epidural spinal cord compression is a portentous complication of metastatic disease, which can result in significant neurological sequela if it is not diagnosed urgently. The patient's neurologic status should prompt imaging evaluation and, although all the imaging modalities may detect ESCC to some degree, MR imaging is the most sensitive study. Various sequences have proven useful in detection of epidural metastasis. In the future, advanced MRI imaging, such as diffusion tensor imaging and functional MRI, may provide functional information about the status of the spinal tract fibers. Though imaging can be helpful in distinguishing between various epidural processes, the radiologic differential diagnosis of more difficult cases can only be narrowed by critical clinical history or biopsy.

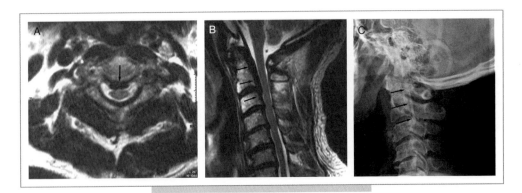

FIG. 52.28. **(A)** Axial T2 -weighted image illustrates a hypointense lesion along the posterior aspect of the cervical vertebral body (arrow), effacing the ventral CSF space and causing mild mass effect on the ventral cord. **(B)** Sagittal T2-weighted image shows a prominent hypointense lesion along the posterior aspect of the dens and C2 body (arrows). **(C)** Lateral radiograph demonstrates prominent density along the posterior aspect of the dens and C2 body (arrows) compatible with ossification of the posterior longitudinal ligament.

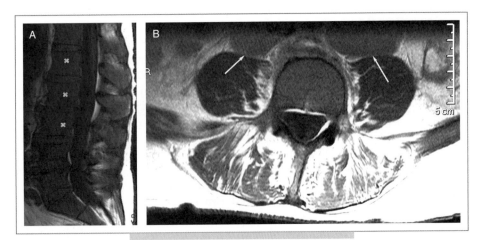

FIG. 52.29. **(A)** Sagittal unenhanced T1-weighted image shows diffuse hypointensity of the vertebral body marrow (⌘), which can be seen in patients with myeloinfiltrative and marrow recruitment processes such as anemia. **(B)** Axial enhanced T1-weighted image illustrates intermediate enhancement of the osseous marrow. There are discrete paravertebral masses (arrows) along the anterior border of the psoas muscles, which are a manifestation of extramedullary hematopoiesis.

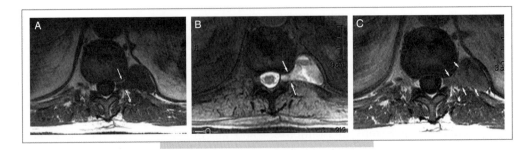

FIG. 52.30. **(A)** Axial unenhanced T1-weighted image demonstrates an isointense lobulated extradural mass extending through the left neural foramen (arrows). **(B)** Axial T2-weighted image illustrates hyperintensity in the lobulated extradural mass (arrows). This was shown to be a schwannoma. **(C)** Axial enhanced T1-weighted image demonstrates intense enhancement of the schwannoma. Courtesy of Matthew R. Hyde, M, University of Virgina Health System.

REFERENCES

1. Barron KD, Hirano A, Araki TS, Terry RD (1959). Experiences with metastatic neoplasms involving the spinal cord. Neurology 8:91–106.

2. Bach F, Larsen BH, Rhode K et al (1990). Metastatic spinal cord compression: occurrence, symptoms, clinical presentations and prognosis in 398 patients with spinal cord compression. Acta Neurochir 170:37–43.

3. Gilbert RW, Kim JH, Posner JB (1978). Epidural spinal cord compression from metastatic tumor: diagnosis and treatment. Ann Neurol 3:40–51.

4. Boogerd W, van der Sande JJ, Kroger R (1992). Early diagnosis and treatment of spinal epidural metastasis in breast cancer: a prospective study. J Neurol Neurosurg Psychiatr 55:1188–1193.

5. van der Sande JJ, Kroger R, Boogerd W (1990). Multiple spinal epidural metastases; an unexpectedly frequent finding. J Neurol Neurosurg Psychiatr 53:1001–1003.

6. Helweg-Larsen S, Hansen SW, Sorensen PS (1995). Second occurrence of symptomatic metastatic spinal cord compression and findings of multiple spinal epidural metastases. Int J Radiat Oncol Biol Phys 33:595–598.

7. Weismann DE, Gilbert M, Wang H, Grossman SA (1985). The use of computed tomography of the spine to identify patients at high risk for epidural metastases. J Clin Oncol 3:1541–1544.

8. Heldmann U, Myschetzky PS, Thomsen HS (1997). Frequency of unexpected multifocal metastasis in patients with acute spinal cord compression. Evaluation by low-field MR imaging in cancer patients. Acta Radiol 38:372–375.

9. Kornblum MB, Wesolowski DP, Fischgrund JS, Herkowitz HN (1998). Computed tomography-guided biopsy of the spine: a review of 103 patients. Spine 23:81–85.

10. Black P (1979). Spinal metastasis: current status and recommended guidelines for management. Neurosurgery 5:726–746.

11. Schiff D, O'Neill BP, Wang CH O'Fallon JR (1998). Neuro-imaging and treatment implications of patients with multiple epidural spinal metastases. Cancer 83:1593–1601.

12. van der Sande JJ, Boogerd W, Kröger R, Kappelle AC (1999). Recurrent spinal epidural metastases: a prospective study with a complete follow up. J Neurol Neurosurg Psychiatr 66:623–627.

13. Huddart RA, Rajan B, Law M et al (1997). Spinal cord compression in prostate cancer: treatment outcome and prognostic factors. Radiother Oncol 44:229–236.

14. Portenoy R, Lipton RB, Foley KM (1987). Back pain in the cancer patient: an algorithm for evaluation and management. Neurology 37:134–138.

15. Adams KK, Jackson CE, Rauch RA, Hart SF, Kleinguenther RS, Barohn RJ (1996). Cervical myelopathy with false localizing sensory levels. Arch Neurol 53:1155–1158.

16. Zevallos M, Chan PYM, Munoz L, Wagner I, Kagan AR (1987). Epidural spinal cord compression due to extradural metastatic tumor. Int J Radiat Oncol Biol Phys 13:875–878.

17. Young RF, Post EM, King GA (1980). Treatment of spinal epidural metastases: a randomized prospective comparison of laminectomy and radiotherapy. J Neurosurg 53:741–748.

18. Helweg-Larsen S, Sorensen PS, Kreiner S (2000). Prognostic factors in metastatic spinal cord compression: a prospective study using multivariate analysis of variables influencing survival and gait function in 153 patients. Int J Radiat Oncol Biol Phys 46:1163–1169.

19. Kim RY, Spencer SA, Meredith RF et al (1990). Extradural spinal cord compression: analysis of factors determining functional prognosis – prospective study. Radiology 176:279–282.

20. Helweg-Larsen S, Johnsen A, Boesen J, Sorensen PS (1997). Radiologic features compared to clinical findings in a prospective study of 153 patients with metastatic spinal cord compression treated by radiotherapy. Acta Neurochir (Wien) 139:105–111.

21. Maranzano E, Latini P, Checcaglini F et al (1991). Radiation therapy in metastatic spinal cord compression. A prospective analysis of 105 consecutive patients. Cancer 67:1311–1317.

22. Barcena A, Lobato RD, Rivas JJ et al (1984). Spinal metastatic disease: analysis of factors determining functional prognosis and the choice of treatment. Neurosurgery 14:820–827.

23. Harrington KD (1981). The use of methyl methacrylate for vertebral body replacement of anterior stabilization of pathological fracture: dislocations of the spine due to metastatic malignant disease. J Bone Joint Surg 63:36–46.

24. Khan FR, Glickman AS, Chu FCH, Nickson JJ (1967). Treatment by radiation therapy of spinal cord compression due to extradural metastases. Radiology 89:495–500.

25. Schiff D (2003). Spinal cord compression. Neurol Clin 21:67–86.

26. Grant R, Papadopoulos SM, Sandler HM, Greenberg HS (1994). Metastatic epidural spinal cord compression: current concepts and treatment. J Neurooncol 19:79–92.

27. Gabriel K, Schiff D (2004). Metastatic spinal cord compression by solid tumors. Semin Neurol 24:375–383.

28. Findlay GF (1984). Adverse effects of the management of malignant spinal cord compression. J Neurol Neurosurg Psychiatr 47:761–68.

29. Sze G (2002). Neoplastic disease of the spine and spinal cord. In *Magnetic Resonance Imaging of the Brain and Spine*, Vol. 2, 3rd edn. Atlas S (ed.). W.B. Saunders, Philadelphia. 1715–1767.

30. Mut M, Schiff D, Shaffrey M (2005). Metastasis to nervous system and intramedullary metastases. J Neurooncol 75:43–56.

31. Arguello F, Baggs R, Duerst R et al (1990). Pathogenesis of vertebral metastasis and epidural spinal cord compression. Cancer 65:98–106.

32. Algra PR, Heimans JJ, Valk J, Nauta JJ, Lachniet M, Van Kooten B (1992). Do metastases in vertebrae begin in the body or the pedicles? Imaging study in 45 patients. Am J Roentgenol 158:1275–1279.

33. Tsukada Y, Fouad A, Pickren J, Lane W (1983). Central nervous system metastasis from breast carcinoma: autopsy study. Cancer 52:2349–2354.

34. Fukui MB, Meltzer CC, Kanal E, Smirniotopolous JG (1996). MR imaging of the meninges Part II. Neoplastic disease. Radiology 201:605–612.

35. Ahmadi J, Hinton D (1993). Dural invasion by craniofacial and calvarial neoplasms: MR imaging and histopathologic evaluation. Radiology 188:747–749.

36. Haddad P, Thaell JF, Kiely JM, Harrison EE Jr, Miller RH (1976). Lymphoma of the spinal extradural space. Cancer 38:1862–1866.

37. Li MH, Holtas S, Larsson EM (1992). MR imaging of spinal lymphoma. Acta Radiol. 33:338–342.

38. Kokkoris C (1983). Leptomeningeal carcinomatosis: how does cancer reach the pia-arachnoid? Cancer 51:154–160.

39. Ushio Y, Posner R, Posner JB, Shapiro WR (1994). Experimental spinal cord compression by epidural neoplasm. Neurology 27:422–429.

40. Westbrook JL, Renowden SA, Carrie LE (1993). Study of the anatomy of the extradural region using magnetic resonance imaging. Br J Anaesth 71:495–498.

41. Blomberg RG, Olsson SS (1989). The lumbar epidural space in patients examined with epiduroscopy. Anesthes Analges 68:157–160.

42. Blomberg R (1986). The dorsomedian connective tissue band in the lumbar epidural space of humans: an anatomical study using epiduroscopy in autopsy cases. Anesthes Analges 65:747–752.

43. Hutzelmann A, Palmie S, Freund M (1997). Abstract. Vertebral body metastases: characteristic MRI findings in epidural infiltration. Aktuelle Radiol 7:169–172.

44. Graus F, Krol G, Foley KM (1985). Early diagnosis of spinal epidural metastasis correlation with clinical and radiological findings. Proc Am Soc Clin Oncol 4:269.

45. Rodichok LD, Harper GR, Ruckdeschel JC et al (1981). Early diagnosis of spinal epidural metastases. Am J Med 70: 1181–1188.

46. Rodichok LD, Ruckdeschel JC, Harper GR et al (1986). Early detection and treatment of spinal epidural metastases: the role of myelography. Ann Neurol 20:696–702.

47. Carmody RF, Yang PJ, Seeley GW, Seeger JF, Unger EC, Johnson JE (1989). Spinal cord compression due to metastatic disease: diagnosis with MR imaging versus myelography. Radiology 173:225–229.

48. Beltram J, Noto AM, Chakeres DW et al (1987). Tumors of the osseous spine: staging with MR imaging versus CT. Radiology 162:565–569.

49. Hollis PM, Malis LI, Zapulla RA (1986). Neurological deterioration after lumbar puncture below complete spinal subarachnoid block. J Neurosurg 64:253–256.

50. Hungerford GD, Powers JM (1977). Avulsion of nerve rootlets with the Cuatico needle during Pantopaque removal after myelography. Am J Roentgenol 129:485–486.

51. Abla AA, Rothfus WE, Maroon JC, Deeb ZL (1983). Delayed spinal subarachnoid hematoma; a rare complication of C1–C2 puncture in a leukemic child. Neurosurgery 12:230–231.

52. Rengachary SS, Murphy D (1974). Subarachnoid hematoma following lumbar puncture causing compression of the cauda equine: case report. J Neurosurg 41:252–254.

53. Rogers LA (1983). Acute subdural hematoma and death following lateral cervical spinal puncture. J Neurosurg 58:284–286.

54. O'Rourke T, George CB, Redmond J III et al (1986). Spinal computed tomographic metrizamide myelography in the early diagnosis of metastatic disease. J Clin Oncol 4:576–583.

55. Algra PR, Bloem JL, Tissing H et al (1991). Detection of vertebral metastases: comparison between MR imaging and bone scintigraphy. Radiographics 11:219–232.

56. Carroll KW, Feller JF, Tirman PF (1997). Useful internal standards for distinguishing infiltrative marrow pathology from hematopoietic marrow at MRI. J Magn Reson Imaging 7:394–398.

57. Kosuda S, Kaji T, Yokoyama H et al (1996). Does bone SPECT actually have lower sensitivity for detecting vertebral metastasis than MRI? J Nucl Med 37:975–978.

58. Metser U, Lerman H, Blank A, Lievshitz G, Bokstein F, Even-Sapir E (2004). Malignant involvement of the spine: assessment by 18F-FDG PET/CT. J Nucl Med 45:279–284.

59. Jadvar H, Cham D, Gamie S, Henderson RW (2004). Fusion positron emission tomography-computed tomography demonstration of epidural metastases. Clin Nucl Med 29:39–40.

60. Sze G, Krol G, Zimmerman RD, Deck MDF (1988). Malignant extradural spinal tumors: MR imaging with Gd-DTPA. Radiology 167:217–223.

61. Masaryk TJ (1991). Neoplastic disease of the spine. Radiol Clin N Am 29:829–845.

62. Petren-Mallmin M (1994). Clinical and experimental imaging of breast cancer metastases to the spine. Acta Radiol Suppl 391:1–23.

63. Kienstra GEM, Terwee CB, Dekker FW et al (2000). Prediction of spinal epidural metastasis. Arch. Neurol 57:690–695.

64. Sodickson DK, Manning WJ (1997). Simultaneous acquisition of spatial harmonics (SMASH): fast imaging with radiofrequency coil arrays. Magn Reson Med 38:591–603.

65. Pruessmann KP, Weiger M, Scheidegger MB, Boesiger P (1999). SENSE: sensitivity encoding for fast MRI. Magn Reson Med 42:952–962.

66. Roemer PB, Edelstein WA, Hayes CE, Souza SP, Mueller OM (1990). The NMR phased array. Magn Reson Med 16:192–225.

67. Ruggieri PM (1999). Pulse sequences in lumbar spine imaging. Magn Reson Imaging Clin N Am 7:425–437.

68. Henkleman EM, Hardy PA, Bishop JE et al (1992). Why fat is bright in RARE and fast spin-echo imaging. J Magn Reson Imaging 2:533.

69. Sevick RJ, Wallace CJ (1999). MR imaging of neoplasms of the lumbar spine. Magn Reson Imaging Clin N Am 7:539–553.

70. Negendank WG, Al-Katib AM, Karanes C, Smith MR (1990). Lymphomas: MR imaging contrast characteristics with clinical-pathologic correlation. Radiology 177:209–216.

71. Sze G, Abramson A, Krol G et al (1988). Gadolinium-DTPA in the evaluation of intradural extramedullary spinal disease. Am J Neuroradiol 9:153–163.

72. Castillo M, Malko JA Hoffman JC Jr (1990). The bright intervertebral disk: an indirect sign of abnormal spinal bone marrow on T1-weighted MR images. Am J Neuroradiol 11:23–26.

73. Ricci C, Cova M, Kang YS et al (1990). Normal age-related patterns of cellular and fatty bone marrow distribution in the axial skeleton: MR imaging study. Radiology 177:83–88.

74. Dunhill MS, Anderson JA, Whitehead R (1967). Quantitative histological studies on age changes in bone. J Pathol Bacteriol 94:259–291.

75. Dooms GC, Fisher MR, Hricak H et al (1986). Bone marrow imaging: magnetic resonance studies related to age and sex. Radiology 155:429–432.

76. Daffner RH, Lupetin AR, Dash N, Deeb ZL, Sefczek RJ, Shapiro RL (1986). MRI in the detection of malignant infiltration of bone marrow. Am J Roentgenol 146:353–358.

77. Ross JS, Delamarter R, Hueftle MG et al (1989). Gadolinium-DTPA-enhanced MR imaging of the post operative lumbar spine: time course and mechanism of enhancement. Am J Roentgenol 152:825–834.

78. Enzmann DR, Rubin JB (1988). Cervical spine: MR imaging with a partial flip angle, gradient-refocused pulse sequence. Part I. General considerations and disk disease. Radiology 166:467–472.

79. Enzmann DR, Rubin JB (1988). Cervical spine: MR imaging with a partial flip angle, gradient-refocused pulse sequence. Part II. Spinal cord disease. Radiology 166:473–478.

80. Baur A, Stabler A, Bruning R et al (1998). Diffusion-weighted MR imaging of bone marrow: differentiation of benign versus pathologic compression fractures. Radiology 207:349.

81. Castillo M, Arbelaez A, Smith KJ, Fisher LL (2000). Diffusion-weighted MR imaging offers no advantage over routine non-contrast MR imaging in the detection of vertebral metastases. Am J Neuroradiol 21:948–953.

82. Eastwood JD, Vollmer RT, Provenzale JM (2002). Diffusion-weighted imaging in a patient with vertebral and epidural abscesses. Am J Neuroradiol 23:496–498.

83. Sugahara T, Korogi Y, Kochi M et al (1999). Usefulness of diffusion-weighted MRI with echo-planar technique in the evaluation of cellularity in gliomas. J Magn Reson Imaging 9:53–60.

84. Castillo M, Smith JK, Kwock L, Wilber K (2001). Apparent diffusion coefficients in the evaluation of high-grade cerebral gliomas. Am J Neuroradiol 22:60–64.

85. Delfaut EM, Beltran J, Johnson G, Rousseau J, Marchandise X, Cotton A (1999). Fat suppression in MR imaging: techniques and pitfalls. Radiographics 19:373–382.

86. Mehta RC, Marks MP, Hinks RS, Glover GH, Enzmann DR (1995). MR evaluation of vertebral metastases: T1-weighted, short-inversion-time inversion recovery, fast spin-echo, and inversion-recovery fast spin-echo sequences. Am J Neuroradiol 16:281–288.

87. Feckenstein JL, Archer BT, Barker BA et al (1991). Fast short-tau inversion-recovery MR imaging. Radiology 179:499–504.

88. Rahmouni A, Divine M, Mathieu D et al (1993). Detection of multiple myeloma involving the spine: efficacy of fat suppression and contrast enhanced MR imaging. Am J Roentgenol 160:1049–1052.

89. Melhem ER, Israel DA, Eustace S, Jara H (1997). MR of the spine with a fast T1-weighted fluid-attenuated inversion recovery sequence. Am J Neuroradiol 18:447–454.

90. Levy LM (1999). MR imaging of cerebrospinal fluid flow and spinal cord motion in neurologic disorders of the spine. Magn Reson Imaging Clin N Am 7:573–587.

91. Levy LM, Di Chiro G (1990). MR phase imaging of cerebrospinal fluid flow in the head and spine. Neuroradiology 32:399–406.

92. Facon D, Ozanne A, Fillard P, Lepeintre JF, Tournoux-Facon C, Ducreux D (2005). MR diffusion tensor imaging and fiber tracking in spinal cord compression. Am J Neuroradiol 26:1587–1594.

93. Edelstein WA, Glover GH, Hardy CJ et al (1986). The intrinsic signal-to-noise ratio in NMR imaging. Magn Reson Med 3:604–661.

94. Nagae-Poetscher LM, Jiang H, Wakana S et al (2004). High-resolution diffusion tensor imaing of the brainstem at 3T. Am J Neuroradiol 25:1325–1330.

95. Barbier EL, Marrett S, Danek A et al (2002). Imaging cortical anatomy by high resolution MR at 3.0T: detection of the stripe of Gennari in visual area 17. Magn Reson Med 48:735–738.

96. Phalke VV, Gujar S, Quint DJ (2006). Comparison of 3.0T versus 1.5T MR: imaging of the spine. Neuroimaging Clin N Am 16:241–248.

97. Lufkin RB, Gronemeyer DHW, Seibel RMM (1997). Interventional MRI: update. Eur Radiol 7:S187–S200.

98. Broaddus WC, Grady MS, Delashaw JB Jr, Ferguson RD, Jane J (1990). Preoperative superselective arteriolar embolization: a new approach to enhance respectability of spinal tumors. Neurosurgery 27:755–790.

99. Lyons MK, O'Neill BP, Marsh BWR, Kurtin PJ (1992). Primary spinal epidural non-Hodgkin lymphoma. Report of eight patient's and review of the literature. Neurosurgery 30: 675–689.

100. Ross J, Brant-Zawadzki M, Chen M, Moore K, Salzman K (2004). *Diagnostic Imaging: Spine*. Amirsys Inc, Altona.

101. Li MH, Holtas S, Larsson EM (1993). MRI of extradural spinal tumors at 0.3 T. Neuroradiology 35:370–374.

102. Laurat E, Cazalets C, Sebillot M, Bernard M, Caulet-Maugendre S, Grosbois B (2003). Localized epidural and bone amyloidosis, rare cause of paraplegia in multiple myeloma. Amyloid 10:47–50.

103. Belber C, Graham D (2004). Multiple myeloma-associated solitary epidural amyloidoma of C2–C3 without bony connection or myelopathy: case report and review of the literature. Surg Neurol 62:506–509.

104. Hovi I, Lamminen A, Salonen O et al (1994). MR imaging of the lower spine. Differentiation between infectious and malignant disease. Acta Radiol 35:532.

105. Calderone RR, Larson JM (1996). Overview and classification of spinal infections. Orthop Clin N Am 27:1–8.

106. Mut M, Schiff D, Miller B, Shaffrey M, Larner J, Shaffrey C (2005). Osteoradionecrosis mimicking metastatic epidural spinal cord compression. Neurology 64:396–397.

107. Smith WK, Pfleidere AG, Millet B (2003). Osteonecrosis of the hyoid presenting as a cause of intractable neck pain following radiotherapy and the role of magnetic resonance image scanning to aid diagnosis. J Laryngol Otol 117:1003–1005.

108. Stabler A, Reiser MF (2001). Imaging of spinal infection. Radiol Clin N Am 39:115–135.

109. Dagirmanjian A, Schils J, McHenry MC (1999). MR of spinal infections. Magn Reson Imaging Clin N Am 7:525–538.

110. Adachi M, Hayashi A, Ohkoshi N et al (1995). Hypertrophic cranial pachymeningitis with spinal epidural granulomatous lesion. Intern Med 34:806–810.

111. Colmenero JD, Jinenez-Mejias ME, Sanchez-Lora FJ et al (1997). Pyogenic, tuberculous, and brucellar vertebral osteomyelitis: a descriptive and comparative study of 219 cases. Ann Rheum Dis 56:709–715.

112. Barret K, Miller ML, Wilson JT (2001). Tophaceous gout of the spine mimicking epidural infection: case report and review of the literature. Neurosurgery 48:1170–1173.

113. Fenton P, Young S, Prutis K (1995). Gout of the spine: two case reports and a review of the literature. J Bone Joint Surg Am 77A:767–771.

114. Bonaldi VM, Duong H, Starr MR, Sarazin L, Richardson J (1996). Tophaceous gout of the lumbar spine mimicking an epidural abscess: MR features. Am J Neuroradiol 17:1949–1952.

115. Murshid WR, Moss TH, Ettles DF, Cummins BH (1994). Tophaceous gout of the spine causing spinal cord compression. Br J Neurosurg 8:751–754.

116. Duprez TP, Malghem J, Van de Berg BC, Noel HM, Munting EA, Maldague BE (1996). Gout in the cervical spine: MR pattern mimicking diskovertebral infection. Am J Neuroradiol 17:151–153.

117. Nawashiro H, Reiko H (2001). Contrast enhancement of a hyperacute spontaneous spinal epidural hematoma. Am J Neuroradiol 22:1445.

118. Chang F, Lirng J, Chen S et al (2003). Contrast enhancement patterns of acute spinal epidural hematomas: a report of two cases. Am J Neuroradiol 24:366–369.

119. Boukobza M, Guichard JP, Boissonet M et al (1994). Spinal epidural hematoma: report of 11 cases and review of the literature. Neuroradiology 36:456–459.

120. Bone DG, Bone GE, Aude F et al (2003). Lumbosacral epidural lipomatosis: MRI grading. Eur Radiol 13:1709–1721.

121. Chourmouzi D, Pistevou-Gompaki K, Plataniotis G, Skaragas G, Papadopoulos L, Drevelegas A (2001). MRI findings of extramedullary haemopoiesis. Eur Radiol 11:1803–1806.

122. Heffez DS, Sawaya R, Udvarhelyi GB, Mann R (1982). Spinal epidural extramedullary hematopoiesis with cord compression in a patient with refractory sideroblastic anemia. Case report. J Neurosurg 57:399–406.

123. Abbassioun K, Amir-Jamshidi A (1982). Curable paraplegia due to extradural hematopoietic tissue in thalassemia. Neurosurgery 11:804–807.

124. Sato N, Sze G (1997). Extradural spinal meningioma: MRI. Neuroradiology 39:450–452.

125. Milz H, Hamer J (1983). Abstract. Extradural spinal meningiomas. Report of two cases. Neurochirurgia (Stuttg) 26: 126–129.

126. Friedman DP, Tartaglino LM, Flanders AE (1992). Intradural schwannomas of the spine: MR findings with emphasis on contrast-enhancement characteristics. Am J Roentgenol 158:1347–1350.

CHAPTER 53

Imaging of Plexopathy in Oncologic Patients

Eric Davis and George Krol

INTRODUCTION

The general term 'plexopathy' is often used to describe symptomatology related to plexus involvement caused by a variety of conditions, including tumor infiltration, inflammatory, traumatic, post-surgical and post-radiation changes. In oncologic patients, such involvement may be due to tumor arising from the nerve fibers (e.g. neurofibroma), direct spread of primary neoplasm from the adjacent organs (breast, lung), compression/infiltration by regional metastases (metastatic nodes) or iatrogenic factors, particularly treatment by surgical intervention or radiation. Patients may present with pain, paresthesias, focal weakness, autonomic symptoms, sensory deficits and muscle atrophy [1–3]. Physical examination fares rather poorly in assessment of the process. Conventional radiologic methods have been utilized all along in an attempt to localize and further characterize the causes of plexopathy, although with rather dismal results. A 'quantum step' progress was made with the introduction of computed tomography (CT) and magnetic resonance imaging (MRI). Technical improvements in recent years made it possible to visualize individual nerves directly [4–7]. A reliable method of visualization of diseased nerve/plexus seems more difficult to find. As new techniques are introduced, improving resolution, depicting more detail and chemical composition of tissue, there arises a need for more thorough knowledge of imaging utility in normal and diseased state.

NORMAL ANATOMY OF PLEXUSES

Plexus is defined as a network of connections of nerve roots, giving rise to further interconnecting or terminal branches. As ventral (motor) and dorsal (sensory) roots leave the spinal cord, they soon unite within the spinal canal to form a spinal nerve. After exiting neural foramina, they form a network of interconnections (plexus), from ventral roots, trunks, cords to individual nerves. Although there are many such stations in the body, the three main plexuses are cervical, brachial and lumbosacral. Plexopathy may result when

any of the above segments of the plexus becomes involved. Since it may not be possible to visualize directly the abnormality within the nerve, one may have to rely on altered adjacent tissue to make a diagnosis. Thus, the knowledge of normal configuration, adjacent tissue characteristics and spatial relationships of plexus components in reference to bony landmarks, vascular structures and muscles is very important in detection of abnormality and interpretation of plexus disease.

Cervical and Brachial Plexus

The cervical plexus lies on the ventral surface of the medial scalene and levator scapulae muscles. It is formed by ventral rami of the cervical nerves C1 through C4. Each ramus at C2, C3 and C4 levels divides into two branches, superior and inferior. These in turn unite in the following way: superior branch of C2 with C1; inferior branch of C2 with superior branch of C3; inferior branch of C3 with superior branch of C4 and inferior branch of C4 joins C5, to become part of the brachial plexus. Terminal cutaneous, muscular and communicating branches supply skin and muscles in the occipital area, upper neck, supraclavicular, upper pectoral region and diaphragm.

The brachial plexus is formed by ventral rami of spinal nerves exiting through the neural foramina of the cervical spine at C5 to T1 levels (dorsal rami innervate posterior paravertebral muscles). Inconsistent contributions may arise from C4 and T2 segments. As the spinal nerves leave the foramina between the vertebral artery anteriorly and facet joint posteriorly, they soon create the first station of connections between anterior and middle scalene muscles: C5 and C6 nerves unite to form the superior trunk; C7 becomes the middle trunk and C8 and T1 form the inferior trunk. The subclavian artery proceeds with the brachial plexus components within the triangle, anteriorly to the trunks and the subclavian vein courses in front of the anterior scalene muscle. The trunks divide just laterally to the lateral margin

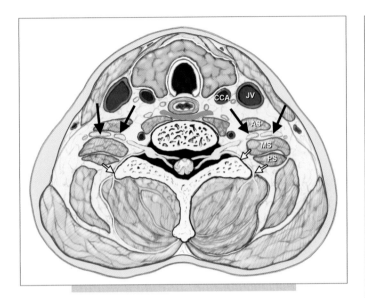

FIG. 53.1. Diagram of anatomical section through lower cervical spine in axial plane. Ventral rami of spinal nerves exiting through neural foramina unite between anterior and middle scalene muscles to form brachial plexus (arrows). Posterior rami innervate posterior paraspinal musculature (open arrow). AS, MS, PS: anterior, middle and posterior scalene muscles; CCA: common carotid artery; JV: jugular vein.

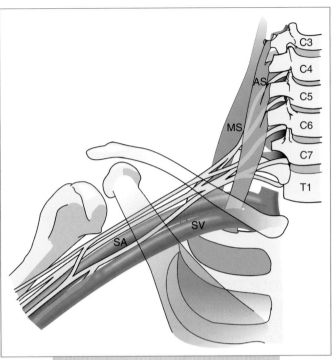

FIG. 53.2. Brachial plexus proceeds laterally towards axillary fossa through the scalene triangle (bordered by anterior and middle scalene muscles anteroposteriorly and first rib inferiorly), where it is located posteriorly to subclavian artery. Subclavian vein travels anteriorly to the anterior scalene muscle. SA: subclavian artery; SV: subclavian vein.

of scalene muscles into three anterior and three posterior divisions. Pectoralis major and serratus anterior muscles constitute anterior and posterior boundaries, respectively. Anterior divisions of superior and middle trunks join to form the lateral cord; anterior division of the inferior trunk becomes the medial cord and posterior divisions of all three trunks unite to form the posterior cord. The medial cord, which receives fibers from inferior C8–T1 trunk, gives off the ulnar nerve. The lateral cord, containing contributions from the superior and middle trunks (C5–C7) becomes the largest nerve of the upper extremity, the median nerve. The posterior cord contains fibers from all three trunks (C5–T1); its main pathway is the radial nerve. Other terminal branches include suprascapular, musculocutaneous, axillary, thoracodorsal, medial cutaneous and long thoracic nerves (Figures 53.1, 53.2 and 53.3) [8,9].

Lumbosacral Plexus

The lumbosacral plexus is formed by ventral rami of the lumbar and sacral nerves, T12 through S4. The lumbar part is formed by roots from T12 to L4 and the sacral component by L4–S4 roots. These divide into anterior and posterior divisions, which give rise to anterior and posterior branches, respectively. Anterior branches of the lumbar plexus include (in craniocaudal direction): iliohypogastric, ilio-inguinal, genito-femoral and obturator nerves; the same of sacral plexus are: tibial component of sciatic nerve, posterior femoral cutaneous and pudendal nerves. Posterior

branches of the lumbar plexus include lateral femoral cutaneous and femoral nerves and those of sacral plexus are: peroneal component of sciatic nerve, superior and inferior gluteal and piriformis nerves. The roots of the lumbar plexus lie on the ventral surface of the posterior abdominal wall, proceeding in diagonal fashion anterolaterally, between fibers of psoas and iliacus muscles. The largest femoral nerve continues behind the inguinal ligament, supplying anterior and medial aspects of the thigh. The sacral plexus proceeds laterally along the posterior wall of the pelvis, where it lies between iliac vessels anterolaterally and piriform muscle posteromedially (Figure 53.4). Terminal branches innervate pelvic organs and the sciatic nerve, the largest nerve of the body, proceeds through the greater sciatic notch to supply regions of posterior thigh and below the knee [8].

RADIOGRAPHIC METHODS OF IMAGING

Evaluation of a plexus with conventional radiography is difficult and yields little information [10]. However, it may be used in preliminary evaluation of plexopathy, mainly to exclude major abnormality, such as bone destruction, fracture, lung infiltration or ligamentous calcifications.

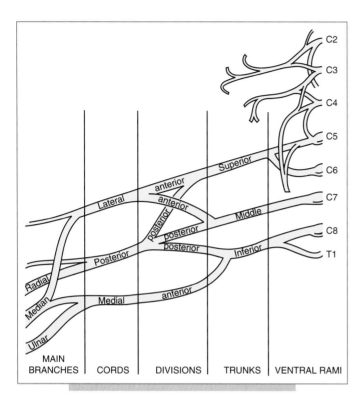

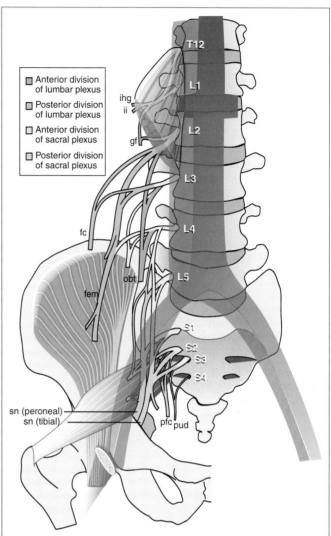

FIG. 53.3. Schematic representation of cervical and brachial plexus network. Cervical plexus (C1–C4): ventral rami C2, C3 and C4 divide, giving off superior and inferior branch at each level. Superior branch of C2 and C1 ramus unite to form ansa cervicalis. Adjacent inferior and superior branches of C2, C3 and C4 merge and give off lesser occipital, greater auricular, transverse cervical, supraclavicular and phrenic nerves (unmarked). Brachial plexus (C5–T1): superior, middle and inferior trunks are formed by ventral rami of C5/C6, C7 and C8/T1, respectively. Each trunk divides into anterior and posterior divisions. Anterior divisions of superior and middle trunks form lateral cord; anterior division of inferior trunk becomes medial cord and posterior divisions of all three trunks unite to form posterior cord. Major nerves of the arm: median, ulnar and radial receive contributions predominantly from lateral, medial and posterior cords, respectively.

FIG. 53.4. Simplified coronal diagram of lumbosacral plexus, depicted on a background of psoas, iliacus and piriformis muscles. Anatomy of the lumbosacral plexus is much less intricate and more variable than that of brachial plexus. Anterior and posterior divisions unite and/divide to form terminal branches (the trunks and cords are not distinguished). Anterior divisions of lumbar plexus unite to form iliohypogastric (ihg), ilioinguinal (ii), genitofemoral (gf) and obturator (obt) branches, whereas posterior divisions give rise to femoral cutaneous (fc) and femoral (fem) nerves. Of sacral plexus, anterior divisions divide to give rise to tibial component of sciatic nerve (sn), posterior femoral cutaneous (pfc) and pudendal nerves (pud), while posterior divisions give off peroneal component of sciatic nerve (sn), part of femoral cutaneous, gluteal and piriformis nerves.

Development of CT heralded major progress in imaging and its utility in detection of plexus disease has been realized by early investigators [11–16]. Introduced in recent years, the new technique of multichannel scanning allows for uninterrupted data acquisition during continuous tube rotation and table advance [17]. Further anatomical detail and tissue characteristics have been provided by MRI. Special sequences have been developed for selective imaging of nerve tissue (neurography). Sonography and positron emission tomography (PET) scanning also have been reported to provide valuable contributions [18–20]. Development of the picture archiving system (PACS) has revolutionized the way studies are viewed and interpreted by radiologists. Perhaps

the best example is a cine mode option, which creates a three-dimensional perception, allowing for better understanding of the extent and configuration of lesions and their relation to adjacent normal structures, particularly vessels.

However, despite valuable contributions from the above imaging methods, the assessment of plexus regions frequently presents a challenging problem for clinicians, as well as radiologists.

Technique – CT

Because of its small size, the cervical plexus is rarely evaluated radiographically as a separate entity. Rather, it is included as a part of head, cervical spine or neck examinations. The cervical plexus extends from C1 down to C4, thus scanning from the skull base down to C5, utilizing small field of view (FOV) (25 cm), is sufficient. The anatomical brachial plexus extends approximately from C5 down to T2 vertebral levels. Adequate coverage of the plexus is provided by scanning the region from C4 to T3. We extend lower range with larger FOV down to T6, to include more peripheral components. Contiguous 4–5 mm spaced axial sections are obtained, perpendicular to the table top. Coned-down views of the area in question may be added to the study. The elements of normal plexus are small and are depicted as nodular or linear areas of soft tissue density. They are difficult to identify and may not be outlined at all [21], particularly on inferior quality examination. Contrast injection is recommended [12], not only for identification of normal vascular structures and differentiation from lymph nodes, but also for more complete information on the enhancement pattern of the lesion. For this purpose, an intravenous injection in dynamic mode is preferred, using an initial bolus of 50 ml, followed by contiguous infusion at the rate of 1 ml/second to a total of 100 ml of non-ionic contrast (Omnipaque 300 or equivalent). The infusion should be administered on the site opposite to the suspected pathology, since high concentrations of intravenous contrast may produce streaking artifacts, obscuring detail [21].

Adequate coverage of the lumbosacral plexus includes axial sections from T12 down to the tip of the coccyx, to visualize greater sciatic notches (GSN). Axial images with 4–5 mm slice thickness and FOV large enough to include both sacroiliac joints are usually sufficient. Intravenous contrast administration (100 ml of Omnipaque 300 or equivalent in dynamic mode or bolus injection) is recommended, unless contraindicated. An intra-oral contrast (Gastrographin, given 2 h in advance) may be beneficial in assessment of plexus pathology, mainly to delineate distal urinary tract and colon, respectively.

Technique – MR

The main advantages of MR imaging over CT are the ability to perform multiplanar scanning without needing to change the patient's position, and superior resolution and tissue characterization [22–24]. Adequate anatomical coverage of the plexus must be assured. Thus, for the brachial plexus, axial sections from levels of C3 down to T6, coronal sections including both gleno-humeral joints and sagittal

sections to cover both axillary fossas should be obtained. For lumbosacral plexus, the coverage in axial plane needs to extend from T12 to the coccyx. Sections of 5 mm thickness/0.25 gap are obtained in axial, sagittal and coronal planes. We prefer scanning in direct axial, sagittal and coronal planes [25], although oblique scanning planes have been advocated for the brachial [26] and sacral plexus [27], to optimize visualization of plexus components and the sciatic nerve. T1, fast spin-echo T2 and STIR (fat suppression) should be obtained. Use of phased-array coils is recommended for greater resolution of detail [28,29]. Intravenous contrast (Magnevist or equivalent) is utilized routinely, in a dose of 0.1 mmol/kg of body weight.

Visualization of Normal Plexus Components on CT and MR

Conventional (non-contrast) CT offers rather poor definition of intraspinal anatomy. Nerve roots exiting through foramina are more consistently seen, particularly in lumbar spine and sacrum, because of larger size and more abundant epidural fat (Figure 53.5). They are depicted as punctate or linear structures of muscle density, contrasting against the adjacent darker fat within the foramen or vicinity (Figure 53.6). Trunks and cords of the brachial plexus are usually blended with muscle fibers and major vessels more distally. With abundant fat tissue, individual nerve components may be visible as discrete, linear areas of soft tissue density, proceeding posteriorly along the subclavian artery (Figure 53.7). Those of the lumbar plexus enter paraspinal musculature (psoas) and are difficult to identify with

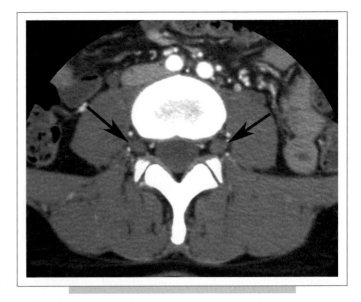

FIG. 53.5. Axial CT section through inferior plate of L3. Nerve roots are clearly seen bilaterally (arrows) contrasting against low background density of fat tissue.

certainty. Femoral and obturator nerves may be visible, contrasted against the intrapelvic/intra-abdominal fat. The largest sciatic nerve (Figure 53.8) is more consistently identified in the lateral aspect of GSN, posteriorly to the ischial spine [12–15].

MR depicts the plexus components with much greater accuracy [22–24,30] and can be further improved when phased-array coils are used [29,31]. Conventional T1-weighted sequences are routinely utilized for anatomical detail. Although individual nerves down to 2 mm in diameter can be seen [29], much of the success depends upon relaxation properties of adjacent tissues. Thus, even a smaller branch surrounded by fat or fluid (i.e. cerebrospinal fluid (CSF)) can be seen clearly, while a larger neural trunk encompassed by infiltrative process will blend with the abnormal tissue and may not be recognized at all. T2-weighted sequences (FSE) provide best detail of normal nerve when contrast interphase of fluid (CSF) or abnormal tissue (e.g.

edema) is present. Thus, spinal nerves within the canal are demonstrated in good detail on T1 and T2 sequences, because they contrast against fat (epidural) and fluid (CSF), respectively.

Special sequences have been designed to create a 'myelographic effect', depicting the cord and roots within the thecal sac throughout the spine [32]. Fat suppression

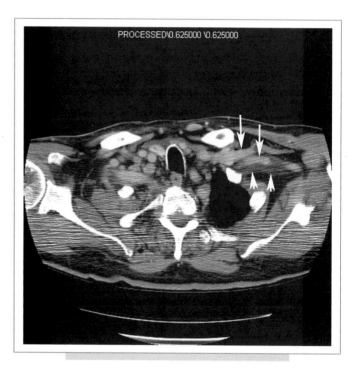

FIG. 53.7. Axial CT image, depicting spatial arrangement of neurovascular bundle (arrows from front): subclavian vein, artery and fibers of brachial plexus.

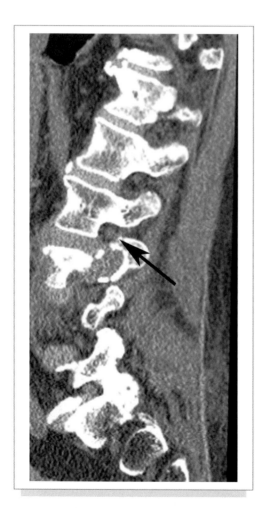

FIG. 53.6. Sagittal reconstruction CT image through the plane of foramina. Nerve root (arrow) surrounded by epidural fat. Note lytic bone lesion below, not compromising the lumen of the foramen.

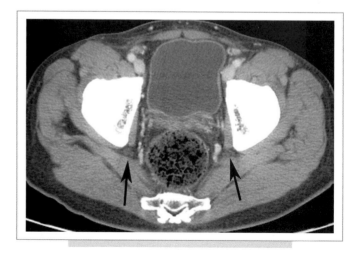

FIG. 53.8. Axial CT section through the greater sciatic notch. Sciatic nerves (arrows) are visualized in the lateral aspect of GSN, posteriorly to the ischial spine.

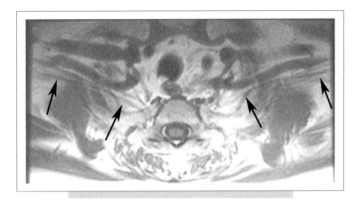

FIG. 53.9. MRI of the brachial plexus. Long segments of plexus fibers (arrows) are demonstrated on axial T1-weighted image. Subclavian arteries and veins visualized anteriorly.

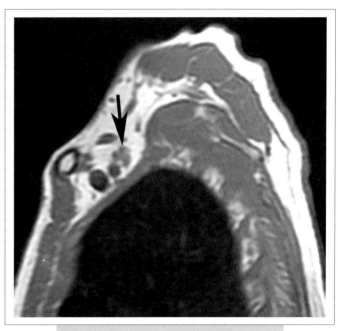

FIG. 53.10. Sagittal T1-weighted image through the axillary fossa on the same patient as Figure 53.9. Brachial plexus components (arrow) depicted as nodular, somewhat irregular soft tissue densities, posteriorly to the artery and vein (rounded, lower intensity structures anteriorly).

sequences (STIR being most commonly used) null fat signal, thus rendering background fat tissue darker and nerve more clearly visible [33,34]. The option can be applied to both T1- and T2-weighted sequences and is most valuable in conjunction with contrast enhancement.

The normal neural components of the plexus can be identified with variable success on a good quality MR scan. Nerve roots within the foramina and immediate vicinity are routinely seen. As they descend with the muscle fibers, trunks and cords of the brachial plexus can be identified on axial images, proceeding between the anterior and middle scalene group laterally, to exit through the scalene triangle. Extended segments of the plexus may be demonstrated on one well placed axial and/or coronal T1-weighted anatomical image (Figure 53.9). Sagittal sections through and laterally to the scalene triangle outline the individual divisions as punctate areas of soft tissue intensity within the fat background, with subclavian artery and vein anteriorly (Figure 53.10). Similarly, the lumbar and sacral roots within the subarachnoid space and neural foramina are seen in great detail, while the interconnecting network within the psoas complex is more difficult to identify. On axial sections, femoral nerve may be visualized as a single trunk proceeding anterolaterally along the ventral aspect of the posterior abdominal wall. The greater sciatic nerve within the notch is more consistently seen, either as a single, oval in shape structure or as a cluster of several smaller individual nerves (Figure 53.11) [7,23,24].

Abnormal Nerve

When involved by a disease, nerve tissue may exhibit swelling, focal or diffuse infiltration, edema, cyst formation or necrosis. In the early stage, when there is no enlargement of peripheral nerve(s), these changes may not be appreciated on imaging modalities but become more apparent as the process progresses. On CT, abnormal nerve or plexus may appear locally or diffusely enlarged, or become indistinguishable from adjacent structures because of infiltrative process or fibrosis. In the brachial plexus, this may be manifested as general thickening of neurovascular bundle (Figure 53.12). Greater tissue discrimination allows MR to assess the character and extent of disease more precisely. T1-weighted images may show segmental or diffuse enlargement and increased T2 intensity may be seen within the nerve in case of involvement by infiltrative process or edema. The extent of the abnormal T2 signal may be demonstrated to the better advantage on fat suppression (STIR) sequences (Figure 53.13). Contiguity (or disruption) of the nerve, increased diameter, change of course or contour (compression), altered intrinsic intensity or enhancement within involved or a compressed segment of the nerve may be observed. Neurography, utilizing combination of T1, T2, STIR, short inversion recovery (IR) sequences and phased-array surface coils depict these findings with greater accuracy [5,35–37]. In a study of 15 patients with neuropathic leg pain and negative conventional MR conducted by Moore et al, it proved definitely superior, revealing casual abnormality accounting for clinical findings in all cases [38].

An interesting phenomenon of increased T2 intensity within normal nerve(s), mimicking a diseased tissue was reported recently by Chappell et al [39]. They observed raising intensity within the peripheral nerve as the orientation

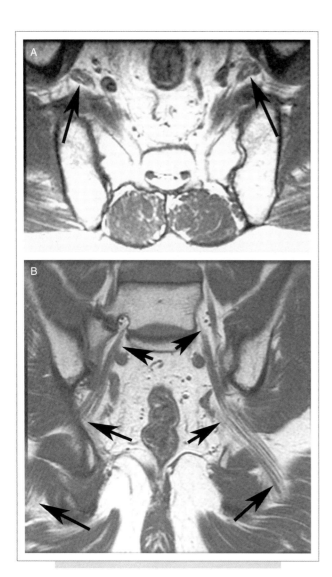

FIG. 53.11. Sciatic nerves (arrows) depicted on axial **(A)** and coronal **(B)** T1-weighted MR images through the pelvic region, as a conglomerate of individual fibers.

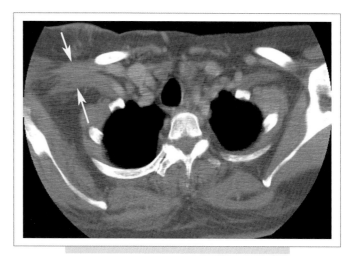

FIG. 53.12. Recurrent breast cancer. Axial CT image demonstrates general thickening of the neurovascular bundle (arrows).

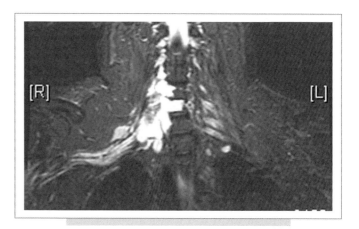

FIG. 53.13. Coronal STIR MR non-contrast sequence enhances visualization of the tumor tissue (bright areas on the right side).

of longitudinal axis of the nerve approached 55 degrees to main magnetic field Bo. As the brachial plexus is usually scanned close to this 'magic angle', T2 hyperintensity within the nerve or plexus should thus be interpreted with caution, to avoid a false positive reading. Administration of contrast is essential in proper evaluation of the extent of disease. Generally, enhancement of intraspinal or peripheral nerve after administration of a conventional dose of gadolinium-based contrast is considered pathological.

Imaging Method of Choice

As x-rays yielded very limited information on plexus involvement, early reports praised the ability of CT to demonstrate anatomical detail of normal and diseased plexus and its advantages over conventional radiography [13,14]. As early as 1988, Benzel et al considered CT a method of choice in evaluation of location and extent of nerve sheath tumors of sciatic nerve and sacral plexus, important in determination of resectability of these lesions [11]. Hirakata et al found CT to be helpful in assessment of patients with Pancoast tumor [16]. They reported obliteration of the fat plane between scalene muscles on CT to be an indication of brachial plexus involvement. The addition of sagittal, coronal and oblique reformatted images improved visualization of brachial plexus and helped to diagnose tumor recurrence [40]. Soon after introduction of MR and as early as 1987, Castagno and Shuman anticipated that the new modality may have substantial clinical utility in evaluating patients with suspected brachial plexus tumor [41]. The advantages of MR over CT in assessment of plexus were reported in comparative studies by several investigators [42–44]. In the

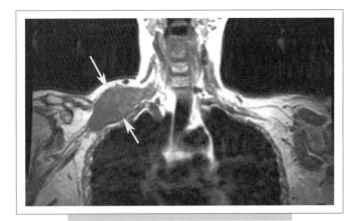

FIG. 53.14. Brachial plexus schwannoma, depicted on coronal T1-weighted MR image. Fusiform mass is orientated along the longitudinal axis of the plexus.

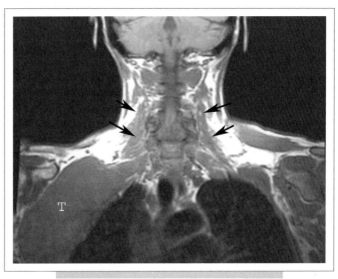

FIG. 53.15. Patient with neurofibroma. Large mass is demonstrated in axillary fossa (T). There is also marked thickening of all components of the plexus bilaterally (arrows).

evaluation of 64 patients with brachial plexopathy of diverse cause by Bilbey et al, the sensitivity of MR was 63%, specificity 100% and accuracy 77% [45]. In a subgroup of patients with trauma and neoplasm these were even higher (81%, 100% and 88%, respectively). Better anatomical definition provided by MR was considered to improve patient care in the study by Collins et al [46]. More recently, in the study of patients with plexopathy following the treatment of breast cancer, Qayyum et al reported high reliability of MR, with specificity of 95%, positive predictive value 96% and negative predictive value of 95% [47]. MR without and with contrast enhancement is also a method of choice in evaluating a patient with plexopathy at our institution.

PRIMARY PLEXUS NEOPLASMS

Primary tumors of peripheral nerve (plexus) are rare, constituting approximately 1% of all cancers. According to the WHO classification proposed in 2000, four groups are distinguished:

1 schwannomas
2 perineuriomas
3 neurofibromas
4 malignant peripheral nerve sheath tumors (MPNST).

Neurofibromas and schwannomas are most prevalent. On MR, these tumors usually present as well defined, rounded or oval masses, oriented along the longitudinal axis of the nerve (Figure 53.14). Larger size or plexiform appearance favors neurofibroma, particularly in patients with neurofibromatosis (Figure 53.15). Both are iso- or slightly hyperintense on T1- and hyperintense on T2-weighted sequences, showing homogeneous or inhomogeneous enhancement. Marked T2 hyperintensity may be seen in some patients [48–52]. Capsule can be identified in approximately 70% of schwannomas and 30% of neurofibromas

[49]. A 'target sign' consisting of a central low intensity area within the lesion on T2-weighted sequence was found to be much more frequent in neurofibromas [53,54]. On CT, low density (in reference to the muscle) was a common feature of plexus tumors and contrast enhancement varied from moderate to marked, as reported by Vestraete [55]. MPNST are rare, arising either as spontaneous mutation or within pre-existing neurofibroma, usually as a transformation to spindle cell sarcoma [56]. It may be difficult to distinguish these tumors from their benign counterparts. Larger size, internal heterogeneity, poor definition of the peripheral border, invasion of fat planes and adjacent edema favor the malignant variant [57–60]. Extremely rarely other malignancies can arise from the nerve proper [61]. There are no pathognomonic radiographic features which could be totally attributed to a particular group; thus thorough knowledge of clinical information, such as age, gender, duration of symptoms, history of von Recklinghausen's disease, etc., is very helpful while interpreting the imaging studies.

EXTRINSIC PLEXUS TUMORS

Plexus may be compressed or infiltrated by an extrinsic neoplasm arising from adjacent structures, with breast, neck, lung and lymph node malignancies being most common offenders for brachial region and pelvic tumors for lumbosacral plexus (Figures 53.16 and 53.17). Clinically, pain is a prominent feature of residual or recurrent tumor [1]. Although local extent of the lesion may be depicted adequately by CT, infiltration of the plexus is difficult to diagnose, except for advanced disease with gross infiltration of the plexus region (Figure 53.18). In a study of 14 patients

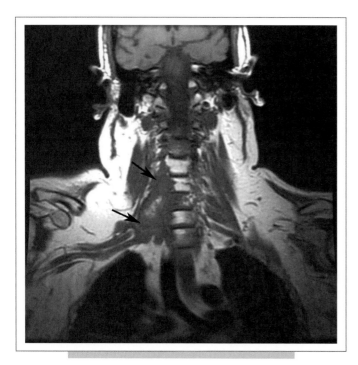

FIG. 53.16. Malignant PNST, Coronal T1-weighted MR image (the same patient as in Figure 53.13). The tumor (arrows) is poorly defined and infiltrates adjacent cervical vertebrae.

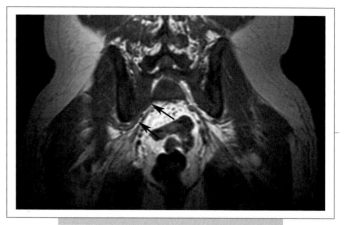

FIG. 53.18. Coronal T1 MR image through the pelvis of patient with extensive bone metastases from prostate cancer and right hip and extremity pain. The fat plane separating nerve fibers of the plexus from abnormal bone obliterated on the right, as compared to the left (arrows).

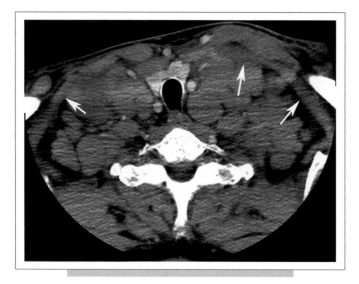

FIG. 53.17. Axial CT section through thoracic inlet of patient with lymphoma. Neurovascular elements (arrows) buried in massive lymphadenopathy.

with Pancoast tumor, obliteration of fat planes between scalene muscles on CT was found to be suggestive of plexus involvement [16]. The potential of MR in evaluation of the extent of thoracic malignancies and brachial plexus involvement was realized as early as 1989 [62]. In the study of chest wall tumors by Fortier, MR clearly delineated the margins and revealed evidence of muscle, vascular or bone invasion [63]. T1 and T2 signal characteristics were non-specific (apart for lipomas), not allowing for confident distinction of benign from malignant process. Qayyum considered MR to be a reliable and accurate tool in evaluation of brachial plexopathy due to tumor, reporting sensitivity and positive predictive value of 96% and specificity and negative predictive value of 95% [47]. Currently, MR without and with contrast remains a method of choice for the assessment of local tumor extent and plexus involvement. PET scanning is considered useful in the evaluation of patients with plexopathy, mainly by excluding recurrent tumor [19,20].

RADIATION INJURY

Plexopathy is a recognized complication of radiation therapy, occurring most commonly in the brachial plexus [1], following regional treatment of breast, lung or neck cancers. The clinical presentation of radiation-induced fibrosis (RIF) is that of a protracted course with low-grade pain, as opposed to recurrent tumor, which progresses more rapidly, with pain being a dominant symptom. Three forms are recognized: acute ischemic, transient and delayed (radiation fibrosis), the latter being most common [64–66]. Radiation-induced fibrosis occurs usually within the first few years after completion of treatment, although latent periods as long as 22 years have been reported [67,68]. It is dose dependent, more likely to occur above 6000 cGy and with larger fraction sizes. Younger patients and those receiving cytotoxic therapy are more vulnerable [1,69]. Changes are confined to the radiation port with a clearly demarcated margin from non-irradiated tissue. The main role of imaging is to distinguish this chronic iatrogenic process from a recurrent

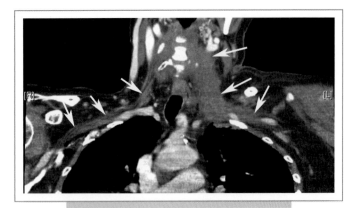

FIG. 53.19. Coronal CT reconstruction image of patient with extensive infiltration of neurovascular structures and scalene muscle complex due to metastatic cholangiocarcinoma. Note normal neurovascular bundle on the right side (arrows).

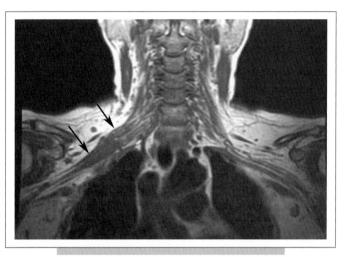

FIG. 53.20. Patient with breast cancer and suspected post-radiation right brachial plexopathy. MR reveals matting and thickening of the right neurovascular bundle (arrows) without recognizable soft tissue mass, consistent with clinical diagnosis of RIF.

tumor, which generally carries an unfavorable prognosis. CT may demonstrate poor definition of neurovascular bundle and increased density of regional fat, without recognizable soft tissue mass [70] and MR may reveal thickening and indistinct outline of plexus components, again without identifiable focal mass (Figures 53.19 and 53.20). There is no uniform agreement as to signal intensity patterns and contrast enhancement of RIF. While some investigators reported low intensity of both T1- and T2-weighted sequences [71], others described variable signal changes [72–74]. Parascalene and interscalene T2 hyperintensity was reported by Bowen [75] and the degree of T2 hyperintensity was found to correspond with severity of fibrosis by Hoeller [74]. Positive contrast enhancement was reported by most investigators, even in very delayed cases [72].

SUMMARY

Clinical assessment of plexus involvement by primary neoplasm, metastases or conditions related to treatment is limited and imaging studies (CT, MR) are routinely requested to characterize further the nature and extent of the process. With its many advantages over CT, MR with and without contrast is presently considered the imaging modality of choice in evaluation of plexus disease. Utilizing high-resolution devices, we are now able to visualize individual nerves (MR neurography). However, there are still many limitations of MR technique, such as inhomogeneous fat suppression or vascular flow artifacts, making the interpretation difficult. Although abnormal signal or enhancement within the individual nerve can be depicted, the process cannot be further characterized (e.g. benign or malignant). Recently introduced into clinical practice, higher field strength (3 T) MR may offer superior resolution and improved options for functional imaging. As for the 'state of the art' approach at present, it includes thorough clinical examination, followed by high-resolution phase-array MR, with and without contrast. Dynamic CT should be reserved for patients unable to have MR and instances when additional information on bone detail or vascular anatomy is also needed. PET scanning plays a complementary role to MR in assessment of neoplastic involvement.

REFERENCES

1. Kori SH et al (1981). Brachial plexus lesions in patients with cancer: 100 cases. Neurology 31(1):45–50.
2. Aguilera Navarro JM et al (1993). Neoplastic lumbosacral plexopathy and 'hot foot'. Neurologia 8(8):271–273.
3. Jaeckle KA (2004). Neurological manifestations of neoplastic and radiation-induced plexopathies. Semin Neurol 24(4): 385–393.
4. Filler AG et al (1993). Magnetic resonance neurography. Lancet 341(8846):659–661.
5. Filler AG et al (1996). Application of magnetic resonance neurography in the evaluation of patients with peripheral nerve pathology. J Neurosurg 85(2):299–309.
6. Aagaard BD et al (1998). MR neurography. MR imaging of peripheral nerves. Magn Reson Imaging Clin N Am 6(1): 179–194.
7. Freund W et al (2007). MR neurography with multiplanar reconstruction of 3D MRI datasets: an anatomical study and clinical applications. Neuroradiology 49:335–341.

8. Clemente DC (1985). The peripheral nervous system. In *Gray's Anatomy*, 30th American edn. Lea & Febiger, Philadelphia. 1149–1282.

9. Castillo M (2005). Imaging the anatomy of the brachial plexus: review and self-assessment module. Am J Roentgenol 185(6 Suppl):S196–204.

10. Posniak HV et al (1993). MR imaging of the brachial plexus. Am J Roentgenol 161(2):373–379.

11. Benzel EC et al (1988). Nerve sheath tumors of the sciatic nerve and sacral plexus. J Surg Oncol 39(1):8–16.

12. Cooke J et al (1988). The anatomy and pathology of the brachial plexus as demonstrated by computed tomography. Clin Radiol 39(6):595–601.

13. Dietemann JL et al (1987). Anatomy and computed tomography of the normal lumbosacral plexus. Neuroradiology 29(1):58–68.

14. Gebarski KS et al (1986). The lumbosacral plexus: anatomic-radiologic-pathologic correlation using CT. Radiographics 6(3):401–425.

15. Gebarski KS et al (1982). Brachial plexus: anatomic, radiologic, and pathologic correlation using computed tomography. J Comput Assist Tomogr 6(6):1058–1063.

16. Hirakata K et al (1989). Computed tomography of Pancoast tumor. Rinsho Hoshasen 34(1):79–84.

17. Rydberg J et al (2004). Fundamentals of multichannel CT. Semin Musculoskelet Radiol 8(2):137–146.

18. Graif M et al (2004). Sonographic evaluation of brachial plexus pathology. Eur Radiol 14(2):193–200.

19. Hathaway PB et al (1999). Value of combined FDG PET and MR imaging in the evaluation of suspected recurrent local-regional breast cancer: preliminary experience. Radiology 210(3):807–814.

20. Ahmad A et al (1999). Use of positron emission tomography in evaluation of brachial plexopathy in breast cancer patients. Br J Cancer 79(3–4):478–482.

21. Krol G, Strong E (1986). Computed tomography of head and neck malignancies. Clin Plast Surg 13(3):475–491.

22. Carriero A et al (1991). Magnetic resonance imaging of the brachial plexus. Anatomy. Radiol Med (Torino) 81(1–2):73–77.

23. Gierada DS, Erickson SJ (1993). MR imaging of the sacral plexus: abnormal findings. Am J Roentgenol 160(5):1067–1071.

24. Gierada DS et al (1993). MR imaging of the sacral plexus: normal findings. Am J Roentgenol 160(5):1059–1065.

25. Blake LC et al (1996). Sacral plexus: optimal imaging planes for MR assessment. Radiology 199(3):767–772.

26. Panasci DJ et al (1995). Advanced imaging techniques of the brachial plexus. Hand Clin 11(4):545–553.

27. Almanza MY et al (1999). Dual oblique MR method for imaging the sciatic nerve. J Comput Assist Tomogr 23(1):138–140.

28. Kichari JR et al (2003). MR imaging of the brachial plexus: current imaging sequences, normal findings, and findings in a spectrum of focal lesions with MR-pathologic correlation. Curr Probl Diagn Radiol 32(2):88–101.

29. Maravilla KR, Bowen BC (1998). Imaging of the peripheral nervous system: evaluation of peripheral neuropathy and plexopathy. Am J Neuroradiol 19(6):1011–1023.

30. Blair DN et al (1987). Normal brachial plexus: MR imaging. Radiology 165(3):763–767.

31. Bowen BC et al (2004). The brachial plexus: normal anatomy, pathology, and MR imaging. Neuroimaging Clin N Am 14(1):59–85, vii–viii.

32. Gasparotti R et al (1997). Three-dimensional MR myelography of traumatic injuries of the brachial plexus. Am J Neuroradiol 18(9):1733–1742.

33. Tien RD et al (1991). Improved detection and delineation of head and neck lesions with fat suppression spin-echo MR imaging. Am J Neuroradiol 12(1):19–24.

34. Howe FA et al (1992). Magnetic resonance neurography. Magn Reson Med 28(2):328–338.

35. Dailey AT et al (1996). Magnetic resonance neurography for cervical radiculopathy: a preliminary report. Neurosurgery 38(3):488–492; discussion 492.

36. Erdem CZ et al (2004). High resolution MR neurography in patients with cervical radiculopathy. Tani Girisim Radyol 10(1):14–19.

37. Lewis AM et al (2006). Magnetic resonance neurography in extraspinal sciatica. Arch Neurol 63(10):1469–1472.

38. Moore KR et al (2001). The value of MR neurography for evaluating extraspinal neuropathic leg pain: a pictorial essay. Am J Neuroradiol 22(4):786–794.

39. Chappell KE et al (2004). Magic angle effects in MR neurography. Am J Neuroradiol 25(3):431–440.

40. Fishman EK et al (1991). Multiplanar CT evaluation of brachial plexopathy in breast cancer. J Comput Assist Tomogr 15(5):790–795.

41. Castagno AA, Shuman WP (1987). MR imaging in clinically suspected brachial plexus tumor. Am J Roentgenol 149(6):1219–1222.

42. Rapoport S et al (1988). Brachial plexus: correlation of MR imaging with CT and pathologic findings. Radiology 167(1):161–165.

43. Thyagarajan D et al (1995). Magnetic resonance imaging in brachial plexopathy of cancer. Neurology 45(3 Pt 1):421–427.

44. Taylor BV et al (1997). Magnetic resonance imaging in cancer-related lumbosacral plexopathy. Mayo Clin Proc 72(9):823–829.

45. Bilbey JH et al (1994). MR imaging of disorders of the brachial plexus. J Magn Reson Imaging 4(1):13–18.

46. Collins JD et al (1995). Compromising abnormalities of the brachial plexus as displayed by magnetic resonance imaging. Clin Anat 8(1):1–16.

47. Qayyum A et al (2000). Symptomatic brachial plexopathy following treatment for breast cancer: utility of MR imaging with surface-coil techniques. Radiology 214(3):837–842.

48. Baba Y et al (1997). MR imaging appearances of schwannoma: correlation with pathological findings. Nippon Igaku Hoshasen Gakkai Zasshi 57(8):499–504.

49. Cerofolini E et al (1991). MR of benign peripheral nerve sheath tumors. J Comput Assist Tomogr 15(4):593–597.

50. Soderlund V et al (1994). MR imaging of benign peripheral nerve sheath tumors. Acta Radiol 35(3):282–286.

51. Saifuddin A (2003). Imaging tumours of the brachial plexus. Skeletal Radiol 32(7):375–387.

52. Hayasaka K et al (1999). MR findings in primary retroperitoneal schwannoma. Acta Radiol 40(1):78–82.

53. Bhargava R et al (1997). MR imaging differentiation of benign and malignant peripheral nerve sheath tumors: use of the target sign. Pediatr Radiol 27(2):124–129.

54. Burk DL Jr et al (1987). Spinal and paraspinal neurofibromatosis: surface coil MR imaging at 1.5 T1. Radiology 162(3):797–801.

55. Verstraete KL et al (1992). Nerve sheath tumors: evaluation with CT and MR imaging. J Belge Radiol 75(4):311–320.

56. Antonescu C, Woodruff J (2006). Primary tumors of cranial, spinal and peripheral nerves. In *Russell and Rubinstein's Pathology of Tumors of the Nervous System*, 7th edn. McLendon RE, Rosenblum KM, Bigner DD (eds). Hodder Arnold, London. 787–835.

57. Levine E et al (1987). Malignant nerve-sheath neoplasms in neurofibromatosis: distinction from benign tumors by using imaging techniques. Am J Roentgenol 149(5):1059–1064.

58. Fuchs B et al (2005). Malignant peripheral nerve sheath tumors: an update. J Surg Orthop Adv 14(4):168–174.

59. Geniets C et al (2006). Imaging features of peripheral neurogenic tumors. Jbr-Btr 89(4):216–219.

60. Amoretti N et al (2006). Peripheral neurogenic tumors: is the use of different types of imaging diagnostically useful? Clin Imaging 30(3):201–205.

61. Descamps MJ et al (2006). Primary sciatic nerve lymphoma: a case report and review of the literature. J Neurol Neurosurg Psychiatr 77(9):1087–1089.

62. Templeton PA, Zerhouni EA (1989). MR imaging in the management of thoracic malignancies. Radiol Clin North Am 27(6):1099–1111.

63. Fortier M, Mayo JR, Swensen SJ, Munk PL, Vellet DA, Muller NL (1994). MR imaging of chest wall lesions. Radiographics 14(3):597–606.

64. Gerard JM, Franck N, Moussa Z, Hildebrand J (1989). Acute ischemic brachial plexus neuropathy following radiation therapy. Neurology 39(3):450–451.

65. Salner AL et al (1981). Reversible brachial plexopathy following primary radiation therapy for breast cancer. Cancer Treat Rep 65(9–10):797–802.

66. Maruyama Y, Mylrea MM, Logothetis J (1967). Neuropathy following irradiation. An unusual late complication of radiotherapy. Am J Roentgenol Ther Nucl Med 101(1):216–219

67. Fathers E et al (2002). Radiation-induced brachial plexopathy in women treated for carcinoma of the breast. Clin Rehabil 16(2):160–165.

68. Nich C et al (2005). An uncommon form of delayed radioinduced brachial plexopathy. Chir Main 24(1):48–51.

69. Olsen NK, Pfeiffer P, Johannsen L, Schroder H, Rose C (1993). Radiation induced brachial plexopathy: neurological follow-up in 161 recurrence free breast cancer patients. Int J Radiat Biol Phys 26(1):43–49.

70. Cascino TL et al (1983). CT of the brachial plexus in patients with cancer. Neurology 33(12):1553–1557.

71. Wittenberg KH, Adkins MC (2000). MR imaging of nontraumatic brachial plexopathies: frequency and spectrum of findings. Radiographics 20(4):1023–1032.

72. Wouter van Es H et al (1997). Radiation-induced brachial plexopathy: MR imaging. Skeletal Radiol 26(5):284–288.

73. Dao TH et al (1993). Tumor recurrence versus fibrosis in the irradiated breast: differentiation with dynamic gadolinium-enhanced MR imaging. Radiology 187(3):751–755.

74. Hoeller U et al (2004). Radiation-induced plexopathy and fibrosis. Is magnetic resonance imaging the adequate diagnostic tool? Strahlenther Onkol 180(10):650–654.

75. Bowen BC et al (1996). Radiation-induced brachial plexopathy: MR and clinical findings. Am J Neuroradiol 17(10):1932–1936.

CHAPTER 54

Imaging of Peripheral Neurogenic Tumors

Tudor H. Hughes

INTRODUCTION

Normal peripheral nerves develop from axons migrating from the neural tube and elements from the neural crest. They are supported by connective tissue stroma and Schwann cells [1]. Myelinated fibers have one axon encased by one Schwann cell. Unmyelinated fibers have many axons encased by one Schwann cell. Each individual axon is surrounded by an endoneurium. Groups of axons are bundled together and encased by perineurium. Multiple groups of axons running together produce a peripheral nerve and this is encased by epineurium. This produces a fasicular appearance of peripheral nerves on imaging with magnetic resonance imaging (MRI) or ultrasound [2–4]. Nerve signal conduction is energy dependant and hence there is a profuse blood supply in the immediately adjacent soft tissues.

The presentation and clinical or imaging findings of neurogenic tumors will depend on the nerve involved. Most often these will be painless masses. Sometimes they will become tender to palpation and shooting pains in the nerve distribution will develop. When produced clinically this is Tinel's sign. If nerve transmission is affected, then secondary effects on muscles supplied by the nerve may become apparent with weakness and wasting.

The location of the mass may be a useful indicator as to the type of lesion. If a plantar digital nerve is involved, a Morton neuroma may be suspected, if a mass is seen beneath the fingernail a glomus tumor is a consideration, if the median nerve is enlarged a neural fibrolipoma may be the cause. PNSTs tend to involve large nerve trunks and nerve sheath ganglia are often seen about the knee.

With imaging, the shape of the lesion can lead to a diagnosis of nerve sheath tumor. The 'string sign' refers to vertical soft tissue extending from both ends of the mass and represents the entering and exiting nerve. The 'split fat sign' refers to the displaced fat surrounding the neurovascular bundle and tumor. With MRI, a target pattern that is the reverse of that normally seen in tumors may be apparent. This has low signal centrally and high signal peripherally on fluid sensitive

imaging. Following intravenous gadolinium, the opposite signal pattern may occur which is also the reverse of most tumors and this combination is virtually diagnostic of a nerve sheath tumor. Enhancement is more often non-specific. Should imaging show central necrosis or hemorrhage in a tumor otherwise resembling a PNST, then ancient schwannoma should be considered.

Although biopsy of the mass can be diagnostic, this is often painful with smaller lesions.

BENIGN PERIPHERAL NERVE SHEATH TUMORS

Benign peripheral nerve sheath tumors (PNSTs) are divided into neurilemoma (schwannoma) and neurofibroma. They are similar lesions clinically and on imaging, but differ histologically and in their frequency of association with NF1. Both contain elements related to the Schwann cell. Neurilemoma is composed entirely of Schwann cells, but neurofibromas contain all elements of peripheral nerves, including fibroblasts, Schwann cells and neurites.

Neurilemoma

Neurilemoma is also known as benign schwannoma and neurinoma. These tend to occur in the 20–30-year age group with an equal sex distribution. They represent approximately 5% of all benign soft tissue tumors [5,6]. Common sites are the flexor surfaces of the upper and lower extremities, particularly the ulnar and peroneal nerves. Classically, the tumor lies eccentric to the native nerve, with the normal nerve components (fascicles) maintained. Histologically, they are composed of two types of tissue – Antoni A and B tissue types. Antoni A tissue is central in location and of high cellularity with bundles or fascicles of spindle cells. If the Antoni A tissue predominates, then the tumor is referred to as a cellular schwannoma. It is this part of the tumor that enhances strongly. Antoni B tissue is peripheral, relatively hypocellular and contains myxoid tissue and is of high water content, accounting for the high signal on T2-weighted

(T2W) images. These tissue types are often intermixed with varying amounts of the two tissues. They are immunohistochemical S100 protein positive [5–7].

Neurilemoma is usually solitary and shows slow growth. At presentation, they are usually an asymptomatic mass less than 5 cm in diameter [6,7]. Fusiform on small nerves and eccentric on larger nerves, they grow within the epineurium. Because of the nerve to which they are attached, they are mobile side to side, but fixed longitudinally in the line of the nerve. When they are multiple, approximately 5% of patients will be found to have NF1 [6].

Large neurilemomas may degenerate and, when they do so, are referred to as 'ancient schwannomas'. In this case, they may show cystic change, areas of calcification, hemorrhage or fibrosis [6,7].

Treatment can be complete surgical resection if they can be separated from the underlying nerve or are superficial and the nerve can be spared [8]. If not, partial resection may suffice and recurrence is rare. Malignant transformation is rare.

Neurofibroma

Like neurilemoma, these also occur in the 20–30-year age group with equal sex distribution [9]. They also represent approximately 5% of benign ST tumors [5]. Although similar in many ways to neurilemoma when imaged, they are histologically different with no Antoni cellular histology, but rather have abundant central collagen [5–7]. They are immunohistochemical S100 protein positive, but less so than neurilemoma [5]. Degeneration is less common than with neurilemoma. There are three classic types: localized (90%), diffuse (typically subcutaneous in the neck) and plexiform (pathognomonic of NF1). The localized form is not associated with NF1 in 90% of cases, in which case, are referred to as solitary. When solitary, malignant change is rare. They are usually superficial and, at presentation, are less than 5 cm in diameter and painless [5,8]. They are fusiform on the native nerve and surgically inseparable. On a large nerve, they remain within the epineurium and are encapsulated. On a small nerve, they extend beyond the epineurium, but are still well defined [5,8].

The diffuse form is usually seen in children and young adults involving the head and neck. Ninety percent of these are not associated with NF1. They are poorly defined subcutaneous, plaque-like infiltrates.

The plexiform type of PNST is essentially pathognomonic for NF1. It has been described as having a 'bag of worms appearance'. These tend to follow the course of a nerve and to be a diffuse, multinodular, coalescent sheet-like mass consisting of diffuse soft tissue or nerve enlargement. They may involve the skin, subcutaneous tissue or deeper nerves and cause distortion of bone growth. In addition, they may cause disfigurement and/or functional disability and cause massive enlargement of extremities (elephantiasis neuromatosa) [10].

Imaging of Benign PNSTs

There are three good review articles on the imaging of PNSTs [11–13]. Radiographically, they usually have either a normal x-ray or non-specific soft tissue mass, rarely with surrounding fat outlining the mass (Figure 54.1). Bony erosion may be seen in longstanding cases, and this has a sclerotic margin compatible with the slow growth of the tumor (Figure 54.2). Calcification is uncommon and, when present, may suggest an 'ancient schwannoma' with degeneration (Figure 54.3). If the tumor is associated with NF1, then some of the associated osseous abnormalities of this condition may be seen, such as pseudoarthrosis, or elephantiasis neuromatosa with bony overgrowth.

Angiography may show displaced vessels or corkscrew vessels at either end of the tumor [14,15]. Scintigraphy with methylene diphosphonate (MDP) will only show mild increased uptake at all phases unless the tumor is calcified. Gallium-67 (Ga-67) shows increased uptake in malignant PNSTs, but not those that are benign [16,17].

With sonography (ultrasound), both neurofibroma and neurilemoma are usually solid non-compressible, well defined hypoechoic soft tissue masses (Figure 54.4) that have faint distal acoustic enhancement. Occasionally, they may have a coarse echotexture, or have focal hyperechoic areas thought to be due to collagen rich regions within the tumor. Ultrasound can also have a target-like appearance with a ring of increased echoes within the tumor, and when present is virtually pathognomonic of a neurogenic tumor. As with other imaging, fusiform shape and string sign suggest a neurogenic tumor (Figure 54.5) [2,3,18,19].

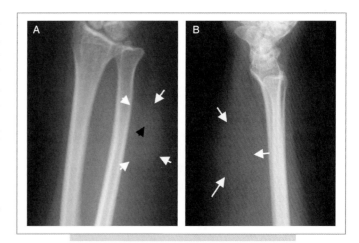

FIG. 54.1. A 40-year-old female with neuroma of the right forearm. The PA **(A)** and lateral **(B)** radiographs of the forearm show a well defined fusiform mass (outlined by white arrows). Note the 'split fat' around the mass and the apparent entering and exiting nerve on the lateral view. A small fleck of calcification could indicate early degeneration (black arrowhead).

Because of the lipid in Schwann cells and relatively high water content, benign PNSTs are low attenuation on unenhanced computed tomography (CT), of the order of 5–25 Hounsfield units (HU). Should this become increased or heterogeneous, then transformation to a malignant PNST should be considered. As with plain films, calcification is uncommon and when present may suggest an 'ancient Schwannoma' with degeneration (see Figure 54.3A).

CT may also show central necrosis, again suggesting an 'ancient Schwannoma'. CT may also show a target-like appearance with increased density centrally due to increased collagen [20–22].

MRI is considered to be the superior imaging modality for demonstrating the classic findings of PNSTs. These include fusiform shape, visualization of the entering and exiting nerve, target signs, split-fat sign, fascicular sign and

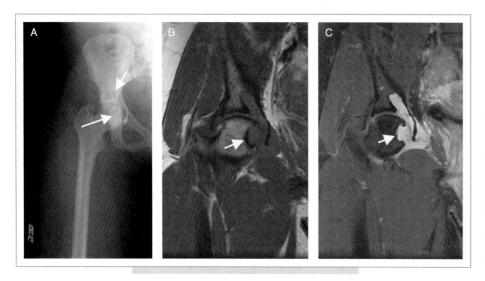

FIG. 54.2. A 35-year old-female with NF1. The frontal radiograph **(A)** of the right hip shows well defined erosions with thin sclerotic margins involving the femoral head and superomedial acetabulum (white arrows). The coronal T1 **(B)** and coronal T1 fat saturation post-intravenous gadolinium **(C)** images demonstrate the intra-articular neurofibroma which strongly enhances and is causing the mechanical bony erosion (white arrows).

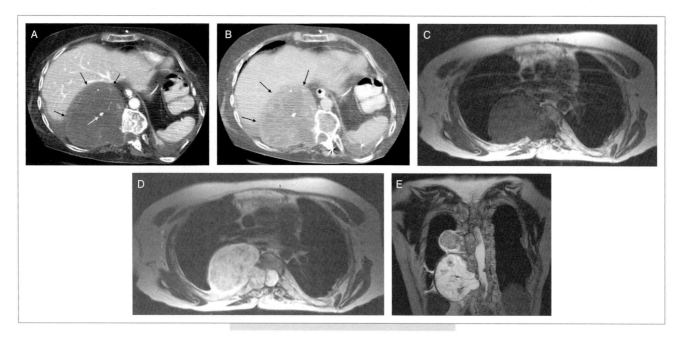

FIG. 54.3. A large right paravertebral ancient schwannoma is seen in this 37-year-old female with neurofibromatosis type 1. **(A)** The arterial phase and **(B)** the equilibrium phase of abdominal CT scan. The tumor is outlined (black arrows). Note the areas of calcification (white arrow) and the early arterial (septal) enhancement with later heterogeneous enhancement. Note also the vertebral erosion. **(C)** Axial T1 images show the homogeneous intermediate signal and **(D)** axial T2 images the heterogeneous pattern often seen in neurogenic tumors due to their fasicular pattern. The coronal T2 image **(E)** shows other additional tumors arising from the spine.

associated muscle atrophy. However, it is not usually possible to separate neurofibroma from neurilemoma by imaging. MR signal is T1 isointense to muscle and T2 variable increased signal [23,24].

The fusiform shape of the tumor and the visualization of the entering and exiting nerve giving a tail to the tumor are considered the most important features that distinguish neurogenic tumors from non-neurogenic tumors when using MRI (Figure 54.6) [23]. Normally, the margins of the tumor are distinct and, if this becomes indistinct, then malignant PNST or diffuse and plexiform NF should be considered. The 'target sign' refers to increased peripheral signal with fluid sensitive imaging from high water content and myxoid tissue [25]. The low signal center is due to the Antoni A tissue high cellularity with bundles or fascicles of spindle cells in the case of neurilemoma (Figure 54.7) and central fibrosis with high collagen content in the case of neurofibroma (Figure 54.8). This is the reverse of most other tumors which, due to central necrosis, have increased fluid signal centrally. The 'fascicular sign' refers to the pattern of the signal within the tumor being similar to a normal peripheral nerve [8], with multiple ring-like structures

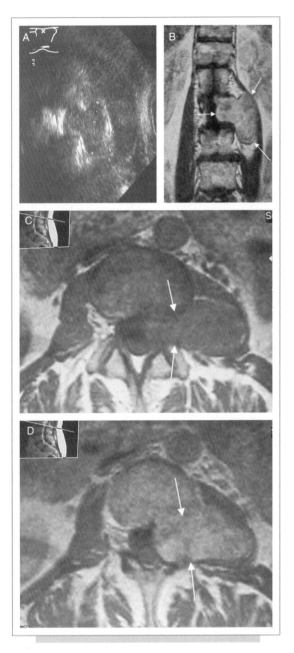

FIG. 54.4. A 55-year-old female having ultrasound for deep vein thrombosis evaluation who was found to have a neurilemoma emerging from the left L2–3 neural foramen. Ultrasound scan of the left flank **(A)**, coronal plane, rotated 90 degrees to match the coronal T1W MRI of the lumbar spine **(B)**. Tumor shown by calipers on the ultrasound and by 3 white arrows on the MRI. Additional axial T1W **(C)** and axial T1 post-intravenous gadolinium **(D)** images show the enlargement of the foramen by the tumor and the mild enhancement (white arrows).

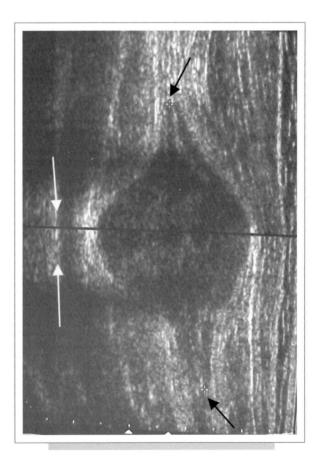

FIG. 54.5. Common peroneal nerve neurofibroma. Longitudinal ultrasound rotated 90 degrees shows a well defined, fusiform, hypoechoic solid mass with posterior acoustic enhancement (white arrows). Note the entering and exiting nerve (black arrows) lies centrally, suggesting neurofibroma rather than neurilemoma.

which are most conspicuous on T2-weighted (T2W) imaging (Figures 54.7 and 54.9). The 'split fat sign' [21] refers to the way the fat of normal neuronal fascial plane wraps around the mass, producing a thin high signal rim with

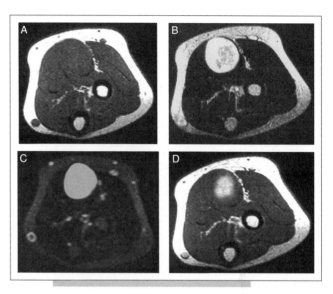

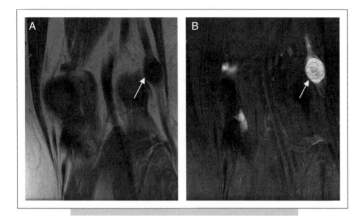

FIG. 54.6. Common peroneal nerve neuroma (white arrows) just above the knee in a 54-year-old female. Coronal T1W **(A)** and coronal protein density with fat saturation **(B)** clearly show the fusiform shape and entering and exiting nerve. In addition, the protein density images show a mild target sign with peripheral high signal.

FIG. 54.8. Neurofibroma of the anterior compartment of the mid forearm. Axial images – T1W **(A)**, T2W **(B)**, STIR **(C)** and T1 fat saturation post-intravenous gadolinium **(D)** – show the pathognomonic signal characteristics of a PNST: isointense to muscle on T1W, high peripheral signal on T2W, very bright with STIR and central enhancement following gadolinium.

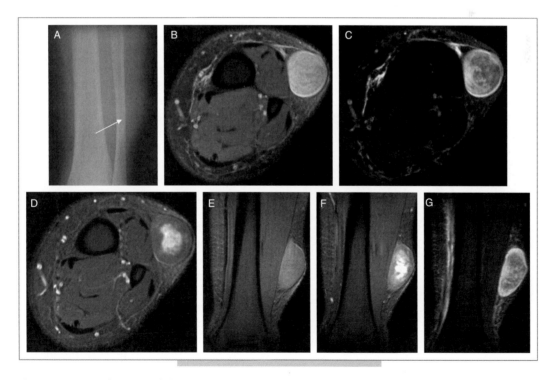

FIG. 54.7. Subcutaneous neurilemoma of the anterolateral aspect of the distal left calf in a 23-year-old man. The frontal radiograph **(A)** shows a focal soft tissue fullness of the lateral calf (white arrow). The three axial MRI images – protein density with fat saturation **(B)**, T2W with fat saturation **(C)** and T1W fat saturation post-intravenous gadolinium **(D)** – and the three coronal images – T1W with fat saturation pre-gadolinium **(E)** and post-gadolinium **(F)** and protein density with fat saturation **(G)** – show some of the characteristic image findings. The fasicular pattern seen best on (B), the peripheral high signal on fluid sensitive images producing the target sign (C and G) and central enhancement seen on (D, E and F combined).

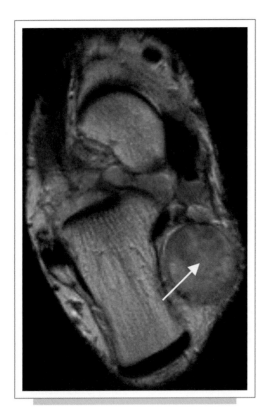

FIG. 54.9. A 26-year-old man with NF1 and a large neurilemoma of the tibial nerve extending into its medial and lateral plantar nerve branches, which shows a partial fasicular pattern on this axial T2W image (white arrow).

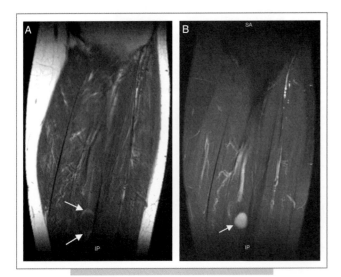

FIG. 54.10. Neurilemoma within the flexor hallucis longus muscle of the calf in a 46-year-old female. Coronal T1W **(A)** and T1 fat saturation post-intravenous gadolinium **(B)** show the split fat sign with triangles of fat at either end of the tumor, high signal on T1W (white arrow) and suppressing on the fat saturation image. Note the central enhancement (white arrow).

triangles at either end on T1W imaging (Figure 54.10). Following intravenous gadolinium (Gd), a useful diagnostic feature is enhancement that follows a reverse target sign of that seen with T2W (center/cellular area enhances more than periphery) (Figures 54.7 and 54.8). This is again the reverse of most non-neurogenic tumors which enhance at the growing periphery rather than the necrotic center and, as such, is almost pathognomonic of benign PNSTs. However, this is not the norm and enhancement is variable for both benign and malignant tumors. Malignant PNSTs usually have more enhancement [8].

Plexiform neurofibromas appear as large lobulated masses following the course of peripheral nerves (Figure 54.11), particularly the sciatic nerve (Figure 54.12). They are usually high signal on T2W (occasionally low due to high collagen content) and low signal on T1W images with variable enhancement.

Neurofibromatosis (NF1)

This is one of the commonest genetic diseases which, in addition to the findings of multiple PNSTs, also shows many other relatively distinctive features due to accompanying mesodermal dysplasia. It is one of the neurocutaneous syndromes or phakomatoses, along with tuberous sclerosis, Von Hippel Lindau, ataxia telangiectasia and Sturge Weber Dimitri syndromes. NF1 was the first phakomatosis to be recognized, described by von Recklinghausen [26] and has become known as von Recklinghausen's disease of bone. There are eight variants numbered NF I–VIII, but 90% of these are NF1 and 9% NF2. NF1 has an incidence of one in 3000 and when hereditary is autosomal dominant, due to a defect at the pericentric location of chromosome 17, but 50% are spontaneous mutations [27,28]. Increased paternal age increases the number of cases of new mutations. The gene at the above location codes for the production of the protein neurofibromen, which appears to be a tumor suppressor. The PNSTs of NF1 may be localized, diffuse or plexiform. NF1 presents with the classic triad of cutaneous lesions, skeletal deformity and mental retardation. The diagnostic criteria for NF1 are shown in Table 54.1. In addition, there may be manifestations of the accompanying mesodermal dysplasia, which can present as a range of musculoskeletal deformities of which scoliosis is the commonest. These are shown in Table 54.2.

All of the three types of neurofibroma mentioned above (localized, diffuse and plexiform) can be associated with NF1 [29]. However, most of the localized and diffuse variants are not associated with NF1, but the presence of a plexiform neurofibroma is virtually pathognomonic of NF1. Neurofibromas in patients with NF1 tend to involve the skin and subcutaneous tissues, but also the peripheral nerve trunks. Neurofibromas in patients with NF1 tend to be multiple and more often involve deep large nerves than do solitary neurofibromas.

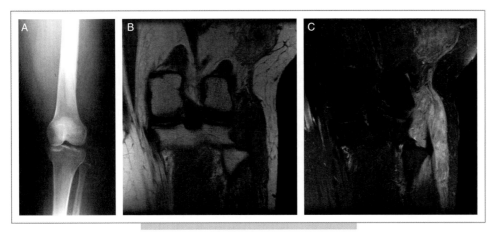

FIG. 54.11. A 35-year-old female with NF1, demonstrates a classic plexiform neurofibroma along the course of the common peroneal nerve. The frontal radiograph **(A)** shows an ill defined soft tissue mass laterally and erosion of the proximal fibula. The coronal MRIs – T1W **(B)** and T1 fat saturation post-intravenous gadolinium **(C)** – show the large heterogeneous serpentine mass resembling a 'bag of worms' with strong enhancement and additional bony erosion of the adjacent tibia.

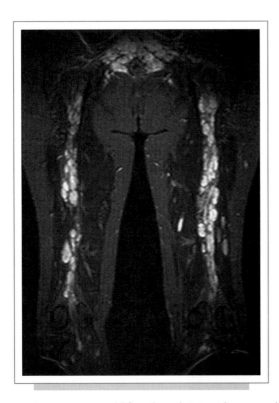

FIG. 54.12. A 29-year-old female with NF1. The coronal T2W fat saturation image shows extensive plexiform neurofibromas following down the course of the sciatic nerve bilaterally.

TABLE 54-1 Diagnostic criteria for NF1
Diagnostic criteria for NF1 are 2 or more of the following:
• 6 or more café-au-lait macules
• >5 mm prepubertal
• >15 mm post-pubertal
• Two or more neurofibromas or one plexiform neurofibroma
• Axillary or inguinal freckling
• Optic glioma
• Two or more Lisch nodules (hamartoma of the iris)
• Distinctive bone lesion
• Sphenoid dysplasia
• Pseudoarthrosis
• 1st degree relative with NF1.

TABLE 54-2 Manifestations of the mesodermal dysplasia of neurofibromatosis type 1
• Scoliosis (short or long segment)
• Kyphosis (often predominates)
• Facial or orbital dysplasia
• Lambdoid suture defects (left sided)
• Pseudoarthrosis (tibia + congenital)
• Periosteal abnormalities (reaction or cyst)
• Multiple NOF or fibroxanthomas
• Rib deformity (ribbon ribs)
• Posterior vertebral scalloping (dural ectasia)
• Elephantiasis neuromatosa
• Mesodermal dysplasia pseudofracture.

Neural Fibrolipoma

This has also been referred to as fibrolipomatous hamartoma, perineural lipoma or intraneural lipoma, but neural fibrolipoma is the preferred name. As the name suggests, the pathology is that of fatty (fibrofatty) infiltration of the nerve and this usually presents in a child or young adult, with equal sex distribution, as a slowly enlarging soft mass.

Approximately 90% involve the upper extremity and 85% the median nerve. Hence the mass is often felt in the palmar region or may present as carpal tunnel syndrome [30–32].

In 30–70% of cases, the neural fibrolipoma is associated with macrodystrophia lipomatosa (Figure 54.13). When involving the distal nerves, this presents as bony and soft tissue overgrowth in the associated sclerotomes (bones supplied by specific nerve root) of the affected nerve as macrodactyly, and hence is often in the second + third digits of the hand (Figure 54.14). This diffuse increase in

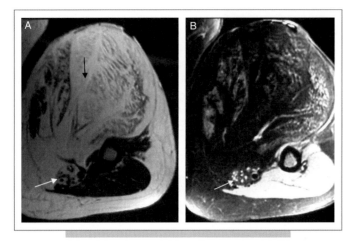

FIG. 54.13. A 47-year-old man with a long history of left arm overgrowth that spares the fourth and fifth digits. Axial T1W **(A)** and protein density with fat saturation **(B)** images of the upper arm demonstrate fatty infiltration of the median nerve (white arrows) and fatty overgrowth with muscle atrophy of the anterior compartment of the arm (black arrows) in this patient with median nerve neural fibrolipoma and accompanying macrodystrophia lipomatosa of the biceps and brachialis muscles. Although the median nerve does not supply the anterior compartment of the arm musculature, it has the same nerve root distribution as the musculcutaneous nerve of the arm which does supply these muscles. Courtesy of Dexter Witte MD, Baptist Memorial Hospital, Memphis.

fibroadipose and osseous overgrowth ceases at puberty, but can lead to secondary osteoarthrosis (Figure 54.15) [33].

Radiographs of the affected area may show a soft tissue mass due to the neural fibrolipoma and/or the macrodystrophia lipomatosa, or the bony overgrowth and secondary osteoarthrosis. Ultrasound shows a cable-like appearance to the enlarged nerve, with alternating hyper- and hypoechoic strands. MRI is nearly always diagnostic and shows the enlarged nerve with low signal fascicles and surrounding fat.

Traumatic Neuroma

Traumatic neuroma is not a neoplasia, but rather an overgrowth of normal nerve components trying to regenerate the injured nerve, particularly the axons [8,34,35]. This represents a normal pattern of healing and is often asymptomatic. These can be divided into two major groups: the spindle type, which are due to chronic friction to the nerve at a point of irritation or microtrauma, due to stretching or compressing of the nerve by localized scar tissue. These are internal focal fusiform swellings (Figure 54.16). Secondly, the lateral or terminal type, which are secondary to partial or total transection of a nerve and produce a bulbous swelling at the site of injury if partial, or on the proximal nerve stump if there has been complete transection [34–36].

They present 1–12 months post-injury with pain and shooting sensations in the nerve distribution if bumped or palpated (Tinel's sign). Pain that is associated with a neuroma does not always have a precise topography and is often difficult to distinguish from phantom limb pain. They are most common in the lower extremity post limb amputation or surgery and are occasionally seen in the brachial plexus and radial nerve [37]. Histologically, they are tangled

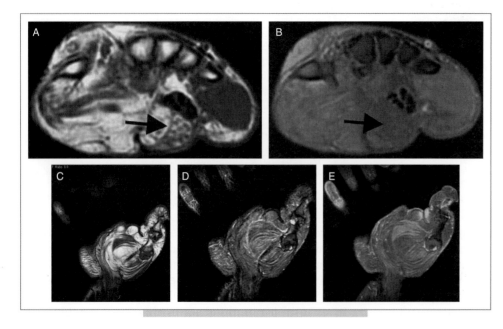

FIG. 54.14. A 39-year-old man with neural fibrolipoma of the median nerve and accompanying macrodystrophia lipomatosa of the distal median nerve distribution. Axial T1W **(A)**, axial T1W with fat saturation post-intravenous gadolinium **(B)**, coronal T1W **(C)**, coronal T2 with fat saturation **(D)** and coronal T1W with fat saturation post-intravenous gadolinium **(E)**. This is the classic location and appearance of these conditions. Note the fatty infiltration of the median nerve which loses signal with fat saturation and does not enhance (black arrows) and fatty overgrowth of the thumb seen best on the coronal images.

masses of axons and other peripheral nerve components, sometimes with partial formation of a fasicular pattern. They have no malignant potential.

Ultrasound imaging shows a non-specific hypoechoic mass, unless the history is noted and the native nerve is observed. With CT scanning, the attenuation is that of muscle and hence non-specific [38]. With MRI, there is intermediate signal on T1W images and intermediate to high signal on T2W imaging, occasionally with a heterogeneous ring (fascicular) pattern (see Figure 54.16). When surrounded by fat, as is often the case due to accompanying muscle atrophy (Figure 54.17), the lesions are well defined. Often, if a larger nerve is involved, the association of the tumor to the severed nerve can be made (Figure 54.17), but with smaller nerves this will be difficult. There is variable enhancement following intravenous gadolinium (Figure 54.18) [39–41].

Nerve Sheath Ganglion

A ganglion is a myxoid collection of inspissated fluid surrounded by a fibrous capsule of compressed tissue. These are often multiloculated. They are thought to arise secondarily to a very minor injury to fibrous connective tissue which then produces a loculated fluid collection.

Nerve sheath ganglia are usually seen in the nerves about the knee (peroneal, tibial). Their origin is debated as to whether they originate within the nerve or arise in the proximal tibiofibular joint and extend to involve the nerve. They may present as a mass if the ganglion is of a large size, but commonly present with foot drop (peroneal nerve) even when of a small size. This is because of the proximity to the osseoligamentous peroneal tunnel between the two heads of peroneus longus which arise from the proximal fibula,

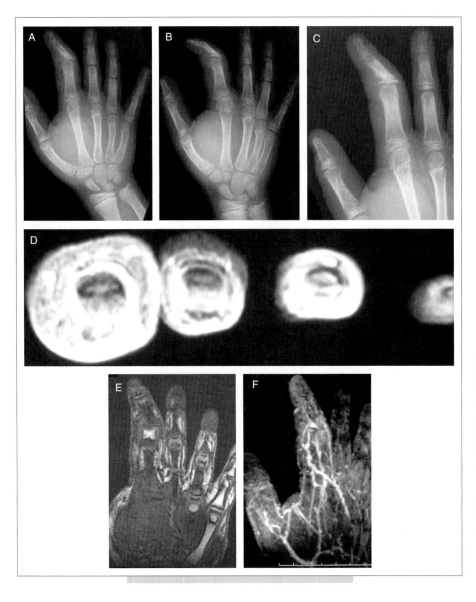

FIG. 54.15. A 14-year-old boy with a median nerve neural fibrolipoma and accompanying macrodystrophia lipomatosa involving the index finger. AP **(A)**, oblique **(B)** and close up oblique **(C)** radiographs of the right hand show the bony and soft tissue overgrowth of the index finger and secondary osteoarthrosis of the proximal interphalangeal joint. The axial **(D)** and coronal **(E)** T1W images show the significant soft tissue and bony overgrowth and the coronal T1W with fat saturation post-intravenous gadolinium magnetic resonance angiogram shows the accompanying increase in blood **(F)** flow to that ray.

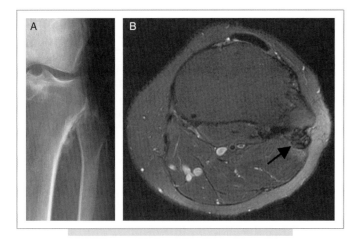

FIG. 54.16. Spindle type traumatic neuroma of the common peroneal nerve. **(A)** AP radiograph of the lateral aspect of the left knee shows post-traumatic dystrophic ossification in the region of the expected course of the common peroneal nerve. **(B)** Axial proton density image at the level of the proximal fibula shows fusiform enlargement of the common peroneal nerve with maintenance of the fasicular pattern (black arrow) secondary to the chronic irritation of the nerve by the post-traumatic changes to the adjacent tissues.

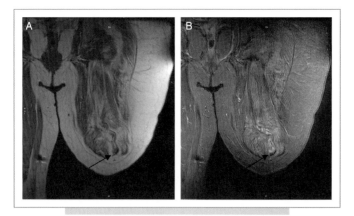

FIG. 54.17. Terminal type traumatic neuroma of the sciatic nerve stump in a 60-year-old female who is 4 years post amputation and has had 1 month of pain. Coronal T1W **(A)** and coronal T2W with fat saturation **(B)** image of the left thigh show the tortuous sciatic nerve surrounded by fat ending in a bulbous tip with increased fluid signal (black arrows).

forming a tunnel through which the nerve passes. Here the nerve becomes compressed by the ganglion [42,43].

Imaging of nerve sheath ganglia will reveal a well defined cystic mass. Hence, on ultrasound, there will be an anechoic structure, possibly with some septation, with increased through transmission [44]. With CT, the density will be slightly above water due to inspissation and proteinaceous

material [45]. With MRI, the mass will have low signal on T1, occasionally with slight increase over that of water due to the proteinaceous material, high signal (light bulb lesion) with T2W imaging and show a thin rim of enhancement following intravenous Gd due to the compressed tissue at the periphery [46]. The other major finding with MRI will be the secondary effects upon the muscles supplied by that nerve (Figure 54.19). For the fibular tunnel this will be both the peroneal and anterior compartments of the leg. Initially, there is seen to be muscle edema and subsequently fatty atrophy.

Morton's Neuroma

Morton's neuroma is actually a perineural fibrosis induced by friction on the plantar nerve at the level of the metatarsal heads [5]. It is far more common in women (up to 18:1) which has led to the speculation that it is induced by the wearing of high heels [47,48]. These cause dorsiflexion at the metatarsophalangeal joints and stretch the plantar nerve at its point of bifurcation into the digital nerves as it passes over the intermetatarsal ligament. Other provocative factors include arthritis, ganglion cysts, running, ballet or an enlarged intermetatarsal bursa [49] as may be seen in rheumatoid arthritis [50].

Another theory is that it is induced by ischemia [51]. There is often an associated inflammatory response with fluid seen in the adjacent intermetatarsal bursa and surrounding soft tissue hyperemia. Morton neuromas are most commonly seen at the third interspace between the third and fourth metatarsal heads. This is thought to be the commonest site because the medial and lateral plantar nerves combine in this interspace to produce a slightly larger interdigital nerve, or that there is more tethering of the nerve due to the two nerves combining. The second interspace is the next most common. They are uncommon at the first interspace, where they go by the name of Joplin neuroma, and rare in the fourth interspace.

Histologically, there is thickening of the epineural fascicles, perineural fibrosis with high collagen content (Renaut bodies) and loss of the myelinated fibers.

Clinically, there may be localized pain or pain radiating to the toes [52]. This is brought on by exercise, or by palpation. A unilateral distribution predominates. The actual neuroma is not usually palpable, but an accompanying inflamed bursa may be. The size of the lesion relates to symptoms with many smaller lesions seen at imaging being found to be asymptomatic.

Radiographs are invariably normal but, sometimes, increased intermetatarsal distance can suggest a large neuroma (Figure 54.20). They are principally used to exclude other pathology. Ultrasound shows a round or ovoid hypoechoic mass with internal echoes in the intermetatarsal space. Being a dynamic interactive investigation, the mass can easily be correlated with the patient's symptoms by pressure on the probe. Any accompanying bursitis is seen as an anechoic adjacent lesion. Color or power Doppler is a

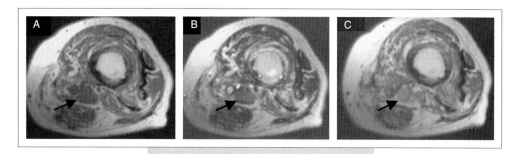

FIG. 54.18. Terminal type traumatic neuroma of the sciatic nerve stump in a 71-year-old man whose status is post above knee amputation of the left leg. Axial T1W **(A)**, axial T2W **(B)** and axial T1W post-intravenous gadolinium **(C)** images show a well defined mass (black arrows) of low signal on T1W, intermediate signal on T2W and with moderate enhancement post-gadolinium.

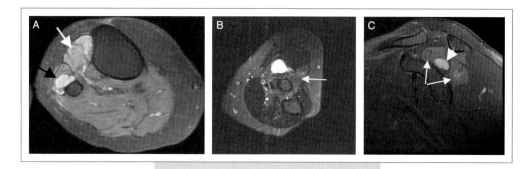

FIG. 54.19. Three different patients with ganglia at osseoligamentous tunnels causing nerve entrapment syndromes. Common peroneal nerve ganglion **(A)**: axial T2W with fat saturation shows the ganglion (black arrow) and the accompanying muscle edema in the peroneal and anterior compartments of the leg (white arrow). Posterior interosseous nerve of the forearm entrapment by a ganglion arising in the annular periradial recess and compressing the nerve beneath the arcade of Frohse at the proximal supinator muscle **(B)**: axial T2W with fat saturation. The edema is seen in the posterior compartment of the forearm and in supinator muscle (white arrow). Suprascapular nerve ganglion compressing the nerve in the suprascapular notch in a 26-year-old man **(C)** (white arrowhead): sagittal proton density with fat saturation and causing edema of both the supra and infraspinatus muscles (white arrows).

useful adjunct to assess for the associated inflammation of the bursitis [53,54].

MR signal characteristics (see Figure 54.20) are of low signal on T1W images, best seen on coronal imaging. With T2W imaging, the signal is variable, but generally remains relatively low or equal to fat. Hence conspicuity on T2W images is less good. Because of this, the tumor is best seen with fat saturation and this also brings out any accompanying intermetatarsal bursitis which will be high signal intensity with T2W imaging. Enhancement with intravenous gadolinium will often demonstrate peripheral enhancement and will increase sensitivity over T2W imaging [55]. It is also helpful in separating fibrosis (enhances solidly) from bursa (enhances peripherally). Intermetatarsal bursitis often coexists with, or causes, a Morton neuroma. The presence of small asymptomatic neuromas can be normal when 3 mm or less in the first to third intermetatarsal spaces. Zanetti and colleagues [56] have shown that MRI is 90% accurate, positive predictive value (PPV) 100%, negative

predictive value (NPV) 60% and introduced the following diagnostic criteria:

1 centered in NV bundle at the intermetatarsal space on the plantar side of intermetatarsal ligament
2 well demarcated
3 signal intensity equal to muscle on T1W.

Erickson and co-workers [57] showed intermetatarsal bursal fluid in 67% of patients with neuroma. With prone imaging, the neuroma is plantar to virtual plantar cortical line. With supine and weight bearing imaging, the neuroma migrates dorsally. The shape and transverse measurement change. Neuromas are most conspicuous on prone images. This is important since neuromas larger than 5 mm have a much better surgical outcome [58,59].

Glomus Tumor

This benign, usually solitary tumor, most often presents in adults with equal sex distribution, unless in the

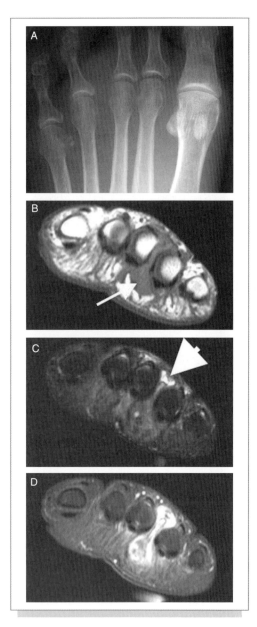

FIG. 54.20. A 51-year-old female with a third interspace Morton neuroma. **(A)** AP radiograph of the left foot, **(B)** coronal T1W, **(C)** coronal T2W with fat saturation and **(D)** coronal T1W with fat saturation post-intravenous gadolinium. The radiograph suggests widening of the third interspace. The MRI images show a well defined mass of low T1 signal (white arrow) which stays of relatively low signal on T2W images apart from accompanying bursal fluid seen dorsally (white arrowhead). Following gadolinium, there is intense enhancement of the neuroma and periphery of the bursa.

commonest location beneath the nail (subungual glomus tumor) when it is more common in women. Seventy-five percent occur in the hand, with the second most common site being the sole of the foot. It arises from the

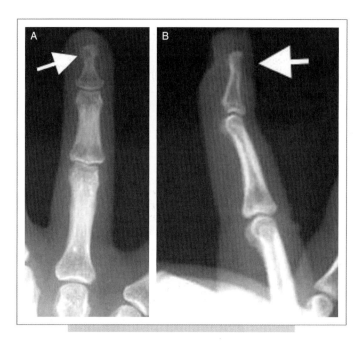

FIG. 54.21. Subungal glomus tumor of the finger. **(A)** AP and **(B)** lateral radiographs of the ring finger show a subungual lucency (arrows) due to bony erosion by the tumor.

neuromyoarterial glomus. Symptoms are out of proportion to the small size of the lesion, which is usually less than 1 cm in size, and include localized tenderness, intense intermittent pain and cold sensitivity.

Histologically, there are three types: vascular, myxoid and solid, ranging from more vascular to more glomus cells.

Imaging with plain films (Figure 54.21) and CT may show well defined bony erosion with a thin sclerotic margin of the adjacent bone if the lesion is subungual. Ultrasound shows a hypoechoic mass and MRI a lesion of low signal on T1W and high signal on T2W [60,61].

MALIGNANT PNST

Malignant PNST is now the accepted name for tumors that previously may have been called neurosarcoma or malignant schwannoma [62]. These account for 5–10% of all soft tissue sarcomas [62,63]. They tend to occur in the 20–50 year age range. There is a nearly even sex distribution overall, but when the tumor occurs in a patient with NF1, there is a strong male predominance of up to 80% [8,62]. They are associated with NF1 in 25–70% of cases and, when they do so, occur a decade earlier than otherwise. The potential for malignant degeneration of a benign PNST in a patient with NF1 is approximately 5% of patients. Malignant PNSTs may also be secondary to radiotherapy, in which case there is usually a 10–20 year lag period from radiotherapy to presentation. This accounts for 11% of malignant PNSTs [62,64].

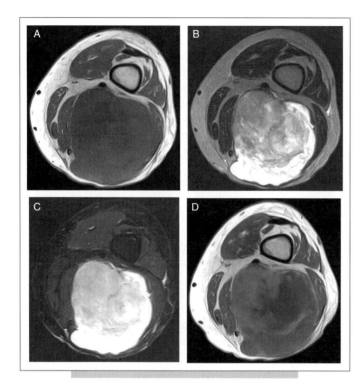

FIG. 54.22. Malignant PNST of the sciatic nerve just above the knee occurring in a 47-year-old man with known history of NF1. The mass has been present for 10 months and is painful and enlarging. **(A)** Axial T1W, **(B)** axial T2W, **(C)** axial STIR and **(D)** axial T1W post-intravenous gadolinium. Note the large size of the lesion, its location along the line of the sciatic nerve, its heterogeneous signal and its heterogeneous enhancement. Courtesy of Robert Downey Boutin, MD, Medical Director, MedTel International, Davis.

Malignant PNSTs commonly involve large nerves such as the sciatic nerve and sacral and brachial plexus [65]. They present with a painful, rapidly enlarging mass, usually larger than 5 cm, often with neurologic symptoms and weakness. The tumor spreads along the nerve and hence may still have a fusiform shape.

Imaging is useful in helping to differentiate benign from malignant PNSTs. Benign PNSTs on MRI will show a well circumscribed homogeneous low T1 signal and variable high T2 signal and may show a target sign of a central zone of low signal with surrounding increased signal on T2W images [66]. Malignant PNSTs (Figure 54.22) show a heterogeneous mass [67] with areas of hyperintensity due to hemorrhage and necrosis and with infiltrative margins. Calcification is seen in 10–15% of tumors. The target sign on T2W imaging and post-Gd is less often seen. When a Ga-67 citrate nuclear medicine study of PNST is positive, it is usually sarcomatous [16,17].

These are usually high-grade sarcomas. As with other sarcoma metastases, metastases from malignant PNSTs

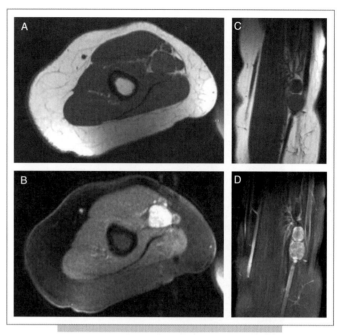

FIG. 54.23. Metastatic melanoma to the median nerve of the arm in a 74-year-old female. **(A)** Axial T1W, **(B)** axial STIR, **(C)** coronal T1W and **(D)** coronal T1W with fat saturation post-intravenous gadolinium. Note the location of the lesion in relation to the exiting nerve trunk, the high signal of the lesion on STIR images and heterogeneous peripheral enhancement.

go principally to the lung and bone. Local nodes are only seen in 9% of patients. Surgical excision and adjuvant chemotherapy and/or radiotherapy are the treatment of choice, but still 40% recurs and 60% metastasizes. The 5-year survival is approximately 44%, but is worse in the older patient with larger and more central lesions. Having underlying NF1 does not affect prognosis.

METASTASES TO NERVES

Hematogenous metastases to nerves are rare and, when seen, involve the larger nerve trunks, such as the sciatic nerve (Figure 54.23). Direct invasion of peripheral nerves tends to occur to the brachial plexus, particularly with apical lung tumors (Pancoast tumors) and to the lumbosacral plexus with colorectal and gynecological malignancy [68–70].

SECONDARY SIGNS

In addition to the direct visualization of a neurogenic tumor, there may also be times when the secondary effects of nerve compression will be visualized by changes in the muscles supplied by that nerve [71]. This most often occurs when the tumor causes nerve entrapment at a fibroosseous tunnel. These tunnels are most commonly adjacent

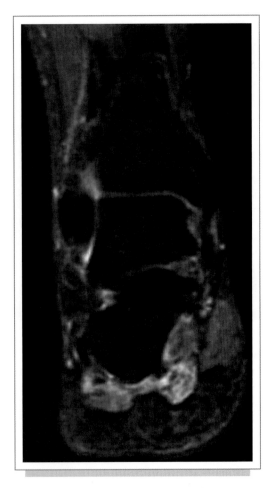

FIG. 54.24. Coronal proton density with fat saturation image of the hind foot. Note the high signal in all of the muscles of the plantar aspect compatible with either tibial nerve or combined medial and lateral plantar nerve compression, likely in the tibial tunnel.

to joints, when a nerve passes between a fixed structure such as the bone and a movable object such as the attachment of a tendon or ligament. Examples of such sites in the upper extremity include the suprascapular nerve in the suprascapular notch or spinoglenoid notch, the posterior interosseous nerve as it passes deep to the arcade of Frohse at the proximal aspect of the supinator muscle, the anterior interosseous nerve as it passes between the aponeurotic two heads of flexor digitorum superficialis and the median nerve in the carpal tunnel.

Within the lower extremity such sites include the common peroneal nerve in the fibular tunnel between the two heads of peroneus longus and the tibial nerve or its branches, the medial and lateral plantar nerves within the tibial tunnel on the medial aspect of the ankle. Examples of the patterns of muscle edema and atrophy are shown in Figures 54.19 and 54.24.

Muscle edema occurs 2–4 weeks after an acute injury and more gradually with nerve compression. This is because of water shift from intracellular to extracellular compartments, probably due to the loss of the energy dependant pump. If the muscle is re-innervated at an early stage, eventually it will return to normal. Once fatty atrophy develops after several months some muscle damage will be irreversible.

SUMMARY

The combination of location, shape and signal characteristics can usually lead to a diagnosis of nerve sheath tumor or, less specifically, neurogenic tumor. The separation of neurilemoma from neurofibroma is problematic. High resolution ultrasound may now be able to separate these in some cases when the nerve can be seen in relation to the tumor. The larger and more heterogeneous a lesion is in the presence of pain suggests malignant nerve sheath tumor.

REFERENCES

1. Damjanov I, Linder J (1996). Anderson's pathology. In *Peripheral Nervous System*, 10th edn. Kissane JM (ed.). Mosby, St Louis. 2799–2803.
2. Fornage BD (1988). Peripheral nerves of the extremities: imaging with US. Radiology 167:179–182.
3. Graif M, Seton A, Nerubai J, Horoszowski H (1991). Sciatic nerve: sonographic evaluation and anatomic-pathologic considerations. Radiology 181:405–408.
4. Ikeda K, Haughton VM, Ho KC, Nowicki BH (1996). Correlative MR-anatomic study of the median nerve. Am J Roentgenol 167:1233–1236.
5. Kransdorf MJ (1995). Benign soft-tissue tumors in a large referral population: distribution of specific diagnoses by age, sex, and location. Am J Roentgenol 164:395–402.
6. Harkin JC, Reed RJ (1969). Tumors of peripheral nervous system. In *Atlas of Tumor Pathology*, 2nd series, fascicle 3. Armed Forces Institute of Pathology, Washington, DC. 29–120.
7. Stout AP (1935). The peripheral manifestations of the specific nerve sheath tumor (neurilemmoma). Am J Cancer 24:751–796.
8. Kransdorf M, Murphey MD (1997). Neurogenic tumors. In *Imaging of Soft Tissue Tumors*. Saunders, Philadelphia. 235–273.
9. Resnick D (ed.) (2002). Tumors and tumor like lesions of soft tissues. In *Diagnosis of Bone and Joint Disorders*, 4th edn. Saunders, Philadelphia. 4219–4222.
10. Harris WC, Alpert WJ, Marcinko DE (1982). Elephantiasis neuromatosa in von Recklinghausen's disease. J Am Podiatr Assoc 72:70–72.
11. Lin J, Martel W (2001). Cross sectional imaging of peripheral nerve sheath tumors. Characteristic signs on CT, MR imaging and sonography. Am J Roentgenol 176:75–82.
12. Murphey M, Sean Smith W, Smith SE, Kransdorf MJ, Temple HT (1999). Imaging of musculoskeletal neurogenic

tumors: radiologic-pathologic correlation. Radiographics 19:1253–1280.

13. Pilvaki M, Chourmouzi D, Kiziridou A, Skordalaki A, Zarampoukas T, Drevelengas A (2004). Imaging of peripheral nerve sheath tumors with pathologic correlation. Pictorial review. Eur J Radiol 52:229–239.

14. Berlin O, Stener B, Lindahl S, Irstam L, Lodding P (1986). Vascularization of peripheral neurilemomas: angiographic, computed tomographic, and histologic studies. Skeletal Radiol 15:275–283.

15. Stener B, Angervall L, Nilsson L, Wickbom I (1969). Angiographic and histologic studies of the vascularization of peripheral nerve tumors. Clin Orthop Rel Res 666:113–124.

16. Levine E, Huntrakoon M, Wetzel LH (1987). Malignant nerve sheath neoplasms in neurofibromatosis: distinction from benign tumors by using imaging techniques. Am J Roentgenol 149:1059–1064.

17. Hammond JA, Driedger AA (1987). Detection of malignant change in neurofibromatosis (von Recklinghausen's) by gallium-67 scanning. Can Med Assoc J 119:252–253.

18. Beggs I (1999). Pictorial review: imaging of peripheral nerve tumours. Clin Radiol 52:8.

19. Beggs I (1999). Sonographic appearance of nerve tumors. J Clinical Ultrasound 27:363–368.

20. Kumar AJ, Kuhajda FP, Martinez CR, Fishman EK, Jezic DV, Siegelman SS (1983). Computed tomography of extracranial nerve sheath tumors with pathologic correlation. J Comput Assist Tomogr 7:857–865.

21. Cohen LM, Schwartz AM, Rockoff SD (1983). Benign schwannomas: pathologic basis for CT inhomogeneities. Am J Roentgenol 147:141–143.

22. Thiebot J, Laissy JP, Delangre T, Biga N, Liotard A (1991). Benign solitary neurinomas of the sciatic and popliteal nerves: CT study. Neuroradiology 33:186–188.

23. Cerofolini E, Landi A, DeSantis G, Mairorana A, Canossi G, Romagnoli R (1991). MR of benign peripheral nerve sheath tumors. J Comput Assist Tomogr 15:593–597.

24. Soderlund V, Goranson H, Bauer HC (1994). MR imaging of benign peripheral nerve sheath tumors. Acta Radiol 35:282–286.

25. Suh JS, Abenoza P, Galloway HR, Everson LI, Griffiths HJ (1992). Peripheral (extracranial) nerve tumors: correlation of MR imaging and histologic findings. Radiology 183:341–346.

26. von Recklinghausen F (1882). Ueber die multiplen fibrome der haut und ihre beziehung zu den multiplen neuromen. August Hirchwald, Berlin.

27. National Institutes of Health (1988). Consensus development. Neurofibromatosis. Arch Neurol 45:575–578.

28. National Institutes of Health Conference (1990). Neurofibromatosis 1 (Recklinghausen disease) and neurofibromatosis 2 (bilateral acoustic neurofibromatosis): an update. Ann Intern Med 113:39–52.

29. Hillier JC, Moskovic E (2005). The soft tissue manifestations of neurofibromatosis type 1. Clin Radiol 60:960–967.

30. Silverman TA, Enzinger FM (1985). Fibrolipomatous hamartoma of nerve: a clinicopathologic analysis of 26 cases. Am J Surg Pathol 9:7–14.

31. Cavallaro MC, Taylor JAM, Gorman JD, Haghighi P, Resnick D (1993). Imaging findings in a patient with fibrolipomatous hamartoma of the median nerve. Am J Roentgenol 161:837–838.

32. Evans HA, Donnelly LF, Johnson ND, Blebea JS, Stern PJ (1997). Fibrolipoma of the median nerve: MRI. Clin Radiol 52:304–307.

33. Ban M, Kamiya H, Sato M, Kitajima Y (1998). Lipofibromatous hamartoma of the median nerve associated with macrodactyly and port-wine stain. Pediatr Dermatol 15:378–380.

34. Enzinger FM, Weiss SW (1995). Benign tumors of peripheral nerves. In *Soft Tissue Tumors*, 3rd edn. Mosby, St Louis. 821–888.

35. Spencer PS (1974). The traumatic neuroma and proximal stump. Bull Hosp Joint Dis 35:85–102.

36. Huber GC, Lewis D (1920). Amputation neuromas: their development and prevention. Arch Surg 1:85–113.

37. Gupta RK, Mehta VS, Banerji AK, Jain RK (1989). MR evaluation of brachial plexus injuries. Neuroradiology 31:377–381.

38. Singson RD, Feldman F, Slipman CW, Gonzalez E, Rosenberg ZS, Kiernan H (1987). Postamputation neuromas and other symptomatic stump abnormalities: detection with CT. Radiology 162:743–745.

39. Boutin RD, Pathria MN, Resnick D (1998). Disorders in the stumps of amputee patients: MR imaging. Am J Roentgenol 171:497–501.

40. Singson RD, Feldman F, Staron R, Fechtner D, Gonzalez E, Stein J (1990). MRI of postamputation neuromas. Skeletal Radiol 19:259–262.

41. Henrot P, Stines J, Walter F, Martinet N, Paysant J, Blum A (2000). Imaging of the painful lower limb stump. Radiographics 20:S219–S235.

42. Yamazaki H, Saitoh S, Seki H, Murakami N, Misawa T, Takaoka K (1999). Peroneal nerve palsy caused by intraneural ganglion. Skeletal Radiol 28:52–56.

43. Nucci F, Artico M, Santoro A et al (1990). Intraneural synovial cyst of the peroneal nerve: report of two cases and review of the literature. Neurosurgery 26:339–344.

44. Leijten FS, Arts W-F, Puylaert JBCM (1992). Ultrasound diagnosis of an intraneural ganglion cyst of the peroneal nerve. J Neurosurg 76:538–540.

45. Gambari PI, Giuliani G, Poppi M, Pozzati E (1990). Ganglionic cysts of the peroneal nerve at the knee: CT and surgical correlation. J Comput Assist Tomogr 14:801–803.

46. Leon J, Marano G (1987). MRI of peroneal nerve entrapment due to a ganglion cyst. Magn Reson Imaging 5:307–309.

47. Alexander IJ, Johnson KA, Parr JW (1987). Morton's neuroma: a review of recent concepts. Orthopedics 10:103–106.

48. Mann RA, Reynolds JC (1983). Interdigital neuroma: a critical analysis. Foot Ankle 3:238–243.

49. Theumann NH, Pfirrmann CWA, Chung CB et al (2001). Intermetatarsal spaces: analysis with MR bursography, anatomic correlation, and histopathology in cadavers. Radiology 221(2): 478–484.

50. Zanetti M, Weishaupt D (2005). MR imaging of the forefoot: Morton neuroma and differential diagnoses. Semin Musculoskelet Radiol 9(3):175–186.

51. Zanetti M, Strehle JK, Zollinger H, Hodler J (1997). Morton neuroma and fluid in the intermetatarsal bursae on MR images of 70 asymptomatic volunteers. Radiology 203:516–520.

52. Wu KK (2000). Morton neuroma and metatarsalgia. Curr Opin Rheumatol 12(2):131–142.

53. Redd RA, Peters VJ, Emery SF, Branch HM, Rifkin MD (1989). Morton neuroma: sonographic evaluation. Radiology 171: 415–417.

54. Kaminsky S, Griffin L, Milsap J, Page D (1997). Is ultrasonography a reliable way to confirm the diagnosis of Morton's neuroma. Orthopedics 20:37–39.

55. Terk MR, Kwong PK, Suthar M, Horvath BC, Colletti PM (1993). Morton neuroma: evaluation with MR imaging performed with contrast enhancement and fat suppression. Radiology 189(1):239–241.

56. Zanetti M, Ledermann T, Zollinger H, Hodler J (1997). Efficacy of MR imaging in patients suspected of having Morton's neuroma. Am J Roentgenol 168:529–532.

57. Erickson SJ, Canale PB, Carrera GF et al (1991). Interdigital (Morton) neuroma: high-resolution MR imaging with solenoid coil. Radiology 181:833–836.

58. Weishaupt D, Treiber K, Kundert H-P et al (2003). Morton neuroma: MR imaging in prone, supine, and upright eight-bearing body positions. Radiology 226(3):849–856.

59. Zanetti M, Strehle JK, Kundert HP, Zollinger H, Hodler J (1999). Morton neuroma: effect of MR imaging findings on diagnostic thinking and therapeutic decisions. Radiology 213(2):583–588.

60. Drape JL, Idy-Peretti I, Goettmann S et al (1995). Subungual glomus tumors: evaluation with MR imaging. Radiology 195(2):507–515.

61. Theumann NH, Goettmann S, Le Viet D et al (2002). Recurrent glomus tumors of fingertips: MR imaging evaluation. Radiology 223:143–151.

62. Enzinger FM, Weiss SW (1995). Malignant tumors of the peripheral nerves. In *Soft Tissue Tumors*, 3rd edn. Mosby, St Louis. 889–928.

63. Kransdorf MJ (1995). Malignant soft-tissue tumors in large referral population: distribution of diagnoses by age, sex, location. Am J Roentgenol 164:129–134.

64. Wanebo JE, Malik JM, Vandenberg SR, Wanebo HJ, Driesen N, Presing JA (1993). Malignant peripheral nerve sheath tumors: a clinicopathologic study of 28 cases. Cancer 71:1247–1253.

65. Ducatman DB, Scheithauer BW, Piepgras DG, Reiman HM, Ilstrup DM (1986). Malignant peripheral nerve sheath tumors: a clinicopathologic study of 120 cases. Cancer 57:2006–2021.

66. Bhargava R, Parham DM, Lasater OE, Chari RS, Chen G, Fletcher BD (1997). MR imaging differentiation of benign and malignant peripheral nerve sheath tumors: use of the target sign. Pediatr Radiol 27:124–129.

67. Mann FA, Murphy WA, Totty WG, Manaster BJ (1990). Magnetic resonance imaging of peripheral nerve sheath tumors: assessment by numerical visual fuzzy cluster analysis. Invest Radiol 25:1238–1245.

68. Nagao E, Nishie A, Yoshimitsu K et al (2004). Gluteal muscular and sciatic nerve metastases in advanced urinary bladder carcinoma: case report. Abdom Imaging. 29(5):619–622.

69. Jaeckle KA (1991). Nerve plexus metastases. Neurol Clin 9(4):857–866.

70. Hirota N, Fujimoto T, Takahashi M, Fukushima Y (1998). Isolated trigeminal nerve metastases from breast cancer: an unusual cause of trigeminal mononeuropathy. Surg Neurol 49(5):558–561.

71. Allieu Y, Mackinnon S (eds) (2002). *Nerve Compression Syndromes of the Upper Limb*. Martin Dunitz.

CHAPTER 55

Neuroimaging of Cerebrovascular Complications in Cancer Patients

Rajan Jain and Shehanaz Ellika

INTRODUCTION

Cerebrovascular disease is the second most common cause of pathologically proven central nervous system disease in cancer patients and was found in 14.6% cancer patients in a large autopsy series, out of which 51% patients were clinically symptomatic related to the cerebrovascular disease [1]. However, etiopathogenesis and clinical presentation of stroke in cancer patients are significantly different as compared to the general population. Most of the causes of stroke in this unique group of patients are either directly related to the tumor and the paraneoplastic syndrome or are related to the therapy apart from various host factors. Aggressive treatment approaches with multimodality management, such as radiation therapy, chemotherapy or combination therapies, can cause additional stress, leading to unusual or accelerated cerebral vasculature disorders. Imaging evaluation of a patient with a known cancer, presenting with stroke or encephalopathy can help diagnose the cause of stroke. Cerebrovascular event can also be the presenting feature frequently, which can lead to an unsuspected malignancy.

Initial imaging evaluation in a cancer patient suspected of a CNS vascular disorder usually involves computed tomography (CT) for assessment of ischemic or hemorrhagic stroke, followed by further evaluation with magnetic resonance imaging (MRI) which is more sensitive. However, management of these patients depends on the cause of the stroke, which will require evaluation of the cerebral vasculature. With recent advances in non-invasive imaging modalities such as CT and MR angiography/venography, the role of diagnostic catheter angiography for cerebrovascular thrombotic or stenotic disease has become limited. Perfusion imaging using MR or CT perfusion techniques can help in assessment of cerebral blood flow, which further aids in evaluation of the cause and extent of cerebrovascular compromise. Imaging features of the cerebrovascular complications associated with tumors, mostly with central nervous system malignancies, either directly related to the tumor or related to the effects of the therapy, are reviewed in this chapter, with an emphasis on various advanced imaging techniques.

TUMOR-RELATED CEREBROVASCULAR COMPLICATIONS

Coagulopathy

Stroke in cancer patients is often related to systemic coagulopathy, which can be a hypercoagulable state causing ischemic stroke or disseminated intravascular coagulation which can cause hemorrhagic stroke. Thrombotic episodes may precede the diagnosis of malignancy by months or years and can present as migratory superficial thrombophlebitis (Trousseau's syndrome), idiopathic venous thrombosis, non-bacterial thrombotic endocarditis (marantic endocarditis), disseminated intravascular coagulation, thrombotic microangiopathy (e.g. hemolytic-uremic syndrome) or with arterial thrombosis [2]. Non-bacterial thrombotic endocarditis (NBTE) is a manifestation of the hypercoagulable state in cancer patients and develops when sterile platelet-fibrin vegetations form on cardiac valves. The possibility of NBTE should be considered in all cancer patients who develop an acute stroke syndrome, as well as patients with cancer presenting with cerebral embolism of unknown etiology [3]. Multiple and often hemorrhagic infarcts are seen in multiple vascular territories caused by embolization of cardiac platelet-fibrin vegetations or by thrombotic occlusion of medium and small-sized vessels. Cerebral angiography can detect vascular occlusions typically involving middle cerebral artery branches [3]. Diffusion-weighted imaging can show multiple small and medium or large disseminated lesions in NBTE as compared to single lesion, territorial infarction and disseminated punctate lesions seen in infective endocarditis [4].

Venous Thrombosis

Cerebral venous thrombosis (CVT) as a complication of cancer can occur due to 'paraneoplastic phenomenon' which is seen more commonly with hematologic malignancies [5]. Venous thrombosis can also occur due to direct

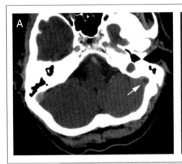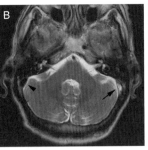

FIG. 55.1. **(A)** Contrast enhanced CT scan in a patient with small cell lung carcinoma showing 'empty delta sign' involving the left sigmoid sinus, filling defect seen due to thrombosis (arrow). **(B)** Axial T2-weighted MR image showing loss of flow void in the left sigmoid sinus suggestive of thrombosis (arrow). Normal flow void is seen in the right sigmoid sinus (arrowhead).

invasion or thrombosis of dural sinus by a brain neoplasm or by dural/calvarial metastases [5]. The usual clinical presentation is with severe headache, convulsions, focal deficits, or rapid progression to stupor and coma of CVT. Initial imaging evaluation of CVT is often done with computed tomography. A non-contrast CT (NCCT) scan can show the dense triangle sign due to occlusion of the superior sagittal sinus by hyperdense clot, hyperdensity along the transverse sinus or rarely as a cord sign due to visualization of a thrombosed cortical vein. NCCT can also show non-specific brain swelling or associated multiple infarcts mostly not conforming to an arterial vascular territory. Multifocal hemorrhagic infarcts are almost as frequent as non-hemorrhagic infarcts in cerebral venous thrombosis [6]. Contrast enhanced CT can show the empty delta sign (Figure 55.1A) due to a filling defect from a non-enhancing clot in the lumen and enhancement of small collateral veins in the walls of the sinus [7]. Prominent tentorial enhancement may be seen due to venous collaterals and congestion. However, in up to 30% of cases, initial CT scan can be non-diagnostic, particularly in patients with isolated intracranial hypertension [6].

Magnetic resonance imaging is very sensitive in diagnosing CVT. Absence of blood flow in the thrombosed sinus results in loss of normal flow-void seen on fast spin-echo images (Figure 55.1B). Various signal abnormalities in the involved dural sinus can be seen on MR images depending on the time interval between onset of thrombosis and imaging. Magnetic resonance imaging also helps better to delineate the associated parenchymal changes secondary to venous drainage obstruction, such as edema, hemorrhagic or non-hemorrhagic infarcts. Diffusion-weighted imaging (DWI) and apparent diffusion coefficient (ADC) pattern seen in venous infarcts is variable and heterogeneous, markedly different from that seen in pure arterial infarcts suggestive of vasogenic edema mixed with cytotoxic edema [8].

This difference probably corresponds in part to the much better recovery observed in non-hemorrhagic venous infarcts with normal or increased ADC, and even areas with initial reduced ADC can be reversible [8]. However, MRI findings can be equivocal in acute thrombosis or in partially thrombosed venous sinuses due to flow artifacts. MR venography (MRV) may assist in establishing the diagnosis and can be done using 2D and 3D phase-contrast and time-of-flight (TOF), as well as 3D contrast-enhanced thin-section gradient echo techniques. Compared with 2D TOF, 3D contrast-enhanced thin-section gradient echo techniques are more sensitive to dural sinus thrombosis and are less affected by saturation artifacts [9]. With the advent of multislice CT scanners, CT venograms can be obtained very quickly with a high yield for diagnoses of venous sinus thrombosis. CT venography has been shown in at least one series to be superior to MRV in visualizing dural sinuses or smaller cerebral veins with low flow and is at least equivalent in the diagnosis of dural sinus thrombosis [10]. CT venograms can be easier to interpret and have fewer artifacts than MR venography and may be especially helpful in the very acute or late stages of CVT, when MRI can be misleading [10]. Catheter angiography, which has been used extensively in the past to diagnose venous sinus thrombosis, has a limited role now with the advent of various non-invasive techniques. The most reliable sign of CVT is the partial or complete lack of opacification of veins or dural sinuses on angiography. Some indirect angiographic signs include dilated collateral veins with a corkscrew appearance and delayed venous emptying.

CVT due to direct tumor invasion shows associated enhancing parenchymal, dural or calvarial metastases encasing, compressing or infiltrating the dural venous sinuses on CT or MRI (Figure 55.2). Differentiation of tumor thrombus from bland thrombus based on imaging findings can be difficult. Enhancement of thrombus filling an occluded venous sinus could be seen with tumor (Figure 55.3), but can also occur in subacute bland thrombus with neovascularization.

Direct Tumor-related Compromise of Arterial Vasculature

Direct tumor-related stroke, which is mostly due to encasement, compression, displacement or invasion of intracranial vessels by tumors, is an uncommon but known event that can cause devastating long-term morbidity. Tumors which have been associated with direct arterial compromise include craniopharyngiomas, suprasellar germinomas, hypothalamic tumors, pituitary adenomas with apoplexy, meningiomas and astrocytomas [11–13]. Meningiomas arising from skull base and cavernous sinus regions can encase and narrow vessels forming the circle of Willis and compromise the cerebral blood flow. MRI and MRA (magnetic resonance angiography) can demonstrate the narrowing or occlusion of the major intracranial vessels by meningiomas (Figure 55.4). Tumors such as high-grade gliomas abutting the major vessels can also lead to vasculopathy and ischemic

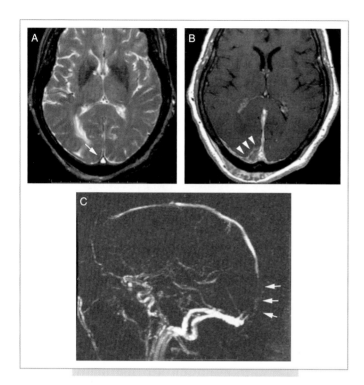

FIG. 55.2. A 59-year-old female with metastatic breast carcinoma. **(A)** T2-weighted images showing loss of normal flow void and hyperintense signal in the superior sagittal sinus (arrow). **(B)** Post-contrast T1-weighted axial image showing enhancing dural metastases encasing and occluding the superior sagittal sinus (arrowheads). **(C)** Phase-contrast MR venogram showing non-visualization of the posterior part of the superior sagittal sinus (arrows).

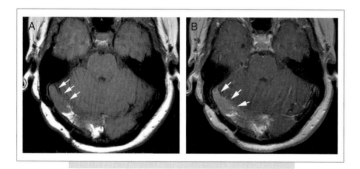

FIG. 55.3. Axial **(A)** pre- and **(B)** post-contrast T1-weighted MR images in a patient presenting with venous sinus thrombosis showing an enhancing filling defect within the distal part of right transverse sinus which revealed tumor thrombus from an undifferentiated metastatic carcinoma on histopathology.

stroke, particularly after radiation therapy (Figure 55.5). Aoki et al described a case of a malignant glioma causing a cerebral infarction by dissection of the middle cerebral artery by invasion of the vessel wall [14] and Züchner et al reported a case of arterial narrowing in a patient with gliosarcoma [15]. Some tumors can also cause slow occlusion of the major

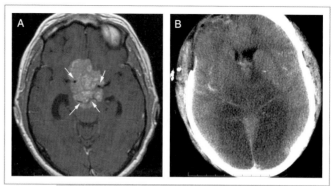

FIG. 55.4. **(A)** Post-contrast T1-weighted axial MR image in a patient with a large suprasellar meningioma showing encasement of bilateral internal carotid artery terminus, basilar tip and bilateral posterior cerebral arteries by the meningioma (arrows). **(B)** Post-operative CT scan showing bilateral posterior cerebral artery territory infarcts due to vascular compression and edema.

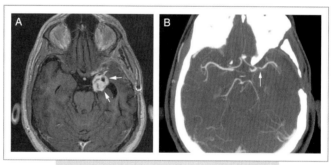

FIG. 55.5. **(A)** Post-contrast T1-weighted axial images in a patient with recurrent anaplastic astrocytoma, showing an enhancing lesion in left temporal lobe, abutting the left middle cerebral artery three years after external beam and stereotactic radiosurgery (60 Gy). **(B)** CT angiogram showing focal narrowing of the M1 segment of left MCA.

intracranial vessels leading to a moyamoya like pattern with development of small immature collateral vessels [16].

Leptomeningeal Metastases

Tumor-related stroke can also occur due to the involvement of the vessels by leptomeningeal metastases. Leptomeningeal metastases or carcinomatous meningitis result from diffuse infiltration of the leptomeninges by malignant cells originating from an extra-meningeal primary tumor site. A rare cause of focal cerebral symptoms in carcinomatous meningitis is ischemic stroke [17]. Cerebral infarction in these patients has been attributed to a vasculopathy resulting from perivascular infiltration and mural invasion of tumor cells and accompanying perivascular inflammation [17]. Angiographic evidence of segmental narrowing and

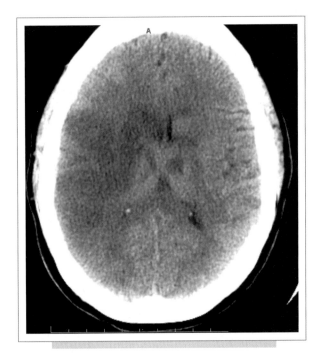

FIG. 55.6. Non-contrast CT scan in a 34-year-old female with left atrial myxoma showing large acute infarct involving right middle cerebral artery territory along with multiple small infarcts in left basal ganglia.

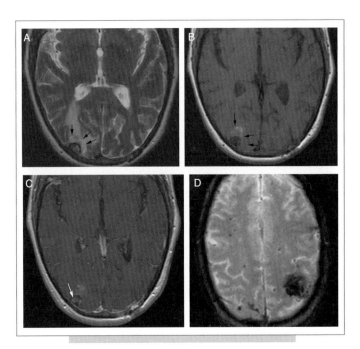

FIG. 55.7. Axial **(A)** T2, **(B)** pre- and **(C)** post-contrast T1-weighted MR images in a patient with atrial myxoma showing hemorrhagic lesion (arrows) with central enhancement (white arrow) in right occipital lobe. **(D)** Axial gradient echo image at supratentorial level showing multiple blooming artifacts due to multiple other small hemorrhagic metastases, mostly at cortico-medullary junction suggesting hematogenous spread and shower of tumor emboli.

wall irregularities in multiple proximal and distal vessels has been documented in the setting of leptomeningeal metastases from carcinoma [17,18].

Tumor Emboli and Neoplastic Aneurysms

Infarction due to tumor emboli presenting with cerebral ischemia can be seen in patients with primary or metastatic neoplasm due to embolization of tumor fragments via the arterial circulation that occurs mostly with lung cancer during surgical resection. Atrial myxomas can also present with systemic embolization in up to 45% of cases, cerebral circulation being involved in one-half of these instances [19,20]. The source of embolus could be the tumor itself or thrombus on the surface of the tumor [19,20]. The middle cerebral artery territory is most frequently affected, presumably because of its dominant flow (Figure 55.6) [21]. Hematogenous spread of the tumor cells with tumor emboli can progress to development of parenchymal brain metastases (Figure 55.7), intracerebral hemorrhage and rarely can also lead to formation of neoplastic aneurysms [22]. Tumor cells usually lodge in smaller, distal, peripheral arteries, attaching to the vessel wall, invading the internal elastic lamina, weakening the endothelium and causing pseudoaneurysm formation similar to mycotic aneurysms [23]. These aneurysms can be diagnosed with catheter angiography and usually appear as irregular, fusiform, distal vessel aneurysms.

Tumor-related Hemorrhagic Stroke

Tumor-related hemorrhage is the most common cause of hemorrhage in patients with solid tumors and occurs more commonly in metastases than in primary brain neoplasms. The mechanism of hemorrhage is likely due to rapid tumor growth with tumor necrosis, rupture of newly formed blood vessels and invasion of adjacent cerebral blood vessels by tumors [24]. A variety of hypervascular metastatic brain tumors such as melanomas, thyroid carcinoma, renal cell carcinoma, hepatocellular and lung carcinomas can present with intratumoral hemorrhage [25–27]. Primary neoplasms like pituitary adenomas, gliomas, meningiomas, schwannomas, oligodendrogliomas and primitive neuroectodermal tumors have also been associated with intratumoral hemorrhage [27]. Hemorrhage usually produces acute symptoms that may be the presenting symptoms or be superimposed on the chronic course of the disease and the patient can present with headache, seizures, and sometimes with associated focal neurologic signs. Imaging evaluation of brain tumor with hemorrhage can occasionally present a diagnostic dilemma on the initial presentation, especially in primary brain tumors or single metastasis. Factors such as atypical location, multiple

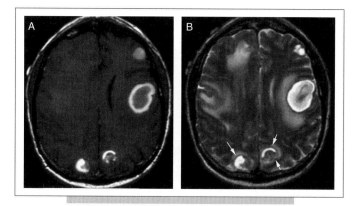

FIG. 55.8. **(A)** Axial pre-contrast T1-weighted MR images showing multiple lesions showing hyperintense signal suggestive of subacute methemoglobin in a patient with metastatic amelanotic melanoma. **(B)** Axial T2-weighted image showing heterogeneous hyperintense lesions with incomplete peripheral hemosiderin rim and surrounding edema.

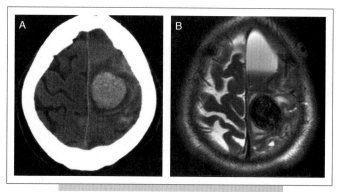

FIG. 55.9. Axial **(A)** non-contrast CT and **(B)** T2-weighted MR images showing intra-tumoral and subarachnoid hemorrhage in a patient with a meningioma.

hemorrhagic sites, disproportionate vasogenic edema and early enhancement may suggest a neoplastic hemorrhage. Hemorrhagic tumors can also show heterogeneous signal characteristics due to blood products in different stages of evolution mixed with tumor (Figure 55.8). Delayed pattern of evolution of the hematoma can be seen in hemorrhagic neoplasms [28,29]. Methemoglobin formation is profoundly influenced by oxygen tension [30,31] and intratumoral hypoxia may significantly prolong persistence of deoxy-hemoglobin and also delay methemoglobin formation. Neoplastic hemorrhage can also be differentiated from non-neoplastic hemorrhage by the absence of the well-defined, complete hemosiderin rim around the hemorrhage [29,32]. This is due to the persistent and marked disruption of the blood–brain barrier that occurs with intracranial tumors and allows more efficient removal of the hemosiderin laden macrophages [29]. Hemorrhage may be purely intratumoral or may extend into the adjacent brain parenchyma depending upon the location of tumor and mechanism of hemorrhage. Similarly, hemorrhage can extend into the extra-axial spaces and tumors can be associated with subarachnoid, intraventricular, subdural or epidural hemorrhage (Figure 55.9). Superficial tumors can rarely present with multiple episodes of subarachnoid hemorrhage and resultant superficial siderosis. Superficial siderosis is caused by deposition of blood products in the leptomeninges, subpial tissue of the brain and spinal cord and cranial nerves as a result of hemorrhage in the subarachnoid space [33]. Various intracranial tumors such as ependymomas, oligodendrogliomas, astrocytomas and vascular spinal tumours can be a source of repeated hemorrhages causing superficial siderosis (Figure 55.10).

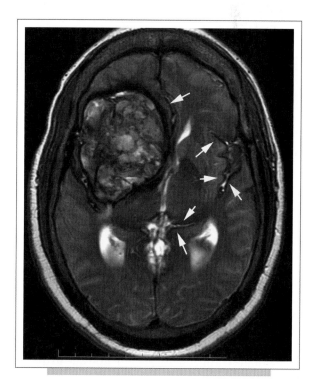

FIG. 55.10. Axial T2-weighted image in a patient with oligodendroglioma showing a large heterogeneous intensity hemorrhagic tumor with hypointense signal along the pial surface of the brain parenchyma suggesting superficial siderosis (arrows).

INFECTION-RELATED CEREBROVASCULAR COMPLICATIONS

Infection in cancer patients occurs due to immunosuppression usually related to antineoplastic therapy or induced by tumor. In the severely immunocompromised host, angioinvasive pathogens like aspergillus gain access to the systemic circulation and disseminate to end organs, such as the brain, causing infarction or hemorrhage as the

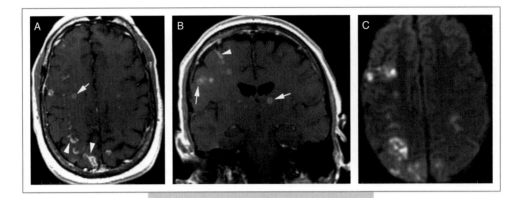

FIG. 55.11. Post-contrast T1-weighted **(A)** axial and **(B)** coronal images in a patient with mycosis fungoides and CNS aspergillosis showing multiple, small, ring enhancing abscesses (arrows) predominantly involving the corticomedullary junction and the basal ganglia. Multiple areas of gyral enhancement are also noted suggesting cerebritis (arrowheads). **(C)** DWI shows corresponding areas of restricted diffusion (low apparent diffusion coefficient – not shown here) suggesting hematogenous spread and septic infarcts.

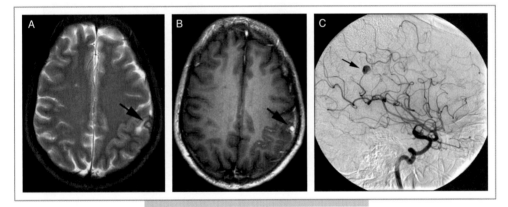

FIG. 55.12. Axial **(A)** T2 and **(B)** post-contrast T1-weighted MR images in a patient with a distal mycotic aneurysm showing a small round peripheral lesion with peripheral hypointense rim and surrounding edema, showing central enhancement (arrow). **(C)** Oblique projection of a left internal carotid angiogram showing a distal left MCA aneurysm filling with contrast (arrow) (courtesy Dr Dheeraj Gandhi).

early radiologic presentation [34]. Aspergillus may cause an infectious vasculopathy, initially leading to acute infarction and hemorrhage, progressing to infectious cerebritis and finally evolving into an abscess. Most of the pyogenic abscesses and metastases to brain are known to have a predilection for the corticomedullary junction due to the hematogenous spread and due to the vascular anatomy of this interface. However, aspergillus lesions are more commonly seen in the basal ganglia and thalami due to involvement of perforating artery territories. MRI can show multiple disseminated small septic infarcts, which will involve basal ganglia, thalami and corpus callosum, in addition to cortical and corticomedullary junction infarcts [34–37]. As these lesions are infarcts, diffusion-weighted images may also be useful for detecting early lesions and can be beneficial in differentiating these from progressive multifocal leukoencephalopathy and tumoral lesions [36]. Associated cortical/gyral enhancement may suggest cerebritis. Further progression to multiple ring-enhancing abscesses, usually measuring less than 3 mm, can be seen commonly in the basal ganglia and at the corticomedullary junction, which may also show restricted diffusion in the necrotic component

on DWI (Figure 55.11). Another fungal organism that can present with encephalopathy and progressive cerebral signs is Candida species. Candida brain abscesses are often associated with candidemia and endocarditis [38].

Mycotic aneurysms are a rare complication of bacterial or fungal infections in immunosuppressed cancer patients. Most cerebral mycotic aneurysms developing in the course of bacterial endocarditis are multiple and are located along the distal branches, particularly of the middle cerebral artery and rarely in the posterior circulation [39,40]. These can rupture causing parenchymal or subarachnoid hemorrhage, but can also be found associated with cerebritis and abscesses. Aspergillus is the most common causative agent of fungal aneurysms because of being highly angioinvasive [41]. Pathologically, there is septic embolization of the vasa vasorum or lumen of the artery with resultant focal arteritis, necrosis and aneurysm formation [42]. CT scan and magnetic resonance imaging are helpful in the diagnosis, however, catheter angiography remains the gold standard. MRI may show a round, mostly peripherally situated lesion with surrounding edema and peripheral hemosiderin rim (Figure 55.12). Rapidly flowing blood through the patent portion of the aneurysmal lumen appears as an area of signal void on MR images [43]. However, slow flowing blood within the patent portion of the aneurysmal lumen may demonstrate variable signal intensity on MR images. In partially thrombosed aneurysms, high signal intensity due to methemoglobin can be seen immediately adjacent to the region of signal void. Many unruptured mycotic aneurysms will resolve on antibiotic therapy alone [44,45]. These patients can be followed up non-invasively with MRA or CTA to demonstrate resolution or progression of the size of the aneurysm. Surgical excision of peripheral aneurysms is made easier after antibiotic therapy, because the aneurysm wall is much thicker from fibrosis and less likely to rupture on handling [46]. The trend towards endovascular treatment of mycotic aneurysms has been applied to only a small number of aneurysms [47,48].

THERAPY-RELATED CEREBROVASCULAR COMPLICATIONS

Radiation-induced Cerebrovascular Complications

The assumed pathogenesis of radiation-induced vascular disease is an acceleration of the atherosclerotic process probably due to endothelial cell damage, fibrosis of the intima-media layer and development of atheromatous plaques, which is increased by occlusive changes in the vasa vasorum leading to ischemia of the arterial wall [49,50]. Radiation therapy administered for head and neck cancers or lymphoma is best known to cause carotid artery stenosis. Carotid duplex ultrasound detected 70% stenosis of internal carotid artery in 11.7% patients irradiated for head and neck cancers in one series [51]. Radiation-induced carotid artery lesions seem to have a disproportionate usually long segment involvement within the irradiated field, and not confined to the bifurcation [52] and thus can be differentiated from the atherosclerotic stenosis. In addition, they are often associated with occlusion of other cervical arteries within the radiation portal [53]. Carotid blow-out syndrome is a rare complication of radiation therapy in the neck. Risk factors include radical tumor resection, flap necrosis, wound infection and recurrent or persistent tumor in addition to radiation therapy. Chaloupka et al have described three stages of carotid blow-out syndrome which are managed differently [54]. Threatened carotid blow-out is when the vessel is exposed but there is no hemorrhage; impending blow-out is when there is pseudoaneurysm formation which may be associated with intermittent hemorrhage. The third stage is carotid blow-out with exsanguinating hemorrhage (Figure 55.13). Emergency surgical ligation of the common carotid artery (CCA) or the proximal internal carotid artery (ICA) without provocative testing has been traditionally the only therapeutic maneuver available for carotid blow-out despite the well-documented risk of cerebral ischemia [55,56]. These poor outcomes have substantially improved with the advent of various endovascular

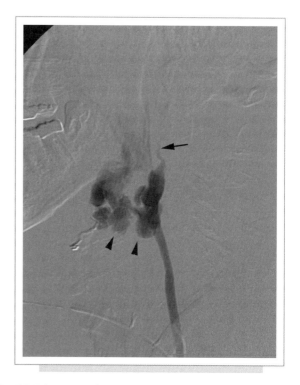

FIG. 55.13. Lateral projection of a left common carotid artery digital subtraction angiogram in a patient with previously irradiated undifferentiated carcinoma of the neck showing marked irregularity of the distal common carotid artery, external carotid and internal carotid artery with contrast extravasation (arrowheads) suggestive of carotid blow-out. Poor contrast opacification of the internal carotid artery suggestive of occlusion (arrow).

techniques, including permanent balloon occlusion of the affected ICA or CCA [57,58]. Endovascular covered stent deployment can be used as a treatment option in carotid blow-out patients thought to be at high risk for surgery or carotid occlusion [59].

Radiation-induced vasculopathy of intracranial vessels is much less common and its incidence has a considerable correlation with radiation dose and the age at the time of radiation therapy. This occurs mostly in a delayed fashion following treatment to the tumors around the circle of Willis [60] and can be accelerated in neurocutaneous syndromes. Catheter angiography usually shows irregularity, narrowing, focal dilatation or even occlusion of the affected vessels (Figure 55.14). Cerebral infarcts can be seen in the distribution of the vessels affected. When large or medium sized vessels are affected, moyamoya-like pattern due to collateral formation can be seen on angiography. Mineralizing microangiopathy is another pathological entity, usually attributed to the combined toxicity of cranial irradiation and intrathecal methotrexate. It is characterized by presence of mineralized deposits in the small vessels. Radiation therapy produces hyalinization and fibrinoid necrosis of

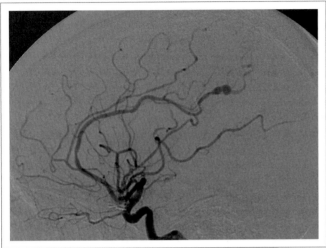

FIG. 55.14. Lateral oblique projection of internal carotid angiogram in a patient who underwent radiation therapy 5 years ago showing irregularity and focal dilatation of one of the distal MCA branches due to radiation-induced vasculopathy.

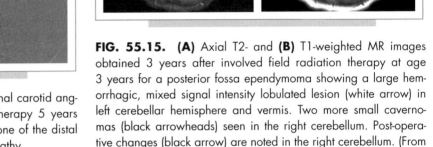

FIG. 55.15. **(A)** Axial T2- and **(B)** T1-weighted MR images obtained 3 years after involved field radiation therapy at age 3 years for a posterior fossa ependymoma showing a large hemorrhagic, mixed signal intensity lobulated lesion (white arrow) in left cerebellar hemisphere and vermis. Two more small cavernomas (black arrowheads) seen in the right cerebellum. Post-operative changes (black arrow) are noted in the right cerebellum. (From Jain et al, Am J Neuroradiol 2005; 26, page 1161 fig. 2 a,b.)

small arteries and arterioles with endothelial proliferation and calcium deposition [61,62]. Dystrophic calcifications in the basal ganglia, subcortical white matter and vascular border zones of the cerebral cortex can be seen in children previously treated with radiation therapy and intrathecal methotrexate [63,64].

Radiation-induced vascular malformations can present with intracranial hemorrhage or can be asymptomatic and found on follow-up imaging [65]. Capillary telangiectasias usually occur 3–9 months after radiation, whereas cavernous malformations can have a latency period ranging from 1 to 26 years [66–68]. Appearance of a new hemorrhagic lesion in a patient previously treated with radiation, especially if treated in childhood, should always raise the possibility of a cavernoma in addition to neoplastic recurrence or metastases [65]. Diagnosis of a radiation-induced cavernoma is usually possible due to typical popcorn-like appearance of a cavernoma nidus with little or no surrounding edema on MRI (Figure 55.15). However, MRI characteristics can be blurred by the presence of acute hemorrhage and follow-up imaging can be useful in such a scenario.

Chemotherapy-related Cerebrovascular Complications

Chemotherapy remains an integral treatment for a wide variety of malignancies. Risk of CNS thromboembolic events increases in patients receiving adjuvant chemotherapy for breast cancer [69] and after chemotherapy in patients with a variety of hematological and solid malignancies [70–72]. A well-recognized cause of ischemic stroke is cisplatin-based chemotherapy [70–72]. The underlying pathophysiology

of these events remains obscure; however, both acute and long-term vascular toxicity may be associated with cisplatin administration [73]. The imaging findings in chemotherapy-associated ischemic stroke are similar to the appearance of ischemic stroke elsewhere. Another commonly used chemotherapeutic drug is methotrexate (MTX) which is thought to induce small-vessel vasculopathy [74]. Areas of hyperintensity on T2-weighted imaging are usually seen located in the periventricular white matter, particularly in the centrum semiovale [75,76] and can be seen in 15–75% of patients during MTX chemotherapy [77]. Despite white matter changes on MR imaging, patients often recover spontaneously from MTX-induced neurotoxicity [63]. DWI has shown low ADC suggesting cytotoxic edema corresponding to T2-weighted lesions; however DWI findings are reversible suggesting a transient metabolic derangement over an angiopathic or demyelinating process [78].

Another commonly used chemotherapeutic agent associated with various neurovascular complications is L-asparaginase. A study of 238 patients treated with L-asparaginase showed cerebral hemorrhages in 2.1%, cerebral thrombosis and thromboemboli in 4.2% of patients [79]. L-asparaginase associated sinus thrombosis occurs usually during the remission/induction period, but has also been observed during the later course of polychemotherapy [80]. Patients usually present with severe headaches and MR or CT venography can be used for the radiological confirmation of the sinus thrombosis.

Posterior reversible encephalopathy syndrome (PRES) consists usually of a reversible encephalopathy, with predominant involvement of the posterior white matter.

It occurs usually in association with acute hypertensive crisis and eclampsia, but has also been seen following treatment with immunosuppressive or chemotherapy agents [81,82]. Intracranial vasospasm has been seen with conventional or MR angiography [83,84] suggesting vasospasm as a possible underlying pathophysiologic mechanism. Hinchey et al initially proposed that the syndrome resulted from sudden elevations in systemic blood pressure exceeding the autoregulatory capability of the brain vasculature [82]. This causes acute disruption of the blood–brain barrier leading to capillary leakage and subsequent vasogenic edema [82]. The selective involvement of structures perfused by the posterior circulation and the rapid reversibility of symptoms indicates loss of autoregulation of the vertebrobasilar system, likely due to its relatively sparse sympathetic innervation [81]. Prompt initiation of antihypertensive treatment or discontinuation of immunosuppressive drugs, if being used, can lead to complete clinical recovery with reversal of MRI lesions in some cases. However, if untreated, permanent neurologic deficit, or even death, may occur as a result of the ensuing cerebral infarctions or hemorrhages [85]. MR findings are typically those of vasogenic edema with T2 and fluid-attenuated inversion recovery (FLAIR) hyperintensities predominantly involving the posterior parieto-occipital regions of the brain bilaterally. FLAIR images can show the lesions of PRES better due to suppression of the adjacent CSF signal [86]. This has allowed for milder cases of PRES to be identified, enabling early initiation of therapy. These lesions usually show higher apparent diffusion coefficient on DWI and ADC maps which is suggestive of vasogenic edema [81,87] unlike cytotoxic edema in ischemic strokes which will have lower ADC.

Endovascular Therapy-related Cerebrovascular Complications

Preoperative embolization of brain tumors, such as meningiomas, is used to induce hemostasis and necrosis of the tumor for easier surgical manipulation and surgical removal of the tumor with less intraoperative blood loss [49,81]. Particulate embolic agents, such as polyvinyl alcohol (PVA) particles [89,90], are generally favored because of the relative ease of use and because the permanence of other agents such as glue or ethanol is not necessary in the operative setting. Extensive tumor infarction has been observed histologically after embolization [89,90] which can also induce spontanous intratumoral or subarachnoid hemorrhage due to rupture of fragile tumor vessels in a large, necrotic meningioma [91]. Apart from hemorrhage, other neurologic complications related to inadvertent embolization of particulate embolic agent to the vessels supplying the normal brain are major concerns which can be prevented by avoiding overly aggressive embolization, because it can cause reflux of embolic material into normal proximal branches. Ischemic stroke or blindness can also occur because of very small extracranial–intracranial arterial anastomoses which open up due to altered flow hemodynamics

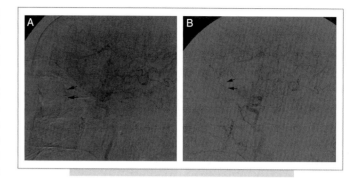

FIG. 55.16. A 49-year-old patient with left frontal meningioma, who underwent preoperative endovascular embolization of the middle meningeal artery branches with PVA particles, complained of complete vision loss in the left eye following embolization. **(A)** Pre- and **(B)** post-embolization lateral projection internal carotid arteriograms showing disappearance of choroidal blush due to occlusion of the retinal artery because of inadvertent embolization of PVA particles through arterial anastomoses.

during embolization (Figure 55.16). These complications can be prevented by superselective angiography performed using microcatheters to confirm the proper position of the microcatheter and to identify any normal branches that might preclude safe embolization [92].

MISCELLANEOUS TUMOR-RELATED CEREBROVASCULAR CONDITIONS

Primary Angiitis of Central Nervous System (PACNS)

Primary angiitis of the central nervous system (PACNS), also known as isolated CNS angiitis or granulomatous angiitis of the CNS, is a non-infectious granulomatous angiitis that has been associated with both Hodgkin's and non-Hodgkin's lymphoma, as well as with other hematologic malignancies. It typically involves small- and medium-sized arteries of the brain and spinal cord and, to a lesser degree, may also involve small veins and venules [93–95]. Associated infarction, hemorrhage as well as demyelination and axonal degeneration can be seen in the adjacent brain parenchyma. The pathogenesis of this condition is unclear, however a hypersensitivity reaction, as well as autoimmune response has been suggested [96,97]. Symptomatology is variable and usually includes headache, focal or global neurological dysfunction, confusion, intellectual deterioration, memory loss and malaise. T2-weighted MR images show hyperintense foci representing ischemia, infarction or inflammatory changes involving both the cortex and white matter [93,98,99]. Post-contrast images show enhancing linear lesions predominantly involving brainstem and white matter which represent inflammation of the perforating arteries and

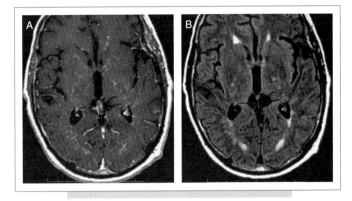

FIG. 55.17. (A) Axial post-contrast T1-weighted images in a patient with Hodgkin's lymphoma and associated PACNS showing extensive linear enhancement through the white matter suggestive of vasculitis. **(B)** Axial FLAIR image showing a few corresponding white matter lesions.

perivascular tissues [93,100] (Figure 55.17). Other less common presentations include hemorrhage [98,99,101] or focal lesion that may simulate a neoplasm [98,100,102]. Patients with Hodgkin's disease who initially present with neurological symptoms due to PACNS had a better outcome in a recent review of literature than those who present initially with Hodgkin's disease and later go on to develop PACNS [13].

Intravascular Lymphomatosis

Intravascular lymphomatosis (IVL) is a rare intravascular subtype of extranodal large cell lymphoma causing multifocal vascular occlusion and resulting in diffuse thrombosis, a high incidence of neurological and cutaneous involvement;

usually showing poor response to therapy and an extremely aggressive clinical course [103–105]. The affected blood vessels show reactive endothelial cells adjacent to lymphoma cells, and complete occlusion of vessel lumen can be seen [106]. Isolated CNS involvement with IVL may pose a diagnostic challenge as it may be difficult to distinguish these from other vasculitides such as primary angiitis of the CNS. MRI usually shows multifocal hyperintense lesions in the deep white matter on T2-weighted images which correlate with edema and gliosis seen on biopsy [107–110]. Small vessels (arterioles, capillaries, postcapillary venules) are more commonly involved than large vessels. Due to the ischemic nature of the lesions in IVL, the imaging findings are more pronounced on DWI in the acute phase.

SUMMARY

Cerebrovascular complications associated with CNS malignancies occur due to various tumor, host and treatment factors and can significantly increase morbidity and mortality. Any patient with a known cancer presenting with new neurological deficit should be evaluated to rule out cerebrovascular compromise, in addition to tumor progression, recurrence or metastases. In addition, a thromboembolic or hemorrhagic stroke as the initial presentation can also lead to diagnosis of an occult malignancy in paraneoplastic syndromes. Neuroimaging studies, particularly evaluating the cerebral vasculature, can help define the cause of stroke and thus help manage these patients. Recent advances in various functional techniques have expanded the diagnostic capabilities of MRI and CT beyond the morphologic evaluation and can be useful in early detection of cerebrovascular disease processes.

REFERENCES

1. Graus F, Rogers LR, Posner JB (1985). Cerebrovascular complications in patients with cancer. Medicine (Baltimore) 64:16–35.
2. Levine M (1997). Treatment of thrombotic disorders in cancer patients. Haemostasis 27(Suppl 1):38.
3. Rogers LR (2003). Cerebrovascular complications in cancer patients. Neurol Clin 21(1):167–192.
4. Singhal AB, Topcuoglu MA, Buonanno FS (2002). Acute ischemic stroke patterns in infective and nonbacterial thrombotic endocarditis. A diffusion-weighted magnetic resonance imaging study. Stroke 33:1267–1273.
5. Raizer JJ, DeAngelis LM (2000). Cerebral sinus thrombosis diagnosed by MRI and MR venography in cancer patients. Neurology 54:1222–1226.
6. Crassard I, Bousser MG (2004). Cerebral venous thrombosis. J Neuro-Ophthalmol 24(2):156–163.
7. Trommer BL, Homer D, Mikhael MA (1988). Cerebral vasospasm and eclampsia. Stroke 19:326–329.
8. Ducreux D, Oppenheim C, Vandamme et al (2001). Diffusion weighted imaging patterns of brain damage associated with cerebral venous thrombosis. Am J Neuroradiol 22:261–268.
9. Liang L, Korogi Y, Sugahara T et al (2002). Normal structures in the intracranial dural sinuses: delineation with 3D contrast-enhanced magnetization prepared rapid acquisition gradient-echo imaging sequence. Am J Neuroradiol 23(10):1739–1746.
10. Ozsvath RR, Casey SO, Lunstrin ES, Alberico TLA, Hassankhani A, Patel M (1997). Cerebral venography: comparison of CT and MR projection venography. Am J Roentgenol 169:1699–1707.
11. Lath R, Rajshekhar V (2001). Massive cerebral infarction as a feature of pituitary apoplexy. Neurology India 49(2):191–193.
12. Mori K, Takeuchi J, Ishikawa M, Handa H, Toyama M, Yamaki T (1978). Occlusive arteriopathy and brain tumor. J Neurosurg 49:22–35.
13. Rosen CL, DePalma L, Morita A (2000). Primary angiitis of the central nervous system as a first presentation in Hodgkin's disease: a case report and review of literature. Neurosurgery 46(6):1504–1508.
14. Aoki N, Sakai T, Oikawa A, Takizawa T, Koike M (1999). Dissection of the middle cerebral artery caused by invasion of malignant glioma presenting as acute onset of hemiplegia. Acta Neurochir 141:1005–1008.

15. Zuchner S, Kawohl W, Sellhaus B, Mull M, Mayfrank L, Kosinski CM (2003). A case of gliosarcoma appearing as ischaemic stroke. J Neurol Neurosurg Psychiatr 74:364–366.

16. Rajakulasingham K, Cerullo L, Raimondi A (1979). Childhood moyamoya syndrome – postirradiation pathogenesis. Child Brain 5:467–475.

17. Klein P, Haley EC, Wooten GF, VandenBerg SR (1989). Focal cerebral infarctions associated with perivascular tumor infiltrates in carcinomatous leptomeningeal metastases. Arch Neurol 46:1149–1152.

18. Wall JG, Weiss RB, Norton L et al (1989). Arterial thrombosis associated with adjuvant chemotherapy for breast carcinoma: a Cancer and Leukemia Group B study. Am J Med 87:501–504.

19. Branch CL, Laster DW, Kelly DL (1985). Left atrial myxoma with cerebral emboli. Neurosurgery 16:675–680.

20. Price DL, Harris JL, New PFJ, Cantu RC (1970). Cardiac myxoma – a clinicopathologic and angiographic study. Arch Neurol 23:558–567.

21. Hofmann E, Becker T, Romberg-Hahnloser R, Reichmann H, Warmuth-Metz M, Nadjmi M (1992). Cranial MRI and CT in patients with left atrial myxoma. Neuroradiology 34:57–61.

22. Jean WC, Walski-Easton SM, Nussbaum ES (2001). Multiple intracranial aneurysms as delayed complications of an atrial myxoma: case report. Neurosurgery 49(1):200–203.

23. Murata J, Sawamura Y, Takahashi A, Abe H, Saitoh H (1993). Intracerebral hemorrhage caused by a neoplastic aneurysm from small-cell lung carcinoma: case report. Neurosurgery 32(1):124–126.

24. Kondziolka D, Bernstein M, Resch L et al (1987). Significance of hemorrhage into brain tumors: clinicopathological study. J Neurosurg 67(6):852–857.

25. Davis JM, Zimmerman RA, Bilaniuk LT (1982). Metastases to the central nervous system. Radiol Clin N Am 20:417–435.

26. Mandybur TI (1977). Intracranial hemorrhage caused by metastatic tumors. Neurology 27:650–655.

27. Rudoltz MS, Regine WF, Langston JW, Sanford RA, Kovnar EH, Kun LE (1998). Multiple causes of cerebrovascular events in children with tumors of the parasellar region. J Neuro-Oncol 37:251–261.

28. Atlas SW, Grossman RI, Gomori JM et al (1987). Hemorrhagic intracranial malignant neoplasms: spin-echo MR imaging. Radiology 164:71–77.

29. Grossman RI, Gomori JM, Goldberg HI et al (1988). MR of hemorrhagic conditions of the head and neck. Radiographics 8(3):441–454.

30. Neill JM, Hastings AB (1925). The influence of the tension of molecular oxygen upon certain oxidations of hemoglobin. J Biol Chem 63:479–492.

31. Pauling L, Coryell CD (1936). The magnetic properties and structures of hemoglobin, oxyhemoglobin and carbonmonoxyhemoglobin. Proc Natl Acad Sci 22:210–216.

32. Gomori JM, Grossman RI, Goldberg HI, Hackney DB, Zimmerman RA, Bilaniuk LT (1986). Occult cerebral vascular malformations: high-field MR imaging. Radiology 158:707–713.

33. Koeppen AH, Dickson AC, Chu RC, Thach RE (1993). The pathogenesis of superficial siderosis of the central nervous system. Ann Neurol 34:646–653.

34. DeLone DR, Goldstein RA, Petermann G et al (1999). Disseminated aspergillosis involving the brain: distribution and imaging characteristics. Am J Neuroradiol 20: 1597–1604.

35. Guermazi A, Gluckman E, Tabti B, Miaux Y (2003). Invasive central nervous system aspergillosis in bone marrow transplantation recipients: an overview. Eur Radiol 13:377–388.

36. Keyik B, Edguer T, Hekimoglu B (2005). Conventional and diffusion-weighted MR imaging of cerebral aspergillosis. Diagn Interv Radiol 11(4):199–201.

37. Miaux Y, Ribaud P, Williams M et al (1995). MR of cerebral aspergillosis in patients who have had bone marrow transplantation. Am J Neuroradiol 16:555–562.

38. Hagensee MH, Bauwens JE, Kjos B, Bowden RA (1994). Brain abscess following marrow transplantation: experience at the Fred Hutchinson Cancer Research Center, 1984–1992. Clin Infect Dis 19:402–408.

39. Frazee JG, Cahan LD, Winter J (1980). Bacterial intracranial aneurysms. J Neurosurg 53:633–641.

40. Meena AK, Sitajayalakshmi S, Prasad VS, Murthy JM (2000). Mycotic aneurysm on posterior cerebral artery: resolution with medical therapy. Neurology India 48(3):276–278.

41. Clare CE, Barrow DL (1992). Infectious intracranial aneurysms. Neurosurg Clin N Am 3:551–561.

42. Molinari GF, Smith L, Goldstein MN, Satran R (1973). Pathogenesis of cerebral mycotic aneurysms. Neurology 23:325–332.

43. Bradley WG, Waluch V, Lai K, Fernandez EJ, Spalter C (1984). Appearance of rapidly flowing blood on magnetic resonance images. Am J Roentgenol 143:1167–1174.

44. Bingham WF (1977). Treatment of mycotic intracranial aneurysms. J Neurosurg 46:428–437.

45. Bohmfalk GL, Story JL, Wissinger JP, Brown WE (1978). Bacterial intracranial aneurysm. J Neurosurg 48:369–382.

46. Roach MR, Drake CG (1965). Ruptured cerebral aneurysms caused by microorganisms. N Engl J Med 273:240–244.

47. Chapot R, Houdart E, Saint-Maurice JP et al (2002). Endovascular treatment of cerebral mycotic aneurysms. Radiology 222:389–396.

48. Khayata MH, Aymard A, Casasco A, Herbreteau D, Woimant F, Merland JJ (1993). Selective endovascular techniques in the treatment of cerebral mycotic aneurysms. Report of three cases. J Neurosurg 78:661–665.

49. Hieshima GB, Everhart FR, Mehringer CM et al (1980). Preoperative embolization of meningiomas. Surg Neurol 14: 119–127.

50. Murros KE, Toole JF (1989). The effect of radiation on carotid arteries. A review article. Arch Neurol 46:449–455.

51. Cheng SWK, Wu LLH, Ting ACW, Lau H, Lam LK, Wei WI (1999). Irradiation induced extracranial carotid stenosis in patients with head and neck malignancies. Am J Surg 178:323–328.

52. Loftus CM, Biller J, Hart MN, Cornell SH, Hiratzka LF (1987). Management of radiation induced accelerated carotid atherosclerosis. Arch Neurol 44:711–714.

53. Melliere D, Becquemin JP, Berrahal D, Desgranges, P, Cavillon, A (1997). Management of radiation-induced occlusive arterial disease: a reassessment. J Cardiovasc Surg (Torino) 38:261–269.

54. Chaloupka JC, Putman CM, Citardi MJ, Ross DA, Sasaki CT (1996). Endovascular therapy of the carotid blowout syndrome in head and neck surgical patients: diagnostic and managerial considerations. Am J Neuroradiol 17:843–852.

55. Coleman JJ (1985). Treatment of the ruptured or exposed carotid artery: a rational approach. South Med J 78: 262–267.

56. Maran AGD, Amin M, Wilson JA (1989). Radical neck dissection: a 19 year experience. J Laryngol Otol 103:760–764.

57. Citardi MJ, Chaloupka JC, Son YH, Ariyan S, Sasaki CT (1995). Management of carotid artery rupture by monitored endovascular therapeutic occlusion (1988–1994). Laryngoscope 105:1086–1092.

58. Morrissey DD, Andersen PE, Nesbit GM, Barnwell SL, Everts EC, Cohen JI (1997). Endovascular management of hemorrhage in patients with head and neck cancer. Arch Otolaryngol Head Neck Surg 123:15–19.

59. Lesley WS, Chaloupka JC, Weigele JB, Mangla S, Dogar MA (2004). Preliminary experience with endovascular reconstruction for the management of carotid blowout syndrome. Am J Neuroradiol 24:975–981.

60. Omura M, Aida N, Sekido K, Kakehi M, Matsubara S (1997). Large intracranial vessel occlusive vasculopathy after radiation therapy in children: clinical features and usefulness of magnetic resonance imaging. Int J Radiat Oncol Biol Phys 38(2):241–249.

61. Chen CY, Zimmerman RA, Faro S, Bilaniuk LT, Chou TY, Molloy PT (1996). Childhood leukemia: central nervous system abnormalities during and after treatment. Am J Neuroradiol 17:295–310.

62. Parker BR (1997). Leukemia and lymphoma in childhood. Radiol Clin N Am 35:1495–1516.

63. Keime-Guibert F, Napolitano M, Delattre JY (1998). Neurological complications of radiotherapy and chemotherapy. J Neurol 245(11):695–708.

64. Sengar AR, Gupta RK, Dhanuka AR, Roy R, Das K (1997). MR imaging, MR angiography, and MR spectroscopy of the brain in eclampsia. Am J Neuroradiol 18:1485–1490.

65. Jain R, Robertson PL, Gandhi D, Gujar SK, Muraszko KM, Gebarski S (2005). Radiation-induced cavernomas of the brain. Am J Neuroradiol 26(5):1158–1162.

66. Hassler O, Movin A (1966). Microangiographic studies on changes in the cerebral vessels after irradiation. 1. Lesions in the rabbit produced by 60Co gamma-rays, 195 kV and 34 MV roentgen rays. Acta Radiol Ther Phys Biol 4(4):279–288.

67. Heckl S, Aschoff A, Kunze S (2002). Radiation-induced cavernous hemangiomas of the brain: a late effect predominantly in children. Cancer 94(12):3285–3291.

68. Reinhold HS, Hopewell JW (1980). Late changes in the architecture of blood vessels of the rat brain after irradiation. Br J Radiol 53(631):693–696.

69. Virapongse C, Cazenave C, Quisling R, Sarwar M, Hunter S (1987). The empty delta sign: frequency and significance in 76 cases of dural sinus thrombosis. Radiology 162:779–785.

70. Czaykowski PM, Moore MJ, Tannock IF (1998). High risk of vascular events in patients with urothelial transitional cell carcinoma treated with cisplatin based chemotherapy. J Urol 160:2021–2024.

71. El Amrani M, Heinzlef O, Debroucker T, Roullet E, Bousser MG, Amarenco P (1998). Brain infarction following 5-fluorouracil and cisplatin therapy. Neurology 51:899–901.

72. Licciardello J, Moake J, Rudy C, Karp D, Hong WK (1985). Elevated plasma von Willebrand factor levels and arterial occlusive complications associated with cisplatin-based chemotherapy. Oncology 42:296–300.

73. Gerl A (1994). Vascular toxicity associated with chemotherapy for testicular cancer. Anticancer Drugs 5:607–614.

74. Kishi S, Greiner J, Cheng C et al (2003). Homocysteine, pharmacokinetics, and neurotoxicity in children with leukemia. J Clin Oncol 15:3084–3091.

75. Asato R, Akiyama Y, Ito M et al (1992). Nuclear magnetic resonance abnormalities of the cerebral white matter in children with acute lymphoblastic leukemia and malignant lymphoma during and after central nervous system prophylactic treatment with intrathecal methotrexate. Cancer 70(7):1997–2004.

76. Matsumoto K, Takahashi S, Sato A et al (1995). Leukoencephalopathy in childhood hematopoietic neoplasm caused by moderate-dose methotrexate and prophylactic cranial radiotherapy – an MR analysis. Int J Radiat Oncol Biol Phys 32(4):913–918.

77. Mahoney DH Jr, Shuster JJ, Nitschke R et al (1998). Acute neurotoxicity in children with B-precursor acute lymphoid leukemia: an association between intermediate-dose intravenous methotrexate and intrathecal triple therapy: a Pediatric Oncology Group study. J Clin Oncol 16:1712–1722.

78. Haykin ME, Gorman M, van Hoff J, Fulbright RK, Baehring JM (2006). Diffusion-weighted MRI correlates of subacute methotrexate-related neurotoxicity. J Neurooncol 76(2):153–157.

79. Fleischhack G, Solymosi L, Reiter A, Bender-Gotze C, Eberl W, Bode U. (1994). Bildgebende Verfahren in der Diagnostik zerebrovaskulärer Komplikationen unter L-Asparaginase-Therapie. Klin Pädiatr 206:334–341.

80. Corso A, Castagnola C, Bernasconi C (1997). Thrombotic events are not exclusive to the remission induction period in patients with acute lymphoblastic leukemia: a report of two cases of cerebral sinus thrombosis. Ann Hematol 75: 117–119.

81. Ay H, Buonanno FS, Schaefer PW et al (1998). Posterior leukoencephalopathy without severe hypertension. Neurology 51:1369–1376.

82. Hinchey J, Chaves C, Appignani B et al (1996). A reversible posterior leukoencephalopathy syndrome. N Engl J Med 334:494–500.

83. Scott M (1975). Spontaneous intracerebral hematoma caused by cerebral neoplasms: report of eight verified cases. J Neurosurg 42:338–342.

84. Slivnick D, Ellis T, Nawrocki J, Fisher R (1990). The impact of Hodgkin's disease on the nervous system. Semin Oncol 17:673–679.

85. Wasserstrom WR, Glass JP, Posner JB (1982). Diagnosis and treatment of leptomeningeal metastases from solid tumors: experience with 90 patients. Cancer 49:759–772.

86. Casey SO, Sampaio RC, Michel E, Truwit CL (2000). Posterior reversible encephalopathy syndrome: utility of fluid-attenuated inversion recovery MR imaging in the detection of cortical and subcortical lesions. Am J Neuroradiol 21:1199–1206.

87. Provenzale JM, Petrella JR, Cruz LCH, Wong JC, Engelter S, Barboriak DP (2001). Quantitative assessment of diffusion abnormalities in posterior reversible encephalopathy syndrome. Am J Neuroradiol 22:1455–1461.

88. Gruber A, Killer M, Mazal P, Bavinzski G, Richling B (2000). Preoperative embolization of intracranial meningiomas: a 17-year single center experience. Minim Invasive Neurosurg 43:18–29.

89. Bendszus M, Klein R, Burger R, Warmuth-Metz M, Hofmann E, Solymosi L (2000). Efficacy of trisacryl gelatin microspheres

versus polyvinyl alcohol particles in the preoperative embolization of meningiomas. Am J Neuroradiol 21:255–261.

90. Bendszus M, Monoranu CM, Schutz A, Nolte I, Vince GH, Solymosi L (2005). Neurologic complications after particle embolization of intracranial meningiomas. Am J Neuroradiol 26(6):1413–1419.

91. Williams RL, Meltzer CC, Smirniotopoulos JG, Fukui MB, Inman M (1998). Cerebral MR imaging in intravascular lymphomatosis. Am J Neuroradiol 19:427–431.

92. Lasjaunias P, Berenstein A (1987). *Surgical Neuroangiography: Endovascular Treatment of Craniofacial Lesions*, Vol 2. Springer-Verlag, Berlin.

93. Campi A, Benndorf G, Filippi M, Reganati P, Martinelli V, Terreni MR (2001). Primary angiitis of the central nervous system: serial MRI of brain and spinal cord. Neuroradiology 43(8):599–607.

94. Ferris EJ, Levine HL (1973). Cerebral arteritis: classification. Radiology 109:327–341.

95. Rogers LR, Cho E, Kempin S, Posner JB (1987). Cerebral infarction from non-bacterial thrombotic endocarditis. Am J Med 83:746–756.

96. Kolodny EH, Rebeiz JJ, Caviness VS Jr, Richardson EP Jr (1968). Granulomatous angiitis of the central nervous system. Arch Neurol 19:510–524.

97. Shoemaker EI, Lin ZS, Rae-Grant AD, Little B (1994). Primary angiitis of central nervous system: unusual MR appearance. Am J Neuroradiol 15:331–334.

98. Greenan TJ, Grossman RI, Goldberg HI (1992). Cerebral vasculitis: MR imaging and angiographic correlation. Radiology 182:65–72.

99. Pomper MG, Miller TJ, Stone JH, Tidmore WC, Hellmann DB (1999). CNS vasculitis in autoimmune disease: MR imaging findings and correlation with angiography. Am J Neuroradiol 20:75–85.

100. Sheibani K, Battifora H, Winberg C et al (1986). Further evidence that 'malignant angioendotheliomatosis' is an angiotropic large-cell lymphoma. N Engl J Med 314:943–948.

101. Alhalabi M, Moore P (1994). Serial angiography in isolated angiitis of the central nervous system. Neurology 44:1221–1226.

102. Calabrese LH, Duna GF (1995). Evaluation and treatment of central nervous system vasculitis. Curr Opin Rheumatol 7:37–44.

103. Glass J, Hochberg F, Miller D (1993). Intravascular lymphomatosis: a systemic disease with neurologic manifestations. Cancer 71:3156–3164.

104. Shanley DJ (1995). Mineralizing microangiopathy: CT and MRI. Neuroradiology 37:331–333.

105. Weingarten K, Barbut D, Filippi C, Zimmerman RD (1994). Acute hypertensive encephalopathy: findings on spin-echo and gradient-echo MR imaging. Am J Roentgenol 162:665–670.

106. Chapin JE, Davis LE, Kornfeld M, Mandler RN (1995). Neurological manifestations of intravascular lymphomatosis. Acta Neurol Scand 91:494–499.

107. al Chalabi A, Sivakumaran M, Holton J, West KP, Wood JK, Abbott RJ (1994). A case of intravascular malignant lymphomatosis (angiotropic lymphoma) with raised perinuclear antineutrophil cytoplasmic antibody titres – a hitherto unreported association. Clin Lab Haematol 16:363–369.

108. Hashimoto H, Naritomi H, Kazui S et al (1998). Presymptomatic brain lesions on MRI in a patient with intravascular malignant lymphomatosis. J Neuroimaging 8:110–113.

109. Liow K, Asmar P, Liow M et al (2000). Intravascular lymphomatosis: contribution of cerebral MRI findings to diagnosis. J Neuroimaging 10:116–118.

110. Wick M, Mills S (1991). Intravascular lymphomatosis: clinicopathologic features and differential diagnosis. Semin Diagn Pathol 8:91–101.

CHAPTER 56

Neuroimaging in Paraneoplastic Syndromes

Steven Vernino and E. Paul Lindell

INTRODUCTION

Imaging studies are an important part of the evaluation of paraneoplastic neurological syndromes. Many CNS paraneoplastic disorders are associated with normal magnetic resonance imaging (MRI). In these cases, imaging of the neuraxis is still necessary to evaluate for alternative (and more common) neurological diagnosis such as inflammatory demyelinating disease, mass lesions, vascular abnormalities and so forth. Characteristic MRI abnormalities are found in several paraneoplastic disorders of the central nervous system. In general, these consist of abnormalities seen on fluid attenuated inversion recovery (FLAIR) and T2 sequences with little or no contrast enhancement. The lack of contrast enhancement helps to distinguish paraneoplastic disorders from malignant disease. Similarly, diffusion-weighted imaging and angiography may be used to help distinguish ischemic vascular disease. In general, paraneoplastic disorders restricted to the peripheral nervous system are not associated with radiological abnormalities.

When evaluating possible paraneoplastic neurological syndromes, diagnostic imaging is also critical for detection of systemic malignancies. Most patients present with neurological symptoms prior to a diagnosis of cancer. In patients with known cancer, the onset of neurological symptoms may herald the recurrence of tumor. Tumors in these patients tend to be small and localized. Serological studies and assessment of cancer risk factors can help identify possible sites of malignancy. In most cases, imaging studies are necessary to identify sites of malignancy for diagnostic biopsy. Utilization of multiple imaging modalities can improve the early detection of small malignancies [1].

PARANEOPLASTIC CEREBELLAR DEGENERATION (PCD)

Despite the dramatic and disabling symptoms of this disorder, imaging of the cerebellum is usually unremarkable early in the disease course (Figure 56.1A). The cerebellum and associated brainstem nuclei appear normal on cranial MRI [2]. Abnormal MRI signal or contrast enhancement in the cerebellum is distinctly unusual, but may be seen in rare cases with rapid progression [3]. In a few isolated case reports of fulminant cerebellar degeneration, FLAIR abnormalities in the cerebellar parenchyma may represent vasogenic edema. Signal abnormalities in the brainstem or mesial temporal lobes may be seen when PCD coexists with other paraneoplastic disorders (such as brainstem or limbic encephalitis). Even when conventional imaging is normal, PET imaging may show hypermetabolism in the cerebellar cortex.

Much later (months after the development of symptoms), cranial MRI and computed tomography (CT) will begin to show diffuse cerebellar atrophy corresponding to permanent neuronal loss. The cerebellum becomes diffusely atrophic, although the loss of tissue is most notable in the midline structures (see Figure 56.1).

PARANEOPLASTIC LIMBIC ENCEPHALITIS

Cranial MRI is abnormal in most cases of paraneoplastic encephalitis (PLE). The usual pattern is bilateral, but asymmetric, T2/FLAIR increased signal in the mesial temporal lobes (involving the amygdala and hippocampal formation)(Figure 56.2). Unilateral involvement can be seen as well. Besides the mesial temporal areas, other limbic structures may also become involved, especially the insular cortex (Figure 56.3). Several reports have shown that the same anatomical areas appear hypermetabolic on cranial PET imaging (Figure 56.3) [4,5]. Because of the location of these changes and the normal mildly increased signal in these areas, the abnormalities may be missed on standard axial MRI series. Coronal T2/FLAIR sequences with thin sections through the mesial temporal lobes optimize the detection of these abnormalities (Figure 56.2). Gadolinium enhancement is minimal, if present at all.

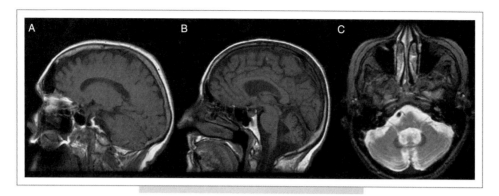

FIG. 56.1. Paraneoplastic cerebellar degeneration. **(A)** Off midline sagittal T1 spin-echo image demonstrates the relative lack of atrophy often seen early in the course of PCD. Despite severe neurological disability, initial brain imaging studies are usually normal. **(B)** Sagittal T1 spin-echo MRI shows prominent cerebellar atrophy. This is the most common finding in more advanced cases. **(C)** Axial T2-weighted image through the cerebellum demonstrates atrophy without signal abnormality.

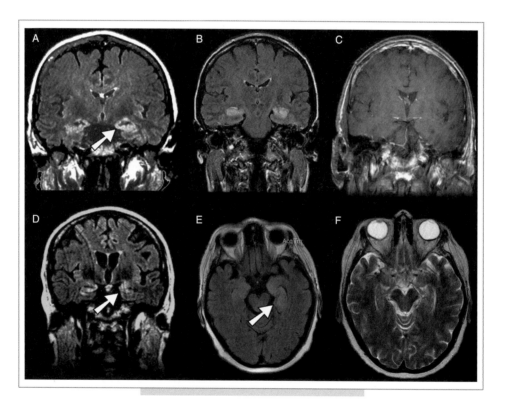

FIG. 56.2. Paraneoplastic limbic encephalitis. **(A)** Coronal FLAIR image through the mesial temporal lobes shows increased signal involving the hippocampus (arrow) and amygdala. These signal abnormalities are often present bilaterally but may be asymmetric. Early in the disease course, there is minimal cortical or mesial temporal atrophy. These changes can be similar to those seen in herpes encephalitis. **(B)** Nearly identical findings can be seen in non-paraneoplastic autoimmune limbic encephalitis. Coronal FLAIR image from a patient with limbic encephalitis and voltage-gated potassium channel antibodies. **(C)** The mesial temporal abnormalities in PLE generally show no enhancement on gadolinium-enhanced images. **(D,E,F)** MRI images from a patient with PLE show that the imaging abnormalities are best appreciated on coronal FLAIR images (**D**, arrow). They can also be seen on axial FLAIR images (**E**, arrow), although the mesial temporal area is also prone to artifact on the axial FLAIR sequence. PLE abnormalities are usually not conspicuous on the axial T2 scan **(F)**.

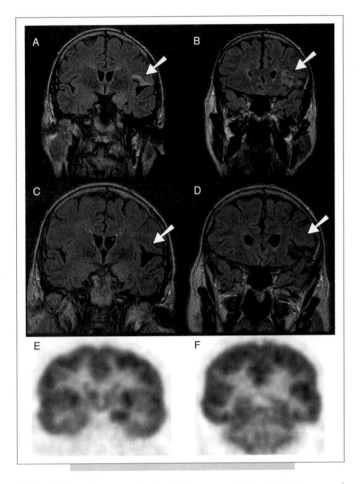

FIG. 56.3. Paraneoplastic limbic encephalitis. **(A)** 3T coronal FLAIR image through the mesial temporal lobes shows increased signal in the left mesial temporal lobe and in the left frontal operculum (arrow). **(B)** Adjacent coronal FLAIR image anterior to **(A)** showing the extension of the abnormal signal to involve the anterior left frontal operculum and anterior insular cortex (arrow). **(C,D)** Coronal FLAIR images demonstrating improvement in the opercular and insular signal (arrows) after immunosuppression treatment. Signal abnormality in the left amygdala persists. **(E,F)** PET imaging with 18-FDG demonstrates hypermetabolism in the areas of abnormal T2 signal seen in panels **(A)** and **(B)**. The PET abnormalities are presumed to represent active inflammation.

The imaging appearance and typical regions of involvement of paraneoplastic limbic encephalitis has significant overlap with other often more aggressive pathologies. Differentiation from herpes encephalitis is an important primary consideration. Herpes encephalitis tends to have a greater degree of enhancement and greater local mass effect, although the signal changes and areas of involvement have significant overlap (Figure 56.4). Mesial temporal sclerosis, gliomatosis cerebri and non-paraneoplastic limbic encephalitis can have identical imaging abnormalities to paraneoplastic limbic encephalitis [6].

In a study of PLE at the Mayo Clinic, cranial MRI was abnormal in 89% of the cases. FLAIR sequences were not performed in the two patients with normal scans [7]. Nearly all cases showed increased T2 and FLAIR signal in the mesial temporal lobe. Bilateral temporal lobe abnormalities were present in 69%, while abnormal contrast enhancement was seen in only 21%. In those cases, the enhancing tissue comprised only a small volume of the total signal abnormality. Extratemporal cortical signal abnormalities were observed in 37%. These changes ranged from diffuse hemispheric cortical and subcortical signal abnormalities involving both cerebral hemispheres to involvement of extratemporal changes in the cortical and subcortical insular region.

Over time, there is typically a partial resolution of the mesial temporal lobe T2 hyperintensity associated with the development of focal temporal and generalized cerebral cortical atrophy. In untreated cases, mesial temporal atrophy can be profound (Figure 56.5).

PARANEOPLASTIC BRAINSTEM ENCEPHALITIS

Various patterns of imaging abnormalities have been reported with paraneoplastic brainstem encephalitis. Often the MRI is normal, however, varying patterns of abnormal T2/FLAIR signal in the brainstem have been reported (Figure 56.6) [8–10]. Enhancement in this cohort of patients is uncommon. Imaging in patients with isolated opsoclonus/myoclonus, however, is generally normal.

PARANEOPLASTIC CHOREA

Identifiable MRI abnormalities have been reported in paraneoplastic chorea and include non-enhancing increased T2/FLAIR signal in the striatum (Figure 56.7) [11]. Occasionally, these may be seen in conjunction with typical changes of paraneoplastic limbic encephalitis. The increased T2 and FLAIR signal in the caudate and putamen may resolve in concert with clinical improvement. The MRI abnormalities in paraneoplastic chorea, non-enhancing signal changes in the striatum, are reminiscent of those seen in sporadic Creutzfeld-Jacob disease (CJD) and acute disseminated encephalomyelitis (ADEM), and there may be some overlap in the clinical presentation of these disorders. However, diffusion imaging in paraneoplastic chorea is normal, which differentiates it from CJD radiographically (Figure 56.7).

PARANEOPLASTIC MYELOPATHY

Several cases of cancer associated myelopathy have been reported [8]. Abnormal increased T2/FLAIR signal and enhancement of long tracts in the cord as well as gray matter has been reported.

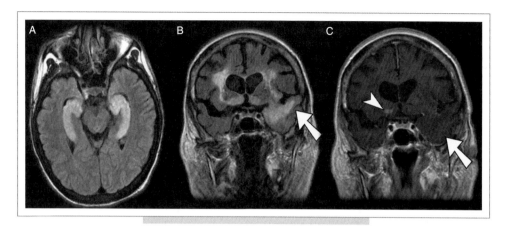

FIG. 56.4. Limbic encephalitis. **(A)** Axial FLAIR image showing extensive signal abnormality which is highly restricted to the mesial temporal lobe in a patient with PLE associated with Hodgkin's lymphoma. Gadolinium-enhanced T1 images were unremarkable in this case. **(B)** Coronal FLAIR image in a case of herpes simplex encephalitis shows signal abnormalities extending throughout the left temporal lobe, insular cortex and bilateral medial thalamus. The findings in herpes encephalitis involve the limbic areas of the brain but are typically more extensive. **(C)** Gadolinium-enhanced T1 coronal image shows areas of contrast enhancement in the thalamus (arrowhead). The left temporal lobe shows low signal and increased volume consistent with vasogenic edema. This degree of contrast enhancement and swelling would be very unusual for PLE.

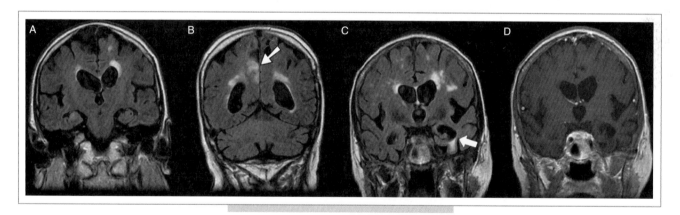

FIG. 56.5. Paraneoplastic limbic encephalitis. Later in the course of PLE, prominent mesial temporal atrophy typically develops. **(A)** Coronal FLAIR image demonstrates moderate bilateral temporal atrophy and abnormal increased right mesial temporal lobe signal. **(B)** Coronal FLAIR image more posteriorly demonstrates abnormal increased signal in the posterior right cingulate gyrus (arrow). **(C)** Coronal FLAIR MRI from a patient with untreated PLE (2 years after onset of symptoms) shows severe asymmetric temporal atrophy with relative preservation of frontoparietal cortex. An area of signal abnormality persists in inferior temporal cortex. **(D)** T1-weighted coronal image after gadolinium shows no enhancement of these abnormalities.

IMAGING OF OCCULT MALIGNANCY IN PATIENTS WITH PARANEOPLASTIC DISORDERS

One of the greatest challenges in evaluating patients with suspected paraneoplastic disorders (PND) is cancer diagnosis. The majority of new diagnoses of PND occur in patients without a history of cancer. In most circumstances, conventional body imaging studies (computed tomography, ultrasonography, etc.) will reveal the malignancy and identify a site for diagnostic biopsy. However, tumors in patients with PND may be missed on initial imaging studies since they are often small and have limited metastases. Yet, patients with well recognized paraneoplastic syndromes and onconeural antibodies are very likely (85–90% in most series) to harbor a malignancy, and early tumor diagnosis and treatment is important to improve outcome. Several recent studies have shown that ^{18}F-fluoro-deoxyglucose

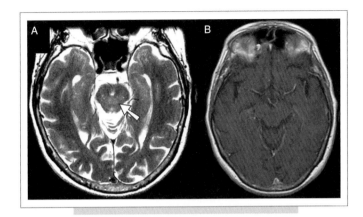

FIG. 56.6. Paraneoplastic brainstem encephalitis. **(A)** Thin section axial fast spin-echo T2 sequence through the midbrain shows abnormal increased T2 signal in the tegmentum and tectum (arrow). Many cases of brainstem encephalitis have no obvious imaging abnormalities. **(B)** Axial post-gadolinium T1 image at the same level demonstrates no corresponding enhancement. The lack of enhancement helps differentiate this entity from other infectious and inflammatory disorders.

whole body positron emission tomography (PET) can improve tumor detection when conventional imaging studies are normal (Figure 56.8) [12]. When limited to patients with PND associated with well-defined antibodies, the sensitivity of PET was reported to be 80–90% [13,14]. It appears that PET has a definite role in evaluating patients with definite PND when conventional oncological screening studies are negative.

SUMMARY

Neuroimaging has a definite role in the diagnostic evaluation of paraneoplastic disorders of the central nervous system. Characteristic imaging findings are seen in PLE and paraneoplastic chorea. In other disorders, imaging is important to exclude other diagnoses. In patients with suspected paraneoplastic disorders (based on serology and clinical presentation), imaging of the body including FDG-PET imaging is important for early detection of small malignancies.

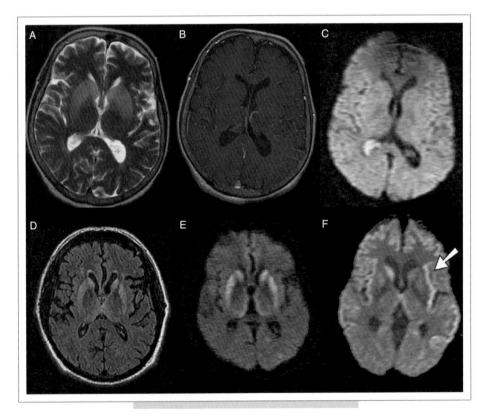

FIG. 56.7. Paraneoplastic chorea. **(A)** Axial spin-echo T2 image shows increased signal in the caudate and anterior putamen consistent with 'striatal encephalitis'. Similar findings are present on FLAIR images. **(B)** Axial post-gadolinium T1 image at the same level shows no enhancement. **(C)** Diffusion-weighted images show normal signal in the basal ganglia. The lack of restricted diffusion is characteristic of paraneoplastic chorea and effectively eliminates Creutzfeld-Jacob disease (CJD) and ischemia as alternative diagnoses. **(D)** In CJD, axial FLAIR image demonstrates similar abnormal T2 signal in the basal ganglia. **(E)** However, on diffusion-weighted imaging, these abnormalities are associated with restricted diffusion. **(F)** In a different patient with CJD, a diffusion-weight image shows abnormality in both the right caudate and along the cortical ribbon of the left insula. Paraneoplastic disorders may show abnormalities of the striatum (chorea) or insular cortex (PLE) but diffusion abnormalities are not seen.

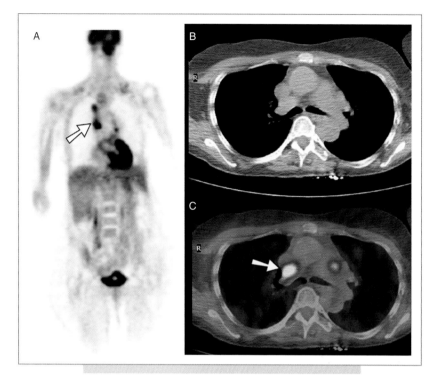

FIG. 56.8. PET imaging for cancer diagnosis. **(A)** Coronal whole body 18-FDG PET image shows abnormal hypermetabolism in right paratracheal and right hilar regions suggesting a lung primary. This patient had PLE and antibodies against both Hu and CRMP-5. **(B)** Non-contrast axial CT scan of the chest in this patient was interpreted as showing no abnormalities. **(C)** Coregistration of the PET images with the CT allows identification of hypermetabolic mediastinal lymph nodes. Biopsy of the most prominent lesion (arrow) confirmed a diagnosis of small cell lung carcinoma.

REFERENCES

1. Berner U, Menzel C, Rinne D et al (2003). Paraneoplastic syndromes: detection of malignant tumors using [18F]FDG-PET. Q J Nucl Med 47(2):85–89.
2. Scheid R, Voltz R, Briest S et al (2006). Clinical insights into paraneoplastic cerebellar degeneration. J Neurol Neurosurg Psychiatr 77(4):529–530.
3. de Andres C, Esquivel A, de Villoria JG et al (2006). Unusual magnetic resonance imaging and cerebrospinal fluid findings in paraneoplastic cerebellar degeneration: a sequential study. J Neurol Neurosurg Psychiatr 77(4):562–563.
4. Kassubek J, Juengling FD, Nitzsche EU, Lucking CH (2001). Limbic encephalitis investigated by 18FDG-PET and 3D MRI. J Neuroimaging 11(1):55–59.
5. Fakhoury T, Abou-Khalil B, Kessler RM (1999). Limbic encephalitis and hyperactive foci on PET scan. Seizure 8(7):427–431.
6. Thieben MJ, Lennon VA, Boeve BF et al (2004). Potentially reversible autoimmune limbic encephalitis with neuronal potassium channel antibody. Neurology 62(7):1177–1182.
7. Lawn ND, Westmoreland BF, Kiely MJ et al (2003). Clinical, magnetic resonance imaging, and electroencephalographic findings in paraneoplastic limbic encephalitis. Mayo Clin Proc 78(11):1363–1368.
8. Taraszewska A, Piekarska A, Kwiatkowski M et al (1998). A case of the subacute brainstem encephalitis. Folia Neuropathol 36(4):217–220.
9. Dalmau J, Graus F, Villarejo A et al (2004). Clinical analysis of anti-Ma2-associated encephalitis. Brain 127(Pt 8):1831–1844.
10. Barnett M, Prosser J, Sutton I et al (2001). Paraneoplastic brain stem encephalitis in a woman with anti-Ma2 antibody. J Neurol Neurosurg Psychiatr 70(2):222–225.
11. Vernino S, Tuite P, Adler CH et al (2002). Paraneoplastic chorea associated with CRMP-5 neuronal antibody and lung carcinoma. Ann Neurol 51(5):625–630.
12. Rees JH, Hain SF, Johnson MR et al (2001). The role of {18F}fluoro-2-deoxyglucose-PET scanning in the diagnosis of paraneoplastic neurological disorders. Brain 124(11): 2223–2231.
13. Linke R, Schroeder M, Helmberger T, Voltz R (2004). Antibody-positive paraneoplastic neurologic syndromes: value of CT and PET for tumor diagnosis. Neurology 63(2):282–286.
14. Younes-Mhenni S, Janier MF, Cinotti L et al (2004). FDG-PET improves tumour detection in patients with paraneoplastic neurological syndromes. Brain 127(10):2331–2338.

Index